CRYSTAL GROWTH 1971

PROCEEDINGS OF THE THIRD
INTERNATIONAL CONFERENCE ON CRYSTAL GROWTH
MARSEILLE, FRANCE, 5–9 July 1971

CRYSTAL GROWTH 1971

PROCEEDINGS OF THE THIRD
INTERNATIONAL CONFERENCE ON CRYSTAL GROWTH
MARSEILLE, FRANCE, 5–9 July, 1971

EDITED BY

R. A. LAUDISE
Bell Telephone Laboratories, Incorporated, U.S.A.

J. B. MULLIN
Royal Radar Establishment, Malvern, U.K.

B. MUTAFTSCHIEV
Université de Provence, Marseille, France

1972

NORTH-HOLLAND PUBLISHING COMPANY – AMSTERDAM

Library of Congress Catalog Card Number: 70-183279

ISBN: 0 7204 0240 9

Reprinted from:

JOURNAL OF CRYSTAL GROWTH 13/14 (1972)

Printed in The Netherlands

CONFERENCE COMMITTEE

On the initiative of M. Schieber (Jerusalem) Secretary General of C.I.C.C. and R. Kern (Marseille), the Groupe Français de Croissance Cristalline (G.F.C.C.) was created in Paris on the 15th January 1968. The founder members are:

S. Amelinckx (*Belgium*)	F. Forrat (*Grenoble*)	V. Luzzati (*Orléans*)
R. Aumont (*Paris*)	J. Friedel (*Orsay*)	H. Makram (*Bellevue*)
J. Bénard (*Paris*)	S. Goldsztaub (*Strasbourg*)	J. C. Monier (*Caen*)
M. Brunet (*Caen*)	R. Hocart (*Paris*)	B. Mutaftschiev (*Marseille*)
B. Chakraverty (*Grenoble*)	R. Kern (*Marseille*)	M. Rodot (*Bellevue*)
J. Chapelle (*Orsay*)	P. Lacombe (*Paris*)	A. M. Vergnoux (*Montpellier*)
H. Curien (*Paris*)	M. Lantieri (*Paris*)	B. Vodar (*Bellevue*)
W. Dekeyser (*Belgium*)	J. Loriers (*Paris*)	A. Weill (*Paris*)
M. Drechsler (*Marseille*)		J. Wyart (*Paris*)

G.F.C.C. co-opted W. Dekeyser as President, A. M. Vergnoux and B. Mutaftschiev as co-secretaries. The assembly expressed general agreement on the proposal of M. Schieber to organise I.C.C.G.3 in Marseille, France in 1971.

The general assembly of C.I.C.C. which was held during I.C.C.G.2 at Birmingham in 1968 decided to organise I.C.C.G.3 in Marseille in 1971. W. Dekeyser and R. Kern were appointed joint Co-Chairmen with B. Mutaftschiev as Secretary General.

FRENCH HONORARY COMMITTEE

P. Aigrain	J. Friedel	L. Royer
J. Bénard	R. Hocart	B. Vodar
H. Curien	P. Lacombe	J. Wyart
	L. Néel	

INTERNATIONAL ADVISORY COMMITTEE

W. Bardsley (*U.K.*)	F. C. Frank (*U.K.*)	H. S. Peiser (*U.S.A.*)
H. Bethge (*G.D.R.*)	P. Hartman (*Netherlands*)	G. M. Pound (*U.S.A.*)
K. J. Button (*U.S.A.*)	R. R. Hasiguti (*Japan*)	C. S. Sahagian (*U.S.A.*)
N. Cabrera (*Spain*)	R. Kaischew (*Bulgaria*)	N. N. Sheftal (*U.S.S.R.*)
J. W. Cahn (*U.S.A.*)	E. Kaldis (*Switzerland*)	I. N. Stranski (*G.F.R.*)
B. Chalmers (*U.S.A.*)		W. A. Tiller (*U.S.A.*)

ORGANISING COMMITTEE

W. Dekeyser (*Co-Chairman*)	M. Drechsler (*Exhibition*)
R. Kern (*Co-Chairman*)	D. Colclough (*Secretary*)
B. Mutaftschiev (*Secretary General*)	M. Dougny (*Secretary*)
A. Masson (*Executive Secretary*)	F. Huillier (*Secretary*)
M. Bienfait (*Treasurer*)	M. C. Marchetti (*Secretary*)
L. Capella (*Local Arrangements*)	L. Stock (*Secretary*)
M. Gillet (*Local Arrangements*)	

PROGRAMME COMMITTEE

R. Kern (*Chairman*)	F. Durand	H. W. Newkirk
A. M. Anthony	P. Gibart	J. Oudar
A. Authier	R. Lacmann	M. Paulus
E. Bauer	R. A. Laudise	M. Pierrot
P. Bennema	J. M. Le Duc	R. B. Roy
J. M. Bermond	H. Makram	K. F. Seifert
R. Boistelle	M. Marais	B. Simon
B. Chakraverty	J. B. Mullin	R. F. Strickland-Constable
	B. Mutaftschiev	

PUBLICATIONS COMMITTEE

B. MUTAFTSCHIEF (Chairman)

R. A. LAUDISE

J. B. MULLIN

M. SCHIEBER (J. Crystal Growth)

FINANCE COMMITTEE

M. BIENFAIT (*Chairman*)

S. ALBINET

C. ALLOARD

R. BOISTELLE

R. CADORET

LOCAL ARRANGEMENTS COMMITTEE

L. CAPELLA (*Co-Chairman*)
M. GILLET (*Co-Chairman*)
G. ALBINET
A. BARONNET
F. BEL
P. BELLANDI
J. BERTALMIO
J. P. BIBERIAN
P. BLIEK
A. BONISCHAT
A. BONISSENT
M. CAHOREAU
C. CARANONI
C. CHAPON
C. CLAEYS
J. P. COUDERT
A. COULET

D. DELAY
B. FELTS
J. L. FIRPO
M. GAUCH
M. GAY
E. GILLET
G. GOUJON
A. GRASSI
P. GUEGEN
R. HASER
J. C. HEYRAUD
J. N. JULLIEN
G. LELAY
J. LE QUERRE
J. L. LERAY
M. MANNEVILLE
M. MATHIEU

J. J. MÉTOIS
B. MINARI
F. MINARI
D. MOYA
M. NAPOLEONI
A. NEMETH
A. OUDIN
M. PASSEREL
G. PEPE
J. PERRIN
B. PICHAUD
G. QUENTEL
A. RENOU
F. SECCIA
D. SUZANNE
E. TRAUGAN
J. F. VIDAL

EXHIBITION COMMITTEE

M. DRECHSLER (*Chairman*)
M. AUDIFFREN

S. BLANCHETON
M. LENOIR

M. PICHAUD
R. ROCHE

Acknowledgements

The French Organising Committee of the Marseille Conference would like to thank all those who contributed to the successful running of the Conference. They are particularly grateful for the support which they received from their British colleagues who organised Birmingham I.C.C.G.2.

This Conference was sponsored by:

Air Force Cambridge Research Laboratories (through the European Office of Aerospace Research, O.A.R.) United States Air Force under grant number EOOAR-70-0079.
Direction de la Recherche et des Moyens d'Essais (D.R.M.E.) auprès du Ministère Français des Armées.
Direction de la Coopération, Sous-Direction des Accords Internationaux et des Echanges Universitaires, auprès du Ministère Français des Affaires Etrangères.
Délégation Générale de la Recherche Scientifique et Technique (D.G.R.S.T.) (Section Métallurgie).

Financial assistance from the following organisations is gratefully acknowledged:

Alusuisse France S.A.
Association des Amis de l'Université de Marseille
Centre Technique des Industries de la Fonderie
CdF Chimie, Société Chimique des Charbonnages
Compagnie des Salins du Midi et des Salines de l'Est
Conseil Général des Bouches du Rhône
Kodak-Pathé
Laboratoires de Marcoussis, C.G.E.
La Radiotechnique – Compelec

L'Electro-Réfractaire
Métallurgie Hoboken-Overpelt, Belgium
N.V. Centrale Research K.N.Z., The Netherlands
Péchiney
Procter et Gamble, France
Thomson C.S.F.
Trindel, Les Milles
Siemens, France
Ugine Kuhlmann

INTRODUCTION

This volume is a selection of papers which were presented at the Third International Conference on Crystal Growth in Marseille (I.C.C.G.3). The papers represent to some extent a cross section of the situation in the field of Crystal Growth throughout the world at a particular point in time – July 1971. However, one cannot expect it to be a perfect representation of all the dimensions of the subject. It was not possible to include all the 321 papers presented at the Conference. A staff of 19 Subject Chairmen of the Programme Committee and more than 60 referees from seven nations spent many days of hard work in order to propose for publication some 200 of the initially submitted papers. A final refereeing of manuscripts through the combined efforts of the Editors, the Subject Chairmen and additional referees resulted in the selection of the some 160 papers published in this volume.

The principle guide lines used in selecting contributed papers were to include only unpublished original research, which was clearly written and documented, whilst striving to keep some equilibrium between different fields of interest in order to reflect as truly as possible the diversity of the Conference and the field of crystal growth.

Since the Proceedings is a refereed issue of the Journal of Crystal Growth it was not possible to include either papers based on older or bibliographic results, or those which could not easily be shortened to the requisite length or those giving new ideas or results that as yet have not been adequately substantiated as finished work. We look forward to seeing the further development and publication of the ideas and results which are not reported here, but which we hope I.C.C.G.3 may have played a role in stimulating and encouraging.

We are happy to acknowledge the work done by the Subject Chairmen and Referees, without whose aid this publication could not have been completed. We would like especially to thank also Dr. R. B. Roy, Dr. A. Bonissent, Dr. P. Tufton and Mr. B. W. Straughan for helping us in the editing as well as Misses L. Stock and D. Colclough, also Mr. A. Bonischot and Mr. J. N. Jullien.

Perhaps the best trend revealed to us by our association with this work, is the truly international aspect of the field of Crystal Growth today, one particularly pleasant demonstration of this for us being the fact that the three Editors represent three different nationalities.

We are certain that in addition to the increased understanding of theory and practice of Crystal Growth which we bring home from Marseille, we will return with increased bonds of friendship with scientists of all nations.

THE EDITORS

Marseille, 20 July 1971

PREFACE

Le volume qui suit donne une bonne idée, bien que partielle, des travaux présentés au 3ème Congrès International de Croissance Cristalline qui s'est tenu du 5 au 9 juillet 1971 au Campus Universitaire de Marseille–Luminy (France). Le Comité de Publication de I.C.C.G.-3 a su mettre en bonne forme ce volume, travail énorme et délicat, avec le doigté et les garanties scientifiques nécessaires. Que ses membres en soient remerciés vivement et à travers eux leurs employeurs qui ont accepté de les rendre disponibles pendant plus de deux mois.

A travers ce volume n'apparait évidemment pas l'activité d'organisation des autres Comités du Congrès. Plus d'une centaine de personnes se sont mises à la tâche pendant plusieurs mois; collègues, chercheurs et techniciens ont donné le meilleur d'eux mêmes pour que l'organisation scientifique, technique, les secrétariats, l'hébergement bon marché, etc. puissent apparaitre convenables aux participants du Congrès. La polyvalence de nos collaborateurs a rendu le Congrès accessible à 650 participants (420 venant de l'étranger) car il était financièrement hors de question de faire appel à des organisateurs professionnels de Congrès et ceci malgré l'aide généreuse dont nous avons profité de la part de nombreux organismes publics et privés, français et étrangers, cités en début du volume.

Parmi les manifestations générales qui n'émergent pas du volume nous devons retenir:

La Session d'Ouverture au Grand Amphithéâtre de l'Ecole des Beaux Arts lors de laquelle Monsieur B. Mutaftschiev a souhaité la bienvenue aux participants tous présents en essayant de justifier d'une manière historique le choix du site de Marseille. Le Professeur H. Curien, Directeur Général du C.N.R.S., a ouvert le Congrès en rappelant la mémoire de Sir Lawrence Bragg décédé le 3 juillet, et dont l'œuvre de cristallographie permet aujourd'hui des réunions interdisciplinaires comme I.C.C.G. Il a ajouté: "I.C.C.G. a tout pour plaire aux organisateurs de la Science du fait que s'y côtoient toutes les disciplines et que les recherches fondamentales et appliquées se fructifient mutuellement".

Monsieur I. N. Stranski (Berlin) a souligné quelques points d'ordre historique qui ont conduit après les années 30 aux conceptions atomistiques de la cristallogenèse. Monsieur F. C. Frank (Bristol) a donné son sentiment sur le devenir de cette science, mais il n'a pas voulu se laisser entrainer à des considérations philosophiques. Monsieur L. Néel (Grenoble) a rappelé ce que les magnéticiens connaissent sur les domaines magnétiques, la complexité des phénomènes usuellement observés dans les couches minces, la nécessité d'aborder les phénomènes volumiques certainement plus simples. Monsieur K. Nassau (Bell Tel.) a relaté d'une manière passionante l'œuvre d'historien faite sur A. V. L. Verneuil, en replaçant l'homme et la méthode dans le cadre si actif de la chimie française de l'époque et de l'industrie naissante des monocristaux à Paris et aux Etats Unis. La présence du neveu de A. V. L. Verneuil et de sa famille a donné une note intime à cette cérémonie.

L'Exposition du Congrès réunissait trente firmes, venant d'Europe et d'Outre Mer, dans le cadre de l'Aula de l'Université. L'ensemble du matériel exposé, les collections de monocristaux formaient un panorama fort agréable et esthétique pui pouvait être admiré par les participants à partir du podium de la buvette-restaurant surplombant l'ensemble. Les congressistes ont passé de nombreuses heures de détente, de visite et de discussions dans cette enceinte. En passant, il faut rappeler aussi que les participants les plus jeunes et donc entreprenants ne négligeaient pas les promenades, les bains de mer dans les Calanques de Luminy où toutes les formes de la nature sauvage pouvaient être appréciées.

La session de films scientifiques réunissant dix documents originaux a été un autre point culminant à côté du Congrès. Certains d'entre eux ont littéralement fasciné l'audience comme cette production presque alimentaire de trichytes d'argent sortant par action électrolytique d'un cristal de sulfure d'argent; leur disparition aisée est aussi étonnante que leur apparition (T. Ohachi). Les expériences sur la stabilité de zones flottantes (J. R. Carruthers) méritaient d'être filmées vu leur difficulté de réalisation. L'admiration a été forcée par le haut degré de technicité atteint par H. E. LaBelle Jr. dans le tirage à grande vitesse de cristaux de saphir sous des formes variées bien définies, directement utilisables par l'industrie.

L'ensemble des films présentés possédait un pouvoir éducatif élevé; le fondamentaliste y trouvait les vitamines stimulant son imagination, le chercheur appliqué le glycogène qui lui permettra de se surpasser.

Les deux Assemblées Générales de C.I.C.C. ont attiré 30 et 20% des participants. L'ésotérisme des sujets abordés, constitution, nominations, selon les usages parlementaires Américains a laissé pantois le monde non Anglo-Saxon. Le signataire reste admiratif devant l'unanimité des décisions déduites à partir des Aye. Le choix du site de I.C.C.G.-4 – Tokyo 1974 a été consolidé par acclamation.

L'organisation scientifique de I.C.C.G.-3 a été faite par les membres du Comité du Programme. Ceux ci s'étaient fixés comme but de suivre les lignes directrices de I.C.C.G.-1 et I.C.C.G.-2. Cependant, une place plus importante, soit 30%, devait être donnée aux aspects fondamentaux de la croissance cristalline. Ce comité a eu raison de faire cette percée car les auteurs ont répondu à la proposition en envoyant leurs papiers fondamentaux. Les amphithéâtres traitant de ces problèmes étaient parmi les plus fréquentés et ceci non seulement par les spécialistes mais aussi par les praticiens. Il a été tenté d'introduire également de nouvelles rubriques comme l'électrocristallisation, la cristallisation industrielle massique, la caractérisation de la perfection cristalline, les hauts polymères. Mis à part la dernière rubrique qui n'a pu aboutir à une session, les autres rubriques ont été un succès, mention particulière devant être faite pour celle concernant la perfection cristalline car elle a nécessité l'aménagement de trois sessions. Le Comité du Programme a tenu compte du voeu exprimé par les participants de I.C.C.G.-2 de multiplier le nombre des conférenciers invités. Sur 23 conférences prévues, 21 ont eu lieu.

Le revers de la médaille a été l'obligation d'organiser cinq sessions parallèles pour les 300 papiers dits "contribués". La disposition en enfilade des amphithéâtres, les tableaux lumineux donnant en temps réel l'état d'avancement des travaux dans chacun d'eux, a minimisé l'inconvénient usuel des sessions parallèles.

Le choix des conférenciers invités a été fait par vote de l'ensemble des membres du Comité du Programme, en se basant sur les données d'une consultation faite auprès de 50 personnalités extérieures à l'organisation du Congrès. Cette manière de procéder s'est révélée intéressante et efficace car des travaux originaux et importants ont pu être mis au jour.

Passons aux acquisitions scientifiques de I.C.C.G.-3 telle qu'elles peuvent apparaître en fin de Congrès.

En ce qui concerne les recherches fondamentales les Congrès de Boston et de Birmingham ont révélé l'importance des travaux sur la stabilité morphologique. Des techniques avancées de physique mathématique avaient permis de formuler des phénomènes précédemment incompris (dendrites, eutectiques).

Lors de I.C.C.G.-3 ces travaux sont venus en retrait. Deux tendances s'opposent cependant, celle (R. F. Sekerka) pensant que les considérations macroscopiques peuvent être développées davantage pour résoudre les dernières obscurités, celle de W. A. Tiller qui essaye d'introduire le discontinu dans le problème sous la forme de mécanismes élémentaires de croissance par couches. Pour le moment le domaine est calme, mais un rebondissement se produira.

Il en est tout autrement des nombreux travaux fort intéressants considérant la croissance comme phénomène atomistique.

Lors du Congrès de Boston, N. Cabrera avait attiré l'attention sur ce domaine alors négligé par certaines écoles. Un exercice particulièrement attrayant consiste actuellement à simuler sur ordinateurs la marche aléatoire des atomes lors de la croissance. A. A. Chernov avait initié cette voie nouvelle en 1965 par l'étude des cristaux mixtes de composition AB. Son travail n'est pas encore dépassé car il doublait son étude expérimentale de simulation d'une analyse littérale en termes de probabilités. A chaque moment il était ainsi en mesure de surveiller son ordinateur. Un congressiste a fait remarquer que ces machines sont des amplificateurs de bêtises si on n'y prend garde. D'où certaines controverses entre les chercheurs hollandais et américains, pratiquant ces exercices. F. L. Binsbergen s'est volontairement borné avec raison à une étude de simulation sur le cristal de Kossel tel qu'il a été traité par Becker et Döring en fixant les paramètres tels qu'ils soient représentatifs de la croissance ignée. La simulation restitue bien le caractère fonctionnel de la croissance avec la sursaturation tel que la théorie explicite le prévoit (nucléation bi- et tridimensionnelle). Il n'en est pas de même pour la grandeur énergie figurant dans les équations. Cette discordance entre théorie et "expérience" suggère de revoir à fond la théorie dans laquelle seulement les configurations non excitées sont prises en considération. Les travaux de P. Bennema, G. H. Gilmer et de l'école de K. A. Jackson se rapportent à des modèles déjà plus complexes et certaines anomalies rencontrées pourraient être dues à des "erreurs expérimentales systématiques". Ce domaine se développera probablement beaucoup ces prochaines années.

Des progrès significatifs ont été réalisés dans les traitements de mécanique statistique, qu'il s'agisse de l'évaporation cristalline (T. Surek) ou de la nucléation homogène (K. Nishioka). La structure des germes stables de très petites dimensions (B. Mutaftschiev) est apparue pouvoir s'écarter des symétries cristallines; leur croissance en un cristal pose un problème (H. S. Peiser), à moins que ces édifices non cristallographiques préfèrent se macler en édifices complexes, réellement observés (S. Ogawa) et dont le mécanisme de formation peut être compris qualitativement (M. Gillet).

Les études expérimentales sur l'électrocristallisation ont vu deux études exemplaires (E. Budevski, J. Corish). Le premier travaillant sur des cristaux de qualité contrôlée a passé en revue toutes les acquisitions profondes qui peuvent en être tirées pour vérifier les différents modes de croissance d'un cristal. Le second travaillant sur des trichytes complexes d'argent produites par électrolyse contrôlée à l'état solide après initiation mécanique ou un prédépôt localisé d'argent, les surtensions nécessaires à la croissance ou la dissolution sont telles que le transfert des atomes d'argent d'une phase à l'autre dicte la cinétique.

Les phénomènes hydrodynamiques paraissent avoir reçu aussi la faveur des fondamentalistes et la liaison avec les oscillations de température, de composition, amène à des considérations pratiques immédiates (D. T. J. Hurle, G. S. Cole, E. O. Schulz-DuBois).

La caractérisation de la perfection cristalline, en particulier par la méthode de Lang, est devenue courante et amène avec d'autres observations telles que figures de croissance et de corrosion, à reconstituer l'histoire d'un cristal (A. Authier, F. C. Frank). L'étude d'une espèce cristalline comme CaF_2 produite par différentes méthodes de croissance fournit des renseignements appréciables en particulier, si différents types de défauts sont considérés (K. Recker).

A partir de ce stade on aborde déjà le domaine des matériaux. Il est heureux de constater que là aussi se fait sentir le souci d'approfondir les problèmes. Un exemple constitue l'étude de la croissance par transport en phase gazeuse pratiquée soit sur le silicium (J. Nishizawa) soit sur GaAs (L. Hollan) sous forme d'hémisphères. La croissance latérale et normale peut être mesurée et en particulier lorsqu'un marquage chronologique est effectué par injection instantanée d'un dopant dans le système.

Les grenats magnétiques sont des matériaux ayant reçu une attention particulière (10 communications); leur intérêt et les exigences cristallochimiques ayant été soulignés par R. A. Laudise. Peu de matériaux nouveaux ont été présentés, mis à part, le ferroélectrique $Sr_2Nb_2O_7$ (M. Takahashi), les nitrures semi conducteurs à large bande interdite (M. Ilegems, J. P. Dismukes, D. K. Wickenden). Par contre des applications nouvelles ont été indiquées pour certains matériaux. Citons les gros monocristaux mixtes présentant un gradient de paramètre cristallin, tel que Cu–Ge, constituant un nouveau type de monochromateur à grand rendement pour diffraction neutronique (F. Forrat).

Du côté de la technologie une activité intense s'est déployée. Dans la pratique industrielle le tirage par la méthode Czochralski sur de grandes longueurs et à diamètre constant, l'automatisation par détection infra rouge couplée à l'ordinateur (D. F. O'Kane) s'avère importante. La méthode du creuset froid a reçu des améliorations considérables grâce à la conception d'induits bien symétriques et de grand rendement. La contamination par le creuset des matériaux tels que Nb, La, Mo, Cu, Al, Si et Ge n'est guère décelable (D. A. Hukin).

Le tirage Czochralski des composés III–V et II–VI a été réalisé sous pression excédentaire du composé volatil grâce à une source secondaire de ce composé, l'équilibrage de la pression se faisant par la mesure de la pression que transmet un gaz inerte vers l'extérieur (J. B. Mullin).

La technologie des conduits de chaleur (heat pipes) venant du domaine de l'énergie nucléaire est introduite maintenant dans les laboratoires de croissance cristalline. Leur conductibilité thermique très élevée, dépassant de quelques ordres de grandeur celle du cuivre, leur pouvoir de volant thermique, permettent des applications insoupçonnées; réalisation de régions sans gradient thermique, de gradients très abrupts. Les formes des conduits thermiques peuvent être facilement adaptées ainsi que leur domaine de température de fonctionnement (J. Steininger).

Indubitablement le clou du savoir faire en matière de cristallogenèse a été présenté par une équipe des Tyco Laboratories. Le principe du tirage des monocristaux de forme bien déterminée à travers des buses se rapproche de la technique des verriers et remonte en ce qui concerne les cristaux à Stepanov (1959) et Gol'tsman (1962). Dans la technique présentée à I.C.C.G.-3 par H. E. LaBelle Jr. la forme du monocristal n'est pas déterminée par l'orifice de la buse comme cela a été le cas auparavant. Ici l'orifice sert seulement à amener par capillarité le bain liquide à la surface supérieure de la buse sur laquelle il s'étend. Un germe mis en contact avec cette couche liquide arrive à tirer un monocristal dont la section est définie précisément par la section pleine de la buse. Ces auteurs ont donc joué d'une manière subtile avec les forces capillaires. Des cen-

taines de mètres de cristaux réfractaires à forme parfaitement définie et souvent sophistiquée, peuvent être tirés à des vitesses atteignant plusieurs centimètres à la seconde. L'assistance a été subjuguée par la technicité apparemment simple. Il n'y a pas de doute, une telle performance aura des retentissements considérables dans l'industrie et des retombées certaines dans les laboratoires de recherche sur la croissance cristalline.

Le domaine est en pleine marche et des efforts sont faits pour attirer de nouveaux adeptes. Ainsi les deux semaines précédant I.C.C.G.-3 ont été consacrées à une Ecole d'Eté de croissance cristalline organisée par P. Hartman à Leyden. L'Union Internationale de cristallographie et I.C.C.G.-3 ont soutenus cette entreprise qui réunissait 120 participants autour d'une brochette de jeunes professeurs.

En résumé, on peut dire que I.C.C.G.-3 bien que n'ayant pas mis au jour la grande découverte, a montré que le domaine de la croissance cristalline est plus actif que jamais. I.C.C.G.-4 à Tokyo, sous la présidence de R. R. Hasiguti nous montrera si cette ardeur peut se maintenir.

R. KERN
Marseille, le 16 juillet 1971

CONTENTS

Section I

Invited papers

I. N. STRANSKI

F. C. FRANK

K. NASSAU

T. SUREK

R. A. LAUDISE

A. AUTHIER

D. T. J. HURLE

F. L. BINSBERGEN

S. OGAWA

S. INO

K. T. AUST

J. CORISH

C. D. O'BRIAIN

M. E. GLICKSMAN

R. J. SCHAEFER

J. A. BLODGETT

C. L. VOLD

R. ROY

W. B. WHITE

B. CHALMERS

H. E. LaBELLE, Jr.

A. I. MLAVSKY

R. C. DeVRIES

J. F. FLEISCHER

D. BUDEVSKI

M. L. TURPIN

J. STEININGER

T. B. REED

M. AVIGNON

Journal of Crystal Growth **13/14** (1972) 3–8 © *North-Holland Publishing Co.*

EINE PLAUDEREI ÜBER GLEICHGEWICHTSFORMEN VON KRISTALLEN

I. N. STRANSKI

Fritz-Haber-Institut der Max-Planck-Gesellschaft, 1 Berlin 33, Germany

A review of the determination of the equilibrium form of very small and infinite size crystals is given, together with calculations of the free energy of formation of a nucleus using the molecular work of separation, the work of separation of a crystal unit from the half-crystal position and the mean work of separation. Various examples are discussed.

Mein Beitrag zur heute beginnenden Tagung sollte eine Plauderei sein. Da ich die Freiheit der Wahl zu haben glaube, so ist es nicht schwer zu erraten, daß es sich um Gleichgewichtsformen von Kristallen handeln wird. Dies ist nicht allein durch Vorliebe sondern auch sachlich bedingt, da die Gleichgewichtsform tatsächlich einen bequemen Schlüssel zum Verständnis und zur Einordnung der Vorgänge der Kristallkeimbildung und des Kristallwachstums darstellt.

Bekanntlich ist die Ableitung der Gleichgewichtsform von Kristallen in Analogie zu der von Flüssigkeitströpfchen entstanden. Sie geht auf Thomson und Gibbs[1]) zurück. Hierher gehören noch die Namen von Curie[2]), Wulff[3]), von Laue[4]), Dinghas[5]) und insbesondere der von Volmer[6]). Letzterem verdanken wir folgende instruktive Formulierung der Gibbs–Thomson'schen Gleichung:

$$\frac{kT}{2V_\mathrm{m}} \ln \frac{p_h}{p_\infty} = \frac{\sigma_1}{h_1} = \frac{\sigma_2}{h_2} = \cdots . \tag{1}$$

(V_m = Molekülvolumen der festen Phase, p_h = aktueller Dampfdruck, p_∞ = Gleichgewichtsdampfdruck des unendlich großen Kristalls, $\sigma_1, \sigma_2, \ldots$ = spez. Freie Oberflächenenergien der Flächen $1, 2, \ldots, h_1, h_2, \ldots$ = Zentraldistanzen der Flächen.) Der Zusammenhang mit der bekannten Wulff'schen Konstruktion der Gleichgewichtsform ergibt sich direkt aus der Gleichung (1), denn sie gibt für jeden beliebigen Wert p_h die zugehörigen Zentraldistanzen an. Daraus folgt auch, daß nicht alle Flächen zur Gleichgewichtsform gehören müssen, sondern nur diejenigen, die die kleinste geschlossene Form begrenzen. Sie entspricht der Gibb'schen Forderung, daß bei konstantem Volumen die Summe der Freien Oberflächenenergien ein Minimum darstellen muß.

Im Prinzip ist damit das Problem der Gleichgewichtsform gelöst, wenn man von den Ungenauigkeiten absieht, die das Vorhandensein von Kanten und Ecken bedingt. Hierbei seien die Berechnungen der Oberflächen-, Kanten- und Eckenenergien des NaCl-Kristalls von Born und Stern[7]) erwähnt.

Ein gewisses Unbehagen bleibt hierbei zurück, weil die Größen spez. Freie Oberflächenenergien (σ_i) bzw. Zentraldistanzen (h_i) nur in einem indirekten Zusammenhang mit den molekularen Vorgängen des Wachstums und der Auflösung stehen. Dies war die eigentliche Veranlassung für Kossel[8]) und auch für mich[9]), die Abtrennungsarbeiten einzelner Kristallbausteine einzuführen. Hierbei spielten die damals aktuellen Berechnungen von Madelung[10]) für NaCl-Gitter auch eine wesentliche Rolle, und so war auch hier das erste Anwendungsobjekt das NaCl-Gitter. Als ein wesentliches Ergebnis stellte sich die sogenannte Halbkristallage (bei Kossel wiederholbarer Schritt) heraus. Sie ist dadurch gekennzeichnet, daß der Baustein in dieser Lage halb so fest gebunden ist wie im Kristallinneren. Die Wahrscheinlichkeiten für eine Anlagerung oder Abtrennung sind bei einem unendlich großen Kristall, der im Gleichgewicht mit seiner Umgebung steht, gleich groß. Wir werden auch hier die Abtrennungsarbeit von dieser Lage mit $\varphi_{\frac{1}{2}}$ bezeichnen.

Mich interessierte insbesondere die Frage, ob sich

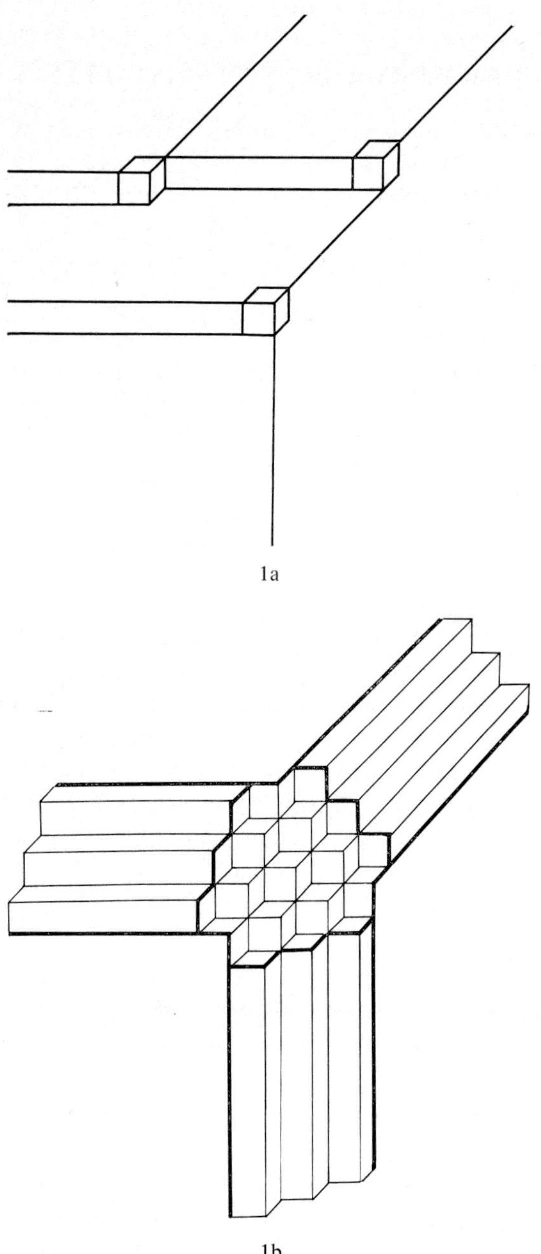

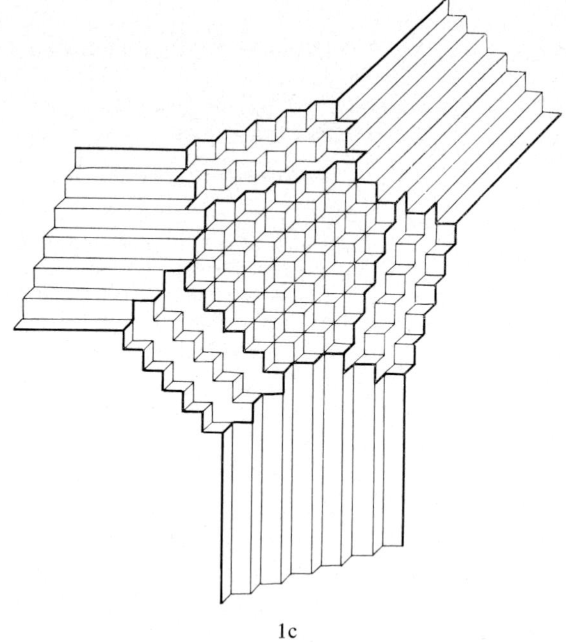

1a

1b

1c

Abb. 1. Gleichgewichtsform eines unendlich großen Kristalls mit einfach kubischem Gitter, wenn nur erstnächste (a), erst- und zweitnächste (b) und erst- bis drittnächste Nachbarn (c) wirksam sind.

hiermit die Gleichgewichtsform des unendlich großen Kristalls direkt angeben läßt. Dazu war es nötig, die Abtrennungsarbeit des Eckbausteins des NaCl-Würfels zu kennen, denn dieser Baustein ist der am schwächsten gebundene innerhalb der gesamten Würfeloberfläche. Es zeigte sich, daß sie beträchtlich größer als $\varphi_{\frac{1}{2}}$ ist. Daraus wurde geschlossen, daß die Würfelecke in diesem Fall atomar scharf ist und die Gleichgewichtsform nur aus {001}-Flächen besteht. Ähnliche

Ergebnisse fand man auch bei anderen Ionenkristallen [vgl. auch Honigmann[11]) und Hartman[12])].

Als ein bequemes Modell für viele Spekulationen ergab sich der von Kossel[8]) eingeführte Kristall mit einfach kubischem Gitter, bei dem nur nichtpolare Kräfte mit begrenzter Reichweite zwischen den Bausteinen wirken. Ich benutzte dieses Gitter, um den Zusammenhang zwischen der unendlichen Gleichgewichtsform und der Reichweite der Gitterkräfte in Kristallen mit nichtpolaren Kräften zu zeigen[13]). Es stellte sich heraus, daß die Gleichgewichtsform nur aus Würfelflächen besteht, solange die Reichweite der Kräfte auf erstnächste Nachbarn beschränkt bleibt. Im Gegensatz zum NaCl-Würfel sind hier die Eckbausteine gerade noch mit $\varphi_{\frac{1}{2}}$ gebunden. Sind daneben auch die Kräfte zwischen zweitnächsten Nachbarn wirksam, so ist das nicht mehr der Fall, und es erscheinen zusätzlich die {111} und die {011} Flächen. Wirken auch drittnächste Nachbarn, so kommt noch die {112} Fläche hinzu (vgl. Abb. 1). Von Kaischew, Krastanow und mir[14]) konnte auch als bei vielen anderen Gittertypen festgestellt werden, daß die Zahl der Gleichgewichtsformflächen mit der Reichweite der Kräfte zunimmt.

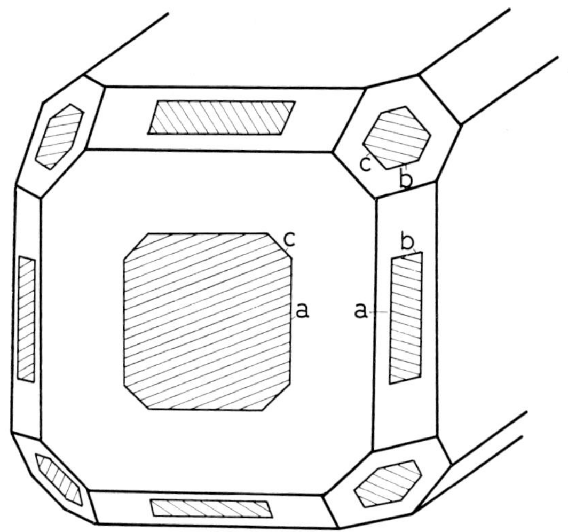

Abb. 2. Schematische Darstellung der zweidimensionalen Keime (schraffierte Flächen) auf den Flächen {001}, {011} und {111} des einfach kubischen Gitters bei Wirksamkeit von erst- und zweitnächsten Nachbarn. Bei kleiner Übersättigung hat der Keim auf {001} die Ränder a und c, der auf {011} die Ränder a und b, und der auf {111} die Ränder b und c. Mit zunehmender Übersättigung verschwinden alle Ränder außer a. Damit ist der zweidimensionale Keim auf {011} eindimensional und der auf {111} nulldimensional geworden, d.h. es bleiben nur Würfelecken und -kanten übrig.

Hierbei sei erwähnt, daß Straumanis[15]) als erster unsere Ergebnisse auf Wachstumsformen von Metallen übertrug und bestätigt fand.

Freilich waren diese Überlegungen rein qualitativ, denn so konnte nur die Art der Flächen, nicht aber ihre relativen Ausdehnungen, angegeben werden. Um weiterzukommen, mußte zunächst die Funktion gefunden werden, die bei endlichen Kristallen die Rolle von $\varphi_{\frac{1}{2}}$ übernimmt, um bei sehr großen Kristallen mit $\varphi_{\frac{1}{2}}$ identisch zu werden. Um dies zu klären, begannen Kaischew und ich[16]) mit einer systematischen Diskussion des Gleichgewichts kleiner Kriställchen. Der Kürze halber beschränken wir uns hier auf das einfach kubische Gitter. Auf die sehr aufschlußreichen Ergebnisse beim NaCl sei hier nur hingewiesen[16]).

Im Gleichgewicht muß an jeder Oberfläche eines Kriställchens die Wahrscheinlichkeit für die Entstehung einer ganzen Netzebene (durch Kondensation) ebenso groß sein wie die für das Verschwinden der ganzen Netzebene (durch Verdampfung). Beide Vorgänge gehen über die Bildung des gleichen zweidimensionalen Keims, so daß dessen Entstehung auf beiden Wegen ebenfalls gleich wahrscheinlich sein muß.

Gibbs[1]) hatte bereits die Notwendigkeit der Annah-

me eines solchen Keims an den Kristalloberflächen angedeutet, doch erst Volmer[6]) und Mitarbeiter definierten ihn in Analogie zum dreidimensionalen Keim mit Hilfe der Freien Randenergien (ρ_i) und der Randlängen (l_i) ohne aber daraus weitere Konsequenzen zu ziehen. Bei den zweidimensionalen Keimen muß in gleicher Weise die Wahrscheinlichkeit für die Entstehung einer Randreihe ebenso groß sein wie die für ihr Verschwinden. Beide Vorgänge gehen diesmal über eindimensionale Keime, und die Wahrscheinlichkeit ihrer Bildung bestimmt auch die Wahrscheinlichkeit ihrer Anlagerung oder Abtrennung. Das Resultat ist sehr einfach und gilt sowohl für den zwei- als auch für den dreidimensionalen Keim. Die von uns gesuchte Größe, die an Stelle von $\varphi_{\frac{1}{2}}$ treten muß, ist $\bar{\varphi}_h$. Es ist der Mittelwert für die Abtrennarbeiten der einzelnen Bausteine beim schrittweisen Abtragen einer ganzen Randreihe des zweidimensionalen Keims bzw. einer ganzen Netzebene beim dreidimensionalen Keim. Er muß für alle Flächen und Ränder der Gleichgewichtsform den gleichen Wert haben und der folgenden modifizierten Gibbs–Thomson'schen Gleichung genügen:

$$\bar{\varphi}_h = \varphi_{\frac{1}{2}} - kT \ln(p_h/p_\infty). \qquad (2)$$

Zu der gleichen Beziehung kommt man auch mit Hilfe einer thermodynamischen Ableitung, indem man die Freie Enthalpie des Keims nach der Zahl der Bausteine differenziert [vgl. Lacmann[17]) und Volmer[6])].

Freilich hat Gleichung (2) ihrerseits nur eine indirekte, wenn auch aufschlußreiche, Beziehung zur Zentraldistanz (h_i) der Gleichung (1). Dafür ermöglicht sie uns, viele neue und ergänzende Schlußfolgerungen zu ziehen. Mit steigender Übersättigung wird $\bar{\varphi}_h$ für die Randreihen kleiner und damit werden auch die Ränder kürzer. Sind mehrere Randarten am zweidimensionalen Keim vorhanden, so wird bei einem Teil der Ränder ihre Länge so kurz, daß aus zwei Ecken eine neue Ecke entsteht. Dies tritt bei der Übersättigung ein, bei der $\bar{\varphi}_h$ die Abtrennarbeit des neuen Eckbausteins unterschreitet. Analog wird die Ausdehnung der Flächen des dreidimensionalen Keims kleiner (Abb. 2). Einzelne Flächen schrumpfen daher auf eine Ecke oder eine Kante zusammen. Dies tritt bei der Übersättigung ein, bei der $\bar{\varphi}_h$ kleiner wird als die Abtrennarbeit des Eckbausteins bzw. die mittlere Abtrennarbeit der Kantenreihe. Während mit der Zunahme der Reichweite der Kräfte die Gleichgewichtsform, wie wir gesehen haben, flächenreicher wird, nimmt mit steigender Übersättigung die Zahl der Gleichgewichtsformflächen ab. $\bar{\varphi}_h$ ist somit das Werkzeug, das uns die Konstruktion der endlichen Gleichgewichtsform erlaubt. Man geht von einer beliebigen Form aus, entfernt von dieser alle Bausteine, deren Bindungsenergie weniger als $\bar{\varphi}_h$ beträgt, und fügt an alle Plätze mit $\varphi_i \geq \bar{\varphi}_h$ Bausteine an. Dann variiert man alle Flächeninhalte so lange, bis sie den gewünschten $\bar{\varphi}_h$-Wert besitzen. Damit ist der Kristallkeim nach Form und Größe bestimmt. Es läßt sich für jede Übersättigung ein dreidimensionaler Keim, auf dessen Flächen zweidimensionale Keime entstehen können, angeben, wobei zwei- und dreidimensionale Keime gleichseitig im Gleichgewicht mit der Mutterphase stehen.

Mit steigender Übersättigung kann der zweidimensionale Keim so klein werden, daß er schließlich zu einem nulldimensionalen entartet, wenn $\bar{\varphi}_h$ kleiner wird als die Abtrennarbeit eines einzelnen Bausteins von dieser Fläche. Dies tritt bei der gleichen Übersättigung ein, bei der die Fläche, auf der sich der zweidimensionale Keim befindet, zu einer Ecke zusammenschrumpft. Auf den Flächen, die mit steigender Übersättigung zu einer Kante zusammenschrumpfen, entartet der zweidimensionale Keim – bei der gleichen Übersättigung – zu einem eindimensionalen (Abb. 2).

Daraus wurde der Schluß gezogen, daß nur die bei der jeweiligen Übersättigung zur Gleichgewichtsform gehörenden Flächen über zweidimensionale Keime und damit Netzebene für Netzebene wachsen können. Auch an einem großen wachsenden Kristall können nur die Flächen über zweidimensionale Keime wachsen, die dies bei der herrschenden Übersättigung am (kleinen) im Gleichgewicht befindlichen Keim tun.

Die zweidimensionale Keimbildung bewirkt eine wesentliche Verlangsamung der Verschiebungsgeschwindigkeit der Fläche im Vergleich zu einer Fläche, deren Wachstum nicht über zweidimensionale Keime verläuft. Aus diesem Grunde können in der Wachstumsform eines Kristalls nur Flächen der Gleichgewichtsform vertreten sein. Da die zweidimensionale Keimbildungsarbeit von Fläche zu Fläche verschieden ist, und die Keimbildungshäufigkeit exponentiell von ihr abhängt, unterscheiden sich die Wachstumsgeschwindigkeiten der einzelnen Flächen sehr stark. Daher müssen nicht alle Flächen der Gleichgewichtsform in der Wachstumsform sichtbar auftreten.

An gerundeten (metallischen) Kristallen erscheinen beim Wachstumsversuch nur solche Flächen, die auch die zugehörigen Gleichgewichtsformen aufweisen, d.h. bei größerer Übersättigung nimmt deren Zahl ab. Hier sei nur auf einige Versuchsreihen und deren Ergebnisse hingewiesen[18], insbesondere auch auf Wachstumsversuche dicht unterhalb des Schmelzpunkts, bei denen interessante Abweichungen beobachtet und gedeutet werden konnten[19].

Die Beziehung (2) liefert uns mit dem $\bar{\varphi}_h$-Wert auch die Möglichkeit, alle denkbaren Keimbildungsarbeiten anzugeben. Sie läßt sich folgendermaßen zusammenfassen:

$$A_k = N\bar{\varphi}_h - \sum_1^N \varphi_i . \qquad (3)$$

Hierin bedeutet N = Zahl der Bausteine des Keims, $\sum_1^N \varphi_i$ = Summe der Energien, die beim schrittweisen Aufbau des Keims gewonnen werden.

Kaischew und ich[20] benutzten das gleiche einfache Kristallmodell unter Berücksichtigung nur erstnächster Nachbarn, um eine Beziehung für die Kristallkeimbildungshäufigkeit zu finden. Das Ergebnis war, wie sich später zeigte, nur qualitativ richtig, da es noch Integrationskonstanten enthielt. Dadurch wurden aber Becker und Döring[21] angeregt, sich des Problems unter Benutzung des gleichen Modells anzunehmen und es eleganter zu lösen. Ihnen folgten noch weitere Arbeiten, so z.B. von Frenkel[22], Kaischew[23], Lothe und Pound [vgl. Hirth und Pound[24]] und von Dunning[25].

Als interessante Anwendungsbeispiele für unsere Methode der mittleren Abtrennungsarbeiten auf die heterogene Keimbildung sei auf einige Arbeiten von Kaischew und Mitarbeiter[26], Mutaftschiev[27] und Lacmann[17,28] hingewiesen. Diese Arbeiten bestätigen die Gleichung (2) für Keime auf glatten Unterlagen, in Hohlkanten, Hohlecken, an Stufen und für solche, die sich auf Unterlagen befinden, die kleiner sind als der Keim selbst.

Erwähnenswert wäre auch noch die Änderung von Kristalltrachten als Folge einer (reversiblen) Adsorption. Solche Vorgänge sind meist ziemlich verwickelt, insbesondere wenn die Adsorption ganz oder zum Teil irreversibel verläuft. Auch hier kann ein Modell besonders instruktive Dienste leisten, um die Grundvorgänge herauszuschälen[29,11]. Wegen der Kürze der Zeit beschränke ich mich darauf, die wichtigsten Veröffentlichungen anzuführen und hier nur die Hauptergebnisse anzugeben[30,31]):

Die Kristallflächen, die keine Adsorption aufweisen, werden trotzdem von ihr beeinflußt. An den Rändern der zweidimensionalen Keime erfolgt nämlich eine Adsorption und damit eine Erniedrigung der Randenergie, was eine Erniedrigung der zweidimensionalen Keimbildungsarbeit zur Folge hat. Demgegenüber wird die Randenergie und damit die zweidimensionale Keimbildungsarbeit auf adsorbierenden Flächen erhöht, weil hier nicht adsorbierende Ränder auftreten. Insgesamt werden die adsorbierenden Flächen gleichzeitig in der Gleichgewichts- und Wachstumsform begünstigt. Es läßt sich auch hier zeigen, daß für die Flächen und für die Ränder der richtige $\bar{\varphi}_h$-Wert herauskommt[32,33,17]).

Prinzipiell ist es auch möglich, in ähnlicher Weise neben den eigentlichen Abtrennungsarbeiten entsprechende Größen für andere Einflüsse (z.B. Schwingungsentropie, Oberflächenlöcher und arteigene Adatome) zu berücksichtigen.

Hier möchte ich noch die experimentelle Seite der Beobachtung von Gleichgewichtsformen streifen. Im Normalfall haben sie nur eine kurze Verweilzeit und erreichen während dieser Zeit nur submikroskopische Größe. Es sei denn, ihre Mutterphase ist räumlich stark begrenzt. Eine Möglichkeit dazu hatte man schon früher bei den sogenannten negativen Kristallen erreicht. Das Verdienst, entsprechende Bedingungen für konvexe Formen geschaffen zu haben, gebührt Lemmlein[34]) und Klija[35]). Diese Methode hat eine erstaunliche Vervollkommnung gefunden, hier in Marseille, bei den Herren Bienfait und Kern[36]). Vielleicht werden wir einige Exemplare davon zu Gesicht bekommen. Hierbei möchte ich auch die Arbeiten von Drechsler und Mitarbeitern[37]) erwähnen.

Bisher war nur von ideal gebauten Kristallen die Rede. Daß Baufehler, insbesondere Versetzungen, eine entscheidende Rolle beim Wachstum spielen, wissen wir seit den grundlegenden Untersuchungen von F. C. Frank und Mitarbeitern[38]). Ich kann mich nur mit einem kurzen Hinweis begnügen, da ich dessen sicher bin, daß hierüber von berufener Seite berichtet wird. Eigentlich liefern gerade die Versetzungen eine entscheidende Unterstützung der These des zweidimensionalen Keimbildungsmechanismus bei Gleichgewichtsformflächen ideal gebauter Kristalle. Denn letztere liefern die Vorbedingung dafür, daß Versetzungen sich an diesen auswirken können. Hierbei möchte ich nicht unterlassen, den hervorragenden Film von Kaischew und Budewski[39]) zu erwähnen, der elektrolytisch geleitete Wachstumsvorgänge zeigt. Hierher gehören natürlich auch die interessanten Arbeiten von Bethge und Mitarbeitern[40]), die es ermöglichten, die Struktur der Wachstums- und Abdampfspiralen mit monomolekularen Stufenhöhen zu untersuchen.

Literatur

1) J. W. Gibbs, Trans. Conn. Acad. **3** (1876/78) 343; *Scientific Papers*, Vol. 1 (Dover Publ., New York, 1961).
2) P. Curie, Bull. Soc. Mineral France **8** (1885) 145; *Oeuvres*, S. 153.
3) G. Wulff, Z. Krist. **34** (1901) 449.
4) M. von Laue, Z. Krist. **105** (1944) 124.
5) A. Dinghas, Z. Krist. **105** (1943/44) 304.
6) M. Volmer, *Kinetik der Phasenbildung* (Steinkopff Verlag, Dresden und Leipzig, 1939).
7) M. Born und O. Stern, Sitzber. Preuß. Akad. Wiss., Physik.-Math. Kl. **48** (1919) 901;
Vgl. auch I. N. Stranski, Sitz. Ber. Wiener Akad. **8** IIb (1936) 840.
8) W. Kossel, Die molekularen Vorgänge beim Kristallwachstum, in Quantentheorie und Chemie, Leipziger Vorträge, Leipzig 1928, S. 1; Nachr. Ges. Wiss. Göttingen, Math.-Physik. Kl. (1927) 135.
9) I. N. Stranski, Z. Physik. Chem. **136** (1928) 259.
10) E. Madelung, Physik. Z. **19** (1918) 324; **20** (1919) 494.
11) B. Honigmann, *Gleichgewichts- und Wachstumsformen von Kristallen* (Steinkopff Verlag, Darmstadt, 1958).
12) P. Hartman, Dieser Kongress.
13) I. N. Stranski, Z. Physik. Chem. B **11** (1931) 342.
14) I. N. Stranski und R. Kaischew, Z. Krist. **78** (1931) 373;
I. N. Stranski und L. Krastanow, Z. Krist. **83** (1932) 155;
I. N. Stranski, R. Kaischew und L. Krastanow, Z. Krist. **88** (1934) 325.
15) M. Straumanis, Z. Physik. Chem. B **13** (1931) 316; **19** (1932) 63; **26** (1934) 246; **30** (1935) 132.
16) I. N. Stranski und R. Kaischew, Z. Physik. Chem. B **26** (1934) 100, 114, 312; Ann. Physik [5] **23** (1935) 330; Physik. Z. **36** (1935) 393;
O. Knacke und I. N. Stranski, Ergebn. Exakt. Naturwiss. **26** (1952) 383.
17) R. Lacmann, Springer Tracts in Modern Physics (Ergebn. Exakt. Naturwiss.) **44** (1968) 1.
18) R. Kaischew, L. Keremidtschiew und I. N. Stranski, Z. Metallk. **34** (1942) 201;
A. Eisenlöffel und I. N. Stranski, Z. Metallk. **41** (1950) 10;
B. Mutaftschiev, Phys. Status Solidi **8** (1965) 223;
R. Kaischew und C. Nanev, Phys. Status Solidi **10** (1965) 779;
St. Budurov und N. Stojcev, Ann. Univ. Sofia, Fac. Chimie **56** (1961/62) 87; Compt. Rend. Acad. Bulg. Sci. **15** (1963) 529, 397;
St. Budurov, E. Ruseva und N. Stojcev, Compt. Rend. Acad. Bulg. Sci. **16** (1963) 653.
19) I. N. Stranski, W. Gans und H. Rau, Z. Elektrochem. **67** (1963) 965;
H. Heyer, F. Nietruch und I. N. Stranski, Kristall u. Technik **3** (1968) 561.

20) R. Kaischew und I. N. Stranski, Z. Physik. Chem. A **170** (1934) 295; B **26** (1934) 317.

21) R. Becker und W. Döring, Ann. Physik [5] **24** (1935) 719.

22) J. Frenkel, *Kinetic Theory of Liquids* (Clarendon Press, Oxford, 1946); *Kinetische Theorie der Flüssigkeiten* (Akademie Verlag, Berlin, 1957).

23) R. Kaischew, Z. Elektrochem. **61** (1957) 35.

24) J. P. Hirth und G. M. Pound, *Condensation and Evaporation*; *Nucleation and Growth Kinetics* (Pergamon, Oxford, 1963).

25) W. J. Dunning, in: *Chemistry of the Solid State*, Ed. W. E. Garner (Academic Press, New York, 1955) S. 159.

26) R. Kaischew, Bull. Acad. Bulg. Sci., Ser. Phys. **1** (1950) 100.

27) R. Kaischew und B. Mutaftschiev, Ber. Chem. Inst. Bulg. Akad. Wiss. **7** (1959) 177; Ber. Physik.-Chem. Inst. Bulg. Akad. Wiss. **3** (1963) 5.

28) R. Lacmann, Z. Krist. **116** (1961) 13.

29) I. N. Stranski, Bull. Soc. Franc. Mineral. Crist. **79** (1956) 359; O. Knacke und I. N. Stranski, Z. Elektrochem. **60** (1956) 816; R. Lacmann und I. N. Stranski, in: *Growth and Perfection of Crystals*, Eds. R. H. Doremus, B. W. Roberts and D. Turnbull (Wiley, New York, 1958) S. 427.

30) G. Bliznakov, Bull. Acad. Bulg. Sci., Ser. Phys. **3** (1952) 23; Comp. Rend. Acad. Bulg. Sci. **6** (1953) 13; G. Bliznakov und E. Kirkova, Bull. Acad. Bulg. Sci., Ser. Phys. **4** (1954) 153; Z. Physik. Chem. **206** (1957) 271.

31) R. Kern, Compt. Rend. (Paris) **234** (1952) 970, 1379, 1696; **236** (1953) 830; Bull. Soc. Franc. Mineral Crist. **76** (1953) 325, 391; **78** (1955) 461, 497; Nova Acta Leopoldina, Suppl. Nr. 1, **34** (1968) 84; M. Bienfait, R. Boistelle und R. Kern, Compt. Rend. (Paris) **256** (1963) 2189; **258** (1964) 880.

32) R. Kaischew und B. Mutaftschiev, Ber. Chem. Inst. Bulg. Akad. Wiss. **7** (1959) 145.

33) R. Lacmann, Z. Krist. **112** (1959) 169.

34) G. Lemmlein, Ber. Akad. Wiss. UdSSR (N.S.) **98** (1954) 973.

35) M. O. Klija, Ber. Akad. Wiss. UdSSR **100** (1955) 259.

36) M. Bienfait und R. Kern, Bull. Soc. Franc. Minéral. Crist. **87** (1964) 604.

37) M. Drechsler und J. F. Nicholas, J. Phys. Chem. Solids **28** (1967) 2597, 2609; M. Drechsler und M. Müller, J. Crystal Growth **3**, **4** (1968) 518; M. Müller und M. Drechsler, Surface Sci. **13** (1969) 471.

38) W. K. Burton, N. Cabrera und F. C. Frank, Phil. Trans. Roy. Soc. (London) A **243** (1951) 299.

39) R. Kaischew und E. Budewski, Contemp. Phys. **8** (1967) 489; Nova Acta Leopoldina, Suppl. Nr. 1, **34** (1968) 105.

40) H. Bethge und M. Krohn, Colloq. Intern. Centre Natl. Rech. Sci. **152** (1965) 389; H. Bethge, Nova Acta Leopoldina, Suppl. Nr. 1, **34** (1968) 50.

Journal of Crystal Growth **13/14** (1972) 9–11 © *North-Holland Publishing Co.*

THE FUTURE OF CRYSTAL GROWTH

F. C. FRANK

University of Bristol, Tyndall Avenue, Bristol BS8 1TL, England

The twin incentives of scientific progress are use and curiosity. Most of the numerous chariots of science are drawn by a pair of horses of these two breeds. Commonly only one of them is pulling at a time, and the real spurts of progress are made on the comparatively rare occasions when they run in double harness. Any attempt to foresee the future must be made only in the broadest of terms to have a chance of not being upset by a totally unforeseen change of circumstances, which still may happen. So let us consider these two prime influences.

Let us first consider use. Industries of very great money value have crystalline products, nowadays crystallized under scientific control. Examples range from salt or sugar to diamonds. For these single product industries, however, the basic principles are simple and well understood, the complications almost too subtle and complex for science: the problems tend to remain the same for years, and this provides the kind of conditions in which the empirical arts of the cookery book can achieve successes which outstrip scientific understanding. That is not to say that industrial crystallization does not still produce phenomena worth the attention of the curious, and that curiosity cannot still contribute to the solution of industrial problems. Industrial Crystallization has to some extent tended to go its own independent way, and I hope that it will maintain a regular recognized place in future International Conferences on Crystal Growth. Nevertheless it does not and I think will not dominate our scene.

Metallurgy has been the source of much of our knowledge of crystal growth, and the exploitation of that knowledge has a new incentive in the use of controlled eutectic textures to obtain desired mechanical properties. Metal crystal growth has also shown a certain tendency to go its own way, and we should welcome the substantial number of papers on crystallization of metals at this conference as correcting that state of affairs.

I imagine though that, as now is the case, the principal incentives from use for the study of Crystal Growth will continue to come from the Electronics Industry, which became the Solid State Electronics Industry some time ago, and is rapidly becoming the Opto-acoustico-magneto-electronics Solid State Industry. This Industry generates demand for crystals of ever-increasing variety. The reasons are easy to see. One is the Bloch theorem. It is so much easier to calculate in principle the behaviour of any wave, such as an electron wave, with a wavelength comparable with the scale of the microstructure of the material in which it travels, if that material has a crystal lattice than if it is amorphous. The lattice may not be essential to the existence of the properties, but it is almost essential for their understanding. Liquid metals and liquid insulators are just about as genuine metals and genuine insulators as are the crystalline metals and insulators, but it has taken a generation longer than it has for crystals to create a theory which can roughly account for these properties in amorphous substances. For this reason we have had many examples in solid state electronics where theoretical understanding has been in the lead: has foreseen the possible existence of properties capable of exploitation for use. The transistor was one of them. The converse happens as well of course, but when a new phenomenon is observed in crystals we seldom have to wait long for a theoretical explanation. I think that waiting time is bound to remain longer, on average, with amorphous substances.

Besides this, theory can tell us that a particular potential device application is available to us in a material furnishing particular symmetry properties, or particular kinds of anisotropy. Commonly it is only a crystal which is capable of satisfying the requirements. It may be that no known crystal does satisfy them, but theory can tell us what kind of crystal might. So the requirements of the electronics industry continue to place demands on crystal growers for the production of entirely new crystals, of widely different kind. If symmetry determines the choice, the melting point may with about equal likelihood be 300 K or 3000 K. The crystal grower who will serve the electronics industry must be very versatile. He must be prepared to turn his hand to quite new practical problems, and to do that with any chance of success he must start with a full understanding of basic principles. He will probably still need some luck and some cookery art before he can finish the job, but the more he understands the fewer will be the attempts he makes which were bound to be fruitless from the outset.

What is there for the curious in this? We understand a great deal about growth from the melt, and the phenomena implied by the key phrases "constitutional supercooling" and "morphological stability" form the core of a subtle and sophisticated science, but it is comparatively recently that we have recognized the range of our ignorance in hydrodynamics relevant to these processes – difficult hydrodynamics at medium Reynolds numbers, magneto-hydrodynamics and hydrodynamics of rotating systems. This is a field for some very worth-while science. G. I. Taylor showed us most of 50 years ago what elegant concepts can be created to aid thought in this area, what unexpected and what surprisingly beautiful conclusions can be drawn, but there are certainly further phenomena to be understood, and that understanding is needed. This I think is the indicated area in crystal growth for the work of an applied mathematician. What is there for the chemical physicist or the physical chemist? You

appreciate that I am trying to identify areas in which science can be both interesting and useful. I think the time has come for us to achieve more rigour, more accuracy, more detail in our basic understanding of deposition from vapours on to surfaces, and I have sufficient hope that we may still discover something unexpected on the way – if only cleverer ways of treating the problems – that I do not think this need be a dull matter of dotting i's and crossing t's. In any case, the practical importance of thin films and epitaxial deposition asks for perfection of the science of the subject.

In the most basic question of the aggregation of deposited atoms on a crystal surface, I believe we have the right theory in outline, but we seldom know the basic facts well enough for quantitative precision. There are, naturally, various degrees of satisfaction with a theory. I can be satisfied with a theoretical result that a nucleation rate is 10^{-1000} m^{-2} s^{-1} even if a better theory would make it 10^{-800} m^{-2} s^{-1} because both are practically zero though one is 10^{200} times the other. We tolerate discrepancies of many powers of 10 in nuclation theory, often because the basic data hardly allow one to do better. That is why I find the detailed and quantitative work of Kaischew and his co-workers on isolated crystal faces so very valuable.

In matters of more immediate practical application, epitaxial deposition by chemical transport through the vapour, it is time we put ourselves in a position to be sure of the facts. To take an extreme example one author can believe that cadmium sulphide is deposited from cadmium sulphide molecules in the vapour while others believe it is transported as cadmium and sulphur. In other practical cases with four or five elements present the number of possible molecular species present can be very large. The equations of chemical equilibrium used to be too difficult for us in a case such as that. Now the computer can solve us the equations promptly, provided we give it the right equations to solve: and there's the rub. Equilibria may or may not be attained in the vapour, and if not, the rate of attainment of equilibrium may be greatly effected by catalytic surfaces. Metal halides may be sublimed as metal halides through hydrogen, or under what seem to be the same conditions they may grow metal whiskers with production of hydrogen halide. I think we should not tolerate remaining in ignorance about such facts. Multiple passage optical absorption cells can tell us what molecular species are present in any vapour space, even in trace quantities. What is present on the crystal surface may be harder to ascertain, but surely easier when we can be sure of the gas phase.

Now let me turn to pure curiosities, without obvious application: though it is seldom the case that solving a scientific puzzle is not useful in the end. Of these, let me pick the curvilinear fern-like forms of vapour-deposited ice on the window pane. In these days of central heating one does not see them so often, but when I was a boy they were common enough on the bedroom window, either at home or at school, and I suppose everyone has seen them sometime. We ought to be ashamed of ourselves for not having explained them yet. The explanation in principle, so I imagine, is related to the curling growth of zinc sulphide crystals when it initiates in the form of thin a-axis whiskers. Any sufficiently thin crystal is elastically bent by its own surface forces if its structure is polar, so that opposite side faces are non-equivalent. When a thin crystal grows on a glass surface the molecular field of the glass will in general destroy the symmetry, unless the surface normal is parallel to a symmetry axis or mirror plane of the crystal. I suppose this to be part of the phenomenon, and I am sure it is not all of it. The regular twist about the radius in spherulitic growth is another such problem, unless it turns out to be in essence the same one. It is time we solved these puzzles, and we must await the solution before we know whether it has consequences.

Journal of Crystal Growth **13/14** (1972) 12–18 © *North-Holland Publishing Co.*

DR. A. V. L. VERNEUIL: THE MAN AND THE METHOD

K. NASSAU

Bell Telephone Laboratories, Incorporated, Murray Hill, New Jersey 07974, U.S.A.

This account is the first detailed biography of Professor A. V. L. Verneuil (1856–1913) to appear in the 58 years since the death of the well-known discoverer of the flame-fusion process for the synthesis of ruby and sapphire. Many facts, never previously published have been brought together with rarely seen photographs to give fresh insight into the history of an important technology and an interesting scientist.

Until recently only two techniques have been of significance in the synthesis of ruby: growth from the flux and growth from the melt. The forerunner of the flux growth was the multiphase $Al_2O_3 \cdot BaF_2 \cdot KOH \cdot K_2Cr_2O_7$ growth system worked out by Professor E. Frémy, at first with C. Feil, and later with Verneuil. This was the first practical synthesis of clear ruby crystals, but it was found impossible to obtain crystals larger than a few millimeters in size.

While this work was proceeding Verneuil was shown some "Geneva rubies" in 1886 and this initiated his interest in the melt-growth concept. Within a few years he revealed his process in the sealed notes of 1891 and 1892, and in 1902–1904 he published the full details of the flame fusion process, the "Verneuil" process.

The scope of Verneuil's professional work is indicated by some 70 publications, several prizes, and extensive consulting and teaching over a 33 year period. Yet when all else has been said, Verneuil will always be remembered as "the Father of Synthetic Ruby".

1. Introduction

In many processes and products the actual originator is often difficult to locate. Various concepts of different origins are frequently involved, with perhaps many improvements after these concepts have been assembled. Yet in the synthesis of red ruby and blue sapphire Verneuil was clearly "the father".

Others performed significant work on the synthesis of ruby both before and after him. Some had worked as long and some with more acclaim. There is M. Gaudin whose work covered at least thirty-three years from 1837 to 1870; there is E. Frémy who wrote the book "Synthèses du Rubis" in 1891; there is always the unknown originator of the "Geneva" synthetic ruby process used in the period 1886 to about 1905; and many others.

Yet after all this is said, one inevitably returns to Professor A. V. L. Verneuil, Doctor of Science. He was involved in the final stages of Frémy's flux work. Although perhaps not the first to use a flame-fusion type of process for the growth of ruby, he discovered the efficient technique known under his name which has been used for over 75 years essentially without change. Photographs of Verneuil are shown in figs. 1 and 2.

This account gives an outline of Verneuil's life. Further details may be found elsewhere[1]). A detailed bibliographical listing of Verneuil's 19 published items on ruby and sapphire is given in table 1; these will be referred to as (A1) to (A12).

Auguste Victor Louis Verneuil was born in Dunkirk, France, on November 3, 1856, the youngest of three brothers. Verneuil's father learned Daguerrotype photography from Daguerre himself, and the young Auguste became interested in chemistry from helping his father with this process in his Paris studio.

2. Verneuil and Frémy

At the age of 17, Verneuil was accepted as laboratory assistant by Dr. Edmond Frémy (1814–1894), Professor of Chemistry and head of the chemistry laboratory at the Museum of Natural History in Paris. It was Frémy's practice to employ promising young men as laboratory assistants. There seems to have been considerable fellowship in this group, shown by the existence of the "Association of the Students of Frémy". It was this group that published a monthly bulletin [in which appeared an account of Verneuil's funeral, including several obituary speeches[2])].

While working for Frémy at the Museum of Natural

Fig. 1.   A. V. L. Verneuil at the age of 55 (from ref. 2).

Fig. 2.   A. V. L. Verneuil at the age of 19 (from ref. 2).

History, Verneuil also engaged in a series of studies which gained him the Bachelor's degree in 1875, the Master's degree in 1880, and finally the Doctor of Science degree in 1886.

Professor Frémy was a chemist of widely varied interests, among which was the synthesis of ruby. Frémy, with his personal assistant C. Feil, succeeded in the first synthesis of clear red ruby. When Feil died, Verneuil became Frémy's personal assistant in 1876, with a private laboratory of his own. The Frémy-Verneuil ruby work continued for some 16 years, with three joint publications in 1887 (A1), 1888 (A2), and 1890 (A3) (figs. 3 and 4). Frémy summarized his extensive work with Feil and with Verneuil in his 1891 book[3]). Despite much effort, the size of these crystals remained too small to be useful. The final process worked out by Frémy and Verneuil involved the recrystallization of alumina with a small amount of added potassium dichromate, by the use of potassium hydroxide and barium fluoride at about 1500 °C. The diffusion of humid air through the porous wall of the crucible was found to be an essential part of the process. Crucibles as large as 50 liter (13 gallon) were used, but crystal size remained limited: a 12 liter run might yield 24000 crystals averaging 0.05 g each. A photograph of some of these crystals and a view of a storage cabinet in Frémy's laboratory are given in figs. 3 and 4.

The mechanism can be assumed to involve the reaction of alumina with the barium fluoride to give gaseous aluminum fluoride, which then reacts with moist air to re-form alumina. Since all of this occurs in a multiphase medium, the resultant vapor-phase nucleation and growth occurs in many small local cavities scattered throughout the porous mass filling the crucible thus producing only small crystals. The growth technique would not be classed as flux-growth today, although it is often referred to in this way.

Frémy thought most highly of Verneuil. In his book he speaks of Verneuil's ardor, talent, perserverance, and remarkable observational talents. He states that he

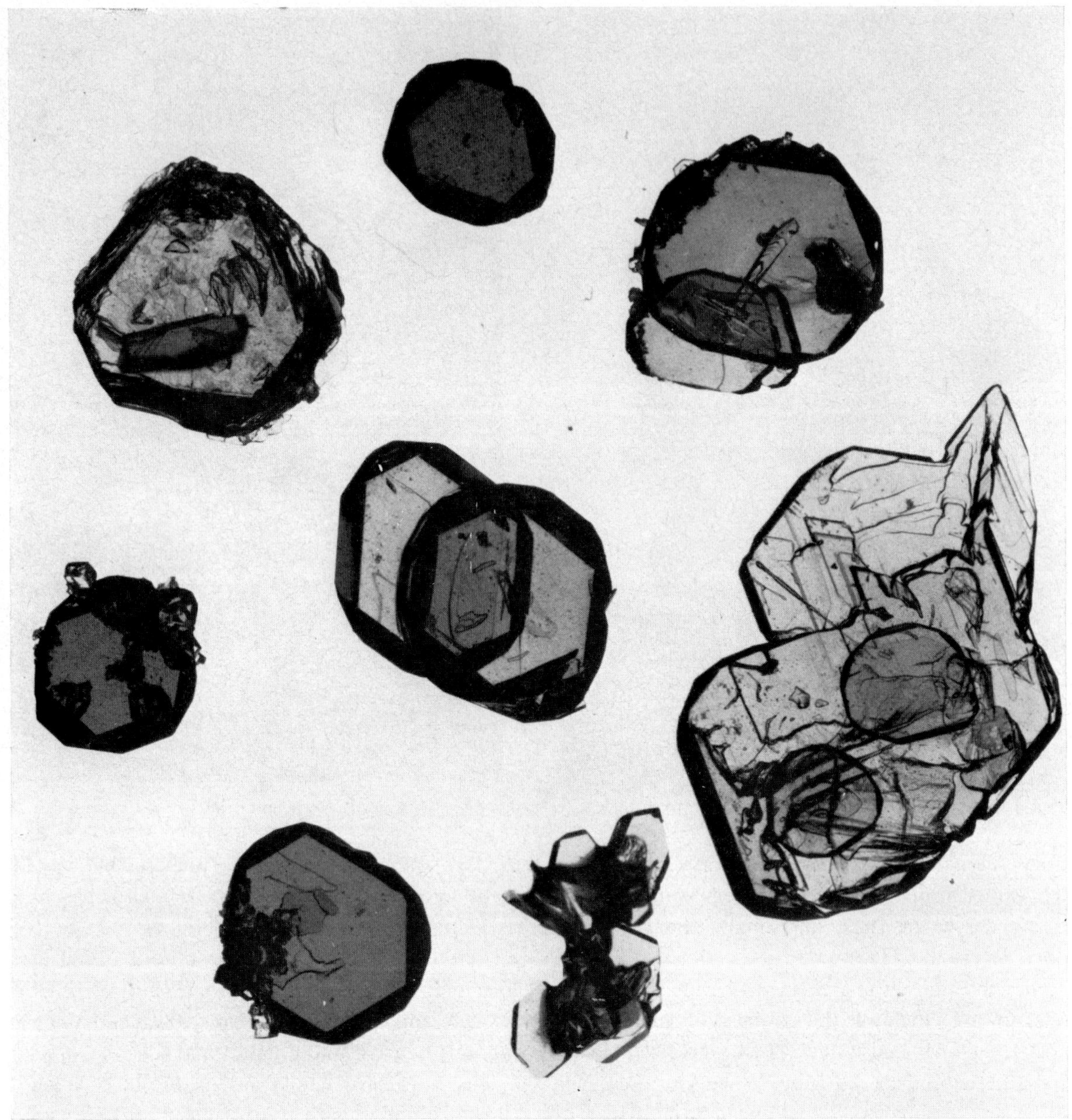

Fig. 3.   Ruby crystals grown by Frémy and Verneuil (largest is 2 mm in diameter).

had been "truly fortunate to find a co-worker such as Mr. Verneuil".

In 1892 Frémy retired and his laboratory disbanded. Verneuil was now appointed to the chair of Applied Chemistry at the Museum of Natural History in Paris, where he remained for 13 years. He also began his teaching career which was to continue until his death.

Besides the ruby work, Verneuil also published on minerals, phosphorescence, charcoal, rubber, the chemistry of selenium and the rare earths, as well as working on industrial consulting projects. If we take a single point in time, the year 1887 for example, we find Verneuil engaged in an astounding number of projects:

(i) work with Frémy on ruby;
(ii) work on the flame fusion growth of ruby;
(iii) work on the phosphorescence of zinc blende;

Fig. 4.   Storage Cabinet in the laboratory of Frémy and Verneuil (from ref. 3).

Fig. 5. Prof. Verneuil (left) and Mr. Spec-Torsky in front of a flame fusion apparatus, about 1910 (from ref. 2).

Fig. 6. View of the laboratory in Paris about 1910 (after an old post-card, courtesy of Mr. H. W. Friedland).

(iv) work on glycerine vor Clolus, Viandey, Linget & Co.;

(v) work on glass for Feil, Mantois, Parra Mantois & Co.;

(vi) teaching at the Polytechnic Association.

3. The Flame Fusion Technique

We know now that the first usable synthetic gemstone was the "Geneva Ruby", sometimes mislabeled "reconstructed" ruby, first manufactured in 1886 and marketed in small quantities until about 1905[4,5]). Since the claim that the "Geneva" product was made by the "reconstruction" of chips of natural ruby without destroying their identity was accepted by many at the time, these rubies were not recognized to be man-made. So that when Verneuil announced his flame fusion synthesis of ruby in 1902, an excited world hailed this as the remarkable achievement it was.

In his 1902 to 1904 publications (A5, A6, and A7), Verneuil gave a report of the production of ruby by a fusion process utilizing a blow-torch. The process is sufficiently well known to make a description unnecessary.

It might however be mentioned that his largest ruby boules were 5–6 mm in diameter and weighed 2–3 g. The origin of his contact with the fusion concept is, however, not generally known. In 1886 Jannettaz examined some "Geneva" rubies and decided they had probably been made by fusion. He reports[6]) that he discussed this with Verneuil who immediately melted some Al_2O_3 in a blow-torch for comparison. This chance contact set the direction for the rest of Verneuil's life. He solved the problems within six years, revealing the essential features of his process in the sealed notes of 1891 and 1892 (A4).

In 1905, at the age of 49, Verneuil was appointed Professor at the National Conservatory of Arts and Sciences in Paris. Much of his time was now taken up with teaching.

In 1909 Verneuil added to his various activities by becoming chief chemist to the firm of L. Heller and Son, of New York and Paris, to investigate the production of blue sapphire. He discovered (A8, A11, and A12) that it was necessary to add about 1.5 % iron oxide and 0.5 % titanium oxide or their equivalent to the alumina feed powder. Views of Verneuil in the Heller laboratory are shown in figs. 5 and 6.

4. The last years

Throughout his life Verneuil had been interested in the arts as well as in the sciences. He had a profound interest in music, playing the piano, in painting, and in the plastic arts, particularly antique pottery. One of Verneuil's last achievements, published only in abstract form, was the successful duplication[7]) of the iron-containing black glaze, showing green reflections, which had been used on old Greco-Italian pottery.

With his many activities of teaching, research, and consulting, exhaustion set in, complicated by diabetes. For several months he continued to serve as consultant from the couch in his living room. He died on April 13, 1913, at the age of 57, in the arms of his brother Ernest.

During his 40 years of professional work, Verneuil's name appeared on some 70 publications and he received several awards. Particularly if one makes allowance for much unpublished work performed for industrial concerns, a body of achievements of most respectable proportion is evident. In addition we must not forget his teaching, at which he was excellent; he inspired both respect and affection in his students. Yet when all else has been said, Prof. Verneuil will always be remembered as *"The Father of Synthetic Ruby"*.

Acknowledgments

I am particularly grateful to Mr. A Verneuil, Professor Verneuil's nephew, of Marseille, France, for much unpublished information; for information, photographs, etc. to Mr. E. S. Heller, Mr. G. Heller, Mr. J. Heller, Mr. H. W. Friedland and the many others acknowledged in ref. 4; and to my wife Julia without whose assistance this study could not have been completed.

TABLE 1

Bibliography of publications
by A. V. L. Verneuil on ruby and sapphire

I. *The Frémy–Verneuil work*:
- (A1) Action des fluorures sur l'alumina; E. Frémy and A. Verneuil,
  - (a) Paris Acad. Sci., Comptes Rendus **104** (1887) 738–740.
  - (b) J. Pharm. **15** (1887) 401–403.
- (A2) Production artificielle des Cristraux de rubis rhomboédriques; E. Frémy and A. Verneuil, Paris Acad. Sci., Comptes Rendus **106** (1888) 565–567.
- (A3) Nouvelles recherches sur la synthése du rubis; E. Frémy and A. Verneuil, Paris Acad. Sci., Comptes Rendus **111** (1890) 667–669.

II. *The flame fusion discovery*:
- (A4) Plis Cachetés; Sur un nouveau procédé de fusion et d'affinage de l'alumine chromée et la production d'une matiere possedant la composition, la dureté et la densité du rubis, No. 4752, 23 décembre 1891 et No. 4849, 19 décembre 1892, ouverts 11 juillet 1910; A. Verneuil, Paris Acad. Sci., Comptes Rendus **151** (1910) 131–132.
- (A5) Production Artificielle du rubis par fusion; A. Verneuil,
  - (a) Paris Acad. Sci., Comptes Rendus **135** (1902) 791–794;
  - (b) Rev. Ind. **33** (1902) 469–470 (with minor changes);
  - (c) Cosmos **936** (1903) 11–12 (with minor changes).
- (A6) Reproduction artificielle du rubis par fusion; A. Verneuil,
  - (a) La Nature **32** (1650) (1904) 177–178;
  - (b) Rev. Ind. **35** (1904) 448–449 (with minor changes);
  - (c) Scientific American Supplement **1535** (1905) 24594 translated into English under the title "The Artificial Production of Rubies", but without the author's name!).
- (A7) Mémoire sur la reproduction du rubis par fusion; A. Verneuil, Ann. Chim. Phys. [8] **3** (1904) 20–48.

III. *The blue sapphire investigations*:
- (A8) Observation sur une Note de M. L. Paris, sur la reproduction de la coloration blue du sapphir oriental; A. Verneuil, Paris Acad. Sci., Comptes Rendus **147** (1907) 1059–1061.
- (A9) Sur la reproduction synthétique du sapphir par la méthode de fusion; A. Verneuil, Paris Acad. Sci., Comptes Rendus **150** (1909) 185–187.
- (A10) Sur la nature des oxydes qui colorent le saphir oriental; A. Verneuil, Paris Acad. Sci., Comptes Rendus **151** (1910) 1063–1066.
- (A11) Process of Producing Synthetic Sapphires. A. V. L. Verneuil, U.S. Patent No. 988,230, March 28, 1911; Application May 10, 1910; Assigned to L. Heller and Son, New York, N.Y.
- (A12) Synthetic Sapphire. A. V. L. Verneuil, U.S. Patent No. 1,004,505, September 26, 1911; Application June 28, 1911; Assigned to L. Heller and Son, New York, N.Y.

References

1) K. Nassau and J. Nassau, Lapidary J. **24** (1971) 1284, 1442, 1524.
2) "Auguste Verneuil, 1856–1913"; Extrait du Bulletin de l'Association des Élèves de Frémy (about 1914) 39 pages and 3 plates. Contents: Auguste Verneuil p. 3, Discours de M. Fleurent p. 7, Discours de M. Léon Lindet p. 15, Discours de M. Maquenne p. 21, Discours de M. Barthélemy p. 25, Notice etc. par M. L. Maquenne p. 29, Liste Chronolique des Travaux d'Auguste Verneuil, p. 37 (the last is incomplete).
3) E. Frémy, *Synthèse du Rubis* (Libraire des Corps Nationaux, Paris, 1891)
4) K. Nassau, J. Crystal Growth **5** (1969) 338.
5) References to the history of the synthesis of ruby will be found in the series: K. Nassau and R. Crowningshield, Lapidary J. **23** (1969) 114, 313, 334, 440, 621.
6) E. Jannettaz, Bull. Soc. Franc. Mineral. Paris **9** (1886) 321.
7) A. Verneuil, Chem. Ztg. **35** (1911) 259.

I – 3

Journal of Crystal Growth **13/14** (1972) 19–26 © *North-Holland Publishing Co.*

MOLECULAR PHENOMENA DURING CRYSTAL EVAPORATION

THOMAS SUREK*

Department of Materials Science, Stanford University, Stanford, California, U.S.A.

The terrace–ledge–kink model of crystal evaporation is reviewed. A computer simulation technique enables the simultaneous treatment of ledge nucleation and subsequent ledge motions and interactions. Transient and steady-state solutions of the evaporation rate and the surface topography are obtained. Conditions for observing low values of the evaporation coefficient are described.

1. Introduction

The relevance of a discussion on crystal evaporation to the subject matter of this conference lies in the similarities between the atomic or molecular mechanisms involved in crystal evaporation and those occurring in crystal growth processes such as growth from the vapor, electrocrystallization or growth from solution or melt. From an experimental point of view, as a means of testing or verifying the proposed mechanistic models, there are both advantages and disadvantages in carrying out evaporation or growth experiments; it is not the intent of this paper to discuss these, however.

Historically, experimental interest in crystal evaporation has centered on two areas: studies of surface morphology or topography and studies of rates of vaporization. The morphology that develops during evaporation of a crystal in vacuum or some gaseous environment provides characteristic information about the nature of the evaporating surface, such as crystallography, defect structure, and the nature of the interactions with impurities in the crystal or the environment. Topographical studies typically include observations of shapes and slopes of etch pits, spacings and heights of macroscopic ledges, and evaporation or growth spirals. The information gained from etch pitting techniques[1], for example, can lead to an understanding of basic dislocation behaviour in the plastic deformation of materials, or in an entirely different application, to the

growth of nearly perfect single crystals required by solid-state electronics technology.

Measurements of the rate of vaporization involve determinations of the evaporation coefficient, α_v, defined by

$$J_v = \frac{\alpha_v(p_e - p)}{(2\pi mkT)^{\frac{1}{2}}}, \tag{1}$$

where J_v is the net rate of vaporization, p_e is the equilibrium vapor pressure at absolute temperature T, p is the actual pressure, m is the molecular mass and k is Boltzmann's constant. Rate measurements are typically of free evaporation rates in Langmuir experiments ($p = 0$) or of equilibrium exchange rates in Knudsen experiments ($p = p_e$). The evaporation coefficient reflects deviations from ideal behaviour, i.e., $\alpha_v = 1$.

In this paper, the atomic mechanisms involved in crystal evaporation are reviewed briefly, and some experimental confirmations of the proposed models are cited. General disagreements with the quantitative aspects of the theory are also noted. Following that, a computer simulation treatment of the evaporation problem is presented in which the actual time dependence – transient and steady state – of both the surface topography and the rate of vaporization is obtained.

2. Review of terrace-ledge-kink model

The role of monomolecular ledges and kinks along ledges in growth and evaporation processes of crystals was recognized by Kossel[2] and Stranski[3]. The importance of spiral dislocations (i.e., dislocations that

---

* Presently Postdoctoral Fellow, Department of Metallurgical Engineering, McGill University, Montreal, Quebec, Canada.

have a component of their Burgers' vector normal to the surface) as a source of ledges at low supersaturations has been shown by Frank[4]). These considerations led to the development of the theory of crystal growth by Burton, Cabrera and Frank[5]), which was later extended to evaporation by Knacke, Stranski and Wolff[6]) and Hirth and Pound[7]).

For evaporation, the model describes the vaporization of vicinal surfaces (i.e., low index terraces separated by monatomic ledges) of pure metals with monatomic vapors. Similar considerations apply to the vaporization of simple ionic and molecular crystals with monomolecular vapors. The model consists of a series of atomic mechanisms (fig. 1): (i) dissociation of atoms from kink positions in ledges to adsorbed sites; (ii) diffusion of adatoms along the low index surface; and (iii) desorption into the vapor phase. The contribution of other atomic processes (such as kink position to vapor) to the vaporization rate is found to be negligible[7]). The rate controlling step is the surface diffusion of adatoms, hence evaporation proceeds by the adatom diffusion-controlled movement of ledges across the low index surface. The topography of the evaporating surface is determined by the heights and spacings of these ledge fronts.

Of immediate concern, therefore, are the sources of ledges in crystal evaporation. Frank's[4]) spiral dislocation source, mentioned earlier, can lead to a self-perpetuating spiral-shaped ledge front which rotates at a constant angular velocity when steady state is reached. Disc-shaped holes, or "Lochkeime" in Bethge's[8]) terminology, which nucleate randomly on a perfect low index surface or at a dislocation (edge or screw)-surface intersection[9,10]) can also provide ledges in evaporation. More importantly, unlike in crystal growth, ledges can form easily at crystal edges in evaporation even at very small undersaturations[7]). For crystal edges comprised of the intersection of two singular surfaces, a larger critical undersaturation may be required, as noted by Cabrera and Coleman[11]).

The key to the terrace–ledge–kink (TLK) model of evaporation lies in the dynamics of monatomic ledges, yet there is no complete theory that describes the entire time dependent dynamics of ledge creation at some source followed by the adatom diffusion-controlled motion of the ledges along the low index surface. The continuum kinematic theory[12–14]) of crystal growth or evaporation

describes, in a macroscopic manner, the time development of an arbitrary initial surface orientation. As its name suggests, the theory is not restricted to a layer mechanism (TLK) of crystal growth, but in essence, the equations represent the conservation conditions for monatomic ledges at a point on the surface. The basic assumption of the kinematic theory (i.e., ledge velocity is a function of ledge density only) is shown to break down near the sources of ledges, such as dislocation spirals[15]).

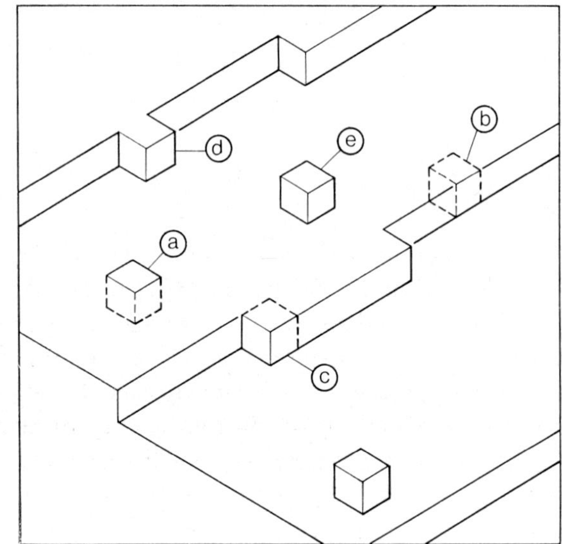

Fig. 1. Terrace-ledge-kink model of a crystal surface showing low index terraces separated by monatomic ledges and atoms in the following positions: (a) in surface, (b) in ledge, (c) kink, (d) at ledge and (e) adsorbed on the surface.

A theoretical prediction for the rate of vaporization in the TLK model was made by Hirth and Pound[7,16]). They described an asymptotic steady state approached by ledges emanating from crystal edge or dislocation spiral sources. Maximum steady-state velocity (i.e., no further acceleration) is approached by such ledges when their spacing becomes

$$\lambda_0 = 6[(D_s/v) \exp (\Delta G_{\mathrm{des}}/kT)]^{\frac{1}{2}} = 6/Q, \qquad (2)$$

where D_s is the surface self-diffusion coefficient, v is the vibrational frequency of the adatoms and ΔG_{des} is the free energy of activation for desorption of the adatoms. The quantity Q is related to the mean free path for random walk, $\overline{X}$, of an adatom on the low index surface. In fact, for straight, parallel ledges $Q = (2)^{\frac{1}{2}}/\overline{X}$, while for concentric, circular ledge fronts $Q = 2/\overline{X}$ [17]). The limiting value of the evaporation coefficient, α_v in eq. (1), is shown to be $\frac{1}{3}$ for the asymptotic steady-state behaviour described above.

Surface diffusion kinetics are important in most instances of crystal evaporation, but other factors may also influence the ledge dynamics during the vaporization of some substances. Entropy constraints, caused by the restricted rotational degrees of freedom of the molecules in the condensed phase, for example, can affect the kinetics of transfer between kinks and adsorbed molecules. Alternatively, the transfer from kink to adsorbed state may have a large activational potential energy and hence be partially rate controlling.

Dissociation of polymeric species during various stages of the stepwise kinetics [6,18]), diffusion in the vapor phase and thermal accommodation effects have also been considered theoretically, as reviewed by Hirth and Pound [9]). Finally, the effects of impurities on the evaporation kinetics and morphology have received considerable attention. Most noteworthy are Frank's [12]) treatment of time dependent impurity adsorption and Cabrera and Vermilyea's [13]) treatment of large, immobile impurity molecules impeding the ledge motions. The effect of impurity poisoning of kink sites has been considered by Chernov [19]).

3. Experiments

The observation of the basic units of the TLK theory, such as adatoms, monatomic ledges and kinks in ledges, presents considerable experimental difficulties. Nevertheless, experimental techniques such as field ion microscopy [20,21]) and field electron microscopy [22]), sometimes used in conjunction with decoration or imaging techniques [23]), can provide resolution on an atomic scale. Alternatively, one can observe structure sensitive macroscopic properties of the surface, such as surface self-diffusion coefficients or the crystallographic dependence of the surface free energy [24]).

There are numerous good experiments which confirm the TLK model for evaporation. The induction [7]) that crystal edges should serve as sources for ledges in evaporation has been demonstrated by the experiments of Sears on p-toluidine [25]) and on zinc whiskers [26]). Observations of growth and evaporation spirals [27-29]) give qualitative support for the TLK model, while measurements of macroscopic ledge velocities versus

ledge height [30,31]) provide semiquantitative agreement with the TLK theory. The continuum kinematic theory has been successfully tested in the dissolution of germanium by Frank and Ives [32]) and the dissolution of lithium fluoride by Ives [33]). Experiments by Hullett and Young [31]) demonstrated that the dissolution shape of a crystal is governed by the motions and interactions of atomic ledges. Bethge's [8]) observations of monatomic ledge motions during the growth and evaporation of NaCl crystals are noteworthy, as these methods may ultimately provide a means of verifying quantitative aspects of the TLK theory.

A great deal of experimental attention has focused on attempts to verify the Hirth-Pound limiting law [7]) (i.e., $\alpha_v = \frac{1}{3}$) in evaporation, but none of the observations were conclusive because of the possible influence of factors, such as impurity effects, during the experiments. Recent experiments by Winterbottom [34]) on the vaporization of silver and by Mar and Searcy [35]) on zinc single crystals both found evaporation coefficients of unity for the respective metals, in apparent discord with the Hirth-Pound limiting law. Finally, this brief review would not be complete without mentioning some excellent reviews on the subject of crystal growth and evaporation, notably those by Hirth and Pound [9]), Heyer [36]) and Strickland-Constable [37]).

4. Kinetic model for ledge dynamics

It is apparent from the foregoing considerations of the TLK model for evaporation that a complete theoretical study should include the process of ledge creation at some source on the evaporating surface, to be treated simultaneously with the motions and interactions of the ledges as they move away from the various sources. Hullett and Young [31]) have recognized that the kinetics of ledge motion based on Mullins and Hirth's [38]) treatment of ledge interactions is amenable to simulation on the digital computer. The sources producing the ledges were neglected in their treatment, however. We have used the same kinetic description of the ledge motions, but included in our treatment [17,39]) a simulation of the ledge sources which have been experimentally verified as operative, and also considered impingement of ledges from different sources. The main features of our model are summarized below:

(i) The ledge system shown in fig. 2 depicts the problem of evaporation proceeding by the motion of

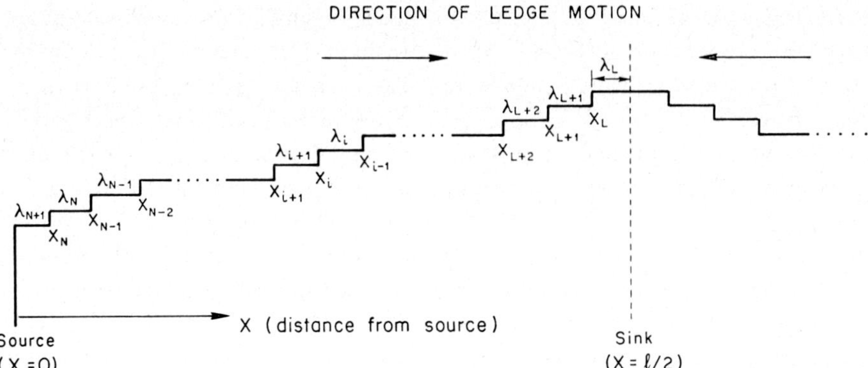

Fig. 2. Schematic cross section of crystal surface evaporating by the motion of straight, parallel ledges.

straight, parallel ledges. The source of ledges is the crystal edge, but similar considerations apply to other sources capable of producing these ledge fronts. The sinks for the ledges, in the most general case, lie along lines of annihilation with ledges from other sources. In fig. 2, ledges nucleate at the source ($X = 0$) and are annihilated at the sink ($X = \frac{1}{2}l$) by ledges from an equivalent source a distance l away. The subscript L is used to denote the leading ledge, while N denotes the last ledge nucleated at the source. Therefore, at any instant of time, ($N-L+1$) ledges are traversing the surface, and ($L-1$) ledges have already been annihilated at the sink (initially, $N = L = 1$).

(ii) The velocity of the ledge at X_i in fig. 2 is given by

$$v_i = \frac{dX_i}{dt} = \frac{D_s a_0 Q n_e}{n_y}\left(1 - \frac{p}{p_e}\right)\left[\tanh\left(\frac{Q\lambda_{i+1}}{2}\right)\right.$$
$$\left. + \tanh\left(\frac{Q\lambda_i}{2}\right)\right], \tag{3}$$

where λ_i is the width of the terrace spacing ahead of the ith ledge, a_0 is the width of one row of atoms, n_y is the number of atoms per unit length of ledge, n_e is the concentration of adatoms in equilibrium with the crystal and p/p_e is the undersaturation ratio. Eq. (3) is valid when the root-mean-square diffusion velocity of the adatoms is greater than the ledge velocity, which is always likely to be the case, and it also assumes that the adatom concentration at the ledge fronts is n_e.

(iii) A new ledge nucleates at the source when the preceding ledge has moved some distance λ_I. This "ledge creation criterion" obviates the need for describing the nucleation problem at the source in detail, and clearly, it can parametrically represent any nucleation condition, in that any value for the critical undersaturation for ledge nucleation can be represented by the proper choice of the parameter λ_I. The ledge creation criterion can be written in the form of an initial condition for eq. (3), viz., $X_i(0 \leq t \leq \tau_i) = 0$, where $X_{i-1}(\tau_i) = \lambda_I$. The time τ_i represents the time of nucleation of the ith ledge.

(iv) The leading ledge annihilates when $2\lambda_L \leq \lambda_A$, where the parameter λ_A is usually taken as one atomic spacing. As in the case of ledge nucleation, the above ledge annihilation criterion can parametrically represent the effects of any physical processes, such as attractive and repulsive interactions between impinging ledges, that may occur during annihilation.

(v) At any instant of time, the net vaporization flux from the surface between source and sink can be written in the form of eq. (1), with α_v replaced by an average evaporation coefficient, $\langle \alpha \rangle$, given by

$$\langle \alpha \rangle = \sum_i \alpha_i \lambda_i \Big/ \sum_i \lambda_i, \tag{4}$$

where $\alpha_i = (2/Q\lambda_i)\tanh(Q\lambda_i/2)$, and the summation is over the ($N-L+2$) terraces in fig. 2.

(vi) The calculations consist of "computer experiments" in which the parameters $\frac{1}{2}l$, λ_I and λ_A are specified, and the surface profile [i.e., $\lambda(X)$ versus X, where X is the distance measured from source to sink] and the rate of vaporization [i.e., $\langle \alpha \rangle$ of eq. (4)] are monitored as a function of time. The use of dimensionless variables, viz., distances measured in multiples of $1/Q$ and time in multiples of $1/K$, where

$$K = \frac{D_s a_0 Q^2 n_e}{n_y}\left(1 - \frac{p}{p_e}\right) = \frac{a_0}{n_y}\frac{(p_e - p)}{(2\pi m k T)^{\frac{1}{2}}}, \tag{5}$$

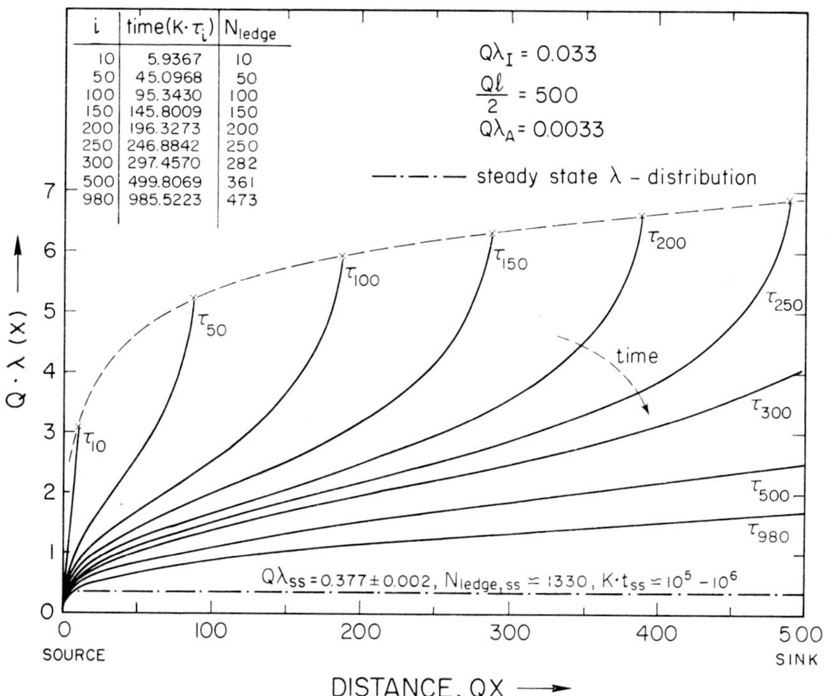

i	time(K·τ_i)	N_ledge
10	5.9367	10
50	45.0968	50
100	95.3430	100
150	145.8009	150
200	196.3273	200
250	246.8842	250
300	297.4570	282
500	499.8069	361
980	985.5223	473

Axes: $Q \cdot \lambda(x)$ (vertical, 0 to 7); DISTANCE, QX (horizontal, 0 to 500)

Fig. 3. Surface profiles at times τ_i when the ith ledge nucleates at the source. $\lambda_I = 0.033/Q$, $\lambda_A = 0.0033/Q$ and $\frac{1}{2}l = 500/Q$. Here, $\times$ denotes first ledge nucleated at source (impinges at sink at $t = 252.3/K$).

obviates the need for specifying any of the system constants in eq. (3). Note that the constant K represents the number of monolayers of the low index surface that would evaporate ideally (i.e., $\alpha_v = 1$) per second.

5. Results and discussion

Fig. 3 shows the time development of the surface profile during a typical simulation run. The parameter values chosen are representative of a real system, e.g., for $Q \simeq 10^5$ cm^{-1}, $\lambda_I \sim 10$ atomic spacings, $\lambda_A \sim 1$ atomic spacing, and the source density, i.e. $1/l^2 \sim 10^4$ cm^{-2}. The terrace spacing behind the leading ledge (denoted by $\times$) increases with time as the leading ledge accelerates away from the source. The surface profile is shown at times τ_i when the ith ledge nucleates at the source. The times τ_i and the number of ledges on the surface (between source and sink) at that instant, N_{ledge}, are given in fig. 3. As one would expect, the time development of the surface profile is consistent with the qualitative predictions of the continuum kinematic theory[12–14]). The ledge flux, i.e., the number of ledges passing a given point on the surface per unit time, decreases monotonically with distance X from

the source for this case of surface diffusion-controlled kinetics. The kinematic theory thus predicts that the ledge density, i.e., the inverse of the ledge spacing, should increase with time at every point on the surface. This is seen to be the case in fig. 3. The following estimates are made regarding the steady-state behaviour of the ledge system in fig. 3: time to reach steady state, $t_{ss} \simeq 10^5$ to $10^6/K$; steady-state ledge spacing, $\lambda_{ss} \simeq 0.377/Q$; and the number of ledges on the surface at steady state, $N_{\text{ledge,ss}} \simeq 1330$ (approx. $\frac{1}{2}l/\lambda_{ss}$). The way in which t_{ss} and λ_{ss} are obtained is described later.

Typical plots of the average evaporation coefficient versus time are shown in fig. 4. Line A corresponds to the case examined in fig. 3. Typically, $\langle \alpha \rangle$ increases linearly with time until the leading ledge impinges at the sink, after which $\langle \alpha \rangle$ approaches its steady-state value at an ever decreasing rate. For $\lambda_I \geq 6/Q$ (cf. line B), steady state is reached by the time the leading ledge impinges at the sink. Correspondingly, the surface profile is nominally invariant with time once impingement of the leading ledge occurs.

The following general conclusions, as illustrated in

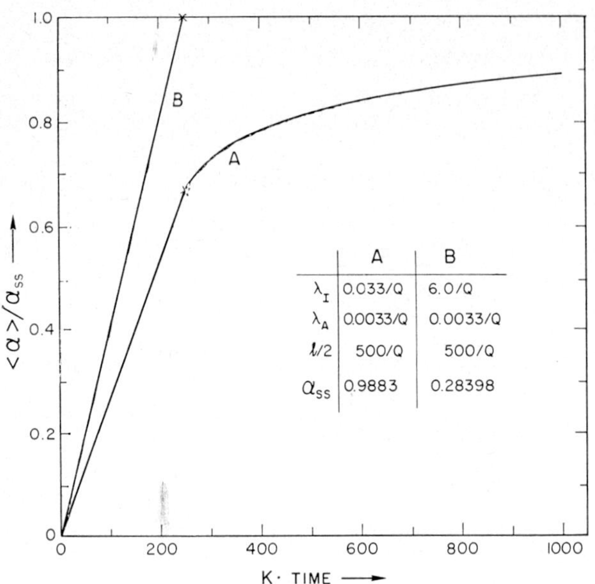

Fig. 4. Plots of the average evaporation coefficient, normalized with respect to its steady-state value, versus time for selected values of λ_I. Here, $\times$ denotes impingement of first ledge at sink.

$$\alpha_{ss} = (2/Q\lambda_{ss})\tanh(Q\lambda_{ss}/2). \tag{7}$$

Table 1 summarizes the steady-state predictions based on simulated experimental values of $\Delta\tau_{N,ss}$ for selected values of λ_I. The quantity $\Delta\lambda = (\lambda_{ss}-\lambda_I)$ represents the effect of a "diffusion induced acceleration"[7]) near the source of ledges.

(d) The time to reach steady state can be estimated by the empirical formula $t_{ss} = n(\frac{1}{2}l/\lambda_I)\Delta\tau_{N,ss}$, where the factor n is between 1 and 10.

(e) The presence of the sink is imperative insofar as reaching steady state is concerned (as well as being a realistic boundary condition). The value of λ_A, however, has no effect on the ledge dynamics (transient or steady state) with the exception of 2 or 3 ledges near the sink. The trivial condition $\lambda_A \ll \frac{1}{2}l$ assures the validity of these results.

TABLE 1

Steady-state predictions for selected values of λ_I for straight ledge sources. Uncertainties are $\pm$ one unit in the last significant digit unless otherwise indicated

$Q\,\lambda_I$	$K\,\Delta\tau_{N,ss}$	$Q\,\lambda_{ss}$ (eq. (6))	$Q\,\Delta\lambda$	α_{ss} [eq. (7)]
0.0033	1.0011	$0.115 \pm .005$	0.112	0.9989
0.033	1.0118	$0.377 \pm .002$	0.344	0.9883
0.33	1.0849	$1.0180 \pm .0006$	0.688	0.9217
1.0	1.2757	$1.8699 \pm .0003$	0.870	0.7839
6.0	3.5213	$7.0302 \pm .0002$	1.030	0.28398
12.0	6.5173	13.0345	1.035	0.15344
60.0	30.5193	61.0386	1.039	0.032766

part by the results in figs. 3 and 4, can be made from these simulation experiments:

(a) Steady-state conditions, i.e., steady-state surface profile and evaporation rate, obtain for a given λ_I independent of $\frac{1}{2}l$ and λ_A, as long as: (i) $\frac{1}{2}l \geq 6/Q$ [i.e., there is no appreciable overlap of diffusion fields from ledges near the source and sink, respectively – cf. eq. (2)]; and (ii) $\frac{1}{2}l \geq 10\lambda_{ss}$ (i.e., the steady-state profile contains at least 10 ledges). Furthermore, the steady-state conditions are uniquely characterized by the value of λ_I, which in turn, is characteristic of a given source.

(b) Steady state is approached near the source first, then it "propagates" toward the sink. Related to this effect is the result that steady-state nucleation rate at the source is reached in a time which is much less than that required to obtain a steady-state surface profile.

(c) Based on the results in (a) and (b), we can estimate the steady-state conditions for a given λ_I by carrying out the simulation experiments until some steady-state nucleation interval, $\Delta\tau_{N,ss}$, is approached at the source. The steady-state ledge spacing is then the solution of

$$\Delta\tau_{N,ss} = \frac{\lambda_{ss}}{(2K/Q)\tanh(Q\lambda_{ss}/2)}, \tag{6}$$

while the evaporation coefficient at steady state is given by [cf. α_i in eq. (4)]

The results of the present calculations have important consequences regarding the theoretical expectations for the evaporation coefficients of solids. Firstly, the asymptotic steady-state behaviour proposed by Hirth and Pound[7,16]) is only a very special case of the dynamic ledge system in fig. 2. Because of a usually rapid nucleation of ledges at the source, and less importantly, impingement of ledges at the sink, a dynamic situation in which an appreciable portion of the surface is covered by ledge spacings of approx. $\lambda_0 = 6/Q$ [see eq. (2)] does not occur, in general. The trivial exception is a ledge source with $\lambda_I \simeq 5/Q$, for which $\lambda_{ss} \simeq 6/Q$ (cf. table 1). A simulation experiment with a single ledge source ($\lambda_I < 1/Q$) and a semi-infinite crystal surface (very large value of $\frac{1}{2}l$), which corresponds to the boundary conditions in the Hirth–Pound solution, showed that the percentage of the surface with ledge spacings $\geq 5/Q$ never exceeds 10%. From this we con-

I – 4

clude that the Hirth–Pound value of $\alpha_v = \frac{1}{3}$ has no general significance as a limiting law in evaporation.

Finally, we list the various experimental conditions, based on the results of the present calculations, which can lead to low values of the evaporation coefficient (i.e., $\langle\alpha\rangle$ appreciably less than unity).

(i) When the evaporation kinetics are controlled by ledge sources characterized by large values of λ_l, low $\langle\alpha\rangle$ or α_{ss} can result (see table 1).

(ii) A low value of $\langle\alpha\rangle$ can be measured in any experiment during the transient stage of the ledge dynamics (cf. fig. 4). The real time-scale of the transient stage is increased for a low density of ledge sources (i.e., large value of $\frac{1}{2}l$), but more importantly, for small values of the system constant K [see eq. (5)].

(iii) Insofar as the temperature and the adatom diffusivity enter the constant Q (cf. equation (2)), the value of $Q\lambda_l$ in table 1 probably increases as the temperature increases, hence decreasing $\langle\alpha\rangle$.

(iv) When other factors, such as impurity effects, entropy constraints, dissociation of polymeric species, etc. affect the ledge kinetics, the conclusions reached in earlier treatments[5,14,40] regarding the lowering of $\langle\alpha\rangle$ still apply. The surface diffusion contribution (if any) to the overall evaporation coefficient multiplies all other effects (represented by a parameter $\bar{n}_s/n_e$, where $\bar{n}_s < n_e$ is some effective adatom concentration at the ledge fronts), such that $\alpha_v = \alpha_{ss}(\lambda_l)\,(\bar{n}_s/n_e)$.

6. Summary

The computer simulation technique described in this paper enabled us to obtain an exact numerical solution to the transient and steady-state behaviour of a ledge system (fig. 2) which is assumed to be representative of a real system. Two consequences of this work are immediately apparent. Firstly, there is a need for a critical re-evaluation of some key evaporation experiments in light of the quantitative predictions of the present calculations. Secondly, the effects of impurities on the ledge kinetics, the presence of macroscopic ledges and the role of impurities in macroscopic ledge formation should receive further theoretical attention. We believe that the computer models proposed here can be extended readily to treat these more complicated problems.

Acknowledgements

The author wishes to express his gratitude to his

thesis advisors, Professor G. M. Pound of Stanford University and Professor J. P. Hirth of Ohio State University, for their guidance and support during the course of the author's dissertation research, a portion of which this paper describes, and also for their help with the preparation of this manuscript. The author is indebted to Professor W. A. Tiller of Stanford University for his encouragement and suggestions for the oral presentation of this paper. Support for this work from the National Science Foundation and the Stanford Center for Materials Research is gratefully acknowledged.

References

1) W. H. Robinson, in: *Techniques of Metals Research*, Vol. II, Part 1, Ed. R. F. Bunshah (Interscience, New York, 1968) p. 291.
2) W. Kossel, Nach. Ges. Wiss. Göttingen (1927) 135.
3) I. N. Stranski, Z. Physik. Chem. **136** (1928) 259.
4) F. C. Frank, Discussions Faraday Soc. **5** (1949) 48, 67.
5) W. K. Burton, N. Cabrera and F. C. Frank, Phil. Trans. Roy. Soc. London A **243** (1950) 299.
6) O. Knacke, I. N. Stranski and G. Wolff, Z. Elektrochem. **56** (1952) 476; Z. Physik. Chem. **198** (1951) 157.
7) J. P. Hirth and G. M. Pound, J. Chem. Phys. **26** (1957) 1216.
8) H. Bethge, in: *Molecular Processes on Solid Surfaces*, Eds. E. Drauglis, R. D. Gretz and R. I. Jaffee (McGraw-Hill, New York, 1968) p. 569.
9) J. P. Hirth and G. M. Pound, Progr. Mater. Sci. **11** (1963) 1.
10) W. H. Robinson and J. P. Hirth, J. Crystal Growth **7** (1970) 262.
11) N. Cabrera and R. V. Coleman, in: *The Art and Science of Growing Crystals*, Ed. J. J. Gilman (Wiley, New York, 1963) p. 3.
12) F. C. Frank, in: *Growth and Perfection of Crystals*, Eds. R. H. Doremus, B. W. Roberts and D. Turnbull (Wiley, New York, 1958) p. 411.
13) N. Cabrera and D. A. Vermilyea, ref. 12, p. 393.
14) A. A. Chernov, Dokl. Akad. Nauk SSSR **117** (1957) 983.
15) M. B. Ives and J. P. Hirth, J. Chem. Phys. **33** (1960) 517.
16) J. P. Hirth and G. M. Pound, Acta Met. **5** (1957) 649.
17) T. Surek, Ph.D. Thesis, Stanford University, 1971.
18) L. Brewer and J. S. Kane, J. Phys. Chem. **59** (1955) 105.
19) A. A. Chernov, Soviet Phys.-Usp. **4** (1961) 116.
20) E. W. Muller, ref. 8, p. 67.
21) G. Ehrlich and F. G. Hudda, J. Chem. Phys. **44** (1966) 1039.
22) R. D. Gretz, ref. 8, p. 425.
23) G. A. Bassett, Phil. Mag. **3** (1958) 1042.
24) N. A. Gjostein and W. L. Winterbottom, in: *Fundamentals of Gas-Surface Interactions*, Eds. H. Saltsburg, J. N. Smith, Jr. and M. Rogers (Academic Press, New York, 1967) p. 42.
25) G. W. Sears, J. Chem. Phys. **27** (1957) 1308.
26) J. B. Hudson and G. W. Sears, J. Chem. Phys. **35** (1961) 1509.
27) E. D. Dukova, Soviet Phys.-Dokl. **3** (1958) 703.
28) A. R. Verma, *Crystal Growth and Dislocations* (Butterworth London, 1953).
29) A. J. Forty, Advan. Phys. **3** (9) (1954).

30) G. G. Lemmlein, E. D. Dukova and A. A. Chernov, Kristallografia **2** (1957) 428.

31) L. D. Hullett, Jr. and F. W. Young, Jr., J. Electrochem. Soc. **113** (1966) 410.

32) F. C. Frank and M. B. Ives, J. Appl. Phys. **21** (1960) 1996.

33) M. B. Ives, J. Appl. Phys. **32** (1961) 1534.

34) W. L. Winterbottom, J. Appl. Phys. **40** (1969) 3803, 3810.

35) R. W. Mar and A. W. Searcy, J. Chem. Phys., **53** (1970) 3076.

36) H. Heyer, Angew. Chem. Intern. Ed. **5** (1966) 67.

37) R. F. Strickland-Constable, *Kinetics and Mechanism of Crystallization*, (Academic Press, London, 1968).

38) W. W. Mullins and J. P. Hirth, J. Phys. Chem. Solids **24** (1963) 1391.

39) T. Surek, G. M. Pound and J. P. Hirth, J. Chem. Phys., in press.

40) J. P. Hirth and G. M. Pound, J. Phys. Chem. **64** (1960) 619.

Journal of Crystal Growth **13/14** (1972) 27–33 © *North-Holland Publishing Co.*

SINGLE CRYSTALS FOR BUBBLE DOMAIN MEMORIES

R. A. LAUDISE

Bell Telephone Laboratories, Incorporated, Murray Hill, New Jersey 07974, U.S.A.

A description of the search for materials to meet the requirements of the bubble domain memory is given. This search is used to illustrate the importance of symbiosis between the crystal grower and the solid state physicist and the important role which the grower should play in the discovery of new materials. Single crystals of flux grown rare earth orthoferrites were among the first materials examined for bubble applications, but the stable bubble sizes proved too large. Decreasing the bubble diameter by reducing the magnetocrystalline anisotropy was possible in $Sm_{0.55}Tb_{0.45}FeO_3$ where Sm^{+3} brought the magnetic spin reorientation temperature close to room temperature. However, the properties of such compositions were too temperature sensitive to be useful. Other noncubic ferromagnetic materials were investigated but success was achieved only when it was discovered that the ordinarily cubic rare earth garnets could be made noncubic by appropriate substitutions. A description of the growth and reasoning which was used in tailor making garnets to desired magnetic requirements is given. The problem of developing economic techniques for the growth of thin garnet films is discussed and recent results on chemical vapour deposition, liquid phase epitaxy and hydrothermal epitaxy are given.

1. Introduction

The concept of magnetic domains was originated by Weiss[1] and has been used to explain magnetic phenomena for sixty years. Recently Bobeck[2] suggested the use of individual domains in single crystal uniaxial materials to make fast high information density stores. In 1969 Bobeck, Fischer, Perneski, Remeika and Van Uitert[3] demonstrated the use of these "bubble" shaped magnetic domains in memory and logic devices. This demonstration followed the successful search by Van Uitert and co-workers for materials possessing unique magnetic properties to fulfill the requirements for bubble devices. We will use this search to illustrate our thesis that the materials scientist and particularly the crystal grower is especially fitted to find the new materials of the future. It is our belief that a thoughtful symbiosis of crystal growers with an interest in structure, bonding, and properties and solid state physicists with a feel for the necessities of devices can lead in many cases to results such as we will describe here.

We assume that the reader is familiar with the functioning of bubble memories (see, for instance, refs. 2–6). The requirements for an economically attractive bubble memory are that it store $\sim 1.6 \times 10^5$ bits/cm² and that it operate at least at a megabit data rate. This leads to the requirement that bubble diameter be from 2.5–7.5 μm and that the domain wall velocity be between 100

and 1000 cm/sec·Oe. Larger bubble size results in inefficient information storage, smaller size bubbles would be difficult to manipulate and detect. Low wall velocities lead to low bit rates, thus, the higher velocities are advantageous. An additional requirement for bubble stability is that the material be magnetically noncubic. It is also required that the bubbles be stable near ambient temperatures without elaborate temperature control and that the material supporting the bubbles be mechanically strong enough for ease in processing, chemically and thermally stable and reasonably easy to grow as crystals. In our magnetic bubble work crystal growth efforts were divided into two phases:

(1) Find a material to meet the above requirements,

(2) Find an economic growth method for the material.

It was shown by Thiele[5,6] that the bubble diameter (B) is related to the magnetic properties such that

$$B \sim \sqrt{K}/M_s^2 \, ,$$

where K is the magnetocrystalline anisotropy (a function of domain wall energy, σ_W) and M_s is the magnetic moment (a function of the number of unpaired electrons).

2. The materials search

To find a useful material a number of magnetic materials had to be prepared. Thus a comparatively rapid growth method generally applicable to magnetic materials was required. Flux growth was the obvious choice, although it was felt that once the material had been found, flux growth might not be an economic growth method for production. Among the first crystals investigated were the rare earth orthoferrites, $R(FeO_3)$ (where R is a rare earth) which are uniaxial canted spin ferromagnetic materials[7-9]). The flux used typically was $PbO–PbF_2–B_2O_3$ and typical growth conditions for the growth of, for instance, $Sm_{0.55}Tb_{0.45}$-FeO_3, in a large (20 cm diameter × 20 cm high, ~3 liter volume) covered platinum crucible were as shown in table 1.

TABLE 1

Crystal components			Flux		
Sm_2O_3	5.5 mole	1920 g	PbO	25 mole	5536 g
Tb_2O_3	4.5 mole	1650 g	PbF_2	27 mole	6600 g
Fe_2O_3	10 mole	1590 g	B_2O_3	4 mole	280 g

Holding temperature 1300 °C
Holding time with crucible rotation 5 hr
Cooling rate 0.5 °C/hr to 900–950 °C

In order to provide crystals suitable for device evaluation, it was necessary to grow from large crucibles.

Following the growth cycle, quenching to room temperature is impractical because:

(1) Often regions of the phase diagram are entered where the crystals will redissolve.

(2) Large thermal strains are produced which sometimes cause cracks and often lead to poor magnetic behavior because of the large magnetostriction of orthoferrites. Such strains are common if the crystals are cooled to room temperature too rapidly.

(3) The frozen flux is hard to remove so that very long leaching in acid to recover the crystals is required. Pouring off the flux following growth is impractical because of crucible size, so that an arrangement for puncturing the bottom of the crucible to let flux run off is made. Crucibles are drained by puncturing a replaceable platinum diaphragm with a steel "spear" following growth. After the flux is drained the grown crystals may be slowly cooled to room temperature *in situ*. The crystals are removed, processed and their magnetic properties are measured.

A study of the orthoferrites[11,3]) showed that in general the bubble size was too large (although $YFeO_3$ may have some applications because of its very high mobility: >2000 cm/sec·Oe). To reduce bubble size several expedients were considered.

(1) Increase $4\pi M_s$. Any ion which might be substituted in hopes of effecting the moment would be likely to distribute itself equally in the two sublattices with no net effect on $4\pi M_s$.

(2) Change the canting angle of the two spin systems in the two sublattices so as to affect their vector sum. This angle is fixed when $[O^=]$ is the anion in the lattice. However, Robins et al.[12]) reported that substitution of, for instance, $0.2\,Ni^{++}$ at Fe^{+3} sites together with $0.2\,(F)^-$ at the $[O^=]$ sites for charge compensation resulted in marked change in the moment which they attributed to a canting angle change[12]). However, reasonable sized crystals have so far eluded preparation.

(3) Reduce the magnetocrystalline anisotropy, K. The easy magnetic axis in orthoferrites undergoes a reorientation from $\langle100\rangle$ to $\langle001\rangle$ [a → c] by a spin reorientation (flipping) process at a particular temperature (T_r), the spin reorientation temperature. Close to T_r, K will be small. T_r for all rare earth orthoferrites except $SmFeO_3$ is below room temperature; T_r for $SmFeO_3$ is 175 °C. Thus, T_r can be made close to room temperature in, for instance, $Sm_{0.55}Tb_{0.45}FeO_3$ and $Sm_{0.60}$-$Er_{0.40}FeO_3$, which gives a small σ_w and B[13]). Unfortunately, it was found that σ_w and B are strong functions of temperature close to T_r so that device requirements could not be met. Another attractive possibility is to make substitutions at the iron site of an element with a high moment thus achieving small B far from T_r. This stratagem was tried by making Co substitutions[14]). A typical composition was $TbFe_{0.95}Co_{0.05}O_3$ where B was about 12 μm. The coercivity of such crystals was found to be too high for usefulness, and it proved difficult to uniformly dope with Co at the very low levels which gave interesting bubble sizes. Thus, orthoferrites were found to be generally unattractive.

The search moved on to other uniaxial ferromagnetic materials. Hexagonal magnetoplumbite compounds such as $PbFe_{12}O_{19}$ were investigated[15]). Crystals were prepared by flux growth. Such compounds generally have such high moments that the resultant bubbles are too small for practical devices. Substitution of nonmagnetic ions such as Al reduce the moment and increase the bubble size to a useful range; unfortunately, the domain wall velocity of the magnetoplumbite compounds was found to be too small to be interesting for devices*. Models for wall mobility are not quantitative, but there are indications that at least in some instances low mobility is correlated with high angular momentum of the magnetic ions (e.g., Co in orthoferrites). Although mobility can be improved by adroit choice of composition and by post growth treatments such as annealing, in no cases could useful mobilities be produced in magnetoplumbites.

Thus it was decided to attempt the preparation of ordinarily cubic materials with appropriate substitution so as to make them uniaxial. The rare earth ferrimagnetic garnets were selected because previous experiments provided a background for modification of properties such as magnetic moment. Ordinarily the rare earth iron garnets are called cubic, they have four equivalent easy $\langle 111 \rangle$ axes. Of course, in a single domain region, due to magnetostriction, a crystal is no longer truly cubic. It was discovered that when two or more appropriate rare earths are substituted at the rare earth site and when growth is under appropriate conditions on certain growth faces, the garnet undergoes a slight distortion so that one of the $\langle 111 \rangle$'s or a direction close to it becomes the unique easy magnetic axis**[19]). The moment can be reduced by the substitution of Al or Ga in the tetrahedral iron sites. Typical conditions for the flux growth of $Er_2TbAl_{1.1}Fe_{3.9}O_{12}$[18]) were as given in table 2.

TABLE 2

Crystal components		Flux	
Er_2O_3	1872 g	PbO	5160 g
Tb_2O_3	936 g	PbF_2	6308 g
Al_2O_3	336 g	B_2O_3	287 g
Fe_2O_3	1870 g		

The flux used was $PbO–PbF_2–B_2O_3$, the same as that used for $RFeO_3$ growth. Holding time was similar as was cooling rate and the temperature at which flux was drained off. Table 3 lists some garnet compositions studied.

All of these compounds have a unique easy axis (small K) and a low magnetization at room temper-

* The manner in which mobilities have sometimes been measured can give unrealistically high results for real device uses. See refs. 16 and 17.
** Although such garnets are often called uniaxial, they are actually orthorhombic.

TABLE 3

Noncubic garnet compositions

I	$Tb_{1.0}Er_{2.0}Al_{1.0}Fe_{4.0}O12$
II	$Gd_{2.34}Tb_{0.66}Fe_5O_{12}$
III	$Er_{1.3}Gd_{0.95}Tb_{0.75}Al_{0.5}Fe_{4.5}O_{12}$
IV	$Er_{1.0}Eu_{2.0}Fe_{4.3}Ga_{0.7}O_{12}$
V	$Y_{1.5}Gd_{1.5}Fe_{4.4}Al_{0.6}O_{12}$
VI	$Y_{1.85}Eu_{0.18}Gd_{0.50}Tb_{0.47}Fe_{4.4}Ga_{0.6}O_{12}$
VII	$Eu_{1.9}Gd_{1.1}Al_{0.5}Fe_{4.5}O_{12}$

ature ($4\pi M_s \approx 100–200$ gauss). Thus, their bubble diameters are in the 7.5 μm range. In addition, their domain wall velocities can be as large as 500 cm/sec · Oe, so that they can meet device requirements. Temperature stability for garnets near room temperature can also be achieved. In garnets the net magnetizations of the iron sublattice and the rare earth sublattice are opposed. At the temperature where sublattice magnetizations exactly balance each other, the net moment becomes zero. This temperature is the compensation point (T_{cp}). This temperature depends upon the rare earth present in the garnet as shown in fig. 1[20]). Beyond T_{cp}, $4\pi M_s$ is usually rather temperature insensitive. Thus, for a temperature stable bubble one wants a system when T_{cp} is below room temperature (so the flat region

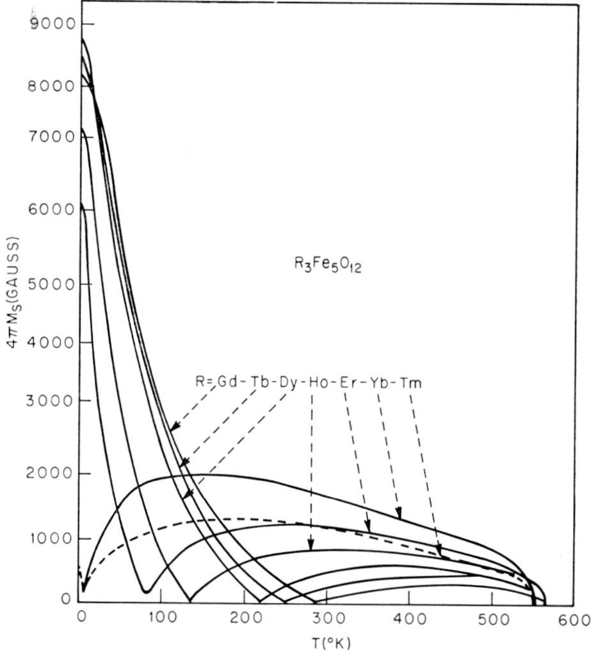

Fig. 1. Magnetization as a function of temperature for some rare earth garnets.

of fig. 1 is at room temperature) and where $4\pi M_s$ is such as to give appropriate B at room temperature. T_{cp} is adjusted by choice of rare earth and by employing solid solutions of two or more rare earths (a requirement for noncubicity anyhow). Finally, $4\pi M_s$ is reduced to the neighborhoud of 150 gauss (required for $\sim 7.5\ \mu m$ B given the usual K of the garnets) by Al or Ga additions.

To avoid the effect of growth and processing strains on bulk garnets, it is convenient to grow a zero magnetostriction composition. In general λ_{111} (the magnetostriction constant in $\langle 111 \rangle$) is negative (see table 4)[21].

TABLE 4

Magnetostriction constants at room temperature

R in $R_3Fe_5O_{12}$	$\lambda_{111} \times 10^6$	$\lambda_{100} \times 10^6$
Sm	-8.5	21.0
Eu	$+1.8$	21.0
Gd	-3.1	0
Tb	$+12.0$	-3.3
Dy	-5.9	-21.5
Ho	-4.0	-3.4
Er	-4.9	2.0
Tm	-5.2	1.4
Yb	-4.5	1.4
Lu	-2.4	-1.4
Y	-2.4	-1.4

Only Eu and Tb have a positive λ_{111}. While λ_{100} is positive for Sm, Eu, Er, Tm, and Yb; zero for Gd and negative for Tb, Dy, Ho, Lu, and Y. Appropriate solid solution compositions to minimize magnetostriction can thus be easily found (to make magnetostriction zero Eu or Tb substitution is necessary).

All of the above criteria can successfully be combined in one material and composition IV of table 3 is an example. Furthermore, such crystals have been grown from the flux without undue difficulty.

The final important question to be answered is, "Why are the garnets noncubic?" It should be pointed out that this noncubicity is not the commonly observed magnetostrictive non-cubicity often seen in ordinary garnets. It occurs in zero magnetostrictive compositions and it can be annealed out by treatment at $\sim 1200\ °C$. The orientation of the easy axis is different for different compositions and different growth faces. Two common examples called Type I and Type II[23] are shown in fig. 2. It has been shown that a model

based on ion pairing between the rare earth and iron ions gives fair quantitative agreement with room temperature magnetic torque measurements[24,25]. It has also been shown that a site model based on the possibility of preferential occupation of sites in the rare earth has the same analytic form as the pair model and requires fewer preference parameters[26,27]. Very recent analysis has shown that all models basically reduce to what are essentially two parameters[28]. In the simplest case, values for these two parameters are fitted to experimental data (e.g., magnetocrystalline anisotropies and direction of easy axis) on say, material grown on a

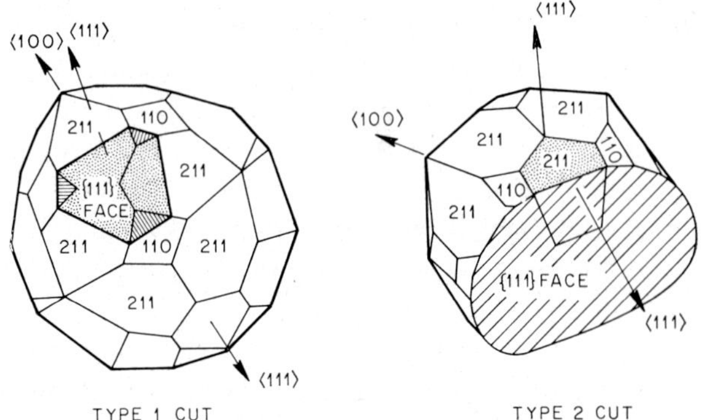

Fig. 2. Orientation of easy magnetic axes in flux grown garnets. Type 1 – On a {211} growth facet; the easy axis is close to the $\langle 111 \rangle$ which is normal to the facet, so that a slice cut $\perp$ to that $\langle 111 \rangle$ and lying across the facet is used. Type 2 – On a {211} growth face, the easy axis is close to the $\langle 111 \rangle$ lying on the {211} face so that a slice cut $\perp$ to that $\langle 111 \rangle$ and lying $\perp$ to the {211} is used.

(211) face, these parameters operated on by appropriate symmetry operators are used to predict the magnetocrystalline anisotropy and easy axis direction for growth on a (110) face. Agreement shows that the symmetry operations were performed correctly but does not necessarily prove the validity of a particular model. In addition to magnetic proofs of noncubicity, very careful lattice parameter measurements also show slight deviation from cubicity[19,29], however, X-ray measurements have so far not revealed any sign of ordering. It would be expected that the amount of order (say $\sim 10^{-3}$ of the rare earths) needed to give the ordered magnetic effects would be unlikely of detection by other means.

It is interesting to point out that since the anisotropy is annealed out at about 1200 °C, the ordered state may be either thermodynamically stable below the anneal out temperature or it may be metastable at all temperatures and result from the growth process. The fact that the direction of the easy magnetic axis is dependent upon the growth face makes it tempting to postulate a growth incuded anisotropy. For instance, if the order in an adsorbed layer (which might be different on different faces) were frozen in during growth, the observed anisotropies and anneal out behavior would be possible. It is interesting to point out that the anisotropy in hydrothermal garnets* is usually larger, as might be expected for a lower temperature process where a higher degree of order in an adsorbed layer would be favored.

Clearly, however, the noncubic garnets fulfill device requirements and provide a system with enough degrees of freedom to allow the magnetic device designer to vary material properties over wide ranges. The crystal growers first job is done. His second and equally challenging job remains.

3. Developing the growth method

An ideal device configuration might require sheets of garnet $\sim 6\ cm^2$ with a thickness of about 7.5 μm. Cutting bulk garnet to this thickness does not appear attractive. Thus, heteroepitaxial processes using nonmagnetic substrates appear to be the appropriate growth methods. A detailed review of work on the preparation of thin films of noncubic garnets would not be appropriate here, especially since much of this research is currently in progress. However, growth methods will be mentioned and some flavor of current progress will be given. Three heteroepitaxial methods have been used (1) chemical vapor deposition (CVD)[31,32]; (2) liquid phase epitaxy (LPE)[33,34]; (3) hydrothermal epitaxy (HE)[30].

It might be pointed out that growth from the melt in terms of simplicity ought to receive careful consideration. Indeed although oxygen pressures are required, $Y_3Fe_5O_{12}$ (albeit of rather poor quality) has been melt grown in the past**[35]. Further, it is interesting to point out that of the orthoferrites $YFeO_3$ is comparatively stable to reduction and can be melted (MP > 1680 °C) without decomposition in one atmosphere of O_2. Oka-

da et al.[37]) float zoned the material and Blank et al.[38]) grew Bridgman crystals. Garnets will not exhibit noncubicity by pair ordering or similar mechanisms if grown above ~ 1200 °C. They can, of course, be made uniaxial by means of strain and strain is easy to obtain in heteroepitaxial growth if the lattice parameter and/or coefficient of expansion of substrate and film do not match. However, there may be some advantages of uniformity and reproducibility in producing the desired magnetic properties by processes not involving strain. Thus, lower temperature heteroepitaxial processes seem dictated.

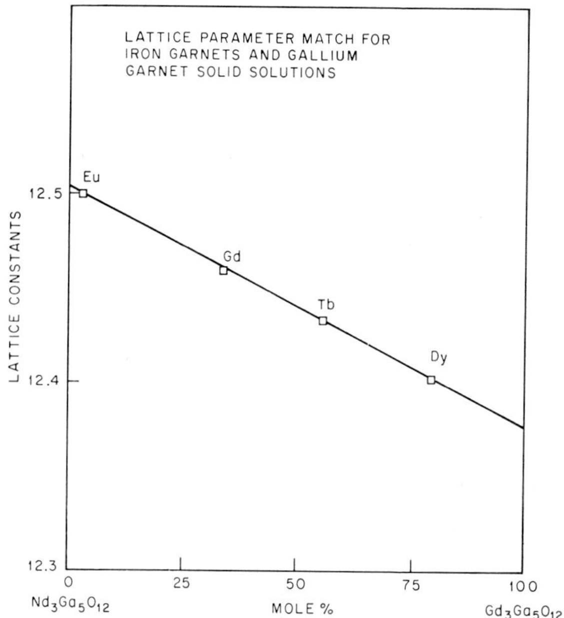

Fig. 3. Lattice parameters in the $Nd_3Ga_5O_{12}$–$Gd_3Ga_5O_{12}$ system.

Lattice parameter and coefficient of expansion match of the substrate and film have received considerable recent attention. Rare earth gallium and aluminum garnets are comparatively easy* to grow by the Czochralski technique and are generally used for substrates. Substrate decomposition in the milieu used for growth can be a problem. However, combinations (such as $Nd_xGd_{3-x}Ga_5O_{12}$) whose lattice parameters match particular magnetic garnets can be devised from Vegard's law plots of lattice parameters[39]) such as fig. 3, and have been grown[40]).

---

* Large noncubic garnets of compositions II–VI have been grown hydrothermally in 20 m KOH, 87% fill, 385 °C crystallization temperature, and a ΔT of 40 °C (ref. 30).

** The Y_2O_3–Fe_2O_3 phase diagram was carefully determined by Van Hook[36]).

* In the perfection ultimately required for device use the growth of these materials is proving a demanding challenge.

We will now briefly discuss the status of the three most used hetero-methods.

(1) CVD – Fig. 4 shows Robinson's apparatus [after Nielsen[32])]. The reaction is:

$$3\,(R)Cl_3 + 5\,FeCl_2 + 6\,O_2 \rightarrow (R)_3Fe_5O_{12} + \tfrac{19}{2}\,Cl_2 .$$

Films are deposited in the neighborhood of 1200 °C and uniaxial garnets have been deposited by Mee et al.[31]) and more recently by Robinson[32]). Ga doping to

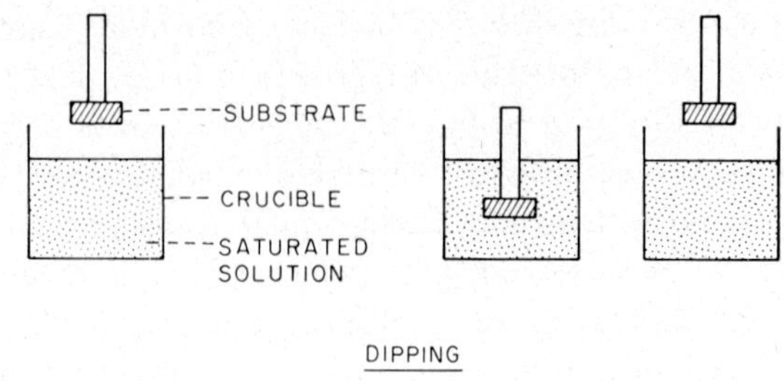

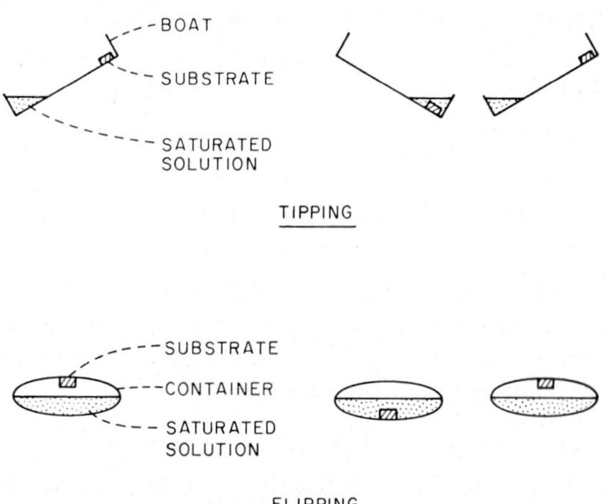

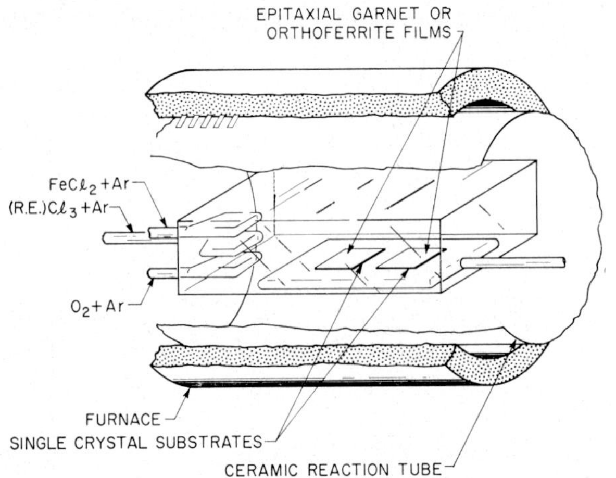
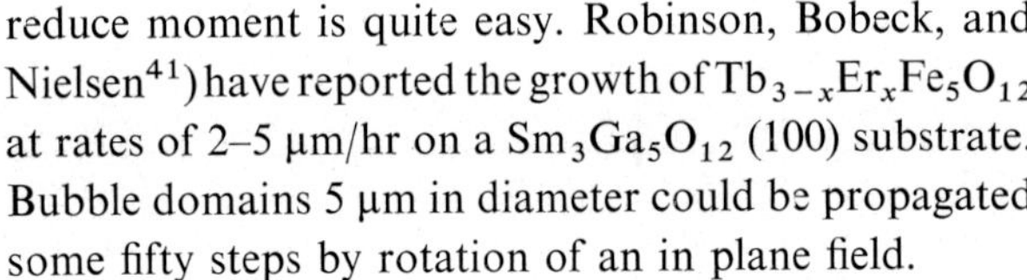

Fig. 4. Schematic of CVD apparatus.

Fig. 5. Geometries for LPE.

reduce moment is quite easy. Robinson, Bobeck, and Nielsen[41]) have reported the growth of $Tb_{3-x}Er_xFe_5O_{12}$ at rates of 2–5 µm/hr on a $Sm_3Ga_5O_{12}$ (100) substrate. Bubble domains 5 µm in diameter could be propagated some fifty steps by rotation of an in plane field.

(2) LPE – Fig. 5 shows a schematic of the three geometries which have been used in semiconductor work. Tipping and dipping have been used for garnet growth. The solvents are similar to those used in flux growth, although more emphasis has been placed upon low volatility components. A variety of films on several substrates have been reported by Shick et al.[42]). Composition IV of table 4 and $Er_2Eu_1Ga_{0.7}Fe_{4.3}O_{12}$ have been deposited as 5–20 µm thick films free of cracks on the (110) and (111) faces of $Gd_3Ga_5O_{12}$ substrates. The flux used was $PbO–B_2O_3$. Growth was by tipping. Solutions were saturated (by equilibration at higher temperatures) at 865–920 °C. Following tipping the temperature was immediately lowered at about 300 °C/hr to 850 °C. For (110) growth the cooling rate was 44 °C/hr and cooling continued to 832 °C. Films con-

taining several thousand bubbles have been made. Bubble size was 6–14 µm.

(3) HE – Early work was on homoepitaxial growth of orthoferrites[30]). Heteroepitaxial growth of garnet under conditions to reduce substrate attack has proven moderately successful. Noncubic films of composition IV have been deposited on $Gd_3(Ga_{5-x}Sc_x)O_5$ and bubbles have been observed. Conditions are given by Kolb and Laudise[30]).

The present problems with all three methods are reproducibly obtaining large area high perfection films. The principal imperfections are:

(1) Those associated with gross decomposition of the substrate – voids in the hetero film – common in hydrothermal material, rare in other methods.

(2) Low angle grain boundaries – several/cm² in CVD and LPE.

(3) Dislocations – probably present in all material, probably not important in affecting magnetic behavior.

(4) Coercivity variations due to Ga gradients – can occur in LPE, insufficient study to evaluate in CVD.

Thus, the noncubic garnets are shown to meet device requirements and the feasibility of growing films has been demonstrated, no unsurmountable obstacles have been revealed, but much hard crystal growth work remains to be done.

Acknowledgments

I would especially like to thank L. G. Van Uitert and A. H. Bobeck, whose work to a large degree comprised the materials search, for discussions and permission to quote from their unpublished work. L. J. Varnerin and J. W. Nielsen and E. M. Gyorgy have provided the author with many stimulating discussions of growth and properties.

References

1) P. Weiss, J. Physique 6 (1907) 667.
2) A. H. Bobeck, Bell System Tech. J. 46 (1967) 1901.
3) A. H. Bobeck, R. F. Fischer, A. J. Perneski, J. P. Remeika and L. G. Van Uitert, IEEE Trans. Mag. **MAG-5** (1969) 544.
4) A. J. Perneski, IEEE Trans. Mag. **MAG-5** (1969) 544.
5) A. A. Thiele, Bell System Tech. J. **48** (1969) 3287.
6) A. A. Thiele, J. Appl. Phys. **41** (1970) 1139.
7) M. A. Gilleo, J. Chem. Phys. **24** (1956) 1239.
8) R. M. Bozorth, Phys. Rev. Letters **1** (1958) 362.
9) R. M. Bozorth, V. Kramer and J. P. Remeika, Phys. Rev. Letters **1** (1958) 3.
10) L. G. Van Uitert, R. C. Sherwood, W. A. Bonner, W. H. Grodkiewicz, Miss L. Pictroski and G. Zydzik, Mater. Res. Bull. **5** (1970) 153.
11) D. Treves, J. Appl. Phys. **36** (1965) 1033.
12) M. Robbins, R. D. Pierce and R. Wolfe, J. Appl. Phys. **42** (1971) 1563.
13) R. C. Sherwood, L. G. Van Uitert, R. Wolfe and R. C. LeCraw, Phys. Letters **25A** (1967) 297.
14) L. G. Van Uitert, R. C. Sherwood, E. M. Gyorgy and W. H. Grodkiewicz, Appl. Phys. Letters **16** (1970) 84.
15) L. G. Van Uitert, D. H. Smith, W. A. Bonner, W. H. Grodkiewicz, and G. J. Zydzik, Mater. Res. Bull. **5** (1970) 455.
16) A. H. Bobeck, A Summary of Magnetic Bubble Technology, in: *Magnetic Oxides* (Wiley, New York, to be published).
17) G. Asti, M. Colombo, M. Guidici and A. Levialdi, J. Appl. Phys. **36** (1965) 3581.
18) L. G. Van Uitert, W. A. Bonner, W. H. Grodkiewicz, Miss L. Pictroski and G. J. Zydzik, Mater. Res. Bull. **5** (1970) 825.
19) A. H. Bobeck, E. G. Spencer, L. G. Van Uitert, S. C. Abrahams, R. L. Barns, W. H. Grodkiewicz, R. C. Sherwood, P. H. Schmidt, D. H. Smith and E. M. Walters, Appl. Phys. Letters **17** (1970) 131.
20) F. Bertaut and R. Pauthenet, Proc. IEEE B **104** (1957) 261.
21) S. Iida, J. Phys. Soc. Japan **22** (1967) 1201.
22) R. C. LeCraw and F. B. Hagedorn, private communication.
23) A. H. Bobeck, presented at the Intern. Magnetics Conf., Washington, D.C., 1970.
24) A. Rosencwaig and W. J. Tabor, J. Appl. Phys. **42** (1971) 647.
25) A. Rosencwaig, W. J. Tabor, F. B. Hagedorn and L. G. Van Uitert, unpublished.
26) A. Rosencwaig, W. J. Tabor and R. D. Pierce, unpublished.
27) H. Cullen, Appl. Phys. Letters **18** (1971) 711.
28) E. M. Gyorgy, A. Rosencwaig, E. I. Blount and W. J. Tabor, Appl. Phys. Letters, to be published.
29) R. L. Barns, private communication.
30) E. D. Kolb and R. A. Laudise, J. Appl. Phys. **42** (1971) 1552.
31) J. E. Mee, G. R. Pulliam, J. L. Archer and P. J. Besser, IEEE Trans. Mag. **MAG-5** (1969) 717; see also D. M. Heinz, P. J. Besser, J. M. Dwens, J. E. Mee and G. R. Pulliam, J. Appl. Phys. **42** (1971) 1243.
32) Mac D. Robinson, reported by J. W. Nielsen, Conf. on Magnetic and Electronic Materials for Computer Memories, AIME, New York, August 31, 1970.
33) R. C. Linares, J. Crystal Growth 3, 4 (1968) 443.
34) L. K. Shick and J. W. Nielsen, J. Appl. Phys. **42** (1971) 1554.
35) M. Jones, Texas Instruments Co., Dallas, Texas, private communication.
36) H. J. Van Hook, J. Am. Ceram. Soc. **44** (1961) 208.
37) T. Okada, K. Matsumi and H. Makimo, presented at the Intern. Ferrite Conf. Kyoto, Japan, 1970.
38) S. Blank, L. K. Shick and J. W. Nielsen, J. Appl. Phys. **42** (1971) 1557.
39) R. L. Barns, private communication.
40) C. D. Brandle, Jr., private communication.
41) Mac D. Robinson, A. H. Bobeck and J. W. Nielsen, to be published.
42) L. H. Shick, J. W. Nielsen, A. H. Bobeck, A. J. Kurtzig, P. C. Michaelis and J. P. Reekstin, Appl. Phys. Letters **18** (1971) 89.

X-RAY TOPOGRAPHY AS A TOOL IN CRYSTAL GROWTH STUDIES

A. AUTHIER

Laboratoire de Minéralogie-Cristallographie, associé au CNRS, 9, Quai Saint-Bernard, Paris 5, France

X-ray topography is a non-destructive method revealing extended imperfections such as dislocations, stacking faults, growth horizons, twins, inclusions and so on; in non too imperfect crystals, it has rapidly proved a very useful tool in studying problems related to crystal growth. The configuration, nature and density of lattice defects usually vary considerably depending on the method and conditions of growth as well as on the chemical nature of the crystal. One very frequent observes very different degrees of perfection for the same crystal grown by different methods, but similar configurations of defects for different crystals grown by the same method.

The dislocation configuration arising in crystals grown from the solution will be studied in more detail. Dislocations in this case tend to be linear and their orientation closely related to that of the growth faces. They usually constitute bundles originating at inclusions and along the interface between the seed and the new crystal, which is where the majority of dislocations are created. The importance of the surface of the seed and of the initial stages of the growth is therefore primordial for the quality of the crystal to be grown. Any fluctuation of the conditions of growth will usually produce a growth horizon along which dislocations are sometimes formed.

1. Introduction

The determination of the nature and distribution of defects is not only necessary for the characterization of crystal perfection, but also useful for a better understanding of the growth mechanism.

Many defects are formed during growth. They may be actively involved in the growth process, such as dislocations with a screw component, but are usually a consequence of the growth conditions, impurity segregation, inclusions, dislocations due to lattice closure, errors around an inclusion, stacking faults, twins etc. They are also often formed after growth and are due to thermoeleastically activated dislocation sources, coalescences of vacancies during rapid cooling, or simply handling.

X-ray topography[1]) is a non-destructive method for ervealing strains in a nearly perfect crystal. Its sensitivity to small distortions and variations of lattice parameters is excellent and can be further improved by using two crystal settings[2]), X-ray interferometry[3]) or moiré techniques[4]). Its geometrical resolution is of the order of one micron in the best conditions and cannot be improved since there are no optics for X-rays and topographs are not enlarged with respect to the crystal. All extended defects such as long range internal stresses, dislocations, precipitates, inclusions, impurity segrega-

tion bands, stacking faults, twins etc... give rise to strains or phase shifts which are at the origin of contrast effects on the topographs[5,6]). The anisotropy of the strains and the Burgers vector of dislocations can be determined by taking various topographs with different reflections.

X-ray topography has been in use for twelve years and has been applied to many different types of crystals: elements and compounds, metallic, covalent, ionic, molecular and so on. In many cases, it has brought very valuable information not only about crystal perfection, but also about the growth mechanism. This is due to the fact that knowledge of both the geometrical shape and Burgers vectors of dislocations for instance often makes it possible to distinguish between defects formed *during* or *after* growth, between edge and screw dislocations and so on.

2. Crystals grown from the melt

The distribution and nature of defects depends highly on the conditions of growth, often more than on the chemical nature of the crystal. One may for instance distinguish between crystals grown from the melt at high temperatures and crystals grown from the solution at temperatures not very different from room temperature.

Dislocations formed during growth at high temperatures are usually subject to climb and have curved shapes[7]). Stresses of thermal origin activate many dislocation sources. During annealing, polygonization may take place, giving rise to subgrains[8]).

During cooling, coalescence of vacancies gives rise to loops as has been observed in the case of aluminium[9,10]). Michell[11]) and coworkers have shown in the case of copper that when single crystals are submitted to a heat treatment in the presence of hydrogen, hydrogen segregates in the vacancy clusters, its accumulation building up an internal pressure which gives rise to rows of punched out loops.

The degree of perfection varies greatly according to the method of growth as has been shown for example by Authier and Sauvage in the case of corundum[12]) and by Wang and McFarlane in the case of magnesium aluminium spinels[13]). The crystal quality improves greatly from Verneuil or flame-growth techniques to Czochralski and flux techniques.

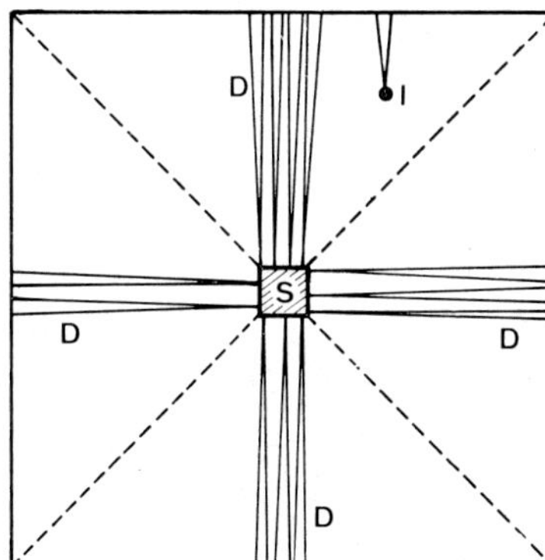

Fig. 1. Defects arising in a crystal grown from the solution S: seed; D: dislocations; I: inclusion.

3. Crystals grown from the solution

Fig. 1 describes schematically a crystal grown from solution. Broken lines show the intersections between neighbouring growth pyramids. Two types of defects can usually be observed: planar defects, growth bands, striae on one hand, dislocations on the other. The former will be described in a later section.

Dislocations run usually in straight lines, approximately normal to the growth faces. Most of them originate at the seed, dividing the crystal in regions which are practically dislocation-free[14,15]) – see for example figs. 2c and 2d. A certain number will be generated along a growth phantom accompanying an abrupt

change from a slow growth by two-dimensional nucleation to fast growth by a dislocation mechanism[16]) – see for example fig. 2c. These growth phantoms occur because of a temperature or a concentration fluctuation. Many dislocations are created because of lattice closure errors around solid inclusions[17]) either isolated or distributed along the seed or a growth phantom; they usually form bundles (fig. 2) and are a consequence of the growth conditions, only sometimes are they actively engaged in the growth process. When a dislocation line leaves one growth pyramid to enter another one, its direction changes abruptly, remaining always normal to the growth face (fig. 2a).

Fig. 2 compares the distribution of defects in triglycine sulphate crystals of various purities and growth conditions. Fig. 2a is an X-ray topograph of an iron doped sample. The seed has a square cross section and can be seen in the lower left hand corner of the picture. A fast horizontal face, normal to b-axis, was present at the beginning of the growth and disappeared rapidly as the slow dominant {110} faces grew. The intersection between the corresponding growth pyramids is revealed by the bends in the dislocation lines. Its slope enables one to calculate the ratio between the growth rates of the fast and slow faces, about four in this particular case. Dislocation bundles may be seen to arise at inclusions along the intersections of {110}, growth pyramids. Some dislocations from the seed continue in the new crystal and can be seen crossing the right hand limit of the seed. This only occurs when the transition between seed and new crystal is very smooth, without inclusions or many defects. The overall dislocation density of this crystal is much higher than that of the much purer samples observed in figs. 2b, 2c, 2d.

Topographs 2b and 2c represent crystals grown from pure solutions at constant supersaturation. The dislocation density is much smaller than in the iron-doped sample. It is the smallest for crystal 2c grown with minimum supersaturation. In the latter case, a minor face normal to the b-axis remained present a long time as a slow face until an accident occurred which induced a concentration fluctuation. The growth mechanism then abruptly changed with many dislocations appearing. The growth rate increased significantly and the face disappeared rapidly.

Small precipitates can be seen in both crystals, some decorating dislocation lines.

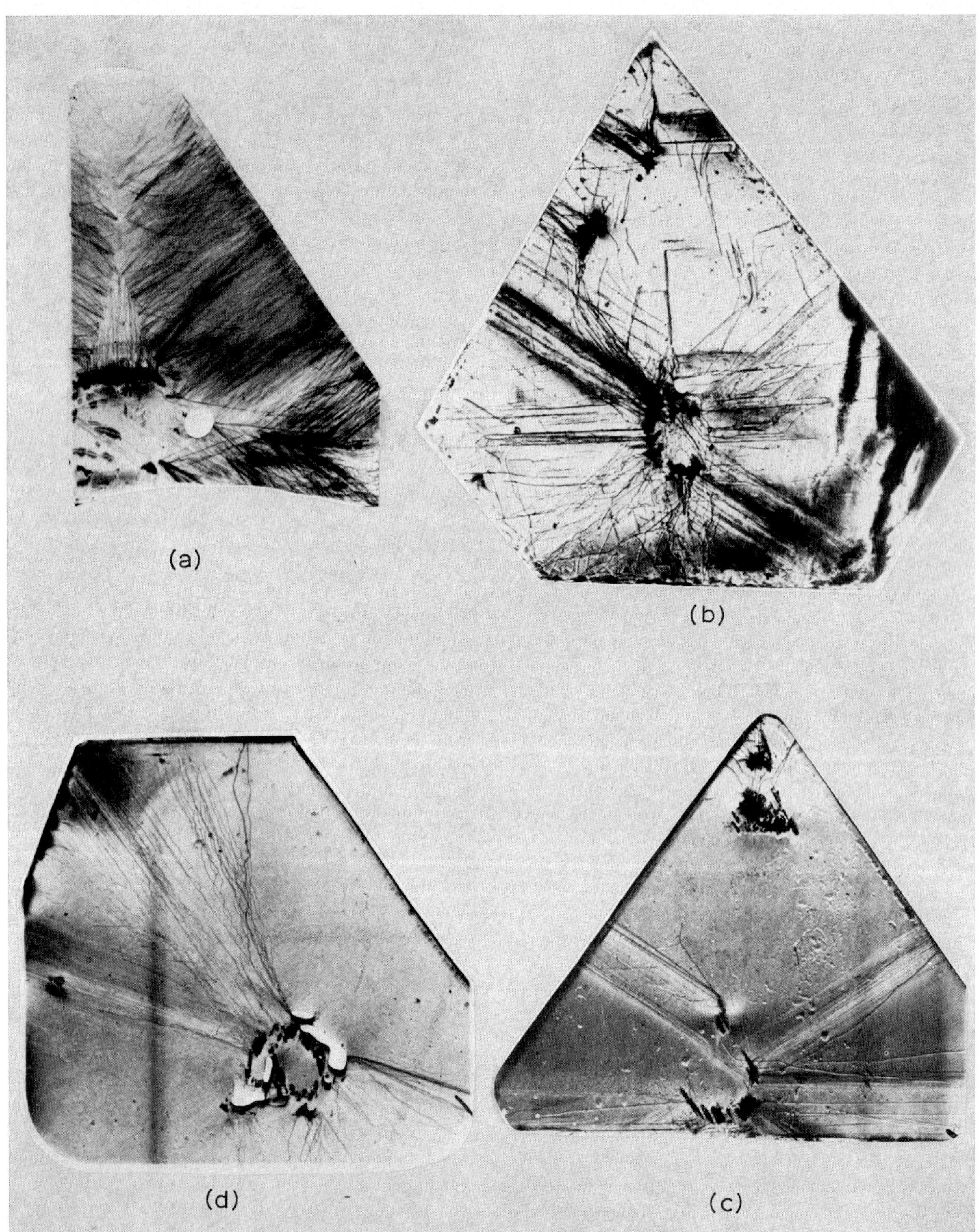

Fig. 2.   X-ray topographs of triglycine sulphate single crystal slabs cut normal to the *c*-axis. Thickness about 4 mm. Observe dislocations radiating from the seed and inclusions. (a) Iron-doped sample – Mo Kα radiation; 040 reflection; ×5.6. (b) pure sample grown at constant supersaturation – Mo Kα radiation; $\bar{2}20$ reflection; ×4.8. (c) pure sample grown at a lower constant supersaturation than sample (b) – Mo Kα radiation; 040 reflection; ×4. (d) very pure sample grown with decreasing temperature from a four times recrystallized solution – Mo Kα radiation; 040 reflection; ×4.

Fig. 3. X-ray topograph of an hexamethylenetetramine crystal. Thickness: 1.25 mm; orientation: (112); Mo Kα radiation; 01$\bar{1}$ reflection; growth bands parallel to {110} growth faces; ×14.5.

Topograph 2d represents a small crystal grown by decreasing the temperature from a four times recrystallized solution. The dislocation density is low and precipitates occur along one growth horizon only.

4. Impurity segregation and growth bands

Growth bands, striae, growth phantoms occur very frequently and can practically always be observed if a sensitive enough method is used. They are usually due to impurity segregation, thermal or concentration fluctuations and are sometimes decorated by inclusions. They may be relatively thick or extremely thin. If they are associated with a phase shift, such as a translational twin, they are visible on X-ray topographs even if they are one atomic layer thick. Planar growth faults are usually accompanied on X-ray topographs by thin fringes which are due to interferences between the various types of X-ray wave-fields having crossed the fault. One example is given on fig. 3 representing a crystal of hexamethylenetetramine with a very low dislocation density. When the growth band is associated with impurity segregation, their contrast on X-ray topographs is related to atomic displacements normal to the band.

5. Conclusion

X-ray topography, by revealing the nature and density of defects formed during growth enables one to characterize crystal perfection and to determine the influence of growth conditions on this perfection. Furthermore, it shows which defects are involved in the growth mechanism and which are a consequence of the growth conditions. Its use has been proved in the past by many examples and it will in the future remain one of the tools to be used in any crystal growth study for the assessment of crystal perfection and a better understanding of the growth processes.

Acknowledgements

I am endebted to Drs. Malek, Moravec and Novotny for the triglycine sulphate crystals, to Mrs. A. Izrael for the corresponding topographs, and to Dr Di Persio for the topograph of a hexamethylenetetramine crystal.

References

1) A. R. Lang, J. Appl. Phys. **30** (1959) 1748.
2) U. Bonse, Z. Physik **153** (1958) 278.
3) U. Bonse and M. Hart, Z. Physik **188** (1965) 154.
4) A. R. Lang and V. F. Miuscov, J. Appl. Phys. Letters **7** (1965) 214.
5) A. Authier, Advan. X-ray Analysis **10** (1967) 9.
6) A. Authier, Phys. Status Solidi **27** (1968) 77.
7) A. E. Jenkinson and A. R. Lang, in: *Direct Observations of Imperfections in Crystals*, Eds. J. B. Newkirk and T. H. Wernick (Interscience, New York) p. 471.
8) J. F. Petroff and A. Authier, Physica Status Solidi **13** (1966) 373.
9) A. Authier, C. B. Rogers and A. R. Lang, Phil. Mag. **4** (1965) 878.
10) C. G'Sell, B. Baudelet and G. Champier, J. Crystal Growth **13/14** (1972) 252.
11) L. M. Clarebrough, D. Michell and A. P. Smith, Phys. Status Solidi **21** (1967) 369.
12) A. Authier and M. Sauvage, Revue Intern. Hautes Temp. (1970).
13) C. C. Wang and S. H. McFarlane III, J. Crystal Growth **3, 4** (1968) 485.
14) S. Ikeno, H. Maruyama and N. Kato, J. Crystal Growth **3, 4** (1968) 683.
15) F. Lefaucheux, this conference (1971).
16) F. Mussard and S. Goldsztaub, J. Crystal Growth, to be published.
17) A. Zarka, Bull. Soc. Franc. Mineral. Crist. **92** (1969) 160.

Reprinted from:
Journal of Crystal Growth **13/14** (1972) 39–43 © *North-Holland Publishing Co.*
Printed in the Netherlands

HYDRODYNAMICS, CONVECTION AND CRYSTAL GROWTH

D. T. J. HURLE

Royal Radar Establishment, Great Malvern, Worcestershire, U.K.

Because the existing literature in this field is not well documented, the author has attempted to use the limited space available for this review to provide a bibliography (though by no means an exhaustive one) of the field rather than to treat one aspect of the subject in detail. The subjects reviewed are the effects of steady forced and natural convection on segregation in melt growth, the role of non-steady convection in forming striations and the effects of gas-phase convection on crystal growth.

1. Introduction

The existence of natural convection in the fluid phase from which crystal growth is taking place, and the value of mechanically-induced mixing as a means of reducing the diffusional barrier to crystal growth have, of course, both been known for a long time. The use of crystal and/or crucible rotation in the crystal pulling technique and of rotating "trees" in aqueous solution crystallisers are examples of the latter.

The role of convection in controlling solute distributions was highlighted by the development of zone refining. Tiller et al.[1] had shown that, in the absence of convective mixing, directional solidification at constant velocity (v) leads to the formation of a solute boundary layer ahead of the growing crystal whose thickness increased, during an initial transient period, to a final, steady value such that the solute concentration in the solid (C_S) became equal to that in the bulk liquid (C_L), so that no macro-segregation occurred. Pfann[2] pointed out that since zone refining worked this could not represent the true situation and convection must be important.

At about the same period, numerous studies of impurity striations in melt-grown semiconductor crystals were being made[92,93]. The helical distribution in rotated, pulled crystals (with pitch equal to the amount of crystal grown per rotation cycle) was recognised as being due to thermal asymmetry in the melt[3–6]. A variety of explanations were put forward to explain the "non-rotational" striations: for example Landau[7] and

Milevskii[8]) invoked the periodic build up and subsequent rapid solidification of a zone of constitutionally-supercooled melt ahead of the interface, whilst Ueda[9]) and Gatos et al.[10]) proposed the periodic build up of thermal supercooling followed by rapid growth which destroyed that supercooling by latent heat evolution.

Komarov and Regel[11]) visually observed periodic growth and re-melting at the interface of a bismuth crystal and suggested that this was caused by turbulent convection. As we now know, non-steady convection is the true cause of non-rotational striations. In the following two sections we review the effects of steady and non-steady convection on segregation in melt growth.

2. Segregation and time-independent flow

The gulf between the "no-mixing" approach of Tiller et al.[1]) and the complete-mixing of Pfann[2,12]) was bridged by the Burton, Prim and Slichter[13]) boundary layer model. In terms of a boundary layer adjacent to the solid–liquid interface of thickness δ within which diffusion is the dominant solute transfer mechanism, they showed that the effective distribution coefficient is:

$$k_{\text{eff}} = \frac{C_S}{C_L} = \frac{k^*}{k^* + (1 - k^*) \exp(-v\delta/D)}, \qquad (1)$$

where k^* is the interface distribution coefficient and D the solute diffusion coefficient. Effective use of this formula requires specification of δ in terms of the degree of motion in the melt. In the case of flow caused by the rotating crystal, Burton et al.[13]) used the Cochran[14]) and von Karman[15]) treatments of flow at a rotating disc to show that:

$$\delta \sim 1.6\, S^{-\frac{1}{3}}(v/\omega)^{\frac{1}{2}}, \qquad (2)$$

where $S = v/D$ in the Schmidt number, v the kinematic viscosity and ω the angular velocity of the crystal. In the general case, δ is expressable in terms of a characteristic dimension and some powers of the Schmidt and Reynolds numbers for the flow. For convective flow, the Reynolds number will be related to the Rayleigh number (non-dimensional temperature gradient). The rotating disc has a unique property which the rotating crystal shares approximately; namely a uniformly-accessible boundary layer. That is, the component of flow normal to the surface of the disc is independent of radial position and hence δ and therefore k_{eff} are constant over the whole crystal interface. It is this property which results in the high degree of radial uniformity of solute concentration attainable with the pulling technique and leads to the use of rotating discs in chemical and electrochemical transport rate studies[16]). Heat and mass transfer in fractional solidification has been reviewed by Wilcox[85]).

By reducing the thickness of the solute boundary layer for $k^* < 1$, convective mixing reduces the degree of constitutional supercooling of the melt (solvent or flux), should it exist. The gradient of constitutional supercooling at the interface ($x = 0$) is[17])

$$\left(\frac{\mathrm{d}S}{\mathrm{d}x}\right)_{x=0} = -\frac{mvC_{\mathrm{L}}(1-k^*)}{D\{k^* + (1-k^*)\exp(-v\delta/D)\}} - G_{\mathrm{L}}, (3)$$

where m = slope of liquidus curve and G_{L} is the temperature gradient in the melt at the interface. Decreasing δ decreases the magnitude of the first term on the RHS of (3). However, combination of (1) and (3) shows that, if the object is to achieve a given solid concentration C_{S}, then the magnitude of $(\mathrm{d}S/\mathrm{d}x)_{x=0}$ is independent of δ.

The effect of fluid flow on the marginal state and critical wavelength for planar interface breakdown has been considered by Delves[18,19]) and by Hurle[20]). The more complex problem of the interaction between the cellular pattern of solute rejection caused by interface breakdown and the natural convection due to the density gradients so set up has not received study. However Hämäläinen[21,22]) has obtained a complex pattern of cellular segregation in thin layers of alkali halide solid

solutions which he ascribes to the interaction between a cellular interface and an array of Bénard convection cells[23]) in the melt.

The combined effects of crystal and crucible rotation have been studied in model crystal-pulling systems using transparent liquids with dye injection to delineate the flow[24-28]). An excellent account of the main effects due to rotational flow has been given by Carruthers and Nassau[28]). Numerical computations of flow patterns resulting from thermal convection and crystal and crucible rotation have been performed by Kobayashi and Arizumi[29,30]). Rotation of the crucible relative to the crystal leads to the formation of a column beneath the crystal [a "Taylor–Proudman cell"[28])] within which motion is controlled essentially by the vertical variation in the radial centrifugal pressure gradient. Outside this cell thermal convective flow usually dominates. The two regions are separated by a stagnation surface which effectively isolates the solute in the outer region from mixing with that in the inner. Changes in the relative rotation rates disrupt the stagnation surface and carry fluid from the outer region into the region adjacent to the crystal, and this can produce marked changes in solute concentration. If the two rotations are counter to each other additional stagnation surfaces can appear[28]). The radial spoke patterns seen on oxide melts[59]) have not been reproduced in the model systems.

Other possible effects which appear not to have been considered are the baroclimic instabilities discovered by meteorologists in models of atmospheric circulation[31,32]) and the non-axisymmetric shear layers reported by Hide and Titman[33]).

The influence of solute concentration gradients near the growth interface (produced by rejection at the interface) on convective motion in the melt has not been extensively examined but may well be very significant. Wagner[44]) calculated the flow velocity and the resulting degree of segregation in a horizontal directional solidification system where the convection was driven by density gradients due to the rejected solute. More recently experimental studies of the effects of both thermal and solutal convection on segregation and interface shape have been made for the solidification of lead[45,46]). It is known that, where, as could occur in this case, an overall density gradient is composed of opposing contributions from two entities which diffuse at different rates (e.g. heat and solute, or a fast and a slow diffusing solute), then curious mixing phenomena can occur[34-39]). It is conceivable, for example, that the phenomenon of "salt fingering"[35]) (in which long, thin solute-rich "fingers" descend into colder fluid) could occur during Stockbarger growth in which a solute which is less dense than the solvent is rejected at the interface but such possibilities appear not to have been investigated. In castings, on the other hand, "A" and "V" segregates[40-42,90]) and "freckles"[43]) have recently been shown to be due to thermosolutal convection in the "mushy" zone of the casting.

3. Striations and time-dependent convection

Several workers have demonstrated temperature fluctuations in the melt which have been ascribed to turbulent free convection[47-50]). The literature on striations in crystals has been reviewed by Muller and Wilhelm[51]). These authors demonstrated rather regular variations in temperature in the molten zone of an InSb charge undergoing zone-melting and they obtained some correlation between the spacing of Te-solute striations in the crystal and the frequencies of temperature oscillation in the melt. They found that they could eliminate this periodic variation by the use of sufficiently short zones and low temperature gradients.

The present author and his colleagues[52-56]) have made a detailed study of thermal convection in rectangular volumes of liquid gallium and have shown that sinusoidal temperature oscillations of large amplitude can be obtained in both "heated from below" and "heated from the side" configurations. The period of the oscillation depends on the melt dimensions. Oscillations have also been demonstrated in cylindrical columns of mercury heated from below by Verhoeven[57]), in molten $NaNO_3$ by Carruthers and Pavilonis[58]) and in molten CaF_2 and $CaWO_4$ by Cockayne and Gates[59]). The first observation of regular temperature oscillations in liquid metals was probably made by A. J. Goss who obtained oscillations in molten Sn, Ge and Bi in 1955 but this work was never published.

Utech and Flemings[60,61]) and Chedzey and Hurle[55,62]) have independently shown that, for conducting melts such as InSb, the oscillations and fluctuations can be damped by the application of a magnetic field normal to the horizontal growth axis. When this is done, there are no striations in the crystal. The basic

frequencies present in puller melts are about an order of magnitude lower than those in horizontal-growth melts of similar volume and, in consequence, the fields required to damp the fluctuations are proportionaly higher. Added to this, the configuration of a puller makes it difficult to locate pole pieces sufficiently close to the melts, so that, overall, the use of magnetic damping in pulling does not look so promising as in horizontal growth. In non-conducting melts, for which of course magnetic damping is not possible, attempts to minimise striations have involved the use of shallower melts and lower temperature gradients to reduce the Rayleigh number[63]) and careful centering of the crucible to minimise thermal asymmetry.

It is then, now well established that low Prandtl number fluids exhibit periodic components to their convective motion under conditions of quite low Rayleigh number (Ra). The number of such modes increases with increasing Ra until, finally, at several orders of magnitude above the threshold value of Ra, the flow can be said to be fully turbulent. It has been demonstrated that these periodic components are the direct cause of solute striations in metal[61]), semiconductor[60,62,86,87]) and oxide crystals[59,63,84,91]) and of "banding" in uni-directionally solidified eutectics[89]). Lateral motions of the radial spoke pattern in oxide melts are a possible further source of striations[59]). The factors which give rise to this periodic motion are not well understood. A consideration of possible mechanisms is beyond the scope of this review but a survey of this field will be published elsewhere[64]). The one form of periodic ("overstable") convective motion that is well understood is that observed in the rotating Bénard cell[23,65]). Carruthers[66]) has attempted to correlate the observed spacings of striae in Si crystals grown with crucible rotation with the frequencies predicted for a Bénard cell rotating at the same rate.

Having established that temperature fluctuations occur in crystal growth melts, we must next ask how the crystal–melt interface responds to such fluctuations and how the concentration of solute incorporated varies. This has been considered theoretically by Carruthers[67,68]) and by Hurle, Jakeman and Pike[69,70]) and elegantly demonstrated experimentally by Gatos and co-workers[71-74]). The last-named authors have introduced very fine striations into pulled crystals of Te-doped InSb by vibrating the crucible at a chosen frequency (which could be varied). Variations in the spacing of these "rate" striations revealed variations in the rate of growth of the crystal on longer time scales. They showed that there were marked changes in the instantaneous growth rate of nonfacetted portions of the interface during one rotation of the crystal and that re-melting could occur during part of the cycle. Only very small variations in normal growth rate occurred on the facetted regions due presumably to the fact that the normal growth rate on the facet is controlled by the maximum undercooling which remains fairly constant although the position on the interface at which it occurs will vary periodically. Nonetheless, striations do occur on the facetted region even though the normal velocity is sensibly constant. (The velocity of steps parallel to the interface cannot of course be constant.)

Hurle and Jakeman[70]) have shown theoretically that the average degree of segregation is reduced by the presence of striations and this has been demonstrated by Barthel and Eichler[75]) in experiments on the floating-zone melting of W doped Mo. The effect is particularly marked for solutes with $k > 1$.

The dependence of the amplitude of the concentration fluctuations on the amplitude and temporal frequency of the temperature oscillations, is complex. At low frequencies, it tends to a constant value, whilst strong damping occurs above a critical frequency which is either the thermal relaxation frequency ($\omega_T = 2K_L/\delta_T^2$) or the solutal equivalent ($\omega_c = 2D/\delta^2$) depending on whether the response rate is controlled by thermal or solute diffusion processes. K_L is the thermal diffusivity of the liquid and δ_T the thermal boundary layer thickness. If the striae spacing (v/ω) is very small solid state diffusion will subsequently anneal out the striations[61]). Clearly some bands of frequencies are particularly to be avoided. It has been claimed[76]), although as yet there is no independent confirmation, that anomalous segregation due to the "facet effect"[77]) disappears if all fluctuations are absent so that no striations are formed.

4. Convection in vapour growth

Apart from two studies[78,79]) on the flow patterns in epitaxial Si reactors, the author is not aware of any detailed studies of the role of natural convection in vapour growth. In these two studies flow patterns were delineated using TiO_2 smoke. The interaction between the streamline flow through the reactor and the vortical motion due to thermal convection above the heated substrate were clearly demonstrated. Bradshaw[80]) has demonstrated the existence of a thermal boundary layer in a Si reactor and successfully interpreted the temperature and concentration dependence of the growth rate on a diffusion-limited boundary layer model.

Reed et al.[88]) have considered the use of forced convective flow as a means to enhance the growth rate in diffusion-limited vapour growth systems and have specified conditions which prevent the occurrence of the analogue of constitutional supercooling under flow conditions.

It is likely that convective temperature fluctuations in vapour growth are of lesser importance than those in melt growth because the higher frequencies obtained in gases[81]) and the lower growth rates reduce the striation spacing to sub-diffusion distances.

The Rayleigh number of a gaseous convecting system at constant temperature difference increases as the square of the total pressure and, in consequence, the violence of the convection becomes a problem in pressurised crystal-growth systems. On the other hand the gaseous diffusion coefficient falls as the reciprocal of the pressure. The overall effects on the mass and heat transfer rates have been studied by Haggerty et al.[82]) and Chesswas et al.[83]).

References

1) W. A. Tiller, K. A. Jackson, J. W. Rutter and B. Chalmers, Acta Met. **1** (1953) 428.
2) W. G. Pfann, Acta Met. **1** (1953) 763.
3) W. D. Edwards, Can. J. Phys. **38** (1960) 439.
4) J. A. M. Dikhoff, Solid-State Electron. **1** (1960) 202.
5) A. C. English, J. Appl. Phys. **31** (1960) 1498.
6) K. Morizane, A. F. Witt and H. C. Gatos, J. Electrochem. Soc. **114** (1967) 738.
7) A. I. Landau, Phys. Metals Metallog. **6** (1958) 132.
8) L. S. Milevskii, Soviet Phys.-Cryst. **6** (1961) 193.
9) H. Ueda, J. Phys. Soc. Japan **16** (1961) 61.
10) H. C. Gatos, A. J. Strauss, M. C. Lavine and T. C. Harman, J. Appl. Phys. **32** (1961) 2057.
11) G. V. Komarov and A. R. Regel, Soviet Phys.-Solid State **5** (1963) 563.
12) W. G. Pfann, *Zone Melting* (Wiley, New York, 1966).
13) J. A. Burton, R. C. Prim and W. P. Slichter, J. Chem. Phys. **21** (1953) 1987.
14) W. G. Cochran, Proc. Cambridge Phil. Soc. **30** (1934) 365.
15) T. von Karman, Z. Angew. Math. Mech. **1** (1921) 244.
16) V. G. Levich, *Physicochemical Hydrodynamics* (Prentice Hall, New York, 1962).

17) D. T. J. Hurle, Solid State Electron 3 (1961) 37.
18) R. T. Delves, J. Crystal Growth 3, 4 (1968) 562.
19) R. T. Delves, J. Crystal Growth 8 (1971) 13.
20) D. T. J. Hurle, J. Crystal Growth 5 (1969) 162.
21) M. Hämäläinen, J. Crystal Growth 1 (1967) 125.
22) M. Hämäläinen, J. Crystal Growth 2 (1968) 131.
23) S. Chandrasekhar, *Hydrodynamic and Hydromagnetic Stability* (Clarendon Press, Oxford, 1961).
24) A. J. Goss and R. E. Adlington, Marconi Rev. 22 (1959) 132.
25) B. M. Turovskii and M. G. Mil'vidskii, Soviet Phys.-Cryst. 6 (1962) 606.
26) D. S. Robertson, Brit. J. Appl. Phys. 17 (1966) 1047.
27) J. R. Carruthers, J. Electrochem. Soc. 114 (1967) 959.
28) J. R. Carruthers and K. Nassau, J. Appl. Phys. 39 (1968) 5205.
29) N. Kobayashi and T. Arizumi, Japan. J. Appl. Phys. 9 (1970) 361.
30) N. Kobayashi and T. Arizumi, Japan. J. Appl. Phys. 9 (1970) 1255.
31) R. Hide, Phys. Fluids 10 (1967) 56.
32) R. Hide and R. J. Mason, Phil. Trans. Roy. Soc. London A. 268 (1970) 201.
33) R. Hide and C. W. Titman, J. Fluid Mech. 29 (1967) 39.
34) J. S. Turner and H. Stommel, Proc. Natl. Acad. Sci. U.S. 52 (1964) 49.
35) J. S. Turner, Deep Sea Res. 14 (1967) 599.
36) G. Veronis, J. Marine Res. 23 (1965) 1.
37) T. G. L. Shirtcliffe, J. Fluid Mech. 35 (1969) 677.
38) D. A. Nield, J. Fluid. Mech. 29 (1967) 545.
39) D. T. J. Hurle and E. Jakeman, Phys. Fluids 12 (1969) 2704.
40) R. J. McDonald and J. D. Hunt, Trans. AIME 245 (1969) 1993.
41) R. Mehrabian and M. C. Flemings, Met. Trans. 1 (1970) 455.
42) R. Mehrabian, M. Keane and M. C. Flemings, Met. Trans. 1 (1970) 1209.
43) S. M. Copley, A. F. Giamei, S. M. Johnson and M. F. Hornbecker, Met. Trans. 1 (1970) 2193.
44) C. Wagner, J. Metals 6 (1954) 154.
45) J. R. Carruthers and W. C. Winegard, Can. Met. Quart. 6 (1967) 233.
46) J. Szekely and P. S. Chhabra, Met. Trans. 1 (1970) 1195.
47) G. S. Cole and W. C. Winegard, Can. Met. Quart. 1 (1962) 29.
48) G. S. Cole and W. C. Winegard, J. Inst. Metals 93 (1964–5) 153.
49) W. R. Wilcox and L. D. Fullmer, J. Appl. Phys. 36 (1965) 2201.
50) K. Morizane, A. F. Witt and H. C. Gatos, J. Electrochem. Soc. 113 (1966) 51.
51) A. Muller and M. Wilhelm, Z. Naturforsch. 19a (1964) 254.
52) D. T. J. Hurle, Phil. Mag., 13 (1966) 305.
53) D. T. J. Hurle, J. Gillman and E. J. Harp, Phil. Mag. 14 (1966) 205.
54) E. J. Harp and D. T. J. Hurle, Phil. Mag. 17 (1968) 1033.
55) D. T. J. Hurle, in: *Crystal Growth*, Ed. H. S. Peiser (Pergamon, Oxford, 1967) p. 659.
56) D. T. J. Hurle, E. J. Harp, C. P. Johnson and C. Purcell, to be published.
57) J. D. Verhoeven, Phys. Fluids 12 (1969) 1733.
58) J. D. Carruthers and J. Pavilonis, J. Appl. Phys. 39 (1968) 5814.
59) B. Cockayne and M. P. Gates, J. Mater. Sci. 2 (1967) 118.
60) H. P. Utech and M. C. Flemings, J. Appl. Phys. 37 (1966) 2021.
61) H. P. Utech and M. C. Flemings, in: *Crystal Growth*, Ed. H. S. Peiser (Pergamon, Oxford, 1967) p. 651.
62) H. A. Chedzey and D. T. J. Hurle, Nature 210 (1966) 933.
63) B. Cockayne, M. Chesswas, J. G. Plant and A. W. Vere, J. Mater. Sci. 4 (1969) 565.
64) E. Jakeman and D. T. J. Hurle, Phys. in Technology, to be published.
65) D. Fultz and Y. Nakagawa, Proc. Roy. Soc. (London) A 231 (1955) 211.
66) J. R. Carruthers, J. Electrochem. Soc. 114 (1967) 1077.
67) J. R. Carruthers, Can. Met. Quart. 5 (1966) 55.
68) J. R. Carruthers and K. E. Benson, Appl. Phys. Letters 3 (1963) 100.
69) D. T. J. Hurle, E. Jakeman and E. R. Pike, J. Crystal Growth 3, 4 (1968) 633.
70) D. T. J. Hurle and E. Jakeman, J. Crystal Growth 5 (1969) 227.
71) A. F. Witt and H. C. Gatos, J. Electrochem. Soc. 113 (1966) 808.
72) A. F. Witt and H. C. Gatos, J. Electrochem. Soc. 114 (1967) 413.
73) A. F. Witt and H. C. Gatos, J. Electrochem. Soc. 115 (1968) 70.
74) K. Morizane, A. F. Witt and H. C. Gatos, J. Electrochem. Soc. 115 (1968) 747.
75) J. Barthel and K. Eichler, Kristall und Technik 2 (1967) 205.
76) A. F. Witt and H. C. Gatos, J. Electrochem. Soc. 116 (1969) 511.
77) K. F. Hulme and J. B. Mullin, Phil. Mag. 4 (1959) 1286.
78) R. Takahashi, K. Sugawara, Y. Nakazawa and Y. Koga, in: *Chemical Vapour Deposition*, 2nd Intern. Conf. (The Electrochemical Society, 1970) pp. 695, 713.
79) F. C. Eversteyn, P. J. W. Severin, C. H. J. v.d. Brekel and H. C. Peek, J. Electrochem. Soc. 117 (1970) 925.
80) S. E. Bradshaw, Intern. J. Electron. 23 (1967) 381.
81) W. T. Mitchell and J. A. Quinn, A.I.Ch.E.J. 12 (1966) 1116.
82) J. S. Haggerty, J. L. O'Brien and J. F. Wenckus, J. Crystal Growth 3, 4 (1968) 291.
83) M. Chesswas, B. Cockayne, D. T. J. Hurle, E. Jakeman and J. B. Mullin, J. Crystal Growth 11 (1971) 225.
84) J. C. Brice and P. A. C. Wiffin, Brit. J. Appl. Phys. 18 (1967) 581.
85) W. R. Wilcox, in: *Fractional Solidification*, Vol. 1, Eds. M. Zief and W. R. Wilcox (Edward Arnold, London, 1967) pp. 47, 157.
86) W. M. Yim, Trans. Met. Soc. AIME 236 (1966) 474.
87) G. Baralis and M. C. Perosino, J. Crystal Growth 3, 4 (1968) 651.
88) T. B. Reed, W. J. LaFleur and A. J. Strauss, J. Crystal Growth 3, 4 (1968) 115.
89) D. T. J. Hurle and J. D. Hunt, *The Solidification of Metals* (Iron and Steel Institute Publication 110, 1968).
90) R. J. McDonald and J. D. Hunt, Met. Trans. 1 (1970) 1787.
91) B. Cockayne, J. Crystal Growth 3, 4 (1968) 60.
92) *Transistor Technology*, Vol. 1, Eds. H. E. Bridgers, J. H. Scaff and J. N. Shive (Van Nostrand, 1958) p. 113.
93) J. A. Burton, E. D. Kolb, W. P. Slichter and J. D. Struthers, J. Chem. Phys. 21 (1953) 1991.

Journal of Crystal Growth **13/14** (1972) 44–47 © *North-Holland Publishing Co.*

STUDY OF MOLECULAR PHENOMENA IN THE NUCLEATION OF CRYSTALLISATION BY COMPUTER SIMULATION AND APPLICATION TO POLYMER CRYSTALLISATION THEORY

F. L. BINSBERGEN

Koninklijke/Shell-Laboratorium Amsterdam (Shell Research N.V.), Amsterdam, The Netherlands

The classical nucleation theory, as set up by Volmer and based on the ideas of Gibbs, considers the nucleus both as an equilibrium configuration and as being in equilibrium with the parent phase. The existence of real equilibrium is doubtful, however, since a nucleus is not large as compared with the fluctuations that establish a thermodynamic equilibrium between phases. Moreover, the size of the nucleus, as calculated according to classical theory, is in some cases below molecular dimensions.
Monte Carlo simulation of nucleation and crystal growth according to a model similar to that used by Volmer shows that the initiation of a stable crystal proceeds via a multitude of possible paths of growth involving all types of non-equilibrium configurations rather than via a nucleus of equilibrium shape. A well-defined nucleus does not appear to exist and, therefore, the "work of formation of nucleus" as used for the calculation of nucleation frequency or rate of crystal growth is a misleading concept. The latter two quantities viz, nucleation frequency and crystal growth, can be determined more justifiably from the parameters governing the attachment of molecules to and their separation from the growing crystal. This object is achieved e.g. through the Monte Carlo simulation by means of a computer. All this has important consequences for a number of theories where the equations of nucleation frequency and rate of crystal growth are based on a activation-free energy of nucleation (work of formation of a nucleus), the latter being calculated on the basis of a model nucleus. The implications for some current polymer crystallisation theories are discussed. The crystallisation of polymers is supposed to proceed via attachment of segments of one or a few repeating units rather than by steps of a whole "fold length" at once. Chain folding as a prerequisite for crystallisation must be rejected; it may be the result of crystallisation.
Unfortunately, one cannot decide from data on crystallisation kinetics alone which theory is right or wrong, since it is the derived interfacial free-energy parameters that are actually the adjustable parameters of either theory.

The present paper may as well have a different title: "Do nuclei exist? Negative evidence from computer simulation". This sounds rather provocative. It has to be that way, because we will have a hard time to change the generally accepted concepts of the nucleation theory as it stands today. This generally used theory, which I will refer to as the "classical nucleation theory", is the one based on the ideas of Gibbs[1]) and Volmer[2]).

In order to understand well the essentials of the revision of the classical concepts, we have to be very precise in our definitions. An *embryo* is a precursor of a new crystal having any size (and shape). The *nucleus* is a very specific embryo: the one of critical size. The *embryo*, is characterized, for instance by its size or by the free energy of its formation, ΔG. It arises as a result of fluctuations in the parent phase and grows as a result of successive advantageous fluctuations. The embryo may also disappear altogether, because the free

energy of its formation is initially positive as a result of the specific interfacial free energy, σ, associated with the interface, Ω, between embryo and parent phase. Therefore, a plot of ΔG versus the size of the embryo (m, in moles) shows a maximum.

$$\Delta G = -m\,\Delta\mu + \Omega\sigma, \tag{1}$$

in which $\Delta\mu$ is the difference in thermodynamic potential of the bulk phases A and B at ambient pressure, p_1:

$$\Delta\mu = \mu_{A,1} - \mu_{B,1}. \tag{2}$$

The embryo corresponding to this maximum is called the *nucleus*, and its free energy, ΔG^*, is regarded as an activation free energy for nucleation (the asterisk denotes the nucleus).

In the case of condensation from the vapour, for example, the embryo is a spherical droplet of radius, r. Mechanical equilibrium requires that the pressure inside the droplet, p_2, should be higher than the ambient pressure:

$$p_2 - p_1 = 2\sigma/r. \tag{3}$$

From this condition it can be derived that:

$$\mu^*_{B,2} = \mu_{A,1}, \tag{4}$$

which is in agreement with the definition according to which the nucleus is the embryo at the free energy maximum: A small change in the size of the nucleus, dm, does not lead to any change in the free energy:

$$\left(\frac{\partial G}{\partial m_B}\right) dm - \left(\frac{\partial G}{\partial m_A}\right) dm = 0, \tag{5}$$

which again leads to condition (4), since by definition $\mu_A = (\partial G/\partial m_A)$, etc.

The conclusion is: *the nucleus is in equilibrium with the parent phase*, be it a labile equilibrium.

The condition $d\Delta G/dm = 0$ allows the calculation of the size of the nucleus as well as the free energy of its formation from an equation like eq. (1). In crystallization, from the vapour or the melt, this is done by assuming that *the nucleus is a crystallite of equilibrium shape*, considering that embryos of any other shape are less probable.

However, if use is made of a molecular model, the contribution of embryos other than equilibrium-shaped can be taken into account as well. This was done indeed when the Kossel crystal[3]) was introduced, a packing of molecules in a primitive cubic lattice, taking only nearest-neighbour interaction into account. We illustrate this for the case of two-dimensional nucleation, i.e. the initiation of a new monomolecular layer on the

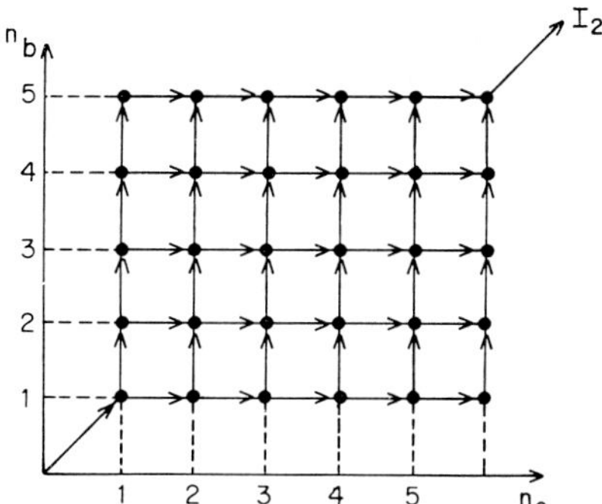

Fig. 2. Network of paths of growth according to Becker and Döring.

growth face of a crystal (fig. 1). The free energy of formation of the two-dimensional embryo is

$$\Delta G_2 = -bca_0 \Delta\mu_v + (2b + 2c)a_0\sigma, \tag{6}$$

in which $\Delta\mu_v$ is $\Delta\mu$ per unit volume. This equation can be written in molecular quantities by introducing

$$\Delta\mu_0 = a_0^3 \Delta\mu_v, \tag{7}$$

and

$$\varphi = 2a_0^2\sigma, \tag{8}$$

so that eq. (6) becomes

$$\Delta G_2 = -n_b n_c \Delta\mu_0 + (2n_b + 2n_c)\varphi, \tag{9}$$

in which n_b is the number of molecules in edge b.

Becker and Döring[4]) (in 1935) took all rectangular configurations into account by calculating the net flux of transition from one rectangular embryo to the next (fig. 2). They did a similar job for three-dimensional nucleation. They concluded that the activation free energy for nucleation was equal to the sum of the free energy of formation of the three-dimensional nucleus, the free energy of formation of the two-dimensional nucleus plus the free energy of attachment of the first molecule of a new row of a monomolecular layer, together forming the highest free energy of formation of any configuration in the lowest path of their network.

Becker and Döring already pointed out that all possible configurations (non-rectangular, distorted edges) should be taken into account, instead of only the transition of one rectangular configuration to the next. This would, however, involve an immensely complicated calculation already up to very small configurations.

Summarizing, the classical nucleation theory uses a few idealizations: (1) a nucleus; (2) the nucleus is in

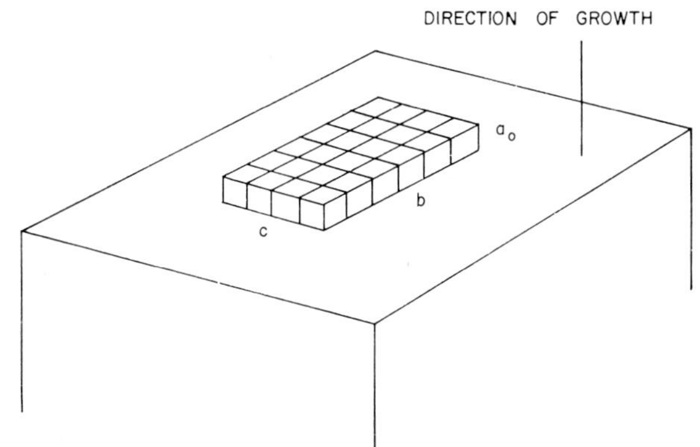

Fig. 1. Two-dimensional embryo on growth face of growing crystal.

equilibrium with its surroundings and in equilibrium shape. What are the consequences?

(1)a. At high degrees of supercooling – which frequently occur, for instance, in polymers – the dimensions of the nucleus are only a few times the thickness of a molecule or polymer repeating unit. Further supercooling would cause a stepwise change in the activation free energy, for instance, of secondary nucleation. However, no stepwise change in the rate of crystal growth with temperature has ever been observed for materials nicely following the growth equation according to secondary nucleation.

b. Isotactic polystyrene allows a supercooling of 100 °C and more. There the calculated nucleus would even be smaller than the thickness of the polymer molecule[5]), although the growth rate of the polymer still follows the equation of secondary nucleation.

(2) A phase equilibrium rests on fluctuating exchange processes between the phases. The fluctuations in question are of the same order of magnitude as the nucleus, especially at higher degrees of supercooling. A thermodynamic potential $\mu_{B,2}$ [see eq. (4)] can therefore not be established. An *equilibrium* with the parent phase is not relevant, neither is an equilibrium shape. Continuous and important distortions of the shape must be expected. In many textbooks the "stability" of the nucleus is considered. This is in fact a misnomer since, even in the classical theory, the nucleus is of all embryos the least stable one.

The present availability of fast digital computers enabled us to carry out a process of embryo growth via all possible configurations. To this end we used the same Kossel model as used in the classical theory. In this model we attached molecules to the growing embryo and simultaneously allowed the molecules to detach. We will now briefly depict such a process for the case of two-dimensional nucleation.

A new monomolecular layer is to be deposited on the top face of a crystal. As we want to know the nucleation frequency per site, we try to deposit a molecule on a selected site. Since we have to use a go-no go scheme in computer simulation, we chose the time-unit, u_t, so small that the probability of such an event taking place within one time unit is very low. We try again and again until one molecule attaches.

We have now reached a rather unstable situation, so that the molecule is prone to detach. In a few of these

cases a second molecule may attach before this actually happens, and so on, until a stable, large configuration is obtained. This case is then counted as a successful trial.

Since we are doing $1/u_t$ trials per second (by definition), the nucleation frequency is related to the fraction of successful trials, f_s, by

$$I = \frac{f_s}{u_t} \equiv Z \exp\left(-\frac{\Delta G_2^*}{kT}\right)_{\text{simulated}}. \qquad (10)$$

In a similar way three-dimensional nucleation was simulated.

In order to be able to compare the simulated activation free energies with the classical ones we introduced dimensionless parameters:

$$P = \varphi/2kT\,, \qquad (11)$$

$$R = \Delta\mu_0/2kT\,. \qquad (12)$$

Table 1 shows the differences.

TABLE 1

	Classical theory	Simulated, eq. (10)
$\dfrac{\Delta G_2}{kT}$	$2\,\dfrac{P^2}{R}$	$1+\tfrac{4}{3}\,\dfrac{P^{5/2}}{R}$
$\dfrac{\Delta G_3}{kT}$	$8\,\dfrac{P^3}{R^2}$	$6.9\,\dfrac{P^{15/4}}{R}$

Remarkably, the nucleation frequency is dependent on the same power of the supercooling in the theory and in the simulation. But fair differences were found in the relation with interfacial free energies. Especially if the interfacial free energy parameter is low, the simulation shows considerably higher frequencies. This has to be attributed to the increasing number of available paths of growth at low P.

Configurations, if printed out, deviate strongly from the ideal equilibrium shape. They even hardly approach rectangular shapes (fig. 3). A full account of the computer simulation method used and of the evaluation of

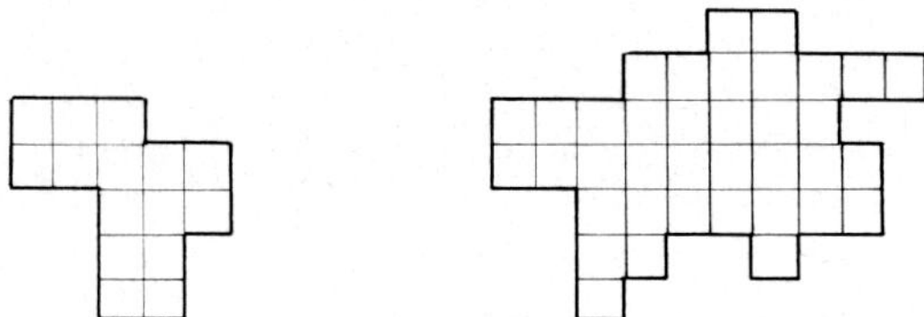

Fig. 3.   Examples of configurations of two-dimensional embryos

the computer results to empirical equation will be submitted to J. Crystal Growth.

In conclusion we may say:

(1) Nuclei do not exist.

(2) The use of a model-nucleus for the calculation of nucleation frequency or rate of crystal growth is invalid.

(3) One is not entitled to calculated interfacial energies or free energies from experiments on nucleation rate or crystal growth rate.

These conclusions affect all the theories making use of such model, e.g. current polymer crystallization theories. The theory of Mandelkern[6]) is merely an adaptation of the classical theory to polymer crystallization. Hoffman and Lauritzen[7]) consider the deposition of one fold length a nucleation event. Moreover they assume the crystallizing polymer molecule to fold back onto itself. This might be the case in crystallization from very dilute solution, but is improbable in crystallization from the melt.

In segmental crystallization, i.e. the attachment and separation of one or a few polymer repeating units as the elementary step, chain folding may be a result of crystallization but will not be a prerequisite. This has recently been illustrated by Fulcher[8]), who showed that after the completion of the primary – i.e. volume-filling – crystallization the crystalline lamellae in the polymer spherulites keep growing at the cost of the amorphous layers between them without altering the sum of thicknesses of crystalline and amorphous layers.

Chain folding with adjacent re-entry is in fact 10^6 to 10^8 times less probable than random segmental crystallization from the melt.

In conclusion, we must stress that it is the kinetic probability of reaching certain configurations that determines the initiation of crystallization and the crystal growth rather than the thermodynamic stability of any configuration.

Experimental evidence such as the relation between rate of crystal growth and degree of supercooling follows both the classical equation based on secondary nucleation and the empirical equation based on computer simulation. The interfacial free energy parameters are the adjustable parameters of either theory. However, we can now reach this result without using forced assumptions such as nuclei. Moreover, the morphological consequences of the two pictures are quite different.

References

1) J. W. Gibbs, *Collected Works*, (Yale Univ. Press, New Haven, 1948) p. 253.
2) M. Volmer, *Kinetik der Phasenbildung* (Steinkopff, Dresden, 1939).
3) W. Kossel, Nachr. Ges. Wiss. Göttingen, Math.-physik. Kl. (1927) 135.
4) R. Becker and W. Döring, Ann. Physik [5] **24** (1935) 719.
5) J. Boon, J. Polymer Sci. C **16** (1967) 1739.
6) L. Mandelkern, *Crystallization of Polymers* (McGraw-Hill, New York, 1963) ch. 8.
7) J. I. Lauritzen and J. D. Hoffman, J. Res. Natl. Bur. Std. **64A** (1960) 73; **65A** (1961) 297.
8) K. U. Fulcher, J. Polymer Sci. **A2** (Polymer Physics) (in press).

    Journal of Crystal Growth **13/14** (1972) 48–56 © *North-Holland Publishing Co.*

FORMATION OF MULTIPLY-TWINNED PARTICLES ON ALKALI HALIDE CRYSTALS BY VACUUM EVAPORATION AND THEIR STRUCTURES

SHIRO OGAWA and SHOZO INO

The Research Institute for Iron, Steel and Other Metals, Tohoku University, Sendai, Japan

During studies on the epitaxy and film growth of metals evaporated onto alkali halide crystals curious particles causing anomalous 111 electron diffraction spots and showing unusual shapes and contrasts on electron micrographs have been found. They have mainly been studied for Au on NaCl. As a result of analysis it has become clear that they have multiply twinned structures and are divided into three kinds, i.e., a pentagonal decahedron particle composed of five distorted tetrahedral twins, an icosahedral particle composed of twenty distorted tetrahedral twins and a rhombic shape particle consisting of five distorted twins. Stability of these multiply-twinned particles that have inherently strained structures is discussed, the total energies of them being calculated and compared with those of a fcc tetrahedron and a fcc Wulff-polyhedron. There is a good agreement in sizes of multiply-twinned particles between calculation and observation. Behavior of the particles during film growth of various metals on NaCl and KCl is described. Finally, the origin of the particles is discussed, most of them being presumed to form by nucleation at the earliest stage of film growth.

1. Introduction

Allpress and Sanders[1]), and Mihama and Yasuda[2]) found abnormal 111 spots on electron diffraction patterns of epitaxial Au films, and the latter researchers ascribed them to some compound particles which showed curious contrasts on electron micrographs, but they did not elucidate the structure of such particles. In our laboratory these particles were studied in detail mainly using dark field electron microscopy, and it was revealed that they are composed of distorted tetrahedral twins, and are called "multiply-twinned particles"[3,4]). In the present paper we describe observations on three kinds of multiply-twinned particles by electron diffraction and electron microscopy, detailed analyses of their structures, discussion on the stability of the particles based on the calculation of energies, behavior of the particles during film growth in various metals, and finally some considerations about the origin of the particles.

2. Electron diffraction patterns and electron micrographs in the initial growth stages of gold films formed by vacuum evaporation on cleavage face of rocksalt

In the present work NaCl and KCl were used as substrates, and were cleaved in an ultrahigh vacuum of pressure 10^{-9}–10^{-10} mm Hg. Evaporated films that were composed of many small discrete particles at early

stages of film growth were backed by carbon evaporation and isolated from the substrates. As the most reliable data concerning multiply-twinned particles have been accumulated for Au on NaCl, results of observations on this system will be described.

Fig. 1 is a typical diffraction pattern obtained from a film 25 Å thick formed at 290 °C. Four {200} spots represent (001) orientation, and twenty-four {220} spots arise from two sets of (111) orientation. However,

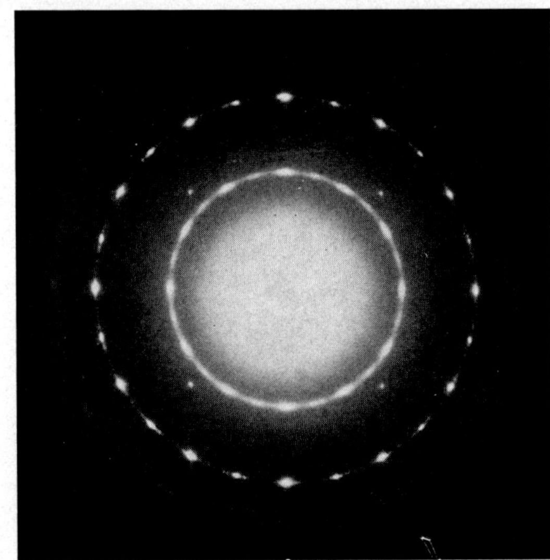

Fig. 1. A typical electron diffraction pattern taken from a Au film 25 Å thick deposited at 290 °C on NaCl (001) face cleaved in ultrahigh vacuum. Abnormal twenty-four 111 spots are seen.

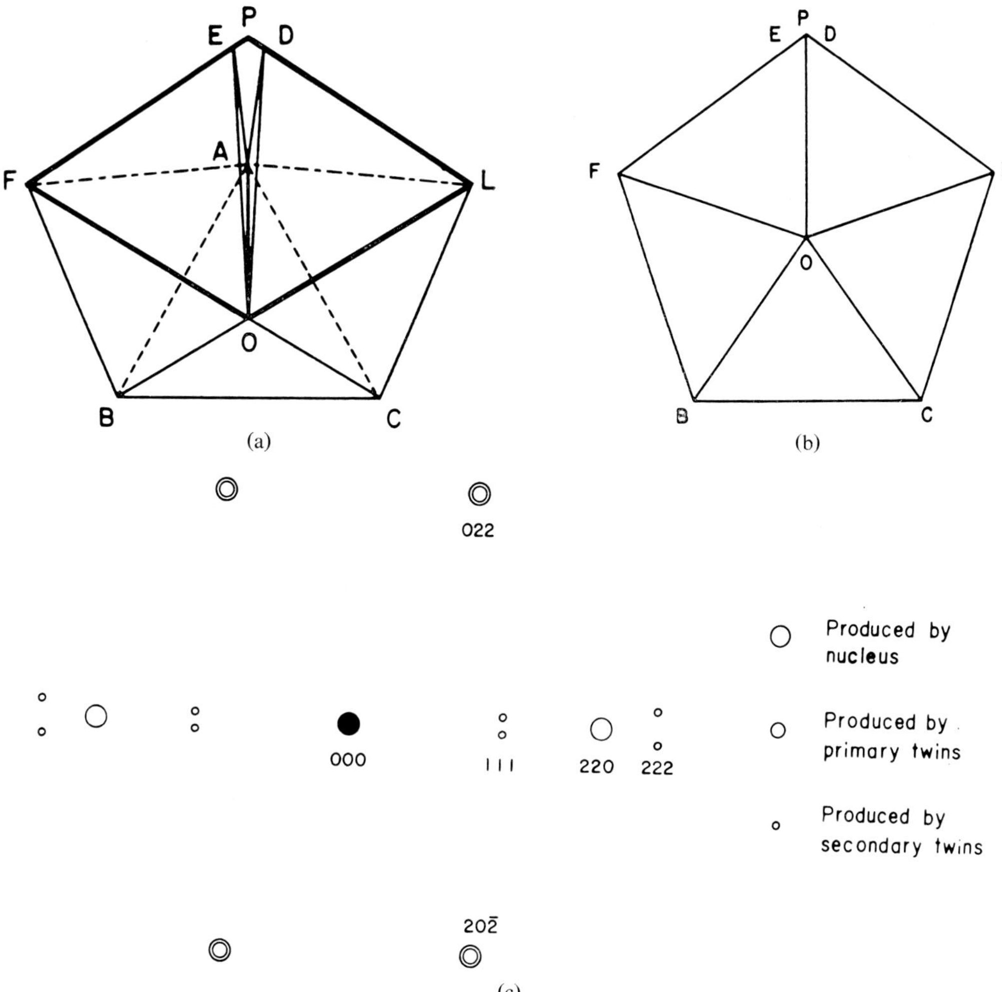

Fig. 6. A structure model for a pentagonal particle. (a): Piling-up of five tetrahedral twins. (b): A five-fold symmetry of one set of five distorted tetrahedra without the gap, viewed along OA direction. (c): A diffraction pattern to be expected from the structure in (a).

used. Also in this case abnormal 111 spots are related to the structure of rhombic particles.

3. Structure of multiply-twinned particles

The shapes and dark field images of pentagonal and hexagonal particles, especially the triangular contrast in fig. 5a, suggest that these particles may be composed of tetrahedra. Based on this suggestion, multiply-twinned particle models were proposed for these particles, and also for a rhombic particle.

3.1. STRUCTURE OF PENTAGONAL PARTICLES

A structure model in fig. 6a is proposed for a pentagonal particle. On a tetrahedron OABC whose ABC face adheres to the substrate surface two tetrahedral twins OABF and OACL pile up primarily, and on these further two tetrahedral twins OAFE and OALD pile up secondarily, resulting in the formation of a gap OAED. If this gap is filled, the pile-up of five tetrahedra will look like a pentagon just as observed in fig. 5a. Actually no gap has ever been observed. A lattice imperfection resulting from disappearance of the gap should distribute inside the particle and give rise to a straining of the structure. Fig. 6b is a shape of such a particle viewed along OA direction. Bagley has once considered such a decahedron consisting of five distorted tetrahedra[5]).

In fig. 6c an electron diffraction pattern resulting

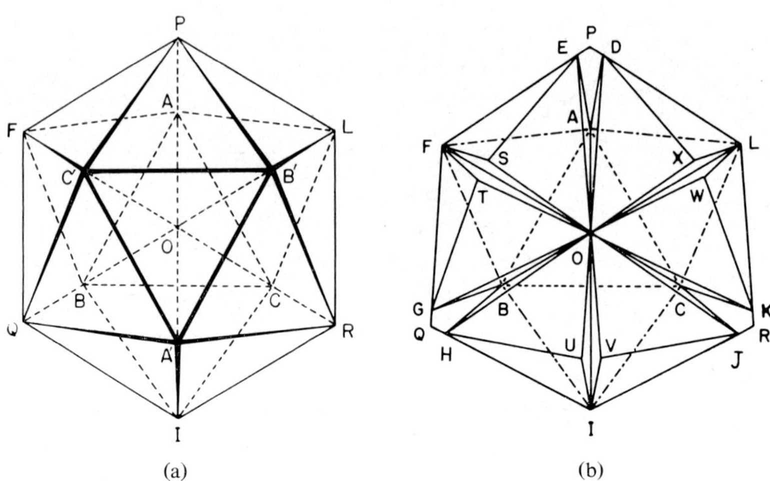

(a)   (b)

Fig. 7. A structure model for a hexagonal particle. (a): An icosahedron composed of twenty distorted tetrahedral twins. (b): A structure resulting from a simple pile-up of thirteen tetrahedra.

from a decahedral particle when the electron beam is incident perpendicularly upon the substrate surface is shown. It is clear that 111 reflections occur from secondary twins and two adjacent 111 spots, separated because of the gap, merge into a diffuse elongated spot as a result of the strain left by the disappearance of the gap. A consideration of possible orientations of the present structure results in the diffraction pattern actually observed. The dark field contrasts in fig. 4 are satisfactorily explained by the present model consisting of five twins of distorted tetrahedra.

3.2. STRUCTURE OF HEXAGONAL PARTICLES

The pile-up of twinned tetrahedra is further continued until a structure composed of twenty tetrahedra is completed, as shown in fig. 7a. Many more gaps come from a simple pile-up of tetrahedra than in the case of a decahedral particle, but as before when these gaps disappear they leave strain in the particle. Thus this icosahedron is a structural model based on a hexagonal particle. A structure composed of thirteen tetrahedra, as shown in fig. 7b, is used for explanation. The icosahedral structure can explain all the observed facts related to hexagonal particles. 111 diffraction spots are diffused and elongated owing to the strained {111} planes as a result of the disappearance of the gaps. The characteristic contrast change observed in fig. 3 is clearly understood. By using one of 111 spots to form a dark field image, for instance, secondary twins OAEF and

OADL and tertiary twins OHIU and OJIV in fig. 7b become bright at the same time producing butterfly-like contrasts, because they cause the 111 reflection, and use of other 111 spots causes rotations of the contrasts. Contrasts corresponding to a head and a tail of a butterfly can be explained by double diffraction of a secondary and a tertiary twin neighboring along the incident beam direction. In fig. 3e many extra diffraction spots are seen, and all of these extra spots are also satisfactorily explained by double diffraction from neighboring twins.

A diffraction pattern to be expected from the above model is in good agreement with the observed diffraction pattern in fig. 1, as is the case with a pentagonal particle.

The strained structure of an icosahedral particle is considered to be no more than the one already described by Mackay as a dense non-crystallographic packing of equal spheres[6]).

3.3. STRUCTURE OF RHOMBIC PARTICLE

The structure of a rhombic particle is not as simple as it appears. A structural model proposed in fig. 8 explains well the external shape and the observed electron microscopic contrasts of a rhombic particle. A cube O'BA'CODAE in (a), truncated at two (111) faces, adheres with its (001) faces to the substrate, and four tetrahedra pile up on it, as shown in (c). A gap which should result from a simple pile-up disappears as in the

I – 9

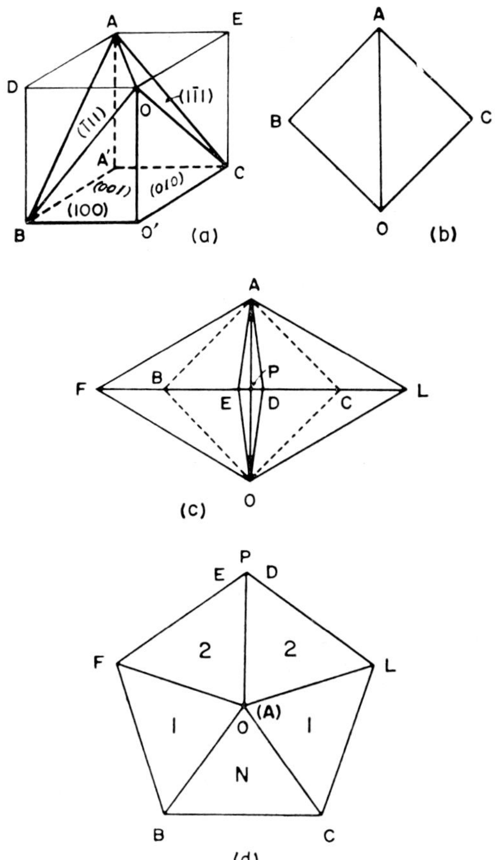

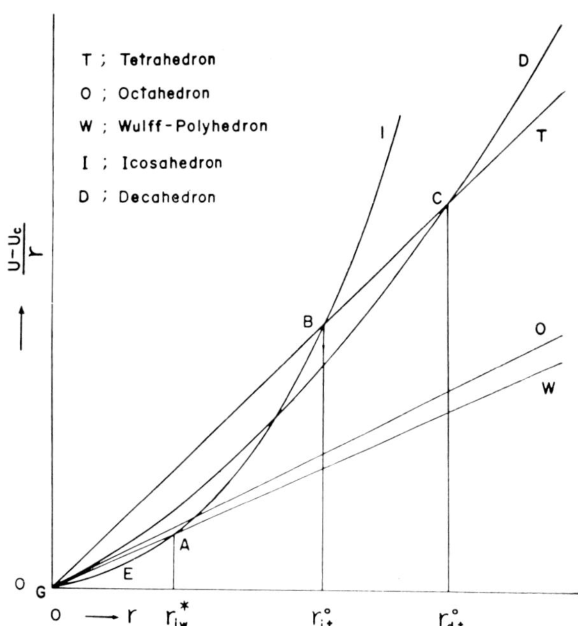

Fig. 9.   Depiction of the stability of an icosahedral and a decahedral particle.

Fig. 8.   A structure model for a rhombic particle. (a): An external form of a nucleus with (001) orientation. (b): A projection of the nucleus on the film plane. (c): A top view of a particle consisting of five twins showing a rhombic shape. (d): A view of the particle showing a pentagonal shape.

cases of a pentagonal and an icosahedral particle, leaving the strain inside the particle. The shape viewed along the OA direction appears as a pentagon in (d). There is thus a close resemblance between a pentagonal and a rhombic particle. A diffraction pattern to be expected from the present structure agrees well with the observation. The present structure explains well the dark field images in fig. 5 which were formed by using 200, 220 and 111 spots.

The above-mentioned particles of three kinds with characteristic structures are given a general name "multiply-twinned particles" in consideration of their twinned structures. This name has been given only because their structures are composed of units having a twinning relation with one another. It does not necessarily mean that they are formed by some real twinning mechanism. The term "nucleus" was used here only for an easier understanding of the structures.

4. Stability of multiply-twinned particles

Some theoretical treatments have been made in order to explain the stability of multiply-twinned particles in comparison with normal fcc particles[7,8]. Here, Ino's investigation will be introduced[9].

The total free energy U including the cohesive energy, the surface energy, the adhesive energy, the elastic strain energy and the twin boundary energy was calculated for a decahedral and an icosahedral particle in comparison with a fcc tetrahedron, a fcc octahedron and a fcc Wulff-polyhedron. The elastic strain energy is characteristic of multiply-twinned particles that have heavily strained structures. In this calculation various energy values characteristic of each metal were estimated from experimental data. Fig. 9 depicts results of the calculation. The abscissa r is the edge length of a polyhedron and a measure of the particle size, and in the ordinate $U - U_c$ is the sum of the energies less the cohesive energy U_c. T, O, W, I and D mean a tetrahedral, an octahedral, a Wulff-polyhedron, an icosahedral and a decahedral particle, respectively. A Wulff-polyhedron particle has the lowest energy per volume of all the crystalline polyhedra, but for $r \leq r_{iw}^*$ an icosa-

TABLE 1

Calculated critical diameters $2r_{iw}^*$, $2r_{it}^0$ and $2r_{dt}^0$ for essentially stable and quasi-stable states

	$\gamma_a = 0$ (in free space)			$\gamma_a \neq 0$ (on NaCl)		
	$2r_{iw}^*$ (Å)	$2r_{it}^0$ (Å)	$2r_{dt}^0$ (Å)	$2r_{iw}^*$ (Å)	$2r_{it}^0$ (Å)	$2r_{dt}^0$ (Å)
Au	106.8	436.3	3961	102.1	409.4	3550
Ag	75.6	306.9	2905	69.8	273.5	2373
Cu	67.6	279.1	2833	63.2	259.2	2404
Ni	43.1	212.6	3385	39.0	189.4	2765
Pd	49.9	263.9	2447	44.8	234.6	1992
Pt	56.3	238.4	3784	51.9	213.4	3118
Pb	97.5	446.1	2258	89.2	398.5	1852
Al	29.3	486.1	2199	15.4	406.4	1602
Si	27.6	133.7	830	25.1	119.2	679
Ge	15.5	109.8	725	13.3	96.9	583
γ-Fe	44.0	200.6	1727	40.2	179.2	1417
β-Co	97.6	375.9	1574	90.9	337.8	1305

hedral particle is more stable than a Wulff-polyhedron particle. The reason for this lies in the abnormal structure of the former. The condensation process of evaporated films is not considered to be in equilibrium, and a tetrahedral particle is frequently found in evaporated films. Therefore, icosahedral particles with $r \leq r_{it}^0$ are considered to be able to be quasi-stable. A decahedral particle is not stable for any size, but quasi-stable for $r \leq r_{dt}^0$. It is suggested that a decahedral particle can grow larger than an icosahedral particle.

Table 1 shows values of calculated critical diameters $2r_{iw}^*$, $2r_{it}^0$ and $2r_{dt}^0$ for particles of various metals grown in free space and on NaCl. γ_a means the adhesive energy. There is no large difference in diameter values between those in free space and those on a substrate. Calculated values of r_{it}^0 and r_{dt}^0 should be compared with observed values of maximum sizes. There seems a good agreement between observation and calculation. For instance, the largest diameter of icosahedral particles of Au formed on NaCl is about 400 Å and diameters most frequently observed converge into about 290 Å, and these values correspond to calculated values for $2r_{it}^0$ and $2r_{it}^m = \frac{4}{3}r_{it}^0$. There is also an agreement for a decahedral particle between the largest observed diameter and $2r_{dt}^0$.

5. Behavior of multiply-twinned particles during film growth

The formation of multiply-twinned particles can be known from abnormal 111 diffraction spots, even if they are not directly observed on electron micrographs.

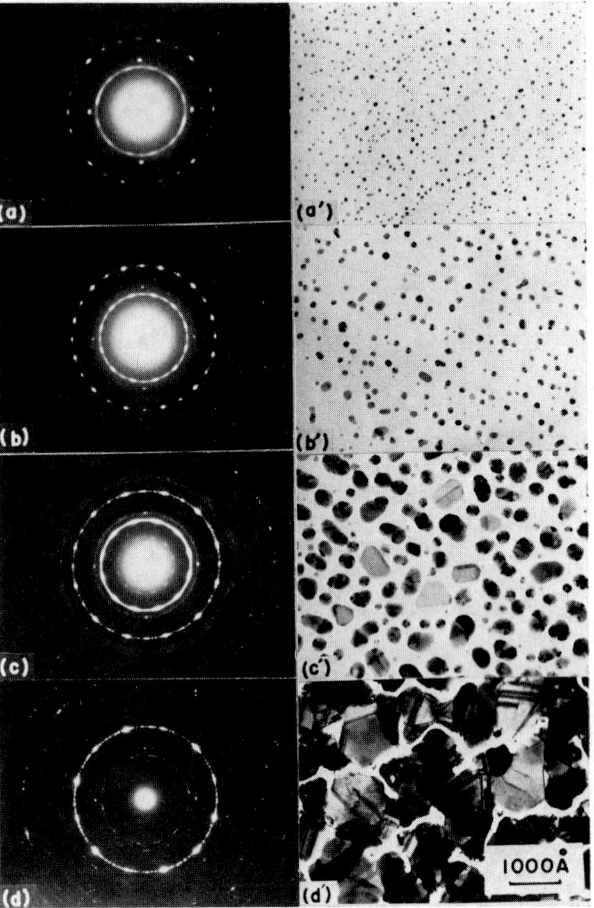

Fig. 10. A sequence of electron diffraction patterns and bright field electron micrographs showing various stages of growth of Au films deposited on NaCl cleavage face at 300 °C. The mean thicknesses are 5 Å, 20 Å, 100 Å and 500 Å for (a), (b), (c) and (d), respectively.

Fig. 10 shows diffraction patterns and corresponding micrographs of Au films formed on NaCl at 300 °C, 5 Å, 20 Å, 100 Å and 500 Å in mean thickness. Multiply-twinned particles appear already in earlier stages of deposition, many more at 100 Å and they almost disappear at 500 Å as a result of coalescence and unstability of these particles.

{111} spots arising from multiply-twinned particles were found also from Ag films formed on NaCl as well as KCl. In the case of Ni and Pd films formed on NaCl as well as KCl diffraction spots originated from rhombic particles were observed. From Al films grown on NaCl as well as KCl diffraction spots originated from multiply-twinned particles have never been observed.

Fig. 11 represents the relation between the density of

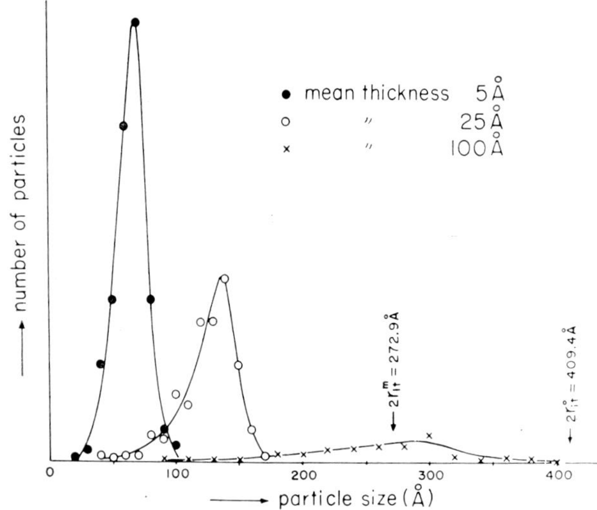

Fig. 11.   A histogram of particle size of icosahedral particles in Au films 5 Å, 25 Å and 100 Å in the mean thickness.

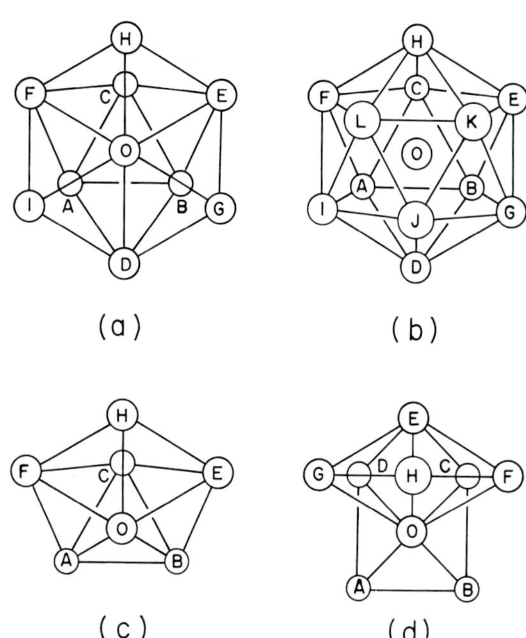

Fig. 12.   Structures of nuclei of multiply-twinned particles. (b): A nucleus of an icosahedral particle, the process of its formation being shown in (a). (c): A nucleus of a decahedral particle. (d): A nucleus of a rhombic particle.

the number of icosahedral particles observed on electron micrographs and the particle size for Au films 5 Å, 25 Å and 100 Å thick formed on NaCl at 350 °C. In the thinnest film the greater part of the isolated particles are icosahedral and the mean particle size is less than 100 Å. With increasing film thickness the mean particle size increases but the total number of the particles decreases. The largest and the mean particle size in the film 100 Å thick are about 400 Å and about 290 Å, respectively. The curves in fig. 11 show approximately the way the number of icosahedral particles decreases by coalescence.

6. Origin of formation of multiply-twinned particles

Fig. 11 shows that a large number of icosahedral particles are formed at a very early stage of film growth, the smallest one being about 20 Å in size. This fact suggests that multiply-twinned particles may be formed by nucleation from deposited atoms. On the other hand, similar observations by cine films have been reported, and here the multiply-twinned particles are formed by growth or coalescence of normal particles[10,11]). Such examples were, however, very few, and most of multiply-twinned particles usually observed are not considered to form by such processes. It will be natural to consider that most of them nucleate at the earliest stage of film growth.

Fig. 12 explains the formation processes of an icosahedral nucleus in (a) and (b), decahedral nucleus in (c)

and a rhombic particle in (d). The plane of illustration is parallel to the substrate surface. OABC in (a) is a tetrahedral nucleus, but positions D, E and F are those which are not permitted in a fcc lattice, and when the adhesive force between metal atoms and substrate ions is smaller than the cohesive force between metal atoms, these positions will be occupied. Positions G, H and I are also abnormal positions. Thus a nucleus forms which is composed of O and twelve atoms surrounding it, as shown in (b). This nucleus has two values of bond length, one of which connects O with the surrounding atoms and the other of which connects, e.g., A with B and is about 5% longer. An icosahedral nucleus possesses, therefore, essentially a strained structure. This is due to the fact that an energy increase caused by the straining is overcome by an surface energy decrease caused by its shape. This nucleus can grow up to an icosahedral particle having the spontaneously strained structure. In (c) a nucleus of a decahedral particle is shown. This nucleus has also a strained structure with three kinds of bond length represented by CO, OA and AB. In (d) the growing process of a nucleus of a rhombic particle is shown. This has also a strained structure.

It is very interesting that there are those anomalous atomic distances in the structures of multiply-twinned particles which are not found in the normal lattice, as can be understood from the above. This may be the reason for the unusual atomic distances that have sometimes been observed[12]).

References

1) J. G. Allpress and J. V. Sanders, Phil. Mag. **10** (1964) 645.
2) K. Mihama and Y. Yasuda, Read before the Meeting of Phys. Soc. Japan held October 1964; J. Phys. Soc. Japan **21** (1966) 1166.
3) S. Ino, J. Phys. Soc. Japan **21** (1966) 346.
4) S. Ino and S. Ogawa, J. Phys. Soc. Japan **22** (1967) 1365.
5) B. G. Bagley, Nature **208** (1965) 674.
6) A. L. Mackay, Acta Cryst. **15** (1962) 916.
7) Y. Fukano and C. M. Wayman, J. Appl. Phys. **40** (1969) 1656.
8) J. G. Allpress and J. V. Sanders, Australian J. Phys. **23** (1970) 23.
9) S. Ino, J. Phys. Soc. Japan **26** (1969) 1559; **27** (1969) 941.
10) G. Honjo and K. Yagi, Read before the Intern. Conf. on Thin Films held in Boston on 28 April–2 May, 1969.
11) Private communication.
12) H. Morimoto and H. Sakata, J. Phys. Soc. Japan **17** (1962) 136.

Journal of Crystal Growth **13/14** (1972) 57–61 © *North-Holland Publishing Co.*

PRINCIPLES OF CRYSTAL GROWTH FROM THE SOLID STATE IN RELATION TO THE PREPARATION OF LARGE CRYSTALS

K. T. AUST

Department of Metallurgy and Materials Science, University of Toronto, Toronto, Canada

The principles involved in the preparation of large crystals by solid state recrystallisation methods are briefly discussed. The intrinsic properties of interfaces such as their energy and mobility, and the effect of impurities on these properties, play a predominant role in our understanding of crystal growth in a solid. The general approach to preparing large crystals by recrystallisation techniques is to inhibit the nucleation and/or growth of most of the grains in a polycrystalline matrix and to induce the nucleation and/or growth of a single grain. The controlling factors consist of (I) the type and amount of driving energy for interface motion, (II) stabilisation of the matrix by orientation texture, inclusions, solutes or grain boundary grooving, and (III) variables such as prior treatment of the specimen, purity of the starting material and annealing conditions.

The various recrystallisation methods for producing single crystals are: strain-anneal grain growth, inclusion-inhibited growth, texture-inhibited growth, surface energy, lineage-induced growth, and sintering. Several of these solid state methods, e.g. strain-anneal, lineage and sintering, have also been utilised for the production of bicrystals and tricrystals having controlled orientations.

1. Introduction

This short discussion is concerned with the principles of nucleation and growth of large crystals in a solid. These principles are related to the energy and mobility of interfaces and the effects of impurities on these properties.

2. Formation of grain boundaries

The nucleation problem in the preparation of single crystals deals with the formation of large-angle grain boundaries in a deformed matrix. This differs from concepts of classical nucleation theory in the sense that there is no collective and simultaneous reorganization of atoms into a newly oriented lattice, but a gradual increase in perfection of a region of distorted matrix until it becomes perfect enough thermodynamically to grow larger. The nucleus appears to have the following characteristics: a different orientation from the deformed matrix, larger than a critical size, relatively strain-free and originates at a preferred site in the matrix, i.e. in a localized region of maximum misorientation. At present there are several conflicting views as to how a nucleus originates. In one theory[1], the so-called "bulge" mechanism, a crystal may form by strain-induced boundary migration in which a difference in dislocation density between two adjacent regions provides the driving energy. When this process results from the growth of existing grain boundaries in a polycrystalline matrix, no new orientations are formed. However, this is not necessarily so if the growth occurs from individual subgrains. The bulge mechanism appears to be important in moderately deformed fine-grained metals and is consistent with observations that nuclei under such conditions form only along the boundaries and with a zero induction period.

In another theory a subgrain, formed by polygonization, increases in size and misorientation by means of boundary migration[2-4]. Again, a low dislocation density region grows into a high dislocation density region obtaining a new unstrained crystal with an orientation different from the matrix, i.e. a large-angle grain boundary is produced. An alternative theory considers nucleation to occur by subgrain coalescence and rotation[5,6]. In this model, two adjacent subgrains may coalesce resulting in the disappearance of the common subboundary by the motion of dislocations from the disappearing boundary into the connecting boundaries around the subgrains. This produces a rotation of the subgrain lattice, yielding an orientation of the coalesced subgrain which is different from the original orientation of the subgrains before coalescence.

Experimental observations based on thin-film electron microscopy have provided support for both the "subgrain coarsening" theory, e.g. ref. 7, and the "subgrain rotation" theory, e.g. ref. 5. Recent observations[8] using transmission electron microscopy indicate that the growth of subgrains in an Al–Cu solid solution is due to the migration of subboundaries and not due to subgrain rotation. The choice between these models, however, rests upon the interpretation of some weak contrast between adjacent fields in electron micrographs. Even closely similar micrographs obtained under identical conditions have produced divergent interpretations.

3. Migration of grain boundaries

The elementary process in grain boundary migration consists of the transfer of single atoms. This process appears to be closely related to that of grain boundary self diffusion, having an activation energy similar to that for the motion of vacancies[9]. In the absence of impurity-drag effects, the velocity of grain boundary motion may be estimated from absolute reaction rate theory for a single atom transfer process[10,11].

It is proposed, based on electron microscope observations, that a migrating grain boundary has a step-like or a ledge structure[13]. The ledge model of the boundary is an alternative description of the boundary coincidence structure of Bishop and Chalmers[14]. In fcc metals, these boundary steps are formed by the {111} crystal planes of both grains where these planes meet the boundary. During boundary migration, it is postulated that atoms are emitted from steps on {111} planes of the grain in which the boundary is advancing and these atoms are absorbed onto steps on {111} planes of the growing grain. Therefore, the mobility of a grain boundary is determined essentially by the density of steps in the boundary. The smaller the inclination of the {111} planes relative to the boundary, the larger is the distance between the steps and the lower is the mobility of the boundary.

Observations that coherent twin boundaries in a fcc lattice have zero or very low mobility support this model since such an interface has no growth steps. The traces of {111} planes on any surface of a non-coherent segment of a twin boundary are perpendicular to the boundary and such an interface should have good mobility. Direct, measured velocities of non-coherent twin boundaries in fcc organic camphor show that these boundaries are highly mobile[15], in agreement with the step or ledge model of boundary motion. Observations on high-purity lead indicate that an abrupt transition occurs at a critical temperature for both the temperature-dependence of grain boundary migration[16,17] and grain boundary energy[18]. These results suggest that the step density or the boundary structure varies with temperature in a manner analogous to a phase transformation[18,19].

Impurities exert a retarding effect on the motion of a grain boundary as a result of solute adsorption at the boundary. The effects of solute on the kinetics of boundary migration depend upon the kind and amount of solute as well as upon the structure of the grain boundary[9]. The motion of the boundary is limited by the velocity of the solute atmosphere which is less than the velocity of the boundary without impurity[20,21]. Similar theories of impurity-controlled boundary motion advanced by Cahn[22] and by Lücke and Stüwe[23] provide a suitable explanation for the case of impurity drag in the grain lattice adjacent to a boundary. In these theories the boundary is represented as a planar discontinuity, i.e. no ledges or steps, characterized by an interaction energy $E(x)$ between an impurity atom and the boundary, and by a diffusion coefficient $D(x)$ of the impurity atoms. Both $E(x)$ and $D(x)$ are functions only of the distance x normal to the boundary, at constant temperature. Experimental results for recrystallization of aluminum containing copper[24] and grain growth of tin containing antimony[25] are in quantitative agreement with the predictions of these theories[26].

Gleiter[27] has also considered the impurity profile along the core of the boundary containing steps and deals with the impurity drag in the boundary core. Since the diffusion coefficient in the boundary core is much higher than in the grains, this theory is believed to describe impurity drag of a boundary at higher velocities than those considered by Cahn or by Lücke and Stüwe. The results obtained for the motion of grain boundaries in zone refined lead containing different solutes[9] are in good agreement with the predictions of impurity drag in the boundary core containing steps[27].

4. Driving energy considerations for crystal preparation

A driving energy for the growth of a large crystal may be obtained by a variation in the free energy of the boundaries or in the free energy of the crystals adjacent to the boundary. In the first case which is applicable to grain growth methods, a boundary will tend to diminish its area, and therefore its energy, when subjected to no other driving force. Andrade and co-workers[28,29]) prepared crystals in wires of several bcc metals by controlling the process of grain growth. With the use of a movable temperature gradient, the number of possible growth centers was limited and crystals several centimeters in length were prepared.

It is also possible to obtain large crystals from a polycrystalline cast metal of zone-refined purity by simply annealing the cast specimen at a temperature near its melting point[30,31]). The driving force for such growth is largely due to the reduction of interfacial energy. In addition, conditions for the growth of a few large grains may be established during the sintering and grain growth of powder materials, e.g. in single phase ceramic systems[32]).

The second type of driving force, namely variation in the free energy of the crystals, is utilized in single crystal techniques such as the strain-anneal[33]) and lineage-induced growth[34]). In the strain-anneal method, a critical strain is required just sufficient to obtain recrystallization in order to allow as few grains as possible to grow into the deformed material when annealed. It is usually assumed that the limiting condition of critical strain is nucleation and not growth; this appears to be true at least for polycrystalline materials[35]). A small and uniform grain size is needed to obtain homogenous deformation during the critical straining in order to reduce the probability of nucleation of stray crystals and to avoid the presence of small, included grains on the surface of the crystal. A steep temperature gradient during the annealing of a critically-strained specimen decreases the probability of forming many new grains and is an important requirement for the preparation of alloy single crystals by this method[36,37]). Deviating orientation are observed in some materials as a result of twinning, which may occur during critical straining or during subsequent annealing.

Single crystals have been prepared from high-purity materials using a combined solidification and recrystallization method[34,38]). This method is based on the fact that lineage or striation substructure, which is obtained in a melt-grown crystal, provides a source of driving energy for the growth of a lineage-free crystal. The main limitation is the requirement of high-purity

material in order to obtain a high velocity of grain boundary migration under the small driving force from the lineage structure.

Large crystals in thin sheets may also be promoted by an additional driving force which is derived from surface energy and its dependence on orientation of the crystal. For example, large grains in which the surface orientation is nearly parallel to low-energy planes have been grown in thin sheets of high-purity iron containing 3% silicon[39,40]). Impurities in the annealing atmosphere or in the specimen may change the surface energy such that crystals are formed with a different surface orientation[41,42]).

5. Matrix stabilization

The stabilization of a polycrystalline matrix is essential in order to promote the growth of as few grains as possible. A stabilized matrix may be achieved by a strong orientation texture, particles, porosity, gas bubbles, suitable solutes, or grain boundary grooves.

The introduction of a strong preferred orientation stabilizes the matrix since the resulting general low and medium angle boundaries have sufficiently low mobility to prevent normal grain growth. This texture-inhibited growth is helpful in attempts to grow single crystals of zone-refined materials by the strain-anneal method. However, the resulting single crystals may be more restricted in orientation.

A strong preferred orientation may form during primary recrystallization of a heavily deformed material. On annealing at a higher temperature this primary texture is removed by relatively few, large secondary grains of different orientation, i.e. by secondary recrystallization. This method for obtaining large grains is dependent on the following factors: (1) the growth of a few crystals is promoted by the presence of a strong preferred orientation and (2) the resulting large secondary grains have a preference for certain orientation relationships with the primary matrix. For example, in the case of copper the secondary grains are related to the primary cube recrystallization texture by rotations near $38°$ or $22°$ about a $\langle 111 \rangle$ axis[43,44]). Grain boundaries with these types of orientation relationships, termed coincidence site orientations, have lower interfacial energies, and greater mobility due to a relative absence of interaction with solute, than do other large-angle grain boundaries[45]). The physical criterion for these special properties of a coincidence boundary may not be coincidence per se, but due to small, periodically repeating units associated with such a structure[46,47]).

Large grains are produced by exaggerated grain growth in which the growth of most of the grains in a fine-grained specimen is inhibited by a fine, uniform dispersion of second-phase particles. This inclusion-inhibited growth is more effective when the driving force for growth is small and due to a reduction of boundary energy. The grain boundaries are also better anchored by particles which are crystalline at the annealing temperature since crystalline particles appear to be immobile and, therefore, are not dragged by moving grain boundaries[48]).

Pores present during sintering of powder compacts serve the same role as growth-inhibiting particles during exaggerated grain growth[32]). In addition, gas bubbles may form preferentially at grain boundaries of a metal annealed after irradiation with neutrons or other nucleons. Such bubbles, if small enough, can inhibit recrystallization[42]) and grain growth[50]). However, the efficiency of bubbles in slowing boundary motion is complicated by the observation that bubbles can migrate bodily through the metal (e.g. refs. 51 and 52).

Matrix stabilization and subsequent growth of a few large grains may also derive from solutes segregated at grain boundaries. Although it is not always easy to distinguish between the role of fine particles and solutes segregated at grain boundaries, a definitive example of segregated solute inhibition of normal grain growth is given by the work of Grenoble and Fiedler[53]). They found that large secondary grains are obtained by a small solute content of sulfur and nitrogen in a primary weak texture of vacuum melted 3% silicon iron.

Thermal grooves formed by the intersection of grain boundaries with the surface of a material may inhibit the motion of grain boundaries and thereby stabilize the matrix. Mullins[54]) has shown that the boundary is essentially stuck in the groove at the surface if the magnitude of the angle the boundary makes with the surface normal is less than a critical value (about $3°$). When grains have diameters larger than about one or two times the sheet thickness, normal grain growth stops due to the formation of thermal boundary grooves. It is then possible to promote the growth of large grains by means of surface energy differences, provided these differences are at least several percent.

Acknowledgement

The author would like to thank the National Research Council of Canada for a grant (NRC A 4611) during the preparation of this paper.

References

1) J. E. Bailey and P. B. Hirsch, Proc. Roy Soc. (London) A **267** (1962) 11.
2) P. A. Beck, J. Appl. Phys. **20** (1949) 633.
3) R. W. Cahn, Proc. Phys. Soc. (London) A **63** (1950) 323.
4) A. H. Cottrell, *Dislocations and Plastic Flow in Crystals* (Clarendon Press, Oxford, 1953).
5) H. Hu, *Recovery and Recrystallization of Metals* (Interscience, New York, 1963) p. 311.
6) J. C. M. Li, J. Appl. Phys. **33** (1962) 2958.
7) J. L. Walter and E. F. Koch, Acta Met. **10** (1962) 1059.
8) H. Gleiter, Phil. Mag. **20** (1969) 821.
9) K. T. Aust and J. W. Rutter, *Recovery and Recrystallization of Metals* (Interscience, New York, 1963) p. 131.
10) D. Turnbull, Trans. AIME **191** (1951) 3.
11) D. G. Cole, P. Feltham and E. Gilliam, Proc. Phys. Soc. (London) **67** (1954) 131.
12) H. Gleiter, Acta Met. **17** (1969) 565.
13) H. Gleiter, Acta Met. **17** (1969) 853.
14) C. H. Bishop and B. Chalmers, Scripta Met. **2** (1968) 133.
15) C. J. Simpson, K. T. Aust and W. C. Winegard, Scripta Met. **3** (1969) 171.
16) K. T. Aust, Can. Met. Quart. **8** (1969) 173.
17) C. J. Simpson, K. T. Aust and W. C. Winegard, Trans. TMS-AIME, **2** (1971).
18) H. Gleiter, Z. Metallk. **61** (1970) 282.
19) E. W. Hart, Scripta Met. **2** (1968) 179.
20) R. L. Fullman, *Metal Interfaces* (Am. Soc. Metals, Cleveland, Ohio, 1952) p. 179.
21) K. Lücke and K. Detert, Acta Met. **5** (1957) 628.
22) J. W. Cahn, Acta Met. **10** (1962) 789.
23) K. Lücke and H. P. Stüwe, *Recovery and Recrystallization of Metals* (Interscience, New York, 1963) p. 171.
24) P. Gordon and R. A. Vandermeer, Trans. AIME **224** (1962) 917.
25) E. L. Holmes and W. C. Winegard, Trans. AIME **224** (1962) 945.
26) P. Gordon and R. A. Vandermeer, *Recrystallization, Grain Growth and Textures* (Am. Soc. Metals, Metals Park, Ohio, 1966) p. 205.
27) H. Gleiter, to be published.
28) E. N. da C. Andrade, Proc. Roy Soc. (London) A **163** (1937) 16.
29) E. N. da C. Andrade and Y. S. Chow, Proc. Roy. Soc. (London) A **175** (1940) 290.
30) J. W. Rutter and K. T. Aust, referred to by K. T. Aust, in: *Techniques of Metals Research*, Vol. 1, part 2 (Wiley, New York, 1968) p. 991.
31) C. J. Beingessner and W. C. Winegard, J. Inst. Metals **93** (1964–65) 480.
32) J. E. Burke, in: *Conf. on Kinetics of High Temperature Pro-*

cesses (MIT Technical Press, and Wiley, New York, 1959) p. 109.

33) H. C. H. Carpenter and C. F. Elam, Proc. Roy. Soc. (London) A **100** (1921) 329.

34) J. W. Rutter and K. T. Aust, Acta Met. **6** (1958) 375.

35) R. W. Cahn, in: *Textures in Research and Practice* (Springer, Berlin, 1969) p. 150.

36) G. K. Williamson and R. E. Smallman, Acta Met. **1** (1953) 487.

37) N. Gane, Bull. Inst. Met. **4** (1958) 94.

38) J. W. Mitchell, *Growth and Perfection of Crystals* (Wiley, New York, 1958) p. 386.

39) J. L. Walter and C. G. Dunn, Trans. AIME **215** (1959) 465.

40) K. Detert, Acta Met. **7** (1959) 589.

41) J. L. Walter and C. G. Dunn, Acta Met. **8** (1960) 497.

42) G. W. Wiener, J. Appl. Phys. **35** (Part II) (1964) 856.

43) M. L. Kronberg and F. H. Wilson, Trans. AIME **185** (1949) 501.

44) C. A. Verbraak, Acta Met. **8** (1960) 65.

45) K. T. Aust, in: *Textures in Research and Practice* (Springer, Berlin, 1969) p. 160.

46) M. Weins, B. Chalmers, H. Gleiter and M. Ashby, Scripta Met. **3** (1969) 601.

47) B. Chalmers and H. Gleiter, Phil. Mag. to be published.

48) M. F. Ashby and R. M. A. Centamore, Acta Met. **16** (1968) 1081.

49) C. E. Ells, Acta Met. **11** (1963) 87.

50) C. E. Ells and W. Evans, Trans. AIME **227** (1963) 438.

51) R. S. Barnes and D. J. Mazey, Proc. Roy. Soc. (London) A **275** (1963) 47.

52) A. P. Greenough, Nature **166** (1950) 904.

53) H. E. Grenoble and H. C. Fiedler, J. Appl. Phys. **40** (1969) 1575.

54) W. W. Mullins, Acta Met. **6** (1958) 414.

THE GROWTH AND DISSOLUTION OF SILVER WHISKERS

J. CORISH and (the late) C. D. O'BRIAIN

Department of Chemistry, University College, Belfield, Stillorgan Road, Dublin 4, Ireland

The growth and dissolution of a small number of silver whiskers which had been preferentially nucleated on a specific site on the surface of a pressed α-Ag$_2$S tablet have been studied photomicrographically. The rates of growth and dissolution may be controlled very conveniently using solid-state electrochemical cells. Silver crossed the silver whisker/α-Ag$_2$S interface with very small overpotentials and the shape of the growth and dissolution curves of the whiskers indicated the operation of a dislocation mechanism. Attempts were made to further elucidate the part played by dislocations in the whisker dissolution process.

1. Introduction

The remarkable ability of α-Ag$_2$S to act as a substrate for silver whiskers has been known for a long time[1]. The growth of these whiskers can be promoted in a variety of ways[2] and has been explained by Wagner[3]. Our interest in such whiskers arose initially from their possible use as a new source of silver in an extension of studies on the transfer of silver across the silver/α-Ag$_2$S phase boundary[4,5]. This transfer process is important because of its role in the unidimensional growth rate of silver sulphide in the reaction $2\text{Ag(s)} + \text{S(l)} \rightarrow \text{Ag}_2\text{S(s)}$[5,6]. This paper describes the nucleation, on a predetermined location on an α-Ag$_2$S substrate, of a small number of well defined whiskers. These whiskers could then be grown and redissolved in an electrochemically controlled process which also allowed their growth and dissolution rates to be measured.

2. Experimental and results

A detailed description of the apparatus and experimental method has been given elsewhere[7,8]. The solid state electrochemical cell[9]

$$\text{Pt} \,|\, \text{Ag} \,|\, \text{AgI} \,|\, \text{Ag}_2\text{S} \,|\, \text{graphite}$$

was used to make the basic thermodynamic and kinetic measurements. The α-Ag$_2$S growth substrate, in the form of a pressed tablet, was made a component in two such cells; the complete circuitry is shown in fig. 1.

Cell A – thermodynamic application: when polariza-tion effects are negligible or under open circuit, the e.m.f. of this cell, E, as measured by the recorder, is related to the chemical potential of the silver in the silver sulphide substrate $\mu_{\text{Ag}}(\text{Ag}_2\text{S})$ by the equation[9]

$$\mu_{\text{Ag}}(\text{Ag}_2\text{S}) - \mu_{\text{Ag}}^0(\text{Ag}) = -EF, \tag{1}$$

where $\mu_{\text{Ag}}^0(\text{Ag})$ is the chemical potential of silver in the pure metallic state and F the Faraday constant. The relationship

$$\mu = \mu^0 + RT \ln a \tag{2}$$

gives the thermodynamic activity $a_{\text{Ag}}(\text{Ag}_2\text{S})$ of silver in the α-Ag$_2$S as

$$a_{\text{Ag}}(\text{Ag}_2\text{S}) = \exp\left(-EF/RT\right), \tag{3}$$

and hence cell A measures the Ag/S ratio in the sulphide tablet.

Cell B – kinetic application: at the operating temperatures of the experiments AgI is a pure ionic conductor[10] and thus the passage of a direct current between the platinum and graphite electrodes results in the transfer of silver to or from the sulphide. When there are silver whiskers on the Ag$_2$S surface and when this silver transfer process is accompanied by no change in the e.m.f. of cell A (steady state condition) then it will be shown that the current through cell B is a measure of the rate of transfer of silver to or from the whiskers.

If silver is added electrochemically to the α-Ag$_2$S tablet using cell B so that the Ag/S stoichiometric limit[9] of the sulphide is exceeded the excess silver is

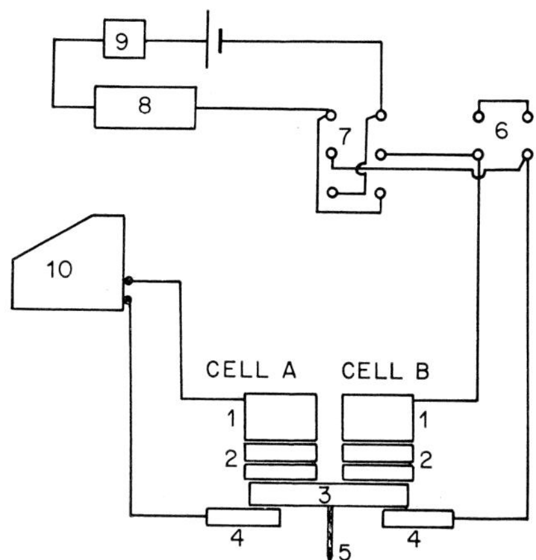

Fig. 1. Schematic diagram of substrate assembly showing the arrangement of the cells and their circuitry. (1) Silver blocks, (2) silver iodide tablets, (3) silver sulphide substrate, (4) graphite, (5) silver whisker, (6) short switch, (7) reversing switch, (8) resistance box, (9) microammeter, (10) Honeywell electronik 19 recorder.

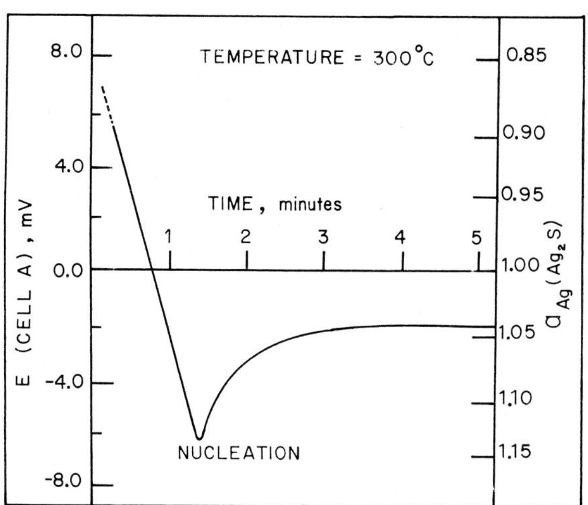

Fig. 2. Time dependence of the potential E of cell A during nucleation of whiskers. The cell B current was 1.0 mA and the curve is from a recorder trace.

quickly precipitated to relieve the supersaturation. The time dependence of the e.m.f. of cell A during this process is similar to that observed when silver whiskers are formed on an α-Ag_2S surface following the removal of sulphur by a stream of hydrogen gas[3]) and is shown in fig. 2. It is likely that the silver precipitated following the electrochemical transfer process also forms incipient whiskers but these were never observed on the exposed sulphide surface. The deposits always occurred in the region of maximum supersaturation resulting from such an addition of silver, i.e. near the Ag_2S/AgI interface. Here any subsequent growth or dissolution of the whiskers could occur by the simultaneous migration of silver ions through the iodide to, or from their tips and sides and electrons to, or from their bases. Silver aggregates in this location were thus of no use in the study of silver transfer processes at the silver whisker/α-Ag_2S interface. Neither could their growth behaviour be observed and so an attempt was made to induce whisker nucleation on the exposed surface of the α-Ag_2S tablet. Earlier experiments[11]) had shown that under certain conditions the required surface nucleation might be accomplished by the use of a beam of silver vapour. Fig. 3 is a scanning electron micrograph showing silver whiskers which resulted when a beam of intensity $\sim 5.0 \times 10^{14}$ atoms cm^{-2} sec^{-1} impinged on the α-Ag_2S surface for a short time either before or during the electrochemical addition of the silver. The nuclea-

tion pattern (fig. 2) was virtually unchanged and the minimum in E was observed to occur exactly at the time of whisker nucleation. When the silver beam was confined to a small area (0.3 mm diameter) of the tablet surface nucleation occurred in that area only and a small number, usually between three and seven, of well-formed whiskers was produced. Similar preferential nucleation was also obtained when a tablet which had been supersaturated by the electrochemical addi-

Fig. 3. Silver whisker growth on an α-Ag_2S substrate without preferential nucleation (40$\times$).

Fig. 4. Group of whiskers grown after prefential nucleation (40×).

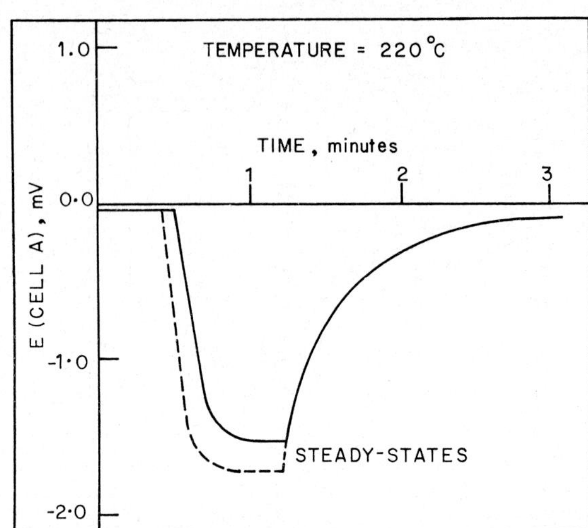

Fig. 5. Curves showing typical behaviour of the potential E of cell A during the short term growth, or regrowth, of a group of silver whiskers. Solid line: growth current 0.12 mA, broken line: growth current 0.16 mA. The curves are from recorder traces and have been drawn to show the superposition of the relaxation curves.

tion of silver to a degree insufficient to cause nucleation was exposed to the 0.3 mm diameter silver beam for two to three minutes. Preferentially nucleated whisker groups were used to obtain all the data presented here because only for these could cross-sectional areas be determined accurately[8,12]) and growth and dissolution behaviour observed satisfactorily. Fig. 4 shows a group of whiskers grown from a preferentially nucleated site. All photographs shown here are scanning electron micrographs of samples which have been removed from the vacuum system as the conventional photomicrograph, with which the in situ observations were made, had too shallow a depth of focus to photograph complete whiskers.

The galvanostatic addition of silver to an α-Ag$_2$S substrate which had been successfully nucleated resulted in the growth of the whiskers, while the potential of cell A varied with time as shown in fig. 5. Steady state potentials dependent on the magnitude of the cell B currents were obtained without nucleation minima and the relaxation curves on switch-off were characteristic of the sample under study. The whiskers were observed to grow from the base and when weighed in the microanalytical laboratory agreed to within 0.5% with a simple application of Faraday's Law to prove that the cell B current was a direct measure of the whisker growth rate. Reversing the cell B currents gave mirror image curves to those of fig. 5 with the steady states

now indicating that the electrochemical withdrawal of silver was being exactly compensated by silver entering the sulphide from the whiskers. The whiskers could be seen dissolving from their bases in a process exactly analogous to the phase boundary studies which have been made[4,5]) on the block silver/α-Ag$_2$S interface.

The data for the current vs. potential curves shown in fig. 6 for the silver transfer process were obtained by the following procedure. Nucleated whiskers were grown galvanostatically, the usual current being 1.0 mA, until 2 to 3 mm long. They were then partially redissolved using a series of reverse currents which were varied stepwise in the range 0.02 mA to 0.60 mA. Each current was used only long enough to characterize the steady state potential attained and the relaxation traces were recorded. The partially dissolved whiskers were then regrown using a similar series of currents and the complete process observed microscopically to ensure that it involved only the regular straight portions of the whiskers. This sequence of operations was used because it provided the best chance of dissolving the whiskers across a uniform interface and because by always maintaining the regrowth currents smaller than the initial growing current the danger of producing new nuclei was minimized. In contrast to other studies[4]) all the

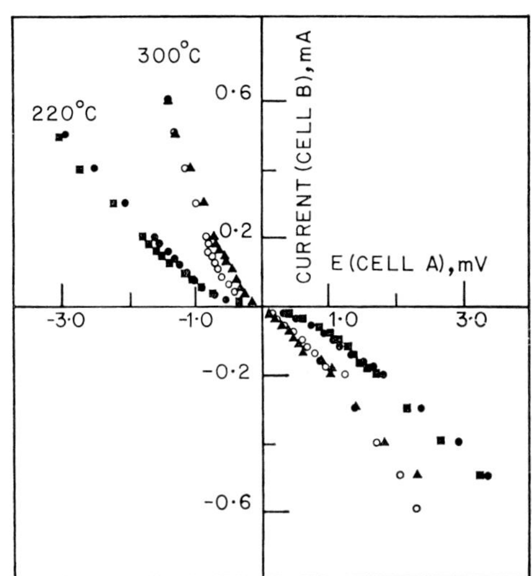

Fig. 6. Current-potential curves for the transfer of silver in both directions across the silver whisker/α-Ag$_2$S interface. The data are from four groups of whiskers; two at 220 °C and two at 300 °C.

Fig. 7.   Silver whisker tip in detail (1300×).

relaxation curves pertaining to any sample were found to be exactly superimposable (fig. 5) indicating a non-changing silver whisker/α-Ag$_2$S interface. In earlier unsuccessful experiments observable changes at this interface (e.g. complete dissolution of a whisker or production of a new one) had invariably been accompanied by noticeable changes in the shapes of relaxation curves.

The fine structure and form of the whiskers are shown in figs. 7 and 8. Their tips appear to be examples of the previously observed phenomenon[13] of the joining together of a number of smaller fibres. Extension of the technique used to measure their cross-sectional area[8,12] showed that they quickly developed roughly cylindrical shapes of uniform cross-section. Voids were sometimes encompassed (fig. 8) and when present were found to extend right through the straight portion of the whisker. The average cross-section of the bases of whiskers grown at 300 °C was 1.5×10^{-5} cm^2, which corresponds to a diameter of ~ 40 μm, while those grown at 220 °C had diameters larger than this by factors of 2 or 3. In the course of the work three anomalously large whiskers with blunted heads and cross-sections of $\sim 70 \times 10^{-5}$ cm^2 were produced but they behaved in the same manner as the more usual kind. The transfer of silver across the silver whisker/α-Ag$_2$S interface at a current density of 0.5 A/cm^2 was accomplish-

ed with overpotentials of ~ 2.9 mV and 2.2 mV at 220 °C and 300 °C respectively. The same silver transfer rate from block silver required overpotentials of the order of 220 mV and 160 mV at these same temperatures[4]). The largest silver transfer rate observed at the silver whisker/α-Ag$_2$S interface was 8.7 A/cm^2 and this required an overpotential of <5 mV at 300 °C.

3. Discussion

The process used here for growing the whiskers on the α-Ag$_2$S may be expected to produce an almost ideal

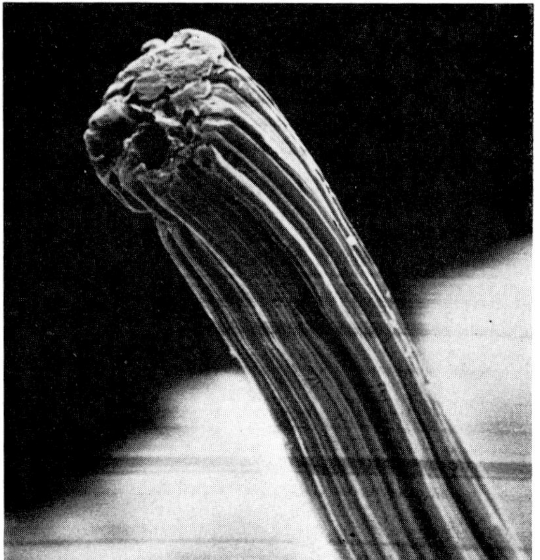

Fig. 8.   Base and lower section of a whisker which had been broken from its α-Ag$_2$S substrate (400×).

contact between the two solids. If this is so then it appears that the large difference in the overpotentials for the transfer of silver across the block silver/α-Ag$_2$S and silver whisker/α-Ag$_2$S interfaces might be interpreted in terms of differences in the quality of the microcontacts achieved in each case. Indeed it has been pointed out[14]) that the pressure dependent effects at the block silver/α-Ag$_2$S boundary as well as the non-reproducible relaxation curves[4]) are typical of microcontact problems. A more recent investigation, however, has shown that this explanation can provide no more than a partial answer since the nature of the block silver, and in particular its grain size, has been found to have large effects in determining the extent of the polarization[5]). The importance of the mosaic structure of the silver is not surprising since the process under study involves the dissolution of the silver into the silver sulphide. Furthermore because the extremely large interdiffusion coefficient of silver in α-Ag$_2$S[9]) and the abundance of available silver sites[15,16]) make the sulphide an ideal environment to accept silver[11]), the process should depend largely on the facility with which silver atoms are removed from the dissolving silver crystals. Frank[17]) has pointed out that crystal imperfections providing perpetual steps in a crystal surface allow dissolution and growth to take place easily whereas atomically smooth surfaces require continuous two-dimensional nucleation[18]) and their growth and dissolution curves show critical supersaturation characteristics[19]).

The electrochemical measurements made in this study readily provide the data for the growth and dissolution curves of the whiskers as shown in fig. 9. The cell B current has been shown to be a direct measure of the growth rate of the whiskers and, using Faraday's Law, current density values may be converted to rates of change in whisker length. The activity of silver in the α-Ag$_2$S, as given by cell A using eq. (3) is analogous to the saturation ratio $\alpha = p/p_0$ usually used to define the magnitude of the driving force in the growth of crystals from their own vapour, and following Burton et al.[20]) the supersaturation σ is defined as

$$\sigma = \alpha - 1. \tag{4}$$

The curves show no critical super or undersaturation characteristics and indicate the operation of a dislocation mechanism. The rates of growth increase with

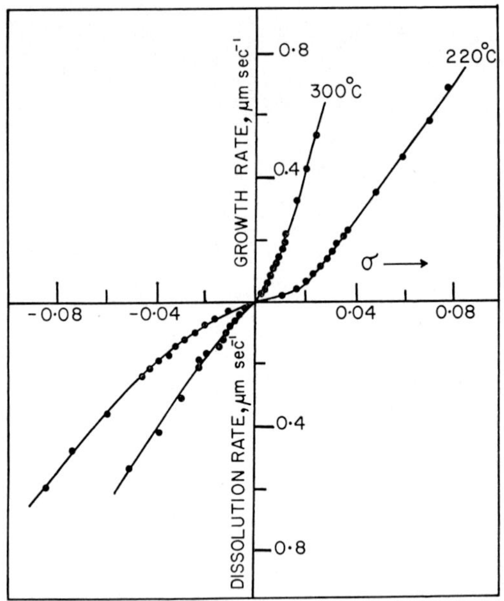

Fig. 9.   The growth and dissolution rates of silver whiskers on an α-Ag$_2$S surface against the supersaturation σ at temperatures of 220 °C and 300 °C as measured electrochemically.

temperature and the curves, which become linear at larger σ values, are qualitatively very similar to those obtained[21]) for the growth and evaporation of polycrystalline benzophenone and interpreted in terms of the Frank theory. Screw dislocations have been observed[22]) in silver whiskers similar to those studied here and it is proposed that it is the presence of these imperfections in the whiskers which is mainly responsible for their growth and dissolution behaviour and also for the extremely low overpotentials observed for silver transfer at the silver whisker/α-Ag$_2$S interface. The latter effect is analogous to that discussed by Vermilyea[23]) for the electrolytic growth and dissolution of copper whiskers in solution. Because of their size the whiskers must contain multiple dislocations[13]) which are presumably introduced during growth by dislocation ends existing in the crystallites which form the surface of the pressed α-Ag$_2$S substrate. Such imperfections are known to act as sites for the congregation of mobile adatoms[24]) and this type of mechanism for whisker nucleation has been proposed by Sears[25]) for the growth of mercury whiskers from the vapour on glass. It would explain the observation (fig. 4) that whiskers do not grow on the total area exposed to the silver beam.

Attempts have been made to probe further the special character of the whisker/α-Ag$_2$S interface. Whiskers were removed from their growth sites and replaced (i) vertically but with their bases in contact with a new area of the α-Ag$_2$S tablet and (ii) horizontally with their sides pressed against the silver sulphide surface. Under no circumstances did the whiskers redissolve exactly as if they were at their original growth sites but the overall results must, unfortunately, be regarded as inconclusive since pressure effects, aggravated by the small size of the whiskers, proved to be the dominant factor in determining the magnitudes of the observed overpotentials.

The extent of the control over the nucleation site and the subsequent rates of growth and dissolution given by the method described here makes it very attractive for the production and study of this type of whisker. It should be capable of refinement to produce just one whisker of a particular size and at a desired location. It has been extended[7] to the growth of silver whiskers on α-Ag$_2$Se and, in principle, should be adaptable to study all of the limited number of systems for which suitable solid-state electrochemical cells exist.

Acknowledgements

Thanks are due to Prof. J. M. Thomas, Dr. C. J. Warde and Dr. E. P. O'Neill for helpful discussions and to Dr. D. L. M. Cansfield and Mr. D. L. Scott for the scanning electron micrographs.

References

1) J. F. Henckel, *Pryitologia*, translated from the German (Miller and Linde, London, 1757).
2) H. K. Hardy, Progr. Metal Phys. **6** (1956) 45.
3) J. Schmalzried and C. Wagner, Trans. Met. Soc. AIME **227** (1963) 539.
4) H. Rickert and C. D. O'Briain, Z. Physik. Chem. **31** (1962) 71.
5) C. J. Warde, Ph.D. Thesis, Natl. Univ. Ireland, Dublin (1968).
6) H. Rickert and C. Wagner, Z. Physik. Chem. **31** (1962) 32.
7) J. Corish, Ph.D. Thesis, Natl. Univ. Ireland, Dublin (1969).
8) J. Corish and C. D. O'Briain, J. Mater. Sci., in press.
9) C. Wagner, J. Chem. Phys. **27** (1953) 1819.
10) C. Tubandt, in: *Handbuch der Experimentalphysik*, Vol. 12 (1) (1932) p. 383.
11) J. Corish, M. T. Duffy and C. D. O'Briain, Trans. Faraday Soc., in press.
12) S. S. Brenner and C. R. Morelock, Rev. Sci. Instr. **28** (1957) 652.
13) S. S. Brenner, Acta. Met. **4** (1956) 62.
14) D. O. Raleigh, Progr. Solid State Chem. **3** (1967) 83.
15) H. Okazaki, J. Phys. Soc. Japan **23** (1967) 355.
16) P. Junod, Helv. Phys. Acta **32** (1959) 567.
17) F. C. Frank, Discussions Faraday Soc. **48** (5) (1949) 67.
18) M. Volmer, *Kinetic der Phasenbildung* (Steinkopff, Dresden und Leipzig, 1939).
19) G. W. Sears, J. Chem. Phys. **24** (1956) 868.
20) W. K. Burton, N. Cabrera and F. C. Frank, Phil. Trans. Roy. Soc. London A **243** (1951) 299.
21) S. A. Kitchener and R. F. Strickland-Constable, Proc. Roy. Soc. (London) A **245** (1958) 93.
22) H. Rickert, private communication to C. D. O'Briain.
23) D. A. Vermilyea, J. Chem. Phys. **27** (1957) 814.
24) G. A. Bassett, Phil. Mag. **3** (1958) 1042.
25) G. W. Sears, Acta Met. **3** (1955) 367.

Journal of Crystal Growth **13/14** (1972) 68–72 © *North-Holland Publishing Co.*

HOLOGRAPHIC STUDY OF CRYSTAL GROWTH*

M. E. GLICKSMAN and R. J. SCHAEFER

Metallurgy Division, Naval Research Laboratory, Washington, D.C. 20390, U.S.A.
and

J. A. BLODGETT

Optical Sciences Division, Naval Research Laboratory, Washington, D.C. 20390, U.S.A.

Holography has two main advantages over conventional microscopy for studies of crystal growth. First, the hologram records the object in great depth, so that events occurring at unpredictable locations can be studied in the reconstructed image. Second, the hologram records the phase as well as the amplitude of the light scattered from the object, thus making possible several interferometric techniques.

A holographic microscopy system has been constructed to study the growth of transparent crystals. Holograms were made of initially planar solid/liquid interfaces developing into cellular or dendritic morphologies. Several interferometric techniques were used to measure the development of interface features. The techniques included double exposure holography, which revealed the crystal growth between the two exposures, and interference of the reconstructed image with a plane wave, which revealed surface contours. The use of a relatively high-powered argon laser permitted exposure times as short as 30 microseconds, so that rapidly growing crystals could be studied.

Interferograms of reconstructed holograms are shown, illustrating the development of micromorphology on solid/liquid interfaces. The measured shape of a dendrite tip is compared to that predicted by a theoretical model.

1. Introduction

Studies of the micromorphology of growing crystals have usually been qualitative or, at best, limited to measurements of the grosser features of such crystals. Experimental techniques for the detailed mapping of surface morphology have simply not been available. Thus, while several detailed theories[1-8] have been developed to describe the steady-state or transient morphologies of solid/liquid interfaces, very little experimental data exist with which to compare the theoretical results. It is thus difficult to judge the validity of the several competing theories.

Two areas of crystal growth in which there is a particular need for experimental measurements are the determination of the exact shape of the tip of growing dendrites, and the observation of the breakdown of an unstable planar solid/liquid interface. Experimental data in these areas have until now been too crude to differentiate between the various competing theories which try to explain these phenomena.

The new technique of holography now provides a tool whereby extremely detailed quantitative data can be obtained from a single image of a solid/liquid interface. In order to make clear the advantages of holography over conventional photography, several of the principles and techniques available for making and reconstructing holograms will be briefly described here.

2. Holographic techniques

A hologram records the amplitude and phase of coherent light scattered from an object illuminated by a laser. A high-resolution film may be used as the recording medium. After development, the hologram is illuminated by coherent light to produce, by diffraction, a reconstructed image of the original object. When the morphological features of interest occur on a microscopic scale, it is important that this reconstructed image have a spatial resolution commensurate with the scale of these features. The attainment of high resolution is possible only with very careful attention to mechanical stability of the holographic apparatus, quality of its optical components, exposure and processing of the film, and alignment of the hologram for reconstruction. These requirements combine to make holography considerably more exacting (and expensive) than con-

---

* This paper was presented with the following one as a combined invited paper by M. E. Glicksman.

ventional photography. However, the image reconstructed from a hologram has several unique characteristics which, for certain purposes, make it more useful than an image recorded by conventional photographic means. The three-dimensional qualities of a holographic image are well-known – a microscope can be focused at any depth within the image to reveal features in sharp detail. Less widely known, but perhaps even more useful, is the fact that the hologram reconstructs the *phase*, as well as the *amplitude*, of the light scattered from the original object. This phase information can be extracted by a variety of interferometric methods, to reveal precise dimensional measurements of the original object. Only if one takes advantage of these characteristics is holography advantageous over conventional photography.

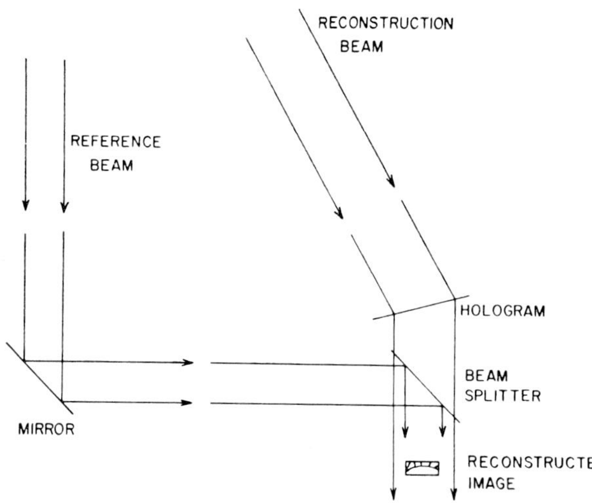

Fig. 1. Optical arrangement for producing interference fringes between reconstructed image of growing crystals and plane wave reference beam.

One useful technique for extracting phase information from a hologram is double-exposure holography. Two holograms are recorded in sequence on the same piece of film, without moving the film or the object. If the shape of the object changes slightly in the interval between the two exposures, two images of the object will be reconstructed, and optical interference effects will be seen between these two images. This technique enables an observer to measure minute changes of bodies with highly irregular shapes, and, thus double-exposure holography can be used to map out the inhomogeneities in growth rate on the surface of a crystal.

A second useful technique involves comparing a reconstructed image from a single-exposure hologram with a coherent optical plane wave. A beam splitter is used to combine the reconstructed image and the plane wave (fig. 1), and fringes then appear along lines of constant optical path difference. By changing the angle of incidence of the plane wave, one can align the fringes in such a way as to facilitate the measurement of interesting features.

Both of these techniques were used in a study of the growth of crystals of two low-entropy-of-fusion organic materials, camphene and succinonitrile. These materials have solidification behavior similar to that of metals[9], with cubic crystal structures and unfaceted solid/liquid interfaces. The transparency of these organic materials makes them superior to metals for optical crystallization studies.

3. Experimental procedure and results

A glass specimen chamber, with flat windows on the top and bottom surfaces[10], was filled with the test material. A nichrome wire heater warmed the top of the chamber, melting all of the material except a thin layer on the bottom of the chamber. Freezing and melting were then controlled by varying the heater power. The specimen was illuminated by a collimated "object beam" passing vertically down through the specimen chamber, then reflecting from a 45° mirror onto the film. A collimated "reference beam" forming an angle of approximately 25° with the object beam, simultaneously struck the film to form the hologram. Exposure times were in the range of 30 to 150 µsec, using the 5145 Å line of an argon ion laser. For reconstruction, the hologram was illuminated to produce a real image, which was examined through a conventional microscope. The reconstructed image is seen as if one were looking up through the bottom surface of the specimen chamber, i.e., through the solid.

Fig. 2 shows a reconstruction of a double-exposure hologram of solidifying succinonitrile. The first exposure was made just as the power to the specimen chamber heater was switched off and the second exposure after one minute of crystal growth. The growth during this interval has resulted in changes in the optical path length for light transmitted through the chamber, since solid and liquid have different indices of refraction. The two reconstructed images, therefore, show inter-

Fig. 2. Reconstruction of double-exposure hologram of solidifying succinonitrile crystals. The growth cell, which contains two bubbles, is ¾ inch in diameter, and is crossed by a fine thermocouple.

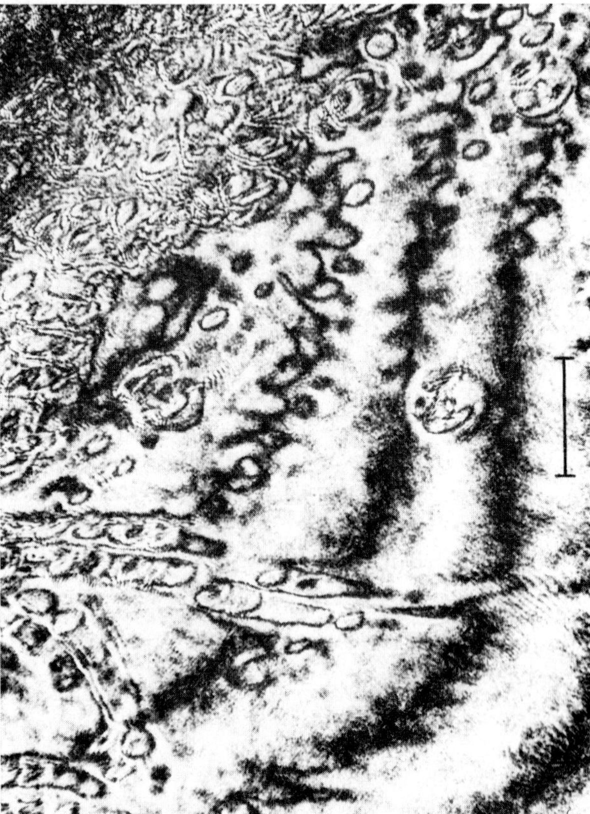

Fig. 3. Enlargement of a small section of the reconstruction shown in fig. 2. The fringes are distorted by microstructural features, but the exact shape of the features is not revealed. Marker indicates 250 μm.

ference fringes, with each fringe representing an integral number of wavelengths in optical path change. Each fringe here represents a motion of 35 μm in the solid/liquid interface (for camphene each fringe represents a motion of 18.5 μm). Growth has been greatest near the edges, and here the initially smooth solid/liquid interface has broken down into a highly irregular structure. Fig. 3 is a more detailed view of one area of the interface, and here it can be seen that the large fringes are distorted by the micromorphological features developing on the interface. Elongated ridges are developing beside crystal sub-boundaries, and isolated knobs, the precursors of dendrites, are appearing in the faster growing regions.

Although double-exposure holography gives a very clear picture of macroscopic growth patterns (fig. 2), on a microscale it is often difficult to interpret the fringes (fig. 3). An annular fringe may represent an elevation or a depression. In most cases it has been found that interferometry between a plane wave and a reconstructed single-exposure hologram (fig. 1) is more useful than the double exposure technique. Fig. 4a is a recon-

struction of a single-exposure hologram of a solid/liquid interface, including within its field several crystal sub-boundaries, with morphological features developing near them. In fig. 4b, a plane reference wave has been added, and fringes have been produced to reveal the exact shape of surface features. Elevations can now be distinguished from depressions by the direction of bending of the fringes. This technique reveals features with heights corresponding to a small fraction of a fringe interval, whereas the same features would be very difficult to recognize by the double-exposure technique. It can be seen in figure 4b that, although protrusions of the interface are developing near the sub-boundaries, large areas of the interface remain smooth.

Fig. 5 illustrates a dendrite advancing across the field of view. The fringes have been aligned to run perpendicular to the dendrite axis. Fringe contours in fig. 5 traced from the photograph are compared with theo-

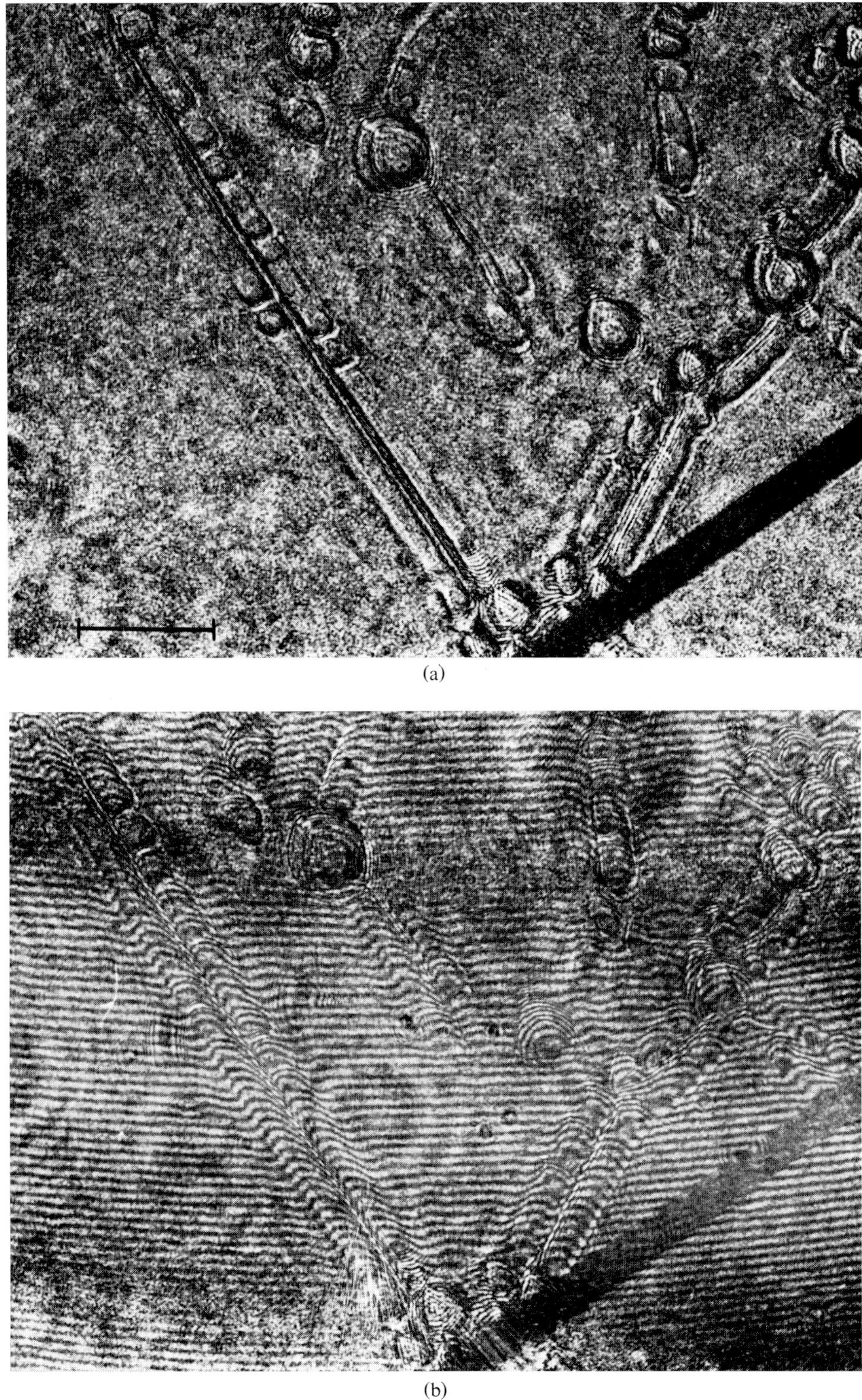

(a)

(b)

Fig. 4. (a) Reconstruction of a section of solid/liquid interface, showing several microstructural features developing adjacent to crystal sub-boundaries; marker indicates 250 µm. (b) The same reconstruction shown in (a), with a plane wave reference beam added to produce contour fringes.

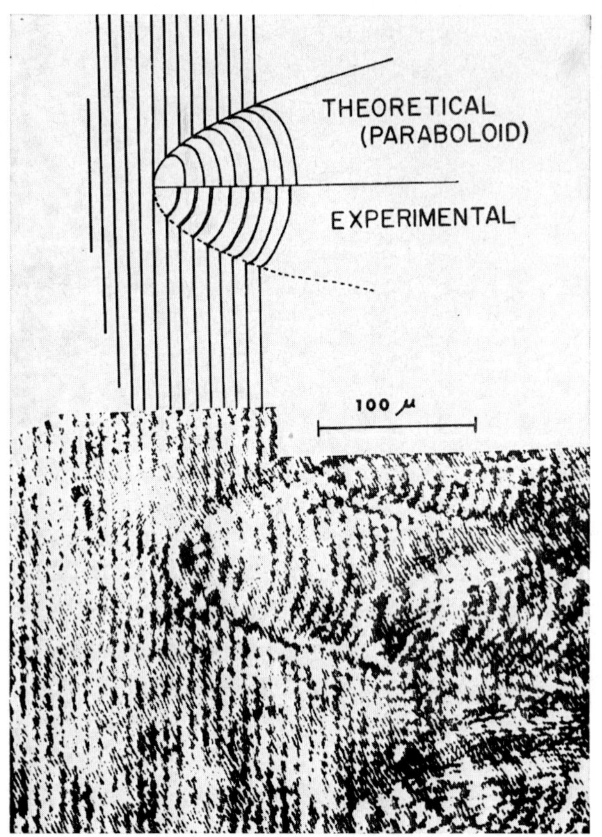

Fig. 5.   Fringes on tip of growing dendrite, and comparison of fringes traced from photograph with fringes calculated by assuming dendrite is a paraboloid of revolution.

retical contours calculated by assuming that the dendrite tip is a paraboloid of revolution. The experimental contours are difficult to follow where they bend sharply, as at the edge of the dendrite, nonetheless, the overall agreement between the theoretical model and these observations is satisfactory. There is no suggestion that branching has started anywhere within ten times the tip radius from the dendrite tip.

4. Discussion

The results shown here illustrate some of the measurements which can be made only through holography. Conventional interference microscopy does not allow comparison of an object with itself at a later time, as in double-exposure holography, nor does it allow alignment of fringes with rapidly changing features. Moreover, because the hologram can produce a real image

when reconstructed, one can examine features within bulk liquid with a microscope objective of short working distance and high numerical aperture, and thus achieve best resolution. The holographic image suffers, however, from granularity and diffraction effects from out-of-focus particles. Since these "optical noise" effects often make it difficult to follow fringes when they are crowded together, the holographic techniques work best on relatively isolated features.

These experiments illustrate the use of holographic techniques to measure dendrite tip shapes and the breakdown of unstable planar solid/liquid interfaces. The studies of dendrites have shown that, near the tip, the dendrite does not differ greatly from a paraboloid of revolution. The crystal anisotropy is not strong enough to produce a detectable four-fold axial symmetry of the crystal in this region. Subsequent studies will measure the initial phases of side-branch development. The observations of ridges on the solid/liquid interface adjacent to grain boundaries and sub-boundaries have shown no indication that this disturbance propagates outward from the boundaries as a wave, other than as a very slight secondary ridge. Thus, the boundaries do not generate periodic structures in adjacent regions.

The primary advantage of holography as a tool for crystal growth studies lies in its use for the detailed measurement of transient crystal shapes. These measurements are providing the first experimental data of sufficient detail to provide a rigorous test of many theories of crystal growth.

References

1) G. P. Ivantsov, Dokl. Akad. Nauk USSR **58** (1947) 567.
2) D. E. Temkin, Dokl. Akad. Nauk USSR **132** (1960) 1307.
3) M. E. Glicksman and R. J. Schaefer, J. Crystal Growth **1** (1967) 297.
4) G. R. Kotler and L. A. Tarshis, J. Crystal Growth **3,4** (1968) 603.
5) W. W. Mullins and R. F. Sekerka, J. Appl. Phys. **34** (1963) 323.
6) R. F. Sekerka, in: *Crystal Growth*, Ed. H. S. Peiser (Pergamon, Oxford, 1967) p. 691.
7) S. R. Coriell and R. L. Parker, ref. 6, p. 703.
8) G. R. Kotler and W. A. Tiller, ref. 6, p. 721.
9) K. A. Jackson and J. D. Hunt, Acta Met. **13** (1965) 1212.
10) R. J. Schaefer and M. E. Glicksman, Met. Trans. **1** (1970) 1973.

Reprinted from:
Journal of Crystal Growth **13/14** (1972) 73–77 © *North Holland Publishing Co.*
Printed in the Netherlands

GRADIENT HOT-STAGE ELECTRON MICROSCOPY OF BINARY METALLIC SOLID/LIQUID INTERFACES*

M. E. GLICKSMAN and C. L. VOLD

Metallurgy Division, Naval Research Laboratory. Washington, D.C. 20390, U.S.A.

Temperature gradient hot-stage electron microscopy of solid/liquid interfaces has been utilized in the study of metallic binary systems. Solute additions were found to have a marked effect on the micromorphology of the solid/liquid interfaces. For example, in a dilute alloy of tin in bismuth, melting always occurred at grain boundaries, leaving isolated "islands" of bismuth-rich solid in a "sea" of binary liquid. A number of binary systems representing the common types of crystallization reactions have been surveyed, and typical results are presented.

1. Introduction

The use of *in situ* methods of imaging the solid/liquid interface has been recognized as a progressive step in the study and analysis of interfacial processes occurring during crystal growth or solution. The usefulness of optical microscopy in studying solidification and melting has been clearly demonstrated for transparent organic compounds[1,2] and for opaque metallic materials[3,4]. However, achieving adequate contrast effects to image the position and shape of the solid/liquid interface has generally been a serious experimental problem in metallic systems, and, moreover, the limited resolution obtainable by optical microscopy has proved to be a severe restriction when investigating crystallization phenomena occurring on the scale of a few microns or less.

The application of transmission electron microscopy to the study of solidification and melting has extended the resolution capabilities of direct observational methods to about 50 Å. It has thus become possible to view micromorphological structures at a solid/liquid interface on a size scale where dislocations, sub-boundaries, and grain boundaries play a major role in determining the solidification structure. Also, the problem of securing adequate contrast between the solid and liquid phases has been virtually eliminated.

Pure bismuth has been studied quite extensively[5–7] using the techniques developed by the authors. These studies have yielded detailed information concerning the micromorphology and the interaction of defects during solidification. Quantitative applications of gradient hot-stage electron microscopy have yielded an absolute determination of the solid/liquid interfacial free energy in pure bismuth[8], and quantitative information concerning the thermodynamic behavior of dislocations near the melting temperature.

The purpose of the present investigation was to extend the usefulness of the basic electron microscope technique into the area of alloy solidification. To this end, a survey of a number of alloy systems representing the common types of crystallization reactions was initiated, the results of which are described in this paper.

2. Experimental procedure

The details of the technique used to image the solid/liquid interface have been adequately described in earlier papers[5,6]. Briefly, the stringent requirements for positional stability of the solid/liquid interface required for high-resolution photography at high magnifications were achieved by utilizing the microscope's electron beam *both* for inducing melting and producing the transmission electron image. The alloy specimens for microscopic examination were generally prepared by sequentially evaporating the two component metals from separate crucibles onto carbon-film substrates. The carbon films were supported by 200-mesh microscope grids. An alternate technique sometimes used in preparing the specimens was simultaneous evaporation of the two components from a single crucible onto the

* This paper was presented with the preceding one as a combined invited paper by M. E. Glicksman.

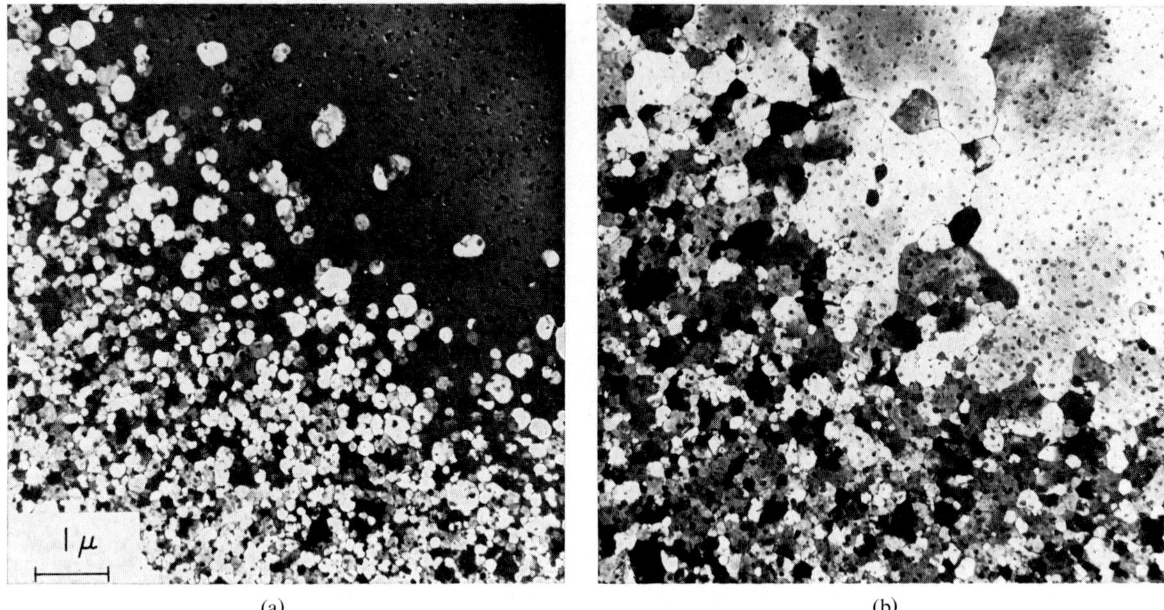

(a)   (b)

Fig. 1.   Melting and solidification of a highly bismuth rich bismuth–tin alloy. (a) "Mushy" zone, characteristic of alloys which melt over a range of temperatures. (b) Structure of the re-solidified alloy of (a). Note occlusion of many grains by the much larger resolidified grains.

substrate. Although the specimen composition could not be determined accurately using either of these methods of preparation, this did not prove to be a serious handicap, because the compositional limits of the specimen could usually be determined from knowledge of the phases present, as identified from the electron diffraction pattern.

Certain general requirements must be met for the successful utilization of the technique in the investigation of multi-component systems:

(a) At least one phase must melt at a reasonably low temperature (within the temperature limitation of the hot stage).

(b) The alloy must exhibit low reactivity with the residual gases in the electron microscope.

(c) The vapor pressures of the component elements must be sufficiently low to prevent rapid evaporation and compositional change at or near the liquidus temperature.

(d) The alloy must exhibit sufficiently high creep strength to retain a continuous film around the partially melted area.

Using the experimental procedures described above a number of binary alloy systems have been surveyed. These include the systems Bi–Sb, Bi–Sn, Bi–Pb, Au–Pb,

Au–Bi, Ag–Bi, Au–Sn, and Au–Sb. It is noted that each binary satisfies to a greater or lesser degree the requirements stated above. The major features displayed by certain of these binary systems are described in the following section.

3. Results

Bi–Sn: The melting of a bismuth–tin alloy, highly rich in bismuth, is presented in fig. 1a. This alloy melts over a range of temperatures, which results in a "mushy" zone, shown here to consist of islands of bismuth-rich solid surrounded by binary liquid. Here, the line of complete melting (the liquidus isotherm) and the line of negligible melting (the eutectic isotherm) are separated by about 5×10^{-4} cm, consistent with a temperature gradient of roughly 4×10^4 °C/cm in the plane of the film. It should be noted that despite these high temperature gradients, thermal convection is absent because of the extreme thinness of the specimen (specimens were seldom thicker than about 2000 Å), and any redistribution of solute occurs by diffusive flow only. This situation is quite unlike the conventional case that occurs in bulk castings. The individual "island" grains can be held stationary, but after freezing, fig. 1b, the resolidified microstructure shows that the previously

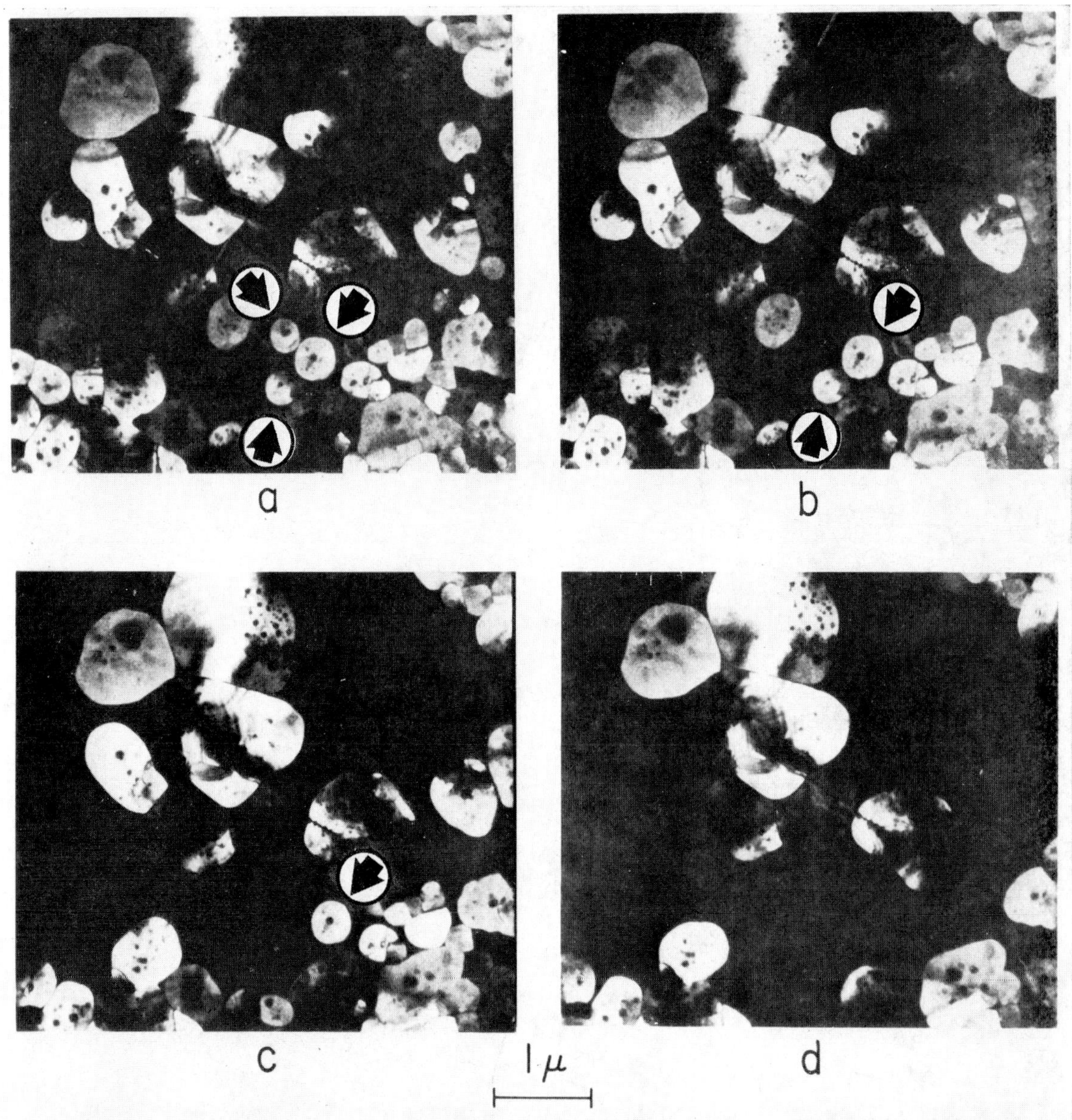

Fig. 2. Melting of a bismuth–lead alloy. This sequence was photographed at time intervals of about 5 to 10 sec. The isolated grains melted slowly until their diameters approached a critical value of about 2.5×10^{-5} cm, whereupon very rapid melting occurred. Note the melting of the arrowed grains on the succeeding frames.

isolated "island" grains have been occluded by the large re-solidified grains, both on the grain boundary and within the grain interiors.

Bi–Pb: A "mushy" zone was also observed in this alloy series, fig. 2. Here the isolated grains shrank very slowly as the temperature rose, and as their diameters approached about 2.5×10^{-5} cm, very rapid melting ensued (figs. 2b–2d). This abrupt increase in the melting

rate was probably caused by the depression of the solidus temperature at small radii of curvature, rather than by a change in the kinetic mechanism governing the melting rate.

Au–Pb: The peritectic decomposition of the intermetallic compound Au_2Pb into a nearly pure gold phase plus a binary gold–lead liquid is illustrated in fig. 3. Here the specimen, the average composition of

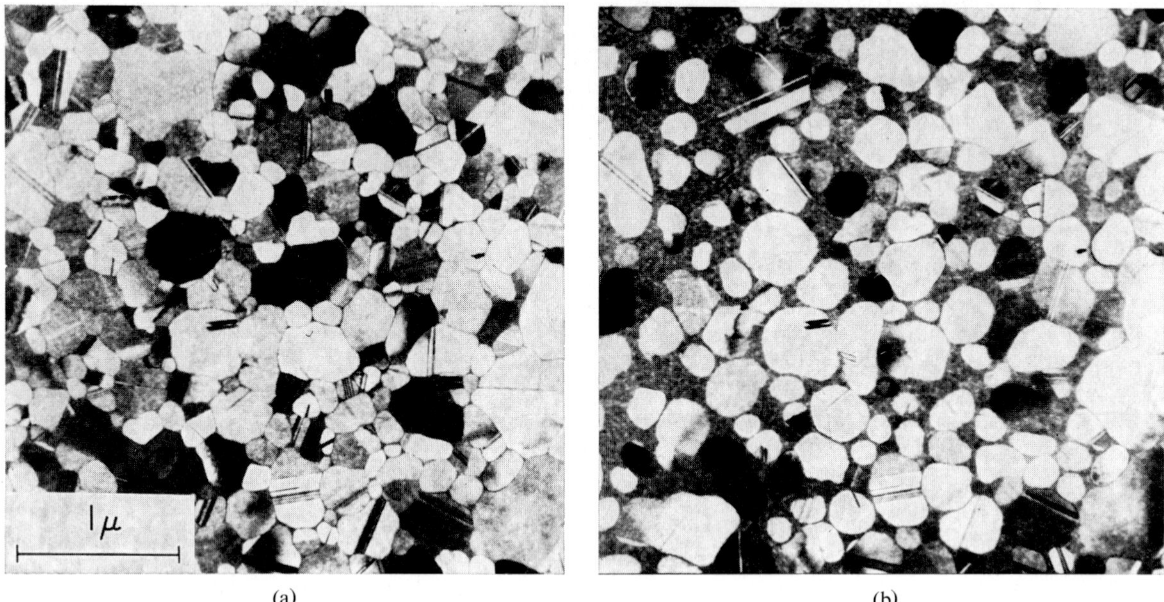

(a)   (b)

Fig. 3. Peritectic decomposition of Au_2Pb into solid gold plus a gold–lead binary liquid. (a) Decomposition at an early stage. (b) Decomposition at an advanced stage, where many of the smaller grains have decomposed completely.

which was on the gold side of the compound Au_2Pb, was slowly raised to the peritectic temperature. At a temperature slightly above the peritectic temperature, a trace of binary liquid was observed to form at the grain corners, fig. 3a, imaging as very small and dark triangular areas. A later stage of the decomposition is depicted in fig. 3b. Here the liquid phase formed at the grain boundaries is much more evident, the peritectic reaction having gone to completeness on many of the smaller Au_2Pb grains. The resulting microstructure consists of islands of gold surrounded by the binary liquid.

Au–Sb: The average composition of this alloy is only slightly to the eutectic side of the compound $AuSb_2$, and the phases in equilibrium are, therefore, predominantly $AuSb_2$ plus a small amount of gold solid solution. As the eutectic temperature (360 °C) is exceeded, grain-boundary liquation was observed, fig. 4, and the resulting microstructure is nearly pure $AuSb_2$, plus a binary liquid of nearly eutectic composition. It was not possible to raise the temperature of the film to the decomposition temperature of the compound, because of a strong tendency for the alloy film to de-wet from its carbon film substrate.

Au–Sn: This alloy was prepared by sequential evaporation of the component elements from separate cru-

cibles; the average composition was between that of the intermediate phase designated as ζ in Hansen[9]) (approximate composition 14 at% Sn) and that of the congruently melting compound AuSn. Fig. 5a is a micrograph of the alloy at a temperature slightly below the eutectic temperature (280 °C). As this temperature is exceeded, fig. 5b, grain-boundary liquation is ob-

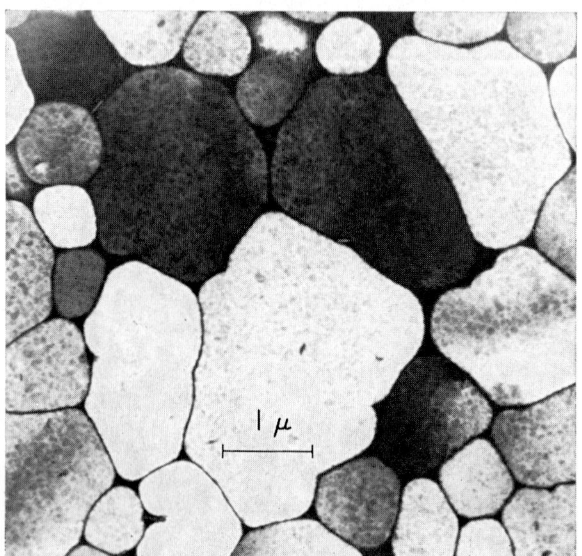

Fig. 4. Phase equilibrium between the intermetallic compound $AuSb_2$ and a liquid of nearly eutectic composition.

I – 13

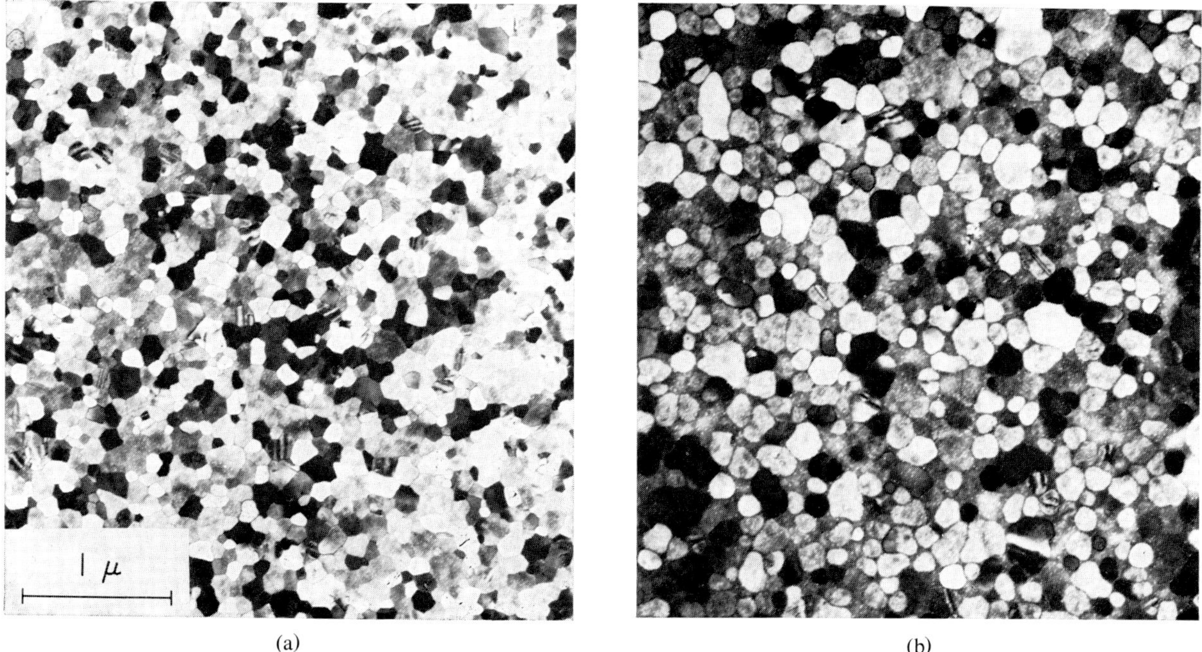

(a)   (b)

Fig. 5. Melting of a gold–tin alloy. (a) Electron micrograph of an alloy composed of ζ phase and AuSn at a temperature just below the eutectic temperature. (b) Grain-boundary liquation as eutectic temperature is exceeded. Note faint grain boundaries in some of the liquid regions, probably caused by a compositional gradient normal to specimen surface.

served. Also, there is an indication of very faint grain boundaries amidst some of the liquid regions, suggestive of an incomplete reaction between the sequentially deposited layers, resulting in a composition gradient normal to the specimen surface and retention of some solid below (or above) the liquid layer.

4. Conclusion

These experiments have shown that the electron microscope is a very useful tool for the study of microscale solidification processes in binary alloy systems. Subject to the restrictions discussed earlier, it provides the only method known at present for direct observation of solidification beyond the limits of optical resolution. The ability to utilize this technique in the study of alloys considerably enhances the applicability of gradient hot-stage electron microscopy.

References

1) L. J. Thomas and J. W. Westwater, Chem. Engr. Progr. Symp. Ser. **59** (1963) 155.
2) K. A. Jackson and J. D. Hunt, Acta Met. **13** (1965) 1212.
3) V. de L. Davies, J. Inst. Metals **93** Part 1 (1964–1965) 10.
4) M. E. Glicksman and R. J. Schaefer, Acta Met. **14** (1966) 1126.
5) M. E. Glicksman and C. L. Vold, Acta Met. **15** (1967) 1409.
6) M. E. Glicksman and C. L. Vold, *The Solidification of Metals* (British Iron and Steel Institute Publication 110, London, 1968) p. 37.
7) M. E. Glicksman and C. L. Vold, Acta Met. **17** (1969) 1.
8) M. E. Glicksman and C. L. Vold, Scripta Met. **5** (1971) 493.
9) M. Hansen, *Constitution of Binary Alloys* (McGraw-Hill, New York, 1958) p. 233.

Journal of Crystal Growth **13/14** (1972) 78–83 © *North-Holland Publishing Co.*

GROWTH OF TITANIUM OXIDE CRYSTALS OF CONTROLLED STOICHIOMETRY AND ORDER

RUSTUM ROY and WILLIAM B. WHITE

Materials Research Laboratory, The Pennsylvania State University, University Park, Pennsylvania 16802 U.S.A

Growth of single crystals of the intermediate oxides of titanium, important because of the unique electrical and magnetic properties of these materials, is difficult to achieve in a controlled manner. The difficulties arise for several reasons: the large number of phases in the Ti–O system, their defect structure and potential for both equilibrium and nonequilibrium disorder of the defects and the change of the oxygen fugacity of the co-existing equilibrium vapor phase with temperature. Precise phase equilibria including solid-vapor equilibria are required and these were first determined in detail.

Once these are known several common techniques can be applied to the crystal growth. Reduced rutile has been grown by flame fusion methods for years. We have grown fully oxidized TiO_2 at a much lower temperature in hydrothermal highly acidic environments. The higher titanium oxides – Ti_2O_3, Ti_3O_5, and the Magneli phases from Ti_4O_7 to $Ti_{10}O_{19}$ have been grown from borate fluxes by precise control of the oxygen fugacity in the furnace atmosphere. Extensive characterization of the oxides indicates an unexplored region of possible ordered Magneli phases between $Ti_{10}O_{19}$ and approximately $Ti_{100}O_{199}$. Growth of these phases from fluxes with ultra-precise temperature and atmosphere control does not seem impossible.

The lower titanium oxides cannot be grown in open systems because of the extremely low oxygen fugacities. The arc-fusion technique of T. B. Reed has proved the best method for preparing high TiO and other phases stable at the melting point. Low TiO remains a crystal-growth problem, as do the oxygen-rich titanium metal phases.

1. Introduction

The phase diagrams for the transition metal oxide systems of which Ti–O is one of the more complicated examples provide excellent illustrations of the difficulties in principle in the crystal growth of *good* crystals of these materials. We find high melting oxides with complicated defect character, we find intermediate oxides with very restricted stability in terms of temperature and oxygen partial pressure that melt incongruently, and we find lower oxides stable only in an extremely reducing environment. Our objectives in the present paper are to review the fundamental data needed, the techniques that have been used for growing crystals of the intermediate oxides of titanium, to show the diversity of methods that are necessary, and to point out their applicability to other transition metal oxide systems.

2. Review of phase equilibria

The condensed *T–x* diagram for the system Ti–O has been determined in our laboratories by Porter[1]) and is presented in fig. 1. A diagram differing only in minor detail has been published by Wahlbeck and Gilles[2]).

The highest oxidation state oxide in the Ti–O system is rutile, TiO_2. The corundum sesquioxide structure has a rather wide range of solid solution. Ti_3O_5 has a unique structure and exists in several polymorphic forms[3,4]), among them one with the pseudobrookite structure. The compounds from Ti_4O_7 to $Ti_{10}O_{19}$ (general formula Ti_nO_{2n-1}) constitute the prototype series, of closely related shear structures which have come to be known as the Magneli Phases[5]). The melting points for these compounds shown in fig. 1 are largely conjectural. The compounds are so closely spaced that precise melting point determinations have not yet been achieved. There are a number of solid–solid phase transitions within the Magneli series. Many of these are rapid and occur just above or below room temperature. Some of these are Mott transitions from metallically conducting to semiconducting states. The magnetic properties and electrical properties have constituted the chief reason for the recent interest in the Magneli phases.

It will be noted that there is a rather wide composition gap between the highest resolved Magneli phase, $Ti_{10}O_{19}$ and the lower stability limit of the oxygen deficient rutile. The lower limit of the rutile stoichio-

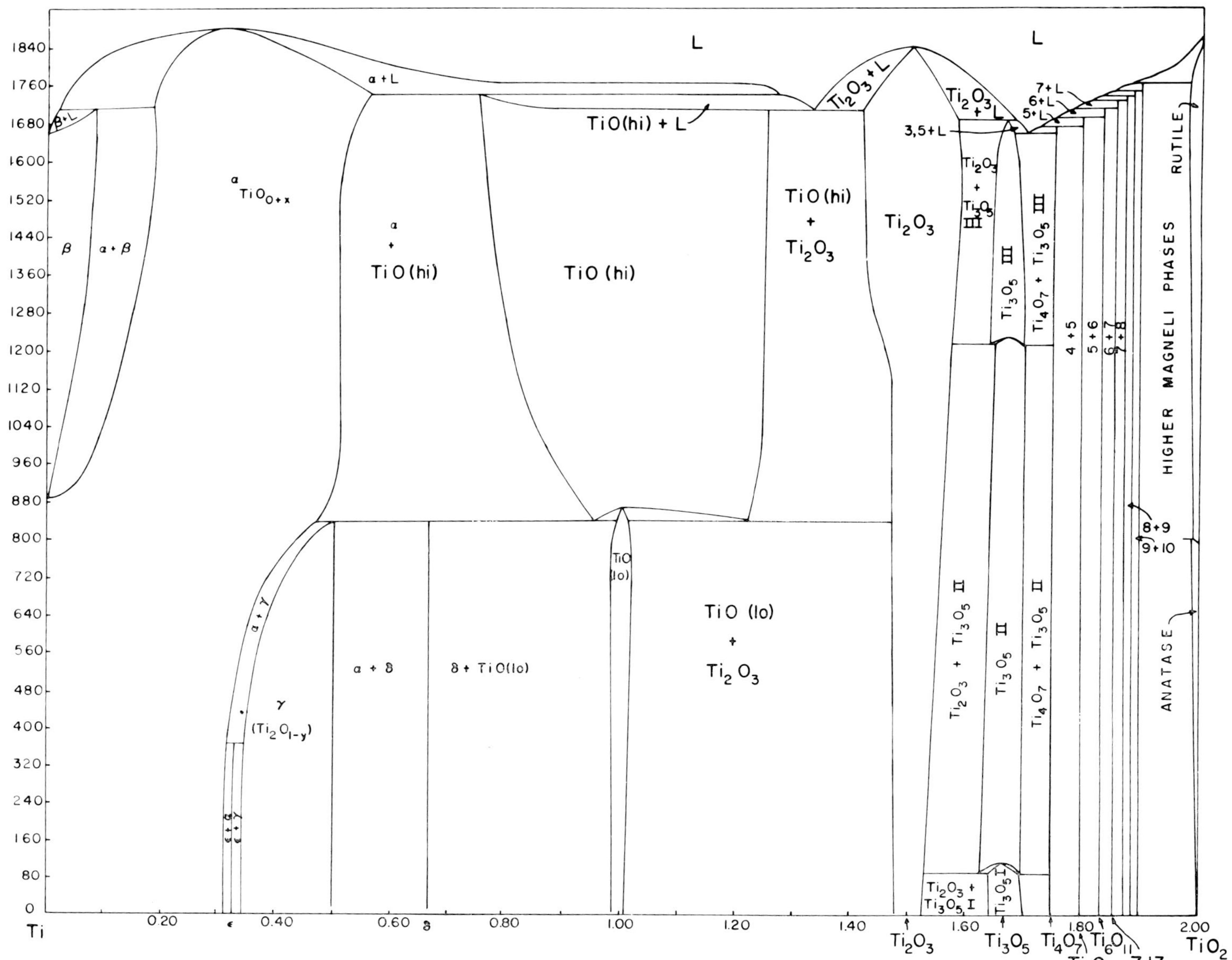

Fig. 1. *T–x* condensed phase diagram for system Ti–O.

metry is approximately $TiO_{1.992}$ as separately determined by Porter[1]) and by Blumenthal and Whitmore[6]). It is now generally agreed that the rutile structure will only tolerate an oxygen deficiency of about 8 parts per thousand before the defects order to form the "shear-structure" phases. We therefore believe, as Anderson and Khan[7]) have recently suggested also, that a rather wide range of undiscovered Magneli phases must exist. Measurements of a shift in band frequency in the diffuse reflectance spectra of these phases by Porter, White and Roy[8]) imply that the highest Magneli phase is $Ti_{100}O_{199}$ and that, in principle, all the phases intermediate between this and $Ti_{10}O_{19}$ should exist.

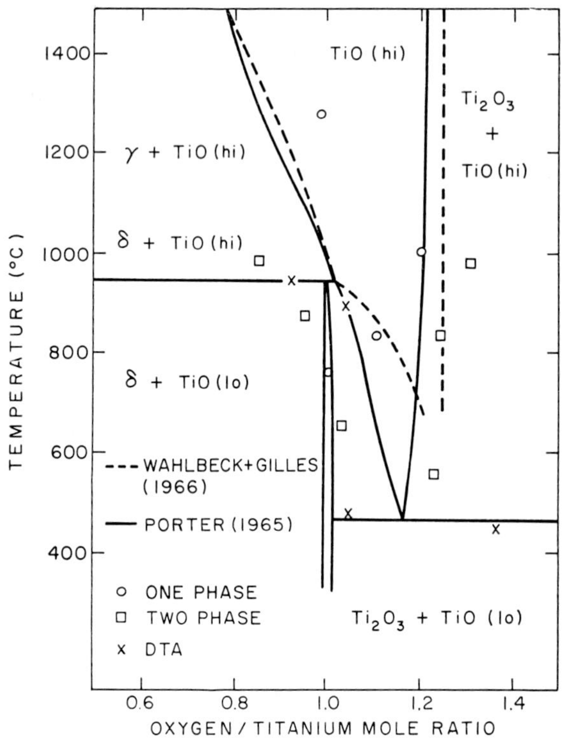

Fig. 2.   Portion of the Ti–O system near the TiO stoichiometry.

Below the sesquioxide stoichiometry are several oxygen-rich titanium metal phases which do not concern us in this paper, and the monoxide, TiO. The low form of TiO has an unknown structure and a relatively narrow range of solid solution. The transition is not polymorphic in the rigorous sense and there is a wide temperature range where the two forms co-exist (fig. 2). High TiO is a yellow brassy compound and is a metallic conductor. It is a rather remarkable compound – first

because of the wide range in stoichiometry as shown in fig. 2, and secondly because of the very large concentration of Schottky defects (up to 15% according to ref. 9) that it accommodates a situation that was referred to by Roy[10]) as a solid solution with "vacuum". These high-defect concentrations must be adequately characterized with respect to concentration, nature (i.e. interstitial, vacancy, etc.) and degree of clustering or ordering, if one intends to determine the properties of crystals of this phase.

The condensed phase diagram presentation discussed so far leaves out a most important variable. Since this is a variable valence system the stability of all phases depends on the oxygen partial pressure, the universal variable in most high-temperature chemistry on earth. One needs therefore to determine the p–T relationship by equilibrating the various compounds and assemblages at known $p(O_2)$ conditions, usually obtained with atmospheres of CO/CO_2 or CO_2/H_2 which have known and buffered oxygen fugacity. Fig. 3 shows the stability of the oxygen-rich phases as a function of $p(O_2)$. These data show that one can synthesize all of the Magneli phases in an open system with oxygen-buffered atmospheres. However, at the sesquioxide limit the equilibrium oxygen pressure is near 10^{-30} quite unattainable in a controlled manner in open systems. Ti_3O_5 can be prepared in extremely reducing atmospheres but even reduction in highly purified hydrogen has failed to produce Ti_2O_3.

It is interesting to find that the controlled oxygen fugacity experiments show quite clearly that the Magneli phases are discrete entities and not members of a continuous series of crystalline solutions. The oxygen isobars stair-step very precisely from one compound to the next. In the regions between the homologous series ratio's, two Magneli phases coexist. Anderson and Khan[7]) have recently pointed out that double shear structures are also possible with such ratios as $Ti_{11}O_{20}$ which would be an intergrowth of Ti_5O_9 and Ti_6O_{11} on an atomic scale. There is no *evidence* for any solid solution. The closeness of spacing in the oxygen isobars implies that oxygen pressure must be controlled very carefully in any crystal growth experiments. Most ordinary crystal growth methods would produce either a mixture of Magneli phases or an apparent single crystal that would contain not only compositional zoning but also phase zoning.

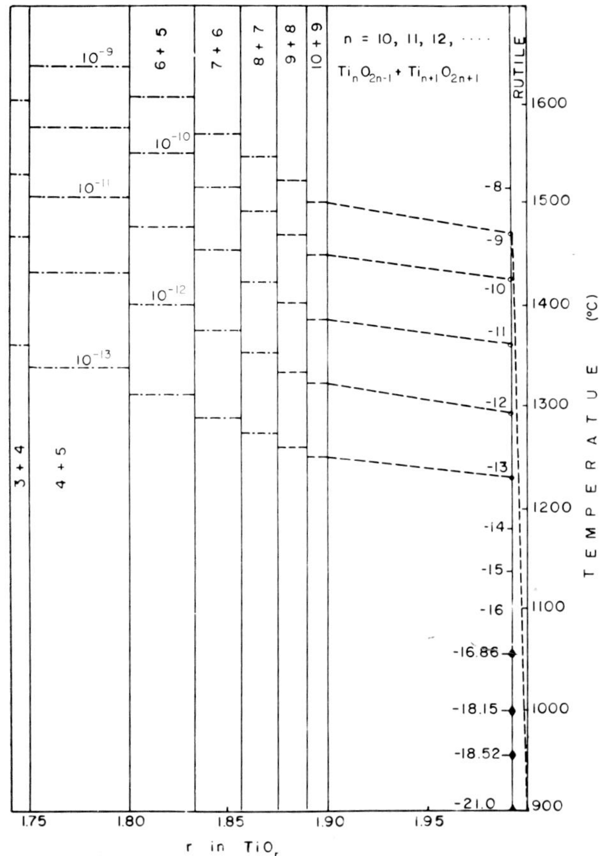

Fig. 3.   Stability of the Magneli phases at various oxygen fugacities.

3. Growth of rutile

Single crystals of rutile have been grown for a long time either by Verneuil or Czochralski methods. Rutile grown by flame fusion is always substantially reduced and is dark blue or black in color. Crystals must be re-oxidized by annealing at about 1000 °C in oxygen. The oxygen is replaced and the crystals become pale yellow and are of reasonable quality, chemically. Typical dislocation contents, however, are in the range of $10^5/cm^3$. Much the same can be said for pulled rutile since it has not been possible to maintain the high oxygen pressures required at the 1800 °C temperatures.

Harvill and Roy[11]) were first successful in growing stoichiometric rutile directly by high pressure solution methods using the highly oxidizing 9 M H_2SO_4 as a solvent. In addition there was a drop in dislocation density from 10^5 in the flame-grown seed to about 10^3 in the hydrothermal overgrowth. Berkes, White and

Roy[12]) grew rutile of controlled and different stoichiometry in alkali borate fluxes. Pale yellow acicular highly insulating crystals were formed when the flux growth experiment was carried out in the most oxidizing atmosphere [$p(O_2) = 1$ atm]. Rutile that is reduced only to the extent of about 1 part per thousand in oxygen is already a pale blue color and semiconducting. At the limit of the true rutile-structure stability field, the material is black and a fairly good electrical conductor. Thus, single crystals of the entire series of reduced rutiles can be grown under controlled conditions from the alkali borate fluxes. Basically the same slow cooling process is used except that the oxygen fugacity of the gas phase above the flux is carefully controlled. The experiments show that a wide range of compositions can be prepared by this technique. By coincidence, the stability range of rutile as a function of $p(O_2)$ and temperature parallels the changing oxygen fugacity of the gas at constant mixing ratio but varying temperature. Thus the temperature of the flux can be lowered over a range of several hundred degrees without extensive compositional zoning. In the absence of this coincidence, in the general case it would be necessary to program a change in mixing ratios of the gases to provide the required change in $p(O_2)$ with temperature during the cooling process.

The crystals grown from alkali borate fluxes were 5 to 10 mm in length and usually a fraction of a mm in cross-section. The advantage to this particular technique is that physical property measurements can be made on specimens of constant composition obtained without any further heat treatment. They provided, for example, an ideal geometry for measurement of dielectric constant by non-contact methods[13]).

4. Growth of Magneli phases

The stability region of most of the Magneli phases can also be reached with ordinary gas mixing techniques. Bartholomew and White[14]) were able to grow from sodium tetraborate solvents, crystals with stoichiometrics all the way from Ti_4O_7 to $Ti_{10}O_{19}$ by controlling the fugacity of oxygen in a CO/CO_2 atmosphere. Crystals were grown of sufficient size to enable the measurements of the transport properties[15]). In this particular situation PbO, PbF_2 and Bi_2O_3 fluxes would be unsuitable because the liquids are unstable at the low oxygen pressures. Growth was initiated by slow cooling of the melt and causes some difficulty. The range of stability for any given Magneli phase spans only a short interval of oxygen fugacity so that only a short cooling interval is possible before the systems slip outside the stability range. This happens because of the changing oxygen fugacity in the gas at constant mixing ratio. In lieu of using complicated automatic equipment to vary the mixing ratio so as to keep the oxygen fugacity constant, only short cooling intervals could be used and this limited somewhat the size of crystals that could be grown.

Crucible materials pose a distinct problem in such systems. Platinum is very rapidly attacked both by the melt and by the titanium oxide at the low oxygen fugacities. For the lowest fugacity range, graphite crucibles could be used because the graphite-CO equilibrium forms a buffer system. The various situations for crucible materials are represented in fig. 4. Ti_3O_5 and Ti_4O_7 can be grown in graphite crucibles. It was found that the higher Magneli phases can be grown from liquids contained in tungsten or molybdenum. As can be seen in fig. 4, the oxygen fugacities are so low that neither of these refractory metals oxidize and they did not seem to be perceptibly attacked by the melts.

An alternative approach to the cooling problem would be to use flux evaporation as used for VO_2 by Aramaki and Roy[16a,b]). The difficulty with flux evaporation is the additional supercooling near the surface from the loss of latent heat of vaporization. In fact, in many borate flux experiments *some* of the growth is by flux evaporation which is unavoidable in the open system which is mandated. The titanium oxide crystals then tend to form on the surface and are much less perfect than those grown at depth in the liquid. The liquid acceleration–deacceleration technique recently published by Scheel[17]), may be an answer to this problem. Flux evaporation growth has the tremendous advantage of both constant temperature and constant oxygen pressure and should be re-investigated with the improved techniques now available.

5. Growth of the lower oxides

Growth of the oxides below Ti_3O_5 is very difficult. One cannot use controlled atmosphere solution growth because of the unattainably low oxygen fugacities. Melt techniques are difficult because of high temperatures, although Reed et al.[18]) have successfully pulled Ti_2O_3

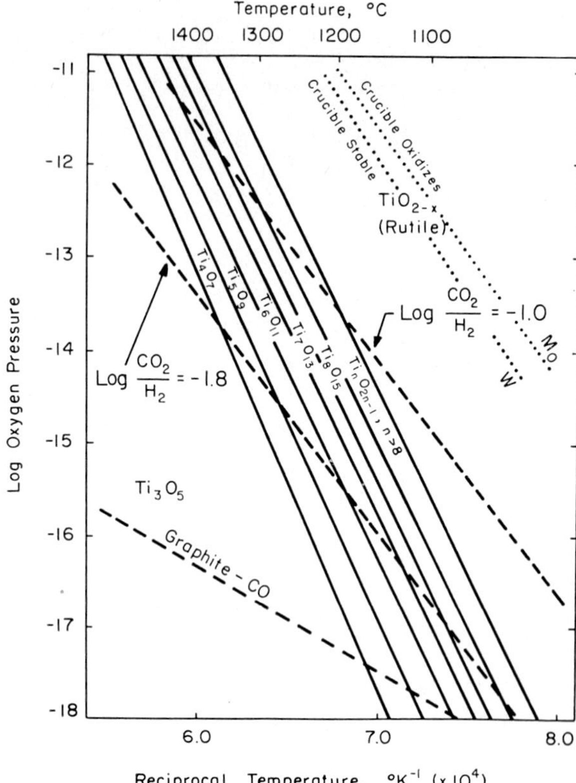

Fig. 4. p–T projection for Ti–O system showing various conditions for borate solvent growth of Magneli phases.

from the melt using a tantalum resistance furnace. The high temperature form of TiO which does melt congruently, can be grown using the arc melting methods of Reed[19]). The materials are contained on a water-cooled copper hearth and melted directly in a high current, low voltage electric arc. Melting takes place in a flowing argon atmosphere to avoid oxygen uptake. The oxide melt does not wet the copper and rolls around on it to form a small single crystal button. Low temperature TiO has not been grown to the authors knowledge. There is no suitable flux and because of the conversion to high TiO, no melt technique is suitable. Hydrothermal techniques are not suitable because the oxygen potentials are below the stability limits of water.

6. Conclusions

This eight year long study started from the basic p–t and t–x phase equilibrium data, and by utilizing a *variety* of methods which provide controlled oxygen fugacities varying from roughly 10^2 to 10^{-30} one can prepare all the higher titanium oxides with at least the *stoichiometry* controlled.

The degree of sophistication of our methods in recording the nature and degree of the order–disorder of defects is such that one can only claim that very crude levels of reproducibility have been attained. Thus we can in fact produce single crystals of a single Magneli phase on demand. However, no one has even attempted to describe, leave alone control, the degree of equilibrium defect clustering as a function of temperature in any of the suboxides. In rutile itself there is still controversy as to the nature of the defect in the slightly reduced materials. Under their stated conditions, Porter and Roy[1] have established unequivocally that interstitial titanium defects are predominant: obviously this may not be true for a sample with different heat treatment.

We have also noted that these single crystals of the titanium oxides have been studied in detail with respect to various properties as follows:

Optical spectra by Porter and Roy[8];

Magnetic susceptibilities showing antiferromagnetism in many by Keys and Mulay[20];

Electronic transport properties by Bartholomew and Frankl[15];

Detailed dielectric loss studies by Groner and Cross[13].

Acknowledgements

This work was supported by the Advanced Projects Agency under Contract DA-49-083 OSA 3140.

References

1) V. R. Porter, Ph.D. Thesis, Pennsylvania State University (1965).
   See also Porter and Roy, Bull. Am. Ceram. Soc. **43** (1964) 263; R. Roy, CNRS Colloque Internationaux No. 149 (1964) pp. 27–32.
2) P. G. Wahlbeck and P. W. Gilles, J. Am. Ceram. Soc. **49** (1966) 180.
3) S. Asbrink and A. Magneli, Acta Cryst. **12** (1959) 575.
4) H. Iwasaki, N. F. H. Bright and J. F. Rowland, J. Less Common Metals **17** (1969) 99.
5) S. Andersson and L. Jahnberg, Arkiv Kemi **21** (1963) 413.
6) R. Blumenthal and D. Whitmore, J. Electrochem. Soc. **110** (1963) 92.
7) J. S. Anderson and A. S. Khan, J. Less Common Metals **22** (1970) 219.
8) V. R. Porter, W. B. White and R. Roy, J. Solid State Chem. (in press).
9) M. E. Straumanis and H. Li, Z. Anorg. Allgem. Chem. **305** (1960) 143.
10) R. Roy, in: *Reactivity of Solids*, Ed. G. M. Schwab (Elsevier, Amsterdam, 1965) p. 321.
11) M. L. Harvill and R. Roy, in: *Crystal Growth*, Ed. H. S. Peiser (Pergamon, Oxford, 1967) p. 563.
12) J. S. Berkes, W. B. White and R. Roy, J. Appl. Phys. **36** (1965) 3276.
13) L. E. Cross and C. F. Groner, J. Appl. Phys. **40** (1969) 126.
14) R. F. Bartholomew and W. B. White, J. Crystal Growth **6** (1970) 249.
15) R. F. Bartholomew and D. R. Frankl, Phys. Rev. **187** (1969) 828.
16) a. R. Roy, Mater. Res. Bull. **1** (1966) 294.
    b. S. Aramaki and R. Roy, J. Mater. Sci. **3** (1968) 643.
17) H. Scheel and E. O. Schulz-Dubois, J. Crystal Growth **8** (1971) 304.
18) T. B. Reed, R. E. Fahey and J. M. Honig, Mater. Res. Bull. **2** (1967) 561.
19) T. B. Reed, Mater. Res. Bull. **2** (1967) 349.
20) L. K. Keys and L. N. Mulay, Phys. Rev. **154** (1967) 453.

     Journal of Crystal Growth **13/14** (1972) 84–87 © *North-Holland Publishing Co.*

EDGE-DEFINED, FILM-FED CRYSTAL GROWTH

BRUCE CHALMERS

Division of Engineering and Applied Physics, Harvard University, Cambridge, Massachusetts 02138, U.S.A.
and
H. E. LaBELLE, Jr. and A. I. MLAVSKY

Tyco Laboratories, Inc., Waltham, Massachusetts 02154, U.S.A.

Crystal formation from the melt by the edge-defined, film-fed growth process is discussed quantitatively in relation to the range of conditions under which continuous growth is possible. It is shown that two distinct modes of growth can occur, in one of which the extraction of latent heat is by radiation into the growing crystal. This is a relatively slow process, and because a planar interface is inherently stable in this thermal situation, it can produce crystals of high quality. Also, because of the thermal geometry, there is no upper limit to the cross section that can be grown in this way. The second mode is that in which heat extraction is largely by conduction into and through the liquid film. Crystals can be grown much faster by this process; the interface is inherently unstable, with the result that imperfections arise from cellular or dendritic growth. In the fast mode, the cross section for continuous growth is limited because the position of the interface is stabilized by edge effects, which become inoperative at large cross sections. The quantitative results of these analyses are in close accord with the results of experiments and with observations made in routine production of sapphire crystals.

1. Introduction

In edge-defined, film-fed crystal growth[1]) a crystal is grown from a liquid film which is supplied from a reservoir by capillary action. It has the advantages of permitting very high speeds of growth with very precise control of the cross-sectional shape and size. The technique is illustrated schematically in fig. 1; a selection of shapes grown by this method is shown in fig. 2.

The purpose of this paper is to discuss the conditions under which continuous growth by this method is possible. In general, it is necessary for liquid to reach the solid–liquid interface at the rate at which solidification takes place, and for the linear rate of growth of the crystal V_g to equal, at least when time-averaged, the rate of upward motion V_p imposed on the crystal. The rate of growth of the crystal is determined by the rate at which latent heat is removed from the vicinity of the interface. It is also necessary for the interface to be stable against excursions from its steady state position. The following analysis applies to the case of a solid cylinder; it will, in a later publication, be developed for the case of tubes and other shapes. We will consider first the supply of liquid to the film.

2. Fluid flow

The flow of liquid to and in the film requires a pressure difference p which is the sum of three components: p_1, required to raise the liquid through the height h is given by: $p_1 = Dgh$, where D is the density of the liquid. p_2, required to force the liquid through the capillary at the required rate, is given approximately by

$$p_2 = 8R_2^2 V_g \mu L D_s / R_4^4 D,$$

where
D_s = density of the solid,
μ = viscosity of the liquid,
L = length of capillary,
R_4 = radius of capillary,
R_2 = radius of crystal.

p_3, required for radial flow in the film, is given approximately by

$$p_3 = \frac{12\mu(R_2^2 - R_4^2)V_g(R_2 - R_4)D_s}{S^3(R_4 + R_2)D}$$

$$= \frac{12\mu V_g(R_2 - R_4)^2 D_s}{S^3 D},$$

where S is the film thickness.

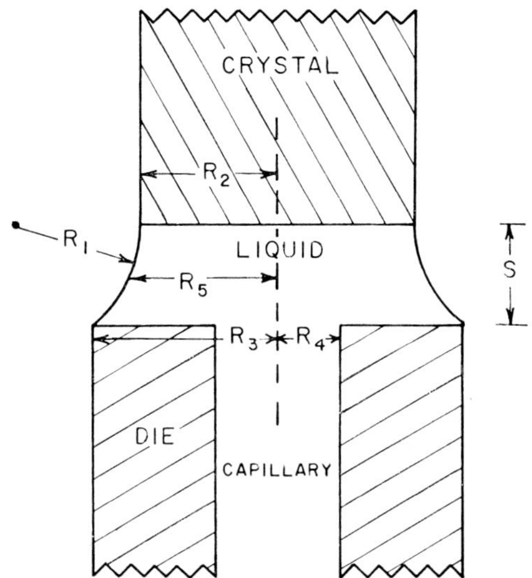

Fig. 1. Edge-defined, film-fed growth.

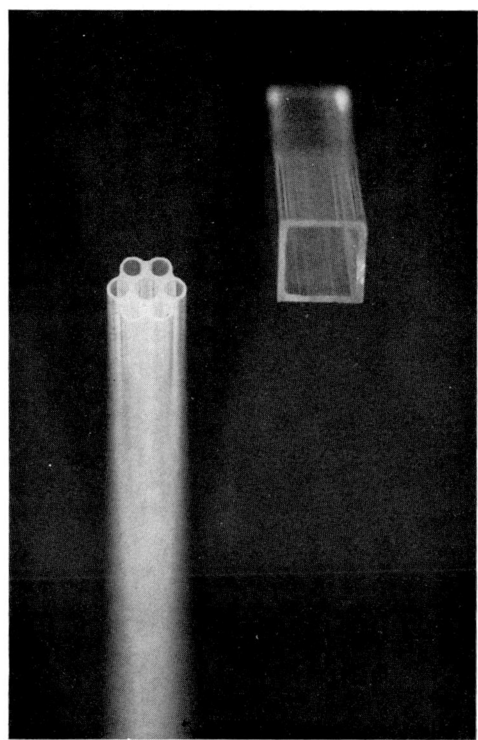

Fig. 2. Sapphire shapes (square tube has 0.65 cm edges).

The numerical values for p_2 and p_3 depend on the values of R_4, R_2, V_g and S; an illustrative case is that of high speed growth of a filament, for which let $R_4 = 10^{-2}$ cm, $R_2 = 2 \times 10^{-2}$ cm, $\mu = 1$ poise, $L = 5 \times 10^{-2}$ cm, $V_g = 2 \times 10^{-1}$ cm/sec, $h = 2$ cm, $D = 3$ g/cm^3, $D_s = 4$ g/cm^3, $S = 4 \times 10^{-3}$ cm. Then

$$p_1 = 6 \times 10^3 \text{ dyne/cm}^2,$$
$$p_2 = 4.3 \times 10^3 \text{ dyne/cm}^2,$$
$$p_3 = 5 \times 10^3 \text{ dyne/cm}^2,$$

from which $p = p_1 + p_2 + p_3 = 1.5 \times 10^4$ dyne/cm^2.

The pressure difference due to surface tension γ, given by

$$p = \gamma \left(\frac{1}{R_1} - \frac{1}{R_2} \right)^*$$

must be equal to this. If $\gamma = 600$ dyne/cm, then $1/R_1 - 1/R_2 = 25$ cm^{-1}, from which $R_1 = 0.013$ cm.

The case shown in fig. 3 is one in which the supply of liquid can be maintained; the value of S is between the upper limit of R_1, at which, as demonstrated below, thermal stability is lost, and the minimum value, of about 10^{-3} cm, at which p_3 would be of such a value that R_1 would be no greater than S.

In terms of surface tension, the film would be stable so long as $S < 2R_1$. It should be noted that this is an approximation, since it derives from the assumption that the profile of the film edge is an arc of a circle of radius R_1; because R_5 varies from point to point, R_1 must also vary. For the extreme case of R_2 infinite, R_1 is circular.

The conditions for R_2 large, and slow growth (so that viscosity can be neglected) give $\gamma/R_1 = Dgh$, or $R_1 = 0.1$ cm, from which the greatest possible mechanically stable thickness of the film is 0.2 cm.

There is no lower limit, from this point of view, on the value of R_2, the radius of the filament, although R_2

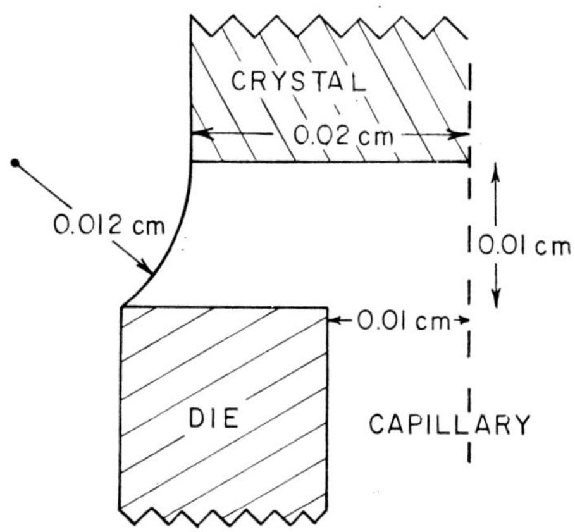

Fig. 3. One set of conditions for stable growth.

* More accurately $p = \gamma(1/R_1 - 1/R_5)$; see fig. 1.

must always be larger than R_1; R_1 is reduced by increasing V_g.

3. Thermal conditions

The rate at which latent heat, L cal/g, is generated is equal to $\pi R_2^2 V_g D_s L$, where V_g is the rate of growth of the crystal. For steady state growth, V_g must equal V_p, the rate of withdrawal of the crystal; hence $\pi R_2^2 V_p D_s L$ must be equal to the rate of extraction of latent heat (Q) from the region of the interface. Q is equal to the difference between J_1, the heat leaving the interface by radiation into the crystal, and J_2, the heat reaching the interface by conduction in the liquid. J_2 can be positive, zero or negative. J_1 is equal to $\pi R_2^2 \sigma T^4$. This is equal to about 40 cal/cm^2/sec for Al$_2$O$_3$ at its melting point. When J_2 is zero (i.e., zero temperature gradient in the liquid) the rate of growth must be given by $LD_s V_g = 40$; LD_s is about 1000 for Al$_2$O$_3$, and the rate of growth for the case of $J_2 = 0$ is therefore 0.04 cm/sec. If J_2 is positive (i.e. heat flows towards the interface) the rate of growth is reduced. We will consider the stability of this case, in which the positive temperature gradient in the liquid opposes the dendritic and the cellular modes of growth, and therefore is conducive to good optical quality. Assume first that the film is cylindrical and that the die is maintained at a fixed temperature. Steady state growth is established when $J_1 - J_2 = LD_s V_g$ which can occur only for a unique value of J_2 (for a given velocity V_g), which implies a unique value of S, since J_2 is the heat which reaches the interface by conduction through the film ($K\Delta T/S$ per unit area). If S has a value greater than that required for steady state growth, J_2 is decreased, $J_1 - J_2$ is increased, which tends to increase the rate of growth and so to bring the interface back towards the steady state position. This mode of growth is, therefore, stable against fluctuation of S. It is also, to a limited extent, stable against fluctuations of V_p because an increase in V_g is achieved by increasing S, and thereby reducing J_2, and vice versa. For useful production rates, J_2 must be small compared with J_1, and the stable range of conditions is, therefore, small.

When J_2 is negative, the rate of growth can be very high, because a high rate of heat extraction can be achieved by conduction through the film in addition to radiation and conduction into the solid crystal. Heat leaves the liquid film by conduction to the die and/or radiation from the free surface of the film. The limitation in this case is that the temperature of the liquid must not anywhere be low enough for nucleation of new crystals to take place. In the case of Al$_2$O$_3$ it is believed that the maximum supercooling that can be permitted is about 200 °C.

For $V_g = 2 \times 10^{-1}$ cm/sec, $\Delta T/S = 1.3 \times 10^4$ deg/cm from which the maximum value of S should be 1.5×10^{-2} cm. It follows that the high speed of 0.2 cm/sec is compatible with the use of the die as a heat sink.

It is observed that good optical quality can be obtained by slow growth, but not for speeds above about 0.04 cm/sec. This corresponds to the J_2 condition. Higher speeds require J_2 negative, which permits cellular or dendritic growth with the consequent imperfections.

In this case, the position of the interface would be inherently *unstable* in a cylindrical film, because any increase of S would reduce J_2 and V_g would decrease. This conclusion is valid only when (a) radiation from the film surface can be neglected, (b) radius of the crystal (R_2) does not depend on S, and (c) changes in die temperature can be ignored. When radiation *cannot* be neglected, that is, when R_2 is small, so that the area of the film surface (approximately $2\pi R_2 S$) is not small compared with πR_2^2, increase of S allows greater heat loss by radiation, which tends to cause V_g to increase with S. Secondly, when an increase in S causes a decrease in R_2, i.e. when $S < R_1$, the linear rate of growth V_g increases in the proportion R_2^2/R_1^2 for a constant rate of formation of solid. Thus an increase in S from the steady state value tends to be self-correcting. On the other hand, if an increase in S causes an increase in $R_2(S > R_1)$, the linear rate of growth decreases; this is unstable. It follows that, for stability, S must be less than R_1.

It is concluded, also, that these stabilizing effects are operative only at relatively small radii; this explains why very rapid growth, which requires heat flow from the interface to the liquid, is a stable mode only for the growth of filaments for which R_2 is not much larger than R_1. This condition is met for a radius of 10^{-2} cm. For crystals of much larger cross section (greater than about 0.4 cm radius) for which the radiation and the variation of R_2 with S are negligible, stable growth can occur only when heat reaches the interface by conduc-

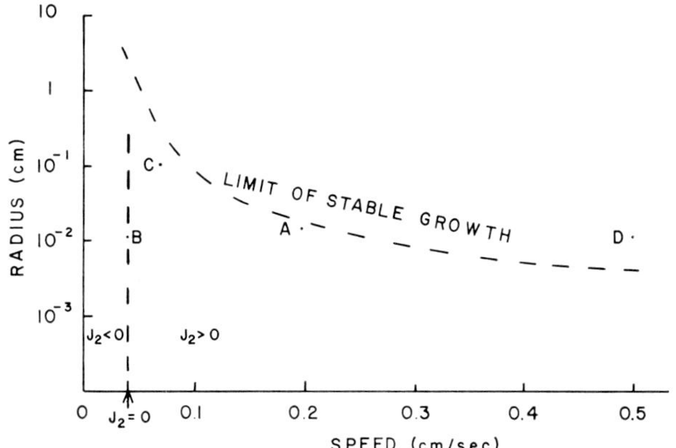

Fig. 4. Stable growth limit as a function of crystal radius and growth speed.

tion (i.e. J_2 is positive) for which the limiting speed is 0.04 cm/sec (for $J_2 = 0$), or 2.4 cm/min. This agrees well with the observed maximum rate for large cross sections of 2.0 cm/min.

These conclusions are summarized in schematic form in fig. 4 in which the limiting radius for stable growth is shown. The point A corresponds to the conditions envisaged in the numerical example given above, while B, C and D correspond to empirically observed conditions as follows. B represents growth of good optical quality filament ($J_2 \geq 0$); C represents growth just within the limit of stability; D represents continuous growth beyond the stability limit in that the crystal profile did not conform closely to the shape of the die. It is concluded that the cross section is limited only by the stability conditions, and not by the existence of a steady state value of S. The criterion is whether S is in stable or unstable equilibrium. The precise shape of the limiting curve of fig. 4 is not known.

There are three other aspects of stability; namely stability against variations in rate of pulling, in the temperature of the die, and in the level of the melt outside the die. It can readily be seen that a range of values of V_g is possible for a given die temperature, because a value of S can be found to give the required heat flow to or from the interface. A similar statement applies to the temperature of the die, while changes in liquid level have a direct effect only on p_1 which causes only minor changes in R_1. Subject to the restrictions

detailed above, S will be self-adjusting, but S must lie between the limits imposed by surface tension and by viscosity; it must be less than R_1 and greater than a value at which the viscous drag in the film would be limiting. This usually provides a considerable available range for S; therefore precise control is not required either for speed or for temperature of for liquid level; this is confirmed by the experience that it is possible to grow sapphire filament continuously for many hours.

Finally it must be pointed out that the foregoing discussion contains the implicit assumption that there is no thermal inertia in the system; hence that a change in S produces an instantaneous change in heat flow, and that the die temperature either is constant or responds instantaneously to changes in the conduction path in the liquid. The die does, in fact, possess a finite heat capacity, with the result that its temperature does not respond instantaneously to a changing heat demand or supply. This results, under some conditions, in an oscillatory mode of growth, of which the various stages are as follows; J_2 must be negative; (1) let V_g be less than V_p (V_g = velocity of growth, V_p = velocity of pulling), then the heat reaching the die from the interface decreases, and the die temperature falls. This causes V_g to increase to a value greater than V_p. The interface approaches the die, the temperature gradient in the liquid increases and the rate of heat flow to the die increases. However, because of the thermal capacity of the die, its temperature rises relatively slowly and the interface position overshoots the steady state position. At some position of the interface, however, the die temperature reaches a value that causes V_g to drop below V_p, after which the cycle is repeated. An oscillatory mode has been observed during high speed growth.

Acknowledgments

We would like to thank our various colleagues at Tyco for useful discussions. This work was supported by the U.S. Air Force Materials Laboratory, Nonmetallic Materials Division, Wright-Patterson Air Force Base, Ohio.

Reference

1) H. E. LaBelle, Jr., Mater. Res. Bull. 6 (1971) 581.

PHASE EQUILIBRIA PERTINENT TO THE GROWTH OF CUBIC BORON NITRIDE

R. C. DeVRIES and J. F. FLEISCHER

Inorganic and Structures Branch, Physical Chemistry Laboratory, General Electric Company, Corporate Research and Development, Schenectady, New York 12301, U.S.A.

The practical synthesis of cubic boron nitride (cubic BN) requires the simultaneous application of high temperatures and pressures. Since the discovery of cubic BN more detailed understanding of the chemistry of the process has been acquired from equilibria or "approach-to-equilibria" studies in the Li-B–N system. This paper describes some of these results and considers their application to the growth of borazon.

The principal compound that coexists with cubic BN at high pressures and temperatures is a high pressure modification of Li_3BN_2 (called $Li_3BN_2(W)$). It has been shown that cubic BN does not form when this compound melts or decomposes. When BN is added to Li_3BN_2 in excess of the composition 53 mole% BN: 47 mole% Li_3N, then cubic BN precipitates from solution. At 55 kb, the system Li_3BN_2–BN has a eutectic at 6% BN and 1610 °C and the upper limit of the cubic BN plus liquid region is at 1750 °C. The solubility curve is a very steep function of temperature and consequently the system is not too favourable for crystal growth by slow cooling.

These new results can be compared with previous results by plotting a series of *P–T* curves defining the stability region for cubic BN under different conditions. Good agreement is found between the solidus determined in the Li_3BN_2–BN study and the lower limit of the catalyst growth area. In effect, this lower limit is directly related to the initial formation of liquid in the Li_3BN_2–BN portion of the system. The lower temperature limit of growth of cubic BN reported from Li–BN compositions is below that of the present study and the catalyst growth area previously defined. This result would have some fundamental basis in fact if a lower temperature reaction involving cubic BN occurred in the ternary system Li–B–N.

1. Introduction

The practical synthesis of cubic boron nitride (cubic BN) requires the simultaneous application of high temperatures and pressures. Wentorf[1,2]) has shown that under these conditions mixtures of Li_3N plus BN can be used to produce crystals of cubic BN. Since the discovery of cubic BN[3]), more understanding of the chemistry of the process has been acquired from equilibria or "approach-to-equilibria" studies in the Li–B–N system and in other systems. This paper describes some of these results and considers their application to previous investigations on the growth of cubic BN.

2. Experimental

The experiments were conducted in internally heated high pressure cells in which the experimental mixture was contained in a Ta-liner inside a BN sleeve (fig. 1). Temperature was determined from a previously calibrated watts versus temperature plot and the pressure calibration was made at room temperature at the fixed points for the phase transformation in Bi, Tl, and Ba. The products of the runs were identified optically and by X-ray techniques.

3. Results and discussion

3.1. The system Li–B–N

Binary relationships. In the Li_3N–BN part of the system the principal phase that can coexist with cubic BN at high pressures and temperatures (besides liquid) is a high pressure modification of Li_3BN_2, called

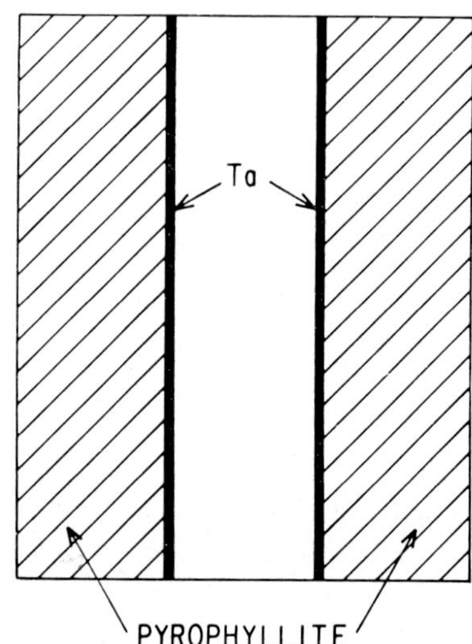

Fig. 1.   Type of cell used for high pressure experiments.

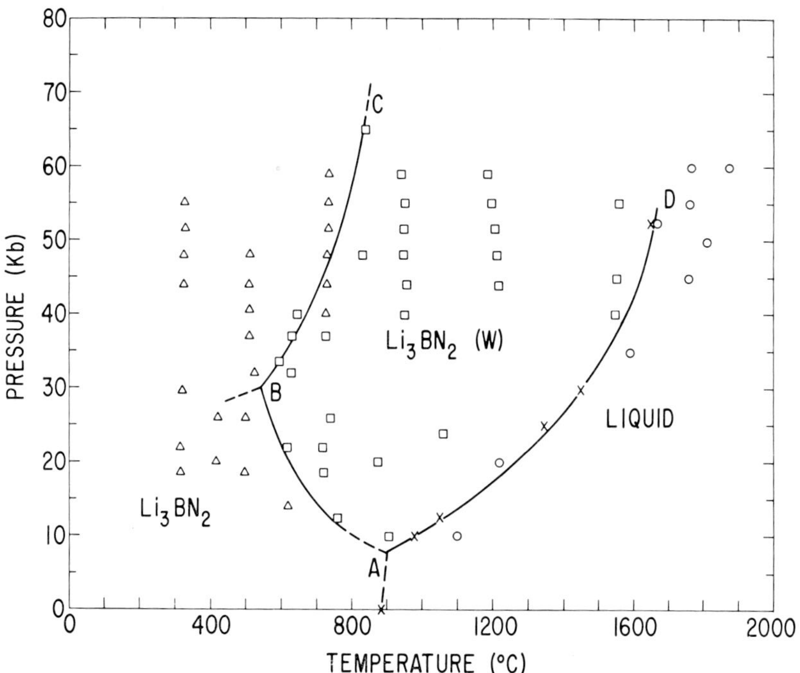

Fig. 2. *P–T* diagram for the system Li_3BN_2.

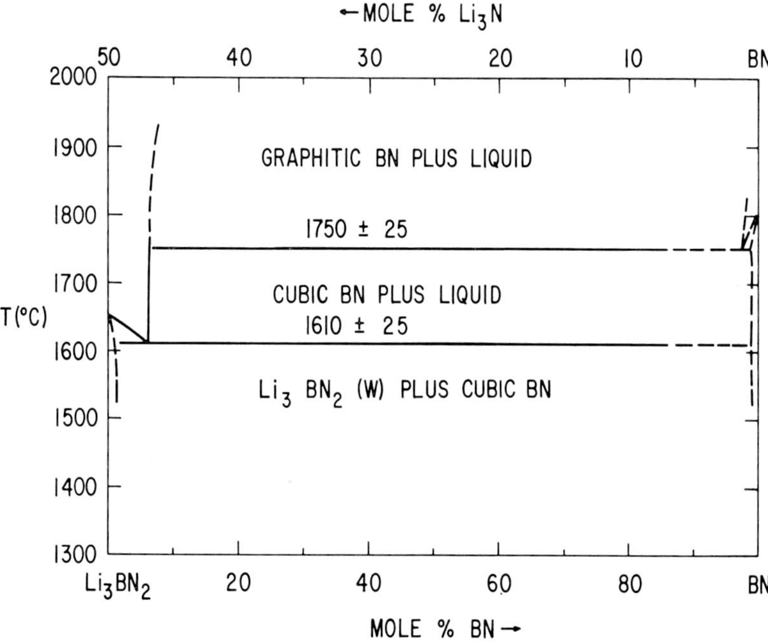

Fig. 3. The system Li_3BN_2–BN at 50 kb.

$Li_3BN_2(W)$[4]). The *P–T* relationships for this compound were determined (fig. 2), and it has been shown that cubic BN does not form when this compound melts or decomposes[4]). However, when the graphite form of BN is added to Li_3BN_2 in excess of the composition 53 mole % BN : 47 mole % Li_3N, then cubic BN precipitates from solution. At 55 kb, the system Li_3BN_2–BN has a eutectic at 6 % BN at 1610 °C and the upper

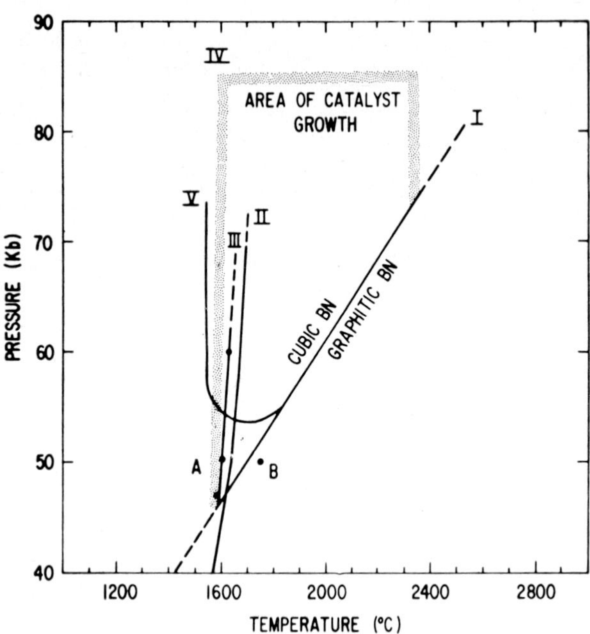

Fig. 4.   Summary of P–T data on stability of cubic BN. (I) direct conversion curve (ref. 5); (II) melting curve of $Li_3BN_2(w)$ (ref. 4); (III) solidus in Li_3BN_2–BN; (IV) catalyst growth area (ref. 5); (V) lower temperature limit of cubic BN growth from Li (ref. 6).

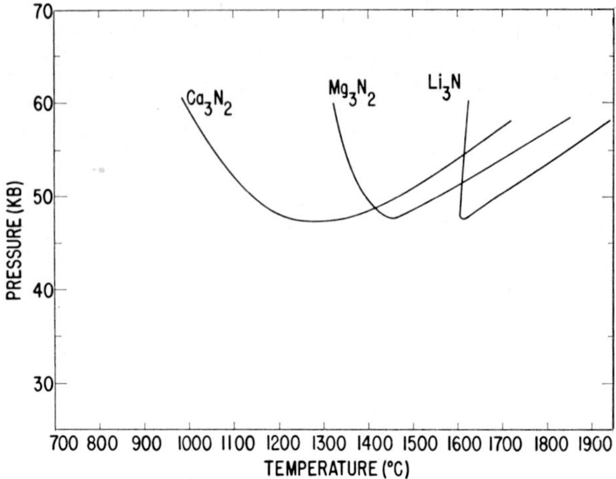

Fig. 5.   Cubic BN stability region for several nitride systems. Cubic BN exists in the region above each curve.

limit of the cubic BN plus liquid region is at 1750 °C (fig. 3). This means a crystal growth range of about 140 °C. However, the solubility is a very steep function of temperature and consequently the system is not too favorable for crystal growth by slow cooling or by thermal gradient.

These new results can be put into the perspective of previous studies by plotting a series of P–T curves defining the stability region for cubic BN under different conditions. When the direct conversion curve of Bundy and Wentorf[5]), the catalyst growth area[5]), and the lower limit of synthesis reported by Kudaka et al.[6]) are compared with the solidus determined in the Li_3BN_2–BN study, good agreement is found between this solidus and the lower limit of the catalyst growth area (fig. 4). The new binary data provide a quantitative understanding of the lower temperature limit of catalyst growth as previously determined from crystal growth studies in the same system. In effect, this lower limit is directly related to the initial formation of liquid in the Li_3BN_2–BN portion of the system.

Ternary relationships. It has been confirmed under our experimental conditions that the lower temperature limit of growth of cubic BN from Li–BN com-

positions as first reported by Kudaka et al.[6]) is below that of the Li_3BN_2–BN system and the catalyst growth area previously defined. This result would have some fundamental basis if a lower temperature reaction involving cubic BN occurred in the ternary system Li–B–N. Studies of the phase relationships in the Li_3N–BN–B–Li portion of the Li–B–N system were made at high pressures and temperatures, and a ternary reaction involving cubic BN, Li_3BN_2(W) and a boron-containing phase and liquid was found to occur at about 1550 °C. This temperature corresponds closely to the lower limit determined from the reaction of metallic Li with BN[6]). In the ternary system, as in the binary, the temperature of first appearance of cubic BN is also related to the lowest temperature of formation of liquid in that part of the ternary system.

In summary, the results from the study of the Li–B–N system show that at 50–55 kb cubic BN is stable over about 140 °C for mole ratios of $Li_3N:BN$ less than 47:53. Ternary phase relations indicate an even lower initial temperature for cubic BN formation.

4. Results from other nitride systems

The Wentorf patent[1]) describes many other successful chemical systems for synthesis of cubic BN. Of particular interest for comparison with the Li–B–N system are those which combine alkaline earth nitrides with BN. Some of these systems also have been studied, and it is sufficient for present purposes to give only a summation of results in so far as they may help in drawing some general conclusions about the formation of cubic BN. When the *P–T* stability regions for cubic BN from such systems are plotted together as in fig. 5, two features are apparent: (1) the marked differences in the minimum temperature for appearance of cubic BN; and (2) the essentially constant pressure minimum.

There is an approximate relationship between the stability of the "complex" or compound formed in the nitride–BN system and the lower temperature limit of the cubic BN stability region. Thus although Li_3BN_2 is quite a stable compound as shown in fig. 1, our results show that $Ca_3B_2N_4$ decomposes or melts at a much lower temperature. There is also some evidence of compound formation between Mg_3N_2 and BN (about at the 1:2 mole ratio) from high pressure experiments, and this "complex" disappears at temperatures between those of the Ca- and Li-compounds. For synthesis of cubic BN from these nitride systems, one interpretation of these data suggests a mechanism of growth which is dependent in part upon the break up of a complex or compound. The lower the stability of the complex, the lower will be the temperature of formation of BN. Or, stated differently, since the minimum temperature for formation of cubic BN is related to the formation of a liquid phase, and these systems appear to be of the eutectic type, the lower the stability of the compound in the binary system with BN, the lower the eutectic or solidus temperature will tend to be.

Kudaka et al.[6]) reported that the minimum temperature of the cubic BN stability region was lower for Li–BN mixtures than for Mg–BN, and this result was correlated with the relative melting temperatures of the two metals. It is clear from fig. 5 that this kind of correlation is consistent with our view of the relative stability of Li_3BN_2 and a binary compound in the Mg_3N_2–BN system. However, the situation is more complex than simple consideration of relative melting point of the two metals when they are reacted with BN since the reactions are ternary rather than simple binary. As pointed out above in the Li–BN case a ternary reaction occurs at a lower temperature than on the binary Li_3N–BN, i.e. the liquidus temperatures are lower on the boron rich side of that binary in the Li–B–N system. In contrast to this our results indicate that in the alkaline earth–BN systems the liquidus temperatures rise immediately as the composition is changed toward the boron rich side from the nitride–BN

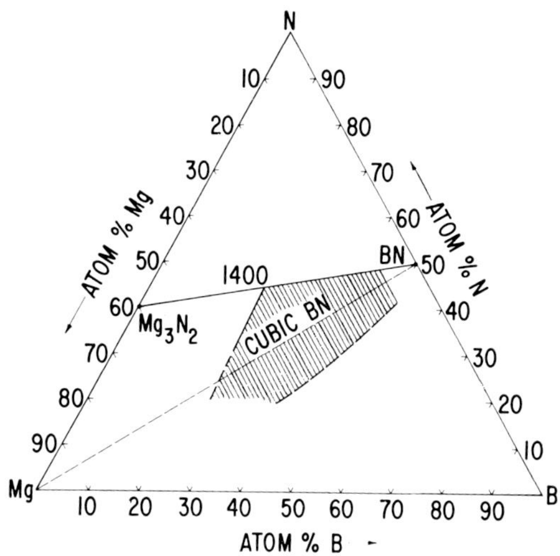

Fig. 6. Portion of stability region of cubic BN in the Mg–B–N system.

join. Although the detailed ternary phase relationships are not known for the Mg–B–N system, it is possible from analogy with known relationships plus some extrapolation to define roughly part of the cubic BN stability region as shown in fig. 6. In this context a line joining Mg with BN is not a join in the phase equilibrium sense but simply a composition line intersecting the primary phase region of cubic BN. From the data of Kudaka et al.[6]) it can be stated that the ternary reaction point defining the minimum temperature for cubic BN formation from the Mg–B–N system would have a temperature of about 1700 °C, and the probable phase assemblage is cubic BN, boron, Mg borides, and liquid. The presence of Mg borides with cubic BN has been reported by Filonenko et al.[7]) from high pressure synthesis of cubic BN using Mg.

The constancy of the minimum pressure for the stability of cubic BN from the nitride–BN systems (see fig. 5) deserves some comment. As discussed earlier the temperature dependence of the lower limit of cubic BN formation correlates with the thermodynamic stability of each of the nitride–BN systems, and it would be expected that the minimum pressures might also vary among different chemical systems. Lower pressure minima have been reported from certain alloy systems[8]), and there is a small but apparently significant difference in the pressure minimum when Li–BN and Mg–BN are compared[6]). Little or no variation in the minimum pressure suggests that the role of kinetics is more important than thermodynamics in the nitride–BN systems. Perhaps this means a common reaction mechanism for formation of cubic BN from these systems. On the basis of the structure of Li_3BN_2, we speculate that there might be a common structurally-based mechanism for formation of cubic BN from these kinds of systems. It is assumed that the structure of Li_3BN_2 is closely related to the known antifluorite lattice of Li_3AlN_2[9]), and that the structure of the melt at high pressures will be somewhat closely related to that structure. From a study of a crystal model of the Li_3AlN_2 lattice it is apparent that besides the usual way of describing such a structure in terms of packing of ions, it also can be considered as made up of B–N–B–N–... chains which are held together by Li ions. In this context the Li_3N–BN melt also might be thought of as a BN polymer crosslinked by Li ions. Perhaps the common step in the transport and precipitation of cubic BN from nitride systems depends primarily on the making and breaking of the metal bonds which link the BN chains, and this process is not very sensitive to either pressure or ion charge.

Acknowledgments

The helpful teaching and encouragement of R. H. Wentorf, Jr., John Birle and R. E. Hanneman are gratefully acknowledged.

References

1) R. H. Wentorf, Jr., U.S. Patent 2,947,617, 8/2/60, assigned to General Electric Co., Schenectady, N.Y.
2) R. H. Wentorf, Jr., J. Chem. Phys. **26** (1959 956; **34** (1961) 809.
3) R. H. Wentorf, Jr., J. Chem. Phys. **26** (1957) 956.
4) R. C. DeVries and J. F. Fleischer, Mater. Res. Bull. **4** (1969) 433.
5) F. P. Bundy and R. H. Wentorf, Jr., J. Chem. Phys. **38** (1963) 1144.
6) K. Kudaka, H. Konno and T. Matoba, Kogyo Kagaku Zasshi, (J. Chem. Soc. Japan, Ind. Chem. Section) **69** (1966) 365.
7) N. E. Filonenko, V. I. Iranov, L. I. Fel'dgun, M. I. Sokhor, and L. F. Vereshchagin, Dokl. Akad. Nauk SSSR **175** (1967) 1266.
8) H. Saito, M. Ushio and S. Nagao, Yogyo-Kyokai Shi **78** (No. 893) (1969).
9) R. Juza and F. Hund, Z. Anorg. Chem. **257** (1948) 13.

Journal of Crystal Growth **13/14** (1972) 93–96 © *North-Holland Publishing Co.*

ON THE MECHANISM OF ELECTROCRYSTALLIZATION

E. BUDEVSKI

*Division of Electrochemical Power Sources, Institute of Physical Chemistry,
Bulgarian Academy of Sciences, Sofia, Bulgaria*

The technique for the preparation of electrodes having one crystallographic plane only was applied in the investigation of the mechanism of electrolytic metal deposition. Impedance and galvanostatic transient methods, as well as steady state measurements were used in the case of silver deposition on dislocation-free silver crystal planes and on planes possessing a limited number of screw dislocations. The spreading out kinetics of mono-atomic layers on dislocation-free planes and the current-square overvoltage law found on planes with dislocations, show that current is proportional to step length. This fact was used for the preparation of electrodes with exactly known step density.

An interesting fact observed on electrodes with known step density is that the exchange current density, obtained from impedance measurements at high frequencies, is proportional to step length. This shows that deposition on a step is much easier than on a plane region. It was found that the exchange current density of ions in solution with ions in kink position is about four orders of magnitude higher than the exchange current density with adions. A simple calculation shows that up to the highest step densities (step distance of about 450 Å) used in the present investigations the direct transfer mechanism is prevailing by at least two orders of magnitude over the surface diffusion mechanism.

1. Introduction

The development of the technique for the preparation of electrodes presenting one single crystal plane only offers the possibility for the investigation of a wide range of problems in the electrochemistry of solid surfaces. A significant feature of these electrodes is that only one crystallographically well defined plane takes part in the electrochemical process. Their most remarkable advantage is that their planes can be obtained with an exactly known surface topography and number of defects, or even without any defects. Some of the electrochemical problems which were investigated with such electrodes will be discussed in this paper.

2. Experimental technique

The experimental technique consists in letting an appropriately oriented silver single crystal grow in a glass capillary as a single crystal filament. By superimposing an alternating current on the direct current of growth the front plane gradually fills up the inner diameter of the capillary exposing in the last stage only one crystallographic plane to the growth process. According to the orientation of the seed crystal this plane may be cubic, octahedral or rhombododecahedral. It was established that the number of emergence points of screw dislocations diminished steadily during the growth process. This enabled us to prepare single crystal planes exhibiting only a limited number of screw dislocations or at a more advanced stage – dislocation-free planes[1,2]).

3. Behaviour of dislocation-free planes and kinetics of two dimensional nucleation

At constant current of growth a very characteristic oscillation of the cathode potential is observed at the relatively high overvoltage of about 10 mV (fig. 1). The period of these oscillations depends on current density, so that the product of the current and the period remains constant, giving an amount of electricity equivalent to a monoatomic layer of silver. This wavering of

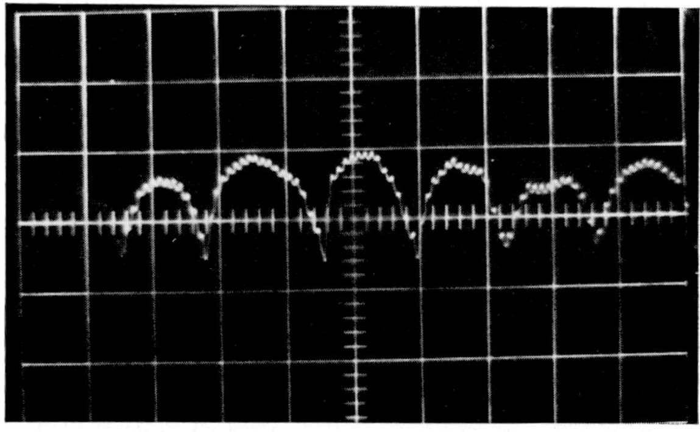

Fig. 1.   Fluctuation of the overvoltage at constant current growth of a dislocation-free plane.

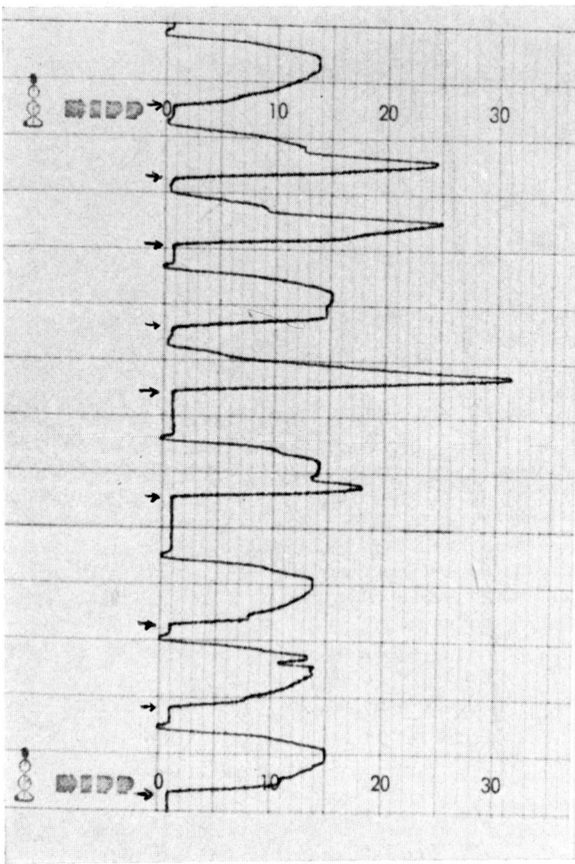

Fig. 2. Current–time curves at constant potential (6 mV over-potential) following a voltage pulse (amplitude 13 mV duration 1 msec) during which a two-dimensional nucleus is formed. The moments of pulse release are marked with arrows.

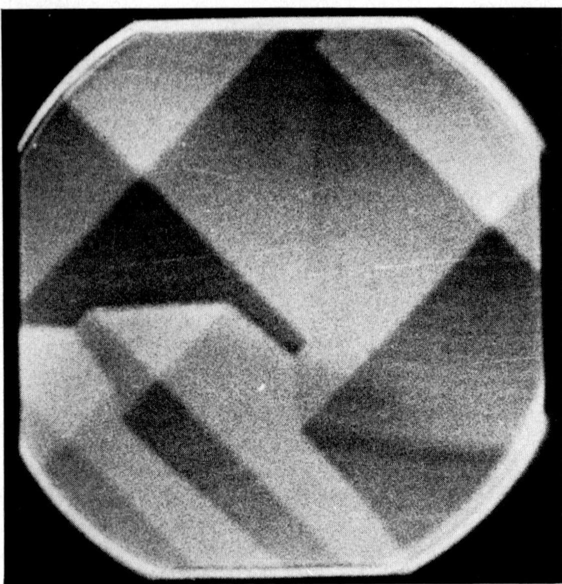

Fig. 3. Pyramidal growth pattern of a cubic plane with low screw dislocation density.

the potential is evidently connected with the nucleation of every new layer for which a critical overvoltage should be exceeded[1,2]).

Under constant potential conditions at potentials which do not exceed this critical overvoltage the cell with a dislocation-free plane remains electrically cut off, i.e. no current flows, because of the very small probability for nucleus formation. A properly chosen electric pulse can trigger the formation of a new layer. After this a current is observed which rises rapidly, reaches a maximum and falls to zero (fig. 2). Here again the area closed under the current–time curve gives an amount of electricity equivalent to a monolayer.

The course of the current–time curve is evidently connected with the change of the periphery length of the growing monolayer. A linear dependence of the current on the length of the periphery of the spreading monolayer can be assumed to explain the shape of the current–time curves[1,2]). The time lag of the nucleus formation after a triggering pulse gives a possibility for the investigation of the kinetics of this process[1,2]). The energy of nucleation and the specific edge energy implied in the Volmer equation can be easily estimated.

4. Role of dislocations on metal deposition

In presence of screw dislocations a crystal plane grows according to the mechanism proposed by Frank[3]). A spiral shaped front is formed at the emergence point of the dislocation, the distance between successive turns of the spiral being inversely proportional to the overvoltage. As a result, a flat pyramid of growth, fig. 3, appears at the dislocation emergence point. The slope of such a pyramid should be proportional to the overvoltage. The coefficient of proportionality permits the evaluation of the edge energy of the step belonging to the spiral-shaped front[2,3]).

This affirmation, which follows directly from the spiral growth theory, was tested using a cubic plane possessing only a limited number of screw-dislocations. The studies on the spreading-out velocity of single monolayers after two dimensional nucleation showed that the current is proportional to overvoltage and to the periphery length of the monolayer. In the case of spiral growth the total length of the spiral front is in-

TABLE 1

Preparation conditions		Step length L (m/cm^2)	Overall exchange current density i_0 (A/cm^2)	Ecd due to steps $i_{0,st} = i_0 - i_{0,ad}$ (A/cm^2)	Ecd for unit step length $i_{0,st}/L$ (μA/cm)
Current density (mA/cm^2)	Corrected overvoltage (mV)				
8.8	1.6	570	0.97	0.91	16
17.6	2.2	790	1.10	1.04	13
35.2	3.2	1140	1.68	1.62	14
70.4	4.5	1890	2.12	2.06	11
132.2	6.1	2160	2.37	2.31	11
Dislocation-free plane		0	$i_{0,ad} = 0.06$	0	

versely proportional to the distance between successive turns of the spiral and hence, directly proportional to the overvoltage. It follows then that the current is proportional to the square of the overvoltage. The factor of proportionality contains the specific edge energy of the growth front which can be calculated from experimental data. The values found lie between 4 and 7×10^{-6} erg cm^{-1} (ref. 4) and coincide well with the values obtained by other methods.

5. Mechanism of metal deposition

There are two possible ways of ion incorporation into the crystal lattice in the process of metal deposition: (i) Ions are discharged at any site of the crystal surface and are transported afterwards by surface diffusion to the sites of growth where they are incorporated – surface diffusion mechanism; (ii) ions are directly discharged on a growth step or site of growth with simultaneous incorporation – direct transfer mechanism.

The experiments described with single planes are of major importance for the investigation of the mechanism of metal deposition. They allow the preparation of cathodic planes with a perfectly known surface topography. Three very important surface configurations can be achieved:

(i) Dislocation-free atomically smooth plane.

(ii) Dislocation-free plane with a monoatomic step of known length. On a dislocation-free plane a new monoatomic layer can be produced by the described constant voltage-pulse technique. The propagation of the layer can be stopped at any desired position by interrupting the current.

(iii) Stepped plane with perfectly known uniform step density. A plane with few dislocations (low dislocation density) can be grown at a constant overvoltage until a regular pyramidal growth pattern is obtained. The slope of the pyramids, i.e. the step density, is related to the overvoltage. A plane of perfectly known step density, i.e. of known total step length, is thus produced.

5.1. Information obtained from dislocation-free planes[5])

A dislocation-free atomically smooth plane behaves as an ideally polarizable electrode in the region between 0 and 10 mV overvoltage. The double layer capacity is easily found using impedance or galvanostatic transient measurements. A value of 30 μF cm^{-2} was obtained for the cubic plane of silver in 6 N silver nitrate solution.

Impedance measurements at high frequencies allow the estimation of the resistance of the charge transfer reaction (ions in solution to adions on a close packed plane) as a resistance parallel to the double layer capacity. A value of 0.45 ohm cm^2 was found for this resistance. The corresponding exchange current density $i_{0,ad} = 0.06$ A cm^{-2} of the charge transfer reaction with adions is easily obtained from this value[5,6]).

5.2. Information obtained from stepped planes[7,8])

Impedance measurements on stepped planes have shown that the charge transfer resistance depends strongly on step density. We can therefore assume two charge transfer reactions taking place in parallel: (i) ion in solution to adion on a close packed plane and (ii) ion in solution to ion on a step line or site of growth, with the exchange current densities $i_{0,ad}$ for the first and $i_{0,st}$ for the second reaction. The overall exchange current density will then be the sum of both.

Table 1 shows a linear dependence of $i_{0,st}$ from the

overall step length L, the ratio $i_{0,\text{st}}/L$ being constant over a wide range of step densities. The true exchange current density of the ions on a step can be calculated from this constant and the height of the steps $h = 2.04 \times 10^{-8}$ cm. A value of 650 A cm^{-2} was found which is four orders of magnitude higher than the i_0 value for adions. This fact might be considered as contradictory to the theoretical calculations of Conway and Bockris[9]. Considering the case of ion discharge on the edge of a growth step Despic[10]) showed recently that this discharge reaction proceeds almost activationless.

5.3. STEADY-STATE DEPOSITION MECHANISM

The current-square overvoltage law found on a plane with screw dislocations has shown that the current is proportional to step length. This is consistent with the direct transfer mechanism, and also agrees with the surface diffusion mechanism under the condition that the diffusion distance λ_0 is smaller than the step distance d in the whole range of investigation. The highest step density used in our case corresponds to a step length of 2160 m cm^{-1} (the last line in table 1) for which $d = 460$ Å.

The resistance of the surface diffusion mechanism at small overpotentials can be calculated from the relation[11,12])

$$R_{\text{SD}} = \frac{RT}{F} \frac{1}{i_{0,\text{ad}}} \frac{d}{2\lambda_0} \text{tgh}^{-1} \frac{d}{2\lambda_0}$$

Assuming a value of $\lambda_0 = \frac{1}{2}d$ for the case of highest step density, where the contribution of the surface diffusion would be the highest, $R_{\text{SD}} = 0.59$ ohm cm^2.

The resistance of the direct transfer reaction depends on step length and can be calculated from

$$R_{\text{DT}} = \frac{RT}{F} \frac{1}{i_{0,\text{st}}^* L},$$

where $i_{0,\text{st}}$ denotes the exchange current density for unit step length which is according to table 1 about 13 µA cm^{-1}. For the case of highest step density is $R_{\text{DT}} = 0.0094$ ohm cm^2.

It can readily be seen that even in the case of highest step density the resistance of the direct transfer reaction is about 60 times lower than the resistance of the surface diffusion reaction. For lower step densities this ratio should be even higher. It follows, at least in that case of silver deposition from concentrated silver nitrate solutions, that the contribution of the surface diffusion mechanism is negligible, or in any case lower than 2%.

References

1) E. Budevski, T. Vitanov, W. Bostanov, Z. Stoinov, A. Kaischeva and R. Kaischew, Phys. Status Solidi **13** (1966) 577; Electrochim. Acta **11** (1966) 1697; Electrokhimia **3** (1967) 856.
2) R. Kaischew and E. Budevski, Contemp. Phys. **8** (1967) 489.
3) F. C. Frank, Discussions Faraday Soc. **5** (1949) 48.
4) W. Bostanov, R. Roussinova and E. Budevski, Commun. Dept. Chem., Bulg. Acad. Sci. **2** (1969) 885.
5) T. Vitanov, E. S. Sevastianov, W. Bostanov and E. Budevski, Electrokhimia **5** (1969) 451.
6) T. Vitanov, E. S. Sevastianov, Z. Stoinov and E. Budevski, Electrokhimia **5** (1969) 238.
7) T. Vitanov, A. Popov and E. Budevski, to be published.
8) E. Budevski, Trans. SAEST (India) **5** (1970) 55.
9) B. E. Conway and J. O'M. Bockris, Electrochim. Acta **3** (1961) 340.
10) A. R. Despic, Croatia Chem. Acta **42** (1970) 265; c.f. also K. J. Vetter, ref. 12, ch. 2, section 76.
11) W. Lorenz, Z. Physik. Chem. **202** (1953) 275; Z. Elektrochem. **57** (1953) 382.
12) K. J. Vetter, in: *Elektrochemische Kinetik* (Springer, Berlin, 1961) ch. 2, section 77.

Journal of Crystal Growth **13/14** (1972) 97–105 © *North-Holland Publishing Co.*

LES MATÉRIAUX COMPOSITES À FIBRES

M. L. TURPIN

Centre des Matériaux, Ecole Nationale Superieure des Mines de Paris, 91 – Corbeil–Essonne, France

Development of composite materials involves the solution of a complex system involving many design parameters.

A fairly large effort is still devoted to the development of B, SiC and Graphite continuous fibres to produce regular, high quality and low cost materials. Some examples of the problems encountered in the CVD growth of SiC and of the development of diffusion barriers on B for use in metal matrices at higher temperatures are given.

The technology of composite fabrication with resin and metal matrices has progressed faster than the scientific understanding of the basic mechanisms. For instance, adhesion between C fibres and resin is low and techniques to improve it is developing slowly. For hot pressing with metal matrices tendencies are to use temperatures and pressures as low as possible.

In situ solidification of eutectic is discussed in some length. Some systems show promise as high temperature materials for turbine blades, others are of interest for their physical properties.

Characterisation of the mechanical properties of composites is a difficult problem, but many data and excellent reviews are available. As the lack of toughness appears as a crucial problem, more work is necessary to understand the role of the interface on the fracture of composites.

Presently, basic science is lagging the technology and effort must be continued in this field to improve our basic knowledge of the mechanisms and to guide our action in finding better combinations of composite properties.

1. Introduction

Dès leur apparition, les matériaux composites à fibres ont suscité un très vif intérêt. En peu d'années s'est effectué un effort considérable de recherche et dès à présent plusieurs éléments de structure ont été ou sont réalisés industriellement dans l'aéronautique et l'aérospatiale. C'est de l'idée à l'utilisation, un délai extrèmement court si on le compare aux lenteurs qui ont marqué les progrès de matériaux, comme le PVC, l'aluminium ou les aciers inoxydables.

Pourtant d'un premier bilan qui comporte succès et échecs, on a actuellement tendance à tirer des conclusions plutôt pessimistes pour ce qui concerne l'avenir de ces nouveaux matériaux, bien qu'on s'accorde à prévoir leur développement. Deviner ce que sera leur place exacte est très aléatoire, d'autant qu'il est pratiquement impossible, pour cause de secret militaire ou industriel, d'avoir accès aux informations les plus pertinentes et qu'il règne une grande incertitude sur l'avenir des programmes aéronautiques et spatiaux, seuls consommateurs importants aux prix actuels des produits[1–3]).

Dans le cadre de ce court exposé, nous soulignerons quelques problèmes importants encore mal résolus qui gouvernent les propriétés des matériaux composés à fibres. Ces nouveaux venus se révèlent d'un usage plus malcommode qu'on l'avait souvent voulu dire et il importe de consolider les bases scientifiques sur lesquelles repose actuellement un développement technologique assez anarchique.

Si les améliorations des matériaux usuels et leur pénétration ont été lentes et obtenues de façon généralement empirique, l'expérience a montré que cette démarche était très mal adaptée à la conception et à la mise en œuvre des composites à fibres car leurs caractéristiques rompaient brutalement avec la chronique antérieure. La substitution brutale du nouveau matériau à l'ancien sans reconcevoir la pièce ou la structure conduit à des déboires ou à des gains très faibles. Conception, fabrication et utilisation des matériaux interagissant dans un système complexe qu'il est indispensable d'étudier en bloc et dont la fig. 1 présente un ébauche de représentation.

Un problème essentiel et bien connu des systèmes interactifs est la définition pour chaque partie de sous objectifs compatibles avec l'objectif global. L'effort de recherche qui s'est porté sur les points 1, 2 et 3 s'est sans doute heurt'à cet écueil et nous essaierons d'in-

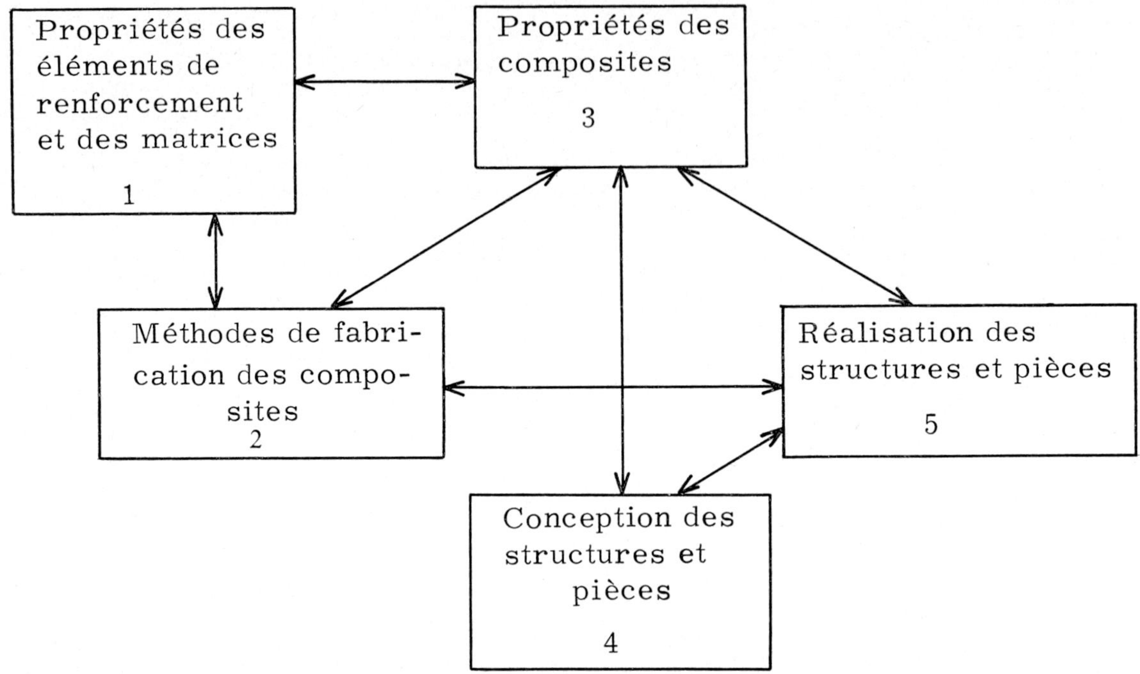

Fig. 1.   Interaction des différentes étapes de réalisation d'une structure.

diquer les inflexions qui apparaissent actuellement dans ces trois domaines qui feront seuls l'objet de cet exposé.

2. Éléments de renforcement

2.1. DES VICTIMES

Au début de l'effort sur les composites à fibres, on vit foisonner des recherches sur la production d'éléments de renforcement. Que ce soit pour obtenir des trichites ou des fibres plus longues, toutes les combinaisons des éléments légers (B, C, N, Al, Mg, Si, O) susceptibles de convenir ont vraisemblablement été étudiées au laboratoire[4]). Citons parmi ceux qui firent naître le plus d'espoir Si_3N_4, SiC, Al_2O_3. Bien peu de ces produits sont aujourd'hui commercialisés.

Les victimes les plus notoires sont les trichites. Une première raison en fut leur prix prohibitif, une seconde leur difficulté d'utilisation, une troisième a été entourée de plus de discrétion et est la fragilité congénitale du composé à fibre obtenu.

Les principales fibres disponibles commercialement sont actuellement :
– la fibre de carbone ($d \simeq 8$ µm, $\bar{R} \simeq 200$–300 HBar, prix $\simeq$ 1500 F/kg) en écheveaux discontinus ou en fils tressés;

– la fibre de bore sur substrat de W ($d \simeq 100$ µm, $\bar{R} \simeq 300$–400 HBar, prix $\simeq$ 3000 F/kg;
– la fibre de BorSiC (bore revêtu de SiC) ($d \simeq 100$ µm, $\bar{R} \simeq 250$–300 HBar, prix $\simeq$ 4500 F/kg);
– la fibre de SiC (sur substrat de W) ($d \simeq 100$ µm, $\bar{R} \simeq 300$–400 HBar, prix non publié).

Tous ces produits restent très chers et on est beaucoup moins optimiste qu'il y a peu sur les possibilités de pouvoir abaisser leur prix dans de fortes proportions (par un facteur 10).

2.2. LES FIBRES DE CARBONE (fig. 2)

Par suite de la très haute température de sublimation du carbone, la structure d'un matériau carboné, dépend fortement de son histoire[5]). La structure des fibres de graphite est maintenant bien connue[6]). Les cristallites de graphite sont de petites tailles ($\sim$40 à 100 Å), orientées avec leur plan basal parallèle à l'axe de la fibre. Le module des fibres croît lorsqu'augmente la taille des cristallites. Un trait caractéristique est l'existence de défauts internes alignés dans l'axe de la fibre[7]. Ce sont des vides de grande taille ($\sim$1 µm) qui se développent lors de la graphitisation à haute température ($>$1800 °C). Ces défauts expliquent l'observation à première vue déroutante que dans le procédé

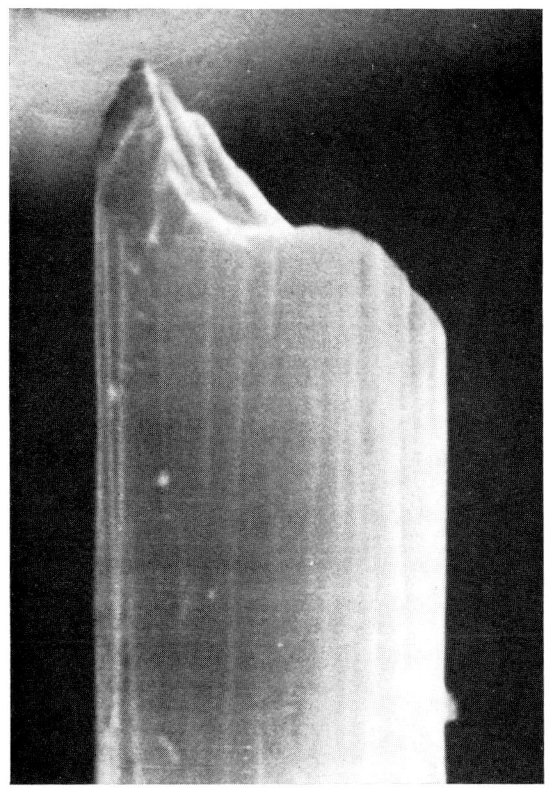

Fig. 2. Fibre de carbone Courtauld haute résistance – microscope à balayage, ×5000.

RAE la résistance des fibres passe par un maximum lorsque la température de graphitisation est d'environ 1500 °C et décroît ensuite. L'origine exacte du dégagement gazeux qui selon toute vraisemblance provoque ces trous reste incertaine.

Deux démarches différentes sont utilisées pour parvenir à la structure recherchée;

(a) *Une technique* (américaine et japonaise) utilise des fibres peu coûteuses du type cellulose qu'on peut aisèment pyroliser sans fusion. Mais on doit, pour obtenir la graphitisation orientée, opérer à haute température sous tension. On peut ainsi utiliser une grande variété de précurseurs[8]) au prix d'un traitement délicat et coûteux).

(b) *La procédé RAE* utilise comme précurseur le polyacrylonitrile qui se graphitise correctement sans tension. Mais il faut éviter une fusion prématurée avec rupture des orientations privilégiées en opérant une oxydation contrôlée vers 200–300 °C au cours de laquelle les fibres sont maintenues sous tension[9]).

2.3. LA FIBRE DE SiC

En collaboration avec le service des poudres, un gros effort de recherche a porté, au Centre des Matériaux, sur la mise au point d'un procédé faible de fabrication

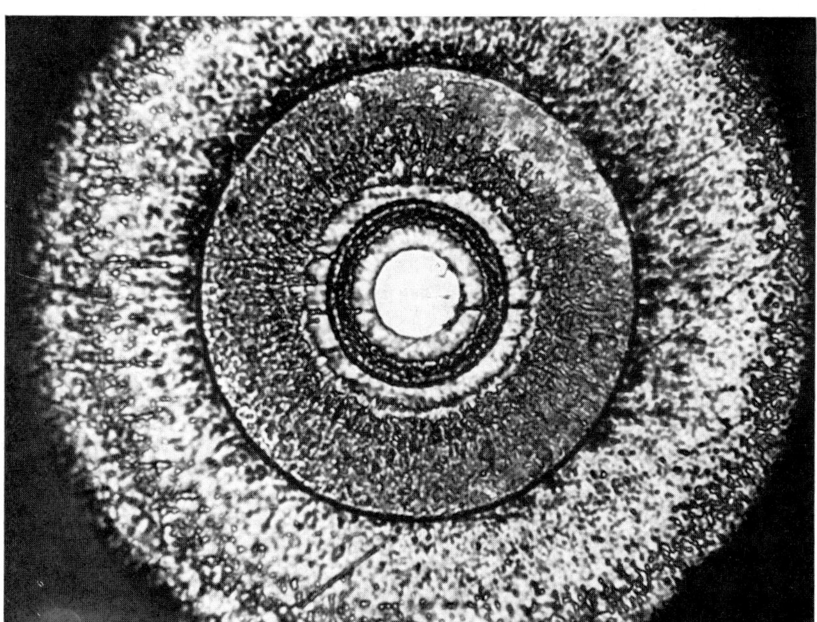

Fig. 3. Coupe transversale d'une fibre de SiC. Réacteur à 3 chambres de dépôt, ×1890.

de ces fibres. Bien que plus lourd que le bore, SiC présente des propriétés qui le rendent intéressant pour des utilisations dans des matrices métalliques à moyenne température.

Le principe du procédé est simple: réaction de type Van Arkel de décomposition sur un fil chaud d'un mélange CH_3SiCl_3 (MTS) et d'hydrogène. Mais des différences profondes sont apparues avec la fabrication de la fibre de Bore.

Fig. 4. Défaut de croissance sur fibre de SiC, microscope à balayage.

La croissance de la fibre reste environ 5 fois plus lente que celle du bore bien que le MTS se dissocie entièrement[10]. L'utilisation de parois chaudes[11] permet d'accélérer le processus. La phase gazeuse est alors pratiquement en équilibre thermodynamique mais en déséquilibre avec SiC, on pense que cette étape lente est la recombinaison de Si et C sur la surface.

Pour obtenir la structure microcristalline (~ 1000 Å) qui présente la résistance souhaitée, il importe de ne pas dépasser une température de filament de 1200 °C. Plusieurs chambres de dépôt étaient initialement nécessaires et la fig. 3, représente une coupe caractéristique des fibres obtenues.

Chaque nouvelle couche germe sur la précédente sans qu'il y ait apparemment épitaxie. Il en résulte une très forte probabilité d'apparition de défauts de croissance. Certaines particulièrement spectaculaires (fig. 4) sont apparues lorsqu'on a utilisé des réacteurs très longs à parois chaudes et nous les attribuons à un effet électrostatique.

Actuellement nous utilisons un réacteur à paroi chaude à une seule chambre qui marche à 150 m/heure

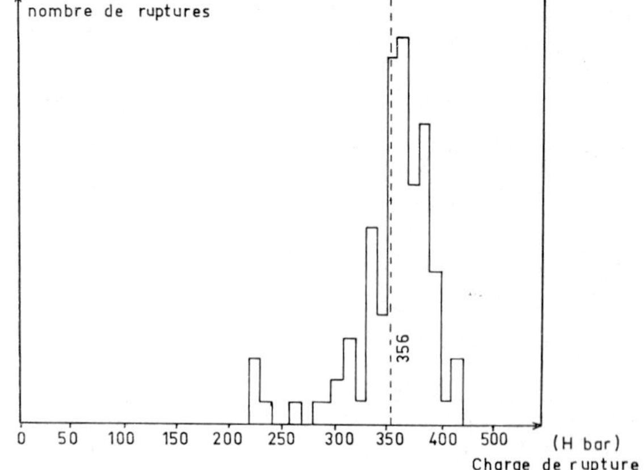

Fig. 5. Histogramme de fibres de SiC. Réacteur E. N. S. Mines de Paris (1971) (100 essais, échantillons de 2.5 mm).

et produit une fibre de résistance moyenne comprise entre 350 et 400 HBar, fig. 5.

Malgré ces résultats encourageants les mécanismes de croissance de SiC restent encore mystérieux et mériteraient un effort de recherche fondamentale.

3. Méthodes de fabrication des composites

Il existe d'excellentes synthèses de la littérature très abondante qui traite de ces problèmes[12,13]. Les comptes rendus annuels du Congrès du SAMPE sont une autre source régulière et importante[14]. Mais chaque producteur s'efforce de conserver le secret des tours de mains qui font sa supériorité et seule l'expérience personnelle permet de constituer le fond d'acquit technologique nécessaire pour élaborer de façon reproductible des matériaux composites de bonne qualité. Certaines difficultés ont suscité un effort particulier et nous verrons successivement:

(1) l'élaboration des composites résine–carbone;

(2) les conditions de pressage à chaud des composés à matrice métallique;

(3) la croissance des eutectiques orientée.

3.1. LES COMPOSITES RÉSINE–CARBONE

La transposition aux fibres de bore et de SiC des techniques de fabrication mises au point pour les fibres de verre n'a pas présenté de difficultés majeures. Par contre les fibres de carbone à haute caractéristique (type RAE) présentent à l'état brut de fabrication une très médiocre adhésion avec la matrice. L'interface matrice–fibre ne supporte que de très faibles efforts de cisaillement (environ 2 HBar). Il en résulte de mauvaises caractéristiques en traction, flexion et compression[15]). La structure de la fibre explique ce phénomène auquel on s'est efforcé de remédier par différents traitements superficiels.

C'est l'oxydation ménagée par exemple par HNO_3 bouillant[16]) qui a donné les meilleurs résultats. On trouvera dans la littérature[15-17]) une comparaison des divers traitements, y compris la whiskérisation en notant que les méthodes de mesure de la cission critique n'étant pas les mêmes, les résultats ne sont pas rigoureusement comparables. Actuellement on s'efforce d'éviter ce traitement ultérieur en agissant sur les paramètres de fabrication. La panoplie des caractéristiques des fibres de carbone proposées s'agrandit ainsi de façon très intéressante.

3.2. LE PRESSAGE À CHAUD DES COMPOSITES À MATRICE MÉTALLIQUE

Parmi les nombreuses techniques essayées[18]) le pressage à chaud reste le plus utilisé. Cette importance est due en partie à la réactivité des fibres de bore qui ne supportent pas le contact avec les métaux légers fondus. Mais c'est également une technique bien adaptée à la fabrication de plaques, coques, tubes et profilés[19]). La nécessité de passer par un semi-produit: le ruban monocouche est souvent un avantage plutôt qu'un inconvénient. Ces feuilles sont maniables et peuvent être découpées, pliées à la forme, empilées avant le repassage. On effectue ainsi la mise en forme et on peut orienter les couches successives en fonction des contraintes que doit subir la pièce.

Le pressage à chaud s'effectue soit sous presse soit en autoclave. Les conditions de temps, température et pression doivent être soigneusement contrôlées. Avec des métaux purs il faut opérer à haute pression pendant de longues durées (tableau 1). Fabriquer de grandes pièces dans ces conditions exige des appareils puissants très coûteux et d'une mise en œuvre extrèmement délicate. On s'est efforcé de diminuer pressions et temps de contact.

TABLEAU 1

Conditions de pressage à chaud de composites à matrice métallique (métaux purs)

Paramètres d'élaboration Composites	Température (°C)	Pression (kg/mm²)	Temps (hr)
Al–B	550	3.5	1
Mg–B	525	7	1
Ti–SiC	900	6	1

Des résultats intéressants sont obtenus en utilisant une phase liquide soit en projetant un alliage de coulée du type ASG (ASTM 60 61), soit en déposant une fine couche d'un métal qui assure une brasure. On descend ainsi à des pressions de 0.1 HBar.

Malheureusement on accroît la fragilité de la matrice ce qui détériore la résistance. Disposer de fibres qui puissent rester assez longtemps en contact avec Al et Mg purs, près ou au-dessus de leur point de fusion reste donc particulièrement intéressant.

3.3. LES EUTECTIQUES ORIENTÉS

La solidification biphasée est une méthode séduisante pour fabriquer en une seule opération un matériau composite. La réalité de ce renforcement fut d'abord montré à l'aide de systèmes simples $(Al–Al_3Ni)$[21]). Un gros effort de recherche suivit, qui parallèlement à des essais systématiques à caractère exploratoire, a comporté une grande partie théorique qui a considérablement accru notre connaissance des mécanismes de solidification de ces systèmes.

(a) *La quête systématique de systèmes utilisables*: Nous ne citerons que pour mémoire les études entreprises pour utiliser des propriétés physiques de ces eutectiques orientée[21]). Il est apparu assez vite que c'était dans les utilisations à haute température que les potentialités des eutectiques orientés étaient les plus fortes, en particulier pour obtenir un meilleur matériau pour aube de turbine.

Vu les contraintes sévères auxquelles sont soumises ces aubes, les eutectiques binaires ne pouvaient convenir. Ils ont servi de systèmes de base à partir desquels on s'est efforcé d'obtenir les propriétés requises par addition d'éléments selon des règles empruntées à la somme des connaissances accumulées dans le développement des superalliages. Parmi tous les systèmes étudiés les plus prometteurs sont les eutectiques à carbures (TaC, TiC) dans une matrice à base cobalt[22]).

Mais l'addition d'éléments a compliqué le problème d'élaboration en accentuant les causes d'instabilité et en allant même jusqu'à supprimer prématurèment eutectique.

(b) *La solidification eutectique monovariante: stabilité et ségrégation:* Ajouter un troisième élément à un eutectique on a un pseudoeutectique binaire à deux conséquences dont les effets se combinent:
– Il accentue l'instabilité de croissance qui supprime la structure régulière.
– Il ségrège, ce qui modifie la composition au fur et à mesure de l'avancement de la solidification.

Dans les systèmes binaires on ne peut obtenir des structures régulières que dans certaines conditions[23,24]). L'addition d'un troisième constituant provoque d'abord la formation de cellules. Pour rendre compte de ce phénomène, a été développée une théorie copiée sur celle des systèmes monophasés et qui s'accorde plus au moins bien avec l'expérience[25]). Pour des compositions qui s'éloignent de la vallée eutectique un autre type d'instabilité apparait plus fréquemment: la croissance de cristaux primaires d'une des deux phases[26]). Aucune théorie ne rend compte actuellement de ce phénomène de façon satisfaisante. Les structures lamellaires sont beaucoup plus sensibles que les structures fibreuses[26,27]) (fig. 6).

Or la présence de cellules comme celle de dendrites provoque généralement une chûte brutale des caractéristiques mécaniques des eutectiques. Par suite de la désorientation des éléments de renforcement ou de leur dégénérescence, les joints de cellule sont des zones de faiblesse à la fois en traction et en fluage. On a pu également sur un eutectique binaire montrer que la présence de certains cristaux primaires modifiait profondément l'allongement à la rupture. Eviter ces irrégularités est un problème essentiel. Or par suite de la ségrégation la composition du liquide n'a, en solidification dirigée, aucune raison de suivre la vallée eutectique et un régime permanent ne pourrait être maintenu[27]). Par contre la fusion de zone permet de conserver un régime permanent et est beaucoup plus favorable. L'expérience confirme ces conclusions mais la technique de fusion de zone s'adapte malaisèment à la réalisation de pièces un peu complexes.

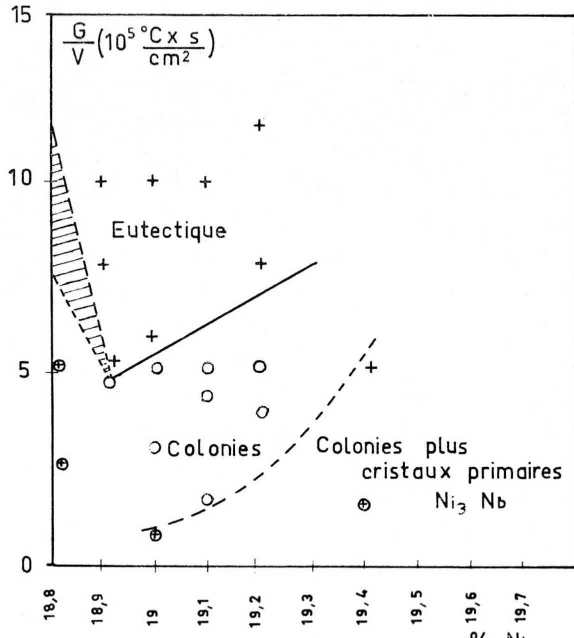

Fig. 6. Limite de stabilité des différentes structures dans le ternaire Ni+15% Cr–Nb.

4. Les propriétés mécaniques, le problème de la fragilité

Au stade de la caractérisation mécanique on peut considérer le composé à fibres comme un matériau homogène anisotrope. Le problème de la caractérisation des diverses propriétés est régulièrement mise à jour par l'ASTM[30]).

De très nombreux calculs théoriques étendent les résultats de l'élasticité et de la résistance des matériaux aux structures composites. Le développement est actuellement hors de proportion avec la base expérimentale qui lui serait nécessaire. Actuellement des produits aussi courants que les résines armées de fibres de verre sont utilisés avec des facteurs de sécurité et des fautes de conception des structures qui font perdre une grande partie de l'avantage potentiel de ces matériaux.

La prévision des caractéristiques du composite à partir de celle des composants a d'abord utilisé la loi des mélanges. Des améliorations successives ont été apportées qui ont fait apparaître des notions importantes comme celle de longueur de transfert. Diverses théories ont tenté d'intégrer le caractère aléatoire de la rupture des fibres soit en utilisant une loi statistique qui représente la répartition des niveaux de résistance[29]) soit par simulation[30]). Cette dernière méthode montre que la notion de valeur moyenne de résistance n'est pas une donnée satisfaisante pour caractériser un lot de fibres. La proportion des basses valeurs est beaucoup plus significative.

Mais les efforts entrepris pour utiliser les composites

TABLEAU 2

Résilience de matériaux composites à fibres; essais Charpy, éprouvettes non entaillées

Matériau	Energie absorbée (daJ/cm²)	Référence
Verre+tissu 181 Narmco 570	1.2	Essais Snecma
Verre – résine époxy	3	éprouvette 4×2
Carbone H–R+résine époxy Narmco 5.505	0.1	
Carbone H–R+résine époxy Den 438	0.5	
CarboneH–R+résine époxy Erla 438	0.7	
Carbone H–R+polyimide	0.5	
Carbone matrice Al	0.06	
Bore+résine époxy	0.5	
Bore+matrice Al	0.25	
Alliage magnésium (Mg+1.5% Mn)	1.63	Morris et Smith
Ti–6 Al–4 V	13.56	éprouvette
Verre+polyester	1.12	2.79×2.79 mm
Al+B (18% en volume)	0.27	

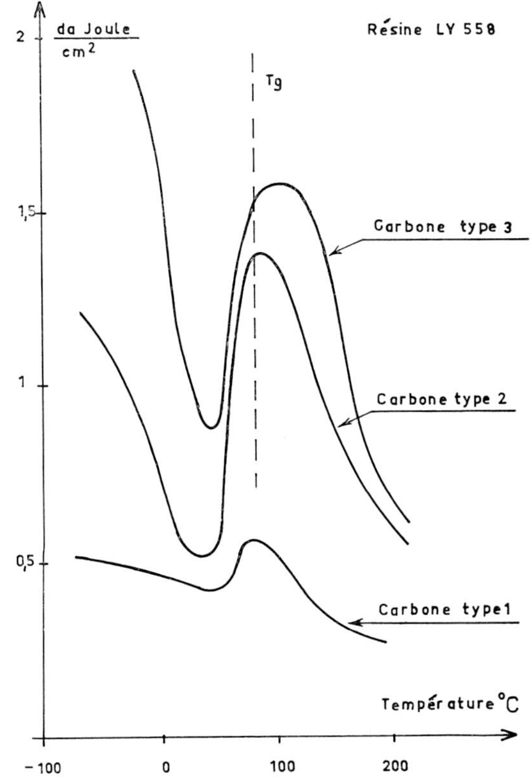

Fig. 7.   Variation avec la température de la résilience de différents composites résine–fibres de carbone.

comme aubes de compresseur ou de turbine ont jusqu'à présent achoppé sur le problème de la fragilité. Les données du tableau 2 situent le problème. Les niveaux de résilience obtenus sont nettement inférieurs à ceux des bons alliages de Titane. Toutefois, dans les meilleurs cas, on obtient des valeurs comparables à celles des alliages légers à haute caractéristique. L'étude plus approfondie des mécanismes de fracture revêt donc une importance toute particulière puisqu'il est possible d'obtenir des gains significatifs. Les composés Al–B tiennent beaucoup mieux que les composés Al–C. Egalement remarquables sont les résultats représentés sur la figure 7[31]): la résilience est nettement fonction de la nature des fibres mais aussi des propriétés, hélas mal caractérisées, de la résine. Des essais[32]) ont montré également sans ambiguïté sur des composites carbone-résine, l'influence sur la résilience de la cission critique de l'interface. Schématiquement la résilience:
– croît lorsque la cission critique décroît
— croît lorsque la taille des fibres croît et que leur module décroît

– croît lorsque la matrice s'écrouit

Pour expliquer ce comportement des modèles ont été développés par Kelly et son école[33–35]).

Ces auteurs évaluent le travail nécessaire pour propager une fissure perpendiculaire à la direction de traction. Le premier résultat important est que la déviation d'une fissure parallèle à l'interface ne peut avoir lieu que lorsque la résistance de cette interface est très faible. En effet l'anisotropie augmente fortement le rapport σ_x/τ_{xy} en avant de la fissure le portant (en régime élastique) de 5 pour un matériau isotrope à 11 pour un matériau anisotrope du type résine–carbone[35]). En pratique cette faiblesse est observée dans les composites résine–carbone fabriqués avec des fibres non traitées[36,39]) et qui présentent effectivement une bonne tenacité.

Matrices résines et métalliques ont un comportement différent. Pour ces dernières la propagation de la fissure se fait par rupture de fibres dans une section à peu près perpendiculaire à la direction de traction. La dissipation d'énergie provoquée par l'avancement de la fissure est due essentiellement à l'écrouissage de la matrice. La zone affectée par cette déformation est proportionnelle au rayon de la fibre[33]), donc la dissipation d'énergie est également proportionnelle à ce rayon. Toutefois si ces fibres sont trop grosses la fissure ouverte par une rupture risque de dépasser la taille critique pour une propagation catastrophique.

Les matrices résines sont fragiles et leur rupture n'absorbe que peu d'énergie. Kelly[35]) considère deux formes de dissipation. La rupture de la liaison entre fibre et matrice et le déchaussement de la fibre. Il prévoit que le renforcement par fibres discontinues est plus efficace et optimal lorsque la longueur de la fibre est égale à la longueur critique de transfert. Dans ce cas également la dissipation d'énergie est proportionnelle au rayon de la fibre.

Trichites et fibres de carbone constituent donc, pour la tenacité de mauvais éléments de renforcement. Dans l'élaboration on a intérêt à éviter des liaisons trop fortes entre matrice et fibre (ce qui augmente la longueur de transfert) et à utiliser des matrices métalliques ductiles et fortement écrouissables. Dans tous les cas un compromis doit être accepté entre résistance et résilience.

5. Conclusions

Les matériaux composites n'ont qu'un marché limité qui leur soit acquis dans combat. Un effort important est nécessaire pour conquérir de nouvelles utilisations.

Il est tout d'abord indispensable de les utiliser au mieux de leur possibilité. On ne peut substituer un composé à fibre au matériau isotrope sans reconcevoir les structures et les tests d'évaluation. Les pionniers ont des déboires qui coûtent parfois très chers, mais il faut saluer leur courage car si le banquier ne prête qu'aux riches (ce qu'il avoue) l'ingénieur ne fait confiance qu'à ce que d'autres ont éprouvé (ce qu'il n'avoue pas toujours).

Il importe de diminuer le prix de ces matériaux, c'est-à-dire d'abord celui des éléments de renforcement qui doivent pouvoir être produits en grande quantité, mais peut-être avec des caractéristiques moins extrêmes mais plus régulières. Des fibres légères et bon marché (SiC ou Al_2O_3) qui présenteraient des résistances comprises entre 150 et 200 HBar, seraient certainement appréciées.

Il faut enfin rechercher pour chaque utilisation le meilleur compromis entre les différentes caractéristi-

ques. L'observation de la nature montre qu'elle sacrifie toujours la résistance à la résilience. La même évolution se dessine dans l'utilisation des composés à fibres et les recherches sur les moyens d'améliorer la tenacité sont particulièrement importantes. Toutefois d'autres caractéristiques (résistance à la corrosion ou à l'érosion par exemple) ne doivent pas être négligées.

Les matériaux composés à fibres présentent des avantages certains et on peut être confiant en leur développement. On peut même penser qu'ils pourraient modifier certaines technologies en rendant possible la construction de structures rigides légères et de grande taille nécessaires par exemple pour capter l'énergie solaire ou éolienne.

Remerciements

Nous tenons à remercier les chercheurs du Centre des Matériaux de l'Ecole des Mines de Paris, car c'est des discussions fructueuses que nous avons pu avoir avec eux et grâce à l'aide matérielle précieuse qu'ils nous ont apportée et aux travaux qu'ils ont effectués et dont nous avons présenté certains résultats que cet article a pu être préparé.

Nous voudrions plus particulièrement exprimer nos remerciements à MM. Gilbert Charles, Pierre Combrade, Gérard Lesoult, Alain Robert et Michel Villamayor. Que MM. Brunetaud et Lescop du laboratoire de la Snecma trouvent ici l'expression de notre gratitude pour leur collaboration et la permission d'utiliser certains de leurs résultats.

Références

1) W. M. Powers, Boron and Graphite Fiber Market Competition, in: *26th Annual Technical Conf.*, 1971; Reinforced Plastics/Composites Division (The Society of the Plastics Industry, Inc.) section 86, p. 1.
2) R. Wetter, Kunststoffe **60** (1970) 756.
3) J. Fray, dans: *Intern. Conf. on Carbon Fibres, their Composites and Application*, ICCF (The Plastics Institute, London, 1971) paper 46.
4) M. Turpin, Bull. Soc. Franc. Crist. **92** (1969) 595.
5) D. V. Badami, New Scientist **5** (1970) 251.
6) D. J. Johnson, dans: ICCF (The Plastic Institute, London, 1971) paper 8.
7) J. V. Sharp et G. S. Burnay, dans: ICCF (The Plastics Institute, London, 1971) paper 10.
8) H. M. Ezekiel, Fibre Sci. Technol. **3** (1971) 243.
9) W. Watt et J. Green, dans: ICCF (The Plastics Institute, London, 1971) paper 4.
10) A. Robert et M. Turpin, dans: Congrès Génie Chimique, Paris, avril 1971.
11) Brevet No. 1586 467, Perfectionnement aux procédés et ap-

pareils pour obtenir des filaments continus de SiC. Compagnie Française Thomson, Houston.

12) G. Lubin, Ed., *Handbook of Fiberglass and Advanced Plastics Composites* (Van Nostrand, New York, 1969).

13) J. A. Alexander, R. G. Shaver et J. C. Withers, Critical analysis of accumulated experimental data on filament – reinforced metal matrix composites, NASA report N69-41247 (1969).

14) Society of Aerospace Material and Process Engineers SAMPE (1966).

15) R. A. Simon et F. R. Barnet, dans: *ICCF* (The Plastics Institute, London, 1971) paper 26.

16) B. Harris, D. W. R. Beaumont et A. Rosen, J. Mater. Sci. **4** (1969) 432.

17) S. P. Prosen, Fibre Sci. Technol. **3** (1970) 81.

18) M. D. Weisinger, Metal Matrix Composite structures Tooling and Fabrication processes (ASM, Society of Automotive Engineers) paper 700752.

19) General Technologies Corporation Newsletters, GTC Reston Virginia.

20) F. D. Lemkey, R. W. Hertsberg et J. A. Ford, Trans AIME **233** (1965) 234.

21) F. S. Galasso, J. Metals **19** (1967) 17.

22) H. Bibring, J. P. Trottier, M. Rabinovitch et G. Seibel, Mem. Sci. Revue Met. **68** (1971) 23.

23) A. Chadwick, Progr. Mater. Sci. **12** (1963) 97.

24) F. R. Mollard et M. C. Flemings, Trans. AIME **239** (1967) 1526.

25) H. Cline, Trans. AIME **239** (1967) 1489.

26) G. Pflieger, Thèse, Université de Grenoble, juillet 1970.

27) P. Combrade et M. Turpin, Solidification monovariante d'alliages ternaires, application à la fusion de zone, Congrès ICCG3, Marseille, 1971.

28) Composite Materials: Testing and Design ASTM-STP 460 (1969).

29) V. R. Riley et J. L. Reddaway, J. Mater. Sci. **3** (1968) 41.

30) M. Sindzingre, Compt. Rend. (Paris) C **271** (1970) 686.

31) A. J. Baker, dans: *ICCF* (The Plastics Industry, London (1971) paper 20.

32) B. Harris, P. Beaumont et E. Moncunill de Ferran, J. Mater. Sci. **6** (1971) 238.

33) G. A. Cooper et A. Kelly, J. Mech. Phys. Solids **15** (1967) 279.

34) G. A. Cooper et A. Kelly, Role of the Interface in the fracture of Fibre-Composite Materials, in: *Interfaces in Composites*, ASM STP 452 (1969) p. 20.

35) A. Kelly, Proc. Roy. Soc. (London) A **319** (1970) 95.

36) A. W. H. Morris et R. S. Smith, Fibre Science **3** (1971) 219.

37) G. R. Sidey et F. J. Bradshaw, dans: *ICCF* (The Plastics Industry, London, 1971) p. 25.

Journal of Crystal Growth **13/14** (1972) 106–112 © *North-Holland Publishing Co.*

APPLICATION OF HEAT PIPE TECHNOLOGY TO CRYSTAL GROWTH*

JACQUES STEININGER** and THOMAS B. REED

Lincoln Laboratory, Massachusetts Institute of Technology, Lexington, Massachusetts 02173, U.S.A.

Heat pipes are highly efficient heat exchangers with effective thermal conductivities several orders of magnitude higher than those of the best metallic conductors. The principle, design and construction of heat pipes are discussed together with specific examples of applications to crystal growth for heat transfer and to obtain isothermal and multizone temperature profiles.

1. Introduction

Heat pipes are highly efficient heat exchangers that have recently been developed for specialized heat transfer applications. Much of the fundamental research and development has been conducted at the Los Alamos Scientific Laboratory of the University of California, following the rediscovery[1]) of a principle originally investigated in the automotive industry. The Los Alamos workers have been primarily interested in heat transfer for nuclear power generating systems in spacecraft. Numerous other applications have since been devised for heat pipes. Thermal applications[2,3]) exploit their high thermal conductance, efficient heat flux transformation capability and potential for isothermal operation. In addition, the well-defined isothermal, isobaric vapor zones inside heat pipes have also been utilized in physical measurements, such as vapor pressure studies on corrosive elements[4]) and vapor phase spectroscopy[5]).

Because controlled transfer of heat is one of the primary principles underlying most techniques of crystal growth, heat pipes offer numerous possibilities of application in this field for heat dissipation, isothermal or multizone temperature profiles, and material containment. This paper presents a brief review of heat pipe principle and performance and a discussion of various applications to crystal growth. The results ob-

tained with several experimental heat pipe devices will be described.

2. Principle and characteristics of heat pipes

A heat pipe is essentially a closed vessel lined internally with a wick saturated with a volatile liquid (fig. 1). Heat supplied by an external heat source causes evaporation of the liquid in the high temperature region of the pipe. Due to the pressure differential between hot and cold regions, the vapor is transported toward the colder regions of the pipe and condenses, releasing the absorbed heat of vaporization. The condensed liquid is then recycled to the high temperature evaporator by capillary action in the wick, assisted in some cases by gravity.

The theory, construction and performance of heat pipes have been extensively studied in the last few years[6–10]). The axial rate of heat transfer in a heat pipe is essentially equal[6]) to the latent heat flux in the evaporator:

$$Q = \dot{m}_v L , \tag{1}$$

where Q is heat transfer rate (cal/sec), $\dot{m}_v$ vapor mass flow rate at evaporator exit (g/sec), and L latent heat of vaporization of the liquid (cal/g).

For a tubular pipe[7]), eq. (1) becomes:

$$Q = A_v \rho \bar{v} L, \tag{2}$$

where A_v is vapor passage cross sectional area (cm^2), ρ vapor density (g/cm^3), and $\bar{v}$ average vapor velocity (cm/sec).

* This work was supported by the Department of the Air Force.
** Present address: Arthur D. Little, Inc., Cambridge, Mass. 02140, U.S.A.

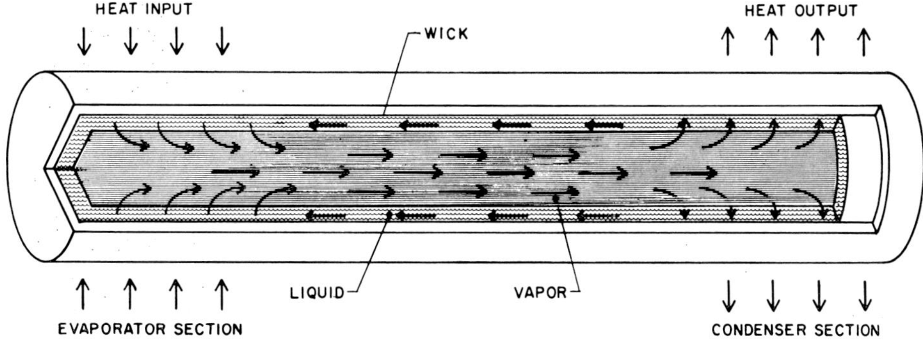

Fig. 1. Heat and fluid flows in heat pipe [from Eastman[2]].

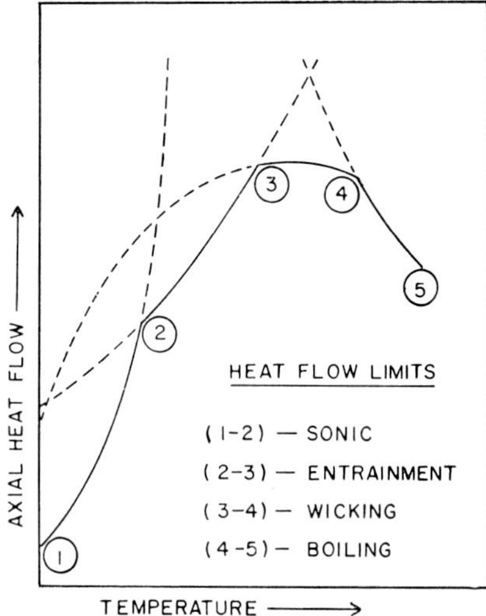

Fig. 2. Theoretical limitations on heat flow [from Kemme[7]].

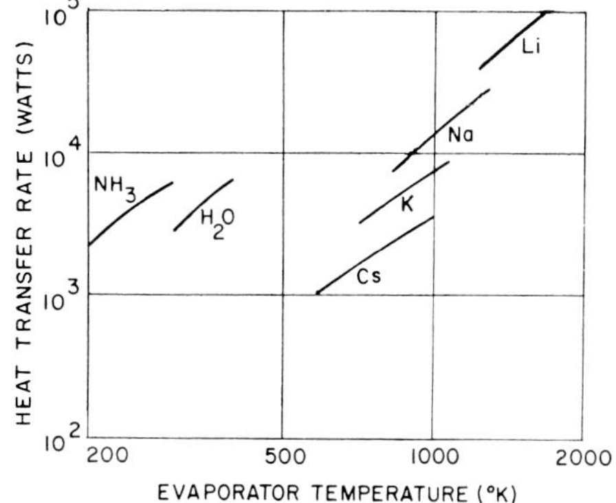

Fig. 3. Maximum heat transfer rates for different working fluids [from Feldman and Whiting[3]].

Because of the high latent heats of vaporization of liquids, effective thermal conductivities several orders of magnitude higher than those of even the best solid or liquid conductors, such as copper or silver, can be obtained with properly constructed heat pipes. There are several limitations on heat transfer however, which are governed respectively (fig. 2) by the sonic velocity of the vapor, by the entrainment of liquid in the wick by the vapor, by the gravity and viscosity forces in the wick, and by boiling of the liquid inside the wick[7]. Of these four, the limitations inherent in the wicking action are usually the most significant. In the absence of gravity, the maximum heat transfer rate limited by the maximum flow rate of the condensed liquid through the wick can be expressed for a tubular pipe[3] by:

$$Q = \sigma A_{\mathrm{w}} DL/bvl. \tag{3}$$

where σ is surface tension of the liquid (dynes/cm), A_{w} free flow area in the wick (cm^2), D wick pore size (cm), v kinematic viscosity (cm^2/sec), b capillary geometric constant (dimensionless), of the order of 20 for wire mesh capillary, and l pipe length (cm).

Fig. 3 shows the maximum theoretical heat transfer rates[3] for a tubular heat pipe 2.5 cm in diameter, 30 cm long, with a woven mesh wick saturated with different liquids. For sodium operating near 900 °C, maximum heat transfer rates of over 10 000 W are indicated, while for lithium at 1400 °C they reach 100 000 W.

From a thermodynamic point of view, evaporation and condensation of a pure liquid constitute a monovariant, reversible phase transformation. Heat pipes therefore tend to assume nearly isothermal temperature profiles. In practice, a small pressure drop is needed to maintain the flow of vapor from the evaporator to the condenser. This pressure drop however can be as small as 1 torr and the temperature gradient along the axis of the pipe can therefore be quite small.

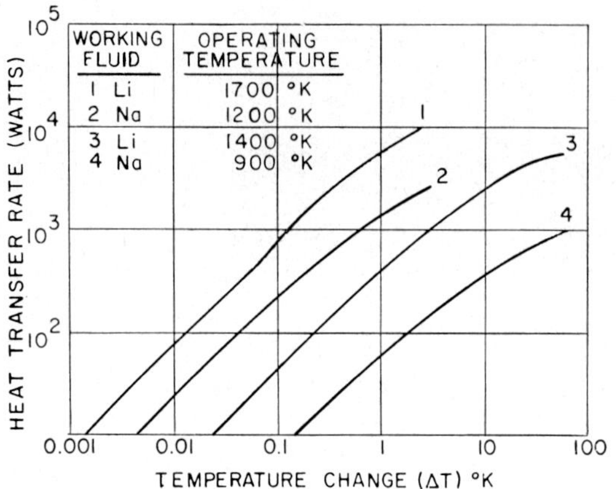

Fig. 4. Minimum temperature gradients for different working fluids [from Feldman and Whiting[3])].

Fig. 4 shows the theoretical temperature gradients[3]) for the same pipe as in fig. 3 as a function of heat transfer rate for different liquids. For sodium at 900 °C, a temperature gradient of 0.01 °C/cm can transfer nearly 2500 W.

It should be noted that heat pipes can be operated in two modes: constant volume and constant pressure. The usual sealed pipes operate at constant volume and variations in input power are matched by variations in internal vapor pressure and temperature. It is also possible however to operate an open pipe where the vapor is contained by an inert gas pressure[4,5]). In this case, variations in power input are matched by variations in the length of the vapor zone and the pipe operates with an isothermal zone of fixed temperature but variable length.

TABLE 1

Heat pipe materials and working fluids

Working fluid	Pipe material	Temperature (°C)	Lifetime (hr)	Ref.
Acetone	Stainless steel	80	>14000	a
Water	Copper	120	>20000	a
Fluorinated hydrocarbons	Copper	240	N.D.	b
Mercury	Stainless steel	350	>28000	a
Potassium	Nickel	600	>35000	a
Sodium	Nickel	900	N.D.	b
	Stainless steel	780	10000	a
	Hastelloy	750	>30000	a
Zinc	Quartz	1100	N.D.	b
Lithium	TZM alloy	1500	10000	a
	Tungsten	>1500	N.D.	c

(a) ref. 8; (b) this study; (c) ref. 2.
N.D.: no data available.

Heat pipes of various designs and materials can be made to fit particular applications. Operating temperatures ranging from below 0 °C to above 2000 °C have been achieved[8-10]). The working liquids can have vapor pressures at their operating temperatures ranging from a few torr to several atmospheres. The selection of container and wick materials is subject to the requirement that the liquid must wet but not react with the materials. Table 1 shows some combinations of materials and liquids that have been operated in various temperature ranges.

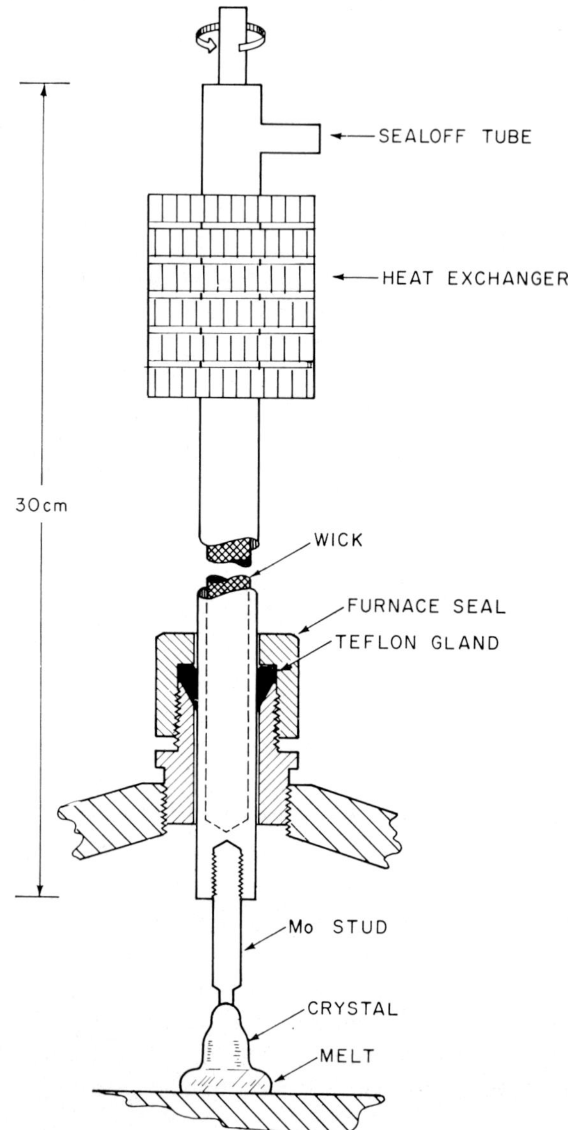

Fig. 5. Heat pipe seed rod for tri-arc crystal puller.

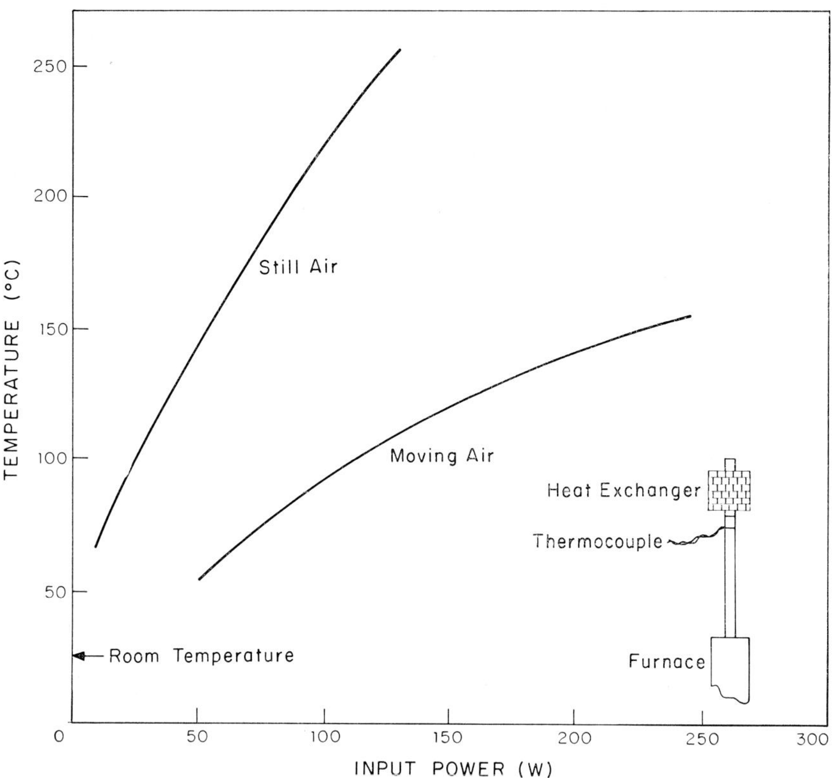

Fig. 6. Heat transfer in heat pipe seed rod.

3. Applications to crystal growth

3.1. HEAT TRANSFER

The initial interest in heat pipes was due to their capacity for achieving very high rates of heat transfer, particularly in confined environments. In a direct application of this property, we have used a commercially available low temperature heat pipe to construct a seed rod for a tri-arc crystal puller. In the original tri-arc furnace[11]), the cooling of the rod was accomplished by circulating water through it, but this was inconvenient because it required a controlled flow of water and the use of rotating seals which tended to leak. The new seed rod is made from a copper heat pipe filled with a fluorinated hydrocarbon with a maximum operating temperature of 240 °C. As shown in fig. 5, the heat liberated at the cold end of the rod outside the furnace is transferred to the room by means of a finned heat exchanger. Because of the relatively high operating temperature, a teflon seal is used at the furnace gland.

Before installing the seed rod in the furnace, its heat transfer properties were measured in still and moving air. Fig. 6 shows the variation of temperature at a point just below the finned heat exchanger as a function of power input to the lower end of the pipe. In moving air, the highest available power input (250 W) gave a temperature at this point of only about 150 °C, and extrapolation of the experimental curve indicates that at 240 °C the power dissipated would be about 600 W, probably more than required for any normal crystal growth experiment. At a power input of 250 W, the temperature drop along 10 cm of the heat pipe was only 9 °C. Under the same conditions, the temperature drop along a solid copper cylinder of the same diameter would be 480 °C.

The new seed rod has been used to grow a number of single crystals of VO, V_2O_3, TiO, FeO, MnO, Gd and Ni with melting temperatures ranging up to 2075 °C for VO. Very high temperature gradients were therefore induced in the short (1 cm) Mo seed holder between crystal and heat pipe.

3.2. Isothermal and multizone temperature profiles

Annular heat pipes can be used as liners, or muffles, inside furnaces to obtain isothermal or multizone temperature profiles suitable for material preparation, crystal growth or annealing. With several heat pipes in series, one can obtain well-defined short zones with high temperature gradients for crystallization without constitutional supercooling, long zones with smooth low temperature gradients for annealing, or narrow high temperature zones suitable for zone melting or solution zoning. Even if the heat source is not uniform, the region inside each heat pipe can be quite isothermal. Thermal instability caused by power transients in the

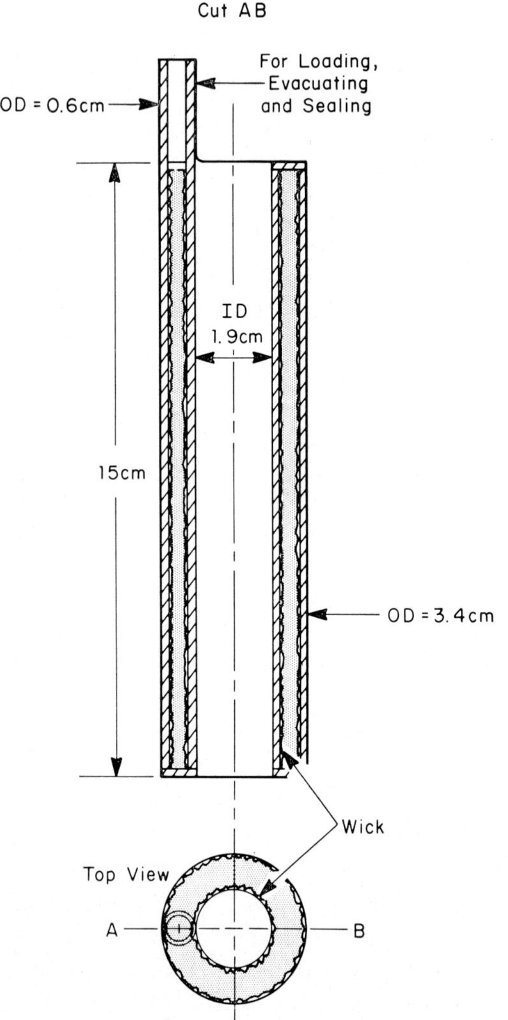

Fig. 7.   Annular crystal growth heat pipe.

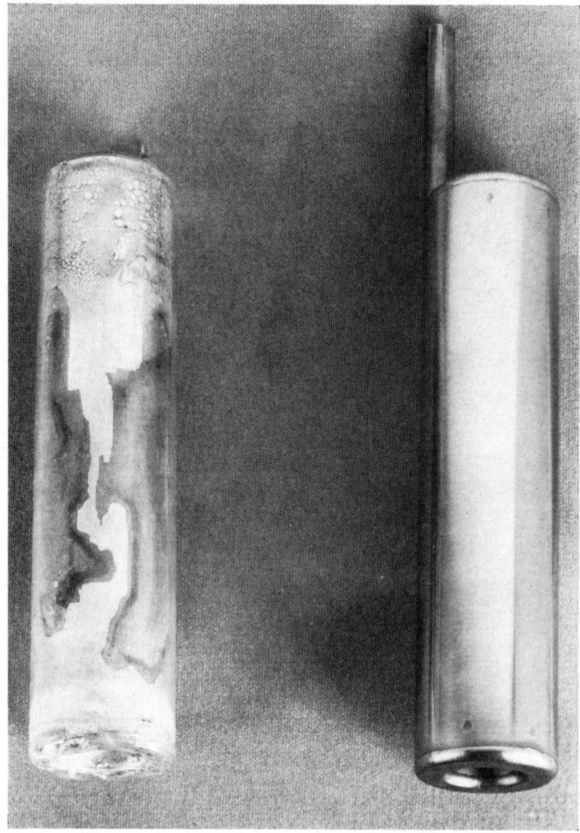

Fig. 8.   Experimental annular heat pipes. Left: quartz-(zinc) pipe. Right: nickel-(sodium) pipe.

heat source can also be much reduced inside the heat pipe.

Several annular heat pipes have been constructed to evaluate their characteristics in different furnace configurations. The basic design shown in fig. 7 consists of a double-walled pipe, 3.4 cm in outside diameter, 1.9 cm in inside diameter and 15 cm long. The inside and outside walls are lined with wicks made up of several layers of fine mesh screen. Fig. 8 shows two of the experimental heat pipes, a quartz-(zinc) pipe and a nickel-(sodium) pipe.

Quartz was selected for the construction of the first experimental annular pipes because of its workability and transparency which made observations possible. The wicks were made of quartz woven cloth and the working liquid was either zinc or cadmium. Fig. 9 shows two temperature profiles measured along the axis of a quartz-(zinc) pipe placed inside a vertical resistance heated furnace, together with the two conventional profiles, obtained without the heat pipe. The

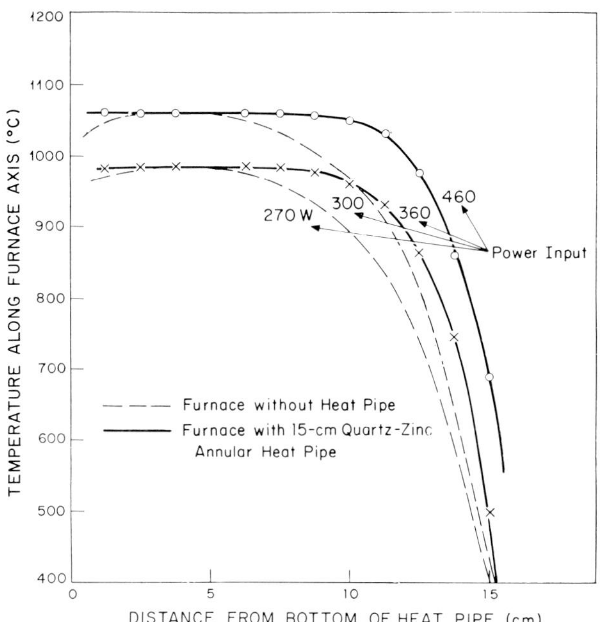

Fig. 9. Temperature profiles inside quartz-(zinc) annular heat pipe.

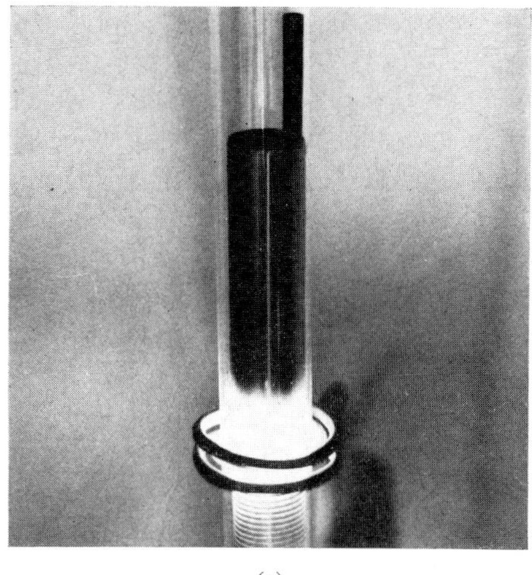

(a)

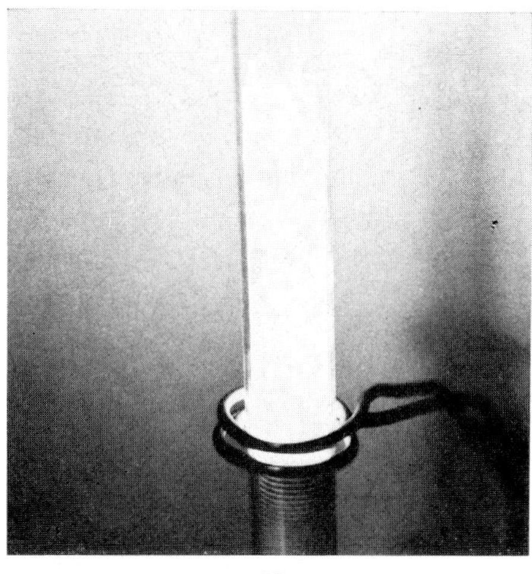

(b)

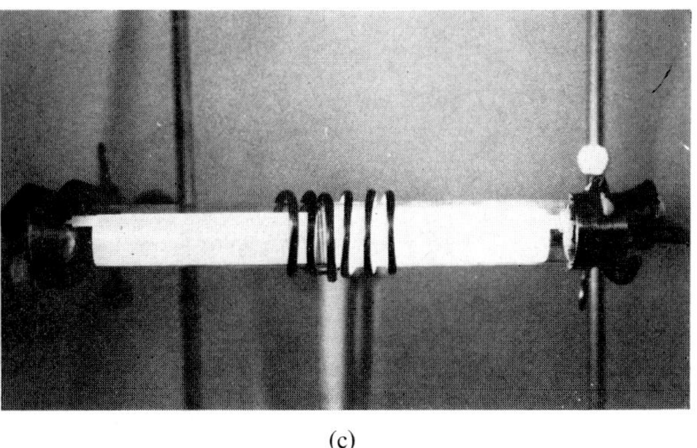

(c)

Fig. 10. Nickel-(sodium) heat pipes at operating temperatures: (a) dummy pipe without sodium; (b) single heat pipe in vertical position; (c) system of two horizontal heat pipes at two different temperatures.

effect of the heat pipe is to raise the temperature profile in the lower part of the furnace to extremely uniform temperatures over several centimeters. The higher temperature profile remained within 0.1 °C of 1065.0 °C for a distance of 8 cm, as measured with a Pt–Pt, 13 % Rh thermocouple inserted along the furnace axis. With this pipe, however, it was not possible to obtain a completely uniform profile, even at the higher power levels. There was always a cold zone at the upper part of the pipe which was attributed to the presence of non-condensable residual gases inside the pipe. As observed by previous investigators[1]), these gases are swept into the colder section of the pipe by the pumping action of the vapor and form a low conductivity vapor lock.

The principal disadvantage of quartz heat pipes is their tendency to break during cooling due to solidification of the liquid metals. Quartz-(mercury) pipes however can be used in a lower temperature range without damage.

Sturdier heat pipes with the same basic design were subsequently constructed of nickel. The wicks were made of several layers of 150×150 mesh nickel wire screen. The working liquid was high purity sodium which was loaded into the pipe by vacuum distillation. The feed tube on the top of the pipe was then pinched and electron beam welded under high vacuum.

The nickel-(sodium) pipes were tested under stringent conditions using RF heating to supply highly localized sources of heat input. Fig. 10a shows a dummy pipe of similar geometry but without working fluid, being tested under the same conditions. The triple loop RF coil brings the temperature at the bottom of the pipe to a cherry red color corresponding to about 850 °C, but the rest of the pipe remains cold. In fig. 10b, a nickel-(sodium) pipe is in operation with the same source of RF heating. This pipe shows a remarkably uniform coloration in the 850 °C temperature range. It should be noted that this uniform coloration extends to the top of the pipe, which indicates that residual gases are practically absent. The very narrow upper tip of the pipe is solid nickel and remains colder because the sodium vapor cannot penetrate it. The longitudinal heat flux in the evaporator section of this pipe was calculated to be in excess of 500 W/cm^2. Fig. 10c shows a combination of two pipes held at temperatures of 850 °C and 750 °C and separated by an alundum spacer 0.8 cm thick. The temperature gradient was less than 1 °C/cm along each pipe but at the junction it was estimated to reach as high as 150 °C/cm.

These pipes are now being studied in the preparation and crystal growth of different types of ternary semiconductor alloys[12]).

3.3. OTHER APPLICATIONS

Heat pipe applications in materials preparation and crystal growth are not limited to the conventional tubular or annular configurations described above. Small but uniform temperature gradients over relatively large areas for epitaxial growth of homogeneous films are being investigated either in off-axis annular heat pipes or in sandwich configuration systems with two flat heaters maintained at slightly different temperatures.

The efficient pumping action and segregation of vapor species in heat pipes[4,5]) can also be used for the containment of vapors of volatile materials.

4. Conclusion

A recently published review article[13]) shows that the number of applications suggested for the exceptional heat transfer characteristics of heat pipes seems to be growing continuously. It is also believed that many of these new applications will be in the field of crystal growth.

Acknowledgements

The authors would like to acknowledge the valuable assistance of A. R. Leyenaar in the fabrication of heat pipes and of R. E. England and R. E. Fahey in their evaluation and testing.

References

1) G. M. Grover, T. P. Cotter and G. F. Erickson, J. Appl. Phys. **35** (1964) 1990.
2) G. Y. Eastman, Sci. Am. **218** (5) (1968) 38.
3) K. T. Feldman and G. H. Whiting, Mech. Eng. **89** (2) (1967) 30.
4) J. Bohdansky and H. E. J. Schins, J. Appl. Phys. **36** (1965) 3683.
5) C. R. Vidal and J. Cooper, J. Appl. Phys. **40** (1969) 3370.
6) T. P. Cotter, Los Alamos Sci. Lab., Los Alamos, N. Mex., Rept. LA-3246-MS (1965).
7) J. E. Kemme, IEEE Trans. Electron Devices **ED-16** (1969) 717.
8) W. E. Harbaugh and G. Y. Eastman, Heating, Piping and Air Cond. **42** (10) (1970) 92.
9) J. E. Deverall, Los Alamos Sci. Lab., Los Alamos, N. Mex., Rept. LA-4300-MS (1970).
10) F. J. Lavoie, Mach. Design **42** (19) (1970) 86.
11) T. B. Reed and E. R. Pollard, J. Crystal Growth **2** (1968) 243.
12) J. Steininger and A. J. Strauss, J. Crystal Growth **13/14** (1972) 657.
13) E. R. F. Winter and W. O. Barsch, in: *Advances in Heat Transfer*, Vol. 7, Eds. T. F. Irvine, Jr. and J. P. Hartnott (Academic Press, New York, 1971).

Journal of Crystal Growth 13/14 (1972) 113–120 © *North-Holland Publishing Co.*

SHAPE STABILITY OF A TWO-DIMENSIONAL NUCLEUS

M. AVIGNON

Laboratoire d'Electrostatique et de Physique du Métal, Section Physique des Couches Minces, C.N.R.S., Cedex 166, 38 – Grenoble-Gare, France

Stability of the lateral interface of a disk shaped nucleus is reviewed when such a nucleus is growing parallel to the basal plane by surface diffusion of ad-atoms and their eventual attachment at the interface. Two physically distinct systems have been considered: (a) a conservative one in which the nucleus grows without addition or removal of matter from the system and (b) a non-conservative one, in which the nucleus is growing in the presence of a vapour source. The system is idealised by considering a single nucleus surrounded by an infinite matrix of ad-atoms and quantities like edge-free energy, interface attachment coefficient as isotropic.

Behaviour of a small fluctuation represented by circular harmonics has been determined in the usual way of Mullins and Sekerka, by solving the steady state diffusion equation (Laplace equation for conservative case, Helmholtz equation for non-conservative system) with appropriate boundary conditions.

For the conservative system, the results are similar to the 3-dimensional case, a perturbation will grow as soon as the nucleus size is greater than a limiting value. The effect of non infinite interface attachment coefficient is to push this radius at which instability sets in at a higher value. For the non-conservative system the situation is radically different: the instability is governed by the parameter R^*/x_s (R^* critical nucleation radius, x_s mean diffusion distance) and the interface attachment coefficient β. When the interface attachment mechanism is too slow (β smaller than a critical value β_c) the nucleus is stable, but when it becomes sufficiently rapid ($\beta > \beta_c$) instability may occur provided that R^*/x_s is smaller than a critical value $[R^*/x_s]_c$; in this case amplification of the perturbation is possible only when the size of the nucleus lies between two limiting values. The interface attachment mechanism appears to be an important stabilizing factor.

The influence of temperature and impurity concentration at the interface on stability are discussed. In connection with experimental evidence, a model is developed for the instability of thin discs of ice crystals growing on the surface of slightly undercooled water.

1. Introduction

Since the "classic" work of Mullins and Sekerka[1], shape stability – instability of the interface between two phases during the growth of the new phase from the parent one – has attracted the attention of many workers[2,3], mainly for 3-dimensional systems. Although not a rare occurrence, morphological stability of a 2-dimensional nucleus has received comparatively little interest.

We have undertaken a study of the stability of the lateral interface of a disc-shaped nucleus on a substrate growing from the vapour by surface diffusion[4] and interface attachment of ad-atoms[5].

However, previous to the theoretical study, the instability of a disc-shaped nucleus of ice growing from undercooled water had been observed by Arakawa[6] in a laboratory experiment, and by Schaefer[7] in the formation of "frazil ice". These observations show clearly how the dendrites start and then develop into 6-sided star shapes; they may ultimately form beautiful "ice flowers" at the surface of freezing water or under a sheet of ice[8]. Glen[9] has suggested that interface instability is the precurser of dendrite development in a comment on Arakawa's paper. This sort of instability is directly amenable to our analysis of a two-dimensional system.

For a 3-dimensional system, the growth of the particle occurs by diffusion or heat flow obtainable, in the steady state approximation, from the solution of the Laplace equation. In the 2-dimensional case, the main difference is that the system, now in a plane, may exchange material or heat with the infinite medium above it. This leads to a non-conservative system governed in the linear exchange approximation by an Helmholtz equation.

2. Statement of the problem: boundary conditions

The purpose of morphological stability analysis is to show that "shape preserving solutions" of the diffusion equation are not necessarily kinetically stable shapes. This means that a small perturbation of the interface shape may be amplified during growth. Evidently, the boundary condition at the interface must play a predominant role in the analysis. When the boundary is slightly deformed, the boundary condition is also modified; this will lead in general to one of the more difficult boundary value problems[10]). Fortunately, to test the instability of an interface a first order perturbation, i.e. a linear analysis, is sufficient; but in this case, only initial behaviour of the perturbation may be predicted and nothing or at least little can be said about its subsequent development.

The correspondence between the growth in a diffusion field and in a thermal field is well known[11]), so we shall state the boundary conditions for the former case. Let us consider a nucleus of height h (h being small with respect to the lateral dimensions) (fig. 1) and represent the interface by an equation $r = \rho(\theta)$ in plane polar coordinates (r, θ). The number of ad-atoms per unit area is $n(r, \theta)$.

When perturbed, the equation of the interface becomes

$$r = \hat{\rho}(\theta) = \rho(\theta) + \delta(\theta), \quad \delta(\theta) \ll \rho(\theta).$$

The nucleus is growing by surface diffusion of atoms of effective volume Ω and their eventual attachment at the nucleus. It is the competition between these two processes – surface diffusion and attachment at a nucleus – which fixes the concentration of ad-atoms at the interface n_I, so that the equation of mass conservation may be written as follows:

$$V_\mathrm{n} = \frac{\Omega D_\mathrm{s}}{h}(\nabla_\mathrm{N} n)_\mathrm{I} = \beta\Omega(n_\mathrm{I} - n_\mathrm{K}),$$

$$V_\mathrm{n} = \frac{\partial\hat{\rho}}{\partial t}\frac{\hat{\rho}}{(\hat{\rho}^2 + \hat{\rho}'^2)^{\frac{1}{2}}},$$

               (1)

V_n being the normal rate of growth of the interface. Here

$$j = D_\mathrm{s}(\nabla_\mathrm{N} n)_\mathrm{I}$$

is the normal flux of ad-atoms at the interface, where

$$(\nabla_\mathrm{N} n)_\mathrm{I} = \left(\frac{\partial n}{\partial r} - \frac{\hat{\rho}'}{\hat{\rho}}\frac{1}{r}\frac{\partial n}{\partial\theta}\right)_{r=\hat{\rho}}\frac{\hat{\rho}}{(\hat{\rho}^2 + \hat{\rho}'^2)^{\frac{1}{2}}}.$$

D_s is the surface diffusion coefficient of the ad-atoms on the surface, n_K is the equilibrium concentration around an interface element of curvature K, as given by the Gibbs–Thomson relation[13]):

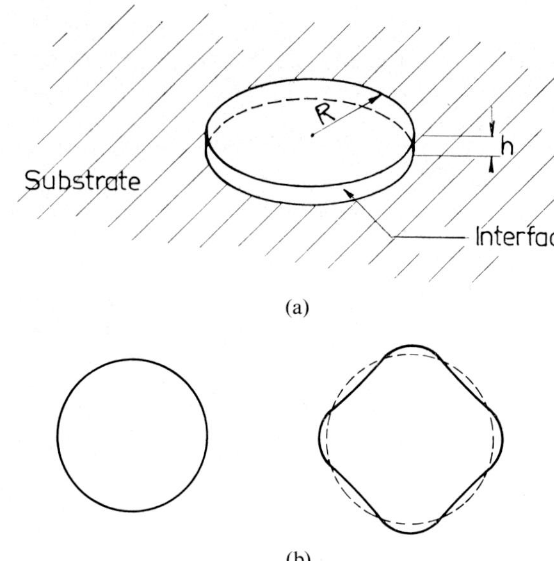

Fig. 1. (a) The disc-shaped nucleus on a substrate. (b) Plane view of the nucleus unperturbed (left and perturbed by harmonic $k = 4$ (right).

$$n_\mathrm{K} = n_\mathrm{eq}(1 + \Gamma K),$$

n_eq being the equilibrium concentration of a flat interface, $\Gamma = \gamma\Omega/k_\mathrm{B}T$ the capillarity constant for an edge-free energy γ and temperature T (k_B the Boltzmann constant), and

$$K = \frac{\hat{\rho}^2 + 2\hat{\rho}'^2 - \hat{\rho}\hat{\rho}''}{(\hat{\rho}^2 + \hat{\rho}'^2)^{\frac{3}{2}}}.$$

For a circular unperturbed nucleus,

$$\hat{\rho} = R + \delta(\theta).$$

β is the interface attachment coefficient[13,14]) which characterizes the rate of exchange of atoms through the interface. When β is infinite, the growth is controlled by diffusion and the condition of local equilibrium $n = n_\mathrm{K}$ is obtained.

The edge-free energy γ and the interface attachment coefficient β are supposed to be isotropic; the effect of slight anisotropy is to provide an initial perturbation that will be enhanced by the diffusion field[15]). A second boundary condition is necessary to determine completely the diffusion field. If the nucleus is considered to be isolated or does not interact with the surrounding nuclei, we may consider that at a certain distance from the nucleus the concentration has a given value. From the exact expression of the normal gradient $\nabla_\mathrm{N} n$, and of the curvature K of an arbitrary interface, we immediately realize the difficulty of the problem since an *exact* solution of the diffusion equation must be obtained for the unperturbed boundary in order to have a realistic perturbation result. This limits the interface geometries for which analytic treatment is tractable without too much difficulty.

We have taken a circular geometry $\rho(\theta) = R$, which is also the shape of the minimum energy for isotropic γ, and then to the first order in the perturbation $\delta(\theta)$,

$$V_{\mathrm{n}} = \partial\rho/\partial t = \dot{R} + \dot{\delta},$$

$$(V_{\mathrm{N}}n)_{\mathrm{l}} = \left(\frac{\partial n}{\partial r}\right)_{r=\rho},$$

$$K = \frac{1}{R} - \frac{\delta(\theta) + \delta''(\theta)}{R^2},$$

where the dot denotes the derivative with respect to time, while the prime denotes derivatives with respect to the angle θ.

An arbitrary perturbation may be Fourier-analyzed in circular harmonics $e^{ik\theta}$, and because of the linear approximation all harmonics will superpose. The study of a single harmonic $\delta_k\, e^{ik\theta}$ is thus appropriate:

$$V_{\mathrm{n}} = \dot{R} + \dot{\delta}_k\, e^{ik\theta},$$

and

$$K = \frac{1}{R} + \frac{(k^2-1)}{R^2}\,\delta_k\, e^{ik\theta}.$$

The rate of growth or decay of the perturbation $\dot{\delta}_k$ is found to be proportional to the perturbation itself, δ_k, so that one initial perturbation is necessary to trigger the instability.

When $\dot{\delta}_k/\delta_k > 0$, the perturbation will grow and the interface is unstable, but only "relatively unstable" until the perturbation grows faster than the nucleus, i.e.

$$\frac{\dot{\delta}_k/\delta_k}{\dot{R}/R} > 1;$$

this latter condition is the condition of "absolute instability"[†].

† Even when the perturbation may grow, if the nucleus radius R grows still faster then the infinitesimal perturbation remains infinitesimal, and so does not produce any change in shape. The nucleus shape is really unstable only then the perturbation grows faster than R [i.e., $(\dot{\delta}/\delta)/(\dot{R}/R) > 1$], thus we have called this condition the condition of "absolute instability" of the nucleus shape.

3. Vapor growth

At low supersaturation, we may assume that dynamic equilibrium is realized and the steady state is valid.

3.1. CONSERVATIVE SYSTEM

The nucleus is growing on a substrate without transfer of matter from the system (the system being the nucleus and the surrounding ad-atoms). The problem is quite similar to the right-circular cylinder[16]), without perturbation along the z-direction, and only the results will be given.

The concentration of ad-atoms is the solution of the two-dimensional Laplace equation $\nabla^2 n = 0$, with the boundary condition (1) and $n = \bar{n}$, at a distance $R' = lR$, l being a constant, so that

$$\dot{R} = \frac{D_{\mathrm{s}}\Omega(\bar{n} - n_{\mathrm{eq}})}{h\,(\ln l + D_{\mathrm{s}}/h\beta R)}\,\frac{1}{R}\left(1 - \frac{R^*}{R}\right), \qquad (2)$$

where

$$R^* = n_{\mathrm{eq}}\Gamma/(\bar{n} - n_{\mathrm{eq}})$$

is the critical nucleation radius, and

$$\frac{\dot{\delta}_k}{\delta_k} = \frac{D_{\mathrm{s}}\Omega(\bar{n} - n_{\mathrm{eq}})\,(k-1)}{hR^2(1 + kD_{\mathrm{s}}/h\beta R)\,(\ln l + D_{\mathrm{s}}/h\beta R)}$$
$$\times \left\{1 - \frac{R^*}{R}\left[1 + k(k+1)\left(\ln l + \frac{D_{\mathrm{s}}}{h\beta R}\right)\right]\right\}. \qquad (3)$$

This result is standard. The perturbation of order $k \geq 2$[†] may grow as soon as a critical radius $R_{\mathrm{c}}(k)$ is exceeded (fig. 2). For absolute instability a higher value $R_{\mathrm{c}}(k)$ must be reached. Compared to the diffusion controlled growth ($\beta \to \infty$), the effect of interface attachment kinetics is simply to increase the value at which instability sets in.

3.2. NON-CONSERVATIVE SYSTEM

This system represents the particularity of the two-dimensional analysis and completely different behavior must be emphasized.

The nucleus is now growing in the presence of a vapor source, and there is a continuous exchange of matter between the substrate and the vapor. There is the possibility of having ad-atoms on the nucleus, for which the mean diffusion distance x_{s} is different from that on the

† To the first order δ_1 simply translates the nucleus, $\delta_1 = 0$.

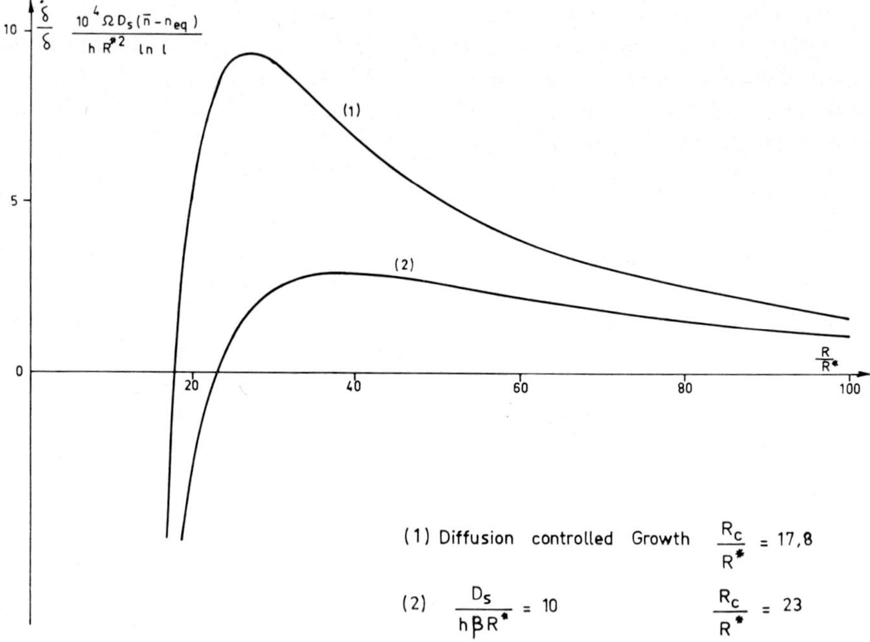

Fig. 2. Amplification rate versus R/R^* for the conservative case. (1) Diffusion controlled growth. (2) $2D_s/hx_s\beta = 10$.

substrate, and two different equations are needed. To simplify, we take a substrate of the same nature as the nucleus, i.e. the case of a singular crystal surface grown by "two-dimensional nucleation". The ad-atom concentration is thus governed by the Burton, Cabrera and Frank equation of the Helmholtz type[14]

$$x_s^2 \nabla^2 \Psi = \Psi,$$

where

$$\Psi = \frac{p}{p_{eq}} - \frac{n}{n_{eq}},$$

p is the actual vapour pressure, p_{eq} the equilibrium vapour pressure. Solutions to the Helmholtz equation are

$$\Psi_1(r) = A_0 K_0\left(\frac{r}{x_s}\right) + A_k K_k\left(\frac{r}{x_s}\right) e^{ik\theta}, \quad r > \rho, \quad (4a)$$

$$\Psi_2(r) = B_0 I_0\left(\frac{r}{x_s}\right) + B_k I_k\left(\frac{r}{x_s}\right) e^{ik\theta}, \quad r < \rho, \quad (4b)$$

where the I and K functions are modified Bessel functions.

The condition $n = n_{eq}p/p_{eq}$ is satisfied at infinity. Two fluxes are contributing to the growth, one from

the inside and one from the outside; hence

$$j = D_s n_{eq}\left(\frac{\partial \Psi_2}{\partial r} - \frac{\partial \Psi_1}{\partial r}\right)_{r=\rho}.$$

After some algebra, using relations among Bessel functions[17], the rate of growth of the nucleus is found to be:

$$\frac{\dot{R}}{R} = \frac{\Omega D_s n_{eq}(\alpha - 1)}{hx_s^2(1 + \beta_0/\beta)}\left(1 - \frac{x^*}{x}\right) y_1(x), \quad (5)$$

with

$$\alpha = \frac{p}{p_{eq}}, \quad x = \frac{R}{x_s}, \quad x^* = \frac{R^*}{x_s}, \quad \left(R^* = \frac{\Gamma}{\alpha - 1}\right),$$

$$y_1(x) = 1/x^2 I_0(x) K_0(x),$$

$$\beta_j = \frac{D_s}{hx_s}[xI_j(x)K_j(x)]^{-1},$$

and for the perturbation

$$\frac{\dot{\delta}_k}{\delta_k} = \frac{\Omega D_s n_{eq}(\alpha - 1)}{hx_s^2(1 + \beta_k/\beta)(1 + \beta_0/\beta)}$$

$$\times \left[\left(1 - \frac{x^*}{x}\right) y_2(k, x) - x^*\left(1 + \frac{\beta_0}{\beta}\right) y_3(k, x)\right], \quad (6)$$

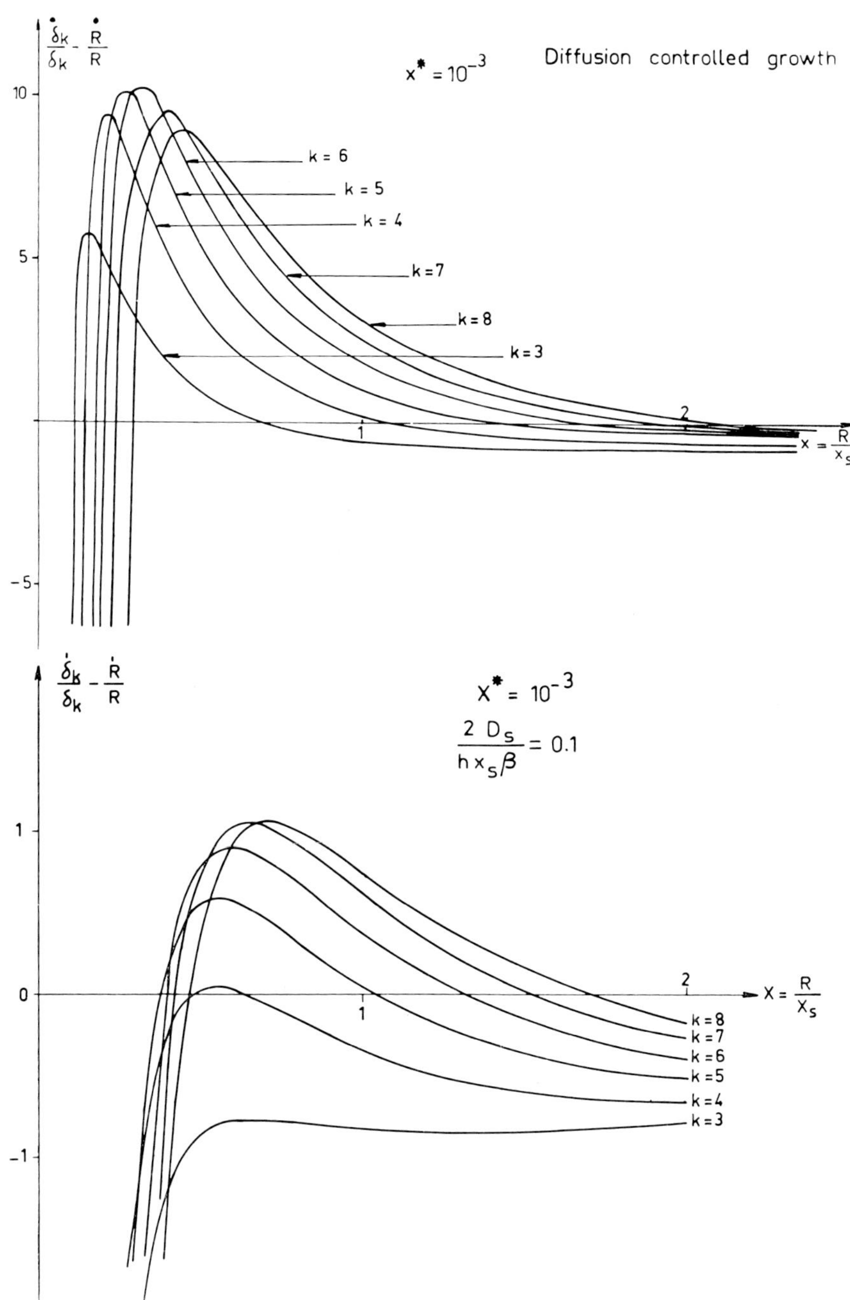

Fig. 3. Absolute amplification rate versus $x = R/x_s$ for the non-conservative case for diffusion controlled growth (upper curves and $2D_s/hx_s\beta = 0.1$ (lower curves).

with

$$y_2(k, x) = \frac{K_1(x)K_{k+1}(x)}{K_0(x)K_k(x)}$$
$$- \frac{I_1(x)I_{k+1}(x)}{I_0(x)I_k(x)} - (k+1)y_1(x),$$

$$y_3(k, x) = (k^2 - 1)/x^3 I_k(x)K_k(x).$$

We first note that the prevailing factor for the instability is no longer the nucleus size R, but the ratio R/x_s. We shall discard the condition of relative instability and consider only the absolute instability because their behaviour is similar[5]) and is represented in fig. 3, for different harmonics ($k = 3$–8), for diffusion controlled growth (β infinite) and $2D_s/hx_s\beta = 0.1$.

We notice that now there exist two threshold values x_{c1} and x_{c2}, between which the perturbation may grow. Outside these values the nucleus is stable. The lower value x_{c1} results, as usual, from the competition between the capillarity which tends to conserve the circular shape and the flux of matter which is more important at a bump than at a trough. But unlike the "conservative" case, the growth front is fed from both sides (internal area as well as outer area of the nucleus); when R is sufficiently important, the internal feed is nearly as important as the external field and a trough is nearly as effective as a bump in enhancing the feed rate, hence the tendency towards disappearance of the instability at x_{c2}.

When interface attachment is finite, we see that the growth amplitude of the perturbation is smaller and the instability range x_{c1}–x_{c2} narrower:

$$\text{if } \frac{R}{x_s} \to 0, \quad \frac{\beta_0}{\beta} \to \infty \quad \text{and} \quad \frac{\beta_k}{\beta} \to \infty .$$

Thus, for very small radius, interface attachment is rate controlling. If x^* is very small, so that the radius at which instability sets in would be sufficiently small for interface attachment to be rate determining, then the values for relative and absolute instability would be

$$x_{c1} \sim \left[k(k+1)\frac{2D_s}{hx_s\beta}x^* \right]^{\frac{1}{2}} ,$$

$$x'_{c1} \sim \frac{2D_s}{hx_s\beta}\frac{k}{k-2} .$$

For the second value to be small, $2D_s/hx_s\beta \ll 1$ is needed. At large radius, β_0/β and β_k/β behave like $2D_s/hx_s\beta$; so, if $2D_s/hx_s\beta \ll 1$, for large radius the growth is controlled by diffusion. In fact, around the limiting radius, diffusion and interface attachment are, in general, competing.

The radii limiting the instability zone are given as a function of $x^* = R^*/x_s$ by the roots of $\dot{\delta}/\delta - \dot{R}/R = 0$ (or $\dot{\delta}/\delta = 0$ for relative instability).

This may easily be solved for x^* versus x_c to obtain

$$x^* = \frac{y_2 - y_1(1 + \beta_k/\beta)}{[y_2 - y_1(1 + \beta_k/\beta)]/x + (1 + \beta_0/\beta)y_3} , \tag{7}$$

which is represented in fig. 4 for absolute instability.

To see the importance of x^*, let us consider the case of "diffusion" controlled growth. As x^* increases, the region of instability narrows down and finally disappears at a critical value of x^*, say $x_c^{*'}$. When $x^* > x_c^{*'}$, the nucleus is stable for all sizes, and below $x_c^{*'}$, the perturbation grows between x'_{c1} and x'_{c2}. When the interface attachment begins to play a role and becomes slower ($2D_s/hx_s\beta$ increases), the critical value $x_c^{*'}$ is decreased until the zero value is reached for $\beta = \beta_c(k)$.

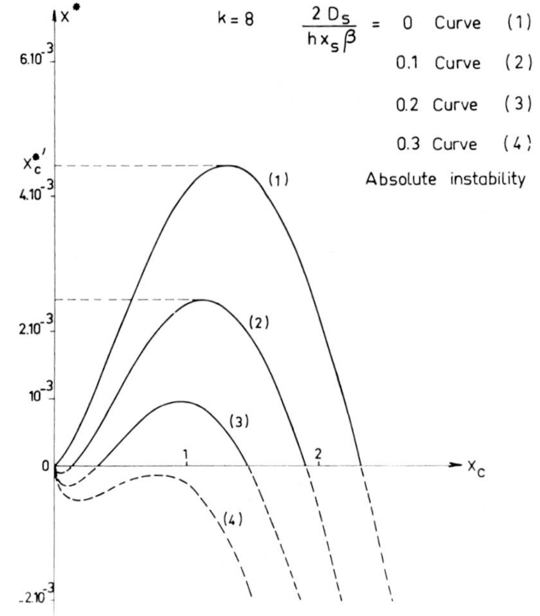

Fig. 4. Absolute stability–instability diagram for harmonic $k = 8$, with $2D_s/hx_s\beta = 0$ (1), 0.1 (2), 0.2 (3) and 0.3 (4).

If the interface kinetics are too slow ($\beta < \beta_c$), the nucleus is always stable with respect to the harmonic considered. When it is rapid enough ($\beta > \beta_c$), instability occurs on condition that $x^* < x_c^{*'}$. This shows the two prevailing factors which prevent the perturbation from starting, $2D_s/hx_s\beta$ on one side and R^*/x_s on the other.

4. Melt growth – ice crystal on water surface

When water is slightly supercooled (generally not more than one tenth of a degree) ice crystals grow on the surface in the form of thin disks up to diameters of about 1–2 mm. These freely floating crystals, of frequent occurrence in cold regions on the surface of lakes or rivers[7]), may then become serrated on the edge. The pictures of Arakawa[6]) are particularly revealing on shape instability, and our two-dimensional analysis may be extended to this.

If we consider at the surface of water a sheet having the disk height h, it exchanges heat by radiation or convection with the air at a temperature T_0 above it.

Heat transport also occurs into the water below, but if the temperature gradient in the direction perpendicular to the considered sheet is very small, this may be neglected.

For small temperature differences, a linear law may be supposed[18]:

$$q = \mu(T - T_0),$$

μ being the "coefficient of superficial heat transfer", so that heat flow is governed by the Helmholtz equations in the "steady state approximation"

$$\frac{K_S h}{\mu} \nabla^2 T_S - (T_S - T_0) = 0 \quad \text{in the solid,}$$

$$\frac{K_L h}{\mu} \nabla^2 T_L - (T_L - T_0) = 0 \quad \text{in the liquid,}$$

K_S and K_L being the thermal conductivities of ice and water, T_S and T_L the temperatures in ice and in the sheet of water.

If we consider the case of local equilibrium at the edge (heat flow controlled growth),

$$T_{LI} = T_{SI} = T_M - \frac{T_M}{L_v} K,$$

where T_M is the melting temperature of a flat surface, and L_v the latent heat of freezing per unit volume;

$$\dot{R} + \dot{\delta}_k \, e^{ik\theta} = \left(\frac{K_S}{L_v}\frac{\partial T_S}{\partial r}\right)_I - \left(\frac{K_L}{L_v}\frac{\partial T_L}{\partial r}\right)_I.$$

Finally, with $K_S \approx 4 \, K_L \approx 5 \times 10^{-3}$ cal cm^{-1} °C^{-1} sec^{-1},

$$\frac{\dot{R}}{R} = \frac{(T_M - T_0)K_L}{L_v X_L^2}\left(1 - \frac{x^*}{x}\right) y_1(x),$$

where

$$X_L^2 = \frac{K_L h}{\mu}, \quad x = \frac{R}{X_L},$$

$$x^* = \frac{R^*}{X_L} \quad \text{with} \quad R^* = \frac{T_M \gamma}{L_v(T_M - T_0)},$$

$$y_1(x) = \left[2\frac{I_1(\tfrac{1}{2}x)}{I_0(\tfrac{1}{2}x)} + \frac{K_1(x)}{K_0(x)}\right] x^{-1},$$

and

$$\frac{\dot{\delta}_k}{\delta_k} = \frac{(T_M - T_0)K_L}{L_v X_L^2}$$

$$\times \left[\left(1 - \frac{x^*}{x}\right) y_2(k, x) - x^* y_3(k, x)\right], \tag{9}$$

where

$$y_2(k, x) = \frac{K_{k+1}(x)K_1(x)}{K_k(x)K_0(x)}$$

$$- \frac{I_{k+1}(\tfrac{1}{2}x)I_1(\tfrac{1}{2}x)}{I_k(\tfrac{1}{2}x)I_0(\tfrac{1}{2}x)} - (k+1)y_1(x),$$

$$y_3(k, x) = \frac{(k^2 - 1)}{x^2}\left[3\frac{k}{x} + 2\frac{I_{k+1}(\tfrac{1}{2}x)}{I_k(\tfrac{1}{2}x)} + \frac{K_{k+1}(x)}{K_k(x)}\right].$$

For the harmonic $k = 5$, the value of x_c^* for "absolute instability" is found to be close to 2×10^{-3}, with values $L_v = 73$ cal cm^{-3}, $\gamma = 20$ erg cm^{-2} (ref. 19), $T_M - T_0 = 0.1$ °C and $\mu \approx 2.5 \times 10^{-4}$ cal cm^{-2} °C^{-1} sec^{-1} (ref. 18). Here, $R^* \approx 1.8 \times 10^{-5}$ cm, while $X_L \approx 1.2$ mm, giving $R^*/X_L \approx 1.5 \times 10^{-4}$, which is below the critical value and thus in the region of possible instability. The radius at which instability sets in is about $R_c/X_L \sim 0.1$, i.e. $R_c \sim 0.12$ mm. This result is consistent with experimental results since instability becomes visible for $R \sim 1$ mm.

5. Discussion

The difference in the results of shape stability–instability for three- and two-dimensional systems appears clearly from the comparison between "conservative" and "non-conservative" cases, since the conservative case is simply the two-dimensional restriction of a 3-dimensional system. It has been seen how the two parameters $2D_s/hx_s\beta$ and R^*/x_s determine instability, and restrict the cases for which instability with respect to infinitesimal fluctuation may set in. These factors are mainly influenced by impurity concentration at the interface through β and by temperature. The interface attachment coefficient β is diminished when the density of active sites on the edge is lowered[13]); thus poisoning of kinks will tend to give a higher value of $2D_s/hx_s\beta$, and to push the system toward the region of absolute stability ($\beta < \beta_c$). The influence of temperature appears in R^*/x_s as well as in $2D_s/hx_s\beta$. It is found that

$$\frac{2D_s}{hx_s\beta} \sim \exp\frac{2\Delta G_{kink} - \Delta G_{des} - \Delta G_{diff}}{2k_BT},$$

$$\frac{R^*}{x_s} \sim \frac{1}{k_BT}\exp\left(-\frac{\Delta G_{des} - \Delta G_{diff}}{2k_BT}\right),$$

where ΔG_{kink}, ΔG_{des} and ΔG_{diff} are respectively energy of formation of a kink, energy of desorption and activation energy for diffusion. Thus $2D_s/hx_s\beta$ increases with temperature (likewise R^*/x_s) up to $k_BT = \frac{1}{2}(\Delta G_{des} - \Delta G_{diff})$; thus lowering of temperature will favour instability.

Let us also note from first order perturbation theory that when the radius at which instability occurs has been attained, the interface will no longer be circular but will have a finite although small distortion; the $\rho(\theta)$ equation of the interface may be used. For example, if the interface has only one harmonic $k = k_1$, of amplitude Δ present, the rate of growth of the harmonic $2k_1$ will be found to be of the form

$$\delta_{2k_1} \sim A(\Delta/R)^2 + B\delta_{2k_1}.$$

An initial perturbation is no longer necessary to trigger instability and the harmonic $2k_1$ will normally appear even if the conditions of its instability are not achieved.

Finally, we note that first order perturbation analysis tells us whether or not something will happen at the interface but very little about sequential events; real behaviour is likely to be influenced by crystal symmetry omitted in our analysis.

References

1) W. W. Mullins and R. F. Sekerka, J. Appl. Phys. **34** (1963) 323.
2) For a review see: R. F. Sekerka, J. Crystal Growth **3, 4** (1968) 71.
3) R. L. Parker, Crystal Growth Mechanisms: Energetics, Kinetics, and Transport, in: *Solid State Physics*, vol. 25 (Academic Press, New York, 1970).
4) M. Avignon and B. K. Chakraverty, Proc. Roy. Soc. (London) A **310** (1969) 277.
5) M. Avignon, J. Crystal Growth **11** (1971) 265. (Part of the paper was presented at A.C.C.G. Conf., Gaithersburgh, Md., 1969).
6) K. Arakawa, J. Glaciol. **2** (1955) 463.
7) V. J. Schaefer, Trans. Am. Geophys. Un. **31** (1950) 885.
8) W. A. Bentley and W. J. Humphreys, *Snow Crystals* (Dover Publ., New York) p. 221, phot. 2–8.
9) J. W. Glen, J. Glaciol. **2** (1955) 483.
10) P. M. Morse and H. Feshbach, *Methods of Theoretical Physics* (McGraw-Hill, New York, 1953).
11) H. S. Carslaw and J. C. Jaeger, *Conduction of Heat in Solids*, 2nd ed. (Oxford Univ. Press).
12) W. W. Mullins, in: *Metal Surfaces* (Am. Soc. Metals, Metals Park, Ohio, 1962) ch. 2.
13) A. A. Chernov, Soviet Phys.-Usp. **4** (1961) 116.
14) W. K. Burton, N. Cabrera and F. C. Frank, Proc. Roy. Soc. (London) A **243** (1950) 299.
15) J. W. Cahn, in: *Crystal Growth*, Ed. H. S. Peiser (Pergamon, Oxford, 1967) p. 681.
16) S. R. Coriell and R. L. Parker, J. Appl. Phys. **36** (1965) 632; G. R. Kotler and W. A. Tiller, in: *Crystal Growth*, Ed. H. S. Peiser (Pergamon, Oxford, 1967) p. 721; S. R. Coriell and S. C. Hardy, J. Res. Natl. Bur. Std. **73** A (1969) 65.
17) M. Abramowitz and I. A. Stegun, Eds., *Handbook of Mathematical Functions* (Appl. Math. Ser. 55) (Natl. Bur. Std., Washington, 1964).
18) Ch. Fabry, *Propagation de la Chaleur*, 3rd ed. (Armand Colin, Paris, 1961).
19) S. C. Hardy and S. R. Coriell, J. Crystal Growth **3, 4** (1968) 569.

Section II

Nucleation, surface diffusion, simulation of nucleation and growth

S. TOSCHEV   H. J. LEAMY
A. MILCHEV   K. A. JACKSON
S. STOYANOV   A. C. ADAMS
D. KASHCHIEV   G. H. GILMER
I. MARKOV   P. BENNEMA
R. A. SIGSBEE   F. C. FRANK

Journal of Crystal Growth **13/14** (1972) 123–127 © *North-Holland Publishing Co.*

123

ON SOME PROBABILISTIC ASPECTS OF THE NUCLEATION PROCESS

S. TOSCHEV, A. MILCHEV and S. STOYANOV

Institute of Physical Chemistry, Bulgarian Academy of Sciences, Sofia 13, Bulgaria

Nucleation is a random phenomenon. The analysis of the expressions giving the time-dependence of the probability of formation of at least one ($P_{\geq 1}$), two ($P_{\geq 2}$) etc. nuclei makes it possible to assess the role of transient effects in nucleation kinetics as well as to evaluate the parameters of the process. Experiments on the electrolytic nucleation of mercury and cadmium on platinum are carried out by means of the double pulse potentiostatic technique ensuring reproducible working conditions and a precise control of the factors governing the deposition process. It is shown that the distribution of the number of the deposited nuclei as well as the $P_{\geq m}(t)$ plots ($t =$ time) obtained are in good agreement with formal probabilistic considerations. The $P_{\geq 1}(t)$ functions are S-shaped over the whole interval of variation of the overvoltage (equivalent to the supersaturation) which is an important indication of the non-steady state character of the process of electrolytic nucleation within the studied time intervals. A semianalytical procedure is developed to calculate stationary nucleation rates and induction times (in the sense of Zeldovich) from the experimental results. The outlined probabilistic approach may prove very useful in situations in which the time-dependence of the mean number of the nuclei formed at constant supersaturation is not available experimentally.

1. Introduction

Nucleation is a random phenomenon. The number of nuclei forming in a supersaturated medium within a given time interval is a random quantity. The kinetic theory of nucleation considers the average number of the nuclei and relates it to the macroscopic parameters and properties of the system including the time. To minimize errors, this average value must be determined on the basis of a large number of experiments. The statistical distribution of the number of nuclei however does not yield any information concerning the kinetics of the process. The situation is different with the time of formation of the first nucleus which is also a random quantity. It will be shown below that the distribution of this time being closely connected with the kinetic characteristics of the nucleation process makes it possible to undertake a quantitative study of nucleation if experimental data are available.

2. Theory

The formation of nuclei at constant supersaturation can be treated as a sequence of random events along the time axis. If these events appear independently from each other, the probability to find m nuclei within the time interval $0, t$ can be calculated by the Poisson formula:

$$P_m = \frac{N^m \exp(-N)}{m!}, \tag{1}$$

where N is the average number of nuclei to be expected in the considered time interval. It is not difficult then to derive the distribution of the time intervals separating the first event, i.e. the first nucleus from the initial moment. In terms of a more general approach to the problem under study we shall be interested in the probability to find the mth nucleus within the time interval t and $t + dt$. The differential distribution of the times elapsing from the initial point and the mth event is then given by (see e.g. ref. 1):

$$dP_{\geq m} = I \frac{N^{m-1}}{m!} \exp(-N) \, dt. \tag{2}$$

Here $I = dN/dt$ is the momentary density of the events in time or in other words, the momentary rate of nucleation. Expression (2) integrates to

$$P_{\geq m} = 1 - \exp(-N) \sum_{k=0}^{m-1} \frac{N^k}{k!}, \tag{3}$$

yielding thus the probability that at least m nuclei are formed after the expiration of the time t. The probability of discovering the first nucleus within the time interval t and $t + dt$, as well as the probability of having at least one nucleus within the interval $0, t$ result from

II – 1

(2) and (3) respectively, setting $m = 1$:

$$dP_{\geq 1} = I \exp(-N)\,dt, \tag{4}$$

$$P_{\geq 1} = 1 - \exp(-N), \tag{5}$$

Expressions (4) and (5) can be further expanded if the time dependence of the average number N is introduced into them. The general formulation of the theory of nucleation proposed by Zeldovich[2]) and Frenkel[3]) makes it possible to derive the required $N(t)$-function. An approximate but still sufficiently accurate solution of the problem reading

$$N = I_0 \tau F(x), \tag{6}$$

where

$$F(x) = x - \frac{\pi^2}{6} - 2 \sum_{n=1}^{\infty} \frac{(-1)^n}{n^2} \exp(-n^2 x), \tag{6'}$$

and

$$x = t/\tau,$$

is given by Kashchiev[4]). Here $I_0 = \text{const}$ is the stationary rate of nucleation at $t \to \infty$, whereas τ is the so-called induction time or non-steady state time lag. Practically when $t \geq 5\tau$, eq. (6) transforms into a simple linear relationship, i.e. the steady state is already attained:

$$N = I_0(t - \tfrac{1}{6}\pi^2 \tau). \tag{7}$$

If the induction time can be ignored

$$N = I_0 t. \tag{7'}$$

Combining expressions (4) and (5) with (7') it follows for steady state nucleation kinetics:

$$dP_{\geq 1} = I_0 \exp(-I_0 t)\,dt, \tag{4'}$$

$$P_{\geq 1} = 1 - \exp(-I_0 t). \tag{5'}$$

Fig. 1 shows schematically the plots of $dP_{\geq 1}/dt$ and $P_{\geq 1}$ against time in the case of stationary nucleation (fig. 1a) and in presence of a perceptible induction time (fig. 1b). It is seen that the time distributions differ markedly in both cases, which in its turn can serve as a criterion for the role of transient effects in nucleation kinetics[5]). A further step in this respect should involve the evaluation of the important parameters I_0 and τ from the experimental $P_{\geq 1}(t)$ or $dP_{\geq 1}(t)/dt$ functions. Particularly when eqs. (4') and (5') hold the quantity

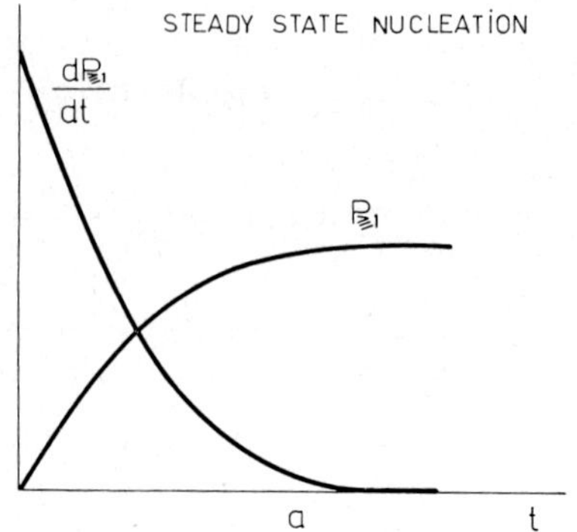

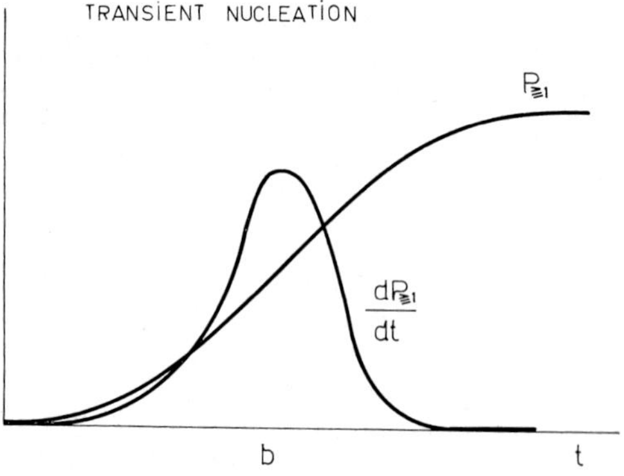

Fig. 1. Plots of $dP_{\geq 1}$ and $dP_{\geq 1}/dt$ versus time for steady state (a) and non-steady state nucleation (b).

I_0 is readily obtained as this has been demonstrated by Skripov and coworkers in a series of statistical experiments on nucleation of gas bubbles and of crystals in overheated liquids[6]) and undercooled melts[7]) respectively. The general case is more complex as the $N(t)$-function appearing in eqs. (4) and (5) is very complicated. A semi-analytical procedure to analyze the statistical distribution $P_{\geq}(t)$ in case of transient nucleation will be described in the second part of this contribution.

It is perhaps worth noting that according to definition the mean time of formation of the first nucleus is

$$\bar{t} = \int_0^{\infty} t\,dP_{\geq 1},$$

which gives by inserting $dP_{\geq 1}$ from (4')

$$\bar{t} = 1/J_0.$$

This rather trivial and well-known relationship is frequently employed, though it is not always clearly realized that its validity is restricted only to cases in which non-steady state effects are of no importance. No physical meaning should be ascribed to the mean time in the case depicted in fig. 1b.

3. Experiment

To illustrate the above outlined probabilistic approach and to assess its merits as a tool for the study of phase formation kinetics, metal nuclei were deposited electrolytically on platinum single crystal electrodes using the double pulse potentiostatic technique. This technique as well as numerous results obtained are described at length elsewhere[8,9]), so that no experimental details will be mentioned. It should only be pointed out that the very nature of the process of electrolytic nucleation enabled us to carry out a large number of experiments under precise control of the overvoltage (equivalent to the supersaturation) and its time-duration. This is a prerequisite to obtaining reliable statistical data within a reasonable period of time.

The first step was to check formally the expressions in the foregoing. Fig. 2 shows the distribution of the number of electrolytically deposited mercury droplets on platinum based on 500 uniform experiments[11]). The discontinuous line represents the plot of eq. (1) with $N = 3.14$ which is an average over all the experimental data. It is seen that the experimental distribution follows the Poisson law with good accuracy.

Another experiment with mercury confirming the applicability of the formal probabilistic treatment is presented in fig. 3. The curves 1, 2 and 3 show the increase with time of the probabilities of forming a minimum of one, two and three nuclei. The circles denote values obtained by multiple repetition of the experiments yielding the relative number of favourable occurrences.

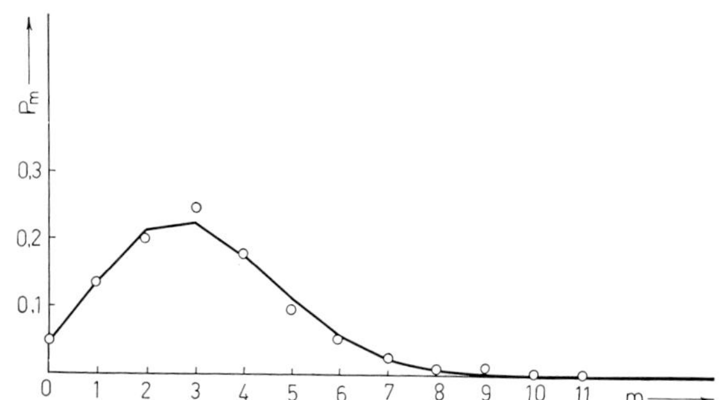

Fig. 2. Poisson distribution of the number of electrolytically deposited mercury nuclei.

The same probabilities are also calculated by means of eq. (3) setting $m = 1$, $m = 2$, $m = 3$ respectively, inserting in (3) experimentally determined values of the average number of the nuclei which are formed within the chosen time intervals (see the points). As is seen, the two methods to determine the probability $P_{\geq m}$ give

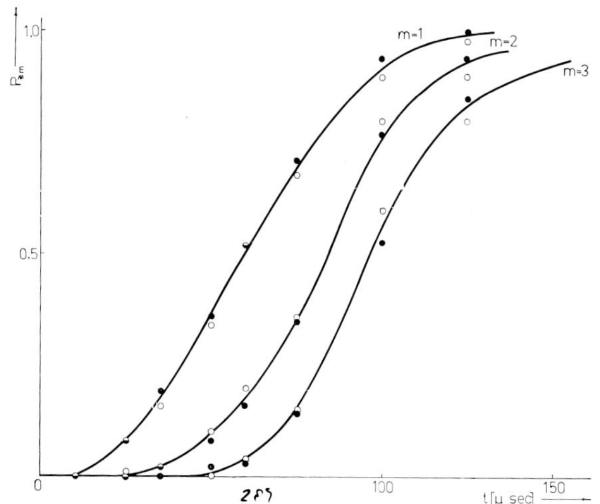

Fig. 3. Plots of the probability to form at least one nucleus, at least two and at least three nuclei in the electrodeposition of mercury.

practically the same results. An important point herewith is that curve 1 is S-shaped which in the light of the considerations in the preceding is evidence for the essentially non-steady state character of the nucleation process in the electrodeposition of mercury on platinum. Particularly in the case of mercury the transient nucleation effects have been very thoroughly investigated by means of the $N(t)$-relationships[10]).

Fig. 4 shows a series of experimental plots of the probability to form at least one cadmium nucleus versus time at various supersaturations[12])*. The curves are also S-shaped and their formal analysis has shown that they can be fairly well approximated by the integral Gaussian distribution. Considering that the $P_{\geq 1}(t)$-plots should satisfy eq. (5) combined with (6), the kinetic parameters of the process – the stationary nucleation rate I_0 and the induction time τ – can be determined by means of the coordinates of the inflection points of the $P_{\geq 1}(t)$-functions. These two quantities are interdependent in the following manner[11]):

* It has been erroneously stated in ref. 12 that the probability to form one nucleus has been measured.

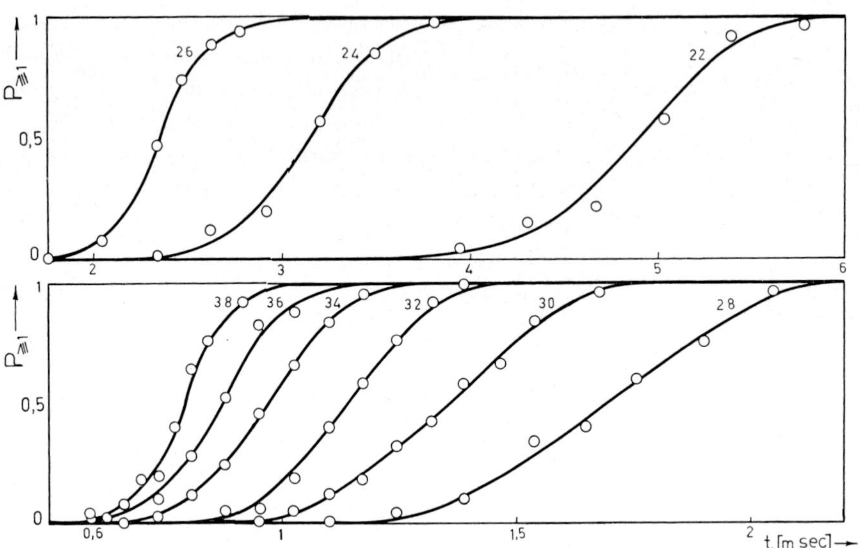

Fig. 4. Experimental $P_{\geq 1}(t)$ plots in the electrodeposition of cadmium on platinium. The attached figures denote the overvoltage in milivolts.

$$\tilde{p} = 1 - \exp\left[-\frac{F(\tilde{x})f'(\tilde{x})}{f^2(\tilde{x})}\right], \tag{8}$$

wherein

$$f(\tilde{x}) = 1 + 2 \sum_{n=1}^{\infty} (-1)^n \exp(-n^2\tilde{x}),$$

$$f'(\tilde{x}) = 2 \sum_{n=1}^{\infty} (-1)^n \exp(-n^2\tilde{x}),$$

$$\tilde{x} = \tilde{t}/\tau,$$

and $F(\tilde{x})$ can be obtained from (6') setting $x = \tilde{x}$. The tabulated function (8) (see ref. 11) makes it possible to evaluate $\tilde{x}$ and hence τ from the experimental values of $\tilde{P}$ and $\tilde{t}$. The steady state rate of nucleation is readily obtained from

$$I_0 = -\frac{\ln(1-\tilde{p})}{\tau F(\tilde{x})}.$$

Since in the case under consideration always $\tilde{P} = 0.5$ the product $I_0\tau$ remains constant irrespective of the supersaturation. The calculations yield $I_0\tau = 135$. Hence all experimental data should be described by the common equation:

$$P_{\geq 1} = 1 - \exp[-135\ F(x)]. \tag{9}$$

Experiment and theory are compared in fig. 5 the full line representing eq. (9). It is interesting to note that for some mathematical reasons eq. (9) is very close to

the integral Gaussian law with mean value $\bar{x} = 0.55$ and dispersion $\sigma^2 = 0.0121$. This is in accordance with the experimental finding.

4. Conclusion

The analysis of the statistical distribution of the time associated with the appearance of the first nucleus at constant supersaturation yields valuable information concerning the kinetics of nucleation. The reliability of the extracted numerical values of the parameters I_0 and τ depends primarily on the adequacy of the theoretical $N(t)$-equation to represent accurately the real situation in the very early stages of the nucleation process. Despite some limitations of the outlined approach which are discussed in detail in another paper[11]), a fairly good description of the basic kinetic features of the

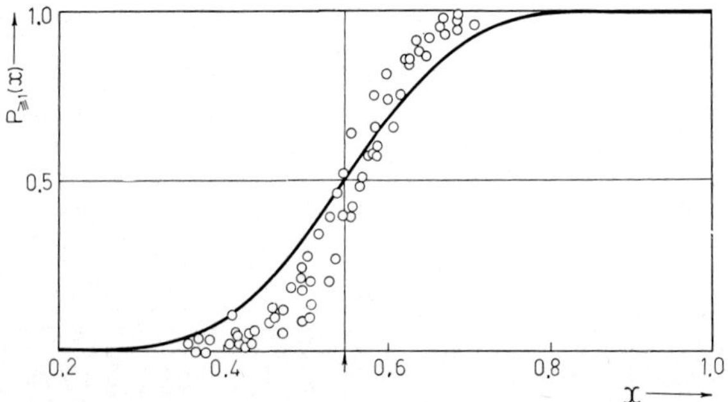

Fig. 5. Comparison of the experimental data from fig. 4 with the theoretical equation (9).

process of nucleation is possible in this way. The significance of these considerations increases particularly when due to some experimental difficulties the number of the nuclei formed cannot be determined as is often the case in threedimensional homogenous nucleation and in twodimensional surface nucleation.

Acknowledgements

The authors are grateful to Professor R. Kaischew and to D. Kashchiev for stimulating discussions.

References

1) E. S. Ventzel, *Teoriya Veroyatnostei* (Goss. Izd. Fiz. Mat. Lit., Moscow, 1962).

2) J. B. Zeldovich, Acta Physicochim. URSS **18** (1943) 1.

3) J. I. Frenkel, *Kinetic Theory of Liquids* (Dover, New York, 1955).

4) D. Kashchiev, Surface Sci. **14** (1969) 209.

5) S. Toschev and S. Stoyanov, Commun. Dept. Chem. Bulg. Acad. Sci. **3** (1970) 295.

6) E. N. Sinitzin and V. P. Skripov, Ukr. Fiz. Zh. **12** (1967) 99.

7) V. P. Skripov, V. P. Koverda and Y. T. Butorin, Kristallografia **15** (1970) 1219.

8) R. Kaischew, Chem. Ing. Tech. **39** (1967) 554.

9) R. Kaischew, S. Toschev and I. Markov. Commun. Dept. Chem. Bulg. Acad. Sci. **2** (1969) 463.

10) S. Toschev and I. Markov, J. Crystal Growth **3, 4** (1968) 436.

11) S. Toschev and A. Milchev, Commun. Dept. Chem. Bulg. Acad. Sci. **3** (1970) 757.

12) S. Toschev and I. Markov, Electrochim. Acta **12** (1967) 281.

NUCLEATION IN EXTERNAL ELECTRIC FIELD

D. KASHCHIEV

Institute of Physical Chemistry, Bulgarian Academy of Sciences, Sofia 13, Bulgaria

Homogeneous nucleation is considered, proceeding in the presence of an externally applied uniform electric field. The dependence of the work required for the formation of a spherically shaped cluster of the new phase on the electric field strength is found. The analysis shows that this work may be either reduced or enhanced by the electric field depending on the ratio of the dielectric constants of the old and the new phases which means that the electric field can either stimulate or inhibit the nucleation process. The dependences of the nucleation rate and of the time-lag on the electric field strength are obtained. It turns out that there exists a critical strength which is connected with the initial supersaturation and which, once reached, causes a strong increase of the nucleation rate. Finally, the possibility is indicated for the direct application of all results obtained to the case of nucleation taking place in an external uniform magnetic field.

1. Introduction

Many experimental investigations show that the electric field exerts a perceptible influence on the nucleation process. Nevertheless, there are only a few theoretical considerations concerning this problem, the most of them being devoted to processes in the Wilson cloud chamber[1-4]).

It is of interest, however, to examine the behaviour of the nucleating system when it is situated in an external electric field and to find expressions for both the nucleation rate and time-lag as functions of the electric field strength. This namely is the purpose of the present paper.

2. Work for cluster formation

Let us consider a parent phase situated in an externally applied uniform electric field E and let p and T be its constant pressure and absolute temperature respectively. Neglecting the effect of electrostriction, we can present $\Delta G(x)$, the isothermal reversible work required for the formation of a cluster of x atoms (or molecules), in the form

$$\Delta G(x) = \Delta G_0(x) + \Delta W_E(x), \tag{1}$$

where $\Delta G_0(x)$ is the same work, but in the absence of the electric field, and $\Delta W_E = W_2 - W_1$ is the free energy change due to the field. According to the known electrostatic formula[5]), the free energy of the electric field before the formation of the cluster and after that will be respectively

$$W_1 = \frac{\varepsilon_m}{8\pi} \int_{\Omega_m} E^2 \, dv,$$

$$W_2 = \frac{\varepsilon_c}{8\pi} \int_{\Omega_c} E_c^2 \, dv + \frac{\varepsilon_m}{8\pi} \int_{\Omega_m - \Omega_c} E_m^2 \, dv, \tag{2}$$

where ε_c and ε_m are the dielectric constants of the new and the old phase, E_c and E_m are the electric fields inside and outside the cluster, Ω_c and Ω_m are the spatial regions occupied by the cluster and by the system and $E^2 = \boldsymbol{E}^2$, $E_c^2 = \boldsymbol{E}_c^2$, $E_m^2 = \boldsymbol{E}_m^2$.

The latter expressions are so general that in order to use them some complementary assumptions must be made. Since in general it is difficult to find $\boldsymbol{E}_c$ and $\boldsymbol{E}_m$ and to calculate the integrals in (2) we shall consider that both the cluster and the system are spheres of radii r_c and R respectively. Then r_c will be connected with x by the relation

$$r_c = (3v_c/4\pi)^{\frac{1}{3}} x^{\frac{1}{3}}, \tag{3}$$

in which v_c is the mean atom volume. All this permits the direct use of the expressions giving the electric potentials inside and outside a dielectric sphere situated in an external uniform electric field. In spherical coordinates r, θ, φ, if the centre of the cluster is chosen as an origin of the coordinates and if the polar axis has the direction of the applied field, the corresponding potentials are[5])

$$U_c = -[3\varepsilon_m/(\varepsilon_c + 2\varepsilon_m)]Er \cos\theta, \quad 0 \le r \le r_c, \tag{4}$$

$$U_m = \{[(\varepsilon_c - \varepsilon_m)/(\varepsilon_c + 2\varepsilon_m)] (r_c/r)^3 - 1\} Er \cos\theta,$$

$$r_c \le r \le R. \tag{5}$$

Finding $E_c^2 = (\mathrm{grad}\ U_c)^2$ and $E_m^2 = (\mathrm{grad}\ U_m)^2$, carrying out the integration in (2) and using $r_c^3 \ll R^3$ and (3), we get

$$\Delta W_E = -k_B T c E^2 x \qquad (6)$$

where $c = \varepsilon_m f(\lambda) v_c / 8\pi k_B T$ is a constant and

$$f(\lambda) = (1-\lambda)/(2+\lambda), \quad \lambda = \varepsilon_c/\varepsilon_m. \qquad (7)$$

Further, by means of (6) and the well-known expression[6,7]) for $\Delta G_0(x)$, taken in the form

$$\Delta G_0(x) = -k_B T s_0 x + a\sigma x^{\frac{2}{3}}, \qquad (8)$$

where $s_0 = \ln(p/p_e)$ is the supersaturation, p_e is the equilibrium pressure, k_B is the Boltzmann constant, σ is the specific surface energy, $a = (4\pi)^{\frac{1}{3}}(3v_c)^{\frac{2}{3}}$ is a constant, we obtain from eq. (1)

$$\Delta G(x) = -k_B T s x + a\sigma x^{\frac{2}{3}}. \qquad (9)$$

The identical structure of (8) and (9) suggests that the quantity

$$s = s_0 + s_E \qquad (10)$$

may be considered as an effective supersaturation at which the cluster formation occurs. The influence of the electric field upon the process is taken into account through the additional term

$$s_E = cE^2. \qquad (11)$$

Thus, as in the conventional nucleation theory, in the presence of an external electric field the sign of s will determine the state of the system: it will be supersaturated, saturated or undersaturated at $s > 0$, $s = 0$ and $s < 0$ respectively.

When $s > 0$ a maximum work $\Delta G_k = \Delta G(x_k)$ must be performed for the formation of a cluster of size x_k, called a nucleus. Directly using the known expressions for the work for nucleus formation and for the nucleus size[6,7]), we may write

$$\Delta G_k = 4a^3\sigma^3/27[k_B T(s_0 + cE^2)]^2, \qquad (12)$$

$$x_k = [2a\sigma/3k_B T(s_0 + cE^2)]^3. \qquad (13)$$

The latter relation is an analogue to the well-known Thomson–Gibbs equation.

3. Nucleation rate

The shortest way of finding the steady state nucleation rate J_{st} in the presence of an external electric field is the direct use of the general expression for its dependence on the supersaturation[7])

$$J_{st} = [(-\mathrm{d}^2\Delta G/\mathrm{d}x^2)_{x=x_k}/2\pi k_B T]^{\frac{1}{2}} D_k N \\ \exp \times (-\Delta G_k/k_B T). \qquad (14)$$

Here N is the total number of atoms in the system and D_k, given by

$$D_k = \alpha a x_k^{\frac{2}{3}}, \qquad (15)$$

is the probability per unit time of an atom joining to the nucleus. In general, this probability depends on E not only through x_k but also through the frequency factor α. Nevertheless, if the possible dependence $\alpha = \alpha(E)$ is weaker than the exponential one, following Volmer[6]) we may write instead of (14) $(s_0 + cE^2 \geq 0)$

$$J_{st} = A \exp[-B/(s_0 + cE^2)^2], \qquad (16)$$

where $B = 4a^3\sigma^3/(3k_B T)^3$ is a constant and A, uniting all pre-exponential factors in (14), may be approximately considered as a constant, too. Eq. (16) shows that at $c > 0$ (i.e. at $\varepsilon_c < \varepsilon_m$), J_{st} is an increasing function of E, which means that the electric field stimulates the nucleation. Conversely, at $c < 0$ (i.e. at $\varepsilon_c > \varepsilon_m$), the field inhibits the process for then J_{st} decreases with an increase of E.

The obvious, extremely strong dependence of J_{st} on E makes it possible to define a critical electric field strength E_{cr} by means of the relation

$$J_{st}(E_{cr}) = 1.$$

It follows from here and from (16) that E_{cr}^2 is a linear function of s_0:

$$E_{cr}^2 = \frac{1}{c}(s_{cr} - s_0), \quad (c \neq 0) \qquad (17)$$

where $s_{cr} = (B/\ln A)^{\frac{1}{2}}$ stands for the critical supersaturation in the absence of an electric field introduced by Volmer[6]).

A more comprehensive description of the nucleation process requires the knowledge of the dependence of the nucleation rate on time t. If the system admits a steady state, the transient nucleation rate $J(t)$ is[8])

$$J(t) = J_{st}\left[1 + 2\sum_{n=1}^{\infty} (-1)^n \exp\left(-n^2\frac{t}{\tau}\right)\right], \qquad (18)$$

where τ, defined by

$$\tau = 8k_B T/\pi^2 D_k(-\mathrm{d}^2\Delta G/\mathrm{d}x^2)_{x=x_k}, \qquad (19)$$

is the time-lag. If J_{st} and τ are known functions of E, eq. (18) will represent the transient nucleation rate when the process takes place in an external electric field. The dependence of J_{st} on E has already been discussed. The connection between τ and E may be obtained by using eqs. (9)–(13), (15) and (19):

$$\tau = 16\sigma/\pi^2\alpha k_B T(s_0+cE^2)^2. \tag{20}$$

Further, on multiplying (14) and (19), we get

$$J_{st}\tau = [32k_B T/\pi^5(-d^2\Delta G/dx^2)_{x=x_k}]^{\frac{1}{2}}N \\ \times \exp(-\Delta G_k/k_B T). \tag{21}$$

This result is different in principle from the previous ones [eqs. (14) and (19)] for the quantity D_k does not enter it. Therefore, the product $J_{st}\tau$ may be found only on the basis of thermodynamic considerations. In view of (9)–(13) and (21) the explicit dependence of $J_{st}\tau$ on E is

$$J_{st}\tau = [8\sqrt{3}\,NB^{\frac{1}{2}}/\pi^{\frac{5}{2}}(s_0+cE^2)^2] \\ \times \exp[-B(/s_0+cE^2)^2].$$

4. Conclusion

In order to derive the basic relation (6) we profited essentially by the simplifying assumption that the cluster is spherically shaped. In principle, the more realistic case of ellipsoidal or polyhedral shape may be treated in the same way but then serious mathematical difficulties have to be surmounted.

Finally, we must draw attention to the fact that all the results obtained also hold good when nucleation takes place in an externally applied uniform magnetic field. In this case it is necessary to replace E, ε_m and ε_c everywhere by H, μ_m and μ_c, the magnetic field strength and the magnetic permeabilities of the old and the new phase respectively.

Acknowledgment

The author would like to thank Professor R. Kaischew for stimulating discussions throughout this work.

References

1) J. J. Thomson, *Elektrizitätsdurchgang in Gasen* (Teubner, Leipzig, 1906).
2) G. Tohmfor and M. Volmer, Ann. Physik **33** (1938) 109.
3) M. A. Leontovich, *Vvedenie v Termodinamiku* (GITTL, Moscow–Leningrad, 1952).
4) K. C. Russell, J. Chem. Phys. **50** (1969) 1809.
5) J. A. Stratton, *Electromagnetic Theory* (McGraw-Hill, New York, 1941).
6) M. Volmer, *Kinetik der Phasenbildung* (Steinkopff, Dresden–Leipzig, 1939).
7) J. I. Frenkel, *Kinetic Theory of Liquids* (Dover, New York, 1955).
8) D. Kashchiev, Surface Sci. **14** (1969) 209.

Journal of Crystal Growth 13/14 (1972) 131–134 © *North-Holland Publishing Co.*

THE ROLE OF ACTIVE CENTERS IN THE KINETICS OF NEW PHASE FORMATION

I. MARKOV and D. KASHCHIEV

Institute of Physical Chemistry, Bulgarian Academy of Sciences, Sofia 13, Bulgaria

A general equation for the time dependence of the number of nuclei is derived in the case of a nucleation process limited by the finite number of active centers and by the existence of zones around the growing nuclei forbidden for nucleation. The particular case of nucleation onto substrate from the vapour phase is considered. A method for determinating the number of active centers and the forbidden zone spreading rate constant with the help of an experimentally-found nucleation rate, a saturation nucleus density and a saturation time is developed.

1. Introduction

The existence of active centers of finite number in a nucleating system may be considered as a factor confining the nucleation process, since the number of nuclei formed until the end of the process cannot exceed the number of active centers. Athermal nuclei[1]) or foreign particles in melts and different surface defects in the case of heterogeneous nucleation could be active centers. On the other hand, spatial regions, to be called "screening zones", formed around the growing nuclei in which no nucleation occurs practically, may interfere as another factor confining the process. The crystallites growing themselves in melts, the diffusion zones with reduced adatom concentration around nuclei formed on substrate from vapours can for instance play the role of such screening zones.

Several papers are devoted to this problem[2-4]). The present investigation attempts a more general treatment including both confining factors.

2. Solution of the problem

We are given a spatial n-dimensional region with a volume V_n and with N_0 randomly located active centers in it ($n = 1, 2, 3$). Let the latter have an equal potency with respect to the nucleus formation. Further, we assume that nuclei can form only on active centers with a frequency $J(t)$ (sec^{-1}) and that a screening zone with a volume V'_n forms around every growing nucleus. The growth rate of this zone may be expressed as a function of direction $\boldsymbol{n}$ and time t:

$$v(\boldsymbol{n}, t) = c(\boldsymbol{n})k(t). \tag{1}$$

The time dependence of the number N of nuclei formed till time t is given by the definition equation

$$N(t) = N_0 \int_0^t J(\tau)\, \theta(\tau)\, \mathrm{d}\tau, \tag{2}$$

where $\theta(t)$ is the ratio of the free active centers and their total number N_0 at the moment t. We consider as free those centers on which nuclei have not yet formed and which are not ingested by the screening zones. Thus, the problem of finding N is reduced to the determination of the ratio θ and the evaluation of the integral (2).

On solving the so-formulated problem we shall use the method of Kolmogoroff[4]). Assuming for simplicity an isotropic growth of the screening zones, i.e. $c(\boldsymbol{n}) = c = \text{const.}$, the volume of such a zone growing freely around a nucleus formed at time t' will be at time t

$$V'_n(t') = \kappa_n c^n \left(\int_{t'}^t k(\tau)\, \mathrm{d}\tau \right)^n, \tag{3}$$

where $\kappa_1 = 2$, $\kappa_2 = \pi$, $\kappa_3 = \frac{4}{3}\pi$.

The ratio $\theta(t)$ is directly given by the probability of ingestion of an arbitrarily chosen active center P at time t by a screening zone. In order for the active center P to be at time t in a screening zone, it is necessary and sufficient that at least one nucleus be formed at time $t' < t$ on another active center P' at a distance from P smaller than

$$c \int_{t'}^t k(\tau)\, \mathrm{d}\tau,$$

or respectively to be formed in the volume $V'_n(t')$ [eq. (3)].

The probability of formation of at least one nucleus in this volume within the time interval $\Delta t'$ is

$$J(t')\bar{N}_0[V'_n(t')]\,\Delta t' + o(\Delta t'),\qquad(4)$$

where the first term expresses the probability of formation of only one nucleus, and $o(\Delta t')$, an infinitesimal compared with $\Delta t'$ is the probability of formation of more than one nucleus. The quantity $\bar{N}_0[V'_n(t')]$ presenting the mean number of active centers in the volume $V'_n(t')$ can be determined as follows:

$$\bar{N}_0[V'_n(t')] = 1 + (N_0 - 1)V'_n(t')/V_n.\qquad(5)$$

This equality accounts for the fact that in the volume $V'_n(t')$ there is one active center with a probability of unity (the center P) and $N_0 - 1$ others with a probability of $V'_n(t')/V_n$. Consequently, the probability of no nucleus formation in $V'_n(t')$ within the time interval $\Delta t'$ is

$$1 - J(t')\,[1 + (N_0 - 1)V'_n(t')/V_n]\,\Delta t' + o(\Delta t').\qquad(6)$$

Then the probability of no ingestion of the arbitrarily chosen active center by the screening zone until the moment t will be

$$\theta(t) = \prod_{i=1}^{s} \{1 - J(t_i)\,[1 + (N_0 - 1)V'_n(t_i)/V_n]\,\Delta t'\} + o(1),\qquad(7)$$

where $t = s\Delta t'$, $t_i = i\Delta t'$, and $o(1)$ is an infinitesimal at infinitesimal $\Delta t'$. Taking the logarithm of eq. (7) and making the limit $\Delta t' \to 0$ we can write finally the expression

$$\theta(t) = \exp\left\{-\int_0^t J(t')\,[1 + (N_0 - 1)V'_n(t')/V_n]\,dt'\right\}.\qquad(8)$$

The time dependence of the number of nuclei can be then obtained by substituting eq. (8) in eq. (2):

$$N(t) = N_0 \int_0^t J(\tau)\exp$$
$$\left\{-\int_0^\tau J(t')\,[1 + (N_0 - 1)V'_n(t')/V_n]\,dt'\right\}\,d\tau.\qquad(9)$$

Let us investigate the extreme cases of eq. (9):
(1) The zone growth rate constant c is zero, i.e. $V'_n(t') = 0$. This means the neglecting of the spatial confinements and nucleation kinetics limited only by

the finite number of active centers. In the stationary case $J(t) = J_0$ the expression

$$N(t) = N_0[1 - \exp(-J_0 t)]\qquad(10)$$

holds which is identical with that derived in refs. 2 and 3.

(2) The number of active centers is large enough. We shall make the limit $N_0 \to \infty$ using the equality $JN_0 = IV_n$, I $(\mathrm{cm}^{-n}\,\mathrm{sec}^{-1})$ being the nucleation rate per unit volume. Substituting J from this equality in eq. (9) and making the limit $N_0 \to \infty$ we get the equation

$$N(t) = V_n \int_0^t I(\tau)\exp\left[-\int_0^\tau I(t')V'_n(t')\,dt'\right]\,d\tau,\qquad(11)$$

which is identical to that of Kolmogoroff[4]). The physical meaning of this result consists in the disappearance of the limiting effect of the active centers on the nucleation kinetics when their number has become sufficiently large. Then the problem is reduced naturally to the one of Kolmogoroff.

3. Nucleation onto substrate ($n = 2$)

As has been mentioned above, different surface defects may be active centers. If we assume a sufficiently small dispersion of the center activity distribution and if we accept that the diffusion zone radius builds up as the square root of time[5,6]), i.e.

$$k(t) = \tfrac{1}{2}t^{-\tfrac{1}{2}},$$

we obtain from eq. (9) in the particular case $n = 2$ and $J(t) = J_0 = \mathrm{const.}$ ($V_2 = S$, S being the substrate area):

$$N(t) = N_s\left[1 - \frac{\mathrm{erfc}(\tfrac{1}{4}A + J_0 t/A)}{\mathrm{erfc}(\tfrac{1}{4}A)}\right]\qquad(12)$$

where

$$N_s = \tfrac{1}{2}\pi^{\tfrac{1}{2}}N_0 A \exp(\tfrac{1}{4}A^2)\,\mathrm{erfc}(\tfrac{1}{4}A)\qquad(13)$$

is the saturation nucleus density and A is given by the expression

$$A = \left[\frac{6J_0 S}{\pi c^2(N_0 - 1)}\right]^{\tfrac{1}{2}} = \left[\frac{6I_0 S^2}{\pi c^2 N_0(N_0 - 1)}\right]^{\tfrac{1}{2}}.\qquad(14)$$

In the case when $A \gg 1$ the asymptotic expansion of eq. (13) is

$$N_s = N_0(1 - 2A^{-2}),$$

from which it follows that

$$N_s \approx N_0, \tag{13a}$$

when $A > 14$. Conversely, upon $A \ll 1$ one gets:

$$N_s = \tfrac{1}{2}\pi^{\frac{1}{2}}N_0 A(1 - A/\pi^{\frac{1}{2}}).$$

It is clear that when $A < 0.02$ and $N_0 \gg 1$,

$$N_s \approx \tfrac{1}{2}\pi^{\frac{1}{2}}N_0 A = (3I_0 S^2/2c^2)^{\frac{1}{2}}. \tag{13b}$$

The time t_s required for reaching the saturation nucleus density can be obtained by the condition $N/N_s = 0.99$. It is not possible to solve exactly the equation

$$\operatorname{erfc}(\tfrac{1}{2}A + J_0 t_s/A) = 10^{-2}\operatorname{erfc}(\tfrac{1}{2}A), \tag{15}$$

but the expression

$$t_s = 7.2A/J_0(4+A) \tag{16}$$

is found to be a fairly good approximate solution as can be verified through its direct substitution into eq. (15). In the corresponding extreme cases t_s becomes:

$$t_s = 1.8A/J_0 \quad \text{for } A < 0.02, \tag{16a}$$

$$t_s = 7.2/(J_0 \quad \text{for } A > 14. \tag{16b}$$

It should be easily shown that eqs. (16a) and (16b) could be derived from (11) and (10) respectively. This means that the value of A may be used as a criterion for determination of which factor dominates as limiting the nucleation process.

In general, we can find from appropriate experimental data the saturation nucleus density N_s, the saturation time t_s and the steady state nucleation rate I_0 from the slope of the linear part of the time dependence of the number of nuclei, since in view of eq. (12)

$$(\mathrm{d}N/\mathrm{d}t)_{t=0} = I_0 S.$$

Provided N_s, t_s and I_0 are known, the following method for determining the number N_0 of the active centers and the zone spreading rate constant c is developed. Dividing eqs. (13) and (16) one obtaines:

$$2N_s/I_0 St_s = F(A) = (1 + \tfrac{1}{4}A)\exp(\tfrac{1}{4}A^2)\operatorname{erfc}(\tfrac{1}{2}A), \tag{17}$$

where the right-hand side $F(A)$ is a known function of the parameter A (fig. 1). By means of the plot $F(A)$ and the ratio $2N_s/I_0 St_s$ calculated from experimental data, the quantity A may be easily determined. It is clear that when $A > 14$ the nucleation process is limited only by the finite number of active centers and when $A < 0.02$, by the spatial confinements.

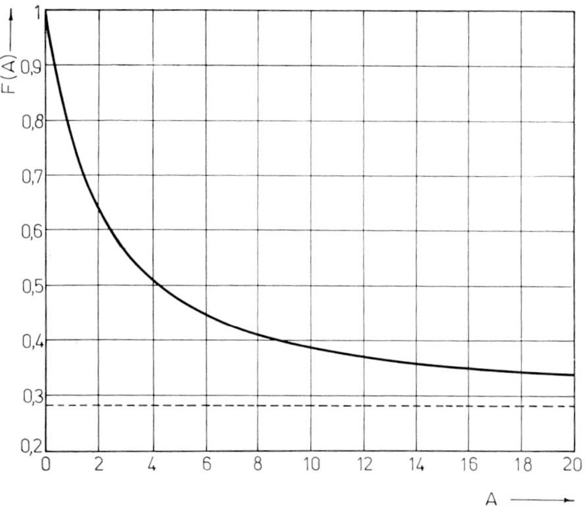

Fig. 1.   Plot of F as a function of A.

Finally, if $N_s \gg 1$ (and therefore $N_0 \gg 1$), with the help of the thus-determined A and using the system of eqs. (14) and (16) we get N_0 and c expressed only through the experimentally found quantities:

$$N_0 = I_0 St_s(4+A)/7.2A, \tag{18}$$

$$c = 7.2(6/\pi)^{\frac{1}{2}}/(4+A)t_s I_0^{\frac{1}{2}}. \tag{19}$$

4. Conclusion

The analysis made shows that the nucleation process onto substrate is simultaneously limited by both the finite number of active centers and the spatial confinements. It is clear that either one or the other will dominate at the ends of a sufficiently large interval of supersaturations. Thus, at high supersaturations leading to great nucleation rates so that $A > 14$, practically all active centers will take part in the process. Conversely, at small supersaturations, i.e. when $A < 0.02$, the spatial confinements will dominate.

It is of interest to discuss some of the results obtained in the present paper. As indicated by eq. (17), the ratio $2N_s/I_0 St_s$ can not be greater than unity and smaller than 0.282 because

$$\lim_{A \to \infty} F(A) = 1/2\sqrt{\pi} = 0.282$$

(the dashed line in fig. 1). For instance the ratio $2N_s/I_0 St_s$ for all curves represented in the paper of Robins and Rhodin[7]) is about 0.46 and $A = 5.5$. The number of active centers is then 3.64×10^{11} while the saturation nucleus density is 3.55×10^{11} which gives

$N_s/N_0 = 0.96$. The same ratio calculated from Robinson and Robins data[8]) is equal to 0.675 so that $A = 1.7$ and $N_0 = 7.4 \times 10^{10}$ while $N_s = 5.36 \times 10^{10}$, i.e. $N_s/N_0 = 0.725$. For the zone spreading rate constant c the value 8.5×10^{-8} cm sec$^{-\frac{1}{2}}$ is obtained.

Obviously, all possible cases of condensation, e.g. crystallization in one- and multicomponent melts, electrolytic nucleation, can be treated in the same way. The problem is then reduced to finding of a concrete functional dependence $k = k(t)$ for every case under consideration.

Acknowledgement

The authors are indebted to S. Stoyanov and M. Georgiev for helpful discussions throughout the present work.

References

1) J. C. Fisher, J. H. Hollomon and D. Turnbull, J. Appl. Phys. **19** (1948) 775.
2) M. Avrami, J. Chem. Phys. **7** (1939) 1103; **8** (1940) 212; **9** (1941) 177.
3) R. Kaischew and B. Mutaftschiew, Electrochim. Acta **10** (1965) 643.
4) A. N. Kolmogoroff, Bull. Acad. Sci. URSS (Sci. Math. Nat.) **3** (1937) 355.
5) M. J. Stowell, Phil. Mag. **21** (1970) 125.
6) I. Markov, Thin Solid Films **8** (1971) 281.
7) J. L. Robins and T. N. Rhodin, Surface Sci. **2** (1964) 346.
8) V. N. E. Robinson and J. L. Robins, Thin Solid Films **5** (1970) 313.

Journal of Crystal Growth **13/14** (1972) 135–139 © *North-Holland Publishing Co.*

ADATOM CAPTURE AND NUCLEI GROWTH

R. A. SIGSBEE

Corporate Research and Development, General Electric Company, Schenectady, New York 12301, U.S.A.

Growth rate and time calculations for heterogeneous cap-shaped nuclei, including curvature effects, are presented. Surface diffusion is dominant until the nucleus size is comparable to the mean adatom diffusion distance. The nucleation exclusion zone around a growing nucleus is initially much smaller than the adatom capture zone.

The results are obtained in a form that is applicable to many experimental cases in which the impinging flux is known. Curvature effects and the critical nucleus size are unimportant when the supersaturation is greater than 100 and the attained nucleus size is greater than 10 times the critical.

1. Introduction

This paper examines the theory of vapor to condensed phase growth of heterogeneous nuclei and resulting adatom depletion effects. Growth of these nuclei involves the following process[1]:

(1) Formation of the adatom concentration n and the embryo size class i concentrations $n(i)$.

(2) Nucleation of supercritical embryos of size class i^*+1 or $r^*(1+\delta)$ and growth of these nuclei which depletes the local adatom concentration.

(3) Concurrent nucleation on areas not depleted of monomer.

We direct our attention here primarily to the growth of nuclei after the critical embryo successfully captures an additional adatom[2].

2. Adatom concentration

The adatom concentration before supercritical nuclei appear, for impingement times greater than τ, is

$$n_\infty = J\tau, \tag{1}$$

where J = impinging flux ($\#\ \mathrm{cm}^{-2}\ \mathrm{sec}^{-1}$), and τ = mean adsorption time of an adatom [$\tau = v^{-1} \exp(\Delta G_{\mathrm{des}}/kt)$].

As shown in fig. 1 a cap-shaped nucleus of radius r_e has a projected radius of $r_e \sin \theta$ or r_e' in the substrate coordinate system r' (ref. 3). Here θ is the embryo to substrate contact angle.

At the nucleus periphery the Gibbs–Thomson effect of curvature on vapor pressure and the assumption of local equilibrium gives the adatom concentration[4]

$$n(r_e') = n_e S^{r^*/r_e}. \tag{2}$$

Here r^* = critical nucleus radius, n_e = adatom concentration in equilibrium with an infinite source maintained at the substrate temperature, and S = supersaturation ratio, n_∞/n_e, or its pressure equivalent, P/P_e.

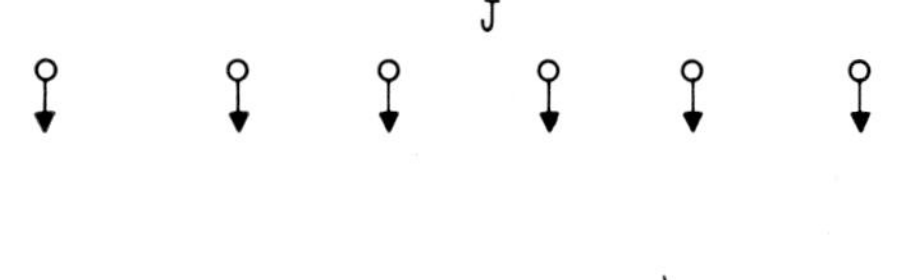

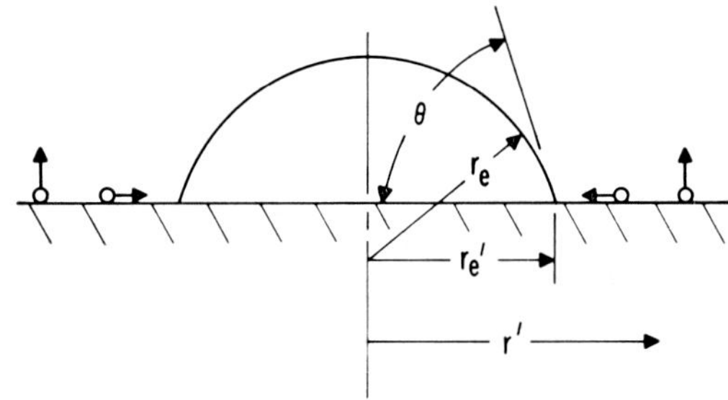

Fig. 1. Cap-shaped nucleus of radius r_e growing on a substrate.

Thus, the adatom concentration at the periphery of an isolated nucleus rapidly approaches n_e as r_e increases beyond r^*. As r_e increases the nucleus volume V grows by the acquisition of surface and vapor beam atoms at a rate

$$\frac{\mathrm{d}V}{\mathrm{d}t} = \underbrace{v\pi r_e'^2\, \frac{n_e(S - S^{r^*/r_e})}{\tau}}_{\text{direct vapor addition}} + \underbrace{v2\pi r_e' D_s\, \frac{\partial n}{\partial r}\Big|_{r\,=\,r'e}}_{\text{surface flux addition}}, \tag{3}$$

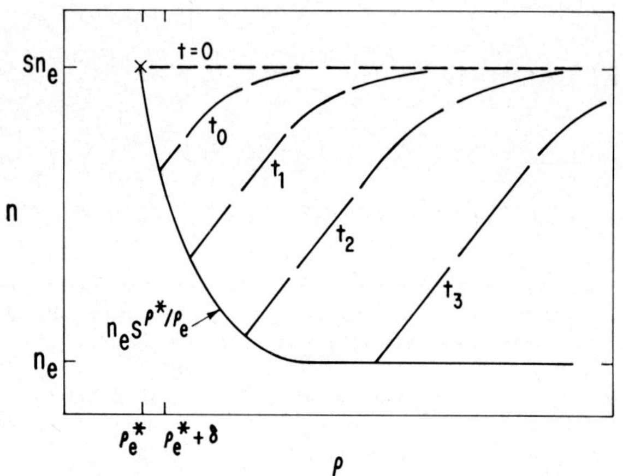

Fig. 2. Adatom concentration versus reduced radius for increasing nucleus size and growth time.

where v is the atomic volume and D_s is the surface diffusion coefficient.

The adatom concentration around a growing nucleus can be found from the steady-state solution of the continuity equation in cylindrical coordinates used by Burton et al.[4] and Chrakravery[3,5,6]. It yields

$$n(r') = n_e \left[S - (S - S^{r^*/r_e}) \frac{K_0(\rho)}{K_0(\rho_e)} \right]. \tag{4}$$

and is plotted in fig. 2 with increasing time and ρ_e as a parameter. Here the K_0 terms are modified Bessel functions of the second kind and the ρ terms are the reduced substrate radii r'/X and r_e'/X etc., where X is $(D_s\tau)^{\frac{1}{2}}$.

We can now evalute the extent and effects of adatom depletion zones around a growing nucleus. The first effect is a capture or sink zone, within which all impinging vapor atoms are captured. A nucleation-exclusion zone surrounding the nucleus will also exist in which nucleation is drastically lowered. A previous simplified nucleation and mass capture model equated both zones to the mean adatom diffusion area[1].

The capture sink zone area A_s, found from eq. (3) for the volume growth rate by considering the total atom current to a growing nucleus, is

$$A_s = \pi \left[2r_e'X \frac{K_1(\rho_e)}{K_0(\rho_e)} + r_e'^2 \right]. \tag{5}$$

For $r_e \ll X(K_1(\rho_e)/K_0(\rho_e))$,

$$A_s = -2\pi X^2/\ln(r_e'/X), \tag{6}$$

which increases slowly with nucleus size, as assumed previously[1]. A_s is similar to Halpern's[7] result obtained from a capture probability function and to that of Mutaftschiev and Chapon[8]. From eq. (6), the capture zone extends to the distance

$$r_s' = X[-2/\ln(r_e'/X)]^{\frac{1}{2}}. \tag{7}$$

A nucleation exclusion zone is more difficult to define since the nucleation rate is only reduced around the nucleus. A description of this region might consider the integrated probability of nucleation to be 0.1 that of regions outside it, similar to Stowell's[9] treatment. This requires an as yet unattainable precision in calculated nucleation rates[10]. Simply defining it as the distance ρ_P where S drops below a cutoff supersaturation S_c, eq. (4) for large S yields

$$K_0(\rho_p)/K_0(\rho_e) = 1 - S_c/S. \tag{8}$$

For $\rho < 0.1$, the exclusion zone radius is

$$r_p' = X(r_e'/X)^{1-S_c/S}, \tag{9}$$

and the nucleation exclusion zone area is then

$$A_p = \pi X^2 [(r_e'/X)^{1-S_c/S}]^2. \tag{10}$$

Values for the capture and exclusion zone areas versus the reduced radius are shown in fig. 3 and table 1 for an S_c/S of one half. These show that nucleation may

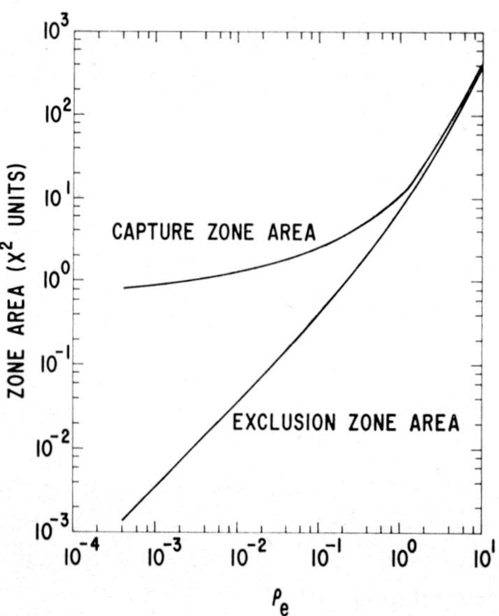

Fig. 3. Adatom capture and nucleation exclusion zones areas versus ρ_e.

TABLE 1

Depletion zone radii and zone areas, calculated for a cutoff supersaturation of $\frac{1}{2}S$, $\rho_e^* = 1.9 \times 10^{-4}$, $S = 10^6$

ρ_e nucleus radius $[10^{(\ )}]$	ρ_p exclusion zone radius $[10^{(\ )}]$	A_p exclusion zone area (X^2 units) $[10^{(\ )}]$	ρ_s sink zone radius $[10^{(\ )}]$	A_s sink zone area (X^2 units) $[10^{(\ )}]$
5 (−4)	2.37 (−2)	1.76 (−3)	5.09 (−1)	8.15 (−1)
1 (−3)	3.35 (−2)	3.52 (−3)	5.34 (−1)	8.93 (−1)
5 (−3)	7.53 (−2)	1.78 (−2)	6.08 (−1)	1.16 (0)
1 (−2)	1.07 (−1)	3.60 (−2)	6.51 (−1)	1.33 (0)
5 (−2)	2.43 (−1)	1.85 (−1)	7.96 (−1)	1.99 (0)
1 (−1)	3.57 (−1)	3.99 (−1)	9.06 (−1)	2.58 (0)
5 (−1)	9.35 (−1)	2.75 (0)	1.43 (0)	6.45 (0)
1.0 (0)	1.51 (0)	7.16 (0)	1.97 (0)	1.02 (+1)
5.0 (0)	5.63 (0)	9.95 (+1)	5.98 (0)	1.13 (+2)
10.0 (0)	10.66 (0)	3.58 (+2)	1.10 (+1)	3.80 (+2)

occur deep within the "capture zone" surrounding a growing nucleus.

3. Nuclei growth

The growth rate for a cap-shaped nucleus can be found from eqs. (3) and (4) which yield (henceforth subscripts are omitted when feasible)

$$\frac{dr_e'}{dt} = \frac{v \sin^3\theta}{2\phi(\theta)} \frac{n_e(S - S^{r*/r_e})}{\tau} \left[\frac{K_1(\rho)}{\rho K_0(\rho)} + \frac{1}{2} \right]. \quad (11)$$

Here $\phi(\theta)$ is the geometric correction term for the nucleus volume $\frac{1}{4}(2 - 3\cos\theta + \cos^3\theta)$ (ref. 2). The growth velocity for a critical nucleus, $r_e = r^*$, is zero by this formulation and re-emphasizes the nature of the nucleation event as a fluctuation in embryo size[2]). Efforts to obtain "corrected" impingement rates for critical nuclei[7,11]) because of adatom depletion are misleading. For the growth of nuclei to sizes comparable to X, the approximation used in Chakraverty's[3,5,6]) and Burton et al.'s[4]) solution for the steady-state gradients,

$$\rho_e \frac{K_1(\rho_e)}{K_0(\rho_e)} = \frac{1}{I_0(\rho_e)K_0(\rho_e)},$$

which gives twice the correct surface contribution for large nuclei, cannot be used. From the growth rate the surface diffusion and vapor contributions are found to be equal when

$$r_e' = 2X \frac{K_1(\rho_e)}{K_0(\rho_e)} \quad (12)$$

Since $K_1/K_0 > 1$ until $r_e'/X \gg 1$, direct vapor addition is unimportant for $r_e < X$.

The time dependence of r_e' for a growing nucleus can be found implicitly by rearranging eq. (11) and integrating from the size $r'^* + \delta r$ and time t_0 at which the nucleation event occurs. In terms of X, J and ρ_e

$$\frac{2\phi(\theta)}{v\sin^3\theta} \frac{X}{J} \int_{\rho*+\delta\rho}^{\rho} \frac{S}{(S - S^{\rho*/\rho})} \frac{2\rho K_0(\rho)\,d\rho}{\rho K_0(\rho) + 2K_1(\rho)} = \int_{t_0}^{t} dt \quad (13)$$

The left hand integral is denoted as the growth time integral W_{sd}, and was integrated numerically for values of ρ^* of 1.9×10^{-4}, $\rho^* + \delta\rho$ of 2×10^{-4}, ρ up to 10 and an S of 10^6. Also, the nucleus growth rate dr_e'/dt can be conveniently expressed in terms of a growth rate factor $w(\rho)$ and the impinging vapor flux

$$\frac{dr_e'}{dt} = \frac{Jv \sin^3\theta}{2\phi(\theta)} w_{sd}. \quad (14)$$

For small nuclei ($r_e < X$) and $S > 100$ the growth time integral W can be approximately expressed as

$$\int_{\rho*+\delta\rho}^{\rho} -\rho^2 \ln\rho\,d\rho = \left[\frac{\rho^3}{9} - \frac{\rho^3}{3\ln\rho} \right]_{\rho*+\delta\rho}^{\rho} = W_a. \quad (15)$$

The nucleus growth rate factor is then approximately

$$w_a = -1/\rho^2 \ln\rho. \quad (16)$$

A tabulation of W_{sd}, W_a, w_{sd} and w_a for values of ρ^* of 1.9×10^{-4}, $\rho^* + \delta\rho$ of 2×10^{-4} and S of 10^6 for ρ up to 10 is given in table 2. W_a and w_a are good approximations for ρ values up to 0.5. A relative insensitivity (less than one part in 10^{-3} for the numerical values) to supersaturation changes when $S > 100$, to the nucleus critical size ρ^* and the initial size $\rho^* + \delta\rho$ when $\rho > 10(\rho^* + \delta\rho)$ was obtained through many calculations. This is due to the rapid approach to one of the $S/(S - S^{\rho*\rho})$ term and the very large growth rates for small nuclei which then require nuclei to spend only a negligible fraction of their total growth time at the small sizes. W_{sd}, W_a, w_{sd} and w_a versus ρ are plotted in fig. 4. In this treatment nuclei have been considered immobile during growth. In some cases small nuclei 20–100 Å in diameter may be mobile[12−14]).

For a disk that grows with a constant height to diameter ratio (also for other constant geometrical shapes such as cubes, etc.) the results for the cap-shaped nuclei are applicable. The proper shape correction factor must

TABLE 2

Growth calculations for isolated nuclei as a function of nucleus size ($\rho^* = 1.9 \times 10^{-4}$, $S = 10^6$)

ρ reduced nucleus size $[10^{()}]$	W growth time integral $[10^{()}]$ W_a analytical approx.	W_{sv} vapor+surface feeding	w growth rate factor $[10^{()}]$ w_a analytic approx.	w_{sv} vapor+surface feeding
2.000 (−4)		0.0		1.444 (+6)
2.100 (−4)		6.048 (−12)		1.933 (+6)
2.200 (−4)		1.048 (−12)		2.052 (+6)
2.300 (−4)		1.575 (−11)		2.024 (+6)
2.400 (−4)		2.080 (−11)		1.939 (+6)
2.500 (−4)		2.610 (−11)		1.833 (+6)
5.00 (−4)	3.070 (−10)	3.168 (−10)	5.263 (+5)	5.182 (+5)
1.00 (−3)	2.39 (− 9)	2.434 (− 9)	1.448 (+5)	1.424 (+5)
5.00 (−3)	2.346 (− 7)	2.395 (− 7)	7.550 (+3)	7.338 (+3)
1.00 (−2)	1.646 (− 6)	1.685 (− 6)	2.171 (+3)	2.118 (+3)
5.00 (−2)	1.887 (− 4)	1.437 (− 4)	1.335 (+2)	1.284 (+2)
1.00 (−1)	8.786 (− 4)	9.204 (− 4)	4.343 (+1)	4.110 (+1)
5.00 (−1)	4.277 (− 2)	5.243 (− 2)	5.771 (0)	4.084 (0)
1.00 (0)	1.111 (− 1)	2.455 (− 1)		1.930 (0)
5.00 (0)		4.543 (0)		7.192 (−1)
1.00 (+1)		1.227 (+ 1)		6.049 (−1)

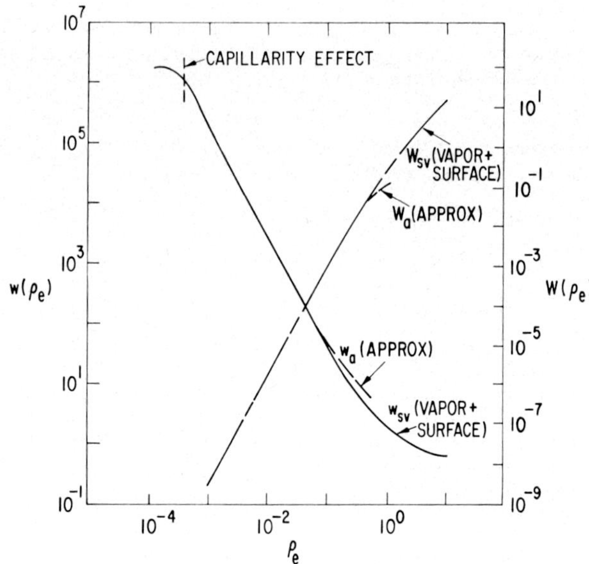

Fig. 4. Growth rate factor w and growth time integral W for cap-shaped nucleus versus reduced nucleus size ρ_e.

then be used instead of $\sin^3 \theta / \phi(\theta)$. A constant shape factor implies that surface and volume diffusion on and in the nucleus are rapid relative to the arrival rate of vapor and adatoms. When a substantial barrier to the nucleation of additional layers on a nucleus disk exists the disk may grow a constant height h. This problem will be treated in detail elsewhere[15]).

4. Summary and conclusions

A modified steady-state solution for the adatom concentration around an isolated nucleus is adequate for describing the actual concentration during growth. The nucleus curvature and critical size have a negligible overall influence on the growth rate for high supersaturations since this effect is quickly dissipated when the nucleus is fractionally larger than the critical radius. Growth rates are dominated by surface diffusion until the nucleus diameter is larger than the mean adatom diffusion distance. For these small cap-shaped nuclei, radii are approximately proportional to the cube root of the growth time. For larger nuclei a numerical integration of the growth rate has yielded generalized values of the growth time as a function of the reduced nucleus size for cap-shaped nuclei. Generalized growth rate values for nuclei and an approximate solution are in-

cluded which exhibit a drastic reduction in growth rates for larger cap-shaped nuclei. These values extend the usefulness of the results to the general evaluation of cap nuclei growth rates and times.

Adatom diffusion to a growing nucleus depletes the surface concentration and effects two zones around the nucleus, a zone in which all impinging vapor atoms are captured and a nucleation inhibition zone in which nucleation is drastically reduced. For small nuclei the adatom capture zone is almost independent of nucleus size while the less extensive inhibition zone retains a large dependence on nucleus size. This allows nuclei formation within the area that is "gettered" by an already existing nucleus. These secondary nuclei may disassociate during later stages of deposition when they are enveloped by the diffusion zones of larger growing nuclei. Growth of an isolated nucleus corresponds to the earliest stages of film formation, to nuclei forming preferentially on dispersed energetic substrate sites, or to nuclei growing at supersaturations too low to nucleate competing neighbors.

References

1) R. A. Sigsbee and G. M. Pound, Advan. Colloid Interface Sci. **1** (1967) 335.
2) J. P. Hirth and G. M. Pound, *Condensation and Evaporation, Nucleation and Growth Kinetics* (Pergamon, Oxford, 1963).
3) B. K. Chakraverty, J. Phys. Chem. Solids **28** (1967) 2401.

4) W. K. Burton, N. Cabrera and F. C. Frank, Phil. Trans. Royal Soc. (London) A **243** (1951) 299.

5) B. K. Chakraverty, J. Phys. Chem. Solids **28** (1967) 2413.

6) B. K. Chakraverty, in: *Basic Problems in Thin Film Physics*, Eds. R. Niedermayer and H. Mayer (Vandenhoek and Ruprecht, Göttingen, 1966) pp. 43–47.

7) V. Halpern, J. Appl. Phys. **40** (1969) 4627.

8) B. Mutaftschiev and C. Chapon, in: *Comptes Rendus des Travaux du Congr. Intern. des Couches Minces*, Cannes 1970 (Societe Francaise des Ingenieurs et Techniciens du Vide, Paris, 1970) p. 197.

9) M. J. Stowell, Phil. Mag. **21** (1970) 125.

10) R. A. Sigsbee, in: *Nucleation*, Ed. A. Zettlemoyer (Marcel Dekker, New York, 1969).

11) B. Lewis and D. Campbell, J. Vacuum Sci. Technol. **4** (1967) 209.

12) R. Kern, A. Masson and J. J. Metois, Compt. Rend. (Paris) B **267** (1968) 142.

13) A. Masson and R. Kern, J. Crystal Growth **2** (1968) 227.

14) A. Masson, J. J. Metois and R. Kern, J. Crystal Growth **3, 4** (1968) 196.

15) R. A. Sigsbee, submitted to J. Appl. Phys.

SURFACE STRUCTURE AND SURFACE DIFFUSION: THE fcc (110)INTERFACE

H. J. LEAMY and K. A. JACKSON

Bell Telephone Laboratories, Incorporated, Murray Hill, New Jersey 07974, U.S.A.

The structure of the (110) crystal–vapor or crystal–melt interface of an fcc crystal has been calculated in zeroth order approximation. The results indicate that structural roughening of the interface occurs in the range $0.03 < kT/L < 0.3$, where both the surface energy (roughness) and surface extent (number of surface atoms) increase rapidly with increasing kT/L. Here L is the enthalpy of vaporization or melting and T is the absolute temperature. The relative population of surface energy levels has also been computed. For $kT/L > 0.3$, the surface approximates an ideally rough configuration. The surface self diffusion coefficient has been computed from the theoretical distribution of atoms at the interface. The results indicate a positive departure from Arrhenius behavior in the temperature interval where roughening occurs. In the L/kT range encountered in metal–vapor systems the theoretically obtained $\log (D_s)$ versus L/kT relationship is linear. The departure from linearity commonly observed in fcc metal–vapor system is therefore not predicted to be the result of structural roughening of the (110) orientation surface.

1. Introduction

Detailed knowledge of the atomic structure of the crystal–vapor or crystal–melt interface is an essential ingredient in the formulation of theories involving mass transport at the interface. For example, the influence of interface roughness on crystal growth has been experimentally[1]) and theoretically[2]) shown to be of great importance. Recently, elegant experimental techniques have been applied to the determination of the surface self diffusion coefficient, D_s, for the (110) surface of several fcc metals. The results of these experiments were reviewed by Gjostein[3]), who noted that a plot of $\ln D_s$ versus T_M/T (T_M = melting point) produced a common curve for all of the available data. For $T \lesssim 0.75\, T_M$, $\ln D_s$ versus T_M/T is linear, while at higher temperatures $\ln D_s$ deviates positively from the low temperature line and D_s approaches a value of 3×10^{-4} cm^2 sec^{-1} at $T = T_M$. Theoretical explanations[3–6]) of this phenomenon rely to varying extent upon the supposition that structural roughening of the interface occurs in the $T_M > T > 0.75\, T_M$ interval. In this paper we summarize the results of a zeroth order approximation calculation of the (110) fcc crystal/vapor (melt) interface structure and assess the extent to which cooperative atomic interactions influence diffusion on this surface.

2. Formulation

Following procedures used in previous calculations[7])

we neglect vibrational and electronic effects and assume that each atom is bound to the crystal by nearest neighbor forces of strength, $\varepsilon = L/6$, where L is the enthalpy of vaporization or melting. In addition, we require that each atom in the ith (110) layer parallel to the interface be situated such that its nearest neighbor site in the $(i-2)$th layer is filled; i.e. the formation of overhanging configurations is not allowed. The equilibrium condition for this interface is:

$$\mathrm{d}X_i/\mathrm{d}t = 0 = (X_{i-2}-X_i)V_i^{+} - (X_i-X_{i+2})V_i^{-} , \quad (1)$$

where X_i is the fraction of filled sites in layer i, V_i^{+} is the condensation flux, and $V_i^{-} = \exp\,[-\varepsilon(1+4X_{i-1}+$

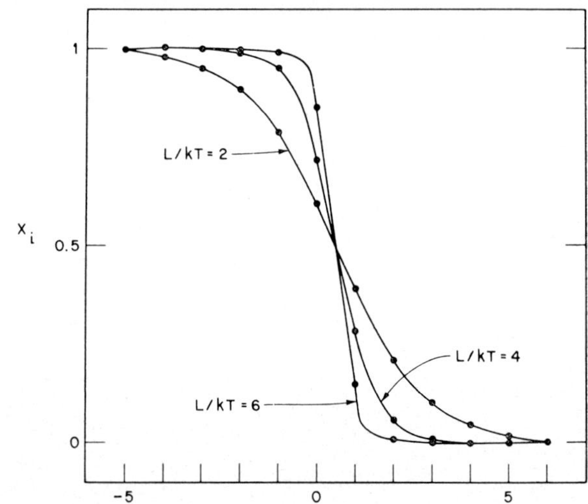

Fig. 1.   (110) fcc crystal–vapor interface profiles for various values of L/kT.

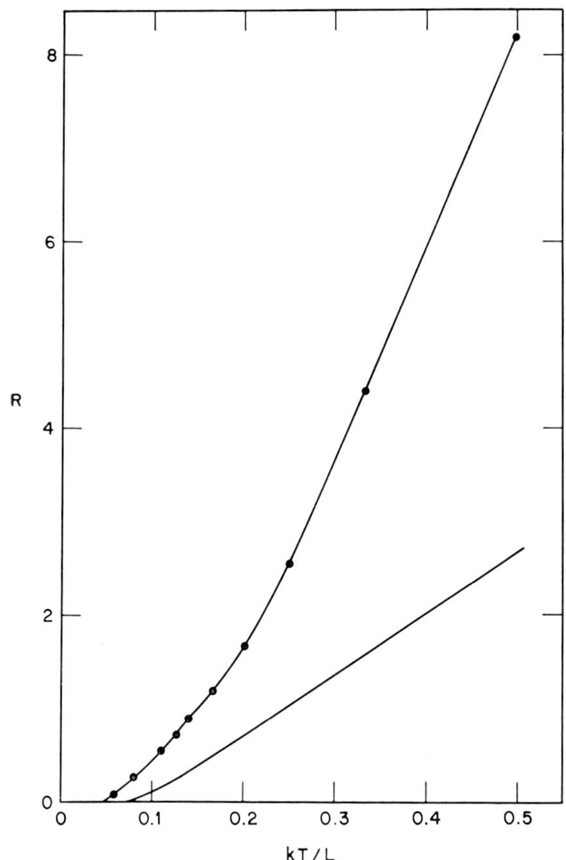

Fig. 2. The kT/L dependence of the roughness of the (110) fcc crystal–vapor interface. The solid curve is that expected from an equilibrium population of adatoms and surface vacancies.

$2X_i+4X_{i+1})/kT]$ is the probability that an atom will evaporate from layer i. The zeroth order approximation enters the calculation via the V_i^- term, in which a random distribution of atoms in each of the (110) planes is assumed. The equilibrium X_i, from which all of the properties of the interface may be calculated, are found by difference methods. Briefly, an initially smooth planar interface is allowed to "evaporate" a number of atoms, $\sum_i (X_i - X_{i+2})V_i^-$, during a time interval Δt. These are uniformly recondensed on the $\sum_i (X_{i-2} - X_i)$ available sites during the next interval. After many iterations the equilibrium state is approached and eq. (1) is satified. Examples of the X_i obtained in this way are given in fig. 1.

3. Results

As previous calculations[7]) have shown, macroscopic characterization of interface structure is best accom-

plished by computation of two parameters; the interface roughness, R, and extent, S. Interface roughness is defined as the energy per unit area of the interface relative to a smooth, flat surface and for this case is given by:

$$R = -6 + \sum_i X_i[1 + 4(1 - X_{i-1}) + 2(1 - X_i) + 4(1 - X_{i+1})] - X_{i+2}. \tag{2}$$

Interface extent is defined as the number of surface atoms per unit area relative to a smooth flat surface and is given by:

$$S = -2 + \sum_i X_i - X_{i+2}X_{i-1}^4 X_i^2 X_{i+1}^4. \tag{3}$$

R and S values computed from eqs. (2) and (3) are compared with those expected from a calculation of the equilibrium population of adatoms and surface vacancies in figs. 2 and 3. For this particular interface, two types of adatom-surface vacancy pairs are geometrically distinguishable; those involving 1 and 5 bonded adatoms. For large L/kT, the equilibrium concentration of surface defects will be simply

$$2 \exp\left[-2\varepsilon/kT\right] + 2 \exp\left[-10\varepsilon/kT\right].$$

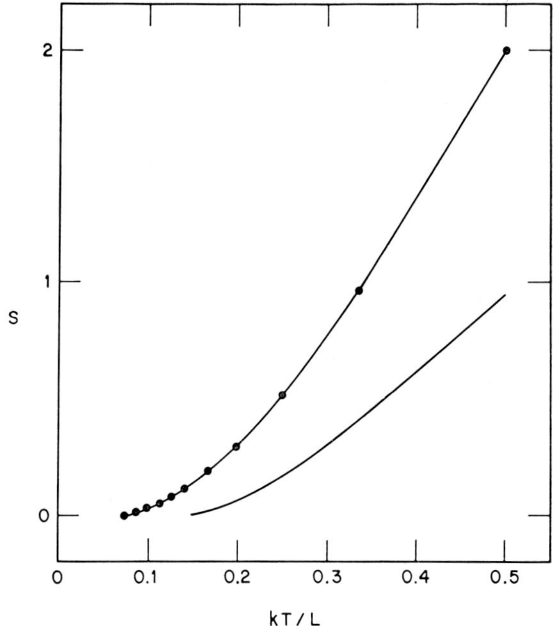

Fig. 3. The kT/L dependence of the extent of the (110) fcc crystal–vapor interface. The solid curve is that expected from an equilibrium population of adatoms and surface vacancies.

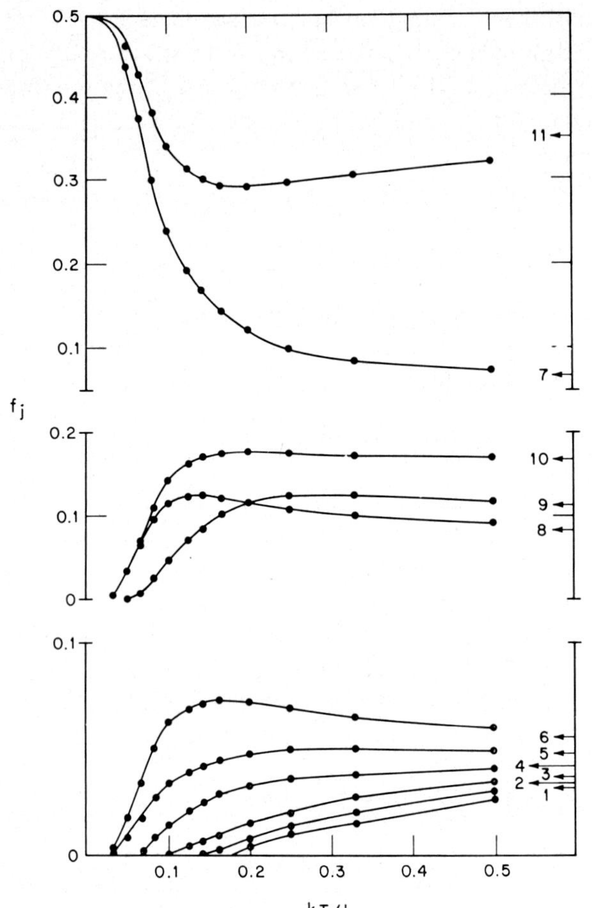

Fig. 4.   The relative population of the j surface energy levels at the (110) fcc crystal–vapor interface.

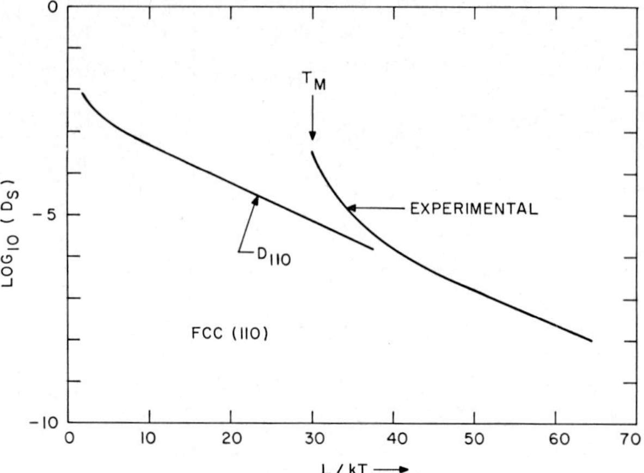

Fig. 5.   Log D_s versus L/kT for the fcc (110) surface. The dashed and solid lines are the theoretical and experimental[3]) results respectively.

4. Discussion

An equilibrium surface configuration as calculated here has the property that when all the surface atoms make statistically probable jumps they produce an equivalent configuration. The surface diffusion coefficient can thus be calculated directly for these configurations. The relationship between surface structure and surface diffusion is embodied in the mechanistic definition of D_s:

$$D_s = 1/4 \sum_{ij} m_{ij} l_{ij}^2 v_{ij} \exp\left[-\Delta G_{ij}/kT\right], \qquad (5)$$

where ΔG_{ij} is the activation energy associated with the transfer of an atom at site i to a vacant site at j, v_{ij} is the attempt frequency for this transfer, l_{ij} is the distance between i and j, m_{ij} is the surface density of ij combinations and the sum runs over all distinguishable ij. For the simple case of adatom migration via interatomic distance jumps on a smooth surface diffusion is a simple singly activated process and a ln D_s versus $1/T$ plot will be linear with slope given by the sum of the adatom formation and migration enthalpies. However, cooperative atomic interactions can produce a population of diffusing surface atoms which deviates from this, so that D_s will be greatly effected[8]). For single atom jumps, D_s will increase since roughening is accompanied by an increase in the number of jumping atoms.

Fig. 5 shows D_s as calculated from eq. (5), assuming that atoms jump singly from one site to the next, and that $\Delta G = \frac{1}{3}\varepsilon$. The calculated D_s agrees reasonably well

Figs. 2 and 3 show that (a) for $L/kT \gtrsim 20$, the interface is considerably rougher than would be expected from a simple adatom-vacancy model, and (b) there exists no easily discernable critical temperature for surface roughening.

In addition to the roughness and extent, the distribution of atoms over the surface energy levels may be obtained from the X_i by combinatorial algebra.

The results of such calculations are given in fig. 4 where the fraction of surface atoms having j neighbors, f_j, is plotted versus kT/L. The arrows on the right hand margin of fig. 4 indicate the values of f_j expected for an ideally rough interface, i.e. an interface for which the difference in energy between the j levels is zero. Fig. 4 shows that for $L/kT \lesssim S$ the interface approximates this ideally rough configuration while for $L/kT \gtrsim 20$ the surface is quite smooth.

with the low temperature experimental values shown in fig. 5. Although the calculated $\ln D_s$ versus L/kT does deviate from linearity, this occurs at temperatures much higher than those encountered in fcc metal-vapor systems. ($L/kT_M \sim 30$ for the systems studied experimentally). Fig. 5 shows that the nearest neighbor central force model employed here leads to an accurate prediction of the activation energy for diffusion at low temperatures.

Surface roughening could be made to account for positive curvature in the experimental data by reducing the energy difference between the 5, 6 and 7 bonded atom configurations. This would produce a rough or "liquid-like" surface similar to that envisioned by Rhead[6]) but would destroy the agreement between theoretical and experimental activation energy at lower temperatures. Alternatively, postulation of long jump distances would increase D_s without affecting this agreement, but would not change the temperature at which cooperative interactions affect changes in D_s. This concept has been extended by Bonzel[5]) who included contributions from atomic pairs and triplets to explain the high temperature behavior of D_s. Neither of these alternatives provides a completely satisfactory reconciliation of theory and experiment. A complete explanation may well require consideration of structural and vibrational effects for both low index and complex surfaces.

5. Conclusions

The equilibrium structure of a (110) surface of an fcc crystal has been calculated for a range of temperatures. The number of surface atoms, the surface roughness, and the number of surface atoms having various numbers of nearest neighbors were determined. From this, the surface self diffusion coefficient was calculated. The low temperature activation energy for diffusion is correctly predicted but the experimental diffusion coefficient is greater than that predicted by the model, and the calculated onset of surface roughening occurs at too high a temperature to account for the enhanced diffusion rate observed experimentally near the melting point.

References

1) K. A. Jackson, *Liquid Metals and Solidification*, (Am. Soc. Metals, Cleveland, Ohio, 1958).
2) H. J. Leamy and K. A. Jackson, to be published.
3) N. A. Gjostein, in: *Surfaces and Interfaces. I. Chemical and Physical Characteristics*, Eds. J. J. Burke, N. L. Reed and V. Weiss (Syracuse Univ. Press, Syracuse, N.Y., 1967).
4) H. P. Bonzel and N. A. Gjostein, in: *Molecular Processes on Solid Surfaces*, Eds. E. Drauglis, R. G. Gretz and R. I. Jaffe (McGraw-Hill, New York, 1969).
5) H. P. Bonzel, Surface Sci. **21** (1970) 45.
6) G. E. Rhead, Surface Sci. **13** (1969) 353.
7) H. J. Leamy and K. A. Jackson, J. Appl. Phys. **42** (1971) 2121.
8) W. K. Burton, N. Cabrera and F. C. Frank, Phil. Trans. Roy. Soc. London A **243** (1951) 299.

    Journal of Crystal Growth **13/14** (1972) 144–147 © *North-Holland Publishing Co.*

COMPUTER SIMULATION OF VAPOR DEPOSITION

A. C. ADAMS and K. A. JACKSON

Bell Telephone Laboratories, Incorporated, Murray Hill, New Jersey 07974, U.S.A.

The initial nucleation which occurs during vapor growth on a crystalline substrate has been simulated in a computer by the use of Monte Carlo methods. The crystalline surface is represented by a 30×30 array and the deposition site is selected randomly as is the alternative between adsorbed atom migration or evaporation. The output was programmed so as to be displayed graphically or tabulated. During each simulation run atoms were given the opportunity of moving between two and three thousand times. The result of simulations at three different temperatures where surface activation energies and binding energies were chosen so as to be on the range reported for silver on silver chloride showed that while good agreement between the simulation and the two-dimensional nucleation theory of Zinsmeister occurred at the start of nucleation, agreement could only be obtained at later stages if direct impingement of depositing atoms is considered.

In summary, the computer simulations were found to be in general agreement with two-dimensional nucleation theory. The most important points of disagreement are: (1) the temperature dependence of the total cluster concentration which is found is to be temperature-dependent at all temperatures investigated; and (2) the cluster size distribution, which is much broader than predicted. The simulation results indicate that the assumption of no direct impingement of depositing atoms is more critical than the assumption of the size independence of the collision cross section. Finally, it has been shown that the nucleation theory can be applied to the co-deposition of two different atomic species.

1. Introduction

Several theories have been developed to describe the initial nucleation that occurs during the vapor deposition of a material onto a crystalline substrate. The development of the theory usually requires that simplifying assumptions be made. Since it is often very difficult to investigate experimentally the effects of these assumptions, we have simulated in the computer the nucleation of a deposited material using Monte Carlo methods to determine the condensation, migration, and evaporation of the atoms on the surface. In these simulations, the effects of different assumptions are easily evaluated by changing the computer program.

2. The computer program

The program is similar to that used by Abraham and White[1]). The crystalline surface is represented by a 30×30 square array with periodic boundary conditions. During the simulation experiments, atoms are deposited on this array with the deposition site selected by two random numbers. After each deposition all atoms on the surface are given the opportunity of migrating to one of the four nearest neighbor sites or of evaporating. The occurrance of either of these processes is determin-

ed by the values of two random numbers. One of these numbers determines if the process is energetically favorable, the other determines if the direction of motion is favorable. For evaporation to occur, the atom must be moving in a direction perpendicular to and away from the surface. Migration can occur only when the atom is moving in one of the four directions parallel to the surface and when the nearest neighbor site in that direction is vacant. The number of times an atom may move between each deposition is determined by the input data. For most of the simulations, atoms were given the opportunity of moving twenty times before another atom was deposited. The program output may be a graphical display of the array at selected times during the simulation. Alternatively, the output may be a tabulation of the number of different sized clusters on the array at various times. Clusters with up to ten atoms were counted; if there were more than ten atoms in the cluster, the total number of atoms in these large clusters was listed. The program also reports the number of atoms that evaporate and the number of times that an atom attempts to deposit on a filled site. The simulations are usually continued until about one hundred atoms have accumulated on the surface. During such a simulation all the atoms are given the opportunity of moving between two thousand and three thousand times.

3. Results and discussion

The results of simulations at three temperatures (800, 1100, and 1400 °C) have been compared to the two-dimensional nucleation theory recently formulated by Zinsmeister[2-4]). In these comparisons, the adsorption energy is taken as 0.4 eV; the activation energy for surface diffusion is 0.08 eV; and the bonding energy of a pair is 1.6 eV. These energies are close to the values reported for silver on sodium chloride[5]). A plot of the adatom concentration versus time (fig. 1) shows good agreement between the simulation and the theory at the start of nucleation. However, by the time ten percent coverage is reached, the adatom concentration is decreasing more rapidly than predicted by the theory (the simulation shows the adatom concentration being nearly proportional to t^{-1}, the theory predicts a $t^{-\frac{1}{3}}$ behavior). This rapid decrease is also observed when the simulations are repeated using a very high adsorption energy so that evaporation of single atoms does not occur.

In developing the theory, it has been assumed that the cross section for the collision of a single atom with a cluster is independent of the cluster size. It has also been assumed that direct impingement of the depositing atoms onto adatoms or clusters does not occur. In the simulations, neither of these assumptions is made and both effects may contribute to the rapid decrease in the adatom concentration. To determine which of these is most important, the simulations have been repeated under conditions in which direct impingement of depositing atoms is not allowed. The results show a much better agreement between the simulation and the theory. This indicates that the theory will be considerably improved when direct impingement of depositing atoms is considered.

Comparisons of the time dependence of the concentrations of the different sized clusters with Zinsmeister's approximative curves[4]) (fig. 2) show several points of qualitative agreement: the small clusters appear before the large clusters; the maximum concentration is reached very quickly; at long times, the different sized clusters approach nearly the same concentration with the concentration decreasing with time; and, the concentration of a particular sized cluster decreases as the

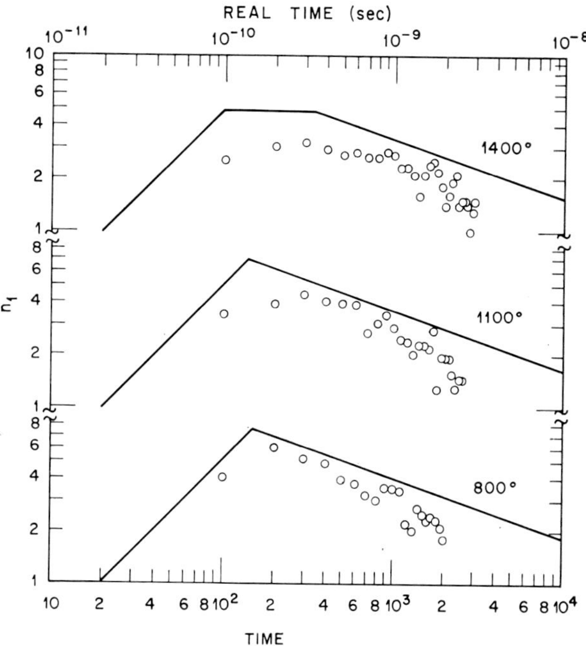

Fig. 1. Adatom concentration versus time for three temperatures, 1400 °C, 1100 °C, and 800 °C. The unit of time is the jump attempt interval. The lines are from Zinsmeister's theory.

temperature increases. However, a quantitative comparison shows a poorer agreement. In the simulation the curves of concentration versus time are much broader than those predicted by theory and the observed maximum concentration is less than expected.

Similarly the size distribution of the clusters on the array at a given time is different than that predicted by the theory. The observed distributions are much broader than the theoretical curves; similar broad distribution curves have been experimentally observed[6]).

Zinsmeister's theory allows the calculation of the total concentration of clusters as a function of time[2-4]). This quantity is very useful since it should correspond to the experimentally observed nuclei density. In terms of this quantity, the simulation results are in good agreement with the theory. Initially the total cluster density increases as t^3, while at long times the increase is proportional to $t^{\frac{1}{3}}$. Both regions are clearly discerned in the simulations (fig. 3). The major discrepancy concerns the temperature dependence of the total cluster concentration. The theory predicts the existence of a critical temperature below which the total cluster concentration becomes nearly temperature independent; the only temperature dependence enters through the

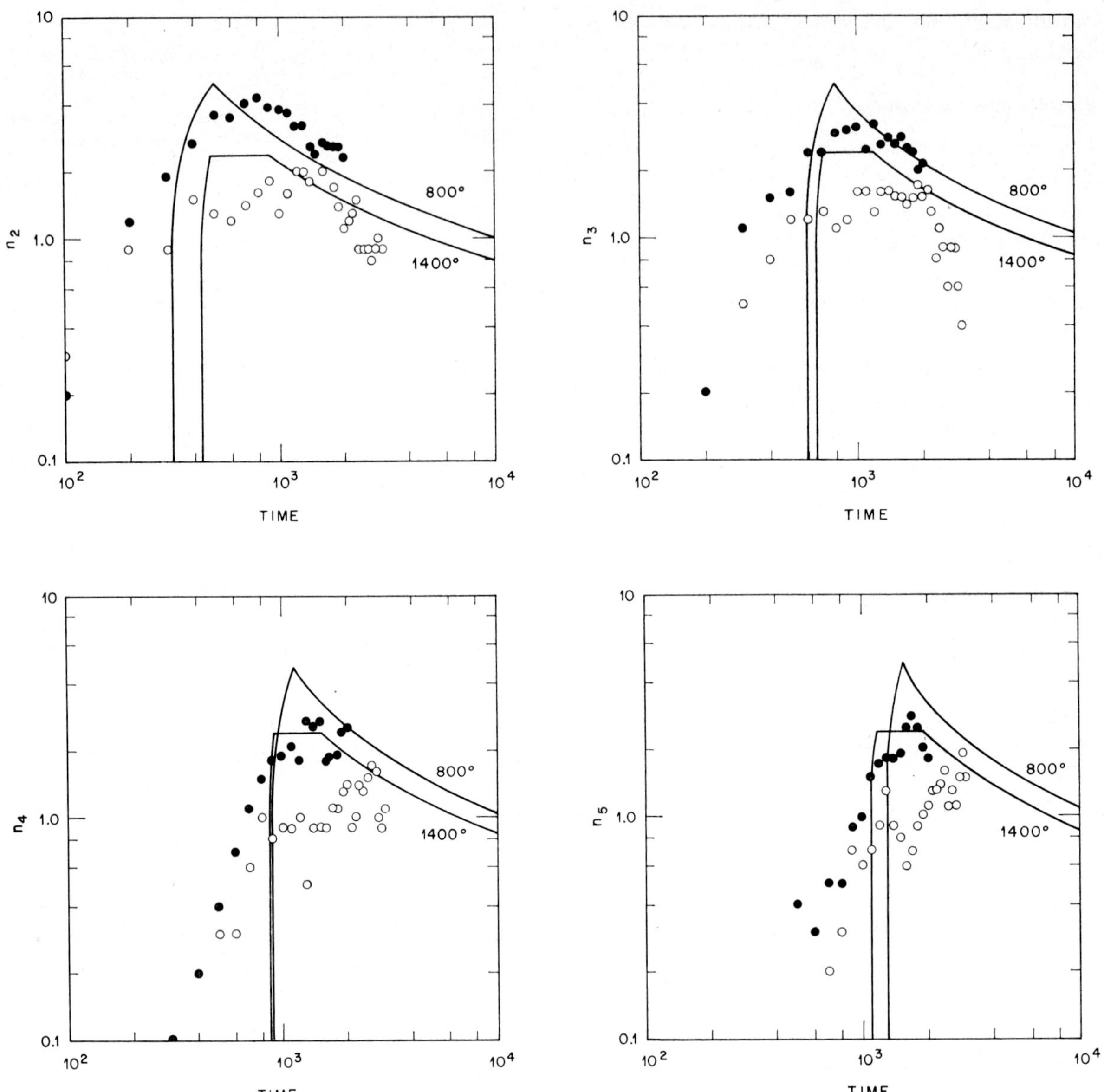

Fig. 2. Time dependence of the concentration of 2, 3, 4 and 5 atom clusters. (●) 800 °C, (○) 1400 °C. The lines are from Zinsmeister's theory.

term for the collision cross section. For the energies used in the simulation, this critical temperature should be near 1100 °C. However, this temperature independence is not observed in the simulations. The discrepancy is probably related to the evaporation of single atoms. Below the critical temperature Zinsmeister's approximate equations assume that evaporation can be neglected. In the simulation, evaporation still occurs at the lower temperature and is clearly temperature dependent. Thus as the temperature decreases below this critical temperature the number of atoms evaporating decreases and the concentration of adatoms increases which in turn allows the formation of more clusters. The net result is an increase in the total cluster concentration as the temperature decreases.

The simulations have also been performed for the case of two different species codepositing simultaneously. For the special case in which there is no inter-

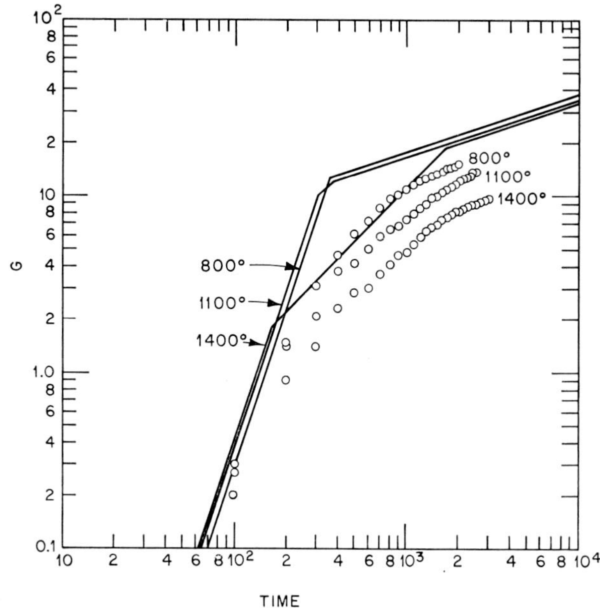

Fig. 3. Total concentration of clusters, G, as a function of time for 1400, 1100 and 800 °C. The lines are from Zinmeister's theory.

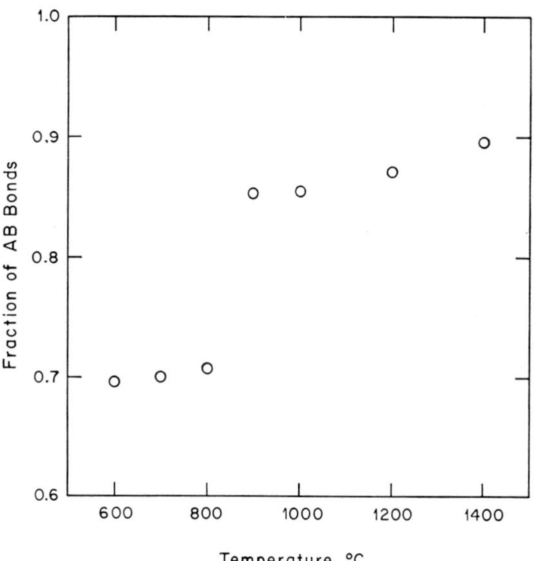

Fig. 4. Ordering versus deposition temperature for an AB alloy with $E_{AA} = E_{BB} = 0.3$ eV, $E_{AB} = 1.6$ eV.

action between unlike atoms, Zinsmeister's equations have been applied by assuming that there are two independent nucleation processes, one process for each species, and that the nucleation for the codeposition is the sum of the two indepedent processes. The time dependence of the total cluster concentration and the adatom concentration which are observed in the simulation are in good agreement with this theory. Similar agreement is observed when there is no interaction between like atoms and a strong interaction between unlike atoms.

For simulations in which there is a weak interaction between like atoms and a strong interaction between unlike atoms, it is observed that the clusters have an ordered AB structure at high temperatures and a disordered structure when nucleation occurs at low temperatures (fig. 4). The transition temperature which is sharply defined, can be calculated by equating the rate of formation and rate of dissociation of an AA or BB pair. The disorder at the lower temperature results because the clusters do not have time to form the most stable configuration.

In summary, the computer simulations are in general agreement with two-dimensional nucleation theory. The most important points of disagreement are: (1) the temperature dependence of the total cluster concentration which is found is to be temperature dependent at all temperatures investigated; and (2) the cluster size distribution, which is much broader than predicted. The simulation results indicate that the assumption of no direct impingement of depositing atoms is more critical than the assumption of the size independence of the collision cross section. Finally, it has been shown that the nucleation theory can be applied to the codeposition of two different atomic species.

References

1) F. F. Abraham and G. M. White, J. Appl. Phys. **41** (1970) 1841.
2) G. Zinsmeister, Vacuum **16** (1966) 529.
3) G. Zinsmeister, Thin Solid Films **2** (1968) 497.
4) G. Zinsmeister, Thin Solid Films **4** (1969) 363.
5) D. Walton, T. N. Rhodin and R. W. Rollins, J. Chem. Phys. **38** (1963) 2698.
6) H. Poppa, J. Vacuum Sci. Technol. **2** (1965) 42.

COMPUTER SIMULATION OF CRYSTAL SURFACE STRUCTURE AND GROWTH KINETICS

G. H. GILMER* and P. BENNEMA

Technical University of Delft, Laboratory of Physical Chemistry, Delft, The Netherlands

A model is described for simulation of a crystal–fluid interface of unlimited thickness. The interface changes from a smooth, flat surface at low temperature to a rough, irregular surface at high temperature. The crystal growth rate is measured versus undercooling or supersaturation for various conditions. The growth rate of a perfect close-packed surface at low temperature shows an exponential dependence on the supersaturation indicating a heterogeneous nucleation barrier. As the temperature increases, the coefficient of the exponent decreases to zero. The growth rates of vicinal surfaces are significantly higher than those of close-packed surfaces only when the temperature is below the point where the nucleation barrier disappears.

1. Introduction

The interface between a crystal and the surrounding fluid changes from a thin transition region at low temperature to a disordered mixture of solid and fluid at high temperature[1]. This structure has an important effect on the growth rate of the crystal since a growth unit can easily adhere to a rough interface at almost any position, but it will have difficulty adhering to a smooth, flat interface. Jackson[2], Mutaftschiev[3], and Temkin[4] have examined the equilibrium and metastable states of interfaces of one, two and an unlimited number of molecular layers respectively. They find that metastable states exist at low supersaturation (or undercooling) for an interface temperature below the point at which appreciable roughening occurs, and conclude that a layer mechanism such as two-dimensional nucleation or spiral growth is required in this region. A rapid increase in growth rate is expected as the supersaturation rises above the limit of existence of the metastable states, and this is called a transition from layer growth to continuous growth[5].

A very important factor in the interface structure is the tendency of molecules of the same phase to cluster together and thereby reduce the surface energy, but the Bragg–Williams approximation used in most of these investigations[2-4] assumes a completely random distribution of solid and fluid molecules in each layer. Clusters of solid molecules are potential nuclei which

can grow out into a new layer in the presence of a supersaturation, and therefore play a critical role in the growth process.

Computer simulation can provide equilibrium averages[6] and the rates of processes occurring in a non-equilibrium environment[7] without the severe assumptions necessary for most analytical solutions. We have developed a crystal-fluid interface simulation model which allows for clustering, an unlimited interface thickness, and unrestricted motion of the interface. The growth rate of the crystalline phase due to a supersaturated fluid can be measured, and any effect of the metastable states should be easily observed. Our computer program is apparently similar in operation to that of Abraham and White[8] for the study of vapor deposition; our program measures the average growth rate after a large displacement of an interface which initially has the equilibrium structure, whereas their program measures the growth of an adsorption monolayer on a flat substrate. The kinetics of both models are so chosen that the probability for a growth unit from the fluid to join the crystal is independent of position. Binsbergen[9] examined nucleation theory with a computer simulation program and included a few runs in which he allowed several layers to be deposited on a flat initial substrate. He measured a growth rate for the process, but a quantitative comparison with his results is impossible since the kinetics he used made it extremely difficult for a growth unit from the fluid to join the crystal as an isolated adsorbed molecule.

The growth processes reported here occur without

* On leave from Washington and Lee University, Lexington, Virginia, U.S.A., under a National Science Foundation grant.

surface diffusion, and the results are probably most appropriate for melt growth. We have performed extensive simulations with surface diffusion and will report this data in another paper[10]). Surface diffusion increases the rate of growth in all cases investigated, but the shapes of the rate of growth versus supersaturation curves are quite similar.

2. Model of the crystal–fluid interface

The molecules of the solid phase occupy a primitive cubic (or tetragonal) lattice and the interface is parallel to the (001) plane. We assume that each growth unit belongs to either the solid or fluid phase (there is no intermediate condition); and solid molecules are only allowed above other solid molecules, where "above" is defined as the direction of increasing fluid concentration.

The surface energy is determined by the number of solid–fluid bonds in the interfacial region since the energy of these bonds is usually not as high as the average of solid–solid and fluid–fluid bonds. ϕ_{SS} is the lateral bond energy between two neighboring solid molecules in the layer of the interface. In the same way ϕ_{FF} represents the bond energy between two fluid molecules*, and ϕ_{SF} the bond energy between a solid and a fluid molecule, and

$$\gamma = 2\frac{(\phi_{SS}+\phi_{FF})-2\phi_{SF}}{kT} \tag{1}$$

is a critical parameter which determines the average number of lateral solid–fluid "bonds" in equilibrium.

The exchange of growth units between the solid and fluid phases is governed by the kinetic rate constants for the transition of a growth unit from the fluid phase to the solid (which we will call a creation), and for the transition from solid back to the fluid phase (an annihilation). We assume that the rate constant k^+ for a creation is independent of the position on the surface. With $\gamma \neq 0$, the potential energy of a growth unit of the solid phase at a surface position and therefore the rate constant for its annihilation depends on the number of its horizontal solid nearest neighbors.

* More precisely, ϕ_{FF} is the interaction between the average contents of two adjacent unit cells of the lattice extended into the region occupied by the fluid, and a similar definition holds for ϕ_{SF}. It is not necessary to assume that the fluid molecules fit into a lattice isomorphous with that of the crystal.

There are five possible sites with 0, 1, 2, 3 or 4 horizontal solid neighbors and rate constants k_0, k_1, k_2, k_3 and k_4 respectively. Since k^+ is independent of position, the free energy of the activated transition state must also be independent of position. Using the theory of rate processes we can show

$$k_1 = k_0\,e^{-\gamma/2}, \quad k_2 = k_0\,e^{-\gamma},$$
$$k_3 = k_0\,e^{-3\gamma/2}, \quad k_4 = k_0\,e^{-2\gamma}, \tag{2}$$

since a site with one horizontal nearest neighbor has a potential energy $\frac{1}{2}\gamma$ lower than a site with none, and so forth. Also, in equilibrium

$$k^+ = k_2, \tag{3}$$

since a face with 100% kink sites must not grow (the rate of creation must equal the rate of annihilation). This determines the rate constants except for a common multiplicative constant, which will only affect the time scale of the process.

When a crystal–melt interface is undercooled, or when a vapor or solution is supersaturated, the effect for a kinked face is to increase the rate of creation as compared with the annihilation rate, and the face will grow. Then

$$k^+ = \exp\,(\Delta\mu/kT)k_2, \tag{4}$$

where $\Delta\mu = \mu_F - \mu_S$ is the difference in chemical potential of growth units in the fluid adjacent to the surface, and in the solid phase. In a dilute solution or vapor,

$$\Delta\mu = kT\ln\frac{c}{c_0} \quad \text{and} \quad k^+ = \frac{c}{c_0}k_2,$$

where c_0 and c are the equilibrium and supersaturated concentrations, respectively. For melt growth, $\Delta\mu = L(\Delta T/T)$, where L is the latent heat of fusion and ΔT is the undercooling.

3. Computer simulation

A computer program in Fortran IV was used with the IBM 360/65 of the Computer Center of the Technical University at Delft. The "solid above solid" assumption implies that the configuration of the interface can be completely specified at any instant of time by giving the height of the solid–fluid boundary at every molecular position on the surface. The interface is therefore represented by a 20×20 (or 40×40) array

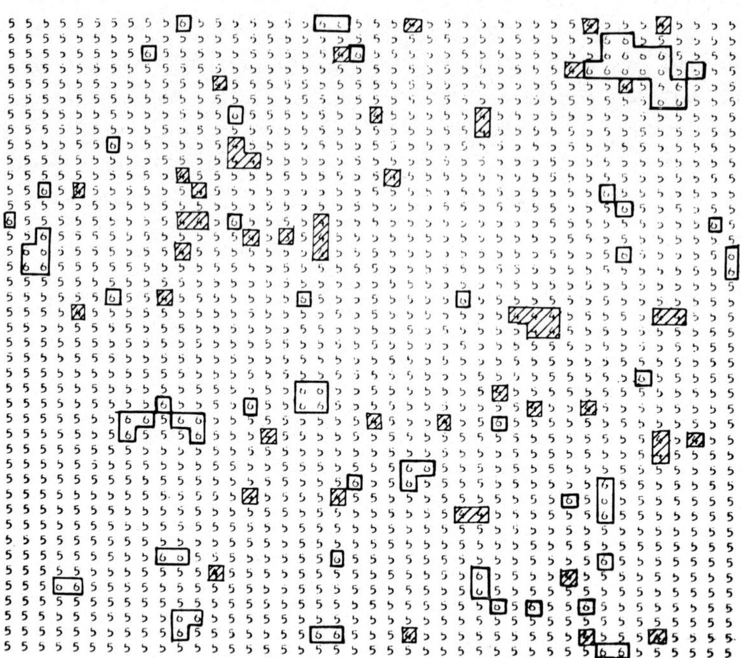

Fig. 1. Equilibrium surface structure for $\gamma = 4$. Cross-hatched areas are lower than the average surface height of approximately five molecular diameters.

of integers with each integer representing the height of the solid–fluid boundary in molecular diameters. Periodic boundary conditions are used to provide neighbors for molecules at the edges of the array.

The exchange between the solid and fluid is simulated by an endless repetition of the following sequence of steps for a 20×20 array. (1) A site on the surface is chosen using a random integer between 1 and 400. (2) A decision on whether to attempt a creation or an annihilation at the chosen site is made using another random number. (3) A decision on whether to execute the creation or annihilation is made with a third random number in such a way that the probability of executing the event is proportional to the rate constant for the event. The creation or annihilation is executed by raising or lowering the surface one unit at the chosen site. One million annihilation and creation attempts are made in two minutes of computer time.

The growth rate is measured by counting the total number of solid molecules at the beginning and at the end of the process, subtracting, and dividing by a time parameter. Since the rate constant for creation is constant, the number of creation events is used as a measure of elapsed time.

The rate of growth R is given by

$$R = \frac{d}{N}\,(J^+ - J),\tag{5}$$

where d is the layer thickness or interplanar distance, N is the number of sites per unit area in a (001) cross-section, J^+ is the flux of molecules being created on the solid surface, and J is the time averaged annihilation flux. J^+ and J are related to the rate constants by

$$J^+ = Nk^+, \quad J = N \sum_{i=0}^{4} \bar{C}_i k_i,\tag{6}$$

where $\bar{C}_i$ is the time averaged fraction of surface sites of type i. Thus

$$R = \frac{d}{N}\,J^+ \left\{ 1 - \sum_{i=0}^{4} \bar{C}_i \exp\left[-(\tfrac{1}{2}i - 1)\gamma - \frac{\Delta\mu}{kT} \right] \right\}.\tag{7}$$

Our program makes regular surveys of the interface structure in order to determine $\bar{C}_i$. Comparison of the actual growth rate with the prediction of eq. (7) provides a convenient and sensitive self-consistency check.

Figs. 1, 2 and 3 show 40×40 arrays representing interfaces at equilibrium. These are all produced from an initially flat interface with each site at a height of five units, and are the results of approximately 10^6 attempted creations and annihilations (600 per site).

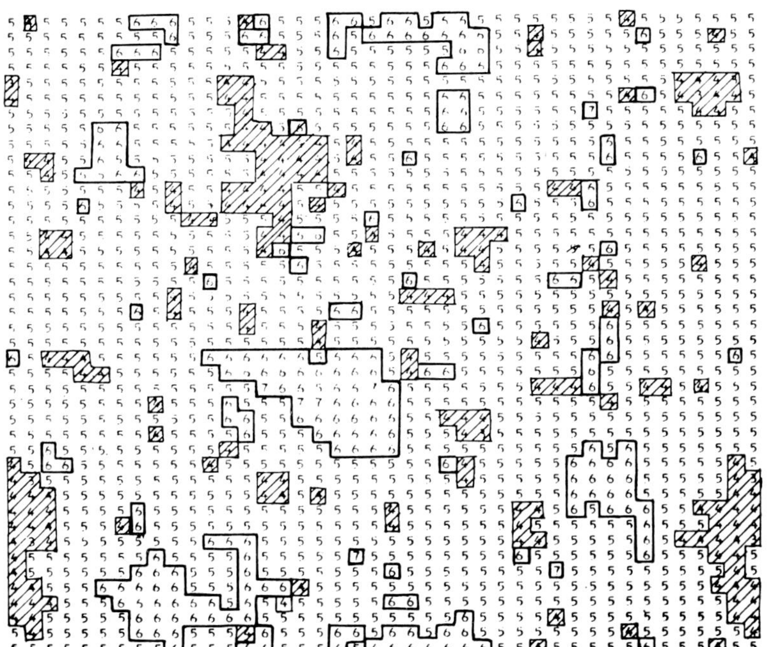

Fig. 2. Equilibrium structure for $\gamma = 3.5$.

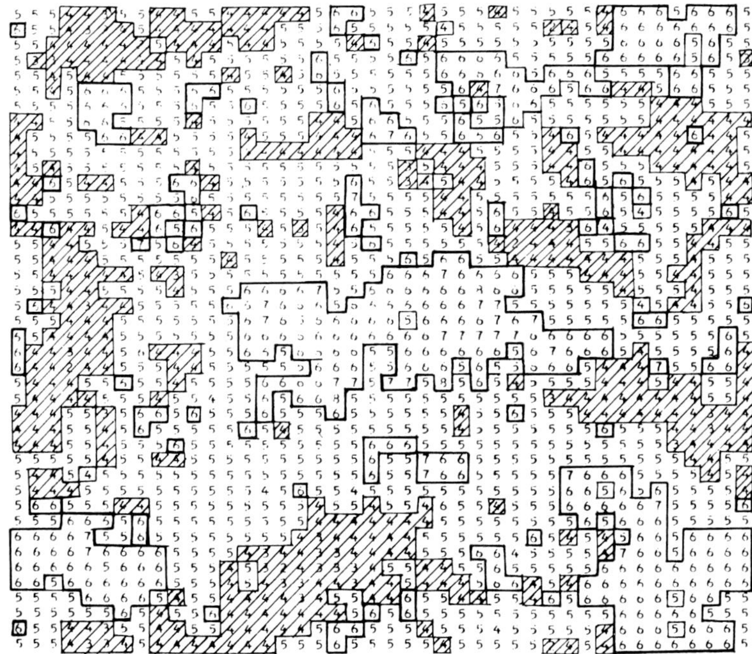

Fig. 3. Equilibrium structure for $\gamma = 3$.

Note the transition from a smooth, flat surface at $\gamma = 4$ to a rough, irregular surface at $\gamma = 3$. One would not expect the nucleation of new layers to be very difficult when large clusters of solid molecules exist in equilibrium as for $\gamma = 3$.

Fig. 4 gives the resulting growth rate when $k^+ > k_2$, or $\Delta\mu > 0$*. The solid line with the steepest slope gives

* Most of the data points represent the average rate during the growth of the equivalent of twenty layers of the solid or a net gain of 8000 growth units.

the rate of growth of a kinked face [from eq. (7)]; and even at $\gamma = 2.5$ the growth rate of the (001) face is far below that of a kinked face.

Nucleation theory is the obvious way to try to explain our growth rate results since the clustering tendency is included. The growth rate was derived by Hillig[11]) for a nucleation rate which is so fast that a new critical nucleus forms on the top of an earlier nucleus before it has covered the surface. This gives

$$R = D_0 \beta^{\frac{5}{6}} \exp(-\pi \gamma'^2 / 3\beta), \qquad (8)$$

where D_0 is a constant independent of $\beta = \Delta\mu/kT$, and $kT\gamma'$ is the excess free energy of a growth unit in the edge of a nucleus.

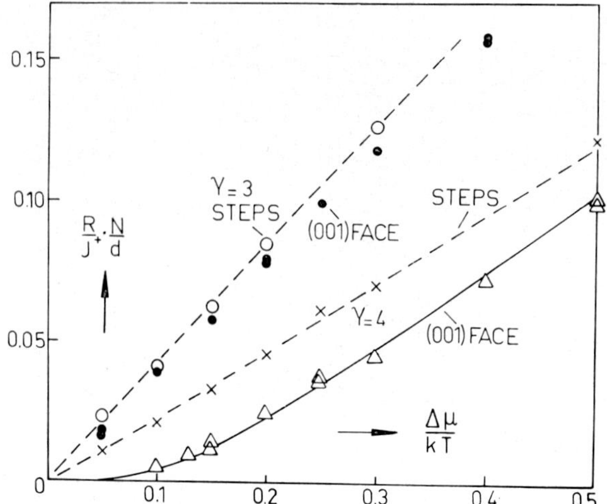

Fig. 5. Growth rate with and without steps for $\gamma = 3$ and $\gamma = 4$.

model[10]). The assumption that nuclei form on top of earlier nuclei is apparently accurate for most of our data since the growth rates of 20×20 and of 40×40 matrices are the same except for high γ and low β ($\gamma = 4$, $\beta = 0.1$) where the 20×20 growth rate was somewhat less.

The effect of steps on the growth rate is also examined. The boundary conditions are adjusted to maintain a difference in level of five molecular diameters between the left and right sides of the surface. This produces five steps in the 20×20 section of the surface. Figs. 5 and 6 show the growth rate at $\gamma = 3$, 3.5 and 4 with and without steps. The stepped growth rate is approximately linear in supersaturation (dashed line), but it is

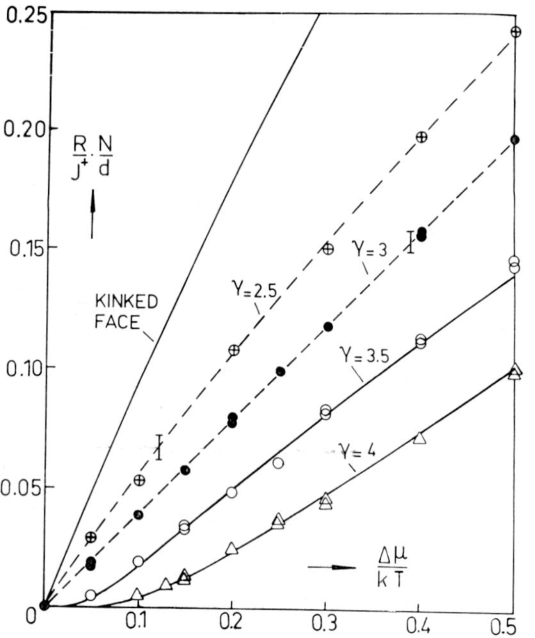

Fig. 4. Growth rate versus $\Delta\mu/kT$ for various values of γ. The growth rate for a kinked face is the solid curve to the left. The vertical lines on the $\gamma = 2.5$ and $\gamma = 3$ curves indicate the upper limit of Temkin's metastable states.

The solid curve through the $\gamma = 4$ data represents eq. (8) with $D_0 = 0.291$ and $\gamma' = 0.49$. This value of γ' is not unreasonable since the excess energy of an edge molecule at $\gamma = 4$ is about 1 kT, and entropy effects will reduce the excess free energy. For $\gamma = 3.5$, $D_0 = 0.314$ and $\gamma' = 0.3$; for $\gamma = 3$ and $\gamma = 2.5$, $\gamma' \cong 0$.

It can be shown that the value of D_0 is consistent with Hillig's expression and a very simple step growth

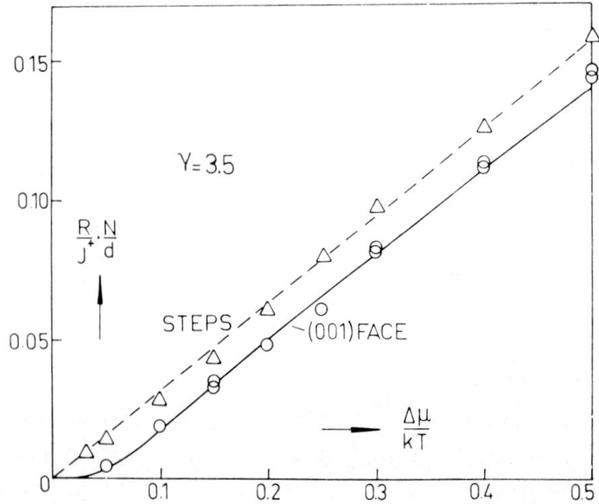

Fig. 6. Identical to fig. 5 except for $\gamma = 3.5$.

still far below the maximum possible growth rate, which would occur if we had a completely kinked face. Even at $\gamma = 3$ the steps cause a small increase over the (001) growth rate.

4. Conclusions

(1) The growth rate curves with and without steps do not show any indication of a change in growth mechanism as $\Delta\mu$ is increased. The analysis of Temkin implies that the metastable states cease to exist at $\Delta\mu \cong 0.38\,kT$ for a $\gamma = 3$ interface, and yet no sudden change in growth rate is observed here or at any other value of $\Delta\mu$ up to $0.5\,kT$. The $\gamma = 3$ equilibrium interface is so rough that any metastable states must be limited to very small $\Delta\mu$ values. At $\gamma = 3.5$ and $\gamma = 4$ there is a smooth increase in the growth rate of the (100) face characteristic of a nucleation curve and we believe that the concept of continuous growth in this case is isomorphous with rapid two-dimensional nucleation.

(2) The edge free energy of steps on the (100) face apparently becomes zero between $\gamma = 3.5$ and $\gamma = 3$ since the $\gamma = 3.5$ surface shows a nucleation barrier to growth and the $\gamma = 3$ surface does not. The surface should become quite rough when the formation of steps involves no increases in its free energy, and the rough $\gamma = 3$ surface in fig. 3 certainly is consistent with a very small edge free energy.

The BCF interface model has a critical temperature for roughening which corresponds to $\gamma = 3.53$, only a slightly lower temperature than that of the simulation model. On the other hand, all of the Bragg–Williams models roughen at a much higher temperature corresponding to $\gamma \cong 2$. The close agreement between the BCF and the simulation models probably results from their proper treatment of clustering.

(3) The fact that steps increase the rate of growth of the surface for γ as low as three implies that the surface may have some stability against macroscopic roughening during growth. If one section has a higher than average $\Delta\mu$ between the solid and adjacent fluid, it will grow faster; but as the neighboring areas lag behind, the steps which form will increase their growth rate and tend to correct the deviation in height.

References

1) W. K. Burton, N. Cabrera and F. C. Frank, Phil. Trans. Roy. Soc. London A **243** (1951) 299.
2) K. A. Jackson, in: *Liquid Metals Solidification* (ASM, Cleveland, 1958) p. 174.
3) B. Mutaftschiev, in: *Adsorption et Croissance Cristalline* (CNRS, Paris, 1965) p. 231.
4) D. E. Temkin, in: *Crystallization Processes* (Consultants Bureau, New York, 1966) p. 15.
5) J. W. Cahn, W. B. Hillig and G. W. Sears, Acta Met. **12** (1964) 1421.
6) D. A. Chesnut and Z. W. Salsburg, J. Chem. Phys. **38** (1963) 2861.
7) R. Gordon, J. Chem. Phys. **49** (1968) 570.
8) F. F. Abraham and G. M. White, J. Appl. Phys. **41** (1970) 1841.
9) F. L. Binsbergen, Kolloid. Z. **237** (1970) 289.
10) G. H. Gilmer and P. Bennema, submitted to J. Appl. Phys.
11) W. B. Hillig, Acta Met. **14** (1966) 1868.

AN OUTLINE OF NUCLEATION THEORY*

F. C. FRANK

H. H. Wills Physics Laboratory, University of Bristol, Bristol, England

I think it is a good thing to re-examine the basis of nucleation theory. To assist that re-examination I wish to offer the simplest of outlines of the theory to provide a background against which one may see where approximations are involved, where refinements are necessary. Let us consider any sequence of processes passing through a sequence of denumerable stages with first order rate constants:

$$(0) \underset{b_1}{\overset{f_1}{\rightleftarrows}} (1) \underset{b_2}{\overset{f_2}{\rightleftarrows}} \ldots \underset{b_n}{\overset{f_n}{\rightleftarrows}} (n) \underset{b_{n+1}}{\overset{f_{n+1}}{\rightleftarrows}} \ldots$$

The system is governed by the equations

$$j_n = q_{n-1} f_n - q_n b_n, \tag{1}$$

and

$$dq_n/dt = j_n - j_{n+1}, \tag{2}$$

where q_n is the occupation number of stage n, f_n and b_n are the forward and backward transition rate constants to and from stage n, j_n is the net transition rate from stage $(n-1)$ to stage n, and t is time.

This is applicable to two-dimensional nucleation on a crystal surface, n denoting the number of atoms (or molecules, etc.) aggregated in a nucleus and q_n the number of such aggregates per unit or other prescribed area (q_0 being then the number of vacant atomic sites, q_1 the number of singly adsorbed atoms) subject to two approximating assumptions:

(a) Transitions between non-adjacent stages have a

* Contributed discussion submitted prior to the meeting.

relatively negligible rate: such transitions are the union of a pair of nuclei of n and m atoms respectively ($n, m > 1$) or its converse.

(b) The rate of transformation between different configurations of nuclei of given n is sufficiently high compared with the transition rate between stages, or alternatively the rate constants f_{ni}, b_{ni}, for the various configurations denoted by i are sufficiently similar to each other, that suitably weighted average rate constants f_n, b_n can be employed in place of a more detailed analysis.

Non-validity of condition (b) would only complicate the problem without changing its nature. Furthermore, it only affects the transient behaviour of the system: it is without any influence on steady state behaviour.

Let us now write

$$f_n/b_n = k_n, \quad \prod_{1}^{n} k_n = C_n; \tag{3}$$

k_n and C_n would be equilibrium constants, if such end conditions were imposed on the system that an equilibrium could exist.

Let us further write

$$V_n = q_n/C_n, \tag{4}$$

$$R_n = 1/C_n b_n. \tag{5}$$

Then equations (1) become

$$j_n = (V_{n-1} - V_n)/R_n, \tag{6}$$

identically those of an electrical resistance–capacity ladder:

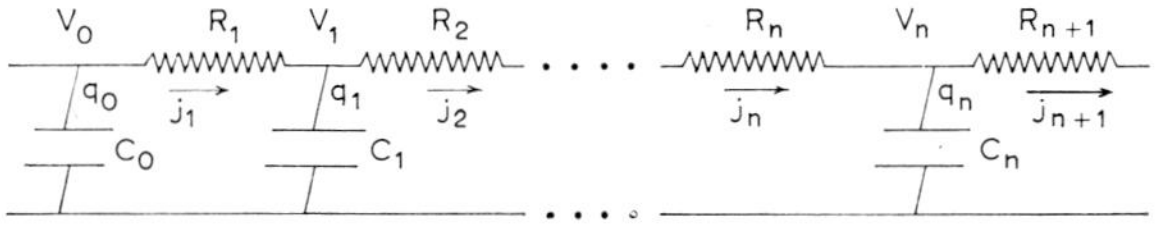

It is not a simple matter to use this analogue experimentally, because a change of conditions, either ambient vapour pressure or temperature, changes all the parameters: but the general behaviour of the system becomes self-evident. In particular, the steady state behaviour becomes simply that of the chain of resistances in series: if those for very large n are very small

$$j = V_0 / \sum_1^\infty R_n. \tag{7}$$

Conveniently, we express the equilibrium constants C_n in terms of free energies g_n:

$$C_n = \exp\left(-g_n/kT\right). \tag{8}$$

If the nuclear aggregate, fluctuating over its various configurations, is compact enough to have a definable periphery l, and the vapour pressure is low enough for use of the ideal gas law, we have

$$g_n = -nkT \ln\left(p/p_0\right) + \int \gamma \, dl, \tag{9}$$

where p is the ambient vapour pressure, p_0 its equilibrium value with the bulk crystal, and γ the specific edge free energy of the two-dimensional nucleus. As further approximations we may treat γ as constant, making the final term $l\gamma$, and estimate l as $4n^{\frac{1}{2}}a$, where a is an atomic length. Eq. (9) thereby becomes

$$g_n/kT = -n \ln\left(p/p_0\right) + 4n^{\frac{1}{2}}\gamma a/kT, \tag{10}$$

from which it follows that for $p > p_0$, g_n/kT has a quasi-parabolic maximum at

$$n = n^* = \left[2\gamma a/kT \ln\left(p/p_0\right)\right]^2. \tag{11}$$

Since b_n^{-1} is a relatively slowly varying function of n in

$$R_n = b_n^{-1} \exp\left(g_n/kT\right), \tag{12}$$

this corresponds to a quasi-Gaussian distribution of high values of R_n in the neighbourhood of n^*. The quantity g_n is within kT of its maximum in a range of order of magnitude $\pm(n^*)^{\frac{1}{2}}$ about n^*. Resistances in this range make the dominant contribution to the total nucleation resistance $\sum R_n$. At the maximum,

$$g_n/kT = g_n^*/kT = n^* \ln\left(p/p_0\right). \tag{13}$$

In further approximation we may replace edge free energy γ by edge energy (neglecting entropy corrections, including the entropy of multiple configurations). On the most closely packed faces of a crystal we may estimate this from a simple model as an energy $\frac{1}{2}\phi$ per atomic length a, where 3ϕ is the lattice energy. The final term of (9) then becomes $2\phi n^{\frac{1}{2}}$ and we obtain in place of (11)

$$n^* = \left[\phi/kT \ln\left(p/p_0\right)\right]^2. \tag{14}$$

If the vapour pressure is less than about 1 atm, Trouton's rule tells us that ϕ/kT is at least 3. A more usual figure would be about twice this. Thus for a supersaturation of about 80% we should have n^* about 100 and g_{n*}/kT about 60: this is a typical order of magnitude for g_{n*}/kT corresponding to nucleation at a measureable rate.

On less closely packed faces γ, and hence n^* and g_{n*}, will be less. The entropy corrections also reduce γ, and consequently n^* and g_{n*}, on the close-packed faces, but not gravely when $\phi/kT \gtrsim 6$, as we shall presently indicate.

Two points may be made at this stage. Irregular variations of g_n with n (e.g. minima when n is a perfect square), for values of $n \ll n^*$ are of no account, because the corresponding values of R_n are in any case very small.

Secondly, such irregularities in the neighbourhood of n^*, insofar as they exist, are of little account, because the total nucleation resistance does not come at a single "critical nucleus" but from a number, a few times $(n^*)^{\frac{1}{2}}$, of large values of R_n in the neighbourhood of n^*. These irregularities are in any case made small by multiple configurations of low excitation energy.

At any one value of n there are w_j distinct configurations for each value ε_j of the total edge energy. Distinguishing these energy levels, which in the simple model increase by steps of ϕ, we have

$$\exp\left(-g_n/kT\right) = C_n = \sum_j w_{nj} \exp\left(-\varepsilon_{nj}/kT\right). \tag{15}$$

Here non-configurational entropy corrections are disregarded, for simplicity. The approximation leading to (14) corresponds to taking only the first term of the series (15), with the "ground state" value ε_{n0} of ε_{nj}, and putting $w_{n0} = 1$. To be accurate this requires the series in (15) to converge strongly. For $\phi/kT = 6$, the Boltzmann factor falls by a factor of 400 at successive terms: but, for $n > \sim 25$, the configuration number for the lower excited states rises even more rapidly, though the Boltzmann factor wins in the end. In consequence, for $n \sim 100$ the sum in (15) may be perhaps a million times its first term, and the nucleation rate larger by a corresponding factor. This could seem serious if ϕ in (14) were regarded as an exact physical parameter rather than, as it is, a model parameter leading to an approximate estimation of γ from first principles. γ is in principle directly measurable from orientation dependence of surface free energy of crystals. An experimental value should be used when possible. When $n^* \sim 100$, the factor of 10^6 in nucleation rate corresponds to a reduction in γ by about 13%: a reduction attributable to configurational entropy.

At sufficiently high temperature the series in (15) is dominated by the terms representing very highly ramified configurations of nuclei. In the simple square-array nearest-neighbour-interaction model this occurs for $\phi/kT < \sim 2$. This condition corresponds to the vanishing of γ, at the critical temperature of surface "melting" when the crystal face becomes non-singular, and there is no nucleation problem at all.

It would nevertheless be desirable to have a more thorough investigation of the numbers of configurations for various excitation energies in nuclei of about 100 atoms, for some reasonably realistic but tractable model, since the use of γ, a quasi-continuum quantity, for processes on this scale, cannot be completely satisfactory.

Section III

Adsorption and heterogeneous nucleation from the vapour phase

A. THOMY
J. REGNIER
J. MENAUCOURT
X. DUVAL
J. SUZANNE
G. ALBINET
M. BIENFAIT

R. J. H. VOORHOEVE
R. S. WAGNER
J. N. CARIDES
NGUYEN TAN TUAN ANH
R. CINTI
B. K. CHAKRAVERTY

Journal of Crystal Growth **13/14** (1972) 159–163 © *North-Holland Publishing Co.*

LES DIFFÉRENTES ÉTAPES DE LA FORMATION D'UN FILM ADSORBÉ DE GAZ RARE OU DE MÉTHANE SUR UNE SURFACE UNIFORME DE GRAPHITE

A. THOMY, J. REGNIER, J. MENAUCOURT et X. DUVAL

Centre de Cinétique Physique et Chimique du C.N.R.S., Route de Vandoeuvre, 54 – Villers–Nancy, France

Many examples show that the growth by adsorption of a crystal on a uniform surface takes place by successive monomolecular layers, as a consequence, the adsorption isothermes present a series of steps.

The data obtained previously could be interpreted by supposing that each layer is formed either by building up of a single phase (steps of sigmoid shape) or by transition from a dilute to a condensed phase (steps with only one vertical part). The adsorption of rare gases or methane on the basal planes of graphite described in this paper, gives an example of a more complex growth of an adsorbed layer.

This is the case for the first layer which passes through three successive two-dimensional states: presumably "gas", "liquid" and "solid"; the two-dimensional solid may undergo rearrangements. The first step is then made up from several "sub-steps".

The comparison of our data with other data obtained by low energy electron diffraction shows that the first layer is different from the others not only by the diversity of the events accompanying its formation, but also by its higher density.

1. Introduction

Il existe actuellement d'assez nombreux travaux relatifs à l'adsorption montrant qu'une surface uniforme (face cristallographique parfaite) impose à un cristal de croître, à plus ou moins grande distance, par couches successives, selon des plans de type déterminé*. Cela se traduit dans les isothermes d'adsorption par une série de "marches", chacune d'elles correspondant à la formation d'une couche d'épaisseur monomoléculaire (cf. fig. 1).

En raison des résultats obtenus, on pouvait penser, jusqu'à ces derniers temps, que les différentes couches se formaient toujours soit par enrichissement progressif d'une seule phase (marches de forme sigmoïde), soit en passant d'un état dilué à un état condensé par l'intermédiaire d'un changement de phase du premier ordre (marches comportant une seule partie verticale). Or, l'étude détaillée des isothermes d'adsorption de gaz rares ou de méthane sur du graphite de surface homogène [presqu'exclusivement constituée de faces (0001)], a montré que l'on pouvait avoir affaire à des phénomènes plus complexes[2-4]). Si les deuxième, troisième, ..., marches ont effectivement une forme simple, ne comportant, à suffisamment basse température, apparement qu'une seule partie verticale (cf. fig. 1), il n'en est

* Parmi les travaux les plus récents, citons celui de Larher[1]) concernant toute une série de corps de structure lamellaire.

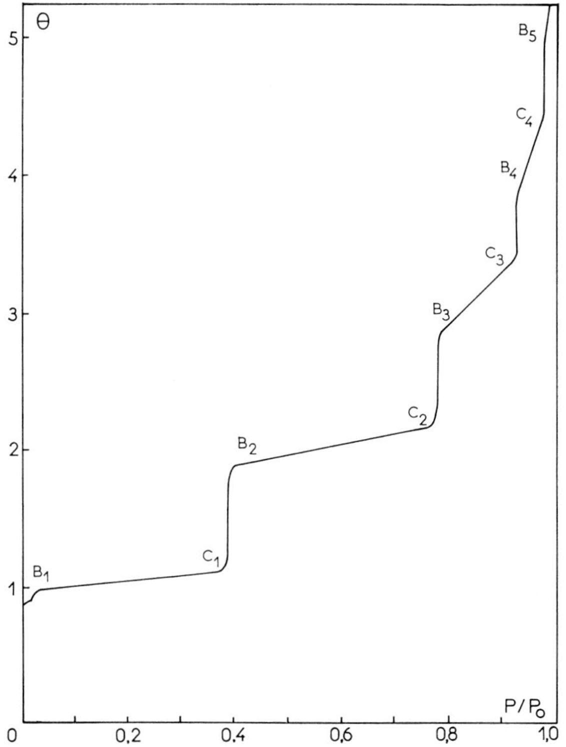

Fig. 1. Isotherme d'adsorption vers 85 °K de krypton sur une surface homogène de graphite. θ = degré de recouvrement de la surface – quantité adsorbée rapportée à la quantité adsorbée au point B_1 (point B habituel); P = pression du gaz en équilibre avec la phase adsorbée; P_0 = pression de vapeur saturante du gaz à la température considérée; OB_1C_1 = première marche; $C_1B_2B_2$ = deuxième marche, etc.

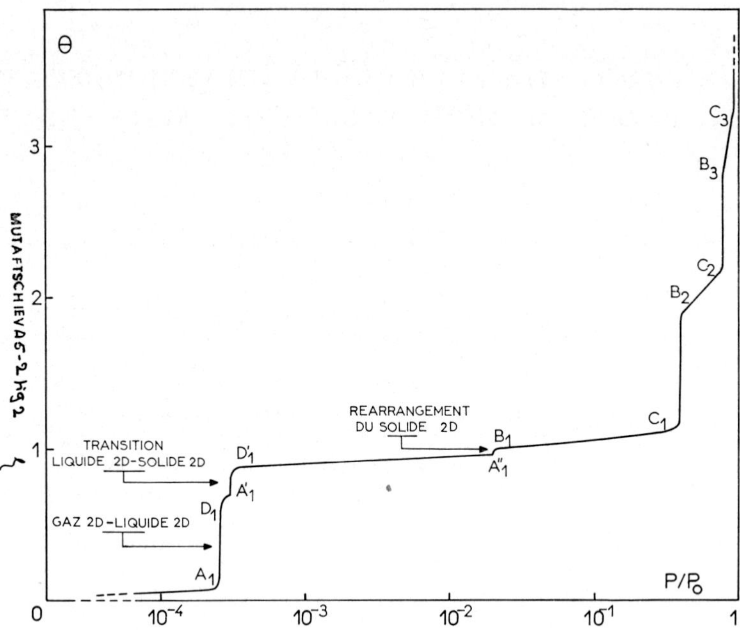

Fig. 2.   La même isotherme que celle de la figure 1, les pressions relatives étant cette fois portées en échelle logarithmique. Les trois "sous-marches" $OA_1D_1A'_1$, $A'_1D'_1A''_1$ et $A''_1B_1C_1$ qui constituent la première marche, se manifestent ici nettement.

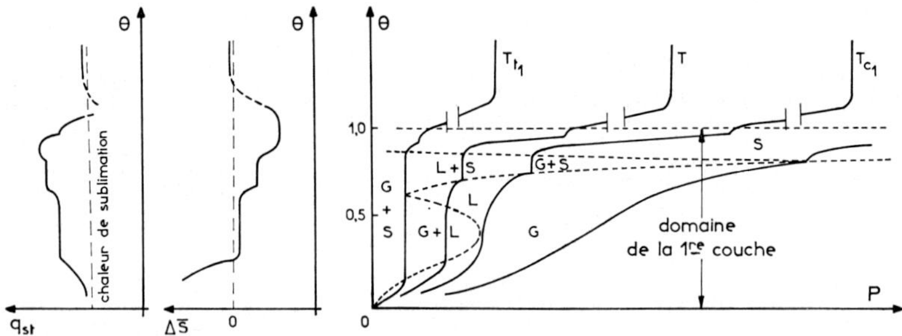

Fig. 3.   Evolution de la première marche en fonction de la température. G — état gazeux 2D; L = état liquide 2D; S = état solide 2D. Sont représentées à titre indicatif, les courbes donnant la variation de la chaleur isostérique q_{st} et de l'entropie différentielle d'adsorption $\Delta\bar{S}$ en fonction de θ, dans le cas le plus complexe où la couche passe successivement par les trois états G, L, S et subit un réarrangement après s'être solidifiée (cas de l'isotherme T).

pas de même de la première marche. En effet, celle-ci peut se composer de plusieurs "sous-marches" nettement marquées qui sont la manifestation de transformations diverses subies par la première couche au cours de sa croissance.

2. Cas particulier de la première couche

2.1. Passage du film par des états bidimensionnels "gazeux", "liquide" et "solide"

Il nous est tout d'abord apparu, dans le cas du krypton[5]) que la première marche pouvait se dédoubler en deux sous-marches: $OA_1D_1A'_1$ et $A'_1D'_1\ldots C_1$ (cf. fig. 2),

le point B_1 nous semblant être alors, un simple "point anguleux". Par la suite, des observations analogues faites notamment avec le xénon et le méthane[2,4]), montraient qu'il s'agissait de phénomènes ayant un certain caractère de généralité.

Telle qu'elle ressort de nos résultats, l'évolution de la première marche est schématisée (fig. 3). Le point essentiel est qu'il existe un domaine de température $(T_{t_1} - T_{c_1})$ dans lequel la première couche subit deux changements de phase du premier ordre, de sorte qu'elle passe alors successivement par trois états. Dans ces états de plus en plus denses, les molécules adsorbées sont de moins en moins mobiles et à des niveaux énergétiques de plus en plus bas, comme l'indiquent les courbes déduites des isothermes, traduisant la variation de l'entropie différentielle $\Delta \bar{S}$ et de la chaleur isostérique d'adsorption q_{st} en fonction du degré de recouvrement θ de la surface (fig. 3)[4]. Nous donnons dans le tableau 1, les valeurs estimées à partir de nos résultats, des températures T_{t_1}

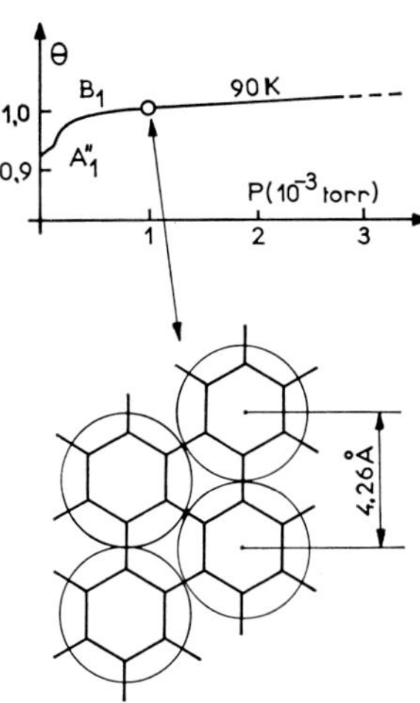

Fig. 4. D'après Morrison et Lander[11]: localisation de la première couche de xénon correspondant aux conditions indiquées. $A''_1 B_1$: "accident" montrant que la première couche subit un réarrangement après s'etre solidifiée.

TABLEAU 1

Températures critiques et témperatures des points triples des phases bidimensionnelles et tridimensionnelles

Adsorbat	T_t (3D) (°K)	T_c (3D) (°K)	T_{t_1} (2D) (°K)	T_{c_1} (2D) (°K)
Néon	24.6	44.4	—	—
Argon	84	151	—	65
Krypton	116	209	77 (environ)	87
Xénon	161	290	99	117
Méthane	91	191	—	75 (environ)

et T_{c_1}, la première étant celle à partir de laquelle se manifestent deux changements de phase (elle définit un véritable point triple), la seconde étant la température critique au-delà de laquelle le premier changement de phase cesse de se produire. Sont également indiquées: la température critique T_c et celle du point triple T_t relatives à la phase tridimensionnelle. Pour des raisons que nous avons déjà exposées[2,4], les changements de phase accompagnant la formation de la première couche sont très vraisemblablement analogues à ceux que subissent les corps purs dans le cas tridimensionnel (3D): entre les températures T_{t_1} et T_{c_1}, le film passerait successivement par des états bidimensionnels (2D) "gazeux", "liquide" et "solide". De tels états ont d'ailleurs été observés par diffraction d'électrons lents dans l'adsorption de brome et césium également sur graphite[6]. L'existence de phases 2D mobiles (gazeuses et liquides) s'explique par la grande uniformité énergétique des faces (0001) du graphite qui se caractérisent par des puits de potentiel peu profonds[7,8].

2.2. RÉARRANGEMENT DE LA PREMIÈRE COUCHE APRÈS QU'ELLE SE SOIT SOLIDIFIÉE

Plus récemment, nous avons découvert dans les isothermes de krypton et de xénon, que le point B_1 n'était pas un simple point anguleux comme nous l'avions d'abord cru, mais qu'il était en réalité précédé d'une partie abrupte $A''_1 B_1$ correspondant à une variation de

θ d'environ 0.05 et portant à trois le nombre de sous-marches pouvant composer la première marche (cf. fig. 2)[2,9,10]. Un tel "accident" traduit un réarrangement de la première couche après qu'elle se soit solidifiée. Il semble bien qu'il s'agisse là d'une manifestation des puits de potentiel qui, bien qu'insuffisamment profonds pour empêcher la formation de phases mobiles, doivent cependant influer sur la structure du solide 2D. C'est en tout cas ce qui ressort des analyses par DEL de Morrison et Lander[6,11] portant sur l'adsorption de divers gaz sur graphite. Plus précisément, la confrontation de nos résultats relatifs au xénon, avec ceux de ces auteurs tend à montrer que les réarrangements observés correspondent à l'établissement d'une corrélation donnée entre le réseau de la première couche et celui du plan graphitique sous-jacent. Dans le cas considéré, à l'issue du réarrangement du solide 2D (au voisinage du point B_1) toutes les molécules adsorbées se trouveraient sur des puits de potentiel selon le schéma de la figure 4.

2.3. PROPRIÉTÉ DU POINT B_1 ET SES CONSÉQUENCES

Tout d'abord, dans les domaines de température que nous avons explorés (de 77 à 109 °K pour le krypton, de 100 à 118 °K pour le xénon et de 77 à 90 °K pour le méthane), pour une surface et un gaz donnés, nous n'avons noté aucune variation sensible de la quantité adsorbée au point B_1. Il en résulte que la concentration

TABLEAU 2

Comparaison des quantités adsorbées au point B_1 avec les données cristallographiques

Adsorbat (C.F.C.)	T (°K)	$\dfrac{V_{B_1}}{V_{B_1}(Kr)} = \dfrac{\Gamma_{B_1}}{\Gamma_{B_1}(Kr)}$ [2]	$\dfrac{\Gamma_0}{\Gamma_0(Kr)}$	a_0 [Réf.] (Å)	d_0 (Å)
Néon	20.4	1.58	1.62	4.50 [15]	3.18
Argon	77.3	1.10	1.11	5.44 [15]	3.85
Krypton	77.3	1	1	5.74 [15]	4.06
Xénon	108.8	0.845	0.84	6.25 [15]	4.42
Méthane	77.3	0.94	0.91	6.01 [16]	4.25

a_0 = paramètre du cristal normal à la température T considérée;

d_0 = distance séparant deux molécules voisines dans un plan (111) du cristal normal ($d_0 = a_0/\sqrt{2}$);

Γ_0 = concentration moléculaire d'un plan (111) du cristal normal ($\Gamma_0 = 2/d_0^2\sqrt{3}$);

V_{B_1} = quantité adsorbée au point B_1 de l'isotherme considérée;

Γ_{B_1} = concentration moléculaire correspondante du film (nombre de molécules par unité de surface) ($\Gamma_{B_1} = 2/d^2_{B_1}\sqrt{3}$, d_{B_1} = distance séparant au point B_1 deux molécules adsorbées voisines).

$V_{B_1}(Kr)$, $\Gamma_{B_1}(Kr)$, $\Gamma_0(Kr)$ sont les mêmes quantités relatives au krypton à 77.3 °K. L'égalité des rapports $V_{B_1}/V_{B_1}(Kr)$ et $\Gamma_{B_1}/\Gamma_{B_1}(Kr)$ résulte du fait que l'aire de la surface ne dépend pratiquement pas de la température (cf. réf. 17).

moléculaire Γ_{B_1} de la première couche au point B_1 peut être considérée comme constante. Comme la concentration moléculaire Γ_0 d'un plan de type donné du cristal normal, soit un plan (111), varie elle-même très peu [pour les gaz rares, elle décroît de moins de 2 % dans les intervalles de température étudiés, cf. Pollack[15]]; on peut également considérer comme pratiquement constant le rapport Γ_{B_1}/Γ_0. De plus, la valeur de ce rapport est approximativement la même quel que soit l'adsorbat. Cela ressort du tableau 2 dans lequel on compare, pour les différents gaz concernés, les concentrations moléculaires Γ_{B_1} et Γ_0 à certaines températures, en prenant comme référence le krypton à 77.3 °K.

La localisation observée par Morrison et Lander dans le cas du xénon implique pour ce gaz une valeur de Γ_{B_1}/Γ_0 égale à 1.07. En vertu de ce qui précède, cette valeur devrait être également celle des rapports Γ_{B_1}/Γ_0 correspondant aux autres gaz. En fait, étant donné la précision avec laquelle est situé le point B_1 et sont connues les quantités adsorbées, nous devons nous contenter d'écrire:

$$\Gamma_{B_1}/\Gamma_0 = 1.07 \pm 0.05,$$

ou:

$$1.02 \leqslant \Gamma_{B_1}/\Gamma_0 \leqslant 1.12.$$

Cependant, cela nous permet de conclure que, dans tous les cas, la première couche est plus dense qu'un plan (111) du cristal normal.

Certes, on peut émettre des doutes sur la température réelle de la surface de l'échantillon dans les expériences de Morrison et Lander, température qui pouvait fort bien être supérieure de plusieurs degrés à celle de l'oxygène liquide utilisé comme fluide réfrigérant. Aussi se peut-il que la localisation observée pour le xénon (fig. 4), corresponde à une valeur de θ sensiblement plus faible que l'unité. Cette remarque ne fait que renforcer notre conclusion selon laquelle, pour tous les gaz considérés, la première couche est effectivement comprimée, et ce dans des proportions pouvant atteindre une dizaine de pour-cent.

L'inégalité précédente entraîne une autre conséquence, à savoir que s'il s'établit effectivement une corrélation au point B_1 entre le réseau de la première couche et celui de la surface, cette corrélation n'est pas nécessairement simple. En effet, dans le cas du krypton par exemple, on peut calculer que la distance d_{B_1} séparant alors deux molécules adsorbées voisines est comprise entre 3.8 et 4.0 Å. De telles valeurs excluent que les molécules de krypton se trouvent toutes, comme celles de xénon, sur des puits de potentiel, les différentes distances séparant ceux-ci étant: 2.46 Å, 4.26 Å, 4.92 Å, etc. Cela montre que les puits de potentiel influent certainement peu sur la densité de la couche. Le fait que la première couche soit comprimée explique que la chaleur isostérique q_{st} commence à décroître déjà avant le point B_1 (à partir de valeurs de θ comprises entre 0.90 et 0.95 qui sont bien en accord avec les conclusions précédentes concernant le rapport Γ_{B_1}/Γ_0[9,12]). Précisons qu'en ce point, q_{st} a encore une valeur élevée (par exemple pour le krypton: 3.9 kcal/mole contre 2.6 pour la chaleur de condensation proprement dite), ce qui montre que le nombre de molécules adsorbées en deuxième couche est encore négligeable, bien que la première couche soit certainement sur le point d'être saturée.

3. Conclusion

Nos résultats font ressortir le caractère particulier de la première couche. Celle-ci se singularise tout d'abord par le fait qu'elle se forme en majeure partie à des sous-saturations environ mille fois inférieures à celles correspondant à toutes les autres couches (cf. fig. 2).

Il s'agit là d'une conséquence de la décroissance rapide du potentiel d'adsorption d'une couche à la suivante [loi en $1/n^3$, n étant le numéro d'ordre de la couche[13])], décroissance telle qu'au niveau de la deuxième couche l'influence de la surface est déjà très faible. Pour cette raison également, si la densité de la première couche diffère de façon sensible de celle d'un plan (111) du cristal normal, ce n'est certainement plus le cas dès la deuxième couche. Ainsi, la première couche assure presqu'à elle seule la transition entre l'adsorbant et le cristal normal de l'adsorbat. Mais la première couche se distingue surtout par la manifestation remarquablement nette des étapes successives de sa formation. En premier lieu, celle-ci peut s'accompagner de changements d'états analogues à ceux auxquels on est accoutumé dans le cas tridimensionnel*. L'existence d'états bidimensionnels gazeux ou liquides serait due en partie à la grande uniformité énergétique de la surface. Pourtant, malgré leur faible profondeur, les puits de potentiel pourraient être responsables des réarrangements que subit également la première couche.

Bibliographie

1) Y. Larher, Thèse, Orsay–Paris (1970).
2) A. Thomy, Thèse, Nancy (1968).
3) A. Thomy et X. Duval, J. Chim. Phys. **66** (1969) 1966.
4) A. Thomy et X. Duval, J. Chim. Phys. **67** (1970) 1101.
5) A. Thomy et X. Duval, dans: *Adsorption et Croissance Cristalline*, Colloq. Intern. du C.N.R.S., Nancy, No. 152 (1965) p. 81.
6) J. J. Lander et J. Morrison, Surface Sci. **6** (1967) 1.
7) A. D. Crowell et D. M. Young, Trans. Faraday Soc. **49** (1953) 1080.
8) E. L. Pace, J. Chem. Phys. **27** (1957) 1341.
9) J. Regnier, Thèse 3ème cycle, Nancy (1969).
10) J. Menaucourt, Thèse 3ème cycle, Nancy (1971).
11) J. Morrison et J. J. Lander, Surface Sci. **5** (1966) 163.
12) A. Thomy, J. Regnier et X. Duval, dans: *Thermochimie*, Colloq. Intern. du C.N.R.S., Marseille, No. 201 (1971).
13) A. Thomy et X. Duval, J. Chim. Phys. **67** (1970) 286.
14) J. Regnier, Résultats non publiés.
15) G. L. Pollack, Rev. Mod. Phys. **36** (1964) 748.
16) S. C. Greer et L. Meyer, Z. Angew. Physik **27** (1969) 198.
17) Y. Baskin et L. Meyer, Phys. Rev. **100** (1955) 544.

* Des résultats récents obtenus au laboratoire tendent à montrer que le nitrure de bore donne lieu à des phénomènes analogues à ceux qui se produisent sur le graphite[14]).

        Journal of Crystal Growth **13/14** (1972) 164–166 © *North-Holland Publishing Co.*

ÉTUDE PAR DIFFRACTION D'ÉLECTRONS LENTS DE L'ADSORPTION DU XÉNON SUR UN MONOCRISTAL DU GRAPHITE

J. SUZANNE, G. ALBINET et M. BIENFAIT

*Laboratoire des Mécanismes de la Croissance Cristalline associé au C.N.R.S.,
Centre Universitaire de Marseille-Luminy, 13 – Marseille 9ᵉ, France*

The first stages of the condensation of xenon on the (0001) face of a graphite crystal are studied by LEED. The specular spot intensity I_{00} for electron energy $V = 32$ V versus xenon pressure shows a step indicating the appearance of the first xenon mono-layer on graphite. This step is defined by (P'_{Xe}, T').
The corresponding phase change obeys the Clausius–Clapeyron law from which the isosteric heat of adsorption of the first xenon monolayer on the (0001) face of graphite can be deduced. i.e. 7.3 ± 0.8 kcal mole^{-1}.

La condensation du xénon sur la face (00.1) du graphite a été étudiée en détail par Thomy et Duval[1–3]) en déterminant les isothermes d'adsorption de ce gaz rare dans des gammes de pression et de température s'étendant de 1 à 10^{-3} torr et de 100 à 120 °K. Les résultats de cette étude indiquent que la surface de graphite se recouvre, à sous-saturation, couche atomique après couche atomique. De plus, au-dessus du point triple bidimensionnel (99 °K), la première couche passe par trois états: deux mobiles ("gaz" et "liquide"), un immobile ("solide").

A priori, il est intéressant de reprendre par une autre technique les mesures déjà obtenues, de les compléter, et d'étendre le domaine d'observation vers les basses pressions ($p < 10^{-3}$ torr). Les électrons lents sont, du fait de leur faible pénétration, un moyen efficace d'étude de la surface du graphite et des premières couches condensées de xénon. De plus, on peut espérer obtenir des renseignements sur la structure de l'adsorbat.

Dans un premier temps, nous avons mis au point un appareillage destiné à réguler la température du porte-cristal au dixième de degré près, dans le domaine des basses températures[4]). Cette constance est nécessaire car nous étudions un équilibre thermodynamique entre le cristal et la couche adsorbée.

L'ensemble cristal–diffracteur se trouve dans une enceinte où on fait initialement un vide de 1×10^{-10} torr. La pression de xénon est ajustée grâce à une vanne à fuite réglable. Elle est contrôlée à l'aide d'une jauge Bayard–Alpert. Afin de limiter les difficultés expérimentales dues à des claquages électriques nous avons

été amenés, dans un premier temps, à étudier la condensation de xénon dans une gamme de pression comprise entre 10^{-10} et 10^{-6} torr. Cela nous oblige, si l'on veut obtenir la formation de la première couche de xénon, de porter l'échantillon de graphite à une température inférieure à celle de l'étude de Thomy et Duval[1–3]). Les mesures sont donc faites en dessous du point triple bidimensionnel et nous devons nous attendre à obtenir la formation directe d'une monochouche solide.

Nous étudions l'intensité de la tache spéculaire de diffraction I_{00}. Elle possède, à faible tension, des maxima lorsque les électrons sont accélérés sous 14, 20, 24 et 32 V. La variation d'intensité de ces maxima est mesurée lorsqu'on fait croître la pression de xénon (p_{Xe}) pour une température donnée de l'échantillon. On trace, pour chaque température, une courbe $I_{00} = f(p_{Xe})$. Pour chaque température, la courbe $I_{00} = f(p_{Xe})$ possède deux parties approximativement horizontales[4]) séparées par une discontinuité. Nous attribuons cette discontinuité à la formation de la première couche atomique de xénon sur la face (00.1) du graphite. Cette interprétation est justifiée par les indications que donne l'extrapolation des résultats de Thomy et Duval aux basses pressions et températures. Les variations de I_{00} sont également étudiées lorsque la pression de xénon dans l'enceinte diminue. Pour une température donnée, le sens de variation de la pression n'influe par sur la position de la discontinuité de la courbe $I_{00} = f(p_{Xe})$. Ceci montre que le phénomène est réversible.

Nous avons retenu comme pression de changement de phase, pour une température donnée, celle qui correspond à la discontinuité des courbes $I_{00} = f(p_{Xe})$. Nous obtenons ainsi des couples (p'_{Xe}, T') caractéristiques de l'apparition de la première couche de xénon sur le graphite. Sur la figure, le logarithme de p'_{Xe} est représenté en fonction de $1/T'$. Nos résultats sont tracés en pointillés; ceux de Thomy et Duval sont reportés en traits pleins. Il apparaît clairement sur la figure que leur domaine d'observation est différent du nôtre.

Cette figure n'est pas un diagramme de phases car la pression p'_{Xe} est celle du xénon en phase vapeur à la température ambiante. Nous pouvons cependant définir, en fonction de p'_{Xe} les domaines de stabilité pour une température donnée des diverses phases bidimensionnelles. C'est ce que nous avons indiqué sur la figure 1. Selon Thomy et Duval, ces diverses phases sont les suivantes: "gaz", première couche "liquide", première couche solide, deuxième couche solide. Le point triple de la première couche atomique ($p'_{Xe} = 2 \times 10^{-4}$ torr, $T' = 99$ °K) se trouve dans un domaine de température et de pression supérieur au nôtre. Ceci nous permet d'affirmer que le changement de phase que nous observons correspond à l'apparition de la première couche atomique solide. La courbe tracée d'après nos mesures est une droite. Nous pouvons écrire:

$$\log p'_{Xe} = \frac{A}{T'} + B,$$

avec $A = 3700$ et $B = 26$. La chaleur isostérique d'adsorption Q_{st} est donnée par l'expression[3]:

$$\left[\frac{\partial \ln p}{\partial(1/RT)} \right]_{\theta} = -Q_{st} \tag{1}$$

où θ est le degré de recouvrement.

Grâce à (1), on peut déduire de la figure la valeur de Q_{st} pour la première couche atomique de xénon, dans un domaine de températures: 80 °K $< T < 90$ °K. Nous trouvons:

$$Q_{st} = 7.3 \pm 0.8 \text{ kcal mole}^{-1}.$$

Cette valeur est voisine de celle trouvée expérimentalement par Thomy et Duval[3] (6.8 kcal mole^{-1}) dans le domaine de température compris entre 100 et 120 °K. Notre résultat est assez imprécis bien que la tempéra-

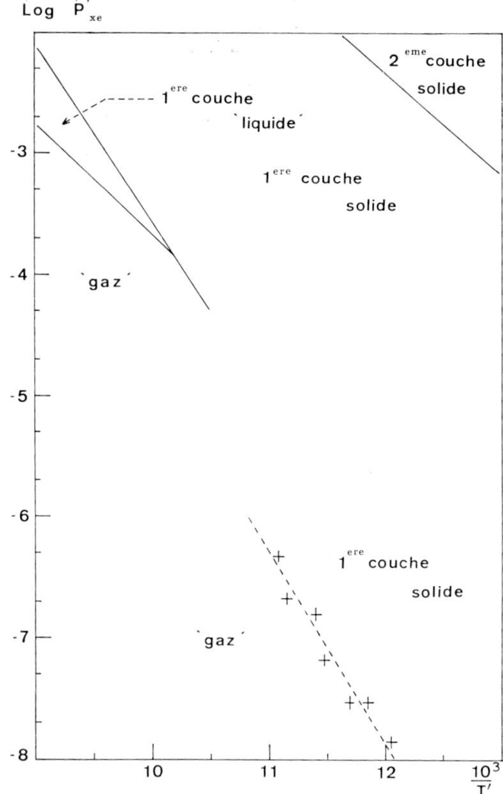

Fig. 1.   Courbes donnant le logarithme de la pression de xénon dans l'enceinte en fonction de l'inverse de la température du cristal. En pointillés, apparition de la première couche solide d'après nos résultats. En traits pleins, changements de phase bidimensionnels successifs d'après Thomy et Duval.

ture du porte-cristal soit connue avec une bonne précision (0.1 °C). Cette imprécision est due à l'incertitude dans la connaissance de la pression dont les causes sont les suivantes: la première tient au fait que la discontinuité de la courbe $I_{00} = f(p'_{Xe})$ n'est pas brutale. Il y a donc une erreur sur p'_{Xe}. Son ordre de grandeur est donné par la dispersion des points expérimentaux. La deuxième est due à l'effusion thermique. Un calcul montre que la pression réelle au voisinage du cristal est environ le double de celle dans l'enceinte. Cette erreur systématique peut être corrigée. Il s'ensuit un déplacement vers les pressions élevées de la droite expérimentale. La pente reste d'ailleurs pratiquement la même. Cependant, ce déplacement est loin d'être suffisant pour expliquer le décalage vers les basses pressions de l'ordonnée à l'origine de notre droite par rapport à celle obtenue par Thomy et Duval[3]. C'est pourquoi nous n'avons pas corrigé la pression mesurée de cette effusion thermique. Il faut chercher l'origine du décalage de notre droite dans une erreur beaucoup plus important due à l'appareil de mesure. La jauge Bayard–Alpert possède une bonne sensibilité dans la mesure des variations de pression mais ne donne pas une bonne mesure absolue de celle-ci. Il nous semble que cette incertitude absolue dans la connaissance de la pression permet de comprendre le décalage de notre droite expérimentale. Seul un étalonnage ultérieur de notre jauge pourra limiter cet inconvénient.

Ce décalage ne crée pourtant pas d'ambiguité dans le choix du changement de phase bidimensionnel observé. En effet, la formation de la deuxième couche se ferait pour des pressions $\sim 10^4$ fois plus grandes.

Nous pensions obtenir des informations sur la structure de la première couche adsorbée par observation de la symétrie et des paramètres du diagramme de diffraction. Nous avons, parfois, obtenu des taches supplémentaires identiques à celles observées par Lander et Morrison[5]. Cela semblerait indiquer que la phase condensée (une seule couche) est ordonnée. Cependant, ces résultats ne sont pas reproductibles et nous préférons attendre de nouvelles mesures avant de conclure.

Nous avons l'intention, d'une part, de poursuivre nos observations et de les étendre à un plus large domaine de température et de pression afin de pouvoir les comparer à l'étude de Venables et Ball[6] réalisée dans des conditions expérimentales tout à fait différentes. D'autre part, nous chercherons à confirmer et compléter ces résultats par l'études des électrons diffusés inélastiquement et en particulier des électrons Auger.

Bibliographie

1) A. Thomy et X. Duval, J. Chim. Phys. **66** (1969) 1966.
2) A. Thomy et X. Duval, J. Chim. Phys. **67** (1970) 286.
3) A. Thomy et X. Duval, J. Chim. Phys. **67** (1970) 1101.
4) G. Albinet, J. Suzanne, J. P. Biberian et M. Bienfait, Compt. Rend. (Paris) **272** (1971) 1155.
5) J. J. Lander et J. Morrison, Surface Sci. **6** (1967) 1.
6) J. A. Venables et D. J. Ball, Proc. Roy. Soc. (London) A **332** (1971) 331.

Reprinted from:
Journal of Crystal Growth **13/14** (1972) 167–173 © *North-Holland Publishing Co.*
Printed in the Netherlands

STUDY OF ADSORPTION AND NUCLEATION OF CADMIUM ON GERMANIUM SINGLE CRYSTALS BY A MOLECULAR BEAM TECHNIQUE

R. J. H. VOORHOEVE, R. S. WAGNER and J. N. CARIDES

Bell Telephone Laboratories, Incorporated, Murray Hill, New Jersey 07974, U.S.A.

A molecular beam mass-spectrometric ultra-high vacuum technique was used for the study of adsorption and nucleation of cadmium on the (111), (211), (110) and (331) faces of 45 Ω-cm germanium single crystals. Extensive cadmium adsorption preceded nucleation on all these faces. On the (111) face, two adsorption states, with heats of adsorption of 25 and 27 kcal/mole were identified, illustrating the barrier for nucleation of bulk cadmium which has a sublimation energy of 26.9 kcal/mole.

The critical supersaturation for nucleation, α^* is found to be related to the atomic roughness of the germanium crystal face, the highest α^* being found on (111), the lowest on (110) and (331). No relation is found between the amount of preadsorbed cadmium and the critical supersaturation. This indicates that preadsorption fills adsorption traps in which cadmium atoms are immobile and unable to participate in statistical clustering.

The effect of electrically active impurities on the nucleation has been investigated for the (111) faces of germanium crystals doped with gallium or arsenic. Both types of dopants enhance nucleation, though only to a small extent.

1. Introduction

Three promising experimental techniques have recently been introduced for the study of heterogeneous nucleation from the vapor. These are *in situ* electron microscopy[1–3], field ion microscopy[4] and ultra high vacuum molecular beam mass spectrometry[5,6].

The latter technique has some important advantages in that it is possible (a) to work with clean, well-defined surfaces, (b) to study quantitatively the adsorption occurring prior to, and also during nucleation, (c) to use unequivocal criteria for the onset of nucleation, provided by the continuous measurement of the deposition and desorption rates, and (d) to measure the thermal accommodation coefficient. The first of these advantages has up till now only partly been exploited with the use of clean, but polycrystalline substrates[5,6].

In the present work, clean single crystals of germanium have been used as substrates, in order to study the influence of the substrate surface structure on the nucleation kinetics. Also, for the first time, the effects of electrically active doping of a semiconductor substrate have been investigated.

2. Experimental

The experimental arrangement (fig. 1) consists of a large germanium single crystal slab (T), a tantalum boat (B) used as cadmium effusion source, and a mass spectrometer (MS). These are mounted in a stainless steel ultra high vacuum chamber, getter-ion pumped to a base pressure of 2×10^{-10} Torr.

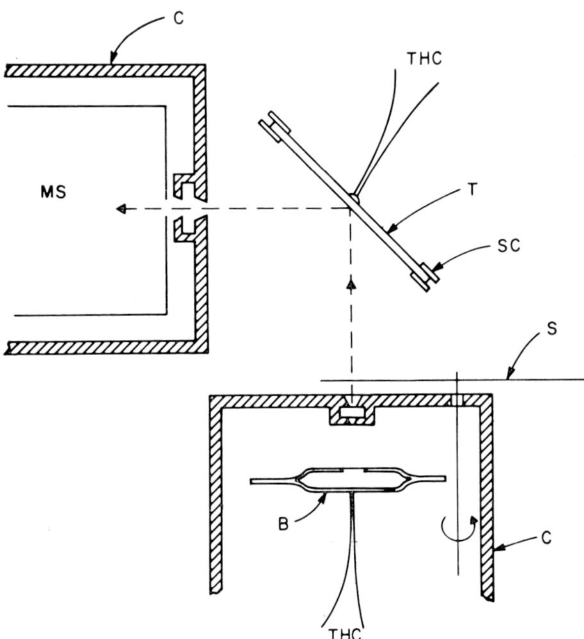

Fig. 1.   Experimental arrangement. (B) Tantalum boat effusion cell, (C) Liquid nitrogen cooled shrouds, (MS) Mass spectrometer, (S) Shutter, (SC) spring contacts, (T) Germanium crystal target, (THC) Thermocouple.

The germanium crystal, with dimensions of $45 \times 12 \times 1.5$ mm, is mounted on molybdenum posts by means of tantalum spring contacts (SC). The crystal is heated electrically, and its temperature is measured and controlled with a chromel–alumel thermocouple cemented to the back of the crystal with a small amount of alumina paste. The alumina also electrically insulates the thermocouple from the germanium.

A rotating shutter (S) is used to intercept the cadmium beam. Liquid nitrogen cooled cans (C) shroud the mass spectrometer and the cadmium source. The collimation of the beam and the field of view of the mass spectrometer were the same as in our earlier work[6]).

Cadmium of 99.9998% purity was used (United Mineral and Chemical Corp.). Germanium slabs were cut from crystals grown by zone-leveling or by the Czochralski method, oriented by X-rays within 3 min of arc of the desired orientation, and Syton*-polished to a mirror finish.

Prior to mounting in the vacuum chamber the crystals were rinsed with a detergent solution, with deionized water and finally with electronic grade methanol. In the vacuum, further cleaning was done by heating to $700°C$, sometimes first in 10^{-7} Torr oxygen for a few minutes and subsequently in better than 5×10^{-10} Torr residual vacuum for up to 20 min. In view of the volatility of GeO and the absence of any stable germanium carbides, this procedure is believed to leave a surface free of carbon and oxygen[7,8]). The final criterion as to the cleanliness of the surface is the reproducibility of the results obtained after various treatments with oxygen, or hydrogen, or in residual vacuum (see section 4). No evidence of (111) facetting on the surfaces has been found to result from these treatments. Direct-replica electron microscopy of the least stable face, the (110), showed a featureless surface down to the resolution limit of 20–50 Å.

The orientation of the germanium crystal and details about doping, resistivity and dislocation density are listed in table 1.

To calibrate the mass-spectrometer for cadmium[6]), a thick cadmium layer (50–100 atomic layers) was deposited on the substrate. The mass spectrometer current, I, which is proportional to the evaporating cadmium flux, was monitored as a function of temperature

* Trade mark of the 3M Company.

during slow heating of the crystal (approximately $0.5 °K/sec$). The calibration relation used was

$$I = \frac{3.52 \times 10^{22}}{(MT)^{\frac{1}{2}}} S_i \beta p_e \left(\frac{T_{\text{ref}}}{T}\right)^{\frac{1}{2}}. \tag{1}$$

Here, S_i is the calibration constant to be determined, T_{ref} is an arbitrary reference temperature, T the target temperature, M the atomic mass of cadmium, β the evaporation coefficient and p_e the equilibrium bulk vapor pressure in Torr obtained from the literature[9]) and given by

$$\log p_e \text{ (Torr)} = 9.067 - 5920/T. \tag{2}$$

The evaporation coefficient was measured separately[6,10]) and was found to be between $\beta = 0.8$ and $\beta = 0.95$ for the orientations studied.

TABLE 1

Properties of germanium crystals

No.	Orientation	Doping (cm^{-3})	Resistivity (Ω–cm)	Dislocation density[a] (cm^{-2})	Growth method[b]
1	(111)	—	45	5×10^3	ZL
2	(331)	—	45	5×10^3	ZL
3	(111)	—	45	5×10^3	ZL
4	(111)	As 10^{19}	0.0015	0	Cz
5	(111)	Ga 2×10^{18}	0.0067	10^3	ZL
6	(211)	—	45	5×10^3	ZL
7	(110)	—	45	5×10^3	ZL
8	(211)	—	45	5×10^3	ZL

[a] Determined by counting etch pits on a (111) face and by X-ray topography.
[b] ZL, zone leveling; Cz Czochralski method.

In calculating the desorption flux from I, incomplete accommodation of cadmium impinging on the germanium target has been neglected. This is permissible since the resulting was found to be 5% or less[10]).

The mass spectrometer current was displayed on an oscilloscope or, more often, on an $X–Y$ recorder.

3. Adsorption

All germanium crystals adsorbed appreciable amounts of cadmium prior to nucleation. The adsorption was initially studied by monitoring the adsorption "spectrum". This is obtained by a measurement of the cadmium flux desorbing from the target while the target is slowly cooled with a cadmium beam impinging. Ad-

III – 3

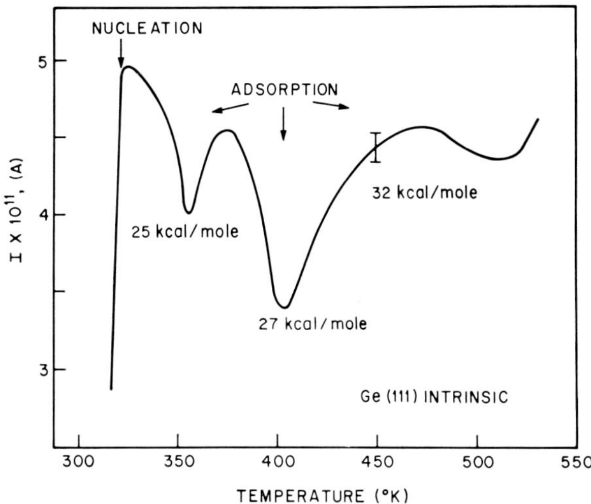

Fig. 2. Adsorption and nucleation on intrinsic Ge (111). The error bar indicates the noise level in the original recording.

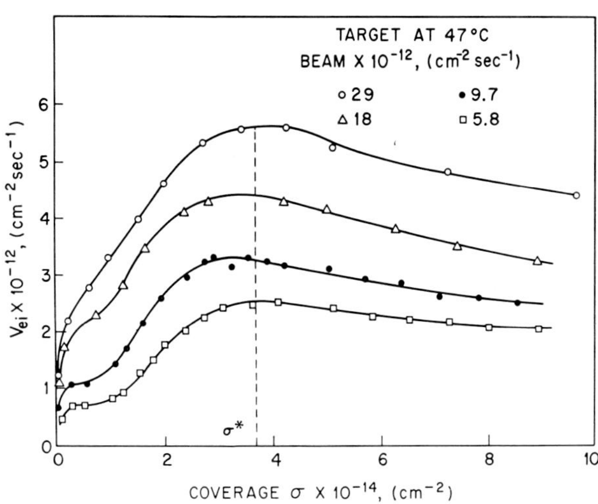

Fig. 3. Nucleation experiments on intrinsic Ge (110) at constant temperature and various beam intensities: cadmium desorption rate versus accumulated coverage of the target.

sorption states show up as minima in the recorded signal. In fig. 2, such a spectrum is shown in a plot of mass spectrometer current versus target temperature for an intrinsic Ge(111) crystal. The depth of high-temperature adsorption dips is exaggerated in comparison with the low temperature adsorptions since natural cooling was employed.

On cooling from 530 °K, two adsorption states at 400 °K and 350 °K can be clearly identified before nucleation takes place at about 325 °K*. A third adsorption state may be suspected at 520 °K, but is of less importance. The adsorption energies indicated near the minima in fig. 2 have been derived from a desorption spectrum by Reynolds' method[11]) of graphical analysis. These energies, 25 and 27 kcal/mole for the two low temperature minima, are close to the heat of sublimation of cadmium, which is 26.9 kcal/mole (refs. 9, 12). This adsorption spectrum is an interesting example of the formation of adsorbed layers in preference to an equally stable bulk-phase which, however, cannot yet form because of a nucleation barrier.

The adsorption spectrum shown was quite reproducibly obtained for two different intrinsic Ge (111) crystals (No. 1 and No. 3, table 1). Gallium-doped Ge (111) showed a qualitatively very similar spectrum, but with the two deep minima shifted somewhat. Intrinsic germanium crystals of orientations other than (111)

showed more diffuse adsorption spectra, without distinctive minima.

Further information about the adsorption preceding nucleation was obtained by measuring the amount of cadmium that had accumulated on the surface before nucleation occurred. This is done through the recording of "desorption transients". The desorption rate of cadmium from the surface, v_{ei}, is measured at constant impinging beam intensity, v_i, and constant temperature as a function of the time, t, elapsed since opening the shutter. Starting with a clean germanium surface, the accumulated coverage with cadmium, σ, is then at any time given by

$$\sigma = \int_0^t (v_i - v_{ei}) \, dt. \tag{3}$$

Examples of such nucleation experiments on intrinsic Ge (110) are given in fig. 3 for various beam intensities at 320 °K. Recordings of v_{ei} versus t have been replotted as v_{ei} versus σ. The maxima in the desorption rates are equivalent to minima in the growth rates and are hence indicative of the crossing of a nucleation barrier. The corresponding coverage, $\sigma^* = 3.7 \times 10^{14}$ cm^{-2}, is almost independent of the beam intensity. In the temperature range studied, 303–353 °K, σ^* is not noticeably dependent on temperature either. Values of σ^* for the various germanium substrates are collected in table 2. The value of σ^* has also been determined as the maximum coverage attained with v_i slightly lower

* Slightly higher supersaturations were needed for nucleation on oxygenated tungsten substrates[10]).

TABLE 2

Adsorption, nucleation and etch rate for intrinsic germanium faces

No.	Orientation	Adsorption parameters		Nucleation parameters[a]			Etch rate[b]
		σ^* (cm^{-2})	σ_S (cm^{-2})	v_i^* (cm^{-2} sec^{-1})	α^*	Δh (kcal/mole)	(μm/min)
1	2	3	4	5	6	7	8
3	(111)	7.5×10^{14c}	7.2×10^{14}	1.2×10^{13}	92	21.1 ± 1.7	4.4
6, 8	(211)	4.0×10^{14d}	$2.6{-}5.3 \times 10^{14}$	6.3×10^{12}	48	25.6 ± 1.8	5.1
2	(331)	4.5×10^{14}	5.7×10^{14}	3.6×10^{12}	28	23.7 ± 1.4	6.1
7	(110)	3.7×10^{14}	4.4×10^{14}	3.6×10^{12}	28	23.2 ± 1.4	6.2

[a] v_i^* and α^* at 325 °K; Δh by least squares fit, with standard error.
[b] Etch rate at 30 °C in H_2O_2–HF (ref. 14).
[c] In a few experiments $\sigma^* = 1.0 \times 10^{15}$ cm^{-2} was obtained.
[d] In a few experiments $\sigma^* = 7.5 \times 10^{14}$ cm^{-2} was obtained.

than v_i^*, the critical impingement rate necessary for extended growth.

An attempt has been made to correlate the σ^* in table 2 with the cadmium coverage needed to satisfy obvious adsorption traps on the surface. Models of germanium surfaces were used for this. Both adsorption sites with high germanium-coordination for adsorbed cadmium, and sites with high overlap between cadmium and a germanium "dangling bond" were counted. Cadmium atoms were kept at a minimum distance of 2.98 Å. The resulting model coverages, σ_S, are also given in table 2.

The σ_S and σ^* appear to agree rather satisfactorily, supporting the belief that cadmium is first filling up the adsorption traps on the surface before nucleation of bulk cadmium takes place. The amount of adsorbed cadmium involved in the statistical clustering envisaged by classical nucleation theory is clearly much less than σ^*.

The preadsorption of the amount of cadmium given by σ_S, one full monolayer, has effectively changed the (111) substrate. Nucleation, therefore, does not occur on a germanium surface but on a cadmium surface, albeit a cadmium surface that does not have any structural resemblence to any face of a cadmium crystal. Preadsorption of cadmium on the other surfaces, (211), (331) and (110), causes nucleation to occur on a substrate that is a hybrid of a cadmium and a germanium surface.

4. Nucleation

Critical impingement rates for nucleation were determined by very slowly cooling a germanium target (about 10 °K/min), with a constant beam of cadmium atoms impinging, and recording the temperature at which the desorbed cadmium flux starts diminishing sharply. That decrease is taken as an indication of the start of nucleation.

The reproducibility of this technique was tested for intrinsic Ge (211). Two different crystals were used. One of them was, in addition to the usual cleaning, subjected to various treatments in the vacuum system, such as heating in oxygen, residual vacuum, or hydrogen, always followed by final heating to 700 °C in better than 5×10^{-10} Torr vacuum. The results for the critical impingement rates as a function of nucleation temperature are shown in fig. 4. The straight line is a least squares fit for the data on target No. 6. The two sets of data on target No. 8 are seen to fit the same line equally well.

The critical supersaturation is given by

$$\alpha^* = v_i^* / (v_e)_{\text{bulk}}, \tag{4}$$

where v_i^* is the critical impingement rate at a temperature T, and $(v_e)_{\text{bulk}}$ is the equilibrium impingement rate for the same temperature, calculated from the kinetic gas theory expression[13])

$$(v_e)_{\text{bulk}} = 3.52 \times 10^{22} p_e (MT)^{-\frac{1}{2}}. \tag{5}$$

For the Ge (211) face (fig. 4), and indeed for the other germanium faces also, rather small critical supersaturations are found, ranging from $\alpha^* = 20$ to $\alpha^* = 100$. The variation of the critical impingement rate with temperature is characterized by an enthalpy of evaporation, Δh, which is close to the sublimation enthalpy of bulk cadmium (table 2). This means that

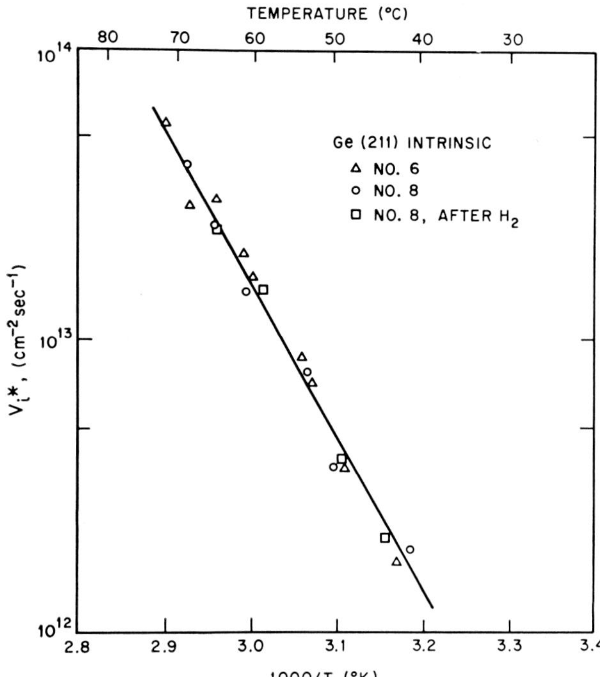

Fig. 4. Critical impingement rates for two intrinsic Ge (211) crystals, as a function of temperature.

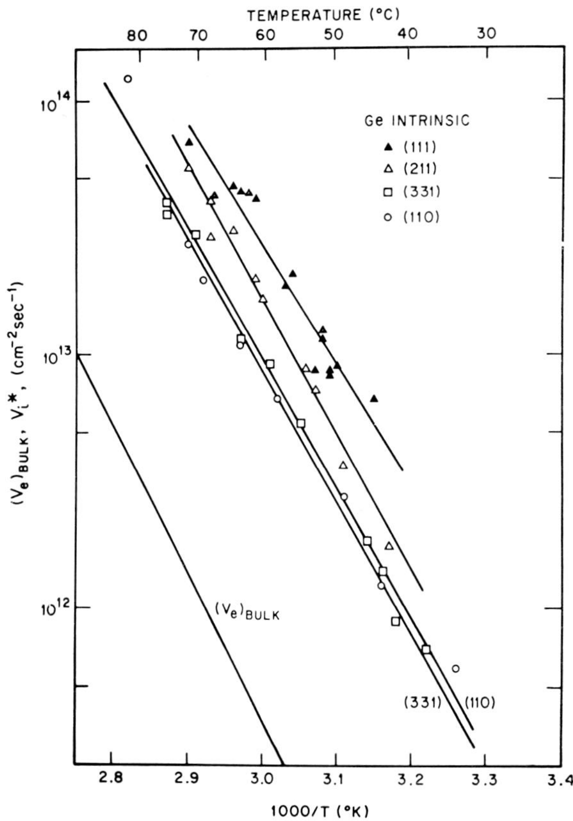

Fig. 5. Critical impingement rates for various intrinsic germanium faces as a function of temperature. $(v_e)_{bulk}$ gives the equilibrium evaporation rate.

the critical supersaturation is only weakly dependent on temperature.

4.1. INTRINSIC GERMANIUM CRYSTALS

In order to determine the influence of the surface structure of germanium on the nucleation kinetics, a series of intrinsic crystals of the orientations (111), (211), (331), and (110) were studied. It is known from bonding configurations and etching studies that the atomic roughness of the surface increases and the stability decreases in this series. This is demonstrated by the etch rates determined in a mixture of hydrogen fluoride and hydrogen peroxide (Superoxol etch), which increases in this same order[14]) (table 2).

The critical cadmium impingement rates are given in fig. 5 as a function of nucleation temperature. It is clear that nucleation is most difficult on the densely packed (111) face and that the critical impingement rate decreases with increasing roughness of the face. In table 2, columns 6 and 8 show that the critical supersaturation, α^*, increases as the etch rate decreases.

It is interesting to see that there is no correlation between the amount of preadsorbed cadmium σ^* and the critical supersaturation α^* (table 2, columns 3 and

6). The adsorbed layer of cadmium does not seem to mask the structural differences in the germanium faces. Comparison of (331) and (110) shows that the increased adsorptive capacity of (331) does not influence the nucleation kinetics.

4.2. EFFECT OF DOPING Ge (111)

The effect of electrically active doping of germanium on its efficacy as a nucleation substrate was determined with gallium and arsenic-doped crystals. The doping levels (table 1) were small enough to prevent precipitation of dopants[15,16]).

Critical cadmium impingement rates for nucleation on (111) faces of intrinsic and doped germanium crystals are collected in fig. 6. Doping is seen to facilitate nucleation, decreasing the critical supersaturation by a factor 1.5 compared with intrinsic germanium. This is equivalent to a decrease in undercooling of about 4 °C. The effect is the same for both p- and n-type germa-

III – 3

nium. It is rather small, but in view of the reproducibility of these measurements (see fig. 4) it may be significant. It appears that the collective electrical properties of the semiconductor substrate are of limited importance for the nucleation kinetics. It is of course possible that the heavy coverage with cadmium in the pre-nucleation adsorption stage is masking the electrical effects, since σ^* is particularly large for both doped and intrinsic (111) crystals.

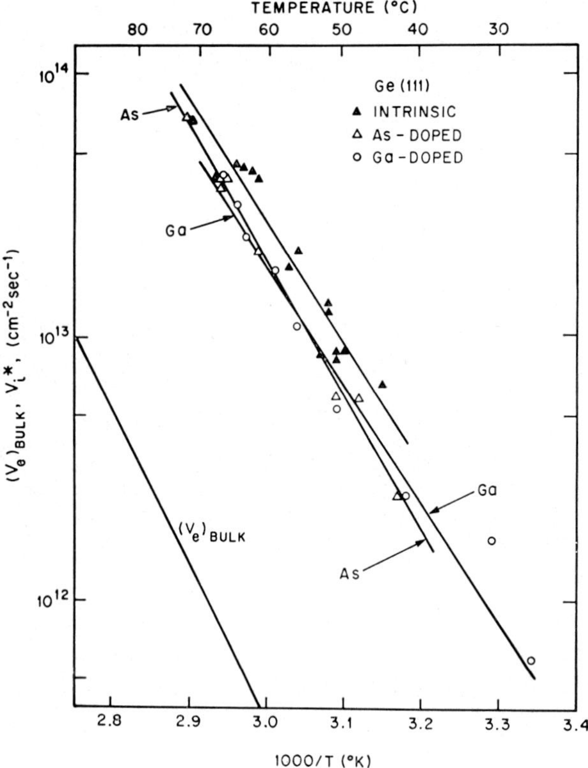

Fig. 6. Critical impingement rates for the (111) face of intrinsic and doped germanium, as a function of temperature. $(v_e)_{bulk}$ gives the equilibrium evaporation rate.

The rather small differences in critical impingement rates for the As-doped crystal (No. 4), which was dislocation free, and the intrinsic crystal (No. 3), with a dislocation density of 5×10^3/cm^2, seems to rule out any significant effect of dislocations on the present nucleation experiments.

5. Summary and conclusions

The study of adsorption of cadmium on a series of germanium crystals of various orientations and electrical doping has revealed that on all these substrates quite substantial adsorption precedes nucleation. The extent of adsorption ranges from 3.7×10^{14} cm^{-2} coverage on (110) to 7.5×10^{14} cm^{-2} on (111). In most cases there are several adsorption states, e.g., for Ge (111) at least two adsorption states (25 and 27 kcal/mole) have been identified. The pre-adsorption appears to saturate adsorption traps on the surface formed by vacant sites of high germanium coordination and by sites where a germanium "dangling bond" is available for bonding with the cadmium.

The surfaces formed by cadmium adsorbed in traps on germanium faces serve as substrates for the nucleation of bulk cadmium. The "real" substrate is therefore a cadmium-rich surface with a structure peculiar to the particular germanium crystal face. The critical supersaturation needed is rather low, smaller than 100, and the critical supersaturation is only weakly dependent on temperature.

The effect of the crystal orientation of germanium on the nucleation of cadmium can be correlated directly with the atomic roughness of the germanium face. Rough faces like (110) and (331) are better nucleation substrates than a smooth face such as (111). There is no relation, however, between the adsorptive capacity of a germanium face and its efficacy as a nucleation substrate. This is clearly shown by a comparison of (331) and (110). These two surfaces show equal α^*, but σ^* differs by a factor 1.2 (table 2).

Doping of germanium with electrically active impurities such as gallium or arsenic only slightly influences the adsorption and nucleation of cadmium. The critical impingement rate for nucleation decreases only by a factor 1.5 for both n- and p-type doping. The possibility of a screening effect by the cadmium adlayer on the (111) faces used for the comparison must be recognized, however. Doping may have a larger effect on germanium faces which show lower pre-adsorption coverage, such as (110).

Acknowledgments

We would like to thank K. E. Benson (BTL, Allentown) for providing the zone-leveled germanium crystals, C. J. Doherty for the skillful orienting and cutting of the crystals, J. R. Patel and P. E. Freeland for providing the Czochralski-grown germanium crystal and for determining the dislocation densities.

References

1) H. Poppa, J. Appl. Phys. **38** (1967) 3883.
2) D. J. Ball and J. A. Venables, J. Vacuum Sci. Technol. **6** (1969) 469.
3) J. A. Venables and D. J. Ball, Proc. Roy. Soc. London A **322** (1971) 331.
4) D. W. Bassett, Surface Sci. **23** (1970) 240.
5) J. B. Hudson and J. S. Sandejas, J. Vacuum Sci. Technol. **4** (1967) 230.
6) R. S. Wagner and R. J. H. Voorhoeve, J. Appl. Phys. **42** (1971) 3948.
7) F. Jona, Appl. Phys. Letters **6** (1965) 205; IBM J. Res. Develop. **9** (1965) 375.
8) R. E. Weber and W. T. Peria, J. Appl. Phys. **38** (1967) 4355.
9) A. N. Nesmeyanov, in: *Vapour Pressure of the Elements*, Ed. J. I. Carasso (Academic Press, New York, 1963) pp. 202, 452.
10) R. J. H. Voorhoeve, R. S. Wagner and J. N. Carides, to be published.
11) T. W. Reynolds, NASA TN D-4789, CFSTI, Springfield, Va. (1968).
12) F. D. Rossini, D. D. Wagman, W. H. Evans, S. Levine and I. Jaffe, *Selected Values of Chemical Thermodynamic Properties* (Circ. 500, National Bureau of Standards, Washington, D.C., 1952).
13) J. H. de Boer, *The Dynamical Character of Adsorption*, 2nd ed. (Cambridge University Press, 1968) p. 7.
14) B. W. Batterman, J. Appl. Phys. **28** (1957) 1236.
15) F. A. Trumbore, Bell System Tech. J. **39** (1960) 205.
16) J. R. Patel, R. F. Tramposch and A. R. Chaudhuri, in: *Metallurgy of Elemental and Compound Semiconductors*, Vol. 12, Ed. O. Grubel (Interscience, New York, 1961) p. 45.

    Journal of Crystal Growth **13/14** (1972) 174–179 © *North-Holland Publishing Co.*

CONDENSATION OF INDIUM FILMS ON SILICON SUBSTRATES OBSERVED BY MASS SPECTROMETRIC TECHNIQUE

NGUYEN TAN TUAN ANH, R. CINTI and B. K. CHAKRAVERTY

Section Physique des Couches Minces, Laboratoire d'Electrostatique et de Physique du Métal, C.N.R.S., Cedex 166, 38 – Grenoble–Gare, France

The nucleation and condensation of indium on silicon substrate has been studied by a mass spectrometric technique. The method permits an exact and unambiguous measurement of the following important quantities:
(a) mean lifetime τ of an ad-atom of indium on silicon,
(b) the sticking as well as the thermal accommodation coefficient,
(c) the critical supersaturation for the onset of nucleation,
(d) the ad-atom population.
We find that both the sticking and the thermal accommodation coefficients are unity in our substrate temperature range. We also find that, for a given substrate temperature, the mean lifetime τ is practically constant when the coverage θ is approximately lower than 0.1. When $0.1 < \theta < 1$, τ depends on the population n of adatoms and can be expressed by:
$\tau = \tau_p \exp\left(-\alpha(T)n/kT\right),$
where $\alpha(T) \simeq 5 \times 10^{-16}$ eV/atom at 909 °K and τ_p is coverage-independent mean lifetime, given by
$\tau_p = 8 \times 10^{-13} \exp\left[(57000 \pm 3000)/RT\right].$
An apparent undersaturation is observed; condensation begins when the equivalent pressure of impinging flux is still lower than the vapor pressure of the massive phase of indium.

1. Introduction

The mass spectrometric method used here consists in following: by means of a mass spectrometer, the flux of reemitted atoms J_r from a surface when a calibrated impinging flux J_i strikes this surface. From the evolution of this flux J_r as a function of time and substrate temperature, one can deduce mean lifetimes τ of ad-atoms, their Debye frequency ν_0, the population of adatoms, the heat of desorption ΔH_{des}, the sticking coefficient α_c, the thermal accommodation coefficient α_T and the critical supersaturation for nucleation.

This method is similar to the pulsed molecular beam technique frequently used in experiments of gas reflection on solid surfaces[1]. It is used by Shelton and Cho[2] and by Cho and Hendricks[3] to measure adsorption mean lifetimes of metallic atoms on tungsten. Hudson and Sandejas[4,5], and Kinawi and Hudson[6] developed and employed it extensively for studying the growth of cadmium films on tungsten and mercury films on pyrex glass.

2. Experimental

The experimental apparatus is outlined in fig. 1. The

Knudsen cells and the substrate heater were regulated to better than ± 2 °C. The response time of the electromagnetical shutter is 3 ms.

Mass spectrometric signals can be either recorded or displayed on an oscilloscope screen and photographed.

A pressure of a few 10^{-9} Torr is maintained during experiments.

The silicon (111) substrate is rectangular 24×14 mm and is 1 mm thick. It is mechanically and chemically ($HCl + H_2$) polished and etched in a CP4 etch before being mounted. Before starting an experiment this substrate is heated for 2 h at 1100 °C and then flashed at

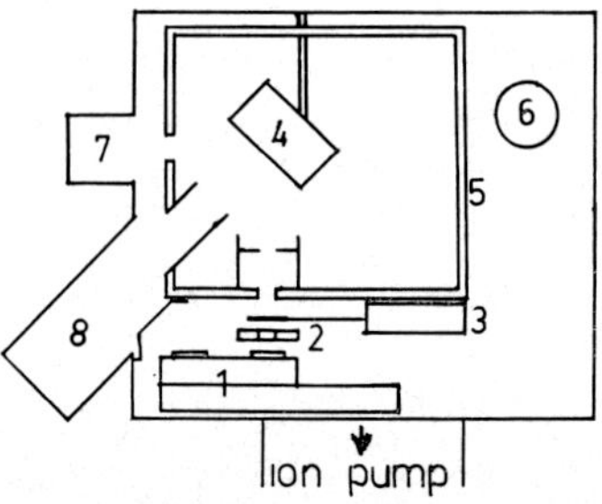

Fig. 1. Experimental apparatus. (1) Knudsen cell, (2) water cooled shield, (3) electromagnetical shutter, (4) substrate heater, (5) liquid nitrogen trap, (6) Bayard–Alpert gauge, (7) ion gun, (8) quadrupole mass spectrometer head.

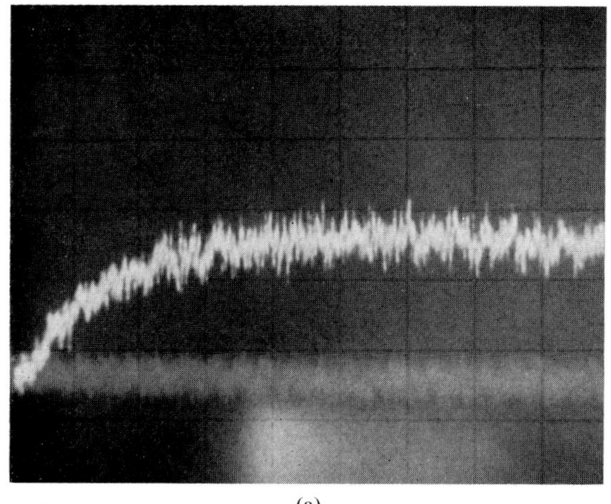

(a)

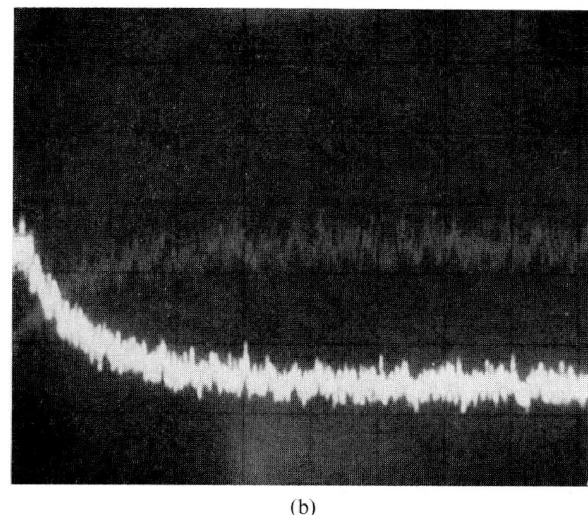

(b)

Fig. 2.   Oscilloscope traces of re-emitted flux J_r upon opening (a), and closing (b) of the shutter. Substrate temperature $T_s = 1151$ °K.

1250 °C for 10 min. Between measurements it is flashed at 1250 °C for several minutes.

The principle of the experimental meaurements is described in detail elsewhere[4,5]). The main features are:

– mean lifetimes τ are obtained from the exponentials of increasing and decreasing curves of the reemitted flux J_r upon opening and closing of the shutter, in a typical adsorption experiment. The heat of desorption ΔH_{des} and the Debye frequency are deduced from the log τ versus $(1/T_s)$ plot, where T_s are substrate temperatures. The fraction of atoms elastically reflected from the substrate, and hence the sticking coefficient α_c, is measured from the portions of J_r curves which do not show time delay.

– the thermal accommodation α_T is measured from the variation of mass spectrometric signal when the substrate temperature is slowly decreased, the incident beam J_i being constant.

– the critical supersaturation for nucleation is measured by determining the substrate temperature at which the re-emitted flux J_r starts to decrease from the equilibrium value $\alpha_c J_i$.

3. Experimental results

3.1. MEAN LIFETIMES τ AND DESORPTION HEAT

Mean lifetimes are measured for fluxes J_i ranging from 6.3×10^{12} to 9.4×10^{13} atoms cm^{-2} s^{-1} and for substrate temperatures T_s from about 800 °K to 1250 °K.

At high temperatures, τ are short and J_r are viewed on an oscilloscope. Typical are oscilloscope traces shown in figs. 2a and 2b, corresponding to the opening and the closing of the shutter for $T_s = 1151$ °K. These traces are pure exponentials, log J_r as function of time being a straight line. As the temperature T_s is gradually decreased, the traces of J_r deviate more and more from exponential forms; the increasing and the decreasing curves are no longer symmetrical.

The measurements of τ are unambiguous at high temperature but more difficult in low temperature range. In the latter case, two values of τ are noted each for temperature, the first $\tau_{\theta \sim 0}$ from the slopes at the beginning of increasing curves and which is related to small surface coverage, the other $\tau_{\theta = \theta_0}$ from the slopes at the origins of decreasing curves and which corresponds to equilibrium surface coverage θ_0. The values of $\tau_{\theta \sim 0}$ for $J_i = 6.3 \times 10^{12}$ atoms cm^{-2} s^{-1} and for $J_i = 1.7 \times 10^{13}$ atoms cm^{-2} s^{-1} are of the same order and the mean lifetimes for a virgin surface is found to be:

$$\tau_P = 8 \times 10^{-13} \exp\left[(57000 \pm 3000)/RT_s\right].$$

Populations of adatoms at adsorption equilibrium are deduced from graphical integration of the curve J_r as a function of time (figs. 2a, 2b).

The curves log τ versus $1/T_s$ for a group of measurements are shown in fig. 3. The corresponding adatom population n as well as the dependence of τ on n are shown in figs. 4 and 5 respectively.

Both the sticking coefficient and the thermal accom-

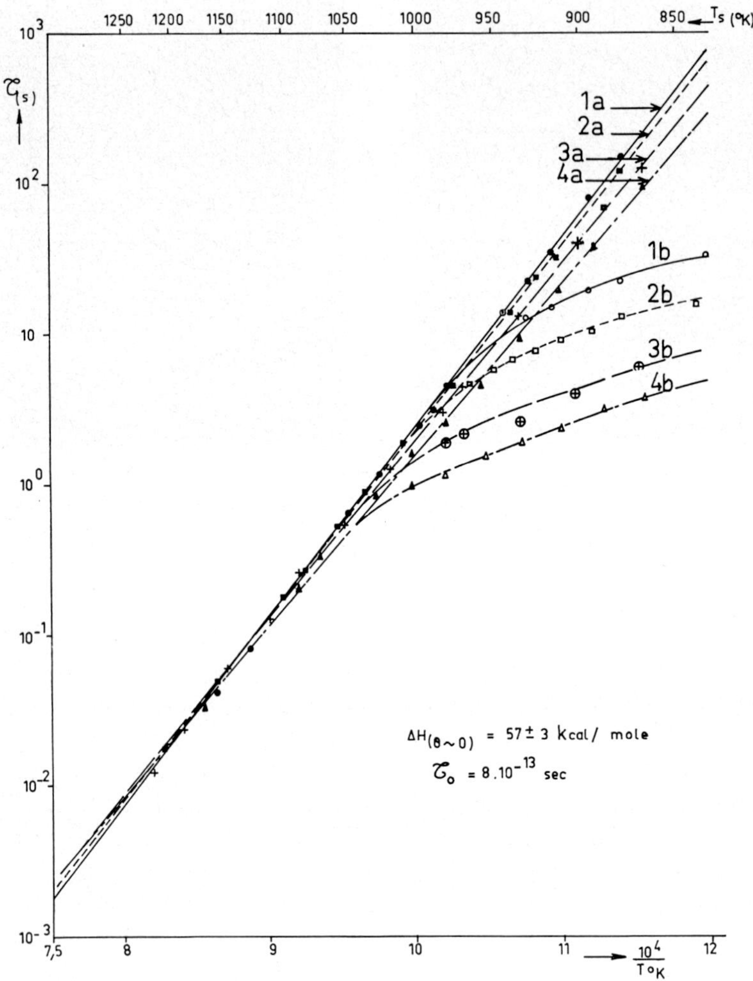

Fig. 3. Mean adsorption lifetimes of indium versus silicon substrate temperatures for a virgin surface (curves 1a, 2a, 3a, 4a) and at equilibrium coverages (curves 1b, 2b, 3b, 4b).

(1a, 1b) $J_i = 6.3 \times 10^{12}$ atoms $cm^{-2}\,s^{-1}$;   (3a, 3b) $J_i = 5.0 \times 10^{13}$ atoms $cm^{-2}\,s^{-1}$;

(2a, 2b) $J_i = 1.7 \times 10^{13}$ atoms $cm^{-2}\,s^{-1}$;   (4a, 4b) $J_i = 9.4 \times 10^{13}$ atoms $cm^{-2}\,s^{-1}$.

modation coefficient were measured and were found to be unity.

3.2. CRITICAL SUPERSATURATION OF NUCLEATION

In these series of experiments a constant impinging flux was maintained. Substrate temperature was slowly decreased and mass spectrometric signals i_s were recorded for every temperature. The temperature T_s was decreased until i_s was zero, all incident atoms having by then condensed on the substrate. A deposit of about a minimum of one hundred monolayers was thus made. After that, the substrate temperature was slowly increased and the corresponding mass spectrometric

signals were recorded. For each experiment, the re-emitted flux curve and the vapor pressure curve were obtained in this manner. Critical supersaturations for nucleation were determined by comparison of these two curves.

Some typical results are plotted in figs. 6a and 6b for impinging fluxes $J_i = 1.6 \times 10^{13}$ and 1.27×10^{14} atoms $cm^{-2}\,s^{-1}$. *The re-emitted flux curves are all below the equilibrium vapor pressure curves. This indicates that condensation begins when the equivalent pressure of impinging flux is still lower than the vapor pressure of the massive phase of indium.* There is an apparent undersaturation. This undersaturation is about 0.9 for the

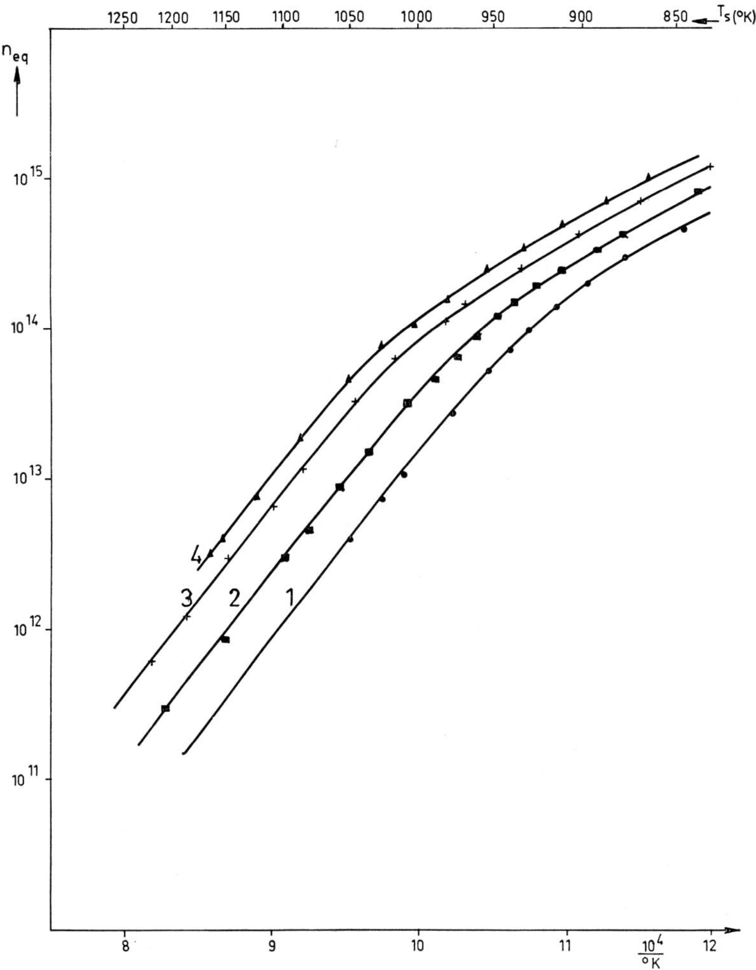

Fig. 4. Equilibrium coverages versus substrate temperatures for different impinging fluxes.
(1) $J_i = 6.3 \times 10^{12}$ atoms cm^{-2} s^{-1};   (3) $J_i = 5.0 \times 10^{13}$ atoms cm^{-2} s^{-1};
(2) $J_i = 1.7 \times 10^{13}$ atoms cm^{-2} s^{-1};   (4) $J_i = 9.4 \times 10^{13}$ atoms cm^{-2} s^{-1}.

four different extensive series of experiments we have done with four different impinging fluxes.

4. Discussion

The experimental results described above can be separated in three parts: the thermal accommodation, the adsorption and the condensation of indium atoms on a silicon substrate.

– In the first one, the results are simple and in good agreement with the theoretical predictions. In fact, in systems having large desorption energies, as the case here, the thermal accommodation must be complete and the sticking coefficient equal to unity in the temperature range studied.

– In the second part, we have seen that the adsorption energy measured on a non-covered surface is

$$\Delta H_{ad} = 57 \pm 3 \text{ kcal/mole}.$$

One can note that this experimental value is very near the sublimation heat of indium given in the literature (58 kcal/mole)[7]. The important point of this part is the demonstration of the dependence of the mean lifetime τ versus the surface coverage θ. This phenomenon, well known in gas on solid chemisorption, is specially clear at low temperature when the coverage becomes important.

Fig. 4 shows the experimental relation between $\log n_{eq}$ and $1/T$. Up to a coverage of about 0.1 monolayer

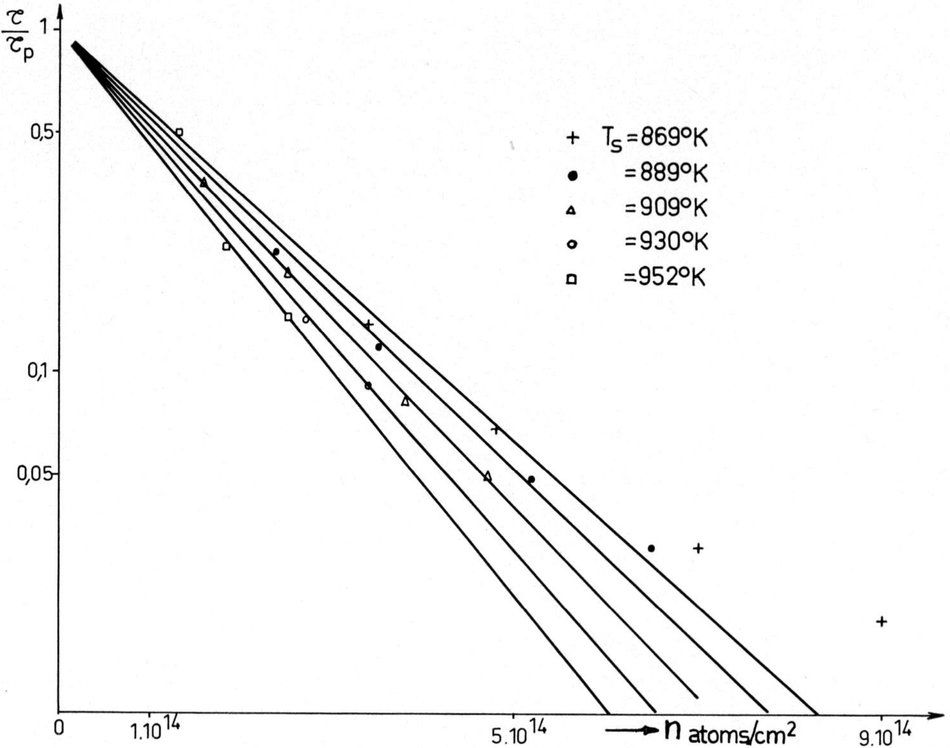

Fig. 5. Variation of mean adsorption lifetimes with adatom population (τ_p mean adsorption lifetimes for a virgin surface, τ at equilibrium coverages).

($\sim 9 \times 10^{13}$ atoms/cm^2), the dependence is linear and in good agreement with the classical expression: $n_{eq} = \alpha_c J_i \tau$, where τ is a constant at a given temperature.

Above this coverage the linearity is not obeyed, the mean lifetime τ is being a function of n, as is seen in fig. 5, and τ can be expressed by the relationship

$$\tau = \frac{1}{v_0} \exp \frac{E_a}{KT} = \frac{1}{v_0} \exp \frac{E_0 - \alpha(T)n}{KT},$$

where v_0 is the Debye frequency, E_a, the adsorption energy when the adpopulation is n, E_0 is the adsorption energy when $n \to 0$.

The coefficient $\alpha(T)$ is slightly dependent with the substrate temperature, it varies from 4.2×10^{-16} eV/atom to 6.2×10^{-16} eV/atom in the temperature range 869–952 °K.

In the series of experiments on condensation, the results are interesting. Nucleation occurs in an adlayer of about a monolayer in each case studied and it is fairly probable that this phenomenon occurs at a

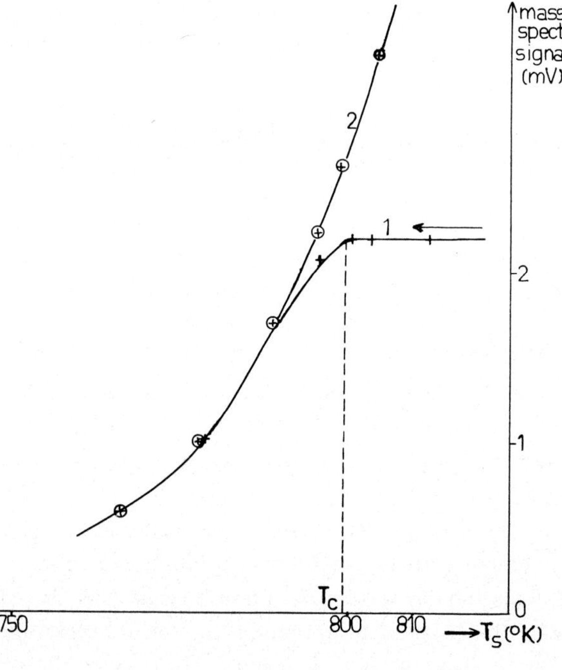

Fig. 6a

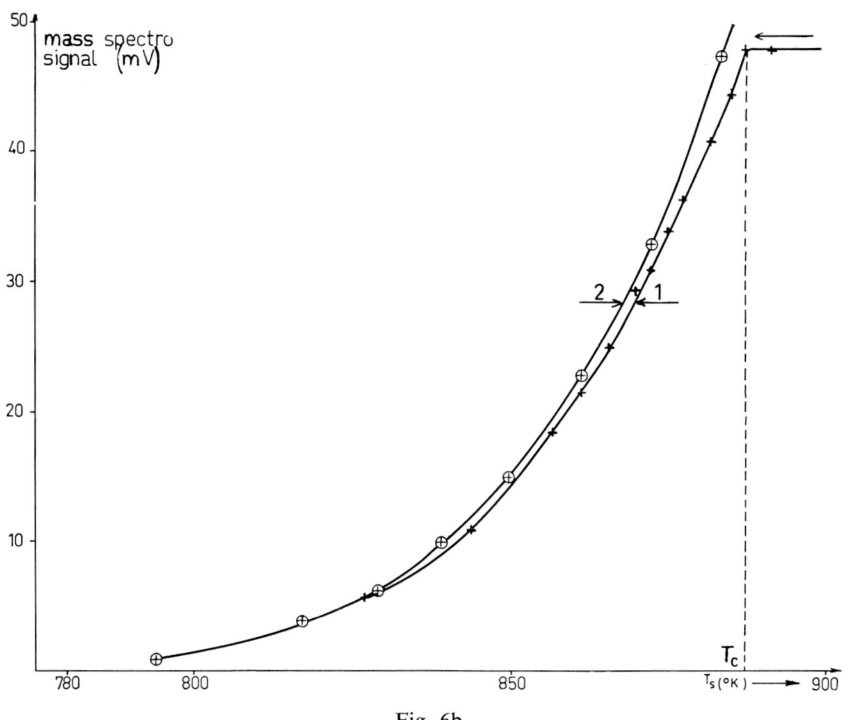

Fig. 6b

Fig. 6. Determination of the critical supersaturation for the onset of nucleation. Curves 1: re-emitted flux J_r. Curves 2: vapor pressure of the massive phase of indium. The critical supersaturation is determined by the point at which J_r starts to decrease. (a) $J_i = 1.6 \times 10^{13}$ atoms cm^{-2} s^{-1}; (b) $J_i = 4.47 \times 10^{14}$ atoms cm^{-2} s^{-1}.

slight undersaturation. This poses a great number of difficulties with both classical and atomistic nucleation theories.

Some of these results can be compared with the cadmium–tungsten studied by the same experimental method by Hudson et al. In both cases, the thermal accommodation of the incident flux is complete but while in our work we observe the nucleation in about a monolayer only with a zero supersaturation or undersaturation, Hudson finds nucleation occurring in several layer thick adatoms with a supersaturation ranging from 1.6 to 1.8.

We are currently working on theoretical interpretation of some of our results.

References

1) D. R. Bates, Ed., *Advances in Atomic and Molecular Physics*, Vol. 3 (Academic Press, New York, 1967) p. 143.
2) H. Shelton and A. Y. Cho, J. Appl. Phys. **37** (1966) 3544.
3) A. Y. Cho and C. D. Hendriks, J. Appl. Phys. **40** (1969) 3339.
4) J. B. Hudson and J. S. Sandejas, J. Vacuum Sci. Technol. **4** (1967) 230.
5) J. B. Hudson and J. S. Sandejas, Surface Sci. **15** (1969) 27.
6) A. A. Kinawi and J. B. Hudson, J. Vacuum Sci. Technol. **6** (1969) 68.
7) R. E. Honig, RCA Rev. **30** (1969) 285.

Crystal morphology, growth forms, epitaxy

D. NENOW	M. LEDESERT
G. A. WOLFF	M. FREY
M. WEINSTEIN	E. GILLET
B. N. DAS	M. GILLET
T. OHACHI	J. S. VERMAAK
I. TANIGUCHI	J. R. ROOS
J. H. KITTERMAN	H. PASCARD
D. R. WINDER	C. QUINTANA
R. HEIMANN	F. HOFFMANN
W. FRANKE	C. SELLA
R. LACMANN	

Journal of Crystal Growth **13/14** (1972) 183–185 © *North-Holland Publishing Co.*

183

ON THE VARIATIONAL FORMULATION OF THE PROBLEM OF SURFACE EQUILIBRIUM OF CRYSTALS

D. NENOW

Institute of Physical Chemistry, Bulgarian Academy of Sciences, Sofia 13, Bulgaria

The Herring formula for the surface equilibrium of crystals is generalised in the two-dimensional case. The generalisation is based on the results obtained by the Born-Stern method as extended by Stranski and depends upon the choice of the position of the dividing surface. The specific edge energy κ may depend not only on the orientation α of the contour at a point but also on its coordinates x and y. Therefore, the treatment corresponds to the use of the differential form of the fundamental equation.
It is shown that the Euler-Lagrange equation of this problem has the form

$$\frac{1}{\rho}\left(\frac{\partial^2 \kappa}{\partial \alpha^2} + \kappa\right) + \frac{\partial^2 \kappa}{\partial n \partial \alpha} \cos (\alpha - \phi) + \frac{\partial \kappa}{\partial n} \sin (\alpha - \phi) - \lambda = 0,$$

where ρ is the radius of curvature of the extremum curve, ϕ is the slope of the normal to the level lines $\kappa = \text{const.}$ at $\alpha = \text{const}$, $\partial/\partial n$ denotes a differentiation after the same normal and λ is the Lagrange multiplier.
In the particular case of a circle this equation corresponds to the generalised two-dimensional Gibbs-Thomson equation for liquids with an equimolecular dividing surface

$$\frac{\kappa}{r} + \frac{d\kappa}{dr} - \lambda = 0.$$

The traditional approach to the rigorous introduction of surface energies of crystals stems from the work of Born and Stern[1]. However, its utilization in strict thermodynamic considerations became possible only after Stranski[2] introduced the concept of "crystalline half-crystal position", which proved essential for the subsequent development of the problem, and generalized the Born–Stern method for finite crystals.

We should like to discuss one fact which up to now has received no attention whatsoever. In the idealized mathematical partitioning of the crystals used in this method the sum of the amounts of substance enclosed in the portions obtained in this way is always equal to the total amount of substance present prior to the operation. The absence of a correction term taking into account the appearance of a surface determines in this case the choice of the dividing surface as an equimolecular one. However, there was no complete comparison of the results obtained by the Born–Stern method as generalized by Stranski with those following from the detailed thermodynamic treatment involving the equimolecular surface in the case of liquids[3]. It is particularly convenient to perform this comparison in terms of the variational formulation of the problem of

the equilibrium form of the crystals as developed by Landau[4], Herring[5], and Burton, Cabrera and Frank[6]. In such a comparison it is essential that the considerations should correspond to the use of the differential form of the fundamental equation in the thermodynamics of surfaces. Therefore, we intend to give here a generalization of the variational formulation of the problem of the surface equilibrium of crystals based on the results obtained by the Born–Stern method as generalized by Stranski and therefore connected with the choice of the position of the dividing surface as an equimolecular one. In the use of this generalization the comparison with the relevant results from the thermodynamics of the liquid surfaces follows in a perfectly natural manner.

At present the exact variational formulation of the problem of surface equilibrium of crystals based on the approach outlined by Stranski is only possible in the two-dimensional case. This is related to the possibility discussed by Nenow, Stoyanov and Kashchiev[7] of using this approach in the general case of a two-dimensional polygonal model. In the three-dimensional case only parallelopipeds can be studied in this way. From this work it follows that for the overall surface

IV − 1

energy ϕ of an arbitrary two-dimensional polygon a distribution of the type

$$\phi = \sum_{i=1}^{N} \kappa_i l_i \tag{1}$$

is always valid, where κ_i are the specific edge energies depending on the size of the crystal and l_i are the lengths of the polygon edges. Upon infinite increase of the number of edges N of the polygonal model, expression (1) changes to the functional

$$\phi = \int_L \kappa \, dl, \tag{2}$$

whose minimum at constant surface of the generalized model under discussion gives the condition for surface equilibrium in the two-dimensional case. The essential difference between the variational problem thus formulated and the problem considered in refs. 3–5 is that here κ from (1) and (2) may depend not only on the orientation of the contour of the model at a given point but on its coordinates, too. Therefore, the treatment in the present paper corresponds to the use of the differential form of the fundamental equation in the thermodynamics of surfaces. Thus the problem of finding the equilibrium form of the two-dimensional model under consideration may be formulated in a parametrical form as

$$\int_L \kappa \left(x, y, \text{arc tg} \frac{\dot{y}}{\dot{x}} \right) (\dot{x}^2 + \dot{y}^2)^{\frac{1}{2}} \, dt = \text{min.},$$

$$\frac{1}{2} \int_L (x\dot{y} - y\dot{x}) \, dt = \text{const.},$$

where x and y are the Cartesian coordinates of a point of the contour, and $\dot{x}$ and $\dot{y}$ are derivatives after the parameter t.

In order to represent this problem in an invariant form, we shall use the Euler–Lagrange equation in the form given by Weierstrass[8]

$$F_{x\dot{y}} - F_{y\dot{x}} + F_1(\dot{x}\ddot{y} - \dot{y}\ddot{x}) = 0, \tag{3}$$

for the expression

$$F = \kappa(\dot{x}^2 + \dot{y}^2)^{\frac{1}{2}} - \frac{1}{2}\lambda(x\dot{y} - y\dot{x}), \tag{4}$$

where

$$F_1 = \frac{F_{\dot{x}\dot{x}}}{\dot{y}^2} = \frac{F_{\dot{y}\dot{y}}}{\dot{x}^2} = -\frac{F_{\dot{x}\dot{y}}}{\dot{x}\dot{y}}, \tag{5}$$

and λ is the Lagrange multiplier. Since

$$\frac{1}{\rho} = \begin{vmatrix} \dot{x} & \dot{y} \\ \ddot{x} & \ddot{y} \end{vmatrix} (\dot{x}^2 + \dot{y}^2)^{-\frac{3}{2}},$$

where ρ is the radius of curvature of the extremum curve, eq. (3) in the form

$$F_{x\dot{y}} - F_{y\dot{x}} + F_1 \frac{(\dot{x}^2 + \dot{y}^2)^{\frac{3}{2}}}{\rho} = 0 \tag{6}$$

is, as is well-known, invariant with respect to the choice of the coordinate system.

Using (4) and (5) we get

$$F_1 \frac{(\dot{x}^2 + \dot{y}^2)^{\frac{3}{2}}}{\rho} = \frac{1}{\rho} \frac{\partial^2 \kappa}{\partial \alpha^2} + \frac{1}{\rho} \kappa,$$

and from (5)

$$F_{x\dot{y}} - F_{y\dot{x}} = \frac{\partial^2 \kappa}{\partial n \, \partial \alpha} \cos(\alpha - \varphi) + \frac{\partial \kappa}{\partial n} \sin(\alpha - \varphi) - \lambda,$$

where α is the slope of the tangent to the extremal given by $\alpha = \text{arc tg}(\dot{y}/\dot{x})$, φ is the slope of the normal to the level lines $\kappa = \text{const.}$ at $\alpha = \text{const.}$, with respect to the x-axis given by

$$\left(\frac{\partial \kappa}{\partial x} \right)_{y,\alpha} = \left(\frac{\partial \kappa}{\partial n} \right)_{\alpha} \cos \varphi,$$

$$\left(\frac{\partial \kappa}{\partial y} \right)_{x,\alpha} = \left(\frac{\partial \kappa}{\partial n} \right)_{\alpha} \sin \varphi,$$

and $\partial/\partial n$ denotes a differentiation after the same normal.

Now (6) acquires the form

$$\frac{1}{\rho} \left(\frac{\partial^2 \kappa}{\partial \alpha^2} + \kappa \right) + \frac{\partial^2 \kappa}{\partial n \, \partial \alpha} \cos(\alpha - \varphi) +$$

$$\frac{\partial \kappa}{\partial n} \sin(\alpha - \varphi) - \lambda = 0. \tag{7}$$

With the exception of the terms containing the derivatives after the normal this expression corresponds to the well-known Herring formula and represents the Euler–Lagrange equation in an invariant form for the problem

$$\int_L \kappa \left(\text{arc tg} \frac{\dot{y}}{\dot{x}} \right) (\dot{x}^2 + \dot{y}^2)^{\frac{1}{2}} \, dt = \min. \, ,$$

$$\tfrac{1}{2} \int_L (x\dot{y} - y\dot{x}) \, dt = \text{const.} \, ,$$

where κ is independent of x and y. The term $(\partial\kappa/\partial n)$ $\sin(\alpha - \varphi)$ appears also in the similar variational problem of a propagation of a light beam according to Fermat's principle in an optically isotropic non-homogeneous medium[9]). The problem discussed here is more general and includes the above two problems as particular cases.

In the particular case of a circle, that is, when κ is a function only of the radius vector r, it can be readily shown that (7) acquires the form

$$\frac{\kappa}{r} + \frac{d\kappa}{dr} - \lambda = 0 \, ,$$

i.e., it corresponds to the generalized two-dimensional Gibbs-Thomson equation for liquids in the case of an equimolecular dividing surface. Therefore (7) can be also considered as a generalization in the case of anisotropic phases.

The author wishes to thank S. Stoyanov and D. Kashchiev for fruitful discussions during the preparation of the manuscript.

References

1) M. Born and O. Stern, Sitzber. Preuss. Akad. Wiss. **48** (1919) 901.
2) I. N. Stranski, Sitzber. Akad. Wiss. Wien, Math.-Naturwiss. Kl. IIb **145** (1936) 840; Z. Kristallogr. A **105** (1943) 278; I. N. Stranski and R. Kaischew, Z. Physik. Chem. B **35** (1937) 427.
3) S. Ono and S. Kondo, Molecular Theory of Surface Tension in Liquids, in: *Handbuch der Physik*, Bd. X (Springer, Berlin, 1960).
4) L. D. Landau, *Sbornik Posvyashtenniy Semidesyatiletiu Akademika A. F. Ioffe* (Izd. Akad. Nauk SSSR, Moscow, 1950) p. 44 (in Russian).
5) C. Herring, in: *The Physics of Powder Metallurgy*, Ed. W. E. Kingston (McGraw-Hill New York, 1951); in: *Structure and Properties of Solid Surfaces*, Eds. R. Gomer and C. S. Smith, (University of Chicago Press, 1953).
6) W. Burton, N. Cabrera and F. C. Frank. Phil. Trans. Roy. Soc. London A **243** (1951) 299.
7) D. Nenow, S. Stoyanov and D. Kashchiev, Surface Sci. **9** (1968) 268.
8) R. Courant and D. Hilbert, *Methods of Mathematical Physics*, Vol. 1 (Interscience, New York, 1953).
9) M. L. Lavrentiev and L. A. Lusternik, *Kurs Variatsionnogo Ischisleniya* (Gostehizdat, Moscow, 1950) (in Russian).

Journal of Crystal Growth **13/14** (1972) 186–190 © *North-Holland Publishing Co.*

A UNIFIED APPROACH TO HABIT AND CRYSTAL GROWTH ANALYSIS

G. A. WOLFF*, MARTIN WEINSTEIN† and B. N. DAS**

Tyco Laboratories, Inc., Waltham, Massachusetts 02154, U.S.A.

Faceting of crystals during growth or solution is directly associated with the occurrence of rate-controlling kinetic steps on the advancing or receding crystal interface. If faceting morphology is characterized by the type of (flat) F-planes, (stepped) S-planes, and (kinked) K-planes comprising the surface, it can alternately be described in terms of a particular combination of specific atomic surface and dislocation sites at these planes. The integration of the atoms of the exposed sites into the underlying crystal lattice, or their removal from the surface, provides the rate-controlling partial reaction step in crystal growth or solution. Since the number of such combinations of the various atomic surface sites, and of the respective crystal habits, is limited, the number of growth and solution mechanisms that are associated with varying matrix environment, composition, and homeotect crystal structures is naturally also limited. All habits associated with possible or actual growth and solution mechanisms can be derived by proper assignment of free energy parameters ("dangling" bond free energy, or equivalent) to the various surface sites that would have to include the effects of adsorption and autoadsorption. Habits of structures of interplaced Bravais lattices can be synthesized from the basic "Bravais habits" by parametric selection rules. Examples are demonstrated for growth and solution morphologies of ZnO, ZnS polytypes, and other materials of adamantine and other structures.

1. Introduction

Although the crystal (equilibrium) habit of various structures and materials has successfully been derived by a number of methods[1-7], the immense difficulties sometimes encountered in their derivation for complex crystal structures, make the development of a more functional method desirable. The main ingredients of the method are the application of Wulff's and Stranski's principles[1] and Hartman's PBC vector concept[2].

2. Definition and analysis of crystal habit

In the following, an intentionally simplified and idealized situation will be described. This is not meant to imply that refinement and sophistication in the applied model have not, or cannot, be used[8].

By their appearance, crystal faces can be classified in three groups, namely, F(flat)-planes, S(stepped)-planes, and K(kinked)-planes. F-planes and S-planes are made up of, and stabilized by, two or more bond chains and one bond chain, respectively[2]. Bond chains are arrays of atoms, molecules, or other crystal building units, with each unit bonded to surface and bulk of the

* Present address: Lamp Division, General Electric Company, Nela Park, Cleveland, Ohio, U.S.A.
† Present address: Turbine Support, Chromalloy-American Corporation, San Antonio, Texas, U.S.A.
** Present address: Naval Research Laboratories, Washington, D.C., U.S.A.

crystal more strongly than a crystal building unit in the equilibrium or growth position. In the latter position the building unit is bonded, or coordinated, to half as many neighbors as the respective unit in the crystal interior. This position is, therefore, also called the half-crystal position; in it a crystal building unit terminates one bond chain which in turn is simultaneously attached to both a F-plane and the terminating step of a half-layer resting on the F-plane.

Since a crystal building unit has an equal chance of attaching to, or leaving, the growth position of a crystal in equilibrium with its matrix, the apex positions of polyhedral crystal forms in equilibrium with their ambient matrix are also equilibrium positions in materials of homopolar bonding. Since the apex units terminate the edges bounding the crystal planes, crystal edges are made up of bond chains. All planar intersections or "cuts" of crystals which truncate the crystal apices therefore result in the formation of an infinite number of K-planes, while all truncating cuts which are parallel to any one edge of the crystal produce an infinite number or set of S-planes.

The total number of S-plane sets which comprise a crystal form can conveniently be drawn as lines in the stereographic projection, as in figs. 1–3. The space enclosed by lines and connecting intersections denotes an infinite number of K-plane positions or their representative crystal apex formed by the intersection of the adjacent F-planes and their common edges. This presentation has the additional advantage that the lines also represent the zonal regions in which the various bond chains are effective.

Bond chains do not have to be continuous and intact along their whole length. It is only necessary that the statistical mean detachment free energy per bond chain building unit at equilibrium be greater than that of a unit in the growth position, or that the "specific" bond chain terminal free energy is of finite value. This is particularly important for adsorption and incongruent phase changes. Small monotonic changes of the concentration in the ambient matrix of one of the constituents may thus produce abrupt habit (and stability) changes in peritectic phases and solid solutions; the influence of adsorbing species on crystal habit has been known for a long time and does not require repeating.

Since bond chains follow specific crystallographic directions, they exhibit translational periodic properties. Their translational period may be equal to, or a sub-multiple of, the corresponding unit cell lattice spacing, while their intra-period configuration and bonding are uniquely identified by $n-1$ [of a total of $\binom{n}{2}$ possible] "inter-cell translation vectors" and composition of the participating n Bravais lattices.

A "PBC-vector"[2]) may thus be a single vector, or else it may be presented by the sum of partial vectors including the aforementioned inter-lattice vectors. Since the PBC-vectors are periodic with respect to all Bravais unit cells which compose the structure unit, the total set of derived crystal equilibrium planes must also be found in every set of crystal planes which can be derived for all habit planes obtained for every Bravais lattice. This is to say that the number of the prospective planes which have been derived for the various Bravais lattices (which ultimately could prove to be equilibrium planes) will mostly be reduced. In the consideration of inter-lattice bonding where the existence of Bravais pseudo-sub-lattices may reasonably be assumed, the identification of a few equilibrium planes is often facilitated. The consideration of lattice interaction is, however, of greatest importance in crystals lacking a center of symmetry where the polar character of interlattice bonding strongly influences the crystal habit.

The practical derivation of crystal morphology may best be described as follows: First, an actual crystal model and/or corresponding two-dimensional projections of the crystal structure investigated are separated or "cleaved" in two parts. The number of specific atomic surface sites of each type per unit mesh area for the obtained "crystal plane" are then listed for the various planes; it is preferable to start with low-index planes associated to low-index zones. Variable free energy parameter values are then assigned to each type of surface site formed by the effected "cleavage" process, simulated adsorption or any other related surface change. Surface energy parameters of newly generated sites may here be identified in terms of "dangling bonds" (i.e. in terms of lattice neighbors removed), specific types or species of surface atoms, ionicity and related terms. The various specific surface free energy sites are then used to derive their most important surface energy PBC vectors $\langle UVW \rangle$ which are obtained by expressing and combining the respective surface free energy values according to the formula

$$\sigma_{hkl} = \frac{Uh + Vk + Wl}{\text{unit mesh area}}$$

$$= \frac{(n_C \varphi_C + n_A \varphi_A)_{hkl}}{\text{unit mesh area}}$$

where n_C, n_A and φ_C, φ_A, respectively, represent number and bond free energy of cation and anion "dangling" free bonds. This specific formula is, for example, used in the calculation for the ZnS diagrams in fig. 2.

In the included figures these principles have been used for the derivation of corresponding habits. The habits are represented in stereographic projections.

It should be realized that F-planes occur in equilibrium as well as in crystal growth, since during growth commencement of each new layer requires two-dimensional nucleation energy barriers; this barrier may be lowered by the catalytic influence of intersecting and self-perpetuating screw dislocations according to the mechanism described by Frank.

3. Influence of Bravais lattice combination, atomic substitution, and unit cell size increase on crystal habit

In the first column of fig. 1 is shown the idealized habit of materials having the structures of the three cubic Bravais types, namely simple cubic (sc), face-

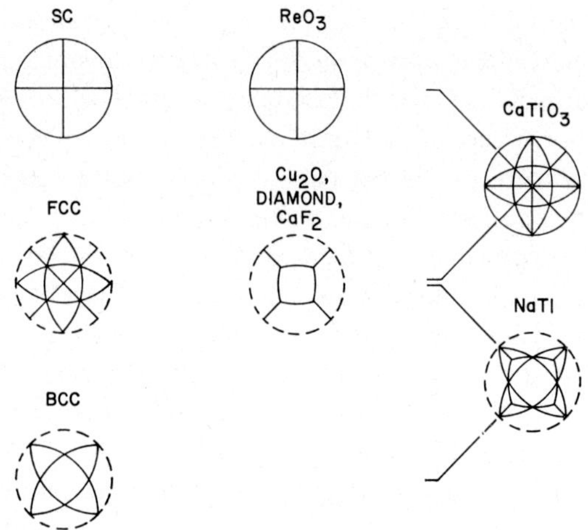

Fig. 1. Stereographic projections of calculated crystal equilibrium forms for materials of cubic structure. Forms are given for the three cubic Bravais lattices, namely the simple cubic (sc), face centered (fcc), the body centered cubic lattice, and their combinations. Intersections of lines denote equilibrium planes (F-planes); connecting lines represent intersections (or crystal edges = S-planes) of two adjoining F-planes. Equilibrium forms have been calculated for nearest neighbour interaction only.

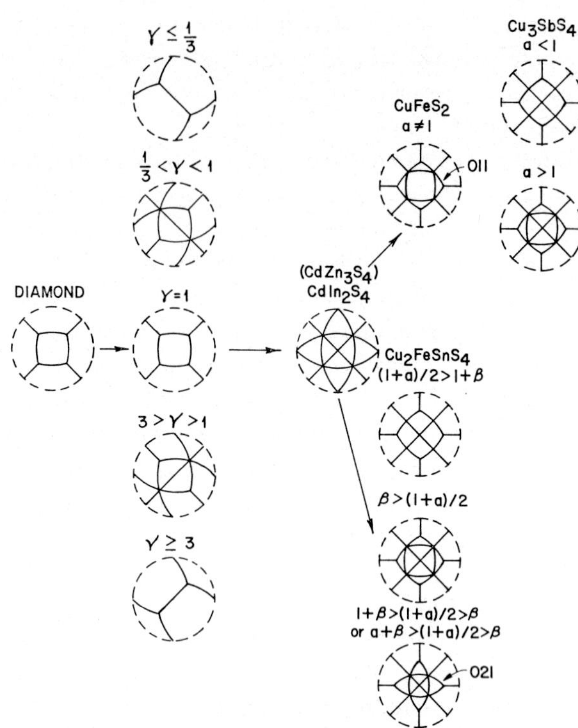

Fig. 2. Stereographic projections of calculated crystal equilibrium forms for materials with diamond structure and with structures (sphalerite, $CdIn_2S_4$, chalcopyrite, stannite, famatinite) derived from the diamond structure by successive partial substitution. All forms of columns 3, 4 and 5 have been derived for the simple case of cleavage (introduction of symmetry center in structure, as "Friedel-condition"). Complete treatment was applied to forms in columns 1 and 2. Perspective drawings of crystal forms column 2 are shown in fig. 4, column 1.

centered cubic (fcc), and body-centered (bcc). The habit has been derived for the first nearest neighbor interaction only, excluding auto-adsorption or deviations of surface structure from the structure in the bulk. Since only straight, linear atomic PBC-vectors due to nearest neighbor interactions are observed, leaving the vectors intact upon "crystal cleavage" parallel to them, the PBC-vector range in all three cases extends over the full zone and 360 °range. In the three stereographic projections on the left of fig. 1 the PBC-vectors due to nearest neighbor interaction appear as uninterrupted straight or circular lines. The lines in the first case intersect in {001} planes which they connect so that a cube of {001} planes with common edges is observed for the sc structure; a cuboctahedron enclosed by {001} and equilateral {111} planes is indicated for the fcc structure; and a rhombic dodecahedron {011} is found for bcc structure. Placing three atoms per sc unit cell on the three equivalent centered cell edge positions $00\frac{1}{2}$, as in ReO_3, does not generate new PBC-vectors, since the ⟨001⟩ vectors which are normally present in the sc structure, as in α-polonium, are only strengthened through the newly introduced positions midway between the atoms in 000 (cell corner) positions.

In cubic $SrTiO_3$ of cubic $CaTiO_3$ perovskite struc-

ture, one Sr atom is introduced in bcc position of the cell, with Ti and O in the previously mentioned positions. Next nearest neighbor interaction φ_2 between Sr and O is thus generated in addition to φ_1 between Ti and O. Sr and three O atoms are now in an ordered fcc Cu_3Au-type arrangement resulting in a superposition of sc and fcc as shown on the right in fig. 1. In perovskite materials one might thus expect {111} in addition to {001}; the observed habit is, however, found to be predominantly cubic. From the derived formula for the specific surface free energy, it can be shown that this will happen for $\varphi_2 + \varphi_3 \geq 0$ when φ_3 (≤ 0) represents the repulsive interaction between the O atoms, and attraction energy is taken positive. Repulsion between O atoms in this case would negate the attraction between Sr and O. Bcc-type interaction between Sr and Ti would produce {011} as shown at the lower left of fig. 1.

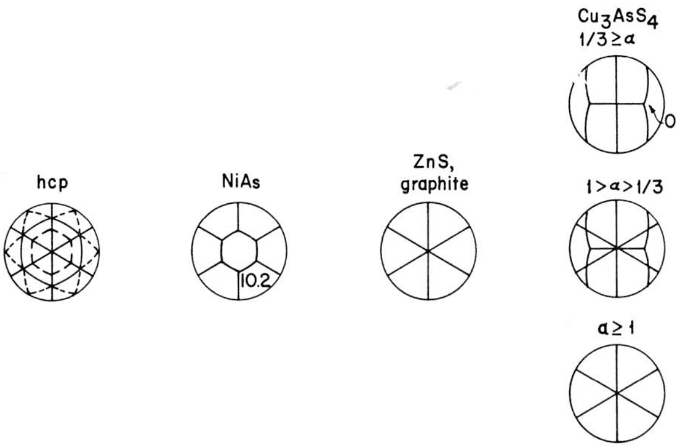

Fig. 3. Stereographic projections of calculated crystal equilibrium forms for materials with hexagonal close-packing (hcp) and derived structures (NiAs, wurtzite, graphite, and enargite).

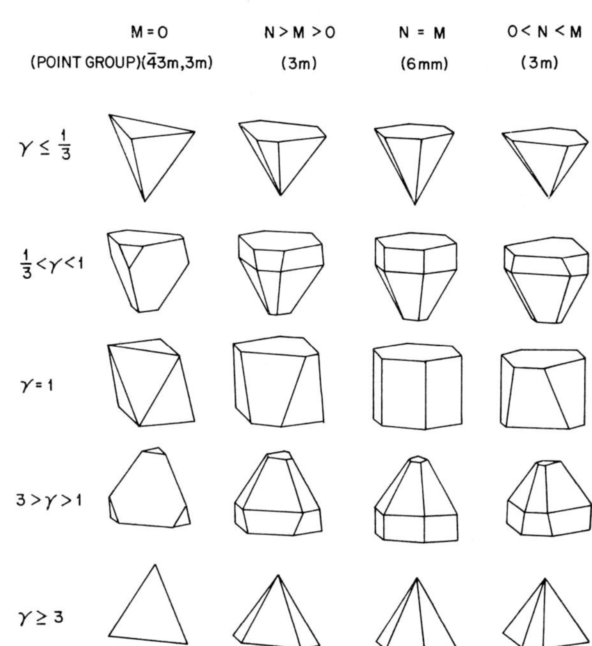

Fig. 4. Calculated equilibrium forms of ZnS structure polytypes for various ratios N/M of layer stacking (columns) and ratios $\gamma = \varphi_C/\varphi_A$ of surface free energy contribution (rows) of cation (φ_C) and anion (φ_A). M and N refer to the sum of odd and even Zhdanov stacking numbers, respectively. In the 15 R polytype, for example, with Zhdanov notation (32)$_3$, $M = 9$ and $N = 6$. The left column relates to column 2 of fig. 2.

For this purpose of demonstrating the type of approach used, the previous example has been given in full detail. Only the less obvious points will be explained for the remaining atoms placed in equivalent $\frac{1}{4}\frac{1}{4}\frac{1}{4}$ and $\frac{1}{2}\frac{1}{2}\frac{1}{2}$ positions. When only the first nearest neighbor interaction φ_1 is considered, the habits of diamond, Cu$_2$O and CaF$_2$ all prove to be octahedral, in spite of the structural differences. This clearly shows that many of the PBC-vectors in the fcc structure have become ineffective. For example the two equivalent $\langle 110 \rangle$ PBC-vectors in {001} of diamond have no effect, since the C atoms in this plane utilize only two bonds for attachment; all C atoms on {001} are thus in equilibrium positions. Inter-lattice interaction in NaTl between the interpenetrating Na- and Tl-lattices, each of diamond structure, produce {011} planes. This appears reasonable, since Na and Tl atoms are placed in a pseudo-bcc arrangement.

In fig. 2 it is shown how the habit changes upon successive substitution of atoms which are all in tetrahedral positions of the diamond structure. The influence of ionicity is indicated by habit variations with $\gamma = \varphi_C/\varphi_A$, while habit changes through bond changes can be followed for the varying conditions given on the right of fig 2, where $\alpha = \varphi_2/\varphi_1$, $\beta = \varphi_3/\varphi_1$; the index number n for the free energy φ_n of cation–anion bonding increases with increasing valency.

In fig. 3 habit changes are demonstrated as they occur upon filling of various positions as interstitials in the hcp lattice. Dotted lines in the habit presentation of hcp denote PBC-vectors corresponding to next nearest neighbor interaction in the lattice. The influence of the substitution of Zn in wurtzite leading to Cu$_3$AsS$_4$ is shown on the right.

In almost all of the investigated cases good agreement is found between derived habits and those observed by our investigators or reported in the literature.

Fig. 4 shows the derived crystal equilibrium forms of the various ZnS polytypes which also relates to column 2 of fig. 2.

Only in these last two presentations is noncentrosymmetrical behavior considered. In all other illustrated examples in figs. 1–3, the crystal habits shown have been calculated under the simplifying assumption of centrosymmetrical behavior even if the structure were not centrosymmetrical. This habit is likely to show in cleavage, however, in all structures. This includes structures lacking a center of symmetry since such a center is introduced during the cleavage process.

The authors wish to acknowledge the support of the Air Force Materials Laboratory under contracts AFML-TR-67-385 and AFML-TR-69-13.

References

1) I. N. Stranski, Discussions Faraday Soc. **5** (1949) 13.
2) P. Hartman, in: *Growth of Crystals*, Vol. 7 (Consultants Bureau, New York, 1969) p. 3.
3) J. D. H. Donnay and D. Harker, Am. Mineralogist **22** (1937) 446.
4) B. Honigmann, Fortschr. Physik. Chem. **4** (1958).
5) R. Kern, in: *Growth of Crystals*, Vol. 8 (Consultants Bureau, New York, 1969), p. 3.
6) R. Lacmann, Ergebn. Exakt. Naturwiss. **44** (1968) 1.
7) G. A. Wolff, in: *Intermetallic Compounds*, Ed. S. H. Westbrook (Wiley, New York, 1967) p. 85.
8) F. H. Cocks, N. B. Das, C. B. Lamport, A. A. Menna, E. A. Trickett and G. A. Wolff, Final Reports AD-666073 (1967) and AD-682518 (1968).

Journal of Crystal Growth **13/14** (1972) 191–197 © *North-Holland Publishing Co.*

CONTROLLED FILAMENTARY GROWTH OF SILVER FROM SILVER COMPOUNDS*

TADASHI OHACHI and ICHIRO TANIGUCHI

Department of Electronics, Faculty of Engineering, Doshisha University, Kyoto, 602 Japan

Growth of filamentary silver was controlled using a solid-state-ionics technique. The controlled growth process was recorded on a 16 mm film, "High-speed growth of silver filaments". The morphology of filaments was examined using a scanning electron microscope.

The growth representing a variety of silver compounds is classified into three groups according to their ionic and electronic conductivity, σ_i and σ_e respectively. (1) Growth from α-Ag$_2$X (X = S, Se, or Te) (σ_i: large, σ_e: large); the maximum growth rate is about 30 μm/sec and corresponds to the maximum nucleation rate. The filament is fluted and thicker at its root and appears to consist of a group of smaller filaments. This growth process is reversible, that is the grown filaments can re-dissolve in the compounds. (2) Growth from α-AgI (σ_i: large, σ_e: small). The maximum growth rate is about 10^3 μm/sec which is also the maximum nucleation rate. This high growth rate may be due to electronic breakdown in α-AgI. A bunch of filaments, which are fin-like and have a $\langle 110 \rangle$ orientation, is cotton-like in appearance. This process is also reversible. (3) Growth from AgBr or AgCl (σ_i: small, σ_e: small); whiskers and dendrites grow very slowly owing to the small ionic conductivity. The dendrites have many steps.

1. Introduction

The phenomenon by which silver filaments grow from silver compounds has been studied by many investigators, therefore many descriptions, such as filament, capillary, filiform, hair, moss, fiber, wire, spike and whisker have been used by individual authors. Recently Hardy[1] and Nabarro[2] briefly reviewed the filamentary growth of metals. The natural growth of silver filaments was well-known by Ercker in the 16th century[1]. It may be not too much of an exaggeration to say that the first scientific approach to crystal growth began with research into silver filamentary growth. It is also of historical interest that the term, "growth" may have been used for the first time in the field of crystal growth by Redwin who reported the growth of moss silver, gold and copper in 1877[3].

In this paper we report the control of growth of silver filaments from silver compounds by using a solid-state-ionics technique and discuss the morphology and growth rate of the filaments. The growth is classified into three groups according to the ionic and electronic conductivity of the compound. The growth process was photographed on a 16 mm film. The

morphology of the filaments was examined by scanning electron microscopy.

2. Growth control

The important factors controlling crystal growth are the nucleation and the supply of atoms to a growth point. These are affected by the following parameters.

The nucleation by: the degree of supersaturation, the supply of atoms to a growth point, surface roughness, and the growth-point temperature.

The supply of atoms to a growth point by: transport of atoms, the conductivity for atoms, dimensions of specimen, the temperature, and the supply of atoms to the compound.

These are not independent, because the degree of supersaturation depends on the supply of atoms. For the nucleation rate the degree of supersaturation is the most important controlling factor. The growth can be controlled by control of the above parameters.

Many methods are available to supply an excess of silver atoms to silver compounds, for example:
(a) removal of the non-metallic component by high-temperature decomposition or hydrogen-reduction,
(b) injecting silver atoms at the surface by causing an electric current to flow in the galvanic cell, Ag/α-AgI/α-Ag$_2$X/Pt,

---

* The colour film, "High-speed growth of silver filaments" was shown during the Conference.

(c) using metastable silver selenide, prepared by abruptly cooling silver selenide containing excess silver from the high-temperature saturated state to room temperature.

These methods were used in the following experiments to control growth.

The filamentary growth is classified into the following three groups according to ionic and electronic conductivity of the compounds concerned.

(I) Controlled growth from α-Ag$_2$X (X = S, Se, or Te) (σ_i: large, σ_e: large).

Method b for the introduction of silver atoms into α-Ag$_2$X was reported by Ohachi and Taniguchi[4]). In order to obtain a supersaturated state and provide a driving force for the atoms, in this case a temperature gradient was maintained across an α-Ag$_2$X specimen. Filaments grew at the top of the low-temperature end as shown in fig. 1. The growth can be considered as the same as solution growth. This growth process is reversible, that is, if the atoms are removed instead of being injected, the grown filaments dissolve into the α-Ag$_2$X. The effect of injected silver atoms was rapidly available at the growth point because of the large ionic and electronic conductivities of α-Ag$_2$X. When the injection of atoms was increased sufficiently (so that the injection rate is larger than the nucleation rate) the

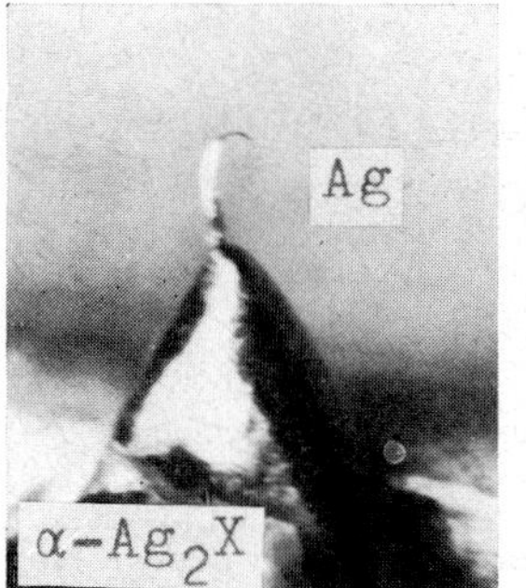

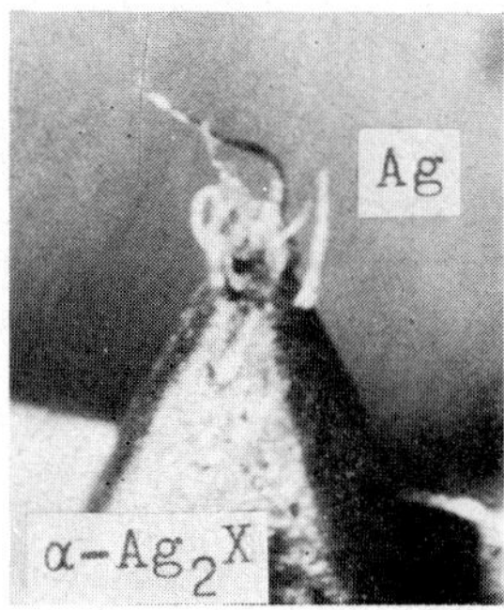

Fig. 1.   A silver filament grown from α-Ag$_2$X ($\times$10).

Fig. 2.   Silver filaments grown from α-Ag$_2$X at high-injection level ($\times$10).

filament diameter or the number of filaments increased, as shown in fig. 2. In this case the growth rate is controlled by the nucleation on the crystal surface – rate-control by nucleation. The maximum growth rate (30 μm/sec) was measured from the film. When the injection rate is small, the growth rate is controlled by the injection – rate-control by injection.

Many other investigators, for example, Kohlshüter[5]) and Wagner[6]), have used method a to obtain filaments.

In the case of method c, the degree of supersaturation is controlled by the excess silver content of the metastable selenide or by a temperature gradient.

Typical shapes of filaments grown from α-Ag$_2$X are shown in fig. 3. The shapes of all filaments are similar

(a)

(b)

Fig. 3.   Scanning electron microscope (SEM) photographs of the parts of filaments grown from α-Ag$_2$X: (a) $\times$330, (b) $\times$3300.

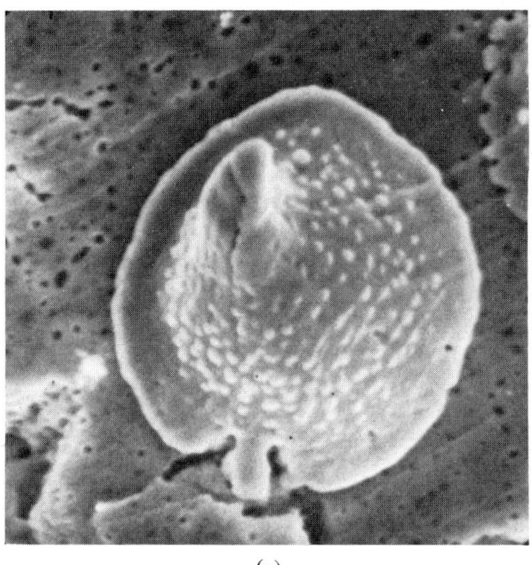

(a)

(b)

Fig. 4.   SEM photographs of (a) a disk-like ($\times$1000), and (b) thick-short ($\times$330) filaments.

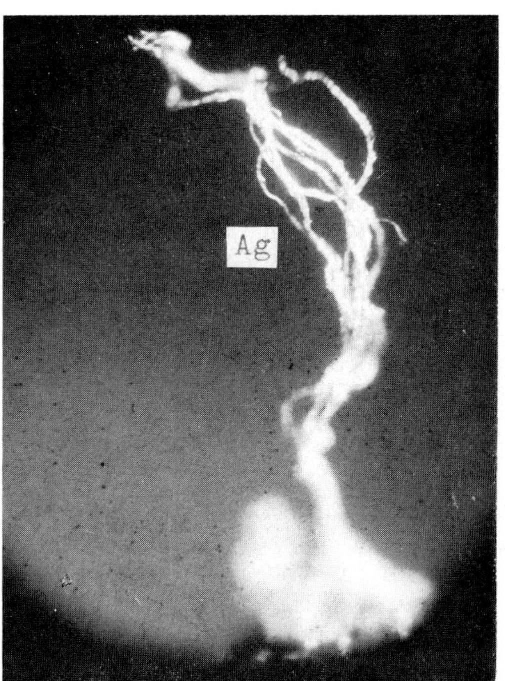

Fig. 5.   A bunch of silver filaments grown from α-AgI using method b ($\times$15).

Fig. 6.   A SEM photograph of silver filaments grown from α-AgI pellet using method c ($\times$330).

and are independent of the compound from which they are formed. The filament is generally curved, thicker at its root, rounded, fluted, and appeared to consist of a group of a smaller filaments. The disk-like or short thick filaments which grow in the case of high temperature and high degree of supersaturation, produced by high frequency induction heating at 2 MHz, are shown in fig. 4. The surface shape of the filament is due to the surface roughness of α-Ag_2X on which nucleation occurred.

X-ray investigation shows that filaments are polycrystals and do not have any specific orientation detectable using an X-ray spot size of 0.5 mm.

(II) Controlled growth from α-AgI (σ_i: large, σ_e: small).

Method b was used for growth from α-AgI, the experimental arrangement was the same as before[4]) except that the growth point was brought into contact

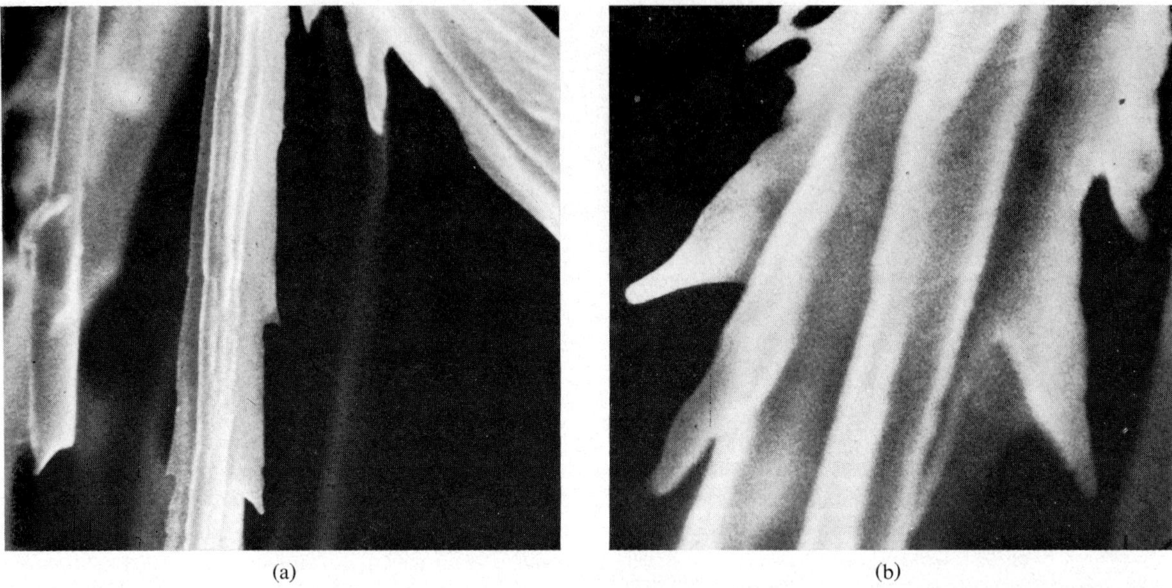

(a) (b)

Fig. 7. SEM photographs of fin-like silver filaments grown from α-AgI. The growth direction is upward. (a) $\times 7500$, (b) $\times 22\,500$.

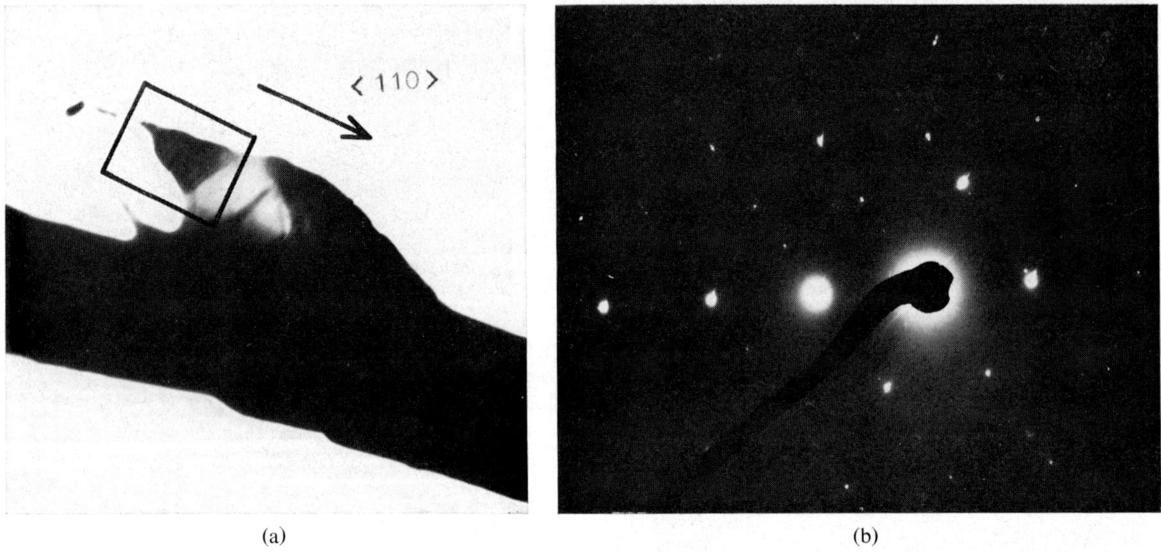

(a) (b)

Fig. 8. An electron microscope photograph of the fin: (a) $\times 6800$. The electron diffraction (110) pattern.

with silver iodide. The cotton-like filaments grew continuously as long as the injection continued (fig. 5.)

Method c was also used as follows; a silver iodide pellet was placed on metastable silver selenide to supply the silver atoms and heated from the bottom. As shown in fig. 6 the cotton-like filaments began to grow immediately from the corner of the silver iodide pellet, and the growth stopped after all the excess silver atoms in the metastable silver selenide were precipitated. The growth rate (about 10^3 µm/sec) which was measured from the film is remarkable high. This method is the most suitable one for mass production of silver filaments. This process is also reversible when the temperature across the AgI gradient is reversed. A cyclic process was observed as the grown filaments dissolved in the metastable silver selenide continuously. The growth is considered to be controlled by nucleation (injection rate is high) or injection (injection rate is low), the same as in the case of growth from α-Ag$_2$X.

One of the cotton-like filaments was thin and like

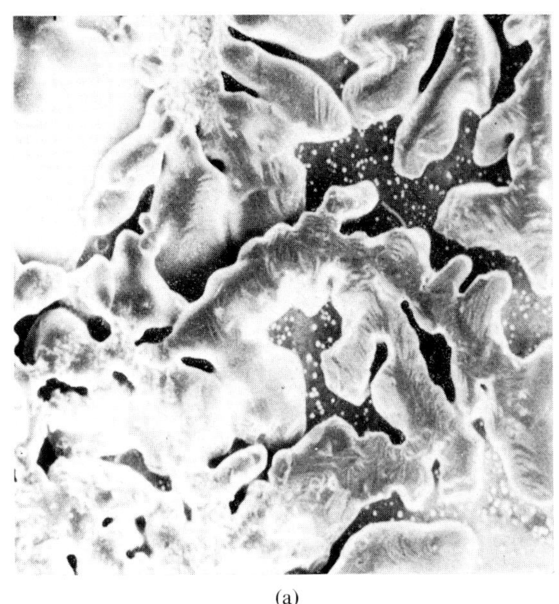
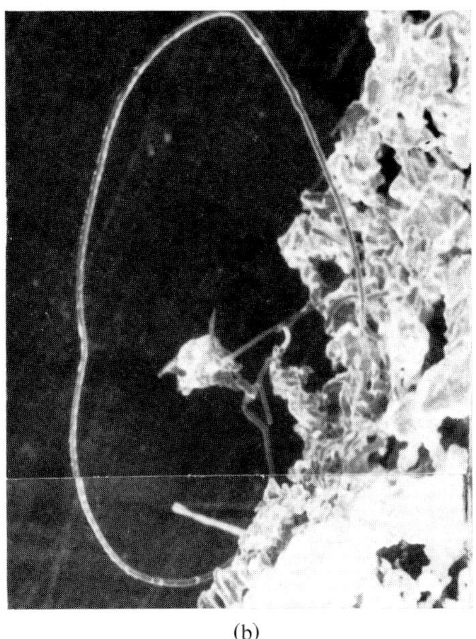

(a)   (b)

Fig. 9. SEM photographs of (a) dendrites ($\times 100$) and (b) a filament ($\times 33$).

the fin of a fish as shown in fig. 7. Selected area diffraction patterns showed that the growth direction of the fin was a $\langle 110 \rangle$ direction and the face of the fin was a $\{110\}$ face. The fin has many faults, for example twins and/or stacking faults, as shown in fig. 8a. The faults seem to be introduced by the high rate of nucleation.

Berry[7]) has also reported the growth of silver filaments from α-AgI by using an α-AgI/Ag contact. In Berry's method, the growth rate was 1 cm/24 hr or 0.12 μm/sec. The difference in the growth rates may result from the difference in the interface resistance for silver atoms.

Brenner[8]) used method a to obtain silver whiskers.

(III) Growth from AgBr or AgCl (σ_i: small, σ_e: small).

Berry's method[7]) was used in this case, replacing the iodide with bromide or chloride. Dendrites were produced mainly, but a filament or whisker could be also obtained. Fig. 9 shows dendrites and a filament grown from AgBr at about 400 °C. The small growth rate is due to the small ionic conductivity of the compound (rate-control by diffusion or injection).

3. Discussion

According to the kind of silver compound involved various shapes and growth rates could be obtained. The growth is classified into three groups as tabulated in table 1.

TABLE 1

The classification of the filamentary growth of silver from silver compounds (at 300 °C); most of these values are from Jost[9])

Types	Compounds	Shapes	Growth rate (μm/sec)	σ_i (mho/cm)	σ_e (mho/cm)	D_{Ag} (cm²/sec)	k_r (eq./cm sec)	Fig. no.
I	α-Ag$_2$X (X = S, Se, Te)	Fluted filament (thicker at its root or same thickness)	max. ~ 30 (large)	1–4 (large)	10–10^3 (large)	2.8×10^{-2}	1.6×10^{-6}	1, 2, 3, 4
II	α-AgI	Fin-like filament	max. $\sim 10^3$ (large)	~ 2 (large)	$< 10^{-7}$ (small)	?	1.4×10^{-9}	5, 6, 7, 8
III	AgBr AgCl	Dendrite or whisker	? ? (both small)	1.5×10^{-3} 1.8×10^{-2} (both small)	3.5×10^{-4} 3.7×10^{-5} (both small)	? ?	1.0×10^{-10} 1.0×10^{-10}	9, 10 9, 10

IV – 3

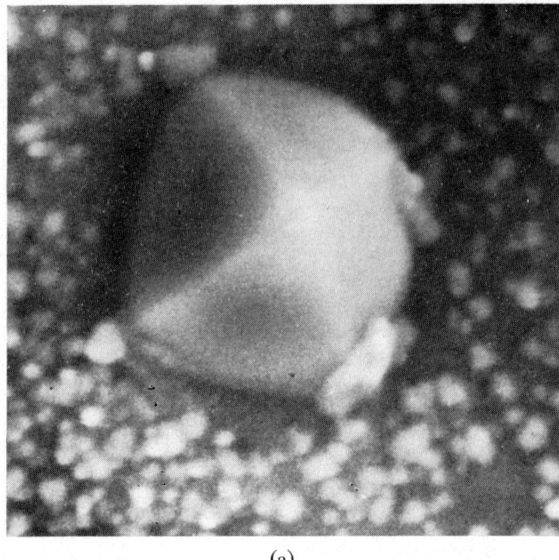

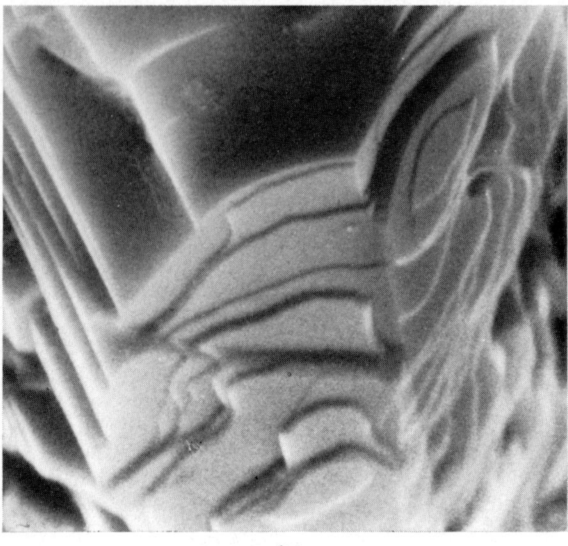

(a)   (b)

Fig. 10.   SEM photographs of the evidence of surface diffusion. (a) a pyramid ($\times 9000$), (b) steps ($\times 3300$). These photographs are the parts of fig. 9a.

The growth rate can be estimated from the rate of supply of the atoms. The best way to determine this rate is by using the diffusion constant for excess silver atoms, D_{Ag}[10,11]) but D_{Ag} for silver halides has not been reported. Therefore, we used the parabolic constant of the tarnishing reaction[10]), which is concerned in the rate of reaction, to determine the growth rate of the compounds. The values of the constant k_r for α-AgI, AgBr and AgCl are nearly of the same order, but their growth rates are very different whereas the constant k_r for α-AgI is 10^{-3} times smaller than that for α-Ag$_2$X, but on the contrary the growth rate from α-AgI is about 300 times larger than that for α-Ag$_2$X. This growth rate anomaly in the case of α-AgI can be explained if it is assumed that electronic breakdown occurs in α-AgI as a result of the electrical field gradient caused by the concentration gradient of the atoms, and the effective k_r of α-AgI thus becomes large.

The reason why the filaments from α-AgI are fin-like and have many faults and a $\langle 110 \rangle$ orientation may be the high growth rate (about 10^3 µm/sec), the low temperature of the growth point (about 180 °C), and the fact that the $\langle 110 \rangle$ direction is a direction along which the available interatomic binding forces are greater than in any other[12]). The nucleation is due to the condensation of atoms resulting from the interatomic binding force. Dendritic growth from AgBr and AgCl is due to

the low growth rate and the high temperature (about 400 °C). The surface temperature, about 400 °C, seems to be enough for silver to diffuse. Some of the evidence for surface diffusion of deposited silver is shown in fig. 10. There are many pyramids near a dendrite and many deposited silver clusters near the pyramids, and many steps are observed on the dendrite.

4. Conclusion

It is concluded that the shapes of the filaments are affected by the growth-point temperature, the nature of the compound, the surface roughness, and the degree of supersaturation. The high-speed nucleation on α-Ag$_2$X and α-AgI, which causes faults such as twins and/or stacking faults, is due to condensation resulting from interatomic binding forces. The growth control technique, which makes it possible to reveal the nucleation stage may become an important tool for studying the nucleation process.

Acknowledgement

The work reported here was made possible by the provision of laboratory facilities by Professor S. Kikkawa, the New Nippon Electric CO., Ltd., and the Sanyo Electric CO., Ltd. This work was partially supported by the Grant-in-Aid from the Science and Engineering Research Institute of Doshisha University.

Note added in proof

After this manuscript had been completed, we received a paper by Corish and O'Briain[13] in which the growth of silver from α-Ag_2S in vacuum has been described. They also reported the growth and dissolution of silver whiskers[14].

References

1) H. K. Hardy, Progr. Metal Phys. **6** (1956) 45.
2) F. R. N. Nabarro and P. J. Jackson, in: *Proc. Intern. Conf. on Crystal Growth*, Cooperstown, New York: *Growth and Perfection of Crystals*, (1958) p. 13.
3) T. A. Redwin, Chem. News **35** (1877) 144.
4) T. Ohachi and I. Taniguchi, Japan. J. Appl. Phys. **8** (1969) 1062.
5) H. W. Kohlschütter, Z. Elektrochem. **38** (1932) 345.
6) C. Wagner, Trans. AIME **194** (1952) 214.
7) C. R. Berry, J. Opt. Soc. Am. **40** (1950) 615.
8) S. S. Brenner, Acta Met. **4** (1956) 62.
9) W. Jost, *Diffusion in Solids, Liquids, and Gases* (Academic Press, New York, 1960, revised ed.).
10) C. Wagner, J. Chem. Phys. **21** (1953) 1819.
11) I. Yokota, J. Phys. Soc. Japan **16** (1961) 2213.
12) J. H. Howey, Phys. Rev. **55** (1939) 578.
13) J. Corish and C. D. O'Briain, J. Mater. Sci. **6** (1971) 252.
14) J. Corish and C. D. O'Briain, J. Crystal Growth **13/14** (1972) 62.

GROWTH OF CADMIUM WHISKERS IN VARIOUS ATMOSPHERES

J. H. KITTERMAN* and D. R. WINDER

Department of Physics, Colorado State University, Fort Collins, Colorado 80521, U.S.A.

The growth of cadmium whiskers from the vapor is described. Growth rates were measured at a supersaturation of approximately 0.1 in atmospheres of argon and of an oxygen–argon mixture. The growth rates are dependent on the radius of the whisker, the chemical nature of the atmosphere and the pressure. In argon, growth rates in excess of the classical linear law for tip impingement support the model of surface diffusion inhibited by the presence of the inert gas. This inhibited surface diffusion is the rate-controlling process. The addition of the active element oxygen increases the number of whiskers and dendritic growth while decreasing the growth rates of whiskers. This decrease is proportional to the partial pressure of oxygen, but the effect seems to be less than that suggested in the previous literature. At the lower supersaturations, growth habits other than whiskers become dominant with small platelets comprising the primary habit at the lowest supersaturation. The effect of a different inert gas was investigated by the substitution of helium for argon.

1. Introduction

The growth of metallic whiskers from the vapor has been investigated experimentally and theoretically for many years. The growth of cadmium crystals in inert gases has been studied by Price[5] (1960), by Shvidkovskii et al.[8] (1964) and by Coleman and Cabrera[1] (1957).

The models usually advanced for growth of whiskers from the vapor consider that one of three mechanisms dominate the growth process: the classical linear equation using tip impingement as the limiting process; the volume diffusion-limited process; and the surface diffusion-limited process. These three predict different dependences of the growth rate on supersaturation and radius of the crystal. Previous work in which volume diffusion was suggested to be the rate controlling process was done by Price[5] (1960), while experimental and theoretical work involving surface diffusion as the rate controlling process has been done by Dittmar and Neumann[2] (1960), Sears[11] (1955), Parker et al.[10] (1963) and Simmons et al.[6,7] (1967, 1968). It is also widely recognized that impurities inhibit the growth rate. This research attempts to distinguish by quantitative measurements between the models of growth when the growth rates based on each model are all of the same order of magnitude, and to measure the effect of different inert gases and of free oxygen when introduced

as an impurity. In this paper only the growth condition of condensation from the vapor is considered.

2. Description of the experiment

The material for the experiment was cadmium of 99.9999% purity, which was doubly vacuum distilled into the growth chamber. The chamber was a square pyrex tube with one face optically polished so that a reasonable optical resolution was obtained for an object the size of the cadmium whiskers to be grown, i.e., approximately 1 micron in diameter. The growth assembly could be alternately evacuated and purged with inert gas. The temperature profile along the chamber during growth was maintained by a hemi-cylindrical reflecting nichrome wire resistance furnace and measured in duplicate runs to the actual growth runs. Fig. 1 shows the completed growth tube assembly and fig. 2 shows the temperature profile along the tube. The microscope (33×) used for direct observation of the growth run was equipped with a traveling stage and a

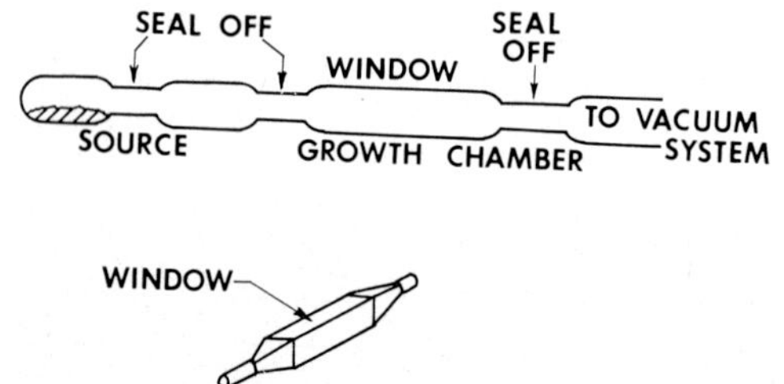

Fig. 1. Growth tube assembly. Growth chamber is 15 cm long and of square cross section: 1 cm × 1 cm i.d.

* Present address: The Timken Company, Canton, Ohio 44706, U.S.A.

bifilar eyepiece. A narrow viewing slot was cut through the reflecting top of the furnace so that an unobstructed viewing path was provided for direct sighting into the chamber. The bifilar allowed growth measurements to be made as growth proceeded, but this method was seldom used because only one growing whisker could be measured at a time. An alternative procedure of time-lapse photography was used extensively. A field of view containing a number of growing whiskers was

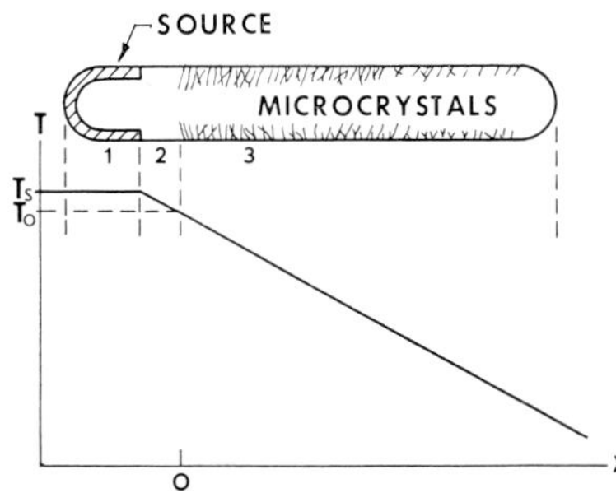

Fig. 2. Growth tube and temperature profile. The source temperature, T_s, and T_0 are 323 °C and 321 °C, respectively. The temperature gradient in the growth region ($x > 0$) is 9.7 °C/cm.

located and the photographic sequence started. After 2 to 4 hours of data were accumulated, the run was stopped. The areas of growth were relocated after photographic prints from the data films were prepared and measurements of the diameters of the actual whiskers were made with another microscope at 500×. Orientations of the whiskers were also measured at this time.

From the temperature and mass deposition profiles the supersaturation ratio is calculated. From the time lapse photographs and the subsequent microscopic observations the growth rate is calculated. These data are reduced and comparisons made with the predictions of the various models of crystal growth.

3. Experimental results

The general observations on habit are similar to those reported in the literature [Price[5]), 1960; Coleman and Sears[12]), 1957; Zakharova and Zinenkova[9]), 1965]. Quantitative growth rates of whiskers for various conditions varied greatly and are summarized in table 1. Also summarized in table 1 are the results of the experiment. The values labeled $\bar{R}$ are the growth rates normalized for radius variation and supersaturation variation, and averaged for the group of equivalent

observations, i.e. for whiskers 2 μm in diameter and a growth tube relative supersaturation of 0.1 [$\bar{R} = R \times$ (0.1 r/σ), σ = relative supersaturation, r = radius of the whisker]. The decrease in growth rate with fill pressure can be seen. This trend makes it reasonable to modify the growth rates to remove the growth rate dependence of the experimental values from equilibrium density changes, c_0, of the cadmium vapor. The column labeled $(R/c_0)_n$ gives the average growth

TABLE 1

Tabulated results for argon and helium growth runs

Oxygen	Inert gas	Argon runs		Helium runs
P	P	$\bar{R}$	$(R/c_0)_n \times 10^3$	$\bar{R}$
(Torr)	(Torr)	(μm/min)	(μm/min)	(μm/min)
0	300	3.0 ± 0.6	8.0 ± 1.5	1.2
0	150	4.0 ± 1.2	14.7 ± 4.2	2.5
Low	150	5.0 ± 1.2	17.0 ± 4.0	
0.2	150	3.5 ± 1.2	11.4 ± 3.8	
2	150	2.0 ± 0.4	8.8 ± 1.8	
0	50	8.4 ± 1.0	27.0 ± 3.2	1.2
0	10	3.6 ± 0.8	7.8 ± 1.8	
0	1	7.8 ± 1.0	13.8 ± 1.8	

rates after removing the variations with equilibrium density. The growth rate now is inversely proportional to the chamber pressure down to 50 Torr and erratic below 50 Torr.

4. Calculation of the supersaturation ratio

The calculation of the supersaturation in the growth tube follows that of Shvidkovskii, Predvoditelev and Zakharova[8]) (1964). In fig. 2 we consider the transport of cadmium from a source at constant temperature T_s down the growth chamber to lower temperatures. Condensation is absent in region 2 and begins in region 3 when the temperature in the chamber reaches the melting point of cadmium, T_0, which is 321 °C. Taking into account that the radial temperature gradient is small compared to the linear temperature gradient, we assume that the problem is one dimensional. The calculation of the supersaturation requires the knowledge of the mass of cadmium deposited at every point in region 3. The mass deposition was determined experimentally by dissolving the cadmium in the growth tube layer-by-layer into 0.1 normal nitric acid and determining the cadmium in solution by atomic absorption. The resulting calculated supersaturation profile is given in fig. 3.

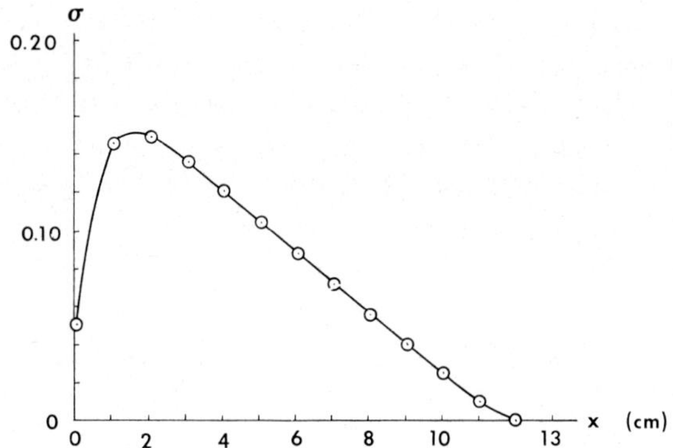

Fig. 3.   Supersaturation, σ, versus distance along growth region.

5. Discussion

The classical linear law, based on the kinetic unbalance of arriving and departing molecules at the tip, predicts the following rate:

$$R_{\text{lin}} = \beta_c \Omega \sigma P_0 / (2\pi mkT)^{\frac{1}{2}}, \tag{1}$$

where β_c is the condensation coefficient ($\beta_c \leq 1$), Ω is the atomic volume of cadmium, P_0 is the equilibrium cadmium vapor pressure, and m, k, and T have their usual meaning. With appropriate values for a growth temperature of 280 °C, assuming $\beta_c = 1$, this rate is 3.24 μm/min, which agrees well with the experimental rate for 300 Torr argon.

In the volume-diffusion domain, the growth rate of a whisker is[5])

$$dl/dt = R_{\text{vol}} = 4D\Omega \sigma c_0 / r, \tag{2}$$

where D is the diffusion coefficient of cadmium in argon calculated from Chapman and Cowling[13]), and $\Omega = 2.165 \times 10^{-23}$ cm^3/molecule. The quantity R_{vol}/c_0 should vary directly with D and thus inversely with growth chamber pressure P. Table 1 indicates this trend; inserting appropriate values of $r = 1$ μm, $D = 0.64$ cm^2/sec, $T = 280$ °C into eq. (2) for 300 Torr argon, a rate of 10.3 μm/min is obtained which is much larger than the observed rate.

In the surface diffusion domain the rate is

$$R = (1 + 2\lambda_s/r)R_{\text{lin}}. \tag{3}$$

The mean free path for surface diffusion, λ_s, is estimated to be 2 μm from the heat of evaporation by a method due to Sears (1956). In this experiment, the rate does not increase over the classical law until the

argon pressure has decreased to between 50 and 100 Torr. At 50 Torr the volume mean free path of argon in argon, given by[4])

$$\lambda = 1/n\pi d^2, \tag{4}$$

where n is the growth chamber gas density (molecules/cm^3) and d is the effective collision diameter [Kennard[4]] 1938], is 2 μm which is the same as the mean free path of surface diffusion. Therefore, a diffusion cadmium molecule on the surface has a good chance of reaching the tip before it suffers a disrupting collision with the inert gas molecules. This model provides the enhanced growth needed at lower pressures and for smaller whiskers. From table 1 the experimental rate, $\bar{R}$, for 2 μm diameter whiskers at 50 Torr argon is approximately 9 μm/min and gives order of magnitude agreement with eq. (3) which would predict a rate of 15 μm/min. Eq. (2) would predict a rate of approximately 50 μm/min at 50 Torr argon.

Finally, the effect of oxygen as an impurity is seen from table 1 which illustrates that the growth rate steadily decreases with increasing oxygen level, until, at 150 Torr Ar and 2 Torr O_2, the rate is below the classical linear law. The effect of oxygen is more severe as the gas fill pressure goes down, if the main oxygen contaminant comes from desorption of the vessel after sealing; in this case the ratio of oxygen flux to cadmium flux onto the whisker surface will increase. If the main oxygen contaminant is due to the inert gas used, then the ratio of oxygen to cadmium flux onto the surface will remain constant. The inert gas may serve to clean the whisker surfaces by bombardment at higher fill pressures. As the inert gas pressure drops to 10 Torr, the effect of residual oxygen becomes more severe, and growth rates of individual whiskers will be lessened. That this is probably the case is corroborated by the appearance of bush growth at all portions of the growth chambers for pressures of 10 Torr and below.

6. Conclusion

Growth of whiskers was observed at all argon pressures down to 1 Torr. The effect of the introduction of oxygen into the system was to induce bush (dendritic) growth. The presence of bush growth was observed to commence at 10 Torr argon and become steadily more pronounced as the chamber pressure decreased, although the appearance of the growth encountered at pressures less than 10 Torr had an appearance at 500× which resembled a conglomeration of tiny (unresolvable) crystals, i.e. a powder, while the dendritic growth induced by the introduction of free oxygen into the system at 150 Torr appeared to be a mass of crumpled whiskers at 500×.

The primary habits observed at the lowest supersaturations were platelets with diameters less than a few microns. A few whiskers were observed which were very short, usually 50 μm or less in length. These observations are for argon pressures above 10 Torr and for helium pressures above 50 Torr. In helium the growth rates for whiskers were essentially constant.

The growth of whiskers in argon exhibited growth rates which varied inversely with radius and with the chamber pressure. The rates increased as chamber pressure decreased until an argon pressure of 50 Torr was reached. At pressures of 10 Torr and lower the rate decreased over the 50 Torr values and became erratic. The mechanism controlling the whisker growth rate down to 50 Torr argon can be explained as surface diffusion modified by the effect of the inert gas. At pressures of 10 Torr and lower, the growth rates decrease and become erratic either because of turbulence in the local environment accompanied by the heavy condensation rate interfering with the surface diffusion process or the possible inhibiting effect of residual oxygen in the system.

The growth rates in helium are explained similarly to the low argon pressure case. The diffusion coefficient of cadmium in helium is about four times that of cadmium in argon at an equivalent pressure and temperature. Thus the deposition rate of cadmium in helium is about four times higher than cadmium in argon at an equivalent pressure. The heavier condensation rate in helium probably interferes with the surface diffusion process.

The growth rate of whiskers varies inversely with oxygen level of the growth chamber at an argon pressure of 150 Torr. This is direct evidence of the inhibiting effect of oxygen on the growth rate of whiskers from the vapor.

References

1) R. V. Coleman and N. J. Cabrera, J. Appl. Phys. **28** (1957) 1360.
2) W. Dittmar and K. Neumann, Z. Electrochem. **64** (1960) 297.
3) R. B. Honig, RCA Rev. **18** (1957) 195.
4) E. H. Kennard, *Kinetic Theory of Gases* (McGraw-Hill, New York, 1938).
5) P. B. Price, Phil. Mag. **5** (1960) 473.
6) J. A. Simmons, H. Oser and S. R. Coriell, in: *Crystal Growth*, Ed. H. S. Peiser (Pergamon, Oxford, 1967) p. 255.
7) J. A. Simmons and S. R. Coriell, J. Appl. Phys. **39** (1968) 3459.
8) B. G. Shvidkovskii, A. A. Predroditelev and M. V. Zakharova, Soviet Phys.-Solid State **6** (1964) 833.
9) M. V. Zakharova and G. M. Zinenkova, in: *Growth and Imperfection of Metallic Crystals*, Ed. D. Ousienko (Consultants Bureau, Plenum Publ., 1965) p. 181.
10) R. L. Parker, R. L. Anderson and S. C. Hardy, Appl. Phys. Letters **3** (1963) 93.
11) G. W. Sears, Acta Met. **3** (1955) 361.
12) R. V. Coleman and G. W. Sears, Acta Met. **5** (1957) 131.
13) S. Chapman and T. G. Cowling, *Mathematical Theory of Non-Uniform Gases*, 2nd ed. (Cambridge Univ. Press, 1952).

DISSOLUTION FORMS OF SINGLE CRYSTAL SPHERES OF RUTILE

R. HEIMANN and W. FRANKE

Institut für Mineralogie der Freien Universität, 1 Berlin 33, Germany

and

R. LACMANN

Fritz-Haber-Institut der Max-Planck-Gesellschaft, 1 Berlin 33, Germany

Spheres of rutile single crystals have been dissolved in H_2SO_4 and in the following melts: $K_2S_2O_7$, KHF_2, KOH, alkali carbonates, PbO and a KHF_2–$K_2S_2O_7$ composite. In all acid environments during the dissolution polyhedra were formed, bounded by more or less rounded faces. These faces were determined. It is assumed the bounding faces are built up from submicroscopic areas of the faces of the equilibrium form. The dissolution form can be described by the shift velocities of a limited number of faces.

1. Experimental

In the course of investigations concerning kinetics and morphology of the dissolution of high refractory oxides[1]) we have dissolved single crystal spheres of rutile with various melts and acids. The polished spheres (4 mm diameter) were made from a synthetic Verneuil-grown rutile. The reaction was carried out in platinum crucibles; the spheres being placed inside a platinum basket which was submersed in the melt. This basket was connected with a 100 Hz vibrator. In this way the spheres were forced to an intensive motion relative to the melt.

2. Results

(I) Using a potassium disulphate melt at 600 °C a ditetragonal bipyramid resulted as the final solution form. The bounding {531} faces were smooth and flat without visible rounding. Fig. 1 shows besides these {531}, a base-like face.

(II) A tetragonal bipyramid was formed in a potassium bifluoride melt at 600 °C. The bounding {331} faces were rough and rounded (fig. 2). The same symmetry was approached in a melt of LiF/NaF/KF with eutectic composition. However, an orientated deposit tetragonal in symmetry was formed. Moreover, a slightly reduction of Ti (IV) visible by blue staining of the sphere occurred. We suppose the failing of a distinct solution form is due to this reduction and/or the deposit. Reduction and the occurrence of a distort-

ed and truncated final solution form was also observed in hydrothermal environments.

(III) Quite another polyhedron was obtained using a melt consisting of 25% KHF_2 and 75% $K_2S_2O_7$. The dissolution form was bounded by a flat base {001}, a ditetragonal prism {210} and a tetragonal bipyramid {221}. All faces were smooth, only a very tiny {$h0l$}

Fig. 1.   Dissolution form obtained by the dissolution of a single crystal sphere of rutile in $K_2S_2O_7$ at 600 °C. Stereo-Scan (Cambridge) (×18).

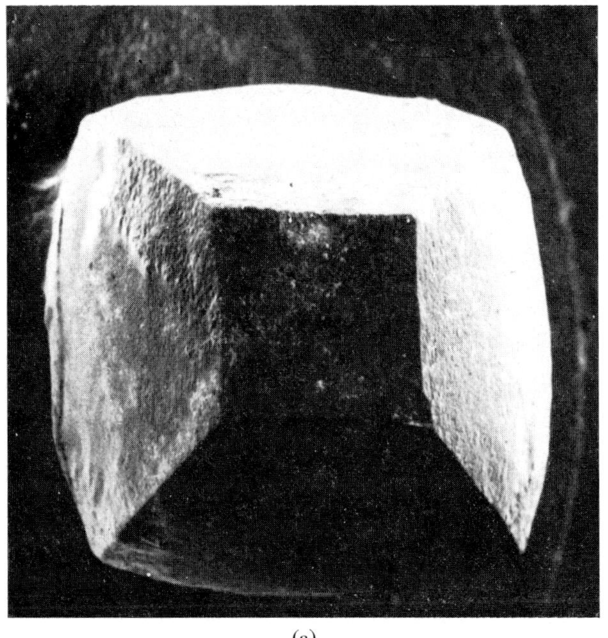

(a)

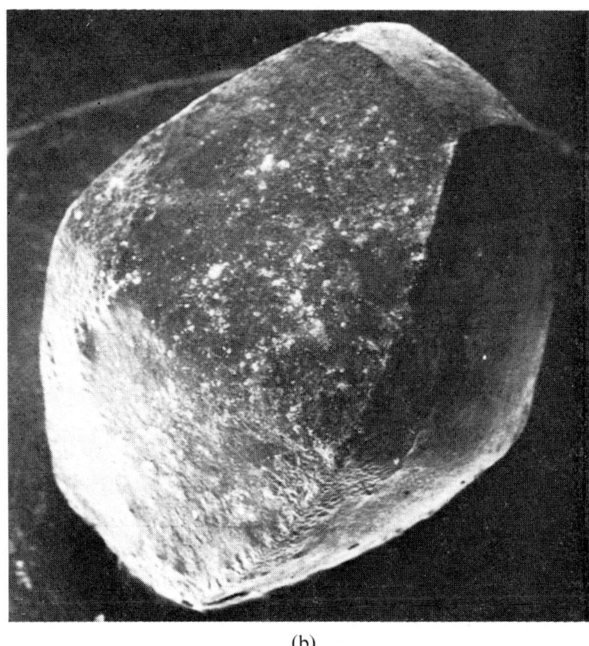

(b)

Fig. 2.   Dissolution form obtained by the dissolution of a single crystal sphere of rutile in KHF_2 at 600 °C. (a) $\parallel$ c-axis; (b) $\perp$ c-axis. Stereo-Scan (Cambridge) ($\times$ 18).

pyramid was rough. The development of the faces in the prism zone is schematically drawn in fig. 3. The same habit of the solution form was attained with a melt of $NaF/K_2S_2O_7$. While this habit is very different from the habits formed by the pure components of the melt, we suggest HSO_3F formed in the composite melt is the attacking agent.

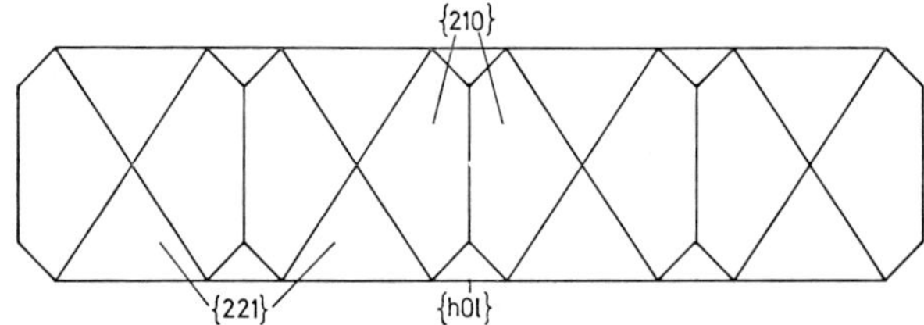

Fig. 3.   Schematic developable surface of the prism zone of the dissolution form obtained by the dissolution of a single crystal sphere of rutile in a $KHF_2/K_2S_2O_7$ composite melt.

(IV) On etching the rutile spheres with boiling sulphuric acid for some time titanium sulphate was deposited orientated on the sphere. This deposit shows a ditetragonal symmetry (fig. 4). The same precipitation can be avoided by using a great excess of acid and adding some potassium sulphate. The bounding faces of the remaining polyhedron were a tetragonal bipyramid and a ditetragonal prism. They were covered with steps and etch pits. Possible indices are {553} and {530} or {703} and {230} respectively. Goniometric measurement by light reflection is very difficult and precarious although these faces are only slightly rounded.

(V) In alkaline melts (KOH, alkali carbonates, PbO) the rutile was dissolved very rapidly, but the final solution form remained a smooth sphere. Obviously all lattice planes have the same velocity of dissolution. This is possibly due to a diffusion controlled mechanism of dissolution.

On the other hand in acid environments the controlling factor of the rate of dissolution is the velocity of the disintegration of the different lattice planes. The resulting faces of the final solution form represent the lattice planes with the highest rate of dissolution. According to the periodic bond chain method by Hartman and Perdok[2]) these faces ought to be K-faces.

Table 1 shows the faces of the dissolution forms obtained by various attacking agents. In the fluoride melt as well as in boiling sulphuric acid orientated deposits were formed. Therefore we suppose rough faces are developed if deposition of reaction products is involved as an intermediate step of dissolution. This

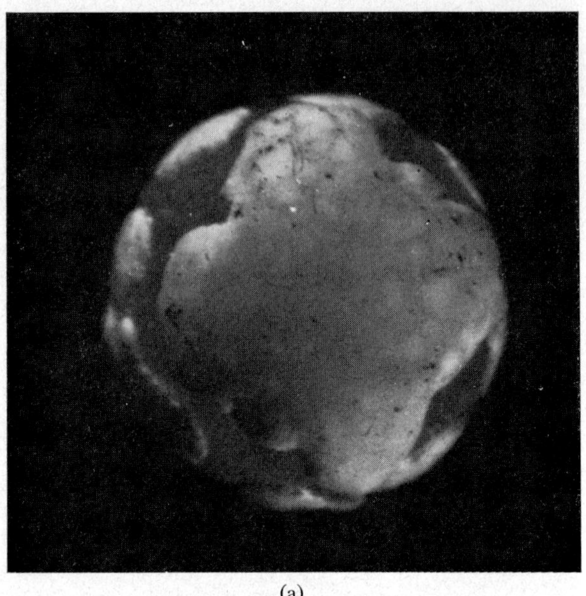

(a)

(b)

Fig. 4. Single crystal sphere of rutile with orientated deposit of titanium sulphate formed by the reaction with boiling sulphuric acid. (a) approximately $\parallel$ c-axis; (b) $\perp$ c-axis ($\times 15$).

is true for pure fluoride melts and sulphuric acid. Smooth faces were obtained when no precipitation effect was detectable.

TABLE 1

Habits of the dissolution forms obtained by the dissolution of single crystal spheres of rutile in various agents

Attacking agent	Dissolution forms
(I) $K_2S_2O_7$	$\{531\}+\{998\}^*+\{001\}^*$
(II) KHF_2	$\{331\}+\{001\}^*$
(III) 25 mol% $KHF_2 +$ 75 mol% $K_2S_2O_7$. or $NaF/K_2S_2O_7$	$\{221\}+\{210\}+\{001\}^*$
(IV) H_2SO_4 (conc.)	$\{553\}+\{530\}+\{001\}^*$ or $\{703\}+\{230\}+\{001\}^*$
(V) KOH, alkali carbonates, PbO	Spheres

* These faces disappear in the course of the dissolution process.

With the exception of the $\{703\}$ all faces obtained by dissolution were situated in the zones $\langle 110 \rangle$, $\langle 350 \rangle$ or $\langle 001 \rangle$. Conforming with investigations of Hartman[3]) all these faces were K-faces, except the $\{hk0\}$ faces.

3. Theoretical considerations

For the interpretation of the experimental results the following statements were made. (1) At the beginning of the dissolution process edges and corners were formed. (2) The areas between these edges being gradu-

ally flattened during the process and some eventually disappearing.

We assume the faces formed by dissolution are built up from submicroscopic areas of the faces of the equilibrium form [Stranski[4]]. These faces can be ablated lattice plane after lattice plane. In this way vicinal faces of the stable faces of the equilibrium form were obtained [see Hirth and Pound[5])]. The radial shift velocity (v) of these stable faces determines the geometrical relations of the dissolution form.

To explain these considerations a twodimensional quadratic pattern is used. The stable ledges should be the $\{10\}$-ledges, leading to the build up of the (11)-ledge by (10)- and (01)-steps (fig. 5a). The shift velocity of the (11)-ledge (v_{11}) is given by $v_{10}/\cos \varphi_{11/10}$, where $\varphi_{11/10}$ is the angle between the [10]- and the [11]-directions. Hence, $v_{11} = \sqrt{2}\, v_{10}$. The shift velocity of all other (hk)-ledges is also determined by v_{11} (fig. 5b). However, it must be considered the shift direction of all these (hk)-ledges also is $[\bar{1}\bar{1}]$.

Fig. 6 shows the dissolution form for $t = 0$, t_0, $2t_0$ and $3t_0$. The dissolution form can be constructed by a simple geometric method. Starting from the periphery of the original circle all points were moved along the directions [11], $[1\bar{1}]$, $[\bar{1}1]$, $[\bar{1}\bar{1}]$ for the distances $v_{10}\sqrt{2}\, t_0$. Points intersecting the coordinate

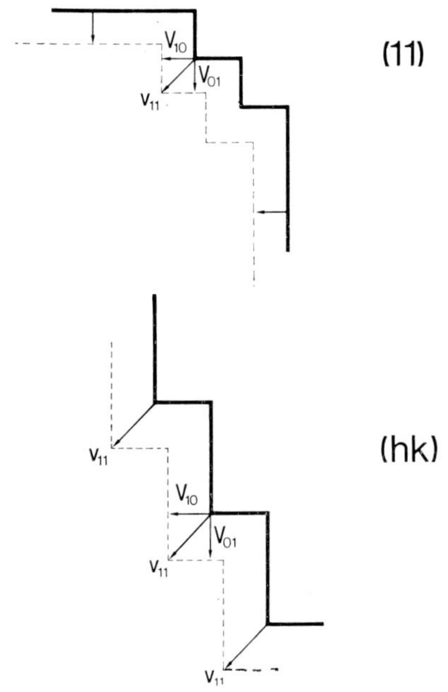

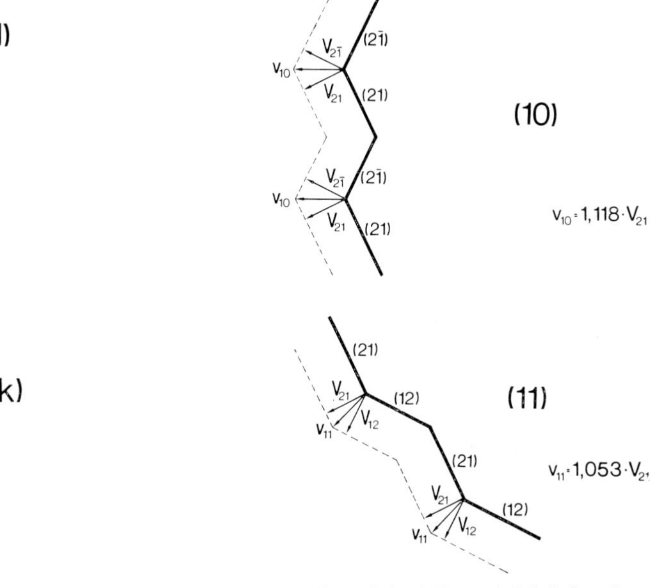

Fig. 5. The shift of the (11)- and (hk)-ledges obtained by the shift velocities v_{10} and v_{01}.

Fig. 7. Formation of the (10)- and (11)-ledges by combination of the (21)- and ($\bar{2}$1)-ledges and (21)- and (12) ledges respectively.

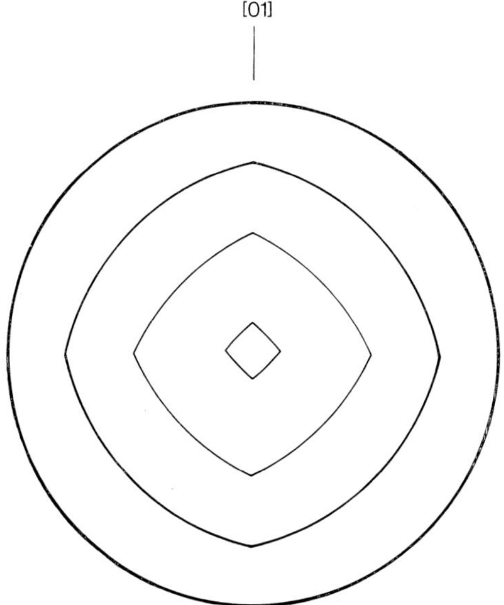

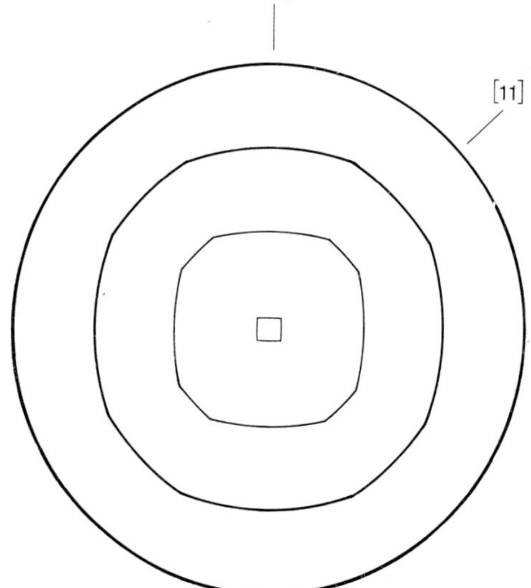

Fig. 6. Dissolution form for $t = 0$, t_0, $2t_0$ and $3t_0$ obtained by the geometric method described in the text.

Fig. 8. Dissolution form for $t = 0$, t_0, $2t_0$ and $3t_0$. The decrease of the (11)-areas in the course of the dissolution process is clearly seen.

axes ($x = 0$, $y = 0$) were eliminated. The figure shows the appearance of the corners on the (10)-poles and the slope of the (11)-areas. The radius of curvature remains constant.

For the interpretation of the appearance of the intermediate faces we assume the {21}-ledge to be the only stable one. In this case the {10}-ledge is not the equilibrium ledge. Moving away from the periphery the (21)-ledges and the (12)-ledges were formed. By combination of these ledges we obtain the (11)-ledge (fig. 7) with

$$v_{11} = v_{21}/\cos \varphi_{21/11} = 1.053 v_{21}.$$

Moreover, the (10)-ledge is formed consisting of

(21)- and (2$\bar{1}$)-ledges (fig. 7). The shift velocity is

$$v_{10} = v_{21}/\cos \varphi_{21/10} = 1.118 v_{21}.$$

At the beginning of the dissolution process the {11}- and {10}-ledges were formed since their shift velocities are faster with respect to the {21}-ledges. Consequently, the {11}-ledges disappear in the final dissolution form because $v_{10} > v_{11}$. Fig. 8 shows the dissolution form at $t = 0$, t_0 $2t_0$ and $3t_0$. The decrease of the {11}-areas is clearly seen.

If the dissolution form is determined by the shift velocities of several faces, higher indexed faces can be formed. When we intend the {11}-ledges are bounding the dissolution form in a rectangular pattern with the coordinates a_0 and b_0 and the ledges of the equilibrium form {10} and {01}, it should be $v_{10}/a_0 = v_{01}/b_0$. This is quite beyond the range of probability. The same is valid in a quadratic pattern and a cubic lattice respectively consisting of more than one kind of ledges and faces in the equilibrium form. In this way we can explain also the observed high indexed faces.

In the investigated systems (I)–(IV) the time-dependence of the dissolution forms can be described by the shift velocities of a limited number of faces. According to Hartman[3]) {110}, {101} and {100} are the faces of the equilibrium form in the system rutile/vapour. Therefore {hkl}, {hhl} and {001} ought to be present in the dissolution form. This is in agreement with the results obtained by $K_2S_2O_7$ melt. The deviations observed in the other systems are probably caused by the interaction between solid and solvent[6]).

The authors are very indebted to the Deutsche Forschungsgemeinschaft for financial aid.

References

1) W. Franke and R. Heimann, J. Crystal Growth **7** (1970) 97.
2) P. Hartman and W. G. Perdok, Acta Cryst. **8** (1955) 49.
3) P. Hartman, Papers and Proc. Fifth General Meeting Int. Mineral. Assoc. Cambridge, 1966, and this Symposium.
4) I. N. Stranski, Z. Physik. Chem. B **17** (1932) 127.
5) J. P. Hirth and G. M. Pound, *Condensation and Evaporation*, (Pergamon, Oxford, 1963).
6) R. Heimann and W. Franke, Neues Jahrb. Mineral. Monatsh. (1969) 161.

Journal of Crystal Growth **13/14** (1972) 207–211 © *North-Holland Publishing Co.*

CROISSANCE DES CRISTAUX DE CYANURE MERCURIQUE ORIENTÉS PAR RAPPORT À LA MATRICE DE CRISTAUX DE LEUR SOLVATE AVEC LE TÉTRAHYDROFURANNE

M. LEDÉSERT et M. FREY

Groupe de Cristallographie et de Chimie du Solide, Laboratoire de Cristallographie et Minéralogie, Université, 14–Caen, France

The growth of mercuric cyanide crystals from a crystal of the mercuric cyanide with tetrahydrofuran solvate $5Hg(CN)_2 \cdot 4T.H.F.$ is obtained through slow decomposition of the solvate crystal to the air. The growth occurs in such a way that mercuric cyanide crystals show an orientation in relation to the matrix constituted by the initial solvate crystal. This phenomenon corresponds to tridimensional structural concordances between solvate and pure salt; they appear in the existence in the solvate of buildings of $Hg(CN)_2$ molecules which are to be found again in mercuric cyanide with the same orientation the same structure and the same respective positions.

1. Introduction

Les cristaux de sels mercuriques peuvent être obtenus à partir de solutions dans de nombreux solvants organiques. Avec plusieurs solvants les cristaux ne se forment que dans un domaine de température bien déterminé; dans les autres domaines on obtient des cristaux d'un ou plusieurs composés d'addition contenant une proportion plus ou moins grande de solvant. A l'air libre les cristaux de certains de ces solvates, en perdant les molécules de solvant, se transforment en cristaux de sel pur qui prennent une orientation bien définie par rapport au cristal de solvate initial[1,2].

Des anomalies de faciès des cristaux de NaCl à partir de solutions aqueuses contenant de l'urée, des cristaux de $Hg(CN)_2$ à partir de solutions dans l'alcool méthylique, ont été interprétées par un mécanisme faisant intervenir, au niveau des faces anormalement développées, des relations structurales entre les composés d'addition NaCl–urée–H_2O ou $Hg(CN)_2 \cdot CH_3OH$ et les sels purs correspondants NaCl et $Hg(CN)_2$[3,4].

On pouvait de façon analogue envisager que le phénomène de croissance des cristaux de sels mercuriques orientés par rapport à la matrice du cristal de solvate recouvrait des correspondances entre les structures cristallines du sel pur et du solvate initial.

Nous exposons ici les résultats qui ont été obtenus pour la croissance de cristaux de cyanure mercurique orientés par rapport à la matrice du solvate cyanure mercurique–tétrahydrofuranne [$5Hg(CN)_2 \cdot 4T.H.F.$].

2. Description du phénomène

Les composés $5Hg(CN)_2 \cdot 4T.H.F.$ et $Hg(CN)_2$ cristallisent respectivement avec les groupes spatiaux $P4_2/mnm$ et $I\bar{4}2d$. Une décomposition ralentie d'un cristal de solvate donne des cristaux de cyanure mercurique ayant tous même orientation; leurs faces sont parallèles à des faces du cristal de solvate initial[1]. Les rangées [001] des deux espèces sont orientées à 90° l'une de l'autre et une face du prisme {100} du cyanure mercurique correspond à une face du prisme {110} du solvate (fig. 1). A cette orientation sont liées des correspondances paramétriques tridimensionnelles rappelées dans le tableau 1; la maille commune ainsi définie correspond à une maille simple pour le solvate et double pour $Hg(CN)_2$.

Des expériences ont été aussi effectuées sur des cristaux très minces obtenus entre deux lamelles et aplatis suivant une face {110}. Après le départ du solvant chaque cristal de solvate est transformé en un seul

TABLEAU 1

Maille commune à $5Hg(CN)_2 \cdot 4T.H.F.$ et $Hg(CN)_2$

$5Hg(CN)_2 \cdot 4T.H.F$	$Hg(CN)_2$	Discordance
[001] = 10.372 Å	[100] = 9.643 Å	−7%
[010] = 13.553 Å	[01$\bar{1}$] = 13.11 Å	−3.3%
[$\bar{1}\bar{1}$0] = 19.167 Å	2[001] = 17.76 Å	−7.3%
[$\bar{1}\bar{1}$0] $\wedge$ [010] = 135°	[01$\bar{1}$] $\wedge$ [001] = 133°	

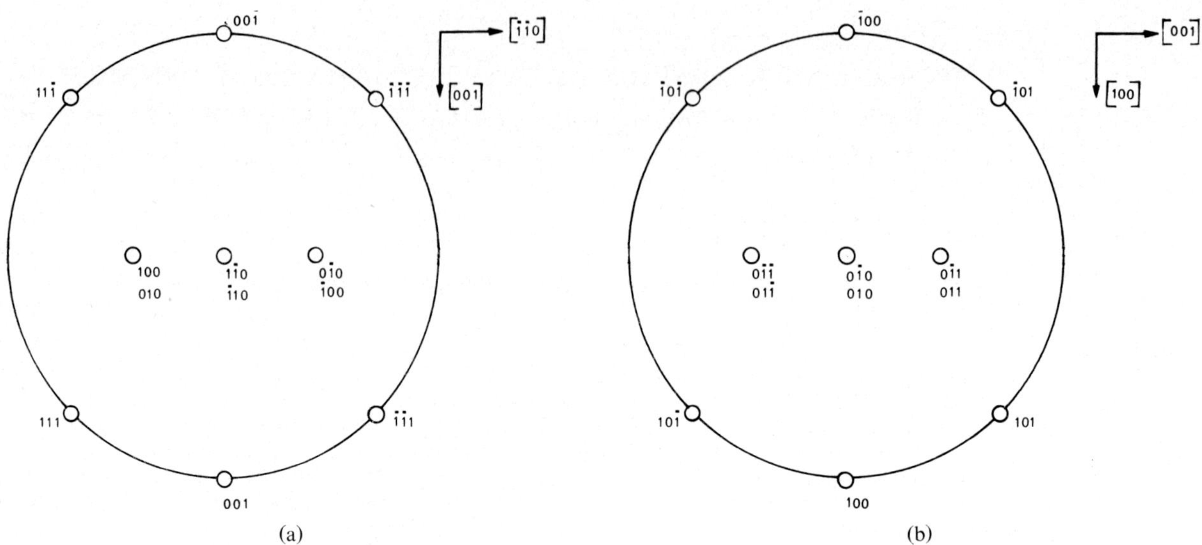

Fig. 1. Correspondance des faces de $5Hg(CN)_2 \cdot 4T.H.F.$ et $Hg(CN)_2$ en fonction de l'orientation observée.

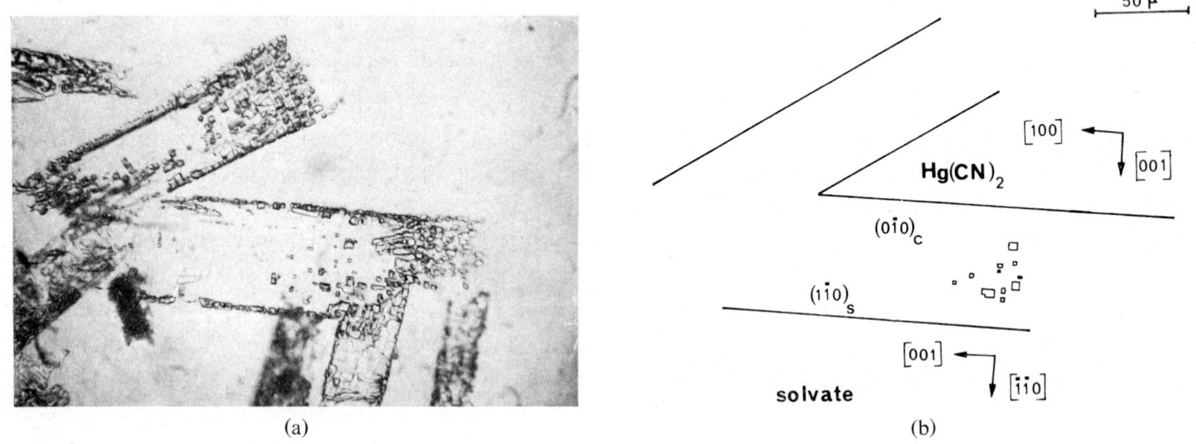

Fig. 2. Cristaux de $Hg(CN)_2$ après décomposition de $5Hg(CN)_2 \cdot 4T.H.F.$ entre deux lamelles. On passe d'un cristal de $5Hg(CN)_2 \cdot 4T.H.F.$ à un cristal de $Hg(CN)_2$ comportant des lacunes.

cristal de cyanure mercurique ayant sensiblement les mêmes contours et présentant des lacunes; celles-ci ont des côtés parallèles aux rangées [100] et [001] du cyanure mercurique (fig. 2).

3. Correspondances structurales

Pour comparer les deux structures cristallines, nous avons utilisé les positions atomiques de Frey et Ledésert[5]) pour $5Hg(CN)_2 \cdot 4T.H.F.$ et de Seccombe et Kennard[6]) pour $Hg(CN)_2$. Les projections ont été effectuées dans un plan perpendiculaire à deux des faces les plus développées qui se correspondent: $(1\bar{1}0)$ du solvate et $(0\bar{1}0)$ du cyanure mercurique. Dans le solvate (fig. 3a) les molécules $Hg(CN)_2$ sont perpendiculaires ou parallèles au plan de projection. Cristallographiquement, elles sont de trois types différents et caractérisées sur la figure par leur atome de mercure (I, II, III). Les cycles tétrahydrofurannes sont représentés par leur atome d'oxygène. Dans le cyanure mercurique (fig. 3b), les molécules $Hg(CN)_2$ sont équivalentes; elles sont parallèles ou presque perpendiculaires au plan de projection.

Tous les atomes de mercure ont, dans les deux structures, un entourage octaèdrique constitué par les

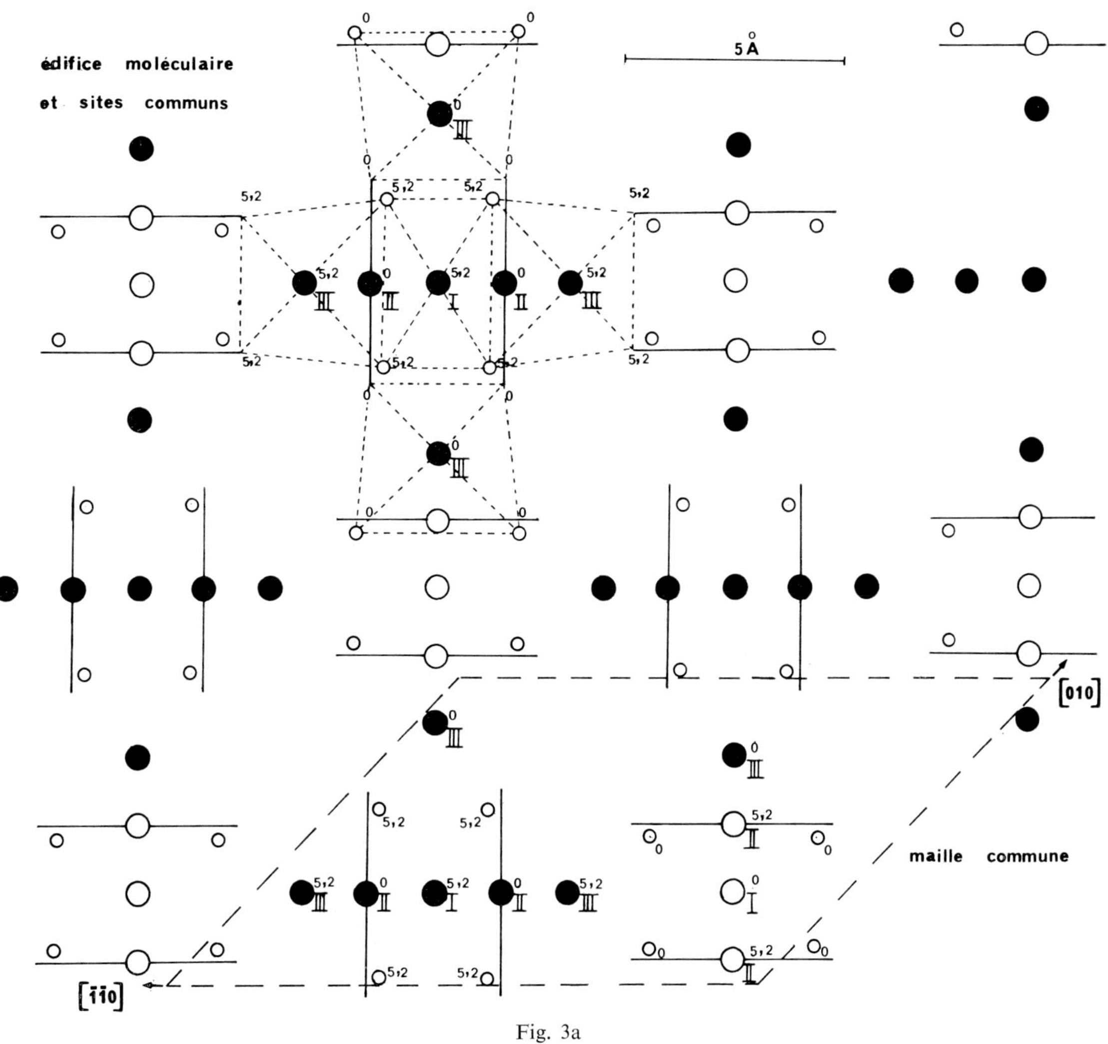

Fig. 3a

deux atomes de carbone de la même molécule $Hg(CN)_2$ et par quatre atomes de molécules voisines. Dans le solvate ces quatre atomes sont: pour les mercures (I), quatre oxygènes des tétrahydrofurannes; pour les mercures (II), quatre azotes de molécules (I) et (III); pour les mercures (III), deux azotes de molécules (II) et deux oxygènes des tétrahydrofurannes. Dans le cyanure mercurique tous les octaèdres sont équivalents.

Les faces $(1\bar{1}0)$ du solvate[7]) et $(0\bar{1}0)$ du cyanure mercurique[4]) sont des faces F au sens de Hartman[8]). La tranche d'épaisseur $d_{1\bar{1}0} = 9.583$ Å du solvate correspond à deux tranches d'épaisseur d_{020} du cyanure mercurique $(2d_{020} = 9.643$ Å$)$ comprises dans la maille commune aux deux espèces. Nous avons donc

recherché dans $2d_{020}$ du cyanure mercurique les molécules et les sites dont les positions relatives sont les mêmes que dans le cristal de solvate initial.

Cinq molécules $Hg(CN)_2$ de la première tranche, deux molécules $Hg(CN)_2$ de la deuxième tranche ont, à quelques faibles écarts angulaires près, les mêmes positions relatives que sept molécules sur dix de la maille du solvate. Il en est de même pour tous les sites d'atomes qui n'appartiennent pas à ces molécules mais font partie des octaèdres de coordination des mercures (I) et (III).

Si nous considérons l'ensemble des deux structures cristallines, les molécules $Hg(CN)_2$ qui se correspondent constituent des édifices perpendiculaires au plan de figure qui présentent entre eux de grandes similitu-

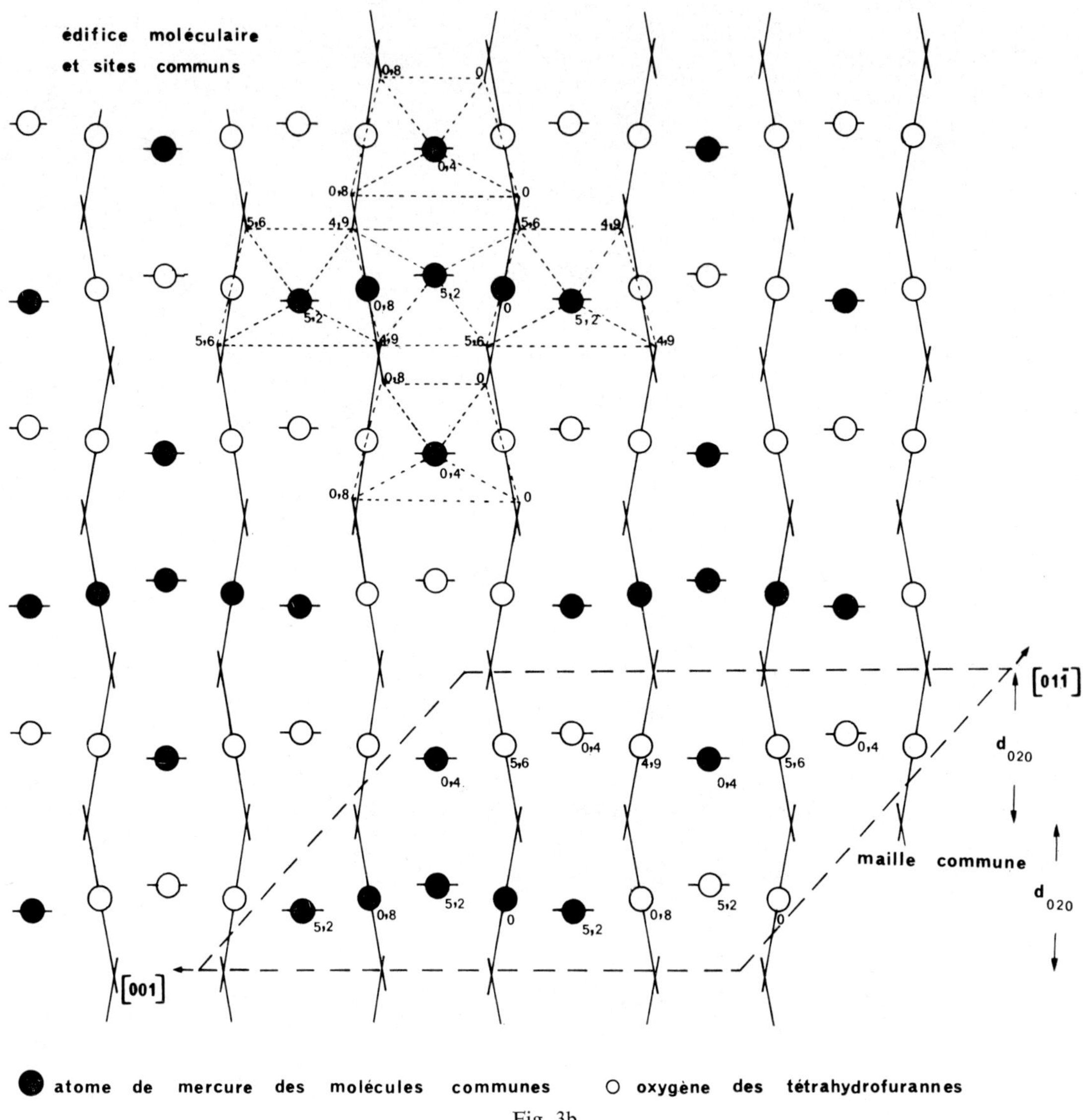

● atome de mercure des molécules communes   ○ oxygène des tétrahydrofurannes

Fig. 3b

Fig. 3. Projection de (a) $5Hg(CN)_2 \cdot 4T.H.F.$ parallèle à [001] et de (b) $Hg(CN)_2$ parallèle à [100]. Les cercles noirs représentent les atomes de mercure des molécules dont les positions relatives sont les mêmes dans $Hg(CN)_2$ et le solvate. Les cotes sont indiquées en Å; pour $Hg(CN)_2$ on a effectué une translation de l'origine de 2 Å suivant [Ī00]. Les octaèdres entourant les mercures des molécules horizontales n'ont pas été matérialisés.

des. Toutes leurs molécules $Hg(CN)_2$ sont liées entre elles ($Hg \cdots N = 2.85$ Å pour le solvate; $Hg \cdots N = 2.74$ et 3.06 Å pour le cyanure mercurique). Ils se retrouvent dans le solvate et le cyanure mercurique avec même orientation, même structure et mêmes positions mutuelles.

Sur chacune des figures 3a et 3b, un de ces édifices est représenté de façon détaillée avec les octaèdres de coordination des mercures (I) et (III) du solvate et de leurs correspondants dans le cyanure. Les sites qui se correspondent dans les deux structures sont aux sommets de ces octaèdres.

Le phénomène de croissance des cristaux de cyanure mercurique orientés par rapport à la matrice du solvate $5Hg(CN_2 \cdot 4T.H.F.$ correspond à des concordances structurales tridimensionnelles remarquables entre le solvate et le sel pur. L'orientation des cristaux de sel pur s'effectue comme si le départ des molécules de solvant perturbait le moins possible l'arrangement des molécules minérales déjà en place dans le solvate.

Références

1) M. Ledésert, M. Frey et J. C. Monier, Bull. Soc. Franc. Minéral. Crist. **90** (1967) 36.
2) M. Ledésert, Compt. Rend. (Paris) **270** (1970) 534.
3) J. H. Palm et C. H. McGillavry, Acta Cryst. **16** (1963) 963.
4) M. Ledésert et J. C. Monier, dans: *Adsorption et Croissance Cristalline* (C.N.R.S., Paris 1965) p. 537.
5) M. Frey et M. Ledésert, Acta Cryst. (1971) (sous presse).
6) R. C. Seccombe et C. H. L. Kennard, Organometal. Chem. **18** (1969) 243.
7) M. Frey, Thèse Caen, No. C.N.R.S. A.O. 4028 (1970).
8) P. Hartman, Z. Krist. **119** (1963) 65.

CROISSANCE CONTINUE, À PARTIR DE GERMES À SYMÉTRIE QUINAIRE, DES CRISTALLITES "MULTIPLES" FORMÉS LORS DE LA NUCLÉATION HÉTÉROGÈNE

E. GILLET et M. GILLET

Laboratoire des Mécanismes de la Croissance Cristalline, associé au C.N.R.S., Université de Provence, Centre de St. Jérôme, 13-Marseille 13e, France

The study of nuclei of gold deposited on alkali halides cleaved in ultra high vacuum shows that the early deposits are multiple particles. The mainly multiple particules have pentagonal or hexagonal contours and consist of five or twenty crystals in twin position. It has been suggested that these particles result from a successive twin formation process based on tetrahedron nucleus. However the particles observed in high resolution electron microscopy (making possible the direct imaging of lattices) did not exhibit defects due to this growth process. An alternative formation mecanism is proposed: the growth is performed by adding atom by atom to a nucleus consisting of seven atoms (pentagonal particles) or twelve atoms (hexagonal particles) packed as closely as possible like pseudo-nuclei which are found in liquids. The pentagonal particles nuclei exhibit fivefold symmetry and further growth layer by layer in a fcc structure slightly distorted (5%) keeps the initial symmetry and form. The hexagonal particles nuclei have also fivefold symmetry axes which are growth-axes. The structure of the multiple particles grown in this manner is in good agreement with the electron patterns and micrographs obtained.

1. Introduction

Lorsque l'on étudie les premiers stades de croissance des couches d'or condensées sur un halogénure alcalin, on remarque, généralement, que les cristallites, dont le diamètre varie entre 20 et 50 Å présentent des formes géométriques relativement bien définies: triangles, rectangles ou carrés, pentagones et hexagones. Leurs structures ont été étudiées par diffraction et microscopie électroniques[1–3]; les auteurs ont montré notamment que les cristallites pentagonaux et hexagonaux sont constitués de particules tétraédriques accolées en position de macle les unes par rapport aux autres. Il est généralement admis que la croissance de ces cristallites se fait par maclages successifs à partir d'un premier germe tétraédrique. Une telle théorie de la formation des cristallites a l'avantage d'expliquer complètement les diagrammes de diffraction et les contrastes des micrographies électroniques obtenus à partir de couches d'or très minces. Cependant, une étude plus approfondie de ce modèle de croissance nous à conduits à formuler quelques remarques:

a) Si la croissance s'effectue de cette façon, elle doit engendrer, entre certaines parties du cristallite des défauts; or, la microscopie à haute résolution[4]) n'en a mis aucun en évidence, exception faite d'une légère déformation des plans réticulaires.

b) Le développement par maclages successifs de certains cristallites, tels que les cristallites pentagonaux (paragraphe 2), suppose nécessairement l'existence de plans de maclages préférentiels qu'il paraît difficile de justifier.

c) La croissance ainsi envisagée ne peut s'effectuer que par la formation de nouvelles macles; or l'observation montre que des cristallites de tailles très différentes ont même structure, c'est-à-dire qu'ils sont constitués par un même nombre de particules tétraédriques, ce qui suggère que ces cristallites croissent d'une manière continue en conservant leur structure initiale.

Le but de cet article est de montrer qu'une telle croissance est possible à partir de groupements d'atomes tels que ceux que l'on trouve dans les liquides et qu'elle permet d'expliquer d'une façon satisfaisante tous les aspects obtenus en diffraction et microscopie électroniques. Ce processus de croissance nous a conduits à considérer un édifice cristallin possédant une symétrie quinaire.

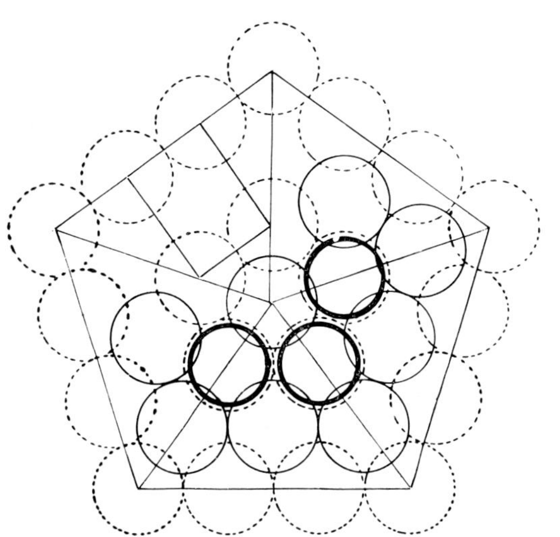

Fig. 1. – – –, atomes de la 1ère couche; —, atomes de la 2ème couche; —, atomes de la 3ème couche.

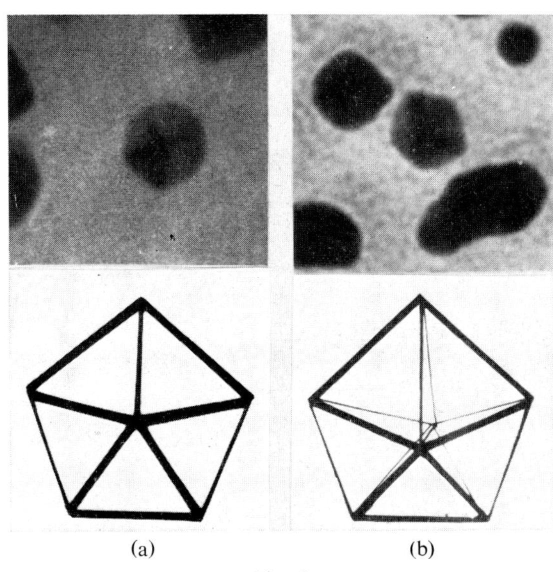

(a)   (b)

Fig. 2.

2. Existence de structures cristallines à symétrie quinaire

On trouve dans la bibliographie quelques exemples de cristaux présentant une symétrie d'ordre 5, tels que les "wiskers" de Ni, Fe, Pt observés au microscope à émission de champ par Melmed et Hayward[5]), ou bien les whiskers de cuivre à symétrie pseudo-pentagonale mises en évidence par Ogburn, Paretzkin et Peiser[6]). Récemment Froment et Maurin[7]) observant au microscope électronique les dépôts obtenus par électro-cristallisation du Nickel ont montré l'existence de nombreux cristallites à symétrie quinaire. Bien que dans tous les cas ces croissances puissent être expliquées par des maclages successifs, aucune observation n'a encore permis de localiser les défauts résultant de l'accolement des différentes macles. Le processus de croissance de tels édifices n'est donc pas entièrement connu. Cependant Melmed et Hayward[5]) ainsi que Schwoebel[8]) suggèrent qu'un empilement à symétrie pentagonale, dont nous donnons le schéma (fig. 1.), peut conduire à un édifice infini possédant un axe de croissance d'ordre 5 perpendiculaire au plan de la figure. L'édifice est constitué par cinq tétraèdres possédant une structure cubique à faces centrées légèrement déformée ayant une direction [110] parallèle à l'axe de croissance. Bagley[9]) a montré que cette déformation conduit à une structure orthorombique centrée.

3. Croissance des cristallites pentagonaux

Les observations au microscope électronique montrent qu'il existe des cristallites ayant un contour pentagonal mais présentant deux contrastes différents schématisés sur les figures 2a et 2b.

Le contraste de la figure 2a provient d'une édifice semblable à celui de la figure 1, c'est-à-dire possédant un axe de croissance pentagonal, normal au support.

Le contraste de la figure 2b correspond à une édifice cristallin qui peut être construit à partir d'un groupement de sept atomes disposés de façon à réaliser le germe le plus compact possible: un décaèdre pentagonal. On peut considérer que le décaèdre pentagonal se forme à partir d'un germe tétraèdrique ABCD auquel se joignent les atomes E, F, G (fig. 3a). La croissance de cet édifice peut se poursuivre de deux façons différentes. Dans le premier cas, les atomes H, I, J, K, L se fixent sur le décaèdre de la figure 3a dans les sites tels que BCE formant ainsi autour de l'atome C un nouveau pentagone qui se déduit du pentagone A, B, E, G, F par une rotation de 30 degrés autour de l'axe CD (fig. 3b). L'édifice complété par l'atome M contient alors treize atomes qui forment l'ensemble le plus compact possible. Dans le deuxième cas, les atomes H, I, J, K, L sont considérés comme fixés au-dessus de B, E, F, A, dans la direction parallèle à l'axe quinaire CD, l'atome M se place alors dans le prolongement de CD. Les groupements de 3 b et 3C possèdent respective-

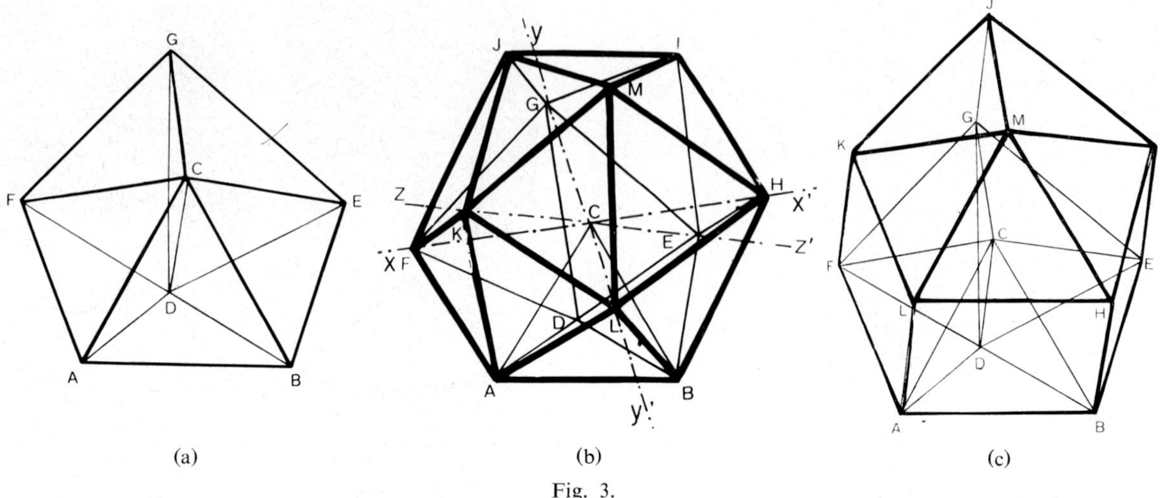

(a)   (b)   (c)

Fig. 3.

ment quarante deux et trente sept liaisons entre les premiers plus proches voisins. La configuration 3b, à ce stade de croissance, serait donc, en première approximation, plus stable que celle de 3c. Cependant, il est possible qu'un déplacement d'ensemble des atomes du pentagone H, I, J, K, L transforme la configuration 3b en la configuration 3c par une rotation de 30 degrés autour de CD, en effet 3c comporte des sites à quatre liaisons alors que 3b ne comporte que des sites à trois liaisons. L'adjonction de cinq atomes respectivement à 3c dans des sites tels que B, H, I, E et à 3b dans des sites tels que M, L, H conduit à des édifices qui comportent l'un et l'autre cinquante sept liaisons entre les atomes plus proches voisins. Ces deux configurations auront donc des probabilités égales d'existence. L'édifice 3b engendrera en croissant un cristallite à contour hexagonal que nous étudierons au paragraphe suivant alors que la configuration 3c conduit à un cristallite pentagonal. Dans ce cas, l'édifice présente, sur sa partie supérieure, cinq plans (111) tels que M, H, I. Nous allons montrer que la croissance par empilement de plans (111) dans la structure cubique à faces centrées conserve la symétrie pentagonale de l'édifice. Supposons que le plan M, H, I (fig. 3c) soit complété par l'atome N placé dans le site B, H, I, E, le plan M, H, I, N (fig. 4) est alors un plan (111) que nous appelons couche b, la couche a étant formée par les atomes C, B, E. Lors de la croissance, un atome se placera dans le site H, I, M qui est un site de la couche c. Ce plan se complètera nécessairement de façon à ce que des atomes se placent au-dessus de I et H dans la direction de l'axe

pentagonal. De ce fait, la croissance ultérieure conserve la symétrie pentagonale de l'édifice qui grossit de façon continue, plan par plan, en restant homothétique à lui-même. La figure 5 représente un modèle de ce, cristal réalisé avec des sphères rigides. Remarquons que le cristal ainsi formé possède dans les parties A, C B et C, F, G, E des plans (110) et (111) perpendiculaires au support, conformément aux observations effectuées en microscopie et diffraction électroniques.

4. Croissance d'un cristallite à contour hexagonal

Comme nous l'avons dit au paragraphe précédent, les cristallites hexagonaux sont issus du modèle de la figure 3b. Nous supposons que les atomes de ce groupement se disposent de façon à réaliser le maximum d'isotropie tout comme les aggrégats élémentaires que certains auteurs signalent dans les liquides[10-12]).

Un tel édifice possède vingt faces (icosaèdre) assimilables à des plans (111), l'une d'entre elles (ADB) étant en contact avec le support. Il possède en outre six demi-axes quinaires, CH et CF, CG et CL, CK et CE, issus respectivement de l'atome central C et portés par les trois droites XX', YY', et ZZ'. Chacun de ces axes peut être considéré comme l'axe de croissance d'un décaèdre qui se développe dans les mêmes conditions que l'édifice de la figure 5. Une construction avec des sphères rigides (figs. 6 et 7) montre que la croissance autour de l'axe CG, par exemple, induit automatiquement une croissance identique autour de CK et qu'il en est de même pour les différents axes pentagonaux. Ainsi la croissance de l'ensemble de

IV – 7

l'édifice s'effectue d'une manière continue par un empilement de plans (111) autour de ces axes, donnant un cristallite dont le contour est hexagonal.

La figure 6 représente une vue de dessus du cristallite posé sur son support et la figure 7 une vue de dessous

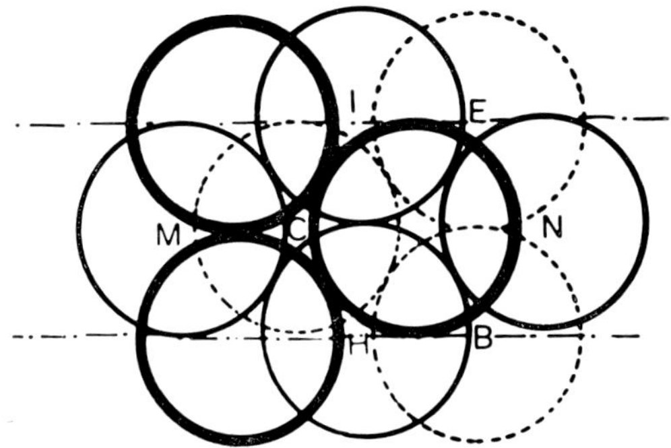

Fig. 4. – – –, couche (a), —, couche (b); —, couche (c).

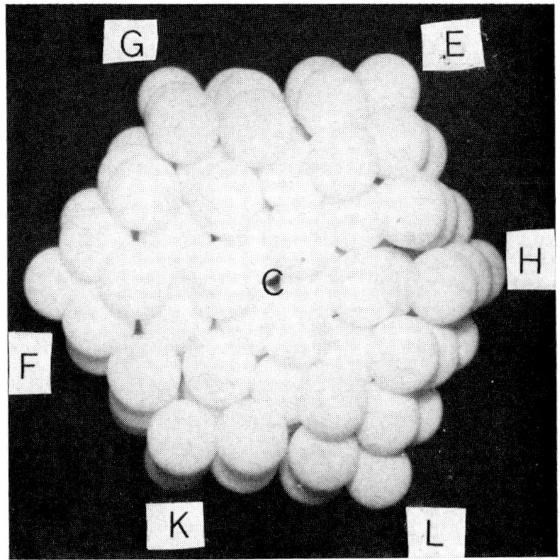

Fig. 6.

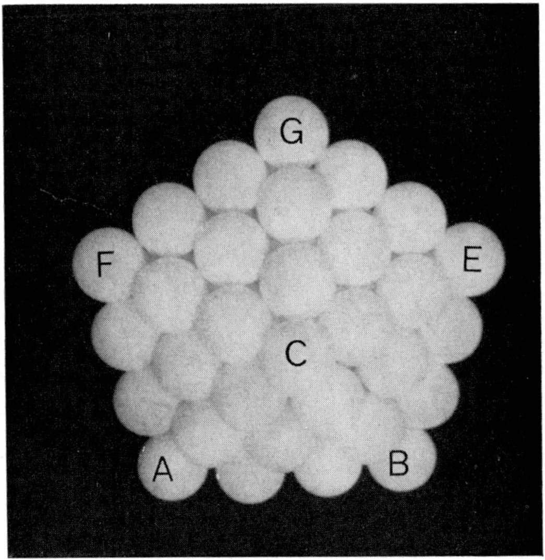

Fig. 5.

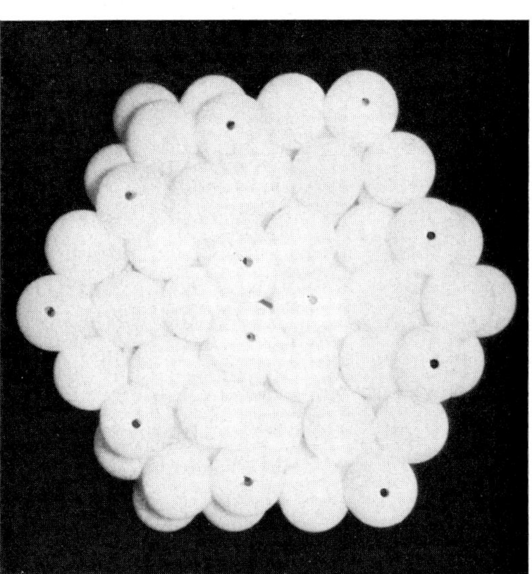

Fig. 7.

sur laquelle les atomes en contact avec le support sont marqués avec des points.

Dès que le cristal atteint la taille correspondant au modèle de la figure 5, on peut considérer qu'il est composé de différentes parties en position de macle les unes par rapport aux autres. Il existe, en particulier, six plans de maclage normaux au support et chacun d'eux contient un axe pentagonal. Par exemple, pour les axes portés par la droite ZZ', ces plans sont CKA pour l'axe CK et CEI pour l'axe CE (fig. 6 et 7); il en résulte que le cristal comporte des plans (111) perpen-diculaires au support dans les parties ayant pour contour CFKG et CEHL. Bien que ces plans ne s'éten-dent pas sur toute l'épaisseur du cristal, ils permettent d'expliquer entièrement les diagrammes de diffraction et sont à l'origine du "contraste en aile de papillon observé sur les micrographies électroniques en fond noir[1–3]).

5. Discussion

Nous avons montré que l'on peut considérer que les cristallites d'or formés par nucléation hétérogène sur un support d'halogénure alcalin peuvent croître d'une manière continue et qu'il n'est pas nécessaire de faire appel à une croissance par multimaclage pour obtenir les structures observées. Les germes envisagés, décaèdre pentagonal et icosaèdre sont les groupement d'atomes les plus compacts et les plus stables que l'on puisse former avec sept ou treize atomes. De tels édifices sont supposés exister dans les liquides, mais on admet alors qu'ils ne peuvent pas donner naissance à un cristal tridimensionnel. Dans le travail nous supposons qu'ils sont à l'origine des cristallites observés, leur formation et leur croissance étant justifiées non seulement par le fait que les observations sont en bon accord avec les modèles proposés mais également parce que les questions qui se posaient dans le contexte d'une croissance par maclages successifs se trouvent résolues par l'existence d'une croissance continue des cristallites.

Il est possible que dans le cas de la nucléation hétérogène ces édifices soient rendus stables par l'action du support. Pour le moment le manque de connaissances quantitatives sur l'interaction dépôt-support ne nous permet pas d'entreprendre une étude de la stabilité de ces cristallites en fonction du support. Cependant, la formation et la croissance de germes initiaux tels que ceux que nous avons envisagés impliquent que l'interaction métal–halogénure alcalin soit faible par rapport à l'interaction métal–métal. Les résultats expériementaux montrent que les cristallites à contour hexagonal sont beaucoup plus nombreux que les cristallites à contour pentagonal; on peut donc penser que la configuration 3c est plus stable que la configuration 3b. Pour ces deux configurations l'interface dépôt–support est identique, il devrait donc être possible, par une étude énergétique tenant compte uniquement des interactions entre atomes métalliques, de confirmer ce résultat.

L'observation microscopique montre d'une part, que la taille des cristallites ne dépasse jamais 80–100 Å, d'autre part que lorsque l'épaisseur du dépôt croît il se forme des aggrégats composés de plusieurs de ces cristallites, dus à leur coalescence par diffusion[13]). On peut en déduire que la structure des cristallites hexagonaux est instable au-delà d'une certaine taille et qu'ils sont très mobiles sur leur support, ce qui facilite leur coalescence. Cette mobilité peut être favorisée par le fait que, le plan de contact avec le support ne contient qu'un très petit nombre d'atomes par unité de surface.

Bibliographie

1) S. Ino, J. Phys. Soc. Japan **21** (1966) 346.
2) J. G. Allpress et J. V. Sanders, Surface Sci. **7** (1967) 1.
3) E. Gillet et M. Gillet, Thin Solid Films **4** (1969) 171.
4) T. Komoda, dans: *European Conf. on Electron Microscopy*, Rome, 1968, p. 131.
5) A. J. Melmed et D. O. Hayward, J. Chem. Phys. **31** (1959) 545.
6) F. Ogburn, B. Paretzkin et H. S. Peiser, Acta Cryst. **17** (1964) 774.
7) M. Froment, G. Maurin et J. Thevenin, Métaux (Corrosion-Ind.) **536** (1970) 3.
8) R. L. Schwoebel, J. Appl. Phys. **37** (1966) 2515.
9) B. G. Bagley, Nature **208** (1965) 674.
10) F. C. Frank, Proc. Roy. Soc. (London) A **215** (1952) 43.
11) J. D. Bernal, Nature **183** (1959) 141; **185** (1960) 68.
12) B. Mutaftschiev, Bull. Soc. Franc. Mineral. Crist. **92** (1969) 558.
13) A. Masson, J. J. Métois et R. Kern, Compt. Rend. (Paris) **267** (1968) 64; J. Crystal Growth **3, 4** (1968) 196; Compt. Rend. (Paris) **271** (1970) 235.

Journal of Crystal Growth **13/14** (1972) 217–224 © *North-Holland Publishing Co.*

ON THE GROWTH PROCESS OF Au AND Ag ON VARIOUS AIR- AND VACUUM-CLEAVED ALKALI HALIDE CRYSTALS

J. R. ROOS and J. S. VERMAAK

Department of Physics, University of Port Elizabeth, Port Elizabeth, Republic of South Africa

The influence of air exposure of various alkali halide crystals on the growth process of Au and Ag was investigated. For this purpose the other growth parameters, i.e. substrate temperature, deposition rate and vacuum conditions, were kept constant. Both metals were then evaporated simultaneously in steps of varying thickness onto air- and vacuum-cleaved surfaces of NaCl, NaBr, NaF, KCl and KBr. The films were examined by transmission electron microscopy and electron diffraction. The experimental results are explained in terms of the hard sphere model proposed by Henning and Vermaak.

1. Introduction

One of the most puzzling results of the growth of fcc metal films on alkali halide substrates is the difference in the final orientation of these metals when grown on vacuum-cleaved and on air-cleaved surfaces of the same substrate[1-3]. Matthews[2] explained the enhancement of epitaxial gold films on air-cleaved NaCl empirically in terms of the accompanying increased nucleation density on such a surface as compared to the vacuum-cleaved surface. This empirical explanation, however, did not explain the very basic process which governs the increase in nucleation density. Recently Henning and Vermaak[4] proposed a model to explain this process. They showed experimentally[3] that the epitaxy of Au films on air-contaminated rocksalt substrates actually results by virtue of the conversion of the NaCl surface layer to a NaOH surface layer. Based on this result they gave an interpretation[4] of the growth process in terms of a critical accomodation centre and critical nucleus concepts. This theory was also shown to be applicable to the growth process of various other metals[2] on air- and vacuum-cleaved NaCl substrates under a variety[5,6] of experimental conditions.

The purpose of this study is to investigate the validity of their model to the growth process of both Au and Ag on various other alkali halide surfaces. For this purpose air- and vacuum-cleaved NaCl, NaBr, NaF, KCl and KBr were used as substrates. Air- and vacuum-cleaved surfaces of NaCl were included mainly because they provided a direct means of checking experimental growth conditions[3]).

2. Experimental procedure

It is common knowledge that the experimental parameters which are critical in the growth of fcc metal films on alkali halide substrates are substrate temperature, deposition rate, vacuum conditions and substrate structure. Thus, in order to determine the influence of the latter on the growth process of Au and Ag, the first three parameters were kept constant while the metals were evaporated on the different substrate surfaces.

All films discussed in this paper were grown in an U.H.V. system pumped by vacsorb and vacion pumps, capable of a vacuum of the order of 10^{-10} Torr. Cleavage of crystals in vacuum was accomplished with a linear motion cleavage device. The blade could be stably and reproducibly adjusted to any required setting, thereby exposing various portions of the substrate crystals which were mounted in a stainless steel block furnace. Substrate temperatures were controlled to 340 ± 10 °C. Au and Ag of 99.999% purity were evaporated from a molybdenum basket at a distance of about 5 cm from the substrates. A fixed evaporation rate of 2 Å/s, measured with an oscillating quartz crystal thickness monitor, was maintained throughout all the experiments.

All the crystals, each of dimension 0.2 cm × 1.5 cm × 1.0 cm, obtained from Harshaw Company, were mounted similtaneously in the furnace block. The air-cleaved crystals were cleaved in the atmosphere of relative humidity of at least 70%, $1\frac{1}{2}$ hr prior to pump down, while the vacuum-cleaved crystals were cleaved just prior to evaporation in a vacuum of the order of 10^{-9} Torr. Gold was evaporated on the surfaces in steps of 10, 60, 300, 600 and 1200 Å thickness and Ag in steps of 10, 60, 600, 1500 and 2000 Å thickness. The vacuum during the evaporation was in the order of 10^{-8} Torr. The films were examined by transmission electron microscopy and diffraction.

3. Experimental results

The electron micrographs and transmission electron diffraction patterns of 10 Å mean thickness, and continuous films of the two metals on all the surfaces are shown in figs. 1–5. These electron micrographs and HEED patterns are representative of 360 samples that were studied. The final orientations of the continuous films on vacuum-cleaved surfaces are in agreement with those obtained by Jesser and Matthews[7]. The particle densities of both metals for 10 Å mean thickness films are given in table 1. These values represent average values obtained on each sample.

4. Discussion

According to Henning and Vermaak's[4] model, for each clean (vacuum-cleaved) (100)-substrate surface there exists a critical overgrowth radius, r_{0c}, given by

$$r_{0c} = (r_a^2 - r_c^2 + d^2)/2(r_c + \sqrt{2d} - r_a), \tag{1}$$

where r_a and r_c are the anion and cation radii respectively and $d = r_a + r_c$. For overgrowths with atomic radii $r_0 \geq r_{0c}$ this surface will, at a specific supersaturation, provide *super-critical* or *critical* accomodation centres for (100) parallel growth, respectively. At the same supersaturation, however, the surface is said to be *subcritical* for (100) parallel growth if $r_0 < r_{0c}$. For the substitutional surfaces such as the hydroxide surface (air-cleaved) of rocksalt, the maximum critical overgrowth radius for (100)-parallel growth is given by[4]

$$r_{0c}^{max} = \frac{r_{as}^2 + r_a^2 + 2r_a r_c}{2[\sqrt{2}\, r_a - r_{as} + (\sqrt{2}+1)\, r_c]}, \tag{2}$$

where, r_{as} is the radius of the substitutional anion. Using these two expressions, the critical overgrowth radii for the various substrates were calculated and are

TABLE 1

TABLE 1

The critical atomic radii r_{0c} and r_{0c}^{max} were calculated from eqs. (1) and (2), using the ionic radii given by Wert and Thomson[8]. The values of r_{0c}^{max} are calculated on the assumption that the anion exchange[4]), does take place for all air-cleaved surfaces. The abbreviations, sc. and pc. under the column O (Orientation), mean single crystalline (001)-parallel orientation and polycrystalline (111)-preferred orientations, respectively, while D represents the number of nuclei per cm². The atomic radius $r_0^{Au} = r_0^{Ag} = 1.44$ Å[8].

Substrate	Cleavage	r_{0c} (Å)	r_{0c}^{max} (Å)	Au $D \times 10^{-10} \times cm^2$	O	Ag $D \times 10^{-10} \times cm^2$	O
NaCl	Air		1.26	72	sc.	12	sc.
	Vacuum	1.64		9	pc.	6	pc.
NaBr	Air		1.26	7	pc.	2	pc.
	Vacuum	1.82		3	pc.	0.5	pc.
NaF	Air		1.26	50	pc.	21	pc.
	Vacuum	1.26		54	pc.	13	pc.
KCl	Air		1.15	60	sc.	16	sc.
	Vacuum	1.44		16	sc.	1.4	pc.
KBr	Air		1.15	22	sc.	25	sc.
	Vacuum	1.60		3	sc.	1.0	pc.

shown in table 1. The values r_{0c}^{max} for the air-cleaved surfaces are calculated under the assumption that the anions in the surface layer are displaced by the OH-anions.

From figs. 1 to 5, it is clear that according to the final orientations of the continuous films, two types of substrate surfaces can be distinguished;
(i) those substrates yielding single crystalline films with the (100) parallel orientation – type S-surfaces, and
(ii) those substrates yielding polycrystalline films with (111)-preferred orientations – type P-surfaces.

(i) *Type S-surfaces.* For Au these type of surfaces are air-cleaved surfaces of NaCl, KCl and KBr and vacuum-cleaved surfaces of KCl and KBr, while for Ag they are air-cleaved surfaces of NaCl, KCl and KBr (see figs. 1, 4 and 5). In terms of the model all these surfaces, except vacuum-cleaved KBr, offer either critical or supercritical accommodation centres for the impinging Au and Ag atoms (see table 1). For this type of surface the model predicts the formation of both (100)- and (111)-nuclei during the nucleation process. Furthermore, that at a specific supersaturation a higher sticking coefficient and therefore a higher nucleation density will prevail on the (super)-critical than on the subcritical surface of the same substrate. Also, even though the surface energy of the (100)-nuclei is greater

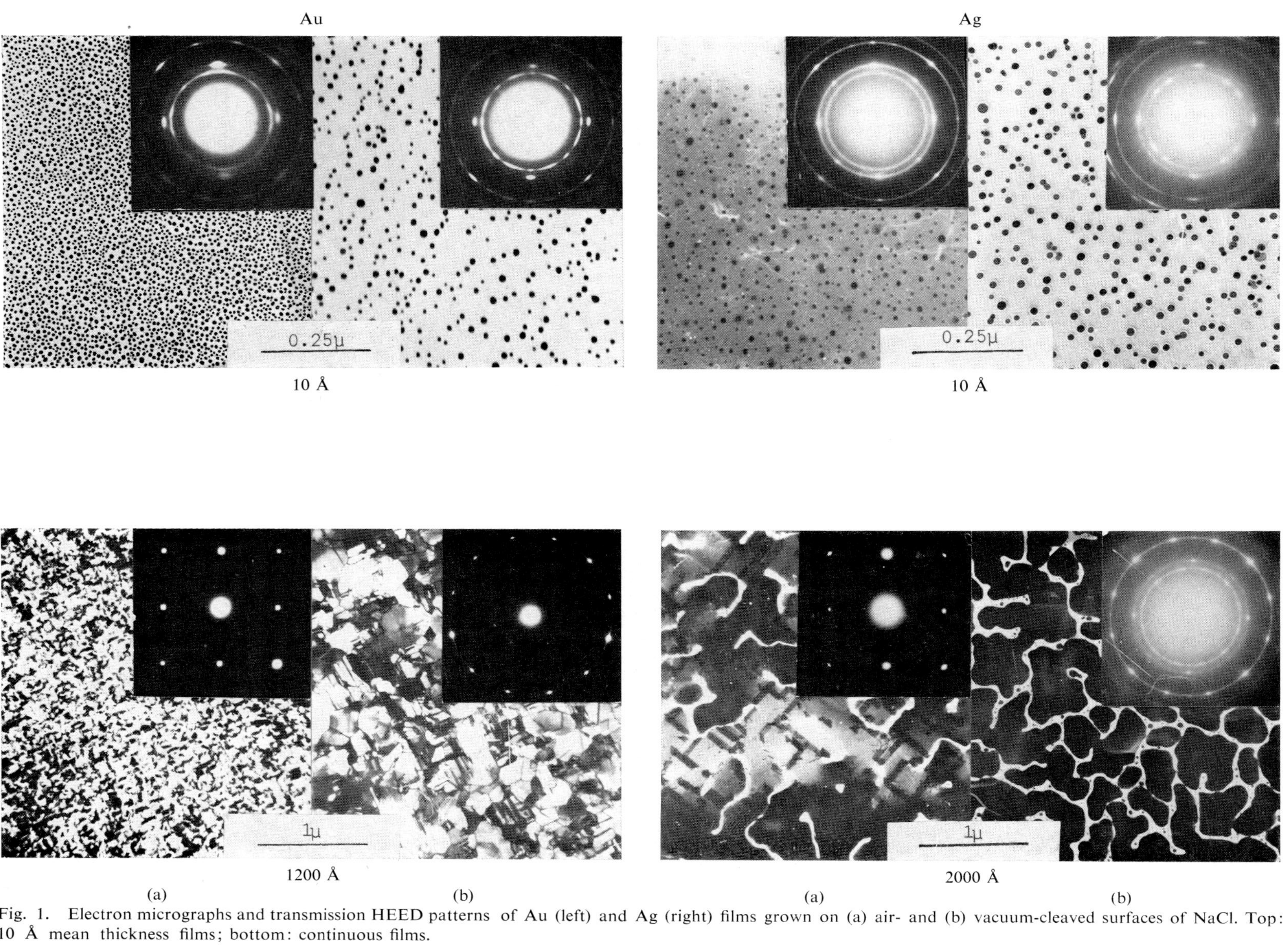

Fig. 1.   Electron micrographs and transmission HEED patterns of Au (left) and Ag (right) films grown on (a) air- and (b) vacuum-cleaved surfaces of NaCl. Top: 10 Å mean thickness films; bottom: continuous films.

J. R. ROOS AND J. S. VERMAAK

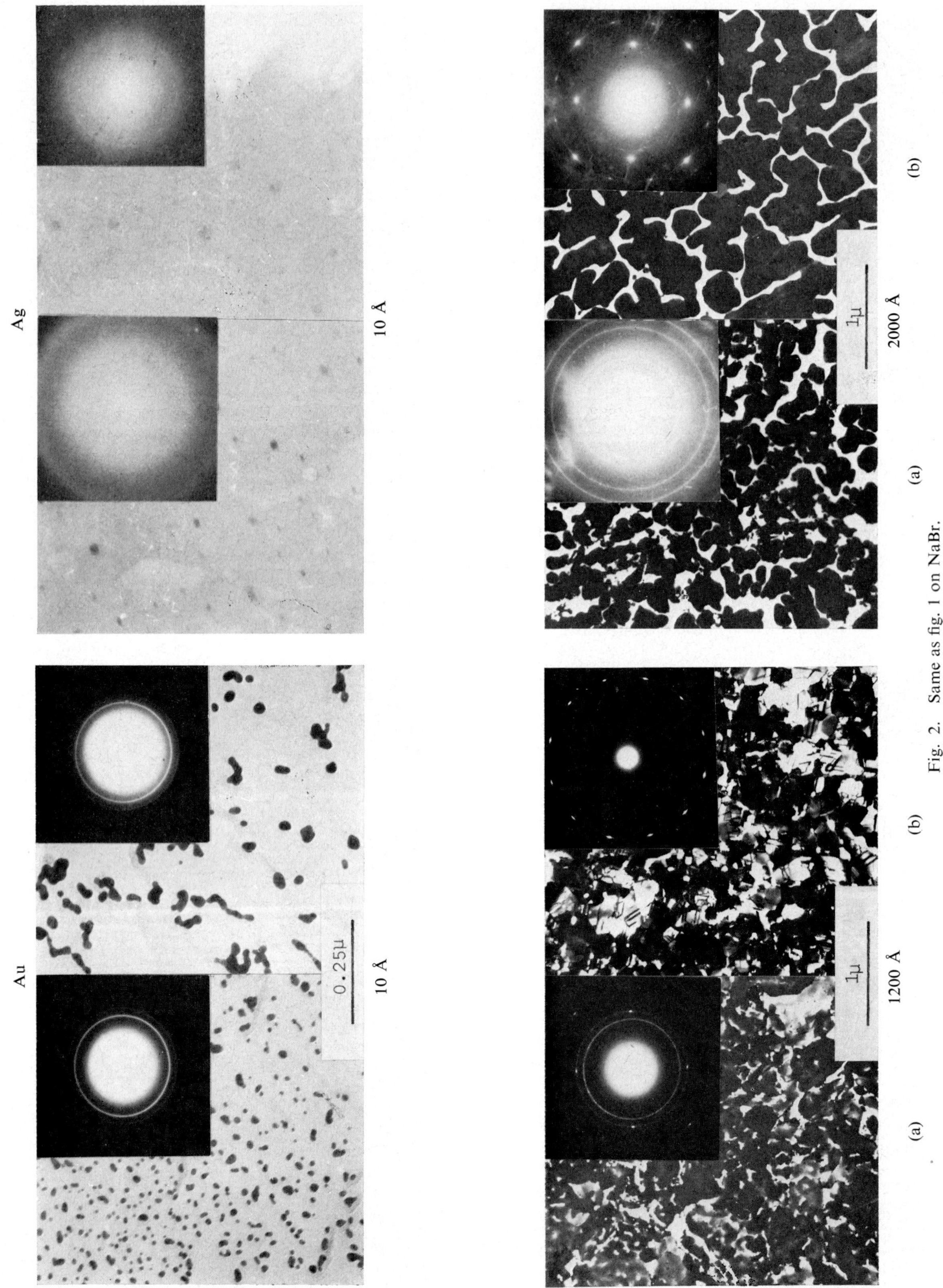

Fig. 2. Same as fig. 1 on NaBr.

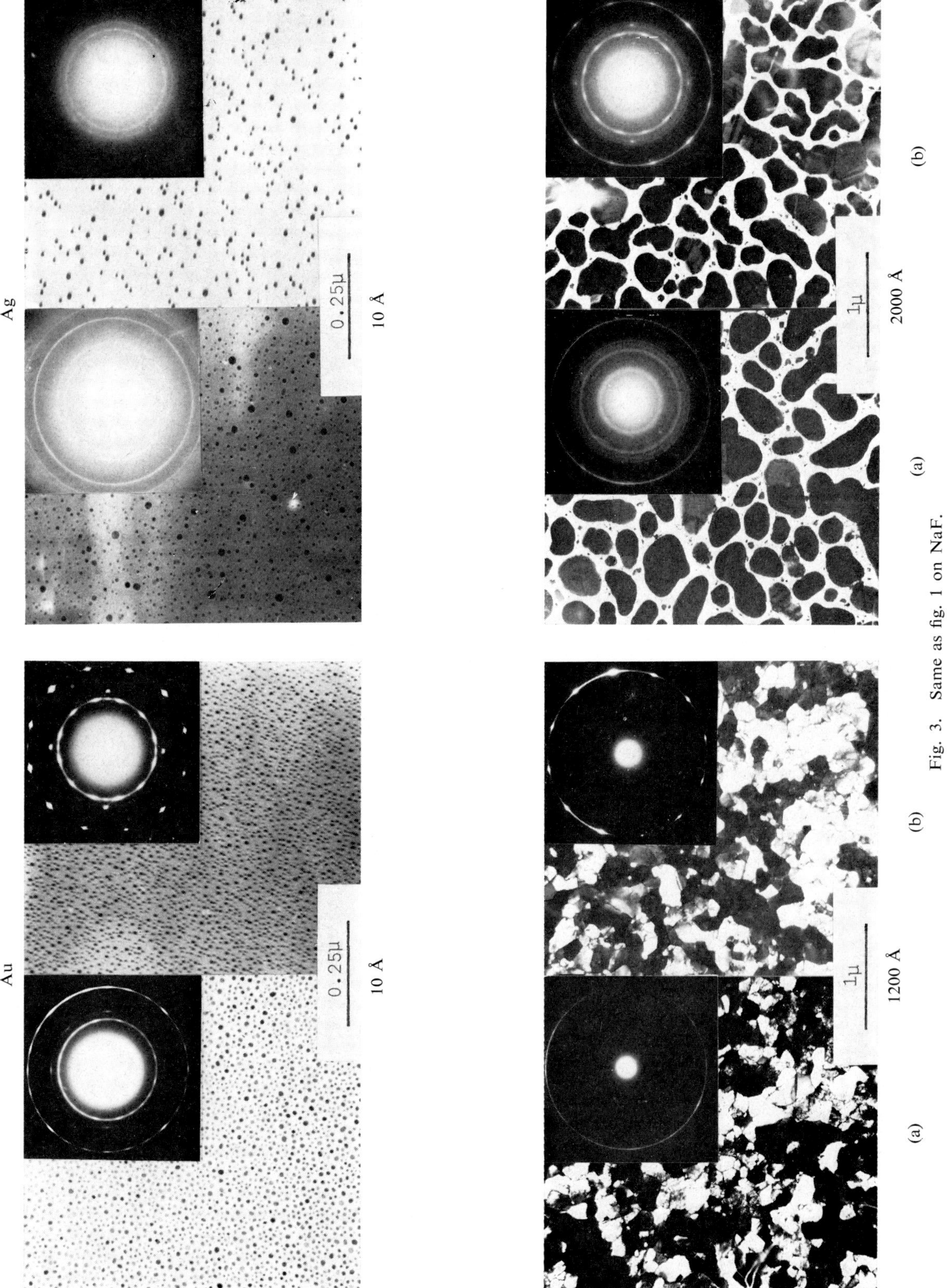

Fig. 3.   Same as fig. 1 on NaF.

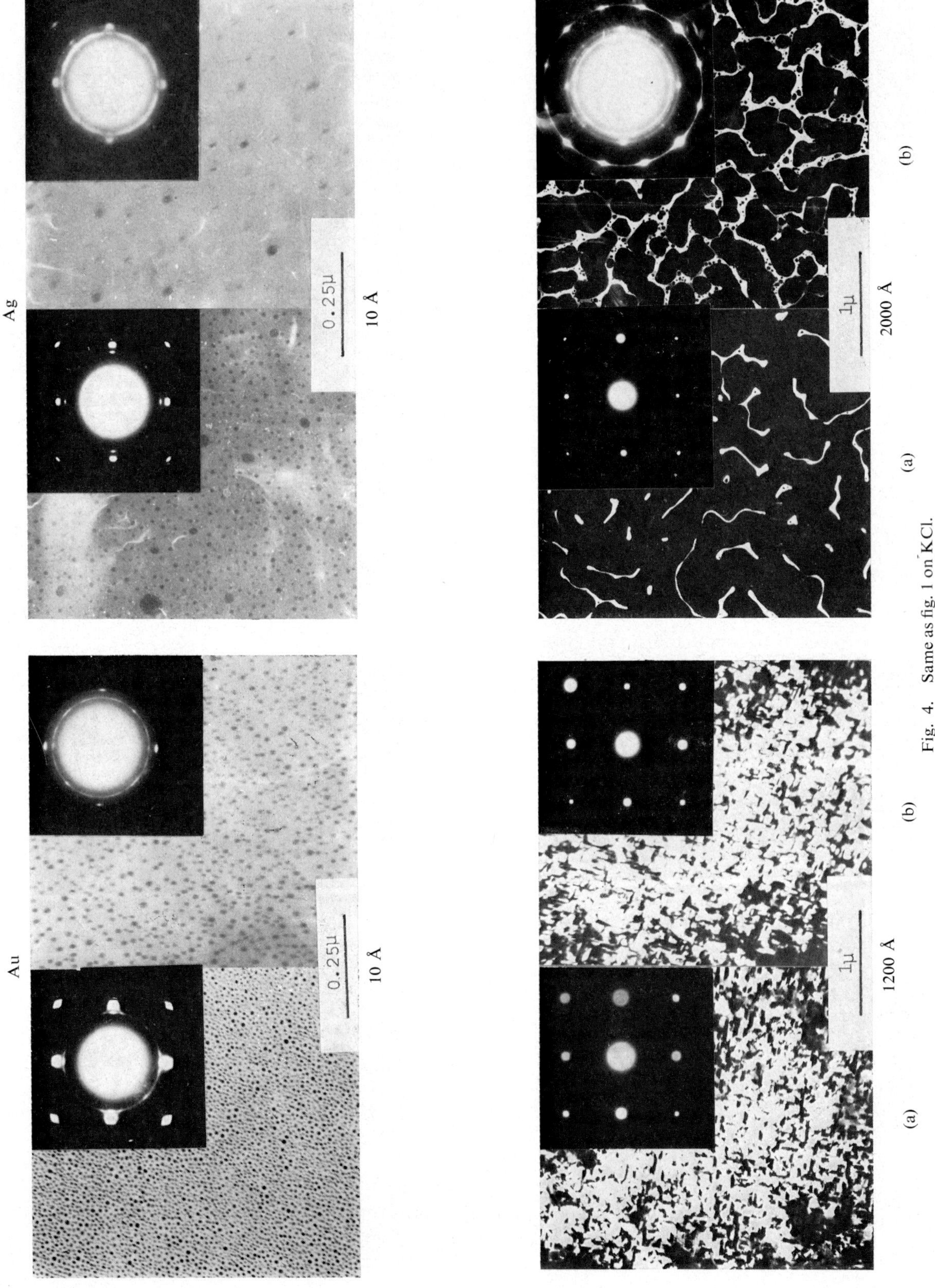

Fig. 4.   Same as fig. 1 on KCl.

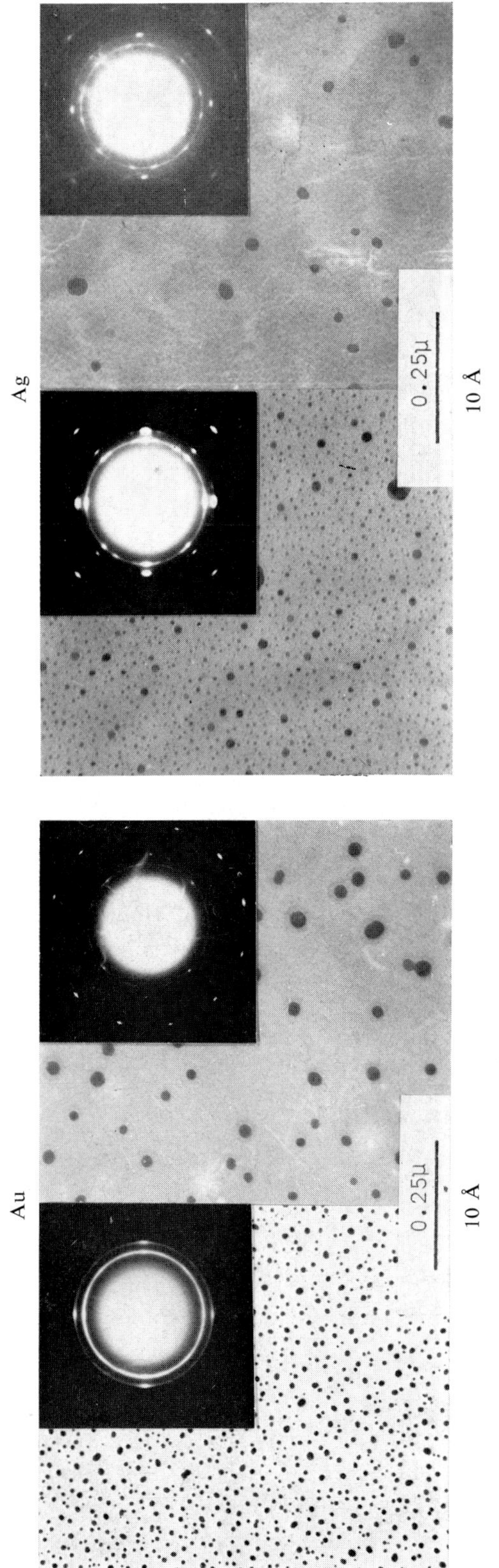

Fig. 5.   Same as fig. 1 on KBr.

than the surface energy of the (111)-nuclei, the growth of the (100)-nuclei will be energetically favoured because they grow on a (super)-critical surface while the (111)-nuclei grow on a subcritical surface[4]). The final orientation of the continuous film thus ought to be of the (100)-parallel type. The results of this experiment are in agreement with these predictions (see figs. 1, 4 and 5 and table 1).

Au grows epitaxially on both vacuum-cleaved KCl and KBr while Ag does not (figs. 4 and 5). Henning and Vermaak[4]) already explained this difference in growth behaviour of Au and Ag on KCl and the same explanation is proposed for KBr.

(ii) *Type P-surfaces.* For Au these type of surfaces are vacuum-cleaved surfaces of NaCl, NaBr, NaF and air-cleaved surfaces of NaBr and NaF, while for Ag they are the vacuum-cleaved surfaces of NaCl, NaBr, NaF, KCl and KBr and the air-cleaved surfaces of NaBr and NaF.

From table 1 it is clear that the vacuum-cleaved surfaces of NaCl, NaBr and KBr are subcritical for (001)-parallel growth. According to the model[4]) such surfaces will yield low nucleation densities which promote polycrystalline growth, as is obtained in tis experiment (see table 1 and figs. 1, 2 and 5).

5. Conclusions

From table 1 it is clear that the model predicts the final orientation of Au and Ag on 12 out of 20 substrates correctly. The model is purely a geometrical description of the fit of the overgrowth atoms on the substrate surface which does not take into account supersaturation conditions and binding between overgrowth atoms and substrate. Furthermore, structural changes in the metal films which occur on removing them from the $NaBr^-$ and NaF-surfaces and preparation difficulties might all be factors that influence the final results.

Acknowledgement

One of us (J. S. Vermaak) is indebted to the C.S.I.R., Pretoria, for financial assistance.

References

1) S. Ino, D. Watanabe and S. Ogawa, J. Phys. Soc. Japan **19** (1964) 881.
2) J. W. Matthews, J. Vacuum Sci. Technol. **3** (1966) 133.
3) J. S. Vermaak and C. A. O. Henning, Phil. Mag. **22** (1970) 269.
4) C. A. O. Henning and J. S. Vermaak, Phil. Mag. **22** (1970) 281.
5) J. W. Matthews, Appl. Phys. Letters **7** (1966) 131.
6) C. A. O. Henning, J. C. Lombaard and J. C. Botha, Appl. Phys. Letters **14** (1969) 109.
7) W. A. Jesser and J. W. Matthews, J. Crystal Growth **5** (1969) 83.
8) C. A. Wert and R. M. Thomson, *Physics of Solids* (McGraw-Hill, New York) pp. 23, 84.

Journal of Crystal Growth **13/14** (1972) 225–228 © *North-Holland Publishing Co.*

225

RECRISTALLISATION DE COUCHES MINCES D'ARGENT SUR LE MICA

H. PASCARD

Laboratoire de Magnétisme, C.N.R.S., 92–Bellevue, France

C. QUINTANA

C.I.F. Torres Quevedo, Madrid, Spain

F. HOFFMANN

Laboratoire de Magnétisme, C.N.R.S., 92–Bellevue, France

et

C. SELLA

Laboratoire de Rayons X, C.N.R.S., 92–Bellevue, France

Thin silver (111) films with a good *flatness of the surface* have been obtained by oriented recrystallization on single crystal substrate: firstly, very thin polycrystalline silver films (300 Å) are evaporated on mica at room temperature: films are continuous, with very small crystals and many defects; then, an annealing at 300 °C in ultra high vacuum produces an *oriented recrystallization*: very large crystals grow with an orientation (111) parallel to the substrate and most defects are eliminated.

The results are qualitatively well explained taking into account the influence of the surface crystalline energy, the thermal annealing energy, the thermal stress energy induced by the substrate, and the surface energy of an adsorbed layer on mica.

These films are utilized to obtain very flat epitaxial films of permalloy in order to study standing spin wave resonances.

1. Introduction

Dans le cadre d'une étude du couplage d'échange entre deux métaux ferromagnétiques, nous avons cherché à réaliser des couches minces métalliques présentant à la fois une orientation cristalline parfaite et une très bonne planéité de surface.

Nous avons choisi d'épitaxier le métal ferromagnétique cubique faces centrées sur un autre métal cfc lui-même déposé sur le plan de clivage (0001) du mica, le mica étant choisi pour sa grande planéité de surface à l'échelle atomique et pour la possibilité de réalisation de plages monocristallines de grandes dimensions[1,2]).

Nous avons d'abord essayé l'épitaxie directe de l'argent sur mica: une orientation (111) parfaite est facilement obtenue (fig. 1) mais la planéité de surface est médiocre (fig. 2A).

Nous avons ensuite étudié l'épitaxie directe de l'or sur le mica: peu d'études avaient été faites dans ce domaine et nous avons trouvé des conditions d'épitaxie en vide classique et en ultra-vide[3]). Nous avons mis en évidence l'influence considérable du traitement thermique du mica sur l'épitaxie: la désorption super-ficielle de molécules d'eau fixées à la surface du mica et la déshydratation des premières couches atomiques de mica joueraient un rôle important dans l'épitaxie. Cependant, l'orientation cristalline obtenue dans les meilleures conditions est moins bonne que dans le cas de l'argent et l'état de surface n'est pas meilleur.

Il semble donc que l'épitaxie à haute température (250 à 300 °C pour l'argent, 400 à 500 °C pour l'or) corresponde à de bonnes conditions énergétiques pour l'orientation cristalline mais non pour la planéité de surface. Ceci est lié au fait que les couches épitaxiques préparées à haute température ne sont continues que pour des épaisseurs assez grandes (de l'ordre de 1500 Å dans le cas de l'argent). On sait en effet que le nombre de germes épitaxiques formés est faible et que ces germes atteignent des dimensions assez importantes avant de coalescer. Il en résulte que les couches ne sont vraiment continues que pour des épaisseurs assez grandes et que la taille des cristaux ainsi formés explique le relief de surface assez important observé sur ces couches. Ces phénomènes sont illustrés par les figures 1 et 2A.

Par contre, le dépôt à température plus faible (à la

IV – 9

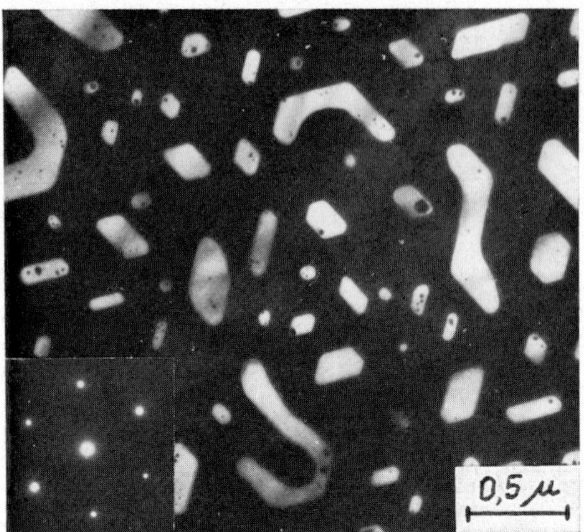

Fig. 1. Dépôt de 800 Å d'argent sur mica, obtenu par épitaxie directe à 300 °C.

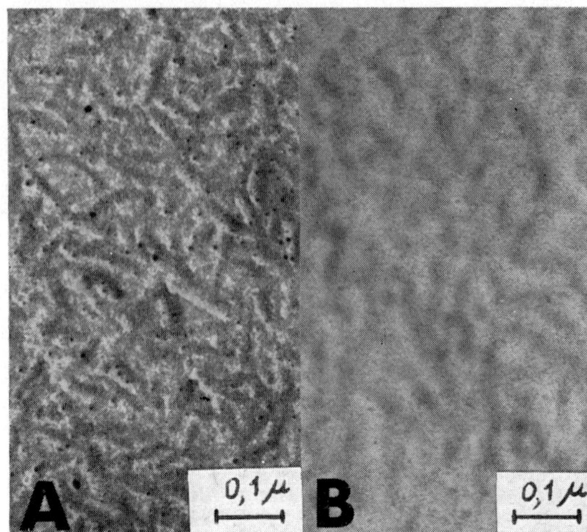

Fig. 2. Réplique au tungstène rhodié de la surface: (A) d'un dépôt de 3000 Å d'argent sur mica obtenu par épitaxie directe à 300 °C; (B) d'un dépôt de 400 Å d'argent sur mica à 20 °C, recuit 8 h à 300 °C.

température ambiante par exemple) permet d'obtenir des couches très minces continues, de bonne planéité de surface, mais d'orientation médiocre (fig. 3).

Nous avons alors recherché, à partir de couches très minces d'argent déposées à la température ambiante sur mica, les conditions de recuit sous vide permettant de réaliser une *recristallisation orientée sur le substrat* et d'obtenir la meilleure orientation cristalline possible tout en conservant une bonne planéité de surface.

2. Conditions expérimentales

Toutes les couches minces d'argent étudiées ont été préparées par évaporation thermique dans un groupe à ultra vide à pompe ionique (400 l/s). Les dépôts étaient effectués soit pendant l'étuvage (10^{-6} Torr pendant l'évaporation), soit après l'étuvage avec pompage cryogénique (10^{-8} Torr pendant l'évaporation).

Pendant le dépôt, le substrat de mica était maintenu à la température ambiante. Le mica utilisé est du type muscovite en provenance des Indes: il est clivé à l'air et placé immédiatement sous vide. La vitesse d'évaporation de l'argent est comprise entre 5 et 10 Å/s.

Tous les recuits ont été réalisés sur substrat et sous vide (entre 10^{-6} et 10^{-8} Torr) aussitôt après l'évaporation sans remise à l'air. Différentes températures (entre 250 °C et 350 °C) et différentes durées (8h ou 24h) de recuit ont été essayées.

Les dépôts ont été étudiés en microscopie et diffraction électronique en transmission. Pour l'observation,

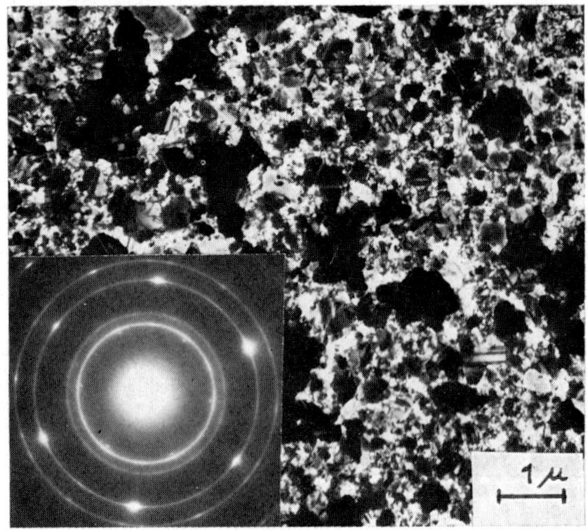

Fig. 3. Dépôt de 400 Å d'argent sur mica à 20 °C à 10^{-6} Torr non recuit. Diffraction sur une surface de 100 μm².

les films sont détachés de leur support par flottaison dans l'acide fluorhydrique dilué.

Résultats

En ce qui concerne l'orientation, il est nécessaire de choisir une durée et une température de recuit suffisantes: des recuits de 8 heures à 300 °C de dépôts d'argent polycristallins (fig. 3) conduisent à la formation de grands cristaux bien orientés (figs. 4 et 5). Par ailleurs nous avons mis en évidence que la recristallisation est plus complète sur des dépôts obtenus à 10^{-6} Torr que

Fig. 4. Dépôt de 400 Å d'argent sur mica à 20 °C à 10^{-6} Torr et recuit 8 h à 300 °C en ultra vide. Diffraction sur une surface de 100 µm².

Fig. 5. Dépôt de 400 Å d'argent sur mica à 20 °C à 10^{-6} Torr et recuit 8 h à 300 °C en ultra-vide.

sur des dépôts à 10^{-8} Torr: de plus grandes zones sont bien recristallisées.

En ce qui concerne la planéité de surface, le paramètre le plus important semble être l'épaisseur: des couches de 200 à 300 Å présentent encore des trous de forme géométrique après recuit (hexagones). A partir de 400 Å environ, les couches sont parfaitement continues. Un autre paramètre important est la température de recuit qui ne doit pas être trop élevée pour ne pas entraîner une trop grande mobilité des atomes superficiels, ou même des phénomènes de sublimation ce qui pourrait accentuer les défauts de surface: la

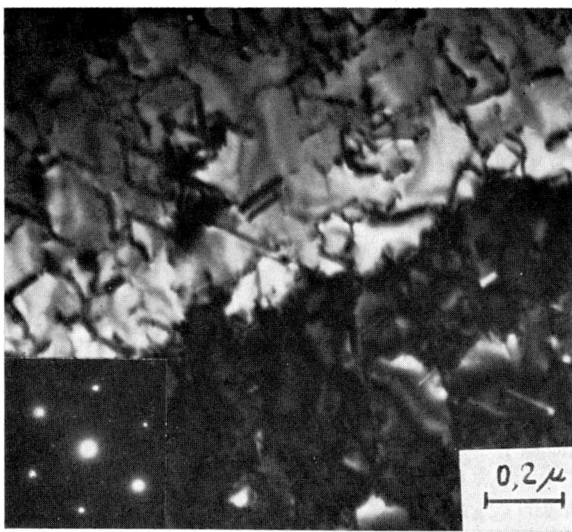

Fig. 6. Dépôt de 400 Å d'or épitaxié à 200 °C sur un dépôt recuit de 400 Å d'argent sur mica.

température de 300 °C semble être suffisante pour obtenir une bonne recristallisation sans détérioriation sensible de la surface.

Les figures 4 et 5 montrent un exemple de couche obtenue dans les conditions optimales: dépôt à 10^{-6} Torr à la température ambiante, épaisseur 400 Å environ, et recuit de 8 heures à 300 °C. Des zones de plusieurs microns carré se sont parfaitement recristallisées. De très nombreux joints de grain ont été éliminés. De plus, la dispersion d'orientation a considérablement diminué. Néanmoins, un certain nombre de défauts subsistent: dislocations, macles.

La planéité de ces couches (fig. 2B) étudiée au moyen d'une technique de réplique à très haute résolution[4]) est meilleure que celle des dépôts obtenus par épitaxie directe sur mica à 300 °C (fig. 2A).

Les couches ainsi orientées par recristallisation peuvent être utilisées comme substrats pour préparer des couches épitaxiques d'autres métaux:

(1) L'épitaxie d'une couche d'or de 400 Å d'épaisseur sur une telle couche d'argent recristallisée sur mica, chauffée à 200 °C, permet d'obtenir un film monocristallin présentant une continuté et une orientation parfaites (fig. 6).

(2) L'épitaxie de 3000 Å permalloy (80 Ni, 20 Fe) sur une telle couche d'argent recristallisée, chauffée à 250 °C conduit à une couche bien orientée aux propriétés magnétiques excellentes: en effet, les couches obtenues par cette méthode présentent de meilleures largeurs de raie d'ondes de spin ΔH que les couches de permalloy

monocristallines déposées sur argent ou sur or épitaxiés sur mica à température élevée (ΔH étant directement relié à la planéité de surface).

4. Discussion et interprétation

De nombreux travaux ont été effectués sur la recristallisation de couches minces métalliques, les uns concernant des couches sans support[5–7], les autres concernant des couches sur support amorphe[5,8,9]. Par contre, peu d'études ont été faites sur la recristallisation sur substrat monocristallin.

Nous allons analyser successivement les mécanismes auxquels on peut s'attendre dans le cas de recristallisation sur support monocristallin, d'une part lors de la cristallisation pendant le dépôt et d'autre part lors de la recristallisation pendant le recuit, et examiner le rôle d'une couche de gaz adsorbée à l'interface mica argent (cas de l'évaporation à 10^{-6} Torr).

Le dépôt d'argent sur mica à la température ambiante conduit à la formation d'une couche à très petits cristaux avec un grand nombre de défauts (fig. 3): joints de grains, macles, dislocations, etc. Ces couches sont polycristallines avec quelques cristaux orientés parallèlement au substrat [orientation (111)], ce qui traduit un début d'épitaxie de l'argent: l'épitaxie serait complète à 250–300 °C. La formation de ces cristaux (111) est aussi favorisée par le fait que pour des épaisseurs très faibles, la croissance d'un métal cfc avec un plan (111) parallèle au plan de la couche, correspond à des conditions énergétiques minimales. Ceci à été vérifié dans le cas de dépôts très minces de métaux cfc sur substrat amorphe, pour lesquels une texture [111] est obtenue[9,10].

Pendant le recuit à 300 °C, de grands cristaux se forment (fig. 4): la plus grande partie du matériau s'oriente parallèlement au substrat, quelques cristaux présentent seulement une texture [111]. On peut penser que pendant la recristallisation, le rôle de l'énergie de surface est prépondérant le matériau ayant un rapport surface sur volume très grand. Comme l'énergie cristalline superficielle pour un métal cfc est minimale pour le plan (111), les cristaux présentant une orientation (111) ou une texture [111] se développent aux dépens de leurs voisins, par migration des joints de grains, la proximité de la surface permettant une élimination facile des défauts. Enfin on peut penser que le rôle de l'énergie de contrainte thermique due à la différence

des coefficients de dilatation du mica et de l'argent est faible, le rôle de l'énergie cristalline superficielle étant prépondérant[10,11]).

La présence d'une couche de gaz adsorbée à l'interface mica–argent introduit une énergie superficielle différente entre dépôt et substrat. Quelques études ont été faites sur l'influence d'une couche contaminatrice sur la recristallisation[5,6]. Dans notre cas, on peut s'attendre lors de l'évaporation à 10^{-6} Torr, à la présence d'une couche de gaz adsorbée (probablement de molécules d'eau) plus importante que dans le cas de l'évaporation à 10^{-8} Torr. L'adhérence de la couche sur le substrat est beaucoup plus forte lorsque le dépôt a lieu à 10^{-8} Torr. En présence d'une telle couche adsorbée (évaporation à 10^{-6} Torr), il est trouvé que d'une part lors du dépôt de l'argent à 20 °C, le nombre de cristaux d'orientation parallèle au substrat est plus faible, et que d'autre part lors du recuit la recristallisation est plus complète. Ces résultats suggèrent que la présence d'une couche de gaz adsorbés entre subtrat et métal, introduit une énergie superficielle correspondant à une liaison argent–mica plus faible. La présence d'une telle couche crée des conditions de surface telles que l'ensemble couche–interface–substrat est intermédiaire entre une couche sans support et une couche rigidement fixée sur substrat: on peut penser que la migration des défauts et des joints de grains pendant la recristallisation peut alors avoir lieu partiellement vers la surface en contact avec le substrat, d'où une recristallisation plus complète.

Références

1) J. C. Bruyère et P. Escudier, Rapport D.R.M.E. No. 65-34-371-00-480-75-01, Grenoble, 1966.
2) H. Jaeger, P. D. Mercer and R. G. Sherwood, Surface Sci. **6** (1967) 309; **11** (1968) 265.
3) H. Pascard, C. Quintana et C. Sella, dans: *Compt. Rend. des Travaux du Congr. Intern. sur les Couches Minces*, Cannes, 1970, pp. 404–408.
4) C. Sella, J. C. Martin, J. Escaig et M. Desforges, dans: *7e Congr. Intern. de Microscopie Electronique*, Grenoble, 1970, pp. 303–304.
5) C. Boulesteix, A. Marraud et R. Rateau, Surface Sci **12** (1968) 75;
C. Boulesteix, Cahiers Phys. No. 197 (1967) 1.
6) R. W. Vook, J. Appl. Phys. **32** (1961) 1557.
7) Hothersall, Phil. Mag. **20** (1969) 433.
8) J. P. Chauvineau, M. Galtier, M. Gandais et Th. Rappeneau, J. Phys. (Paris) **25** (1964) 142.
9) J. P. Chauvineau, G. Devant et M. Gandais, dans: *Proc. Second Colloq. on Thin Films*, Göttingen, 1967, pp. 63–72.
10) F. Witt et R. W. Vook, J. Appl. Phys. **39** (1968) 2773.
11) K. Kinosita, K. Maki et K. Takeuchi, dans: *Proc. Second Colloq. on Thin Films*, Göttingen, 1967, pp. 118–126.

Crystal-melt interface, growth of ice

J. ZELL
B. MUTAFTSCHIEV
B. LACMANN

I. N. STRANSKI
A. G. CROWTHER

Journal of Crystal Growth **13/14** (1972) 231–235 © *North-Holland Publishing Co.*

SIMULATION OF A CRYSTAL-MELT INTERFACE IN A MONOATOMIC SYSTEM

J. ZELL and B. MUTAFTSCHIEV

Laboratoire de Croissance Cristalline, Université de Provence, Centre de St. Jérôme, 13-Marseille 13e, France

This study deals with the simulation of the crystal–melt interface using a hard sphere static model. The liquid, represented by an assembly of randomly distributed 3000 ping-pong balls (Bernal model), is brought in contact with the lattice plane constructed from the same balls. After the maximum density of the assembly is reached, the co-ordinates of all the balls in the liquid and the crystal within 8 atomic planes from the interface, are determined.

Using the Lenard-Jones potential, the work of rupture $|\varepsilon|$ along the planes parallel to the interface as a function of the normal distance (d), has been calculated. This work, is equal to the energy of cohesion of the liquid or of the solid, depending on whether the rupture occurs in the bulk of one of these phases. For the rupture at the interface, $|\varepsilon|$ is approximately equal to the energy of adhesion between crystal and melt. If the crystal plane under consideration is a (0001) face of the hcp lattice, $|\varepsilon| = f(d)$ has a sharp minimum at the interface; the minimum is less pronounced if ordered clusters of a few "liquid" atoms are present at the interface.

In all the cases under consideration, the minimum indicates an imperfect wetting of the crystal face by the melt. The presence of a mono-atomic step on the crystal face enhances the wetting of the face by the melt. A graphical separation of the (linear) energy of adhesion at the step from the adhesive energy of the surface, shows that the higher organisation in the liquid due to the presence of a step propagates to some 5 to 8 atomic distances from the interface.

1. Introduction

A knowledge of the structure and the properties of crystal-melt interfaces is of great importance in understanding nucleation and growth processes in the melt. The first theoretical approaches to the phenomena of melting and growth were developed by Tamman[1]) and Stranski[2]). More recently, the same problem has been variously treated by Skapski[3]), Jackson[4]), Mutaftschiev[5]), Temkin[6,7]), and others. The diverging theoretical analyses are a consequence of the variation in the initial models which have been chosen to describe the crystal–melt interface. For example, in some models the liquid in the immediate neighbourhood of the interface has been considered to be either a crystalline lattice whose molecules are larger than those of the crystal itself[2]) or to be an isomorphic extension of the crystal lattice with a different binding energy[5–7]). In other models, involving the liquid density at the interface, the crystal surface is considered as a mathematical plane[8,9]).

In the case of such important properties of the interface such as the wettability of a dense crystal plane by the melt, some authors[3,4,6]) consider that wetting is excellent whilst others[2,5]) judge it to be poor. Clearly,

the experimental difficulties associated with measuring the contact angle at the crystal-melt interface has helped to promote this general confusion. In view of the unreality of all theoretical models reliable experimental studies are at a premium.

In an earlier experimental investigation, we determined both the contact angle and the interfacial energy between a clean (0001) face of cadmium and its melt[10]). The wettability of the crystal face was found to be relatively poor (contact angle $= 37°$) and the interfacial energy obtained was much larger than some theoretical estimates[3,4]). This prompted us to search for an experimental model which could take into account the structure of the interface because we appreciated that it is impossible to develop even an approximate idea of the structure at the junction of a disordered liquid with an ordered crystal face[11]). In order to study the structure of the interfaces we therefore decided to apply Bernal's random model of rigid spheres[12]).

2. Justification of Bernal model for the crystal-melt interface

The random distribution of rigid spheres, as proposed by Bernal, has already given results which are in semi-quantitative agreement with the expected bulk properties of simple liquids. It predicts quite well the radial distribution curves obtained by neutron or X-ray diffraction. It equally takes into account the volume change on fusion and gives reasonable values for the latent heat of fusion of rare gases[13]). Insofar as the model is purely structural, it is incapable of accounting for dynamic phenomena. However, if as in a crystal lattice the volume of a sphere is assumed to be the average vibrational volume of the atom the Bernal model represents the properties of simple liquids as satisfactorily as the cellular theory of Lennard-Jones and Devonshire[14]) or the hole theory of Eyring et al.[15]).

A criticism often made against this model is that the force of interaction between the spheres is neglected and replaced uniquely by gravitational forces. However if one takes precautions to see that a maximum density is attained, the atomic distribution will be representative of the real situation. A similar kind of criticism could be posed against the equivalent simulation of a crystal–melt interface using a solid sphere model. Furthermore, one can imagine the formation of two-dimensional ordered clusters on the crystal surface whose size and distribution are difficult to simulate. Our purpose then is to study the structure and the specific surface energy in the limiting case of an atomically smooth crystal surface. Measurements have additionally been made on the effect of the presence of ordered two-dimensional clusters.

3. Determination of the adhesion energy of the crystal-melt interface using the hard sphere model

By definition, the energy of adhesion is the sum of the work needed to mechanically separate the two phases of the interface and the work of subsequent re-arrangement at the two surfaces so created.

If one has a randomly distributed hard sphere model of a liquid in contact with a plane crystal surface composed of the same spheres, and if one knows the coordinates of all the spheres, one can determine the adhesion energy between the liquid and the crystal in the following manner (provided the law of intermolecular forces is known):

(i) Consider the model represented by fig. 1a in which a cylinder of "liquid.' of unit base area is in contact with a crystal surface. We can identify the following total potential energies of interaction:

ε_{vv} – between all the spheres inside the cylinder,

ε_{ve} – between all spheres inside the cylinder and all spheres outside it which are above the base of the cylinder but below the imaginary height noted in the figure,

ε_{vc} – between all spheres inside the cylinder and all those in the crystal lattice.

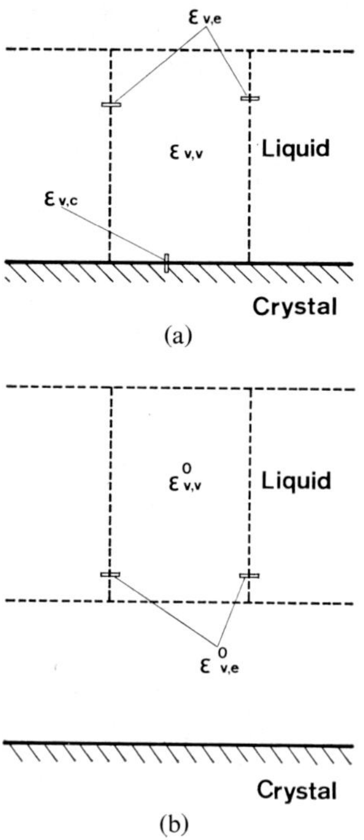

Fig. 1.   Principle of the adhesion energy determination by a hard sphere model.

(ii) Now displace the same cylinder perpendicularly to the interface by a distance sufficiently large to allow a homogeneous distribution of the spheres. Then, the following total potential energies can be identified (fig. 1b):

ε^0_{vv} – between all spheres inside the cylinder,

ε^0_{vc} – between all spheres inside the cylinder and all spheres outside it which are above the infinite plane coincident with the base of the cylinder but below the imaginary height.

The work of adhesion is then equal to:

$$\beta = -(\varepsilon_{vc} + \varepsilon_{vv} - \varepsilon^0_{vv} + \varepsilon_{ve} - \varepsilon^0_{ve}). \tag{1}$$

In this model we assume that the creation of a free liquid surface is not associated with molecular re-arrangement.

An analogous method can be used to calculate the specific edge energy of a step on a crystal–melt interface; the importance of edge energy in crystal growth processes is well known[16]).

Because eq. (1) involves differences between large similar numbers, the experimental determination of the specific interfacial energy by solid sphere models requires accurate measurements on models comprising a very large number of spheres. In this preliminary study, we have confined ourselves to the measurement of another physical parameter which permits an approximate appreciation of the energy of adhesion; in so doing important information regarding the structure at the interface is obtained.

4. Experiments and discussion

A liquid model consisting of 3000 ping-pong balls ($\sim$. 3.8 cm diameter) has been used for experimental measurements. The "liquid" is placed in a cylinder 70 cm in diameter and 50 cm in height. A rigid lattice plane constructed from the same balls represents the (0001) face of an hcp lattice. When the "liquid" and the "crystal face" have been brought into contact, the essembly is vibrated until the compactness reaches a maximum value. The position of the crystal plane happens to be unimportant for the distribution of liquid spheres at the interface. Having fixed the balls in their final positions by glueing, the essembly was fully characterised by determining the co-ordinates of each ball immediately prior to its removal from the assembly. The reference plane for the coordinate measurement was the top hcp lattice plane of the crystal model. The hcp positions of 8 more layers of balls in the "solid" were calculated. So for the determination of the interaction energies the coordinates of some 5000 balls were available.

The calculation consisted in determining the potential energies between all the balls residing within a cylinder A (fig. 2) with a base parallel to the crystal face of diameter 30 cm and a height 15 cm, and the balls situated outside of the cylinder within a half space B limited by the base plane of the cylinder from the side of the crystal. The potential energy between two balls was calculated using the Mie function with the constants $n = 6$ and $m = 12$ (Lennard-Jones potential). The results were subsequently converted into energy per unit area by inserting, for the minimum potential

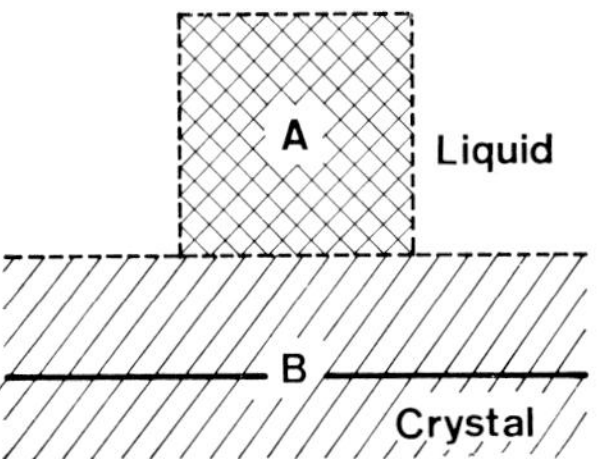

Fig. 2. Principle of the determination of the specific work of rupture by a hard sphere model.

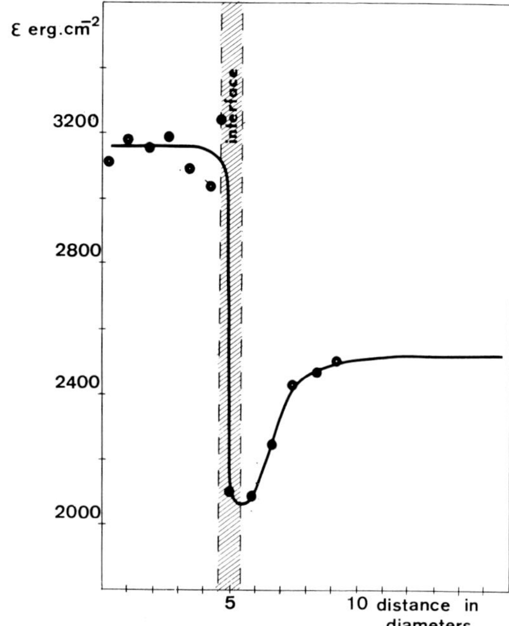

Fig. 3. Specific work of rupture as a function of the normal distance to the interface for a smooth (0001) face.

energy of a pair of free atoms. U_0 and the equilibrium inter-atomic distance r_0 (equal to the diameter of a ball in the model), the values,

$$U_0 = -2 \times 10^{-13} \text{ erg} \quad \text{and} \quad r_0 = 2 \times 10^{-8} \text{ cm}.$$

The absolute energy values, obtained from the calculation, $|\varepsilon|$ is exatcly equal to the cohesive energy of the crystal or of the liquid depending upon whether the plane of separation AB is considered relative to the bulk of the crystal or to the bulk of the melt. If the plane AB is situated just at the interface, the magnitude of $|\varepsilon|$ is approximately equal but always greater than the energy of adhesion, the difference arising from the omission of the terms due to the re-arrangement of the liquid in the neighbourhood of the interface.

Fig. 3 presents the dependence of $|\varepsilon|$ erg/cm^2 as a

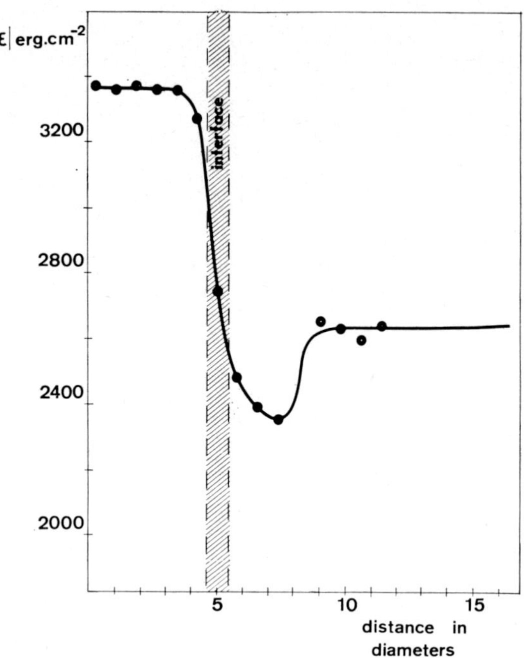

Fig. 4. Specific work of rupture as a function of the normal distance to the interface for a (0001) face containing randomly distributed two-dimensional clusters.

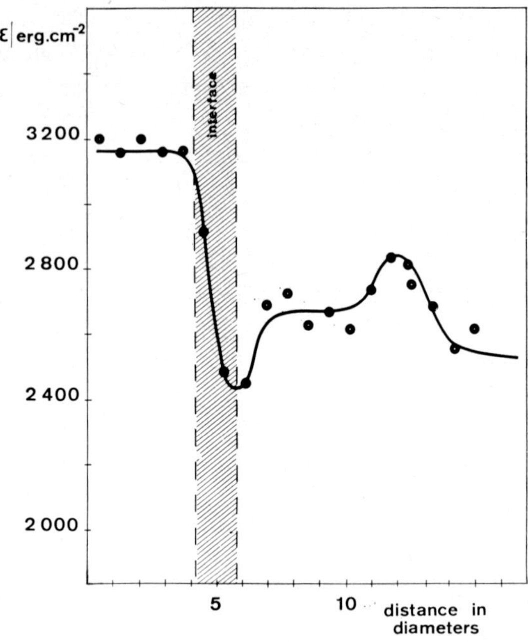

Fig. 5. Specific work of rupture as a function of the normal distance to the interface for a (0001) face including a running monoatomic step.

function of the position of the plane AB. The minimum in the curve at the interface indicates imperfect wetting of the crystal face by the liquid.

It is interesting that the model demonstrates in the case of smooth surface that many balls belonging to the "liquid" phase are situated in adsorption sites on the crystal face (potential wells). From the measured coordinates one finds that the number of such balls represents about 50% of the number of balls belonging to the lattice plane of the crystal. One half of these balls are on the ad-sites a (hcp) and the other half on position b (fcc). Thus the number of holes (point defects) at the interface is very important, it explains the pronounced minimum in $|\varepsilon|$.

Fig. 4 represents the results from another measurement of the model. The decrease in energy beyond the crystal surface extends up to 3 to 4 lattice distances into the liquid. This is due, in this case, to the presence of organised islands of small dimensions on the crystal surface. In view of the existance of the two types of adsorption sites for the atoms at the lattice plane, these small islands frequently cannot meet in a coherent manner, and thereby give rise to a number of linear faults at the close packed plane. This results again in a

zone of small density running parallel to the interface and hence in a minimum in $|\varepsilon|$.

If one identifies the minimum value of $|\varepsilon|$ with adhesion energy β and the value of $|\varepsilon|$ within the liquid with the energy of cohesion $2\sigma_L$, one can estimate the contact angle from the equation of Young–Dupré:

$$\beta/\sigma_L = 1 + \cos\alpha.$$

The values obtained from the curves in figs. 3 and 4 are $\alpha = 49°$ and $\alpha = 36°$ respectively. In spite of the simplicity of the experimental model it is interesting and probably significant that these two values are of the same order of magnitude as the measured contact angle between the melt and the (0001) plane of Cd crystal as quoted previously[10]).

A further experimental study was carried out in a similar way to the one just described but here the model of the (0001) face included a running monoatomic step. Fig. 5 shows the results of one of these measurements.

The maximum in $|\varepsilon|$, which appears within a distance of 4 to 5 ball diameters from the interface on the liquid side is consistently reproduced in all measurements. For a better understanding of its significance, one has to realise that the curve in fig. 5 is a result of two different effects. Firstly, it takes into account the value of $|\varepsilon|$ on two smooth interfaces displaced from one another by a lattice distance whose total area is equal to the base of the cylinder under consideration. Secondly, it contains the linear effect of the step itself on the distribution of spheres in the liquid. On substracting from the curve in fig. 5, the arithmetical mean value of $|\varepsilon|$ obtained from two curves of the type in fig. 3 displaced by a lattice distance, one gets a curve as shown in fig. 6. It is to be noticed that the presence of a

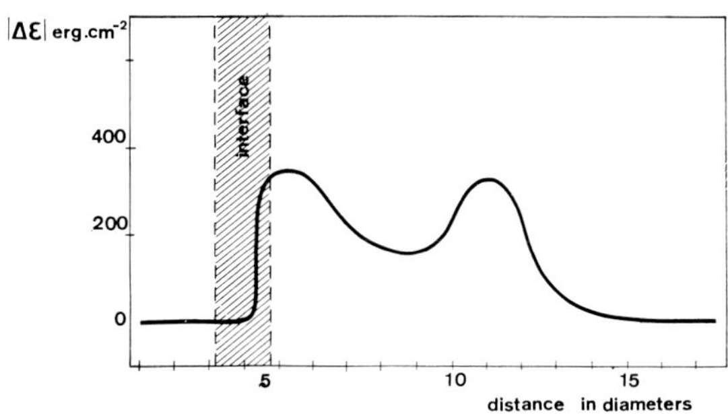

Fig. 6. Specific work of rupture (per unit length) of the liquid from a monoatomic step on the (0001) face.

step contributes positively to the adhesive energy and the cohesive energy of the liquid near the interface. This contribution is not confined to the nearest neighbour atoms to the step, but extends up to several interatomic distances.

This effect must have repercussions on the specific edge energy at the crystal-liquid interface. It appears to have small values which facilitates crystal growth by the formation of two-dimensional nuclei as suggested in various experimental works[17,18]).

If the crystal face in contact with the liquid is a low density plane (non-singular face) it may be identified with more than one system of closely spaced groups of steps. One could, thus, expect an increase of the order in the liquid phase extending to many atomic distances from the interface. In macroscopic terminology, this will mean a perfect wetting of the crystal and an extremely low interfacial energy, in agreement with the ideas originally advanced by Stranski[2]).

References

1) G. Tammann, Physik. Z. **11** (1910) 609.
2) I. N. Stranski, Naturwissenschaften **30** (1942) 425.
3) A. S. Skapski, Acta Met. **4** (1956) 576.
4) K. A. Jackson, *Liquid Metals and Solidification* (Am. Soc. Metals, Cleveland, Ohio, 1958) p. 174.
5) B. Mutaftschiev, in: *Adsorption et Croissance Cristalline* (C.N.R.S., Paris, 1965) p. 231.
6) D. E. Temkin, in: *Crystallization Processes* (Consultants Bureau, New York, 1966) p. 15.
7) D. E. Temkin, in: *Growth and Imperfections of Metallic Crystals* (Consultants Bureau, New York, 1968) p. 11.
8) H. Reiss, H. L. Frisch and J. L. Lebowitz, J. Chem. Phys. **31** (1959) 369.
9) R. J. Greet, as quoted by D. Turnbull, in: *Liquids, Symp. Gen. Mot. Res. Lab.* (Elsevier, Amsterdam, 1965) p. 6.
10) B. Mutaftschiev and J. Zell, Surface Sci. **12** (1968) 317.
11) B. Mutaftschiev, Bull. Soc. Franc. Minéral. Crist. **92** (1969) 558.
12) J. D. Bernal, in: *Liquids, Symp. Gen. Mot. Res. Lab.* (Elsevier, Amsterdam, 1965) p. 25.
13) J. L. Finney and J. D. Bernal, Nature (1967) 1079.
14) J. E. Lennard-Jones and A. F. Devonshire, Proc. Roy. Soc. (London) A **163** (1937) 53; **165** (1938) 1.
15) S. Glasstone, K. J. Laidler and H. Eyring, *The Theory of Rate Processes* (McGraw-Hill, New York, 1941) p. 477.
16) W. K. Burton, N. Cabrera and F. C. Frank, Phil. Trans. Roy. Soc. London A **243** (1951) 299.
17) J. W. Cahn, W. B. Hillig and G. W. Sears, Acta Met. **12** (1964) 1421.
18) G. Goujon, C. Chapon and B. Mutaftschiev, J. Crystal Growth **5** (1969) 317.

THE GROWTH OF SNOW CRYSTALS

R. LACMANN and I. N. STRANSKI

Fritz-Haber-Institut der Max-Planck-Gesellschaft, 1 Berlin 33, Germany

The growth and orientation of ice-dendrites (snow) is explained by means of a stable quasi-liquid film of H_2O-molecules on the ice surface, the unwettability of the edges at temperatures around $-15\,°C$, and the literature values of the evaporation coefficients of ice and water.

1. Introduction

The growth of ice from the vapour-phase is distinguished by the dendritic growth form which is called snow and which is formed under certain conditions. The interpretations existing up to the present explain it only by the influence of diffusion[1]. But the dendrite branchings have a constant spacing depending on the growth conditions, and the angles between the dendritic directions are always 60° respectively 120° (fig. 1).

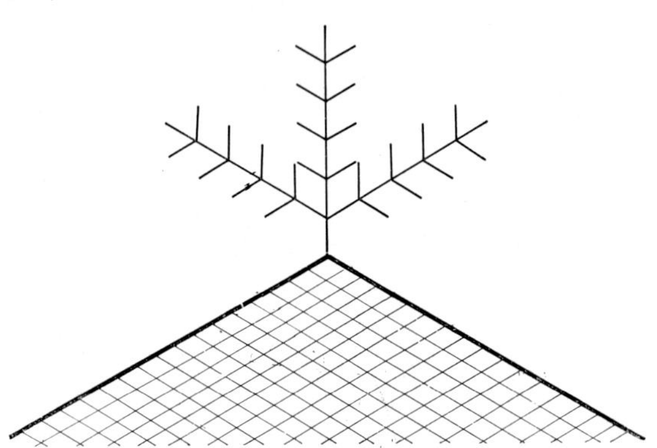

Fig. 1. A snow crystal (schematic).

Because this growth form is so characteristic and is not observed with another substance in the same way, it is logical to explain this not only by diffusion but also by the abnormal properties of water and ice[2].

Compared with this the orientations of dendrites of ionic crystals could be interpreted by means of the different free energies of formation of a two-dimen-

sional nucleus on the surface, at the edge and at the corner of a crystal[3]. That is not possible in the case of snow because the effects of electrostatical binding forces are too small, and the dendritic growth is observed only at temperatures below the melting point, optimally around $-15\,°C$.

2. The quasi-liquid film

In general the planes of the growth shape are not wetted completely by their own melt[4,5]. In the case of ice having a considerable volume contraction during the melting the planes of the growth shape are also wetted completely by water. This consideration is in agreement with other theoretical calculations[1] and with experimental tests, showing an adhesion between ice crystals down to a temperature of about $-25\,°C$[1,6]. Complete wettability means

$$\sigma_E > \sigma_W + \sigma_{E/W}$$

(σ = specific interfacial free energy, E = ice, W = water, E/W = interface ice/water). Then the specific free energy of the surface of an ice crystal (Δf) with a quasi-liquid film of thickness δ below 0 °C is given by (fig. 2)

$$\Delta f = \sigma_W + \sigma_{E/W} + \delta\,\frac{kT}{V_m}\,\ln\,(p_W/p_E)$$

$$= \sigma_W + \sigma_{E/W} + \frac{\delta}{V_m}\,\Delta\mu_{E/W}. \tag{1}$$

p_W, p_E are the equilibrium vapour pressures of water

and ice, p_W is greater than p_E below 0 °C, and equal to p_E at 0 °C; $\ln(p_W/p_E)$ and $\Delta\mu_{E/W}$ are functions of the supercooling $\Delta T = 0\,°C - T$, V_m is the molecular volume in the quasi liquid film. At $\delta = 0$, Δf becomes $\sigma_W + \sigma_{E/W}$. Then the surface has no quasi-liquid film and Δf should be equal to σ_E. The contradiction can be explained by

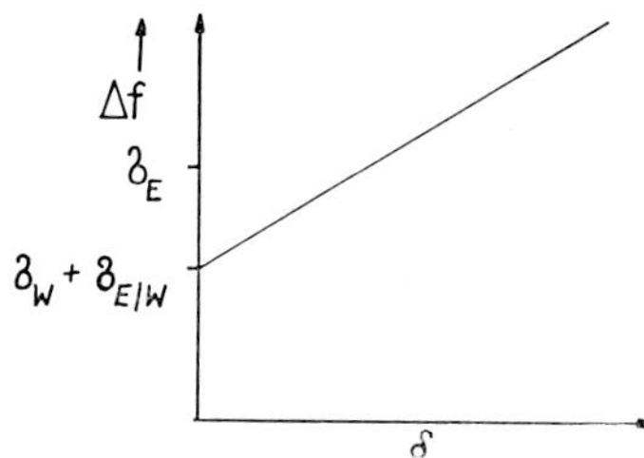

Fig. 2. The specific free energy of an ice surface (Δf) as a function of the thickness of the quasi-liquid film (δ) according to eq. (1).

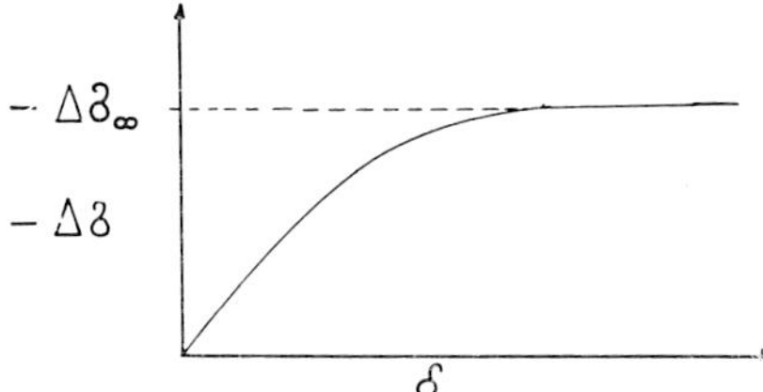

Fig. 3. $-\Delta\sigma = \sigma_E - \sigma_W - \sigma_{E/W}$ as a function of δ according to eq. (2).

the dependence of the σ-values on the film thickness (δ) which was not taken into account. We define $-\Delta\sigma = \sigma_E - \sigma_W - \sigma_{E/W}$. The quantity $\Delta\sigma(\delta)$ must be equal to zero at $\delta = 0$ and it must be constant for great δ-values ($-\Delta\sigma_\infty$). For instance we could assume

$$-\Delta\sigma(\delta) = \frac{\delta\,\Delta\sigma_\infty}{A + \delta},\tag{2}$$

(fig. 3) and we get

$$\Delta f(\delta) = \sigma_E - \frac{\delta\,\Delta\sigma_\infty}{A + \delta} + \frac{\delta}{V_m}\Delta\mu_{E/W},\tag{3}$$

as shown in fig. 4. A is a parameter of the radius of action of the intermolecular forces. We can get other relations for $\Delta\sigma(\delta)$ by integration of the isotherms of multilayer adsorption[7]).

The equilibrium film thickness δ_{G1} is given by the minimum of $\Delta f(\delta)$. We get

$$\delta_{G1} = -A + \left(\frac{A\,\Delta\sigma_\infty V_m}{\Delta\mu_{E/W}}\right)^{\frac{1}{2}}\tag{4}$$

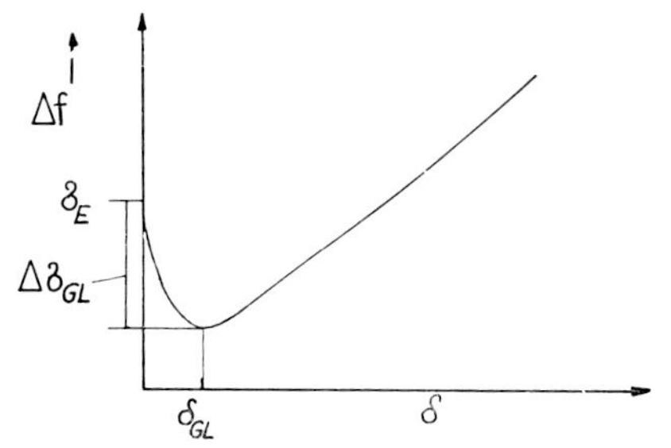

Fig. 4. Δf as a function of δ according to eq. (3).

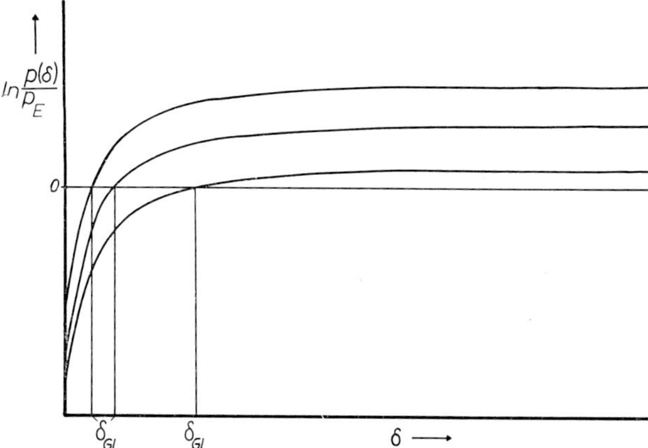

Fig. 5. $\ln(p(\delta)/p_E)$ as a function of δ according to eq. (5) for 3 different temperatures.

At $\Delta\mu_{E/W} = 0$, which means at 0 °C, δ_{G1} becomes equal to ∞, with decreasing temperature δ_{G1} decreases, too.

We get the equilibrium vapour pressure of the quasi-liquid film $[p(\delta)]$ from the differential quotient $\partial\Delta f(\delta)/\partial N$, where N is the number of molecules per unit area in the quasi-liquid film. Here, $p(\delta)$ is given by (fig. 5).

$$\ln\frac{p(\delta)}{p_E} = -\frac{V_m\,A\,\Delta\sigma_\infty}{kT\,(A + \delta)^2} + \ln\frac{p_W}{p_E}.\tag{5}$$

At $\delta = \infty$ for all temperatures $p(\infty)$ is equal to p_W. At $\delta = \delta_{G1}$, $p(\delta_{G1})$ is equal to p_E; indicating equilibrium between ice and the quasi-liquid film.

3. The wettability of the edges

The quasi-liquid film at the edges and at the corners is a special problem. If the film has the same thickness as in the middle of the plane (fig. 6), a curvature with the radius δ_{G1} will develop resulting in an increase of the equilibrium vapour pressure at the edges and corners (p'). It is given by the Thomson equation

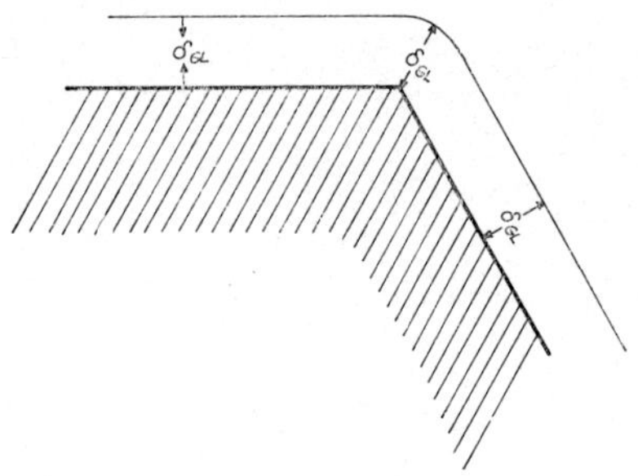

Fig. 6.   The quasi-liquid film at the edge (δ = const.).

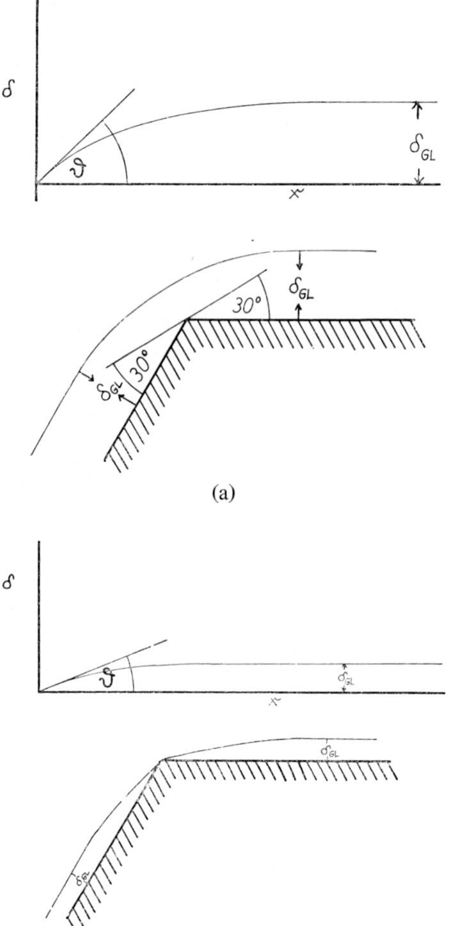

(a)

(b)

Fig. 7. δ as a function of the distance from the edge (x) according to eqs. (5)–(8) and the wettability of the edges. (a) Directly below the melting point. (b) In the region of $-15\ °C$.

$$\ln \frac{p'}{p(\delta_{\mathrm{GI}})} = \ln \frac{p'}{p_{\mathrm{E}}} = B\,\frac{V_{\mathrm{m}}\,\sigma_{\mathrm{W}}(\delta)}{kT\,\delta_{\mathrm{GI}}}. \tag{6}$$

B is equal to 2 for the corner but equal to 1 for the edge. In the latter case a curvature only exists in one direc-

tion. In the case of the snow it needs but to take into account the edge with $B = 1$.

The equilibrium vapour pressure will remain constant, if the film thickness $\delta(x)$ near the edges decreases (fig. 7). This is connected with a fall of the equilibrium vapour pressure, to compensate it, the surface must have a curvature with the radius r. We get the relation for equilibrium from eqs. (5) and (6).

$$-\frac{V_{\mathrm{m}}\,A\,\Delta\sigma_{\infty}}{kT\,(A+\delta(x))^2} + \ln\frac{p_{\mathrm{W}}}{p_{\mathrm{E}}} + \frac{V_{\mathrm{m}}\,\sigma_{\mathrm{W}}(\delta(x))}{kT\,r(x)} = 0. \tag{7}$$

Thus the film thickness (δ) and the radius of curvature (r) are functions of the distance from the edge (x). The differential equation, which is to be solved is given by

$$\frac{1}{r} = \frac{\partial^2\delta/\partial x^2}{[1+(\partial\delta/\partial x)^2]^{\frac{3}{2}}}, \tag{8}$$

and eq. (7)[8]. A complete solution is only numerically possible. The angle ϑ between the surface of the quasi-liquid film and the interface ice/quasi-liquid film is greater than $30°$ at the point $x = 0$ at temperatures directly below the melting point (fig. 7a). At lower temperatures the angle is less than $30°$ (fig. 7b). This means that in the case of fig. 7a the edge is wettable by its own melt; the edge is unwettable in the case of fig. 7b, as shown in the figure.

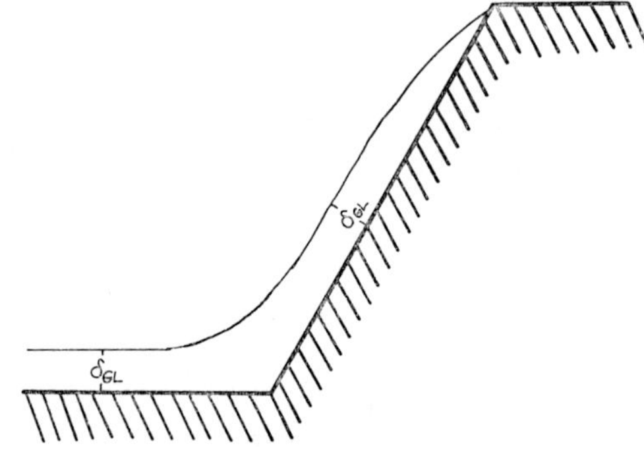

Fig. 8.   The quasi-liquid film at a concave edge.

On the other hand if δ is greater than δ_{GI}, the surface of the quasi-liquid film must show a concave curvature in order that $p(\delta) = p_{\mathrm{E}}$ (fig. 8). Then $\delta(x)$ and $r(x)$, for $r < 0$, can be determined in a concave edge also.

If the dimension of the surface (x_0) is small, then in the middle of the surface the thickness will be smaller than δ_{GI}, if $p(\delta)$ is also equal to p_{E} (fig. 9).

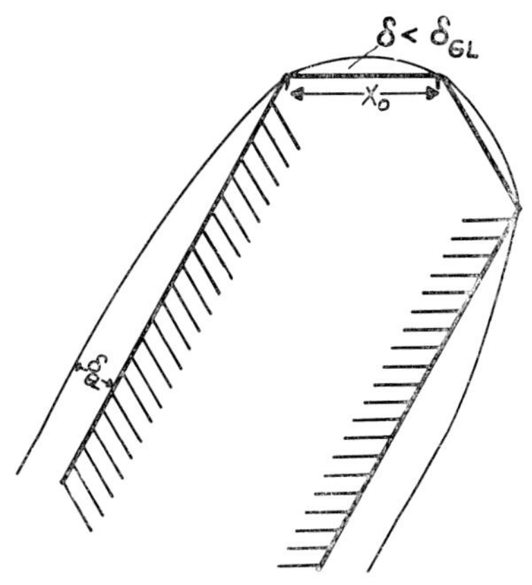

Fig. 9.   The quasi-liquid film on a small face.

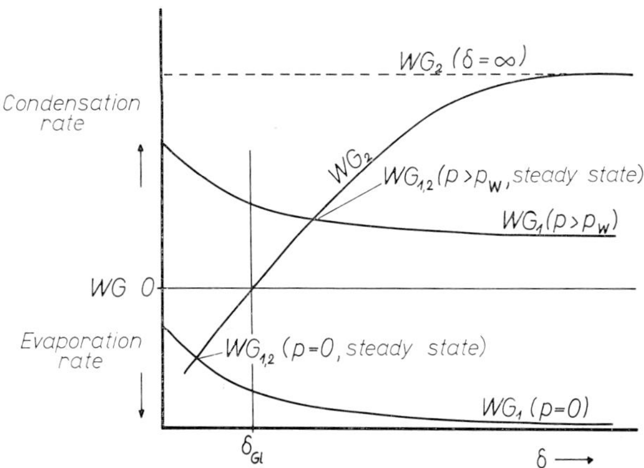

Fig. 10. WG_1 and WG_2 as a function of δ.

4. The rate of evaporation of ice and water

In accordance with the kinetic theory of gases the linear velocity of the interface between a condensed phase and a vapour phase is given by[9]:

$$WG = \frac{\alpha(p - p_{Gl})V_m}{(2\pi mkT)^{\frac{1}{2}}},\tag{9}$$

with p the pressure of the vapour phase and α the condensation or evaporation coefficient. In the case of evaporation ($p < p_{Gl}$) WG is less than zero and in the case of condensation ($p > p_{Gl}$) WG is greater than zero. We get the maximal linear velocity, if α is equal to 1. The following evaporation coefficients of ice and water have been measured for $p = 0$

Ice: $t < -40\ °C$, $\alpha \approx 1$[10,11];
Ice: $-13 \leqslant t \leqslant -2\ °C$, $\alpha = 0.014$[12];
Water: $t \approx 0\ °C$, $\alpha = 0.0415$[12];
Water: $t \approx 100\ °C$, $\alpha = 0.02$[12].

The small evaporation coefficient of water is a result of the hydrogen-bondings and the specific structure of water. This effect will not appear, if ice has no quasi-liquid film, and so α (ice) is equal to 1 at low temperatures. In the temperature region below 0 °C the evaporation and condensation coefficient of ice is determined by the phase transitions vapour $\rightleftharpoons$ (1) $\rightleftharpoons$ quasi-liquid film $\rightleftharpoons$ (2) $\rightleftharpoons$ ice. The phase transition (1) will be retarded similarly to that of water especially, if δ is large and the properties of the surface of the quasi-liquid film are identical with those of water. Fig. 10 shows the dependence of WG_1 on δ for $\alpha = $ const. and for $p = 0$ with $p_{Gl} = p(\delta)$ given by eq. (5). The phase transition (2) is dependent on δ and also dependent on

the supercooling ($\Delta T = 0°C - T$). At $\delta = \delta_{Gl}(\Delta T)$, WG_2 must be equal to zero, because the thickness of the film does not change. For $\delta < \delta_{Gl}$, WG_2 will be less than zero, this is the case of evaporation. If δ is greater than δ_{Gl}, WG_2 must be greater than zero, and at large film thickness the linear velocity [$WG_2(\delta = \infty)$] will be the growth velocity of ice from the melt (water) increasing with ΔT. For instance, the following growth velocities from water have been measured[13,1]:

$$WG_2(\delta = \infty) = 1.67 \exp(-0.35/\Delta T)\ (\text{cm/sec}),$$

$$WG_2(\delta = \infty) = 0.16\ \Delta T^{1.7}\ (\text{cm/sec}).$$

The first relation shows that the growth is retarded by the formation of two-dimensional nuclei. Fig. 10 schematically illustrates the dependence of WG_2 on δ. In the steady state WG_1 and WG_2 must have the same value [$WG_{1,2}\ (p = 0)$]. Because both processes are retarded, it is understandable that the effective evaporation coefficient of ice between $-13\ °C$ and $-2\ °C$ (0.014) is smaller than that of water near 0 °C (0.0415) as shown in fig. 10.

5. The interpretation of the growth of snow crystals

The growth shape of ice from the vapour should be bounded by $\{0001\}$- and $\{10\bar{1}0\}$-faces, if the supersaturation is not too low. These are the faces of the equilibrium shape if we only take into account the interaction between first nearest neighbours[14]. The snow dendrites occur at the edges between two $\{10\bar{1}0\}$-faces. The H_2O-molecules of the vapour-phase can be directly built into the crystal at the edges only. The formation of two-dimensional nuclei is not necessary,

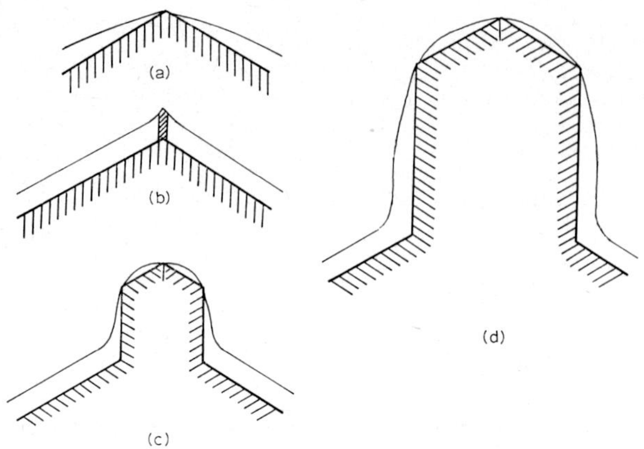

Fig. 11.   The growth of a snow crystal.

because the striking molecules are bound not only to the "ice"-molecules of the edge, but also to the "quasi-liquid" molecules of the film. This is not the case at temperatures near below 0 °C (cf. fig. 7a). There the edges are being wetted by the quasi-liquid film. This explains the fact why snow crystals are formed only at lower temperatures, particularly around -15 °C. The quantity δ must be greater than δ_{GI} for the growth at the faces, because then WG_2 is greater than zero (fig. 10). Therefore the equilibrium vapour pressure of the quasi-liquid film is greater than p_E [eq. (5)] and the super-saturation of the vapour phase with reference to the quasi-liquid film is smaller than that with reference to the ice-phase.

Fig. 10 also shows a graph of WG_1 against δ for $p > p_W$. In the steady state $WG_1 = WG_2 = WG_{1,2}$ (fig. 10) with*

$$WG_1(\delta > \delta_{GI}) = \frac{\alpha[p - p(\delta > \delta_{GI})]V_m}{(2\pi mkT)^{\frac{1}{2}}}. \tag{10}$$

We must take a value of $0.02 - 0.04$ for α, like that of

---

* If $p > p_W$ it should be possible in principle, that WG_1 is greater than WG_2, and the quasi-liquid film must become thicker and thicker and its surface will finally be identical with that of the water. But the calculations show, that this can be only the case for small supercoolings and high supersaturations of the vapour-phase. Under these conditions the edges will be wetted by the quasi-liquid film.

water and it is $p - p(\delta > \delta_{GI}) < p - p(\delta_{GI}) = p - p_E$. On the other hand, α is about 1 and the effective super-saturation is equal to $p - p_E$ at the edges. It appeares also that the growth velocity is essentially greater at the edges than that at the faces, so that the dendrites can originate.

If in course of the growth a dendrite has reached a certain length and width, three new edges will originate and new dendrite branchings can grow at these edges in the same way. Fig. 11 shows the schematic disappearance of growth in course of time.

Acknowledgement

Support of this work by the Deutsche Forschungsgemeinschaft is gratefully acknowledged.

References

1) N. H. Fletcher, *The Chemical Physics of Ice* Cambridge, Univ. Press, 1970).

2) W. Luck, Fortschr. Chem. Forsch. **4** (1964) 653;
D. Eisenberg and W. Kauzmann, *The Structure and Properties of Water* (Clarendon Press, Oxford, 1969).

3) M. Hille, H. Rau and J. Schlipf, Z. Elektrochem., Ber. Bunsenges. Physik. Chem. **63** (1959) 285; in: *Growth and Perfection of Crystals*, Eds. R. H. Doremus, B. W. Roberts and D. Turnbull (Wiley, New York) 1958, p. 325.
M. Hille and I. N. Stranski, Z. Elektrochem., Ber. Bunsenges. Physik. Chem. **65** (1961) 789;
J. Schlipf, Fortschr. Mineral. **38** (1960) 74.

4) M. Volmer and O. Schmidt, Z. Physik. Chem. B **35** (1937) 467.

5) I. N. Stranski, Naturwissenschaften **30** (1942) 425; Z. Physik **119** (1942) 22.

6) H. H. G. Jellinek, J. Appl. Physics **32** (1961) 1793.

7) R. Lacmann, Ber. Bunsenges. Physik. Chem., in preparation.

8) B. Baule, *Die Mathematik des Naturforschers und Ingenieurs*, Vol. 1 (Hirzel, Leipzig, 1953).

9) M. Volmer, *Kinetik der Phasenbildung* (Steinkopff, Leipzig, Dresden, 1939).

10) K. Tschudin, Helv. Phys. Acta **19** (1946) 91.

11) H. Kramers and S. Stemering, Appl. Sci. Res. **3** (1953) 73.

12) L. V. Delaney, R. W. Houston and L. C. Eagelton, in: *Condensation and Evaporation of Solids*, Eds. E. Rutner, P. Goldfinger and J. P. Hirth (Gordon and Breach, New York, 1962) p. 683.

13) W. B. Hillig, in: *Growth and Perfection of Crystals*, Eds. R. H. Doremus, B. W. Roberts and D. Turnbull (Wiley, New York, 1958) p. 350.

14) L. Krastanow, Meteorolog. Z. (1943) 15.

Journal of Crystal Growth **13/14** (1972) 291–243 © *North-Holland Publishing Co.*

PRELIMINARY INVESTIGATION INTO THE GROWTH OF ICE CRYSTALS FROM THE VAPOUR IN AN ELECTRIC FIELD IN THE TEMPERATURE RANGE −11 to −15 °C

A. G. CROWTHER

Physics Department, University of Manchester Institute of Science and Technology, Manchester, England

The growth of ice crystals from the vapour in a continuous diffusion cloud chamber is found to be profoundly influenced by the presence of an electric field. If a certain threshold value of electric field is exceeded growth becomes more rapid and needle-like crystals are formed. These crystals are seen to grow along lines of electric flux with growth velocities up to a hundred times those occurring normally. The author discusses the observed phenomena and argues that the effects occur in the vapour phase and are a consequence of the non-uniformity of the electric field created by the geometry of the growing ice crystal. The problem is shown to be explicable in terms of boundary layers at the ice/vapour interface but analytical solutions lie outside the scope of the paper.

1. Introduction

Experiments have been undertaken in the Laboratory to extend the research of Bartlett, Van den Heuvel and Mason[1]) into the growth of ice crystals from the vapour in an electric field. Work was restricted to the temperature region −11 to −15 °C where the accelerated growth phenomenon reported by previous workers is most pronounced.

2. Results

The diffusion cloud chamber (fig. 1) [Mason[2])]: Normal ice crystal growth within a diffusion cloud chamber between −11 °C and −15 °C is in the form of dendritic structures with higher growth rates than exist in any other temperature region above −25 °C. However, in the presence of an electric field the crystals growing within this range of temperature forfeit their dendritic character for even higher longitudinal growth rates as ice needles. An average linear growth rate of 0.27×10^{-6} m/sec has been measured for normal growth at −12 °C and this has been increased to 15.3×10^{-6} m/sec by the application of a field of 160×10^3 V/m. These average growth rates are shown in fig. 2. The needles follow closely the electric lines of force and appear to only deviate from their paths if a small crystal fragment or droplet is captured, thereby creating a structural discontinuity.

Electric growth appears to override normal growth provided the electric field exceeds 46×10^3 V/m.

Below this value there is no visible indication of disturbance from normal growth. This indicates that a threshold value may exist below which electric growth is impossible. However, also associated with these low electric fields is a time-lag between field application and

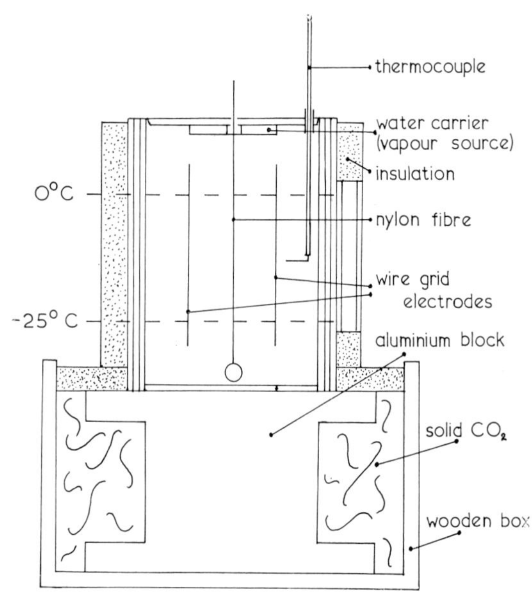

Fig. 1. The diffusion cloud chamber.

commencement of rapid growth. This lag is linked to the field magnitude and the tendency is for increasing time-lag with decreasing field, as shown in fig. 3. Bartlett et al suggest, probably correctly, this is due to the dependence of electric growth on the local field at the growing tip of the ice crystal. Thus, only when a certain geometric configuration has been obtained can electric growth commence and this will be a more exacting condition when the applied field is low.

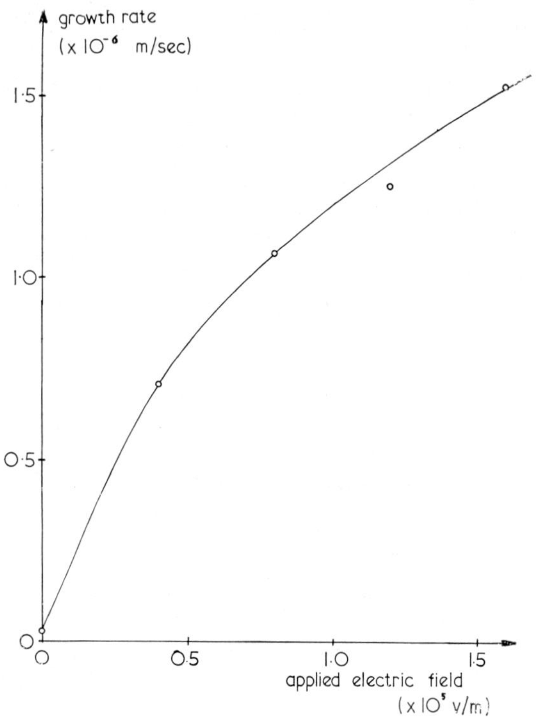

Fig. 2. Average growth rate of ice crystals in an electric field.

The habit of electric crystals tends to be a direct extension of normal growth at -12 °C. The growth direction of 85% of the crystals examined was parallel to the basal plane (1120); the remainder being parallel to the optic axis (0001).

Electric growth of ice crystals is independent of field polarity and any breaking fragments always travel to the nearest electrode, indicating that most of their charge is probably induced in the electric field and is not simply self-charge. These breakages frequently occur during electric growth, especially at high field strengths, and may be due to the accumulation of defects in a small volume of the crystal.

3. Discussion

The observations show that electric growth demands the presence of small radii tips on the affected crystals. The electric crystal has a form typical of normal

growth at high supersaturation and leads to the tentative suggestion that the electric field enhances this quantity.

The nature of the ice crystal lattice is such that its units are hydrogen bonded and the structure is readily polarised. Thus the surface entropy is changed by the proton-jump mechanism in an electric field. If this

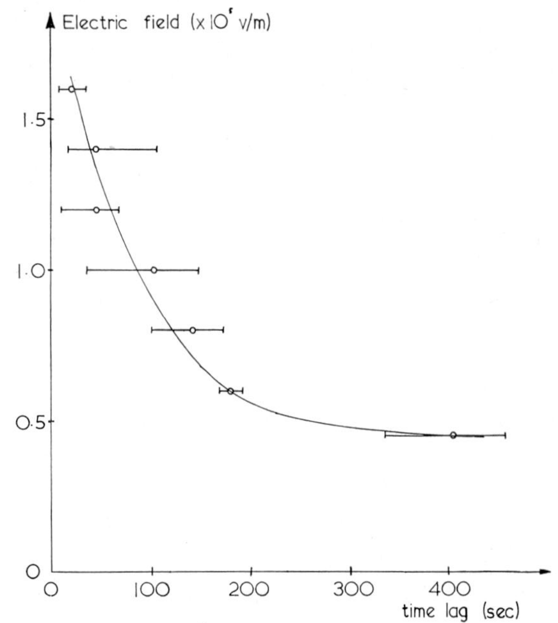

Fig. 3. Time lag between instant of field application and initial appearance of electric crystal.

field is applied in, say, a direction parallel to the c-axis the (0001) face will become polar and the surface energy increased. The crystal would be expected to grow in this direction and the equilibrium form would be a prismatic needle.

To explain the increased growth-rate, consideration of possible processes occurring in the vapour phase must be made. Two effects are plausible: (1) The enhanced electric field close to the crystal edges has an orientative effect upon the dipoles or dipole clusters leading to their being more readily adsorbed. (2) The non-uniform nature of the field at the crystal tip drives molecules along its gradient; the surface in the tip region being the most probable area for deposition to occur.

The first of these effects raises two problems, both of which require further study. One is concerned with the orientation of dipoles in the vapour and the other with the formation and lifetime of molecular clusters. Orientation of the dipoles is, in the case of electric growth of ice crystals, unlikely to be a significant process since it would not account for similarity of growth in fields of opposite polarity. Nor could it explain continuation of growth on field reversal. Bartlett et al reject this phenomenon on the grounds that the energy available to the dipoles from the applied field is very small compared with their own thermal energy and consequently orientation will not occur.

The clustering molecules, however, may be of far greater importance than has been suggested in the literature. Dorsey[3] claims that below $0°$ C less than 0.04% of vapour molecules are double. Nevertheless, it is apparent that ice crystals grow not by the addition of single molecules to their lattices, but by the addition of molecular groups. It is imagined that this clustering will be produced by a mechanism occurring at the vapour/ice interface where molecules will collect prior to their installation into the crystal lattice. At such an interface the molecules will lose one or more degrees of freedom and possess therefore less total energy than in the vapour, whereupon hydrogen bonding may become energetically viable. The accelerated growth within the electric field could then be caused by one of two effects. The mechanism creating clusters may become more favourable or the flux of vapour to the clustering region may be increased. Of these, the latter is most probably dominant.

The increase of the vapour flux to the growing crystal seems to be due to the second of the effects given above; the motion of unchanged molecules in the non-uniform field. This non-uniformity will exist only a short distance into the vapour thereby creating a small volume around the crystal depleted of water molecules. The replenishing of this volume will then control the growth rate of the crystal, providing the crystal geometry remains constant. Thus the growth rate will be a function of the vapour diffusion coefficient and for a particular temperature and supersaturation the growth rate/applied field relationship would be expected to "saturate" (see fig. 2).

Consequently, a situation will develop around a growing ice crystals tip wherein three regions will be manifest. In the immediate vicinity will be a layer of adsorbed molecules awaiting inclusion into the lattice or re-evaporation. Ajoining this will be the region of influence of the non-uniform field and, finally, there will be the normal vapour. The "non-uniform field" region will be characterised by the increased mobility it gives to the vapour molecules but this property will be imparted to any molecule which may be approximated to a dipole. This could lead to the inclusion of alien molecultes in the ice lattice producing a mechanically weaker crystal than is normal. The general polarisation of the crystal together with the movement of L- and D-defects under the influence of the field will produce the charge separation witnessed by observation of crystal fragments travelling to the nearest electrode.

More consideration of the theory concerning this phenomenon is required together with extended experimentation into other ice crystal growth situations. Specifically the growth of crystals in free fall may give particular insights into the physics of the process.

References

1) J. T. Bartlett, A. P. Van den Heuvel and B. J. Mason, J. Appl. Math. Phys. **14** (1963) 599.
2) B. J. Mason, Advan. Phys. **7** (1958) 235.
3) N. E. Dorsey, *Properties of Ordinary Water-Substance* (Reinhold, New York, 1940).

Section VI

Characterization and assessment of crystal perfection

A. FREUND
J. SCHNEIDER
C. G'SELL
B. BAUDELET
G. CHAMPIER
H. FEHMER
W. UELHOFF
C. C. WANG
S. H. McFARLANE III
R. F. BELT
R. C. PUTTBACH
D. A. LEPORE
F. LEFAUCHEUX

R. LECKEBUSCH
K. RECKER
D. E. COX
F. C. MERKERT
I. T. STEINBERGER
E. ALEXANDER
Y. BRADA
Z. H. KALMAN
I. KIFLAWI
S. MARDIX
T. OGAWA
M. UCHIDA

Journal of Crystal Growth **13/14** (1972) 247–251 © *North-Holland Publishing Co.*

TWO NEW EXPERIMENTAL DIFFRACTION METHODS FOR A PRECISE MEASUREMENT OF CRYSTAL PERFECTION*

A. FREUND and J. SCHNEIDER

Institut Max von Laue–Paul Langevin, Avenue des Martyrs, Cedex 156, 38 – Grenoble, France

Two instruments have been developed for the independent measurement of lattice strains, tilts and changes in the lattice constant using diffraction of X-rays at very high (89.5°) and of γ-rays at very low (0.5°) Bragg angles.

The X-ray instrument consists of a double crystal diffractometer on which two crystals are arranged in the $n; - n$-position thus eliminating the high wavelength dispersion occurring at high Bragg angles. Wavelengths from the continuous spectrum are used. By measuring both broadening and position of the reflection curve, lattice strains and changes in the lattice constant can be determined at the same time with an accuracy of 10^{-6} and 10^{-7} respectively which is limited only by the dynamical diffraction properties of the crystal.

The γ-ray instrument represents a single crystal diffractometer. The 412 keV γ-radiation of a gold source $(\lambda = 3.02 \times 10^{-2}\,\text{Å}\ \Delta\lambda/\lambda = 10^{-6})$ is collimated into a beam of a cross section of $0.2 \times 10\,\text{mm}^2$ and an angular divergence of 10 seconds of arc. Because of the low Bragg angle only lattice tilts contribute to the broadening of the rocking curve. Theoretical linewidths calculated for a perfect crystal are of the order of magnitude of only 0.1 seconds. Consequently the sensitivity of detecting lattice defects is very high. However, to determine lattice tilts of less than 10″ the integrated reflection power has to be used. A further advantage of the short wavelength employed is the weak absorption. Therefore, a larger volume will be seen by the γ-rays and thick (up to 20 mm) crystal samples can be studied without cutting and surface preparation thus allowing a rapid control of crystal perfection, perhaps even during the process of crystal growth. Some typical reflection curves are given for both instruments.

1. Introduction

In order to develop techniques for the growth and preparation of single crystals with a controlled perfection one has to find methods for the detection of crystalline defects. In many cases this leads to the production of perfect or nearly perfect crystals. There is further interest in the lattice tilts and strains, and the change in the lattice constant caused by different kinds of defects. Measurements of absolute integrated intensities anomalously diffracted, rocking curves, and Bragg intensity measurements have shown that crystals with dislocation densities of less than $10^4/\text{cm}^2$, vacancy cluster densities of less than $4 \times 10^5/\text{cm}^3$, and crystals irradiated with 10^7 fast neutrons/cm^2 exhibit X-ray diffraction properties equivalent to perfect crystals for which the dynamical theory of X-ray diffraction is valid. For the observation of individual dislocations, stacking faults and impurity clusters there have been employed various methods e.g. the etch pit technique, electron microscopy and X-ray topography. However, these techniques need a careful and lengthy preparation of

the samples. Therefore, it seemed to be useful to develop methods allowing a rapid control of lattice perfection with a high sensitivity at the same time.

2. A useful characterization of imperfect crystals

In order to see the influence of imperfections on the diffraction properties of X-rays it is of advantage to consider the reciprocal space. Let τ be a reciprocal lattice vector with $|\tau| = 1/d$, $d = $ lattice spacing. The existence of imperfections produces both a tilt and a strain of the real lattice. This can be described by a mean misorientation angle η and a mean variation $(\Delta\tau/\tau)_S$ of the length of the vector τ.

In the following it will be shown that when doing diffraction experiments on even perfect crystals, their reciprocal lattice points seem not to be well-defined. This effect can be represented by an artificial variation $(\Delta\tau/\tau)_d$ of the length of the reciprocal lattice vectors.

From the dynamical theory of X-ray diffraction it follows that for a perfect crystal the width $\Delta\theta$ of the Bragg peak is finite and given by

$$\Delta\theta = \frac{2r_e\lambda^2 F_H}{\pi V \sin 2\theta_B}, \tag{1}$$

---

* Dedicated to Professor Dr. H. Maier–Leibnitz, Munich–Grenoble, on the occasion of his 60th birthday.

where r_e, λ, F_H, θ_B, and V have their usual meanings. From the differentiation of Bragg's law we get for monochromatic radiation and a perfect crystal ($\eta = 0$):

$$\Delta\theta = \frac{\Delta\tau}{\tau} \tan \theta_B.$$

Combinings eqs. (1) and (2) we can define

$$\left(\frac{\Delta\tau}{\tau}\right)_d = \frac{4}{\pi} r_e \frac{d^2 F_H}{V}. \tag{3}$$

The quantity $(\Delta\tau/\tau)_d$ depends only on crystallographic constants.

For an imperfect crystal there will be the following relation between the measured halfwidth $\Delta\theta$, the lattice tilt η and the strain $(\Delta\tau/\tau)_S$, which are assumed to have Gaussian distributions:

$$(\Delta\theta)^2 = \left[\left(\frac{\Delta\tau}{\tau}\right)_d^2 + \left(\frac{\Delta\tau}{\tau}\right)_S^2\right] \tan^2 \theta_B + \eta^2. \tag{4}$$

For low Bragg angles only η will give rise to a line broadening, and for high values of θ_B, η becomes negligible in comparison to the first term on the right hand side of eq. (4). The resolution of measuring both lattice strains and tilts will be theoretically limited only by $(\Delta\tau/\tau)_d$.

3. A backscattering instrument for a precise measurement of lattice strains and relative changes in the lattice constant

The instrument presented was first proposed by Bottom[1]), then adapted to neutrons by Alefeld[2]) and to X-rays by Sykora[3]). It represents a considerable improvement of the apparatus of Sykora which was possible by using modern X-ray tubes with a narrow line focus of 8×0.4 mm^2 and a power of 1.2 kW. A double crystal diffractometer was constructed that allows diffraction at Bragg angles from $3°$ up to $89.5°$ ($\tan \theta_B = 115$) and even higher values. For $\theta_B \approx 90° \to \lambda \approx 2d$.

In general there do not exist characteristic wavelengths which correspond to a given d. Therefore, one has to select wavelengths out of the continuous spectrum. In order not to get too long wavelengths (high absorption) and to obtain a good resolution [low value of $(\Delta\tau/\tau)_d$] higher order reflections are preferable. However, with increasing reflection order the intensity decreases.

A schematic setup of the instrument is shown in fig. 1.

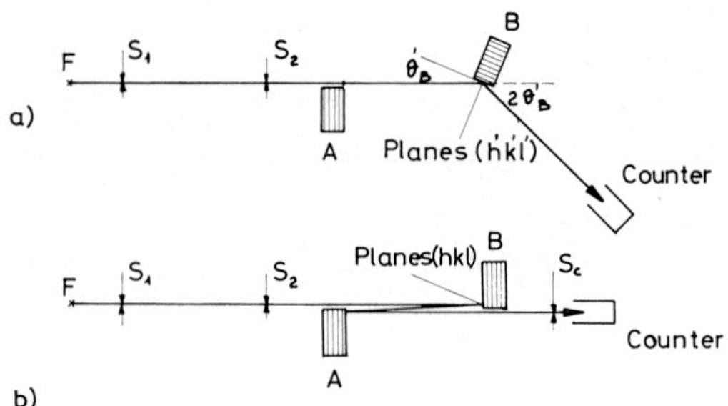

Fig. 1.   Schematic setup of the backscattering instrument. (a) in position for adjustment; (b) in measuring position.

The X-ray beam emerging from the source F is collimated by two slits S_1 and S_2 of variable widths. The vertical divergence is limited by a Soller collimator. The $n;$-n-dispersion-free arrangement of two crystals A (nearly perfect monochromator) and B (sample) with identical or almost identical lattice spacings is used.

In order to find the exact angular position of the sample B corresponding to a Bragg angle of e.g. 89.5° the Bragg peak of another set of reflecting planes ($h'k'l'$)

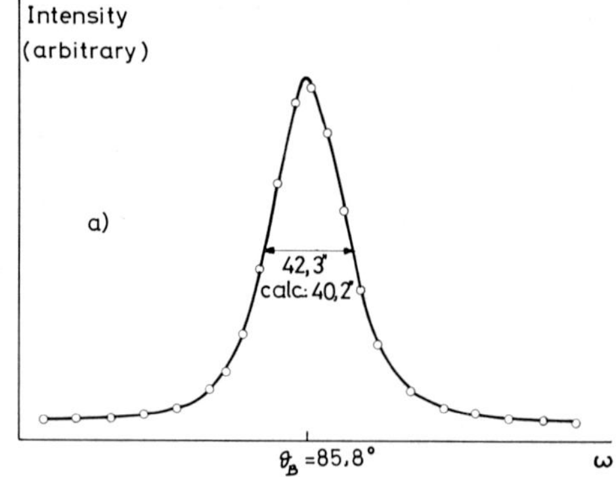

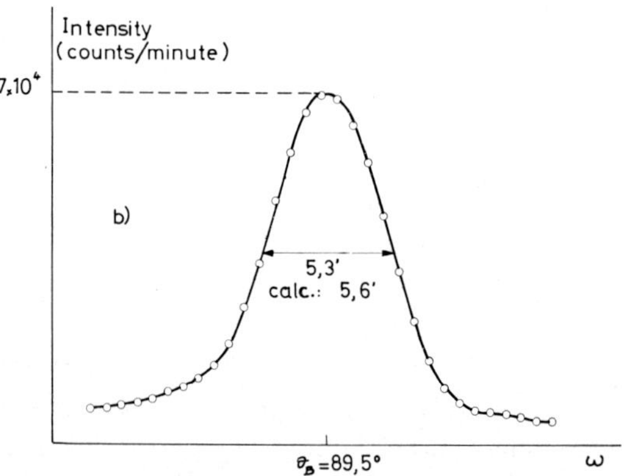

Fig. 2.   Reflection curves obtained with two perfect silicon crystals. (a) at a Bragg angle of 85.8°; (b) at a Bragg angle of 89.5°.

making a known angle with the planes (*hkl*) to be investigated serves as calibration of the angular reading using a characteristic wavelength and a well collimated primary beam (fig. 1a). Then both crystals are set in the backscattering position (fig. 1b) and crystal A will analyse the radiation scattered by the sample performing a ω-scan or a so-called τ-scan by heating the crystal A. The sample can be moved within all translational

The measured values of $\Delta\theta$ corresponding to Bragg angles of 85.8° and 89.5° agreed almost exactly (5%) with the theoretical values calculated for a perfect crystal, (figs. 2a and 2b). Because of the high intensity obtained it is possible to choose still higher order reflections and in this way to improve the resolution which is estimated to be 10^{-6} for $(\Delta\tau/\tau)_s$ and 10^{-7} for the change in the lattice constant.

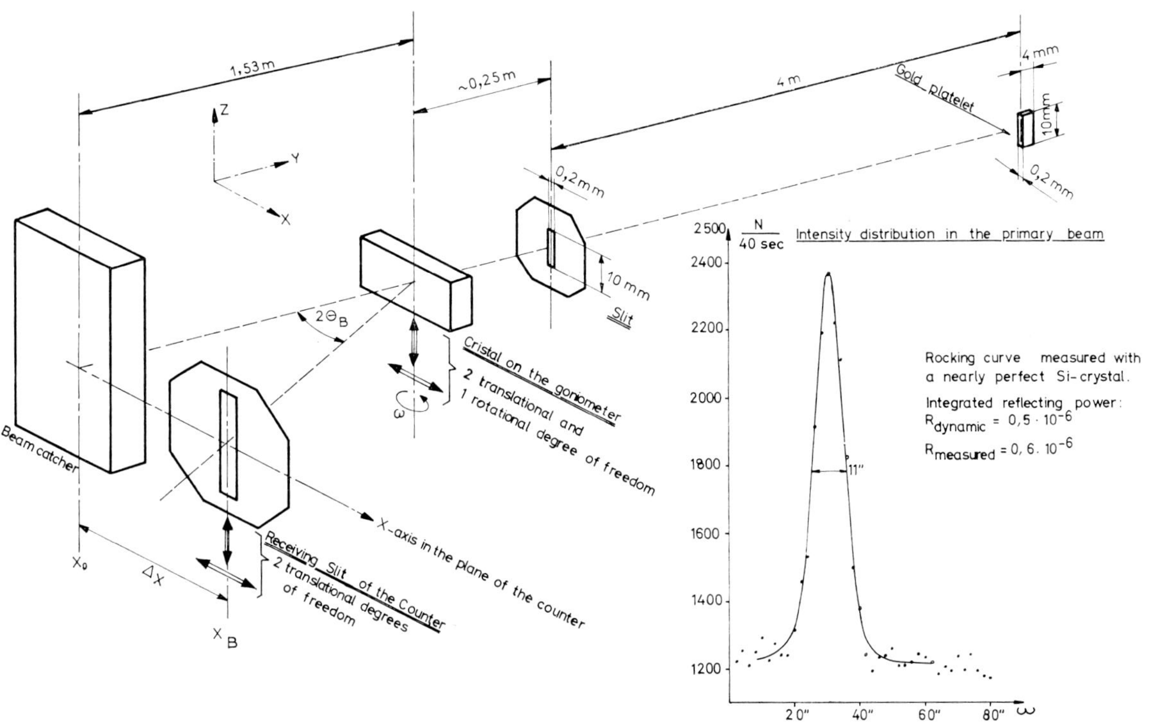

Fig. 3.   Schematic setup of the γ-diffractometer.

and rotational degrees of freedom thus allowing both a fine adjustment of the crystals and a choice of a well defined domain on the surface of the sample to be studied. A temperature control of the crystal holders is provided which works with an accuracy of ± 0.01 °C. Strains and changes in the lattice constant will be obtained at the same time from the broadening and the position of the Bragg peak respectively.

For the test of the instrument two perfect silicon single crystals were mounted on the two axes (distance: 50 mm) of the diffractometer. The 333-reflection $(d = 1.04$ Å$, (\Delta\tau/\tau)_d = 10^{-5})$ was utilized with $\lambda = 2$ Å selected from the spectrum of a Cr-tube operated at 30 kV, 28 mA. After two reflections the peak intensity was 7×10^4 counts per minute (scintillation counter).

4. Gamma-diffractometer

From the characterization of an imperfect crystal given above one can deduce that only lattice tilts contribute to the broadening of the rocking-curve, if the Bragg-angles are very small. This condition could be fulfilled by a single crystal diffractometer working with the 0.03 Å γ-radiation of radioactive gold Au[198] (half-life: 2.7 d).

The Bragg-angles are of the order of magnitude of 30′ and therefore the samples are investigated in the Laue-case. A further important advantage of the short wavelength employed is the weak absorption of the radiation, so that thick crystal samples (≈ 20 mm) can be studied without surface preparation.

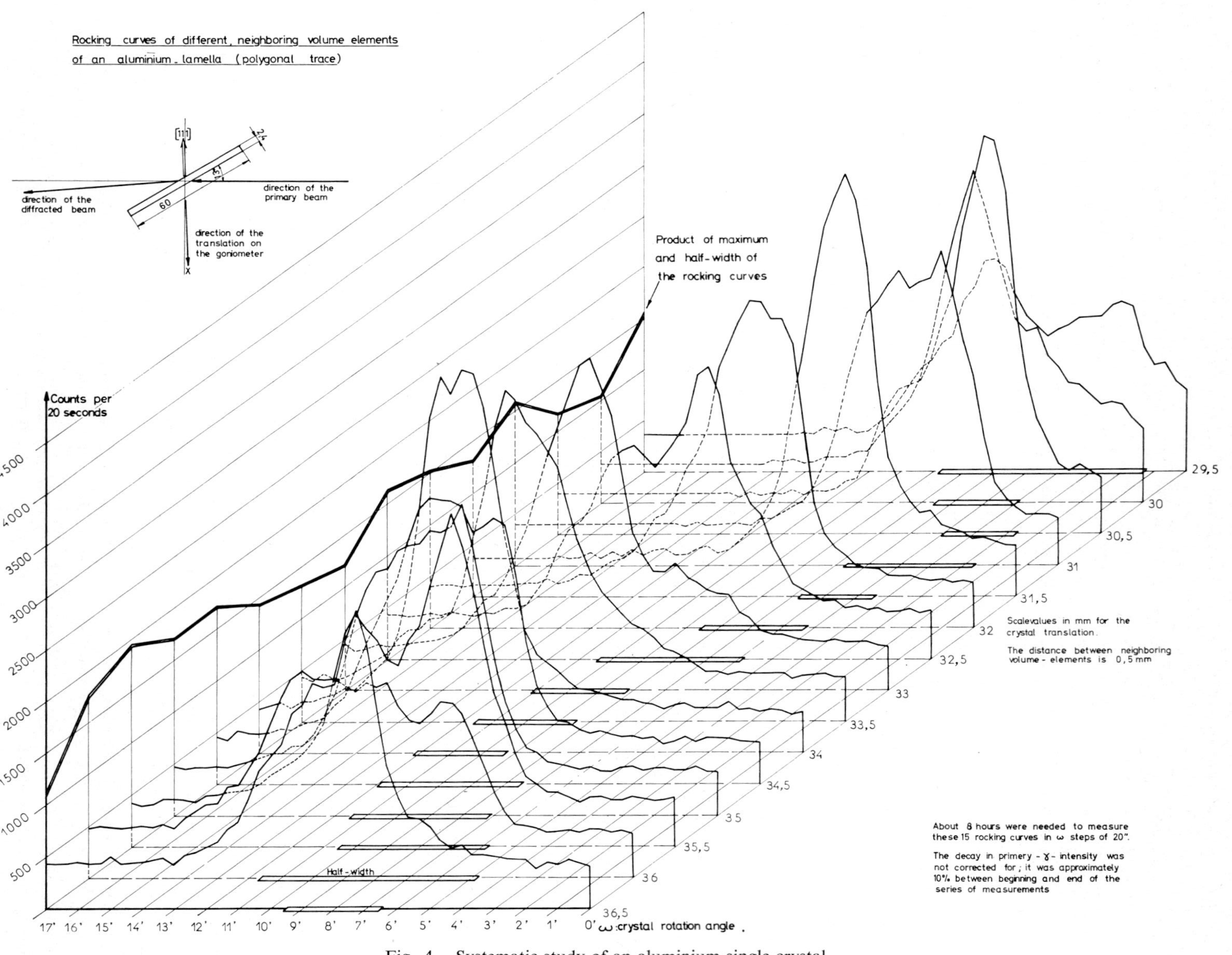

Fig. 4. Systematic study of an aluminium single crystal.

5. Description of the instrument

The gamma-ray diffractometer is composed of a γ-source, 2 slits, a goniometer, a beam catcher and a NaI(Tl) scintillation counter. These parts are mounted on a I-shaped support 6 m in length (see fig. 3).

The total activity of the source is about 60 curies. The slit selects out a narrow beam with following properties: $\lambda = 3.02 \times 10^{-2}$ Å; $\Delta\lambda/\lambda \approx 10^{-6}$; angular divergence $\approx 10''$; beam cross-section behind the slit: width $\Delta x = 0.2$ mm, height $\Delta z = 10$ mm. The primary intensity at the beginning of the measurement was: $P_0 \approx 6500$ counts/sec. Depending on the crystal perfection we can work with one γ-source up to a week. To measure the rocking curve we bring the counter, with its receiving slit large enough to accept all of the scattered radiation, in the position x_B. This corresponds to an angle $2\theta_B$ with respect to the direction of the primary beam, where θ_B is the Bragg-angle. With the goniometer we rotate the crystal, in steps chosen between $2''$ and $30''$, through the range of angles where Bragg-reflection takes place. On the goniometer there is a possibility for translation of the crystal in the x-direction in order to examine successively neighboring volume elements, 0.2 mm in width.

6. Theory

The half width of the diffraction pattern given by the dynamical theory is proportional to λ, the average scattering cross-section per unit volume of the crystal, Q, is proportional to λ^2 for small Bragg-angles, and so the application of Zachariasen's[4]) general diffraction theory may be applied up to relatively large crystals with a high degree of perfection. For the moment this theory is used to calculate a minimum value of the mosaic spread for which the kinematical theory is still applicable, and to estimate from the measured integrated reflecting power the half-width of the rocking curve of the sample, even if it is much smaller than the angular resolution of the instrument.

7. Applications

Example 1: The integrated reflecting power of a brick-shaped Si-crystal ($\Delta x = 5$ mm, $\Delta y = 10$ mm, $\Delta z = 13$ mm) was measured, the resultant value was about 20% higher than that predicted by dynamical theory (0.5×10^{-6}). A distortion of the crystal at the surface due to cutting was assumed to be one of the reasons for this deviation. This could be verified qualitatively. Using the translation possibility on the goniometer we showed the depth of the damaged region due to cutting (diamond saw) to be about 0.15 mm, the measured integrated reflecting power increased up to 3 times the dynamical value.

Example 2: Often the diffraction of neutrons in monochromator crystals is described by Darwin's mosaic model, neglecting primary extinction. The crystals have a mosaic spread much wider than $10''$ and their rocking curve, measured with the gamma-diffractometer, is proportional to the distribution function $W(\Delta)$ which describes the orientation of the blocks. Because of the good angular and spatial resolution of the Gamma-diffractometer detailed information about the homogeneity of the crystal structure can be obtained (see fig. 4). Furthermore it is possible to calculate the exact diffraction pattern for neutrons directly from $W(\Delta)$ and not from any fitted mathematical distribution function.

Acknowledgements

It is a great pleasure for the authors to express their thanks to Professor Dr. H. Maier-Leibnitz for suggesting this work and for his great interest during its performance. We also want to thank Dr. Kalus and Dr. Tippe for useful discussions.

References

1) V. Bottom, Anales Acad. Brasil. Ciéncias **37** (1965) 407.
2) B. Alefeld, Bayer. Akad. Wiss. Math.-Nat. Kl. **11** (1967) 109.
3) B. Sykora, Diplomarbeit, Technische Hochschule, Darmstadt, Germany, 1968.
4) W. H. Zachariasen, Acta Cryst. **23** (1967) 558.

PERFECTION DE CRISTAUX D'ALUMINIUM EN FONCTION DES TRAITEMENTS THERMIQUES SUBIS

C. G'SELL, B. BAUDELET et G. CHAMPIER

Laboratoire de Physique du Solide, ENSMIM, Laboratoire associé au C.N.R.S. n⁰ 155
Parc de Saurupt, 54 – Nancy, France

Dislocation configurations have been studied by X-ray topography in aluminium single crystals cooled to ambient temperature from various temperatures at different cooling rates (10 °C/h to 50 000 °C/h). For coolings from temperatures lower than 500 °C, one observes random dislocations and also rows of coaxial dislocation loops parallel to ⟨110⟩ whose characteristics depend on the cooling parameters. For coolings from 600 °C, one observes "walls" composed of a high density of random dislocations. These "walls" define areas containing a relatively low density of random dislocations and rows of loops in the case of the rapid coolings. The influence of the cooling parameters on the dislocation configurations can be interpreted in terms of the formation and migration of the vacancies in aluminium. The particular configurations observed after cooling from 600 °C can be explained by considering the structural transformation of the aluminium oxide at high temperature. This study allows us to determine the best conditions to prepare crystals of high perfection by the strain-annealing method; the upper temperature of annealing must not be greater than 500 °C and the cooling rate must be as low as possible.

1. Introduction

Nous avons étudié par topographie aux rayons X les configurations de dislocations contenues dans des monocristaux d'aluminium de haute pureté après refroidissement jusqu'à la température ambiante. Nous préciserons l'influence des paramètres du refroidissement sur les configurations observées. Nous pourrons ainsi indiquer quel traitement thermique permet d'obtenir des cristaux de grande perfection.

2. Technique expérimentale

Les monocristaux sont préparés par écrouissage critique et recuit avec de l'aluminium de zone fondue contenant seulement quelques ppm d'impuretés. Leur épaisseur est égale à 450 microns. La densité de dislocations dans les cristaux utilisés est alors inférieure à 1000 cm cm^{-3}. Les échantillons sont ensuite soumis au traitement thermique suivant: chauffage jusqu'à une température supérieure T_s avec une vitesse égale à 30 °C/heure, palier thermique à la température T_s pendant 10 heures et refroidissement jusqu'à la température ambiante avec une vitesse constante dT/dt. Durant le traitement thermique, le cristal est situé dans une enceinte où l'on réalise un vide de 10^{-3} mm Hg. Les cristaux ont été refroidis depuis quatre températures supérieures (300, 400, 500 et 600 °C); quatre gammes de vitesses de refroidissement ont été retenues (10, 10^2, 10^3 et 10^4 °C/heure).

3. Résultats expérimentaux

L'étude des configurations présentes dans les cristaux après refroidissement conduit à classer les résultats expérimentaux en deux groupes: d'une part les refroidissements depuis des températures inférieures ou égales à 500 °C et d'autre part les refroidissements depuis 600 °C.

3.1. Refroidissements depuis des températures inférieures ou égales à 500 °C

Après refroidissement, les cristaux contiennent des dislocations dont la densité est sensiblement constante dans tout le volume cristallin. Leurs caractéristiques dépendent des paramètres du refroidissement.

3.1.1. *Influence de la vitesse de refroidissement*

Les quatre topogrammes de la figure 1 montrent les configurations de dislocations contenues dans des monocristaux ayant subi quatre refroidissements depuis 300 °C jusqu'à la température ambiante avec des vitesses respectivement égales à 10 °C/heure (1a), 150 °C/heure (1b), 3800 °C/heure (1c) et 22000 °C/heure (1d).

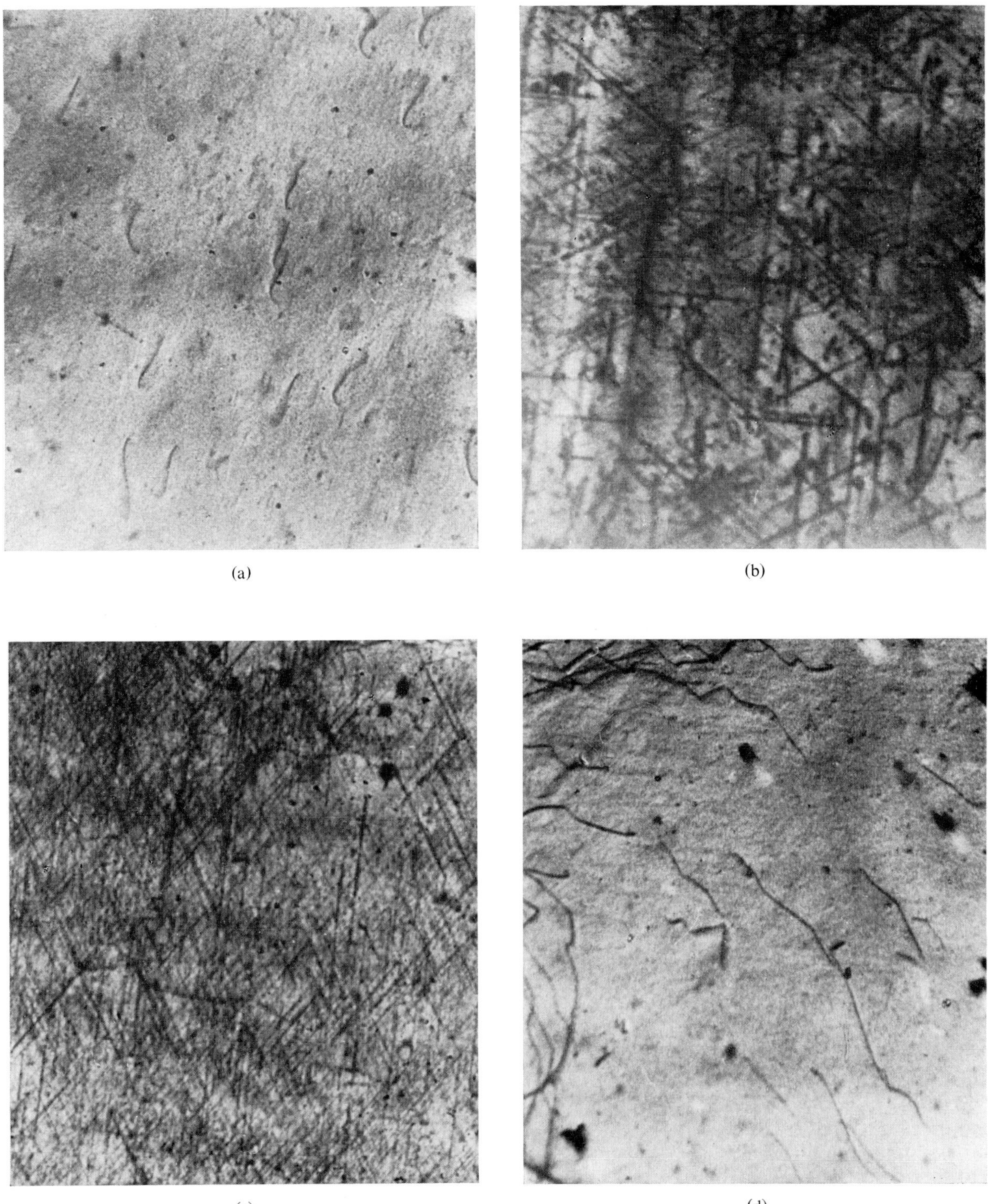

(a)

(b)

(c)

(d)

Fig. 1. (a) Configuration de dislocations d'un cristal refroidi depuis 300 °C à la vitesse de 10 °C/heure. Topogramme {111}. (b) Configuration de dislocations d'un cristal refroidi depuis 300 °C à la vitesse de 150 °C/heure. Topogramme {111}. (c) Configuration de dislocations d'un cristal refroidi depuis 300 °C à la vitesse de 3800 °C/heure. Topogramme {111}. (d) Configuration de dislocations d'un cristal refroid depuis 300 °C à la vitesse de 22000 °C/heure. Topogramme {111}. ($\times$35).

Le topogramme 1a présente des segments indépendants de dislocations courbes dont la densité est égale à quelques centaines de cm cm^{-3}. Le topogramme 1b montre deux types de dislocations: des dislocations courbes dont la densité est aussi faible que précédemment et des alignements de boucles coaxiales de direction $\langle 110 \rangle$ dont la densité est égale à 4000 cm cm^{-3} environ; le diamètre des plus grandes boucles est de l'ordre de 25 microns. Sur le topogramme 1c, on observe également ces deux types de dislocations, cependant la densité des alignements est plus élevée (10 000 cm cm^{-3}) et le diamètre des boucles est inférieur (10 microns au plus, à la limite de résolution de la méthode de topographie aux rayons X). Le topogramme 1d présente des dislocations courbes semblables à celles du topogramme 1a, leur densité est également très faible. Nous avons remarqué de plus que l'intensité des rayons X diffractés par le cristal est beaucoup plus élevée dans ce cas; nous interprèterons plus loin l'origine de cette augmentation.

3.1.2. *Influence de la temperature initiale*

Pour une vitesse de refroidissement donnée, les types de dislocations contenues dans les cristaux après traitement thermique sont les mêmes si la température initiale est égale à 300 °C, 400 °C ou 500 °C. Cependant, dans les cas où l'on observe des alignements de boucles coaxiales, le diamètre des boucles est d'autant plus grand que la température supérieure est plus élevée. Par exemple, un cristal refroidi depuis 500 °C avec une vitesse de 170 °C/heure contient des boucles de dislocations dont le diamètre dépasse 200 microns. Cette influence de la température initiale est d'autant plus faible que la vitesse de refroidissement est plus élevée.

3.2. Refroidissements depuis 600 °C

Les topogrammes réalisés avec des cristaux d'aluminium après refroidissement depuis 600 °C présentent un ensemble de taches noires qui masquent les configurations de dislocations. Lohne et al.[1]) ont montré que ce type de contraste est dû à des agglomérats de lacunes formés au cours du refroidissement sous le film d'oxyde cristallisé à haute température [Doherty et al.[2]). On a dû éliminer par polissage chimique une couche superficielle de quelques dizaines de microns pour obtenir des topogrammes de qualité convenable.

Les cristaux refroidis depuis 600 °C contiennent des parois de dislocations qui délimitent des zones à l'intérieur desquelles la densité de dislocations est plus faible. Si la vitesse de refroidissement est égale à 10 °C/heure ou 10^2 °C/heure, ces parois sont épaisses (100 microns) et limitent des zones dont les dimensions sont égales à quelques centaines de microns (fig. 2). On observe dans ces zones des dislocations courbes dont la densité est de l'ordre de quelques milliers de cm cm^{-3}.

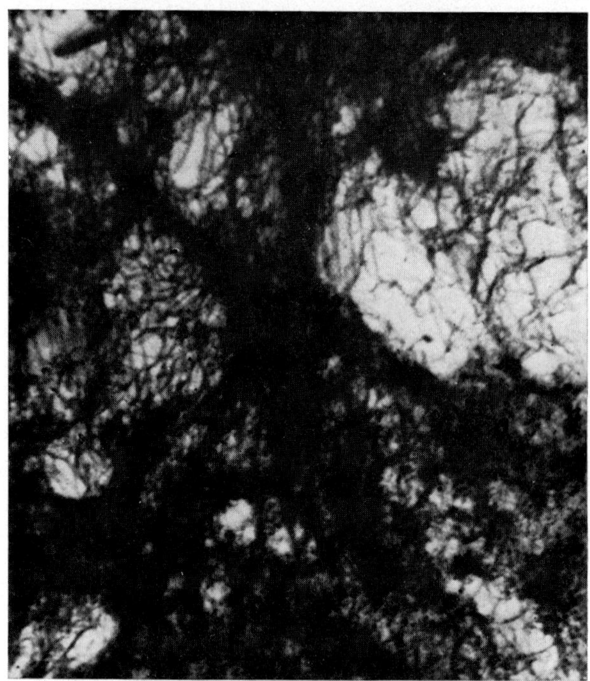

Fig. 2. Structure à "parois" de dislocations dans un cristal refroidi depuis 600 °C à la vitesse de 10 °C/heure. Topogramme $\{111\}$ ($\times 35$).

Si la vitesse est égale à 10^3 °C/heure ou 10^4 °C/heure, les parois sont plus minces. Elles délimitent des zones de plusieurs millimètres qui contiennent des dislocations courbes et des alignements de boucles de densité voisine de 10 000 cm cm^{-3}; le diamètre des boucles est inférieur à 15 microns.

4. Discussion

L'origine des alignements de boucles de dislocations coaxiales peut être interprétée suivant le mécanisme décrit par Bardeen et Herring[3]). Il consiste en une montée de segments de dislocations placés à l'interface de microprécipités et de la matrice cristalline. La nature de ces microprécipités reste encore incertaine, il pourrait s'agir soit de particules d'alumine, soit de précipités d'un composé intermétallique. Les sources de Bardeen et Herring fonctionnent lorsque la sursaturation en lacunes est supérieure à une valeur critique qui est d'autant plus faible que la longueur de la dislocation d'interface est plus grande. Dans les cristaux étudiés, des observations par topographie aux rayons X en cours de refroidissement [Baudelet[4]] ont montré que la sursaturation minimale nécessaire pour créer des alignements de boucles est voisine de 1.1, la longueur maximale correspondante des segments de dislocation d'interface susceptibles de monter étant alors égale à 0.3 micron.

4.1. REFROIDISSEMENT DEPUIS DES TEMPÉRATURES INFÉRIEURES OU ÉGALES À 500 °C

Quand le cristal est maintenu à la température supérieure T_s, la densité des dislocations originelles décroît et se stabilise à une valeur faible égale à quelques centaines de $cm \, cm^{-3}$ [Baudelet et al[5]]. Au début du refroidissement, avant l'apparition des premiers alignements de boucles coaxiales, les lacunes en sursaturation s'éliminent sur deux types de puits: la surface du cristal et les dislocations courbes originelles. Le calcul du nombre de sauts nécessaire pour éliminer les lacunes sur ces puits [Quere[6]] montre que l'on peut négliger l'influence des dislocations courbes par suite de leur faible densité. Dans cette hypothèse, un calcul de diffusion [Shewmon[7]] permet d'évaluer à chaque instant la sursaturation en lacunes en tout point du cristal. Cette sursaturation est maximale dans le plan médian du cristal. Quand sa valeur devient supérieure à 1.1, les premiers alignements de boucles apparaissent. On peut ainsi calculer la température à laquelle cette sursaturation critique est atteinte au cours des refroidissements (voir tableau 1).

TABLEAU 1

T_s (°C) / $-dT/dt$	10 °C/h	10^2 °C/h	10^3 °C/h	10^4 °C/h
300	295	296	296	296
400	333	395	395	395
500	333	469	494	494

Les boucles des alignements croissent par absorption des lacunes et constituent de nouveaux puits. On peut donc s'attendre à ce que les caractéristiques des alignements de boucles observés à la température ambiante dépendent des paramètres du refroidissement.

Si la vitesse de refroidissement est égale à 10 °C/heure, l'influence de la surface est importante. Les premiers alignements sont formés à basse température (voir tableau 1) et leur taille ne peut dépasser la limite de visibilité de la topographie aux rayons X. Les dislocations courbes observées sur les topogrammes sont certainement les dislocations originelles du cristal à la température T_s.

Si la vitesse de refroidissement est de l'ordre de 10^2 °C/heure, des alignements de boucles sont formés à une température plus élevée (voir tableau 1) et peuvent se développer en absorbant des lacunes. La taille des boucles est d'autant plus grande que la température supérieure est plus élevée car la mobilité des lacunes est supérieure à haute température.

Si la vitesse de refroidissement est de l'ordre de 10^3 °C/heure, la sursaturation atteint des valeurs élevées et il apparait un grand nombre d'alignements. La quantité de lacunes pouvant être absorbée par chacun d'eux étant plus faible, leur taille finale est plus petite.

Si la vitesse de refroidissement est supérieure à 10^4 °C/heure, la densité des alignements formés est encore plus grande. Chacun ne peut absorber qu'un nombre de lacunes relativement petit. La taille finale des boucles est inférieure à la limite de visibilité de la topographie aux rayons X et le fond continu intense observé sur les topogrammes serait dû à ces petites boucles alignées en forte densité.

4.2. REFROIDISSEMENTS DEPUIS 600 °C

Dans le cas des refroidissements depuis 600 °C, la surface est recouverte d'une couche d'oxyde cristallisée et elle ne peut plus être considérée comme un puits parfait. Dans le cas extrême où la surface n'absorbe pas de lacunes, la température de formation des premiers alignements est voisine de 593 °C, quelle que soit la vitesse de refroidissement. Ces alignements se développent très rapidement et, par interaction mutuelle des boucles, se transforment pour donner des dislocations courbes [Baudelet[3]] et forment à température élevée des parois de dislocations. De nouveaux alignements peuvent ensuite être formés au cours du refroidissement; ils apparaissent sur les topogrammes si la vitesse de refroidissement est supérieure à 10^3 °C/heure. Pour les vitesses inférieures, les lacunes effectuent un nombre de sauts suffisant pour atteindre les parois de dislocations et ces alignements ne peuvent pas croître.

5. Conclusion

Dans un cristal d'aluminium refroidi depuis une température inférieure ou égale à 500 °C jusqu'à la température ambiante, nous avons observé deux types de dislocations: d'une part des dislocations courbes en faible densité, quelques centaines de cm cm^{-3}, d'autre part des alignements de boucles de dislocations coaxiales dont la densité est d'autant plus grande que la vitesse de refroidissement est plus élevée.

Pour un cristal refroidi depuis 600 °C, la présence d'une couche d'oxyde cristallisée conduit à une configuration en "cellules" limitées par des "parois" de dislocations.

Nous avons interprété l'origine de ces configurations en considérant l'élimination des lacunes en sursaturation sur deux types de puits, la surface et les alignements de boucles.

Cette étude permet en outre de déterminer les conditions optimales pour la préparation de monocristaux d'aluminium à faible densité de dislocations: température de recuit inférieure ou égale à 500 °C et vitesse de refroidissement la plus faible possible, de l'ordre de 10 °C/heure.

Bibliographie

1) O. Lohne et B. Nøst, Phys. Status Solidi **36** (1969) 195.
2) P. E. Doherty et R. S. Davis, J. Appl. Phys. **34** (1963) 619.
3) J. Bardeen et C. Herring, dans: *Imperfections in Nearly Perfect Crystals*, Eds. W. Shockley, J. H. Hollomon and R. Maurer (Wiley, New York, 1952) p. 261.
4) B. Baudelet, Thèse, Nancy, 1970.
5) B. Baudelet et G. Champier, J. Phys. (Paris) **30** (1969) 999.
6) Y. Quere, *Défauts Ponctuels dans les Métaux* (Masson, Paris, 1967).
7) P. G. Shewmon, *Diffusion in Solids* (McGraw-Hill, New York, 1963) p. 17.

Journal of Crystal Growth **13/14** (1972) 257–261 © *North-Holland Publishing Co.*

DIE ZÜCHTUNG VERSETZUNGSFREIER KUPFEREINKRISTALLE

H. FEHMER

Physikalisches Institut der Universität Münster, 44 Münster, Germany

und

W. UELHOFF

Institut für Festkörperforschung der KFA Jülich, 517 Jülich, Germany

Copper single crystals (diameter: 5–10 mm, length: 10 cm) were grown by the Czochralski method. The crystals were free from dislocations, when they were grown under optimal conditions: [100]-growth direction, high vacuum and medium pulling speed (6 cm/h). Lattice defects were caused by the following unfavourable growth conditions: (a) when grown with [110]-orientation of the growth direction or grown under hydrogen the crystals contain subgrains, (b) growth under impure Ar caused prismatic loops around Cu_2O-particles and (c) micro-segregation caused by a high pulling speed.

In the dislocation-free copper single crystals three kinds of defects were observed: (a) vacancy-clusters (volume density: 10^6 cm^{-3}), (2) voids ($\sim 10^5$ cm^{-3}, (3) a kind of defect not identified till now, which generates holes on the surfaces of the crystal samples when they are electropolished.

1. Einleitung

Bei der Züchtung von nahezu perfekten Kupferkristallen gelang es bereits früher[1] beim Czochralski-Verfahren, in vereinzelten Fällen versetzungsfreie Kristallstücke zu erhalten. Nun konnte Dash[2] (1958) als erster zeigen, daß durch Czochralski-Züchtung von [100]- oder [111]-orientierten Si-Kristallen versetzungsfreie Halbleiterkristalle herzustellen sind, wobei für diese Orientierungen das Herauswachsen der im Impfling enthaltenen Versetzungen in optimaler Weise möglich ist. Da das kubisch flächenzentrierte Kupfer die gleichen Gleitelemente aufweist wie Si und Ge, haben wir nach Entwicklung geeigneter Impfling-Präparationsverfahren Kupfereinkristalle dieser Orientierungen hergestellt, um auf diese Weise auch große Kupfereinkristalle auf ihrer ganzen Länge versetzungsfrei zu erhalten.

2. Zur experimentellen Technik

Einige Angaben zur Kristallzüchtung sind in Tabelle 1 zusammengestellt.

Die Impflingherstellung wie auch die vor der Untersuchung erforderliche Probenpräparation (Zerschneiden der Kristalle, Polieren, Folienherstellung) wurde elektrolytisch durchgeführt[3,4].

3. Die Wachstumsparameter zur Züchtung versetzungsfreier Einkristalle

Die Züchtung dieser Kristalle gelang reproduzierbar bei [100]-Orientierung des Impflings. Sie erfolgte mit einer Ziehgeschwindigkeit von 6 cm/h unter Hochvakuum (5×10^{-6} Torr).

Abb. 1 zeigt einen Ausschnitt aus einer Röntgenaufnahme, die mit der Doppelkristalltechnik[5] hergestellt wurde. Bis auf die Abbildung von Löchern in der Oberfläche sind keine weiteren Kontraste, insbesondere

TABELLE 1

Experimentelle Daten zur Kristallzüchtung

Verfahren	Czochralski
Ausgangsmaterial	5N Kupfer (Johnson & Matthey)
Heizung	induktiv
Tiegel	
Material	Graphit
Durchmesser	45 mm
Höhe	60 mm
Kristallgeometrie	
Impfling	5 mm ⌀
Hals	0.5–1.5 mm ⌀
Durchmesser des	
Kristallrumpfes	5–10 mm ⌀
Länge des Kristallrumpfes	10 cm
Steuerung der Kristall-geometrie	durch Programmregelung der Schmelzentemperatur bei konstanter Ziehgeschwindigkeit
Schwankungen der Schmelzentemperatur	±0.15 °C
Rotationsfrequenz	25 R.p.M.
Ziehgeschwindigkeit	s.u.
Atmosphäre	s.u.

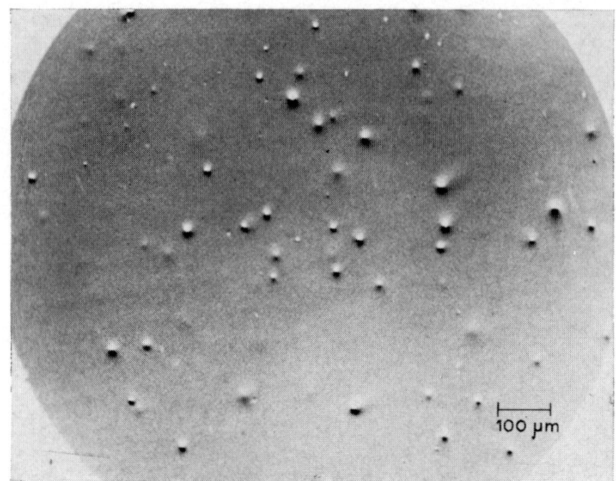

Abb. 1.   Versetzungsfreier Kupfereinkristall [Doppelkristall-aufnahme, Cu–Kα_1-Strahlung, (331)-Reflex].

keine Versetzungslinien zu beobachten (zum Vgl. siehe Abb. 2).

Auch bei Anwendung der Borrmann-Topographie wurden keine Versetzungslinien beobachtet. Die Auf-nahme in Abb. 3 wurde von einer 400 µm dicken Folie, die sich über den gesamten Kristallquerschnitt er-streckte, angefertigt. Ein Vergleich mit Abb. 8 zeigt, daß es sich bei den zahlreichen schwarzen Punkten und Strichen, die auf Borrmann-Aufnahmen von Kristallen dieser Qualität durchweg beobachtet wurden, nicht um Versetzungen handelt. Vielmehr ergab eine Unter-suchung der Folien in allen vier möglichen (111)-Re-flexen, daß die Richtung der Striche stets parallel zur Richtung des Beugungsvektors *g* verläuft. Die Natur der diesen Kontrasten zugrunde liegenden Gitter-defekte wird im letzten Abschnitt diskutiert.

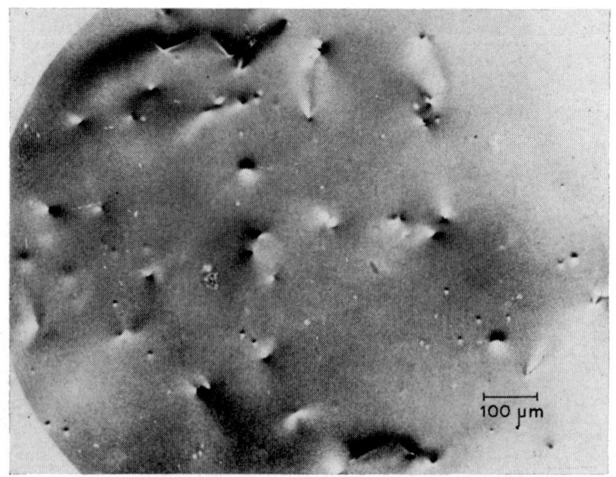

Abb. 2.   Kupfereinkristall mit 3×10^3 cm^{-2} Versetzungen [Dop-pelkristallaufnahme, Cu–Kα_1-Strahlung, (331)-Reflex].

Unter den obengenannten Bedingungen wurden 5 Kristalle gezüchtet. Die röntgenographischen Auf-nahmen zeigten in keinem Fall Versetzungslinien oder Durchstossungspunkte von Versetzungslinien. Die Kristalle dürfen daher in diesem Sinne als versetzungs-frei bezeichnet werden.

Die Versetzungsdichte der Impflinge lag bei diesen fünf Kristallen zwischen null und ca. 10^5 cm^{-2}. Bei den hohen Versetzungsdichten wurden im Impfling Sub-grenzen beobachtet. Zur Untersuchung des Elimina-tionsprozesses wurde ein Längsschnitt aus dem Bereich des Impflings mit folgenden Ergebnissen untersucht. (1) An der Anfangsphasengrenze können beim Anschmel-

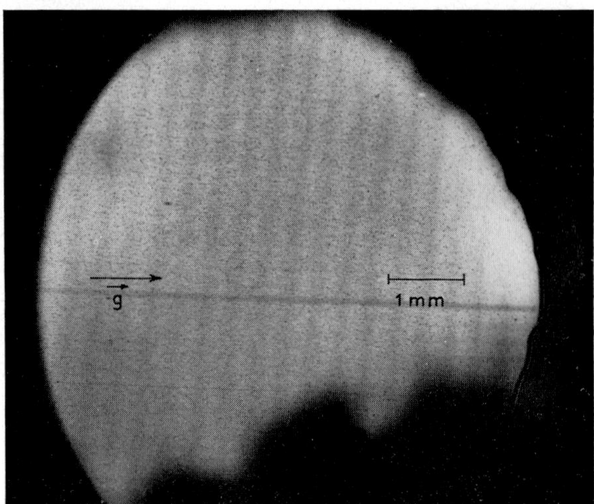

Abb. 3.   Kupfereinkristallfolie ($\approx$400 µm dick) frei von Ver-setzungslinien [Borrmann-Aufnahme, Cu–Kα_1-Strahlung, (111)-Reflex].

zen Poren bis zu 300 µm Durchmesser entstehen; (2) die Versetzungsdichte beträgt im Impfling ca. 10^3 cm^{-2}, geht an der Anschmelzstelle sprungartig auf ca. 10^4 cm^{-2} hoch, bleibt im untersuchten Bereich des Halses nahezu konstant oder nimmt (lokal) noch zu. Dabei kann es zur Bildung von Polygonisationswänden kommen. (3) Wenigstens nach Erreichen des End-durchmessers ist der Kristall versetzungsfrei (das Gebiet der extremen Verjüngung mit ca. 0.35 mm Durch-messer konnte nicht untersucht werden). Aus diesen Beobachtungen kann gefolgert werden, daß die Ver-setzungselimination nicht allein durch den Mechanis-mus des Herauswachsens der Versetzungen erfolgt. In diesem Falle wäre bereits 6 mm unterhalb der An-fangsphasengrenze Versetzungsfreiheit zu erwarten

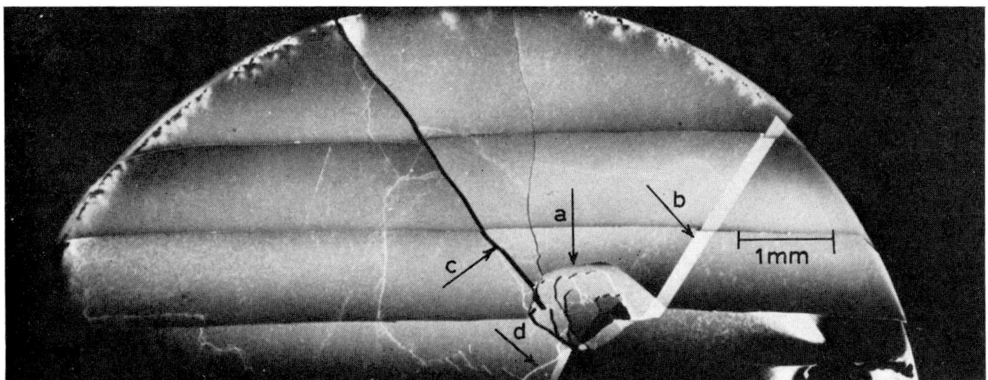

Abb. 4. Substruktur in einem [110]-orientierten Kupfereinkristall (am rechten Bildrand versehentlich verursachte Druckstelle) [Berg–Barrett-Aufnahmen, Cu–Kα_1-Strahlung, (420)-Reflex].

gewesen. Der Bereich hoher Versetzungsdichte in der Verjüngung ist aber wenigstens doppelt so groß. Erst im dünnsten nicht untersuchten Bereich geht sie auf null zurück.

4. Die Auswirkungen von ungünstigen Wachstumsbedingungen

Bei [110]-orientierten Impflingen enthalten die Kristalle Subkörner mit Verschwenkungen bis zu 4°. Die zusammengesetzten Berg-Barrett-Aufnahmen in Abb. 4 zeigen im einzelnen: Ein zentrales Haufwerk (bei a) von Subkörnern ist vom Impfling durch den Kristallhals (ca. 1 mm Durchmesser) in den Kristall hineingewachsen. Von diesem Haufwerk ausgehend durchsetzen drei niedrig indizierte ebene Subgrenzen (bei b, c, d) den gesamten Kristall. Außerdem enthält der Kristall noch weitere Subgrenzen mit geringerer Orientierungsdifferenz. Die Orientierung der erstgenannten Subgrenzen sowie Größe und Richtung der Verkippung läßt den Schluß zu, daß die sie bildenden Versetzungen nicht auf einer Gleitebene liegen und daß sie schräg zur Wachstumsrichtung verlaufen. Es ist daher nicht möglich, daß diese Versetzungen in den Kristall hineingewachsen sind oder durch Gleitung bzw. Gleitungsmultiplikation entstanden sind, wobei durch einen nachfolgenden Kletterprozeß die Subgrenzen gebildet werden. Wir nehmen daher an, daß die Versetzungen dieser Grenzen beim Wachstum aufgrund eines noch unbekannten Prozesses an der Peripherie des Haufwerkes entstanden sind, wobei die Zahl der erzeugten Versetzungen pro Längeneinheit der Subgrenze ständig so groß ist, daß die Verkippung der bereits im Hals entstandenen Subkörner aufrechterhal-

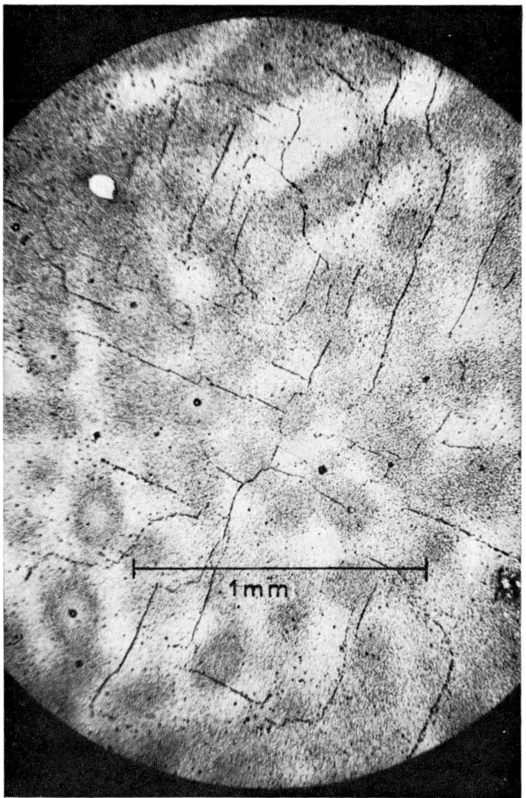

Abb. 5. Ätzmuster auf dem Querschnitt eines [100]-orientierten Kupfereinkristalles (Ziegeschwindigkeit: 60 cm h^{-1}). Orthogonales System von Feinkorngrenzen im Kristallinneren (Richtung der Subgrenzen: [110]).

ten wird. Bei den dichter liegenden feineren Subgenzen ist dagegen nicht auszuschließen, daß sie durch Klettern von Versetzungen entstanden sind, die möglicherweise durch thermische Spannungen gebildet wurden. – Aufgrund dieser Beobachtungen kann nicht erwartet werden, daß bei nicht versetzungsfreien Impflingen versetzungsfreie (110)-Kristalle erhalten werden können.

Auch bei optimaler Impflingsorientierung darf die Ziehgeschwindigkeit nicht beliebig groß sein. (100)-Kristalle, die mit 60 cm/h gezogen wurden, zeigten eine ähnliche Zellstruktur wie Bridgman-Kristalle aus technisch reinem Kupfer[1]). Außerdem enthalten sie ein regelmäßiges System von Subgrenzen (Abb. 5). Die Entstehung dieser Zellstruktur dürfte aufgrund der Arbeiten von Chalmers und Mitarbeitern auf das sehr niedrige Verhältnis G/R (G: Temperaturgradient in der Schmelze, R: Wachstumsgeschwindigkeit) zurückzuführen sein, das bei 60 cm h^{-1} den Wert von 5 °C min cm^{-2} hat, bei der optimalen Ziehgeschwindigkeit von 6 cm h^{-1} um den Faktor 100 größer ist [6]). Die Hauptverunreinigungen des Kupfers sind in Tab. 2 zusammengestellt.

TABELLE 2

Massenspektroskopische Analyse
(Angaben in Atom ppm, absoluter Fehler: etwa Faktor 2)

	Au	Se	Zn	Ni	Fe	K	Cl	Si	Al	C
Regulus	0.4	24	2.4	2.1	4.6	7.9	5	81	3.1	16
Einkristalle*	2.1	16	3.0	3.9	8.0	15	8.0	150	7.4	31

* Mittelwerte von zwei Einkristallen.

Diese Konzentrationen sind offenbar ausreichend, um zur Zellbildung zu führen. Hinsichtlich der Entstehung der Versetzungen kann aufgrund ihrer Anordnung in den Subgrenzen nicht entschieden werden, ob sie durch den Tiller-Mechanismus[7]) oder durch thermische Spannungen (verursacht durch die stark gekrümmte Phasengrenze) entstanden sind.

Kristalle, die unter strömendem Argon (nominelle Reinheit 99.99%) entweder bei 5×10^{-4} oder 0.5 Torr gezogen wurden, enthalten Oxyduleinschlüsse, die zu sternartigen Defekten Anlaß geben (Abb. 6). Sie bestehen, wie im Ätzbild festgestellt und teilweise auch auf Röntgenaufnahmen beobachtet wurde, aus prismatischen Versetzungsschleifen und entstehen durch prismatisches Stanzen[8]). Wegen der beobachteten Konzentrationsabnahme der Defekte von außen nach innen, ist anzunehmen, daß der Sauerstoff nicht in der Schmelze enthalten ist, sondern erst in den wachsenden Einkristall diffundiert. Mit zunehmendem Argondruck nimmt die Größe der Defekte zu. Eine Partialdruckmessung am Rezipienten ergab, daß das

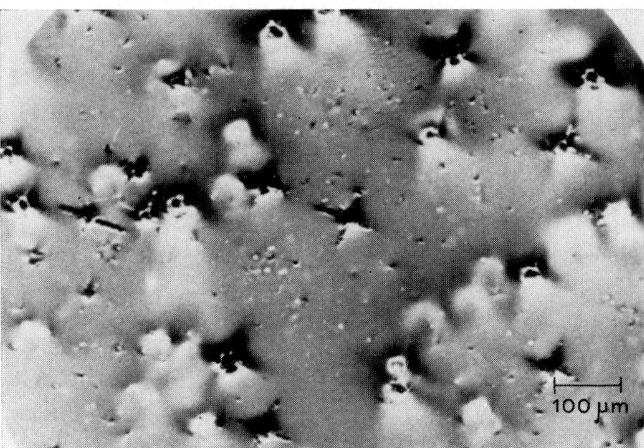

Abb. 6.   Cu$_2$O-Einschlüsse in Kupfereinkristallen ($\simeq 30$ cm^{-2} eingewachsene Versetzungslinien). Von den Einschlüssen ausgehende prismatische Versetzungsschleifen [Doppelkristallaufnahme, Cu–Kα_1-Strahlung, (331)-Reflex].

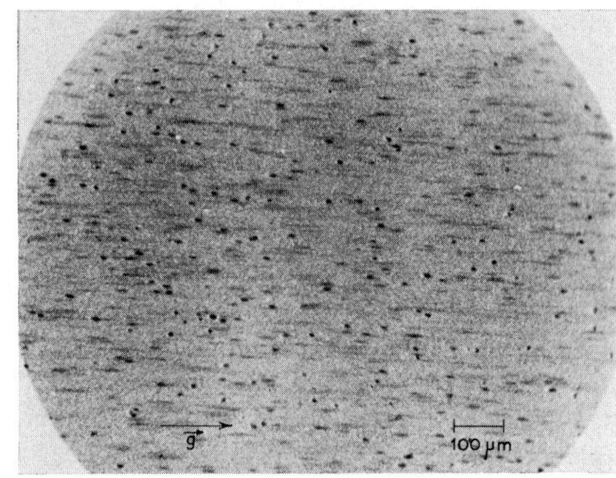

Abb. 7.   Punktförmige Defekte in versetzungsfreien Kupfereinkristallen, Ausschnittvergrößerung von Abb. 3.

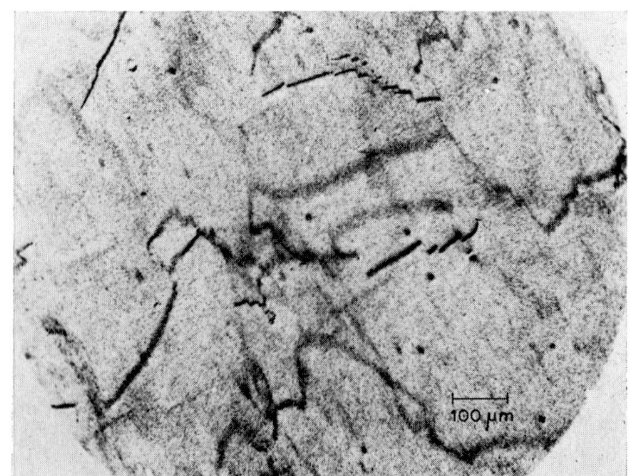

Abb. 8.   Kupfereinkristallfolie mit 2×10^3 cm^{-2} Versetzungen. Die abgebildeten Punkte sind Bilder von Löchern [Borrmann-Aufnahme, Cu–Kα_1-Strahlung, (200)-Reflex].

VI – 3

Verhältnis $P_{O_2}/P_{Ar} \approx 10^{-2}$ betrug, der nominelle Reinheitsgrad also nicht vorlag. Insgesamt gesehen bedeutet dieses Ergebnis, daß zur Herstellung von Einkristallen, die frei von diesen Defekten sind, eine möglichst vollständige Entfernung des Sauerstoffs aus dem Argonschutzgas erforderlich ist. Der Sauerstoff-Partialdruck sollte deutlich kleiner als 5×10^{-6} Torr sein.

Züchtungsversuche unter Wasserstoff bei den gleichen Drücken wie oben angegeben lieferten andere Ergebnisse: Sternartige Defekte wurden nicht beobachtet. Dagegen hatten die Einkristalle Subgrenzen in der Nähe der Mantelflächen. Eine hinreichende Erklärung hierfür steht zur Zeit noch aus.

5. Defekte in versetzungsfreien Einkristallen

Die von uns hergestellten versetzungsfreien Einkristalle dürfen nicht als fehlerfrei betrachtet werden. Die Lang-Aufnahme in Abb. 3 zeigte kleine schwarze Punkte und Stricke (black spots), die etwa 10 μm groß sind (Abb. 7). Ihre Volumendichte beträgt ca. 10^6 cm^{-3}. Es wurden von uns nie die bis zu 150 μm großen Defekte beobachtet, die Young[9]) in Kupferkristallen mit Versetzungsdichten von ca. 10^2 cm^{-2} gefunden und als Frank'sche wie prismatische Versetzungs-schleifen identifiziert hat. Auch die von uns beobachteten Defekte wurden ebenso wie bei Young durch Leer-stellenkondensation verursacht, denn die Kristalle mit höheren Versetzungsdichten ($\approx 10^3$ cm^{-2}) enthalten diese Defekte nicht (Abb. 8). Sie zeigen vielmehr neben Versetzungen ausschließlich die Bilder von Löchern (s.u.). Außerdem ist in Kristallen mit geringer Ver-setzungsdichte die Defektdichte in der Versetzungsum-gebung um den Faktor 10 niedriger als in versetzungs-freien Bereichen. Eine Kontrastuntersuchung in ver-schiedenen Reflexen ergab, daß es sich um Verset-zungsschleifen mit Schraubenkomponenten handeln kann. Auszuschließen sind dagegen Schleifen mit reinem Stufencharakter, Frank'sche Versetzungen sowie Einschlüsse oder Poren.

Derartige Hohlräume wurden dagegen mit optischen Methoden in allen untersuchten Einkristallen unab-hängig vom Herstellungsverfahren in etwa gleicher Anzahl nachgewiesen[10]). Insbesondere wurden sie auch in versetzungsfreien Kristallen gefunden. Ihre Volu-mendichte beträgt 1×10^5 bis 5×10^5 cm^{-3}. Ihre Ursache dürfte ebenfalls auf Leerstellenkondensation zurückzuführen sein. Letztere wird vermutlich durch heterogene Keimbildung eingeleitet, da diese Defekte auch bei höheren Versetzungsdichten beobachtet wurden. Andere Entstehungsmechanismen wie Siede-oder Gasblasen oder Reduktion von Cu_2O-Partikeln können ausgeschlossen werden.

Auf elektrolytisch polierten Flächen von Einkristal-len wurden ebenfalls unabhängig von Herstellungsver-fahren und Versetzungsdichte Löcher mit wesentlich größerer Flächendichte (Faktor 7 bis 30) beobachtet. Nur ein geringer Bruchteil ist auf die o.g. Poren zurück-zuführen. Die Entstehungsursache für den größeren Teil ist bisher nicht bekannt.

Wir danken Herrn Prof. Dr. E. Kappler, Universität Münster, für sein förderndes Interesse sowie Herrn Dr. H. Beske (KFA Jülich) für die Durchführung der massenspektroskopischen Analysen.

Literatur

1) W. Uelhoff, in: *Realstruktur und Eigenschaften von Reinst-stoffen*, Herausgeber J. Kunze, B. Pegel, K. Schlaubitz und D. Schulze (Akademie-Verlag, Berlin, 1967) S. 261–273.
2) W. Dash, J. Appl. Phys. **29** (1958) 705.
3) H. Fehmer und W. Uelhoff, J. Sci. Instr. [2] **2** (1969) 771.
4) H. Fehmer und W. Uelhoff, J. Sci. Instr. [2] **2** (1969) 767.
5) U. Bonse, in: *Direct Observation of Imperfections in Crystals*, Eds. J. B. Newkirk and J. H. Wernick (Interscience, New York, 1962) S. 431.
6) W. A. Tiller, K. A. Jackson, J. W. Rutter und B. Chalmers, Acta Met. **1** (1953) 428;
W. A. Tiller und J. W. Rutter, Can. J. Phys. **34** (1956) 96.
7) W. A. Tiller, J. Appl. Phys. **29** (1958) 611.
8) F. W. Young, Jr., in: *Direct Observation of Imperfections in Crystals*, Eds. J. B. Newkirk and J. H. Wernick (Interscience, New York, 1962) S. 103.
9) F. W. Young, Jr., T. O. Baldwin, A. E. Merlini und F. A. Sherrill, in: *Advances in X-Ray Analysis*, Eds. G. R. Mallet, M. Fay and W. M. Mueller (Plenum, New York, 1966) S. 1;
F. W. Young, Jr., T. O. Baldwin und F. A. Sherrill, in: *Lattice Defects and their Interactions*, Ed. R. R. Hasiguti (Gordon and Breach, New York, 1967) S. 544.
10) E. Kappler, W. Uelhoff, H. Fehmer und F. Abbink, *Herstel-lung von Kupfereinkristallen kleiner Verzetsungsdichte* (West-deutscher Verlag, Opladen, 1971) S. 33.

    Journal of Crystal Growth **13/14** (1972) 262–267 © *North-Holland Publishing Co.*

EPITAXIAL GROWTH AND CHARACTERIZATION OF GaP ON INSULATING SUBSTRATES*

C. C. WANG and S. H. McFARLANE III

RCA Laboratories, Princeton, New Jersey, U.S.A.

Heteroepitaxy of gallium phosphide on insulating substrates has been studied. Undoped epitaxial films with thicknesses up to 25 µm have been successfully grown on (0001) and ($\bar{1}$012) sapphire and on (111) spinel by the reaction between trimethyl gallium and phosphine using hydrogen as the carrier gas. Gallium phosphide has also been grown on thin (2000–3000 Å) silicon grown on sapphire, forming a multilayer structure. The epitaxial film-substrate composites have been characterized by X-ray diffraction, electron microscopy, and optical methods. Information on the growth characteristics, epitaxial orientation relationships, and surface structures have been obtained. The crystalline perfection of the epitaxial gallium phosphide has been examined by transmission X-ray topography. It has been found that the epitaxial GaP deposits are composed of crystallites which are misoriented by $\pm 0.1°$ from the nominal orientation of the layer. Electrical properties of selected films have been assessed by Hall measurements. Electron mobilities up to 70 cm²/V-sec have been measured in unintentionally doped films grown on (0001) sapphire.

1. Introduction

The achievement of single crystal growth of large area semiconductor films on oxide insulating substrates is of scientific interest as well as technical importance. Extensive studies[1-4] on the epitaxial growth and properties of elemental semiconductors (silicon and germanium) on sapphire and spinel substrates have been made in the past few years. More recently, the successful epitaxial growth of several III–V compounds on insulators was reported[5-7], and the properties of GaAs-on-sapphire have been characterized[6]. Results are presented in this paper on the heteroepitaxial growth of GaP on insulating substrates, and on insulating substrates covered with a thin film of epitaxial silicon. The crystalline nature of the composites was examined in detail by X-ray diffraction topography.

2. Experimental

Epitaxial growth of GaP films was carried out using the reaction between trimethyl gallium [$(CH_3)_3Ga$] and phosphine (PH_3). Substrates (0.07 cm × 1 cm²) employed for most growth runs are Czochralski grown sapphire and thin silicon (2000–3000 Å) grown on sapphire, although Al-rich spinel grown by the Verneuil flame

fusion method[8]) was also used. The growth system, as shown in fig. 1, consists of a water cooled vertical quartz reactor (6.25 cm diameter × 40 cm long), a trimethyl gallium source reservoir (kept at 0 °C), a phosphine (10% by volume in H_2) supply, and carrier gas flow controls. Hydrogen purified by a Pd-diffuser was used as the carrier gas. RF induction heating and pyrolytic graphite or SiC coated graphite susceptors were employed for the deposition. The substrate temperature was measured by an infrared pyrometer. In a typical growth run, the reactor is first thoroughly flushed by H_2 to purge the system of air. Predetermined and equilibrated PH_3 and $(CH_3)_3Ga$ gas flows are then sequentially admitted to the reactor in which the substrate is heated to the growth temperature. The growth parameters studied in order to achieve optimum growth include growth temperature, flow rates, substrate orientation, and surface preparation.

The orientation relationship and a preliminary assessment of crystalline perfection were obtained using the Laue X-ray back-reflection and glancing angle electron diffraction methods. In addition, the GaP layers and sapphire substrates were characterized by transmission X-ray topography and X-ray diffraction methods using a Lang camera[9]) and scintillation counter with Ag radiation. The GaP/sapphire combination provides the opportunity to examine the epitaxial layer by transmission topography while the film remains on the substrate. The product of the linear absorption

---

* Research reported in this paper was jointly sponsored by the Air Force Materials Laboratory, Wright-Patterson Air Force Base, under Contract No. F33615-70-C-1536 and RCA Laboratories, Princeton, N.J.

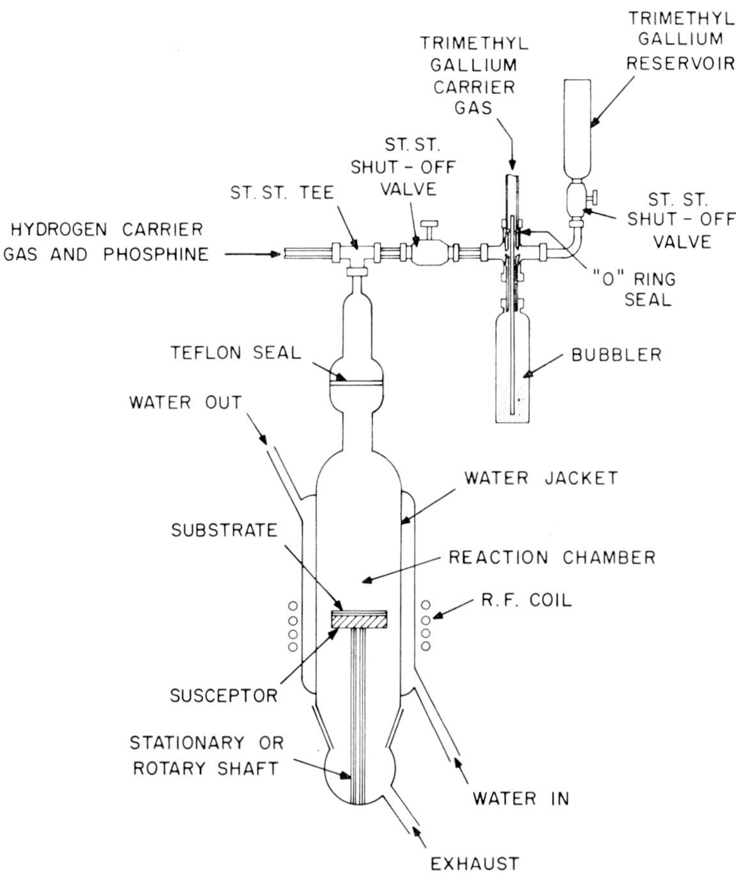

Fig. 1.   Epitaxial growth apparatus.

coefficient and thickness (μt) of the 0.7 cm thick sapphire substrate is 0.44, and μt for a 20 μm GaP layer is 0.18. Thus, both μt values are on the same order and meet the criterion of $\mu t < 1$ for transmission topography.

The films have been examined by optical spectroscopy and scanning electron microscopy. Information on the surface characteristics and film thickness was obtained. The electrical properties of selected epitaxial films were determined by Hall measurements[10]) using six-terminal Hall bar samples. The Hall mobility was derived from the Hall coefficient and resistivity.

3. Results and discussion

3.1. EPITAXIAL GROWTH AND FILM PROPERTIES

Single crystal (111) GaP films with thicknesses up to 25 μm have been grown on (0001) and ($\bar{1}$012) sapphire, and on spinel substrates. The substrate orientation plays a critical role in the ease with which the epitaxial layers grow. The ease of growth of GaP on sapphire increases in the order: (0001)sapphire > (111)spinel > ($\bar{1}$012)sapphire. Films grown on ($\bar{1}$012) sapphire exhibit misorientations of up to 6° from (111). The epitaxy of GaP on spinel is more sensitive to substrate surface preparation than that of GaP on sapphire. (111) GaP films have been consistently grown on (111) spinel substrates which had a high quality surface. The optimum growth temperature has been found to be around 800 °C. Films exhibit poor crystallinity at lower growth temperatures and tend to be inhomogeneous at higher temperatures. The flow conditions which yield the best results for the particular apparatus geometry used are:

H_2 carrier gas flow: 3.0 l/min,
PH_3 (10 % in H_2) flow: 450   ml/min,
H_2 carrier gas for $(CH_3)_3Ga$:   34   ml/min.

Under these conditions, the film growth rate is about

0.3 µm/min. Higher flow rates than these favor heavy deposition at the center of the substrate, while slower flow rates favor deposition at its periphery. It has been found that the film quality improves with increasing thickness. Films less than 10 µm in thickness exhibit a high degree of imperfection. Twinning about the $\langle 111 \rangle$ direction has been commonly observed in the (111) films from electron diffraction. The sterographic projection of (111) GaP on (0001) sapphire is presented in fig. 2 showing the orientation relationship.

Electrical measurements revealed that the as-grown GaP films on sapphire are generally n-type with net carrier concentrations of 10^{16}–$10^{18}/cm^3$. The highest electron mobility measured was 70 cm^2/V-sec (63% of bulk value) for a film 6 µm thick with a net carrier concentration of $1.4 \times 10^{18}/cm^3$.

In addition to films grown directly on the oxide substrates, (100) GaP has also been grown on thin (100) silicon (2000–30000 Å) grown on ($\bar{1}$012) sapphire. The same conditions were used as those employed for grow-

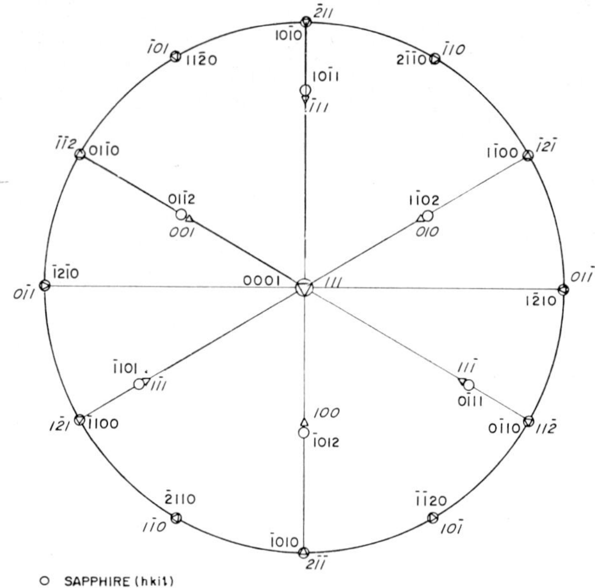

Fig. 2.   Stereographic projection of (111) GaP on (0001) sapphire.

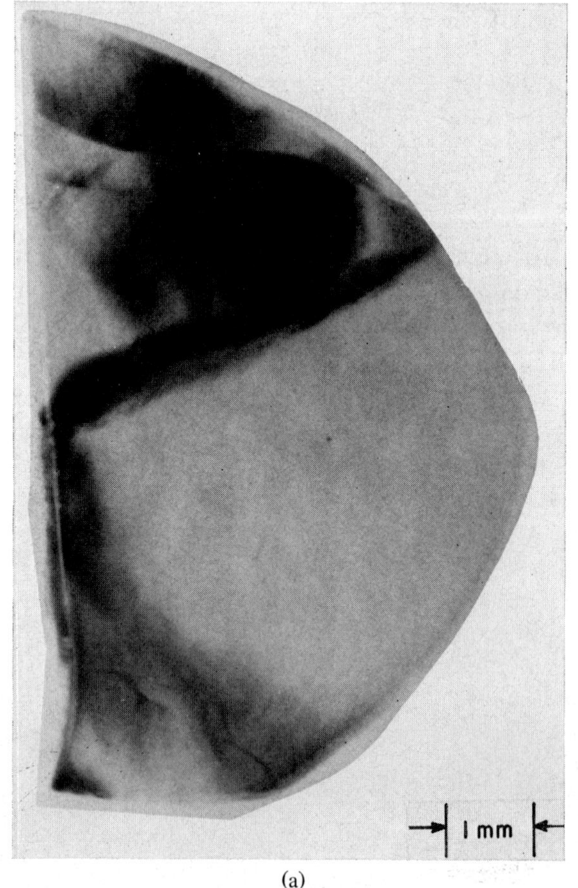

(a)

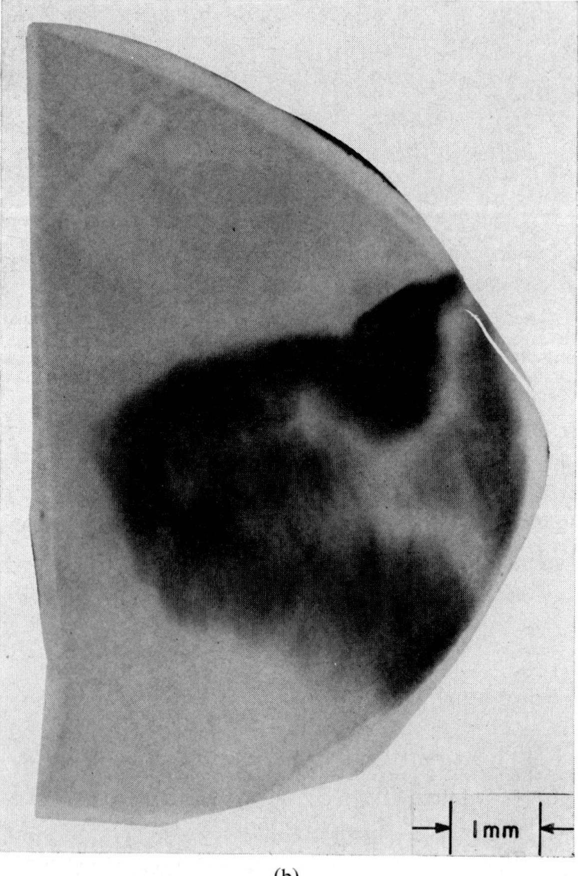

(b)

Fig. 3.   X-ray topograph of (0001) sapphire substrate. Silver radiation, diffracting from the ($1\bar{2}10$) planes. The crystal was rotated 0.1° between the two topographs.

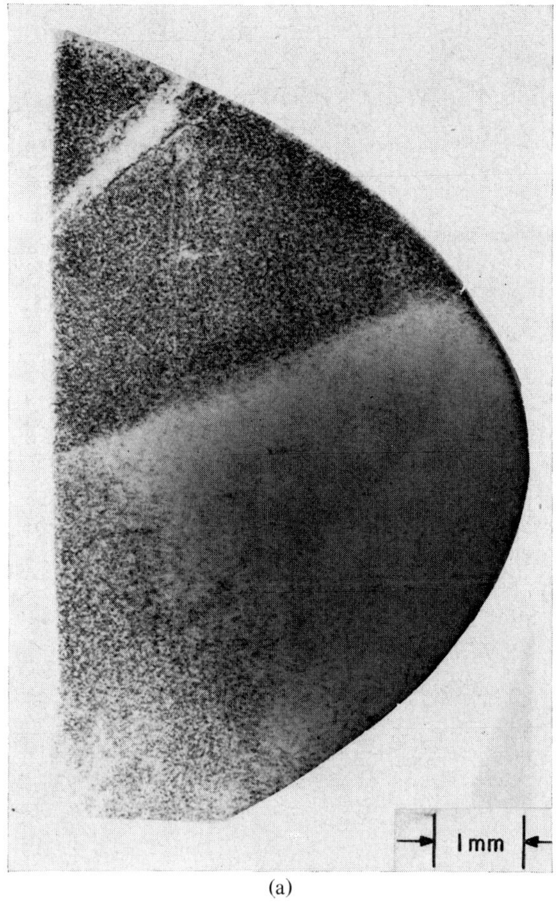

(a)

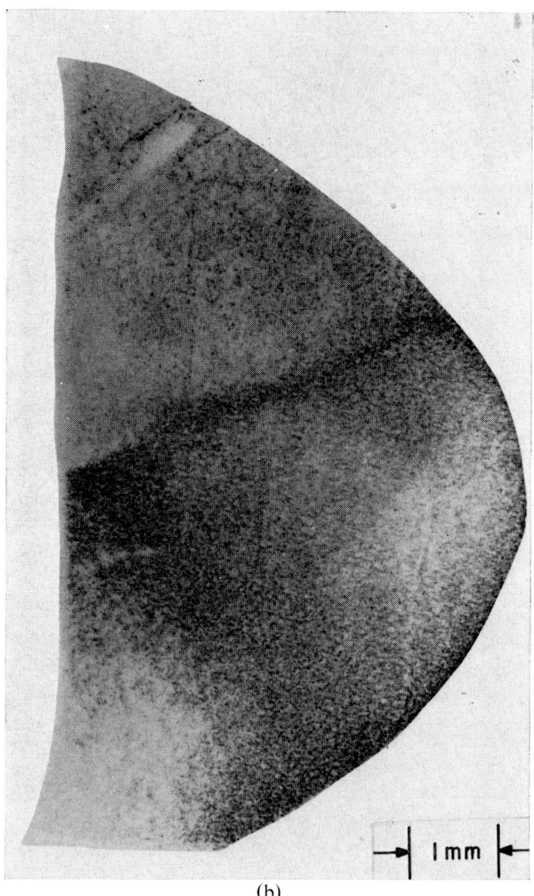

(b)

Fig. 4. X-ray topograph of (111) GaP layer grown on the substrate shown in fig. 3. Silver radiation, diffracting from the $(02\bar{2})$ planes. The crystal was rotated 0.1° between the two topographs.

ing films on (0001) sapphire. The GaP films grown on the Si/sapphire composite substrate generally showed better crystalline perfection than films grown directly on the sapphire substrates.

3.2. CHARACTERIZATION OF GaP FILMS BY X-RAY TOPO-GRAPHY

Examination by X-ray topography reveals that the layers are composed of small crystallites which are mis-oriented in the range of ±0.1° from the nominal orientation of the layer. Further, the general orientation of the layers follows the orientation of large grains in the substrate. Figs. 3 and 4 are topographs taken with silver radiation of (111) GaP grown on a (0001) sapphire substrate. The two topographs of the sapphire substrate shown in fig. 3 were taken at crystal settings 0.1° apart and clearly show the presence of at least two subgrains in the crystal. Fig. 4 shows topographs ob-

tained from the GaP film, again at a crystal setting difference of 0.1° between the two topographs. It is evident that the GaP layers in figs. 4a and 4b follow the orientation of the sapphire grains shown in figs. 3a and 3b, respectively. Thus the typical grain size is about 1–10 μm with misorientations in the range of ±0.1°. Note also in fig. 3 the clear distinction between the areas where the sapphire grain is present or absent. However, in fig. 4, the GaP film orientation is much less sharply defined. That is, in the top part of fig. 3a, a sapphire grain exhibits strong diffraction contrast, and there is no diffraction from this same region in fig. 3b. The GaP film in the top part of fig. 4a also exhibits strong diffracted contrast, and the image appears to be due to a high density of small crystallites. In fig. 4b, this same region is still exhibiting diffracted intensity from a small number of crystallites. Thus, the X-ray topographs show that the GaP film is composed

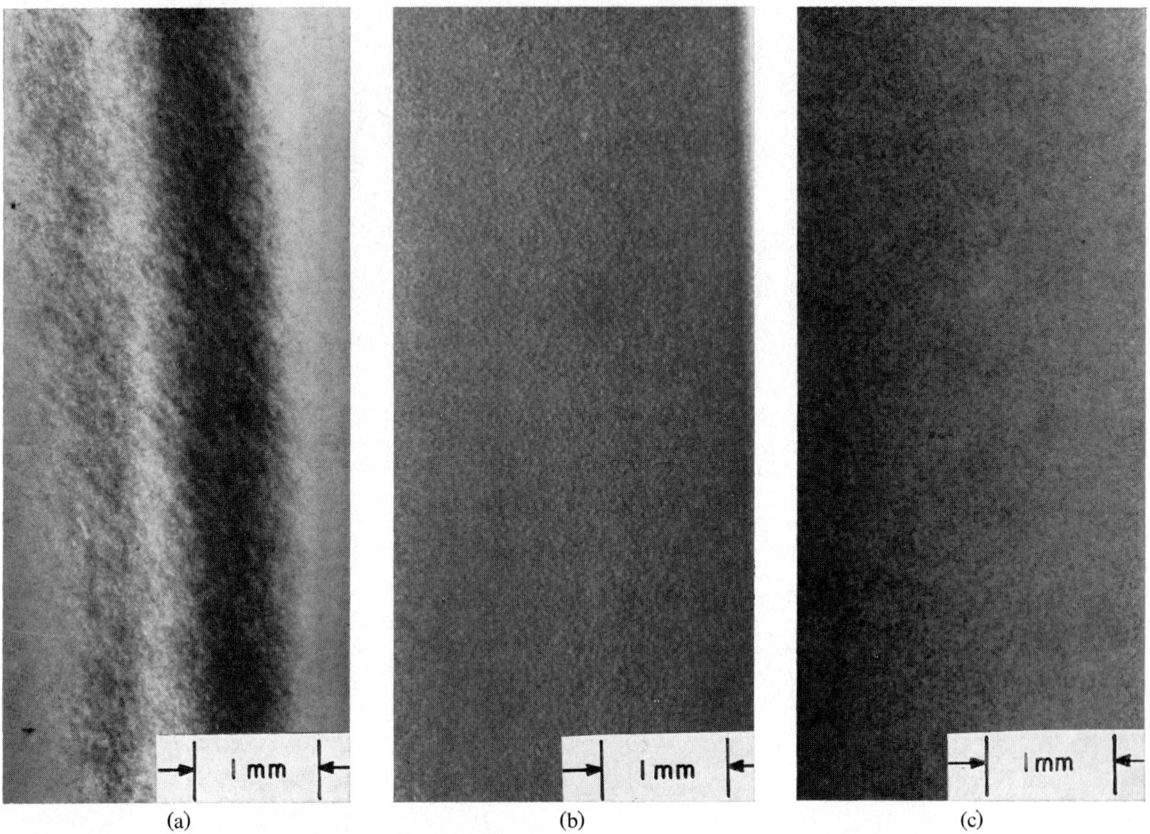

Fig. 5.   X-ray topographs of ($\bar{1}012$) sapphire substrate and the (100) GaP layer grown on it. Silver radiation was used. (a) Sapphire substrate, ($0\bar{1}1\bar{2}$) reflection, (b) and (c) GaP layer, ($02\bar{2}$) reflection. Crystal was rotated 0.1° between the two topographs.

of small crystallites which generally follow the orientation of the grains in the substrate, but are misoriented from each other by $\pm 0.1°$.

A similar situation is shown in fig. 5, where (a) is a topograph of a ($\bar{1}012$) sapphire substrate and (b) and (c) are topographs of the (100) GaP layer [on (100) silicon] taken at a crystal setting difference of 0.1°. Since the film-substrate composite is bent, the crystal does not remain in proper alignment for diffraction from the $K_{\alpha 1}$ component of the incident X-rays as the crystal is translated in the beam. Thus, the two vertical bands of strong diffracted intensity in fig. 5a are due to diffraction from both the $K_{\alpha 1}$ and $K_{\alpha 2}$ components of the incident beam. No grains were observed in topographs taken of this wafer, but the general dislocation density is too high to resolve individual dislocations. The topographs of the GaP layer in figs. 5b and 5c again show contrast due to small crystallites. This (100) layer appears, however, to have less spread in the range of misorientation than was observed on the (111) layer.

The departure of the film from parallel epitaxy can be determined by measuring the angular separation of

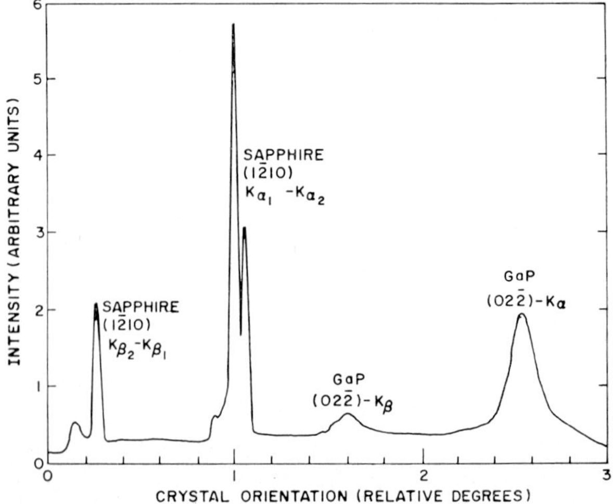

Fig. 6.   Trace of diffracted intensity versus crystal orientation for (111) GaP layer grown on (0001) sapphire. Peaks from both the substrate and layer are observed. (Silver radiation.)

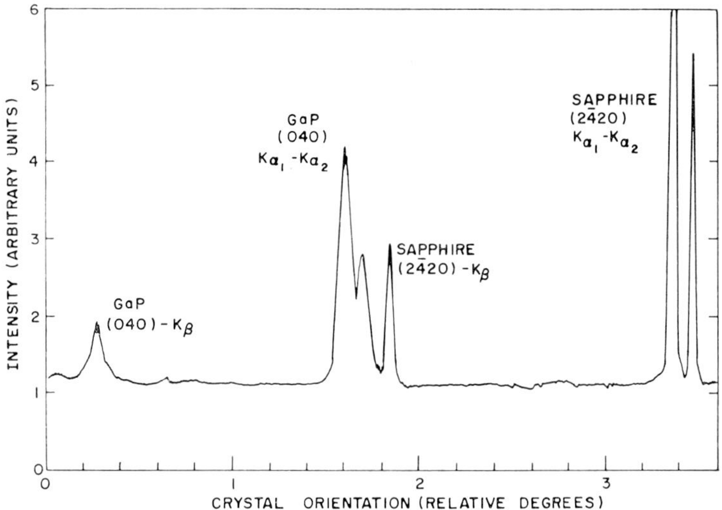

Fig. 7.   Trace of diffracted intensity versus crystal orientation for (100) GaP layer grown on ($\bar{1}$012) sapphire. Peaks from both the substrate and layer are observed. (Silver radiation.)

diffracted peaks from the layer and substrate and comparing this value with that expected for exactly parallel epitaxy. Fig. 6 shows a trace of intensity versus angular setting of the crystal obtained using the Lang camera as a diffractometer, where the crystal is the (111) GaP on (0001) sapphire combination shows in figs. 3 and 4. The separation of the K_{α_1} sapphire ($1\bar{2}10$) peak and K_3 GaP ($02\bar{2}$) peak is 1.55°, a difference of 0.12° from the calculated value of 1.67°. An indication of the error in this measurement can be seen in that the observed separation of the K_{α_1} and K_3 peaks for ($1\bar{2}10$) sapphire and ($02\bar{2}$) GaP agree to within $\pm 0.01°$ of the value calculated. A similar trace is shown in fig. 7 for the (100) GaP on (100) Si/($\bar{1}$012) sapphire shown in fig. 5. The difference between the observed and calculated separation of the K_{α_1} lines for (040) GaP and ($2\bar{4}20$) sapphire is 0.09°. The orientation of the GaP layer on both (0001) and ($\bar{1}$012) sapphire substrates is around a tenth of a degree off exact parallel epitaxy. Note also in fig. 7 that the GaP crystallites are sufficiently well oriented to permit resolution of the silver K_α doublet (the doublet separation is 0.095).

4. Conclusions

The successful growth of epitaxial GaP on insulating substrates provides a new composite system which is of scientific and technical interest. The epitaxial layers are composed of misoriented subgrains with sizes on the order of 1–10 μm and misorientations in the range of

$\pm 0.1°$. The film perfection and growth characteristics depend inherently on the substrate perfection and surface preparation. It is anticipated that the future improvement on substrate quality will lead to improved heteroepitaxial films.

Acknowledgment

The authors are grateful to N. Pastal for film deposition work. They also wish to thank J. T. McGinn and D. A. Kramer for experimental measurements, and J. F. Corboy for growing thin film silicon on sapphire. Review of this work by G. W. Cullen and D. Richman is appreciated.

References

1) T. S. LaChapelle, A. Miller and F. L. Morritz, in: *Progress in Solid State Chemistry*, Vol. 3, Ed. H. Reiss (Pergamon, London, 1967).
2) G. W. Cullen, G. E. Gottlieb, C. C. Wang and K. H. Zaininger, J. Electrochem. Soc. **116** (1969) 1444.
3) C. C. Wang, G. E. Gottlieb, G. W. Cullen, S. H. McFarlane III and K. H. Zaininger, Trans. Met. Soc. AIME **245** (1969) 441.
4) D. J. Cumin, J. Electrochem. Soc. **117** (1970) 95.
5) H. M. Manasevit, Appl. Phys. Letters **12** (1968) 156.
6) H. M. Manasevit and W. I. Simpson, J. Electrochem. Soc. **116** (1969) 1725.
7) P. Rai-Choudbury, J. Electrochem. Soc. **116** (1969) 1745.
8) C. C. Wang, J. Appl. Phys. **40** (1969) 3433.
9) A. R. Lang, J. Appl. Phys. **30** (1959) 1748.
10) E. H. Putley, *The Hall Effect and Related Phenomena* (Butterworths, London, 1960).

Journal of Crystal Growth **13/14** (1972) 268–271 © North-Holland Publishing Co.

CRYSTAL GROWTH AND PERFECTION OF LARGE Nd:YAG SINGLE CRYSTALS

ROGER F. BELT, RICHARD C. PUTTBACH and DONALD A. LEPORE

Airtron, Division of Litton Systems, Inc., 200 East Hanover Avenue, Morris Plains, New Jersey 07950, U.S.A,

Single crystals of Nd:YAG are [111] grown in production quantities to a size of 4 cm diameter and 20 cm long. The principal boule defects are a 2 mm diameter {211} faceted core, strain lines along ⟨110⟩, and impurity striae characteristic of the growth interface. The initial distribution coefficient of Nd was found to be 0.15. The Nd concentration along the boule length increases about 20–30% in 20 cm. Lang X-ray topographs of slices cut along [111] clearly reveal the defective core and the striae distribution. Excessive strain in the boules is visible under crossed polarizers and in Twyman–Green interferograms. The cause of strain is probably a growth rate larger than the equilibrium rate required to incorporate a fixed amount of Nd. Efforts to increase the amount of Nd in the crystal by Lu additions have been fruitful but the same defects remain in Nd, Lu:YAG when examined by X-ray topography.

1. Introduction

Single crystals of neodymium doped yttrium aluminum garnet (Nd:YAG) have been grown[1,2]) in the laboratory and the main features were characterized[3,4]). Unfortunately most of the grown boules were 1–2 cm in diameter and a limited yield of high quality laser rods is obtained. The widespread utilization of the Nd:YAG laser now demands a quality rod at low cost. Progress in the growth has continued to where the diameter of rough boules approaches 4–5 cm. The rod yield is increased proportionately but each growth variable must be controlled to a greater degree. In this paper some of the persistent growth problems are discussed along with their effects on boule quality and laser rod operation.

2. Experimental

Single crystals of Nd:YAG are grown by the Czochralski method with RF apparatus designed for optimum thermal configuration, automatic pulling and diameter control, and excellent stability over a three week period. Starting materials are purified oxides of at least 99.999% assay. The Nd_2O_3 is controlled to have a low Dy content. Growth is performed in 7–11 cm diameter iridium crucibles under an enclosed atmosphere of N_2–Ar–O_2. For the largest boules (5 cm diameter × 25 cm long), special starting methods and annealing cycles are used to prevent thermal strain.

Nd analyses are performed by X-ray emission and neutron activation. Sample preparation for Lang topography consisted of slicing, mechanical diamond polishing to a thickness of 0.25 mm, and a final chemical polish in 85% H_3PO_4. The thickness of the slice was 0.1–0.2 mm. Topographs are recorded by means of a Rigaku–Denki Lang Camera using AgK$_\alpha$ radiation. Exposures were made on Kodak Type M Film at scan rates of 1 mm/hr.

Cylindrical laser rods are core drilled, cut, and polished to specifications of $\lambda/10$ on flatness, ± 4 sec of arc on parallelism, and ± 5 min on perpendicularity. A Perkin–Elmer Model 723 Twyman–Green interferometer is used to evaluate rod quality.

Fig. 1. Large boule of single crystal Nd:YAG (scale in inches).

3. Results and discussion

3.1. BOULE GROWTH

Large boules (fig. 1) have been grown under relatively high thermal gradients and at growth rates of ~ 1 mm/hr. The growth rate is dependent on the amount of Nd_2O_3 initially incorporated into the melt. At a high boule content of Nd (1.5 at%) the growth rate is lowered to get a high quality material. Likewise the undoped YAG can be grown at a faster rate. While diameter control is important, it is imperative to maintain the liquid solid interface shape at a reproducible curvature. Most of our boules are grown along a [111] direction although other directions have been used with no deleterious effects on growth rate or core structure. The core diameter remains small (1–2 mm) even when the boule diameter increases. Thus in accordance with the analysis of Brice[5]) the supercooling or interface curvature changes.

Large optically clear boules of Nd:YAG grown along [111] exhibit a roughly hexagonal cross section due to faceting of {211} planes. Ridges are directed towards {110} planes. Fig. 2a is a typical example of a section viewed along [111]. Even though the core is but a few mm in diameter, the strain pattern may extend to several times this. Fig. 2b shows the strain exhibited by the core. Laser rods cannot be cut from such an area. Fig. 2c is the corresponding Twyman–Green interferometer pattern of the boule. The hexagonal strain pattern is highly visible and divides the boule into six nearly equivalent pie-shaped areas. Depending on the diameter one or more laser rods of the best quality can be cut from the strain-free areas. High areas of strain are directed along $\langle 110 \rangle$ in the (111) plane while $\langle 211 \rangle$ are much more strain free. Any boule which contains cracks, precipitates, or strained areas visible under crossed polars is generally not suitable for laser rods.

3.2. NEODYMIUM DISTRIBUTION

Primarily because of the larger radius of the Nd^{+3} compared to Y^{+3}, the dopant distributes itself unevenly between the grown crystal and the melt. From measurements on more than 50 boules, the initial distribution coefficient, $k_i = C_S/C_L$, was found to be 0.12–0.15 or slightly smaller than that reported by Cockayne[2]). The Nd concentration along a 20 cm boule continually increases until crystal quality becomes very

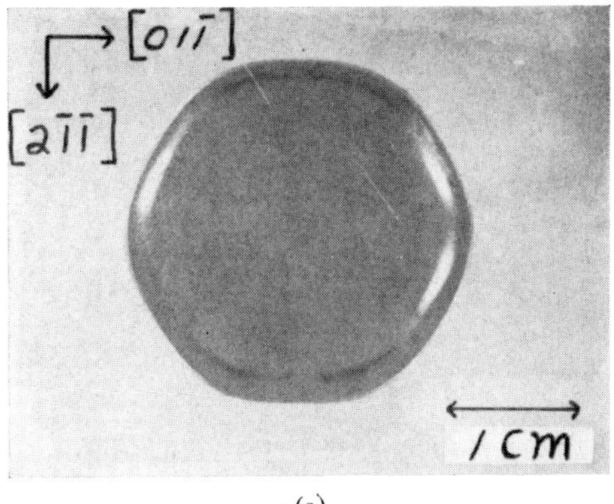

(a)

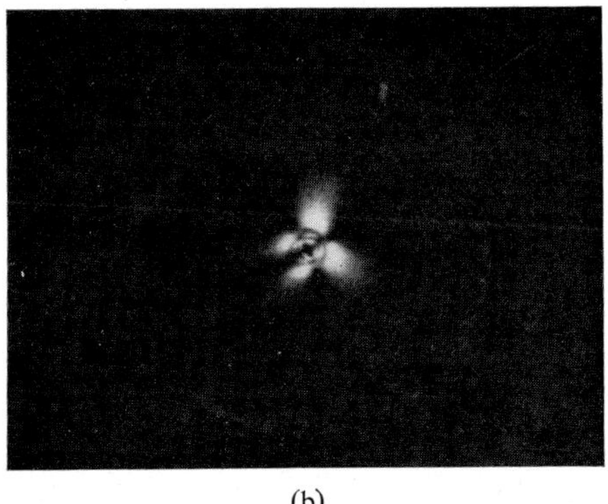
(b)

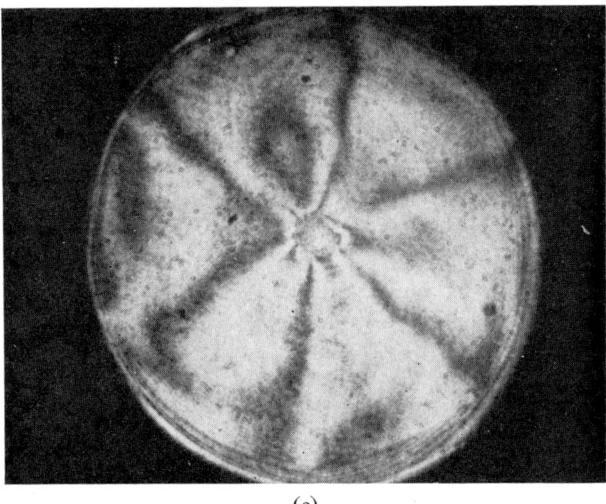
(c)

Fig. 2. (a) Cross section of [111] grown boule in ordinary light. (b) Same boule under crossed polarizers. (c) Twyman–Green interferometer pattern of polished boule.

poor. The axial gradient corresponds to a 20–30% Nd increase over the length measured from the seed to the terminus. Thus a 30–50 mm long laser rod may have an absolute end to end variation of about 0.05–0.10 at% if the initial solid concentration, C_S, is about 1 at%. The maximum rod to rod variation, caused by cutting

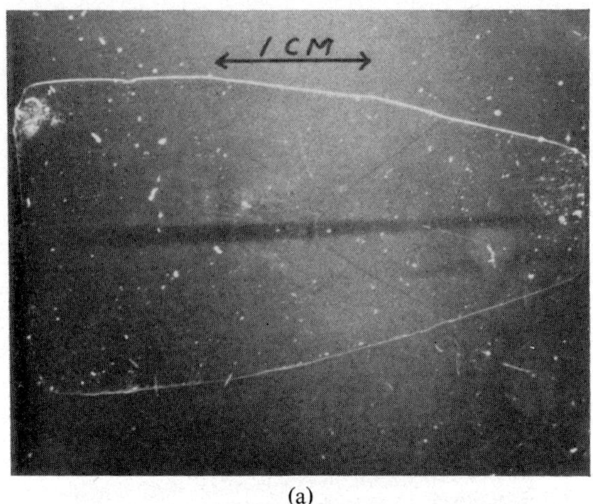

(a)

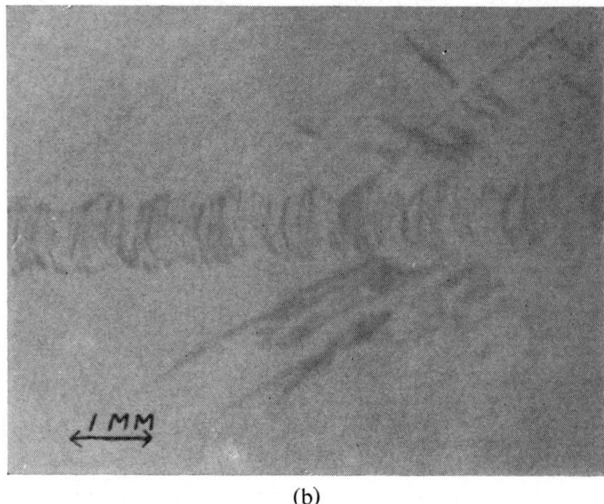

(b)

Fig. 3.   (a) Section of boule near seed; slice cut parallel to [111] growth axis; crossed polarizers. (b) Lang X-ray topograph showing core structure, striae, and strained areas. Ag K_α radiation, (444) reflection.

of rods from different areas of a boule, is no more than 0.1–0.15 at%. Radial gradients are less than $\pm 0.03\%$ except for the core area which is about 20% higher in Nd than the remaining area.

The theoretical and experimental distributions of Nd along a 20 cm boule of Nd:YAG were also compared. Portions of the boule were sampled along the length and the Nd found by X-ray emission. The theoretical curve was calculated from the equation

$$C_l = C_0 k_i (1-g)^{k_i - 1}.$$

In this equation C_l is the Nd concentration at length l, C_0 is the initial melt concentration, k_i is the distribution coefficient, and g is the fraction of the melt pulled. The agreement between the measured and calculated values was excellent. The experimental variations of Nd in different laser rods discussed above is evident. This probably causes some differences in energy for identical rods examined in the same laser cavity.

3.3. STRIAE FORMATION AND DISTRIBUTION

X-ray topographs and etching studies on many large boules grown in our production laboratory have proved that dislocation densities are less than 100/cm². The nost common residual defects are impurity striae precipitating on the growing interface. In Nd:YAG the likely impurity is Nd_2O_3 although striae still occur in undoped crystals. Thus either the Y_2O_3 or Al_2O_3 plays a supporting role. The striae are present in all high quality boules and laser rods. Sometimes they can be

viewed visually but X-ray topographs give the best record of their shape and distribution. The strain arising from striae does not affect the laser quality of rods and is not visible on a Twyman–Green interferogram. Striae are thought to be caused by temperature fluctuations in the melt which arise from convection or external heat variation.

Fig. 3a is a boule slice cut parallel to [111] and photographed under crossed polarizers. Areas of strain were present but not cracks or visible precipitates. Good quality laser rods could not be cut from the boule because of the strain. Fig. 3b is the Lang X-ray topograph from the same slice taken using (444) reflection. The core region consists of periodic striae due to facet formation. The screw like form[6]) is probably due to a slight asymmetry of the [111] growth and the thermal axis coupled with a change in the crystal diameter. The latter is computed to be about 1 mm/hr. The screw is not observed in crystal boules of uniform diameter. The period of the screw is about 1 mm and bears a relation to the pull rate (0.6 mm/hr) but not the rotation rate (40 rpm). The angle between the core striae and (111) was measured to be 20–25°. Thus the principal planes of the striae appear to be {211} and not {110}. Outside the core the striae are convex into the melt and are irregular in period, width, and intensity on an X-ray topograph. Highly strained lines are bunched together and overlap many striae. The strained areas frequently contain micro cracks.

Fig. 4 is an X-ray topograph of a section cut from a

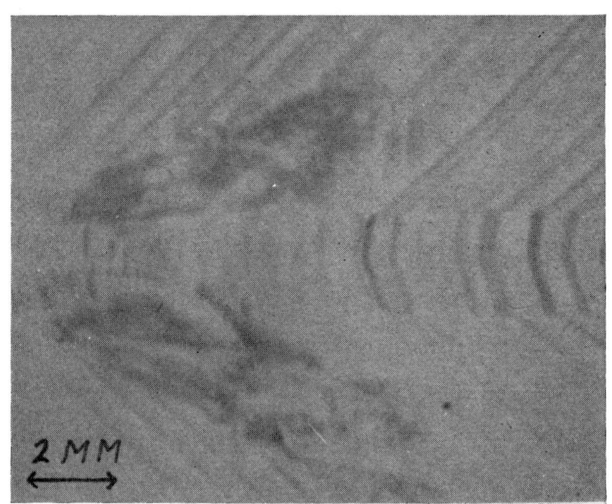

Fig. 4. X-ray topograph of constant diameter boule; slice cut parallel to [111], Ag K_α radiation, (444) reflection.

uniform diameter boule. The (444) reflection was again used. The core diameter is almost constant but small fluctuations can be noted. At the point where core striae are absent or very irregular, strained areas and micro cracks appear outside the core and lead to a damaged boule. Subsequently the core heals, facets appear again, and strain is absent outside the core. It is not known what initiates the strained areas although thermal fluctuations and Nd_2O_3 solubility are the most likely contributors. If the growth rate is very low and constant, the interface supercooling should be equal both on and off $\{211\}$ facets. Obviously for Nd:YAG, a growth rate must be chosen where quality remains high. From our experiments the suggested rates will depend on the level of Nd required in the crystal. Solutions to the problem include growth at a constant composition or programming the rate to reflect composition changes.

3.4. LUTETIUM ADDITIONS TO Nd:YAG

Methods to increase the concentration of Nd in YAG have been discussed previously[7]. Size compensation by Lu has been suggested as one of the most efficient procedures to retain high optical quality of the rod. Production growth of Nd,Lu:YAG in large boules has shown a negligible increase of the Nd distribution coefficient with additions of up to 10 mole% of Lu_2O_3.

Higher Lu levels may seriously impair the quality and lower growth rates must be used. In identical laser cavity geometries, the Lu doped rods show an increase of 30% in output energy for equivalent input energy. The effects of Lu are real and beneficial but it is very difficult to quantitatively assess the important parameters of each rod. Lang X-ray topographs of Nd:YAG and Nd,Lu:YAG boule sections show identical defects and strain distributions. It has also been suggested[8] that some of the effects of Lu may be attributed to a better matching of lamp spectral output and pump bands of the crystal or spectral absorption changes in the ultraviolet region. We feel that the benefits attributed to Nd,Lu:YAG might also be obtained with crystals containing a high Nd concentration and no Lu.

4. Conclusions

The production growth of Nd:YAG laser crystals has progressed to a reliable state of 4×20 cm boules. Core, strain, and striae are the remaining defects. Nd levels vary along a boule and eventually lead to observable strain. High quality boules come only from stringent control of growth rate, thermal configuration, and interface shape. Lu additions give some advantages to Nd:YAG but do not promote better growth or change defects.

Acknowledgements

The authors wish to thank T. DeStefano for growth of the crystals, J. Moss for the X-ray topographs, and R. Thornton for rod testing. A portion of this work was also performed under a U.S. Army Electronics Command Contract No. DAAB05-71-C-2611.

References

1) M. Kestigian and W. W. Holloway, Jr., in: *Crystal Growth*, Ed. H. S. Peiser (Pergamon, New York, 1967) p. 451.
2) B. Cockayne, Phil. Mag. **12** (1965) 943.
3) J. Basterfield, M. J. Prescott and B. Cockayne, J. Mater. Sci. **3** (1968) 33.
4) B. Cockayne, J. Crystal Growth **3, 4** (1968) 60.
5) J. C. Brice, J. Crystal Growth **6** (1969) 205.
6) K. Marizane, A. Witt and H. C. Gatos, J. Electrochem. Soc. **114** (1967) 738.
7) M. Kestigian and W. W. Holloway, Jr., J. Crystal Growth **3, 4** (1968) 455.
8) J. Strozyk, private communication, Nov. 1970.

OBSERVATIONS DE DISLOCATIONS DANS DES CALCITES SYNTHÉTIQUES

F. LEFAUCHEUX

Laboratoire de Minéralogie–Cristallographie, Université Paris VI, 9 quai St Bernard, Paris V°, France

Hydrothermally grown single crystals of calcite have been examined. They are of two types: crystals grown from a rhomboedral seed and spontaneously nucleated plates. Lattice defects have been studied by X-ray Lang topography. The dislocation densities, their distribution and preferred orientations are described. An interpretation is proposed for the origin of the many dislocation bundles observed.

1. Introduction

L'étude entreprise au laboratoire a pour objet la comparaison des défauts présentés par des monocristaux de calcite synthétique avec ceux déjà mis en évidence dans de nombreuses calcites naturelles[1,2]). Il était intéressant de choisir une méthode de croissance se rapprochant le plus du processus naturel, introduisant un nombre limité de défauts, et permettant de travailler avec ou sans germe; la méthode hydrothermale nous a semblé préférable. Elle a déjà été mise en œuvre par Ikornikova[3]) et Pogodin[4]).

2. Détails expérimentaux

Les cristaux ont été obtenus par voie hydrothermale dans des autoclaves de 1 litre avec des solutions de chlorure d'ammonium (5 % en poids) et de chlorure de lithium (7 % en poids) à une température de 200 °C, la différence de température entre la zone de dissolution et la zone de cristallisation étant de 10 °C)[5]. La pression totale du remplissage est de 200 bar. Le temps de croissance est de l'ordre de 1 mois. On dispose dans l'autoclave d'une série de germes obtenus par clivage $(10 \times 3 \times 2 \ mm^3)$ et fixés sur une tige de téflon. On observe d'une part, la croissance sur ces germes de cristaux rhomboédriques avec des troncatures (111), avec triplement dû au poids environ, soit une vitesse de 150 µm par jour, d'autre part des plaquettes (111) d'environ un cm^2 nucléées spontanément. Nous avons étudié par topographie aux rayons X des cristaux obtenus dans ces conditions, soit au laboratoire de minéralogie cris-

tallographie de l'Université Paris VI, soit à l'institut de cristallographie de Moscou. Leurs répartitions de défauts sont analogues. Ce sont principalement des dislocations, des inclusions et des bandes de croissance. Avant examen, les échantillons sont attaquées pendant 30 sec dans une solution à 10 % d'acide chlorhydrique dans l'éthanol.

3. Résultats

3.1. CRISTAUX POUSSÉS SUR GERMES

Les lames examinées ont été clivées à partir du cristal selon les orientations représentées sur la figure 1.

3.1.1. *Distribution des dislocations*

Les dislocations se présentent en général sous forme de gerbes prenant naissance principalement à la surface du germe et, en nombre beaucoup moins grand'à partir d'inclusions solides au sein du nouveau cristal. On peut ainsi distinguer dans le cristal deux types de régions: les unes, situées à l'aplomb du germe et contenant une densité de dislocation comprise entre 10^3 et 10^4 lignes-cm^2, les autres où la densité reste inférieure à 500 dislocations par cm^2 et la qualité cristalline reste très bonne, sauf à l'intersection des pyramides de croissance des différentes faces. Les topographies 2 et 5 de lames de type b clivées transversalement, 3 d'une lame de type a clivée longitudinalement et contenant le germe, 4 d'une lame de type c, clivée longitudinalement, très prés du germe mais ne le coupant pas, illustrent cette répartition. Les régions A, B, C, D de la figure 2 montrent des zones sans dislocations ou presque.

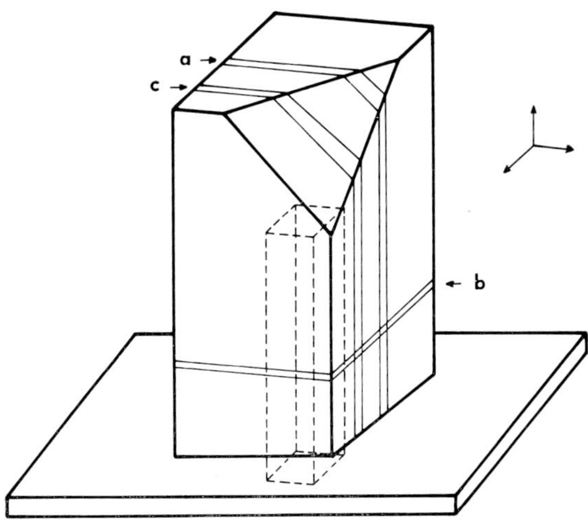

Fig. 1.   Un cristal synthètique sur son support.

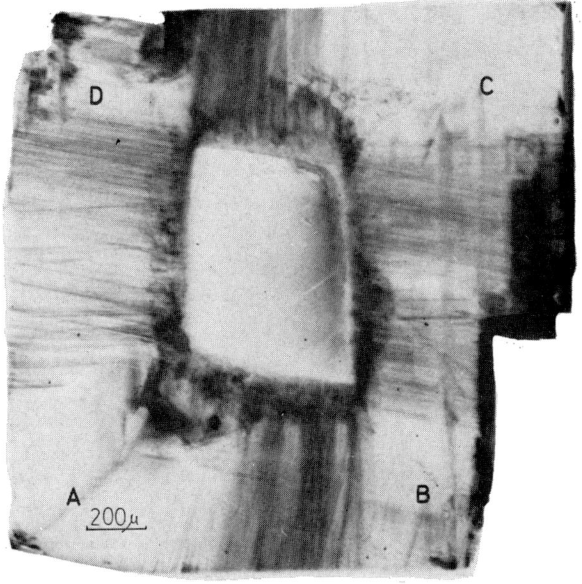

Fig. 2.   Cristal I$_3$; lame 1 de type b. Plan réflecteur (100) 5×4.5 mm².

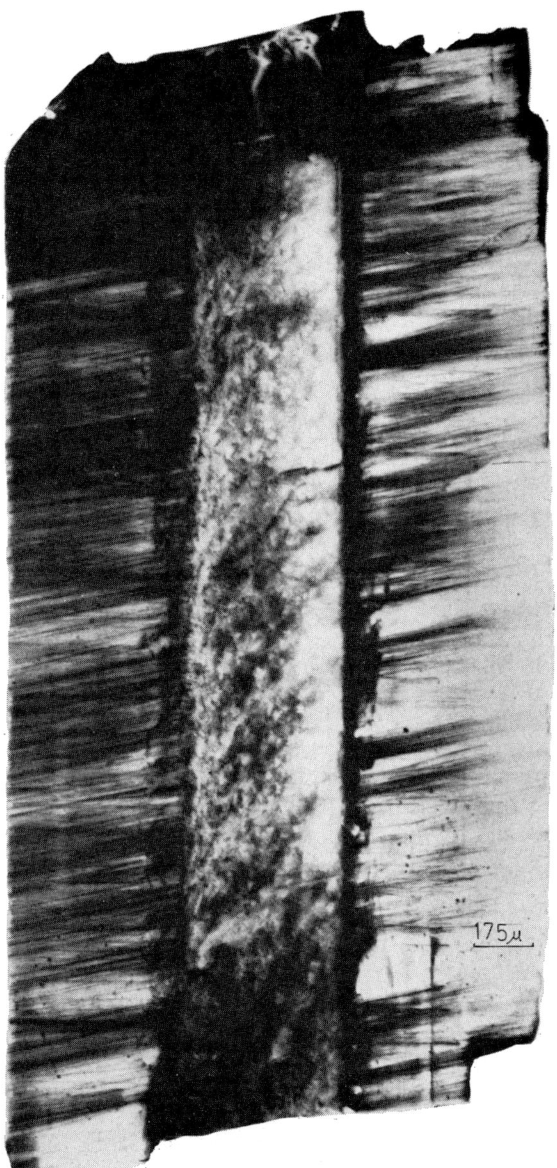

Fig. 3.   Cristal I$_3$; lame 2 de type a. Plan réflecteur (010) 5.5×11 mm²; barre d'échelle: 1 mm.

3.1.2. *Orientation*

Les dislocations le plus souvent rectilignes, partent en faisceaux suivant deux directions principales: l'une située dans un plan perpendiculaire aux faces de croissance (100), l'autre parallèle à une arête de rhomboèdre de clivage ⟨100⟩. Jusqu'à présent la détermination du vecteur de Burgers de ces dislocations n'a pu être faite car nous n'avons par observé d'extinction avec les réflexions utilisées qui sont les plus intenses et les plus commodes du point de vue géométrique (100) et (1$\bar{1}$0). Le même effet a été constaté par Ikeno[6]) dans le chlorure de sodium qui a pourtant une structure de plus haute symétrie.

3.1.3. *Origine probable des gerbes de dislocations*

La présence d'inclusions solides induit dans le cristal des contraintes qui sont, en général, supérieures à la limite élastique et sont relachées par la formation de

Fig. 4. Cristal I_2; lame de type c. Plan réflecteur (010) 4×8 mm²; barre d'échelle: 1 mm.

Fig. 5. Cristalpoussé au laboratoire; lame de type b. Plan réflecteur (010) 5×8 mm²; barre d'échelle: 1 mm.

gerbes de dislocations. Cet effet est particulièrement visible sur la figure 4 et également a été observé dans des cristaux naturels de dolomite[7]).

La majorité des dislocations présentes prennent naissance à la surface du germe. Dans les expériences effectuées jusqu'à présent, elle sont indépendantes de la qualité cristalline du germe mais dépendent essentiellement des conditions initiales de la croissance et de l'état de surface du germe. Les premières couches déposées ont, en général une très mauvaise qualité cristalline et sont, dans certains cas, accompagnées d'un dépôt d'inclusions solides visibles ensuite optiquement. Pour que cette zone soit le moins imparfaite possible, la montée en température de l'autoclave doit se faire lentement avec une température identique dans les zones de dissolution et de croissance. Ce n'est que lorsque la température de cristallisation est atteinte que l'on peut augmenter la température de la zone de dissolution de façon à amorcer le transport de matière.

Des modifications des conditions de croissance en cours d'expérience peuvent, elles aussi, induire des zones de mauvais cristal à l'origine de gerbes de dislocations.

Fig. 6.   Plaquette (111). Plan réflecteur (11̄0) 10×6 mm²; barre d'échelle: 1 mm.

3.2. PLAQUETTES (111)

Les plaquettes observées ont une épaisseur d'une à plusieurs centaines de microns. Optiquement très claires, on ne distingue que des ressauts de croissance selon les directions $\langle 11̄0 \rangle$, intersections des faces de croissance $\{100\}$ avec le plan (111).

La figure 6 révèle un grand nombre de dislocations dont on peut dégager certaines caractéristiques:

– dislocations partant de précipités, au niveau de bandes de croissance, par exemple en A.

– dislocations perpendiculaires aux faces de croissance.

Ces deux propriétés avaient déjà été notées sur les précédentes topographies.

– dislocations se courbant au niveau de fautes (en B puis en C) pour prendre finalement une direction perpendiculaire à la face (010) de croissance rapide.

– dislocations très grossièrement rectilignes, parallèles au front de croissance (010).

Ces courbures de dislocations ainsi que des dislocations perpendiculaires aux faces de croissance avaient déjà été observées dans des calcites naturelles par Sauvage[1]).

Ces plaquettes ont, d'autre part, permis la mise en évidence de bandes de croissance et la détermination de leur vecteur faute[5]).

4. Conclusion

Les cristaux de calcite obtenus par synthèse hydrothermale ont une bonne qualité cristalline: faible densité de dislocation et désorientations très faibles. L'origine des dislocations est principalement associée aux stades initiaux de la croissance à partir du germe. La répartition des défauts présente beaucoup d'analogies avec celle trouvée dans les cristaux de calcite naturelle, ce qui n'est pas le cas de ceux obtenus par la méthode de tirage à partir de la zone fondue[8]). Les expériences en cours ont pour objet de préciser les conditions d'apparition des défauts et de les appliquer à l'interprétation des conditions de croissance des cristaux naturels.

Bibliographie

1) A. Authier et M. Sauvage, J. Phys. (Paris) **27** (1966) 137.
2) M. Sauvage et A. Authier, Bull. Soc. Franc. Minéral. Crist. **88** (1965) 879.
3) N. I. Ikornikova, dans: *Crystal Growth*, Vol. 4, Ed. N. N. Sheftal (Engl. Transl.) p. 76.
4) Yu. V. Pogodin et V. M. Sergeev, dans: *Crystal Growth*, Vol. 7, Ed. N. N. Sheftal (Engl. Transl.) p. 163.
5) F. Lefaucheux, Bull. Soc. Franc. Minéral. Crist. **94** (1971).
6) S. Ikeno, H. Maruyama et N. Kato, J. Crystal Growth **3, 4** (1968) 683.
7) A. Zarka, Bull. Soc. Franc. Minéral. Crist. **92** (1969) 160.
8) J. J. Brissot, Congrès de Croissance Cristalline, ICCG-3, Marseille, 1971.

       Journal of Crystal Growth **13/14** (1972) 276–281 © *North-Holland Publishing Co.*

PERFEKTION VON CaF$_2$-EINKRISTALLEN IN ABHÄNGIGKEIT VON DER ZUCHTTECHNIK

R. LECKEBUSCH und K. RECKER

Mineralogisch-Petrologisches Institut der Universität, Poppelsdorfer Schloß, 53 Bonn, Germany

To investigate the dependence of crystal perfection on growth conditions, a number of calcium fluoride crystals were grown with different growth techniques and growth parameters. The crystals were examined for stress induced birefringence with the polarizing microscope, for dislocations with chemical etches and for anion vacancies with X-irradiation. The following growth techniques were used: (1) Growth from the melt by the (a) Bridgman- and (b) Czochralski-methods, (2) growth by sublimation, (3) flux growth (CaCl$_2$/NaF), and (4) gel growth (starch). These techniques involve both different growth temperatures (melt growth 1405 °C, vapour phase growth ca. 1405 °C, flux growth ca. 1000 °C, gel growth ca. 20 °C) and different growth rates (melt growth ca. 1–10 mm/h, vapour phase growth ca. 1 mm/h, flux growth ca. 0.1 mm/h, gel growth ca. 10^{-4} mm/h). The melt grown and vapour phase grown crystals had sizes of some cm, the flux grown and gel grown crystals of about 1 mm. The crystals were etched with 3 M HCl and X-rayed by copper radiation. Dependent on the growth techniques and different growth parameters the dislocation densities varied from 10^2 (vapour phase grown crystals) up to 10^6 cm^{-2} (Czochralski grown crystals), the anion vacancies from 10^{17} (vapour phase grown crystals) up to 10^{19} cm^{-3} (Czochralski grown crystals).

1. Einführung

Synthetische Einkristalle sind Realkristalle, d.h., sie enthalten mehr oder weniger zahlreich Baufehler und Verunreinigungen aller Art. Ihre Perfektion (Freiheit von Verunreinigungen und Baufehlern) ist naturgemäß von den Bildungsbedingungen abhängig, also vom Zuchtverfahren und den Zuchtparametern, vor allem von der Wachstumsgeschwindigkeit (Übersättigung), der Wachstumstemperatur, dem Milieu, dem Temperungsprogramm nach der Züchtung u.a.m. Um die Abhängigkeit der Perfektion von den Bildungsbedingungen an einem wissenschaftlich wie technisch gleich interessanten Material grundlegend zu untersuchen, wurden CaF$_2$-Einkristalle nach verschiedenen Verfahren mit unterschiedlichen Bedingungen gezüchtet und die Spannungsdoppelbrechung sowie die Anzahl der in ihnen enthaltenen Anionenfehlstellen (Punktdefekte) und Versetzungen (lineare Defekte) bestimmt. Über die ersten Ergebnisse dieser Arbeiten wird berichtet.

2. Experimentelles

2.1. ZÜCHTUNGEN

Untersucht wurden CaF$_2$-Kristalle, die aus Schmelze (Bridgman- und Czochralskiverfahren), Dampfphase,

Fluxmittel und Gel gezüchtet wurden. Ausgangsmaterial für die Schmelz- und Dampfzüchtungen waren hochreine, im Vakuum nach dem Bridgmanverfahren aus CaF$_2$-Pulver (Reinheit: "für optische Zwecke") mit Zusatz von PbF$_2$ vorgezüchtete CaF$_2$-Kristalle. Für die Flux- und Gelzüchtungen wurden p.a.-Substanzen verwendet. Eine Auswahl der gezüchteten und untersuchten Kristalle ist in Tab. 1 zusammengestellt. Die Tabelle enthält außerdem einige Zuchtparameter, Angaben über phänomenologische Eigenschaften und die Untersuchungsergebnisse. Einzelheiten zu den Zuchtverfahren:

(a) Bridgmanverfahren: Benutzt wurde eine Vakuumapparatur eigener Konstruktion[1]) mit einem Rezipienten aus Sinterkorundrohr und Rhodium-Außenheizung. Bei den Züchtungen wurde dem bereits durch Kristallisation vorgereinigten CaF$_2$-Ausgangsmaterial z.T. nochmals ca. 2 Gew.-% PbF$_2$ zugemischt, um eine möglichst große Sauerstofffreiheit zu erreichen. Verwendete Senkgeschwindigkeiten des Tiegels (= Wachstumsgeschwindigkeiten der Kristalle): 1, 5, 10 und 15 mm/h.

(b) Czochralskiverfahren: Zur Züchtung diente eine A.D. Little-Züchtungsapparatur mit HF-Heizung. Die CaF$_2$-Einkristalle wurden aus Graphittiegeln bei 0.5 atm Argon-Überdruck unter Verwendung von

TABELLE 1

Gezüchtete CaF$_2$-Einkristalle mit Zuchtparametern, phänomenologischen Eigenschaften, Spannungsdoppelbrechung, Versetzungsdichten und Anionenfehlstellendichten

Krist. Nr.		1	2[a]	3	4	5	6	7
Zuchtverfahren		Bridgman	Bridgman	Czochralski	Czochralski	Dampf	Flux	Gel
Wachstumstemperatur (°C)		1405	1405	1405	1405	$\leqq$1405	$\sim$1000	25
ΔT im Krist. (°C · cm^{-1})		$\sim$10	$\sim$10	$\sim$100	$\sim$100	$\sim$10	$\sim$0	0
Wachstumsgeschw. (mm h^{-1})		10	10	6	1.2	1.2	$\sim$0.1	$\sim$10^{-4}
Zuchtgefäß		Graphit	Graphit	Graphit	Graphit	Graphit	Platin	Polyacryl
Atmosphäre		Vakuum 10^{-3} Torr	Vakuum 10^{-3} Torr	Argon 1.5 atm	Argon 1.5 atm	(Vakuum 10^{-6} Torr) Eigendampfdruck	(Luft)	Gel
Krist.- Größe (mm)	l	30	30	25	25	25	1[b]	1[b]
	$\varnothing$	15	15	15	15	15		
Wachstumsform		zylindrisch	zylindrisch	$\sim$zylindrisch	$\sim$zylindrisch	zylindrisch	{111}	{100}
Transparenz		klar durchsichtig	klar durchsichtig	klar durchsichtig	klar durchsichtig	klar durchsichtig	klar durchsichtig	klar durchsichtig
Spannungs-Doppelbrechg.[c]		0 bis w	0 bis w	w bis m	w bis m	0 bis w	0	0
Geätzte Flächen		(111)-Spaltflächen	(111)-Spaltflächen	(111)-Spaltflächen	(111)-Spaltflächen	(111)-Spaltflächen	(111)	(100)
Versetzungsdichte (10^5 cm^{-2})		2.9	2.75	6.5[d] 10.5[e]	14.5	<0.001	<0.001	<0.001
Anionenfehlstellendichte[f] (10^{18} cm^{-3})		6.3	0.9	1.3[d] 5.0[e]	7.5	0.45	sehr hoch	sehr gering

Anmerkungen zur Tabelle:
[a] Mit 2 Gew.-% PbF$_2$.
[b] Kantenlänge.
[c] 0 = keine; w = schwache; m = mittlere.
[d] Wachstumsanfang.
[e] Wachstumsende.

[f] Berechnet nach Ref. 5:

$$N = 1.29 \times 10^{12} \frac{n}{(n^2+2)^2} k_{\mathrm{m}} \frac{H}{f}$$

(n = Brechungsindex, k_{m} = Absorptionskonstante in cm^{-1}, H = Halbwertsbreite in eV, f = Oszillatorenstärke).

Impfkristallen gezüchtet. Der Tiegel war thermisch isoliert, der wachsende Kristall nicht. Das Argon-Schutzgas (Reinheit 99.99%) wurde keiner weiteren Reinigung unterworfen. Die Impfkristalle [hergestellt nach dem unter (a) beschriebenen Bridgmanverfahren] wurden mit 4 U/min gedreht. Die Ziehgeschwindigkeiten betrugen 1.2, 6 und 12 mm/h. Es wurden keine besonderen Maßnahmen (Einschnürung, Nachheizung) zur Verringerung der Versetzungsdichten bei der Züchtung getroffen.

(c) Züchtung aus der Dampfphase: Die Dampfzüchtungen erfolgten in vertikaler Anordnung in einer Hochvakuumapparatur eigener Konstruktion[2]). Als Heizung diente eine Graphitwendel, als Zuchtgefäß ein Graphittiegel mit spezieller Impfspitze (Konus + kugelförmige Erweiterung + Kapillare)[3]). Die Verdampfungstemperatur betrug 1500 °C, die Ziehgeschwindigkeit (= Wachstumsgeschwindigkeit der Kristalle) 1.2 mm/h. Verfahrensbedingt lag zu Beginn der Züchtungen im Zuchtgefäß ein Vakuum von ca. 10^{-6} Torr vor. Nach der Keimbildung und dem Verschließen der Impfkapillaren durch die Kristallisation wuchsen die Kristalle unter dem eigenen Dampfdruck.

(d) Züchtung aus Fluxmittel: In Anlehnung an Sinyukova und Stepanov[4]) wurden in verschlossenen Platintiegeln Mischungen aus NaF, CaCl$_2$ (sicc.) und NaCl (z.B. im Molverhältnis 36.3:18.3:45.4) an normaler Atmosphäre aufgeschmolzen, ca. 13 h auf einer Temperatur von 890 °C gehalten, mit 15 °C/h auf ca. 760 °C und danach schnell auf Raumtemperatur abge-

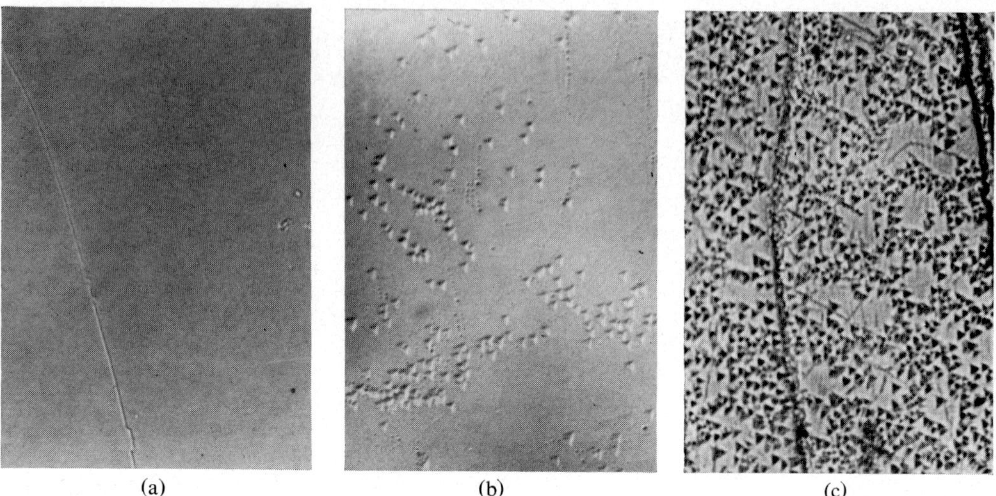

Abb. 1.   Angeätzte (111)-Spaltflächen von CaF$_2$-Kristallen. (a) Kristall Nr. 5, gezüchtet aus dem Dampf; (b) Kristall Nr. 2, gezüchtet nach dem Bridgmanverfahren; (c) Kristall Nr. 3, gezüchtet nach dem Czochralskiverfahren. Vergrößerung 120×.

kühlt. Es bildeten sich zahlreiche, klar durchsichtige CaF$_2$-Oktaeder mit Kantenlängen bis zu 1 mm. Diese wurden durch Lösen des Schmelzkuchens mit Wasser isoliert.

(e) Gelzüchtung: In ein zylindrisches Zuchtgefäß aus Polyacryl, zur Hälfte gefüllt mit Stärkegel, wurde ein zylindrisches Polyacrylrohr getaucht, das den oberen Gefäßteil in zwei konzentrische Räume teilte. Im inneren Raum wurde das Gel vorsichtig überschichtet mit 1-normaler KF-Lösung, im äußeren Raum mit 1-normaler CaCl$_2$-Lösung. Ca^{++}- und F$^-$-Ionen diffundieren in das Gel ein und reagieren hier zu CaF$_2$. Nach 6 Monaten hatten sich im Gel klar durchsichtige CaF$_2$-Würfel mit Kantenlängen bis 1 mm gebildet.

2.2 Versetzungsdichte

Die Versetzungsdichten wurden durch Anätzen der CaF$_2$-Kristalle und Bestimmen der Ätzgrubendichte (Mikroskop) ermittelt. Als Ätzmittel diente 3-normale HCl. Die Ätzzeiten betrugen 1 bis 10 Stunden.

2.3. Fehlstellendichte

Die Kristalle wurden zur Farbzentrenbildung mit Röntgenstrahlen (Cu, 45 kV, 15 mA) bestrahlt, die Farbzentrenabsorption spektralphotometrisch vermessen und aus der Intensität und Halbwertsbreite der Maxima (bei Sättigung) die Zahl der Farbzentren errechnet. Diese Zahl ist ein Maß für die Anionenfehlstellendichte[3]).

3.   Untersuchungsergebnisse

In Tab. 1 sind die Untersuchungsergebnisse (Stärke der Spannungsdoppelbrechung, Versetzungs- und Fehlstellendichten) für eine Anzahl untersuchter CaF$_2$-Einkristalle zusammengefaßt.

3.1. Zu den Versetzungsdichten

Abb. 1 zeigt als Beispiel angeätzte (111)-Spaltflächen der CaF$_2$-Einkristalle Nr. 5, 2 und 3 (Tab. 1).

(a) Dampfgezüchtete Kristalle: Die sehr geringe Zahl der Versetzungen (Tab. 1) ergibt sich zwanglos aus der kleinen Wachstumsgeschwindigkeit, dem relativ kleinen Temperaturgradienten im wachsenden Kristall und dem verfahrensbedingten geringen Angebot und Einbau von Verunreinigungen (besonders oxidischer Phasen).

(b) Fluxgezüchtete Kristalle: Wegen der geringen Größe der Kristalle ließ sich die effektive Zahl der Versetzungen nicht genau bestimmen. Sie liegt etwa in der gleichen Größenordnung wie die der dampfgezüchteten Kristalle. Die geringe Zahl der Versetzungen ist auf die sehr kleine Wachstumsgeschwindigkeit und den praktisch fehlenden Temperaturgradienten zurückzuführen. Die ohne Zweifel vorhandenen Einbauten des Fluxmittels wirken sich auf die Versetzungsdichte anscheinend nicht aus.

(c) Gelgezüchtete Kristalle: Hier gilt Ähnliches wie für die fluxgezüchteten Kristalle.

(d) Kristalle, gezüchtet nach dem Bridgmanverfahren: Die relativ große Zahl der Versetzungen in diesen Kristallen ergibt sich zwangsläufig aus der Kristallisation im Tiegel und der großen Wachstumsgeschwindigkeit.

(e) Kristalle, gezüchtet nach dem Czochralskiverfahren: Der Vergleich zahlreicher Zuchtprodukte zeigte, daß bei diesen weniger die Wachstumsgeschwindigkeit als vielmehr der starke Temperaturgradient im wachsenden Kristall sowie Einbau von Sauerstoff (evtl. submikroskopische Ausscheidung von CaO[7]) primäre Ursachen der Versetzungen sind.

3.2. ZU DEN FEHLSTELLENDICHTEN

Voraussetzung für die Bildung von Farbzentren sind Anionenfehlstellen, die durch reine Schottky- oder Frenkeldefekte oder (bei CaF$_2$) durch Einbau einwertiger Fremdkationen bzw. zweiwertiger Fremdanionen hervorgerufen werden. In CaF$_2$-Kristallen ist die Zahl der Schottky- und Frenkelfehlstellen vernachlässigbar gering gegenüber den durch Fremdionen verursachten Fehlstellen. Einwirkung weicher Röntgenstrahlung führt i.allg. nur zur Besetzung vorhandener Fehlstellen mit Elektronen und nicht zur Erzeugung neuer Fehlstellen. Die Farbzentrenbildung im CaF$_2$ ist deshalb ein Maß nicht nur für die Zahl der Anionenfehlstellen, sondern auch für die Menge der vorhandenen Verunreinigungen (insbes. O^{--}, Na$^+$). Abb. 2 zeigt als Beispiel die Bestrahlungsspektren der Kristalle Nr. 5, 2 und 3 (Tab. 1).

Die Spektren zeigen zwei Maxima (370 und ca. 600 nm), die zwei verschiedenen Zentrenarten angehören, welche durch Sauerstoffeinbau hervorgerufen werden. Das kurzwellige Maximum wird sog. F-Zentren zugeordnet[6]) (Elektron in Anionenfehlstelle, Übergang 1s→2p), während die Zuordnung des langwelligen Maximums noch offen ist. Die Spektren zeigen bereits qualitativ, daß die Farbzentrendichte in den verschiedenen Kristallen sehr unterschiedlich ist. Entsprechend der Stärke der Maxima zeigen die bestrahlten Kristalle eine schwache bis starke Blaufärbung. In Kristallen, die nach dem Czochralskiverfahren gezüchtet wurden, ist die Zahl der Fehlstellen in den verschiedenen Kristallteilen unterschiedlich groß (im Gegensatz zu den Kristallen Nr. 2 und 5).

Abb. 3 zeigt die Abhängigkeit der Intensitäten des F-Zentrenmaximums (370 nm) von der Bestrahlungs-

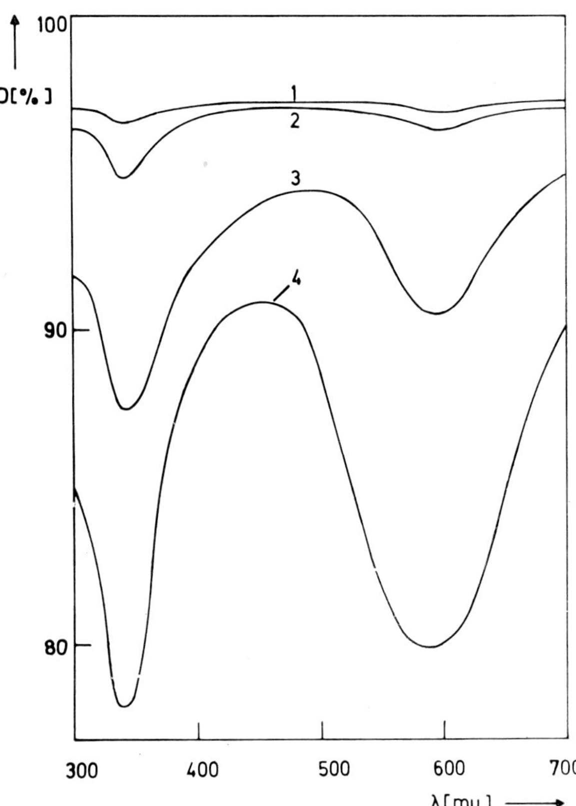

Abb. 2. Absorptionsspektren (300–700 nm) röntgenbestrahlter CaF$_2$-Einkristalle (Bestrahlungszeit 2 min). (Kurve 1) Kristall Nr. 5, gezüchtet aus dem Dampf; (Kurve 2) Kristall Nr. 2, gezüchtet nach dem Bridgmanverfahren; (Kurve 3) Kristall Nr. 3, gezüchtet nach dem Czochralskiverfahren, Wachstumsanfang; (Kurve 4) Kristall Nr. 3, Wachstumsende (D = Durchlässigkeit).

zeit. (Ein analoges Verhalten zeigt das Maximum bei 600 nm).

(a) Dampfgezüchtete Kristalle: Sie enthalten die geringste Zahl von Anionenfehlstellen und damit auch die kleinste Sauerstoffmenge. Selbst nach längerer Bestrahlung zeigen sie keine visuelle Farbe und nur sehr geringe Farbzentrenbildung. Ursache ist die Züchtung im abgeschlossenen System aus reinem CaF$_2$-Dampf.

(b) Kristalle, gezüchtet nach dem Bridgmanverfahren: Mit Zusatz von Fluorierungsmitteln (PbF$_2$) gezüchtete Kristalle enthalten erwartungsgemäß wesentlich weniger Fehlstellen und damit auch Sauerstoff als Kristalle, die ohne PbF$_2$-Zusatz gezüchtet wurden. Eine Änderung der Fehlstellendichte in Abhängigkeit von den anderen Zuchtparametern (z.B. der Wachstumsgeschwindigkeit) wurde nicht festgestellt.

(c) Gelgezüchtete Kristalle: Die Farbzentrendichte ließ sich hier wegen der geringen Größe der Kristalle nicht quantitativ bestimmen. Die Kristalle zeigten aber auch nach langer Bestrahlungszeit keine sichtbare Verfärbung, d.h., sie enthalten nur wenige Anionenfehlstellen.

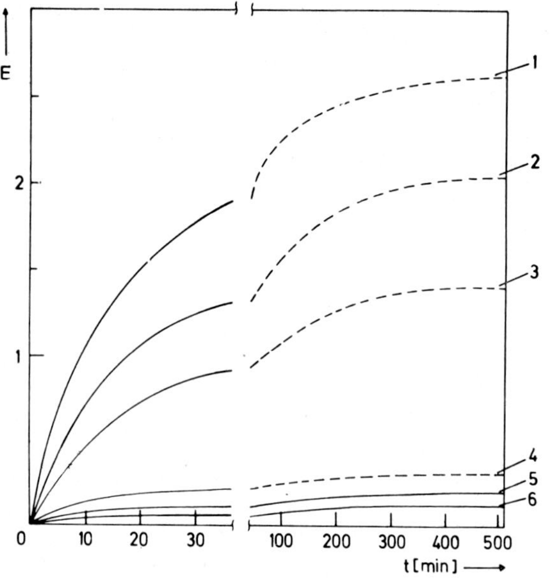

Abb. 3. Abhängigkeit der Intensität des F-Zentrenmaximums von der Bestrahlungsdauer. (Kurve 1) Kristall Nr. 4, gezüchtet nach dem Czochralskiverfahren (WG = 1.2 mm/h); (Kurve 2) Kristall Nr. 1, gezüchtet nach dem Bridgmanverfahren; (Kurve 3) Kristall Nr. 3, gezüchtet nach dem Czochralskiverfahren (WG = 6 mm/h), Wachstumsende; (Kurve 4) Kristall Nr. 3, Wachstumsanfang; (Kurve 5) Kristall Nr. 2, gezüchtet nach dem Bridgmanverfahren (+PbF$_2$); (Kurve 6) Kristall Nr. 5, gezüchtet aus dem Dampf (E = Extinktion, t = Bestrahlungszeit, WG = Wachstumsgeschwindigkeit, – – – = extrapolierte Kurventeile).

(d) Fluxgezüchtete Kristalle: Auch hier ließ sich die Farbzentrendichte nicht quantitativ bestimmen. Die Kristalle zeigten aber bereits nach kurzer Bestrahlungszeit eine starke Blaufärbung, ein Beweis für eine große Fehlstellendichte. Diese kann durch hohe Sauerstoffgehalte als auch durch Substitution von Ca^{++}- durch Na^+-Ionen (Fluxmittel) hervorgerufen sein.

(e) Kristalle, gezüchtet nach dem Czochralskiverfahren: Die hohen Fehlstellendichten werden durch große Sauerstoffgehalte (stammend aus dem Ausgangsmaterial oder der Argonatmosphäre) hervorgerufen. Entsprechend zeigen die Zuchtkörper am Wachstumsende größere Fehlstellendichten als am Wachstumsanfang und langsam gezüchtete Kristalle

(lange Züchtungszeit) größere Fehlstellendichten als schnell gezüchtete Kristalle.

Der Zusammenhang zwischen Farbzentrenentstehung und O^{2-}-Gehalten der Kristalle läßt sich durch die UV-Spektren belegen. Die UV-Absorptionskante des völlig reinen CaF$_2$ liegt bei 130 nm. Kleine Mengen

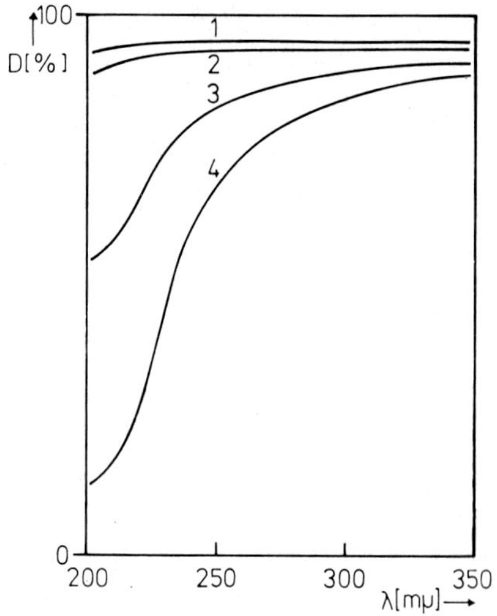

Abb. 4. Absorptionsspektren (200–350 nm) von CaF$_2$-Einkristallen. (Kurve 1) Kristall Nr. 5, gezüchtet aus dem Dampf; (Kurve 2) Kristall Nr. 2, gezüchtet nach dem Bridgmanverfahren; (Kurve 3) Kristall Nr. 3, gezüchtet nach dem Czochralskiverfahren, Wachstumsanfang; (Kurve 4) Kristall Nr. 3, Wachstumsende (D = Durchlässigkeit).

von O^{2-}-Verunreinigungen im CaF$_2$ erzeugen bei $\lesssim$ 200 nm ein starkes Absorptionsmaximum. Die Stärke der Absorption bei 200 nm (Meßgrenze) ist deshalb ein Maß für den O^{2-}-Gehalt der Kristalle. Als Beispiel zeigt Abb. 4 die Absorptionsspektren der CaF$_2$-Einkristalle Nr. 5, 2 und 3 (Tab. 1) im Bereich von 200–350 nm.

Die Spektren (Abb. 4) weisen unterschiedlich starke Absorption bei 200 nm auf. Der Spektrenvergleich aller Kristalle zeigt, daß die dampfgezüchteten Kristalle und die schmelzgezüchteten Kristalle mit Fluorierungszusatz weitgehend sauerstofffrei sind, während alle übrigen Kristalle mehr oder minder Sauerstoff enthalten. Das UV-Absorptionsverhalten der CaF$_2$-Kristalle (Tab. 1) entspricht ungefähr dem gefundenen Verhältnis der Farbzentrenkonzentrationen. Hierdurch wird bestätigt, daß die Farbzentrenbildung weitgehend vom Sauerstoffgehalt der Kristalle abhängt.

Der Deutschen Forschungsgemeinschaft möchten wir an dieser Stelle für die Unterstützung des Forschungsvorhabens und die Bereitstellung einer A. D. Little-Kristallzuchtapparatur sowie eines Spektralphotometers Cary 14R herzlich danken.

Literatur

1) K. Recker und R. Leckebusch, Feinwerktechnik **74** (1970) 506.

2) K. Recker und R. Leckebusch, J. Cryst. Growth **9** (1971) 274.

3) K. Recker und R. Leckebusch, J. Cryst. Growth **5** (1969) 125.

4) I. A. Sinyukova und I. V. Stepanov, in: *Growth of Crystals*, Vol. 2, Eds. A. V. Shubnikov und N. N. Sheftal (Consultants Bureau, New York, 1959).

5) J. H. Schulman und W. D. Compton, *Color Centers in Solids* (Pergamon, Oxford, 1963).

6) J. Arends, Phys. Status Solidi **7** (1964) 805.

7) M. S. Abrahams und P. G. Herkart, J. Appl. Phys. **36** (1965) 274.

 Journal of Crystal Growth **13/14** (1972) 282–284 © *North-Holland Publishing Co.*

THE PREPARATION, CRYSTAL GROWTH AND PERFECTION OF DOUBLE HALIDES OF CsNiCl$_3$ TYPE*

D. E. COX and F. C. MERKERT

Brookhaven National Laboratory, Upton, New York 11973 U.S.A.

Several double chlorides and bromides of CsNiCl$_3$ type have been prepared by dehydration of aqueous solutions. Crystals of RbFeCl$_3$, RbNiCl$_3$, and CsMnBr$_3$, 0.5–1 cm^3 in volume have been grown from the melt by the Bridgman technique. The mosaic spreads determined by neutron diffraction techniques range from 15 to 30 min of arc.

1. Introduction

There has been a rapidly growing interest in the past two or three years in the physical properties of mixed halides containing 3d transition metal ions. One particular topic which is currently receiving a good deal of attention is the study of magnetic behavior in one- or two-dimensional systems. Two-dimensional behavior is exhibited in a particularly striking fashion in mixed fluorides isostructural with K$_2$NiF$_4$, in which there are widely separated layers of magnetically coupled Ni ions[1]). The principal experimental technique used in these investigations is neutron diffraction, which depends, however, on the availability of suitably large and perfect crystals (around 0.5–1 cm^3 in volume).

More recently, systems which might be expected to exhibit one-dimensional properties have been sought, and the most promising candidates appear to be compounds with the hexagonal CsNiCl$_3$ type of structure[2]). Comparatively little is known about the preparation, phase relationships and crystal growth in systems of this type, as previous work on ABX$_3$ compounds has been confined mainly to oxides and fluorides with the simple cubic perovskite structure, in which there is cubic close-packing of the A and X ions. When A is Rb or Cs, B a divalent transition metal ion, and X is Cl, the CsNiCl$_3$ type of structure predominates, however. In this structure, the Cs and Cl ions are in hexagonally close-packed layers, and the Ni ions are in

octahedrally coordinated chains along the c axis. Adjacent chains are about 7 Å apart, and it is this feature which is conducive to one-dimensional magnetic behavior.

The lack of data for these compounds is undoubtedly due in part to the fact that they are markedly hygroscopic and inconvenient to handle. Moreover, previous syntheses have generally required as an initial step the preparation of the appropriate transition metal dihalides, themselves also hygroscopic[2,3]). The present paper describes an alternative and simpler method of synthesis, together with the crystal growth and characterization by neutron diffraction techniques of some mixed chlorides and bromides of this type. As far as is known, the bromides have not been previously reported in the literature.

2. Synthesis and crystal growth

The preparation of anhydrous transition metal chlorides and bromides by dehydration of hydrates in a stream of the hydrogen halide is a useful technique for elements on the right hand side of the first row[4]), and a similar technique was utilized in the present study. For the Mn, Co, and Ni compounds, appropriate amounts of Rb$_2$CO$_3$ or Cs$_2$CO$_3$ (Varlacoid Chemical Company, nominally 99.9%) and the transition metal carbonate (Baker Analyzed grade) were dissolved in dilute HCl or HBr. The solutions were evaporated to dryness under an infrared lamp. Care was taken to avoid overheating and thus prevent the formation of hydrolysis products as much as possible. The coarsely

* Work performed under the auspices of the U.S. Atomic Energy Commission.

powdered products were transferred to vitreous graphite boats, heated for three hours just above their melting points in an atmosphere of anhydrous HCl or HBr which had been purified by passage through cold traps maintained at $-78\,°C$ or $-60\,°C$ respectively, and cooled in a stream of dry helium. In the case of the iron compounds, Fe_2O_3 was used as a starting material, and after evaporation the products were first heated in H_2 until reduction was essentially complete, then in an approximately 1:1 mixture of H_2 and HCl, and finally cooled in a stream of helium.

TABLE 1

Hexagonal lattice constants and melting points ($\pm10\,°C$) of some ABX_3 compounds

Compound	Lattice constants (Å)	Melting point (°C)
$RbFeCl_3$	$a = 7.05_4, c = 6.02_1$	525
$RbNiCl_3$	$a = 6.95_2, c = 5.90_5$	760
$CsMnBr_3$	$a = 7.61_6, c = 6.51_6$	565
$RbFeBr_3$	$a = 7.42_5, c = 6.31_7$	440
$RbCoBr_3$	$a = 7.34_3, c = 6.26_7$	375
$CsCoBr_3$	$a = 7.52_9, c = 6.32_4$	470

TABLE 2

Chemical analysis figures (%) for some ABX_3 compounds; theoretical values are in parentheses

	$RbFeCl_3$	$RbNiCl_3$	$RbCoBr_3$	$CsCoBr_3$
A	34.5(34.5)	34.5(34.1)	23.5(22.3)	30.9(30.8)
B	22.7*(22.6)	23.4(23.4)	15.4(15.3)	13.4(13.7)
X	42.9(42.9)	42.4(42.5)	63.0(62.4)	55.6(55.5)
Total	100.1	100.3	101.9	99.9

* Total Fe. Separate determination of Fe^{2+} gave 22.5%.

The final products were found to be hygroscopic in varying degrees, and were handled in a plastic glove bag containing P_2O_5 as a desiccant. X-ray powder patterns were of quite good quality and showed a single hexagonal phase in all cases with the lattice parameters listed in table 1. In particular, no traces of any hydrolysis products were observed. Wet chemical analyses were performed on some of the compounds, and the results (see table 2) were generally quite satisfactory with the exception of a rather high value for Rb in $RbCoBr_3$.

In a number of cases, the pseudo-binary phase diagrams of ACl-BCl_2 systems of this type have been reported, and in most instances the $ABCl_3$ compounds melt congruently at relatively low temperatures (in the region of 500–600 °C)[5]). The Bridgman technique was therefore chosen for crystal growth. An approximate melting point was determined in each case (see table 1), the compounds were sealed in suitably pointed quartz ampoules about 1 cm in diameter, and lowered at rates between 1–3 mm/h through a temperature gradient of about 10 °C/cm. The quartz did not appear to be attacked to any significant extent, and the resulting boules, about 1.5–2 cm long, could be removed very easily. Owing to the ease of hydration and likelihood of surface attack, conventional Laue X-ray techniques are not very useful for characterization, and an examination was made with the neutron diffraction technique instead. The samples were found to cleave quite readily, and were mounted on a cleavage face in a cylindrical thin-walled aluminum holder having an O-ring flange. Neutrons are very advantageous in this respect, as they have high penetrating power, and rocking curves are therefore representative of the whole sample. A measure of the crystal perfection is given by the rocking curve, which ideally is Gaussian in shape with a full-width at half-maximum Γ (the "mosaic spread"). This

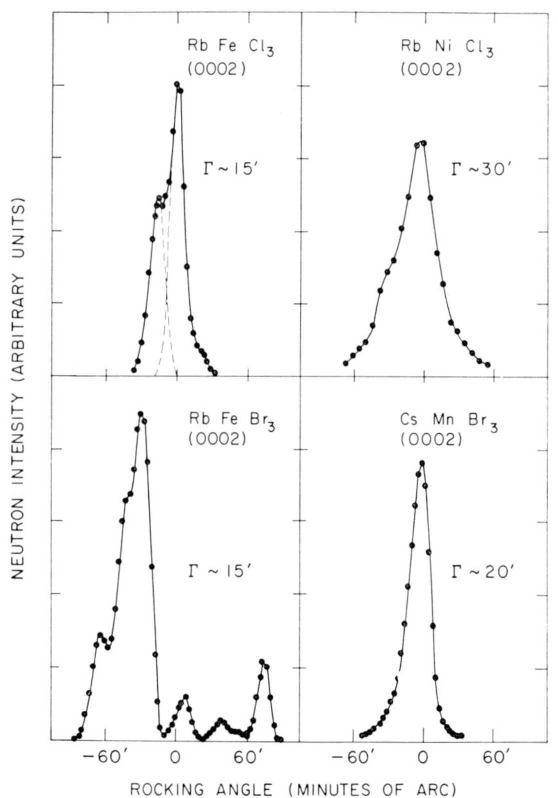

Fig. 1. Neutron rocking curves for some ABX_3 compounds. Γ is the mosaic spread (full-width at half-maximum).

quantity (appropriately corrected for instrumental resolution) is related to the standard deviation η of the mosaic blocks by the expression

$$\Gamma = 2\eta(2 \ln 2)^{\frac{1}{2}}.$$

Large, suitably cleaved pieces from each boule (0.5–1.0 cm^3 in volume) were mounted as described above and the resulting rocking curves for $RbFeCl_3$, $RbNiCl_3$, and $CsMnBr_3$ were quite encouraging (fig. 1). The crystals have mosaic spreads of 15–30 min of arc, typical of fairly imperfect mosaic crystals. Some asymmetry of the curves is evident in the case of $RbFeCl_3$ and $RbNiCl_3$, which is indicative of low angle boundaries. This is also a very common finding in fluoride systems. The results for $RbFeBr_3$, $RbCoBr_3$ and $CsCoBr_3$ were less satisfactory. A typical curve is shown in fig. 1 for $RbFeBr_3$, in which there is clearly a much more pronounced substructure over an angular range of some 2°, although the individually resolved peaks have a relatively narrow half-width of about 15 min of arc.

Further work, including vapor transport growth, is currently in progress in an effort to improve the quality of some of the crystals.

Acknowledgement

We wish to gratefully acknowledge the assistance of J. K. Rowley in performing the chemical analyses.

References

1) R. J. Birgeneau, H. J. Guggenheim and G. Shirane, Phys. Rev. B **1** (1970) 2211.
2) N. Achiwa, J. Phys. Soc. Japan **27** (1969) 561.
3) L. H. Pierce, J. Crystal Growth **8** (1971) 295.
4) R. Colton and J. H. Canterford, *Halides of the First Row Transition Elements* (Wiley-Interscience, New York, 1969) pp. 7, 11.
5) E. M. Levin, C. R. Robbins and H. F. McMurdie, *Phase Diagrams for Ceramists* (American Ceramic Society, 1964), and *Supplement* (1969).

Journal of Crystal Growth **13/14** (1972) 285–291 © *North-Holland Publishing Co.*

GROWTH AND PERFECTION OF ZnS CRYSTALS GROWN FROM THE VAPOR PHASE

I. T. STEINBERGER, E. ALEXANDER, Y. BRADA, Z. H. KALMAN, I. KIFLAWI and S. MARDIX*

The Racah Institute of Physics, The Hebrew University, Jerusalem, Israel

The mechanism of polytype formation in vapor-phase grown ZnS crystals is reviewed. The role of $60°$ dislocations in the determination of the photovoltaic properties of ZnS is stated. It is demonstrated that in the crystals there exist well-defined uniform regions virtually free of stacking faults and dislocations, except for the generating screw.

1. Introduction

Vapor-phase grown ZnS crystals seem to be the only system in which the mechanism of polytype formation is well understood at present. The same crystals are also distinguished by the occurrence of uncommon photoelectric effects; among them the most spectacular one is the appearance of photovoltages which may reach hundreds of volts[1,2] ("Anomalous Photovoltaic Effect", abbreviated as APE). These effects imply the existence of built-in electric fields. In the present paper experimental results will be presented to illustrate that there exist crystallographically uniform regions (cubic, hexagonal or polytypic) in the crystals which are virtually free of dislocations and faults. It will be shown that faults and dislocations are concentrated on planes separating uniform regions. The internal fields responsible for the occurrence of the APE will be attributed to the occurrence of $60°$ dislocations at these boundaries.

2. Results

2.1. Crystal growth and polytype formation

ZnS crystals were grown from ZnS vapor in the presence of H_2S, HCl or a mixture of HCl and Ar, by conventional techniques[3,4]. The temperature of the charge was usually about $1250\,°C$, and the crystals grew in a region which was kept at about $1150\,°C$.

Observation of the growth through a quartz window showed that at first a polycrystalline substrate was formed and subsequently thin needles grew in the direction of the c-axis. The needles often broadened to platelets with the main faces parallel to a (10.0) or $(2\bar{1}.0)$ plane (fig. 1).

The crystals obtained are usually not uniform but consist of parallel strips, all perpendicular to the c-axis. The various strips differ from each other in birefringence and often in the tilt of external faces as well (fig. 2a). Many strips are wide enough (≈ 0.05 mm) for X-ray diffraction studies. Oscillation photographs[5-14] revealed that uniform strips are structurally homogeneous and usually of high crystallographic perfection. They may have the hexagonal wurtzite structure, the cubic sphalerite structure or one of many possible polytypic structures.

A large number of experiments, using different techniques led to the following mechanism of polytype formation in the crystals[15,16]: the crystals grow as hexagonal 2H (wurtzite) needles around a screw dislocation having a Burgers vector which is either equal to the repeat distance c of the 2H structure, or to an integral multiple of $c*$. The latter cases – which are much less common – are the only ones conducive to polytype formation. While the crystals cool down in the growth tube, stacking faults bounded by Shockley

---

* Present address: University of Rhode Island, Kingston, Rhode Island 02881, U.S.A.

* Instead of a single "large" screw dislocation one may visualize the growth as taking place around several cooperating screws, each having a Burgers vector of strength c. The needles in question have a hollow croe along the c-axis.

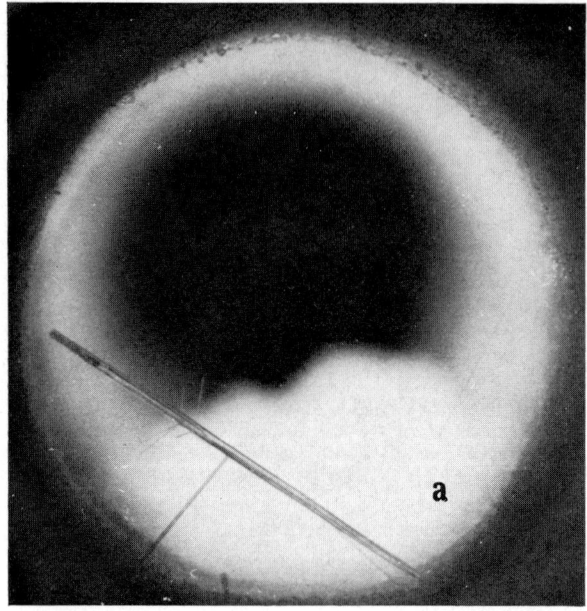

1 a

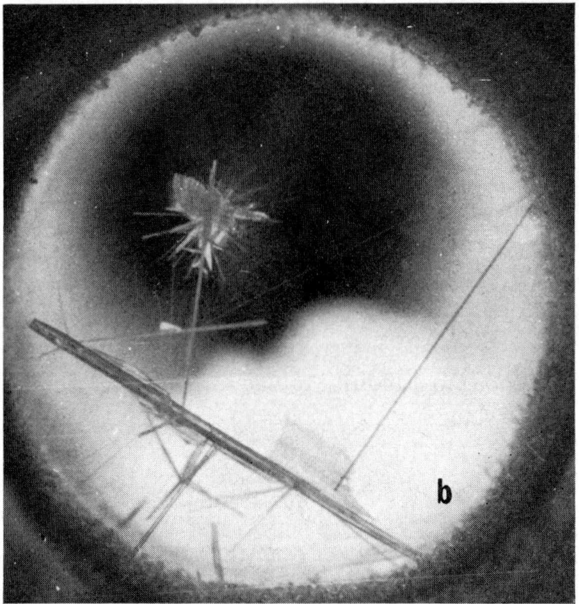

1 b

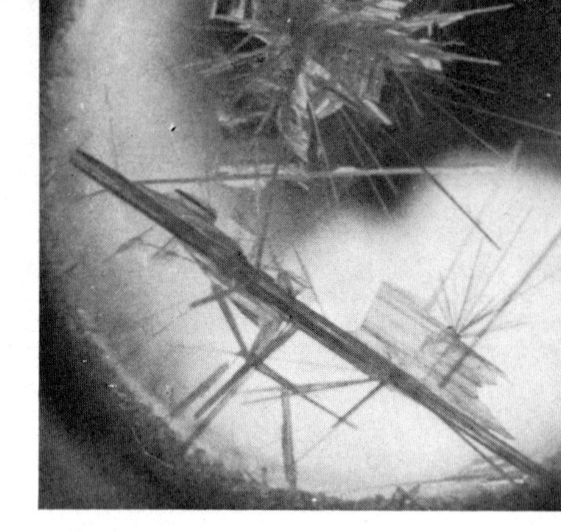

1 c

Fig. 1. (a), (b) and (c) three stages of growth of ZnS crystals.

partials are introduced, probably by mechanical or thermal stresses. Since the low temperature structure of ZnS is that of sphalerite, the stacking faults in wurtzite have negative energies in a wide temperature range and thus they tend to expand, pushing the partials in front of them. On the other hand, the presence of the [00.1] screw dislocation implies a topology simi-lar to that of a multiple spiral ramp. Thus the stacking faults expand along the spiral ramp surfaces to large distances along the c-axis. Accordingly, the transformation creating the polytype occurs by means of a periodic slip process[17]) of close packed planes, thereby introducing a plastic shear in the entire transformed region manifested (in addition to the structure change) by a tilt of the crystal faces[18]). It should be noted that expansion of stacking faults bounded by dislocation dipoles is also a very common feature. This process changes the structure without involving tilt[19]).

2.2. FAULTS AND DISLOCATIONS

There are several kinds of experimental data indicating that the wider strips of uniform birefringence are of uniform structure*, and are usually free of stacking faults and dislocations. "Wider" means here having a width of more than 1 micron; at narrower strips the difficulties of optical resolution prevent us from making a definite statement.

Fig. 3a represents an X-ray oscillation photograph of a typical uniform region of a ZnS crystal grown by the above method. It should be emphasized that the reflection spots are very sharp and well-defined. This is the experience with all the polytypes hitherto identi-

* "Double polytypes" are exceptions to this rule.

2 a

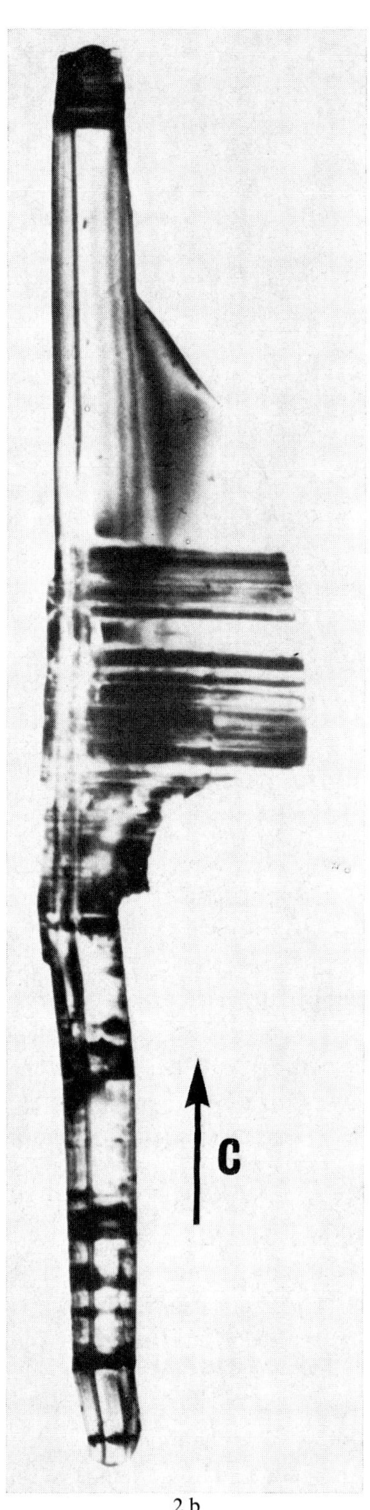

2 b

Fig. 2. (a) Microphotograph between partially crossed polarizers of a ZnS crystal containing several polytypic regions; enlargement: ×20. (b) Lang topograph of the same crystal using (00.2) reflection (×20).

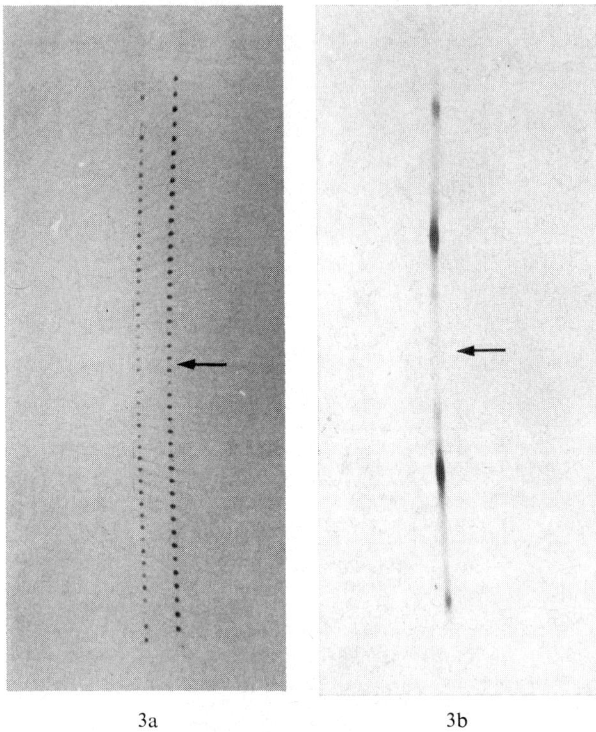

3a   3b

Fig. 3. (a) (10.l) row line obtained from the oscillation photograph (CuKα and Kβ) of a ZnS polytype 84R (5 5 5 3 3 3 2 2)₃. (b) (10.l) row line of a heavily faulted region of a ZnS crystal.

fied (more than 160) in the crystals grown in Jerusalem[5-14]). This result shows conclusively that the most common crystal faults expected to exist in the polytypes, namely stacking faults, are absent from the uniform regions. In contrast, fig. 3b contains the oscillation photograph of a heavily faulted part of a crystal, which included many different narrow regions; here the smearing out caused by the stacking faults is apparent. Both the uniformity of structure within each strip and the absence of stacking faults indicate that there are no partial dislocations within the strip.

Another indication for the quality of the uniform crystal regions concerned can be obtained by considering the etch pit pattern of (10.0) and (2$\bar{1}$.0) faces of ZnS crystals. The etching was performed in a dilute H_2O_2 solution[4]). It is seen that triangular etch pits often appear in rows at the interfaces of two adjoining regions (fig. 4). On the other hand, experiments on hundreds of crystals showed that etch pits within uniform regions are very rare. Since there is no exact and consistent relationship between structure change and the etch pits, they cannot be attributed to points of

emergence of partials; moreover, such partials should probably have an energy which is too low for etch pit formation.

The most probable perfect dislocation in these types of structure has a Burgers vector equal to the nearest neighbor distance and parallel to the close-packed direction within the basal planes. The dislocation line also lies in one of the close-packed directions within the basal plane. The etch pits mark, in all probability, the points of emergence of these "60° dislocations".

Lang X-ray diffraction topographs utilizing the basal reflection common to all polytypes show a dislocation line along the entire long axis of the needles. The direction of the dislocation line changes slightly at the boundary between various polytype regions, in accordance with the tilt angle between these regions. The trace of this dislocation disappears in prismatic plane topographs, thus establishing that its Burgers vector is perpendicular to the basal planes. Since the tilt of the various regions occur only after the crystals have grown[18]), it follows that the dislocation in question was originally a pure screw. Mardix et al.[20]) have shown that the dislocation has a large Burgers vector in crystals which contain polytypes: the length of the Burgers vector is in full accord with that expected from the periodicity of the polytypes involved. The dislocation in question can thus be identified with the generating screw dislocation, around which the original growth took place.

Apart from occasional single 60° dislocations within uniform regions, the only other faults hitherto observed by Lang topography occur at boundaries between polytypes. Contrast appears at such boundaries both in basal plane topographs (fig. 2b) and prismatic reflection topographs (fig. 5). Because of the approximately uniform darkening it is rather difficult to assess their identity. The main contribution to the contrast observed in the basal plane topographs is probably a consequence of the difference of the interplane spacings of the various structures. Regarding the prismatic-plane contrast, two kinds of dislocations can be safely assumed to exist at the boundaries which may contribute to it, namely Shockley partials and 60° perfect dislocations. Moreover, differences of nearest neighbor distances within the basal planes should contribute to it as well. It should be noted in this respect that no precise determinations of unit cell dimensions were made either

VI – 9

4a

4 b

Fig. 4.   Etch pits on prismatic faces of ZnS crystals. (a) heavily faulted region ($\times 300$); (b) border region between a cubic region and a polytype ($\times 200$).

in cubic or in hexagonal vapor-grown pure ZnS crystals. No precision measurements were made of the cell dimension in polytypes. The only precision data available refer either to natural crystals or to powders.

It should be noted that the absence of faults and dislocations from uniform regions is in full accord with the results of Farkas–Jahnke and Dornberger–Schiff[21]; in a similar specimen these authors found relative X-ray reflection intensities as expected for nearly perfect crystals.

2.3. INTERNAL ELECTRIC FIELDS AND DISLOCATIONS

Edge dislocations lying wholly within a basal plane in wurtzite, sphalerite* or one of the polytypic structures, are charged electrically because of the partially polar character of the chemical bonds. The sign of the charge depends on whether the dislocation line lies in a Zn-plane or in an S-plane. Now, considering an edge

* "Basal planes" in the sphalerite structure refer in these specimens to the basal plane perpendicular to the direction of growth.

Fig. 5. Lang topographs of a hexagonal crystal containing a few basal plane faults, using (11.0) reflection.

dislocation (or a dislocation having an edge component like the 60° dislocation) as being the last row of an extra (prismatic) plane, its sign will be dependent on the position of the edge of the extra plane with respect to the crystallographic polarity along the c-axis, namely whether the edge borders the extra plane on its (00.1) (case a) or its (00.$\bar{1}$) side (case b). Since the same energy considerations of broken bonds should apply to the interior of the crystal as to its surface, the charge is positive in case "a" and negative in case "b". Thus any "extra plane" will have opposite electrical charges at its two edges and a set of extra planes (introduced, e.g., by mechanical pressure or by mismatch of basal plane dimensions of adjoining structures) will involve two sets of charged dislocation lines, lying on different basal planes, preferentially along boundaries between different structures.

60° dislocations are very likely candidates for these

charged edge-type dislocations. Counting the etch pits gives then an indication of the charge and field involved. Taking – as a typical figure – 100 etch pits per millimeter, one finds fields of the order of 10^4 V/cm, in good accord with estimates based on the asymmetric Franz–Keldysh effect[22]). The appearance of the APE following the application of slight mechanical pressures[23]) and its dependence on crystallographic polarity[4]) give further support to this model.

3. Conclusions

Faults and dislocations in vapor-phase grown ZnS crystals are concentrated at boundaries between uniform regions differing in structure. Indirect evidence shows that the triangular etch pits on prismatic faces should be identified with points of emergence of electrically charged dislocations; these are, in turn, responsible for the internal fields which cause the APE. Further work on ZnS crystals, employing X-ray topography and precision X-ray diffraction, is under way in this Laboratory.

Acknowledgement

The authors are pleased to thank Mr. I. Natanson for growing the crystals and Mr. A. Kessar for technical assistance.

References

1) S. G. Ellis, F. Herman, E. E. Loebner, W. J. Merz, C. W. Struck and J. G. White, Phys. Rev. **109** (1958) 1860.
2) G. Shachar and Y. Brada, J. Appl. Phys. **41** (1970) 3127.
3) K. Patek, M. Skala and L. Souckova, Czech. J. Phys. **12** (1962) 313.
4) O. Brafman, E. Alexander, B. S. Fraenkel, Z. H. Kalman and I. T. Steinberger, J. Appl. Phys. **35** (1964) 1855.
5) O. Brafman, E. Alexander and I. T. Steinberger, Acta Cryst. **22** (1967) 347.
6) S. Mardix, O. Brafman and I. T. Steinberger, Acta Cryst **22** (1967) 805.
7) S. Mardix, E. Alexander, O. Brafman and I. T. Steinberger, Acta Cryst. **22** (1967) 808.
8) S. Mardix and O. Brafman, Acta Cryst. **23** (1967) 501.
9) S. Mardix and O. Brafman, Acta Cryst. B **25** (1968) 258.
10) I. Kiflawi, S. Mardix and I. T. Steinberger, Acta Cryst. B **25** (1969) 1581.
11) S. Mardix, I. Kiflawi and Z. H. Kalman, Acta Cryst. B **25** (1969) 1586.
12) I. Kiflawi, S. Mardix and Z. H. Kalman, Acta Cryst. B **25** (1969) 2413.
13) I. Kiflawi and S. Mardix, Acta Cryst. B **26** (1970) 1192.
14) S. Mardix and I. Kiflawi, Cryst. Lattice Defects **1** (1970) 129.
15) S. Mardix and I. T. Steinberger, Israel J. Chem. **3** (1966) 243.

16) E. Alexander, Z. H. Kalman, S. Mardix and I. T. Steinberger, Phil. Mag. **21** (1970) 1237.

17) S. Mardix, Z. H. Kalman and I. T. Steinberger, Acta Cryst. A **24** (1968) 464.

18) S. Mardix and I. T. Steinberger, J. Appl. Phys. **41** (1970) 5339.

19) I. Kiflawi, S. Mardix and I. T. Steinberger, to be published.

20) S. Mardix, A. R. Lang and Y. Blech, Phil. Mag. **24** (1971 683.

21) M. Farkas-Jahnke and K. Dornberger-Schiff, Acta Cryst. A **26** (1970) 35.

22) G. Shachar, Y. Brada, E. Alexander and Y. Yacobi, J. Appl. Phys. **41** (1970) 723.

23) G. Shachar, S. Mardix and I. T. Steinberger, J. Appl. Phys. **39** (1968) 2485.

    Journal of Crystal Growth **13/14** (1972) 292–294 © *North-Holland Publishing Co.*

GROWTH AND TEXTURES OF ZnS CRYSTALS

TOMOYA OGAWA and MASAHARU UCHIDA

Department of Physics, Gakushuin University, Mejiro, Tokyo, Japan

The stacking order in ZnS is studied continuously along the polar axis with a noval X-ray rotation camera. Various sorts of fundamental polytypic ordering exist, for example 3C, throughout the crystals on a macroscopic scale, while there are many other modifications as shown by the birefringent striations and Berg–Barrett topographs.

It is well known that there are many stacking faults and polytypes in ZnS crystals. In order to study the stacking sequences continuously along the polar axis, a new rotation camera was prepared (fig. 1). The rotating crystal moves up and down along its rotation axis and the horizontal motion of the film is synchronized with the vertical movement of the rotating crystal. When the rotation axis is adjusted to be parallel to the polar axis and X-rays diffracted from the crystal are selected by the slits, the periodicity of stacking sequences is recorded on the film continuously along the polar axis.

In the figure, the upper and lower parts represent the schematic drawings of the horizontal and vertical projection, respectively. Here, C is the crystal, G the goniometer head, S_1 the pin hole collimator for incident X-ray beams, S_2 the slit for diffraction beams, T the beam trap, F the X-ray film, and "Motor" is the motor for rotation of the crystal.

The reflectivities of the net planes in the ideal wurtzite and zincblende are respectively represented by the hexagonal and trigonal systems in fig. 2. The phase angles of the structure factors and Miller indices are also shown in the figure.

Since the reflection from the net planes with indices $h-k = 3n \pm 1$ is very sensitive to the stacking sequence of basal planes along the polar axis, the photographs were taken with the reflection planes along the column of $h-k = 1$.

The two photographs in the middle part of fig. 2 were taken with a new rotation camera with a micro-

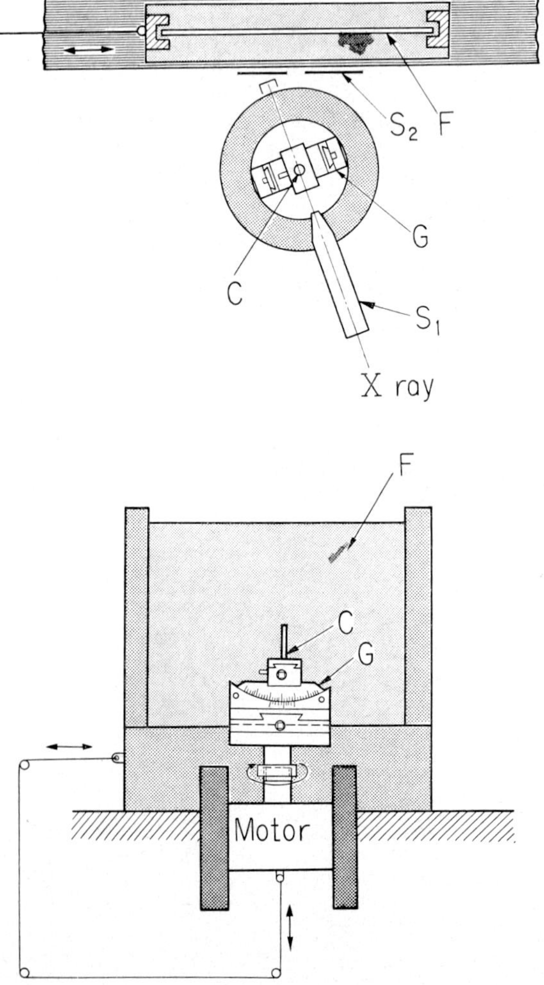

Fig. 1. An X-ray rotation camera for continuous observation of stacking sequences.

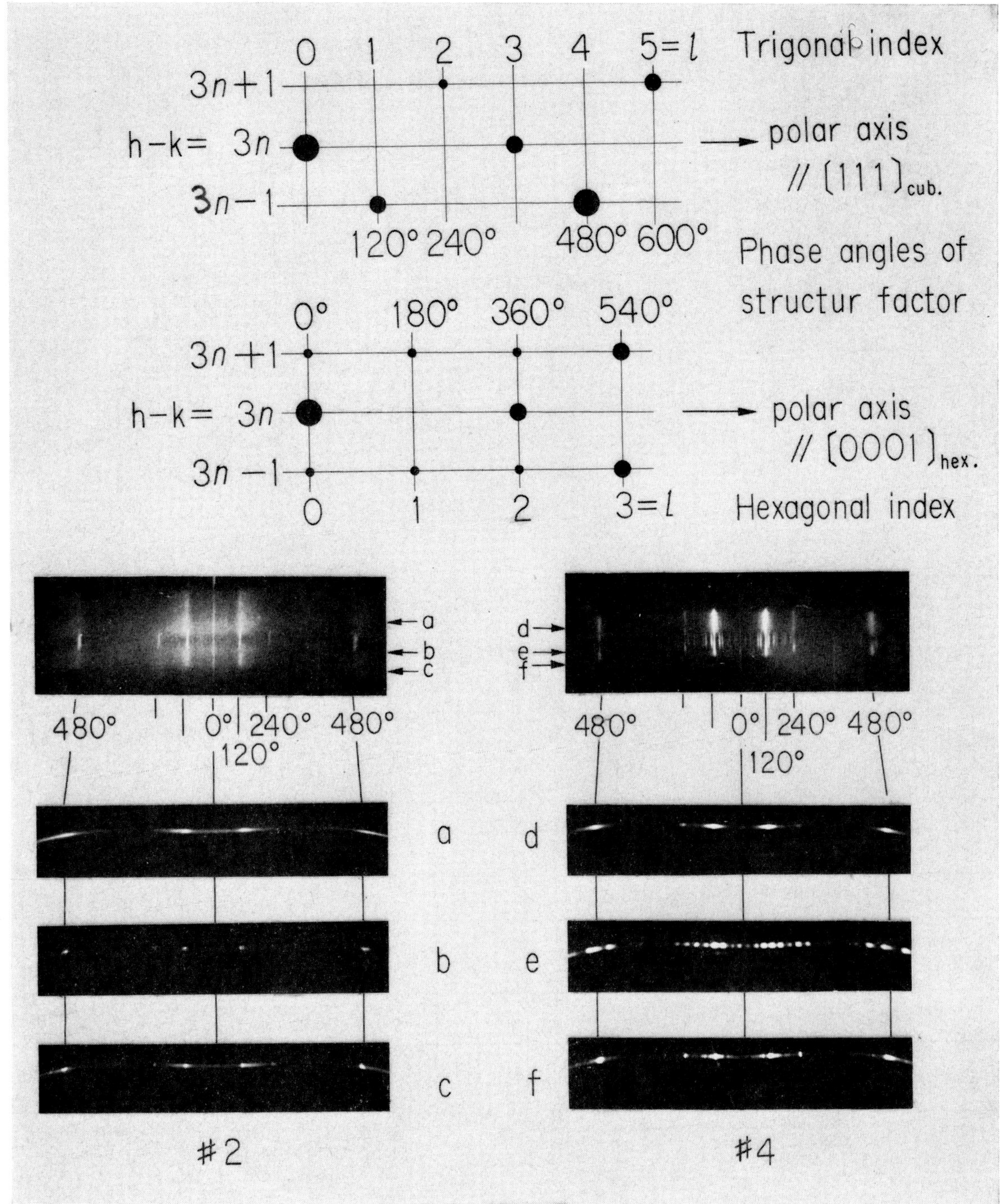

Fig. 2. The upper two diagrams indicate the X-ray intensities from zincblende and wurtzite on their reciprocal lattices. The size of the circle is proportional to the square of the structure factor. The middle two photographs were taken with the camera shown in fig. 1, along the column with $h-k = 1$. The two sets of the lower three photographs were taken by an X-ray rotation camera at the positions indicated by arrows on the middle photographs.

focus X-ray source, and the others were taken with an ordinary rotation camera.

The ordinary photographs, indicated by a, b, c and d, e, f, correspond to the positions indicated by the arrows in the middle two photographs. The picture e shows a superstructure, while the other pictures show stacking faults.

It is clear from these two photographs that there is a dominant periodicity throughout the crystal over a macroscopic distance, even if the crystal is faulted. Of course, the traces of the diffraction spots are smeared out due to the stacking faults.

It is concluded that the dominant periodicity is cubic 3C in most crystals, while there are many other modifications[1-6] as shown by the birefringent striations and Berg–Barrett topographs.

References

1) T. Ogawa, J. Phys. Soc. Japan **16** (1961) 2360.
2) T. Ogawa, Y. Fukuse and Y. Kawai, Japan. J. Appl. Phys. **4** (1965) 948.
3) T. Ogawa, J. Phys. Soc. Japan **25** (1968) 1126.
4) T. Ogawa, J. Phys. Soc. Japan **25** (1968) 913.
5) O. Brafman and I. T. Steinberger, Phys. Rev. **143** (1966) 501.
6) O. Brafman, E. Alexander and I. T. Steinberger, Acta Cryst. **22** (1967) 347.

Section VII

Vapour growth

J. NISHIZAWA
T. TERASAKI
M. SHIMBO
J. BLOEM
H. M. MANASEVIT
J. K. KENNEDY
W. D. POTTER
L. HOLLAN
C. SCHILLER
J. P. CHANE
P. MERENDA
S. IIDA
K. ITO
CHR. BELOUET
T. NISHINAGA
H. OGAWA
H. WATANABE
T. ARIZUMI

K. PLOOG
P. RAUH
W. STOEGER
H. SCHMIDT
M. ILEGEMS
J. P. DISMUKES
W. M. YIM
V. S. BAN
P. LILLEY
P. L. JONES
C. N. W. LITTING
K. NASSAU
J. W. SHIEVER
A. KUHN
A. CHEVY
E. LENDVAY
S. I. RADAUTSAN

F. G. DONIKA
I. G. MOUSTIA
C. PAPARODITIS
R. SURYANARAYANAN
H. WIEDEMEIER
E. A. IRENE
A. K. CHAUDHURI
M. P. CALAGHAN
R. W. BRANDER
H. K. BOWEN
W. D. KINGERY
M. KINOSHITA
C. A. GOODWIN
J. OUVRARD
R. WANDJI
B. ROQUES
H. S. SHIM
J. G. BYRNE

Journal of Crystal Growth **13/14** (1972) 297–301 © *North-Holland Publishing Co.*

LAYER GROWTH IN SILICON EPITAXY

JUN-ICHI NISHIZAWA

Research Institute of Electrical Communication, Tohoku University, Sendai, Japan

TAKESHI TERASAKI

Semiconductor Research Institute, Sandai, Japan

and

MASAFUMI SHIMBO

Research Institute of Electrical Communication, Tohoku University, Sendai, Japan

Silicon epitaxial growth of perfect crystals using hydrogen reduction of $SiCl_4$ has been investigated from the viewpoints of step motion and surface morphology. From experiments using semi-spherical substrate, it is found that very flat surface (we call this surface O.F. ("on facet") surface, evenness <10 Å) is formed exactly on some special crystallographic planes such as (111), (110), (112), (113), ... and rarely on (100). On O.F. surface growth rate (vertical growth rate) was very small and was about 0.1–0.05 compared with the usual growth rates mentioned by other authors where the substrate is inclined a few degrees from the exact (111) surface or non-perfect crystals. The lateral growth rate was measured, using circular small steps with height of about 1500 Å on the (111) O.F. surface, and was very large (about 150 µm/min at 1200 °C) under growth condition in which the usual growth rate is about 0.5 µm/min.

From these experiments, we concluded that two-dimensional growth was very dominant in silicon epitaxial growth using hydrogen reduction of $SiCl_4$. This conclusion is also supported by other results concerning to surface morphology and growth with screw type stacking fault as nucleus center and so on. These large values of lateral growth rate and small value of vertical growth rate have not yet been reported, and can also be seen in GaAs, and so on. Nearly all values reported before about $SiCl_4$ were those for non-perfect crystals. But it was not obtained in silicon epitaxy using pyrolysis of SiH_4.

Many authors have investigated silicon epitaxial growth using hydrogen reduction of $SiCl_4$. These investigations related mainly to thermodynamic treatments of gas reactions and growth kinetics with growth rate, mole fraction, flow rate and so on, or to crystallographical studies with growth rate, orientation-dependency and defects, or to impurity distributions with various growth conditions. But those were about non-perfect crystals.

Recently, a method of obtaining highly perfect crystal layers on substrates variously doped with different impurities has been reported[2]. These phenomena are of course relevant to device fabrication and integrated circuit definition[3].

In this paper we report the growth mechanism of silicon epitaxy from the viewpoint of crystallography with particular reference to the observation of step motion and surface morphology. The experimental conditions are as follows: mole fraction of $SiCl_4$ is 0.5–3 %; the velocity of gas flow in a reaction tube is 2–6 cm/sec; the growth temperature is normally 1150–1200 °C although sometimes it is less in order to obtain the

activation energy during induction or resistance heating. The substrates doped with phosphorus, antimony and boron which had a 10^{-2}–300 Ω cm resistivity were used. The grown specimen was observed in the "as-grown" condition by optical microscopes using various methods such as interference contrast (Nomarski), phase contrast and multiple beam interference. The observations show quite different results from those of non-perfect crystals.

From experiments using semi-spherical substrates it is found that very flat surfaces which show unevenesses less than about 10 Å are formed exactly on some special crystallographic planes such as (111), (110), (112), (113), ... and rarely on (100). These flat surfaces are shown on stereographic projection in fig. 1. The areas of (111) or (113) flat surfaces are usually greater than those of other flat surfaces. These very flat surfaces are inclined at no more than few seconds from the crystallographic planes. These facetted surfaces are called O.F. surfaces ("on facet" surfaces). Vertical growth rates on these O.F. surfaces are very slow compared with other surfaces. The diameters of these "facets" are in the order of a mm for layer growth of about 10 μm thickness. This means that growth rates for any crystallographic direction change significantly on "facet" development. And we can very often observe microscopically the existence of regions where growth was obviously promoted by the expansion of layers having layer thickness

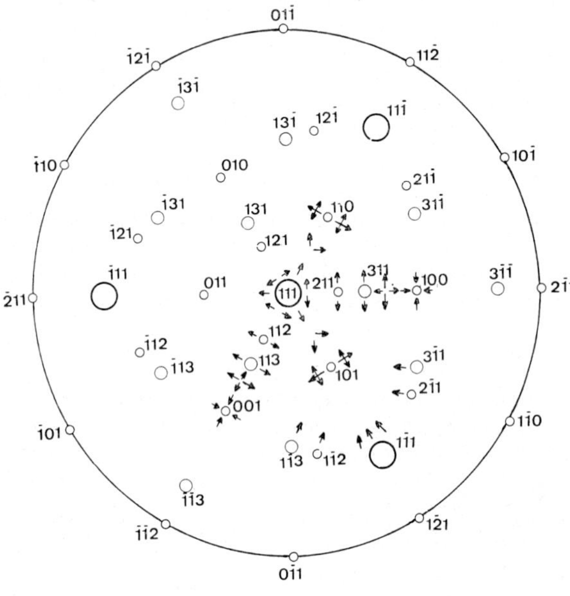

Fig. 1.   Stereographic projection showing several "on facet" (O.F.) surfaces (○), and directions of shadows.

Fig. 2.   Circular step on O.F. surface before growth.

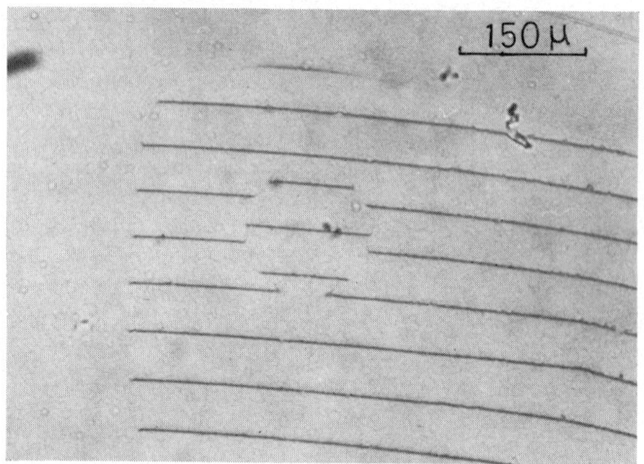

Fig. 3.   Fringes of multiple beam interference for fig. 2.

of about 50–500 Å. We concluded that two-dimensional growth, namely expansion of two-dimentional layer, was a marked feature of silicon epitaxy using hydrogen reduction of $SiCl_4$ as explained below.

To measure the two-dimensional growth rate (expansion of layer growth) we formed a circular step-like mesa, whose step height was about 1000–2000 Å and whose diameter was 50–500 μm, on perfect a (111) O.F. surface using conventional thermal oxidation and photo-etching, and re-oxidation techniques. The surface of the top of the circular mesa was a perfect O.F. surface but its sides had many steps and kinks. There are no visible nucleating sites on the O.F. surface. Photographs of these surfaces are shown in figs. 2 and 3. Until now the growth rate for a ⟨111⟩ type direction was measured on a substrate inclined a few degrees and more from a perfect (111) plane, and referred to as the growth rate of a ⟨111⟩ type direction. For these inclined sub-

strates the growth rate is not a sensitive function of orientation for small inclinations of the crystallographic plane.

After growth, we observe the morphological change of the O.F. step and measure the expansion of the O.F. step. The examples are shown in figs. 4 and 5. One of these results is shown in fig. 6, curve A. The absolute value of the growth rate depends on the gas composition, the temperature, the flow conditions and also on the substrate orientation. Therefore the growth rate observed on the inclined surface was considered as a vertical axis, because it gives rise to the growth not sensitive to surface orientation, and was used to represent the gas composition, the temperature and the flowing condition. Under the growth conditions in which the growth rate normal to the surface is about 0.4 μm/min, the expansion speed of the growing layer is extraordinarily fast and is about 150 μm/min at the

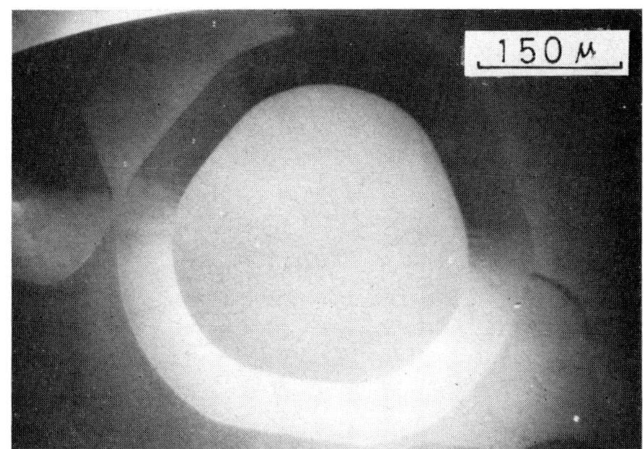

Fig. 4.   Step motion after 1 min growth at 1200 °C.

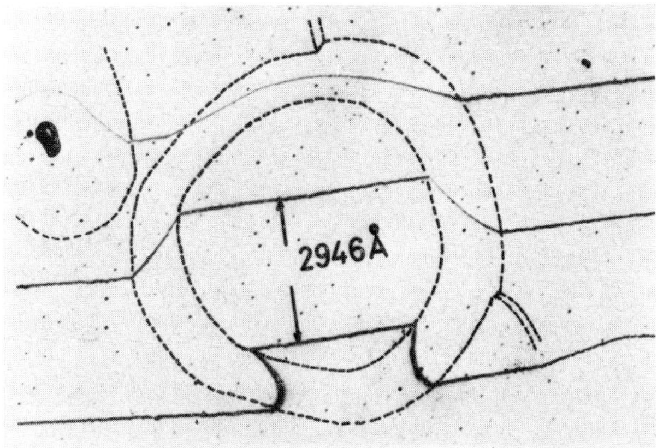

Fig. 5.   Fringes of multiple beam interference for fig. 4.

almost wash out the O.F. circular step during growth even in a short time. Growth twins especially enhance the growth rate.

The growth rate at the bottom of an O.F. mesa is larger than that at the top and is typically more than about 40 μm/min, also shown in fig. 6 (curve B). Compared with O.F. growth, this phenomena may be due to enhanced surface migration, an increase in the number of stable sites and/or a decrease in the time interval for initial growth. If the growth in promoted by the supply of mono-atomic atoms from the gas phase and the sticking probability of atoms is nearly a constant at any surface, although this is an approximation, the small value of the normal growth rate and the

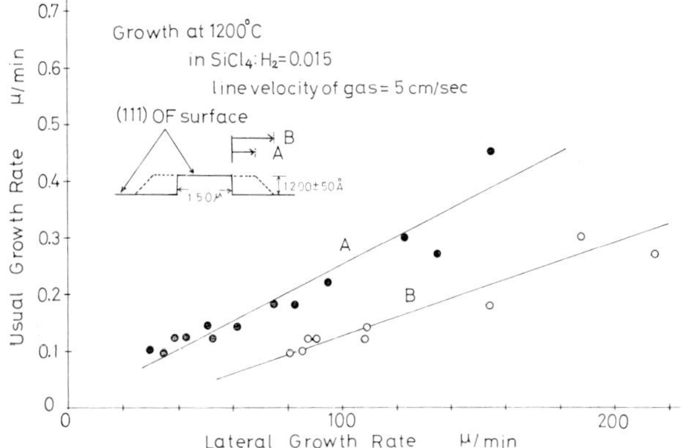

Fig. 6.   Lateral growth rate measured from spread of circular step on O.F. surface.

top surface of an O.F. step, namely about 5000 atomic layers are arranged regularily in the lattice for 1 sec from the measurement of the change of the mesa shape. On the other hand, the vertical growth rate on the perfect (111) surface is very slow and less than about 500–200 Å/min under the same growth conditions. The growth rate noted by the other authors[1]) is not that on a perfect (111) surface. Thus on a perfect (111) plane it is found that the lateral growth rate is very large, and that the vertical growth rate is very small. On other O.F. surfaces such as (110) O.F. the same surface phenomena is also observed. In these experiments the growth rates must be measured very carefully under clean conditions because the lateral growth rate is extraordinarily fast.

Enhancement of the growth rate by very small particles like dust on an O.F. surface can occur by creating nucleation centers; so causing many other steps, which

large value of the lateral growth rate on a perfect (111) O.F. surface may be explained as a result of a large surface migration of adsorbed atoms, because there are a small number of steps and kinks for a stable lattice arrangement. This is in accord with the result by Kinoshita[6]).

If these assumptions are correct, the lateral growth rate must increase linearly as a function of the increase of diameter of the circular mesa, under constant environmental conditions. Decisive results to support the above assumptions could not be obtained within experimental error, though the lateral growth rate does appear to slightly increase with increase of the diameter beyond 10 μm. The slope of curve B in fig. 6 is not changed within experimental error by the increase of the area of the O.F. circular mesa adsorbing atoms which can be supplied to stick to many kinks at the growing mesa step. This suggests that the result is that the growth rate seems to be limited mainly by the probability being caught by the kinks of the step, or by the migration velocity through diffusion length toward the kink, and in the same time by the total amount of sticking atoms on the migrating diffusion zone at the surface. This second conclusion may also be supported from the fact that generation of pyramids also initiates at the center part of an O.F. surface, not only the lateral growth of the circular step at low temperatures below 1200 °C. Figs. 7, 8 and 9 show the grown surfaces at 1140 °C, at 1050 °C, and at about 1000 °C respectively. The formation of a pyramid becomes a stochastic distribution following the decrease of temperature below 1200 °C. This can be understood as follows: the sticking probability on flat surfaces is increased very rapidly with the decrease of temperature, and these figures show the adsorbed atoms at peripheries diffused toward the stable kink at the step edge. Therefore there is no pyramid there and the diffusion length is less than 10 μm in fig. 9. The extraordinary large value of this lateral growth rate has been obtained only in the case of hydrogen reduction of $SiCl_4$, and has not yet been obtained for the epitaxial growth method using pyrolysis of SiH_4 at the same growth temperature and same normal growth rate using the same O.F. substrate. In epitaxial growth by pyrolysis of SiH_4, the lateral growth rate on an O.F. surface is not larger than the usual growth rate and even the normal growth rate on an O.F. surface seems not to be significantly smaller than the usual growth rate. This seems to indicate that the growth processes between the two methods are quite different. The expansion of layer growth in hydrogen reduction of $SiCl_4$ is characterized by a crystal orientation dependence; namely the circular shape develops into a triangular shape in the course of growth. The dependence is sufficiently marked as to enhance the average growth rate. One of the easy growth directions is the $\langle 110 \rangle$ direction for expansion of layer growth.

On an as-grown surface, a surface depression is always associated with a stacking fault which is wedge-shaped and which is from about twice to a hundred times longer than the length of the stacking fault, which is called a "Stacking Fault Shadow" by the authors of ref. 4. It is a result of blocking of the expansion of layer growth by the stacking fault which is thought as a twinned layers, so that the direction of the stacking fault shadow is approximately parallel near to the easy

Fig. 7.   Step motion after 1.5 min growth at 1140 °C.

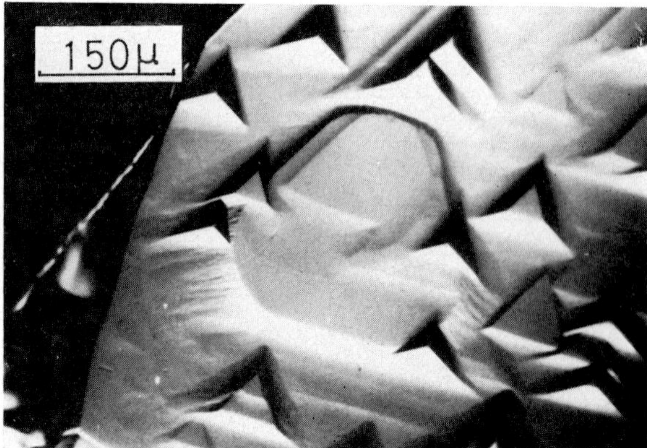

Fig. 8.   Step motion after 1.5 min growth at 1050 °C.

Fig. 9.   Step motion after 1.5 min growth at 1000 °C.

growth direction of layer growth. The arrows in fig. 1 show the experimental directions of stacking fault shadows at peripheries of some special crystallographic planes on which O.F. surfaces, such as (111), (110), (113), ..., are formed. Pattern deviation and washing out of buried layer in intergrated circuit technology during growth on near (111) surface is thus understandable and can be explained and analyzed quantitatively from these experimental results. On the other hand, the same phenomena for SiH_4 does not occur under corresponding conditions of $SiCl_4$ reduction. It is suggested that combinations between the existence of screw type defects as nucleus centers, as was published before[5]), and the very large value of the two-dimensional growth rate, give possible promises of important developments in our understanding of the growth mechanism in the new features.

References

1) S. Mendelson, J. Appl. Phys. **35** (1964) 1570.
2) J. Nishizawa, T. Terasaki, K. Yagi and N. Miyamoto, Seoul Intern. Conf. on Electrical and Electronics Eng. (Seoul. Sept. 1970).
3) C. M. Drum and C. A. Clark, J. Electrochem. Soc. **115** (1968) 664; **117** (1970) 1405.
4) H. Sunami, T. Terasaki, N. Miyamoto and J. Nishizawa, J. Appl. Phys. **40** (1969) 4670.
5) M. Shimbo, T. Terasaki and J. Nishizawa, J. Appl. Phys. **42** (1971) 487;
K. Kijima, N. Miyamoto and J. Nishizawa, J. Appl. Phys. **42** (1971) 486.
6) K. Kinoshita, Paper given at the Intern. Conf. of Crystal Growth (1967).

  Journal of Crystal Growth **13/14** (1972) 302–305 © *North-Holland Publishing Co.*

DOPING IN CHEMICALLY VAPOUR DEPOSITED EPITAXIAL SILICON

J. BLOEM

Semiconductor Development Laboratory, N.V. Philips Gloeilampenfabrieken, Nijmegen, The Netherlands

The incorporation of phosphorus in silicon has been studied, using PH_3 as dopant during epitaxial growth from SiH_4 in hydrogen. Interesting variations in doping level as a function of growth rate (variation of p_{SiH4}) were found: at 1100 °C the concentration of donors increases with silicon growth rate, at 1200 °C the concentration is constant and independent of growth rate, at 1300 °C the donor concentration decreases for higher growth rates.

The high temperature results can be explained by a diffusion controlled incorporation of phosphorus giving a constant flux of dopant atoms towards the growing interface. For a constant value of PH_3 with increasing growth rate the same amount of phosphorus is incorporated in increasingly thicker silicon layers, leading to the observed results.

The increase of dopant concentration with growth rate at lower temperatures is explained by an additional effect, viz. increasing the growth rate or lowering the temperature increases the chance of trapping dopant atoms in the growing layer.

The temperature dependence of the gas–solid segregation coefficient for phosphorus also depends on growth rate. Extrapolation to zero growth rate gives a temperature dependence of the equilibrium segregation coefficient of ~ 0.3 eV.

1. Introduction

Doping in the epitaxial growth of silicon has been the subject of several studies[1,2]), but the mechanism of dopant incorporation is a still not well understood. Apart from the thermodynamic considerations kinetic factors are often of importance. To study these different aspects an experimental study has been performed, using PH_3 as a dopant and SiH_4 as silicon source in a conventional R.F. heated horizontal reactor[3]). Special attention has been paid to the influence of the silicon growth rate on the doping level.

2. Experimental

Growth rates and resistivities were measured by the infrared interference technique and the four-point probe method on layers grown on n^+ and p-type silicon substrates, oriented 3° off the (111) plane in the direction of the nearest (110)[4]. The temperature of the substrates was measured with an optical pyrometer, the values being corrected for the emissivity of silicon and the absorption of quartz[5]). The temperature used ranged from 1350–1500 °K.

The concentration of free electrons in the epitaxial layers, measured at room temperature, is a measure of the concentration of phosphorus introduced in the layers.

Fig. 1 shows the concentration of phosphorus in layers grown at 1400 °K, as a function of the partial pressure of PH_3 in the gas phase. The curve is more or less linear up to concentrations of 10^{18} cm^{-3}, for higher concentrations the slope decreases.

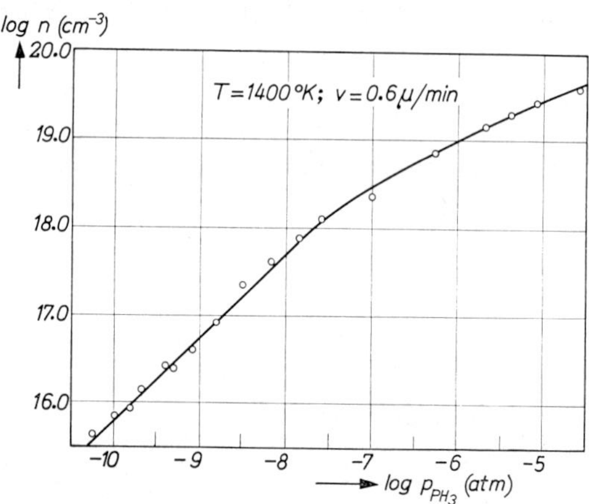

Fig. 1. Donor concentration in epitaxial silicon layers grown at 1400 °K as a function of partial pressure of PH_3 in the gas phase. The slope of the curve changes from 1.0 to 0.5 for concentrations greater than 10^{18} cm^{-3}.

A number of detailed measurements were performed on the influence of the silicon growth rate on the kinetics of the incorporation of dopant. Fig. 2 shows the results for some partial pressures of PH_3 at temperatures of 1350 °K, 1400 °K, 1420 °K and 1500 °K. At

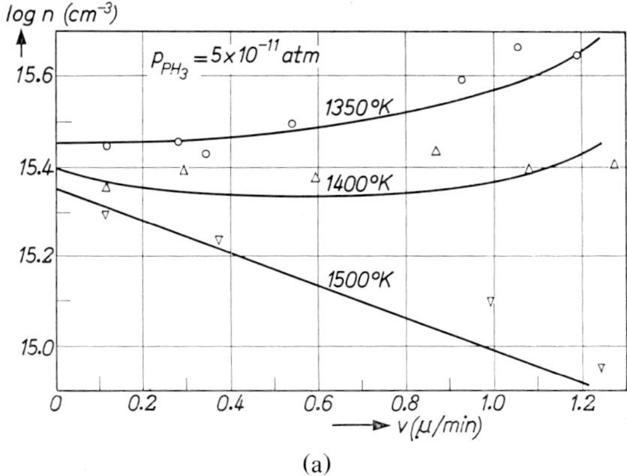

(a)

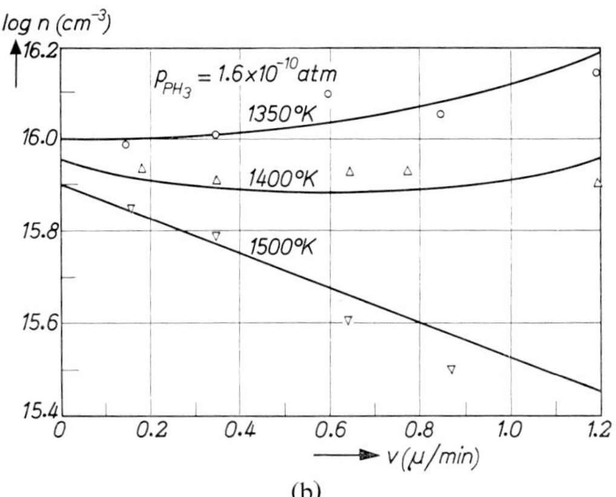

(b)

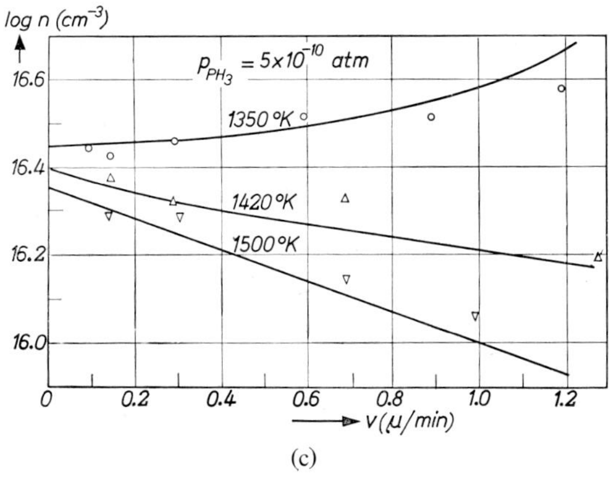

(c)

Fig. 2. Variation of donor concentration in silicon layers with silicon growth rates (v): (a) for $p_{PH3} = 5 \times 10^{-11}$ atm, (b) $p_{PH3} = 1.6 \times 10^{-10}$ atm, and (c) $p_{PH3} = 5 \times 10^{-10}$ atm. The data points are mean values of at least three runs at temperatures of 1350 °K ($\circ$), 1400 and 1420 °K ($\triangle$) and 1500 °K ($\triangledown$). v varies linearly with p_{SiH4} ($v = 1.0$ μm/min for $p_{SiH4} = 1.4 \times 10^{-3}$ atm). The solid curves give values calculated with eq. (12).

the lower temperature the phosphorus concentration increases with growth rate, whereas at the higher temperature the concentration decreases with growth rate.

The implication of this behaviour is that the temperature dependence of the incorporation also depends on the growth rate. Fig. 3 gives this temperature dependence for a growth rate extrapolated to zero, together with the apparent temperature dependencies for growth rates of 0.6 and 1.2 μm/min.

3. Discussion

In fig. 1 a linear relation is found between the partial pressure of PH_3 and the donor concentration in silicon. This indicates that a simple reaction scheme involving adsorbed phosphorus atoms (P_{ad}) which are subse-

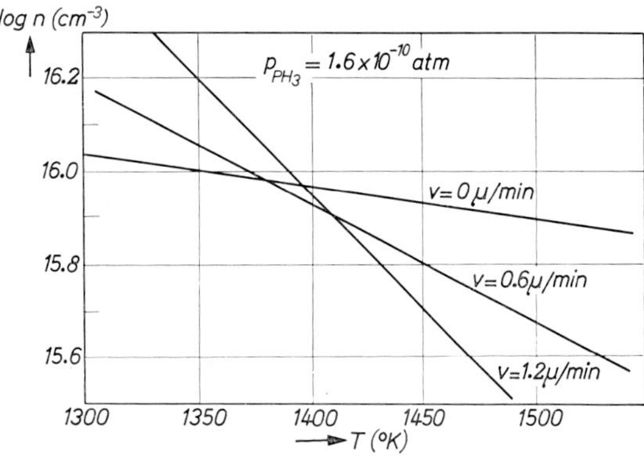

Fig. 3. Temperature dependence of the incorporated donor concentration for $p_{PH3} = 1.6 \times 10^{-10}$ atm. The line $v = 0$ μm/min gives values found by extrapolating the growth rate of silicon to zero.

quently incorporated into the silicon lattice (P_{Si}), can describe the incorporation

$$PH_{3g} \rightleftharpoons P_{ad} + \tfrac{3}{2} H_2 , \tag{1}$$

$$P_{ad} \rightleftharpoons P_{Si} , \tag{2}$$

$$P_{Si} \rightleftharpoons P_{Si}^+ + e . \tag{3}$$

Eq. (3) denotes the ionization of the incorporated atoms. At the growth temperature the intrinsic carrier concentration in silicon (n_i) is equal to $2–5 \times 10^{19}$ cm^{-3}, which explains the change in slope near these concentrations in fig. 1[6,7].

Fig. 2 indicates that depending on the silicon growth rate the concentration of phosphorus can be constant, increasing or decreasing, a situation which can be compared with doping in crystal growth from the liquid phase[8-10].

A combination of diffusion control and adsorption kinetics is found to give a reasonable explanation of the experimental results. Diffusion controlled kinetics in epitaxy has been studied by Eversteyn et al.[11]), who gave evidence for the existence of a stagnant layer of constant thickness (δ) over the length of the susceptor.

The number of dopant atoms transported towards the interface (n_a per cm^2 per sec) depends on diffusion over the stagnant layer, giving[3])

$$n_a = \frac{D_1}{RT} \frac{p_{\text{dope}} - p_{\text{eq}}}{\delta}, \tag{4}$$

where D_1 is the diffusion coefficient of the dopant in hydrogen (cm^2/sec) and p_{dope} and p_{eq} are the partial pressures of dopant in the main gas and the equilibrium dopant pressure near the silicon interface (atm), respectively. The growing layer, having a growth rate v, incorporates a number of dopant atoms (n_g) equal to

$$n_g = vC_s = \frac{D_{\text{SiH}_4}}{RT} \frac{p_{\text{SiH}_4}}{\delta} \frac{p_{\text{eq}}}{\gamma}. \tag{5}$$

Here the silicon growth rate is expressed as in eq. (4); C_s is the dopant concentration in the layer (atom/atom Si), p_{SiH_4} is the silane vapour pressure in the main gas (atm) with a diffusion coefficient D_{SiH_4}; and γ is the gas-phase segregation coefficient defined as the ratio of the mole fractions of monoatomic dopant in the vapour and ionized dopant atoms in the solid

$$\gamma = \frac{x_{\text{gas}}}{x_{\text{Si}}} = \frac{p_{\text{eq}}}{p_{\text{tot}}} \frac{1}{C_s} = \frac{p_{\text{eq}}}{C_s}. \tag{6}$$

In the 1 atm gas ($p_{\text{tot}} = 1$) the mole ratio of the dopant is equal to its partial pressure.

Combining (4) and (5) gives in the steady state condition ($n_a = n_g$)

$$\frac{p_{\text{dope}}}{p_{\text{eq}}} = 1 + \frac{D_{\text{SiH}_4}}{D_1} \frac{p_{\text{SiH}_4}}{\gamma} \approx 1 + \frac{p_{\text{SiH}_4}}{\gamma}. \tag{7}$$

p_{eq} can be eliminated [eq. (6)] to obtain

$$C_s = p_{\text{dope}}/(\gamma + p_{\text{SiH}_4}). \tag{8}$$

To convert C_s into concentration per cm^3 one has to multiply by 5×10^{22}, the number of Si atoms per cm^3. For low values of γ it is seen that C_s depends on p_{SiH_4}, i.e. on the silicon growth rate; p_{eq} becomes smaller than p_{dope} and goes to zero in the ideal diffusion controlled situation. Then

$$C_s = p_{\text{dope}}/p_{\text{SiH}_4}, \tag{8}$$

and C_s decreases with increasing growth rate (fig. 2, 1500 °K). For low growth rates and high values of γ eq. (8) reduces to

$$C_s = p_{\text{dope}}/\gamma, \tag{9}$$

the equilibrium value of C_s being maintained independent of growth rate since only a small amount of dopant is introduced into the layer ($p_{\text{dope}} \approx p_{\text{eq}}$).

Phosphorus has a low γ value ($\gamma \sim 10^{-3}$, $p_{\text{SiH}_4} \sim \sim 10^{-3}$–$10^{-4}$), so that the prediction of eq. (8), a decrease of C_s with increasing growth rate, found at the higher temperatures, can be expected. Extrapolation of the experimental curves to zero growth rate gives the equilibrium value of the segregation coefficient; which is found to be rather independent of temperature. At the lower temperatures, therefore, the same behaviour as at 1500 °K is expected, viz. a decrease of donor concentration with increasing growth rate. This is not found experimentally, however, and another effect must be responsible for this different behaviour.

According to Hall[8]), Trainor et al.[9]) and Holmes[10]), adsorption of donor atoms on the surface may give rise to a surface concentration higher than the concentration in the bulk; both concentrations being in equilibrium with the co-existing gas-phase. Increasing the growth rate will give rise to an increased probability of the trapping of excess donor atoms in the bulk phase.

The time needed to grow one monolayer of silicon at 0.6 μm/min is typically 4×10^{-2} sec (monolayer thickness $\sim 4 \times 10^{-8}$ cm). This time has to be compared with the characteristic time needed for a buried donor atom to diffuse to the surface. The diffusion coefficient of phosphorus in silicon is $D = 10^{-12}$ cm^2 sec^{-1} at 1350 °K and $D = 10^{-11}$ cm^2 sec^{-1} at 1500 °K[12]); hence with $D \approx x^2/t$ and $x \approx 4 \times 10^{-8}$ cm, we obtain $t \approx 10^{-4}$ sec at 1500 °K.

The order of magnitude is such that for the lower temperatures and higher growth rates, an increased donor concentration can be expected because of insufficient time to exchange with the surface. The donor concentration in the bulk (C_s') can now be given as[8])

$$C_s' = C_s + (C_{\text{surf}} - C_s) \exp(-v_c/v), \tag{10}$$

where C_s is the bulk concentration without trapping, C_{surf} is the surface concentration of dopant, v is the growth rate and v_c the growth rate for which the characteristic time for the growth of one monolayer equals the relaxation time of the out-diffusion process. For $v = 0$ the value of C_s' equals C_s, whereas for high growth rates C_s' will be equal to C_{surf}.

Hall[8]) also indicates that for small growth rates (and $C_{surf} > C_s$) the variation of C_s' with growth rate can be given in the simple form

$$C_s' = C_s \exp (v/v_i), \tag{11}$$

or, with (7) and (11),

$$C_s' = \frac{p_{dope}}{\gamma + p_{SiH_4}} \exp \frac{v}{v_i}. \tag{12}$$

Analysis of the experimental curves then gives $v_i = 0.75\ \mu m/min$ at $1350\ °K$, $1.5\ \mu m/min$ at $1400\ °K$, $2.2\ \mu m/min$ at $1420\ °K$ and $8.7\ \mu m/min$ at $1500\ °K$. This corresponds with values of t between 3×10^{-2} sec and 3×10^{-3} sec, and leads to an activation energy of 2.8 eV, close to the activation energy for diffusion of phosphorus in silicon of 2.4–2.6 eV[12]). With $D \approx x^2/t$, values of x are found of 4 atom layers as effective distance over which the buried donor atoms have to diffuse in order to exchange with the surface. Extrapolation of the experimental curves to $v = 0$ gives the values of the equilibrium segregation coefficient γ ranging from 9×10^{-4} at $1350\ °K$ to 1.1×10^{-3} at $1500\ °K$ (activation energy ~ 0.3 eV).

From fig. 3, finally, it is obvious that the growth rate dependence of C_s is nearly absent around $1400\ °K$, at which temperature the counteracting influences just balance out.

4. Conclusion

It has been demonstrated experimentally that the concentration of phosphorus incorporated during the epitaxial growth of silicon shows a definite dependence on growth rate. The simplest quantitative model to describe this phenomenon uses a combination of diffusion controlled kinetics and the trapping probability of adsorbed phosphorus atoms, the latter effect being predominant at lower temperatures.

Diffusion in the gas phase as well as in the bulk of the crystal plays a part in the description. The diffusion controlled introduction of dopant via the gas phase reduces the donor concentration for higher growth rates; the trapping effect, on the other hand, is observed because outdiffusion, restoring the equilibrium concentration, can be slow relative to the addition of new silicon layers. This effect increases the bulk concentration of dopant going to higher growth rates.

References

1) W. Shepherd, J. Electrochem. Soc. **115** (1968) 541.
2) P. Rai-Choudhury and E. J. Salkovitz, J. Crystal Growth **7** (1970) 353, 361.
3) J. Bloem, J. Electrochem. Soc. **117** (1970) 1397.
4) K. E. Bean and P. S. Gleim, Proc. IEEE **57** (1969) 1469.
5) F. G. Allen, J. Appl. Phys. **28** (1957) 1510.
6) P. Rai-Choudhury, J. Phys. Chem. Solids **30** (1969) 1811.
7) H. Reiss, J. Chem. Phys. **21** (1953) 1209.
8) R. N. Hall, Phys. Rev. **88** (1952) 139.
9) A. Trainor and B. E. Bartlett, Solid State Electron. **2** (1961) 106.
10) P. J. Holmes, J. Phys. Chem. Solids **24** (1963) 1239.
11) F. C. Eversteyn, P. J. W. Severin, C. H. J. van den Brekel and H. J. Peek, J. Electrochem. Soc. **117** (1970) 925.
12) C. S. Fuller and J. A. Ditzenberger, J. Appl. Phys. **25** (1954) 1439.

306   *Journal of Crystal Growth* **13/14** (1972) 306–314 © *North-Holland Publishing Co.*

THE USE OF METALORGANICS IN THE PREPARATION OF SEMICONDUCTOR MATERIALS: GROWTH ON INSULATING SUBSTRATES

H. M. MANASEVIT

*Autonetics Division of North American Rockwell Corporation
3370 Miraloma Avenue, Anaheim, California 92803, U.S.A.*

Semiconductor films of II–VI and III–V compounds and alloys have been produced on semiconductors and insulators by decomposing appropriate Group II and Group III metalorganic compounds in the presence of Group V hydrides and Group VI hydrides or alkyl compounds. Group II metalorganics are also found to be suitable dopant sources for the III–V compounds. The halide-free growth process is compatible with existing CVD systems for the formation of elemental semiconductors.

1. Introduction

Recently, a process has been developed which appears to overcome the temperature restrictions, flow requirements etc. of CVD[1–3] for the epitaxial growth of compound semiconductors. This process requires only one controlled hot temperature zone for the *in situ* formation and growth of the semiconductor compound on the heated substrate and occurs in an atmosphere probably free of etching species.

In this process semiconductor films of II–VI and III–V compounds and alloys are produced by decomposing appropriate Group II and Group III metalorganic compounds in the presence of the appropriate Group V and Group VI hydrides or alkyls. In addition, the Group II metalorganics and Group VI hydrides are sources for the introduction of dopant impurities.

Earlier published reports in which the metalorganic–hydride process for compound semiconductor formation was described were made by Didchenko et al. in 1960[4] and Harrison and Tompkins in 1962[5]. In both cases the experiments were performed under relatively low pressure conditions in closed systems. Didchenko reported the preparation of InP by first reacting trimethylindium (TMI) and PH_3 at $-125\,°C$ and then decomposing the reaction product at about 275–300 °C into a dark gray powder identified as InP.

Harrison and Tompkins[5] described experiments leading to the production of high-resistivity crystalline films of InSb at $\sim 160\,°C$ from TMI and SbH_3. They also believed they had formed GaAs from trimethyl-

gallium (TMG) and AsH_3 by mixing them at room temperature in 1:1 molecular proportions in the vapor state for 18 hours and then heating the mixture to about 200 °C. The film they obtained displayed high resistivity, but unfortunately was not characterized further so as to be definitely identified as GaAs.

In 1968, Manasevit and coworkers[6,7] described the use of metalorganics and hydrides in CVD for the preparation of thin films of single-crystal semiconductor compounds and alloys. They used triethylgallium (TEG) or trimethylgallium (TMG) and Group V hydrides in producing GaAs, GaP, $GaAs_{1-x}P_x$, and $GaAs_{1-x}Sb_x$ on semiconductors and insulators such as sapphire (Al_2O_3), spinel $(MgAl_2O_4)$, and beryllia (BeO).

Later, Thomas [8] decomposed TEG and triethylphosphine (TEP) and grew layers of GaP on Si at 485 °C. Rai-Choudhury confirmed Manasevit et al. by using the TMG–AsH_3 process to grow GaAs on GaAs and Al_2O_3[9], and found the electrical properties of the films comparable on both substrates. The early stages of growth[10] and properties[11] of GaAs on Al_2O_3 were studied. Manasevit et al.[13] also studied the metalorganic–hydride process for the formation of epitaxial III–V aluminum compounds. AlAs and $Ga_{1-x}Al_xAs$ were prepared on GaAs and (0001) Al_2O_3 using trimethylaluminum (TMA) as the source of Al and TMG as the Ga source.

In 1970, Minden[13] reported some optical properties of AlAs films grown on GaAs by the TMA–AsH_3 process. Also, Manasevit et al. expanded their studies to include the II–VI compounds[14] and demonstrated

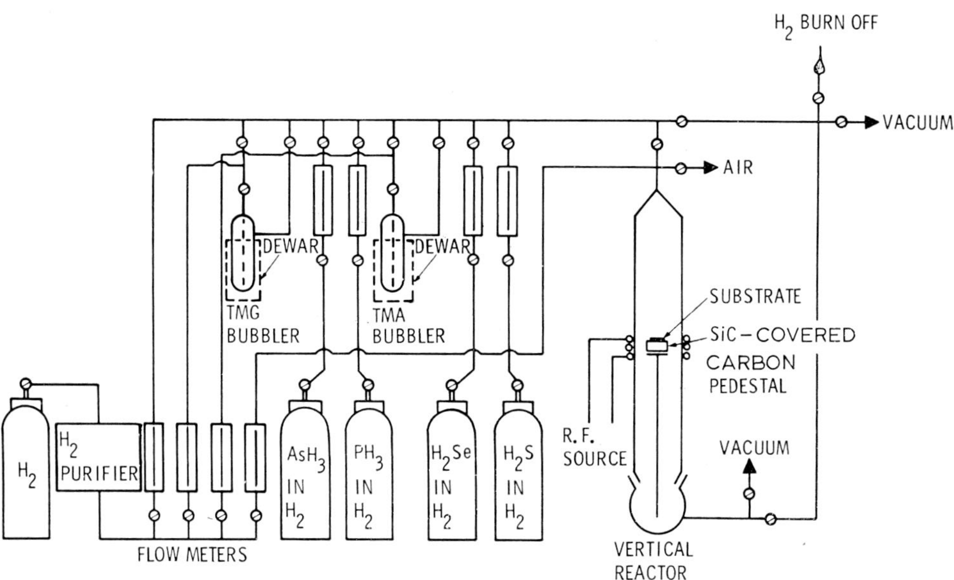

Fig. 1. Schematic of deposition apparatus.

the feasibility of using metalorganics in the preparation of the sulfides, selenides, and tellurides of Zn and Cd, on insulating substrates. The same Group II metalorganics were also found to be effective dopants in the preparation of p-type GaAs in these halide-free systems[15]). Lindeke et al.[16]) used gallium diethylchloride and AsH$_3$ to prepare homoepitaxial layers of GaAs. It was not ascertained if the process could be used to grow epitaxial films on insulating substrates, but it seemed to provide epitaxial films at much lower temperatures than other reported halide processes.

The studies in our laboratories have since been expanded to include single-crystal films of the nitrides of Al and Ga on Al$_2$O$_3$ and other substrates[17]).

The results of the studies of compound semiconductor growth using CVD processes involving metalorganics are reviewed in the following sections.

2. Experimental

The rather simple type of apparatus used in these studies is shown schematically in fig. 1. It is typical of those used in studying the growth of elemental semiconductors. It consists principally of a single vertical 60 mm O.D. quartz tube 38 cm long containing a SiC-covered carbon pedestal which can be inductively heated and rotated; stainless-steel bubblers for containing the liquid metalorganics and SbH$_3$; appropriate flow meters for monitoring the carrier gas, Group V hy-

dride(s), and dopant flows; a H$_2$ burn-off area; and a manifold made from $\frac{1}{4}$-inch stainless-steel tubing. Provisions are made for bypassing the quartz reactor and keeping the reactants separate until the gases are equilibrated and ready to be mixed for film formation.

The metalorganics are introduced into the reactor by bubbling a carrier gas through them and are mixed with the appropriate hydrides – such as AsH$_3$, PH$_3$, AsH$_3$–PH$_3$, AsH$_3$-SbH$_3$, NH$_3$, H$_2$S, and H$_2$Se – controlled independently by simple flow meter adjustments.

The structural nature of the deposits is determined by X-ray diffraction, reflection electron diffraction (RED), and/or electron microscopy techniques. The presence and relative compositions of As, P, and Sb in the films and the compositions of the alloys are determined using an electron microprobe.

The electrical properties of the films are evaluated from measurements of the Hall effect made on specially shaped samples etched in the films by standard photolithographic techniques. The carrier concentrations are deduced from the Hall coefficient R_H according to $n = 1/eR_H$, and the Hall mobility is determined from the product $\mu_H = \sigma R_H$, where σ is the electrical conductivity.

3. Studies of GaAs epitaxy on insulators

Both TEG (b.p. = 143 °C at 760 mm Hg) and TMG (b.p. = 55.6 °C at 760 mm Hg) were successfully used in the preparation of GaAs. Because of its greater volatility, most studies were made using TMG. Growth rates of GaAs of up to 1.5–2 µm/min were easily achieved using TMG. Epitaxial growth has been achieved at 650–750 °C on a number of insulating substrates of various orientations. These are compiled in table 1[18]).

TABLE 1

Orientation relationships between GaAs and single crystal oxides

Substrate	Parallel relationschips
(0001) Al_2O_3	(111) GaAs $\parallel$ (0001) Al_2O_3
$(11\bar{2}3)$ Al_2O_3	(111) GaAs $\parallel$ $(11\bar{2}3)$ Al_2O_3
$(11\bar{2}5)$ Al_2O_3	(111) GaAs $\parallel$ $(11\bar{2}5)$ Al_2O_3
$(11\bar{2}6)$ Al_2O_3	(111) GaAs $\parallel$ $(11\bar{2}6)$ Al_2O_3
$(01\bar{1}2)$ Al_2O_3	(111) GaAs $\parallel$ $(01\bar{1}2)$ Al_2O_3
(100) $MgAl_2O_4$	(100) GaAs $\parallel$ (100) $MgAl_2O_4$
(110) $MgAl_2O_4$	(100) GaAs $\parallel$ (110) $MgAl_2O_4$
(111) $MgAl_2O_4$	(111) GaAs $\parallel$ (111) $MgAl_2O_4$
$(10\bar{1}0)$ BeO	(100) GaAs $\parallel$ $(10\bar{1}0)$ BeO
$(10\bar{1}1)$ BeO	(111) GaAs $\parallel$ $(10\bar{1}1)$ BeO
(0001) BeO	(111) GaAs $\parallel$ (0001) BeO
(100) ThO_2	(100) GaAs $\parallel$ (100) ThO_2

A rather unexpected result was the observed growth of (100) GaAs on both Czochralski (110) $MgAl_2O_4$ and (100) $MgAl_2O_4$[7]). Epitaxy on (100) $MgAl_2O_4$ was found to be very surface-sensitive and not always reproducible; on (110) $MgAl_2O_4$ epitaxy was reproducible, but the electrical properties of the films were sometimes inconsistent. Further work is needed to explain the unusual crystallography. High quality reproducible films were produced on (111) $MgAl_2O_4$.

GaAs films on (0001) Al_2O_3 were found to possess a much lower degree of residual stress than that found for Si on Al_2O_3 or $MgAl_2O_4$. This is not unexpected since the mean linear thermal expansion coefficients for GaAs $(6.4 \times 10^{-6}/\text{deg C})$[19]) and Al_2O_3 ($\perp c$-axis, $7.7 \times 10^{-6}/\text{deg C}$), between 20–500 °C)[20]) are not too far apart at room temperature. Essentially no bowing could be observed for a 35 µm GaAs film grown on a 10 mil Al_2O_3 substrate. However, a 1 µm film did curl a 5 µm Al_2O_3 platelet concave downward, indicating that the residual stress in the GaAs is compressive.

3.1. PROPERTIES OF UNDOPED GaAs FILMS ON INSULATING SUBSTRATES

Studies on thick films deposited by the TMG–AsH$_3$ process have shown that excellent quality GaAs can be grown on (0001) Al_2O_3, (111) stoichiometric $MgAl_2O_4$[11]) and (110) stoichiometric $MgAl_2O_4$. The quality of the material, however, is found to be a function of film thickness, improving as the film is grown thicker. Since the electrical properties of the films change with thickness, the measured electrical parameters are necessarily averages over the thickness of the film. In general, the outermost layers of a film will have properties superior to the average value, and the layers nearer the interface will have inferior values.

Thick nominally-undoped films have been found to be n-type with net (average) carrier concentrations ranging from $\sim 10^{15}$ to 10^{17} cm^{-3} (refs. 9, 11), depending on the purity of the materials used to grow the films and the AsH$_3$–TMG concentration ratio used. Average mobilities greater than 5000 cm^2/V-sec have been obtained[11]).

Angle lapping and staining of n-type films have revealed the presence of a layer of high resistivity near the substrate surface[11]). It was found to be independent of film thickness, and therefore probably not related to diffusion of impurities from the substrate. The thickness of the layer was dependent on the carrier concentration of the donors in the film.

Once the film has converted to n-type, the carrier concentration rises rapidly to the thick-film value. The mobility tends to continue to improve as the film becomes thicker, usually "saturating" at a maximum value by the time the film has grown to ~ 20 µm.

3.2. P-TYPE DOPING STUDIES USING DIMETHYLCADMIUM AND DIETHYLZINC

An all metalorganic-hydride process has been used at Autonetics to produce p-type GaAs films on Al_2O_3 and $MgAl_2O_4$[15]). Dimethylcadmium (DMCd) and diethylzinc (DEZ) have been combined with TMG–AsH$_3$ to produce films with net acceptor carrier concentrations up to about 5×10^{17} cm^{-3} (Cd-doped) and 8×10^{19} cm^{-3} (Zn-doped). The observed doping levels are found to be exponential functions of growth temperature when other parameters remain fixed. Nearly bulk electrical properties have been achieved in Zn-doped films as thin as 1 µm and in thick (> 10 µm) Cd-doped films.

3.3. Electron microscopy and reflection electron diffraction (RED) studies

Studies of the early stages of growth of GaAs, under conditions which produce epitaxial films, indicate the formation of many discrete nuclei which coalesce to form large islands and eventually produce complete surface coverage. Annealing at or near the growth temperature has shown that the nuclei are highly mobile and move in such a fashion as to promote the growth of larger islands, as seen in fig. 2. Several of the crystallites in the figure appear to have merged, and most possess the triangular exterior shape associated with the growth of planes with trigonal symmetry. The RED pattern for this film indicates the presence of some twin structure in the GaAs. In addition, stacking faults could be inferred from the spots of irrational indices clustered closely around the indexed diffraction spots. Ordered structure has been confirmed in a deposit a fraction of a micron thick (~ 3000 Å) by RED patterns, although twinning is quite common in such thin layers. As film thickness increases, the quality of the growth improves, with electron microscope replicas showing a much smoother surface appearance and the RED patterns exhibiting Kikuchi lines and little evidence of twin structure in the upper layers.

3.4. GaAs on GaAs

The Autonetics group has also performed a number of epitaxial growth experiments on (111) Cr-doped GaAs, but an extensive study has not yet been made. However, thick films (~ 35 μm) with measured mobilities exceeding 6500 cm^2/V-sec at room temperature at carrier concentrations of $\sim 5 \times 10^{15}$ cm^{-3} have been produced on GaAs and used to prepare working Gunn-effect devices[21]).

Thickness measurements of thin-film GaAs growth made simultaneously on (110) and (100) GaAs substrates indicated similar growth rates. This is different from what has been reported in halide systems[22]). This seems to indicate that the presence of halides in transport growth processes influences the growth rates on different orientations of the substrate.

Rai-Choudhury[9]) used the metalorganic–hydride method to deposit GaAs on (100) GaAs as well as on (0001) Al$_2$O$_3$. The low mobilities and high carrier concentrations found in films grown on both substrate materials are likely due to impurities in the reactants.

(a)

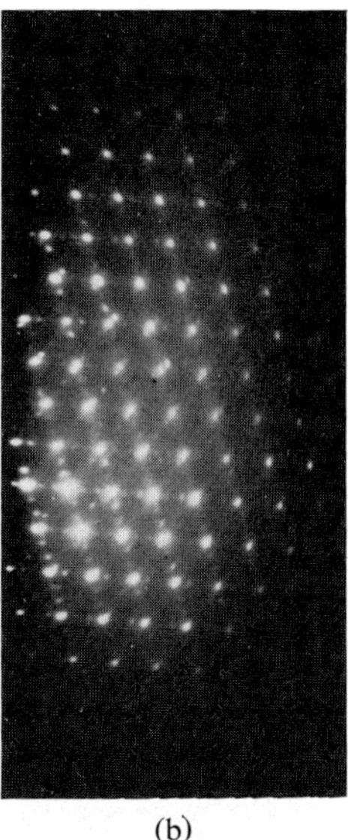

(b)

Fig. 2. GaAs/(0001)Al$_2$O$_3$ produced by reaction of residual TMG with AsH$_3$ for 6 min in an AsH$_3$–H$_2$ atmosphere. (a) Surface structure; (b) (110) RED pattern.

The absence of published film thickness data precludes the suggestion of any other reasonable explanations.

The studies of Lindeke et al.[16]) were limited to GaAs substrates, using gallium diethylchloride and AsH$_3$ to prepare GaAs. Room temperature mobilities of ~ 5000 cm^2/V-sec at carrier concentrations of 4.2×10^{15} cm^{-3} were obtained in films grown on (100) GaAs at deposition temperatures of 575 °C. It was not reported if the process would provide epitaxial films on insulating oxide substrates.

4. Other heteroepitaxial III-V compound semiconductor studies

The metalorganic-hydride growth process has been extended to the preparation of other epitaxial III–V compound semiconductors. Metalorganics were also found to be effective replacements for the hydrides in some growth systems. Table 2 indicates those III–V compounds that have been reported to date.

By reacting PH_3 with TMG, we were able to form GaP. However, it was apparent from the studies that controlling the growth of single-crystal films of GaP on insulators would be somewhat more difficult than the control of GaAs film growth. Whereas GaAs films only 0.3 μm thick are single crystal on Al_2O_3, GaP films of the same thickness exhibited only highly preferred orientation on (0001) Al_2O_3 although single-crystal growth occurred on GaAs. However, the films on Al_2O_3

TABLE 2

Epitaxial III–V compound semiconductors formed from metalorganics and hydrides

Compound	Substrate	Reactants	Growth temperature (°C)	Ref.
GaAs	Ge, GaAs, Al_2O_3, $MgAl_2O_4$	$TMG–AsH_3$ $TEG–AsH_3$	650–750	6, 7, 9
GaP	GaAs, Al_2O_3	$TMG–PH_3$	700–725	7
	Si	TMG–TEP	485	8
$GaAs_{1-x}P_x$ ($x = 0.1$–0.6)	GaAs, Al_2O_3, $MgAl_2O_4$	$TMG–AsH_3–PH_3$	700–725	7
$GaAs_{1-x}Sb_x$ ($x = 0.1$–0.3)	GaAs, Al_2O_3	$TMG–AsH_3–SbH_3$ $TMG–AsH_3–TMSb$	725	7
AlAs	GaAs, Al_2O_3	$TMA–AsH_3$	700	12, 13
$Ga_{1-x}Al_xAs$ ($x = 0.2$–0.9)	GaAs, Al_2O_3	$TMG–TMA–AsH_3$	700	12
AlN	Si, Al_2O_3, α-SiC	$TMA–NH_3$	1250	17
GaN	Al_2O_3, α-SiC	$TMG–NH_3$	925–975	17

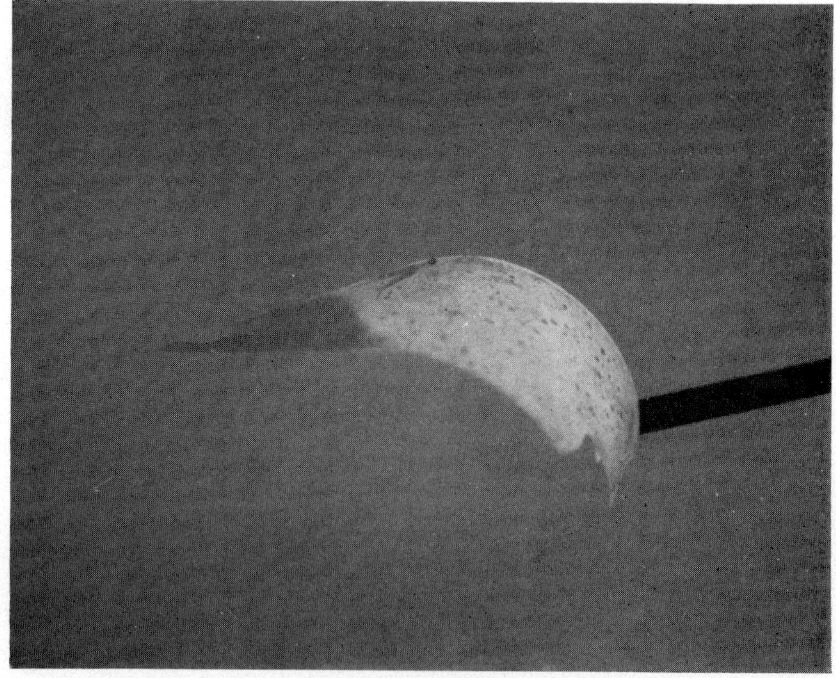

Fig. 3. Bending of a thin sapphire platelet caused by residual stress in the GaP film.

improved in quality with thicker growth; the surfaces were identified as single-crystal (111) orientation by RED.

As was observed for GaAs on a thin Al_2O_3 platelet, GaP growth caused curling of the platelet in a concave downward direction. This indicated the residual stress in the GaP film to be compressive. This effect is shown in fig. 3.

Additions of PH_3 to the AsH_3–TMG mixtures normally used for growing GaAs films led to the formation of single crystal $GaAs_{1-x}P_x$ films. In the preliminary studies, single crystal films with the composition $GaAs_{0.9}P_{0.1}$ and $GaAs_{0.4}P_{0.6}$ were grown on (0001) Al_2O_3 and (111) $MgAl_2O_4$.

Thomas[8] produced single-crystal growth of GaP on Si by the thermal decomposition of a gas-phase mixture of TEG and TEP at 485 °C. Extremely slow growth rates (~ 2000 Å/hr) were used to produce 2700–4500 Å thick films. Unfortunately it was not reported if the success of the process is growth rate, substrate, and/or temperature limited.

Manasevit and Simpson[7] found that adding SbH_3 to AsH_3–TMG mixtures yields $GaAs_{1-x}Sb_x$ alloys. Films with approximate compositions in the range $GaAs_{0.9}Sb_{0.1}$ to $GaAs_{0.7}Sb_{0.3}$ were obtained by arbitrarily changing the AsH_3 flow and/or H_2 flow through the liquid SbH_3. Better growth, as determined by RED, was obtained on (100) GaAs than on (0001) Al_2O_3 at the higher concentration of Sb. Epitaxial alloy films were also achieved by using TMSb in place of SbH_3.

The metalorganic–hydride process was also demonstrated to be feasible for the production of single-crystal films of AlAs and the mixed III–V Al compounds, materials of interest for opto-electronic devices[12]. By pyrolyzing TMA and AsH_3 in a H_2 atmosphere at about 700 °C, single-crystal films of AlAs have been formed on GaAs[12,13] and on single-crystal insulating substrates[12]. A 1–2 μm thick yellow layer of (111) AlAs formed on (0001) Al_2O_3 was found to be unstable in air.

Minden[13] has recently reported on some of the optical properties of AlAs films prepared by the TMA–AsH_3 process on GaAs substrates. The optical absorption edge measured on a polycrystalline layer on quartz indicated a direct bandgap of 3.5 eV. The indirect gap was 2.8 eV; and the high-frequency dielectric constant was determined to be 15.2. The surfaces of the AlAs films grown on GaAs were also reported to be unstable.

We attempted to produce AlP by the pyrolysis of TMA in H_2 at 700 °C in an atmosphere containing excess P formed by decomposing PH_3. The resulting film (~ 1 μm thick) was not stable in the laboratory atmosphere, even during the minute it took to transfer the wafer to the electron microscope chamber. Reid et al.[23] also observed reaction at the AlP surface on exposure to moist air for films formed at 900 °C. Some of the unexpected optical effects observed by Richman for AlP films[24] might also be related to this instability.

Alloys of $Ga_{1-x}Al_xAs$ have also been prepared in the Autonetics laboratories by mixing different quantities of TMA and TMG and reacting these with AsH_3 at the heated pedestal. Simple flowmeter adjustments controlled the TMA and TMG ratios. Alloys over the composition range $Ga_{0.1}Al_{0.9}As$ to $Ga_{0.8}Al_{0.2}As$, as analyzed with the electron microprobe, have been prepared on Al_2O_3 and GaAs substrates.

Nitrides of Al and Ga have also been produced in our laboratories by the metalorganic–hydride process. Pyrolysis of TMA and NH_3 in H_2 formed high-resistivity single-crystal insulating films of AlN on Si, α-SiC

Fig. 4.   AlN growth features on (111) Si.

and Al_2O_3 at temperatures of about 1250 °C[17]). Crazing which was observed in 1 µm thick films grown on (111) Si (fig. 4), did not occur on (01$\bar{1}$2) Al_2O_3. Single-crystal films of GaN have been produced from NH_3 and TMG on α-SiC and Al_2O_3 at temperatures of about 925–975 °C. The only GaN film on Al_2O_3 that was measured was n-type, with a carrier concentration above 10^{19} cm^{-3}, and a mobility ~ 60 cm^2/V-sec[17]).

5. Heteroepitaxial II-VI compound semiconductor films

II–VI compounds have also been grown on insulating substrates using metalorganics and hydrides[14]). The II–VI compounds which have been produced by this

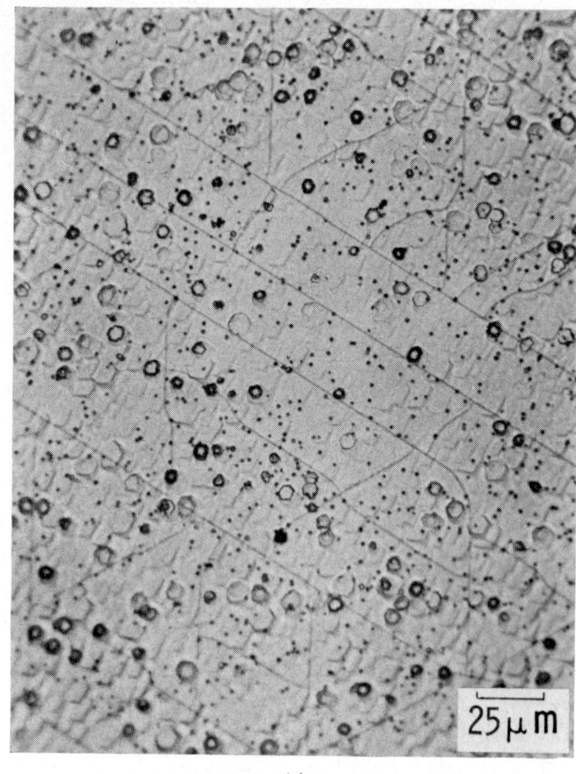

(a)

TABLE 3

Compound II–VI semiconductors formed from metalorganics and hydrides

Compound	Substrate	Reactants	Growth temperature (°C)	Ref.
ZnSe	Al_2O_3, $MgAl_2O_4$, BeO	DEZ–H_2Se	725–750	14
ZnS	Al_2O_3, $MgAl_2O_4$, BeO	DEZ–H_2S	750	14
ZnTe	Al_2O_3	DEZ–DMT	500	14
CdSe	Al_2O_3	DMCd–H_2Se	600	14
CdS	Al_2O_3	DMCd–H_2S	475	14
CdTe	Al_2O_3, $MgAl_2O_4$, BeO	DMCd–DMT	500	14

method are summarized in table 3. We have produced these films by reacting DEZ or DMCd with H_2Se, H_2S or dimethyltellurium (DMT) using essentially the same apparatus previously described for the formation of GaAs and other III–V compounds, but in some experiments a vertical quartz reactor containing a side arm and a center tube extending to about 1–2 inches from the pedestal was used. The hydrides are usually introduced into the reactor through the side arm prior to admitting the metalorganic in its carrier gas via the center tube. The product of their mixing is then decomposed at the heated pedestal to produce the II–VI compound. ZnSe growth on a number of orientations of Al_2O_3, BeO, and $MgAl_2O_4$ supplied the same heteroepitaxial relationships as those found for the growth of GaAs on these substrate planes (see table 1). This, therefore, suggests a similar growth mechanism. Goodman[25]) also obtained (111) ZnSe growth on (01$\bar{1}$2) Al_2O_3 but by an evaporation technique.

The nature of the surface of ZnS growth on Al_2O_3, $MgAl_2O_4$, and BeO is recorded in fig. 5. Some crazing can be observed in the films grown on (0001) Al_2O_3 and (111) $MgAl_2O_4$. The films were grown at about 750 °C from DEZ and H_2S.

Single-crystal films of (111) ZnTe were also produced on Al_2O_3. They were made by reacting DEZ with DMT at about 500 °C. The films possessed a deep orange color and varied in reflectivity with changes in substrate orientation.

Replacing the DEZ with DMCd led to the formation of the corresponding Cd-group VI semiconducting compounds. CdS was obtained on (0001) Al_2O_3 by reacting DMCd with H_2S at 475 °C; the growth of CdSe on (0001) Al_2O_3 from DMCd and H_2Se was achieved at about 600 °C; and single-crystal films of CdTe were produced at about 500 °C from DMCd and DMT on Al_2O_3, $MgAl_2O_4$, and BeO.

6. Conclusions

The results to date using metalorganics in the preparation and doping of compound semiconductors are encouraging. The technique obviates the use of multiple temperature zones for the CVD process, simplifies

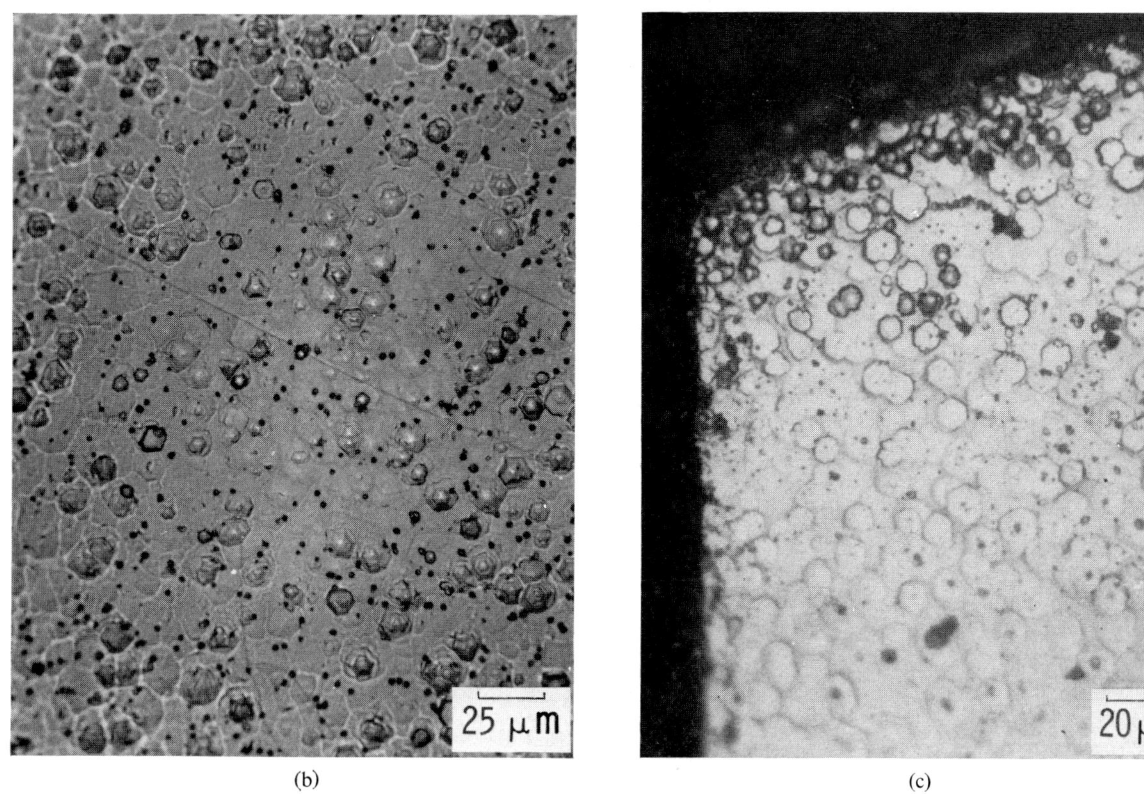

Fig. 5. ZnS growth on (a) (0001) Al_2O_3; (b) (111) $MgAl_2O_4$; (c) $(10\bar{1}1)$ BeO.

the procedure of compound growth considerably, minimizes autodoping which might be caused by the attack of the substrate by transport agents used in other CVD processes, and makes compound semiconductor film growth compatible with the growth of elemental semiconductors. One can expect this new CVD technique to be very instrumental in making different compound semiconductor films more available for device study.

Acknowledgements

The author gratefully acknowledges the efforts of W. I. Simpson and F. M. Erdmann in preparing the heteroepitaxial films, A. C. Thorsen for electrical measurements, and wishes to thank R. P. Ruth for many helpful discussions and a review of the manuscript.

The research was supported in part by the National Aeronautics and Space Administration under Contract No. NAS 12-2010 and by the Army Ballistic Missile Defense Agency through the United States Army Electronics Command under Contract No. DAAB07-69-C-0452.

References

1) J. J. Tietjen and J. A. Amick, J. Electrochem. Soc. **113** (1966) 724.
2) J. R. Knight, D. Effer and P. R. Evans, Solid-State Electron. **8** (1965) 178.
3) M. Michelitsch, W. Kappallo and G. Hellbardt, J. Electrochem. Soc. **111** (1964) 1248.
4) R. Didchenko, J. E. Alix and R. H. Toeniskoetter, J. Inorg. Nucl. Chem. **14** (1960) 35.
5) B. Harrison and E. H. Tompkins, Inorg. Chem. **1** (1962) 951.
6) H. M. Manasevit, Appl. Phys. Letters **12** (1968) 156.
7) H. M. Manasevit and W. I. Simpson, J. Electrochem. Soc. **116** (1969) 1725.
8) R. W. Thomas, J. Electrochem. Soc. **116** (1969) 1450.
9) P. Rai-Choudhury, J. Electrochem. Soc. **116** (1969) 1745.
10) H. M. Manasevit and A. C. Thorsen, Met. Trans. **1** (1970) 623.
11) A. C. Thorsen and H. M. Manasevit, J. Appl. Phys. **42** (1971) 2519.
12) H. M. Manasevit, J. Electrochem. Soc. **116** (1969) 250C; **118** (1971) 647.
13) H. T. Minden, Appl. Phys. Letters **17** (1970) 359.
14) H. M. Manasevit and W. I. Simpson, J. Electrochem. Soc. **117** (1970) 196C; **118** (1971) 644.
15) H. M. Manasevit and A. C. Thorsen, J. Electrochem. Soc. **117** (1970) 407C; **119** (1972) 99.
16) K. Lindeke, W. Sack and J. J. Nickl, J. Electrochem. Soc. **117** (1970) 1316.

17) H. M. Manasevit, F. M. Erdmann and W. I. Simpson, J. Electrochem. Soc. **118** (1971) 1864.

18) H. M. Manasevit and A. C. Thorsen, Heteroepitaxy of III–V Compound Semiconductors on Insulating Substrates, Final Report, Jan. 1970, NASA-ERC Contract No. NAS 12-2010.

19) M. E. Straumanis, J. P. Krumme and M. Rubenstein, J. Electrochem. Soc. **114** (1967) 640.

20) J. B. Austin, J. Am. Ceram. Soc. **14** (1931) 795.

21) J. M. Owens, Proc. IEEE **58** (1970) 930.

22) D. W. Shaw, J. Electrochem. Soc. **115** (1968) 777.

23) R. J. Reid, S. E. Miller and H. L. Goering, J. Electrochem. Soc. **113** (1966) 467.

24) D. Richman, J. Electrochem. Soc. **115** (1968) 945.

25) A. M. Goodman, J. Electrochem. Soc. **116** (1969) 364.

Journal of Crystal Growth **13/14** (1972) 315–318 © *North-Holland Publishing Co.*

EPITAXIAL DEPOSITION OF DEGENERATE n-TYPE LAYERS OF Ge ON GaAs

J. K. KENNEDY and W. D. POTTER

Air Force Cambridge Research Laboratories AFSC,
L. G. Hanscom Field, Bedford, Massachusetts 01730, U.S.A.

The growth conditions required for the deposition of degenerate epitaxial layers of n-type germanium on n-type GaAs via the thermal decomposition of GeH_4 have been established. The growth rate has been determined as a function of the deposition temperature for several germane to hydrogen ratios and as a function of the percentage of germane and arsine in the reaction mixtures. Deposition conditions necessary for the elimination of the formation of an interfering p-layer in the germanium due to the thermal breakdown of the GaAs surface during the deposition were also determined.

1. Introduction

The objective of this investigation was to determine the requisite growth conditions for the preparation of epitaxial layers of degenerate n-type germanium on n-type GaAs for use as a hot-electron type infrared detector. The thermal decomposition of GeH_4 was chosen as the growth technique for several reasons. (1) Both germane and an easily decomposible n-type dopant, arsine, were commercially available in high purity and are stable gases at room temperature. (2) Deposition can be made at relatively low temperatures. (3) There are no reactive by-products of the decomposition reactions.

While no reference has been found concerning the preparation of degenerate n-type Ge layers on GaAs from GeH_4, several authors have reported the preparation of such layers on germanium and GaAs via the iodide disproportionation and the chloride thermal decomposition reactions[1−5]. Two of these authors, when preparing degenerate n-type films, reported a decrease in the growth rate which they attributed to an increase in the etching rate during growth[2,3].

One complication in the deposition of homogeneous degenerate n-type layers of Ge on GaAs was reported by Papazian and Reisman[6]. In their depositions of germanium on GaAs made below 700 °C they found a thin p-type layer in the germanium which started at the GaAs–Ge interface. Beyond this layer their deposits were n-type. For our purposes, the p-layer had to be eliminated. If it is assumed that the Ga atoms responsi-

ble for the p-layer are generated at the GaAs–Ge interface by the thermal breakdown of the GaAs substrate, then the problem is to determine whether the p-layer can be eliminated by lowering the deposition temperature and, if this can be accomplished, to determine if the Ge layers which result are still of sufficiently high quality to be compatible with the photolithographic techniques required for detector fabrication.

2. Experimental

A schematic diagram of the thermal decomposition system used in this investigation is shown in fig. 1. The gas handling system was constructed entirely of pyrex and stainless steel. The water cooled decomposition chamber was constructed of pyrex and quartz. The reactant gases undoped GeH_4, and 10% AsH_3 in H_2, were obtained from the Matheson Company and Precision Gas Products, Incorporated, respectively. The hydrogen carrier gas which comprised the bulk of the reaction mixture was standard quality hydrogen which was purified by passage through an Ag–Pd diffuser. The He gas used to purge the diffuser and the system, when required, was 99.998% He. The flow system is constructed so that the GeH_4 can be doubly diluted while the AsH_3 can be triply diluted before entering the reactor. This arrangement permits a wide range of concentration for both gases. The 1 liter dilution bulbs and the main 2 liter mixing bulb were installed to insure gas phase concentration stability throughout the run.

The substrates which were oriented 3° off the ⟨100⟩ were n-type, Sn doped, GaAs obtained from the Mon-

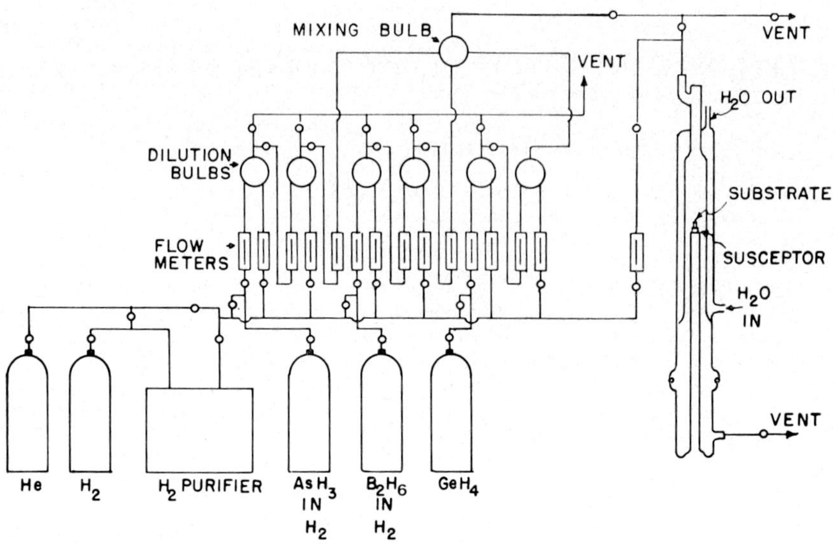

Fig. 1.   Schematic diagram of the thermal decomposition system.

santo Company. The slices were chemically polished with NaClO using the Riesman and Rohr[7]) technique. Immediately before being placed in the reactor, the samples were washed, etched in a solution of 90 H_2SO_4: 5 H_2O_2: 5 H_2O, rinsed in high purity water, and blown dry with filtered He as described by Berkenblit, Reisman and Light[8]). The samples were then placed on the susceptor and positioned in the reactor. The susceptors which were 1 inch diameter, 0.5 inch high and were fabricated by the Ultra-Carbon Corporation of Ultra "F" Purity Graphite Grade UT-6, were baked in H_2 at approximately 1000 °C for 30 min prior to each run.

With the sample in place the decomposition chamber was flushed with H_2 while sufficient time was allotted for the reactant mixtures, which were flowing to vent, to equilibrate in the dilution and mixing bulbs. Substrate temperatures during the run were monitored by an infrared radiation thermometer. Thickness measurements were made interferometrically on samples beveled at a shallow angle. Carrier concentrations were determined from resistivity measurements made with a 4-point probe on layers deposited on p-type GaAs.

3. Results and discussion

In light of the presence of the p-layer found by Papazian and Reisman, the bulk of our depositions were made at 650 °C and below. The growth rate of the films deposited were evaluated as a function of the deposition temperature and the Ge/H_2 and As/H_2 ratios.

Regardless of the experimental conditions used, the growth rate was found to be independent of the deposition time. Fig. 2 shows the growth rate as a function of the deposition temperature for several Ge/H_2 ratios without arsenic doping. The Ge/H_2 ratio was varied by

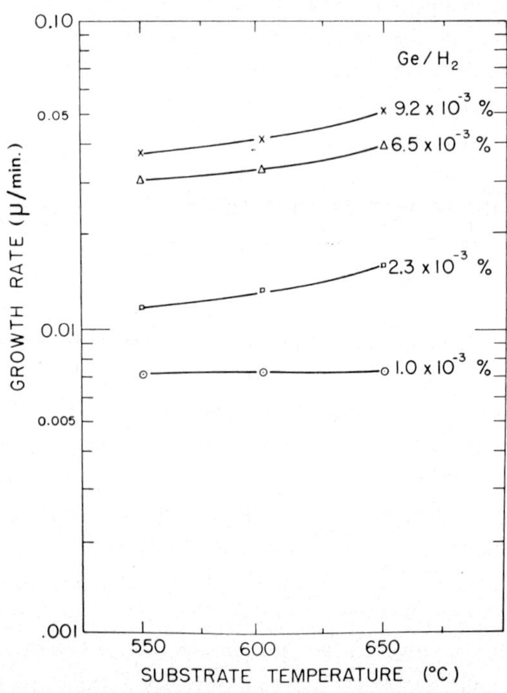

Fig. 2.   Growth rate as a function of substrate temperature at varying Ge/H_2 percentages for a constant linear gas stream velocity of 7.5 cm/sec.

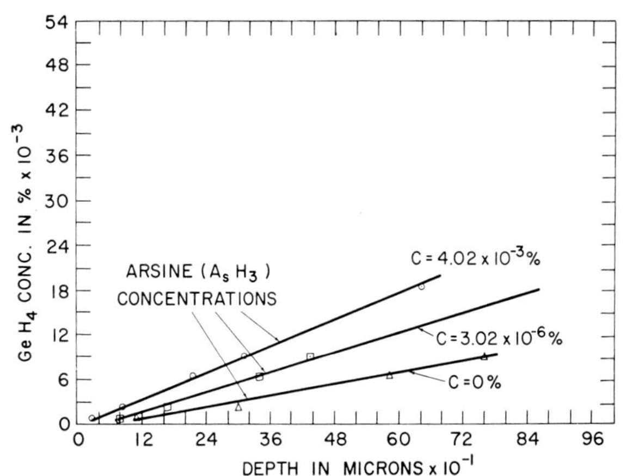

Fig. 3.   Effect of germane concentration on layer depth for several arsine concentrations for 150 min growth time at 650 °C.

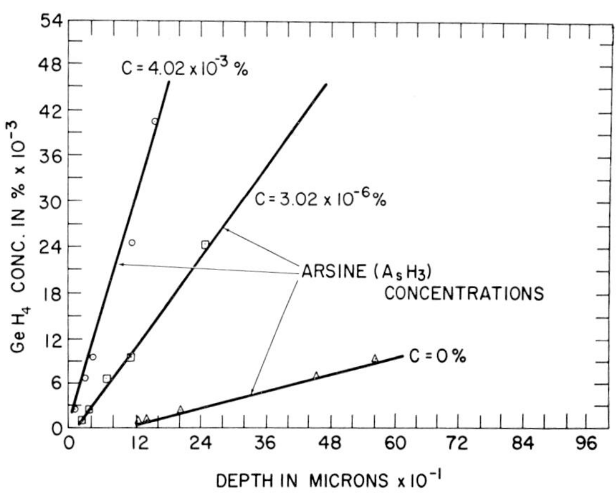

Fig. 4.   Effect of germane concentration on layer depth for several arsine concentrations for 150 min growth time at 550 °C.

changing the amount of GeH_4 added to the gas stream while keeping the total gas flux constant. As shown in fig. 2, the rate of growth increases with increasing temperature.

In figs. 3 and 4, the layer thicknesses found in 150 min runs are shown as a function of the percentage of GeH_4 and AsH_3 in the reaction mixtures at 650 and 550 °C (results at 600 °C were intermediate). These figures clearly indicate that as the deposition temperature is lowered the layer thickness becomes a very strong function of the arsine concentration in the reaction mixture, decreasing with increasing arsine concentration and decreasing temperature. This decrease cannot be attributed to etching since there are no etchants present. It is assumed, therefore, that the effect is due to "poisoning" of the growing surface by arsenic atoms. This trend continues as the deposition temperature is lowered still further. For example, at 450 °C using an AsH_3 percentage of 4.03×10^{-3} and a GeH_4 percentage of 9.2×10^{-3}, the germanium layer deposited in a 150 min deposition, although definitely present, was less than 0.08 µm, the thinnest layer we can measure with any confidence.

The percentage of arsine used in the vapor phase was quite high, being in some cases equal to or greater than the percentage of germane. The use of these high concentrations was found necessary in order to attain the requisite doping levels for degeneracy. The doping levels attained using an arsine percentage of 4.03×10^{-3} were all between 5×10^{18} and 2×10^{19} atoms/cm³ when depositions were made in the 450 °C to 550 °C range

for the germane concentrations shown in fig. 2. The higher doping levels were attained at high AsH_3/GeH_4 ratios; however, throughout the temperature range investigated, the doping concentration did not appear to be a function of the deposition temperature. The resistivity of the germanium layers deposited without intentional addition of dopants was approximately 0.1 ohm-cm n-type. This was determined by depositing a germanium layer on a high resistivity (approximately 3 ohm-cm) p-type Ge substrate and measuring the layer resistivity with a 4-point probe, and corresponded to a carrier concentration of approximately 2×10^{16} donors/cm³.

The presence of a p-layer in a deposit was determined by slowly etching the germanium and determining the conductivity type of the remaining deposit with a standard thermal probe. A 5:1 solution of 30% $H_2O_2:H_2O$ (by volume) heated to approximately 60 °C was used as the preferential etchant. This solution removes approximately 0.1 µm/min in a very uniform manner. The thin p-layer, described by Papazian and Reisman, was found in all samples deposited at 500 °C and above regardless of doping conditions. The only exceptions were samples deposited at 650 °C with no intentional doping. For these samples thin layers (less than 3 µm were p-type throughout. The thickness of the p-layer increased, as expected, with increasing deposition temperature. When the deposition temperature was lowered to 475 °C and then to 450 °C no p-layer could be detected. The growth rates at these temperatures, however were very low. (5.7×10^4 µm/min, and 4.6×10^4

µm/min respectively) but by extending the deposition time useable thicknesses (approximately 0.5 µm) could be deposited and these layers were sufficiently smooth to be compatible with the photolithographic techniques used for detector fabrication. Samples deposited at 450 °C and 475 °C are currently being evaluated as hot electron type infrared detectors.

For the deposition of germanium layers on germanium Silverstri[9]) using the chloride system found that if a smooth, shiny, mirror-like epitaxial layer can be deposited, then in general, the smoothness of such a layer increases with decreasing reactant gas to carrier gas ratio and with increasing deposition temperature. Silvestri considered a deposit structured as opposed to smooth when it "exhibited growth figures characteristic of the particular substrate orientation employed". This trend was found to hold in our depositions of germanium on germanium and GaAs using germane. In the case of undoped depositions on GaAs, increased structuring was observed when the Ge/H_2 ratio was increased. However, in the 450 °C to 650 °C range, the structuring did not appear to decrease with increasing deposition temperature. When heavy arsenic doping was used the structuring due to growth figures gave way to needle formation. The size of the needles followed the same trend as the growth figures, increasing with increasing Ge/H_2 ratio, but remaining about constant with temperature variation.

4. Summary

The growth conditions required for the deposition of smooth, shiny, degenerate epitaxial layers of n-type germanium on n-type GaAs via the thermal decomposition of GeH_4 have been determined. The complication of a p-lyaer formation in the germanium due to the thermal breakdown of the GaAs surface during deposition was eliminated by lowering the deposition temperature to less than 500 °C.

Acknowledgement

The authors wish to express their appreciation to Drs. F. D. Shepherd, Jr., and A. C. Yang for suggesting this investigation and for their helpful discussions and comments.

References

1) T. Arizumi, I. Akasaki and T. Nishinaga, Japan J. Appl. Phys. **2** (1963) 757.
2) G. I. Uspenskaya, V. N. Shabanov and V. A. Tolomasov, Soviet Phys.-Cryst. **12** (1968) 827.
3) A. N. Stepanova and E. I. Givargizov, Fiz. Tverd. Tela **5** (1963) 3034.
4) J. C. Marinace, IBM J. Res. Develop. **4** (1960) 248.
5) R. L. Anderson, Solid State Electron. **5** (1962) 341.
6) S. A. Papazian and A. Reisman, J. Electrochem. Soc. **115** (1968) 961.
7) A. Reisman and R. L. Rohr, J. Electrochem. Soc. **111** (1964) 12.
8) M. Berkenbliit, A. Reisman and T. B. Light, J. Electrochem. Soc. **115** (1968) 966.
9) V. J. Silvestri, J. Electrochem. Soc. **116** (1969) 81.

Journal of Crystal Growth **13/14** (1972) 319–324 © *North-Holland Publishing Co.*

ÉTUDE DE L'ANISOTROPIE DE LA CROISSANCE ÉPITAXIALE DE GaAs EN PHASE VAPEUR

L. HOLLAN et C. SCHILLER

Laboratoires d'Electronique et de Physique Appliquée, 94 – Limeil-Brévannes, France

Thick epitaxial GaAs layers have been grown on GaAs hemispheres. During growth abrupt regular changes were made in the doping level. After growth, sections were cut along a given crystallographic plane; the region of different doping was revealed and observed under an optical microscope. This technique allowed the study of the growth along many crystallographic directions. and the determination of growth rates.

1. Introduction

La méthode de transport en phase vapeur mise en œuvre dans cette étude est celle utilisant le trichlorure d'arsenic et le gallium[1,2]). Bien qu'un grand nombre de chercheurs utilisent cette méthode, les études fondamentales concernant les mécanismes de croissance sont peu nombreuses. Néanmoins, l'aspect thermodynamique de ce transport a été approfondi par le calcul et par l'expérience[3–6]) dans quelques études assez concordantes.

Dans certaines de ces études, l'aspect cinétique a également été abordé. Il apparaît clairement qu'au niveau du dépôt une limitation cinétique de la croissance intervient, devenant dans la plupart des cas le phénomène prépondérant dans le contrôle de cette croissance. La nature exacte de la limitation cinétique n'est pas connue, il s'agit vraisemblablement d'un phénomène d'adsorption-désorption mettant en jeu une énergie d'activation. Le rôle du plan cristallographique dans ce processus est très important; l'anisotropie marquée qui apparaît a surtout été étudiée par Shaw[4,6]). Le but de notre contribution est d'élargir l'étude de l'anisotropie de la croissance, grâce à une nouvelle technique expérimentale qui fournit des renseignements permettant de faire des progrès vers une compréhension exacte des mécanismes dans ce type de croissance en phase vapeur.

2. Techniques expérimentales

Le type de réacteur dans lequel les croissances ont été effectuées, a été décrit précédemment[2]). Pour rendre possible cette étude, nous l'avons doté d'un système de dopage permettant de faire varier le niveau de dopage très rapidement: de 10^{15} à 10^{17} at. cm^{-3} à intervalles de temps réguliers, variant de 5 à 40 minutes.

Les croissances ont été effectuées sur substrat GaAs monocristallin dopé au tellure ($n \sim 10^{17}$ at. cm^{-3}) poli de façon à obtenir des demi-sphères de diamètre 5 ou 10 mm, et de plan de base (001), la couche perturbée par polissage étant enlevée par décapage non sélectif.

Au cours de ces longues croissances, les facettes se développent sur les substrats hémisphériques, comme en témoigne la figure 1, qui montre l'extension des facettes (001), (113), (011) et (111) B. Ensuite, le cristal est soumis à une découpe suivant une orientation cris-

Fig. 1. Macrophotographie d'une demi-sphère épitaxiée avec facettes développées. Grandissement 16.

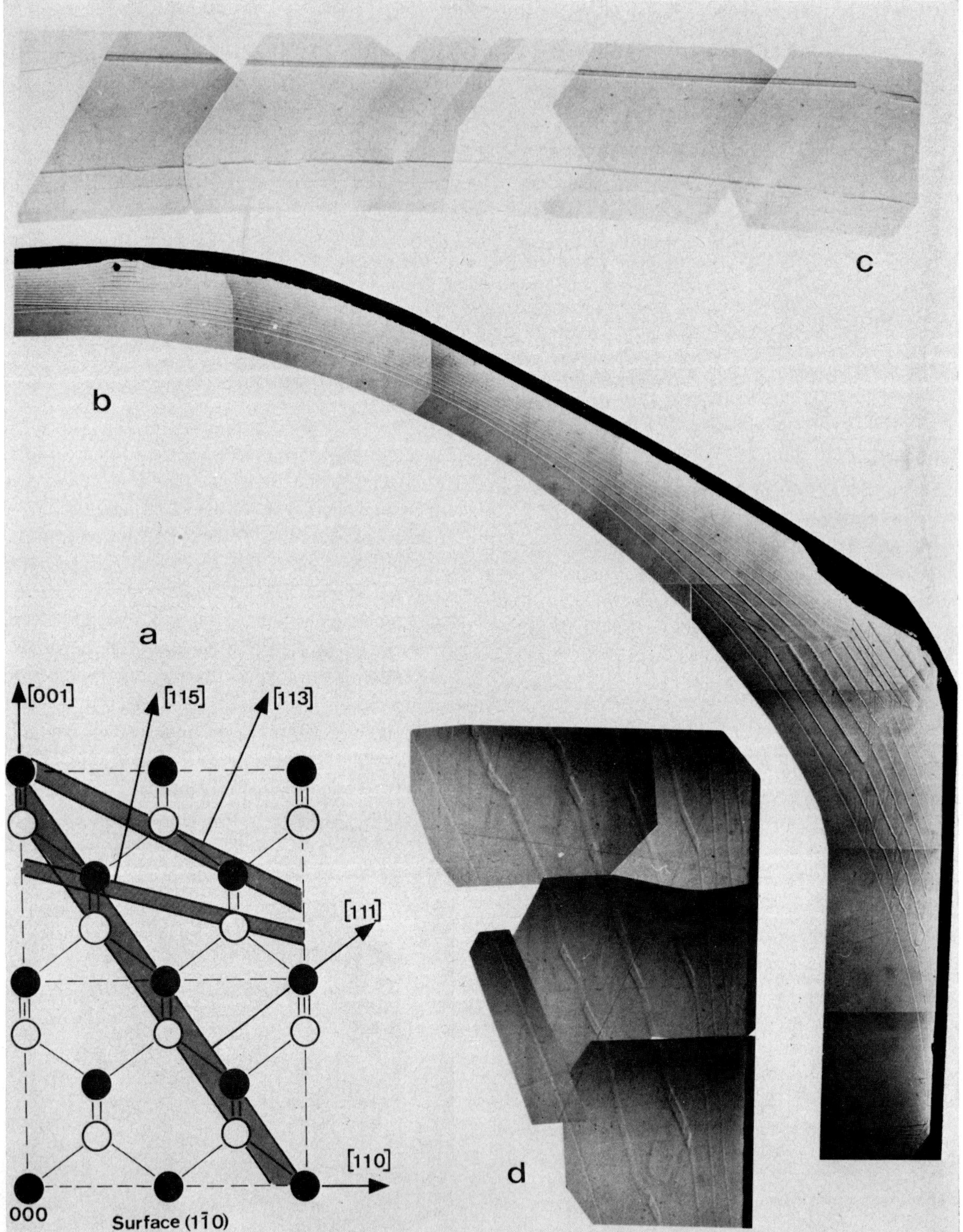

Fig. 2.   Coupe révélée suivant un plan (1̄10) d'une croissance à 750 °C. (a) Schéma cristallographique. Atomes de Ga en cercles pleins et de As en cercles creux, projetés sur le plan (1̄10). Les plans polaires apparaissant en grisé sont normaux au plan de découpe. (b) Vue générale de la coupe, grandissement 57. (c) Détail de la région (111), grandissement 144. (d) Détail de la région (110), grandissement 234.

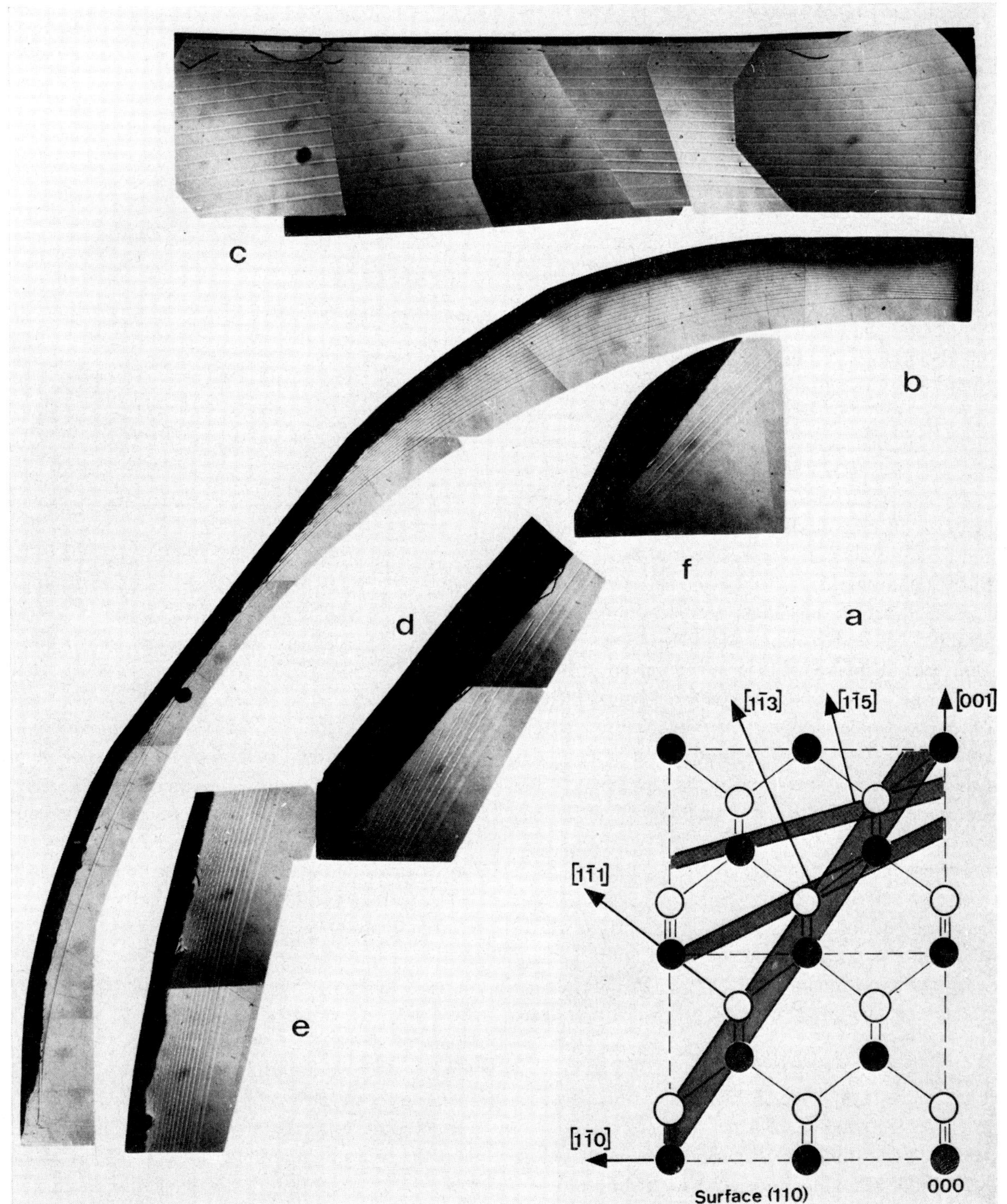

Fig. 3. Coupe révélée suivant un plan (110) d'une croissance à 750 °C. (a) Schéma cristallographique – mêmes conventions que la figure 2a. (b) Vue générale de la coupe, grandissement 57. (c) Détails des régions (001) c, (1̄1̄1) d, (2̄2̄1) e, et (1̄1̄2) f, grandissement 234.

tallographique donnée, de telle façon qu'après enro-
bage et polissage le plan choisi pour la découpe – qui
est perpendiculaire au plan de base de la demi-sphère –
passe exactement par le sommet de celle-ci et par les
pôles (111)A gallium ou (111)B arsenic. On obtient
alors une section perpendiculaire à toute une série de
plans d'indices simples. L'opération peut se répéter
avec une autre découpe. La surface ainsi obtenue est
ensuite plongée dans une solution de révélation chimique qui fait apparaître les zones de niveaux de dopage
différents bien visibles au microscope optique en contraste interférentiel.

Sur les figures 2 et 3 (qui représentent un montage à
partir d'un nombre important de photos), nous avons
donné en exemple la vue globale de deux sections,
effectuées sur une même croissance, mais dans les plans
(1̄10) et (110), ainsi que certains détails à plus fort
grossissement. La croissance avait été effectuée à 755 °C
avec des séquences de dopage, 10 min de type n, puis
5 min de n$^+$, et 30 min de n répétées 14 fois.

Sur ces photomontages, la mesure des taux de croissances en fonction de l'angle devient simple et sans
ambiguité. Pour cela, on procède de la façon suivante:
le centre de la sphère est déterminée par construction
géométrique, et l'orientation (exacte à 1° près) à partir
des facettes bien développées, qui peuvent être indexées
(111)A ou B, (001), (110). Ensuite, pour obtenir des
courbes, il suffit de faire – en fonction de l'angle – des
mesures de longueurs entre une ou plusieurs zones
révélées.

On remarque que, par cette nouvelle technique, deux
causes d'erreurs importantes sont éliminées: celle qui
apparaîtrait du fait de la définition imparfaite de la
surface, et celle qui résulterait de la variation du taux
de croissance en fonction du temps. On constate en
effet une légère variation, mais elle est aisément évaluée.

3. Résultats et interprétations

L'intérêt de la méthode apparaît immédiatement: on
peut observer les différents stades de la croissance dans
le temps, l'apparition et la disparition des facettes et de
défauts ou d'instabilités dans certaines régions.

La figure 4 donne les courbes de taux de croissance.
Dans l'interprétation de ces courbes, nous avons été
amenés à préciser l'orientation des plans polaires en
zones (110) et (1̄10).

Les figures 2a et 3a montrent que, lorsque l'on passe

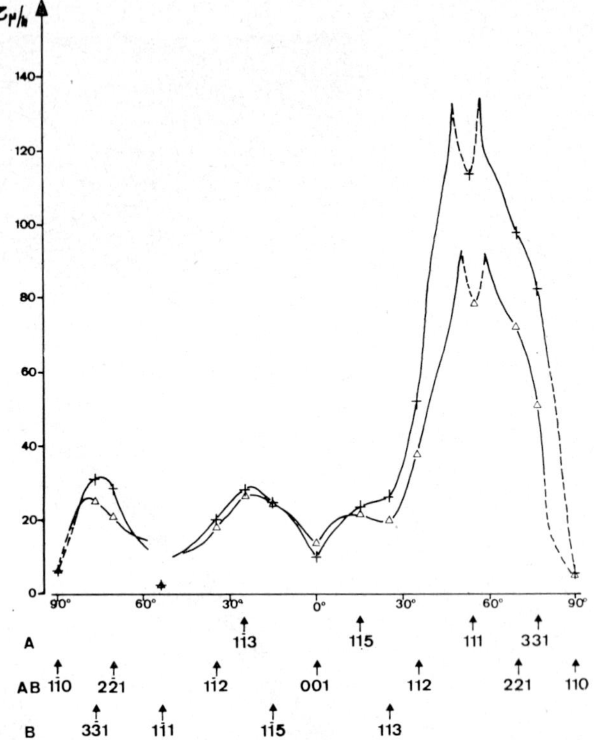

Fig. 4. Courbes des taux de croissance (τ en μm/h) en fonction
de l'angle avec (001) des plans en zones (110) et (1̄10) pour deux
températures d'épitaxie. (+) 725 °C, (△) 755 °C.

du pôle (001) à un pôle (111)A par exemple (A indiquant un plan gallium et B un plan arsenic), la polarité
des plans intermédiaires change alternativement à
chaque passage par un plan non polaire. L'exemple le
plus simple est celui des plans (113)B, (112) et (111)A
pour lesquels la polarité change de sens de (113)B à
(111)A, le plan (112)AB étant non polaire.

Sur la figure 4, nous avons donc représenté ces plans
par des flèches d'une part, et des points situés sur la
courbe d'autre part. Il faut noter que, contrairement
aux études précédentes[4,6]), ces courbes ne sont pas
seulement déterminées par ces points, mais par un
grand nombre de mesures intermédiaires qui nous ont
amenés à les tracer en continu.

Il apparaît assez clairement que ce sont surtout les
plans de polarité gallium A qui correspondent à des
maxima. Le problème des minima discontinus, correspondant en général à des facettes, sera développé
plus loin. Remarquons cependant la dissymétrie de
part et d'autre du plan (001) qui permet d'expliquer les
différences de taux de croissance observées dans le cas
de croissance des plans de surface désorientés de quel-

ques degrés par rapport à ce plan cristallographique, et le minimum (marqué surtout à 755 °C) dans la direction (113)B qui donne des facettes de bel aspect (fig. 1).

Le seul paramètre qui différencie les deux courbes est la température. Contrairement à ce qui pouvait être prévu[2,4]), c'est en général à plus basse température que les taux de croissance sont légèrement plus élevés, sauf dans le cas du plan (001).

Ce fait peut être expliqué par l'utilisation de pressions partielles faibles, ce qui aurait comme résultat d'aplatir le maximum des courbes de $\log \tau = f(1/T)$ (réfs. 2, 4), et de le déplacer vers les plus basses températures. D'autres essais a plus basse température d'une part et à pressions partielles plus élevées d'autre part sont en cours.

3.1. INFLUENCE DES DÉFAUTS DE CROISSANCES ET DES IMPURETÉS

Cette technique de visualisation des différentes étapes de la croissance dans le temps, permet de suivre l'évolution de certains défauts macroscopiques. Ainsi le détail de la fig. 5 montre la section d'une pyramide, défaut typique souvent observé sur les faces (001) non désorientées.

On observe – probablement à cause de la nucléation rendue plus facile par la présence d'un défaut[3]) – que le taux de croissance du plan (001) est nettement accru. Dans ce cas, les plans (113) et (1$\bar{1}$5) limitent la croissance et apparaissent donc comme les flancs de la pyramide. Il est apparu également, que le taux de croissance du plan (110) dépend beaucoup de sa perfection; c'est seulement en l'absence de défaut que l'on retrouve la valeur faible représentée sur la courbe de la figure 4.

La photo a de la figure 5 montre la cicatrisation d'un défaut causé probablement par une poussière accrochée au substrat à cet endroit. Elle révèle les possibilités de cette technique de visualisation dans ce domaine.

Par ailleurs, on peut voir qu'en première approximation, nous n'avons pas noté d'effet de la différence des dopages (qui restent toutefois faibles) sur le taux de croissance. Par contre, le coefficient de ségrégation des impuretés semble souvent être très différent suivant que l'on est sur un plan d'indice simple ou un plan vicinal. Ce phénomène n'apparaît que qualitativement, grâce à la sélectivité de l'attaque. Nous pouvons le noter en particulier sur les détails c et d de la figure 2. Ceci prouve que les mécanismes de la nucléation et de la

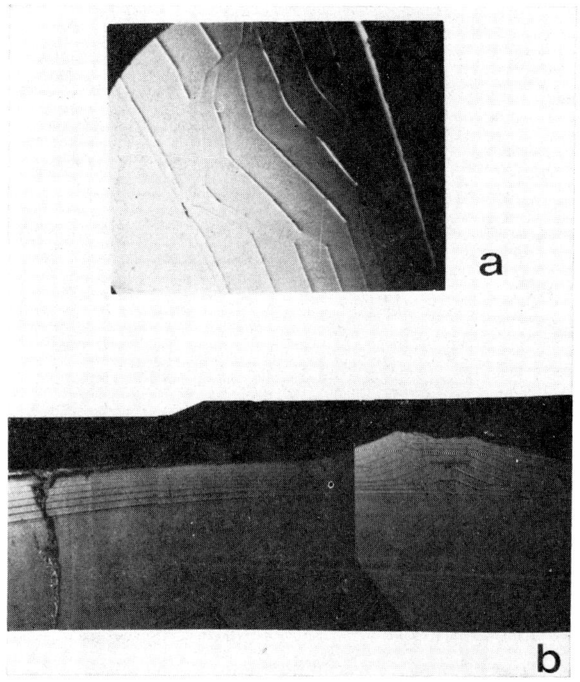

Fig. 5. Détails de défauts de croissance, grandissement 165. (a) région (221), (b) région (001).

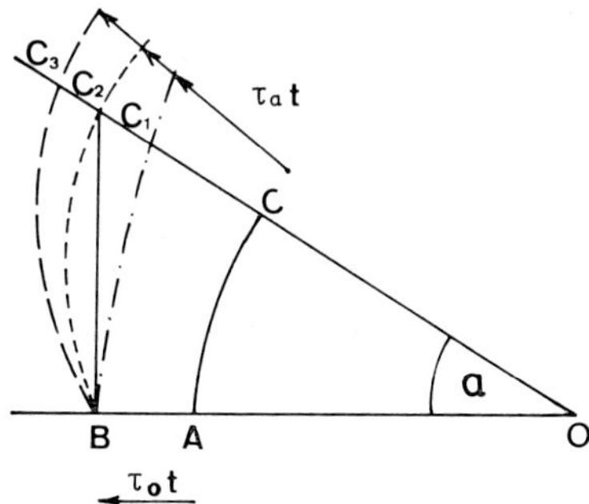

Fig. 6. Construction de Wulff pour une surface sphérique.

croissance sont très différents entre plans d'indices simples – donc sans marches – et plans vicinaux représentant automatiquement des marches.

3.2. FACETTES

L'apparatition et la disparition de facettes correspondant à des plans cristallographiques denses et à des minimums de la vitesse de croissance, peut être expliquée en première approximation par la construction de Wulff.

Représentons une partie de la section du substrat hémisphérique par l'aire $\overset{\frown}{OAC}$ d'angle quelconque α (figure 6) et soit $\overrightarrow{AB}$ la normale au plan considéré. Le segment $\overline{AB}$ représente la vitesse normale τ_0 du plan d'indices simples, multiplié par le temps de croissance t, et $\overrightarrow{OA}$ le rayon de courbure R du substrat en B. Suivant la construction de Wulff menons la perpendiculaire en B à AB. Si en chaque point de l'arc $\overset{\frown}{AC}$ on porte la grandeur $\tau(\alpha)t$ suivant la normale (t étant le temps) on obtient trois types de courbes illustrées par BC_1, BC_2 et BC_3.

Trois cas peuvent se présenter:

$$(1) \quad \tau_\alpha < \frac{\tau_0 + p(1-\cos\alpha)}{\cos\alpha} \text{ avec } p = \frac{R}{t}.$$

La surface apparente sera déterminée par la courbe en pointillés du type $\overset{\frown}{BC_1}$ et la facette n'apparaît pas,

$$(2) \quad \tau_\alpha = \frac{\tau_0 + p(1-\cos\alpha)}{\cos\alpha}.$$

La facette $\overline{BC_2}$ apparaît et limite la croissance jusqu'à C_2 où l'égalité est vérifiée, et au-dela de C_2 la surface de croissance suit la courbe en pointillés,

$$(3) \quad \tau_\alpha > \frac{\tau_0 + p(1-\cos\alpha)}{\cos\alpha}.$$

La facette $\overline{BC_2}$ s'étend au-delà de C_2, et ne peut être limitée plus loin, que par une autre facette. Le même raisonnement géométrique simple peut alors être appliqué pour expliquer le déplacement de la jonction entre deux plans simples (ex. figure 2) ou la disparition – au bout d'un certain temps – d'un plan lent entouré de deux autres encore plus lents.

Pour aller plus loin dans l'étude de l'anisotropie et des mécanismes de la croissance, il est très important d'avoir une connaissance précise, non seulement des taux de croissance des plans simples et de leur évolution en fonction de la température, mais aussi des courbes $\tau(\alpha)$ au voisinage des plans simples. Pour que la mesure de ces dernières devienne possible, il faut être dans le premier cas. Or, dans la formule discutée, on a la possibilité de jouer sur le paramètre p; en particulier, si le minimum est assez marqué pour atteindre l'égalité (2) il faut un rayon de courbure grand et un temps petit.

Dans cette étude, avec un rayon de courbure de 5 mm et un temps de croissance de 30 min (valeur en-dessous de laquelle les mesures deviennent imprécises), seul le voisinage des plans (111)A et B et (110) ne peut être résolu. Nous avons alors représenté les points cor-respondants aux plans simples par leur valeur exacte (mesurable entre deux facettes). Dans le cas des plans (111)A et (110) une valeur limite inférieure de τ_α peut néanmoins être déterminée à partir des figures comme celle représentée par le détail supérieur de la figure 2 C où $\tau_{\text{effective}} \geq \tau_{\text{mesurée}}$.

Pour l'étude du voisinage de ces plans la même méthode reste néanmoins applicable, mais la réalisation de "lentilles sphériques" à grand rayon de courbure apparaît nécessaire.

4. Conclusions

La nouvelle technique de "visualisation" des différents stades de la croissance sur substrat sphérique permet d'effectuer l'étude de l'anisotropie de la croissance dans de bonnes conditions. Appliquée dans cette étude à la croissance du GaAs en phase vapeur, elle pourrait être généralisée à d'autres systèmes.

En ce qui concerne les mesures et les interprétations, l'abondance des renseignements fournis par les coupes n'a pas permis d'être complet. Aussi, d'autres essais sont en cours, notamment pour établir dans une gamme plus large l'influence de la température d'une part, et étudier les morphologies des surfaces d'autre part. Ces études doivent permettre une meilleure compréhension des mécanismes de croissance dans ce type de croissance épitaxiale résultant d'un transport en phase vapeur.

Remerciements

Nous tenons à remercier Monsieur et Madame Oberlin du C.N.R.S. et Monsieur Chane pour leurs discussions fructueuses, Messieurs Daugeron et Dubois pour la réalisation des sphères, Mesdames Valette et Mennechez pour les découpes et les polissages, Madame Lesartre pour le dépouillement et le montage des clichés et Messieurs Marpeau et Legrade pour les montages photographiques.

Bibliographie

1) J. R. Knight, P. Effer et P. R. Evans, Solid State Electron. **8** (1965) 178.
2) A. Boucher et L. Hollan, Onde Electr. **50** (1970). 165.
3) R. R. Fergusson et T. Gabor, J. Electrochem. Soc. **111** (1964) 585.
4) D. W. Shaw, J. Electrochem. Soc. **115** (1968) 405.
5) A. Boucher et L. Hollan, J. Electrochem. Soc. **116** (1970) 932.
6) D. W. Shaw, dans: *Proc. Second Intern. Symp. on Gallium Arsenide*, Dallas, Texas, 1968 (Inst. Phys. and Phys. Soc., London), p. 50.

Journal of Crystal Growth **13/14** (1972) 325–330 © *North-Holland Publishing Co.*

ÉTUDE DE L'ÉPITAXIE LOCALISÉE DU GaAs

J. P. CHANE, L. HOLLAN et C. SCHILLER

Laboratoires d'Electronique et de Physique Appliquée, 94 – Limeil–Brévannes, France

A process has been designed for the observation of the different growth stages. By performing sequential high and low doping levels, the stacking of the successively deposited layers can be clearly revealed on a cleavage plane, and henceforth the growth morphology can be studied. The anisotropy of the etch and of the growth on window edges has been clearly shown. (111), (113)B, and (001) oriented substrates have been used. The influence of the window orientation and of the temperature has been studied in the case of the (001) plane. The corresponding growth features are given.

1. Introduction

L'élaboration de structures coplanaires de GaAs (à partir desquelles certains dispositifs seront réalisés), peut être envisagée par épitaxie localisée. Le substrat monocristallin de GaAs est recouvert d'une couche de SiO_2 dans laquelle sont pratiquées par photogravure des ouvertures de géométrie déterminée; la croissance épitaxiale a uniquement lieu à l'intérieur de ces ouvertures.

L'épitaxie localisée pose, par rapport à l'épitaxie classique, des problèmes spécifiques essentiellement liés aux effets de bord. En effet, au niveau des fenêtres, le masque de SiO_2 définit dans le plan du substrat une limite à partir de laquelle les phénomènes d'anisotropie de décapage et de croissance vont se développer. Ces effets de bord peuvent d'ailleurs interférer lorsque les ouvertures sont étroites. En outre, l'anisotropie doit également intervenir dans le dopage.

Ces phénomènes n'ont pas encore été étudiés en détail, seule l'étude de la morphologie extérieure des dépôts a été abordée[1–3]. Dans ce qui suit, nous nous proposons de compléter l'étude de l'anisotropie par la mise en évidence des différentes étapes de la croissance épitaxiale.

2. Technique d'étude

La technique utilisée peut être résumée selon ses phases chronologiques.

2.1. Préparation des échantillons

Les substrats de GaAs sont soumis à un polissage mecano-chimique puis décapés. Un masque de 2000 Å de SiO_2 est ensuite déposé sur la surface par pyrolyse de $Si(OEt)_4$.

La géométrie des fenêtres photogravées est:
– soit circulaire (diamètre 100 μm)
– soit composée de sillons parallèles de largeur croissante (3, 6, 11, 20 et 50 μm et groupés en faisceaux perpendiculaires (fig. 1).

2.2. Epitaxie et technique de visualisation

La croissance se fait par transport en phase vapeur par $AsCl_3$. Le mode opératoire dérive de celui qui est utilisé en épitaxie classique[4]. Un décapage "in situ" précède systématiquement la croissance. Pour visualiser

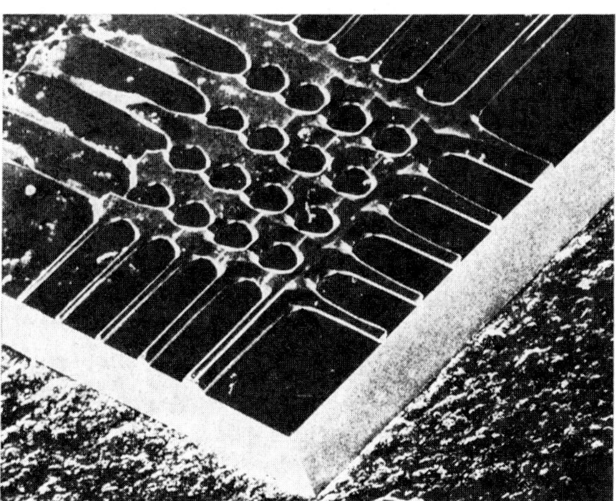

Fig. 1. Photographie au microscope électronique à balayage d'un substrat (001) avec ses motifs orthogonaux orientés suivant [110] et [1Ī0] et les plans de clivage servant à l'observation.

celle-ci le dopage est modifié périodiquement d'abord toutes les 2 min, puis toutes les 4 min. Après l'épitaxie l'échantillon est clivé suivant une des directions cristallographiques contenues dans le plan du substrat, de façon à observer dans le plan de clivage la section des motifs localisés. Cette observation se fait, après révélation chimique, au microscope optique à contraste interférentiel qui fait apparaître l'empilement des couches épitaxiales successives (figs. 2, 4, 6, 7). Les avantages de cette technique apparaissent immédiatement: on voit "dessinée" l'évolution dans le temps des figures de croissance et de multiples effets discutés dans le chapitre qui suit.

3. Étude expérimentale et interprétation

L'étude a consisté à faire varier un certain nombre de paramètres et à dégager l'incidence de ceux-ci sur la morphologie des dépôts.

3.1. INFLUENCE DU DÉCAPAGE

Il ne s'agit, dans cette étude, que des effets de décapage "in situ" effectués dans le réacteur à la température du dépôt. D'autres figures de décapage peuvent également être obtenues par un traitement chimique préalable[5]).

Le décapage "in situ" peut être minimum, donc correspondre à la fraction de micron qui doit être éliminée pour permettre une épitaxie sans aucune perturbation cristallographique à l'interface. Il peut être aussi intentionnellement accentué dans le but de faire des couches enterrées. La figure 2 montre les résultats obtenus dans le cas de deux fenêtres de 5 et 50 µm de largeur, décapées les unes profondément (10 µm) les autres superficiellement ($\sim$1 µm). Dans ce dernier cas, les effets de bord son très faibles; par contre, dans les fenêtres fortement décapées, la croissance se déroule d'une façon anisotrope par une succession de plans différents qui se développent ou s'eliminent suivant leur vitesse relative. De même, sur la figure 3 représentant la surface (001) d'une fenêtre circulaire (diamètre 100 µm) l'anisotropie du décapage apparaît nettement; il se prolonge sous le masque de silice pour donner une forme elliptique, dont le grand axe correspond à la direction [110].

3.2. ORIENTATION CRISTALLOGRAPHIQUE

Nous avons étudié l'influence sur la morphologie de croissance de 3 paramètres cristallographiques:

Fig. 2.   Influence du décapage sur la morphologie de croissance; en haut: croissance après décapage minimum, en bas: croissance après décapage profond (10 µm).

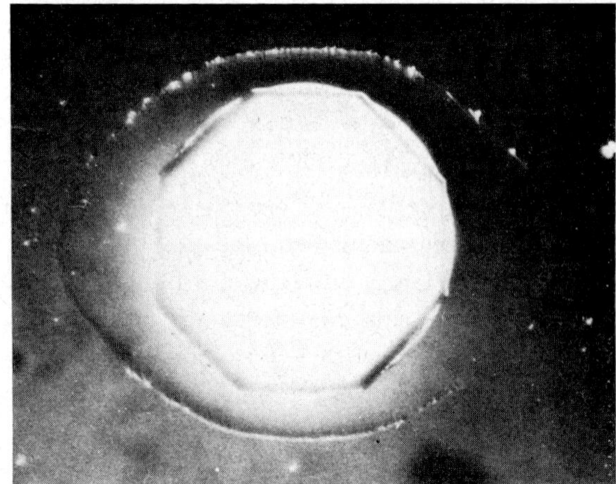

Fig. 3.   Effet de sous-gravure dans le cas d'un motif circulaire (100 µm) par suite de l'anisotropie de décapage dans le plan (001).

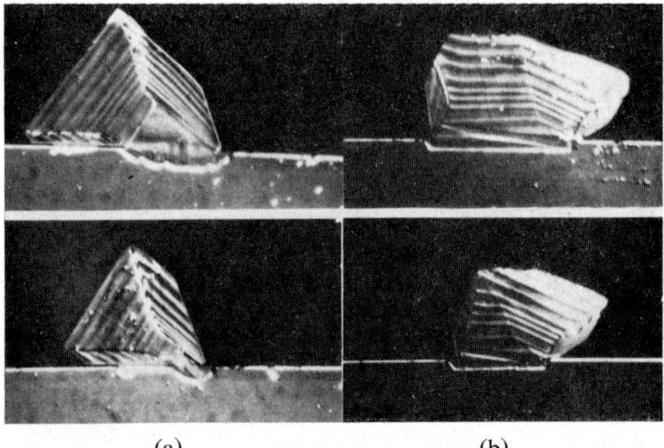

(a)   (b)

Fig. 4.   (a) Figure de croissance sur substrat orienté (111). (b) Figure de croissance sur substrat orienté (113).

– la direction du plan de surface du substrat,
– dans le cas particulier du plan (001), la direction cristallographique suivant laquelle le motif est orienté,
– la direction de désorientation du substrat.

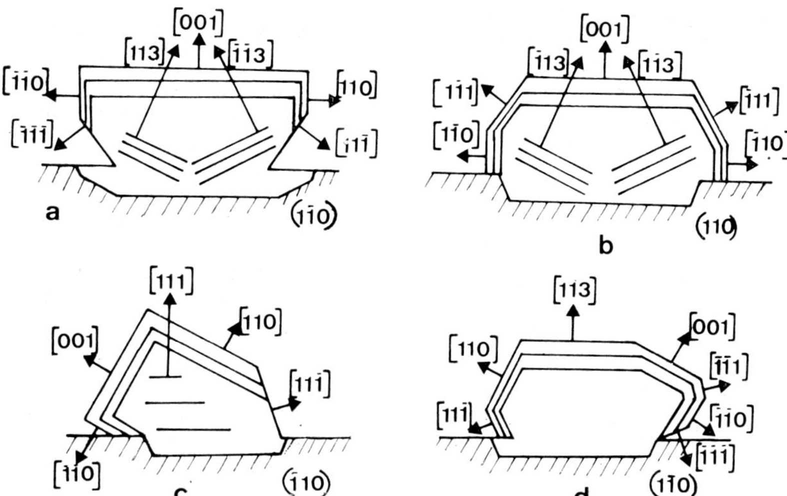

Fig. 5. Schémas de croissance: (a) et (b): sur substrat orienté (001) pour des motifs orientés suivant la direction [1$\bar{1}$0] et [110]; (c) et (d): sur substrat orienté suivant (111) et (113).

– Influence du plan de base. Nous avons utilisé des substrats orientés suivant (111), (113) et (001).

La figure 4a représente la morphologie de croissance sur des substrats orientés suivant le plan (111). Les motifs (largeur: 20 et 5 µm) clivés perpendiculairement étaient désorientés de 30° par rapport à une direction [111].

Dans le but de faciliter la compréhension, nous avons schématisé sur la figure 5c les différents plans qui apparaissent. Précisons que la convention adoptée pour la polarité est de noter les plans de gallium A [par exemple (111) ou ($\bar{1}$13)], et les plans arsenic B, ainsi ($\bar{1}\bar{1}\bar{1}$) ($\bar{1}$11) ou (113).

On constate que le plan de base très rapide est le seul à croître dans le caisson décapé mais qu'il disparait rapidement dès que ce caisson est rempli, au profit des plans (001) (110) et ($\bar{1}\bar{1}$0). Notons que ces trois plans se développent sans que leur vitesse relative ne conduisent finalement à la disparition de l'un d'eux: ceci est dû à l'orthogonalité des plans (110) avec le plan (001).

La figure 4b représente la croissance suivant le plan (113) pour deux largeurs de fenêtre. La croissance se développe suivant les plans prépondérants (113) et (001) limités latéralement par les plans (110) et ($\bar{1}\bar{1}$1) (voir fig. 5d).

Cependant le plan le plus intéressant est évidemment le plan (001), aussi l'avons-nous étudié d'une façon approfondie. Dès le départ, il est apparu nécessaire de considérer l'influence de l'orientation du motif dans ce plan.

– Influence de la direction du motif dans le plan (001). La figure 6a représente la section de sillons orientés parallèlement à la direction [110] tandis que la figure 6b correspond aux sillons orientés suivant [1$\bar{1}$0]. La figure 1 illustre cette disposition des motifs et montre les plans de clivage permettant l'observation. Il faut noter que la distinction entre ces deux directions présentes dans le plan (001) ne peut pas être faite par diagramme de Laüe, et qu'il a fallu mettre au point une méthode basée sur la réflexion anormale des rayons X au voisinage des discontinuités d'absorption λK_{Ga} et λK_{As}[6]), pour lever l'incertitude.

La morphologie des figures de croissance est radicalement différente dans les deux cas, ce qui résulte de l'incidence de la polarité du plan sur les vitesses de croissance[7]). En effet, dans le cas des figures 6a (direction [110]) le décapage fait apparaître sur les flancs des caissons les plans lents ($\bar{1}\bar{1}\bar{1}$)B qui, au début de la croissance, sont éliminés rapidement par suite de la concavité des caissons, au profit de plans vicinaux intermédiaires auxquels les plans ($\bar{1}$13)A vont se substituer (fig. 5b). Ceux-ci sont finalement limités par de nouveaux plans (1$\bar{1}$1)B et (1$\bar{1}$0). Le fait que ces derniers plans apparaissent l'un et l'autre prouve que leurs taux de croissance sont voisins. Dans la phase de croissance convexe, l'apparition du plan (001) dans le cas des

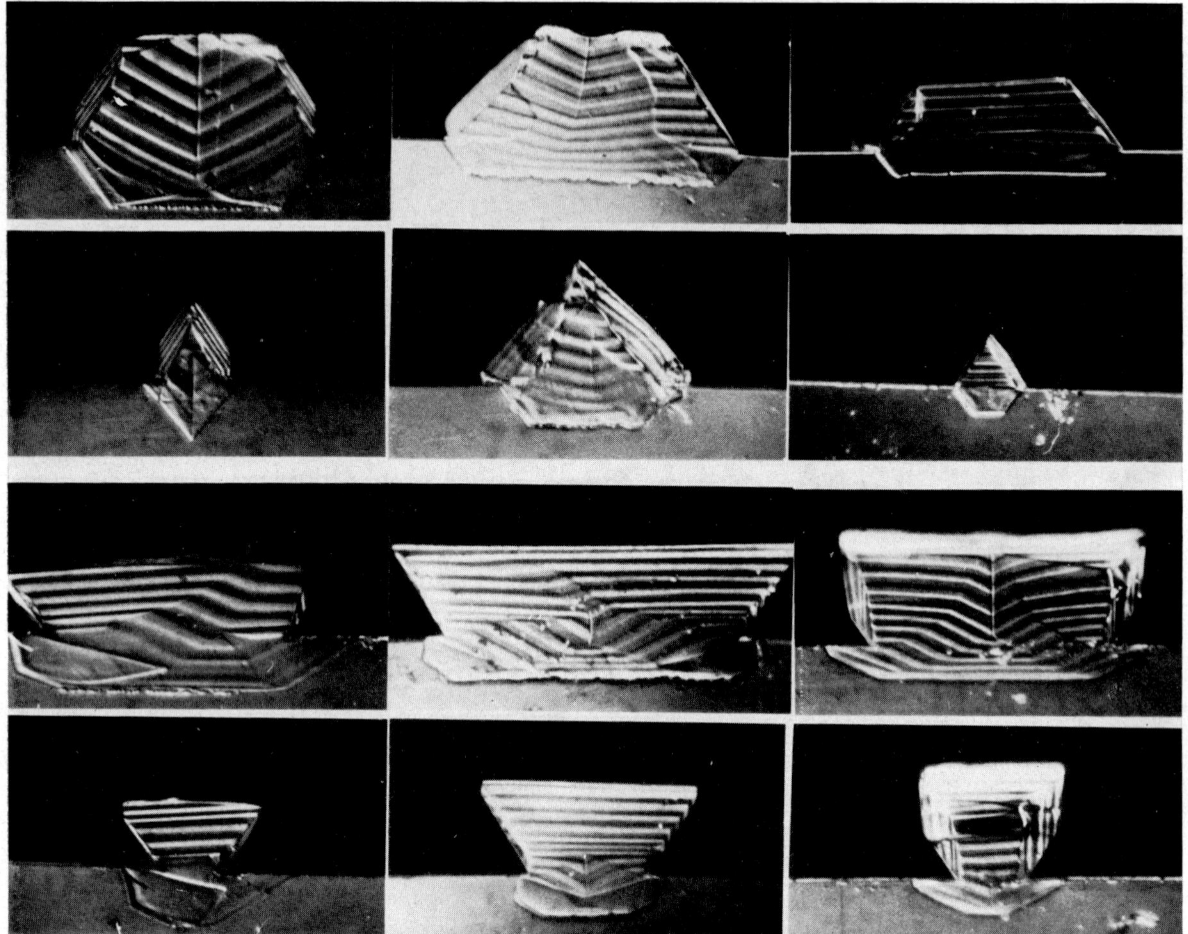

Fig. 6. Figures de croissance sur plan (001). Influence de la direction de désorientation du substrat ([110], [100] et [1$\bar{1}$0]: colonnes de gauche, du centre, de droite). (a) Motifs orientés suivant [110] (les deux lignes supérieures). (b) Motifs orientés suivant [1$\bar{1}$0] (les deux lignes inférieures).

sillons larges est également à noter. Une prolongation du temps de croissance conduirait cependant à sa disparition, et aboutirait à la formation d'un triangle limité par ($\bar{1}\bar{1}\bar{1}$)B. Dans le cas de la figure 6b direction [1$\bar{1}$0]) le décapage fait apparaître les plans (113)B et (551) suivant la désorientation du substrat (voir plus loin). Les plans (111)A qui existent dans cette direction sont immédiatement éliminés par les plans (113), qui, à leur tour sont limités par le plan (001); ce dernier s'élargit, n'étant plus limité que par les plans ($\bar{1}\bar{1}$1) qui dans ce cas, divergent. D'après son développement, la vitesse de croissance du plan (110) est alors comparable à celle du plan (001) et non plus à celle de ($\bar{1}\bar{1}\bar{1}$)B. Il a déjà été observé[7]) que le taux de croissance du plan (110) pouvait varier dans des proportions considéra-

bles, et c'est vraisemblablement les défauts ou les plans voisins qui doivent influencer la nucléation.

Enfin, quelle que soit la direction du motif, on note l'interférence des effets de bord pour des fenêtres étroites, ce qui se traduit en particulier par le remplissage très rapide des caissons décapés.

— Influence de la désorientation. Afin d'éliminer les défauts macroscopiques de la surface des couches épitaxiales, on désoriente toujours les substrats de quelques degrés par rapport au plan (001). Cette désorientation peut se faire suivant l'une des trois directions [110], [100] et [1$\bar{1}$0]. Or, suivant la direction choisie, les morphologies obtenues peuvent différer notablement. Ceci apparait sur les figures 6a et 6b dont les trois colonnes

représentent – dans le même ordre – les sections correspondant aux trois désorientations citées.

La première différence importante concerne le décapage. En effet, dans le cas des sillons orientés suivant [110], les plans des caissons sont toujours limités par les plans $(\overline{1}\overline{1}\overline{1})$B à faible vitesse. Ainsi, ces caissons restent-ils pratiquement symétriques quelle que soit la direction de désorientation du substrat. Par contre, dans le cas des sillons orientés suivant [1$\overline{1}$0] les plans $(\overline{1}\overline{1}\overline{1})$B n'apparaissent pas. Les bords des caissons sont limités, selon la désorientation, soit par des plans (113) faisant un angle de 25° avec la surface, soit par les plans (113) et (551) (ces derniers faisant un angle de 83°). On aboutit à des caissons tantôt symétriques,

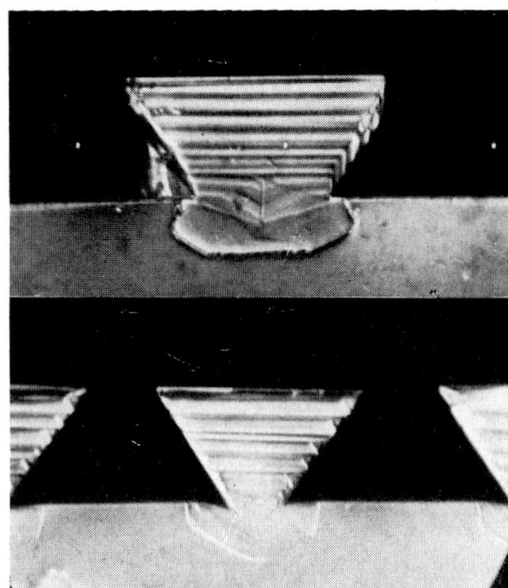

Fig. 7. Influence de la température de dépôt: (a) 750 °C, (b) 725 °C.

tantôt dissymétriques selon la désorientation du substrat.

Dans le cas de la croissance, la technique de marquage permet d'observer le mécanisme suivant lequel le plan de surface [(001) désorienté] est remplacé par le plan (001) exact qui apparaît à la surface de croissance. En effet, comme le plan exact correspond à une vitesse minimum, la désorientation fait apparaître un plan voisin plus rapide qui s'élimine en même temps que le plan (113) situé entre les deux précédents.

3.3. EFFET DES IMPURETÉS

Il est bien établi[8,9]) qu'il existe des différences importantes entre les coefficients de ségrégation des impuretés pour les différents plans cristallographiques, et même, dans certains cas, entre le plan exact et les plans

vicinaux[7]). Ces remarques étant faites, l'examen des figures 6 et 7 montre clairement qu'il parait impossible d'obtenir par épitaxie localisée des couches à dopage homogène. En effet, le taux des impuretés sera différent pour chaque plan cristallographique qui apparait et le dopage sera inhomogène aussi bien latéralement qu'en profondeur, les variations pouvant atteindre un ordre de grandeur. Dans le cas d'un décapage très faible (fig. 2) ce sont seulement les bords du motif ayant crû suivant (111)B qui seront nettement plus dopés, le reste conservant un dopage homogène.

3.4. INFLUENCE DE LA TEMPÉRATURE

La plupart des croissances ont été effectuées à 750 °C (figs. 2, 4, 6) avec cependant quelques expériences à 700 °C et 725 °C (figs. 3, 7). Il apparaît que l'abaissement de la température accentue les effets d'anisotropie: les facettes deviennent plus nettes ainsi que les différences de dopage. Par ailleurs, les énergies d'activation étant différentes suivant les plans[10]), la variation exponentielle des vitesses de croissance avec la température peut entraîner des changements importants dans les rapports de vitesse avec la température et donc des modifications des morphologies. Ainsi dans le cas de la figure 7, les facettes (110) observées à 750 °C (7a) n'apparaissent pratiquement plus à 725 °C (7b): seuls les plans $(\overline{1}\overline{1}\overline{1})$ limitent alors la croissance du plan (001). De même l'influence du dopage devient plus sensible: on observe une variation importante du taux de croissance quand on passe d'un dopage faible à un dopage élevé, ce dernier ayant pour effet d'augmenter le taux de croissance du plan (001), et de diminuer celui du plan (111), ce qui conduit à une ligne de racordement brisée observée au bord du motif 7b.

4. Conclusion

Dans cette étude, nous avons mis au point une technique de visualisation permettant d'étudier dans d'excellentes conditions les problèmes spécifiques concernant l'épitaxie localisée. Dans le cas particulier de l'arséniure de gallium, nous avons mis en évidence l'effet du décapage précédant le dépôt, l'influence de la géométrie des motifs, de l'orientation des motifs dans le plan du substrat du plan cristallographique choisi, de la désorientation du substrat par rapport à ce plan ainsi que le rôle de la température. L'ensemble de ces résultats est nécessaire à la connaissance de la morphologie finale d'un motif de largeur donnée, enterré ou non.

Bibliographie

1) D. W. Shaw, J. Electrochem. Soc. **113** (1966) 904.
2) D. W. Shaw, J. Electrochem. Soc. **115** (1968) 777.
3) Y. Isibashi et M. Yamaguchi, Japan. J. Appl. Phys. **9** (1970) 1007.
4) A. Boucher et L. Hollan, Onde Electr. **50** (1970) 165.
5) Y. Tarni, Y. Komiya et Y. Harada, J. Electrochem. Soc. **118** (1971) 118.
6) C. Schiller, Compt. Rend. (Paris) **272** (1970), 12, 764.
7) L. Hollan et C. Schiller, J. Crystal Growth, à paraître.
8) C. M. Wolf, G. E. Stillman et W. T. Lindley, dans: *Proc. Symp. on Gallium Arsenide*, 1968 (Inst. Phys. and Phys. Soc., London) p. 43.
9) D. W. Shaw, J. Electrochem. Soc. **115** (1968) 405.

Journal of Crystal Growth **13/14** (1972) 331–335 © *North-Holland Publishing Co.*

UNE NOUVELLE MÉTHODE D'ÉPITAXIE EN PHASE VAPEUR D'ARSENIURE DE GALLIUM DE HAUTE PURETÉ

P. MERENDA

Laboratoire Central de Recherches Thomson CSF, 91 – Corbeville, France

High purity gallium arsenide can be prepared by vapour phase epitaxy in many ways, two of which are now well known and widely used, that is, the $AsCl_3/Ga/H_2$ system and the $AsH_3/Ga/HCl$ system. This paper describes a new vapour phase epitaxial process which uses the advantageous features of these two systems in order to give good control of stoichiometry and purity. In this process, gallium is transported by HCl, through the monochloride GaCl, and As_4 is simply evaporated from a solid deposit; the originality of the system lies in the fact that both HCl and As_4 come from the reduction of $AsCl_3$, by pure hydrogen. The results of measurements on the growth rates of epitaxial layers versus $P_{As_4}{}^0$ and $P_{GaCl}{}^0$ are given together with the results and a discussion of the influence of the As/Ga ratio on the free carrier concentration and mobility found in epitaxial layers.

1. Introduction

L'arséniure de gallium de haute pureté peut être préparé en phase vapeur par plusieurs procédés dont deux sont aujourd'hui bien connus et largement utilisés : le système $AsCl_3/Ga/H_2$ et le système $AsH_3/Ga/HCl$.

Cet article concerne un procédé de synthèse épitaxiale, nouveau et original qui se propose de réunir les avantages des deux systèmes cités ci-dessus.

2. Principe de la méthode

Au point de vue thermochimique le système envisagé est une synthèse à partir des éléments As et Ga, le gallium étant transporté par de l'acide chlorhydrique sous forme de monochlorure GaCl.

Il s'agit donc d'un système $As_4/H_2/Ga/HCl$ analogue à celui utilisé par Conrad, Reynolds et Jeffcoat[1], ou Shaw[2] ; son originalité réside dans l'utilisation d'$AsCl_3$ comme source unique d'arsenic et de HCl, cette utilisation se faisant sans manipulation de l'arsenic hors du réacteur.

Les principales réactions thermochimiques qui interviennent sont les suivantes :

$$4\,AsCl_3 + 6\,H_2 \xrightarrow{850\,°C} As_4\,(g) + 12\,HCl\,(g), \tag{1}$$

$$\tfrac{1}{4}\,As_4\,(g) \xrightarrow{15\,°C} As\,(s), \tag{2}$$

$$HCl\,(s) + Ga\,(l) \underset{\longleftarrow}{\xrightarrow{800\,°C}} GaCl\,(g) + \tfrac{1}{2}\,H_2\,(g), \tag{3}$$

$$As_4\,(s) \xrightarrow{450\,°C} \tfrac{1}{4}\,As_4\,(g), \tag{4}$$

$$As_4\,(g) + 6\,GaCl\,(g) \underset{\longleftarrow}{\xrightarrow{T\,<\,700\,°C}} 4\,GaAs\,(s) + 2\,GaCl_3\,(g), \tag{5}$$

$$H_2\,(g) + \tfrac{1}{2}\,As_4\,(g) + 2\,GaCl\,(g) \rightleftarrows 2\,GaAs\,(s) + 2\,HCl\,(g). \tag{6}$$

Les réactions (1), (2) et (3) sont utilisées pour synthétiser GaCl à partir de $AsCl_3$, H_2 et Ga.

L'arsenic évaporé dans la réaction (4) provient de l'arsenic piégé en (2) pendant une manipulation précédente.

Enfin les réactions (5) et (6) sont les réactions qui conduisent au dépôt de GaAs, la réaction (5) n'étant à prendre en considération qu'au dessous de 700 °C[3]). Cette méthode, comme la méthode $AsH_3/Ga/HCl$[4]) permet de doser indépendamment les flux d'arsenic et de gallium entrant dans le système, ce qui autorise un contrôle précis du rapport As/Ga dans la phase vapeur.

D'autre part, le recours à $AsCl_3$ comme source unique d'arsenic et de HCl élimine les aléas dus à l'utilisation de gaz en bouteille ($AsCl_3$ et HCl)[5]) et la haute pureté du produit de départ autorise des dépôts épitaxiaux de pureté comparable à ceux obtenus par la méthode classique ($AsCl_3/Ga/H_2$).

3. Description du réacteur d'épitaxie

La conception de l'appareil découle de la volonté d'éviter toute manipulation de l'arsenic entre la phase de piègeage et la phase d'évaporation.

Ceci implique que le piégage de l'arsenic, lors de la

production de HCl par réduction de AsCl$_3$, et sa réévaporation lors de son utilisation en source d'arsenic soient effectués directement dans le réacteur.

Le procédé étant nécessairement séquentiel on peut obtenir un fonctionnement continu en concevant un appareil symétrique dans lequel une branche sert alternativement de source de HCl (donc de GaCl) et de source d'arsenic, l'autre branche jouant le rôle complémentaire.

La figure 1 montre le schéma de principe de l'appareil utilisé. Cet appareil comprend huit zones de tem-

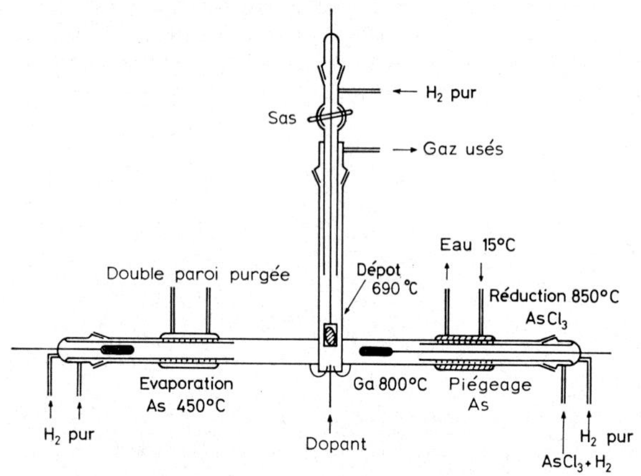

Fig. 1.   Schéma de l'appareil.

pérature régulées séparément. Le réacteur est en forme de T. Dans la configuration représentée la branche droite du T est utilisée en source de Gallium. Une nacelle contenant du gallium (Alusuisse 99.9999) est introduite dans la zone gallium. Du trichorure d'arsenic (pureté 99.9995 provenances diverses), entrainé par un flux d'hydrogène pur est réduit par celui-ci dans la zone de réduction.

L'arsenic produit est piégé dans la zone arsenic maintenue à 15 °C par un manchon refroidi par eau. L'acide chlorhydrique restant passe sur la nacelle de Gallium et entraine ce dernier sous forme GaCl. La branche gauche du T est utilisée en source d'arsenic. La nacelle de gallium, retirée de la zone gallium, est stockée à l'entrée de la branche, sous hydrogène pur. La branche source d'As ne reçoit qu'un flux d'hydrogène pur qui entraine l'arsenic évaporé dans la zone arsenic à la température $T_{As} = 450$ °C.

Lorsque l'arsenic de la branche source d'As est épuisé on inverse le rôle des deux branches en introduisant la nacelle de Ga de gauche dans le réacteur, la

nacelle droite étant ressortie, et en intervertissant les températures des zones Arsenic droite et gauche.

Les flux de GaCl et d'arsenic As$_4$ sont introduits dans une chambre de réaction petite par deux injecteurs de faible diamètre ($\sim$ 1 mm). Deux autres injecteurs permettent d'introduire des gaz de dopage.

Cette disposition entraine une absence quasi totale d'effets de mémoire et une réponse très rapide du système lors de la réalisation de dépôts dopés. La branche latérale du T comporte un sas qui permet d'introduire le substrat sans arrêter l'appareil dont le fonctionnement est ainsi possible "en continu".

4. Conditions expérimentales

Des conditions expérimentales typiques sont les suivantes:

– Branche gallium
T (évaporation AsCl$_3$) = 27 $\pm$ 0.1 °C, P_{AsCl3} = 14 torr,
Débit H$_2$/AsCl$_3$ = 20 à 100 cm^3/min,
T (réduction AsCl$_3$) = 850 $\pm$ 0.5 °C
T (piègage As) = 15 $\pm$ 1 °C
T (formation GaCl) = 800 $\pm$ 0.5 °C
Débit H$_2$ pur = 100 à 20 cm^3/min.

– Branche arsenic
T (évaporation As) 440 $\pm$ 0.5 °C à 460 $\pm$ 0.5 °C;
Débit H$_2$/As = 40 à 120 cm^3/min;
Débit H$_2$ pur = 80 à 0 cm^3/min.

– Conditions de dépôt
Temperature de dépôt = 710 $\pm$ 0.5 °C à 660 $\pm$ 0.5 °C;
Gradient de température = 7 °C/min.

Nous avons vérifié que dans ces conditions la réduction de AsCl$_3$ est pratiquement totale.

Le calcul du flux de GaCl transporté, effectué en supposant la réaction (3) totale et en négligeant la production de GaCl$_3$ (très faible à 800 °C) conduit à des flux compris entre 5×10^{-5} et 2.5×10^{-4} mole/min.

Le flux d'arsenic peut être calculé d'après la loi:

$$\log P_{As_4} = -\frac{6777}{T} + 10.559.$$

qui donne P_{As_4} en Torr, et en supposant complète la saturation de l'hydrogène d'entrainement. Les valeurs trouvées, cohérentes avec le temps nécessaire à évaporer la totalité de l'arsenic piégé se situent entre 7.6×10^{-5} et 6×10^{-4} atom g/min selon les valeurs de T (évaporation As$_4$) et D H$_2$/As.

La gamme de rapports As/Ga [définis comme (flux atomique d'arsenic)/(flux molaire de GaCl)] explorée s'étend donc de 0.3 à 12.

5. Mode opératoire

Le mode opératoire est simple et peut être décrit comme suit:

– l'équilibre thermique des différentes zones du four étant réalisé on règle le débit d'hydrogène sur l'arsenic à la valeur choisie;

– la branche gallium étant maintenue sous un flux d'hydrogène pur on introduit le substrat dans la zone de dépôt.

Lorsque le substrat a atteint la température de dépôt on lui fait subir un léger décapage en admettant par la branche arsenic un flux d'hydrogène chargé d'AsCl$_3$ pendant 1 à 2 min. On arrête alors ce décapage et l'on admet le flux d'AsCl$_3$ sur la branche gallium, ce qui permet le dépôt épitaxial.

Lorsque la couche épitaxiale à atteint l'épaisseur choisie on arrête le dépôt en coupant le flux d'AsCl$_3$ sur la branche gallium.

Un avantage important de ce mode opératoire est que le substrat ou la couche épitaxiale ne se trouvent jamais au contact d'hydrogène pur, le débit d'arsenic étant maintenu pendant la totalité de l'opération.

Ceci évite les effets d'interface ou de couche superficielle isolante dus à la dissociation de l'arseniure de gallium en l'absence d'une pression partielle d'arsenic suffisante.

La préparation du substrat est classique: elle comprend un polissage mécano-chimique au brôme méthanol suivi d'une attaque chimique dans le mélange H$_2$SO$_4$–H$_2$O$_2$–H$_2$O (proportions 5–1–1) effectuée juste avant le dépôt.

Un soin tout particulier est apporté aux opérations de rinçage qui suivent cette attaque chimique.

Les substrats, semi-isolants ou dopés Te, Se, Sn, sont généralement orientés dans des plans voisins de (100), une légère désorientation systématique (de l'ordre de 2°) conduit à une meilleure reproductibilité de l'état de surface.

6. Résultats expérimentaux

Nous nous sommes attachés à déterminer l'influence des pressions partielles $P^o_{As_4}$ et P^o_{GaCl} sur la vitesse de croissance du dépôt ainsi que l'évolution des caracté-

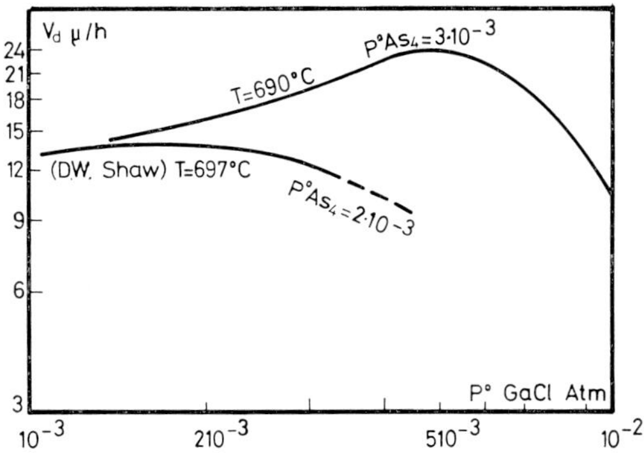

Fig. 2.   Vitesse de croissance à pression d'arsenic constante.

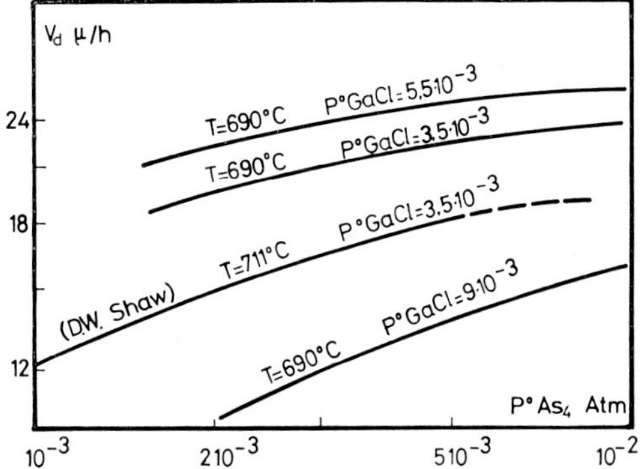

Fig. 3.   Vitesse de croissance à pression de GaCl constante.

ristiques électriques et de la pureté en fonction du rapport As/Ga.

Les études ont été menées sur des substrats orientés (100) et à température de dépôt fixée (690 °C).

7. Étude de la vitesse de croissance

Les figures 2 et 3 montrent les variations de la vitesse de croissance des couches épitaxiales avec les pressions partielles de GaCl et d'As$_4$ dans la chambre de réaction P^o_{GaCl} et $P^o_{As_4}$.

L'allure des lois de variation trouvées est en accord avec les observations de Shaw[6] effectuées dans un système très similaire.

En échelle logarthimique on observe une variation quasi linéaire de la vitesse de dépôt V_d en fonction de la pression partielle de As$_4$ à $P^o_{GaCl} =$ constante. Par contre l'étude de la variation de V_d en fonction de la pression partielle de GaCl à $P^o_{As_4} =$ constante montre que la vitesse de croissance passe par un maximum lorsque P^o_{GaCl} croit.

Cependant les vitesses de dépôts mesurées sont supérieures à celles observées par Shaw. Ceci s'explique probablement par une plus grande valeur du coefficient β traduisant la fraction du flux gazeux qui se met effectivement en équilibre avec le substrat [β = (flux efficace)/(flux total)]. Ce facteur qui dépend beaucoup de la géométrie du système et des écoulements gazeux est ici accru par l'utilisation d'un système d'injecteurs.

Pour $P^{\mathrm{o}}_{\mathrm{As}_4} = 3 \times 10^{-3}$ et $P^{\mathrm{o}}_{\mathrm{GaCl}} = 6 \times 10^{-3}$ la vitesse maximum atteinte est 24µm/h.

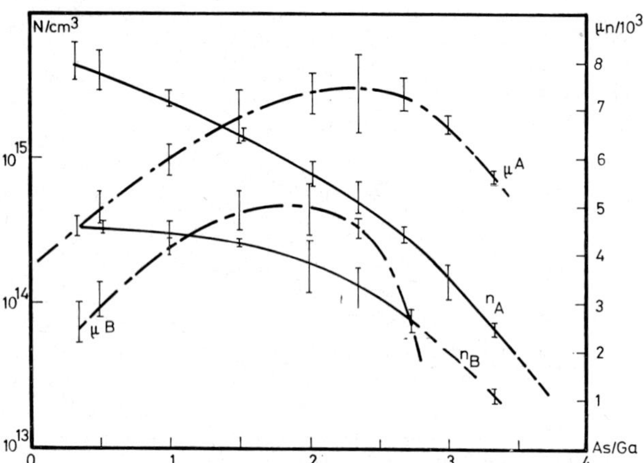

Fig. 4. Variation de n et μ_n en fonction de As/Ga.

La qualité cristalline et l'état de surface des dépôts sont très satisfaisants et peu sensibles à la valeur du rapport As/Ga (qui vaut $4\,P^{\mathrm{o}}_{\mathrm{As}_4}/P^{\mathrm{o}}_{\mathrm{GaCl}}$).

Seuls les dépôts effectués en atmosphère très pauvre en arsenic (As/Ga $\ll 1$) montrent les défauts caractéristiques de la croissance VLS sur des gouttelettes de Ga (formation de pyramides).

Aux très fortes valeurs du flux de GaCl ($> 10^{-2}$ mole/min) des traces d'attaque de la surface apparaissent, liées sans doute à la présence d'un excès de HCl n'ayant pas réagi avec le gallium (temps de contact trop bref).

8. Étude des caractéristiques électriques et de la pureté

Nous avons étudié les variations du nombre de porteurs libres et de leur mobilité en fonction du rapport As/Ga pour deux trichlorure d'arsenic de provenance différente et avons cherché à corréler ces résultats avec des dosages d'impuretés effectués à la sonde ionique (Caméca).

Les figures 4 et 5 représentent les variations du nombre de porteurs libres et de la mobilité à l'ambiante de

couches épitaxiales non intentionnellement dopées déposées à partir d'un trichlorure d'arsenic A, dont la bonne qualité avait été testée dans un système classique $\mathrm{AsCl}_3/\mathrm{Ga}/\mathrm{H}_2$ et d'un trichlorure d'arsenic B n'ayant jamais conduit à des résultats corrects dans le système classique. L'étude de ces courbes permet de dégager plusieurs points intéressants:

(a) Il est possible dans le système considéré de faire varier considérablement le coefficient de séparation des

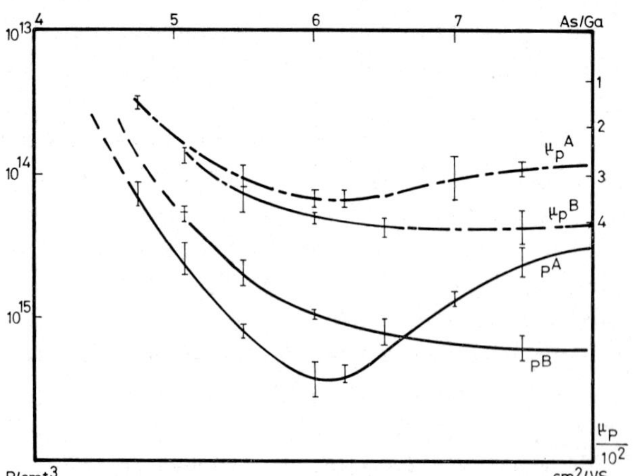

Fig. 5. Variation de p et μ_p en fonction de As/Ga.

impuretés entre la phase vapeur et la phase solide en agissant sur le rapport As/Ga. Pour des valeurs suffisamment élevées de ce rapport, on obtient même un passage du type n au type p.

(b) La mobilité des porteurs libres présente un optimum pour chaque type de conductivité. Ce maximum est évidemment plus net pour le type n, où il se situe entre As/Ga = 2 et As/Ga = 3.

(c) Il existe une plage de variation du rapport As/Ga pour laquelle les dépôts sont très peu dopés et ont de très faibles mobilités. Ce qui les rend fréquemment non mesurables. Le matériau se comporte alors comme s'il était très compensé.

Cette plage est plus large pour le trichlorure d'arsenic B, qui ne permet jamais d'atteindre des mobilités élevées, que pour le trichlorure A qui, pour $n = 5 \times 10^{14}/\mathrm{cm}^3$ mène à des mobilités à l'ambiante de l'ordre de 7500 cm^2/V-sec; les mobilités à l'azote liquide atteignant 70 000 cm^2/V-sec.

Des dosages d'impureté effectués à la sonde ionique dans les couches épitaxiées préparées à partir du trichlorure d'arsenic A pour les rapports As/Ga faibles font apparaitre des traces de soufre et de silicium (quelques 1/10de ppm à 1 ppm).

Les dosages effectués sur les couches préparées à partir du trichlorure d'arsenic B sont assez similaires à deux différences, près :

(1) les concentrations en dopant correspondant à des niveaux peu profonds sont sensiblement plus faibles.

(2) On retrouve dans toutes les couches une concentration en Fer variant entre 0.1 ppm pour As/Ga grand et 3 ppm pour As/Ga petit.

Ceci permet de proposer l'explication suivante au comportement différent des deux trichlorure d'arsenic: Le trichlorure d'arsenic B contient un peu moins d'impuretés, menant à des niveaux peu profonds (Si, Se, etc.) que le trichlorure A, mais est sensiblement plus riche en Fer, qui est connu pour donner un piège dans GaAs. Ceci explique probablement les faibles mobilités des dépôts B et la plus grande largeur de la zone de transition entre le type n et le type p. Cette largeur de transition apparait donc comme un critère de bonne qualité pour un trichlorure d'arsenic donné.

D'autre part, en ce qui concerne les valeurs extrèmes du rapport As/Ga, des niveaux profonds mal identifiés sont détectés par des mesures d'effet Hall et de résistivité en fonction de la température.

On peut tenter d'expliquer la chute des mobilités, pour ces valeurs très faibles ou très élevées du rapport As/Ga, par la formation de complexes associant défauts de stoechiométrie et impuretés.

Ces défauts complexes induits par la non-stoechiométrie seraient responsables de niveaux profonds agissant comme pièges et limitant la mobilité.

Il s'agit là d'une hypothèse dont la vérification demandera la mise en œuvre de nombreux moyens de caractérisation et leur correlation.

9. Conclusion

La nouvelle méthode d'épitaxie d'arséniure de gallium proposée répond bien aux objectifs qui lui avaient été assignés: elle permet d'optimiser les conditions de dépôt de couches épitaxiales de GaAs en fonction du rapport As/Ga avec une grande souplesse. L'utilisation d'un produit source de HCl et d'As unique liée à l'absence de manipulation de l'arsenic, qui sont les points originaux de ce procédé, autorisent une grande pureté des dépôts effectués et une reproductibilité excellente.

Bibliographie

1) R. W. Conrad, R. A. Reynolds et H. W. Jeffcoat, Solid State Electron. **110** (1967) 507.
2) D. W. Shaw, J. Electrochem. Soc. **115** (1968) 405.
3) A. Boucher and L. Hollan, J. Electrochem. Soc. **117** (1970) 923.
4) J. J. Tietjen and J. A. Amick, J. Electrochem. Soc. **113** (1966) 724.
5) B. E. Berson, R. E. Enstrom and J. F. Reynolds, RCA Rev. **31** (1970) 21.
6) D. W. Shaw, J. Electrochem. Soc. **117** (1970) 683.

MORPHOLOGICAL STUDIES ON SELECTIVE GROWTH OF GaAs

SHINYA IIDA and KAZUHIRO ITO

Central Research Laboratory, Hitachi, Ltd., Kokubinji, Tokyo, Japan

Selective growth was conducted in windows on GaAs substrates using a GaAs–AsCl$_3$–H$_2$ system. The substrates used here were oriented in the {001}, {111}B, {110}, and {113} planes. Growth features on these planes were studied morphologically, and the growth processes in the windows were observed by periodic doping during growth.

The following results were obtained: (1) Growth surfaces are closely related to the nature of the walls revealed in the windows. (2) Growth occurs from both the bottom and walls of windows. (3) The dominant growth facets are {110} and {111} B planes which have the slowest growth rates and rectangular windows having a (001) substrate within the ⟨100⟩ direction. (4) Favourable selective growth is obtained in rectangular windows on {001} type substrates where the edge directions are ⟨100⟩.

1. Introduction

Electronic devices such as Gunn oscillators and milli-meter-wave devices have been developed, which employ epitaxial crystal growth from the vapor phase[1]. Based on this development, the demands for integrated-circuit technology based on GaAs are increasing. The importance of selective deposition as an integrated-circuit technology has been realised. This technique involves selective etching of the substrate and selective growth within windows. Studies of this problem have been reported previously by Shaw et al.[2].

In the selective growth of GaAs, interesting features have been observed because of the polarity of its structure. The study of selective etching has been reported elsewhere[3]), and in this paper, morphological studies of selective growth within windows on substrates with different orientations will be described.

2. Experimental

The substrates used here had {001}, {111} As, {110}, {112}, and {113} surfaces. These were sliced within 1° of the above-mentioned principle orientation and polished to a mirror-like finish. After being lightly etched on the surface in a 5 H$_2$SO$_4$:1 H$_2$O$_2$:1 H$_2$O solution, SiO$_2$ masks $\simeq$ 2000 Å thick were deposited on their surface by pyrolysis of tetraethoxy-silane, and circular or rectangular windows cut into the masks.

Window etching was conducted in a solution consisting of 1 H$_2$SO$_4$:10 H$_2$O$_2$, which proved successful in producing flat-bottomed windows for all substrate

orientations. Selective growth was effected in the windows using the GaAs–AsCl$_3$–H$_2$ system. The deposition apparatus and experimental conditions were identical with those previously described[4]. Tin-doped GaAs single crystals were used throughout this work.

Morphological observations were made by using an optical microscope and a goniomicroscope. Intentional dopings were made using SnI$_4$.

3. Results

3.1. EXTERNAL STRUCTURES

3.1.1. *Growth on (001) substrates*

The growth surfaces in circular windows on the (001) plane show two raised facets of (001) planes at opposite positions around the circle edge. These positions correspond to those of the (111) and ($\overline{1}\overline{1}1$) walls in the window as revealed by etching. The development of these facets increases as the growth thickness increases. A typical example is shown in fig. 1.

Using rectangular windows, two experiments with window directions parallel to ⟨110⟩ and ⟨100⟩ were undertaken. In the former, (001) growth facets appear on opposite sides along the [$1\overline{1}0$] or [$\overline{1}10$] directions. The area occupied by the facets in the window increases as the growth thickness increases, and the positions of the facets are crystallographically similar to those in the circular windows. When the growth is effected over the window depth, {110} side facets appear perpendicular to the substrate surface on the (001) facet on one side, and on the {111} As facet on the opposite side.

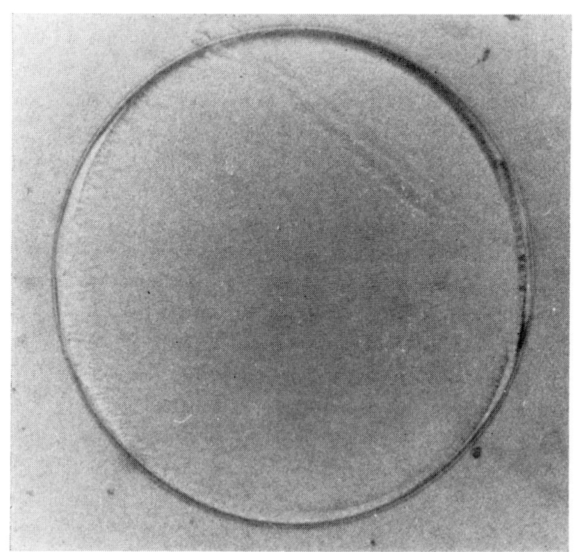

Fig. 1. Growth surface on (001) substrate with circular windows.

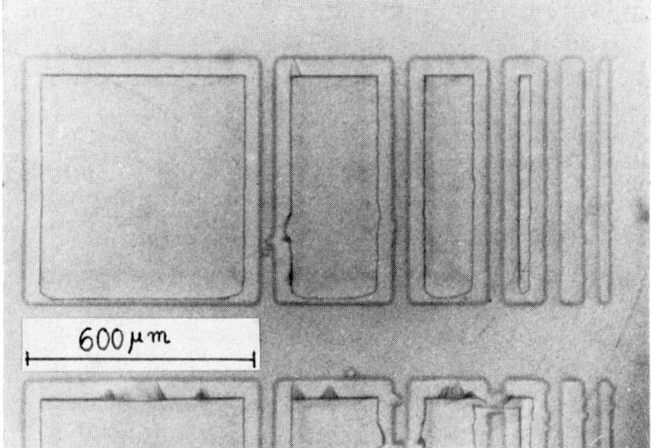

Fig. 3. Growth surface on ($\bar{1}\bar{1}\bar{1}$) As substrate with rectangular windows.

In the latter, which lies at an angle of 45° to the window, flat smooth growth surfaces are obtained, although traces of facets do appear at two corner edges. This favorable result is attributed to the symmetrical window structure since four {100} walls perpendicular to the substrate surface are revealed by etching of the window. Schematic cross-sectional views of the growth status in both experiments are shown in fig. 2.

3.1.2. Growth on ($\bar{1}\bar{1}\bar{1}$) As substrates

In selective growth on the ($\bar{1}\bar{1}\bar{1}$) As plane for circular windows, three raised areas near the circumference in the window are clearly distinguishable from the lower central area. In the rectangular windows parallel to [$1\bar{1}0$] and [$11\bar{2}$] directions similar growth characteristics are observable. Typical growth surfaces are shown in fig. 3. It is evident from this figure that the central area and the raised edge area are clearly distinguished. In this instance, the growth thickness of the central and edge areas of 10–60 μm in width are about 1 and 3 μm respectively. The windows below 50 μm in width are completely filled with growth from the edge area.

3.1.3. Growth on (110) substrates

Selective growth was carried out in the circular windows, and the growth thickness was found to be quite

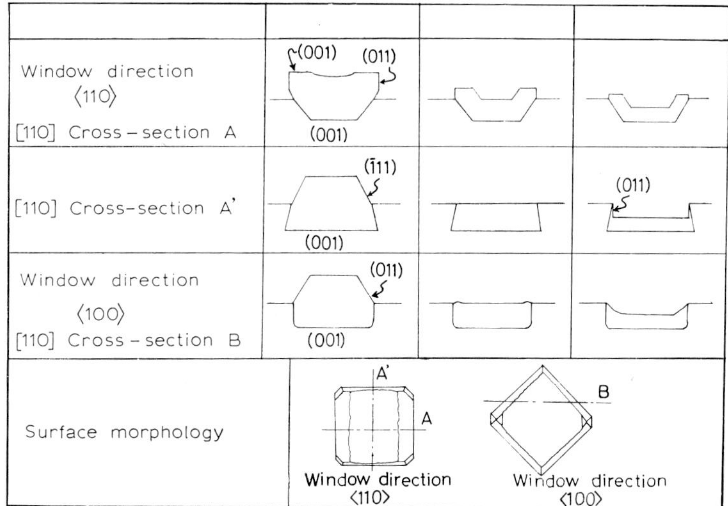

Fig. 2. Schematic cross-sectional views of growth on (001) substrates.

different in the central and circumference areas of the window. The growth height at the circumference is about three times as high as that in the central part. The marginal growth surface is smooth but stepped, and spreads over on to the SiO_2 mask when growth is prolonged, resulting in the octagonal morphology shown in fig. 4.

Growth was also effected in rectangular windows

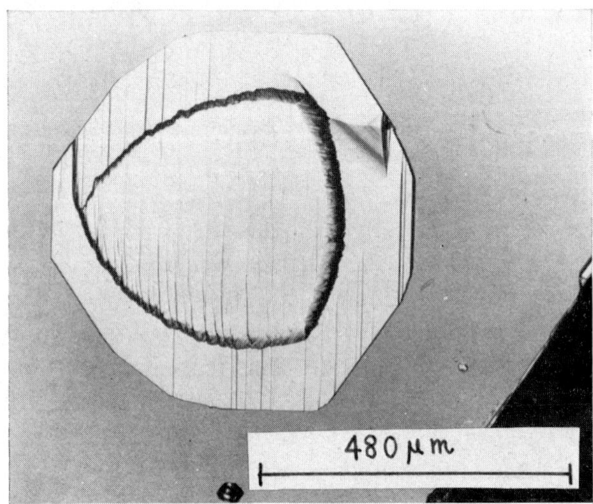

Fig. 4. Growth surface on (110) substrate with circular windows.

whose directions are parallel to [110] and [001]. The growth surface is divided into a raised marginal area and central area similar to the growth on $(\overline{1}\overline{1}\overline{1})$ As substrate.

3.1.4. *Growth on {112} and {113} substrates*

Growth surfaces on (112) Ga, $(\overline{1}\overline{1}2)$ As, (113) Ga and $(\overline{1}\overline{1}3)$ As are uneven, and no faceted growth parallel to the substrate surface was observed although side facets of {110} and {111} As planes are observable.

The growth surface on the (112) Ga plane becomes roughened, especially on the side which is revealed on the (111) Ga wall within the window. Growth surfaces on the $(\overline{1}\overline{1}2)$ As plane, both in the circular and rectangular windows, display ⟨110⟩ striations but become flat as the thickness increases. Further, eccentric selective growth is obtained because of three side facets with a gentle slope to the substrate surface as shown in fig. 5. The growth characteristics on the (113) Ga and the $(\overline{1}\overline{1}3)$ As surfaces will not be described here because

they are very similar to those on the (112) Ga and $(\overline{1}\overline{1}2)$ As planes.

3.2. INTERNAL STRUCTURES OF THE SELECTIVE GROWTH IN WINDOWS

Internal structures were observed on the {110} cleavage plane perpendicular to the substrate surface by employing a staining method. Furthermore, in order to observe the details of the growth process, intentional doping was carried out at periodic intervals during growth.

Fig. 5. Growth surface on $(\overline{1}\overline{1}2)$ As substrate with the rectangular windows.

Cross-sectional views of the growth on the (001) plane with ⟨110⟩ directional windows are shown in fig. 6a, b. It can be seen from fig. 6a, $(\overline{1}10)$ cross-section that growth from the wall ceases when it reaches a vertical $(\overline{1}10)$ plane; thereafter the growth proceeds only from the bottom and forms the $(1\overline{1}1)$ As side facet. On the contrary with the $(1\overline{1}0)$ cross-section, since the growth from the wall is faster than that from the bottom of the window, (001) faceted growth is found.

A cross-sectional view of selective growth on the (001) with ⟨100⟩ rectangular windows is shown in fig. 7. In this case, growth proceeds both from the wall and the bottom, and the surface becomes flat, but with a slightly curved section near the edges because of overlap of the two growth processes. The {110} type side facets are observed during growth over all window depths. During growth on the $(\overline{1}\overline{1}1)$ As plane, a straight line originating from an edge between the $(\overline{1}\overline{1}1)$ Ga wall and the bottom was observed on the stained (110) cross-section. This feature may be explained by difference in

(a)

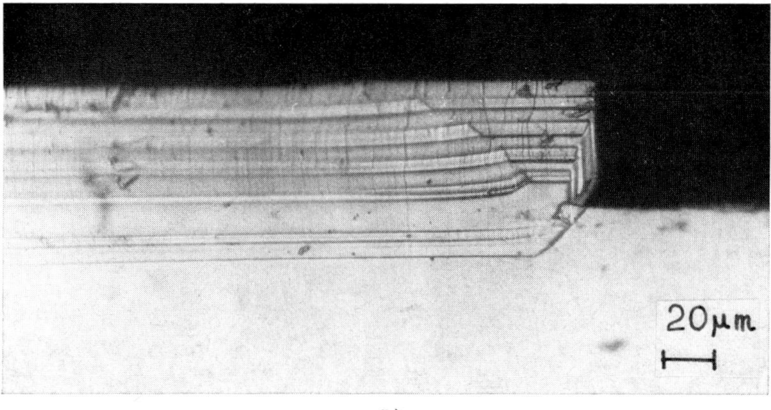

(b)

Fig. 6. Stained cross-sectional views along {110} cleavage planes of growth on (001) substrate with ⟨110⟩ directional windows. (a) (1̄1̄0) cross-section, (b) (11̄0) cross-section.

Fig. 7. Stained cross-sectional view along cleavage plane of growth on (001) substrate with ⟨100⟩ rectangular windows.

growth rates on the wall and base and by the effect of impurity concentration on growth rate which will be described later.

Such lines are observable during growth on other planes near the wall, thus indicating a difference in observed impurity concentration. This effect was not observed for growth on the (001) plane with the ⟨100⟩ direction rectangular window.

TABLE 1

List of observed side facets appearing during selective growth

Substrate	Window direction	I			II			Remarks
		Meas. angle	Ident. plane	Theor. angle	Meas. angle	Ident. plane	Theor. angle	
(001)	[110]	90°	(1$\bar{1}$0)	90°	90°	($\bar{1}$10)	90°	
	[1$\bar{1}$0]	55°	(1$\bar{1}$1)	54°44′	55°	($\bar{1}$11)	54°44′	
	[100]	45–50°	(011)	45°	45–50°	(0$\bar{1}$1)	45°	
	[0$\bar{1}$0]	45–50°	(101)	45°	45–50°	($\bar{1}$01)	45°	
(112)	[110]	54°	(110)	54°44′				
	[111]	59–61°	(2$\bar{1}\bar{1}$)*	60°	59–61°	($\bar{1}$21)*	60°	
($\bar{1}\bar{1}$2)	[110]	20°	($\bar{1}\bar{1}\bar{1}$)	19°28′	84°	(11$\bar{1}$)	90°	
	[111]	29–30°	($\bar{1}$0$\bar{1}$)	30°	29–30°	(0$\bar{1}$1)	30°	
(113)	[110]	65°	(110)	64°46′				
	[33$\bar{2}$]	40–42°	(3$\bar{1}$3)*	40°28′	40–42°	($\bar{1}$33)*	40°28′	
($\bar{1}\bar{1}$3)	[110]	30°	($\bar{1}\bar{1}\bar{1}$)	29°30′	70°	(11$\bar{1}$)	79°58′	
	[33$\bar{2}$]	26°	(023)*	23°06′	26°	(203)*	23°06′	

* Stepped plane.

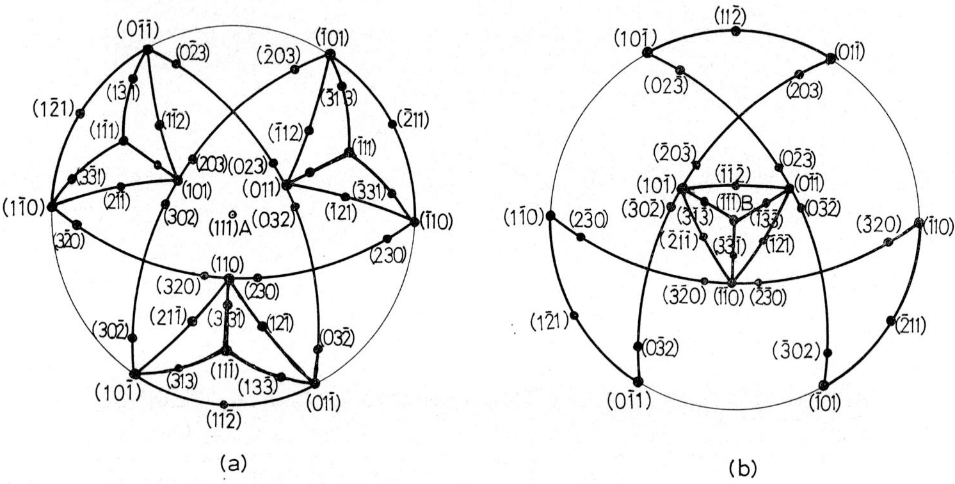

Fig. 8.   Plot of observed side facets of selective growth in standard stereo projection.

4. Discussion

Information on side facets which appear on the growth surfaces is summarized in table 1 and plotted in fig. 8 in stereographic projection. The observed planes are {110} type, {111} As type, and others near them. The relative growth rates as a function of substrate orientation under our growth conditions using large area substrates are in the following order: (111) Ga:(100):(110):(111) As, 20:15:1:1. This agrees qualitatively with Shaw's data[5]) for the Ga–AsCl$_3$–H$_2$ system, and it is felt that observed facets represent the slowest growth rate in the zone.

However, it is expected that the exact facets to appear will depend on the exact nature of the growth conditions, since relative growth rates depend strongly on growth conditions[5,6]).

It is already known that impurity concentrations in the vapor growth layers of GaAs depend on substrate orientation[7]). In the present work, a marked difference is found in the growth layer on (111) As substrate and on the other three orientations, (001), (110), and (111) Ga. Furthermore the differences in doping levels on (111) As and (001) substrates decrease in the order Te > Se > Sn > S. In the case in which we found a line in the tin-doped layer, it originated at an edge between the wall and the bottom which were $\{111\}$ Ga and $\{111\}$ As planes respectively. Thus it was deduced that the growth areas from the wall and the base of the window are divided into two regions by very different impurity concentrations, although they were not measured. Further however, a difference in growth rate on these planes can be obtained, and is found to be about 20:1.

It is undesirable that there should be regions of different impurity content in growth layers within window and also that surfaces should be present. Accordingly, the following conclusions emerged. (1) Sulfur is a favorable donor impurity, and tin ranks next to sulfur for obtaining electrically homogeneous selective layers in a window. (2) Selective growth on the (001) substrate with a $\langle 100 \rangle$ type directional rectangular window is at present the best method of obtaining a flat, smooth growth surface providing that window etching is carried out using H_2SO_4–H_2O_2–H_2O solution.

Acknowledgments

We wish to thank Dr. H. Kusumoto for encouragement during the course of this work. We also express our appreciation to M. Nomura, H. Sato and H. Hirao, for their interest and for stimulating discussions.

References

1) D. P. Brady et al., Proc. IEEE **54** (1966) 1497;
   J. J. Tietjen et al., Solid State Electron. **9** (1966) 1049.
2) D. W. Shaw, J. Electrochem. Soc. **113** (1966) 904;
   P. Rai-Choudhury et al., J. Electrochem. Soc. **118** (1971) 107.
3) S. Iida et al., J. Electrochem. Soc. **118** (1971) 768.
4) S. Iida et al., J. Electrochem. Soc. Japan **37** (1969) 186.
5) D. W. Shaw, J. Electrochem. Soc. **115** (1968) 777.
6) R. E. Ewing et al., J. Electrochem. Soc. **111** (1964) 1266.
7) F. V. Williams, J. Electrochem. Soc. **111** (1964) 886.

KINETICS OF VAPOUR-PHASE EPITAXIAL GROWTH OF GaAs$_{1-x}$P$_x$

CHR. BELOUET*

Philips Research Laboratories, N.V. Philips' Gloeilampenfabrieken, Eindhoven, The Netherlands

Results are presented of a detailed study of the vapour phase epitaxial growth of GaAs$_{1-x}$P$_x$ ($0.35 < x < 0.41$). Layers were grown on (001) faces of GaAs substrates in an open-tube chloride transport system using AsH$_3$ and PH$_3$. The growth rate curve revealed the existence of a strong similarity between the growth kinetics of GaAsP and GaAs. A quasi-equilibrium growth is observed at the high temperature side of the maximum of this curve. At the low temperature side kinetically limited growth is found having about the same activation energy as has been reported for GaAs. A model is presented in which this activation energy originates from a property of the group III element.

The phosphorus content of the layers is found to be a linear function of the deposition temperature. Furthermore, it is shown that the crystal perfection, as judged from microscopic observations, X-ray Berg–Barrett topographs. Hall-effect measurements and optical measurements reaches an optimum for deposition temperatures around the maximum of the growth rate curve.

1. Introduction

GaAs$_{1-x}$P$_x$ ($0.35 < x < 0.41$) layers, unintentionally doped were grown onto GaAs (001) oriented substrates in an open tube system using PH$_3$ and AsH$_3$ and HCl for the transport of the gallium[1]). Growth rate and x were carefully studied as a function of deposition temperature at fixed reactant input partial pressures.

To determine the temperature yielding the optimum crystal perfection, at a fixed set of reactant input partial pressures, the layers were systematically investigated by means of microscopic observations, Berg–Barrett X-ray topography, Hall-effect measurements at 300 °K and optical measurements at 4.2 °K.

2. Experimental

The growth experiments were performed in a 25 mm I.D. quartz tube using a total flow of about 100 ml/min (300 °K). The stability and reliability of all the flows were better than 99.5%. Significant deposits in front of the substrate were prevented by injecting a small stream of HCl between the Ga source (kept at 900 °C) and the substrate. Growth-rates were reproducible within 10%.

As substrates we used mechano-chemically polished semi-insulating GaAs slices obtained from dislocation-free pulled crystals, the surface of deposition being about 2° off the (001) plane tilted around ⟨110⟩. Prior

to insertion in the cold reactor, already under hydrogen, the substrate was etched in CH$_3$OH–Br$_2$. Just before growth, the slice was etched at growth temperature with HCl, which left a mirror-like surface. Growth was then started with a deposition of GaAs and after a well-defined period of time PH$_3$ was mixed with the AsH$_3$ flow ahead of a buffer chamber. After 30 min, during which the "transient" GaAs$_{1-x}$P$_x$ layer ($\sim$15 µm) was grown, GaAs$_{1-x}$P$_x$ with a constant composition was deposited.

3. Results

3.1. Growth kinetics

The growth rate for GaAs$_{1-x}$P$_x$ was studied as a function of temperature, fig. 1. The shape of the growth curve is similar to that reported for GaAs[2]). At the low-temperature side a kinetically limited growth is

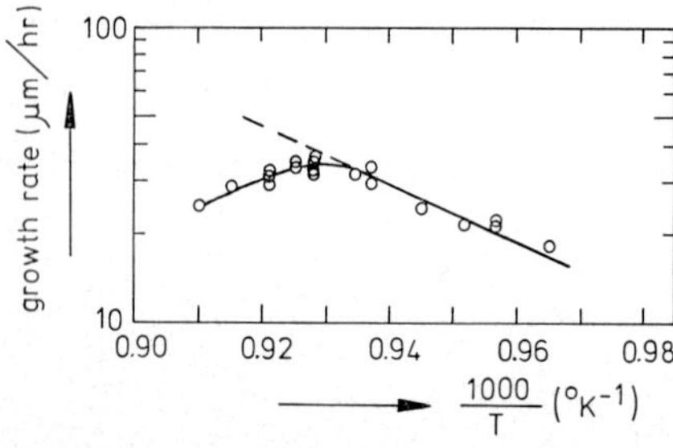

Fig. 1. Growth rate, at fixed flows, as a function of reciprocal deposition temperature for the GaAs$_{1-x}$P$_x$ layer of constant composition. The flows (moles/min) were: GaCl = 4.45 × 10^{-5}, AsH$_3$ = 1.21 × 10^{-5}, PH$_3$ = 0.43 × 10^{-5}, HCl = 0.70 × 10^{-5}.

* On leave from Laboratoires d'Electronique et de Physique, 94 – Limeil–Brévannes, France (present address).

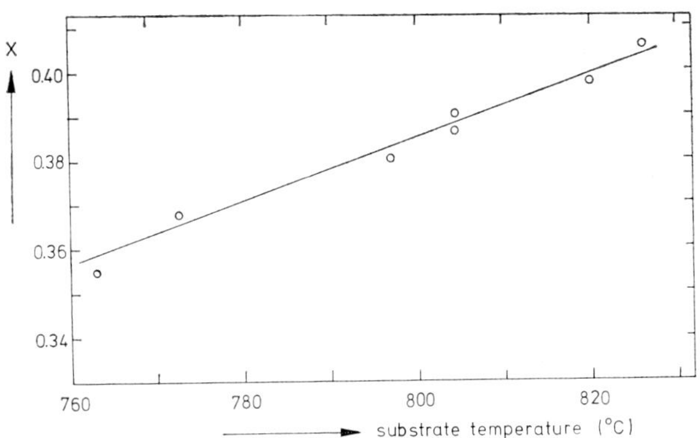

Fig. 2. The phosphorus content x as a function of the deposition temperature at fixed flows (the same as for fig 1).

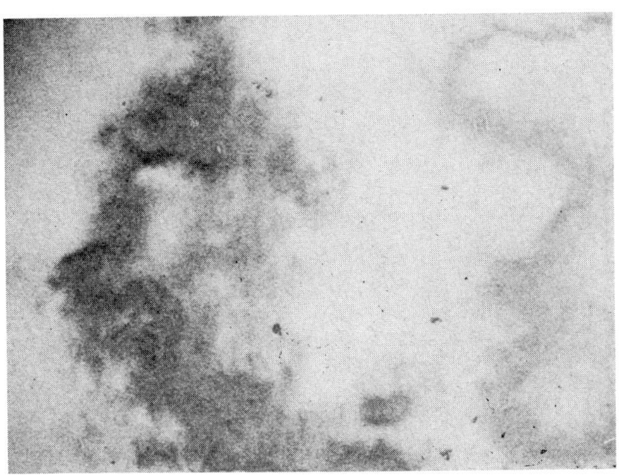

Fig. 3. Berg–Barrett X-ray topograph, magnification $22.5\times$; deposition temperature = 804 °C.

observed, with an activation energy E of ~ 45 kcal/mole, near to the values of 48.5 and 40 kcal/mol, reported for GaAs[2,3]). At the high-temperature side a quasi-equilibrium growth regime is found, as for GaAs[3]).

We will now try to explain the existence of similar activation energies for the growth of both GaAs and GaAs$_{1-x}$P$_x$ on a (001) surface.

The deposition of GaAs$_{1-x}$P$_x$ can be described by the overall reaction (c.f. ref. 3)

$$\text{GaCl (v)} + \tfrac{1}{4}(1-x)\,\text{As}_4 + \tfrac{1}{4}x\,\text{P}_4 +$$

$$\tfrac{1}{2}\,\text{H}_2 \underset{v_2}{\overset{v_1}{\rightleftharpoons}} \text{GaAs}_{1-x}\,\text{P}_x\,(s) + \text{HCl (v)},$$

with $P_{\text{H}_2} \simeq 1$,

$$v_1 = k_1\, P_{\text{GaCl}}\,(P_{\text{As}_4})^{\frac{1}{4}(1-x)}\,(P_{\text{P}_4})^{\frac{1}{4}x},$$

$$v_2 = k_2\, P_{\text{HCl}},$$

where

$$k_1 = A_1 \exp\left(-\Delta E_1/kT\right),$$

$$k_2 = A_2 \exp\left(-\Delta E_2/kT\right).$$

In the quasi-equilibrium growth regime, $v_1 \simeq v_2$ at the surface, while in the kinetically limited regime $v_1 > v_2$ and for $v_1 \gg v_2$ an activation energy E for the growth will be found.

We suggest that the process of binding a GaCl molecule to the surface at a gallium site is the limiting process for both GaAs and GaAsP. E can then be identified with the ΔE needed for the chemical activation of the GaCl molecule, namely the transition

$$\text{s}^2\text{p} \xrightarrow{\Delta E} \text{sp}^2\;;$$

consequently $\Delta E \simeq E \simeq \Delta E_1$. This ΔE is independent of vapour composition and temperature, in agreement with observations on GaAs[2]).

This model requires that the life time of the bound GaCl molecule be large compared with the life times in the subsequent processes involved in GaAsP synthesis. For surfaces other than (001) this requirement may no longer be fulfilled, in particular when the binding of a GaCl molecule at the surface does not decrease the number of dangling bonds of that surface.

3.2. Compositional effects

The phosphorus concentration x was determined by X-ray powder diffraction analysis. The main results are shown in fig. 2. The variation of x with temperature, fig. 2, is independent of the growth rate and probably also of the growth mechanism. This variation may be due to a relative change with temperature of the sticking probability or of the surface migration of the group V elements.

Microprobe analysis revealed no evidence for preferential As-to-P ratios in the transient layer, in contradistinction to literature[4]).

3.3. Quality of the layers

Specific macroscopic surface defects were found for each growth regime. In the region of quasi-equilibrium growth conditions these defects are craters with well-defined crystallographic orientations, whereas pyramids are specific for the kinetically limited growth. Around the growth curve maximum the surface gener-

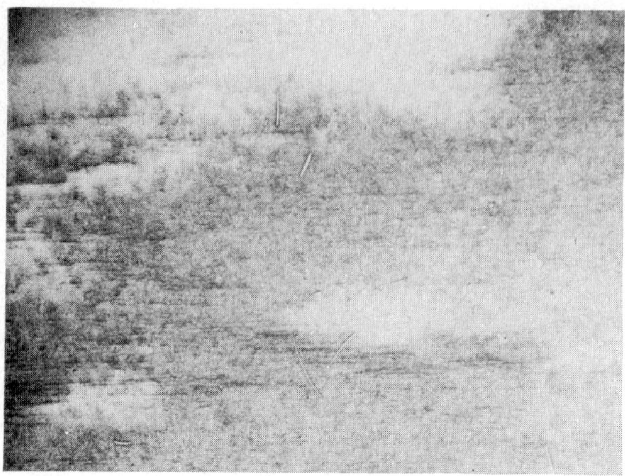

Fig. 4. Berg–Barrett X-ray topograph, magnification $22.5\times$; deposition temperature = 794 °C.

Fig. 5. Berg–Barrett X-ray topograph, magnification $22.5\times$; deposition temperature = 773 °C.

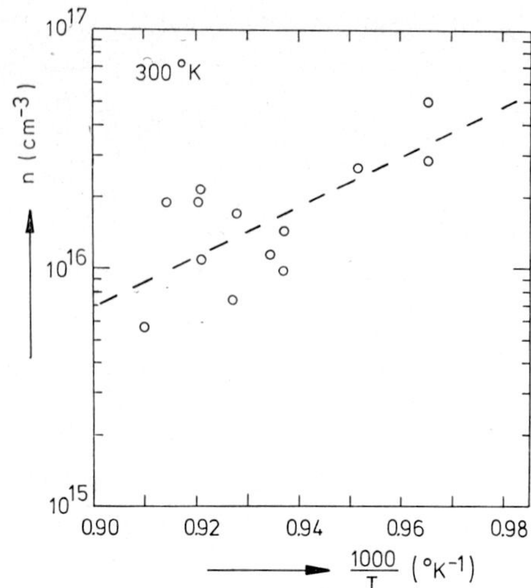

Fig. 6. The concentration of free electrons n at 300 °K as a function of the reciprocal of the deposition temperature. In calculating n from the Hall-effect, a Hall-factor of 1 was used.

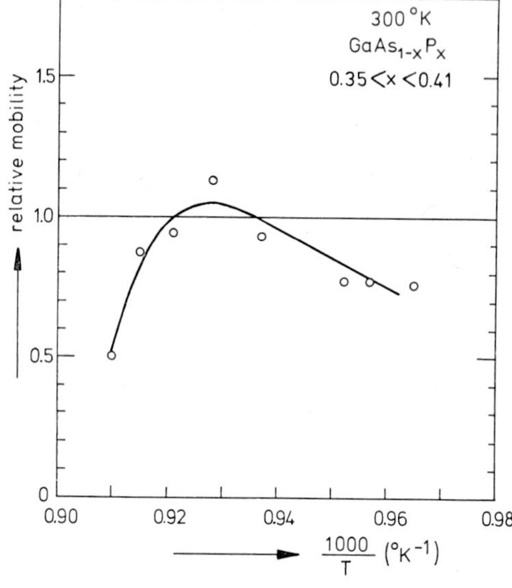

Fig. 7. The Hall mobility for $0.35 < x < 0.41$, taken with respect to that of ref. 1, as a function of the reciprocal of the deposition temperature.

ally shows no macroscopic defects. In the whole temperature range steps aligned along the $\langle 110 \rangle$ directions were observed. These are nearly absent when the surface is almost exactly a (001) plane.

The different growth regimes also show up in Berg–Barrett X-ray topographs. Essentially elastic strains are found around the maximum (fig. 3). At the high-temperature side the crystal quality rapidly degrades when temperature increases. At the low-temperature side, the elastic strains are progressively replaced by networks of dislocations when the temperature decreases (figs. 4 and 5).

From Hall-effect measurements, all layers were found to be n-type, n decreasing with increasing temperature (fig. 6). When the mobilities in the layers with n between 0.5 and 2×10^{16} cm^{-3} are compared with those of ref. 1, it is seen, fig. 7, that the best results are obtained around the maximum of the growth curve.

3.4. OPTICAL MEASUREMENTS

Photoluminescence spectra taken at 4.2 °K on etched samples show one or more narrow lines (half-width $\gtrsim 2.5$ meV) near the band edge. At least some of these emissions will involve shallow donors or acceptors.

Furthermore, at lower photon energies a broader peak (half-width ~ 70 meV) is found often. Extrapolation of its photon energy to GaAs yields ~ 1.15 eV. A luminescence in GaAs having about this energy has been attributed to a recombination involving Ga vacancies[5].

4. Conclusions

It is found that a strong similarity exists between the growth kinetics of GaAs$_{1-x}$P$_x$ and GaAs when grown epitaxially on (001) surfaces using an open-tube chloride-transport system with AsH$_3$ and PH$_3$. The activation energy for the growth rate in the kinetically limited growth region has about the same value for both GaAs$_{1-x}$P$_x$ and GaAs.

The phosphorus content of the layers shows a linear dependence on deposition temperature.

Tentative explanations are offered for this effect.

Furthermore, it is shown that the crystal perfection closely depends on the growth regime; it reaches an optimum for deposition temperatures around the maximum of the growth-rate curve.

Acknowledgement

The author gratefully acknowledges helpful discussions with all members of the group of Dr. A. J. Bosman and the assistance of Mr. J. de Vries in the experimental part of the work.

References

1) J. J. Tietjen and J. A. Amick, J. Electrochem. Soc. **113** (1966) 724.
2) D. W. Shaw, J. Electrochem. Soc. **117** (1970) 683.
3) A. Boucher and L. Hollan, J. Electrochem. Soc. **117** (1970) 932.
4) R. E. Ewing and D. K. Smith, J. Appl. Phys. **39** (1968) 5943.
5) E. g., E. W. Williams and D. M. Blackmoll, Trans. Met. Soc. AIME **239** (1967) 387.

    Journal of Crystal Growth **13/14** (1972) 346–349 © *North-Holland Publishing Co.*

VAPOR GROWTH OF BORON MONOPHOSPHIDE USING OPEN AND CLOSED TUBE PROCESSES

TATAU NISHINAGA, HIROSHI OGAWA*, HISATSUNE WATANABE** and TETSUYA ARIZUMI

Department of Electronics, Faculty of Engineering, Nagoya University, Nagoya, Japan

Single crystal layers of boron monophosphide as large as $1.0 \text{ cm}^2 \times 30 \text{ }\mu\text{m}$ were grown epitaxially on silicon substrates from the vapor phase by thermal reduction of BBr_3 and PCl_3.
The layers grown on $\{111\}$ surfaces were single crystal boron monophosphide of zinc blende structure, while the layers on $\{100\}$ surface were polycrystalline. The undoped single crystal BP layers were always n-type with a resistivity of 5×10^{-3} Ω-cm.
Small crystals of BP were also grown by vertical closed tube method. The transport rate of BP was obtained experimentally as a function of iodine concentration.

1. Introduction

Semiconducting properties of boron compounds have not been fully developed because it is difficult to grow large single crystals. Though boron monophosphide has the longest history among them, the largest BP single crystal grown so far is of the order of several millimeters in size, hence very little work has been devoted to the physical and electrical properties of this material. On the other hand, a rapid growth in the field of the luminescent diode requires the development of the new materials such as the III–V compound alloys and research work on boron compounds is important from this point of view.

Many ways of growing BP crystal have been proposed. Stone and Williams[1] succeeded in preparing polycrystalline films from boron trichloride and elemental phosphorus. Recently, Medvedeva et al.[2,3] carried out closed tube experiments using iodine and sulphur as transport agents. They obtained single crystals, the maximum size being $2 \times 2 \times 1$ mm. By using the solution growth technique, Baranov et al.[4] grew single crystal platelets up to 4 mm in diameter. Williams[5] proposed epitaxial growth on cubic SiC crystals since they have almost the same lattice constant as BP. Very recently Chu et al.[6] obtained epitaxial layers of BP on hexagonal SiC substrates by thermal decomposition of a mixture of diborane and phosphine and by

hydrogen reduction of boron tribromide–phosphorus trichloride mixtures. Though SiC substrates are most suitable from the view point of lattice matching, the difficult etching of SiC prevents the removal of substrates from BP layers; Si substrates are desirable for this purpose, although the lattice mismatch is as large as 16.5%.

This report describes firstly experiments on the growth of epitaxial BP layers on Si substrates and secondly on the growth of BP crystals by the vertical closed tube method.

2. Experimental

2.1. Epitaxial growth of BP on Si substrates

The experimental apparatus used is shown in fig. 1. Hydrogen carrier gas was purified by a palladium diffusion purifier and introduced through two parallel bubblers one containing BBr_3 and the other PCl_3 and into a vertical quartz reaction tube. The reason for employing BBr_3 instead of BCl_3 is that BCl_3 is not easy to handle because of its high vapour pressure at room temperature. Both BBr_3 and PCl_3 had a purity of 99.999% and they were transferred to each bubbler in a dry nitrogen atmosphere. The temperatures of two bubblers were kept at -30 to -10 °C and the hydrogen flow rate through each bubbler was varied from 200 to 350 cm^3/min. The quartz reaction tube employed was of 65 mm OD within which the Si wafers were put on a glassy (vitreous) carbon pedestal of 23 mm OD. The pedestal was heated by rf from the outside of the

---

* Present address: Toyota Central Research and Development Labs., Inc., Nagoya, Japan.
** Present address: Nippon Electric Co., Ltd., Tokyo, Japan.

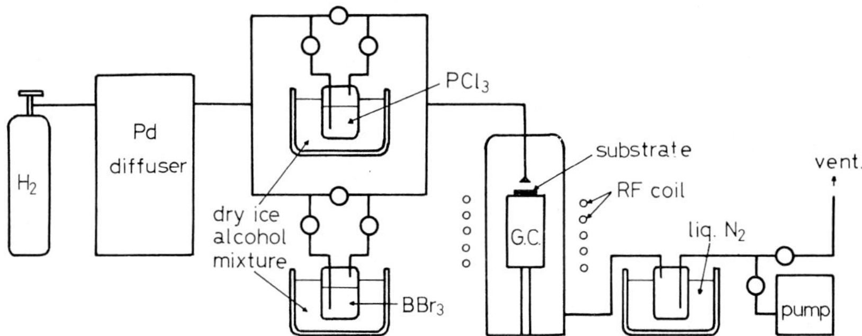

Fig. 1. Experimental apparatus for the epitaxial growth of boron monophosphide.

reaction tube. Before the reaction, the whole system was evacuated to 10^{-3} Torr, then the system was flushed with pure hydrogen. After two or three cycles of this process, a flow of pure hydrogen was maintained for one hour. Prior to the deposition, Si wafers having {111} and {100} faces were treated in hydrogen at 1300 °C for 10 min to remove any oxide layer. Following this treatment, the temperature was lowered to 900–1100 °C to start the deposition. Then PCl_3–H_2 gas mixture was introduced first in order to keep the system under an excess phosphorus pressure. The BBr_3–H_2 mixture was introduced next to begin the deposition. The over-all reaction,

$$BBr_3 + PCl_3 + 3\,H_2 \rightarrow BP + 3\,HBr + 3\,HCl$$

gave BP layers of 10–30 µm thickness after reaction times of 30–60 min. A one hour etch of the grown sample in a mixture of 1 part HF and 4 parts HNO_3 removed the Si substrate completely. Because BP was almost unetched, we could obtain single crystal BP layers without any damage. The BP layers were then examined by X-ray Laue back reflection and the crystals grown on Si {111} surfaces were found to be monocrystalline with a zinc blende structure. The electron reflection diffractions on both surfaces of the layer, the

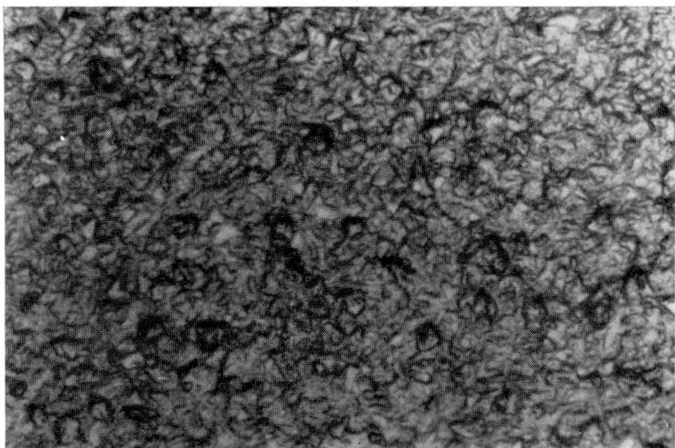

Fig. 3. Surface microphotograph of the single crystal BP layer grown on a Si {111} surface 1100×.

growing side and the substrate side, confirmed that the layer was monocrystalline. Each photograph of the electron reflection diffractions is shown in figs. 2a and 2b. By contrast, single crystal BP layers have not been grown on Si {100} surfaces. The surface and the cross sectional view of a single crystal BP grown on Si {111} surface are shown in fig. 3 and fig. 4 respectively.

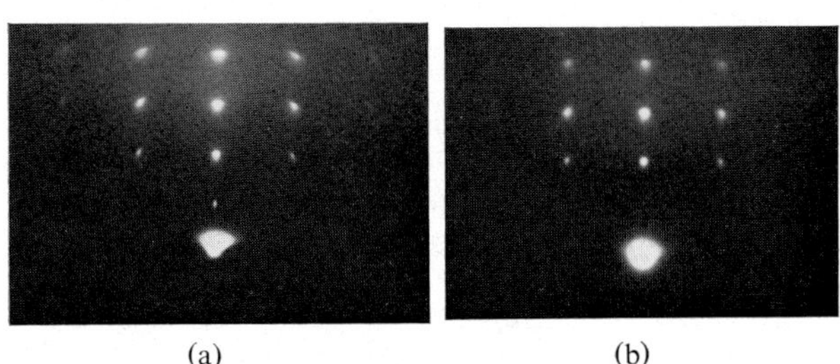

(a)   (b)

Fig. 2. Electron reflection diffraction patterns from single crystal BP layer grown on a Si {111} surface; reflection from (a) growing surface side, (b) Si substrate side.

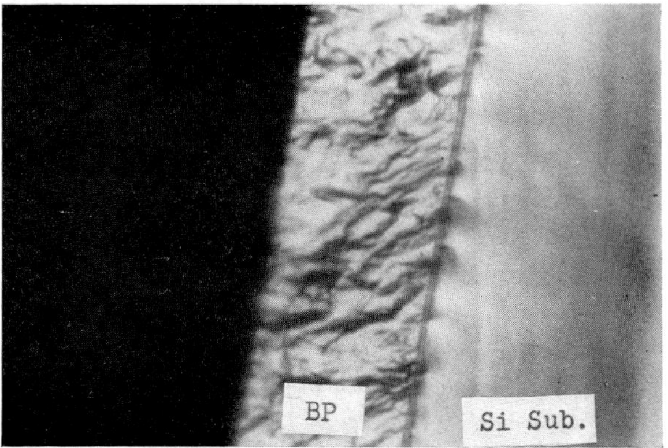

Fig. 4. Cleaved cross section of BP single crystal layer with Si substrate. The layer is 13 µm in thickness.

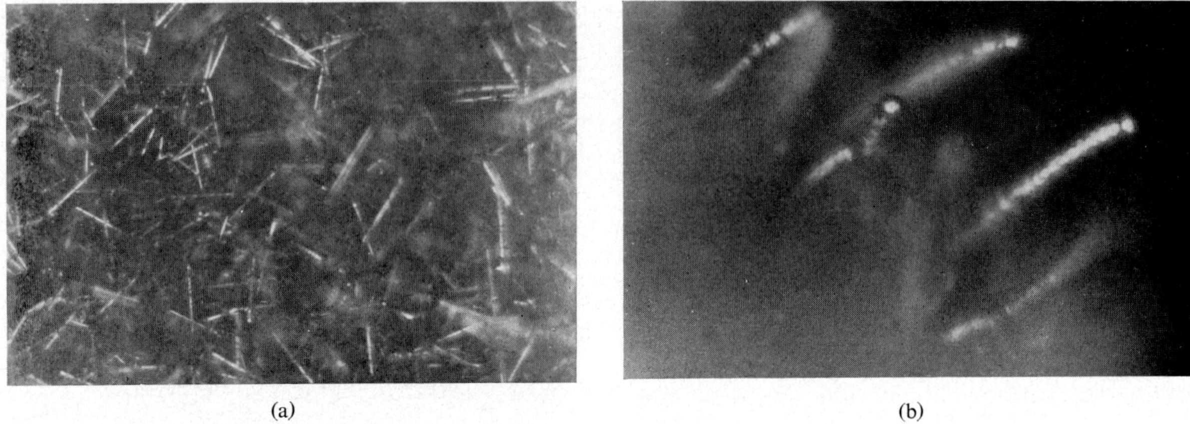

(a)   (b)

Fig. 5.   Dense BP whiskers grown on the low temperature part of a Si substrate; (a) 270×, (b) 1100×.

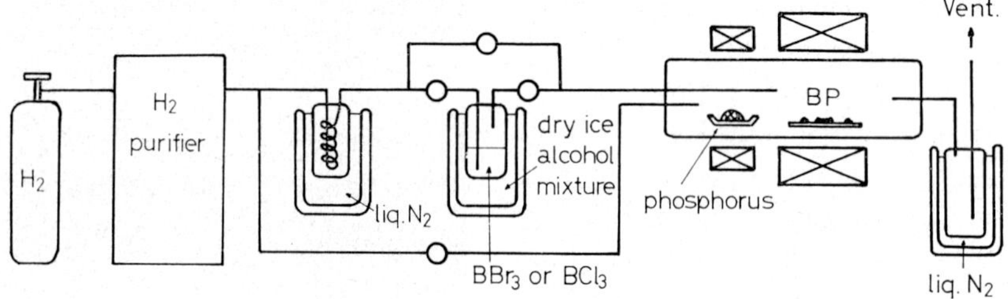

Fig. 6.   Schematic diagram of the apparatus used for BP synthesis.

Because of the differences of the lattice constant and the thermal expansion, grown samples were severely distorted to form a concave shape with the BP layer inside. This deformation was plastic and remained somewhat after the removal of the Si substrate. Thermoelectric power measurements showed the layers were always n-type and the resistivity was found to be of the order of 5×10^{-3} Ω-cm by Van der Pauw's method. On the silicon surface where the temperature was low and hence the supersaturation was thought to be high, dense BP whiskers were observed as shown in figs. 5a and 5b.

2.2. Closed tube experiment

An evacuated quartz tube 10 mm OD and 10 cm long which contained the source polycrystalline BP films at the lower end and 0.8–10 mg of iodine, was inserted into a two-zone vertical furnace. In this system, the deposition of BP occurs[2]) according to the disproportionation reaction

$$2\,BP + BI_3 \rightleftarrows 3\,BI + P_2 .$$

In the source zone where the temperature was kept at 1100 °C, the reaction proceeds towards the right, and towards the left in the seed zone where the temperature was 900 °C. The gas molecules are carried by diffusion, laminar flow and especially in this case by convection. The source BP films were prepared in the apparatus shown in fig. 6. Boron halide (BBr_3 or BCl_3) and phosphorus vapor were carried by hydrogen (480–600 cm³/min) to the high temperature region (950–1000 °C) where the polycrystalline BP films were grown.

The closed tube transport rate obtained experimentally is shown in fig. 7 by the solid circles. The total

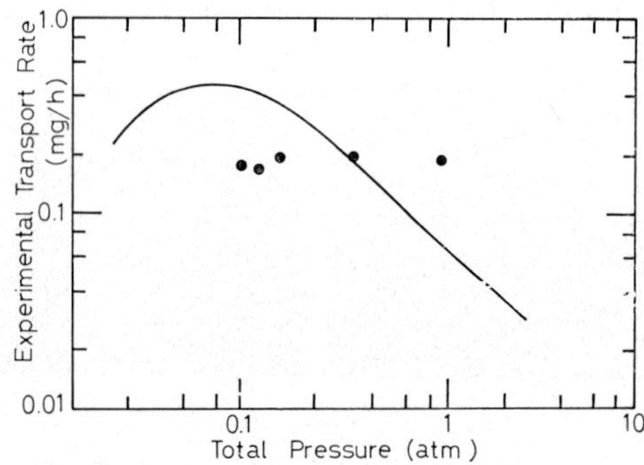

Fig. 7.   Transport rate of BP in closed tube as a function of total pressure. Line: theory; points: experimental.

VII – 10

pressure is readily obtained by the thermodynamical calculation if the total amount of iodine put in the system is known. The theoretical transport rate, n_{Bp}, which is shown by the solid line in the same figure, may be written in the following form employing the two-zone approximation[7]) which took into account only diffusion and laminar flow,

$$n_{\mathrm{BP}} = -\frac{D_0}{L R \overline{T}} \left(\frac{\overline{T}}{T_0}\right)^{\frac{3}{2}} \times$$

$$\sum_{i=1}^{2} \frac{\Delta P_i \sum_j m_j \overline{P}_j - \overline{P}_i \sum_j m_j \Delta P_j}{\sum_j P_j \sum_j m_j \overline{P}_j}$$

where

D_0 diffusion coefficient at T_0 (°K) under 1 atm assuming equal for all gases,
L effective length between the source and seed zones,
R gas constant,
$\overline{T}$ average temperature, $\frac{1}{2}(T_{\mathrm{source}} + T_{\mathrm{seed}})$,
P_j partial pressure of j th gas molecule,
$\overline{P}_j$ $\frac{1}{2}(P_{j\,\mathrm{seed}} + P_{j\,\mathrm{source}})$,
ΔP_j $P_{j\,\mathrm{seed}} - P_{j\,\mathrm{source}}$,
m_j a half of the iodine atom number contained in the jth gas molecule; e.g. if $j = 1, 2$ and 3 denote BI, BI$_3$ and P$_2$ molecules respectively, m_1, m_2 and m_3 are $\frac{1}{2}$, $\frac{3}{2}$ and 0.

To see only the functional dependency on the total pressure, both experimental and theoretical transport rates are plotted in the same figure. The experimental transport rates obtained were larger than those of Medvedeva et al.[2]). The disagreement between theory and experiment suggests the convection current may largely influence the diffusion and the laminar flow. The crystal shape obtained by this process was polyhedral with a diameter of 0.6 mm. The crystals were transparent and glowed reddish when they were illuminated by white light.

Acknowledgement

The authors wish to thank Professor K. Mihama, Department of Applied Physics, Nagoya University, for taking the electron reflection diffraction photographs.

References

1) B. D. Stone and F. V. Williams, in: *Proc. Conf. on Ultra-purification of Semiconductor Materials*, Boston, 1961, p. 645.
2) Z. S. Medvedeva, J. H. Greenberg, E. G. Žukov and N. P. Lužnaja, Kristall und Technik **2** (1967) 523.
3) Z. S. Medvedeva, J. H. Greenberg and E. G. Žukov, Kristall und Technik **4** (1969) 487.
4) B. V. Baranov, Ya. A. Oksman, V. D. Prochukhan and V. N. Smirnov, Optika i Spectroskopiya **19** (1965) 987.
5) F. V. Williams, US pat. 3210624, Oct. 5, 1965.
6) T. L. Chu, J. M. Jackson, A. E. Hyslop and S. C. Chu, J. Appl. Phys. **42** (1971) 420.
7) T. Arizumi and T. Nishinaga, Japan. J. Appl. Phys. **4** (1965) 165.

BINÄRE VERBINDUNGEN DES BORS MIT ELEMENTEN DER V. HAUPTGRUPPE UNTER ERHALTUNG VON IKOSAEDRISCHEN GRUNDGERÜSTEN: TETRAGONALES BORNITRID UND RHOMBOEDRISCHES BORPHOSPHID

K. PLOOG*, P. RAUH, W. STOEGER und H. SCHMIDT

Institut für Anorganische Chemie der Universität München, Germany

Pyrolysis of BBr_3–H_2 mixtures in the presence of N_2 at 1150 to 1500 °C was used to form tetragonal boron nitride of composition $B_{58}N$ to $B_{23}N$ on BN substrates. At increased temperatures the content of nitrogen in the grown material increased from 2.40 to 4.50 wt%; while the density increased from 2.404 to 2.460 g cm^{-3} and the lattice constants a from 8.79 to 9.02 Å and c from 5.08 to 5.26 Å. The structure of the B_xN is derived from the lattice of the so-called "I-tetragonal boron": The basic I-tetragonal framework with 4 B_{12}-icosahedra and 2 single B-atoms is preserved, and the nitrogen atoms occupy one or two of the larger tetrahedral holes. BBr_3–PBr_3–H_2 mixtures were pyrolized on tantalum wires at 1150 °C to prepare rhombohedral boron phosphide of composition $B_{48}P$ to B_9P. The products have variable composition, but can be all derived from the same basic framework: rhombohedrally linked B_{12}-icosahedra as in boron carbide are cross-linked by single B- and/or P-atoms, which occupy sites in the large elongated cavity along [111]. With rising phosphorus content, a three-dimensional lattice expansion is observed, a increases from 5.83 to 5.99 Å and c from 11.71 to 11.98 Å. The a/c ratio remains nearly constant.

1. Einleitung

Im Rahmen unserer Untersuchungen über Verbindungen des Bors mit Elementen der IV. und V. Hauptgruppe interessieren uns besonders solche Phasen, bei denen ikosaedrische Bor-Grundgerüste erhalten oder gar erst durch die Fremdatome induziert werden.

So konnten wir durch Pyrolyse von BBr_3 und H_2 bei Temperaturen von 900 bis 1300 °C unter Zusatz von CH_4 oder $CHBr_3$ die neuen borreichen Boride $B_{48}B_2C$ und $B_{48}B_2C_2$[1]) reproduzierbar darstellen. In diesen beiden Boriden wird das instabile Gitter $(B_{12})_4B_2$ des sogenannten "I-tetragonalen Bors" durch Besetzung von weiteren Tetraederlücken im I-tetragonalen Ikosaedergerüst durch ein bzw. zwei Kohlenstoffatome stabilisiert. Die stabilisierende Wirkung von C-Atomen beim Einbau in das I-tetragonale Gitter erscheint durch folgende grobe Abschätzung plausibel: Kohlenstoff bildet bevorzugt die Tetraederkonfiguration aus, und die Kovalenzradien und Elektronegativitäten von Bor und Kohlenstoff sind ähnlich.

Auch Stickstoffatome sollten in der Lage sein, an Stelle von Kohlenstoff das I-tetragonale Gitter zu stabilisieren, denn in einem N-haltigen I-tetragonalen

Gitter dürfte der kleinere Kovalenzradius des Stickstoffs (0.74 Å) durch Aufnahme von Elektronen aus dem Borgerüst ausgeglichen werden, da der Elektronenmangel von B_{12}-Ikosaedern[2,3]) zweifellos von der großen Elektronegativität des Stickstoffs (3.07 Pauling-Einheiten) kompensiert wird.

Dagegen ist zu erwarten, daß der Einbau von Phosphoratomen in ein ikosaedrisches Grundgerüst zu einer sehr starken Verzerrung der I-tetragonalen Grundstruktur oder zur Induktion von abweichenden Strukturen führt; denn das zu große P-Atom hat wegen seiner mit dem Bor vergleichbaren Elektronegativität von 2.06 Pauling-Einheiten kaum eine Möglichkeit zur Radienverringerung durch Elektronenabgabe. Es sollte daher möglich sein, Bor-Phosphor-Phasen mit niedrigen Phosphorgehalt darzustellen, denen folgende alternative Strukturmerkmale zugrunde liegen: (a) Es werden nur solche, bereits vom elementaren Bor her bekannte Ikosaedergerüste stabilisiert, in denen größere Lücken als z.B. im sogenannten "I-tetragonalen Bor" vorhanden sind; oder (b) der Einbau von Phosphoratomen induziert neue, bisher beim elementaren Bor nicht auftretende Verknüpfungsmüglichkeiten von B_{12}-Ikosaedern und stabilisiert derartige Gerüste.

Für das Arsen mit einem Atomradius von 1.21 Å und einer Elektronegativität von 2.20 Pauling-Einheiten gelten ähnliche Überlegungen wie für den Phos-

---

* Present address: Mineralogisches Institut der Universität Bonn, Abteilung für Kristallstrukturlehre und Neutronenbeugung, Poppelsdorfer Schloß, 53 Bonn, Germany.

phor; Ergebnisse aus eigenen Untersuchungen über das System Bor-Arsen liegen allerdings noch nicht vor.

Die bekanntesten und präparativ sowie strukturell am besten untersuchten Bor-Stickstoff- und Bor-Phosphor-Phasen haben ein B- zu E^V-Atomverhältnis von eins zu eins: das hexagonale Bornitrid BN[4]) und das kubische Borphosphid BP[5]). In beiden Gittern ist jedes Atom der einen Atomsorte von drei bzw. vier Atomen der anderen Atomsorte umgeben. Es treten keine größeren Koordinationspolyeder auf, die im wesentlichen aus Boratomen bestehen, und die als selbständige Baueinheiten entscheidend den Charakter der Struktur bestimmen.

Größere Bor-Koordinationspolyeder treten erst bei solchen in der Literatur beschriebenen Verbindungen zwischen dem Bor und Elementen der V. Hauptgruppe auf, bei denen das Atomverhältnis B zu E^V größer als fünf zu eins ist. Die Zusammensetzung der für die vorliegenden Untersuchungen interessanteren, zweiten bekannten, borreicheren Bor-Phosphor-Phase wird allerdings von verschiedenen Autoren unterschiedlich angegeben.

Matkovich[6]), Spinar und Wang[7]), und Peret[8]) geben die Zusammensetzung mit $B_{13}P_2$ an, während Greene et al.[9,10]) ihre Analysenergebnisse mit der Formel $B_{12}P_2$ interpretieren. Übereinstimmung unter den genannten Autoren herrscht lediglich bei der Diskussion von Teilen der Struktur dieses borreichen Borphosphids: B_{12}-Ikosaeder als dominierende Strukturelemente besetzen wie beim α-rhomboedrischen Bor[11]) und beim Borcarbid[12,13]) die Eckpunkte eines Rhomboeders. Die Phosphoratome sind in einem langgestreckten Hohlraum im Zentrum des Rhomboeders (Richtung [111]) untergebracht. Unklar ist bisher jedoch die genaue Lage und die Funktion der P-Atome in dem Ikosaedergerüst. Auffallend sind besonders bei der Arbeit von Matkovich[6]), der das $B_{13}P_2$ durch thermische Zersetzung von BP im Graphit-Tiegel bei 1400 bis 1700 °C darstellte, die analytisch nachweisbaren beträchtlichen Mengen an Verunreinigungen (u.a. 1% C und 1% Al), die ihrerseits die Anordnung der B_{12}-Ikosaeder und damit die endgültige Kristallstruktur entscheidend beeinflussen können.

Da das Bor bei der Wahl seines Gitters empfindlich auf Fremdatome reagiert[1]), mußte bei unseren Untersuchungen die Darstellung von borreichen Bor-Stickstoff- und Bor-Phosphor-Phasen von vornherein unter sehr reinen Bedingungen erfolgen. Nur dadurch konnte gewährleistet werden, daß die entstehenden Ikosaeder Anordnungen ausschließlich durch den Einbau von N- oder P-Atomen induziert und/oder stabilisiert werden.

2. Arbeitsmethodik

Um die unkontrollierte Diffusion von Fremdatomen aus Tiegelmaterialien beim Schmelzen zu vermeiden, wählten wir zur Synthese von Bor-Stickstoff-Phasen die Pyrolyse von flüchtigen Borverbindungen in Gegenwart von elementarem Stickstoff und Wasserstoff unter Normaldruck bei Temperaturen von 1150 bis 1500 °C und zur Darstellung von Bor-Phosphor-Phasen die Pyrolyse von flüchtigen Bor- und Phosphorverbindungen in Gegenwart von Wasserstoff bei 1150 °C ebenfalls im strömenden System. Diese Temperaturen des Kristallwachstums liegen weit unter dem Schmelzpunkt des reinen Bors (> 2000 °C) und dem der zu erwartenden B–N- und B–P-Phasen. Daher lassen sich mit dieser Methode in kinetisch kontrollierten Reaktionen auch Phasen herstellen, die beim Schmelzpunkt bereits instabil sind.

Besondere Aufmerksamkeit mußte der Wahl einer geeigneten Pyrolysenunterlage geschenkt werden. Als Pyrolysenunterlage zur Darstellung von Bor-Phosphor-Phasen diente hochreiner Tantaldraht von 0.75 mm Durchmesser und 60 mm Länge, der durch elektrische Widerstandsheizung auf die erforderliche Temperatur aufgeheizt wurde. Bei Temperaturen bis 1200 °C ist Tantal als Substratoberfläche besonders vorteilhaft, da es unterhalb dieser Temperatur nur wenig mit dem aufwachsenden Bor und Phosphor reagiert. Zur Darstellung von Bor-Stickstoff-Phasen, die bei höheren Temperaturen erfolgte, scheidet Tantal als Substrat dagegen aus, da ab 1300 °C die Tantalboridbildung sehr stark einsetzt. Als Abscheidungsunterlage, die auch bei höheren Temperaturen nicht mit dem aufwachsenden Material reagiert, erwiesen sich Graphit-scheibchen als geeignet, die mit einer 0.2–0.5 mm dicken Schicht von gut haftendem, röntgenamorphem Bornitrid BN bedeckt waren und die durch ein Hochfrequenzfeld aufgeheizt wurden. Die BN-Überzüge bilden sich bei der Pyrolyse von 2,4,6-Trichlorborazol bei 1200 °C. Nach unseren bisherigen Erfahrungen verhindern diese Überzüge, die bis 1700 °C völlig stabil sind, die Diffusion von Kohlenstoff aus der Graphitunterlage in das abzuscheidende Material. Auch eine Diffusion von Stickstoff aus dem BN in das aufwachsende Bor kann ausgeschlossen werden.

2.1. Ausgangsprodukte zur Darstellung von Bor-Stickstoff-Phasen

Als Ausgangsprodukte für die thermische Zersetzung wählten wir als flüchtige Borkomponente das Bortribromid, das sich mit Wasserstoff als Reduktionsmittel bereits bei unseren Versuchen zur Bildung der Gitter des reinen Bors[14]) bestens bewährt hatte. Als flüchtige Stickstoffkomponente bot sich zunächst der Ammoniak an. Mehrere Vorversuche zeigten jedoch, daß trotz spezieller Versuchsführung NH_3 bereits vor Erreichen der heißen Substratoberfläche vollständig mit dem gasförmigen BBr_3 zu einer festen weißen Additionsverbindung reagierte, die sich ober- und unterhalb der heißen Zone an den Wänden des Zersetzungsgefäßes niederschlug. Damit war eine reproduzierbare Dotierung des aufwachsenden Borgitters mit Stickstoff nicht mehr gewährleistet, und NH_3 schied als Stickstofflieferant aus.

Molekularer Stickstoff, N_2, mit seiner hohen Dissoziationsenergie von 226 kcal/Mol liefert beim Aufheizen der Substratoberfläche durch direkte Widerstandsheizung bis 1200 °C keine zur N-Dotierung des aufwachsenden Bors geeigneten Spezies[15]). Erst bei Verwendung eines Hochfrequenzfeldes zum Aufheizen des Substrats treten bereits ab 1150 °C einbaufähige N-Spezies auf, und es werden N-dotierte Borgitter gebildet. Dabei muß der Stickstoff jedoch in erheblichem Überschuß dem gasförmigen BBr_3-H_2-Reaktionsgemisch beigemischt werden.

2.2. Ausgangsprodukte zur Darstellung von Bor-Phosphor-Phasen

Neben dem BBr_3-H_2-Gemisch als Borkomponente bot sich als Phosphorkomponente das Phosphortribromid an, damit nach Möglichkeit Austauschreaktionen der Art

$$BBr_3 \text{ (g)} + PX_3 \text{ (g)} \rightleftharpoons BBr_{3-n}X_n \text{ (g)} + PX_{3-n}Br_n \text{ (g)},$$
$$(n = 1, 2, 3)$$

vermieden wurden. Die thermochemischen Daten[16,17]) ließen ein gleichzeitiges Aufwachsen von Bor und Phosphor unter Bildung fester Bor-Phosphor-Phasen erwarten.

2.3. Versuchsapparatur

Die ausheizbaren, in Duranglas ausgeführten Zersetzungsapparaturen sind schematisch in Abb. 1a (Darstellung von B–N-Phasen, Hochfrequenzheizung) und Abb. 1b (Darstellung von B–P-Phasen, Widerstandsheizung) skizziert. Sie bestehen aus folgenden, über teflongedichtete Schliffe verbundenen Teilen:

A Hg-Überdruckventil.

B Strömungsmesser.

C Mit Silicagel gefüllte und mit flüssigem Stickstoff gekühlte Fallen zum Ausfrieren von Spuren H_2O aus dem Wasserstoff bzw. Stickstoff.

D_v Thermostatisiertes Vorratsgefäß für BBr_3 (D_h analog für PBr_3). Die im unteren Teil des Vorratsgefäßes austretenden Gasblasen werden in ein spiralförmig aufgewickeltes Glasrohr (1000 × 5 mm) geleitet. Hier erfolgt eine innige Durchmischung der gasförmigen und flüssigen Phase unter Bildung kleiner Gasbläschen, die in der Schlange die Flüssigkeit mit nach oben führen. Dadurch wird eine relativ gute Sättigung des Trägergases mit dem Reaktionspartner entsprechend dem gewählten Dampfdruckverhältnis gewährleistet, die auch nahezu unabhängig von der Höhe des Flüssigkeitsspiegels und in gewissen Rahmen von der Strömungsgeschwindigkeit des Trägergases ist.

E Mit zwei Glasfritten versehenes Zwischenstück, das als Sperre für Staub, Nebel und Druckwellen dient und in dem der mit BBr_3-Dampf beladene H_2-Gasstrom mit den N_2 bzw. mit dem PBr_3-H_2-Gasgemisch vermischt wird.

F Pyrolysegefäß. Das Pyrolysegefäß zur Darstellung von BN-Phasen besteht aus einem Quarzrohr (300 × 30 mm). Ein Graphitstab (150 × 3 mm) hält die Substrat-Scheibe (15 mm Durchmesser, 3 mm Dicke), die durch ein Hochfrequenzfeld geheizt wird, in das Zentrum des Quarzrohres.

F_1 Auf 60 °C thermostatisiertes Pyrolysegefäß zur Darstellung von B–P-Phasen mit dem durch Stromdurchgang erhitzten, U-förmigen Tantaldraht.

F_2 Y-förmiger Aufsatz auf das Pyrolysegefäß, der die mit Hilfe von eingeschraubten kühlbaren Ta-Schliffkernen eingeführten Tantal-Elektroden enthält. Die Befestigung des eigentlichen Abscheidungsdrahtes an den Elektroden geht aus der vergrößerten Abbildung F_3 hervor.

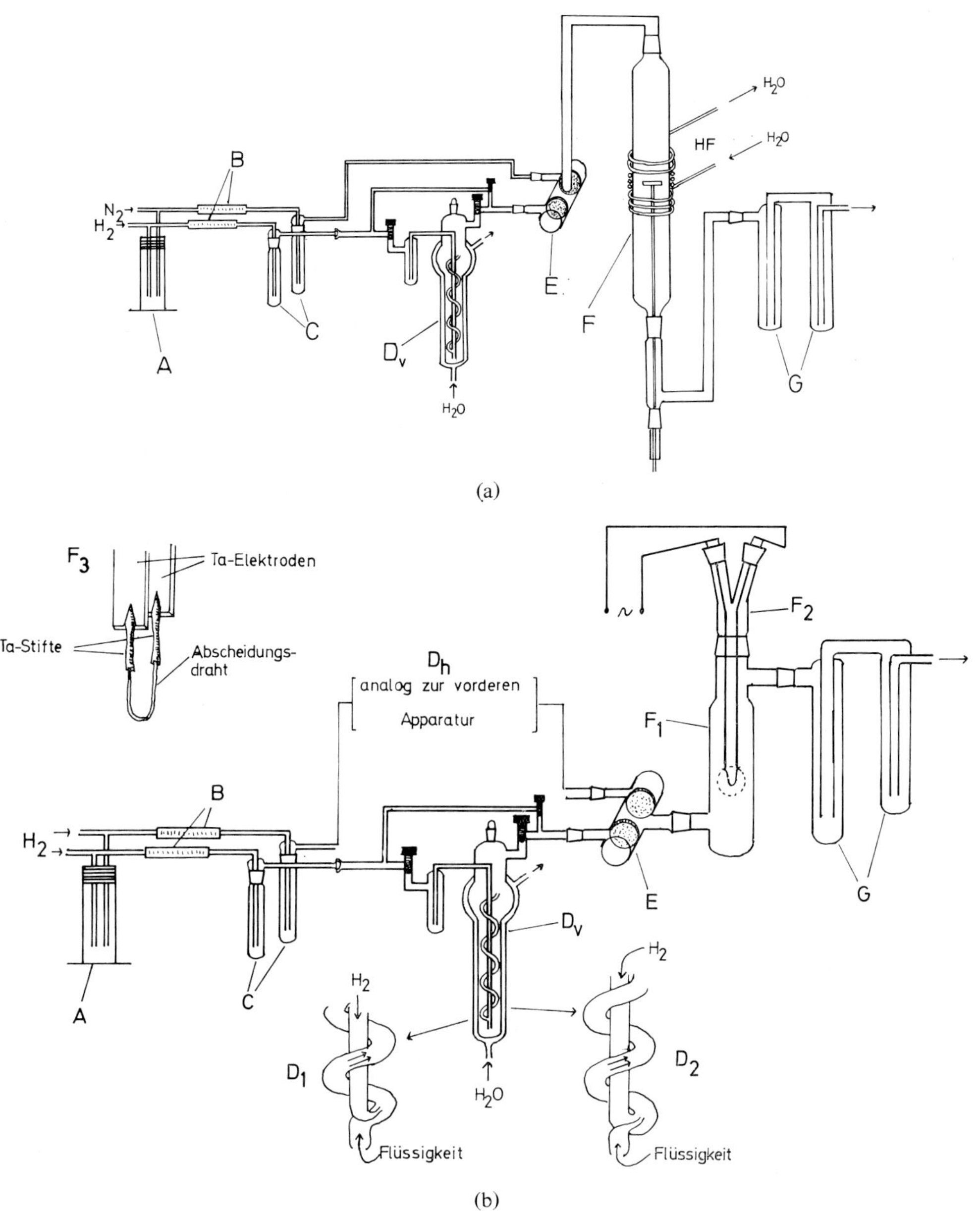

Abb. 1. Apparatur zur pyrolytischen Darstellung von tetragonalem Bornitrid (a) und rhomboedrischem Borphosphid (b).

G Auf $-78\,^\circ$C gekühlte Ausfrierfallen, in denen nicht umgesetzte Reaktionspartner und flüchtige Zersetzungsprodukte aufgefangen werden.

2.4. VERSUCHSFÜHRUNG

Der Wasserstoff dient in unseren Reaktionssystemen gleichzeitig als Trägergas und als Reduktionsmittel für die Ausgangsverbindungen BBr_3 und PBr_3. Er belädt sich auf seinem Weg durch die Vorratsgefäße D_v und D_h mit einer dem Dampfdruck entsprechenden Menge an dampfförmigen Reaktanten. (N_2 wird direkt in das Mischgefäß E eingeleitet.) Um ein Kondensieren des Dampfes nach Austritt aus den Vorratsgefäßen zu verhindern, werden die Rohrleitungen bis zum Reaktionsraum mit Heizdrähten auf eine Temperatur von etwa 60 °C erwärmt. Vor ihrem Eintritt in das Reaktionsgefäß F werden die Reaktionspartner im Zwischenstück E sorgfältig miteinander vermischt, um eine gleichmäßige Zusammensetzung des aufwachsenden Materials zu gewährleisten. Bei der Mischung von BBr_3- und PBr_3-Dampf bildet sich die Additionsverbindung $BBr_3 \cdot PBr_3$ in Form farbloser Kristalle, die aber bei den durch die Zusatzheizung erreichten Temperaturen flüchtig sind und mit dem Gasstrom mitgeführt werden.

Die Reaktionsbedingungen zur Darstellung von Bor-Stickstoff- und Bor-Phosphor-Phasen wurden so gewählt, daß sie ohne Zusatz der N- bzw. P-Komponente bei 1150 °C zur Bildung von kristallinem II-tetragonalem Bor, einer Modfikation des reinen elementaren Bors, führten[14]). Durch sukzessive Erhöhung des N_2- bzw. PBr_3-Anteils im gasförmigen Reaktionsgemisch können dann Pyrolyseprodukte mit steigendem Stickstoff- bzw. Phosphorgehalt entstehen. Das Molverhältnis $BBr_3 : N_2$ bzw. $BBr_3 : PBr_3$ im Gasraum wird durch geeignete Wahl der Dampfdrücke und der H_2- (bzw. N_2-) Strömungsgeschwindigkeiten eingestellt.

Analysen der B–N-Phasen: Zur Stickstoffanalyse wurde das Verfahren von Syavtsillo et al.[18]) für N in Si_3N_4 für das schwerer aufschließbare Borid abgeändert. Das pulverisierte Material wurde mit LiOH im Argonstrom bei 750 °C aufgeschlossen und entstehendes NH_3 volumetrisch bestimmt. Zur Bestimmung des Borgehalts wurde die Substanz mit rauchender HNO_3 aufgeschlossen und die Borsäure mittels Mannit auf pH = 7.8 titriert. Analysen der B–P-Phasen: Das pulverisierte Abscheidungsprodukt wurde in einer Natriumperoxid-Schmelze aufgeschlossen und der Phosphor nach dem Verfahren von Lieb und Kuhn als Ammoniummolybdatophosphat gefällt und gravimetrisch bestimmt. Das Bor ließ sich volumetrisch durch Titration der Borsäure (Zusatz von Mannit) mit Natronlauge bestimmen.

3. Beschreibung und Interpretation der Ergebnisse

3.1. TETRAGONALES BORNITRID

Durch Pyrolyse von BBr_3–N_2–H_2-Gemischen unterschiedlicher Zusammensetzung konnten die in Tabelle 1 aufgeführten kristallinen Bor-Stickstoff-Phasen dargestellt werden. Die untere Temperaturgrenze ergibt sich daraus, daß erst ab 1150 °C für die Dotierung des Borgitters geeignete N-Spezies aus dem molekularen Stickstoff gebildet werden. Bei der oberen Temperatur-

grenze von 1600 °C entsteht neben dem tetragonalen Bornitrid hexagonales Bornitrid BN in Form von weißen Fäden.

Ohne Zusatz von Fremdgas entsteht bei 1150 °C reproduzierbar das II-tetragonale Bor in kristalliner Form. Eine geringe Menge molekularer Stickstoff induziert jedoch das β-rhomboedrische Gitter, B_{105}[19]), wie das analog bei Anwesenheit geringer Mengen CH_4 im Synthesegas der Fall war[1]). Ob das β-Bor statistisch Stickstoff aufgenommen hat, kann mit den bisherigen Analysenmethoden noch nicht entschieden werden.

Bei 1150 °C und höherem N_2-Anteil im Synthesegas setzt die Bildung in sich homogener Abscheidungen von tetragonalem Bornitrid ein. Nach Tabelle 1 nimmt mit steigender Bildungstemperatur der Gehalt an Stickstoff von 2.40 bei 1150 °C bis 4.50 Gew.% bei 1400 °C zu; synchron dazu steigen die Dichten von 2.404 bis 2.460 g cm^{-3} und die Gitterkonstanten: a von 8.79 bis 9.02 Å und c von 5.08 bis 5.26 Å.

Morphologisch besonders gut ausgebildete, metallisch glänzende Kristalle des tetragonalen Bornitrids wachsen bei 1400 °C in Form von Stäbchen auf der heißen Substratoberfläche auf. Die [001]-Stabachse steht dabei etwa parallel zum Temperaturgradienten senkrecht auf der Unterlage. Die bis zu 1 mm langen Kristalle mit einem Durchmesser von 0.4 mm werden von (110)-Flächen als Prismenflächen begrenzt.

Auf Grund der bisher vorliegenden Röntgenuntersuchungen leitet sich die Struktur des tetragonalen Bornitrids vom sogenannten "I-tetragonalen Bor" ab. Das I-tetragonale Grundgerüst (Abb. 2) mit 4 B_{12}-Ikosaedern und 2 B-Einzelatomen pro Elementarzelle [= $(B_{12})_4B_2$] bleibt erhalten, und die Stickstoffatome besetzen ein oder zwei der größten noch freien Tetraederlücken. Nach der Strukturuntersuchung von Hoard et al.[20]) am $B_{48}B_2$ sind die B_{12}-Ikosaeder (zentriert in

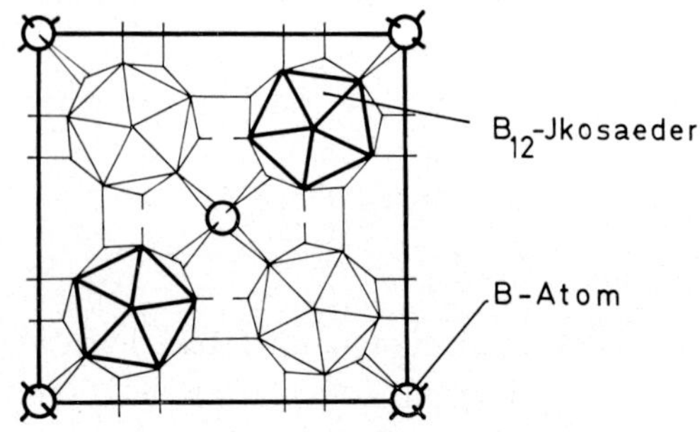

Abb. 2. Die Elementarzelle des "I-tetragonalen Bors" $B_{48}B_2$ nach Hoard et al.[20]) in Richtung der c-Achse projiziert.

TABELLE 1

Präparative und kristallchemische Daten der dargestellten Bor-Stickstoff-Phasen

Molverhältnis $N_2:BBr_3$ im Reaktionsgemisch	Substrattemperatur (°C)	Aufgewachsenes Material			Beobachtete Gitterkonstanten		Tetragonales a/c-Verhältnis	Volumen der Elementarzelle ($Å^3$)	Berechnete Dichte (g cm^{-3})	Kristallchemische Zusammensetzung
		Beobachtete Dichte (g cm^{-3})	Gefundener N-Gehalt (Gew.%)	Analytische Zusammensetzung	a (Å)	c (Å)				
kein N_2	1150	2.36	–	reines B	*tetragonal* 10.12	14.14			2.36	B_{192}
1:5	1150	2.34	<0.2	fast reines B	*hexagonal* 10.94	23.81			2.30	B_{105}
10:1	1150	2.404	2.40	$B_{53}N$	*tetragonal* 8.79	5.08	1.730	392	2.405	$B_{48}B_2(BN)B_{0.2}$
10:1	1200	2.424	3.24	$B_{39}N$	8.80	5.09	1.729	394	2.425	$B_{48}B_2(B_{0.7}N_{1.3})B_{0.8}$
10:1	1300	2.452	4.27	$B_{29}N$	8.87	5.13	1.729	404	2.451	$B_{48}B_2(B_{0.2}N_{1.8})B_{2.6}$
10:1	1400	2.460	4.50	$B_{28}N$	9.02	5.26	1.715	428	2.459	$B_{48}B_2(N_2)B_6$

TABELLE 2

Präparative und kristallchemische Daten der dargestellten Bor–Phosphor-Phasen (Substrattemperatur: 1150 °C)

Strömungsgeschwindigkeit des H_2 in l/Std. durch BBr$_3$	Strömungsgeschwindigkeit des H_2 in l/Std. durch PBr$_3$	Molverhältnis PBr$_3$:BBr$_3$ im Reaktionsgemisch	Aufgewachsenes Material			Beobachtete Gitterkonstanten (Å)	Hexagonales a/c-Verhältnis	Volumen der rhomboedrischen Elementarzelle (Å)	Berechnete Dichte (g cm^{-3})	Kristallchemische Zusammensetzung
			Beobachtete Dichte (g cm^{-3})	Gefundener P-Gehalt (Gew%)	Analytische Zusammensetzung					
9	0	kein PBr$_3$	2.36	–	reines B	tetrag. $a = 10.12; c = 14.14$			2.36	B_{192}
6	3	1:100	2.36	<0.1	fast reines B	hex. $a = 10.94; c = 23.81$ rh. $a = 10.14; \alpha = 65°17'$			2.30	B_{105}
6	3	1:45	2.42	5.6	$B_{48}P$	hex. $a = 5.83; c = 11.71$ rh. $a = 5.15; \alpha = 68.90°$	0.498	114.9	2.43	$B_{12}B_2(B_{0.7}P_{0.3})$
6	3	1:15	2.47	9.8	$B_{26}P$	hex. $a = 5.84; c = 11.75$ rh. $a = 5.17; \alpha = 68.92°$	0.497	115.8	2.48	$B_{12}B_2(B_{0.45}P_{0.55})$
5	4	1:5	2.48	19.3	$B_{12}P$	hex. $a = 5.97; c = 11.94$ rh. $a = 5.26; \alpha = 69.08°$	0.499	122.8	2.49	$B_{12}BP(B_{0.9}P_{0.1})$
3	6	1:2	2.56	24.1	B_9P	hex. $a = 5.99; c = 11.98$ rh. $a = 5.29; \alpha = 69.80°$	0.499	124.1	2.57	$B_{12}BP(B_{0.5}P_{0.5})$

den Lagen $\frac{1}{4}, \frac{1}{4}, \frac{1}{4}; \frac{1}{4}, \frac{3}{4}, \frac{3}{4}; \frac{3}{4}, \frac{1}{4}, \frac{3}{4}$ und $\frac{3}{4}, \frac{3}{4}, \frac{1}{4}$) über je zehn interikosaedrische Bindungen miteinander verknüpft und bilden eine tetragonal verzerrte raumzentrierte Packung. Die restlichen zwei Bindungen pro Ikosaeder werden zu Einzelatomen betätigt, die tetraedrische Lücken (verzerrt zu Biphenoiden der Symmetrie $\overline{4}2$ m) besetzen. In $B_{48}B_2$ sind solche Einzelatome zwei Boratome pro Zelle, die die Positionen $0, 0, 0$ und $\frac{1}{2}, \frac{1}{2}, \frac{1}{2}$ besetzen. In den von uns dargestellten tetragonalen Bornitriden besetzen bis zu zwei Stickstoff- und sechs (zusätzliche) Boratome weitere Tetraederlücken im Ikosaedergerüst und führen zu einer dreidimensionalen Gitteraufweitung (Zunahme des Zellvolumens um etwa 10%). Zu der Frage, welche Plätze die Einzelatome im Ikosaedergerüst besetzen, läßt sich folgendes aussagen: Wären nur die vier größten der insgesamt acht Tetraederlücken pro Elementarzelle ($0, 0, 0; \frac{1}{2}, \frac{1}{2}, \frac{1}{2}; 0, 0, \frac{1}{2}$ und $\frac{1}{2}, \frac{1}{2}, 0$) symmetrisch durch je zwei B- und zwei N-Atome besetzt (ideale Zusammensetzung $B_{48}B_2N_2$), so würde man die Raumgruppe $P4_2/nnm$ erwarten, die auch für das $B_{48}B_2$ gefunden wurde[20]). Zusätzliche schwache Röntgenreflexe bei Einkristallaufnahmen tetragonaler Bornitride – $hk0$, $h+k$ ungerade und $0kl$, $k+l$ ungerade – weisen jedoch auf eine Symmetrieerniedrigung im tetragonalen Gitter zur Raumgruppe $P\overline{4}2m$ hin. Auf Grund der Analysenergebnisse und der Dichtebestimmungen wird diese Verminderung der Symmetrie wahrscheinlich durch unsymmetrische Besetzung eines Teiles der kleineren Tetraederlücken durch zusätzliche Boratome verursacht. Die Besetzung dieser kleinen Lücken dürfte auch entscheidend für die Gitteraufweitung gegenüber dem $B_{48}B_2$ verantwortlich sein.

Die Abb. 2 zeigt deutlich, daß nicht alle Boratome des tetragonalen Ikosaedergerüsts die bevorzugte pentagonal pyramidale Koordination[21]) ausbilden können. Von den von jedem Ikosaeder gebildeten zwölf nach außen gerichteten Bindungen liegen nur acht in Richtung der pentagonalen Ikosaederachse, während die restlichen vier Bindungen um etwa 20° (in $B_{48}B_2$) von der fünfzähligen Achsenrichtung abweichen. Dadurch tritt im I-tetragonalen Ikosaedergerüst eine erhebliche Spannung auf, die durch den Einbau weiterer Einzelatome, besonders in die kleineren Lücken, und die dadurch bedingte starke Gitteraufweitung noch verstärkt wird. Wir nehmen an, daß mit der Zusammensetzung $B_{48}B_2N_2B_6$ und der Ver-

größerung des Zellvolumens um etwa 10% die obere Stabilitätsgrenze für das I-tetragonale Gitter erreicht ist.

Das größere Phosphoratom ($r = 1.06$ Å) dürfte daher abweichende Ikosaedergerüste induzieren, bei denen entweder von vornherein größere Lücken vorhanden sind, oder bei denen eine genügend große Gitteraufweitung ohne allzu großen Stabilitätsverlust möglich ist.

3.2. Rhomboedrisches Borphosphid

Bei einer Substrattemperatur von 1150 °C lassen sich die in Tabelle 2 aufgeführten kristallinen Bor-Phosphor-Phasen darstellen. Die Einkristalle sind allerdings klein (Länge: < 0.1 mm).

Ohne Zusatz von Fremdgasen entsteht durch Pyrolyse von BBr_3-H_2-Gemischen reproduzierbar das II-tetragonale Bor in kristalliner Form. Bereits 1 Mol% PBr_3 im Gemisch der gasförmigen Reaktanten reicht jedoch aus, um ein anderes Gitter des elementaren Bors zu induzieren: das β-rhomboedrische Bor[19]) mit geringfügig erhöhter Dichte. Obwohl diese Bormodifikation auf Grund ihrer Struktur beträchtliche Mengen an Fremdatomen in fester Lösung aufnehmen könnte, wird kein Phosphor eingebaut. Die geringe Menge an analytisch nachgewiesenem Phosphor dürfte höchstens für die Bildung von P-haltigen Kristallkeimen ausreichen, die dann als Matrix für das weitere Kristallwachstum von nahezu reinem β-Bor dienen.

Wie aus Tabelle 2 hervorgeht, nimmt bei sukzessiver Erhöhung des PBr_3-Anteils im gasförmigen Reaktionsgemisch der Phosphorgehalt in den aufgewachsenen B–P-Phasen von 5.6 bis 24.1 Gew.% zu; gleichzeitig steigen die Dichten von 2.42 bis 2.56 g cm^{-3} und die Gitterkonstanten: hexagonal: a von 5.83 bis 5.99 Å und c von 11.71 bis 11.98 Å (rhomboedrisch: a von 5.15 bis 5.29 Å und α von 68.90 bis 69.80°). Das P:B-Atomverhältnis der Pyrolyseprodukte bleibt allerdings weit hinter dem P:B-Verhältnis im Synthesegas zurück, denn mit steigendem PBr_3-Gehalt reagiert ein immer größer werdender Teil des Phosphortribromids bereits vor Erreichen der Substratoberfläche zu polymeren Produkten (siehe unten) und steht damit nicht mehr für den Einbau in das aufwachsende Material zur Verfügung.

Die Röntgenuntersuchungen zeigen, daß die sehr unterschiedlich zusammengesetzten Produkte sich von derselben Grundstruktur ableiten. Bereits etwa 2 Atom% Phosphor reichen aus, um ein rhomboedrisches Gitter, das sich von der Struktur des Borcarbids[12,13]) ableitet, zu induzieren. Mögliche Raumgruppen auf Grund der systematischen Auslöschungen: $R\bar{3}m$, $R\bar{3}$, R3m, R32, R3.

Die Elementarzelle des Borcarbids ist schematisch in Abb. 3 dargestellt. An jedem Eckpunkt der primitiven rhomboedrischen Zelle ist ein B_{12}-Ikosaeder (Kantenlänge etwa 1.8 Å) zentriert und so orientiert, daß genau die Hälfte aller ikosaedrischen Boratome an direkten interikosaedrischen Bindungen beteiligt sind. In dem resultierenden starren dreidimensionalen Gerüst existieren relativ große, langgestreckte Hohlräume im Zentrum der Rhomboederzelle in Richtung der Hauptraumdiagonalen [111] (Positionen 1b plus 2c in Abb. 3), die im Falle des Borcarbids von einer C_3- bzw. CBC-Kette und im Falle der rhomboedrischen Bor-Phosphor-Phase von B- und P-Atomen besetzt sind. Die Besetzung der Positionen 2c durch Bor und/ oder Phosphoratome ist für die Stabilisierung des rhomboedrischen Ikosaedergerüsts von entscheidender Bedeutung, denn nur über sie kann die zweite Hälfte

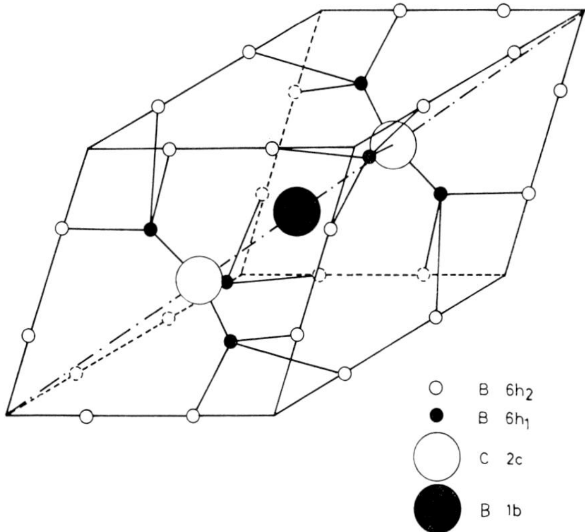

Abb. 3.   Die primitive rhomboedrische Elementarzelle des Borcarbids $B_{12}CBC$. Jede Rhomboeder-Ecke ist durch Ikosaeder-fragmente besetzt.

der ikosaedrischen Boratome miteinander verknüpft werden. Weiterhin enthält die Zelle je eine interstitielle Lücke oberhalb und unterhalb der Positionen 2c auf der Gittergeraden [111], die von B- oder P-Atomen besetzt werden können.

TABELLE 3

Nach dem Debye–Scherrer-Verfahren gewonnene Röntgenbeugungsdaten für die tetragonalen Bornitride der Zusammensetzungen $B_{48}B_2(BN)B_{0.2}$ bis $B_{48}B_2(N_2)B_6$

$B_{48}B_2(BN)B_{0.2}$		$B_{48}B_2(B_{0.7}N_{1.3})B_{0.8}$		$B_{48}B_2(B_{0.2}N_{1.8})B_{2.6}$		$B_{48}B_2(N_2)B_6$		hkl
d (Å)	I_{rel}	d (Å)	I_{rel}	d (Å)	I_{rel}	d (Å)	I_{rel}	
–	–	–	–	6.12	1	6.06	1	110
4.39	10	4.38	10	4.37	10	4.34	9	200
3.92	10	3.90	10	3.91	10	3.90	10	111
3.11	5	3.10	5	3.10	5	3.09	5	121
2.54	7	2.52	8	2.54	8	2.52	10	002
								031
2.43	10	2.42	10	2.41	10	2.41	10	131
2.19	5	2.18	6	2.18	5	2.17	5	231
2.01	1	–	–	–	–	–	–	041
1.963	2	1.965	3	1.945	1	1.939	1	141
1.909	2	1.905	3	1.888	1	1.886	1	331
1.745	1	–	–	–	–	–	–	232
1.657	1	1.654	2	1.643	3	1.649	4	042
								051
1.632	1	1.625	2	1.627	1	–	–	113
								142
1.545	2	1.545	2	1.530	1	–	–	123
								251
1.444	7	1.439	7	1.447	7	1.449	7	133
								342
1.388	3	1.385	2	–	–	–	–	260
								161
1.370	2	1.367	2	1.368	3	1.370	4	252

TABELLE 4

Nach dem Debye–Scherrer-Verfahren gewonnene Röntgenbeugungsdaten für die rhomboedrischen Borphosphide der Zusammensetzungen $B_{12}B_2(B_{0.7}P_{0.3})$ bis $B_{12}BP(B_{0.5}P_{0.5})$

| $B_{12}B_2(B_{0.7}P_{0.3})$ | | $B_{12}B_2(B_{0.45}P_{0.55})$ | | $B_{12}BP(B_{0.9}P_{0.1})$ | | $B_{12}BP(B_{0.5}P_{0.5})$ | | hkl |
d (Å)	I_{rel}	d (Å)	I_{rel}	d (Å)	I_{rel}	d (Å)	I_{rel}	
4.67	4	4.69	2	4.74	1	4.77	2	01.1
3.86	10	3.90	9	3.91	9	3.92	9	01.2
2.95	4	2.96	4	2.98	7	2.99	6	11.0
2.56	10	2.58	8	2.57	10	2.58	9	01.4
2.48	10	2.51	10	2.53	10	2.52	10	02.1
2.35	2	2.40	1	2.39	1	2.39	1	11.3 00.5
2.17	2	2.16	2	2.16	6	2.16	5	02.3 01.5
1.979	2	1.979	1	1.979	1	1.979	1	00.6
1.894	4	1.909	5	1.924	5	1.927	5	12.1
1.734	2	–	–	–	–	–	–	12.3 02.5
1.641	5	1.649	4	1.649	6	1.649	7	03.2 11.6
–	–	1.630	1	1.630	1	–	–	12.4
1.552	5	1.569	2	1.579	1	1.589	4	03.3 02.6
1.491	7	1.502	7	1.507	8	1.506	8	12.5
1.470	7	1.478	7	1.487	5	1.491	4	22.0
1.403	5	1.412	6	1.422	6	1.422	6	13.1
1.372	6	1.383	6	1.393	5	1.399	5	12.6 13.2
1.285	5	1.290	4	1.301	4	1.304	3	03.6
–	–	–	–	1.289	3	1.289	4	04.0 13.4
1.265	1	–	–	–	–	–	–	04.1
1.241	1	–	–	–	–	–	–	04.2
1.212	1	–	–	1.228	3	1.228	3	04.3 13.5
–	–	–	–	1.193	1	1.193	2	22.6
–	–	–	–	1.159	2	–	–	23.2 13.6

In der Elementarzelle des rhomboedrischen Borphosphids stehen also insgesamt fünf Gitterplätze für die Besetzung durch Einzel-Bor- oder-Phosphoratome zur Verfügung. Die Positionen 2c müssen ständig besetzt sein (Stabilisierung der Ikosaeder-Querverbindungen), die Position 1b und die beiden interstitiellen Lücken können je nach den Reaktionsbedingungen besetzt werden oder frei bleiben. Diese vielfältigen Möglichkeiten bedingen die Phasenbreite des rhomboedrischen Borphosphids mit seinem weiten Zusammensetzungsbereich von $B_{48}P = B_{12}B_2(B_{0.7}P_{0.3})$ bis $B_9P = B_{12}BP(B_{0.5}P_{0.5})$ [bzw. bis $B_{12}PBP$ nach Matkovich[6]]. Um dieses Gitter zu induzieren, genügt es schon, wenn analytisch gesehen nur für etwa jede dritte Elementarzelle ein Phosphoratom zum Einbau zur Verfügung steht: $B_{12}B_2(B_{0.7}P_{0.3})$. Steigender Phosphorgehalt führt zu einer dreidimensionalen Gitteraufweitung, das Volumen der Elementarzelle nimmt um etwa 10% zu. Gegenüber dem Borcarbid, das von allen bisher bekannten Ikosaederanordnungen am besten die bevorzugte Koordinationsgeometrie für jedes Boratom erreicht[21]), beträgt die Zunahme des Elementarvolumens etwa 15%. Im Vergleich mit dem I-tetragonalen Gitter scheint der durch die Gitteraufweitung bedingte Stabilitätsverlust bei rhomboedrischer Ikosaederanordnung geringer zu sein, und die großen Phosphoratome können ohne Schwierigkeit eingebaut werden.

Die in Tabelle 2 angeführten kristallchemischen Zusammensetzungen für rhomboedrisches Borphosphid ergeben sich aus den Analysendaten, den Gitterkonstanten und den beobachteten Dichten. Neben dem B_{12}-Ikosaeder sind in jeder Elementarzelle noch weitere drei Gitterplätze besetzt. Welche der oben diskutierten fünf Möglichkeiten dabei realisiert werden, kann erst eine genaue Röntgenstrukturanalyse an genügend großen Einkristallen ergeben.

Nach den Ergebnissen unserer Untersuchungen kann das rhomboedrische Borphosphid als nichtstöchiometrische Verbindung unter Erhaltung seiner wesentlichen Strukturmerkmale in einem weiten Zusammensetzungsbereich existieren. Es ist daher müßig, die Frage nach der "richtigen" Zusammensetzung[6-10] dieser Phase zu stellen. Die vielfältigen Möglichkeiten zur Platzbesetzung und zum Austausch zwischen Bor- und Phosphor-Einzelatomen haben zur Folge, daß die Dichte und die analytische Zusammensetzung der aufwachsenden Reaktionsprodukte sehr empfindlich von den Reaktionsbedingungen, vor allem vom P-Partialdruck im Reaktionsraum abhängen. Letzlich hängt es nämlich vom Phosphordruck nahe der glühenden Substratoberfläche ab, ob einer der fünf verfügbaren Gitterplätze von Bor oder von Phosphor besetzt wird. Es ist anzunehmen, daß bei einem zu geringen P-Partialdruck zunächst eingebaute Phosphoratome aus dem Gitter hinausdiffundieren und durch Leerstellen oder Boratome ersetzt werden. Auf diese Weise kann die Entstehung aller Produkte mit rhomboedrischer Struktur mit weniger als etwa 18 Gew.% Phosphor, also weniger als ein P-Atom pro Elementarzelle, erklärt werden.

Auffallend bei der Gasphasenreaktion zur Darstellung von B–P-Phasen sind die gelben bis braunen Zersetzungsprodukte, die sich an der Wand des Zersetzungsgefäßes niederschlagen und deren Menge bei gleichzeitiger Farbvertiefung mit steigendem PBr_3-Anteil im Reaktionsgemisch zunimmt. Dieses in gebräuchlichen Lösungsmitteln unlösliche Substanzgemisch besteht zum größten Teil aus polymeren Produkten: Polyphosphane der Zusammensetzung P_nH_{n+2}, P_nH_n, P_nH_{n-2}, P_nH_{n-4} usw. bis herab zum elementaren Phosphor, die sich aus intermediär entstehendem P_2H_4 bilden[22]); hochpolymere Halogenide $(P_3Br)_x$, die aus P_2H_4 und Br_2 oder PH_4Br und PBr_5 entstehen; und gemischte Polymere der Zusammensetzung $(H_2PBHBr)_n$, die bei der Addition von B_2H_5Br an PH_3 gebildet werden[23]). Daneben sind geringe Mengen an PH_3, PH_4Br und höheren Borhalogeniden bzw. Borwasserstoffen in dem Gemisch enthalten.

Danksagung

Der Deutschen Forschungsgemainschaft und dem Fonds der Chemischen Industrie danken wir für die wertvolle finanzielle Hilfe.

Literatur

1) K. Ploog und E. Amberger, J. Less-Common Metals **23** (1971) 33.
2) H. C. Longuett-Higgins und M. de V. Roberts, Proc. Roy. Soc. (London) A **230** (1955) 110.
3) W. N. Lipscomb und D. Britton, J. Chem. Phys. **33** (1960) 275.
4) C. E. Frahme, Dissert. Abstracts **28** (1968) 885-B.
5) W. Kischio, Z. Anorg. Allg. Chem. **349** (1967) 151.
6) V. I. Matkovich, Acta Cryst. **14** (1961) 93.
7) L. H. Spinar und C. C. Wang, Acta Cryst. **15** (1962) 1048.
8) J. L. Peret, J. Am. Ceram. Soc. **47** (1964) 44.
9) R. I. Stearns und P. E. Greene, J. Elektrochem. Soc. **112** (1965) 1239.
10) R. A. Burmeister, Jr. und P. E. Greene, Trans. Met. Soc. AIME **239** (1967) 408.
11) B. F. Decker und J. S. Kasper, Acta Cryst. **12** (1959) 503.
12) G. S. Zhdanov und N. G. Sevast'yanov, Compt. Rend. Acad. Sci. USSR **32** (1941) 432.
13) H. K. Clark und J. L. Hoard, J. Am. Chem. Soc. **65** (1943) 2115.
14) E. Amberger und K. Ploog, J. Less-Common Metals **23** (1971) 21.
15) K. Ploog, unveröffentlicht.
16) Yu. D. Chistyakov, Yu. N. Tikkonov und Ya. M. Pinchuk, Tr. Vses. Elektrotekh. Inst. No. 75 (1967) 74.
17) *JANAF Thermochemical Tables* (Dow Chemical Company, Midland, Michigan, 1964).
18) S. V. Syavtsillo, A. M. Nikol'skaya und T. E. Mashko, Chem. Abstr. **64** (1965) 14955b.
19) R. E. Hughes, C. H. L. Kennard, D. B. Sullenger, H. A. Weakliem, D. E. Sands und J. L. Hoard, J. Am. Chem. Soc. **85** (1963) 361.
20) J. L. Hoard, R. E. Hughes und D. E. Sands, J. Am. Chem. Soc. **80** (1958) 4507.
21) J. L. Hoard und R. E. Hughes, in: *The Chemistry of Boron and its Compounds*, Ed. E. L. Muetterties (Wiley, New York, 1967) S. 25ff.
22) E. Wiberg, M. van Ghemen und G. Müller-Schiedmayer, Angew. Chem. **75** (1963) 814.
23) J. E. Drake und J. Sipson, Inorg. Nucl. Chem. Letters **3** (1967) 87.

       Journal of Crystal Growth **13/14** (1972) 360–364 © *North-Holland Publishing Co.*

VAPOR EPITAXY OF GALLIUM NITRIDE

M. ILEGEMS

Bell Telephone Laboratories, Incorporated, Murray Hill, New Jersey 07974, U.S.A.

The epitaxial growth of GaN by vapor phase reaction between GaCl and NH_3 in a He carrier gas ambient is described. Single crystal layers 100–200 μm thick and ~ 1 cm^2 in area, were obtained on (0001) oriented sapphire substrates at deposition temperatures near 1050 °C. The best undoped layers grown had carrier concentrations of 1–2×10^{17} cm^{-3} with electron mobilities near 400 cm^2/V sec at 300 K.

1. Introduction

Gallium nitride, a III-V compound semiconductor with a direct fundamental energy gap near 3.5 eV[1,2], is a presently very promising compound for electroluminescent and lasing applications[3]. Much recent progress has been made in the growth of single crystal material in needle form[4] or epitaxially on sapphire[5-7], although the chemistry of this compound is still incompletely understood. Significant in this respect are the large discrepancies between different observations concerning the stability of GaN at high temperatures[8,9], as well as the fact that the origin of the high n-type conductivity of most nonintentionally doped GaN grown to date has not yet been established. The understanding of this latter problem is of course extremely important as our ability to obtain p-type material will determine to a large extent the technological future of the GaN compound.

This paper presents the experimental technique with which single crystalline layers of GaN on (0001) sapphire substrates, with significantly lower carrier concentrations than previously reported for undoped material[4-6], were obtained. The electrical and optical properties of these layers are related to the experimental conditions during synthesis and the factors that may have a determining influence on the characteristics of vapor grown GaN are briefly examined.

2. Experimental

The vapor deposition apparatus consists of a 75 cm long horizontal single-zone furnace containing SiO_2 reaction and insert tubes with provisions for the inlet and outlet of the reagent and carrier gases. HCl gas[10], diluted in a He carrier, is passed over an Al_2O_3 boat containing liquid Ga and provides for the transport of Ga as GaCl. A second gas stream consisting of NH_3 diluted in a He carrier, bypasses the Ga charge and discharges directly near the center of the reaction apparatus. The reaction of formation of GaN occurs immediately upon mixing of the (GaCl + He) and (NH$_3$ + He) flows; the actual deposition zone is approximately 5 cm long and the temperature over this zone is essentially constant. Several substrates, each 1 cm^2 in area, are used per run and are placed lengthwise along the deposition zone. Considerable variation in physical properties and surface morphology are observed between layers grown in a single run as a function of the exact positioning of the substrates, with the layers closest to the GaCl:NH$_3$ mixing zone invariably exhibiting preferable characteristics as discussed further below.

3. Growth conditions

Growth parameters adopted empirically were (1) a reaction zone temperature in the range 1040–1060 °C, (2) a Ga source temperature of ~ 900 °C and (3) flow rates in the ranges 1.6–1.8 ml/min for HCl, 360–400 ml/min for NH$_3$, ~ 1000 ml/min (HCl carrier) and ~ 400 ml/min (NH$_3$ carrier) for He. The heaviest GaN deposits are always observed in regions closest to the GaCl:NH$_3$ mixing zone. The thickness of these depos-

Fig. 1. Interference contrast micrograph of thin ($\sim$ 25 μm) as-grown GaN surface on (0001) Al$_2$O$_3$. Magnification 160 $\times$

Fig. 2. Interference contrast micrograph of thick ($\sim$ 140 μm) as-grown GaN surface on (0001) Al$_2$O$_3$. Magnification 160 $\times$.

its, after a standard 3 hr growth run, is, for the first substrate, in the 100–200 μm range and is practically uniform over the entire substrate area. For substrates farther downstream in the deposition zone the layer thickness decreases rapidly with distance from the GaCl:NH$_3$ mixing zone and is down to 5–20 μm for the last substrate showing complete coverage. Growth ceases completely at distances from 6 to 10 cm downstream of the mixing zone. Substrates placed immediately upstream of the NH$_3$ outlet also show no growth, indicating that upstream diffusion of NH$_3$ is negligible.

The grown layers are transparent and uncolored when undoped. Upon cooling and as a result of thermal expansion mismatch, the 250 μm thick Al$_2$O$_3$ substrates used generally develop cracks when the GaN layer thickness exceeds $\sim$ 100 μm. In most cases the fracture of the Al$_2$O$_3$ substrate does not extend into the epitaxially grown layer.

Powder X-ray diffraction patterns on material removed from the substrates establish that the material grown is hexagonal GaN having the wurtzite structure. Laue diffraction patterns and high energy electron diffraction patterns of layers grown on (0001) oriented sapphire show that these are single crystal of like orientation. The polarization effects observed in reflectance and photoluminescence studies[1]) on these layers also support the (0001) orientation assignment.

The growth morphology of the epitaxial layers varies considerably as a function of the position of the substrates in the deposition zone as illustrated by figs. 1 and 2. Fig. 1 is typical for the thin (5–50 μm) layers obtained on the substrates far downstreeam from the gas mixing zone. These layers often exhibit a pyramidal growth structure of hexagonal symmetry and with clearly defined growth steps. For deposits of intermediate thicknesses (50–100 μm) obtained typically on substrates in the middle of the reaction zone. the growth steps between successive levels of pyramidal growth become less clearly distinguishable and the surfaces have a flat or conical growth aspect with no visual external evidence of hexagonal symmetry. Finally the thickest layers (100–250 μm) grown on substrates placed immediately adjacent to the GaCl:NH$_3$ mixing zone show the layered aspect of fig. 2 and which is very similar to the structure observed on liquid or vapor epitaxially grown layers of other Ga-V compounds.

While all undoped layers obtained are n-type, the lowest carrier concentrations are invariably observed in the thickest layers grown closest to the $GaCl: NH_3$ mixing zone and having the surface structure shown in fig. 2. Such layers have been grown to date with carrier concentrations as low as $n = 2 \times 10^{17}$ cm^{-3} with mobilities of $\mu \simeq 380$ cm^2/V s at room temperature. The value of n is nearly two orders of magnitude below comparable literature values[5,6] for undoped material while the value of μ is also considerably higher than previously reported[5]. The low temperature reflectance spectra obtained on these layers are well resolved[1] and the photoluminescence spectra show well structured near-bandgap emission[1] and clearly defined donor-acceptor pair recombination bands[11] at lower energies. In contrast, the downstream samples have carrier concentrations that lie usually above 10^{18} cm^{-3} and yield broad poorly resolved reflectance, luminescence and absorption spectra. At present no definite interpretation for these related observations is available.

It has been suggested earlier[5] that the high donor concentrations commonly observed in GaN are related to a native defect, possibly a N vacancy. Our present results, while clearly showing that substantial reductions in carrier concentrations are possible over previously obtained results, are not inconsistent with such hypothesis if the observed changes in carrier concentrations are attributed to differences in stoichiometry between layers grown in different regions of the reaction zone. Other factors that could contribute to the increase in carrier concentration observed are strain effects, resulting from the lattice mismatch between Al_2O_3 and GaN [approximately 13% along a in the (0001) plane[7]], or the presence of donor impurities. The magnitude of the variations in carrier concentrations observed in samples from a same growth run argues, however, against the impurity hypothesis. Also, the impurity concentrations obtained by emission spectrographic analysis were significantly lower than the measured carrier concentrations in these layers. Likewise, while strain effects may contribute significantly to the broadening of the optical spectra taken on very thin samples, they are not expected to have a marked influence on the electrical characteristics of the material.

4. Crystal doping

Preliminary efforts to obtain type conversion in GaN by doping with acceptor impurities have not been successful. The group II elements Zn and Cd are readily introduced from the vapor during growth. Their presence does not, however, appear to alter the carrier concentration of the grown layers appreciably. In photoluminescence, the main feature introduced by these elements is a broad emission band approximately 0.6 eV below bandgap, which may be due to the formation of complex centers. At the highest Zn concentration investigated (Zn transport rate $\simeq 100$ µmole/min), the Zn-doped layers have a yellowish tint. Attempts to introduce Zn by diffusion were not successful and Zn was not detectable in photoluminescence following closed ampoule vapor diffusions into preselected substrates at temperatures up to 1000 °C and times up to 36 hr. Doping with Si and Ge was attempted by introducing these impurities in elemental form in a high temperature region of the growth system. At the doping levels achieved in this manner (transport rates $\simeq 1$–4 µmole/min), no compensation or type conversion effects were observed.

Conductive p-type behavior in GaN using Ge as the dopant has been previously reported[5]. These initial results, which were ill-reproducible, have not however been confirmed and appear presently in doubt. While the evidence is inconclusive, the present doping experiments indicate that, even though significant advances have been made in improving the quality of the material available, the question of whether p-type behavior will be obtainable is still unresolved.

5. Phase equilibria considerations

The dominant reactions prevailing throughout the reaction zone illustrated schematically in fig. 3 are the reaction of formation of GaN:

$$GaCl\,(g) + NH_3\,(g) \rightarrow GaN\,(s) + HCl\,(g) + H_2\,(g), \quad (1)$$

and the dissociation reaction of NH_3:

$$NH_3\,(g) \rightarrow \tfrac{1}{2}N_2\,(g) + \tfrac{3}{2}H_2\,(g). \quad (2)$$

The equilibrium constant for reaction (1) can be calculated from available thermochemical data[12] for HCl and NH_3, and from calculated free energies of formation for GaCl and GaN based upon the pres-

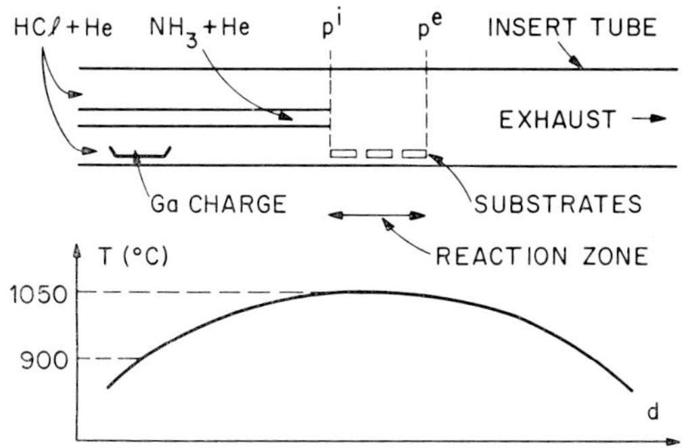

Fig. 3. Schematic of deposition apparatus with indication of temperature conditions (not to scale).

ently accepted values of the thermodynamic properties ref. 12–14) of these compounds and their constituent elements. For GaN, for which no entropy or heat content data are available, the corresponding values tabulated for ZnO, an isoelectronic compound of nearly equal molecular weight (81.38 for ZnO as compared to 83.73 for GaN), were adopted following a previously established procedure[8]). The resulting free energies of formation are, at 1050 °C,

$$\Delta G^f_{GaN} = 6.6 \text{ kcal/mole}, \quad \Delta G^f_{GaCl} = -40.0 \text{ kcal/mole},$$

Combining these values with the tabulated[12]) energies of formation of HCl and NH$_3$ yields

$$\Delta G \,(1050\ °C) = -2.2 \text{ kcal/mole}$$

for reaction (1) and, assuming unit activity coefficients throughout:

$$K_1 = (p^e_{HCl} p^e_{H_2})/(p^e_{GaCl} p^e_{NH_3}) \simeq 2.4, \tag{3}$$

where p^e designates the equilibrium partial pressure.

The dissociation of NH$_3$ is rate limited and equilibrium considerations do therefore not apply to reaction (2). This property precludes an exact solution of the reaction equilibria. However, on the basis of measurements of the degree of NH$_3$ dissociation observed under flow rate and temperature conditions comparable to those prevailing during GaN growth, we believe that the dissociation of NH$_3$ is probably below 1 % at the input end of the reaction zone (fig. 3). This indicates that NH$_3$ is present throughout the reaction zone mainly in its metastable, nondissociated form with a partial pressure close to its input partial pressure.

Under the flow conditions stated in section 3 the Ga transport rate, established for a 15 to 20 g Ga

charge, is in the range of 60 to 70 μmole/min. Taking into account the small fraction of HCl diverted from the reaction zone by the presence of the insert tube, this yields an HCl to GaCl conversion rate equal to unity, within the accuracy of the flow meter calibration.

Assuming negligible NH$_3$ decomposition and complete conversion of HCl, the partial pressures of NH$_3$ and GaCl at the entrance of the reaction zone are thus under our experimental conditions:

$$p^i_{GaCl} \simeq 1 \times 10^{-3} \text{ atm}, \quad p^i_{NH_3} \simeq 220 \times 10^{-3} \text{ atm}.$$

At the output end of the reaction zone, the partial pressures of the reagent species are related by the equilibrium relation (3) and by the mass conservation condition:

$$p^e_{GaCl} + p^e_{HCl} = p^i_{GaCl}. \tag{4}$$

Rewriting (3) under the form

$$p^e_{HCl}/p^e_{GaCl} = K_1 p^e_{NH_3}/p^e_{H_2}, \tag{5}$$

and observing that $p^e_{NH_3}/p^e_{H_2} \gg 1$ throughout the reaction zone one obtains $p^e_{HCl}/p^e_{GaCl} \gg 1$, indicating that the reaction of formation of GaN should go near completion if equilibrium conditions are reached. In practice however, presumably mainly as a result of the high flow rates adopted, the reaction may not be completed and while the total yield is difficult to evaluate because of growth on the insert tube walls, the useful yield of epitaxial GaN is usually below 10 %.

6. Conclusions

The present results demonstrate that no fundamental mechanism exists which prevents the growth of undoped GaN of reasonably low electrical conductivity. It is expected that a systematic investigation of the influence of the different growth parameters will lead to further improvements in the characteristics of the epitaxial layers as well as to an increased understanding of the chemistry of the GaN compound.

Acknowledgments

The author is indebted to R. Dingle and H. C. Montgomery for evaluation of the grown material, to C. D. Thurmond for numerous enlightening discussions and to R. B. Zetterstrom for experimental collaboration.

References

1) R. Dingle, D. D. Sell, S. E. Stokowski and M. Ilegems, Phys. Rev. B **4** (1971) 1211.
2) J. I. Pankove, H. P. Maruska and J. E. Berkeyheiser, Appl. Phys. Letters **17** (1970) 197.
3) R. Dingle, R. F. Leheny, K. L. Shaklee and R. B. Zetterstrom, Appl. Phys. Letters **19** (1971) 5.
4) R. B. Zetterstrom, J. Mater. Sci. **5** (1970) 1102.
5) H. P. Maruska and J. J. Tietjen, Appl. Phys. Letters **15** (1969) 327.
6) D. K. Wickenden, K. R. Faulkner, R. W. Brander and B. J. Isherwood, J. Crystal Growth, **9** (1971) 158.
7) B. B. Kosicki and D. Kahng, J. Vacuum Sci. Technol. **6** (1969) 593.
8) Z. A. Munir and A. W. Searcy, J. Chem. Phys. **42** (1965) 4223.
9) R. C. Schoonmaker, A. Buhl and J. Lemley, J. Phys. Chem. **69** (1965) 3455.
10) Nominal minimum purities of materials and gases used: Ga: 99.9999%, HCl: 99.99%, NH_3: 99.999%, He: 99.995 to 99.999%. The He gas is purified before use by passage through a 77 K cold trap.
11) R. Dingle and M. Ilegems, Solid State Commun. **9** (1971) 175.
12) *JANAF Thermochemical Tables*, U.S. Dept. Commerce Document PB168370 (1965).
13) D. D. Wagman, W. H. Evans, V. B. Parker, I. Halow, S. M. Bailey and R. H. Schum, National Bureau of Standards Technical Note 270–3 (1968).
14) K. K. Kelley, U.S. Bureau of Mines Bulletin 584 (1960).

Journal of Crystal Growth **13/14** (1972) 365–370 © *North-Holland Publishing Co.*

EPITAXIAL GROWTH AND PROPERTIES OF SEMICONDUCTING ScN

J. P. DISMUKES, W. M. YIM and V. S. BAN

RCA Laboratories, Princeton, New Jersey 08540, U.S.A.

Epitaxial growth of ScN has been investigated in the temperature range 750–1150 °C, using the reaction of ammonia with the volatile scandium halides produced by the reaction of HCl, HBr or HI with Sc metal. The optimum temperature range for epitaxial growth of ScN on α-Al$_2$O$_3$ was found to be 850–930 °C. The linear thermal expansion coefficient of ScN was measured to be 8.1×10^{-6}/°C, a relatively good match to α-Al$_2$O$_3$. Mass spectrometric studies suggest that the vapor species formed by the reaction of HCl with Sc metal is ScCl$_2$, and that this reacts further with NH$_3$, which is only 5–10% decomposed at 950 °C, to form ScN. As-grown ScN is n-type with electron concentrations in the range 10^{20}–10^{21} cm^{-3}, and halogen or hydrogen appears to be the principal donor. Doping with C and Si during growth did not give p-type material, nor did post-growth annealing in Mg and Zn vapor. The Hall mobility of ScN is 150 cm^2 V^{-1}sec^{-1} at 300 °K for an electron concentration of 1×10^{20} cm^{-3}, and exhibits a $T^{-1.85}$ dependence between 300 °K and 77 °K. The optical energy gap of ScN is 2.1 eV at 300 °K.

1. Introduction

Recently, we reported epitaxial growth[1]) of ScN using an open-tube reaction of Cl$_2$ or HCl with Sc metal, giving a volatile scandium chloride which further reacted with NH$_3$ to form ScN. These epitaxial layers were all high-conductivity n-type material in which chloride was identified as the most probable donor impurity.

In this work we have further investigated ScN vapor-phase growth using mostly the HCl, HBr, and HI transport of Sc metal, with particular emphasis upon the effect of halogen on the electrical and optical properties and doping behaviour in ScN. We have also studied by mass spectrometry the fundamental chemistry of vapor-phase reactions occurring during growth of ScN.

2. Experimental

Growth of ScN on α-Al$_2$O$_3$ and several other substrates was investigated with the previously described apparatus[1]) over the temperature range 750–1150 °C, principally using HCl, HBr, and HI to transport the Sc metal, but occasionally using anhydrous, freshly prepared ScCl$_3$[2]) transported in He. Representative conditions for growth of ScN are shown in tables 1 and 2.

The vapor-phase compositions during growth were examined using a separate but similar growth system described in detail by Ban[3]), to which was coupled a time-of-flight mass spectrometer.

Vapor-deposited ScN was analyzed chemically for scandium by EDTA titration[4]) and for nitrogen by conversion of the nitride to nitrate in fused NaOH and determination of the nitrate concentration by measuring the electrode potential of the nitrate ion with an ion-

TABLE 1

Effect of temperature on growth and electrical proporties of ScN[a]

Sample No.	Growth temperature (°C)	Growth rate (μm/hr)	n (10^{20} cm^{-3})	μ (cm^2 V^{-1} sec^{-1})
91[b,p]	1150	0.9	3.1	47
68[c,p]	1000	0.6	5.1	38
57[c,s]	930	1.9	1.1	151
66[c,s]	930	1.8	8.3	28
67[c,s]	850	1.4	0.9	158
72[c,p]	850	2.6	4.7	28
73[c,p]	750	2.8	2.7	8

[a] 25 cm^3/min NH$_3$ in 1000 cm^3 /min carrier gas; 1–3 cm^3/min HCl in 1000 cm^3/min carrier gas; growth time 1–3 hr.
[b] He carrier gas.
[c] H$_2$ carrier gas.
[s] Single crystalline layer.
[p] Polycrystalline layer.
[b] Typical impurity concentration by spark-source mass spectrographic analysis in ScN are 3×10^{19} atoms/cm^3 metal impurities (mainly Si, Fe, Al, Mg, Ca, and several rare earth metals), 4×10^{19} atoms/cm^3 oxygen, and 2×10^{19} atoms/cm^3 carbon, in comparison with typical values in Sc metal of 1.8×10^{20} atoms/cm^3 total metal impurities, 2.3×10^{20} atoms/cm^3 oxygen, and 7×10^{19} atoms/cm^3 carbon.

TABLE 2

Effect of ammonia and hydrogen halide flow rates on growth and electrical properties of ScN at 930 °C

Sample No.	Growth rate (μm/hr)	n (10^{20} cm^{-3})	μ (cm^2 V^{-1} sec^{-1})	Flow rates (cm^3/min)			
				NH$_3$	HCl	HBr	HI
21[a,p]	0.5	47	0.3	400	20	–	–
34[a,p]	1.5	8.1	45	400	–	–	20
39[a,s]	0.8	5.4	18	100	–	–	20
42[a,p]	2.6	19	19	100	–	20	–
46[a,s]	2.2	1.9	176	25	–	3	–
57[a,s]	1.9	1.1	151	25	3	–	–
66[a,s]	1.8	8.3	28	25	1	–	–
58[a,p]	7.2	7.6	32	5	3	–	–
70[a,s]	2.8	4.2	70	1	3	–	–

[a] NH$_3$ in 1000 cm^3/min H$_2$ carrier gas; hydrogen halide in 1000 cm^3/min H$_2$ carrier gas.
[s] Single crystalline layer.
[p] Polycrystalline layer.

sensitive glass electrode[5]), manufactured by Orion Research Co., Cambridge, Mass. USA, and then comparing the potential with that of standard nitrate solutions. The analyses were accurate to $\pm$ 0.2 wt% Sc and $\pm$ 0.2 wt% N. Impurities in Sc metal and in ScN were determined by spark-source mass spectrographic analysis. Lattice parameters were measured by the Debye-Scherrer method both at room temperature and as a function of temperature to determine the thermal expansion coefficient. Crystallinity of the ScN layers was examined by electron diffraction and by the back reflection Laue technique. Surface morphology was investigated by scanning electron microscopy.

The optical absorption at 300 °K of as-grown single crystal layers on α-Al$_2$O$_3$ was measured in the wavelength range $\lambda = 0.4$–2.0 μm using a Cary Model 14 spectrophotometer. The absorption coefficient α was calculated, neglecting reflection, as $\alpha = (1/t)\ln (I_0/I_T)$, where t is the thickness, and I_0 and I_T are the initial and transmitted intensities, repectively. Samples for Hall effect measurements were fabricated by sandblasting the required pattern, and the measurements were made both at 300 °K and at 77 °K using an ac bridge technique.

3. Chemistry of ScN vapor growth

By mass spectrometry we studied the vapor-phase reactions involved in the vapor growth of ScN, namely: (1) the thermal decomposition of NH$_3$, (2) the HCl transport of Sc metal, and (3) the reaction resulting in

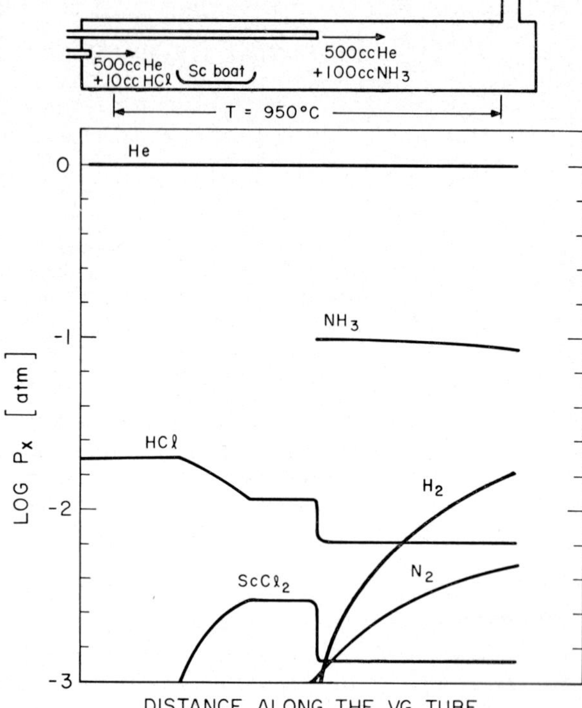

Fig. 1. Mass spectrometric determination of the variation of partial pressures (P) of gases along the growth tube during epitaxial growth of ScN. The apparatus and conditions for growth are schematically represented in the upper part of the drawing. Changes in P_{ScCl2} and P_{HCl} due to the deposition reaction are not apparent in this plot because of the scale used.

the deposition of ScN. At 950 °C the free energy of the reaction[6]),

$$NH_3 \text{ (g)} \rightarrow \tfrac{1}{2} N_2 \text{ (g)} + \tfrac{3}{2} H_2 \text{ (g)}, \tag{1}$$

is -21 kcal/mole, indicating that at thermochemical equilibrium NH$_3$ is completely decomposed. However, when NH$_3$ in He was introduced as shown in fig. 1, only 5–10% of the NH$_3$ did decompose, and thus most of the NH$_3$ was available for reaction with a scandium halide.

At 950 °C only about 40% of the HCl reacted with Sc metal to yield a volatile scandium chloride, while the rest passed downstream unreacted. Mass spectra showed that ScCl$_2^+$ was practically the only scandium chloride ion present. But since this ion could form by ionization of either the dihalide (ScCl$_2$) or the trihalide (ScCl$_3$) parent molecule[7−9]), we analyzed the scandium chloride condensed in the exhaust of the growth tube, and identified only ScCl$_3$ both by wet chemical analysis and by x-ray diffraction. Hence, at first sight, $ScCl_3$ would appear to be the gaseous species formed by reaction of HCl with Sc metal. However, one would not expect to isolate $ScCl_2$ as a pure condensed phase since $ScCl_2$, if formed at the reaction temperature, would disproportionate into $ScCl_3$ and Sc at lower temperatures[2]). Although we did not detect metallic Sc in the condensed phase, it might have further reacted with excess HCl, quartz, or with C from the carbon-coating of the growth tube. The observation only of the dihalide ion in the HCl-Sc reaction is analogous to that previously observed in reaction of Cl_2 with Y metal[8,9]), a metal chemically similar to Sc.

Hence we tentatively identify $ScCl_2$ as the predominant scandium chloride species, and write the following chemical reaction for the deposition of ScN:

$$ScCl_2\,(g) + NH_3\,(g) \rightarrow ScN\,(s) + 2\,HCl\,(g) + \tfrac{1}{2}\,H_2\,(g). \quad (2)$$

The evolution of HCl in the above reaction was confirmed by a slight increase of the HCl signal in mass spectra upon introduction of NH_3 into the system after maintaining a steady-state flow of HCl over Sc metal. At lower temperature ($T < 350$ °C) NH_3 reacts directly with the unreacted HCl to form solid NH_4Cl, but at the normal growth temperatures of 850–930 °C, this reaction is suppressed.

The above results are summarized in fig. 1. The important features are the drop in the P_{HCl} over the scandium boat and the consequent increase in P_{ScCl_2}, and the gradual drop in P_{NH_3} and the accompanying rise in P_{N_2} and P_{H_2} due to thermal decomposition of NH_3.

The abrupt drop of P_{HCl} and P_{ScCl_2} in the middle of the growth tube is a spurious effect due to dilution by the additional volume of NH_3 plus He. The low efficiency of the HCl conversion to $ScCl_2$, the small fraction of $ScCl_2$ reacted to form ScN, and a very slight decomposition of NH_3 show that thermochemical equilibrium was not established in the system.

4. Effect of growth conditions

Of various growth parameters investigated, the temperature (table 1) and the reagent concentrations (table 2) had the most significant effects upon growth rate, crystallinity, and electrical properties of ScN, and this held true whether HCl, HBr, or HI was used to transport the scandium. Single crystal samples were readily obtained in the temperature range 850–930 °C, whereas mostly polycrystalline samples were obtained outside this range, Under given flow rates of reactant gases, the growth rate generally increased with decreasing temperature, for instance, from 0.9 μm/hr at 1150 °C to 3 μm /hr at 750 °C. Large flow rates of ammonia and hydrogen halide usually gave polycrystalline material and, in general, smaller growth rates. Likewise, only polycrystalline deposits with inferior electrical properties were obtained by using anhydrous $ScCl_3$ as the starting source for scandium.

The highest Hall mobility and lowest carrier concentration were obtained for single crystals grown in the temperature range 850–930 °C and with the flow rates of ammonia and hydrogen halide of 25 cm^3/min and 3 cm^3/min, respectively.

5. Properties of ScN

5.1. Chemical and structural characteristics

In striking contrast to the reactivity[1,10,11]) of the mononitrides of Y and the rare earth metals, ScN is inert to air and water, probably due to the adherent film of $Sc(OH)_3$ observed on ScN surfaces by electron diffraction, and dissolves only in strong acids or molten NaOH. The ScN is stable in air up to 550 °C, but oxidizes at higher temperatures to Sc_2O_3, as previously reported[12]).

Spark-source mass spectrographic analysis (table 1) showed that vapor-grown ScN has a smaller concentration both of metal impurities and of oxygen and carbon than the Sc metal starting material. Typical analyses on vapor-grown ScN indicate no appreciable deviation from stoichiometry to within the accuracy of the analysis (calculated for ScN: Sc = 76.4 wt %, N = 23.6 wt % found: Sc = 76.0 ± 0.2 wt %, N = 23.8 ± 0.2 wt %), Thus our vapor grown material appears relatively free of nitrogen vacancies reported in large amounts by previous workers[11,12]).

The room temperature lattice parameter of ScN having the rock-salt structure is 4.5025 ± 0.0005 Å, which is slightly larger than the previously reported values of 4.490 Å[11]) and 4.499 Å[12]). The average linear expansion coefficient of ScN is 8.1×10^{-6}/°C over the temperature range 23–555 °C. Single crystal ScN layers grown on α-Al_2O_3, $MgAl_2O_4$, and TiO_2, which have expansion coefficients[13-16]) close to that of ScN, were adherent and crackfree, whereas layers grown on MgO, which has an expansion coefficient[15]) of $14.4 \times 10^{-6}/°C$, cracked away from the substrates.

The scanning electron micrograph of a typical single crystalline ScN sample on $\langle 1\bar{1}02 \rangle$ α-Al_2O_3 in fig. 2 shows an island-type structure similar to that sometimes observed in the epitaxial growth of silicon[16]). The orientation of this ScN sample, determined from Laue photographs, was $\langle 321 \rangle$ ScN // $\langle 1\bar{1}02 \rangle$ α-Al_2O_3.

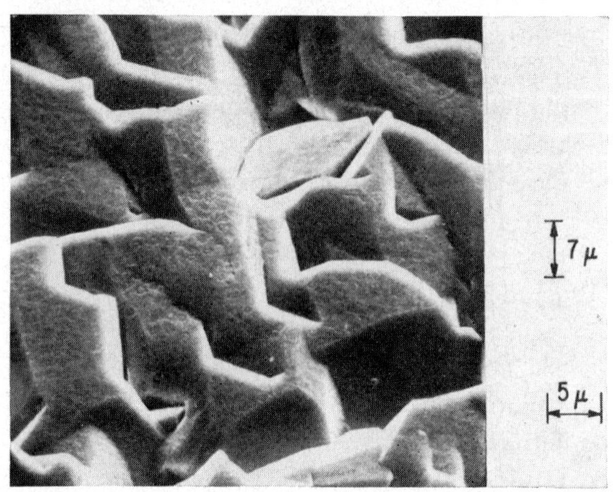

Fig. 2. Scanning electron micrograph of a single-crystalline (321) ScN surface at 45° and 1500×, showing an island-type structure.

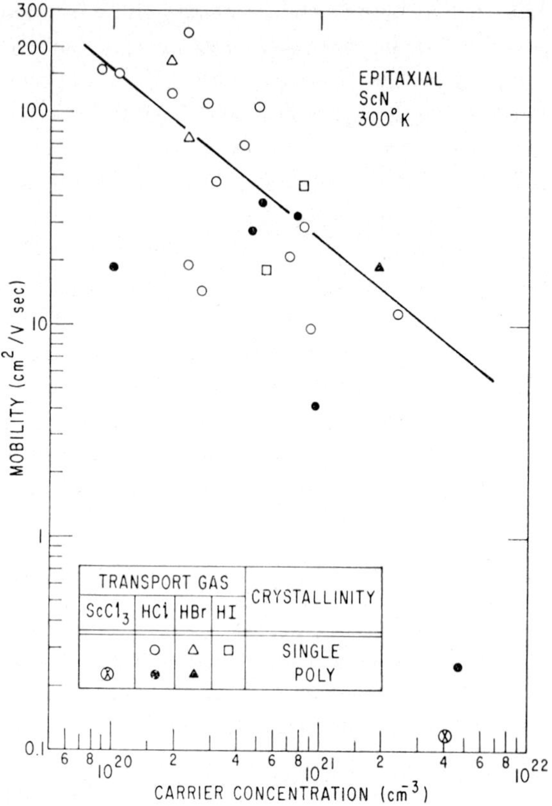

Fig. 3. The dependence of Hall mobility upon electron concentration, at 300 °K, in vapor-grown ScN.

5.2. ELECTRICAL PROPERTIES

The ScN layers grown by the present technique were all highly-conducting n-type material with carrier concentrations in the range 10^{20}–10^{21}/cm³. The dependence of Hall mobility upon carrier concentration was measured both at 300 °K and at 77 °K. The data are shown only for 300 °K, in fig. 3. The difference in carrier concentration between 300 °K and 77 °K is less than 10%, indicating that nearly all of the donor impurities remain ionized down to liquid nitrogen temperature. As seen in fig. 3, the Hall mobility of polycrystalline samples was generally between one and two orders of magnitude lower than the mean of single crystal samples.

Even at the large values of carrier concentration in ScN, for which some contribution from ionized impurity scattering might be expected, the mobility varies from 300 °K to 77 °K as $T^{-1.85}$ at 1×10^{20}/cm³ and as $T^{-1.42}$ at 2×10^{21}/cm³. This temperature dependence is close to the $T^{-1.5}$ variation for a semiconductor in which the mobility is limited by accoustical lattice scattering.

Fig. 4 compares the dependence of Hall mobility upon carrier concentration of ScN with that[17]) of several III–V compound semiconductors, Ge, and Si. The mobility of ScN (~ 150 cm²/V sec) at 1×10^{20}/cm³ is as large as that in the 10^{17}–10^{18}/cm³ range for AlAs and GaP, which have comparable values of energy gap. Further, the mobility of ScN at 1×10^{20}/cm³ is twice that of Si and only 10% smaller than that of Ge. An optimistic extrapolation suggests that the mobility of ScN at lower carrier concentration might lie between that of Ge and Si; hence, ScN could become a useful semiconductor material if controlled doping can be achieved.

5.3. DOPING BEHAVIOUR

The most obvious source of donors in vapor-grown ScN would be nitrogen vacancies. Nevertheless, several of our results suggest that they are not a major contribution. First, the chemical analysis indicates no gross deviation from stoichiometry. Second, no systematic increase in donor concentration occurs over the range of growth temperatures 750–1150 °C. Third, the donor

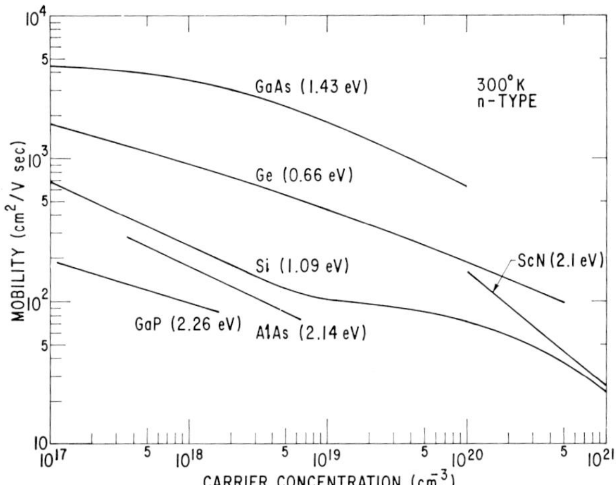

Fig. 4. The dependence of Hall mobility upon electron concentration, at 300 °K, for ScN, several III–V compound semiconductors, Ge, and Si.

concentration decreased with decreasing concentrations of ammonia and hydrogen halide.

In our previous work[1] it was concluded that chlorine, which was the major impurity in vapor-grown ScN, acted as a donor. Since the bond strengths of scandium halides decrease in the series Sc–Cl, Sc–Br, Sc–I, it was hoped that the use of HBr or HI in this work, instead of HCl, to transport Sc metal would reduce the halogen contamination and hence the electron concentration in ScN. However, the present results indicate approximately the same amounts of halogen contamination (10^{20}–10^{22} atoms/cm^3) and the same range of electron concentrations (10^{20}–10^{21} cm^{-3}), irrespective of the hydrogen halide used.

One other donor impurity for which we have not analyzed is hydrogen. There is some hydrogen in the system even using He as a carrier gas due to the decomposition of NH_3, and it could be present in the ScN as amide (NH_2) or imide (NH) substituted for nitrogen. The existence of such type of compounds is

known; for example, the compound CaNH, which is isoelectronic with ScN, has the rock-salt structure and forms continuous solid solutions with $Ca(NH_2)_2$[18,19]. Thus, the complex NH or NH_2 substituted for nitrogen in ScN could render the material n-type.

In an attempt to convert ScN p-type, we have doped it during growth with additions of C as ethylene or Si as silane. However, no p-type conversion occurred, nor was the high electron concentration reduced. Further efforts were made by post-growth annealing of ScN at 1000 °C in atmospheres of p-type dopants such as Zn or Mg, or in vacuum. Annealing in Zn vapor for 1 hr showed no appreciable effect on the electrical properties, while annealing in Mg vapor reduced the carrier concentration from 2.3 to 1.1×10^{20} cm^{-3}, and significantly increased the Hall mobility from 19 to 158 cm^2 V^{-1} sec^{-1}. Annealing in vacuum, on the other hand, increased the carrier concentration and reduced the mobility. In no cases, however, was p-type material obtained.

5.4. OPTICAL ENERGY GAP

The optical absorption spectrum of ScN specimens with high electron concentrations ($n \geq 10^{20}/cm^3$) was characterized by two distinctive absorption processes: the free carrier absorption and the fundamental edge absorption. One such example is shown in fig. 5, where log α at room temperature is plotted as a function of log λ for a 2 μm thick ScN epitaxial layer ($n = 2.9 \times 10^{20}/cm^3$). The absorption at long wavelengths beyond about 1 μm can be ascribed to free carrier absorption, since the absorption in this range was found to increase with increasing carrier concentration and vary approximately as λ^2, which is a functional dependence expected of free carrier absorption upon wavelength[20]. The steeply rising absorption in the short wavelength region near 0.6 μm must be due to a band-to-band transition, since this was characteristic of all samples measured irrespective of their carrier concentrations in the range 1–8 × $10^{20}/cm^3$. The free carrier absorption was subtracted from the measured data by extrapolating[20] it into the region of edge absorption, with the assumption that the λ^2 dependence is still obeyed over this region, thus generating the square points shown in fig. 5.

The absorption coefficients obtained in this manner show an exponential dependence upon photon energy,

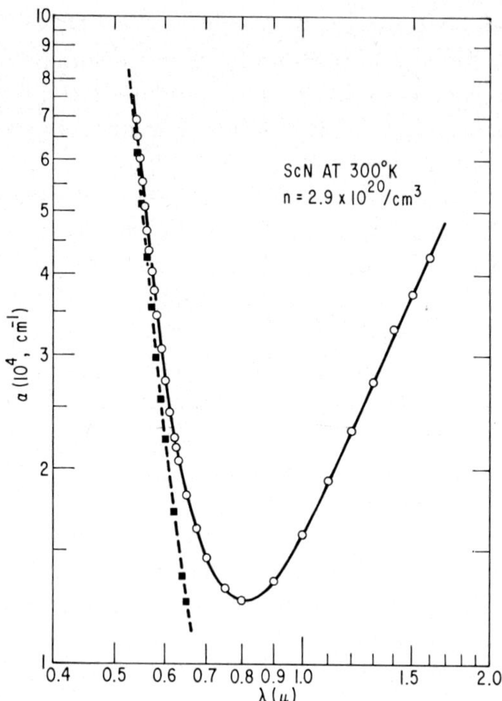

Fig. 5. The dependence of optical absorption coefficient (α) upon wavelength (λ), at 300 °K, for a 2μm thick single-crystalline ScN layer having alectron concentration 2.9×10^{20}/cm³. The circles represent the as-measured α values, and the squares are those for which free carrier absorption was substrated.

and this functional relationship extends to energies below as well as above the apparent absorption edge. The exponential "tail" is indicative of a perturbation of the band structure due to high concentration of charged impurities[21,22]). Additional effects due to high impurity concentration are the possible bandgap *shrinkage* arising from impurity banding and the apparent bandgap *increase* by Burstein shift, the magnitudes of which are not known. Because of the high carrier concentration and the related effects, it is not possible at this time to assign an exact value of the energy gap, nor the type of optical transitions involved, for ScN. However, from the observed absorption edge near 0.6μ, the bandgap of ScN can be estimated to be 2.1 eV. The present value is larger than the previous value reported by Busch et al.[11]) of 1.75 eV, which was obtained by diffuse reflectance measurements on powders containing only 80% of the theoretical nitrogen content of stoichiometric ScN.

Acknowledgments

The authors thank R. J. Paff, R. T. Smith, and W. Roth for X-ray and electron diffraction measurements, Dr. W. L. Harrington and B. L. Goydish for analytical measurements, and R. J. Ulmer, E. J. Stofko and T. Zrebiec for technical assistance. We are grateful to Dr. E. Kaldis for stimulating discussions.

The research reported herein was made possible by the support of the Advanced Research Projects Agency under Order No. 1034, through the United States Army Electronics Command, Fort Monmouth, New Jersey 07703, USA, under Contracts Nos. DAAB07-69-C-0145 and DAAB07-70-C-0155.

References

1) J. P. Dismukes, W. M. Yim, J. J. Tietjen and R. E. Novak, RCA Rev. **31** (1970) 680.
2) O. G. Polyachenok and G. I. Novikov, Russ. J. Inorg. Chem. **8** (1963) 1479.
3) V. S. Ban, J. Electrochem. Soc. **118** (1971).
4) K. L. Cheng, Anal. Chem. **33** (1961) 761.
5) S. S. Potterton and W. D. Shults, Anal. Letters **1** (1967) 11.
6) O. Kubaschewski, E. L1. Evans and C. B. Alcock, *Metallurgical Thermochemistry*, 4th ed. (Pergamon, Oxford, 1967).
7) J. W. Hastie and J. L. Margrave, High Temperature Sci. **1** (1969) 481.
8) J. D. McKinley, J. Chem. Phys. **41** (1964) 2814.
9) J. D. McKinley, J. Chem. Phys. **42** (1965) 2245.
10) R. Didchenko and F. P. Gortsema, J. Phys. Chem. Solids **24** (1963) 863.
11) G. Busch, E. Kaldis, E. Schaufelberger-Teker and P. Wachter, in: *Colloq. Intern. du Centre National de la Recherche Scientifique*, Paris–Grenoble, May 5–10, 1969, Tome I, No. 180, published 1970.
12) M. D. Lyutaya and V. F. Bukhanevich, Russ. J. Inorg. Chem. **7** (1962) 1290.
13) F. A. Mauer and L. H. Bolz, National Bureau of Standards Report WADC TR 55–473 (1955).
14) W. J. Campbell and C. Grain, U.S. Bureau of Mines Report of Investigation 5757 (1961).
15) R. J. Beal and R. L. Cook, J. Am. Ceram. Soc. **40** (1957) 279.
16) D. Richman and R. H. Arlett, in: *Semiconductor Silicon*, Eds. R. R. Haberecht and E. L. Kern (Electrochemical Society, New York, 1969) p. 200.
17) S. M. Sze and J. C. Irvin, Solid State Electron. **11** (1968) 599.
18) H. Hartmann, H. J. Frohlich, and F. Ebert. Z. Anorg. Allg. Chem. **218** (1934) 218.
19) R. Juza and H. Schumacher, Z. Anorg. Allg. Chem. **324** (1963) 278.
20) J. I. Pankove and P. Aigrain, Phys. Rev. **126** (1962) 956.
21) J. I. Pankove, Phys. Rev. **140** (1965) A2059.
22) D. Redfield and M. Afromowitz, Appl. Phys. Letters **11** (1967) 138.

Journal of Crystal Growth **13/14** (1972) 371–374 © *North-Holland Publishing Co.*

THE EPITAXIAL GROWTH OF ZINC SULPHIDE ON SILICON BY FORCED VAPOUR TRANSPORT IN ARGON FLOW

P. LILLEY, P. L. JONES and C. N. W. LITTING

Electronic Engineering Laboratories, Manchester University, Oxford Road, Manchester, England

Thin films of zinc sulphide have been grown epitaxially on silicon (111) substrates using an open tube flow method in which argon was the transport gas. The conditions for epitaxial growth have proved to be very similar to those reported previously for a hydrogen flow system, although a higher source temperature is required with argon.

1. Introduction

Epitaxial growth of zinc sulphide on silicon (111) substrates in a hydrogen gas flow has recently been reported by the authors[1]. The reversible reaction

$$ZnS + H_2 \rightleftharpoons Zn + H_2S \tag{1}$$

was employed with hydrogen acting as both reactant and carrier gas. Important features in the process were found to be the hydrogen flow rate and the substrate temperature in addition to the surface conditions of the silicon substrate, and the pre-cleaning treatment of the zinc sulphide powder source.

This report describes the results obtained using the same apparatus and similar procedure but with argon instead of hydrogen as the transport gas. In this case the reversible reaction

$$ZnS \rightleftharpoons Zn + \tfrac{1}{2} S_2 \tag{2}$$

is employed and the elements are transported from the powder source to the silicon substrate in an argon gas stream.

This technique has been used by Dev[2] to grow single crystal plates of zinc sulphide in a quartz chamber.

2. The growth procedure

The basic features in the design of the quartz growth system (fig. 1) are that separate regions are provided for the cleaning of the silicon wafers and the deposition of the zinc sulphide. The operation of the system may be summarised as follows.

(a) The system is loaded and flushed with hydrogen.
(b) The zinc sulphide powder source is baked in a flow of hydrogen at the temperature to be used in the subsequent growth period. After twenty minutes the furnace is removed and the system allowed to cool to room temperature.
(c) The silicon wafer supported by the graphite susceptor is heated by RF induction to 1260 °C in a gas flow of hydrogen and hydrogen chloride. The resulting clean and etched slice is then allowed to cool to room temperature in a hydrogen flow.
(d) The zinc sulphide is brought to the required source temperature.
(e) By means of the quartz push rod, the slice is moved from the cleaning region of the apparatus into the growth chamber where the required deposition temperature is achieved by careful positioning of the slice.
(f) The argon flow and deposition is commenced when the temperature of the substrate is reasonably stable.

3. Experimental results

From experience with the hydrogen flow system it is apparent that an ideal substrate temperature for successful epitaxial deposition in this system is in the region of 500 °C. A similar result is obtained with argon as the carrier gas and this substrate temperature is adopted as a standard in this system.

To obtain appreciable transport of zinc sulphide a source temperature greater than 1150 °C was found to be necessary.

With a source temperature of 1170 °C, the growth of

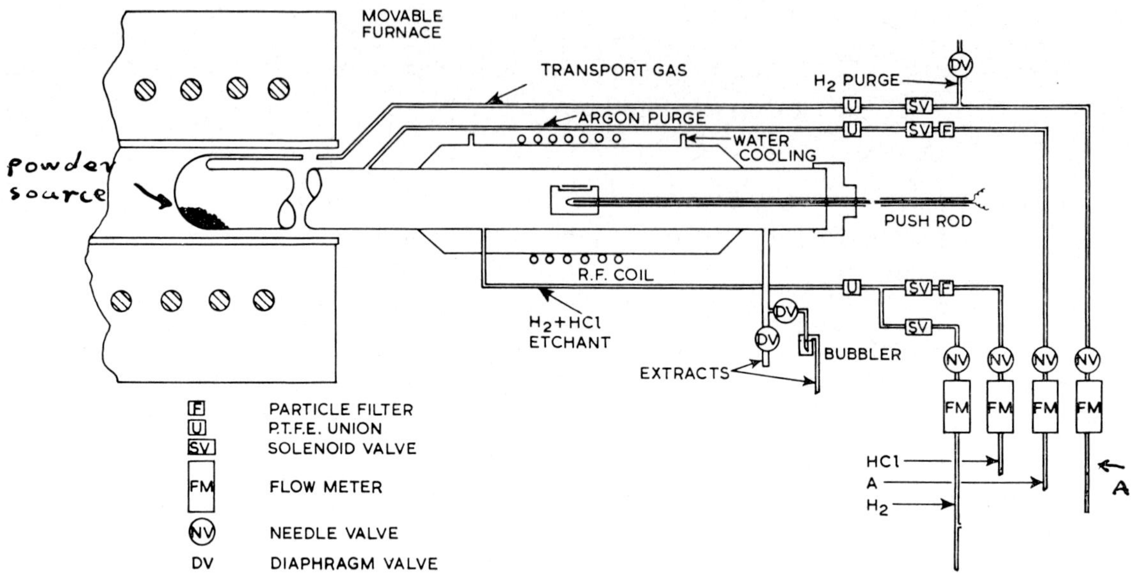

Fig. 1.   Apparatus for the growth of single crystal zinc sulphide on silicon wafers.

deposits was investigated with argon flow rates from $0.5\ cm^3/min$ to $5 \times 10^3\ cm^3/min$. In all cases transparent deposits were obtained, and the faster the flow the more rapidly the growth occurred. With the thicker layers, bands of interference colours were clearly visible, indicating variations of thickness in the direction of gas flow.

3.1. OBSERVATIONS WITH AN OPTICAL MICROSCOPE

The topography of the layers was examined using a Normarski type Interference Contrast attachment. The layers were found to be continuous and matt in appearance without facets and a typical micrograph is shown in fig. 2.

Thickness measurements were made optically using the method previously described[1]), and hence growth rates under various conditions were established. Typically at flow rates of 30 and $170\ cm^3/min$ the growth rates were 20 and $115\ Å/h$ respectively. Increasing the source temperature to $1220\ °C$ resulted in a growth rate of $450\ Å/h$, at a flow rate of $170\ cm^3/min$.

3.2. REFLECTION ELECTRON DIFFRACTION

The diffraction pattern from a film grown with a flow rate of $170\ cm^3/min$ and a source temperature of $1170\ °C$ is shown in fig. 3a. It indicates a good single crystal structure for the bulk of the deposit although the superimposed faint ring pattern does indicate the pre-sence of some polycrystalline material on the surface.

Growth with a flow rate of $30\ cm^3/min$ using the same source temperature gives the diffraction pattern of fig. 3b. In this case the spots are slightly arced indicating a slight degree of misorientation in the layer. Layers grown with flow rates in excess of $300\ cm^3/min$ were completely polycrystalline.

It is apparent that flow rates between about 40 and $250\ cm^3/min$ are necessary for the growth of single crystal deposits, when using source and substrate temperatures of 1170 and $500\ °C$ respectively.

Fig. 2.   Interference contrast optical micrograph of a layer grown on silicon. Source temperature 1170 °C, substrate temperature 500 °C, flow rate 170 cm³/min; 1 cm ≡ 25 µm.

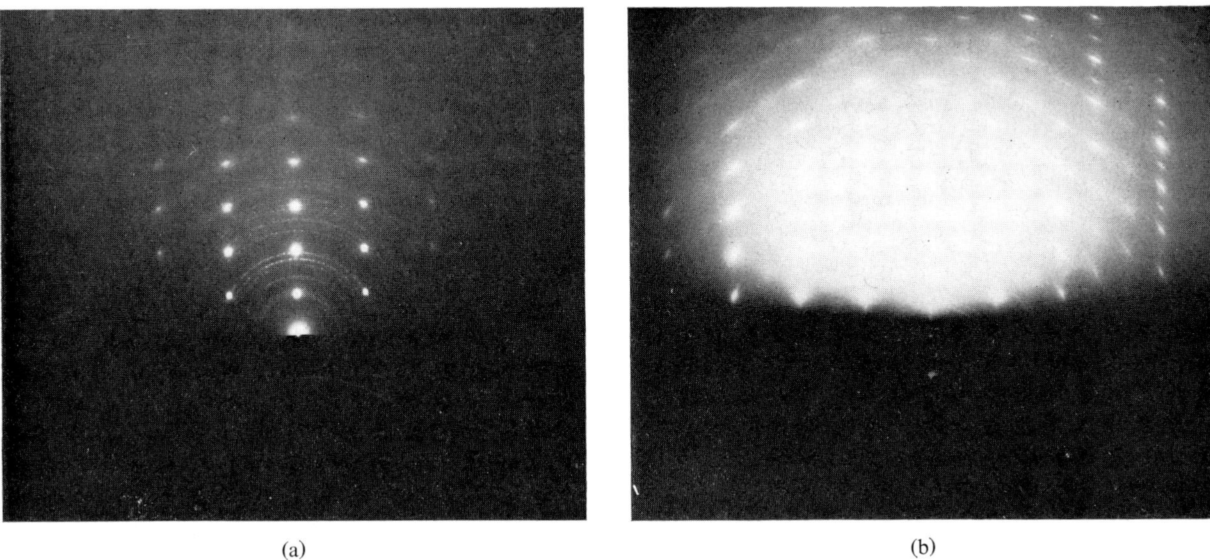

(a)   (b)

Fig. 3. Reflection electron diffraction patterns from layers grown on substrates at 500 °C with argon flow rate: (a) 170 cm^3/min; (b) 30 cm^3/min.

(a)   (b)

Fig. 4. Transmission electron diffraction patterns from: (a) epitaxial growth; (b) silicon substrate.

3.3. TRANSMISSION ELECTRON DIFFRACTION

The specimens were prepared for examination by jet etching the silicon surface of a small chip until a suitably thin area was obtained[3]. using a mixture of nitric and hydrofluoric acids.

The specimens were mounted in the microscope so that the beam was normal to the growth surface, a (111) plane. Fig. 4a shows a typical diffraction pattern from a specimen grown with a flow rate of 170 cm^3/min. For comparison, the spot pattern obtained from a cleaned silicon slice is shown in fig. 4b. It is evident from these patterns that the overgrowth is strongly epitaxial and contains hexagonal material. However as previously explained[1]), the presence of cubic material cannot be excluded.

4. Discussion

It is of interest to compare the growth requirements to give epitaxial deposition in argon flow with those required in the hydrogen flow method.

VII – 14

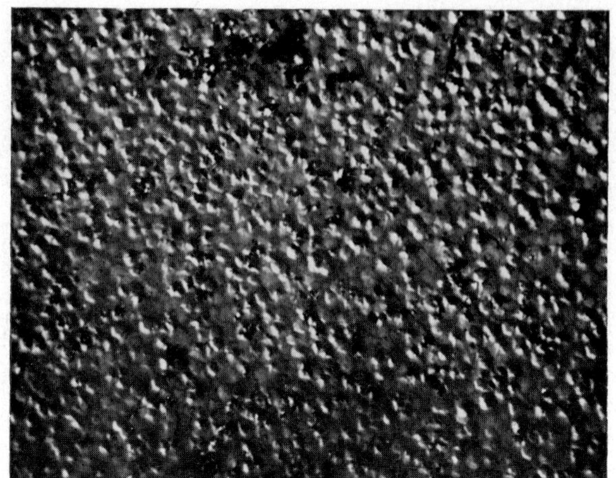

Fig. 5. Interference contrast optical micrograph of a layer grown on silicon in hydrogen flow. Source temperature 1050 °C, substrate temperature 500 °C, flow rate 40 cm³/min; 1 cm ≡ 25 μm.

Flow rates and substrate temperatures are similar and the essential difference is in the source temperature. As argon is non-reactive, a much higher temperature of the source is required to promote an adequate transport of source material [eq. (2)] than when using the reaction with hydrogen [eq. (1)]. Even so growth rates obtained are an order of magnitude lower when argon is the transport gas.

The optical micrographs of specimens grown in argon are significantly different from those grown in hydrogen flow, under similar conditions of flow rates and substrate temperatures. The layers grown in argon (fig. 2) show characteristic fine structure, random in nature, whereas the layers grown in hydrogen (fig. 5) show considerable facetting. Even so, for the two methods of growth, there are no significant differences in the crystal structures revealed by electron diffraction, in either transmission or reflection modes.

A direct comparison of thicker layers, grown by each of the two methods has not been possible because when argon is used, the high source temperatures required are at the limit of the capabilities of the apparatus. An attempt to grow a thick layer in argon, using a source temperature of 1220 °C, was abandoned after several hours and produced a layer of only 2000 Å. Hence, since layers of about 1 μm are necessary for examination by X-rays, it has not been possible to establish the presence of cubic material by this technique.

5. Conclusions

Single crystal growth of zinc sulphide thin films on silicon (111) substrates can be obtained using argon as the transport gas provided the source temperature is sufficiently high. However, as a method of growth for thick layers it is significantly inferior to the hydrogen flow method in that although the crystal structure is as good, the growth rates are inconveniently low.

The experiments indicate that for both hydrogen and argon, the range of flow rates over which epitaxy is observed is similar. The most significant effect of the carrier gas is on the rate of transport of zinc sulphide from the source.

References

1) P. Lilley, P. L. Jones and C. N. W. Litting J. Mater. Sci. **5** (1970) 891.
2) I. Dev, Brit. J. Appl. Phys. **17** (1966) 761.
3) P. L. Jones, C. N. W. Litting, D. E. Mason and V. A. Williams Brit. J. Appl. Phys. (J. Phys. D) [2] **1** (1968) 283.

Journal of Crystal Growth **13/14** (1972) 375–379 © *North-Holland Publishing Co.*

VAPOR GROWTH OF II–VI COMPOUNDS AND THE IDENTIFICATION OF DONORS AND ACCEPTORS

K. NASSAU and J. W. SHIEVER

Bell Telephone Laboratories, Incorporated, Murray Hill, New Jersey 07974, U.S.A.

By a series of improvements in the growth and doping technique it has been possible to prepare single crystals of II–VI compounds such as CdS, CdSe, and ZnSe in a rapid, reproducible and impurity controlled manner. This has resulted in the identification of the chemical nature of the shallow donors and acceptors for the first time.

A sublimation process was used to provide thin strain-free platelets for laser stimulated fluorescence studies in the edge emission region.

Essential features of the technique include the use of high carrier gas flow and rapid growth ($\sim$ 1 hr for CdS) necessary to prevent excessive impurity exchange with the apparatus, and a dilution series technique in which small amounts of impurity are added and successive growth runs are performed without further addition. In this way the very narrow and low concentration region can be reached between the amount always present in starting material (0.1 ppm Na, 1 ppm Cl, etc.) and the amount that produces excessive line broadening and makes the chemical identification impossible ($<$ 20 ppm Li, $<$ 200 ppm Cl, etc.). By this technique the chemical species involved in the edge luminescence (excitons bound at donors and acceptors and sharp line pair apectra) have been identified for the first time. In CdS only Li and Na were found to give shallow acceptors; shallow donors include F, Cl, Br, I, Ga, In, and interstitial Li and Na.

1. Introduction

The so-called edge emission[1]) of II–VI compounds was discovered some 30 years ago. Despite intensive investigation, the chemical nature of the shallow donors and acceptors involved has remained uncertain. The high band-gap II–VI compounds (ZnS, ZnSe, ZnTe, CdS, and CdSe) are direct gap materials and would be expected to provide junction luminescence and lasers in the visible region if they could be made both n and p type. Since this requires low resistivity material for efficient operation, the study of the chemical species producing shallow acceptors and donors becomes significant.

The edge emission features were studied by Henry under laser excitation[2]). Bound excitons (I_1 lines at acceptors, I_2 and I_3 at donors) and pair features (sharp lines and distant pair peaks) involve shallow donors and acceptors.

We can now assign three causes to the previous lack of success in identifying the chemical nature of the edge emission features in CdS (and also in CdSe and ZnSe):

(1) The highest purity II–VI compounds available contain a number of impurities at the 1 ppm level; these produce a number of bound exciton lines;

(2) The addition of a variety of very small amounts of dopants produces broadening so that the features become indistinguishable from one another; and

(3) We have observed that considerable leaching and exchange of impurities can occur with the growth apparatus (e.g., Li, Na, Al, Si) particularly if growth is slow.

We have devised a growth technique to overcome these difficulties and have assigned the two known I_1 lines in CdS to excitons at substitutional Li and Na[3]), and seven I_2 (and I_3) lines to excitons at substitutional F, Cl. Br, I, In, Ga [4]), and interstitial Li. Pair spectra also involve the Li acceptor. Preliminary data show similar behavior in CdSe, and a detailed study with J. L. Merz in ZnSe is well underway.

2. The growth technique

The most widely used technique for the growth of II–VI compounds is that of Piper and Polich[5]). This and most other vapor phase systems use relatively slow growth (1 or more days) in sealed or slow gas flow arrangements, frequently with mullite or alumina as part of the apparatus.

The growth technique used in this work employs sublimation in a fairly rapid gas stream in the all fused quartz apparatus shown in fig. 1. The outer tube is

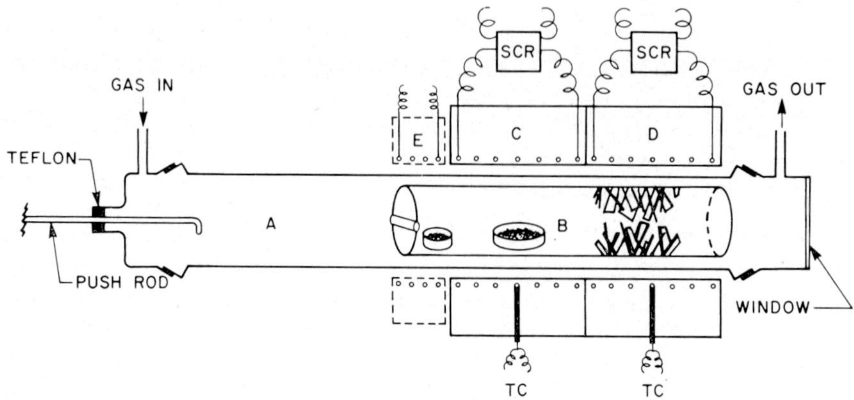

Fig. 1.   Growth apparatus used for II–VI compounds.

Fig. 2.   Split inner tube showing CdS crystal.

$1\frac{1}{2}$ in. O.D. and 62 in. long, 24 in. of it being in the two zone furnace C, D. The inner tube, 1 in. O.D. and 30 in. long, can be moved from the cold position A to the growth position B and back by means of the push rod without atmospheric contamination. An additional lower temperature furnace E with the second boat shown is used when the dopant is very volatile.

Dopant additives are mixed in with the source material (Na_2SO_4, K_2CO_3, etc.), placed in the second boat (Na, Ca, etc.), or volatilized into part or all of the gas stream (HCl, I_2). In the case of HCl and HBr part of the gas stream is first passed thru the aqueous solution and then through conc. H_2SO_4 to remove the water.

Growth conditions are summarized in table 1; growth occurs in a broad temperature region, the value given being an average. In some experiments a high

Fig. 3. Pure and doped CdS crystals showing morphology variation with impurity content. Scale in mm.

purity alumina inner tube was used instead of quartz. It is preferred that the source should not be completely used up.

The II–VI materials used as the source were Eagle Picher Industries ultrahigh purity grade. A typical CdS sample contained 1–5 ppm by weight of Pb, Si, and Fe; 0.5–2 ppm of Mg, Mn, Cl, and K; and 0.1–0.5 ppm of Li, Na, and F. Dopants used were chemical reagent grade. The K_2CO_3 used, for example, contained 11 ppm Na and less than 0.5 ppm Li.

3. Results

Crystals grown by this technique are shown in figs. 2 and 3. The following are the advantages of this growth technique:

(a) with the high gas flow and all quartz system, growth is very rapid (1 hr with CdS) and contamination from the growth apparatus is minimal; and

(b) the use of a dilution series to avoid concentration broadening of spectral features; strain-free platelets are

TABLE 1

Growth conditions

II–VI compound	Gas and flow		Source temperature (°C)	Growth temperature[a] °C)	Growth time (hr)
CdSe	Ar,	0.7 l/min	1070	880	1
CdS	Ar,	0.7 l/min	1120	920	1
ZnSe	N_2/H_2[b],	0.5 l/min	1275	1100	3

[a] Temperature gradient is about 13 °C/cm.
[b] Forming gas: 85% N_2, 15% H_2.

essential to avoid equally serious strain-broadening.

Dopant impurity is added during growth until it interferes with growth, producing dendrites, prisms, etc., instead of the desired flat strain-free platelets (typically 30 μm in thickness). The concentration is reduced somewhat and platelets are examined for edge emission features. If these are broad, as in fig. 4a (about 0.5 g Na_2SO_4 added to 10 g CdS source material), the inner growth tube is now scrapped out and a new source boat is used with no further dopant

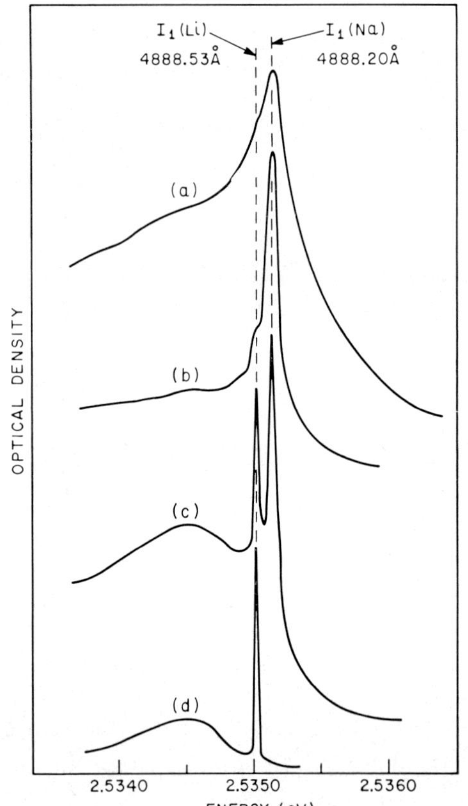

Fig. 4.   Edge emission in the I_1 region at 1.6 °K for Na doped CdS crystals (a to c) and nominally pure CdS (d).

added; the features are still broad (fig. 4b). A light acid wash of the inner tube and a fresh source now gave a satisfactory sharp spectrum (fig. 4c). In addition to the dominant I_1 (Na) line at 4888. 20 Å, the weaker I_1 (Li) line at 4888.53 Å indicates the usual Li contamination present in "pure" material. A prolonged acid wash now removes essentially all traces of dopant, giving the nominally undoped spectrum (fig. 4d). The sample of fig. 4c contained 40 ppm Na and 0.4 ppm Li, that of fig. 4d 0.1 ppm Na and Li. Only Li and Na gave distinct and strong I_1 lines; with K, Rb, and Cs only a weak I_1 (Li) line is seen, derived from the Li present in the source or leached from the quartz tube. The I_1 (Li) line is always accompanied by an I_2 line at 4866.98 Å, which is attributed to a shallow donor caused by interstitial Li. Acting as donors, F, Cl, Br, I, In, and Ga gave I_2 and I_3 lines.

With fluorine addition, reaction with the quartz (or with alumina) appeared to prevent successful doping. (Diffusion with CdF_2 at 500 to 800 °C in evacuated sealed quartz tubes by A. M. Sergent gave satisfactory results.) Reaction with the inner tube was observed in other instances. Thus crystals grown in a high purity alumina ceramic inner tube showed a dominating I_1 (Na) line, and even with a reasonable quantity of added Li, the Na line still dominated; presumably the Li was absorbed and Na liberated from the alumina. Similarly the addition of Al_2S_3 in quartz produced Si doping at the 1000 ppm level (ordinarily the Si concentration is less than 5 ppm) with Al less than 5 ppm.

Other dopants tried included K, Rb, Cs, N, P, As, Sb, Sn, Al, Tl, Ca, Sr, and Ba. None of these gave evidence of forming simple shallow donors or acceptors, although a shallow complex acceptor due to P was observed[6]). A variety of additives were tried in each case; thus, for example, K_2SO_4, K_2CO_3, KF, KI, KCl, KBr, and K metal additions were tried for the introduction of K. The simple shallow donors and acceptors identified in CdS are given in table 2 together with the preferred method of introduction and the concentrations to yield sharp, distinctive edge emission features.

Pair spectra identified include an acceptor-double donor system. In this case each pair line showed a satellite which was attributed to the Li^6 isotope. The Li attribution was confirmed by growth in the presence of Li^6 (added in the form of Li_2CO_3, 95.6 at% Li^6,

TABLE 2

Shallow donors and acceptors in CdS

Dopant impurity	Wavelength (in air, Å, at 1.6 °K)	Dopant added as (B in boat, G in gas)	Concentration in ppm (wt)
Li	$I_1 = 4888.53$[a]	$Li_2SO_4 - B$	0.1–10 (0.1)[b]
Na	$I_1 = 4888.20$	$Na_2SO_4 - B$	10–100 ($<$0.05)[b]
F	$I_2 = 4870.44$	(CdF_2 diffusion)	$-$(0.1)[b]
Cl	$I_2 = 4869.17$	$HCl - 5\%$ G[c]	5–12 (2–4)[b]
Br	$I_2 = 4869.05$	$HBr - 1\%$ G[c]	2–8 ($<$ 0.1)[b]
I	$I_2 = 4868.78$	$I_2 - 100\%$ G[c] 70 °C	6($<$ 0.1)[b]
In	$I_2 = 4869.75$	In – B	$<$ 1 ($<$ 0.1)[b]
Ga	$I_2 = 4869.36$	GaP – B	3 ($<$ 0.1)[b]
Li	$I_2 = 4866.84$	$Li_2SO_4 - B$	0.1–10 (0.1)[b]

[a] I_1 is exciton bound to a neutral acceptor, I_2 is exciton bound to a neutral donor.

[b] Amount present in nominally pure material (0.1 ppm Li $\simeq 10^{16}$ atoms per cm^3).

[c] % figure indicates amount of gas stream saturated.

O.R.N.L., Oak Ridge, Tenn.), which intensified the satellite lines so that they became stronger than the Li[7] lines[3,6]. Simple donor acceptor pair spectra in CdS previously reported by us[2] also involve the Li acceptor. We have given a detailed description of magneto-optic studies of the substitutional donors[4,7] interpreting the so-called two-electron transitions and giving central cell corrections. Lifetimes of the bound excitions were also reported[8].

Preliminary experiments with CdSe using the conditions of table 1 have indicated that there also Li and Na cause bound exciton lines in the edge emission. Pair spectra have been observed and chemical identifications have been made[6]). Work on ZnSe is also underway.

Lack of p-type occurrence in CdS has been attributed to self-compensation by native defects[9]. An additional factor can now be added[6]: only two elements are suitable for providing shallow acceptors (Li and Na) and both of these can self-compensate by also forming shallow donors when entering interstitially.

Acknowledgments

We wish to thank C. H. Henry and A. M. Sergent for extensive interactions in this cooperative endeavor.

References

1) Background data may be found in: *Physics and Chemistry of II–VI Compounds*, Eds. M. Aven and J. S. Prener (North-Holland, Amsterdam, and Wiley, New York, 1967).,
2) C. H. Henry, R. A. Faulkner and K. Nassau, Phys. Rev. **183** (1969) 798.
3) C. H. Henry, K. Nassau and J. W. Shiever, Phys. Rev. Letters **24** (1970) 820.
4) K. Nassau, C. H. Henry and J. W. Shiever, in: *Proc. Tenth International Conf. on the Physics of Semiconductors*, Cambridge, Mass., 1970, Eds. S. P. Keller, J. C. Hensel and F. Stern (USAEC-CONF-700801) (National Bureau of Standards, Springfield, Va., 1970) p. 629.
5) W. W. Piper and S. J. Polich, J. Appl. Phys. **32** (1961) 1278.
6) C. H. Henry, K. Nassau and J. W. Shiever, Phys. Rev. B **4** (1971) 2453.
7) C. H. Henry and K. Nassau, Phys. Rev. B **2** (1970) 997.
8) C. H. Henry and K. Nassau, Phys. Rev. B **1** (1970) 1628; J. Luminescence **1,2** (1970) 299.
9) G. Mandel, F. F. Morehead and P. R. Wagner, Phys. Rev. **136** (1964) A826.

GROWTH OF GaSe SINGLE CRYSTALS FROM THE VAPOR PHASE

A. KUHN and A. CHEVY

Laboratoire de Luminescence II, Université de Paris VI, Equipe de Recherche associée au CNRS, Paris, France

and

E. LENDVAY

Research Institute for Technical Physics of the Hungarian Academy of Sciences, Budapest, Hungary

GaSe crystals were prepared by vapor transport using iodine. On their basal surfaces triangular and non-polygonized spiral growth hills with two or more cooperating spirals and a multiplet structure of the steps were observed. As a result of spiral cooperation and of the interaction of different steps originating from different hills at high transport rates the growth pattern on the surface is very complex.

The revealed non-basal dislocations are probably screw dislocations, at the bottom of the etch pits spiral steps have often been observed. These pits were either in a close packed bunch containing more than 80 pits at high supersaturation or were distributed on the top and along the edges of the hill steps.

The Burton–Cabrera–Frank theory was applied to calculate the supersaturation using the lowest d value between the macrosteps. The main growth directions are $[1\bar{2}10]$.

1. Introduction

Crystals of some $A^{III}B^{VI}$ compounds have a layer structure and that is the reason for their anisotropy. Among these compounds GaSe is very important[1].

Growth mechanism and lattice defects in GaSe have not been studied in detail. Some data on stacking faults have been published[2–4], and non-basal GaSe defects and growth spirals have been found[5–7].

2. Experimental method

The GaSe crystals selected for study were prepared by the vapor transport method[8] using iodine as transport agent. An evacuated quartz tube with a diameter of 20 mm and a length of 200 mm containing the polycrystalline GaSe and the iodine with a concentration of 5 mg/cm³ was heated in a classical two zone furnace when conditions were: $T_1 = 870$ °C, $T_2 = 720$ °C (higher transport rates); or $T_1 = 805$ °C, $T_2 = 700$ °C (lower transport rates), during a period up to ten days.

Plate-like GaSe single crystals up to 2 cm² surface were obtained. To reveal non-basal dislocations diluted chromic sulphuric acid was used[1,9,10].

3. Results

3.1. MORPHOLOGY OF THE CRYSTALS

Crystals generally show hexagonal or dendritic morphology (fig. 1). According to known data [2,11] the lateral planes were (0001) basal surfaces. Triangular and non-polygonized spiral growth hills were found on these covering planes. Two characteristic formations are shown in fig. 2. In fig. 2a, at the triangular corners two cooperating spirals are seen, and in fig. 2b there are two cooperating spiral groups of the same sign and each of them contains at least four or five elementary steps.

At some places ledges or macrosteps are formed, the step distribution on a macrostep being very inhomogenous. On spiral growth hills the step density usually strongly increases at the macrostep edge region.

On the (0001) GaSe surface two different triangular hill orientations have been found, the apexes point in opposite $[10\bar{1}0]$ directions, but the frequency of their appearance is different. The opposite hill orientations have been found sometimes on lamellae grown out from the side of the crystal (fig. 3).

As a result of spiral cooperation and of the interaction of different steps originating from different hills, at high transport rates the growth pattern on the (0001) surface is very complex. Sometimes the development of hills is prevented in one or more directions perpendicular to the hill edges and truncated forms develop. After having stopped a moving triangular front on (0001), often small whisker hexagonal (or trigonal) needle-like spikes shoot out from the front, perpendicular to the surface as is shown in fig. 4. These formations show planes of high indices at the top.

Fig. 1. Surface of GaSe crystals (magnification 10×5).

(a)

(b)

Fig. 2. Cooperating spirals (a) and cooperating spiral groups (b) on GaSe (20×10).

At the large hill sides a periodicity in growth can be observed, well developed, regular triangular fronts are followed by layers with rounded fronts and high kink density at the curved edges (fig. 4). At lower super-saturations great and very smooth (0001) surfaces were grown on which the step height was very low, see fig. 5. The step edges are decorated by some droplets of water as previously noted by Sickafus and Winder[12]).

3.2. DISLOCATION ETCHING OF THE CRYSTALS

The non-basal dislocation structure and the growth hills have been investigated using the selective etchant described in ref. 1. As the emergence points of dis-locations conical etch pits were developed. The dis-locations thus revealed are probably screw dislocations, since at the bottoms spiral steps are often observed.

Fig. 3. Two oppositely oriented hills on GaSe (10×20).

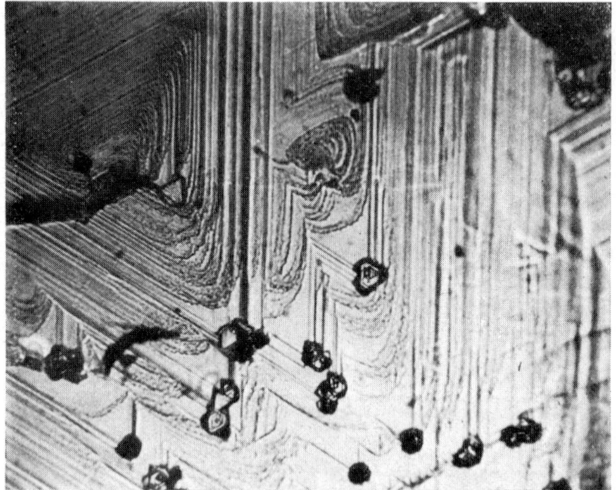

Fig. 4. Needle-like spikes on layers of GaSe with rounded fronts (10×20).

These pits are also seen at the top of the spiral growth hills, where usually screw dislocations operate. The growth steps running in [10$\bar{1}$0] directions were never attacked seriously by the etchant; however, on the truncated forms traces of etching were found and instead of triangular growth fronts there appear curved and rounded forms.

At the hill tops numerous etch pits were observed, either in a close packed bunch containing a large number of pits or distributed on the top and along the edges of the hill steps as is seen in fig. 6. In cases of thin plates the pits revealed at the hill tops could also be observed just under the hill at the other side of the crystal proving that the dislocations pass through the

Fig. 5. Growth surface of GaSe decorated by droplets[12]) (10×20).

crystal. A group of pits under a great hill is shown in fig. 7, the number of the distinguishable pits is more than 80.

The small perpendicular needles were completely dissolved during the etch process and at the corresponding points etch pits were found. No rotation or orientation change such as were described by Williams[11]) have been observed on oppositely oriented growth hill edges.

4. Discussion

4.1. GROWTH OF GaSe IN ⟨0001⟩ DIRECTION

The presence of spiral steps and hills on the (0001) surface of GaSe single crystals shows the importance of non-basal screw dislocations in the growth process. The step height and the spacing between neighboring turns of the spirals have been found to be dependant on experimental circumstances.

As is well known, the size of the critical nucleus and the supersaturation existing in the environment of growing spirals can be evaluated using the measured distance d between the neighbouring spiral arms. The critical radius of the nucleus ρ_0 is given by the Burton–Cabrera–Frank (BCF) theory of crystal growth[13]) as: $\rho_0 = d/4\pi$. At low transport rates d has a value between 10–50 μm which correspond to sizes of critical nuclei 1.6–8 μm. In principle, the value of supersaturation is[6]): $\sigma = \exp\ (a\theta/L_0kT) - 1$, where a is the interatomic distance in GaSe on the (0001) plane ($\sim$ 3.7 Å),

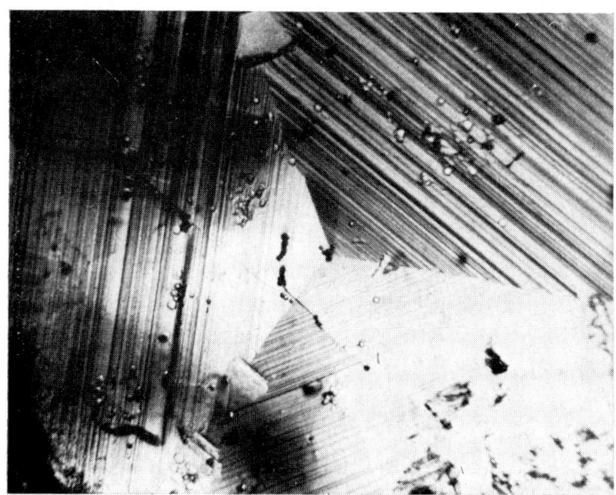

Fig. 6. Etch pits on the top and along the edges of a GaSe-hill (10 × 50).

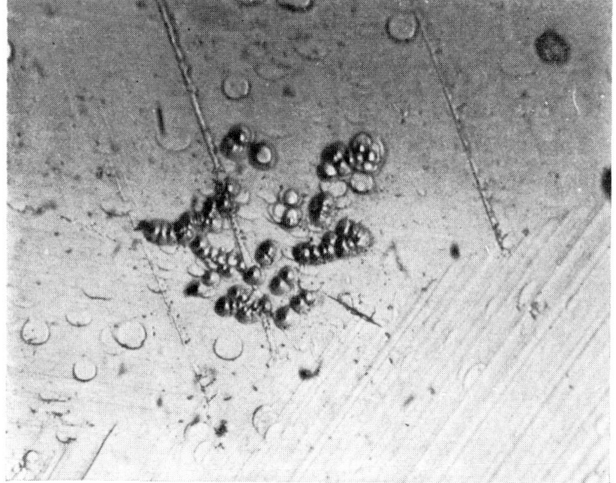

Fig. 7. Group of etch pits under a great hill of GaSe (10 × 20).

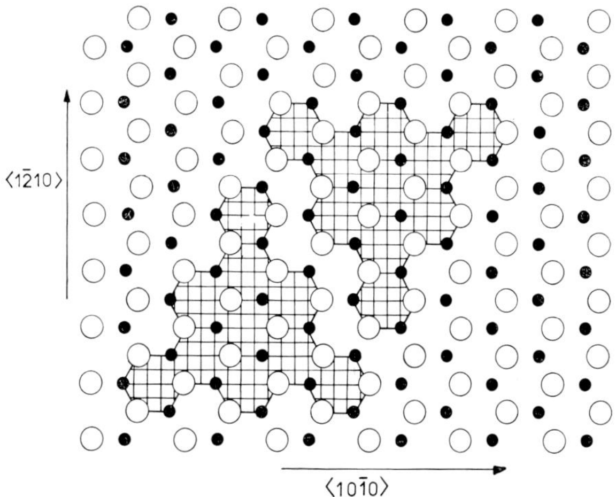

Fig. 8. Growth hills in the lattice of GaSe.

L_0 is the size of critical nucleus ($L_0 = 2\,\rho_0$), and θ is the binding energy between two nearest neighbours in the crystal. We obtain: $\sigma \cong 0.134 - 0.053\%$.

Although this uses a rough approximation for θ/kT, these values show remarkable agreement with those described in ref. 6. At high transport rates, using nearly the same temperature gradient but transport material in concentrations an order of magnitude higher than in the case of low transport rates, spiral hills showing multiplet character were found. Owing to the composite nature of the macrosteps observed the lowest d values between macrosteps were taken into account. In these cases the spacing between the spiral arms was found to be of the order of 2 µm which corresponds to a critical nucleus size of 0.32 µm and a supersaturation value of 0.67%. All of these observations show that the mechanism of GaSe growth from the vapor phase is mainly a dislocation mechanism in the ⟨0001⟩ direction corresponding to the BCF theory of crystal growth[13]).

4.2. THE GROWTH HILLS

The growth hills observed on the GaSe (0001) surfaces were generally polygonized showing triangular pyramidal forms. In undisturbed lattices both apex orientations mentioned before correspond to the lowest configuration energy being parallel with the [101̄0] edges of the crystal; therefore, their appearance is equally probable (see fig. 8). These edges contain both kinds of atoms and represent the most favourable positions on account of the stoichiometry and ionic character. In the presence of screw dislocations the frequency of the formation of one or the other type will depend on the frequency of screw dislocations or bunches with a definite (+ or −) sign. The fact that no polar character of the oppositely oriented hills could be detected by etching[1]) further supports this assumption. In case of CdTe the same triangular hill orientation difference was explained by the supposition that on the surface 60° dislocations emerge with opposite screw components and this produces symmetrical hill orientation[14]). Since on the (0001) surface many pits were observed, this possibility cannot be excluded for GaSe either. At high supersaturation it seems probable that the orientation of triangular hills is rather the result of random cooperation phenomena than of simple effects.

By etching numerous etch pits have been revealed at the top of spiral growth hills having macrosteps, showing that in these cases the hill formation is probably due to the action of dislocation bunches. At low supersaturations only a few pits were revealed at the top region, but at high ones the average number of pits at the top exceeded 40.

The ledge formation observed on the spiral hill sides can be explained either supposing the simultaneous generation of spiral steps, which move altogether on the surface, or a poisoning mechanism which has been involved in the formation of certain types of dendrites[15]).

4.3. GROWTH IN THE MAIN DIRECTION

GaSe grows from the vapor phase as thin plate like crystals. The main growth directions are [1$\bar{2}$10]. Thus the order for velocities is: $v_{[1\bar{2}10]} \gg v_{[0001]}$. Since growth by dislocation mechanism is faster than the two dimensional nucleation, this inequality means that the growth process in the main directions approximates the speed of a dendritic growth.

At the side planes the lamellar structure of GaSe can easily be observed, making the edges and side planes similar to the structure in which parallel twin planes lie in the lateral planes. In the case of PbI_2 which is also a hexagonal layer compound it was found that at high supersaturations dendrites form and in the dendritic arms different polytypes appear[16]). The formation of differently oriented layers and stacking faults can easily be imagined if the forces between the fourfold layers are Van der Waals. X-ray examinations show that the vapor phase grown GaSe crystals really consist of a mixture of ε and γ modifications[7]). In fig. 3 also a misorientation between an inner and the outer layer is shown, therefore the most probable growth mechanism in the main growth directions is the "dendritic" growth. At high supersaturations the development of the dendritic GaSe crystal was actually observed.

References

1) E. Lendvay, A. Kuhn, A. Chevy and T. Ceva, J. Mater. Sci. **6** (1971) 305.
2) Z. S. Basinski, D. B. Dove and E. Mooser, Helv. Phys. Acta **34** (1961) 273.
3) F. Jellinek and H. Hahn, Z. Naturforsch. **16** (1961) 713.
4) K. Schubert, E. Dörre and M. Kluge, Z. Metallk. **46** (1955) 216.
5) H. U. Bölsterli and E. Mooser, Helv. Phys. Acta **35** (1962) 538.
6) G. D. Guseinov and A. M. Ramaranzade, Phys. Status Solidi **23** (1967) 461.
7) R. H. Williams, J. Mater. Sci. **5** (1970) 566.
8) R. Nitsche, H. U. Bölsterli and M. Lichtensteiger, J. Phys. Chem. Solids **21** (1961) 199.
9) J. Woods, Brit. J. Appl. Phys. **11** (1960) 296.
10) M. Harsy and E. Lendvay, J. Mater. Sci **5** (1970) 828.
11) R. H. Williams, Trans. Faraday Soc. **66** (1970) 1113.
12) E. N. Sickfus and D. R. Winder, J. Appl. Phys. **35** (1964) 2541.
13) A. R. Verma, *Crystal Growth and Dislocations* (London, 1953).
14) I. Teramoto and M. Inoue, Phil. Mag. **8** (1963) 1593.
15) D. D. Saratovkin, *Dendritic Crystallization* (Consultants Bureau, New York, 1959).
16) J. I. Hanoka and V. Vand, J. Appl. Phys. **39** (1968) 5288.

Journal of Crystal Growth **13/14** (1972) 385–388 © *North-Holland Publishing Co.*

CROISSANCE DE CERTAINS POLYTYPES DES PHASES TERNAIRES SEMI-CONDUCTRICES DU SYSTÈME ZINC–INDIUM–SOUFRE À PARTIR DE LA PHASE GAZEUSE

S. I. RADAUTSAN, F. G. DONIKA et I. G. MOUSTIA

S. Lazo-Polytechnical Institute, Academy of Sciences of the M.S.S.R., Kishinev, U.S.S.R.

Some new results obtained from the investigation of the complex ternary semiconductor phases $ZnIn_2S_4$, $Zn_2In_2S_5$ and $Zn_3In_2S_6$ in the phase sections of the $ZnS–In_2S_3$ diagram are given and discussed.

Single crystals were grown by the method of chemical transport reactions using I_2 as a transport carrier. Semi-automatic apparatus for temperature programming was used.

It was established that at different temperature gradients various forms grow in the ampoule. In all experiments three-packet polytype forms of the phases $ZnIn_2S_4$ and $Zn_2In_2S_5$ were obtained.

The triangular etch pits, pyramids and hexagonal screw of crystal growth which are characteristic of many polytype forms obtained from the gaseous phases were observed on the (0001) planes.

Hexagonal bonding between layer-like packets in the cells limits the available number of the polytype forms of the phase in the package of N sulphur atom layers.

The mechanism of coalescence of different polytypes during the growth from the gaseous phase may be explained in the same way.

Luminescence investigation of the three-packet polytype form and other polytypes coalesced on its surface shows the difference in their properties.

An investigation of photoconductivity in polarised and non-polarised light showed a defined anisotropy of physical characteristics of the phases in different crystallographic directions.

Les phases ternaires $ZnS–In_2S_3$ ont été étudiées au préalable dans le système ternaire Zn–In–S, en tenant compte des propriétés semi-conductrices prononcées des compositions binaires de départ.

Hahn et Klingler ont, les premiers, étudié la nouvelle phase ternaire $ZnIn_2S_4$[1]). Les travaux ultérieurs[2–5]) ont montré que les monocristaux de $ZnIn_2S_4$ possèdent des propriétés photoélectriques et luminescentes intéressantes. Les études effectuées à l'Académie Moldave des sciences ont permis de révéler encore l'existence de deux phases:

$$Zn_3In_2S_6 \text{[6]}) \text{ et } Zn_2In_2S_5 \text{[7,8]}).$$

L'étude des propriétés physiques a montré une différence importante entre les caractéristiques semi-conductrices de ces phases (tableau 1). Divers auteurs ont remarqué des résultats différents[9]) en ce qui concerne les propriétés des cristaux $ZnIn_2S_4$ qui était expliqué par la diversité des conditions technologiques d'élaboration.

Une analyse plus minutieuse de la structure par rayons X des trois phases a permis de révéler le phénomène du polytypisme[10]) connu auparavant dans les cristaux de ZnS, SiC, CdI_2 et autres[11]).

Dans la présente étude on cite les résultats des recherches sur la croissance de diverses modifications polytypiques des phases ternaires ainsi que l'examen des particularités structurales des monocristaux obtenus. On tente de faire ressortir l'influence des diversités structurales des polytypes sur certaines propriétés physiques des cristaux. Les monocristaux ont été obtenus par la méthode chimique en phase vapeur, l'iode étant utilisé comme agent de transport.

On a réussi à déterminer les conditions optimales de croissance de différentes phases en faisant varier le diamètre et la longueur de l'ampoule, le gradient de température, de même que la quantité d'iode par cm^3.

Les recherches structurales ont montré la croissance de différentes formes polytypiques dans l'ampoule en fonction du gradient de température des zones d'évaporation et de cristallisation. Toutes les expériences donnaient le plus souvent des polytypes à trois empilements.

La vitesse de croissance des plaques monocristallines est maximum dans la direction $[2\bar{1}\bar{1}0]$, ce qui concorde avec la symétrie hexagonale du réseau.

On observe des pyramides de croissance à base

TABLEAU 1

Certaines propriétés des phases de la section $ZnS–In_2S_3$

Composition	Symétrie du réseau	Paramètre du réseau (Å)	Temperature de formation (°K)	Largeur optique de la bande (eV)	Maximum de photoconductibilité		Maximum de photoluminescence 300 °K	
					(nm)	(eV)	(nm)	(eV)
β-$\{\square_{\frac{1}{3}}In_{\frac{2}{3}}\}$ In_2S_4	cubique (spinelle)	$a = 10.76$	1350	2.05	610	2.03	$(1^e)880$ $(2^e)790$	1.4 1.57
$ZnIn_2S_4$	hexagonale	$a = 3.85$ $c = 3.086\ N$	1420	2.6–2.8	590	2.1	780 700	1.59 1.77
$Zn_2In_2S_5$	hexagonale	$a = 3.85$ $c = 3.086\ N$	1470	2.7	460	2.68	720	1.72
$Zn_3In_2S_6$	hexagonale	$a = 3.85$ $c = 18.50$	1500	2.8	412	3.0	690	1.8
ZnS	hexagonale (wurtzite)	$a = 3.82$ $c = 3.13\ N$	2150 (100 atm)	3.4–3.6	337	3.9	460	2.7

Remarques:

(1) La température de formation de $ZnIn_2S_4$ a été déterminée pour le mélange de divers polytypes; le spectre de photoconductibilité et de photoluminescence sont rélévés sur des cristaux du polytype à trois paquets.

(2) Quant aux $Zn_2In_2S_5$ les mesures n'ont pas été effectuées que sur des cristaux du polytype à trois paquets.

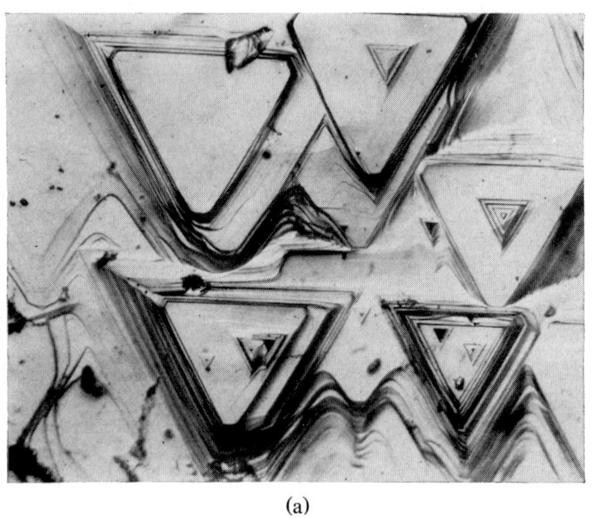

(a)

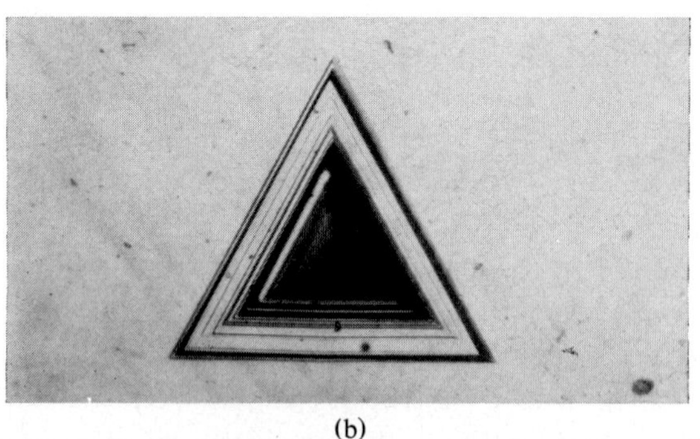

(b)

Fig. 1. Pyramides à base triangulaire de croissance (a) et les cavités à base triangulaire (b) sur les plans (0001) des cristaux de phases ternaires $Zn_mIn_2S_{3+m}$ ($\times 87$).

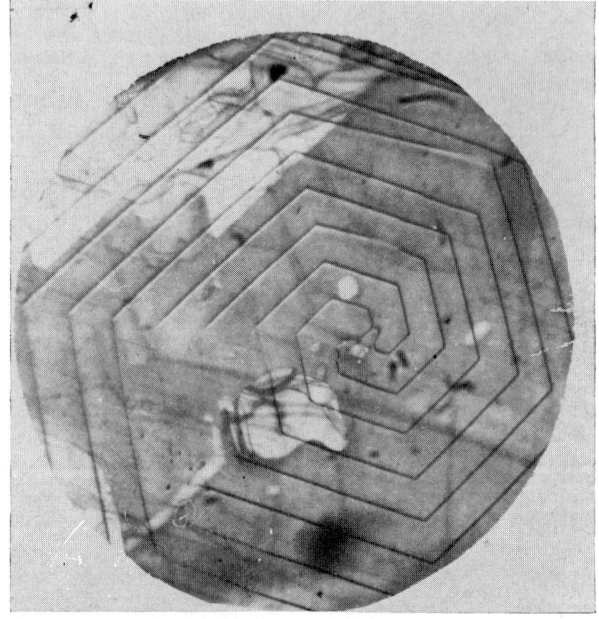

Fig. 2. Spirales hexagonales de croissance sur la face (0001) d'un des polytypes de la phase $Zn_2In_2S_5$ ($\times 270$).

triangulaire et des cavités à symétrie ternaire sur les plans (0001) qui peuvent résulter de l'émergence d'une ou plusieurs dislocations vis (figs. 1a, 1b).

Sur certains échantillons produits à partir de la phase vapeur, on observe des spirales de croissance hexagonales caractéristiques du polytypisme (fig. 2). La hauteur de la spirale mesurée par la méthode interférométrique d'environ 675 Å est conditionnée par les

TABLEAU 2

Structure des polytypes des phases ternaires $Zn_mIn_2S_{3+m}$

Polytype	Paramètres du réseau (Å)		Nombre des couches N	Assemblage compact d'atomes du soufre	Groupe spatial
	a	c			
$ZnIn_2S_4$ (I)	3.85	12.34	4	…(ABAB)…	P3m1
$ZnIn_2S_4$ (II)$_a$	3.85	24.68	8	…(ABABABAB)…	P$\bar{3}$m1
$ZnIn_2S_4$ (II)$_b$	3.85	24.68	8	…(ABACACAB)…	P6$_3$mc
$ZnIn_2S_4$ (III)$_a$	3.85	37.02	12	…(ABCACABCBCAB)…	R3m
$Zn_2In_2S_5$ (II)$_a$	3.85	30.85	10	…(ABABABABAB)…	P6$_3$mc
$Zn_2In_2S_5$ (III)$_a$	3.85	46.27	15	…(ABCBCBCACACABAB)…	R3m
$Zn_3In_2S_6$ (I)	3.85	18.5	6	…(ABABAB)…	P$\bar{3}$m1

paramètres de la maille élémentaire d'un des polytypes $Zn_2In_2S_5$.

Dans certains travaux les spirales de croissance s'expliquent par la présence de certaines dislocations vis qui peuvent expliquer l'apparition de différents polytypes à partir des structures fondamentales[11]).

Nous avons montré[10]) que les phases ternaires ZnS–In$_2$S$_3$ peuvent être exprimées par la formule générale $Zn_mIn_2S_{3+m}$ où $m = 1, 2, 3, ...$ La composition de la phase est déterminée par la valeur m. Suivant le nombre de couches d'atomes de soufre $N = Z(3+m)$ (Z est le nombre de paquets dans une maille élémentaire) pour la phase donnée se forment des familles de structures polytypiques différent par le paramètre c.

Les polytypes identifiés jusqu'à présent sont donnés dans le tableau 2. L'analyse des structures montre que d'une manière générale, les phases ternaires $Zn_mIn_2S_{3+m}$ sont caractérisées par un feuillet de composition suivante:

$$\left[m(SZnS) + (SIn_0SIn_tS\pi S) \right] = m\,ZnS + In_2S_3$$

où In$_0$ est l'atome d'indium dans les lacunes octaèdriques; In$_t$ est l'atome de l'indium dans les lacunes tétraédriques, π est une couche de lacunes de cations.

La liaison hexagonale entre les paquets dans les structures réelles réduit le nombre possible des polytypes compte tenu de l'assemblage compact à N couches d'atomes de soufre.

De la même manière peut être expliqué le mécanisme d'obtention de différents polytypes lors de la croissance à partir de la phase gazeuse. La figure 3 montre la liaison entre les polytypes à un paquet et ceux à 3 paquets $ZnIn_2S_4$.

Le défaut d'arrangement des couches d'atomes du soufre d'un polytype amène la croissance d'autres structures polytypiques possibles.

Les cristaux produits de cette manière ne se distinguent pas extérieurement de ceux des polytypes purs, cependant on observe une différence dans certaines propriétés physiques. La figure 4 montre les spectres de luminescence des monocritaux d'un polytype pur à 3 paquets $ZnIn_2S_4$ (courbe 1) et celui de $ZnIn_2S_4$ (111) imbriqué à d'autres polytypes (courbe 2). En confrontant les courbes on peut voir que le spectre de lumines-

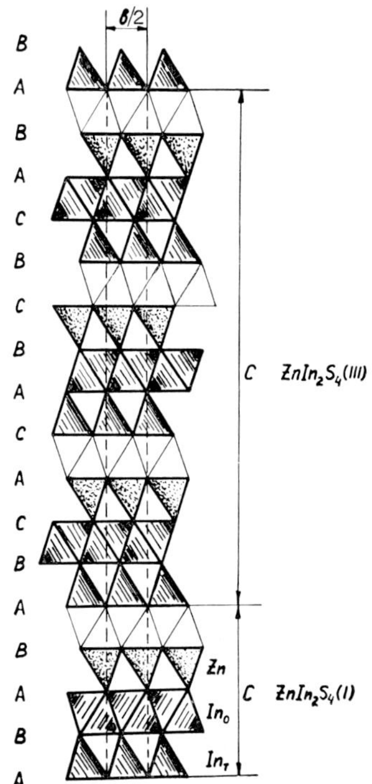

Fig. 3. Modèle du raccord des polytypes à un paquet et à trois paquets des phases $ZnIn_2S_4$ [la projection des structures en polyèdres coordinés sur le plan (2$\bar{1}\bar{1}$0)].

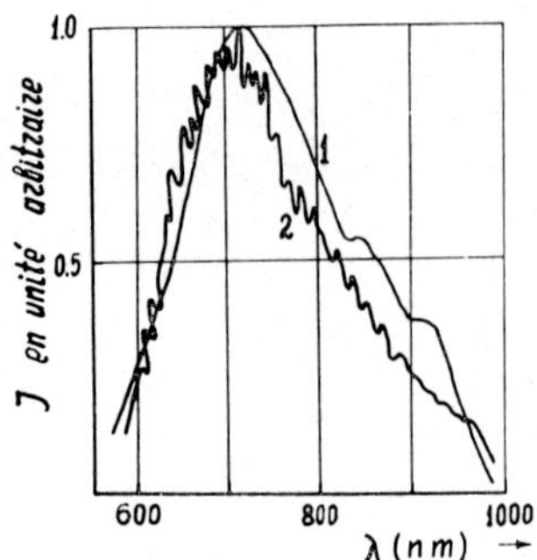

Fig. 4.   Spectres de luminescence des monocristaux du polytype pur à trois paquets $ZnIn_2S_4$ (courbe 1) et celui $ZnIn_2S_4$ (111) raccordé à d'autres polytypes (courbe 2).

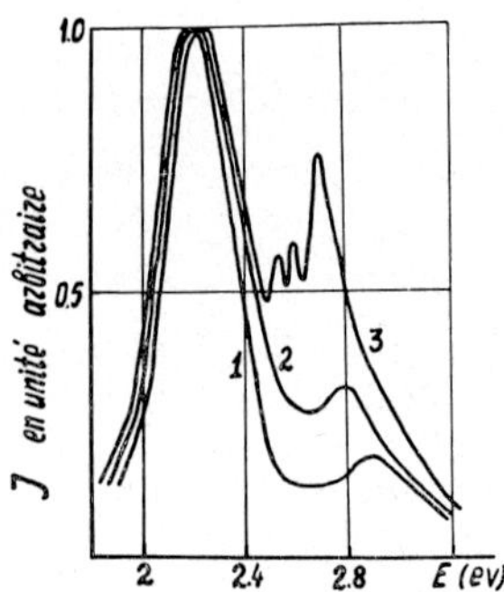

Fig. 5.   La répartition de la photoconductibilité du polytype à trois paquets $ZnIn_2S_4$ (111) dans la lumière naturelle (courbe 1) et polarisée: $E \perp C$ (courbe 2), $E \parallel C$ (courbe 3).

cence des cristaux imbriqués possède une structure fine à période d'oscillations de l'ordre d'énergie du phonon optique local ou discret, c'est-à-dire 0.2–0.4 eV[12]).

En examinant la photoconductibilité $ZnIn_2S_4$ (111) en lumière naturelle et polarisée on distingue nettement l'anisotropie des propriétés des cristaux (fig. 5). En lumière naturelle les pics de photoconductibilité correspondent à des énergies de 2.2 et 2.9 eV (courbe 1). Une position différente des pics énergétiques sur les courbes de photoconductibilité est observée en fonction de la polarisation (courbe 2: $E \perp C$, et courbe 3: $E \parallel C$). Ce phénomène est lié à l'anisotropie importante de la structure dans ces directions cristallographiques.

Le phénomène de l'anisotropie de photoconductibilité a été également révélé sur d'autres phases ternaires et probablement peut être expliqué par le caractère compliqué de la structure de bande de ces matériaux. L'examen des caractéristiques physiques montre leur corrélation étroite avec les caractéristiques structurales des phases étudiées. Les méthodes physiques et structurales de recherche pour le contrôle des modifications

polytypiques permettront de choisir les régimes optimales technologiques de la croissance des cristaux.

Bibliographie

1) H. Hahn et W. Klingler, Z. Anorg. Allg. Chem. **263** (1950) 177.
2) R. Nitsche, J. Phys. Chem. Solids **17** (1960) 163.
3) R. Nitsche, H. Bölsterli et M. Lichtensteiger, J. Phys. Chem. Solids **21** (1961) 199.
4) S. Shionoya et J. Tomoto, J. Phys. Soc. Japan **19** (1964) 1142.
5) A. Andriech, V. Zhitar, I. Lerman, S. Radautsan et V. Sobolev, Neorgan. Mater. **3** (1967) 1812.
6) V. Zhitar, N. Goryunova et S. Radautsan, Izv. Akad. Nauk MSSR **2** (1965) 9.
7) S. Radautsan, F. Donika, S. Semiletov, V. Zhitar et I. Mustya, *Mater. Dokl. v Nauchnoi Tekhnich. Konf.*, Kishinev Polytechnical Institute, 1969.
8 S. Radautsan, F. Donika, G. Kyosse, I. Mustya et V. Zhitar, Phys. Status Solidi **34** (1969) K 129.
9) L. Berguer et V. Prochukhan, *Troinie Almazopodobnie Poluprovodniki* (Izd-vo Metallurgiya, Moskva, 1968).
10) S. Radautsan, F. Donika, G. Kyosse et I. Mustya, Phys. Status Solidi **37** (1970) K 123.
11) A. R. Verma et P. Krishna, *Polymorphism and Polytypism in Crystals* (New York, 1966).
12) I. Damaskine, F. Donika, I. Mustya, S. Pichkine et S. Radautsan, Fiz. i Tekhn. Poluprovodnikov **4** (1970) 2009.

Journal of Crystal Growth **13/14** (1972) 389–392 © *North-Holland Publishing Co.*

PREPARATION OF RARE EARTH CHALCOGENIDE THIN FILMS BY THE CO-EVAPORATION TECHNIQUE

C. PAPARODITIS and R. SURYANARAYANAN

Laboratoire de Physique des Solides, C.N.R.S. 92-Bellevue, France

Co-evaporation technique is shown to be a very powerful and straight forward method for the preparation of rare earth chalcogenide thin films LnX (Ln = Eu, Yb, Sm; X = Te, Se). All the monochalcogenides crystallising in the rock-salt structure have been obtained as single crystal films. For Sm, chalcogen rich compositions $SmX_{1+\delta}$ have been obtained crystallising in the bcc structure. Solid solutions of $Eu_{1-y}\,M_yX$ (M = Pb, Yb, Gd) crystallising in the rock-salt structure have been shown to exist. The optical and magneto-optical properties of these films are studied to characterise them. As an exemple, the optical transitions in YbTe and YbSe are shown to arise from a diamagnetic ground state of Yb^{2+} ion, whereas $YbSe_{1.5}$ has a paramagnetic ground state.

1. Introduction

A wide range of physical properties have recently been reported in the family of rare earth chalcogenides LnX (Ln = Eu, Yb, Sm; X = Se, Te). The europium chalcogenides are magnetic semiconductors possessing interesting optical, magneto-optical and electrical properties[1]. Pressure-induced semiconductor–metal transitions in samarium chalcogenides have recently been observed[2]. Except for the Eu chalcogenides, the optical and magneto-optical properties and consequently the electronic band structure of other rare earth chalgogenides are unknown or not well understood. These properties can be better explored on good quality thin films of these compounds. We report here on the preparation of such films by co-evaporation of elements in vacuum. This offers the following advantages: (1) elimination of the need to prepare the solid compound LnX as a starting material; (2) a better control of the film composition by varying the rate of evaporation of individual elements.

2. Experimental details

Knudsen cell type crucibles made of tantalum and graphite were used to evaporate the rare earth and the chalcogen respectively. The temperatures of the crucibles were maintained within $\pm$ 2 °C. In our experimental set up[3] (fig. 1) a large number of films varying in composition can be prepared in a single batch. About a gram of each material was used during a run. The substrates (pyrex, NaCl and CaF_2) held at a con-

trolled temperature between 400 and 500 °C are kept in an enclosure where the residual gases are essentially those of the elements evaporated. The pressure measured outside this enclosure is $\simeq 5 \times 10^{-6}$ mm Hg. The thickness of the films obtained ranges between 100 and 5000 Å. X-ray analysis was done on these films to determine the structure and electron diffraction was carried out to check the epitaxy.

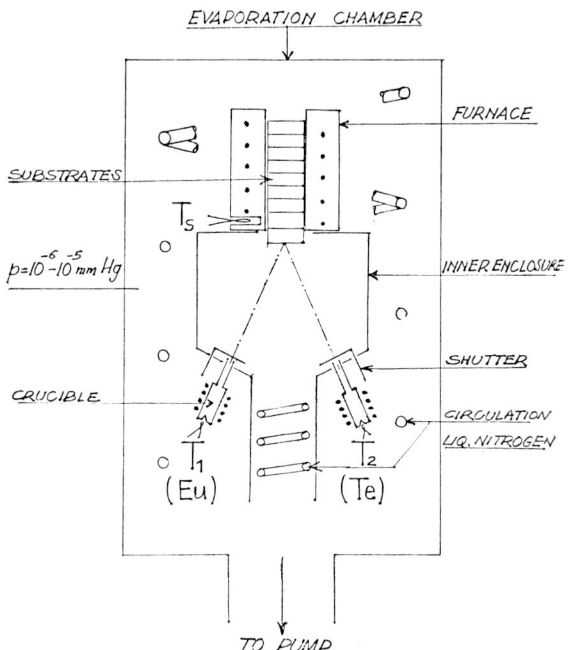

Fig. 1. Schematic representation of the evaporation chamber in the experimental set-up. The actual set-up is equipped with 3 crucibles.

TABLE 1

Film prepared	Lattice constant (Å)	Structure	Epitaxy on	Colour of the film by transmission	Magnetic properties from MCD measurements
EuTe	6.60	NaCl	CaF$_2$	Red	Antiferro
EuSe	6.20	NaCl	CaF$_2$ NaCl	Pink	Metamagnetic
YbTe	6.36	NaCl	CaF$_2$	Purple	Diamagnetic
YbSe	5.93	NaCl	CaF$_2$	Green	Diamagnetic
YbSe$_{1.5}$	5.65	NaCl	CaF$_2$	Amber	Paramagnetic
SmTe	6.60	NaCl	CaF$_2$	Green	
SmSE	6.19	NaCl	CaF$_2$	Green	
SmTe$_{1+\delta}$	9.38	Th$_3$P$_4$		Red	
SmSe$_{1+\delta}$	8.75	Th$_3$P$_4$		Orange	

3. Results

We have prepared the monochalcogenides of Eu, Yb and Sm crystallising in NaCl structure whose lattice parameters agree with those found in the literature (table 1). The films obtained on pyrex showed a (200) preferential orientation. Epitaxial growth was observed for the films deposited on (111) face of CaF$_2$. EuSe also showed epitaxial growth on (100) face of NaCl (fig. 2). All these films had a remarkable adherence to the substrate and were stable as checked by the optical and magneto-optical measurements.

3.1. EUROPIUM CHALCOGENIDES

The absorption spectra (fig. 3) of EuTe and EuSe films in the energy range of 1–5 eV agree well with those reported on the single crystals[4]. Magnetic circular dichroism (MCD) of these films as a function of temperature down to 5 °K has helped to study the effect of magnetic ordering on the optical properties[5]. The MCD signal plotted against temperature goes through a maximum situated very near the Néel temperature as should be expected in an antiferro-magnetic material.

3.2. YTTERBIUM CHALCOGENIDES

The optical density of YbTe and YbSe films is shown in fig. 4. We have observed that slow alteration of the surface resulted in a progressive decrease in reflectance in the ultra-violet. MCD experiments revealed these two compounds to be diamagnetic and the absorption bands located in the vicinity of 2 and 3 eV were assigned to transitions from the 4f^{14} ground state of Yb^{2+} to 4f^{13}(^{2}F$_{7/2}$)5d(t$_{2g}$) and 4f^{13}(^{2}F$_{7/2}$)5d(e$_g$) respectively[6].

In addition to the monochalcogenide YbSe, we have also prepared a Se rich phase which crytallises in the NaCl structure and whose composition was found to be approximately YbSe$_{1.5}$ by electron microprobe analysis. YbSe$_{1.5}$ is transparent below 3 eV (fig. 4) and MCD experiments reveal a paramagnetic ground state[6]).

Fig. 2. Electron diffraction diagram (by transmission) of a EuSe film deposited on (100) face of NaCl.

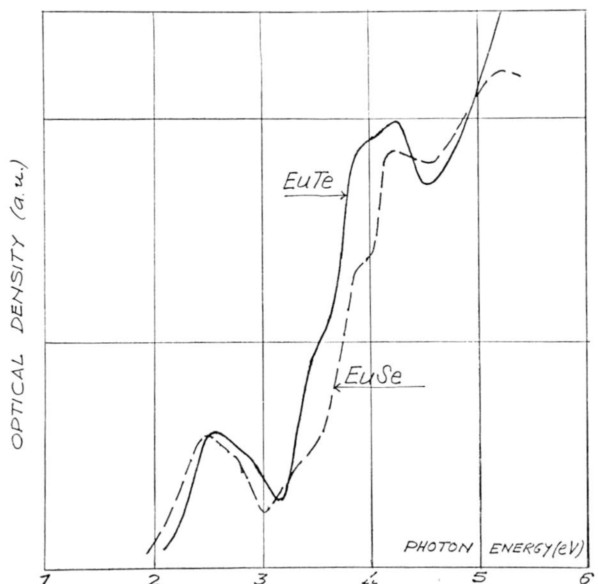

Fig. 3. Energy dependence of optical density for EuTe and EuSe films.

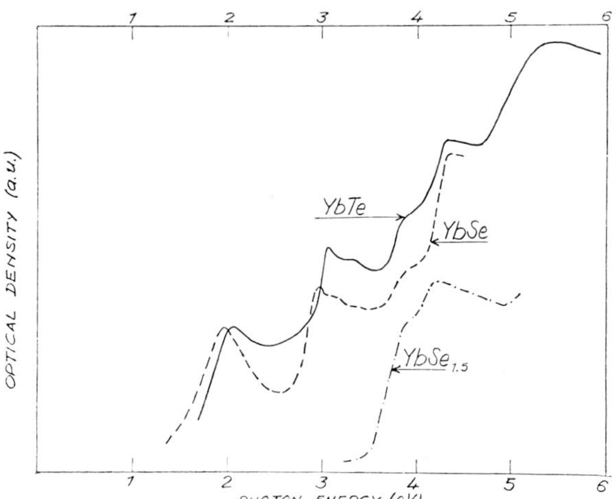

Fig. 4. Energy dependence of optical density for YbTe, YbSe and YbSe$_{1.5}$ films.

3.3. SAMARIUM CHALCOGENIDES

Whereas the Sm monochalcogenides crystallise in a NaCl structure, the chalcogen rich compositions have been found to crystalllise in a bcc structure. In the case of SmS$_{1+\delta}$ as well as SmSe$_{1+\delta}$, it was shown earlier that solid solutions exist which allow δ to vary continuously from 0.33 to 0.50, giving rise to compounds crystallising in a Th$_3$P$_4$ structure[7]. The lattice parameter of our SmSe$_{1+\delta}$ films is found to be 8.75 Å which is slightly smaller compared to the reported value of 8.78 Å for SmSe$_{1.5}$[7]. In the case of SmTe$_{1+\delta}$, it was observed that SmTe$_{1.5}$ crystallising in a Th$_3$P$_4$ structure does not exist and a calculated value of 9.32 Å for the lattice constant has been reported[8]. Our SmTe$_{1+\delta}$ films show a lattice constant of 9.38 Å, leading us to the conclusion that they do belong to a Th$_3$P$_4$ structure. Whereas SmTe and SmSe have prominent absorption bands situated around 1.2 and 2 eV and 1.03 and 1.85 eV respectively[9], SmTe$_{1+\delta}$ and SmSe$_{1+\delta}$ films are transparent in this region and has strong absorption from 2.5 eV[10].

3.4. TERNARY COMPOUNDS

By using three crucibles, we have been able to prepare solid solutions of Eu$_{1-y}$M$_y$X (M = Pb, Yb, Gd) crystallising in the NaCl structure. In particular, single crystal films grown on NaCl were obtained for Eu$_{1-y}$Pb$_y$Te solid solutions. The optical properties of these films as a function of y showed a non variation of the first absorption front for $0 < y \lesssim 0.3$ and hence indicated that the transitions are arising from the ground state of Eu^{2+} ions[11]. In the $0.3 \lesssim y < 1$ range, a regular variation of the absorption front was observed.

4. Conclusion

We have shown that the technique of co-evaporation of elements in vacuum is a powerful straight forward method for the preparation of single crystal rare earth chalcogenide thin films and solid solutions of the type Eu$_{1-y}$M$_y$X. In addition to the monochalcogenides, we have prepared YbSe$_{1.5}$ crystallising in NaCl structure and SmSe$_{1+\delta}$ and SmTe$_{1+\delta}$, both crystallising in a bcc system and belonging to a Th$_3$P$_4$ type structure. The structural, optical and magneto-optical properties of these films confirm their excellent quality.

Acknowledgments

We thank J. Badoz and his collaborators of Laboratoire d'Optique Physique (Paris 5°) for the MCD measurements, C. Sella for the electron diffraction and Rouy for electron microprobe analysis.

References

1) See for example, S. Methfessel and D. Mattis, Magnetic Semiconductors, in: *Handbuch der Physik*, Vol. 18, No. 1 (Springer, Berlin, 1968).

2) E. Bucher, V. Narayanamurti and A. Jayaraman, Annual

Conf. Magnetism and Magnetic Materials, Miami, Nov. 1970; J. Appl. Phys. **42** (1971) 1741.

3) C. Paparoditis and H. Rousseau, Avisem Conf., Saclay Sept. 1966.

4) G. Busch, G. Güntherrodt and P. Wachter, Intern. Conf. Magnetism, Grenoble, Sept. 1970; J. Physique **32** (1971) C1–928.

5) J. Ferre, M. Billardon, J. Badoz, R. Suryanarayanan and C. Paparoditis, Intern. Conf. Magnetism, Grenoble, Sept. 1970; J. Physique **32** (1971) C1–930.

6) R. Suryanarayanan, C. Paparoditis, J. Ferre and B. Briat, Solid State Commun. **8** (1970) 1853.

7) A. Benacerraf and M. Guitard. Compt. Rend. (Paris) **248** (1959) 2012;
M. Picon, L. Domange, J. Flahaut, M. Patrie and M. Guitard, Bull. Soc. Chim. France (1960) 221.

8) M. Prado, Thèse Université de Paris, 1963 (unpublished).

9) R. Suryanarayanan, C. Paparoditis, J. Ferre and B. Briat, Colleg. Intern. Couches Minces, Cannes, Oct. 1970.

10) C. Paparoditis, R. Suryanarayanan and J. Ferre, to be published.

11) R. Suryanarayanan and C. Paparoditis, in: *Les Eléments des Terres Rares*, Colloq. Intern. C.N.R.S., Grenoble, No. 180 (1969) p. 149.

Journal of Crystal Growth **13/14** (1972) 393–396 © *North-Holland Publishing Co.*

CRYSTAL GROWTH BY VAPOR TRANSPORT OF GeSe, GeSe$_2$, AND GeTe and TRANSPORT MECHANISM AND MORPHOLOGY OF GeTe*

HERIBERT WIEDEMEIER, EUGENE A. IRENE and ASIM K. CHAUDHURI

Department of Chemistry, Rensselaer Polytechnic Institute, Troy, New York 12181, U.S.A.

Vapor transport experiments with iodine were performed on the systems GeSe, GeSe$_2$ and GeTe to define optimum growth conditions in terms of crystal size and surface perfection. Mass transport rate studies on the GeTe–iodine system were carried out in the temperature gradients 560 → 445 °C and 420 → 350 °C with iodine pressures ranging from 0 to about 2 atm (based on diatomic iodine and assuming a perfect gas). Evaluation of experimental data in terms of flux (moles/cm^2 sec) as a function of iodine pressure revealed the predominant transport mode in different pressure ranges. Comparison of experimental results with calculated material fluxes based on a diffusion model was used to substantiate the transport mechanism given by the reaction

$$\text{GeTe (s)} + \text{GeI}_4 \text{ (g)} = 2 \text{ GeI}_2 \text{ (g)} + \tfrac{1}{2} \text{ Te}_2 \text{ (g)}.$$

Diffusion controlled transport yields predominantly octahedral type crystals and convective flow causes mainly platelet type crystals. The as-grown surfaces of the platelets are (111) planes. Surface imperfections increase with increasing predominance of the convectional contribution to the transport.

1. Introduction

Theory, experimental techniques and application of the "vapor transport" or "chemical transport reaction" have been discussed in detail by Schäfer[1]. A distinct advantage of this method is the possibility of using high purity elements as source material and of operating at temperatures below the melting points of the compounds. These advantages apply to the systems investigated in the present work.

Germanium selenides (GeSe and GeSe$_2$) and germanium telluride (GeTe) can be prepared by fusion of the elements at about 700 °C. Prior to this work, single crystals of GeSe[2], GeSe$_2$[3,4], and GeTe[5,6] have been obtained by the Czochralski and Bridgman techniques. The disproportionation of germanium diiodide[7–9] suggests that elemental mixtures of germanium and the appropriate chalcogen could be transported with iodine at relatively low temperatures. The work reported here is concerned with the growth of single crystals of GeSe, GeSe$_2$, and GeTe by vapor transport with iodine, with the mass transport rate and transport mechanism of the GeTe-iodine system, and

with the morphology of GeTe crystals as a function of growth conditions.

2. Experimental procedures

The crystal growth and rate studies were performed in sealed tubes of fused silica of 15 mm inner diameter and about 14 cm in length. The cleaning of the ampules, the introduction of iodine, the sealing of the tubes, and the transport procedures were described previously[10].

High purity elemental germanium (99.999%) and selenium (99.999%) and tellurium (99.999%), respectively, were used as starting materials in stoichiometric ratios corresponding to the formulas GeSe, GeSe$_2$, and GeTe. For some of the low temperature rate experiments preannealed GeTe was also employed as starting material. The identification of the transport products was established by X-ray diffraction techniques. Lattice parameters of transported materials and of reference samples, prepared in the absence of iodine by annealing stoichiometric ratios of the elements at about 700 °C, agreed with literature values for GeSe[11], GeSe$_2$[12], and GeTe[13], respectively.

3. Results and discussion

3.1. CRYSTAL GROWTH EXPERIMENTS

The results of systematic transport studies in terms

---

* Based on part of a thesis to be submitted by Eugene A. Irene to the Graduate School of Rensselaer Polytechnic Institute as partial fulfillment of the requirements for the degree of Doctor of Philosophy.

of optimum temperature gradient and iodine pressure with respect to crystal size and surface quality are summarized below. The disproportionation of germanium diiodide [9]) according to

$$Ge\ (s) + GeI_4\ (g) = 2\ GeI_2\ (g) \qquad (1)$$

suggests that analogous reactions could occur in the transport of the germanium chalcogenides. At temperatures above 450 °C, GeI_2 (g) is the predominant species in reaction (1), below 450 °C the tetraiodide species is predominant [9]).

Germanium selenide (GeSe). Large and well developed single crystals are obtained by transporting germanium selenide in a temperature gradient 570 → 450 °C with iodine concentrations ranging from 0.8–1.7 mg/cm^3 tube volume. GeSe single crystal platelets of several mm edge length with very smooth surfaces and metallic luster are obtained. In view of reaction (1) the transport of GeSe can be represented by the equation

$$GeSe\ (s) + GeI_4\ (g) = 2\ GeI_2\ (g) + \tfrac{1}{2}\ Se_2\ (g). \qquad (2)$$

Sublimation of GeSe without iodine occurs in the above temperature gradient. However, there is a decisive difference in the morphology of the sublimed and transported products. Sublimation yields microcrystalline material over an extended region of the tube whereas transport yields well defined single crystals of macroscopic dimensions.

Germanium diselenide ($GeSe_2$). Germanium diselenide is transported from low to high temperature in the gradient 400 → 550 °C with iodine concentrations of 0.9–1.2 mg/cm^3 tube volume. Transparent single crystal platelets of orange-yellow color are obtained.

The above temperature limits are in a range where GeI_2(g) or GeI_4(g) is predominant in reaction (1). Therefore, it appears that the transport of $GeSe_2$ is best represented by two individual reactions:

Low temperature: Source region

$$GeSe_2\ (s) + 2\ I_2\ (g) = GeI_4\ (g) + Se_2\ (g); \qquad (3a)$$

High temperature: Condensation region

$$GeI_2\ (g) + Se_2\ (g) = GeSe_2\ (s) + I_2\ (g). \qquad (3b)$$

This implies that GeI_4 (g) decomposes at higher temperatures to GeI_2 (g) and I_2 (g). The iodine reenters the process by means of reaction (3a).

Sublimation of $GeSe_2$ (without iodine) becomes measurable at temperatures above 550 °C. The upper temperature of the deposition region for optimum crystal size is affected by the onset of sublimation. The cold → hot transport of $GeSe_2$ can be used for a simultaneous transport of GeSe and $GeSe_2$ whereby both phases condense at opposite ends of the temperature gradient. This suggests a method of purification of these compounds.

Germanium telluride (GeTe). Germanium telluride is transported from high to low temperature in the gradient 590 → 440 °C with iodine concentrations in the range 1.0–1.5 mg/cm^3 tube volume. Large sized single crystals of octahedral habit and platelets with bright metallic luster and smooth surfaces are obtained. By analogy to the GeSe system the GeTe transport can be described by the equation

$$GeTe\ (s) + GeI_4\ (g) = 2\ GeI_2\ (g) + \tfrac{1}{2}\ Te_2\ (g). \qquad (4)$$

3.2. Transport mechanism and crystal morphology of the system GeTe-iodine

Under optimum growth conditions the sublimation rate of GeTe (without iodine) is considerably less important than in case of GeSe. Therefore, the GeTe–iodine system was selected for quantitative rate studies and as a model for thermodynamic calculations of transport rates to confirm the above suggested reactions.

3.2.1. *Mass transport rate studies*

Material transport was studied as a function of total pressure in the temperature gradients 560 → 445 °C and 420 → 350 °C. The transport rate was determined from the weight difference of the starting material and/or the weight of crystals recovered and from the total elapsed time. It is assumed that the transport rate is constant for the duration of an experiment. The transport times were long enough (18–70 hr) to average out initial transient variations in the transport rate.

The iodine pressure was computed from the initial iodine concentration for the average temperature of the tube, assuming ideal behavior and the exclusive presence of diatomic species. This pressure is proportional to the actual total pressure in the tube. The material flux (moles/cm^2 sec) is plotted as a function of the iodine pressure (atm) in fig. 1. Both curves show essentially the same trend with three apparently different

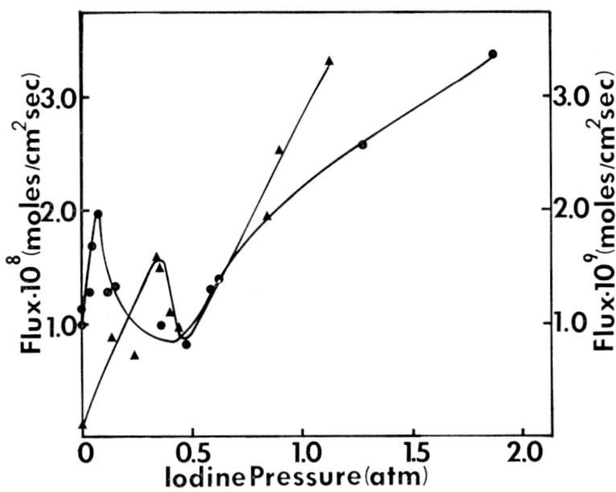

Fig. 1. The rate of transport in the system GeTe–iodine as a function of iodine pressure, based on diatomic iodine. Left hand side ordinate: (●) 560 → 445 °C temperature gradient. Right hand side ordinate: (▲) 420 → 350 °C temperature gradient.

regions. (1) At low total pressures the transport rate increases with increasing pressure. This indicates that the heterogeneous solid-gas phase equilibria in the source and/or condensation region are rate limiting. This region extends to higher total pressures for the lower temperature gradient than for the higher gradient as expected. (2) At intermediate pressures the material flux is inversely proportional to the total pressure indicating that the transport is diffusion controlled. (3) At higher pressures the material transport increases considerably with increasing total pressure revealing the predominance of the convectional transport mode. These observations are in agreement with theoretical considerations[1] concerning diffusive and convective transport.

3.2.2. Transport mechanism

Assuming heterogeneous equilibria in the source and condensation region of the reaction vessel and diffusion to be the rate determining process for the material transport between equilibrium spaces[1] the material flux can be approximated by the equation

$$F_j \propto \frac{\Delta P_i}{\sum P}, \tag{5}$$

where F_j is the flux of compound j, ΔP_i is the difference in equilibrium partial pressures of the transporting species at the high and low temperature, and $\sum P$ is the total pressure at the mean transport temperature. The proportionality constant contains an average diffusion coefficient for the gas mixture, stoichiometric coeffi-

cients of the transport reaction, and dimensional parameters. Applied to reaction (4) and with experimental data from fig. 1 one obtains

$$F_{\text{GeTe}} \propto \frac{(P_{\text{GeI}_2})_{560} - (P_{\text{GeI}_2})_{445}}{(\sum P)_{503}}. \tag{6}$$

With known thermochemical data for GeTe(s)[14], Te$_2$(g)[15,16], GeI$_4$(g)[9,17,18], and GeI$_2$(g)[9]) and estimated values for the heat capacities of GeTe(s)[14]) and GeI$_2$(g) (15 cal/deg mole) and for the absolute entropy of GeI$_2$(g) (83 eu) K_p for reaction (4) is calculated as a function of temperature. From K_p and the initial amount of GeI$_4$(g), corresponding to experimental iodine concentrations of the diffusion range (fig. 1), equilibrium partial pressures of all gaseous species and total pressures are available as a function of temperature. Assuming that the proportionality constants [eq. (6)] are approximately the same for the high and low temperature gradient a flux ratio of GeTe for the two gradients is calculated to be 17. An experimental value of about 10 is found for this ratio. In view of the various assumptions this is a reasonable agreement supporting the validity of reaction (4). In addition, X-ray diffraction patterns of all materials found in transport tubes quenched from various temperatures show only the presence of compounds in reaction (4).

By analogy the transport of GeSe probably follows the same mechanism. Experiments are in progress to verify reaction (2). Due to a lack of sufficient thermochemical data for the GeSe$_2$ system thermodynamic confirmation of reaction (3) is presently not feasible.

3.2.3. Crystal morphology

Previous studies revealed that the growth habit of MnS and MnSe crystals[10]) is affected by the surface state of the silica tubes and that the mass transport mode (diffusion or convection) has a strong effect on the morphology of CdSe crystals[19]). The morphology of GeTe crystals (fig. 2) shows that material transport by diffusion yields predominantly octahedral type crystals and that convective flow causes mainly platelet type crystals. Laue photographs of GeTe platelets indicate that the as-grown surfaces are (111) planes. Under these conditions the fast growing direction is in the (111) plane, and the slow growing direction is in the ⟨111⟩ direction. The variation in crystal habit as a function of transport mode suggests that octahedral

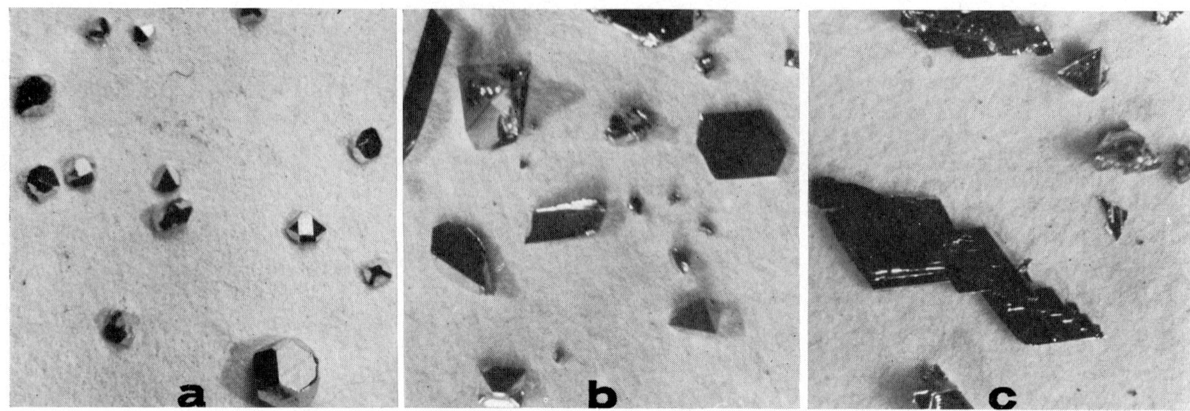

Fig. 2.   Morphology of GeTe crystals as a function of transport mode for the temperature gradient $560 \to 445$ °C. (a) Sublimation (without iodine), diffusion controlled; crystal size; 0.5–1 mm. (b) Transport (with iodine), predominantly diffusion controlled; crystal size: 3–4 mm. (c) Transport (with iodine), predominantly convection controlled; crystal size: 4–6 mm.

type crystals represent the "equilibrium" habit. This would be expected in view of the rhombohedral structure of GeTe with only a small distortion ($\alpha = 88°\ 35'$) from cubic symmetry. The ratio of octahedral to platelet type crystals decreases with increasing iodine pressure and twinning becomes more frequent. Microscopic observations of crystal faces show that diffusion conditions yield extremely smooth surfaces. Crystals grown under convection conditions have faces with random imperfections, growth spirals, and striation patterns. These observations could be explained by bulk flow of matter in the convection range, thus considerably exceeding the critical saturation limit in the condensation region.

4. Summary

Vapor transport experiments with iodine of the systems GeSe, $GeSe_2$, and GeTe have shown that these materials can be obtained in form of large and well developed single crystals at rather low temperatures. From mass transport rate studies the predominant transport mode of the GeTe-iodine system was determined as a function of total pressure. Comparing experimental material flux ratios with computed values based on a diffusion model it was possible to verify the transport mechanism. This approach appears promising for the elucidation of more complicated transport mechanisms. The change in crystal habit as a function of transport mode for GeTe is in accord with structural considerations and elementary diffusion and convection models for gas motion.

Acknowledgements

The authors are pleased to acknowledge the support of this work by the National Aeronautics and Space Administration and by the National Science Foundation.

References

1) H. Schäfer, *Chemical Transport Reactions* (Academic Press, New York, 1964).
2) A. Okazaki, J. Phys. Soc. Japan **13** (1958) 1151.
3) S. A. Dembrovskii and E. N. Loitsker, Izv. Akad. Nauk. SSR, Neorg. Mater. 3 (1967) 2092.
4) N. P. Gavaleshko, M. V. Kurik, and A. I. Savchuk, Fiz. Tekh. Poluprovodnikov 1 (1968) 1099.
5) P. F. Weller, J. Electrochem. Soc. **113** (1966) 90.
6) R. Tsu, W. E. Howard and L. Esaki, Solid State Commun. **5** (1967) 167.
7) R. P. Ruth, J. C. Marinace and W. C. Dunlop, J. Appl. Phys. **31** (1960) 995.
8) J. C. Marinace, IBM J. Res. Develop. **4** (1960) 248.
9) R. F. Lever, J. Electrochem. Soc. **110** (1963) 775.
10) H. Wiedemeier and A. G. Sigai, J. Crystal Growth 6 (1969) 67.
11) Liu Ch'ün-hua, A. S. Pashinkin and A. V. Novoselova, Russ. J. Inorg. Chem. 7 (1962) 496.
12) Liu Ch'ün-hua, A. S. Pashinkin and A. V. Novoselova, Russ. J. Inorg. Chem. 7 (1962) 1117.
13) T. B. Zhukova and A. I. Zaslavskii, Soviet Phys.-Cryst. **12** (1967) 28.
14) C. Hirayama, J. Chem. Eng. Data 9 (1964) 65.
15) D. R. Stull and G. C. Sinke, Advan. Chem. Ser. **18** (1956) 201.
16) K. K. Kelley, U.S. Bur. Mines Bull. **584** (1960) 185.
17) K. K. Kelley and E. G. King, U.S. Bur. Mines Bull. **592** (1961) 42.
18) K. K. Kelley, U.S. Bur. Mines Bull. **584** (1960) 73.
19) A. G. Sigai and H. Wiedemeier, J. Crystal Growth **9** (1971) 244.

Journal of Crystal Growth **13/14** (1972) 397–401 © *North-Holland Publishing Co.*

A NEW SYSTEM FOR THE CHEMICAL VAPOUR DEPOSITION OF SiC

M. P. CALLAGHAN and R. W. BRANDER

The General Electric Company Limited, Semiconductor Laboratories, Hirst Research Centre, Wembley, England

Good quality epitaxial layers of SiC have been grown by the gas phase reaction of trichlorosilane and n-hexane in a reducing atmosphere of hydrogen. Direct radiant heating of the α-SiC substrates using tungsten halogen lamps was employed.

1. Introduction

Equipment for the epitaxial deposition of SiC has been developed to a high degree of sophistication using radio frequency heated susceptors. Limitations inherent in this process have led to a study of deposition onto substrates directly heated by radiant power. The feasibility of growing single crystal layers of polytype identical to the substrate and of good crystal quality has been demonstrated over small areas heated to 1600 °C.

2. Experimental procedure

The standard deposition process reported earlier[1] uses trichlorosilane and n-hexane as the liquid source materials and these are held in a refrigerated bath at -38 °C. The reaction is a hydrogen reduction process and the total flow of hydrogen is 3.5 l/min with the relative silane and hexane mole ratios being 1.7×10^{-3} and 7.0×10^{-5} respectively and the doping is carried out by the addition of diborane or nitrogen. The hydrogen supplied from gas cylinders is further purified by passing it through a Ag/Pd diffusion unit and is handled in stainless steel equipment with connections welded where possible. The normal vertical reaction chamber used with this apparatus is constructed of a double walled quartz glass tube with water cooling; the (0001) seed crystal of SiC is supported on a SiC coated tungsten susceptor which is located in the centre of the tube. The growth temperature of 1600 °C, as measured with an optical pyrometer, is achieved by

eddy current heating. It is with these source materials and gas handling equipment coupled with an alternative design of reaction chamber that this paper is concerned. The design of a new reaction chamber stems from problems encountered with the materials used for the susceptor[1].

For satisfactory rf coupling it is necessary to use either carbon or metallic element susceptors or compounds of these. For normal silicon epitaxy it is satisfactory to use graphite elements usually with a SiC protective layer, however it has been found that the increased temperature required for SiC growth results in the thermal stress cracking of the SiC layer and so variable amounts of carbon are given off into the vapour reaction by the formation of methane and other hydrocarbons. Although pyrolytic graphite and vitreous carbon do reduce the rate of formation of the hydrocarbons it has not been found possible to use other than metallic susceptors. All the refractory metals have low temperature eutectics with silicon but some protection is obtained by using SiC coatings. The standard susceptor material that has given reasonable working life is tungsten and the optical and crystallographic quality of the epitaxial layers grown on (0001) SiC substrates is very good[2]. However, detailed analysis of the device characteristics obtained from diodes made by the c.v.d. process has shown that there is a limitation on luminescent response time and efficiency which can be ascribed to the presence of trapping centres in the vapour deposited layers[1]. The use of graphite susceptors although yielding poorer growth, has

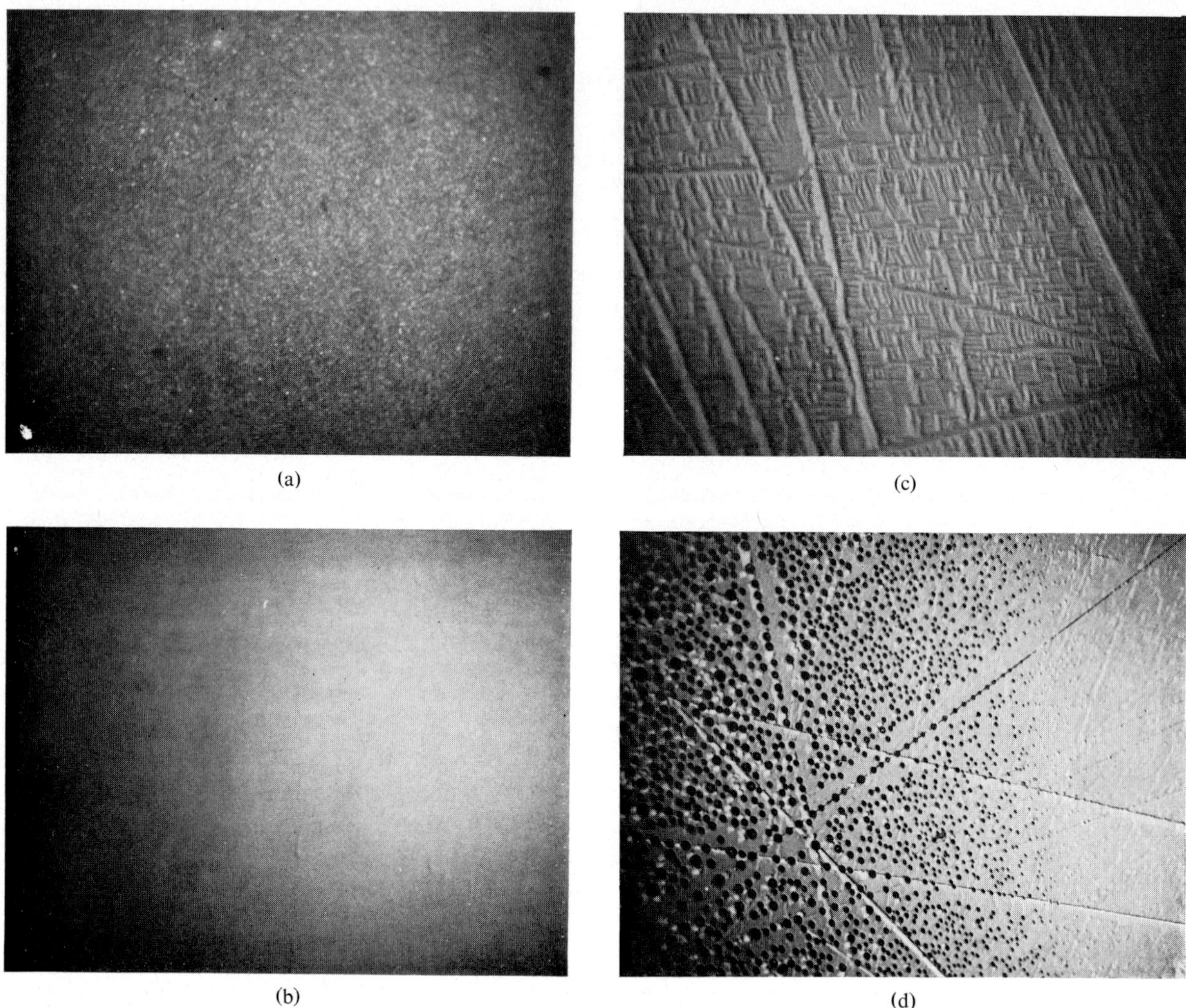

(a)   (c)

(b)   (d)

Fig. 1.   Micrographs ($\times 85$) of epitaxy showing (a) early results, (b) good growth, (c) orientation effects, (d) preferential nucleation of excess silicon.

confirmed that it is the presence of metals which is detrimental to junction properties although the level of contamination is insufficient to be detected by any of the physical analysis methods that have been tried. Luminescence work has enabled the contamination level to be estimated at 10^{16} cm^{-3}. It has therefore been necessary to evolve a direct heating system for the crystal substrate in order to be able to remove sources of metallic contamination from the reaction zone.

The method considered feasible was to use direct radiation onto the substrate surface from either (i) a laser, (ii) a krypton or xenon lamp, or (iii) a tungsten halogen lamp. Consideration of the absorption of SiC (ref. 3–5) and of quartz glass and the possible complica-

tion of ultra violet light on the growth process[6]) together with such factors as equipment size and cost indicated that of the possible sources of radiant power the tungsten halogen lamp with a peak emission at 1 µm would be most suitable. The lamp system chosen consisted of a 1 kw filament lamp mounted in a deep gold-plated aspheric reflector which is water cooled and capable of yielding a focus of 5 mm × 8 mm at a distance of 43 mm from the reflector rim (Conray 71 from Argus Engineering). The lamp is used in conjunction with a quartz glass tube of rectangular cross-section which provides the atmosphere control and support for the specimen carrier. The initial design used one lamp with a resistance heated substrate support and

VII – 20

although some results of such growth showed them to be already monocrystalline (fig. 1a) by X-ray analysis, insufficient power was available. The final form of the equipment consisted of two tungsten halogen lamps in a vertical plane on opposite sides of the reaction tube (fig. 2). These lamps were powered through a voltage stabiliser and auto-transformer, and could be individually adjusted in a vertical line by screw threads and translated along the length of the tube by means of runners. This latter adjustment enabled several depositions to be made sequentially by moving the lamps to an area of the tube uncontaminated by reaction products. It has not been found possible to measure the temperature directly during deposition either optically, because of the scattered radiation, or by means of a thermocouple because of the growth conditions. The temperature attained can be estimated by comparison with a subsidiary experiment in which a thermocouple is used to register the temperature of a dummy crystal made of graphite. Measurements show that temperatures of the order of 1600 °C can be obtained with care over small areas of about 25 mm²). The comparison is not entirely realistic but experiments on the relative growth qualities have shown that it is usuful in practice.

3. Results

The resultant growths are found to be comparable in crystallographic quality in the central region of the substrate to those obtained from the standard reaction chamber system using similar gas flow rates and compositions, with the epitaxy having no discernable features and being of the same polytype as the substrate (fig. 1b). As the substrate is being heated directly the temperature gradient across it due to the edge radiation is estimated to be 100 °C, and so the quality of growth deteriorates towards the edges due to the incorrect Si/C ratio in the vapour phase and the deposition of excess silicon globules occurs in these regions. The general topography of the central areas shows the effects of orientation at small angles off the (0001) plane and also the patterns left after the removal of scratch damage (fig. 1c). In between these two regions the excess silicon nucleates preferentially on the topography of the slice (fig. 1d). The effects of non-stoichiometry can also introduce changes in polytype of the growth layers. Fig. 3a shows the breakdown of the growth of hexagonal material into cubic in the presence

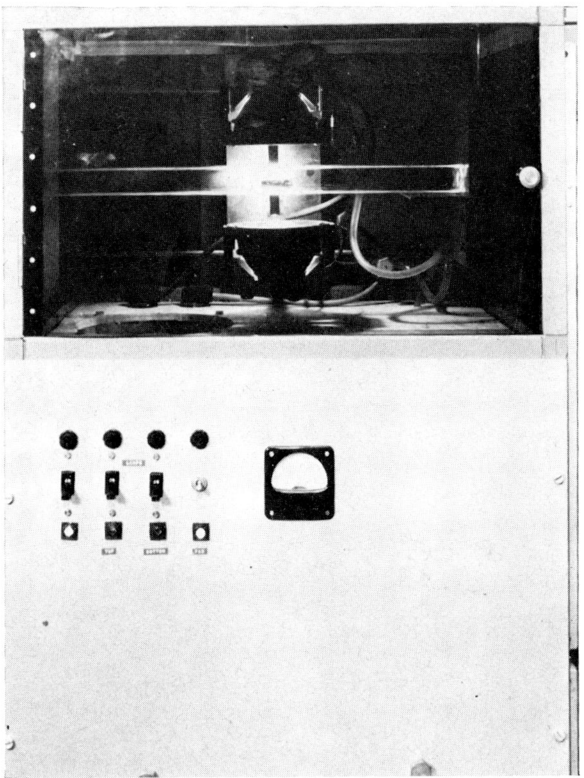

Fig. 2.   The radiant heating equipment and reaction chamber.

of excess silicon in the vapour phase. At first sight this does not occur in every case of an incorrect Si/C ratio (fig. 3b), however further inspection by X-ray topography[7]) or after decoration as in fig. 3c shows that the cubic form is present though not visible optically.

Different forms of substrate supports have been used but all have been based on a silica table in which an aperture has been trepanned. Various materials have been used to bridge the trepanned gap to support the crystals and two points have been noted. The physical support size has to be kept to a minimum in order to reduce as far as possible the re-radiation and conduction of power away from the sample, and the materials used must be both refractory and resistant to reduction in hydrogen. Initially silica filaments were used but were not satisfactory as the reduction products formed fluffy deposits which were carried into the growth zone causing faulting in the layers. Some intermediate work has been carried out using fine tungsten wire and filaments of sapphire and lucalox (Mg stabilised alumina) have been used successfully. All these materials become etched and may introduce some contamination though

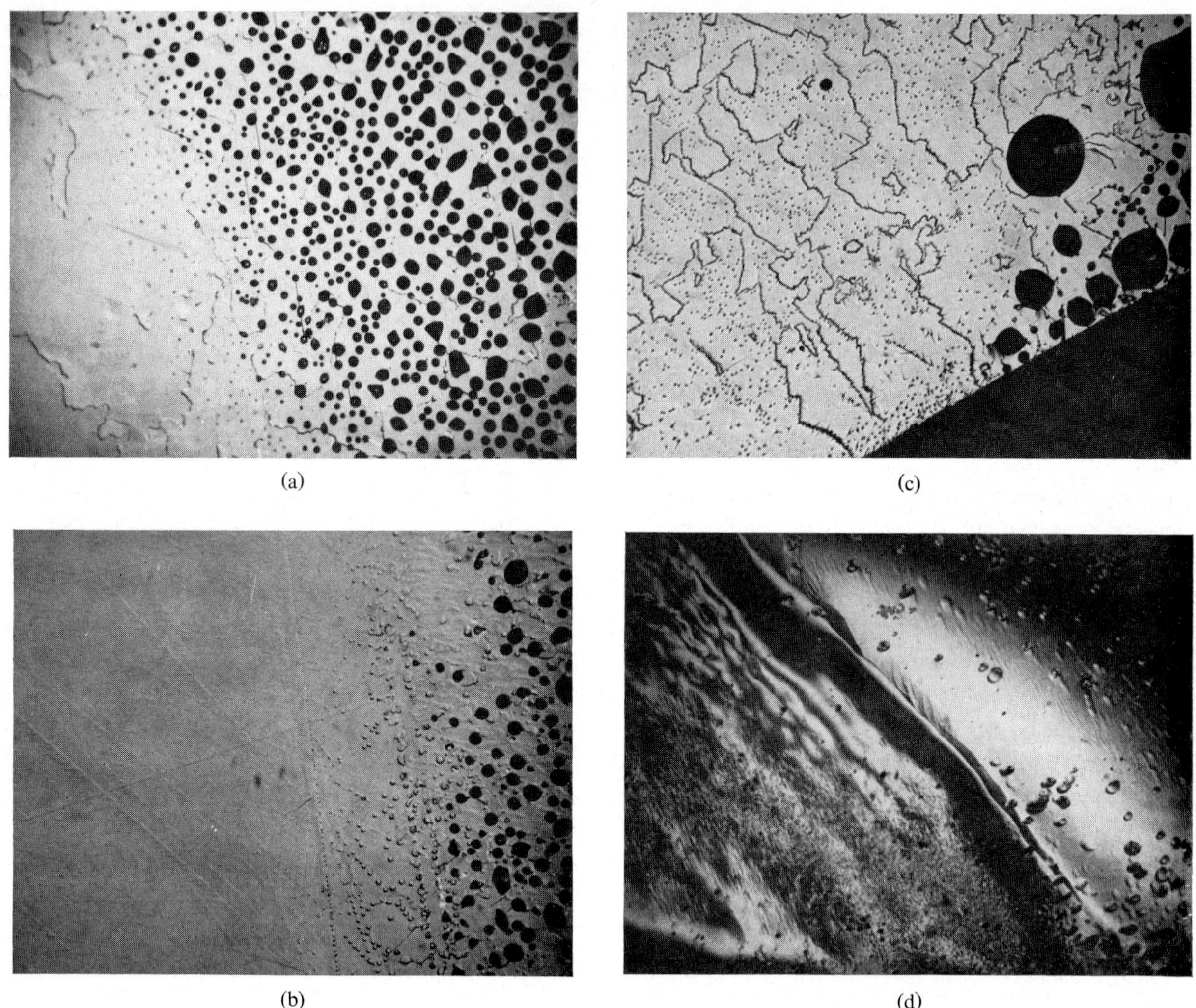

Fig. 3. Micrographs ($\times 85$) of growth in an ambient of excess silicon showing (a) presence of β-SiC, (b) without β-SiC, (c) decoration of the β-SiC mosaics, (d) etching of an α-Al₂O₃ support.

none has been detected. Fig. 3d shows the increased rate of etching of the sapphire where contact with the SiC seed has been made.

Methods of reducing the lateral temperature gradients have been considered and combined reflector/supports have been constructed from silica incorporating various reflecting layers. Gold layers have reflectances of up to 95 % in this part of the spectrum and these were used initially. It was found most expedient to use various liquid metal preparations which could be painted and baked onto the silica at about 750 °C. It was estimated that the temperature of the reflectors would not rise above 300–400 °C and would therefore be stable and inert. In practice the gold was etched away in certain areas on the downstream end of the supports and this has been accounted for by the formation of a chloride compound which volatilizes above 180 °C[8,9]. Similar effects have been found for coatings of platinum and palladium and though some protection has been obtained by sputtering a layer of SiO_2 over the metal coatings, complete coverage is difficult because of the geometry of the reflector. The most satisfactory growth over complete substrates has been achieved using a dielectric coated reflector composed of 21 layers alternately of ThF_4 and ZnS. On a reference plate the coating was measured by a transmission photometer to have a reflectivity at 1 μm of greater than 99 %[10]).

The characterisation of the epitaxial layers from the

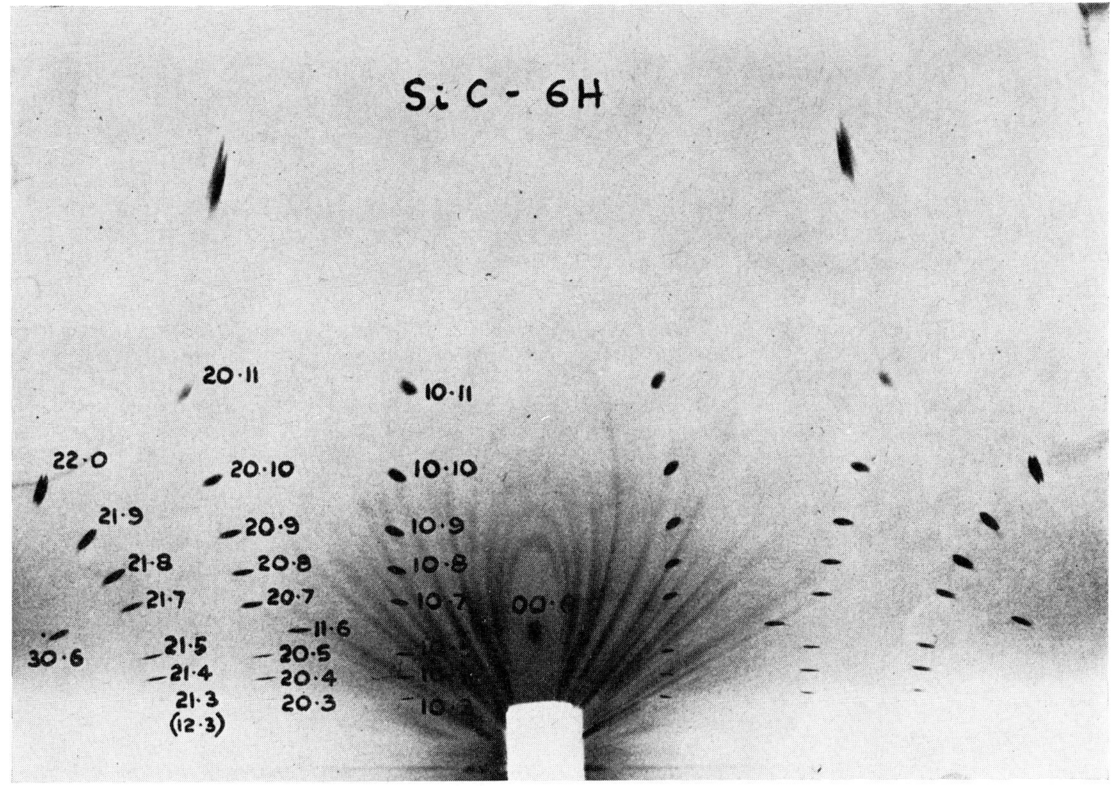

Fig. 4. X-ray inclined beam oscillation photograph of typical epitaxial layer.

point of view of crystallographic perfection is always difficult when the layer thickness is only of the order of a few microns. The rotating crystal method is one of the most powerful tools for X-ray analysis but it is not suitable for examining surface layers. A novel inclined beam oscillation technique has been used to study the perfection of these layers over a significant area[11]. Fig. 4 illustrates an analysis using this technique of typical epitaxial material from this deposition process, the spot twinning is caused by a small misalignment of the crystal relative to the axis of oscillation. From lattice parameter measurements the growth adjacent to the excess silicon globule deposits has been shown to be stoichiometric within the limits of detection of one percent.

The initial success of the growth in this form of reaction chamber has been confirmed both by optical topography and by X-ray analysis of the epitaxial layers. Furthermore, the deposition of alternately doped layers has yielded visually satisfactory injection electroluminescent diodes.

References

1) A. Todkill and R. W. Brander, Mater. Res. Bull. 4 (1969) S293.
2) R. W. Brander and R. P. Sutton, J. Phys. D 2 (1969) 309.
3) M. Namba, J. Phys. Soc. Japan 14 (1959) 228.
4) W. G. Spitzer, D. A. Kleinman, C. J. Frosch and D. J. Walsh, in: *Silicon Carbide* (Pergamon, Oxford, 1960) pp. 347–357.
5) H. G. Lipson, ibid, pp. 371–375.
6) M. Kumagawa, H. Sunami, T. Terasaki and J. Nishigawa, Japan. J. Appl. Phys. 7 (1968) 1332.
7) C. A. Wallace, Z. Krist. 126 (1968) 444.
8) D. M. Liddell, *Handbook of Nonferrous Metallurgy, Recovery of the Metals* (McGraw-Hill, New York, 1945) ch. X.
9) A. Stetefeldt, Eng. Mining J. (29 August 1903) 340.
10) W. G. Freer, private communication.
11) C. A. Wallace, private communication.

Journal of Crystal Growth **13/14** (1972) 402–405 © *North-Holland publishing Co.*

CHEMICAL VAPOR TRANSPORT OF TRANSITION METAL OXIDE SOLID SOLUTIONS

H. K. BOWEN, W. D. KINGERY, M. KINOSHITA and C. A. GOODWIN

Ceramics Division, Department of Metallurgy and Materials Science, Massachusetts Institute of Technology, Cambridge, Massachusetts 02139, U.S.A.

Preparation of solid solution crystals by chemical vapor deposition has been carried out well below the melting temperature. Most of the transition metal monoxides are mutually soluble and are amenable to this process. The composition of the solid solution growth was controlled by the activity of the source components. Solid solutions in the CoO–NiO, wustite, and FeO–MnO systems have been prepared with excellent control of homogeneity, stoichiometry and perfection. The CoO–NiO system is considered in detail. Most compositions showed only isolated dislocations in electron micrographs. One composition, $Ni_{0.66}Co_{0.34}O$, showed a two phase structure with 200–500 Å particles in the matrix. Electron diffraction studies suggest either an interpenetrating two phase structure or an ordered structure showing super-lattice peaks. Electron micrographs and electron diffraction of oxidized specimens showed a completely different structure. These results suggest an apparent miscibility gap below 800 °C in the NiO–CoO system.

1. Introduction

Chemical vapor deposition (CVD) offers unique opportunities for the preparation of compounds and solid solutions at temperatures below the melting point. The transition metal monoxides have rock salt crystal structures, similar lattice parameters (4.30 ± 0.15 Å) and most appear to be mutually soluble. Solid solutions in the CoO–NiO, wustite, and FeO–MnO systems have been prepared with excellent control of homogeneity and composition. The CoO–NiO system is considered in detail.

2. Experimental

A closed vapor transport system similar to those described by Schäfer[1]) and by Nitsche[2]) was used. Reagent grade powders were annealed in controlled oxygen atmospheres. Pellets of specified composition ($Ni_xCo_{1-x}O$) were placed in fused quartz ampoules (2 cm × 15 cm) which were evacuated (10^{-6} torr) and baked out (300 °C) before being back-filled with 1–10 torr electronic grade HCl. The ampoules were placed in two-zone gradient furnaces. Single crystals were allowed to grow at nucleation sites on the wall of

TABLE 1

NiO–CoO single crystal solid solutions

No.	Starting material	Final Crystal composition	Transporting temperature		Conditions time	Crystal size
			T_1 (°C)	T_2 (°C)		
	NiO/CoO	NiO/CoO			(days)	(mm)
1	50/50	45/55	900	800	4	0.8
2	50/50	45/55	900	850	7	1.5
3	50/50	48/52	820	800	7	2.0
4	62.5/37.5	54/46	900	800	4	1.0
5	75/25	66/34	900	800	7	3.0
6	87.5/12.5	84/16	900	800	7	2.5
7	85/15	85/15	900	800	7	1.5
8	60/40		900	800	7	1.0
9	30/70	28/72	900	800	7	1.5
10	15/85	15/85	900	800	7	2.0
11	100/0	100/0	900	800	7	1.0
12	0/100	0/100	900	800	7	3.0

the quartz tube or on MgO substrates. The substrates were either freshly cleaved or chemically polished (25% H_3PO_4). For a deposition temperature of 800 °C and $\Delta T = 100$ °C, the transport rate was 5–10 mg/hr.

Table 1 contains the composition and growth conditions of several crystals. The composition of the grown crystals was determined by wet chemical analysis and was reproducible within $\pm$ 3%. Further investigations of the homogeneity and substructure of the solid solutions were accomplished with transmission electron microscopy (TEM). Specimens were mechanically thinned to $\sim$ 50 μm and subsequently thinned to less

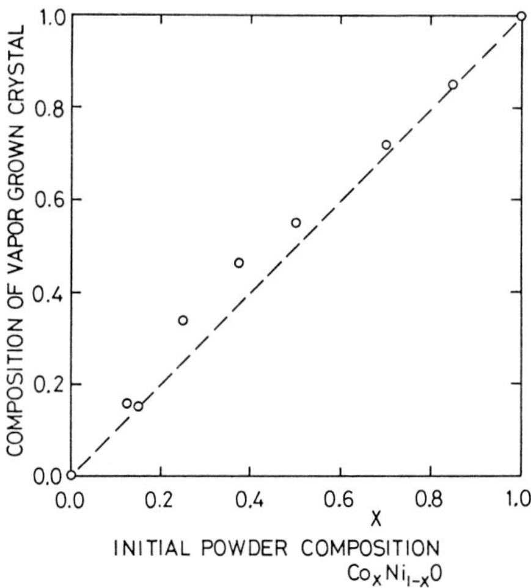

Fig. 1. Plot of source powder composition and vapor deposited crystal composition (samples 1, 4–7, 9–12).

than 800 Å by ion beam thinning (argon sputtering). Argon (99.99%) at 0.2 torr was sputtered against the rotating specimen at a 15° angle. The thinning rate at 6 kV and 200 μA beam current was about 1 μm/hr.

Besides direct observation of dislocations by transmission electron microscopy, etch pits were studied on as-grown and polished surfaces using H_3PO_4 as an etchant.

3. Results and discussion

The deposition kinetics of the NiO–CoO system follow the diffusion limited case discussed by Schäfer[1].

$$\dot{m} = \frac{Dg}{SRT} \Delta P,$$

where D is the gaseous diffusion coefficient, g is the tube cross section, S is the diffusion length, and ΔP the

pressure difference of the vapor products between the hot and cold ends. The pressure difference for each species is reflected by its chemical potential in the solid solution,

$$Co_xNi_{1-x}O \text{ (s)} + 2 \text{ HCl (g)} = x \text{ CoCl}_2 \text{ (g)}$$
$$+ (1-x) \text{ NiCl}_2 \text{ (g)} + H_2O \text{ (g)}.$$

Although no activity-composition data for the NiO–CoO system are available in the literature, comparison with several known systems (e.g. MgO–FeO, MnO–

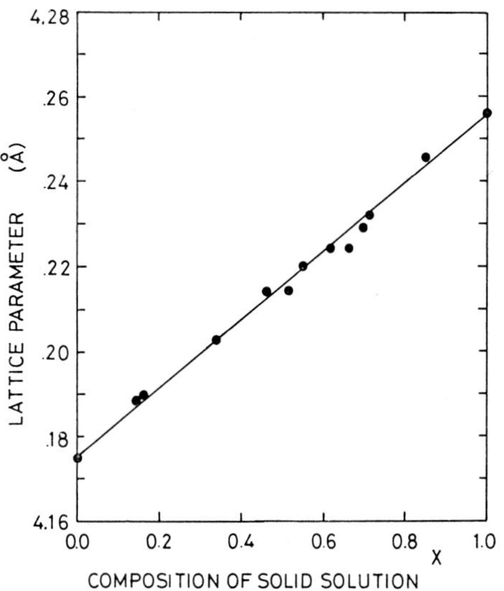

Fig. 2. Lattice parameter vs. composition for Co_xNi_{1-x} O (Debye–Scherrer powder diffraction).

CoO, MnO–FeO, MgO–CoO)[3]) suggests nearly ideal solution behavior. The plot of starting powder composition and vapor deposited crystal composition for crystals deposited at 800 °C is given in fig. 1.

The transported compositions were single crystals of the rock-salt structure as determined by Laue diffraction patterns and Debye–Sherrer data (fig. 2). Vegard's rule holds for the solid solutions in agreement with earlier polycrystalline studies[4]). No extra diffraction lines, e.g., nickel–cobalt spinel, were observed.

The relative quality of the CVD grown crystals was compared to NiO–CoO solid solutions prepared by flame fusion. In general, the flame fusion crystals contained higher dislocation densities, many subgrain boundaries and inclusions. The dislocation etch pits of vapor grown crystals required 20 times the etching time to reveal as the Verneuil crystals (concentrated H_3PO_4

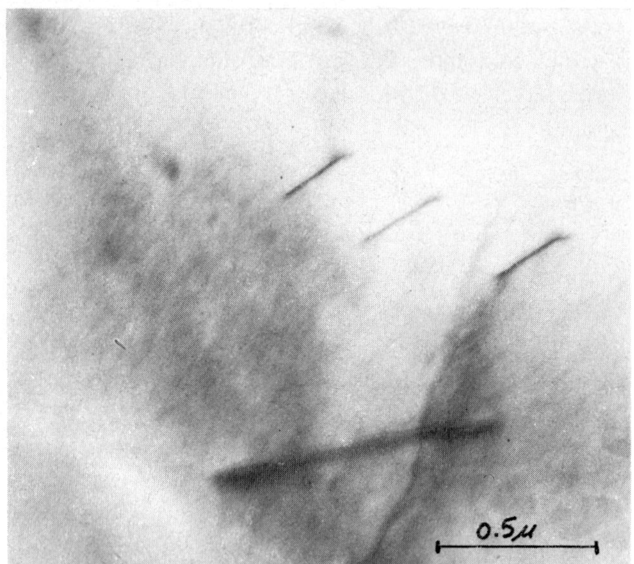

Fig. 3. Transmission electron micrograph of dislocations in CVD $Ni_{0.84}Co_{0.16}O$.

Fig. 4. Electron diffraction pattern of CVD $Ni_{0.84}Co_{0.16}O$ {110}.

at 80 °C). No inclusions or voids were observed in the CVD crystals.

The highest dislocation densities ($10^8/cm^2$) of the vapor grown crystals were at the surfaces where the crystals nucleated on the tube walls. In some cases there were also cracks in these regions apparently due to differences in thermal expansion. Typically the etch pit density within the crystal was less than $10^6/cm^2$; however, large areas could be found in most crystals which were essentially dislocation free. This was especially true for crystals deposited on MgO substrates.

Large decreases in the dislocation density as the crystal grows away from the substrate is consistent with observations of the CVD growth of wustite[5], magnesia[6] and alumina[7]. Fig. 3 shows isolated dislocations typical of most crystals. The number of dislocations observed in transmission agreed well with etch pit counts.

The small scale homogeneity and dislocation structure of most compositions is shown in fig. 3 with a corresponding electron diffraction pattern shown in fig. 4. Examination of crystals of $Ni_{0.66}Co_{0.34}O$ did reveal inhomogeneity on a 100 Å scale (fig. 5). Several areas were scanned and two separate crystals were thinned and observed by TEM. All samples of this composition yielded the grainy texture seen in fig. 5.

The particle sizes ranged from 200–500 Å. Fig. 6 is the electron diffraction pattern of this structure.

The complex diffraction patterns in fig. 6 suggests an ordered structure with superlattice peaks. However, two details in these patterns are contrary to a superlattice structure. First, there are groups of diffraction spots around a main spot but consistently spread between the main diffraction spots. Secondly, the extra diffraction spots are not exactly $1/n$ (n is an integer) values of the distance between main diffraction spots. However, the patterns are similar to wustite patterns which are thought to correspond to ordered vacancies[8]. The grainy submicrostructure suggests a two-phase, interpenetrating structure. This is not inconsistent for phase separation of similar rock salt structures with about the same lattice parameter.

The degree to which the two-phase structure is an artifact of the preparation or observation processes has been investigated. Since cobalt and nickel can be oxidized to plus three oxidation states, one possibility is that either heating in the ion beam or electron beam and subsequent oxidation could precipitate "spinel-type" regions. Zintl[9] has pointed out that the spinel phase is unstable below ~ 900 °C in a vacuum. Calculations of maximum sample temperatures from the beams yielded 250 °C for the argon thinner and 550 °C for the electron microscope. Thus thinned samples

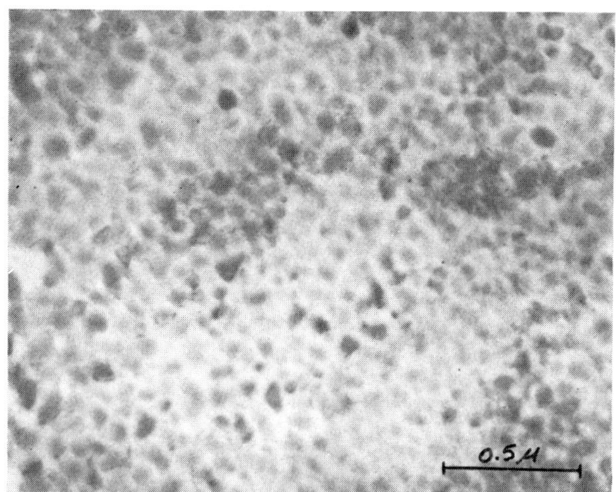

Fig. 5. Two-phase submicrostructure of $Ni_{0.66}Co_{0.34}O$.

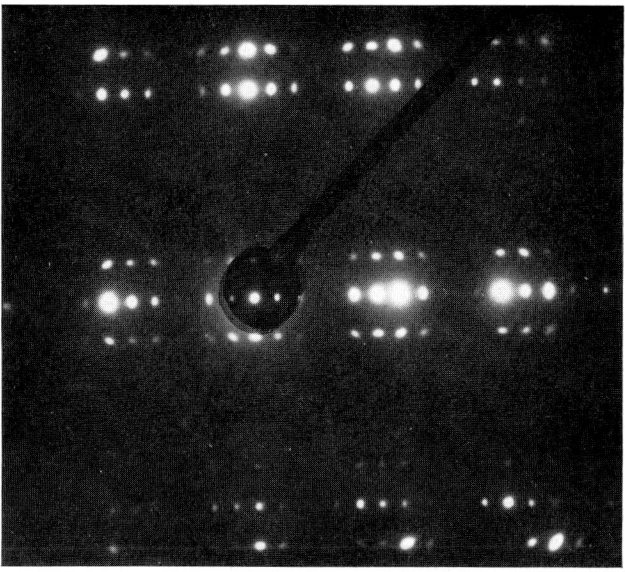

Fig. 6. Electron diffraction pattern of two-phase structure (fig. 5) {211}.

could anneal at reasonable times in the electron microscope. Nevertheless, prolonged heating in the electron microscope ($\frac{1}{2}$ hr) did not produce similar two-phase structure in samples with other compositions. Short time (5 hr) and long time (96 hr) anneals (750 °C) in air caused microregions of spinel; yet, the microstructure and electron diffraction patterns did not correspond to the as-grown crystals in figs. 5 and 6. Annealing studies of a homogeneous composition of the "as-grown" crystal ($Ni_{0.50}Co_{0.50}O$) have revealed a similar two phase structure. At 700 °C and an oxygen partial pressure of 10^{-8} atm, an annealing time of 7 days was required to realize phase separation of 100–400 Å particles. Additional studies are being carried out to determine the region of the immiscibility gap. Similar growth and characterization studies are also under way on the FeO–MnO system.

In conclusion, solid solution crystals of controlled composition can be grown at temperatures well below the melting point. The crystal perfection is much better than Verneuil grown samples which had higher dislocation densities, subgrain boundaries and contained voids and inclusions. Typically, vapor grown crystals were prepared with less than 10^{6} dislocations/cm^2. Much lower densities appear possible with proper selection of a substrate material. An immiscibility gap exists below 800 °C as revealed by the submicron two-phase structure of two compositions which were studied.

References

1) H. Schäfer, *Chemical Transport Reactions* (Academic Press, New York, 1964).
2) R. Nitsche, Fortschr. Mineral. **44** (1967) 231.
3) F. C. M. Driessens, Ber. Bunsenges. **72** (1968) 754.
4) T. Sakata and K. Sakata, J. Phys. Soc. Japan **13** (1958) 675.
5) H. K. Bowen, Ph. D. Thesis, Massachusetts Institute of Technology, January, 1971.
6) T. Vasilos, B. J. Wuensch, P. E. Gruber and W. H. Rhodes, Report of Transport Processes in Ceramic Oxides, No. 1130, Avco Corp., Lowell, Mass., June 1970.
7) H. Hobbs, personal communication, Lexington Laboratories, Cambridge, Mass.
8) F. Koch and J. B. Cohen, Acta Cryst. B **25** (1969) 275.
9) G. Zintl, Z. Phys. Chem. N.F. **48** (1966) 340.

Journal of Crystal Growth **13/14** (1972) 406–409 © *North-Holland Publishing Co.*

CONTRIBUTION DES RÉACTIONS CHIMIQUES DE TRANSPORT À L'ÉTUDE STRUCTURALE DES SILICIURES DE FER

JACQUES OUVRARD, ROLAND WANDJI ET BERNARD ROQUES

Laboratoire de Chimie du Solide, Associé au CNRS, Service de Chimie Minérale B, Université de Nancy I, 1, rue Grandville, 54-Nancy, France

Following a discussion on the control of the chemical transport reactions of binary alloys, this paper describes the use of this crystal growing method in the study of the Fe–Si system.

Vapour transport is shown to be a powerful tool for the preparation of the following types of phases: (1) compounds which decompose thermally below their melting point; e.g., β-FeSi$_2$; (2) phases which are stable at high temperatures but decompose when cooling; e.g., Fe$_5$Si$_3$ and Fe$_2$Si; (3) compounds whose homogeneity region increases with temperature; e.g., FeSi.

1. Introduction

Dans le cadre de plusieurs études consacrées à des siliciures et germaniures de métaux de transition, les auteurs ont été amenés à appliquer la technique de transport par réaction chimique en système fermé. Cette méthode de cristallogénèse leur a permis de réaliser des préparations assez délicates et ils se proposent d'en souligner l'intérêt à l'aide de quelques résultats relatifs au binaire fer–silicium.

2. Remarque sur le choix des conditions opératoires

Dans le cas où la méthode de transport est appliquée à la préparation d'alliages binaires, il est préférable d'utiliser comme source des alliages biphasés et en équilibre, plutôt que de traiter directement le composé recherché. En effet, le dépôt au puits est alors monophasé et sa composition est bien définie, puisque fixée par le maintien à des valeurs constantes de l'activité thermodynamique des constituants de l'alliage à la source.

Ce fait expérimental (tableau 1) s'explique par le raisonnement simple suivant.

Les ampoules utilisées pour la réalisation des transports en système fermé peuvent être schématisées sous la forme de deux enceintes communicantes, S et P, qui sont maintenues à des températures différentes, T_S et T_P, et jouent respectivement le rôle de source et de puits.

Si l'échantillon placé à la source est un alliage binaire à deux phases, les activités de ses constituants sont fixées et leurs valeurs déterminent totalement la composition de la phase gazeuse dans cette enceinte. Il en serait de même en P si le dépôt était également biphasé. Comme la composition du gaz au puits dépend déjà de celle de la source, elle serait alors soumise à deux contraintes difficilement compatibles. Dans un dépôt monophasé, au contraire, les activités sont variables dans un large domaine; elles peuvent donc s'ajuster en fonction des conditions opératoires et conservent ensuite leurs valeurs tant que l'alliage à la source reste biphasé.

Si un alliage monophasé est utilisé comme donneur, sa composition a toutes chances d'évoluer au cours d'une expérience de transport; une variation concomitante de composition doit alors se produire dans le dépôt au puits.

Ces considérations thermodynamiques sont également applicables au cas des systèmes ternaires; elles montrent alors que le choix d'alliages triphasés comme source permet un meilleur contrôle des réactions de transport.

3. Les réactions de transport en tant que méthode de cristallogénèse: contribution à l'étude des siliciures de fer

Grâce à leur effet de séparation et à la bonne définition de leurs produits, les réactions de transport représentent donc un moyen éventuel de purification, dans la mesure où elles sont soigneusement contrôlées.

Il s'agit de plus d'une méthode efficace de cristallogénèse qui complète utilement les procédés métallurgiques classiques pour la détermination des diagrammes

TABLEAU 1

Essais de transport du siliciure FeSi$_2\beta$

Alliage à la source	Températures (°C)		Pression P_0* d'iode (atm)	Durée (jours)	Produit au puits	Remarques sur le dépôt
	Source	Puits				
FeSi$_2\beta$	950	825	$\leqslant 1.5$	25	néant	
FeSi$_2\beta$–Si	750	825	$\leqslant 1.5$	8	Si	
	825	750	$\leqslant 1.5$	8	néant	
			> 1.5	–	Si	
FeSi–FeSi$_2\beta$	750	825	0.1 à 2	10	néant	
	825	750	$\leqslant 1.5$	25	néant	
			> 1.5	–	FeSi$_2\beta$	très faible
FeSi–FeSi$_2\alpha$	1050	900	0.5	8	FeSi	β Dans la zone froide
	1030	900	–	–	FeSi$_2\alpha+\beta$	α dans la zone intermédiaire
	1010	900	–	–	Si	
	1000	900	–	–	néant	
	1030	960	–	–	néant	
	1030	930	–	–	FeSi$_2\alpha$	β dans la zone froide
	1030	910	–	–	FeSi$_2\alpha+\beta$	α dans la zone intermédiaire
	1030	870	–	–	FeSi$_2\beta$	
	1030	830	–	–	FeSi$_2\beta$	
	1000	830	2.0	–	FeSi$_2\beta$	très faible
	–	–	1.5	–	FeSi$_2\beta$	faible
	–	–	1.0	–	FeSi$_2\beta$	moyenne
	–	–	0.5	–	FeSi$_2\beta$	maximale; les dimensions des cristaux atteignent alors 2 à 3 mm
	–	–	0.1	–	FeSi$_2\beta$	très faible
	–	–	0.05	–	néant	nulle

* La pression initiale d'iode P_0 est définie comme la pression d'iode moléculaire qui régnerait dans l'amopule à la température T_S en l'absence de réaction solide–gaz; l'iode est alors assimilé à un gaz parfait.

d'équilibre de certains systèmes binaires. Ses applications à l'étude du système fer–silicium (fig. 1) ont fait ressortir son efficacité dans la résolution des problèmes suivants:
– monocristallisation de phases stables à la température ambiante mais qui se transforment dans l'état solide, à température peu élevée; le disiliciure FeSi$_2\beta$ appartient à cette catégorie;
– préparation de phases stables à haute température mais qui se décomposent au refroidissement; c'est le cas des siliciures Fe$_5$Si$_3$ et Fe$_2$Si;
– obtention à l'état sursaturé de phases dont le domaine d'homogénéité s'élargit en fonction de la température; la préparation du monosiliciure de fer en fournit un exemple.

3.1. Préparation du siliciure FeSi$_2\beta$

Le disiliciure de fer existe sous deux formes différentes, α et β[1]. Cette transformation de phases a retenu l'attention, il y a déjà plusieurs années, car elle s'accompagne d'un changement intéressant de propriétés électriques; la phase β est semiconductrice avec un pouvoir thermoélectrique élevé, alors que la phase α a un caractère métallique.

Il était toutefois difficile d'interpréter cette transition car la structure de la variété β restait inconnue en raison de l'échec des essais de monocristallisation.

Les réactions chimiques de transport avec l'iode ont permis de surmonter les difficultés inhérentes aux propriétés de cette phase; existence dans un domaine de températures inférieures à 970 °C et formation difficile par réaction dans l'état solide. Elles ont fourni des petits monocristaux sur lesquels une étude structurale[2] et des mesures électriques[3] ont pu être effectuées.

La mise au point de cette préparation a été tout de même délicate, comme en témoigne le tableau 1 où sont regroupées les conditions opératoires essayées. En effet, la composition de l'alliage donneur et le gradient de température ont une forte influence sur le résultat du

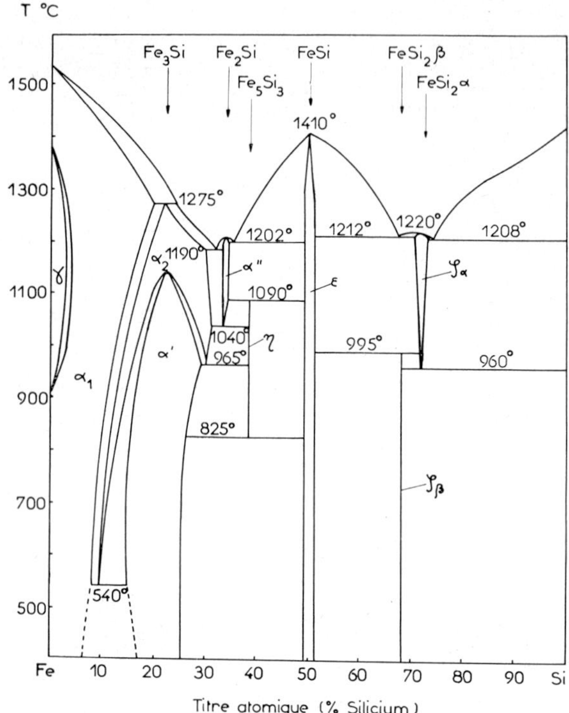

Fig. 1. Diagramme d'équilibre du système fer–silicium d'après Koster et Godecke[1].

transport; leur choix est donc limité et doit être rigoureusement contrôlé.

Il est à remarquer que l'utilisation d'un alliage donneur biphasé: $FeSi$–$FeSi_2\alpha$, est pratiquement impérative dans ce cas précis.

3.2. PRÉPARATION DU SILICIURE Fe_5Si_3

Le problème posé par le siliciure Fe_5Si_3 est différent. Cette phase n'est stable qu'entre 825 °C et 1090 °C, et il est difficile de l'obtenir pure à la température ordinaire; même après une trempe énergique, les échantillons polycristallins sont toujours partiellement décomposés en α' et $FeSi$, comme le montrent en particulier les spectres Mössbauer[2]).

Cette transformation a pu être évitée par refroidissement rapide de monocristaux, probablement en raison de la diminution du nombre de défauts bidimensionnels.

La monocristallisation de ce siliciure est facilement réalisable dans les conditions suivantes:
– alliage à la source: α'–Fe_5Si_3;
– agent activant: I_2 sous des pressions initiales comprises entre 0.5 et 1.5 atm;
– températures: $T_S = 970$ °C; $T_P = 1010$ °C.

3.3. ÉTUDE DE LA PHASE α'' ("Fe_2Si")

Plusieurs faits expérimentaux ont conduit à envisager l'existence d'une nouvelle phase, α'', dans le système fer–silicium; de formule stoechiométrique Fe_2Si, elle serait stable entre 1040 et 1215 °C environ[1]).

Toutefois, cette phase n'a jamais pu être isolée car elle se décompose très vite au refroidissement. Ainsi, les alliages polycristallins de composition Fe_2Si sont toujours polyphasés à la température ordinaire; ils sont constitués en majeure partie des phases α' et Fe_5Si_3 avec un peu de monosiliciure.

Au vu des résultats obtenus dans le cas du siliciure Fe_5Si_3, il a paru utile d'appliquer la méthode de transport chimique.

Les premiers essais ont été effectués dans les conditions suivantes:
– alliages à la source, de compositions variables entre 30 et 50 at% Si;
– températures de la source et du puits: $T_S = 1100$ °C et $T_P = 1160$ °C, choisies dans le domaine d'existence de la phase présumée α'' et où celle-ci devrait être en équilibre avec α_2 et Fe_5Si_3 (fig. 1);
– agent de transport: I_2, sous des pressions initiales comprises entre 0.05 et 1 atm.

Les cristaux formés au puits ont été soit trempés, soit refroidis naturellement, et ils ont été ensuite soumis à une analyse radiocristallographique; d'après celle-ci, la solution solide α_2 est toujours le produit du transport mais sa composition varie en fonction de celle de l'alliage à la source (tableau 2).

TABLEAU 2

Composition de l'alliage à la source (at% Si)	Phases présentes dans les produits	
	Trempés	Refroidis lentement
31	α_2	α' ($+\alpha_2$?)
36 à 40	α_2	$\alpha'+Fe_5Si_3(+\alpha_2$?)

Ces résultats n'étaient pas ceux escomptés mais ils confirmaient l'existence de deux domaines biphasés, et ainsi d'une phase intermédiaire, dans la région étudiée du diagramme d'équilibre. En effet, si tous les alliages donneurs avaient appartenu au même domaine biphasé: α_2–Fe_5Si_3, les activités thermodynamiques de leurs constituants seraient restées constantes et elles auraient fixé la composition du dépôt.

Les essais de transport ont alors été poursuivis dans de nouvelles conditions opératoires:

– alliage à la source, à 34 at% Si, constitué des deux phases α'' et Fe_5Si_3, à la température de la source;
– températures comprises entre 1040 et 1090 °C avec $T_S < T_P$;
– pressions initiales d'iode variables entre 0.1 et 1,5 atm.

Ces conditions permettent donc de se maintenir dans le petit domaine d'équilibre entre la phase α'' et le siliciure Fe_5Si_3.

L'étude aux rayons X indique que les produits ainsi obtenus, une fois trempés, contiennent une nouvelle phase qui est probablement celle cherchée; pour certains cristaux, en effet, les conditions de reflexion ne correspondent pas aux structures des siliciures connus.

Nous ne sommes pas encore en mesure de proposer un groupe d'espace, car l'interprétation des clichés de diffraction présente quelques difficultés, mais ce problème devrait être résolu dans un proche avenir.

3.4. ÉTUDE DU MONOSILICIURE DE FER

Dans le cas du monosiliciure, il serait malvenu d'insister sur les mérites du transport chimique car d'autres méthodes permettent de préparer cette phase sous forme de monocristaux très développés (méthode de Czochralski).

Toutefois, il n'est peut-être pas inutile de mentionner quelques observations qui confirment la conclusion des études précédentes, à savoir que les monocristaux sont plus facilement maintenus hors équilibre que les produits mal cristallisés.

Ce fait s'est à nouveau manifesté dans les essais de transport du monosiliciure et il a conduit à mettre en doute l'opinion suivant laquelle cette phase aurait un domaine de composition étroit et sensiblement constant en fonction de la température.

Des échantillons riches en fer sont en effet obtenus dans les conditions suivantes:
– alliage à la source: α'–FeSi;

– pression initiale d'iode comprise entre 0.1 et 2 atm;
– températures: $T_S = 750$ °C, $T_P = 820$ °C.

A leur sortie du four, ils sont monocristallins et présentent bien les propriétés paramagnétiques du monosiliciure; par contre, leur recuit suivant des cycles thermiques, entre la température ambiante et 600 °C, fait apparaître un ferromagnétisme qui témoigne d'une précipitation de la phase α'.

L'analyse chimique montre d'ailleurs que leur teneur en fer varie entre 50.5 et 51.5 at% environ, l'excès de métal augmentant en fonction de la pression initiale d'iode dans les ampoules de transport.

Ainsi ces résultats indiquent un élargissement du domaine d'homogénéité en fonction de la température et celui-ci mérite un nouvel examen.

4. Conclusion

En définitive, deux qualités des réactions chimiques de transport ressortent de leur application à l'étude des siliciures de fer. Cette méthode permet la cristallisation de phases dont le domaine de stabilité thermique est limité à des températures peu élevées. Elle facilite la préparation de phases métastables à la température ordinaire; leur cristallisation doit en effet réduire le nombre de sites de germination pour les transformations dans l'état solide.

Remerciements

Nous remercions vivement le Professeur J. Protas et ses collaborateurs pour l'aide qu'ils nous ont apportée dans les études cristallographiques.

Bibliographie

1) W. Koster et T. Godecke, Z. Metallk. **59** (1968) 502.
2) R. Wandji, Thèse de Doctorat d'État, Nancy, 1970.
3) Étude effectuée par W. Ruppel et U. Birkholz, Institut für Angewandte Physik, Université de Karlsruhe (communication privée).

Journal of Crystal Growth **13/14** (1972) 410–413 © *North-Holland Publishing Co.*

THE EFFECT OF SUBSTRATE TEMPERATURE AND REACTANT GAS PRESSURE ON THE CHEMICAL VAPOR DEPOSITION OF TUNGSTEN FILMS

H. S. SHIM and J. G. BYRNE

University of Utah, Salt Lake City, Utah 84112, U.S.A.

The growth kinetics, crystal morphology and mechanical behavior of chemically vapor deposited tungsten films are strongly affected by substrate temperature. Lower substrate temperatures result in vastly stronger and harder materials. For example, a temperature of 500 °C resulted in material with a higher yield strength than did a temperature of 700 °C. The difference amounted to 3000 psi* at four test temperatures between 200 °C and 800 °C. Deposition pressure had little effect on film strength or appearance. Pressures of 300 and 500 torr were evaluated for deposits made at 500 °C. The main structural difference observed between films produced at the two substrate temperatures was that at 700 °C the columnar grains grew straight through the film but at 500 °C there was a "fanning out" or a distribution in growth directions through the film thickness. Scanning electron microscopy reveals that the individual tungsten crystals advance with four sided pyramids at the interface, similar in appearance to those found in the electrodeposition of copper single crystals. We find that these pyramids become steeper and more irregular with increasing substrate temperature. The pyramids have {111} surfaces, and propagate by the advance of ledges up the {111} facets. An activation energy of 16 kcal/mole was determined for film growth, indicative of control by surface diffusion, adsorption and desorption processes.

1. Introduction

Chemical vapor deposition is becoming an increasingly important industrial crystallization process for such applications as epitaxial and oriented crystal growth, conducting and insulating films, fibers and composites, and coatings. In order to produce such materials with specific desired properties, it is crucial that we gain a deeper knowledge of the influence of such key growth parameters as temperature and gas pressure.

The current study was aimed at examining the earlier suggestion of Farrell et al.[1]) that low deposition temperature favors higher vacancy concentration and the probability of gaseous impurity-vacancy coalescence to produce bubble nucleation during subsequent annealing. Substrate (or deposition) temperatures of 500 and 700 °C and total gas pressures of 300 and 500 torr were evaluated. The gases involved were hydrogen and tungsten hexafluoride. The reaction at the substrate which produces the solid crystallites of tungsten is:

$$\text{WF}_6 \, (g) + 3\,\text{H}_2 \, (g) \rightarrow \text{W} \, (s) + 6\,\text{HF} \, (g) .$$

Earlier experiments in this laboratory have shown that the radial texture of C.V.D. tungsten is primarily ⟨100⟩.

* psi × 7.04 × 10⁻⁴ = kg/mm².

2. Experimental

The deposition reaction was produced on the inner surface of a copper tube. Two substrate temperatures 500 and 700 °C, and two total gas pressures, 300 and 500 torr, were evaluated by mechanical testing.

Following deposition, rings were spark cut from the composite copper–tungsten tubes. Interior ring surfaces were examined by scanning electron microscopy (SEM). The entire inner surface was then electropolished to the desired radius needed for mechanical testing. The outer copper substrate was then etched away with nitric acid. Knoop microhardness measurements were made on all surfaces of the deposited tungsten and conventional metallography was performed. Mechanical testing was performed in a special high temperature hoop stress testing apparatus[2]) which had been directly calibrated with tensile tests. The test temperatures were 240, 520, 800 and 1000 °C. SEM examinations of all fractured specimens were performed.

3. Results and discussion

Long straight columnar crystals deposited radially and uniformly on the inner substrate surface. Generally, lower temperatures and pressures favored smoother growth surfaces and slower growth rates. Fig. 1 shows the effects not only of temperature and pressure but also flow rate. The maximum growth rates in the current study [designated (s) for Shim in fig. 1] were slightly lower than those of Holman[3]) although the manometer pressure for both was 500 torr. Salt Lake City has a lower average atmospheric pressure than Livermore, California, which would require a higher absolute pressure to attain the same manometer reading as at our laboratory.

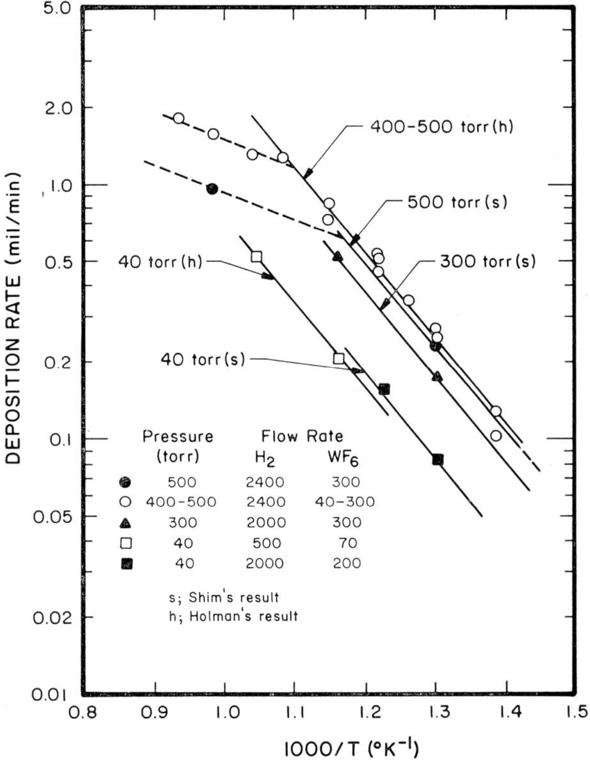

Fig. 1. Deposition rate versus $1000/T$ for various flow rates, temperatures and pressures.

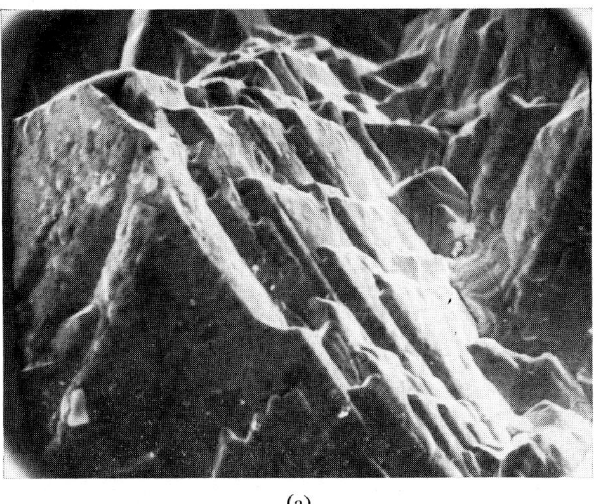

(a)

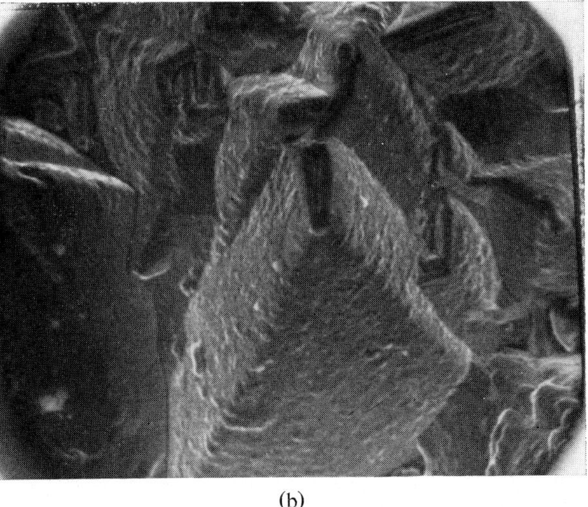

(b)

Fig. 2. Pyramid structures of growth surfaces revealed by scanning electron microscopy: (a) Current material, grown at 500 °C (700 ×); (b) Current material, grown at 700 °C (700 ×).

The larger dependence of growth rate on temperature and pressure suggests that surface processes in which adsorbed species play an important part are rate controlling. The break in the 500 torr (s) curve in fig. 1 implies that the process becomes reactant supply limited at higher temperatures. The growth rate in this high temperature region is controlled by gas phase diffusion of WF_6, H_2, and HF.

The activation energy of approximately 16 kcal/mole (which applies to the lower temperature end of the curve), is in the proper range for surface diffusion, adsorption, and desorption, as was pointed out by Holman[3]). It is too high for gas phase diffusion and too low to represent chemical reactions at the surface.

Fig. 2a shows an oblique view of the four sided pyramid shape typical of C.V.D. tungsten. This structure is similar to the pyramidal structures reported in the electrodeposition of copper single crystals[4]). Fig. 2a, which represents growth at 500 °C, when compared with fig. 2b, which represents growth at 700 °C, suggests that the pyramids developed at 700 °C have steeper sides and less crystallographic facets than those developed at 500 °C. Holman[3]) suggested that the pyramid faces in C.V.D. tungsten are {111}. This would agree with our earlier pole figure results[2]) which indicated largely a ⟨100⟩ fiber axis. Fig. 2a strongly

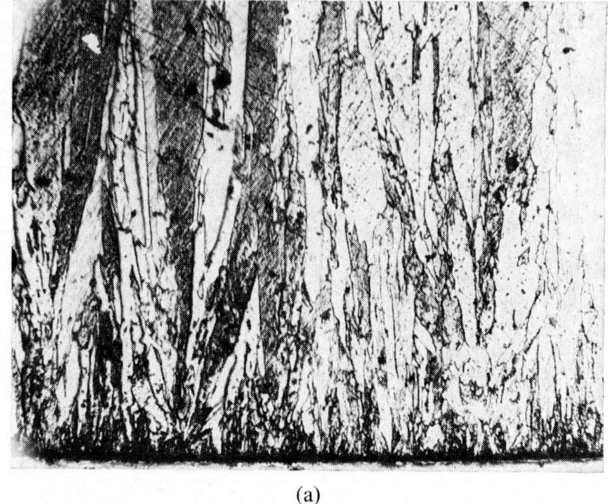

(a)

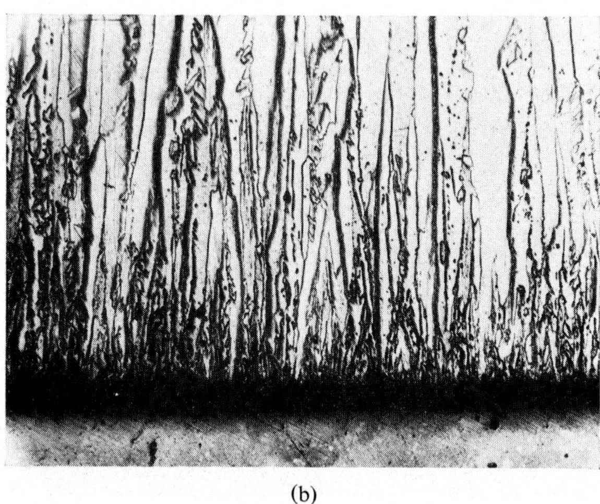

(b)

Fig. 3.   Optical micrographs showing the long dimension of the columnar crystals. The substrate is at the lower edge of each micrograph. (a) 500 °C substrate (35×); (b) 700 °C substrate (35×).

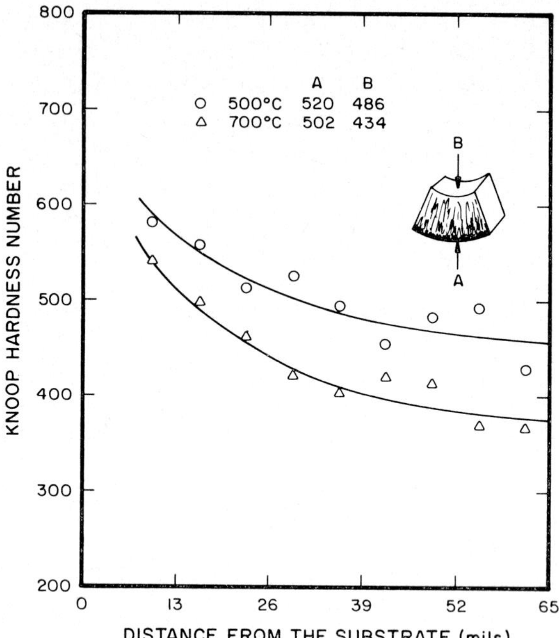

Fig. 4.   Microhardness versus distance from substrate. Substrate side is labeled A in sketch. (1 mil = 10^{-3} inch.)

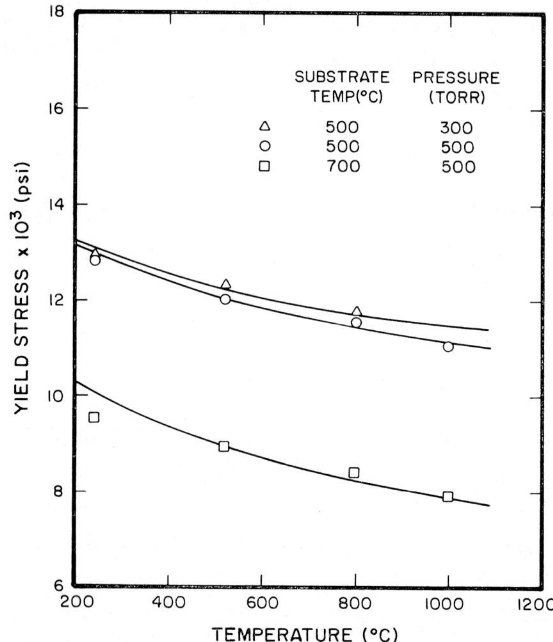

Fig. 5.   Yield strength versus test temperature for films grown at 500 °C and 700 °C.

suggests that the pyramids grow by the advance of ledges on the pyramid faces. The open tops of the pyramids may form when ledges from different pyramid sides intersect. Figs. 3a and 3b are optical micrographs showing the long dimension of the columnar crystals. Note in fig. 3a that there is more "fanning out" of the crystals (at 500 °C) whereas in fig. 3b (700 °C) the crystals are quite straight. They mey be one of the "defects" which impart higher strength to test rings grown at 500 °C, when they are stressed circumferentially.

Fig. 4 shows that the substrate side of each material (ring O.D.) is harder than the last surface to grow. It is also seen that the lower substrate temperature has the higher microhardness. With this as backround, let us now consider fig. 5. This represents the yield stress values as determined with our high temperature hoop stress apparatus, as a function of test temperature. Consistent with the microhardness results, the material grown at 500 °C is stronger than that grown at 700 °C. There is approximately the same temperature dependence for both materials and thus an almost constant 3000 psi difference exists between the curves. Where specimens deposited at 300 torr instead of 500 torr were tested, it is seen that almost no pressure effect on yield strength exists.

The morphologies of the fracture faces for all tests were observed with scanning electron microscopy. The most general conclusion one may draw from observing all of these is that material grown at 500 °C fractures on a finer scale than that grown at 700 °C.

4. Conclusions

(1) Lower growth temperatures lead to high yield strengths in C.V.D. tungsten films. The main reason for this appears to be the more disturbed microstructure, i.e., the columnar crystals are not as uniformly directed as for higher growth temperatures.

(2) Total gas pressure during growth does not have a large effect on film structure or strength.

(3) The activation energy value of 16 kcal/mole, for the growth of C.V.D. tungsten, is indicative of control by surface diffusion, adsorption and desorption processes.

(4) Four-sided pyramids are found in the growing interface. These are believed to propagate by the advance of ledges up {111} pyramid facets. The pyramids seem to increase in steepness with growth temperature.

Acknowledgment

This work was supported by the Advanced Research Projects Agency.

References

1) K. Farrell, J. E. Federer, A. C. Schaffhauser and W. C. Robinson, Jr., in: *Second Intern. Conf. on Chemical Vapor Deposition*, Los Angeles, California (Electrochem. Soc., 1970) p. 263.
2) J. S. Chun, P. S. Nicholson, A. Sosin and J. G. Byrne, in: *Second Intern. Conf. on Chemical Vapor Deposition*, Los Angeles, California (Electrochem. Soc., 1970) p. 263.
3) W. R. Holman and F. J. Huegel, in: *Proc. Conf. Chemical Vapor Deposition Refractory Metals, Alloys and Compounds* (Am. Nuclear Soc. and AIME, 1967) p. 127.
4) R. Sard and R. Weil, Electrochem. Acta **15** (1970) 1977.

Electrocrystallization

I. EPELBOIN
F. LENOIR
R. WIART
W. A. SCHULTZE

U. BERTOCCI
C. BERTOCCI
B. C. LARSON
H. R. SHANKS

Journal of Crystal Growth **13/14** (1972) 417–420 © *North-Holland Publishing Co.*

ÉTUDE PAR ANALYSE DE L'IMPÉDANCE CATHODIQUE DES PROCESSUS ÉLÉMENTAIRES DE LA CROISSANCE CRISTALLINE D'UN DÉPÔT ÉLECTROLYTIQUE

I. EPELBOIN, F. LENOIR et R. WIART

Laboratoire de Physique des Liquides et Electrochimie, Groupe de Recherche du C.N.R.S. associé à l'Université de Paris VI, 11, quai St. Bernard, Paris 5e, France

In the case of copper electrocrystallisation, analysis of the impedance diagram, at very low frequencies, leads to a value of a diffusion coefficient D, in agreement with the value found from the limiting current. This confirms that we are dealing with the transport of Cu^{2+} ions through the diffusion layer.

As for nickel, electrocrystallisation is shown to occur in two successive reactions, where (NiOH) acts as an adsorbed intermediate consumed at the electrode. Addition of inhibitors such as an acetylenic alcohol or an alkali-metal sulphonate to the electrolyte changes the kinetic parameters of these two successive reactions, and decreases the electrochemical double-layer capacity.

1. Introduction

Le but de ce travail est d'apporter des renseignements sur deux processus élémentaires de l'électrocristallisation des métaux: le transport de matière et l'adsorption.

Dans les exemples que nous présentons, le transport de matière par diffusion convective peut toujours être contrôlé au cours de la croissance cristalline en utilisant une électrode à disque tournant à laquelle on applique la théorie hydrodynamique de Levich[1]). Par ailleurs, nous associons étroitement l'analyse de l'impédance cathodique à celle de la courbe de polarisation stationnaire.

Nous commençons par l'étude de la croissance cristalline du cuivre en milieu sulfurique aux fortes tensions cathodiques. Nous montrerons ensuite, que l'électrocristallisation du nickel se prête mieux à une étude quantitative du processus d'adsorption des espèces MeOH (Me = métal) et à celle de l'action de certains inhibiteurs organiques qui modifient de façon spécifique la croissance cristalline. Dans tous les cas, celle-ci est étudiée sur des dépôts polycristallins suffisamment épais afin d'éviter l'influence du support.

2. Transport de matière

L'électrocristallisation du cuivre en milieu sulfurique, aux fortes surtensions, est un exemple type où la valeur du courant limite de diffusion et l'analyse de l'impédance cathodique prouvent, de façon indépendante, que les ions Cu^{2+} sont transportés à travers la couche de diffusion.

Avec la solution $CuSO_4$ 0.5 M + H_2SO_4 0.5 M, et pour des densités de courant d'électrolyse comprises entre 10 et 500 mA/cm², nous avons trouvé que la vitesse de rotation Ω de l'électrode à disque doit dépasser 2000 tr/min pour que le processus de transport de matière ne limite plus le courant d'électrocristallisation[2]). En effet, pour $\Omega < 2000$ tr/min, le courant dépend de la vitesse de rotation de l'électrode à disque et prend une valeur limite quand la polarisation cathodique augmente. Conformément à la théorie[1]), nous avons vérifié que ce courant limite est proportionnel à $\Omega^{\frac{1}{2}}$ en régime de convection forcée ($\Omega \geqslant 35$ tr/min). Du facteur de proportionnalité, on déduit un coefficient de diffusion D de 0.62×10^{-5} cm²/s, en accord avec la valeur que nous avons déterminée par une étude analogue du courant limite, mais avec une solution polarographique ($CuSO_4$ 10^{-2} M + H_2SO 0.5 M). Cette valeur de D correspond au coefficient de diffusion des ions Cu^{2+} (réf. 3).

L'analyse, en fonction de la fréquence, des termes réels (R) et imaginaires (G) de l'impédance cathodique, $Z = R - jG$, relevée à des fréquences relativement élevées, ne renseigne pas sur les processus de transport de matière. En effet, quelle que soit la vitesse de rotation Ω, l'impédance se réduit en très haute fréquence à la résistance de l'électrolyte et, entre 20000 et 100 Hz, à une boucle capacitive (voir fig. 1) qui correspond probablement au transfert d'électron.

En dessous de 100 Hz, aux vitesses de rotation du disque inférieures à 2000 tr/min, on constate l'appari-

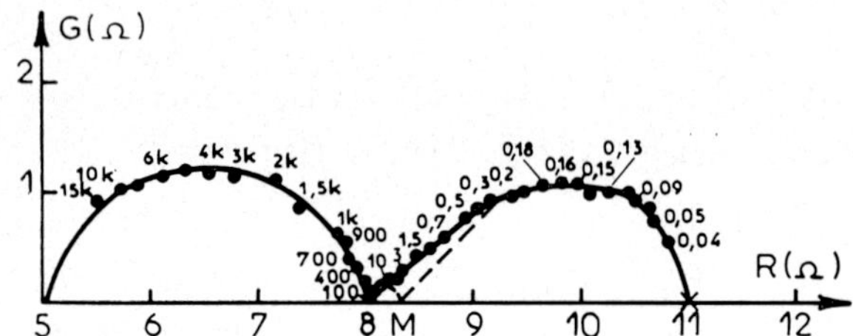

Fig. 1. Diagramme d'impédance $Z = R - jG$, paramétré en hertz, correspondant a l'électrocristallisation du cuivre, obtenu avec une électrode à disque tournant ($\Omega = 115$ tr/min). Composition de l'électrolyte: $CuSO_4$ 0.5 M + H_2SO_4 0.5 M. Température $= 26$ °C. Densité de courant $= 100$ mA/cm².

tion d'une impédance qui est dûe au transport de matière par diffusion, puisqu'elle varie considérablement avec Ω. Le diagramme d'impédance indiqué à titre d'exemple sur la figure 1 est obtenu en régime laminaire de convection forcée ($\Omega = 115$ tr/min) et pour une densité de courant de 100 mA/cm². On constate qu'entre 0.04 et 0.5 Hz le diagramme correspond à l'expression de l'impédance de diffusion Z_d obtenue à partir du modèle de la couche de diffusion de Nernst[4])

$$Z_d = R_d \frac{\text{th } (j2\pi F\delta^2/D)^{\frac{1}{2}}}{(j2\pi F\delta^2/D)^{\frac{1}{2}}}, \tag{1}$$

où R_d est la résistance de diffusion, F la fréquence, δ l'épaisseur de la couche limite de diffusion et D le coefficient de diffusion des ions à travers cette couche. Avec le disque tournant, δ est connu[1]) et l'expression de Z_d permet d'évaluer D à 0.6×10^{-5} cm²/s. Cette valeur est celle trouvée plus haut à partir du courant limite et prouve que le modèle de Nernst peut être appliqué à l'étude de la croissance cristalline d'un dépôt électrolytique d'un métal comme le cuivre.

3. Adsorption des cations

La séparation entre une impédance de diffusion et celle dûe à l'adsorption et au transfert d'électron est souvent très délicate. Ainsi, on observe sur la figure 1 qu'entre 100 Hz et 0.5 Hz environ, la courbe expérimentale s'écarte de la courbe théorique [formule (1)]: en effet, on sait[4]) que si les processus d'adsorption étaient négligeables, l'impédance Z_d devrait, à ces fréquences, être une droite d'origine M et inclinée à 45° (représentée en pointillé sur la figure 1).

On admet souvent dans la littérature[5,6]) que la réaction de transfert de charge s'effectue suivant le mécanisme.

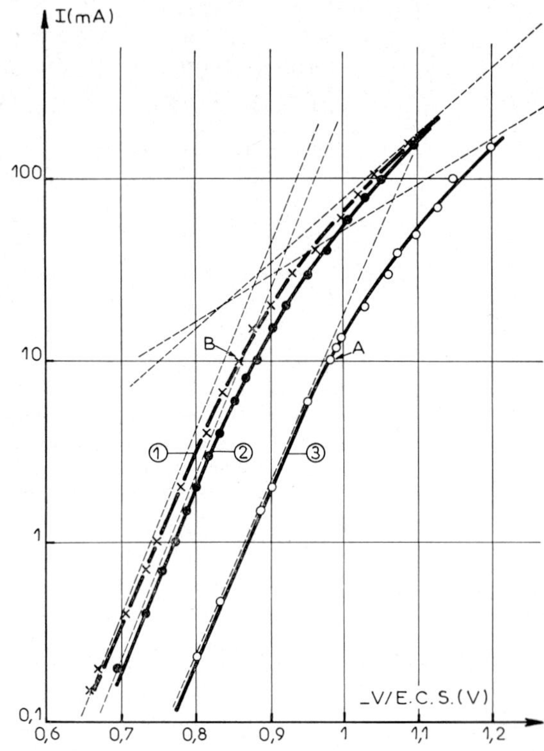

Fig. 2. Influence des inhibiteurs organiques sur la courbe courant–tension $I(V)$ stationnaire, obtenue dans le cas de l'électrocristallisation du nickel avec une électrode à disque tournant ($\Omega = 2000$ tr/min), de surface 0.2 cm² et à partir d'un électrolyte de type Watts (concentration de $Ni^{2+} = 1.22$ M; pH $= 4.5$; température $= 50$ °C. ($\times$) courbe 1: sans inhibiteur organique; ($\bullet$) courbe 2: avec benzène sulfonate de sodium (3×10^{-2} M); ($\bigcirc$) courbe 3: avec butyne-2 diol-1,4 (8×10^{-3} M). Les courbes expérimentales sont en trait plein; les courants partiels, correspondant aux deux réactions de transfert, sont en pointillé. (A) $I = 10$ mA, $V = -0.98$ V/E.C.S.; (B) $I = 10$ mA, $V = -0.86$ V/E.C.S.

$$Cu^{2+} \xrightarrow{(1)} Cu^+ \xrightarrow{(2)} Cu$$

et qu'il est limité par la réaction (1) dès que la surtension dépasse quelques dizaines de millivolts. En fait, l'analyse fine de l'impédance Z de l'électrode à disque, surtout aux vitesses pour lesquelles le transport de matière est négligeable ($\Omega \geqslant 2000$ tr/min), montre que le processus de transfert de charge est bien plus compliqué que le modèle proposé. En effet, nous avons indiqué par ailleurs[2]) qu'outre la boucle capacitive observée en haute fréquence dûe d'après[6]) au mécanisme précité, on constate aux fréquences comprises entre 200 et 0.8 Hz une deuxième boucle capacitive suivie en très basse fréquence d'une boucle inductive

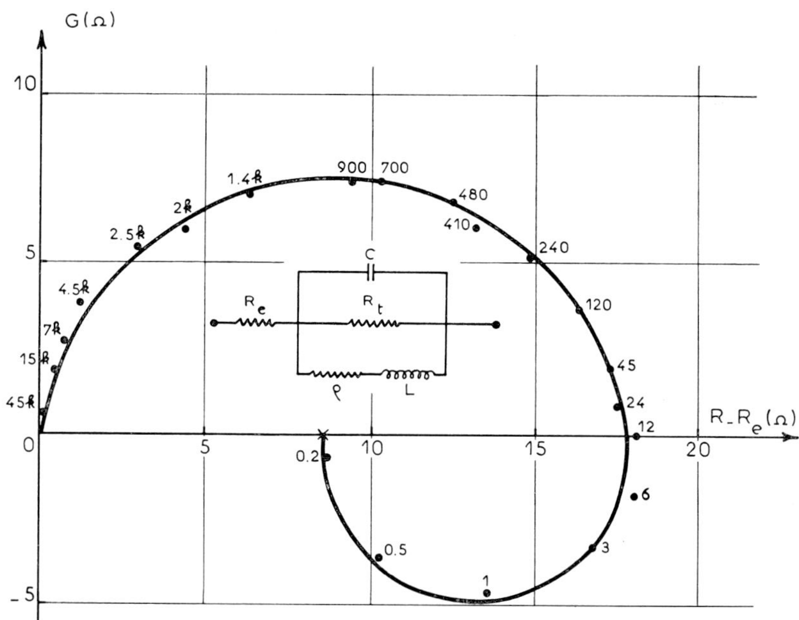

Fig. 3. Diagramme d'impédance $Z = R - jG$, paramétré en hertz, correspondant à l'électrocristallisation du nickel, en présence du butyne-2 diol-1,4 (8×10^{-3} M). Les conditions d'électrolyse sont celles du point A de la figure 2. Le schéma représente le circuit équivalent à l'impédance Z de l'électrode.

($G < 0$) qui laisse supposer l'influence d'un processus d'adsorption.

L'électrocristallisation du nickel à partir d'un électrolyte aqueux usuel de type Watts ($NiSO_4, 7 H_2O$: 300 g/l; $NiCl_2, 6H_2O$: 35 g/l; H_3BO_3: 40 g/l) se prête mieux à l'étude de l'adsorption des ions métalliques et du transfert de charge. En calculant par la méthode des moindres carrés[7]) les éléments du schéma représentatif de l'impédance à différentes tensions cathodiques, nous avons pu séparer l'impédance dûe aux réactions cathodiques de celle dûe à la capacité de double couche électrochimique. Les résultats obtenus avec des vitesses de rotation supérieures à 2000 tr/min, auxquelles le transport de matière n'intervient pas, indiquent que la réaction cathodique s'effectue en plusieurs étapes par l'intermédiaire de $(NiOH)_{ads}$[8])*.

Dans le cas du nickel, l'évolution de la courbe de polarisation et de l'impédance cathodique avec la concentration de Ni^{2+} dans l'électrolyte prouvent que l'électrocristallisation s'effectue selon un processus en deux réactions électrochimiques successives au cours duquel $(NiOH)_{ads}$ ne joue pas le rôle de catalyseur, mais celui d'intermédiaire consommé à l'électrode[9]).

* On trouve, dans ce mémoire, une description des méthodes expérimentales utilisées ici.

4. Action des inhibiteurs sur la croissance cristalline

On sait que la croissance cristalline du nickel est modifiée de façon très différente suivant qu'on ajoute à la solution de Watts un alcool acétylénique ou un sulfonate alcalin[10]). En outre, avec les alcools acétyléniques, l'action inhibitrice et la variation du micro-profil du dépôt sont limités par le transport des molécules organiques vers la cathode[11]). L'analyse de l'impédance, relevée à l'abri de l'influence de la diffusion convective ($\Omega \geqslant 2000$ tr/min), montre, par contre, que les deux types d'inhibiteurs s'adsorbent sur la cathode, car ils diminuent la capacité de la double couche électrochimique d'une façon significative. L'analyse de l'impédance cathodique associée à celle de la courbe de polarisation montre que la présence de ces inhibiteurs ne modifie pas le mécanisme de transfert en deux réactions successives; par contre, elle fait varier de façon spécifique les paramètres cinétiques de ces réactions.

A titre d'exemple, la figure 2 représente les courbes de polarisation stationnaires obtenues dans l'électrolyte de type Watts. Sans inhibiteur, on obtient la courbe 1 dont les deux asymptotes représentent les courants partiels correspondant respectivement à deux réactions de transfert successives au cours desquelles l'intermédi-aire $(NiOH)_{ads}$ est consommé à l'électrode[9]). On con-state que l'addition à l'électrolyte de butyne-2 diol-1,4, 8×10^{-3} M (courbe 3), modifie les courants partiels de ces deux réactions de transfert. Par contre, l'addi-tion de benzène-sulfonate de sodium, 3×10^{-2} M (courbe 2), modifie seulement le courant partiel de la première réaction.

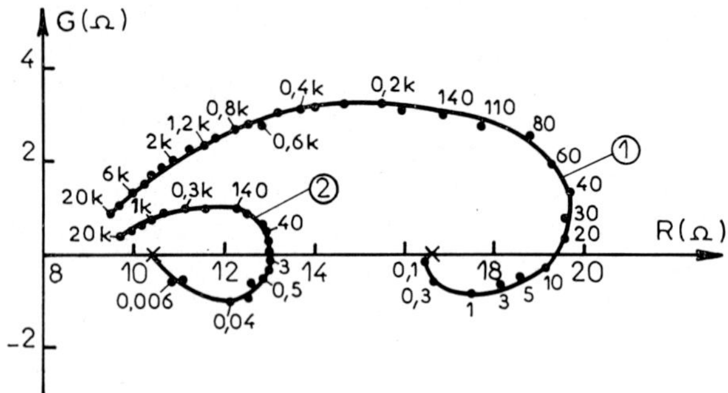

Fig. 4. Diagramme d'impédance $Z = R - jG$, paramétré en hertz, correspondant à l'électrocristallisation de l'argent, obtenu avec une électrode à disque tournant. Composition de l'électrolyte: $AgNO_3$, 100 g/l + HNO_3, 2.7 g/l. Température = 26 °C. Densité de courant: 3.5 mA/cm². Courbe 1: avec sel de Seignette 2 g/l (Ω = 800 tr/min); courbe 2: sans inhibiteur (Ω = 1000 tr/min).

L'impédance de l'électrode est capacitive à toutes les fréquences lorsque la polarisation cathodique est suffisamment forte pour que le courant corresponde pratiquement à celui de la deuxième réaction de trans-fert. Au contraire, dans le domaine des faibles polari-sations cathodiques, en accord avec le processus en deux réactions successives, l'impédance de l'électrode présente un caractère inductif en basse fréquence, que l'électrolyte contienne ou non un inhibiteur. A titre d'exemple, le diagramme d'impédance représenté sur la figure 3 a été obtenu dans les conditions de polarisa-tion du point A de la figure 2. Les éléments du circuit équivalent à l'impédance sont également représentés sur la figure 3 et leurs valeurs sont les suivantes: résistance de l'électrolyte R_e = 9.7 ohm, capacité de double couche électrochimique C = 40 µF/cm², résis-tance de transfert R_t = 9.4 ohm, résistance faradique ρ = 14 ohm, inductance L = 0.3 H. A l'exception de la résistance de l'électrolyte, les valeurs des éléments du circuit équivalent sont modifiées en absence de butyne-2 diol-1,4 dans la solution. Par exemple, au point B de la figure 2 pour lequel le courant est le même qu'en A,

C est multiplié par 2, ρ par 3 et L environ par 2; R_t diminue de 25 %.

Des travaux récents[12]) sur l'électrocristallisation de l'argent dans une solution de nitrate d'argent ($AgNO_3$, 100 g/l + HNO_3, 2.7 g/l) ont montré, que l'addition d'un inhibiteur du type sel de Seignette empêche la croissance dendritique du métal et permet d'obtenir une surface homogène. Avec une telle solution, nous avons pu effectuer des tracés reproductibles de la courbe de polarisation cathodique pour des densités de courant inférieures à 20 mA/cm². Les diagrammes d'impédance obtenus dans ces conditions expérimentales et à l'abri de l'influence du processus de transport par un choix convenable de la valeur de Ω, ont une allure semblable à celui de la figure 3: on observe (figure 4) un phéno-mène inductif ($G < 0$), en présence (courbe 1) et en absence (courbe 2) d'inhibiteur dans l'électrolyte. Nous n'avons pas pu vérifier, jusqu'à présent, si la boucle inductive disparaît aux fortes densités de courant, car les mesures de caractéristique de polarisation et d'im-pédance sont alors perturbées par la croissance den-dritique du métal. Aussi, une étude approfondie de l'adsorption et du transfert d'électron au cours de la croissance du dépôt cathodique est-elle plus difficile dans le cas de l'argent que dans celui du nickel.

Bibliographie

1) V. G. Levich, *Physicochemical Hydrodynamics* (Prentice Hall, Englewood Cliffs, N.J., 1962).
2) I. Epelboin, F. Lenoir et R. Wiart, Compt. Rend. (Paris) **271** (1970) 338.
3) M. F. Fouad et A. M. Ahmed, Electrochim. Acta **14** (1969) 651.
4) C. Deslouis, I. Epelboin, M. Keddam et J.-C. Lestrade, J. Electroanal. Chem. **28** (1970) 57.
5) E. Mattson et J. O'M Bockris, Trans. Faraday Soc. **55** (1959) 1586.
6) Z. A. Tkachik, K. M. Gorbunova et E. S. Seavastyanov, Elektrokhim. SSSR **5** (1969) 351;
   K. M. Gorbunova et Z. A. Tkachik, Electrochim. Acta **16** (1971) 191.
7) J.-P. Badiali, H. Cachet et J.-C. Lestrade, Electrochim. Acta **16** (1971) 731.
8) R. Wiart, Thèse, Paris (1968); Oberfläche-Surface **9** (1968) 213, 241, 275.
9) I. Epelboin et R. Wiart, J. Electrochem. Soc. **118** (1971) 1577.
10) I. Epelboin, M. Froment et G. Maurin, Plating **56** (1969) 1356.
11) M. Froment et R. Wiart, dans: *20e Réunion du CITCE*, Strasbourg, 1969, p. 192.
12) M. Froment, G. Maurin, J. Vereecken et R. Wiart, Compt. Rend. (Paris) **271** (1970) 259.

Journal of Crystal Growth **13/14** (1972) 421–426 © *North-Holland Publishing Co.*

ON THE MORPHOLOGY OF ELECTROCRYSTALLIZED COPPER

W. A. SCHULTZE

Laboratorium voor Metaalkunde, Technische Hogeschool, Delft, The Netherlands

The morphology of copper crystals deposited from an acid copper sulphate electrolyte solution depends on the current density and on the orientation of the copper monocrystal face on which the crystals grow.

At low c.d.'s the growth deposit generally reproduces the orientation of the substrate, and consists of sub-crystals which coalesce with increasing thickness of the deposit forming coarser crystals. The subcrystals show a tendency to be bounded by close-packed planes. There is a relation between the form of the sub-crystals and the shapes of the facets observed on the deposit surfaces.

Under the applied conditions (c.d. 0.1–25 mA/cm^2) the formation of growth twins is the only departure from substrate orientation reproducing growth. This could be confirmed by the use of a microbeam X-ray camera with a minimum beam diameter of 0.03 mm that made it possible to study minor details in the structure of the copper deposits.

In no case could crystals grown from a nucleus with a random orientation to the copper substrate be observed. At the same "overall" c.d. the formation of these growth twins depends strongly on the orientation of the surface plane on which the copper is deposited.

Using the results of the experiments on twin growth it is shown that in a certain c.d. interval repeated twinning can lead to textured deposits. At c.d.'s above this interval polycrystalline deposits are formed.

1. Introduction

Fischer[1]) distinguishes three main types in the morphology of electrodeposited metals. All these forms can be obtained by crystallizing copper out of a purified acid copper sulphate electrolyte solution and changing the current density (c.d.) at room temperature. At low c.d.'s the deposited crystals grow in layers and reproduce under favourable conditions, e.g. copper on a copper substrate, the orientation of the substrate crystals. An increase in c.d. gives outward growing crystals forming a polycrystalline deposit with a preferred orientation. At still higher c.d.'s a randomly orientated polycrystalline deposit is formed. Between the first two types a transitional form can be observed that contains many twin lamellae. There is experimental evidence that the transition between the reproducing type of growth and the other types of deposit on a copper monocrystal does not only depend on a variation in c.d. but is also influenced by the orientation of the growing crystal face[2–6]). The information that is available regards mainly the growth on {111}-, {100}-, and {110}-faces. In most papers on this subject it is reported that with an increase of c.d. the {110}-face shows the least tendency to form textured, twinned or randomly orientated deposits. About the behaviour

of the other two faces contradictory data are found in literature.

Several theories have been proposed to explain the development of a texture in electrodeposited metals. An important contribution was made by Reddy[7]) in 1963 using a suggestion of Wilman[8]). Reddy's theory is based on the assumption that electrodeposits show growth textures only after passing through a random orientation stage in which the crystals are randomly disposed with respect to each other and to the substrate. If that stage is attained a texture is formed if the growth process satisfies the further requirements that the growth of an electrodeposit is associated with the formation of facets of slowly growing faces and that the deposit crystals adopt an outward mode of growth. These latter two stipulations are simultaneously fulfilled only if slowly growing facets are developed normal to the substrate surface.

Another mechanism for the formation of a preferred orientation is given by Pangarov[9]). The main idea of this theory is that the type of two-dimensional nucleus that is formed on an amorphous substrate as a result of the deposit conditions is decisive for the preferred orientation. Extending a method of calculation that Kaischew and Bliznakov[10]) used for a Kossel-crystal to the fcc lattice, Pangarov finds that the nucleation energy for

TABLE 1

Relation between c.d. and the formation of growth twins on differently orientated planes

C.D. (mA/cm²)	Test	Orientation substrate surface						
		{111}	{100}	{110}	{210}	{311}	{331}	{531}
0.1	1	−	−	−	−	−	−	−
1	2	−	++	−	−	−	−	−
5	3	−	+++	−	+	−	−	−
5	4	−	+++	−	+	−	−	−
5	5	−	+++	−	+	−	−	−
10	6	+	+++	−	+++	+	−	+
25	5	+++	+++	+++	+++	+++	+++	+++

− no twins.
+ some "first" twins.
++ some multiple twins.
+++ many multiple twins.

a nucleus with $\{hkl\} \parallel$ substrate depends on the relative overvoltage with which the electrocrystallization process is performed. These calculations indicate that a favourable situation for the formation of fcc-metal nuclei giving rise to a $\{110\} \parallel$ substrate texture only exists for a small interval in overvoltage at a rather high c.d. This is not in agreement with experimental results showing that in a copper deposit a $\{110\} \parallel$ substrate a texture is formed over a wide interval in the higher c.d. range. To explain the mechanism for the real situation that the substrate is not amorphous, as with the deposition of copper on a copper substrate, Pangarov assumes that two-dimensional nuclei with a preferred orientation are formed at passive sites on the surface that act like amorphous bases.

The experiments described in this paper were designed to study the way in which the morphology of electrodeposited crystals change from a substrate-reproducing form to textured or random polycrystalline forms.

2. Materials and procedures

Both coarse crystalline copper (99.999) with crystals of a known orientation and monocrystals were used as a substrate. Monocrystal slices (diam. 5–8 mm, thickness about 0.5 mm) with different surface orientations (within 2° from the poles given in table 1) were cut with an acid saw from a monocrystal rod and electropolished till they were microscopically flat in the centerpart. They were sintered (3 h at 960 °C in vacuo) on a polycrystalline copper bar ($100 \times 10 \times 10$ mm) and simultaneously plated. Every separate specimen (i.e. a bar with seven differently orientated crystal slices) was electroplated with a c.d. of respectively 0.1, 4, 5, 10 and 25 mA/cm² in 1.5 litre electrolyte solution taken from the same purified stock-solution of an acid copper sulphate electrolyte (0.8 m $CuSO_4$ and 0.3 m H_2SO_4) at 20–25 °C, without agitation (except the experiment with 25 mA/cm²).

All electroplating was done underneath a plastic hood to prevent dust contamination. The crystallizing process was repeatedly interrupted to take micrographs and X-ray photographs of the deposit surface. After each interruption the specimens were electropolished for 30 s before replating them. To be quite sure that the electrolyte solution had not changed it was controlled before and after each test in order to ascertain whether it was still capable of giving predominantly substrate orientation reproducing growth on coarse grained polycrystalline copper (1 A h/l; 5 mA/cm²).

After the various tests (see table 1) cross sections of the metal deposits (total thickness between 300 and 800 μm) were analysed both with a reflecting light microscope and by X-ray back-reflection photographs. For the latter purpose a microbeam camera was built with a minimum beam diameter of 0.03 mm to study minor details in the structure of the copper deposits.

3. Results and discussion

3.1. SUBSTRATE ORIENTATION REPRODUCING GROWTH

Below a critical c.d. the orientation of the substrate was reproduced in the deposit. The copper deposit obtained under the aforementioned conditions consists of more or less outward growing subcrystals which coalesce with increasing thickness of the deposit forming coarser crystals. This is shown on the micro-

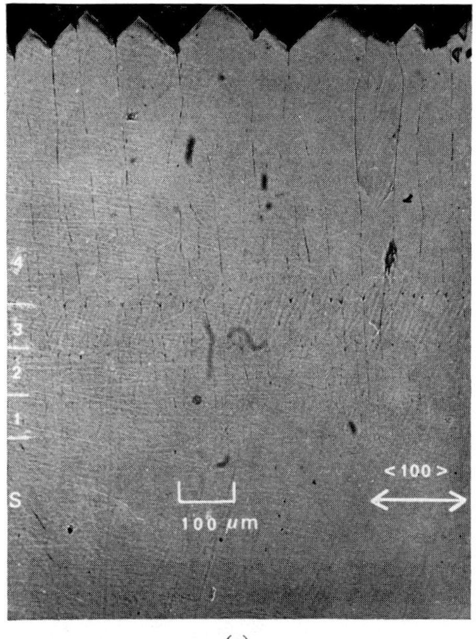

(a)

(b)

Fig. 1. Copper deposit on a {110}-plane (c.d. 5 mA/cm²). (a) Cross section parallel to a {110}-plane; (b) the same as (a), dark field micrograph. S. substrate; 1 to 4: sequence of deposits.

graphs of a cross section of a deposit on a {110}-plane in fig. 1. The size of these subcrystals decrease with an increase in c.d. On a microscopical scale these subcrystals show a tendency to be bound by close-packed planes. The existence of subcrystals may be ascribed to the influence of the electrical field which causes an

outward growth and the tendency of growing crystals to be bounded by slowly growing planes.

This confirms the ideas given by Reddy and Wilman. As a result of small differences in the rate of growth of the surface facets, the subcrystals may coalesce in a way that is schematically drawn in fig. 2 for a ridged deposit surface as shown in fig. 1.

3.2. DEPARTURE FROM SUBSTRATE ORIENTATION REPRODUCING GROWTH

Above the critical c.d. structural components appear in the deposit that do not reproduce the structure of the substrate crystal. The combined techniques of measuring the geometrical configuration with the microscope and the structural orientation with the micro-Laue back-reflection camera made it possible to find the orientation relation between these parts and the substrate. Under the applied conditions (c.d. 0.1–25

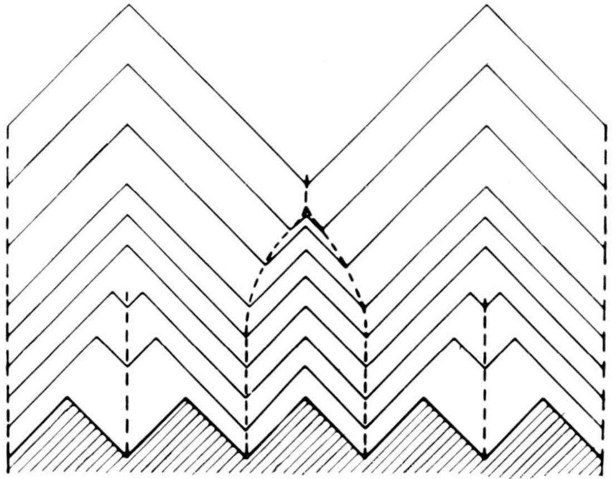

Fig. 2. Schematic representation showing the coalescence of subcrystals in a copper deposit to larger units. Dashed lines: subboundaries; solid lines: growth stages.

mA/cm²) the formation of growth twins is the only departure from substrate orientation reproducing growth. Actually this kind of growth is epitactical, as there is a definite orientation relation, but after repeated twinning this relation between crystal and substrate becomes so complicated that it looks as a random growth. In none of the investigated cases crystals grown from a nucleus with a truly random orientation to the copper substrate were observed.

At the same "overall" c.d. the formation of growth twins depends strongly on the orientation of the surface plane (see table 1). Each plane seems to have its own

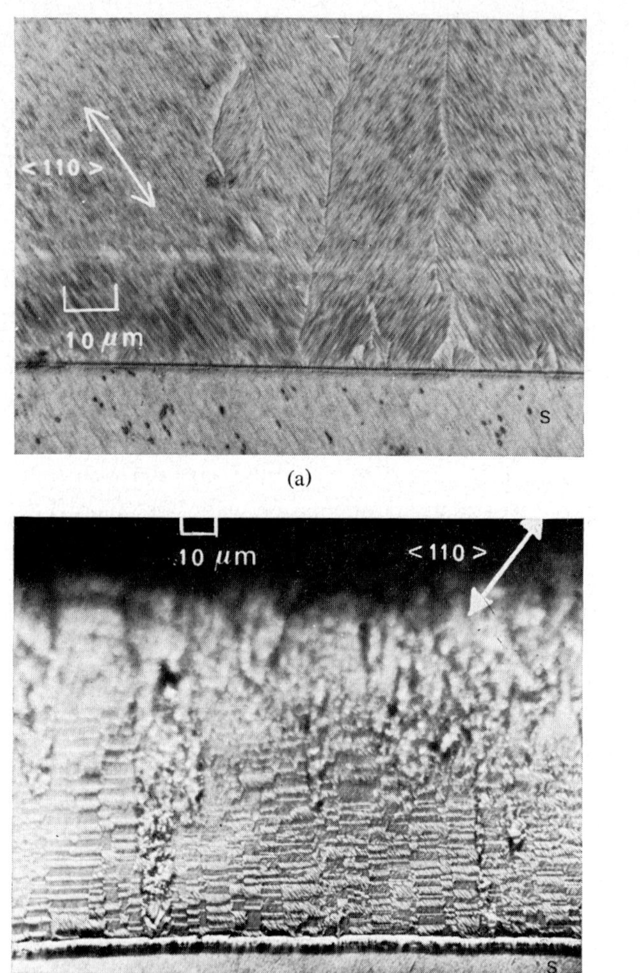

Fig. 3. Formation of growth twins in a deposit on a {111}-face. (a) 10 mA/cm²; (b) 25 mA/cm². Cross section parallel to {110}.

Fig. 4. Twin growth during electrocrystallization on a {100}-plane (c.d. 5 mA/cm²). (a) Isolated form bound by {111}-planes (in the dark part of the micrograph). (b) Twin growth on such a {111}-facet (see arrow). (c) Stereogram belonging to (b) with the plane of projection parallel to the substrate surface. The twin position is indicated by symbols with an accent.

limited c.d. above which the growth of twins starts. An explanation of this phenomenon can be based on two conditions that are thought necessary requirements for the formation of growth twins; firstly, the presence of {111}-facets of a certain extension on which twins can be formed and, secondly, a high local c.d. that promotes on these facets the occurrence of the "ABA-arrangement" of the arriving Cu-atoms. The latter can be concluded from the experiments on crystals with a {111}-surface plane on which twinning begins at an "overall" c.d. of about 10 mA/cm² and occurs more frequently with increasing c.d. (fig. 3). On all the growing deposit surfaces {111}-facets can be observed if a certain thickness is reached. The reason that twins are more abundant on {100} and {210} at a low c.d. has

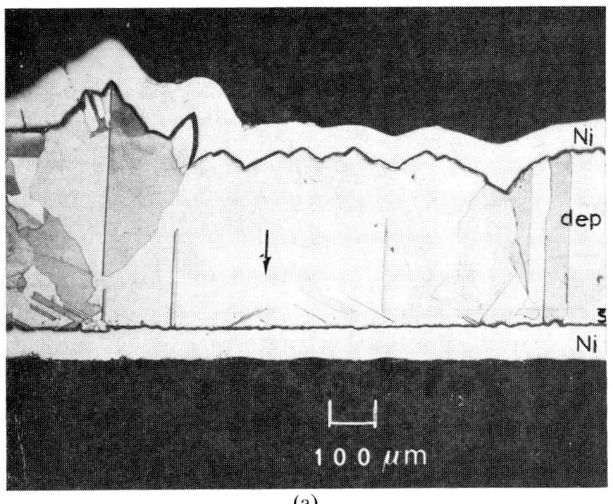

(a)

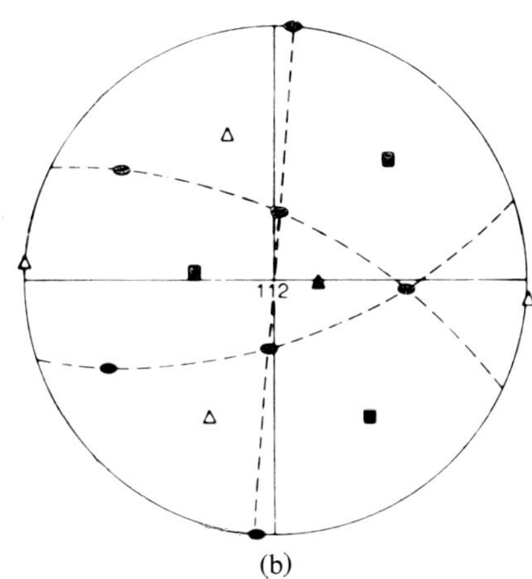

(b)

Fig. 5. Copper deposit on a coarse-grained Cu 99.999 cathode. (a) Cross section. (b) Stereographic projection on the cross section with an indication of the position of the twinning planes. The low-angle twins are not continued into the deposit.

to be sought in the fact that the steepness of the observed {111}-facets is greater on these planes than on the other ones that were investigated. At the top of these steep {111}-facets the limiting c.d. for twinning can be reached even at low "overall" c.d.'s. The observation of growth twins formed on pyramids with octahedral faces on a {100}-face supports this hypothesis (fig. 4).

Another observation is that the inclination of the twinning plane is important for the "survival" of a twinned crystal. The coarse grained copper crystals that were used to control the electrolyte solution quality contained quite a number of twin lamellae in the substrate crystals. Many of the twin boundaries continue rectilinearly into the deposit (fig. 5). This stands in marked contradiction to ordinary crystal boundaries

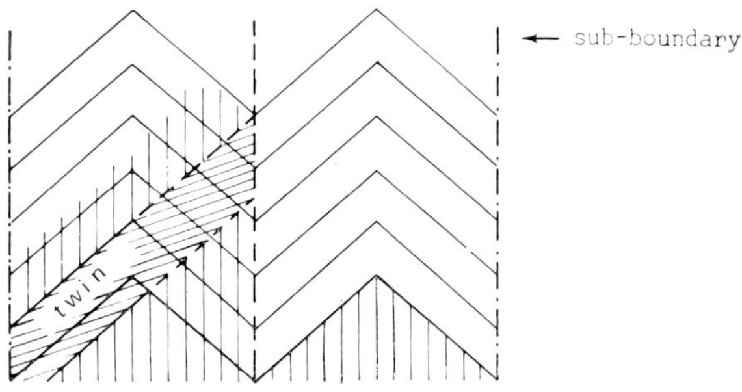

Fig. 6. Schematic representation showing the envelopment of a twin crystal by growing sub-crystals.

(a)

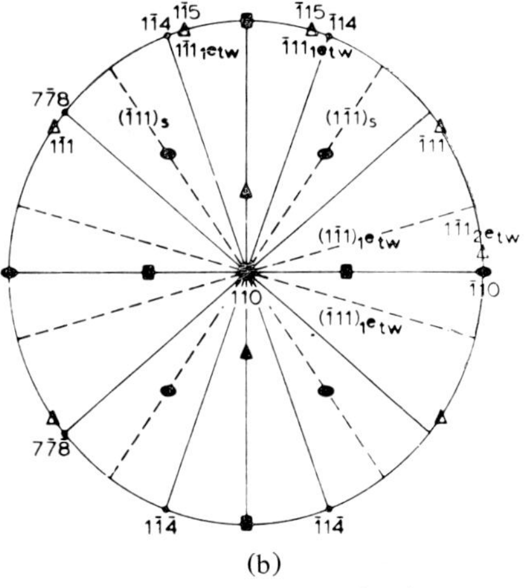

(b)

Fig. 7. Cross section of a deposit on {100}, parallel to {110}, together with a stereographic projection on the section. The etch grooves in a ⟨110⟩-direction agree with the stereographically determined twin positions.

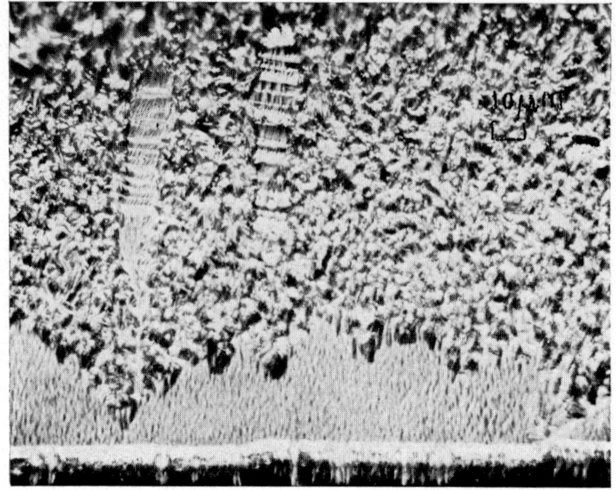

Fig. 8. Cross section of a deposit on a {110}-plane; c.d. 25 mA/cm². Section parallel to {100}.

that continue into the deposit in a direction more or less perpendicular to the substrate surface. From a number of observations in which the orientation of the twinning plane was determined it could be concluded that the continuation of the twinning plane into the deposit depends on the position of the plane with respect to the substrate plane. If the angle is smaller than about 45° the twin tends to disappear (fig. 5). This fact explains the observation that twin growth on a crystal with a {111}-surface begins at a lower c.d. than on a {110}-plane on which {111}-growing facets are formed under 35° with the substrate surface and therefore should be more inclined to form growth twins. In fig. 6 is schematically shown how a twin in a deposit on a facetted {110}-plane can be envelopped by growing subcrystals.

3.3. The formation of polycrystalline deposits with a preferred or random orientation

A closer examination of a deposit on {100} with 10 mA/cm² (a c.d. that finally leads to deposits with a {110} ∥ substrate preferred orientation) shows that in the polycrystalline deposit there is always a twin relation between two neighbouring crystals (fig. 7). Using the transformation matrices:

$$\begin{pmatrix} -1 & 2 & 2 \\ 2 & -1 & 2 \\ 2 & 2 & -1 \end{pmatrix} \quad ; \quad \begin{pmatrix} -1 & -2 & 2 \\ -2 & -1 & -2 \\ 2 & -2 & -1 \end{pmatrix} \quad ;$$

$$\begin{pmatrix} -2 & 2 & -2 \\ 2 & -1 & -2 \\ -2 & -2 & -1 \end{pmatrix} \quad ; \quad \begin{pmatrix} -1 & -2 & -2 \\ -2 & -1 & 2 \\ -2 & 2 & -1 \end{pmatrix}$$

for a rotation over 180° around resp. $[111]$; $[1\bar{1}1]$; $[\bar{1}\bar{1}1]$ and $[\bar{1}11]$ it can be shown that starting from a given crystal orientation after a number of twinning operations a crystal is formed with an orientation close to {110} ∥ substrate surface. Once formed, these crystals grow either without further twinning, because a crystal with {110} ∥ substrate does not twin easily, or twins are formed only on the almost perpendicular {111}-planes. In the latter case the only effect is a rotation around the preferred ⟨110⟩-axis ⊥ substrate. Sometimes this can result in growth forms with pseudo five-fold symmetry.

With a c.d. of 25 mA/cm² even on {110} the deposit changes from monocrystalline to a random polycrystalline structure (fig. 8). From the experimental evidence that in this structure repeated twinning can be observed it can be assumed that in this case growth twins are formed on all possible {111}-facets which leads to random polycrystallinity.

Acknowledgment

The author wishes to thank Prof. Dr. W. G. Burgers for the critical review of the manuscript.

References

1) H. Fischer, *Elektrolytische Abscheidung und Elektrokristallisation von Metallen* (Berlin, 1954).
2) H. Leidheiser, Jr. and A. T. Gwathmey, Trans. Am. Electrochem. Soc. **91** (1947) 95.
3) T. H. Orem, J. Res. Natl. Bur. Std. **60** (1958) 597.
4) L. Peraldo Bicelli and G. Poli, Electrochim. Acta **11** (1966) 289.
5) S. C. Barnes, Acta Met. **7** (1959) 700.
6) Yu. M. Polukarov and Z. V Semenova, Elektrokhimiya **2** (1966) 184.
7) A. K. Reddy, J. Electroanal. Chem. **6** (1963) 141.
8) H. Wilman, Trans. Inst. Metal Finishing **32** (1955) 281.
9) N. A. Pangarov, J. Electroanal. Chem. **9** (1965) 70.
10) R. Kaischew and G. Bliznakov, Compt. Rend. Acad. Bulgare Sci. **1** (1948) 23.

Journal of Crystal Growth **13/14** (1972) 427–432 © *North-Holland Publishing Co.*

427

OBSERVATIONS ON THE ELECTROCRYSTALLIZATION OF COPPER*

U. BERTOCCI, C. BERTOCCI and B. C. LARSON

Solid State Division, Oak Ridge National Laboratory, Oak Ridge, Tennessee 37830, U.S.A.

The defects produced during epitaxial electrodeposition of copper on highly perfect copper single crystals have been studied by correlating various X-ray techniques, including Borrmann and Berg–Barrett topography, with optical microscopy and dislocation etching. In some of the deposits obtained, the measurement of the half-width of the Bragg peak and anomalously transmitted intensity revealed no significant concentration of defects. However, dislocation etching indicated the presence of small loops. Instances of formation of twin crystals were observed. Annealing of the deposits at temperatures up to 1000 °C caused the development of a small number of large strain centers and some indication of aggregation of defects.

1. Introduction

Although the epitaxial electrodeposition of copper on copper has been studied extensively[1]), the crystalline perfection of the deposit is not well established. Divergent back-reflection X-ray topography[2]) and measurements of the half-width of Bragg peaks[3]) have been employed, but these methods give little information about either the specific nature or the spatial distribution of any defects formed in the single crystal deposit. Hence it has not been possible to correlate surface features visible by optical microscopy with defects in the deposit.

The technique of Borrmann X-ray topography has been developed in recent years[4]) and employed for the examination of defects in metals, and copper has been extensively studied[5]). Since this technique utilizes anomalous transmission of the X-rays, specimens several hundred microns thick can be examined. Copper can be electroplated onto highly perfect copper substrates, and defects can be observed in the bulk of the deposit by stereo techniques[6]). Aggregates of point defects that are too small to be imaged topographically may be detected by integrated intensity measurements of anomalous transmission[7]), as shown for neutron irradiated copper where good agreement with electron microscopy observations[8]) was found.

In the present study we have examined epitaxial copper deposits, both as deposited and after annealing at various temperatures, by several X-ray techniques and correlated these observations with those from optical microscopy and dislocation etching on the same crystals. Some overvoltage measurements during galvanostatic transients on low defect density deposits were also carried out.

2. Experimental methods

Copper single crystals of purity 99.999%, in the shape of thin lamellae, 10×10 mm and from 0.2 to 0.5 mm thick were employed as electrodes. The two main surfaces were (111). The electrodepositions were carried out in an acidic $CuSO_4$ solution, mostly at 0.5 mA/cm². The thickness of the deposits ranged from 30 to 150 μm. Details of the procedure have been given elsewhere[9]).

The deposits were characterized by X-ray half-width measurements of the (444) Bragg reflection and were examined by Borrmann and Berg–Barrett X-ray topography. In addition, integrated anomalous transmission intensity measurements were made on some deposits.

The defect structure as shown in the X-ray topographs was correlated with observations by optical microscopy, and additional information about the defects was gathered by electropolishing the surface and etching with a dislocation etch[10]).

Current-potential measurements were taken applying short current pulses and recording the voltage–time curves on a cathodic ray oscilloscope. The measurements were taken after long cathodic depositions so as to achieve a "steady state" surface and defect structure.

* Research sponsored by the U.S. Atomic Energy Commission under contract with Union Carbide Corporation.

VIII – 3

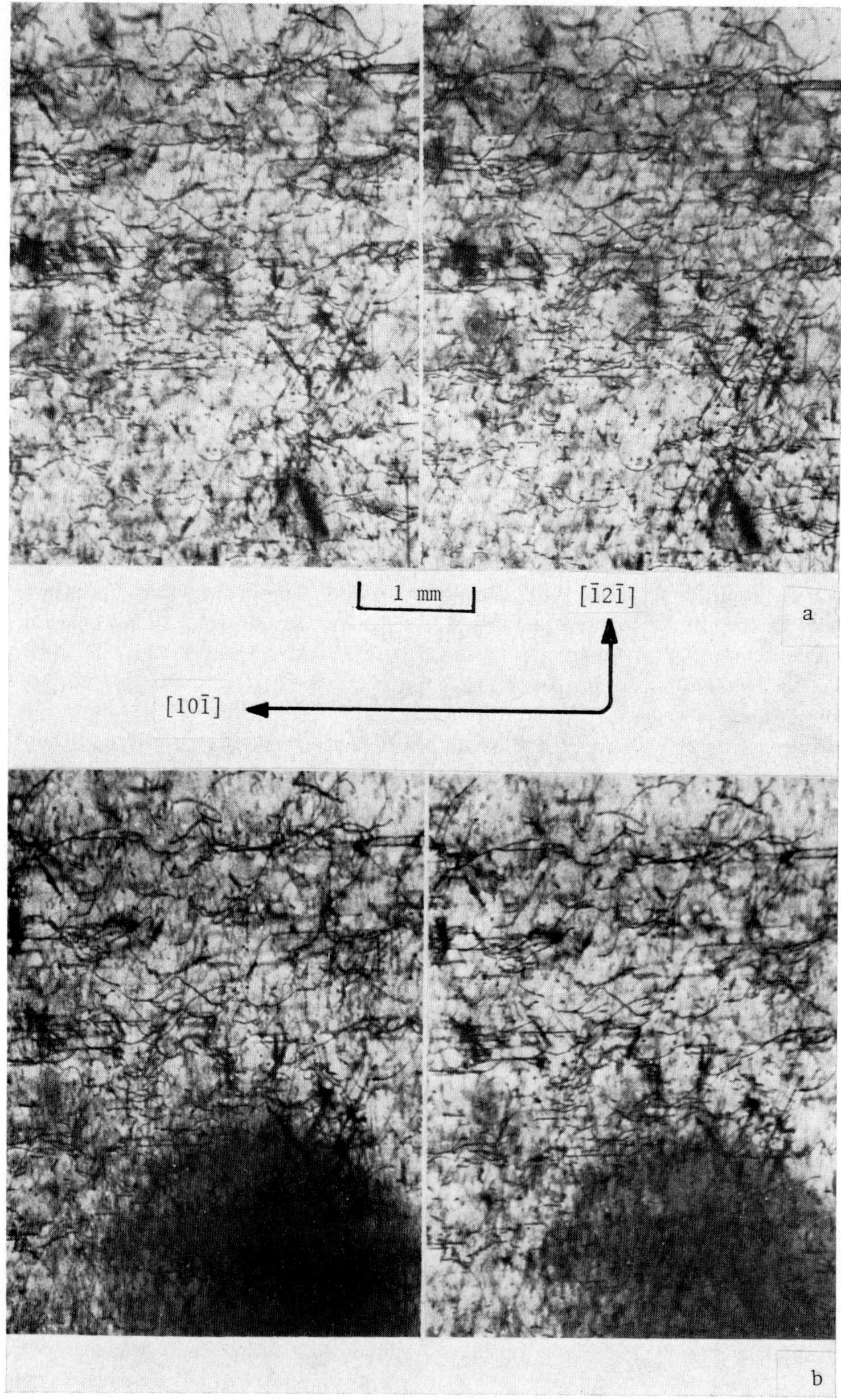

Fig. 1. Borrmann stereo topographs. Diffraction vector [Ī1Ī]. (a) Before deposition. (b) After depositing 60 μm of Cu at 0.5 mA/cm² on each side.

TABLE 1

Integrated intesity measurements for anomalous X-ray transmission

Specimen characteristics	Wavelength	Integrated intensity $\times 10^6$		Annealing
		Measured	Theoretical	
Total thickness 440 $\pm$ 20 μm	CuKα	$R_H = 1.60$	1.63	No
Deposit thickness 260 $\pm$ 30 μm		$R_T = 1.25$	1.23	
c.d. 0.7 mA/cm²		$R_H = 1.64$		400 °C
		$R_T = 1.18$		
		$R_H = 0.86$		600 °C
		$R_T = 0.60$		
Total thickness 600 $\pm$ 20 μm	MoKα	$R_H = 0.66$	0.70	No
Deposit thicknes° 270 $\pm$ 30 μm		$R_T = 0.69$	0.67	
c.d. 0.5 mA/cm²		$R_H = 0.67$		400 °C
		$R_T = 0.69$		
		Not measurable		600 °C

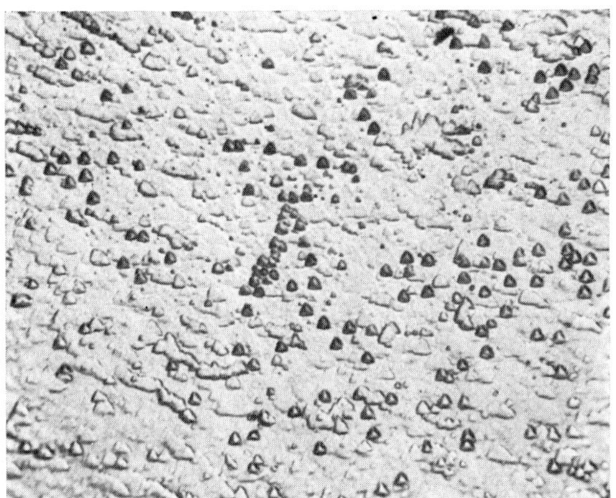

Fig. 2. Micrograph of deposit after slight electropolishing and dislocation etching. Nomarski interference contrast. Magnification ×225.

3. Examination of the perfection of the deposits

For current densities of the order of 0.5 mA/cm², corresponding to an average growth rate of 15 μm/day, epitaxial deposits could be produced for which X-ray examination did not reveal a substantial amount of defects for thicknesses ranging up to 150 μm. The half-width of the (444) Bragg peak was very close to 2″ of arc, the value for a perfect crystal. Also the integrated intensity for anomalous transmission corresponded to that of a perfect crystal within the experimental errors. Data are summarized in table 1.

Stereo pairs of Borrmann topographs, taken before and after depositing about 60 μm of Cu on each side of a 350 μm thick electrode are shown in figs. 1a and 1b. Fig. 1b does not show any appreciable defect con-

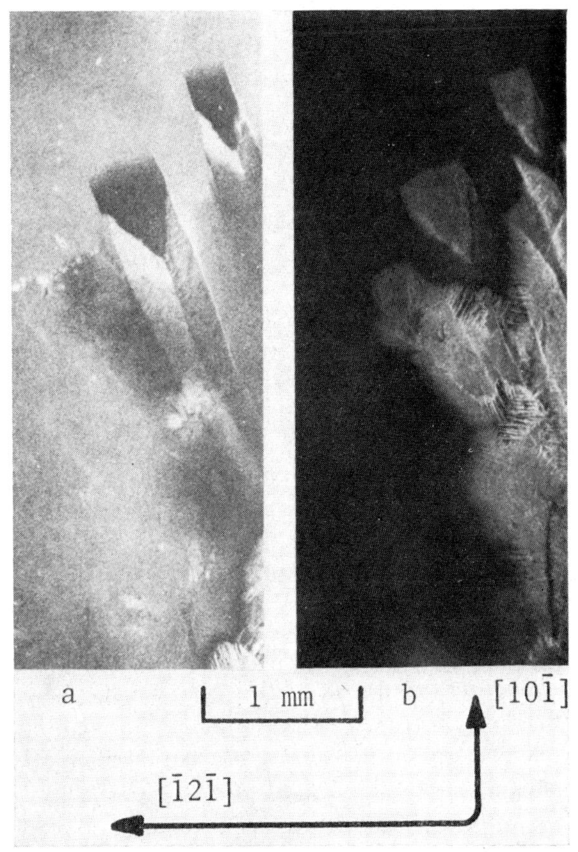

Fig. 3. Berg–Barrett topograph of crystal after depositing 120 μm of Cu. (a) Diffraction vector [131]. Twin crystal appear dark. (b) Diffraction vector rotated 60° with respect to (a). Twin crystals appear light.

centration in the deposit close to the X-ray exit surface (which appears close to the observer in the stereo topograph), whereas some reduction of intensity, due

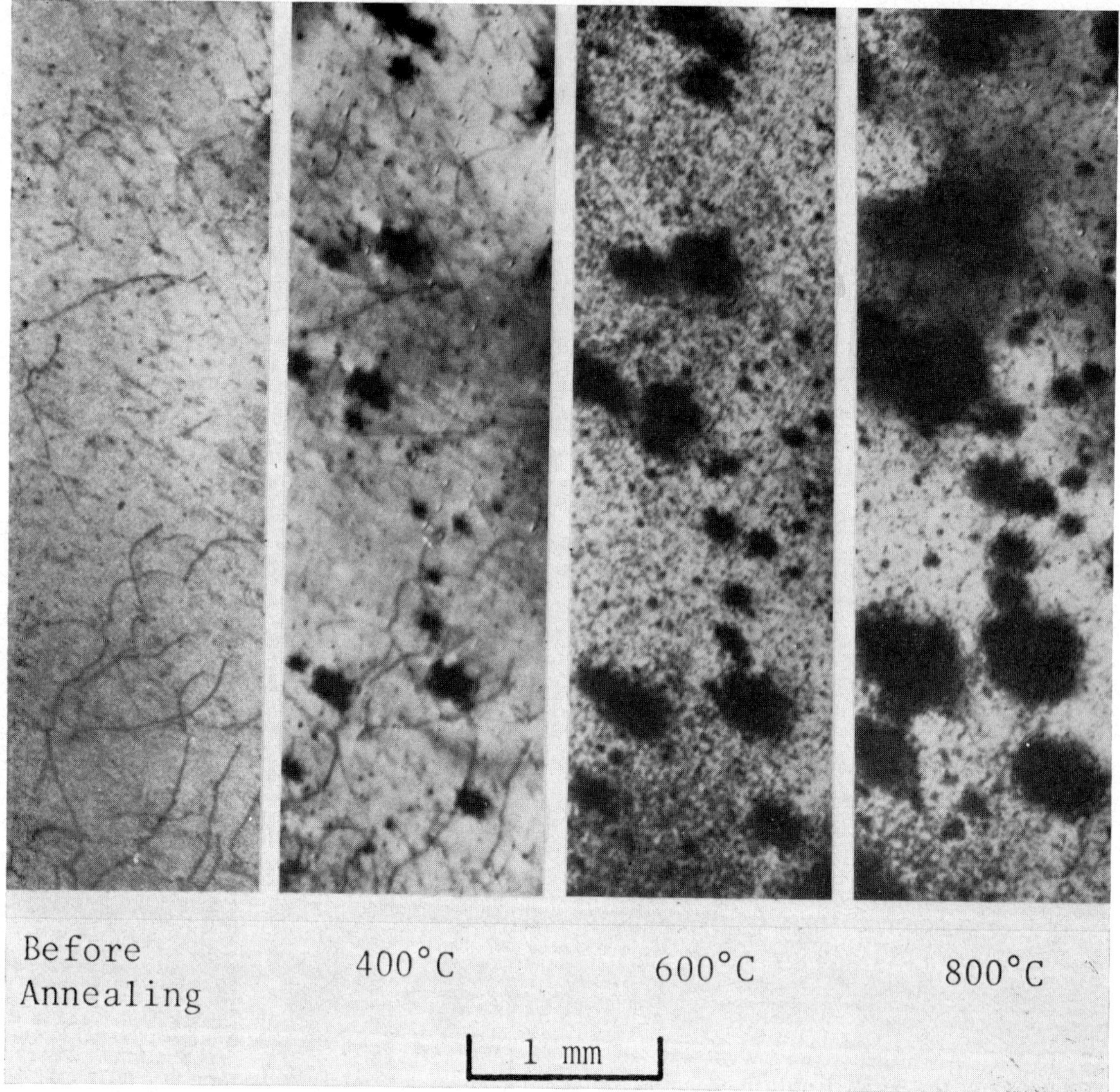

Fig. 4.   Borrmann topographs of crystal with 130 µm deposit on each side before and after annealing at various temperatures.

to defects in the deposit on the other face can be seen in the lower part of the topograph. It was found that a small deviation of the surface orientation from (111), of the order of 3° to 6°, decreased the probability of formation of defects visible on Borrmann topographs. A typical appearance of a surface after electropolishing and etching with a dislocation etch is shown in fig. 2. The pit density varied between $5 \times 10^5/\mathrm{cm}^2$ and $5 \times 10^6/\mathrm{cm}^2$. The flat-bottomed pits are similar to those seen after etching neutron-irradiated copper[11]).

In many instances the defects revealed by Borrmann topography were due to twinning. Patches of twin crystal, extending a few square millimeters, with the twin parallel to the surface could be detected by observing the rotation of the triangular etch-pits or by taking Berg-Barrett X-ray topographs of the surface. The twin crystals were often surrounded by areas of lower perfection. An example of Berg-Barrett topography of a deposit containing twins is shown in fig. 3.

4. Annealing studies

A number of deposits were given three hours annealing in H_2 at temperatures ranging from 200 °C to 1000 °C, with the intent to investigate the stability of the defects revealed by X-ray topography, and to see whether defect aggregation might occur which would allow detection of imperfections which might be present but undetected in the original deposits.

The results showed that regions of low perfection tended to be stable up to 1000 °C. A typical example of the evolution during annealing of areas of high perfection is illustrated in fig. 4. A small number of large strain centers were observed to develop at 400 °C, some of them having prismatic loops punched out in [110] directions. These isolated defects tended to increase in size by annealing at 600 °C and 800 °C.

Moreover, after annealing at 600 °C, the sharpness of the topographs was considerably reduced by the appearance of very small spots, which remained after annealing at 800 °C. The integrated intensity for anomalous transmission did not change after annealing at 400 °C, but a factor of two decrease was observed after annealing at 600 °C, as shown in table 1.

5. Electrochemical measurements

Voltage–time curves during short galvanostatic pulses were recorded on a number of specimens. The measurements were performed after long depositions at 0.5 mA/cm² in order to insure a steady state condition of the surface. Some of the data are shown in fig. 5.

Analysis of the steady state values and of the slopes of the voltage–time curves showed that the results are consistent with a model for which the reaction $Cu \rightleftharpoons Cu^+ + e^-$ occurs reversibly, whereas the rate-determining step is given by the reaction $Cu^+ \rightleftharpoons Cu^{++} + e^-$, with an exchange current density of the order of 1 mA/cm².

No appreciable contribution attributable to crystallization overvoltage could be found. However, since Cu^+ ion diffusion is very important in determining the shape of the voltage–time curve in the first millisecond, the presence of irreversible crystallization effects could easily have been overlooked. The double layer cpacitance was found to be about 22 μF/cm².

6. Discussion

A marked influence of the orientation of the substrate on the perfection of the deposit was found. Small deviations from (111) orientation produced deposits of higher perfection as characterized by the different

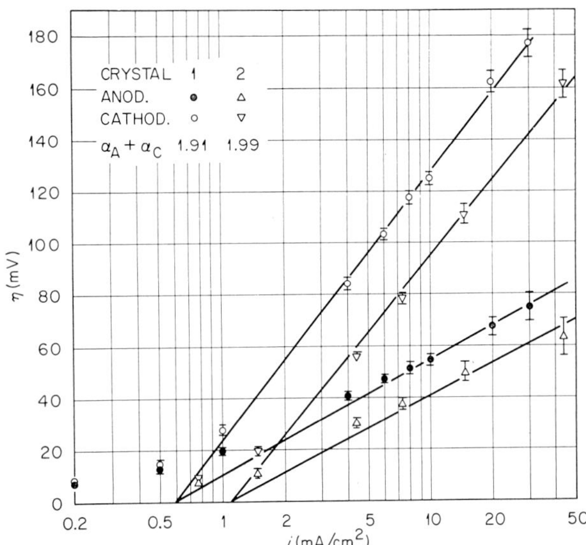

Fig. 5. Overvoltage versus current density curves measured after long deposition at 0.5 mA/cm² on two different specimens.

X-ray methods. This may be related to orderly step motion without need for random nucleation. However, even in the most perfect deposits dislocation etching has consistently revealed the presence of defects too small for imaging in Borrmann topography, and in sizes and concentrations such that the integrated intensity for anomalous transmission was not noticeably affected. The similarity of the etching pattern with that of neutron-irradiated copper suggests that these defects are loops smaller than about 0.2 μm in diameter.

Another type of defect was the twin crystals. They were almost always found on areas of orientation very close to (111), supporting the conclusion that two-dimensional nucleation may be responsible for twin formation.

The study of the defect structures of electrodeposits after annealing has revealed the formation of large strain centers. Although very conspicuous in the Borrmann topographs, their concentration is very small. It is not possible to draw any firm conclusion as to their origin and nature. It can be tentatively proposed that they result from small occlusions due to step overgrowth. The appearance of very small spots uniformly distributed over the specimen might be considered as an indication of defect aggregation producing strain fields of topographic size. The concomitant, observed decrease in anomalous transmission intensity would be in agreement with this interpretation, although the previously discussed large, isolated defects might be responsible to a certain extent for the increased absorption. A quantitative assessment of the significance of the intensity data is therefore not possible, although efforts were made to improve the reliability of the measurements by using a small diameter beam (~ 1 mm) and taking readings at several points. Finally, the stability of the observed defects above 800 °C suggests that impurities, rather than intrinsic defects such as vacancies and self-interstitials, might be involved.

Acknowledgment

The authors wish to thank Dr. F. W. Young, Jr. for numerous helpful discussions regarding this work.

References

1) R. Piontelli, G. Poli and G. Serravalle, in: *Trans. Symp.* on *Electrode Processes*, Ed. E. Yeager (Wiley, New York, 1959) p. 67;
   J. O'M. Bockris and H. Kita, J. Electrochem. Soc. **109** (1962) 928;
   A. Damjanovic, T. H. V. Setty and J. O'M. Bockris, J. Electrochem. Soc. **113** (1966) 429;
   T. Hayashi, S. Higuchi, H. Kinoshita and T. Ishida, J. Electrochem. Soc. Japan **37** (1968) 64;
   L. H. Jenkins, J. Electrochem. Soc. **117** (1970) 630;
   R. Sard and R. Weil, Electrochim. Acta **15** (1970) 1977.
2) P. E. Lighty, D. Shanefield, S. Weissmann and A. Shrier, J. Appl. Phys. **34** (1963) 2233.
3) J. J. O'Connor, B. Rubin and R. Spector, AFCRL-66-538, Office of Aero-space Research (1965).
4) F. W. Young, Jr., T. O. Baldwin, A. E. Merlini and F. A. Sherrill, in: *Advances in X-ray Analysis*, Vol. 9 (Academic Press, New York, 1966) p. 1.
5) F. W. Young, Jr., in: *Crystal Growth*, Ed. H. S. Peiser (Pergamon, Oxford, 1967) p. 789;
   U. Bertocci, C. Bertocci and F. W. Young, Jr., J. Appl. Phys. **40** (1969) 1674.
6) T. S. Noggle, B. F. Day, F. A. Sherrill and F. W. Young, Jr., Bull. Am. Phys. Soc. **10** (1965) 324.
7) P. H. Dederichs, Phys. Rev. B **1** (1970) 1306.
8) F. W. Young, Jr., T. O. Baldwin and P. H. Dederichs, *Vacancies and Interstitials in Metals* (North-Holland, Amsterdam, 1969) p. 619.
9) U. Bertocci and C. Bertocci, J. Electrochem. Soc., **118** (1971) 1287.
10) F. W. Young, Jr., J. Appl. Phys. **32** (1961) 192;
    U. Bertocci, L. D. Hulett and L. H. Jenkins, J. Electrochem. Soc. **110** (1963) 1190.
11) F. W. Young, Jr., J. Appl. Phys. **33** (1962) 749.

Journal of Crystal Growth **13/14** (1972) 433–437 © *North-Holland Publishing Co.*

433

GROWTH OF TUNGSTEN BRONZE CRYSTALS BY FUSED SALT ELECTROLYSIS*

HOWARD R. SHANKS

Institute for Atomic Research and Department of Physics, Iowa State University, Ames, Iowa 50010, U.S.A.

Although tungsten bronzes and related compounds have been prepared by fused salt electrolysis for some time, very little detail about the actual growth methods and results have been reported. This work describes a successful electrolytic cell which has been developed for the growth of a variety of tungsten bronze crystals. This cell has been used to establish the parameters for the growth of Na_xWO_3 crystals with predictable composition and in addition the phases of bronze obtained as a function of melt composition and temperature. Growth rates were measured and found to vary considerably for the different phases. Large crystals of a new triclinic compound, $Na_4W_{14}O_{43}$, were obtained during the study of different melt compositions. Crystals of the lithium isomorph of this compound were also prepared.

1. Introduction

The growth of tungsten bronzes by fused salt electrolysis was first reported by Scheibler[1]) in 1860. Since that time a number of investigators[2–6]) have reported the use of this technique for the preparation of various bronze crystals** but almost no details about the actual growth methods and results have been reported. Recently Banks, Fleischmann and Meites[7]) reported on studies of the nature of the species reduced in the melt to form Na_xWO_3 during electrolysis. Their results correlate strongly with the phase diagram reported for the system[8]). In an effort to help clarify the process by which bronze crystals are grown by electrolysis we would like to report the results of an extended study to establish the parameters for growth of Na_xWO_3 crystals with predictable composition.

2. Experiment

A very successful electrolytic cell was developed for the growth of cubic sodium tungsten bronze crystals. The furnace consisted of a 10 cm i.d. alundum tube on which was wound a Kanthal heating element. This winding was insulated on the outside by a double layer of insulating brick. The crucible was placed on an insulating block to center it in the furnace as shown in fig. 1. The crucibles used for growth of bronze crystals

were made of an electrical ceramic glazed on the inside only with a lead free glaze. The crucibles were nominaly 9 cm in diameter and 9.5 cm high with a wall thickness of 0.3 cm. These crucibles had a lifetime of up to 6 months at the growth temperature.

The crucible with a charge of Na_2WO_4 and WO_3 was heated to the growth temperature at a rate of about 50 °C/hr. Faster heating rates tended to crack the crucible. The amount and composition of the melt in the

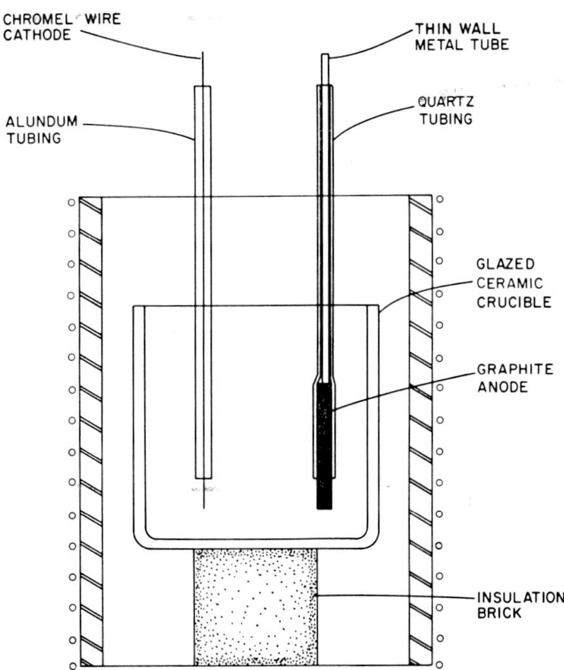

Fig. 1. Electrolytic cell for the growth of tungsten bronze crystals.

---

* Contribution No. 3013. Work was performed in the Ames Laboratory of the U.S. Atomic Energy Commission, Iowa State University, Ames, Iowa.

** The name "bronze" as used in this paper refers to compounds of the general form M_xWO_3.

VIII – 4

crucible could be changed by opening the top of the furnace and slowly adding additional material.

The anode consisted of a small graphite rod connected to a thin wall metal tube with a quartz sleeve. Unprotected graphite rods eroded off at the surface of the melt in a few hours but by extending the quartz sleeve to just below the surface of the melt the lifetime of the anode could be increased to five to seven days. The cathode consisted of a No. 22 chromel wire in an alundum tube. The alundum tube extended to just below the surface of the melt to protect the wire from erosion. The electrodes were supported by metal clips from the top of the furnace. They could be withdrawn from the melt at any time to remove crystals or for replacement.

The melt was held at a constant temperature during the growth of a crystal by means of a proportional control system referenced to a thermocouple placed next to the furnace winding. The temperature did not fluctuate more than 2 °C during the growth of a crystal. An additional thermocouple was placed on the side of the crucible to measure the temperature.

A constant current power supply was used for the electrolysis of the melt. The currents used ranged from 10 to 75 mA. The constant current supply was used in this study in order to measure growth rates. Crystal quality decreased with increasing current. A constant voltage supply was used for routine growth of crystals in order to maintain a constant current density.

The crystals as grown on the bare electrode usually consisted of a group of intergrown single crystals which had nucleated along the wire. Large single crystals could be grown, however, if a small seed crystal was placed in the melt as the electrode. This was done by drilling a small hole from one corner along the diagonal into a cube of bronze with {100} faces of the appropriate x-value for the melt. The size of the cube was usually 3 to 4 mm on a side and was cut from a sample grown in the melt to be seeded. A wire was forced into the hole and then insulated with an alundum tube. Joints in the alundum tube and the joint between the seed crystal and the tube were insulated with refractory cement so that the seed crystal was the only conducting part of the electrode in the melt. The electrode was introduced into the melt and allowed to sit for several minutes before current was turned on. This cleaned the crystal surface so that no additional crystals were nucleated.

(a) Cubic   (b) Tetragonal I

(c) Tetragonal II   (d) $Na_4W_{14}O_{43}$

Fig. 2.   Crystals of sodium tungsten bronzes and $Na_4W_{14}O_{43}$ grown by electrolysis. All photos are to the same scale. The cubic crystal is about 3 cm on a side.

The current was then turned on and the crystal was allowed to grow for several days.

For the growth of large crystals it was found desirable to orient the seed in the melt so that the $\langle 111 \rangle$ direction was vertical. Since the crystals tend to grow as cubes, this left a corner of the cube rather than a face of the cube toward the bottom of the melt. Crystals grown with a flat face down tended to have gas bubbles trapped on the bottom side which became cavities in the crystal as it grew. By orienting the seed with one corner facing down the gas bubbles escaped and did not disturb the growing faces. By increasing the size of the crucible to 11.5×15 cm high in a larger furnace, crystals of cubic Na_xWO_3 as large as 7 cm on a side have been grown.

Single crystals could be grown without seeding by withdrawing the cathode wire a short distance up into the alundum tube. Only one crystal nucleated with this technique but there was no control of orientation. This technique has also been used by Ferretti[9]).

We have been able to seed only the cubic phase of

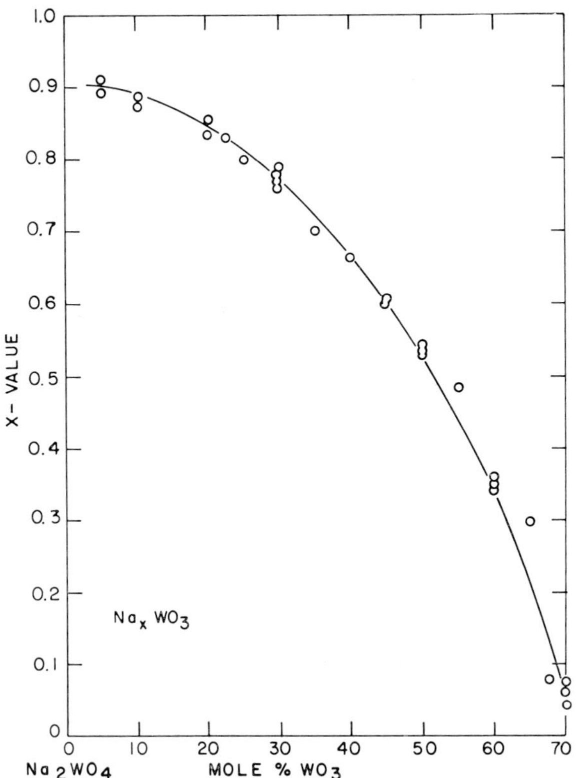

Fig. 3. The composition of the sodium tungsten bronze crystals obtained for different melt compositions when the temperature of the crucible is held just above the melting point of the charge.

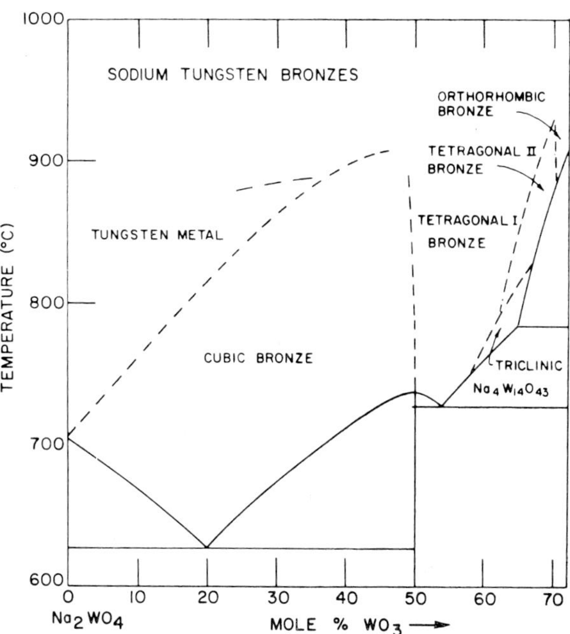

Fig. 4. The phases of sodium tungsten bronze obtained by electrolysis as a function of melt composition. The liquidus curve is taken from the Na_2WO_4–WO_3 phase diagram of Mellor.

Na_xWO_3. Only intergrown clusters of the other phases have been grown as shown in fig. 2. Single crystal samples large enough for most transport measurements could be obtained, however, from the clusters.

3. Results

In order to study the bronzes as a function of x-value, it is important to know what melt composition gives a particular x-value so that the required crystal composition can be obtained predictably. A series of melts were prepared with compositions ranging from 0 to 75 mole% WO_3 in Na_2WO_4. The compositions of the crystals grown were determined by neutron activation analysis. Fig. 3 shows the x-value of the bronze obtained as a function of the melt composition when the temperature of the crucible is held just above the melting point of the charge. The scatter is due to the water of hydration on the WO_3 at room temperature which makes accurate determinations of the crucible charge composition difficult.

The crystal growth rate decreased linearly with increasing temperature from the melting point to some higher temperature where the amount of bronze deposited suddenly decreased rapidly to zero and tungsten metal was deposited. The phases of bronze obtained by electrolysis as a function of melt composition and temperature are shown in fig. 4. The liquidus curve is from the phase diagram reported by Mellor[8]).

Cubic phase material was obtained for all melt compositions with less than 50 mole% WO_3. It was possible to obtain cubic crystals with x-values from 0.49 to 0.93. The highest x-value crytals were obtained from a melt of nearly pure Na_2WO_4 held at the melting point. Tetragonal I crystals were obtained with x-values from 0.25 to 0.49 but composition was strongly dependent on temperature as well as melt composition. Tetragonal II crystals with x-values from 0.25 to 0.07 were obtained from melts with 62 to 70 mole% WO_3 respectively. Orthorhombic crystals were obtained from melts with greater than 70 mole% WO_3. The highest x-value for this structure is 0.07. The lowest x-value obtained was about 0.01 but this limit was due to the high temperature limit of the furnace. Compositions down to 0.005 have been obtained by vapor transport methods.

During the study with different composition melts a new compound, $Na_4W_{14}O_{43}$, was obtained. Large crystals of this black triclinic semiconductor were grown over a narrow temperature range of 10 to 15 °C just above the melting point for melt compositions near 65 mole% WO_3. The growth rate for crystals was near or exceeded those rates observed for the cubic bronze phase.

4. Other bronze systems

We have used the electrolytic cell with little or no modification for the growth of a number of other tungsten bronze crystals. The compositions and structures of the bronzes that have been prepared are shown in fig. 5. The alkali carbonate rather than the tungstate was used for growth of Rb_xWO_3. The appro-

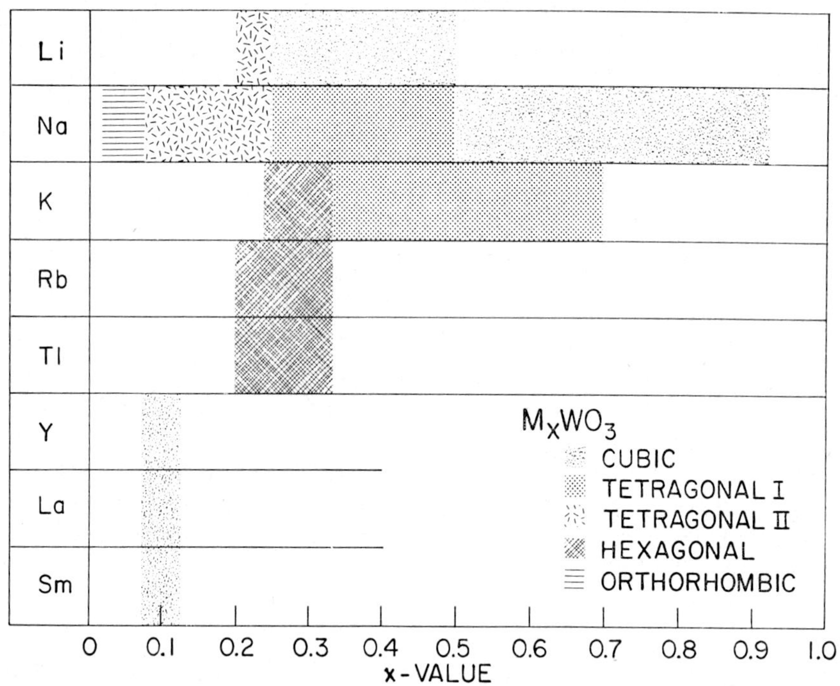

Fig. 5. Dependence of crystal structure on x-value for several tungsten bronzes prepared in the Ames Laboratory by electrolysis. The crystal symmetry increases with increasing x-value. Tetragonal I and hexagonal bronzes are superconductors; tetragonal II and orthohombic bronzes are semiconductors.

Growth rates were determined at the lowest temperature at which the desired x-value could be obtained with a constant current of 25 mA. Since the surface area of the cathode, on which the crystal is growing, changes with time a meaningful current density can not be defined. The growth rates for cubic Na_xWO_3 ranged from 0.3 g/hr in 10 mole% WO_3 melts to 0.6 g/hr in 50 mole% WO_3 melts. A minimum growth rate of 0.25 g/hr occurred at a composition of about 20 mole% WO_3 where Banks et al.[7] reported that the species in the melt begin to polymerize. The rates for tetragonal II and orthorhimbic phases were about 0.5 g/hr. The growth rate for the tetragonal I phase was much lower with values around 0.1 g/hr.

priate fluoride was used for the growth of Y, La[10]) and Sm bronzes.

A lithium isomorph of the compound $Na_4W_{14}O_{43}$ was grown from a melt of 50 mole% WO_3 in Li_2WO_4 over a narrow temperature range just above the melting point.

5. Conclusions

This study shows that large bronze crystals of the desired phase and composition can be prepared predictably by the electrolysis of the appropriate fused salt. In addition a comparison of the phases obtained for various melt compositions with the electrochemical data reported by Banks et al.[7] suggests a correlation between the change in the polymerization of the species in the melt and the phases observed. The transition from the cubic to tetragonal I structure occurs at a melt composition where they report a change from a tetramer to a higher polymer as the predominant species. Such changes can probably also be expected at higher WO_3 concentrations in the melt where lower symmetry bronze structures are observed.

Acknowledgments

I would like to thank A. F. Voigt and his group for the neutron activation analysis of the samples and P. A. Millis for help with the growth of the crystals. I would also like to thank G. C. Danielson and P. H. Sidles for helpful discussions throughout the study.

References

1) C. Scheibler, J. Prakt. Chem. **80**, (1860) 204.
2) L. Kahlenberg and H. H. Kahlenberg, Trans. Am. Electrochem. Soc. **46** (1924) 181.
3) V. Spitzen and L. Kaschtanoff, Z. Anorg. Allgem. Chem. **157** (1926) 141; Z. Anal. Chem. **75** (1928) 440.
4) B. W. Brown and E. Banks, J. Am. Chem. Soc. **76** (1954) 963.
5) L. D. Ellerbeck, H. R. Shanks, P. H. Sidles and G. C. Danielson, J. Chem. Phys. **35** (1961) 298.
6) H. R. Shanks, P. H. Sidles and G. C. Danielson, *Nonstoichiometric Compounds* Advances in Chemistry No. 39, (Am. Chem. Soc., 1963).
7) E. Banks, C. W. Fleischmann and L. Meites, J. Solid State Chem. **1** (1970) 372.
8) J. W. Mellor, *Inorganic and Theoretical Chemistry*, Vol. 11 (1931) ch. 62, p. 673.
9) A. Ferretti, E. I. DuPont Company, private communication.
10) H. R. Shanks and G. C. Danielson, J. Appl. Phys. **38** (1967) 4923.

Section IX

Solution growth

A. GLASNER
M. TASSA
F. MUSSARD
S. GOLDSZTAUB
S. TROOST
G. PILKINGTON
W. J. DUNNING
S. LE BIHAN
M. FIGLARZ
I. F. NICOLAU

M. ITTU
R. DABU
E. ACKER
S. HAUSSÜHL
K. RECKER
H. M. LIAW
J. W. FAUST, Jr.
R. Z. LeGEROS
J. P. LeGEROS

Journal of Crystal Growth **13/14** (1972) 441–444 © *North-Holland Publishing Co.*

THE THERMAL EFFECTS OF NUCLEATION AND CRYSTALLIZATION OF KBr SOLUTIONS

I. EXPERIMENTAL AND MATHEMATICS, ILLUSTRATED

ABRAHAM GLASNER and MENACHEM TASSA

Department of Inorganic and Analytical Chemistry, The Hebrew University of Jerusalem, Israel

The rate of cooling of concentrated KBr solutions containing varying amounts of added Pb^{2+} ions was studied. Internal evolution of heat due to nucleation and to crystallization was observed and measured. This part of the paper includes the mathematical treatment of the continuous temperature measurements, some examples of curves plotted by the computer together with a general interpretation of them, and a discussion of the effect of Pb^{2+} ions on nucleation and the rate of crystallization.

1. Introduction

A thermal method based on Newton's cooling law was employed in a study of the processes taking place in the crystallization of KBr from aqueous solutions. The method was first used by Glasner and Kenat[1]) in the case of KCl. In this study the method has been greatly improved by the use of modern sensitive equipment for temperature measurement and control and also by the extensive exploitation of the computer for calculations and data recording. Observation of the thermal effects due to nucleation was thus made possible.

2. Experimental

The calorimeter was a chrome-nickel plated brass cylinder with an air-tight cover. The salt solution was held in a pyrex glass beaker, gold-plated on the outside to reduce heat loss by radiation and screw-fastened to the inside of the cover of the cylinder. Two water-tight holes were bored in the cover to carry the stirrer and the temperature sensing probe.

The solution was stirred continuously with a pyrex glass rod at a constant rate of revolution controlled by a synchronous motor and flexible transmittance.

A specially purified large batch of KBr was used in the whole series of experiments. Care was taken to minimize the Pb^{2+} content by fractional crystallization.

The solutions were made up to be saturated at 44 °C by weighing out: 193 g KBr $+ (250 - X)$ g $H_2O + X$ g of a standard $PbBr_2$ solution.

The closed calorimeter was then heated in an air oven at ~ 70 °C for several hours, to ensure dissolution of nuclei, and afterwards submerged in a thermostated water bath at 25.000 $\pm$ 0.005 °C controlled by a Tromac (U.S.A.) thermistorized temperature controller, marked PC 1000.

The system may be considered stabilized when the temperature of the outer surface of the calorimeter has become equal to that of the water bath. From this point on, cooling of the solution may be expected to be regular according to Newton's Law.

Readings of the temperature of the solution in the calorimeter were printed out at constant intervals of 60 sec in the temperature range 60–28 °C. Each run took more than five hours.

The thermometric system was based on a digital quartz thermometer supplied by Hewlett–Packard, U.S.A. This apparatus flashes the temperature readings every second with a resolution of 10^{-3} °C. Two scanning probes were used simultaneously, one in the stirred solution and one in the water bath. The quartz thermometer was connected, through an electronic programmer built for the purpose to a digital printer supplied by Takeda Riken, Japan. Thus any one of three temperature values could be printed out at predetermined time intervals: T_i (solution), T_0 (water bath), and $T_i - T_0$ (difference). The corresponding T_i and t_i (time) values were punched out and fed to the computer (CDC 6400), which digested the data according to the expressions and relations in the written program described below. The computer was also instructed to plot the required curves.

3. Mathematical evaluation of heat effects

The cooling law in the case of a wide temperature difference $T_i - T_0$ is given by:

$$dT/dt = -k_0[1 + \alpha(T_i - T_0)](T_i - T_0), \qquad (1)$$

where k_0 is the cooling coefficient of the particular system and α is an experimental correction factor[1]).

The integrated expression is:

$$\ln\{(T_i - T_0)/[1 + \alpha(T_i - T_0]\} = -k_0 t + C,$$
$$C = \text{constant}. \qquad (1a)$$

Hence, when the logarithmic term on the left is plotted versus time a straight line is obtained with the slope k_0 if the cooling law is valid. In practice (fig. 1, curve K) the plot is very nearly linear up to the time, t_c, when crystallization starts; in pure KBr solutions a distinct kink is observed at t_c.

3.1. THE TEMPERATURE DEVIATION

The computer calculates k_0, α and C for each run from the data supplied in the temperature range 60–44 °C (the saturation temperature). When the experimental parameters are known $T_{i\,cal}$, the expected temperature at any time t_i if no reaction of any kind has occurred in the cooling solution, can be calculated by a simple transformation of eq. (1a):

$$T_{i\,cal} = \exp(C - k_0 t_i)/[1 - \alpha \exp(C - k_0 t_i)] + T_0.$$

The temperature deviation D_i at any time is therefore given by:

$$\begin{aligned} D_i &= T_i - T_{i\,cal} \\ &= T_i - \exp(C - k_0 t_i)/[1 - \alpha \exp(C - k_0 t_i)] - T_0. \quad (2) \end{aligned}$$

Curve D in fig. 1 is typical of such a temperature deviation curve, being a plot of the temperature elevation of the solution due to reactions taking place therein. Curve D is sigmoidal in form, starting to rise gently soon after the saturation point and then rising steeply in the region corresponding to the kink in curve K, and finally levelling off, suggesting a constant rate of crystallization as the temperature of the solution approaches that of the water bath.

3.2. HEAT OF REACTION

The heat energy in calories, dH, evolved in time dt can be calculated as follows:

$$dQ = hk_0[1 + \alpha(T_i - T_0)](T_i - T_0)\,dt,$$

where dQ is the flow of energy from the calorimeter to the thermostat [from eq. (1)]; h is the total heat capacity of the calorimeter and the solution in it.

The corresponding equation for the heat flow if no reaction takes place is:

$$dQ_{cal} = hk_0[1 + \alpha(T_{i\,cal} - T_0)][T_{i\,cal} - T_0]\,dt.$$

Considering that α is of the order of 10^{-2} and T_i is larger than $T_{i\,cal}$ only by a fraction of a degree, we write:

$$dH = dQ - dQ_{cal} = hk_0[1 + \alpha(T_i - T_0)](T_i - T_{i\,cal})\,dt. \qquad (3)$$

The curve dH (cal/min), see fig. 1, described by the plotter resembles in shape the devation curve D. The conversion factor is the *nearly* linear variable: $hk_0[1 + \alpha(T_i - T_0)]$. Hence either of these two curves, D or dH, may serve to represent the relative rate of reaction throughout a run.

The differential dH/dt (curve A) and the integral (curve H) of eq. (3) were also plotted.

A distinct feature of the differential curve A is a narrow high peak appearing at the time of the start of crystallization proper. When lead ions are added to the crystallizing solution this peak is shifted gradually along the time axis and simultaneously becomes bell-shaped, like a wide low absorption band.

The integral curve H allows the calculation of the amount of KBr crystallized after the lapse of any time t_i, by dividing H_i by the heat of crystallization L_{KBr}[2]).

3.3. SUPERSATURATION

The supersaturation S_i, in g KBr/1000 g H_2O, is given by the equation:

$$S_i = (T_s - T_i)\gamma - 4H_i/L_{KBr}, \qquad (4)$$

where T_s is the temperature of saturation (44°C) and γ is the solubility gradient g KBr/[1 °C × 1000 g H_2O][3]). Curve S, fig. 1, represents supersaturation throughout an experiment. It has a distorted bell shape, rising quickly and then sinking at a slower rate. When no lead ions, or only a small quantity of lead ions, were added to the solution a hump appears at the point of initial crystallization, indicating the sudden release of *part* of the supersaturation.

However, the supersaturation curves should be treat-

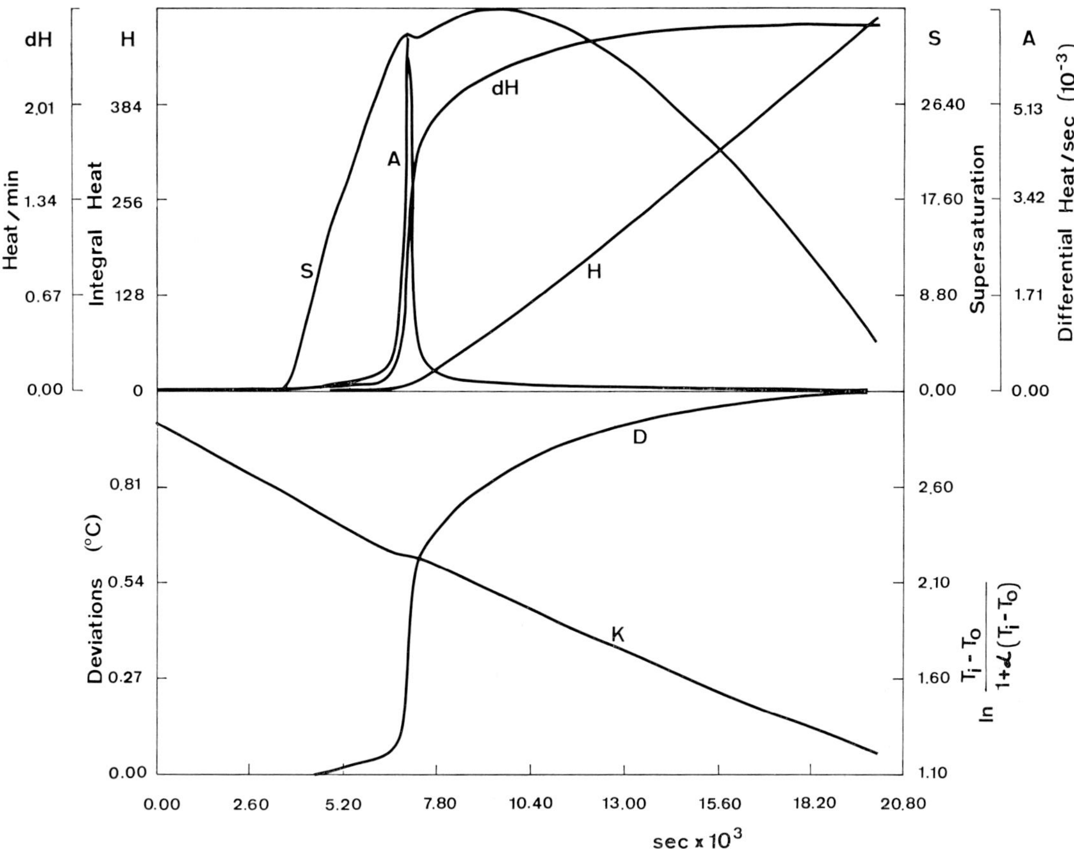

Fig. 1.   Cooling curve K and several derived curves presenting the thermal effects in a cooling KBr solution. (D) Temperature deviation; (A) differential rate of heat evolution; (dH) heat evolution per minute; (H) integral total heat; (S) "supersaturation" curve.

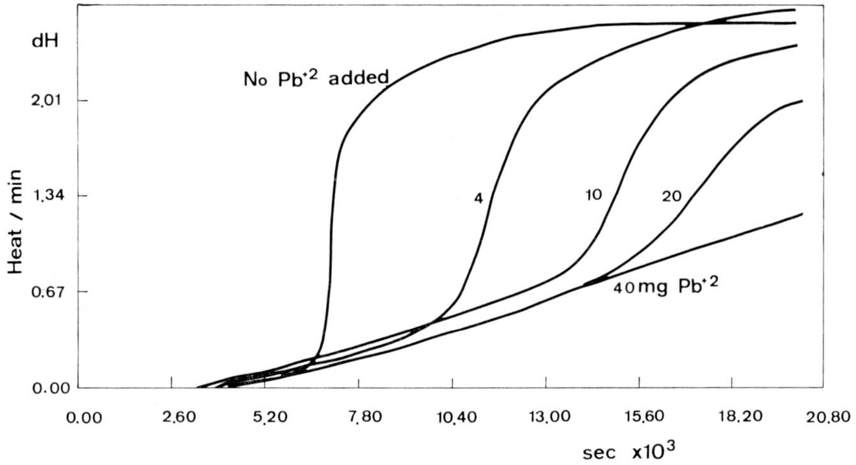

Fig. 2.   Heat evolution, with added Pb^{2+} ions, as marked on the curves.

ed with caution, since H_i in eq. (4) includes the heat of reaction evolved before the crystallization point. The latter fraction of H_i increases appreciably with the addition of Pb^{2+} ions, as shown in fig. 2 and below.

4. Summary of results and interpretation

Fig. 2 shows the dH curves of five runs with increasing amounts of added Pb^{2+} ions. The effect of the lead ions on the processes taking place in the cooling solution are made clearly perceptible by sectioning the sigmoidal curve into three parts: 1 – Nucleation; 2 – Accelerated crystallization; 3 – Uniform rate of crystallization.

Section N (Nucleation). When 40 mg Pb^{2+} ions were added, section N (Nucleation) extends to the very end of the run, and when the calorimeter was opened only a few crystals, weighing about 0.3 g, were found in the solution. Hence the larger part of the total heat evolved, about 150 cal, is due to a reaction which is not crystallization. Since H_i for section N increases proportionately with the amount of lead ions, in parallel with the extension of the supersaturation limit, the reaction must be attributed to the formation of lead-centered nuclei or other nucleators originally present in the "pure" salt. The more lead ions are added to the solution, the smaller are the "nuclei" formed. These "nuclei" are kept in the solution and will precipitate, in the absence of seed crystals, only after their diameter has grown to much more than 100 Å.

Section A (Accelerated crystallization). The rate of evolution of heat increases spontaneously. Crystallization becomes visible at the turning point between the first two sections of the dH curve, and section A therefore represents the passage from the nucleation process to the crystallization process. The total heat evolved in the latter process is found to be proportional to the product of the weight of the crystals formed and the heat of solution in a saturated solution[2]). Hence it is concluded that crystal growth starts with matured "nuclei", i.e., seed crystals, which have grown beyond a certain critical size m_0. Smaller "nuclei" are captured by the seed crystals and are laid down on its surface. The heat evolved in the crystallization process is thus perceived as being related to the lattice energy released when small sub-colloidal crystallites are fitted together into well-grown crystals[4]). The slope of the accelerating section A decreases with the increase in the amount of lead ions added, due to the inverse relation between the size of the "nuclei" and the amount of lead ions.

Section U (Uniform rate of crystallization). In experiments with the "pure" salt, without lead ions added, the last section of the dH curves attained a constant value of 2.62 cal/min, equivalent to 0.094 g KBr crystallizing per minute. At the same time the supersaturation decreases and approaches zero (see fig. 1). Clearly, the rate of crystal growth is *not* a simple function of the supersaturation. However, crystal growth by the adherence of preformed "nuclei" on the surface of seed crystals, may well assume a constant rate, owing to a balance between the diminishing number of "nuclei" in the solution and the increase in the size of these "nuclei" as the solution continues to cool gradually.

When small amounts of lead ions were added, very nearly the same maximal values of dH; namely 2.6 cal/min, were obtained in several experiments. Even if the solutions contained 4, 10, or 20 mg Pb^{2+} the dH curves showed a tendency to attain the very same value. The experiments with large amounts of Pb^{2+} ions could not be continued for longer periods because of technical limitations. However, the above observations indicate that in a series of solutions of similar composition and rate of cooling, when only the amount of lead ions present varies, the rate of crystallization attains the same maximal rate at a gradually decreasing temperature in inverse order of the Pb^{2+} concentration.

This completes the description of the outline of the processes taking place in a crystallizing solution on cooling. A quantitative evaluation of the results will be given in the second part.

References

1) A. Glasner and J. Kenat, J. Crystal Growth **2** (1968) 119.
2) J. Wüst and E. Lange, Z. Physik. Chem. **116** (1925) 196.
3) H. Stephen and T. Stephen, *Solubilities of Inorganic and Organic Compounds* (Pergamon Press, Oxford, 1963) Table No. 252.
4) A. Glasner, Israel J. Chem. **7** (1969) 633.

Journal of Crystal Growth **13/14** (1972) 445–448 © *North-Holland Publishing Co.*

SUR LA CROISSANCE DU CHLORATE DE SODIUM EN SOLUTION

F. MUSSARD et S. GOLDSZTAUB

Laboratoire de Minéralogie, Université de Strasbourg, Strasbourg, France

Frequently, the homologous (100) faces of a sodium chlorate crystal grow from solution with two very different rates V_R and V_L. Interference microscope measurements and X-ray topography show that the slow rate V_L is due to surface nucleation while the fast rate V_R is connected with dislocations.

La croissance en solution aqueuse du chlorate de sodium a été fréquemment étudiée. Plusieurs chercheurs[1,2]) ayant signalé que des faces pourtant homologues {100} d'un même cristal ne se développaient pas à la même vitesse, nous avons tenté d'apporter l'explication de ce phénomène. Les expériences ont été effectuées sous le microscope dans de petites cuves de 0.5 mm de profondeur. On constate que certaines faces ont très peu progressé; c'est pourquoi nous avons qualifié de "blocage" ce phénomène. En fait une observation plus attentive révèle que ces faces "bloquées" croissent malgré tout à une vitesse très lente (V_L), les autres faces étant à croissance rapide (vitesse V_R). En abaissant progressivement la température de la solution mère et en photographiant le cristal, on peut déter-

miner la position des faces en fonction du temps et de la température (fig. 1). On en déduit ensuite l'évolution du rapport V_R/V_L (fig. 2), rapport qui diminue très rapidement avec la température, c'est à dire lorsque la sursaturation augmente.

D'autre part en utilisant soit un micrscope interférentiel Baker, soit un microscope classique muni d'un interféromètre de Michelson, on photographie les courbes d'égale concentration de la solution autour du cristal en voie de croissance. Au voisinage de la face "bloquée" sur la figure 3 n'apparaît qu'une frange d'interférence au lieu de trois pour les autres faces. La

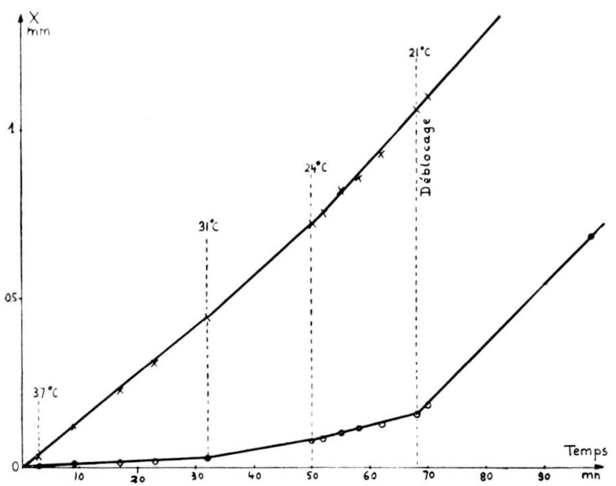

Fig. 1. Evolution d'une face rapide et d'une face bloquée en fonction du temps.

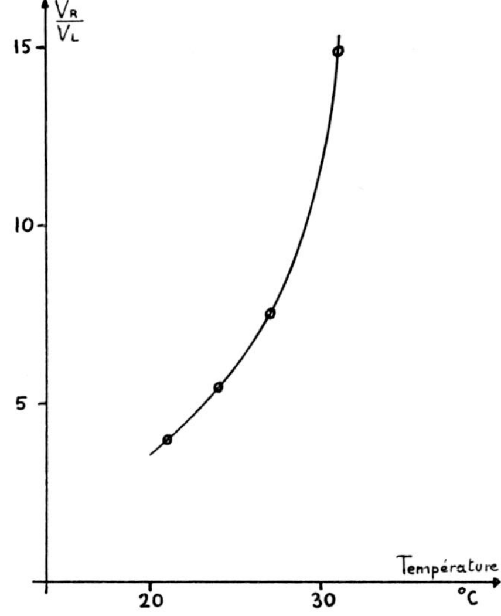

Fig. 2. Rapport des vitesses de croissance d'une face rapide et d'une face bloquée en fonction de la température de la solution.

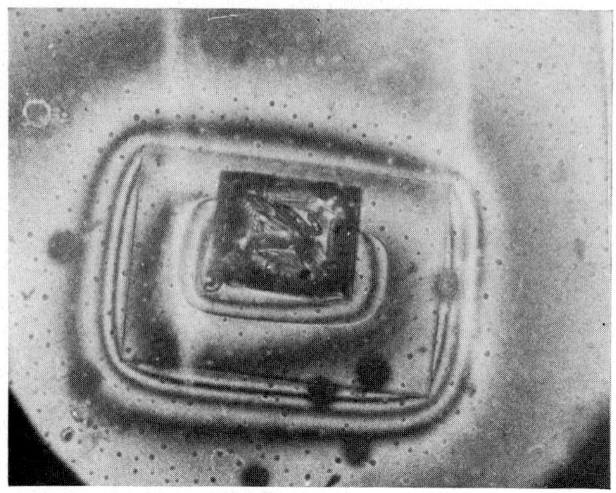

Fig. 3.   Courbes d'égale concentration de la solution autour d'un cristal présentant une face bloquée.

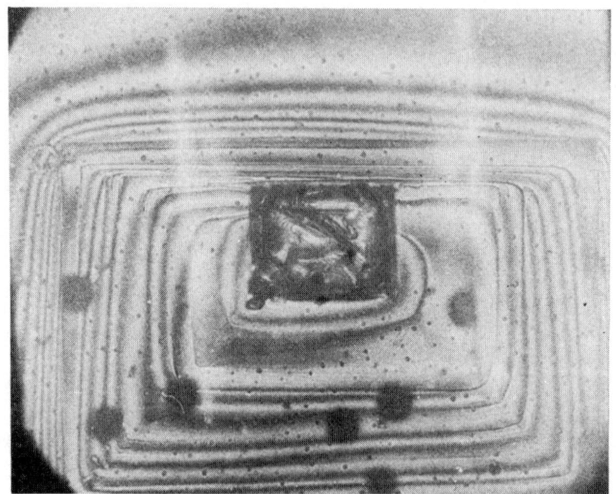

Fig. 4.   Déblocage se manifestant par l'apparition de nouvelles franges.

5a

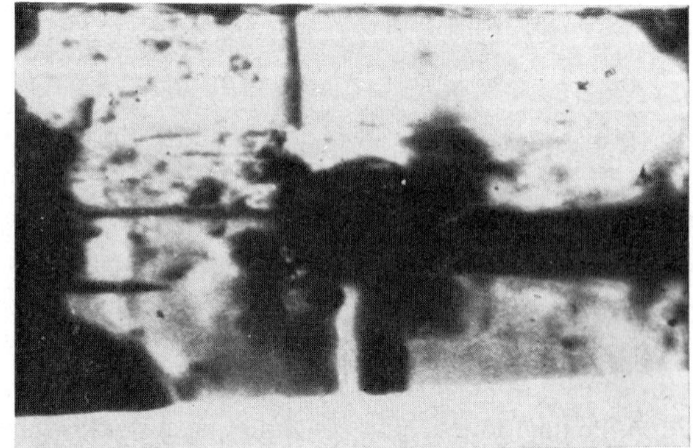

5b

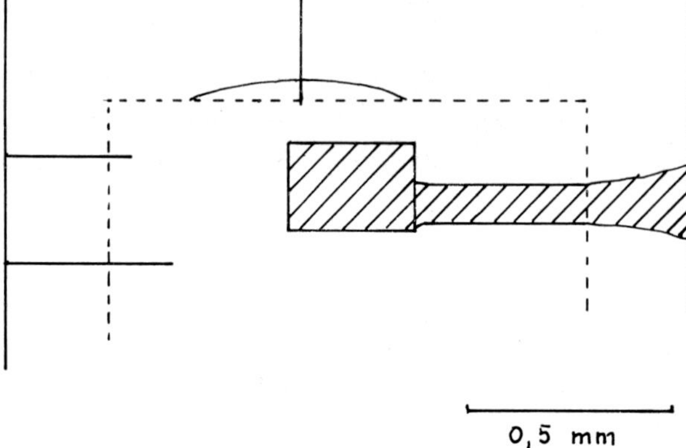

5c

Fig. 5.   (a) Topographie suivant 200; (b) topographie suivant 020; (c) supersposition schématique du déblocage (pointillés) et des topographies. (a) $g \rightarrow$; (b) $g \uparrow$.

concentration étant uniforme loin du cristal, la sursaturation est donc plus faible au contact des trois faces à croissance rapide que de la face "bloquée".

Parfois, lorsque la sursaturation augmente, on observe le "déblocage" de certaines faces à croissance lente; c'est le cas notamment du cristal de la figure 3: des franges supplémentaires apparaissent sur la face bloquée et s'étendent rapidement le long de celle-ci (fig. 4). Simultanément la vitesse de cette face passe de V_L à V_R comme on le voit sur la figure 1: après déblocage, les deux courbes présentent la même pente.

Nous avons pensé que la théorie de Burton, Frank et Cabrera[3]): pouvait expliquer au moins qualitativement ces phénomènes:

— la croissance rapide à faible sursaturation étant due à la présence de dislocations vis,

— la croissance lente à forte sursaturation correspon-

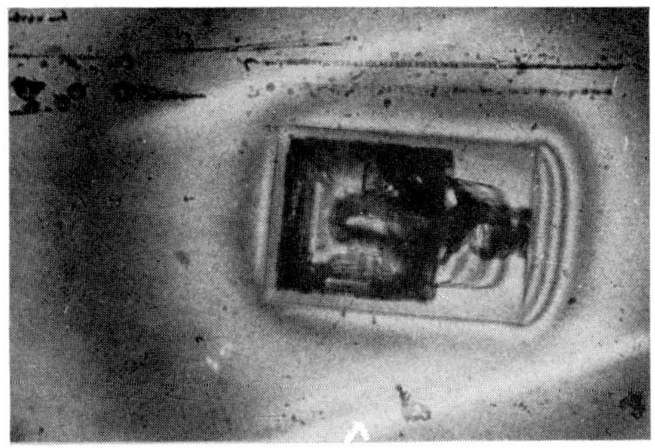

6a

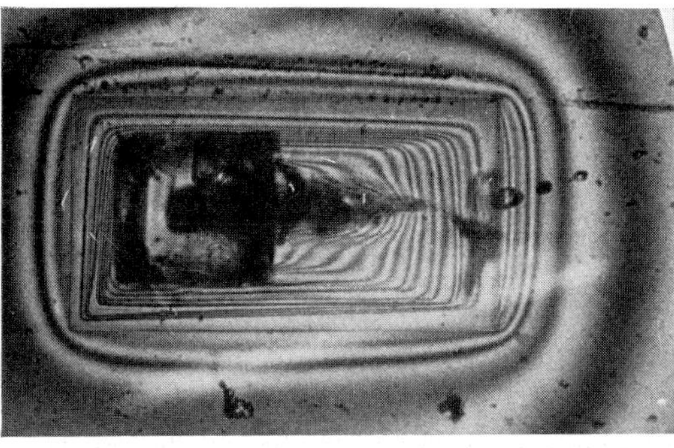

6b

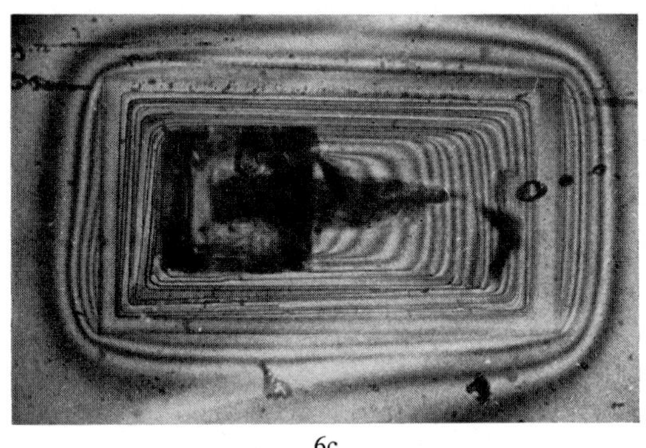

6c

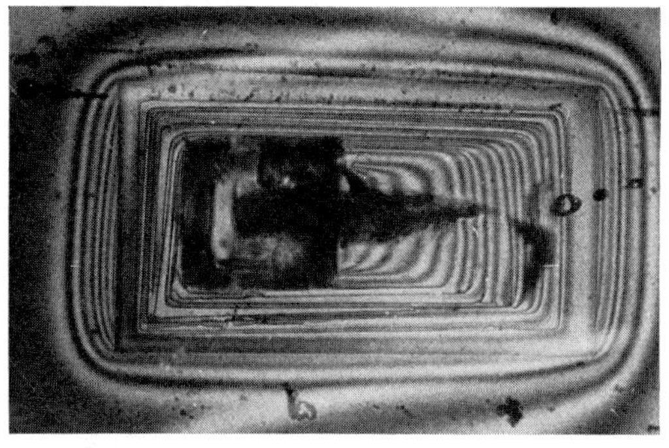

6d

6e

Fig. 6.   Evolution d'un cristal présentant initialement trois faces bloquées. On note duex déblocages successifs (c et e).

5a : le pointillé correspond au cristal lors du déblocage. On constate que dans la partie du cristal engendrée par croissance lente, il n'y a pas de dislocation ; par contre, le déblocage coïncide avec l'origine d'une dislocation.

Plusieurs expériences de ce type ont été effectuées et ont mené au même résultat. L'une d'elles (fig. 6) est plus intéressante, chaque face offrant une particularité :
– à l'origine, une seule face croît rapidement (figs. 6a et 6b, face de droite),
– une première face se débloque (figs. 6c et 6d, face de gauche)
– une deuxième face se débloque (fig.6e, face inférieure)
– la dernière face reste bloquée.

Après avoir immergé le cristal dans le benzène (fig. 7a), pour repérer les inclusions, on fait les topographies 200 et 020 (figs. 7b et 7c). On constate sur le schéma 8, superposition de 7a, 7b et 7c :
– la présence de dislocations à caractère vis partant du germe et émergeant sur la face initialement rapide (face de droite),

dant au mécanisme de germination bidimensionnelle,
– le déblocage serait consécutif à l'apparition de dislocations à caractère vis.

Pour vérifier ces hypothèses, nous avons utilisé la méthode de Lang[4, 5]) pour mettre en évidence les dislocations du cristal étudié. Dès que sa taille est suffisante, il est sorti de la solution et des topographies sont prises avec les réflexions 200 et 020 (figs. 5a et 5b), de telle sorte que les seules dislocations vis susceptibles d'être invisibles émergeraient sur les faces inférieure ou supérieure du cristal qui ne sont pas étudiées. Le schéma 5c représente la superposition de la figure 4 et de

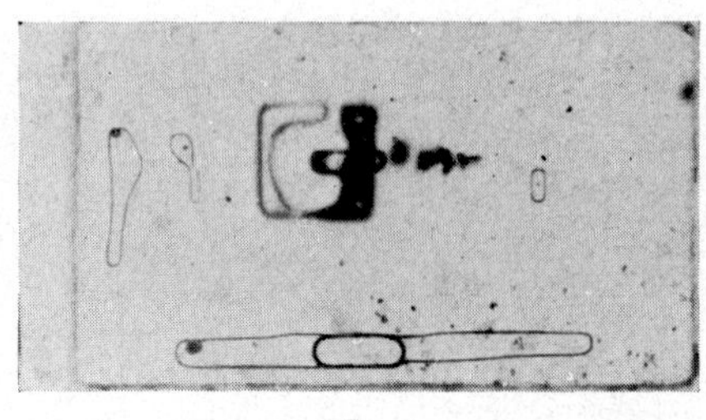

7a

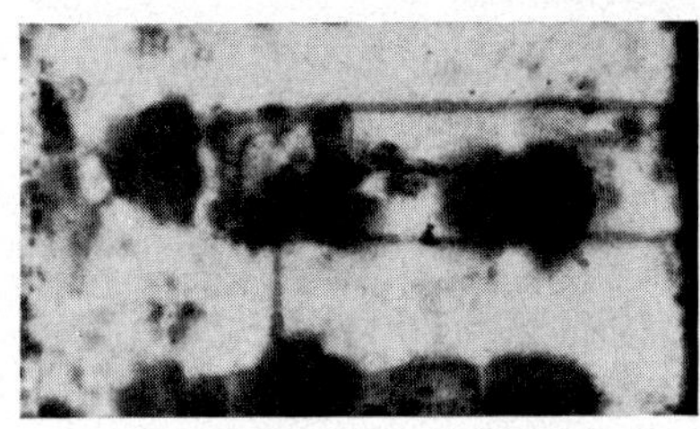

7b

7c

Fig. 7. (a) Mise en évidence des inclusions; (b) topographie suivant 200; (c) topographie suivant 020. (b) $g \rightarrow$; (c) $g \uparrow$.

– à partir des déblocages (pointillés), apparition de dislocations,

– une ligne de dislocation coin (car éteinte sur la figure

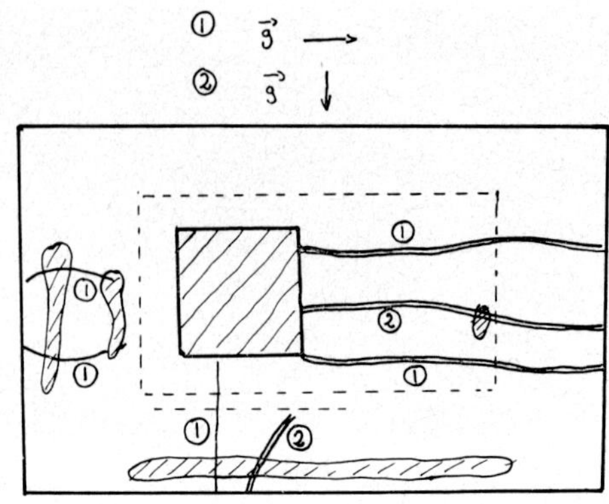

Fig. 8. Superposition schématique des topographies. La position du cristal lors des déblocages est figurée en pointillé.

7c), partant du germe et émergeant sur la face inférieure. On remarquera que son influence sur la vitesse de croissance est nulle.

– pas de dislocation émergeant sur la face supérieure qui est restée bloquée jusqu'à la fin de l'expérience.

De l'ensemble de ces faits, on peut conclure que la croissance lente du chlorate de sodium correspond à la germination bidimensionnelle, tandis que la croissance rapide est liée à la présence de dislocations, ce qui était prévu par la théorie.

Bibliographie

1) W. J. Humphreys-Owen, Proc. Roy. Soc. (London) A **197** (1949) 218.
2) M. Quivy, Rapport C.E.A. (1961), non publié.
3) W. K. Burton, N. Cabrera et F. C. Frank, Phil. Trans. Roy. Soc. London A **243** (1951) 299.
4) A. R. Lang, Acta Cryst. **12** (1959) 249.
5) A. R. Lang, J. Appl. Phys. **30** (Z959) 1748.

Journal of Crystal Growth **13/14** (1972) 449–453 © *North-Holland Publishing Co.*

CRYSTAL GROWTH OF SODIUM TRIPHOSPHATE HEXAHYDRATE FROM AQUEOUS SOLUTIONS

S. TROOST*

Unilever Research Laboratory, Vlaardingen, The Netherlands

The growth rate (R) of the various crystallographic faces of a single sodium triphosphate hexahydrate crystal, $Na_5P_3O_{10} \cdot 6H_2O$, is measured as a function of the supersaturation in solutions prepared from both anhydrous phases: sodium triphosphate Phase I ($Na_5P_3O_{10}$–I) and sodium triphosphate Phase II ($Na_5P_3O_{10}$–II). It is very striking that at the same degree of supersaturation the linear growth rate of the (100) faces and the (001) faces of the hexahydrate in a solution made of Phase I is much higher than in a solution of Phase II. Since it has been shown that the structure of the hexahydrate crystals is the same in both cases there must be a difference in structure of the solute in the solution. Particularly the difference in the structure of the P_3O_{10} group, and in the sodium co-ordination between the crystal structures of Phase I and Phase II, is determinative of the number and shape of the ions in the supersaturated solutions. In view of this we assume that the rupture of the bond between the sodium cation and triphosphate anion is the rate-determining step for crystal growth of hexahydrate.

1. Introduction

Sodium triphosphate, $Na_5P_3O_{10}$, occurs in two anhydrous modifications, $Na_5P_3O_{10}$–Phase I, metastable at temperatures below 410 °C and $Na_5P_3O_{10}$-Phase II, stable below this temperature and as the hexahydrate, $Na_5P_3O_{10} \cdot 6H_2O$. All three forms find extensive application in the detergent industry, and for this reason their properties have been studied by Van Wazer[1]. Many of the observed phenomena in the sodium triphosphate–water system, such as the greater ease of hydration of Phase I, lack a full explanation. Also not yet fully understood is the fact that Phase II can easily be dissolved in water but crystallizes comparatively slowly, even from a highly supersaturated solution.

2. Experimental

The growth rate (R) of the various crystallographic faces of a single hexahydrate crystal as a function of the supersaturation in solutions of both anhydrous phases, is measured in a specially designed growth cell[2,3]. The growth rates of the (100) and ($0\bar{1}0$) faces were measured by time-lapse microphotography through a microscope. The very low growth rate of the (001) face was determined optically by means of the rotary compensator, developed by Ehringhaus, using

the considerable birefringence shown by hexahydrate crystals. The reproducibility of the growth measurements is very satisfactory, considering the fact that a new single crystal of $Na_5P_3O_{10} \cdot 6H_2O$ was used for each experiment.

The degree of supersaturation (σ) was determined by comparing the measured concentration (c) in the solution with that of a saturated hexahydrate solution (c_s) at the same temperature ($\sigma = c/c_s - 1$). All measurements have been done at the temperature of 37.5 °C.

3. Results

From the so determined $R(\sigma)$ curves (see figs. 1 and 2) and complementary experiments the following conclusions can be drawn:

– There is no linear relationship between growth rate and supersaturation. The test whether the growth rate is diffusion-controlled, which seems unlikely, experiments have been made to establish the influence of the flow rate of the solution. The results show that the growth rate is independent of the flow rate. The measured growth rate is also a factor 400 lower than that calculated for diffusion controlled growth. It can therefore be concluded that the diffusion is not growth rate determining.

– It is very striking that at the same degree of supersaturation (e.g. $\sigma = 0.43$) the growth rate of the (100) (107 μm/h) and the (001) (11.20 μm/h) faces of the

* Present addres: Akzo-Zout Chemie-Research, P.O. Box 25, Hengelo (O), The Netherlands.

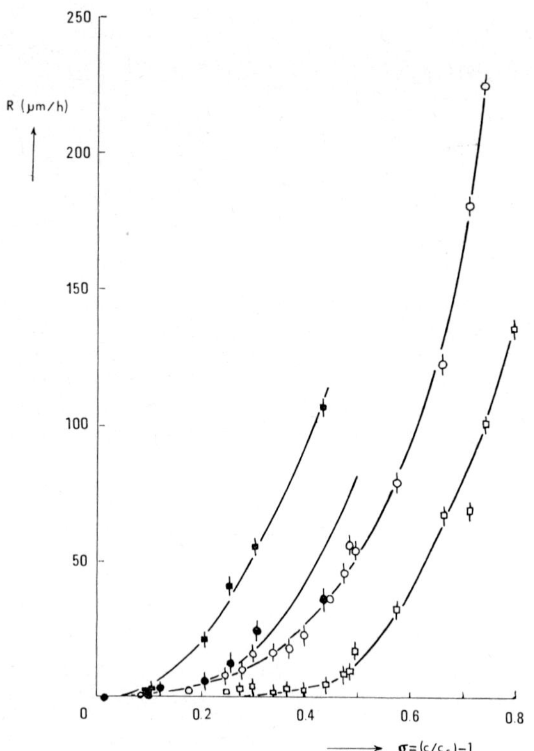

Fig. 1. Growth rate of (100) and (0$\bar{1}$0) faces plotted against degree of supersaturation: (■) (100)-Phase I; (●) (0$\bar{1}$0)–Phase I; (□) (100)–Phase II; (○) (0$\bar{1}$0)–Phase II.

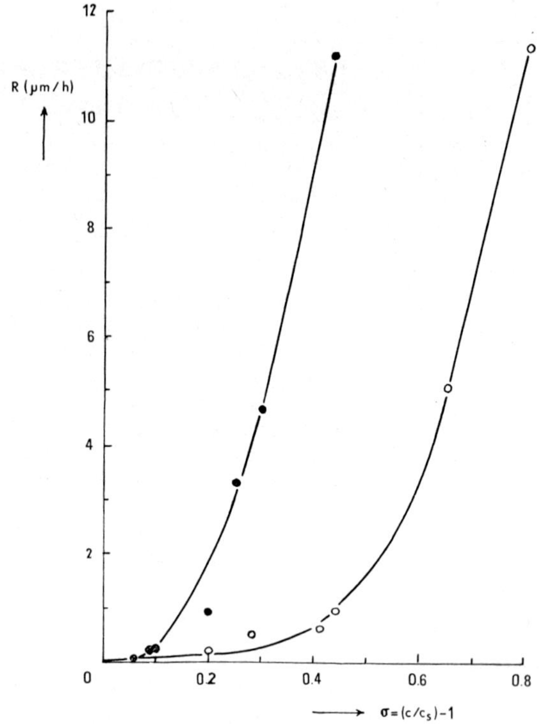

Fig. 2. Growth rate of (001) faces plotted against degree of supersaturation: (●) Phase I; (○) Phase II.

4. Discussion

At the same degree of supersaturation, the growth rate of hexahydrate crystals from solutions made from

hexahydrate in solutions made from Phase I is much higher than in solutions of Phase II (resp. 5.3 μm and 0.94 μm/h). The (100) faces grow about 20 times faster; the growth rate of the (001) face is about 10 times higher. The growth rate of the (0$\bar{1}$0) faces, however, is about the same for both solutions (36 μm/h).

– The growth rate of the (001) face is much smaller than that of the other faces in both types of solution.

– In solutions of Phase II in the ease of a relatively low degree of supersaturation ($\sigma < 0.40$) (see figs. 1 and 2), the growth of the (0$\bar{1}$0) faces is much faster than that of the (100) and (001) faces. This results in the formation of acicular crystals. At $\sigma > 0.40$, the growth rate of the (100) faces increases strongly, resulting in the formation of tabular crystals.

– At low degrees of supersaturation ($\sigma < 0.10$) of solutions of Phase I the growth of the (100) faces is predominant: only at a degree of supersaturation above $\sigma > 0.20$ does the growth rate of the (0$\bar{1}$0) faces increase strongly resulting in the formation of tabular crystals. Ultimately, plate like crystals are formed.

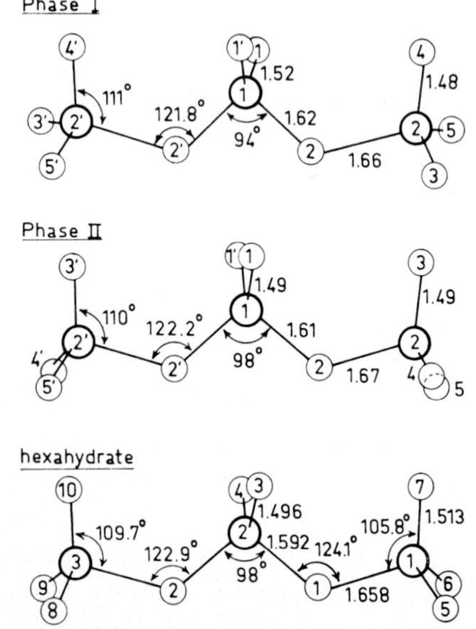

Fig. 3. Configuration of P_3O_{10} groups in Phase I, Phase II and in hexahydrate respectively. Bold-faced circles: phosphor; smaller light-faced circles: oxygen.

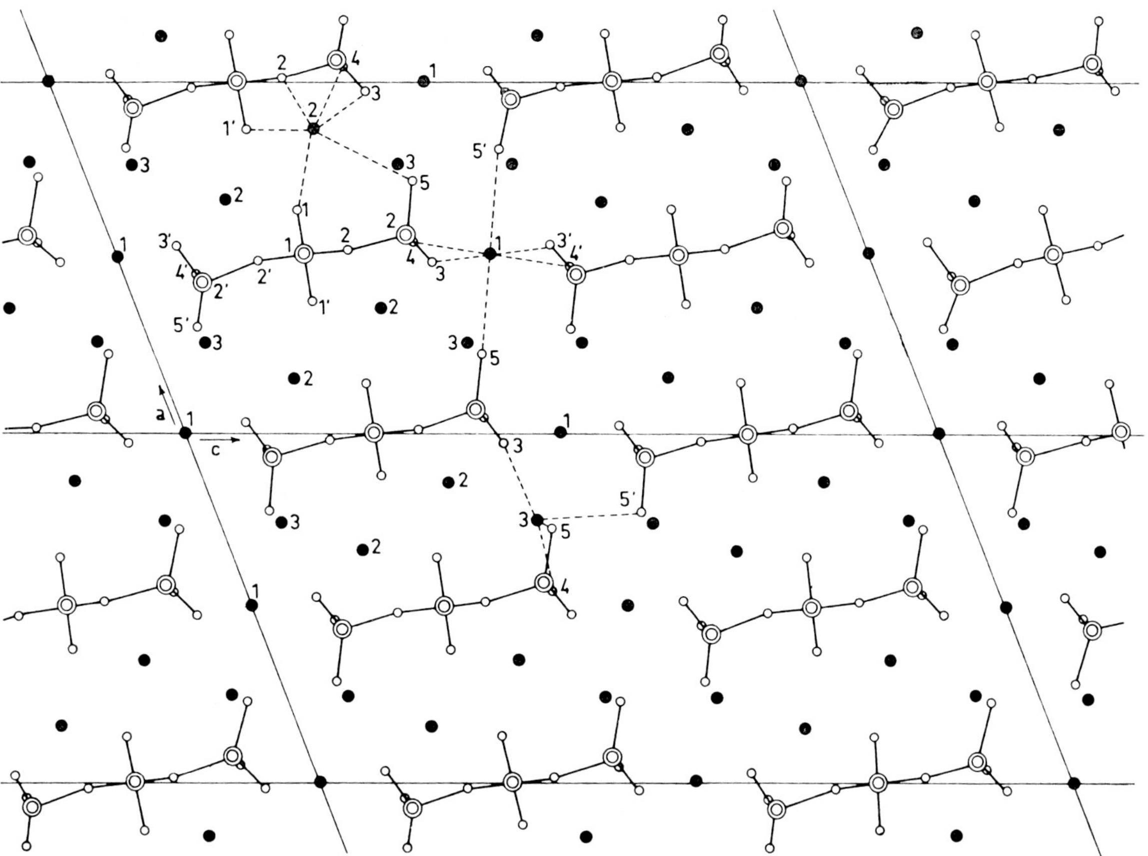

Fig. 4.   Crystal structure of Phase I: projection along b-axis. Solid circles: sodium; concentric circles: phosphor; open circles: oxygen.

Phase I is different from that of solutions of Phase II. Since it has been shown by Dyroff is his thesis[4]) that the structure of the hexahydrate crystals is the same in both cases, there must be a difference in structure of the dissolved ions in the solution. We have carried out osmotic measurements at the same temperature to get an idea of the difference in structure and found that supersaturated solutions of Phase I contain more particles than equimolar solutions of Phase II. According to Wall and Doremus[5]) in their measurements of the fraction of sodiums bound to phosphate ions in unsaturated solutions the $[NaP_3O_{10}]^{4-}$ ion is present, while in saturated solutions the predominant ion is $[Na_2P_3O_{10}]^{3-}$.

The supersaturated solutions were prepared at 37.5 °C by dissolving $Na_5P_3O_{10}$–Phase I or Phase II in water. The difference between the crystal structures of Phase I and Phase II determined by Corbridge[6]) are caused by differences in the structure of the P_3O_{10} group (fig. 3) and the co-ordination of the sodium cation. In the

crystal structure of Phase I and Phase II the anion has a two-fold axial symmetry. The principal difference between Phase I and Phase II is that in Phase I the anion deviates to a greater extent from the trans-symmetrical configurations and has a higher degree of distortion occurring in the terminal PO_4 groups. In the hexahydrate structure the triphosphate ion has no biaxial axis and the anion is more elongated. Since Na^+ ions in solution remain partially linked to the $[P_3O_{10}]^{5-}$ anion, we may assume that this difference in structure of the P_3O_{10} group has great influence on the structure of the dissolved ions in the solutions.

We also assume that only those Na^+ ions which are surrounded by two 0 atoms of the same P_3O_{10} chain remain linked in their original position to the anion. For Phase I solutions there are then only $[Na_2P_3O_{10}]^{3-}$ ion configurations (fig. 4).

We might imagine that the $Na^+(3)$ ion in solution also links two P_3O_{10} groups in the [100] direction. However, this is very unlikely on account of steric

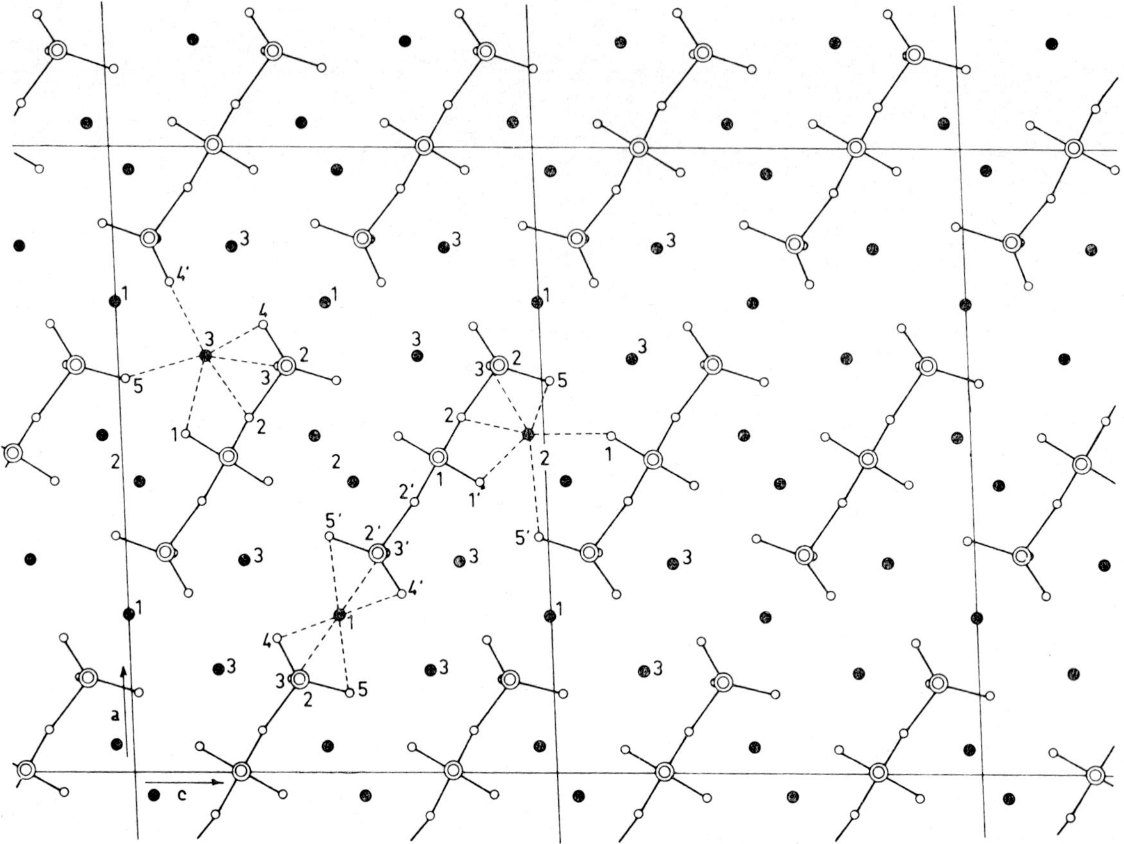

Fig. 5.   Crystal structure of Phase II: projection along b-axis solid circles: sodium; concentric circles phosphor; open circles: oxygen.

hindrance between the P_3O_{10} groups since their axes are parallel.

For Phase II there are more possibilities (fig. 5). One $Na^+(1)$ ion has a very special six-fold co-ordination, for it is on two sides linked to two different P_3O_{10} groups via O(4) and O(5) atoms. We assume therefore – as Corbridge did – that in a supersaturated solution of Phase II, the anions are linked to form associated units

$$- [\, Na_2P_3O_{10} \,]^{3-} - Na^+ - [\, Na_2P_3O_{10} \,]^{3-} .$$

In both cases the original configuration of the P_3O_{10} group is maintained by the Na^+ ions.

The growth rate of the hexahydrate crystals is very low in comparison with most other inorganic salts. This very slow growth must be clearly linked up with the formation of $[Na_2P_3O_{10}]^{3-}$ ions. The structure of the P_3O_{10} group in the hexahydrate crystal lattice is different from that in the two anhydrous modifications. For a change in the configuration to occur, it is necessary that at least one of the two Na–O bonds, with which the Na^+ ion is linked to the anion, be broken. This rupture, followed by a change of configuration of the anhydrous P_3O_{10} group into that of the hexahydrate, will be the rate-determining step. This difference in the configuration of the P_3O_{10} group will also be the reason why the two anhydrous modifications easily yield supersaturated solutions, whereas this will never be the case with hexahydrate; in this latter case, the P_3O_{10} group already has the correct configuration for growth.

As we noted already, the growth rate of the (001) and (100) faces in solutions of Phase II is much lower than that in Phase I solutions whereas the growth rate of the $(0\bar{1}0)$ faces in Phase II solutions is retarded only slightly. These phenomena can be explained as follows. The direction of the P_3O_{10} chain of $Na_5P_3O_{10} \cdot 6H_2O$ runs parallel to the $(0\bar{1}0)$ faces since the direction of the chain is [101]. The directivity of the associated units is unfavourable for the possibilities of fit to the (001) and (100) faces, whereas this directional effect is favourable for the ($0\bar{1}0$) faces. For the P_3O_{10} groups to fit in the hexahydrate crystal, the chain of the associated unit in the Phase II solution must be broken. In saturated solutions, the associated units disintegrate. The concentration of $Na_5P_3O_{10}$ on the surface of the growing crystal is more or less equal to the saturation concentration. This concentration, which is lower than that in the solution, is favourable for the rupture of the sodium ion bridge between the P_3O_{10} chains. The rupture of bonds has no favourable effect at all on the (001) and (100) faces but it has on the ($0\bar{1}0$) faces, since the disrupted P_3O_{10} groups finish up precisely the correct position. Consequently the rate-determining step for growth from Phase II solutions is the disintegration of the associated units. Although the configuration of the anion in the Phase II crystal structure more resembles that in the hexahydrate than that in the Phase I crystal structure, the growth rate in Phase I solutions is nevertheless higher than that in Phase II solutions.

By measurement of the crystal growth of sodium triphosphate hexahydrate we find a "memory effect" in the concentrated sodium triphosphate solution. The solutions remember the original anhydrous modifications, Phase I and Phase II, from which they were prepared.

In our opinion the exact model of the site binding of the sodium ion to the triphosphate anion will be found by means of high resolution nuclear magnetic resonance spectroscopy.

Acknowledgements

The author thanks Prof. Dr. W. G. Perdok (University of Groningen) and Dr. H. L. Spier (Unilever Research Laboratory, Vlaardingen) for their stimulating interest, and wishes to express his thankfulness to the Managment of the Unilever Research Laboratory Vlaardingen for their permission to publish the results in this form.

References

1) J. R. van Wazer, *Phosphorus and Its Compounds*, Vol. 1 (Interscience, New York, 1958) pp. 642–655.
2) S. Troost, J. Crystal Growth **3,4** (1968) 340.
3) S. Troost, Thesis, University of Groningen (1969).
4) D. R. Dyroff, Thesis, Calif. Inst. Technol. Pasadena (1965).
5) F. Wall and R. H. Doremus, J. Am. Chem. Soc. **76** (1954) 868.
6) D. E. C. Corbridge, Acta Cryst. **11** (1958) 315; **13** (1960) 263.

THE GROWTH OF POTASSIUM CHLORIDE CRYSTALS IN THE PRESENCE OF LEAD

G. PILKINGTON and W. J. DUNNING

Department of Physical Chemistry, The University, Bristol, England

Potassium chloride seeds (cleaved from pure, melt grown crystals) were grown from aqueous solutions with supersaturation ratios in the range 1.005 to 1.050 and with concentrations of lead chloride $\geqslant 6 \times 10^{-5}$ M. While the seeds were growing the {100} cleavage faces were studied by using reflection microscopy and two beam interference microscopy.

On the initially flat cleavage faces many growth hills developed during the early stages of growth. The hills were approximately conical and their surfaces were inclined at angles $\gtrsim 3°$ to the close-packed {100} planes. The most steeply inclined hills grew most rapidly, both parallel and perpendicular to the crystal surface, and soon overran their neighbours. Crystal growth occurred solely through the continuing development of these hills.

Profiles of growth hills were determined at time intervals and it was found that the development of the profiles could be described in terms of the kinematic theory of crystal growth. Furthermore, on the otherwise smooth surfaces of the hills, kinematic shock waves of sub-microscopic steps were observed. In many instances, the shock waves had the forms of spirals centered on the apices of the hills. These observations are consistent with the thesis that the hills originate at the points of emergence in the surface of groups of dislocations with large Burgers vectors.

1. Introduction

Crystals of KCl grown from pure aqueous solutions are either polycrystalline or develop as hoppers and ultimately contain inclusions of solution; in fact, it may be impossible to grow single crystals from pure solution[1]. In contrast to this large single crystals free from macroscopic defects can be grown with ease from solution containing small amounts of additives; one of the most effective additives for this purpose is the cation of lead[2].

In this investigation the surface feature of KCl crystals were studied during growth from solutions with various supersaturations and various $PbCl_2$ concentrations. Different types of growth features were observed, each type occurring within a well defined domain of solution conditions. This paper reports the features observed in one of these domains, the domain within which large single crystals can be produced.

2. Experimental

Crystal seeds with {100} faces were cleaved with a razor blade from large melt grown crystals. The seeds were placed in a cell through which solution was continuously circulated from a stock flask; the concentrations of KCl and $PbCl_2$ and the temperature of the solution could be held constant or changed indepen-

dently in a controlled manner when desired. The crystals were studied, while growing, by means of reflection microscopy and a standard metallurgical microscope; the optical arrangement was similar to that used by previous workers[3]. A second, and similar, growth apparatus was also used. For the latter apparatus the crystals could be studied by means of a microscope fitted with an interference objective (a description of which has been given by Terrell[4]), but the properties of the solution could not be varied during growth.

3. Results

3.1. Development of growth hills

First a description is given of the growth features observed when a solution with the following properties was passed over cleaved crystal seeds; supersaturation ratio (actual concentration, expressed in g KCl/g water divided by saturation concentration) 1.005 to 1.050, lead concentration $\geqslant 6 \times 10^{-5}$ M, and temperature 25 °C to 45 °C. A variation of the solution properties within these limits changed only the speed with which the growth features developed.

The features observed by optical microscopy on these freshly cleaved surfaces were low angled grain boundaries, scratches made by the cleaving blade, and the cleavage steps. At the initiation of growth the cleavage

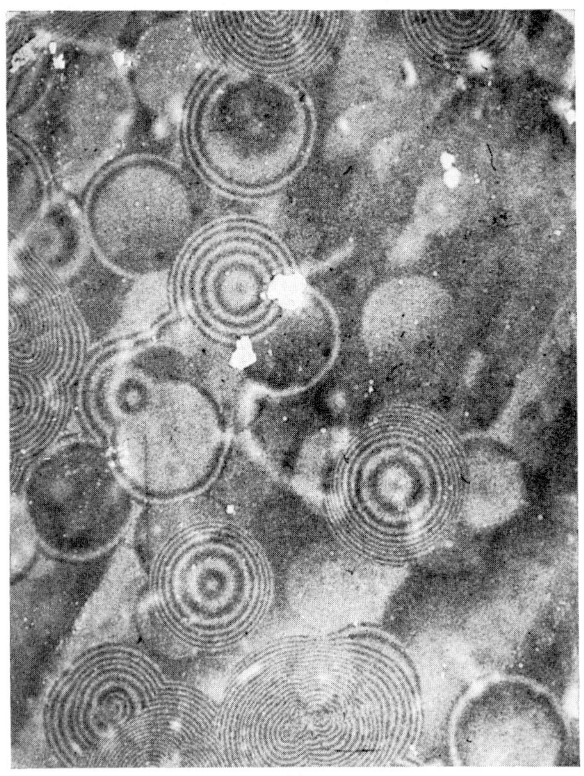

(a)

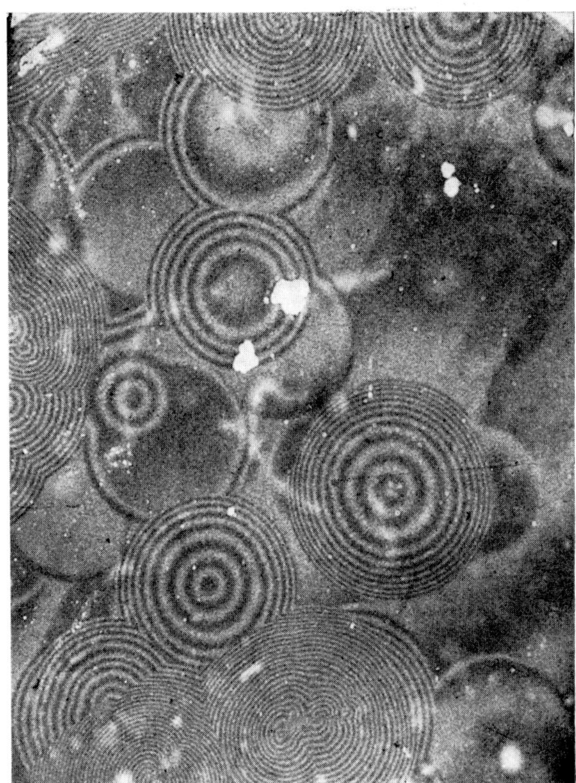

(b)

Fig. 1.   Crystal surface after (a) 24 min, (b) 33 min growth. ×100.

steps, initially steep cliffs, advanced slowly, concomitantly becoming less steeply inclined to the cleavage plane. After a short time growth hills appeared; they formed at scratches, at points along the cleavage steps and at points where no features could previously be seen on the surface. The hills were so densely packed on some parts of the surface that they overlapped before they had grown to a size at which they could be clearly distinguished (density $> 10^5/cm^2$). Elsewhere, areas of about 300 μm)² contained only one or two hills on each of them (density $\sim 10^3/cm^2$). As growth proceeded, the steeper growth hills and the higher cleavage steps overran neighbouring features.

The initial stages in the development of these growth features could be studied in greater detail by the use of two-beam interference microscopy. The fringe patterns obtained by this method represent contour maps of the developing surface. The hills which formed on any given cleavage face had slopes varying over a wide range, see figs. 1a, 1b.

The surfaces of the growth hills are vicinal faces inclined at small angles to the close-packed 100 plane.

These vicinal surfaces consist of trains of submicroscopic steps separated by flat lands of (100) surface. During growth the steps advance, the lateral movement of each contributing a small increment of growth perpendicular to the (100) face. The steps are generated by growth sources at the apices of the hills; the flux of steps generated at any given time is called the activity of the source[5]).

The growth sources originating many of the hills became inactive after a short period of growth; these hills subsequently continued to increase in diameter and developed flat tops. For the early stages of growth it was very common for the activities of growth sources to change unpredictably and abruptly, both increases and decreases being observed. This caused successive rings of the surface of the growth hills to have different slopes.

As growth proceeded the activities changed less often and the number of hills diminished as the steep ones overran their neighbours. After about 50 μm of bulk growth perpendicular to the surface the hills present were steep and their density was about 1/mm². During the following 2 mm of bulk growth, which is the most that has been studied, the density of the growth hills on the surface remained approximately constant. Occasionally, a new growth hill developed on the side of an existing one, presumably because of an increase in activity of a previously dominated growth source; alternatively, a growth hill became less steep and was overrun by a neighbour.

3.2. THE PROFILES OF GROWTH HILLS

As noted above, a vicinal surface consists of a train of steps on a close-packed surface. Frank[6]) examined the way in which such a vicinal surface would develop if the rate of advance of the steps across the close-packed face depended only on local step spacing. He found that the development of the face is described mathematically by a kinematic wave equation. Further, if the step velocity increases with decreasing spacing, smoothly curved concave regions of the vicinal face are unstable with respect to the formation of discontinuities in surface orientation [see fig. 3, the profile in (d) changes to the profile in (e)]. For the mathematical description of the development of the face these discontinuities are analogous to shock waves.

By making measurements on the fringe patterns of the type shown in fig. 1, one can construct profiles of the growth hills. Further, by determining profiles at a series of times one can monitor the step kinetics. Such an analysis for the KCl crystals indicated that the step velocity increased with decreasing spacing, the profiles consisting of smoothly curved convex regions and shock wave discontinuities in concave regions in agreement with the prediction of Frank[6]).

3.3. KINEMATIC SHOCK WAVES

The shock waves have been studied extensively using reflection microscopy; the visibility of the waves becomes less for smaller angles of discontinuity and they cannot be resolved for angles less than about $\frac{1}{2}°$.

Shock waves were seen on growth hills a short time after growth had started. On average, clearly visible ones were observed on 20% of the hills at any particular instant, but this percentage varied from crystal to crystal. The systems of shock waves can be divided into three classes. A system of class (a) consists of a shock wave in the form of a single closed loop on a growth hill; alternatively there may be two or more closed loops in which case they have different degrees of

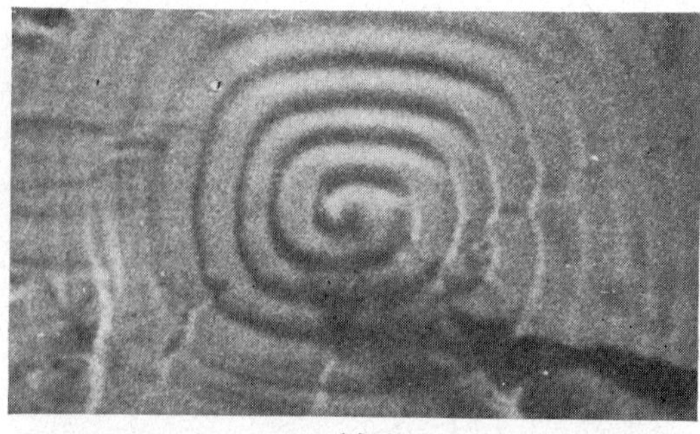

(a)

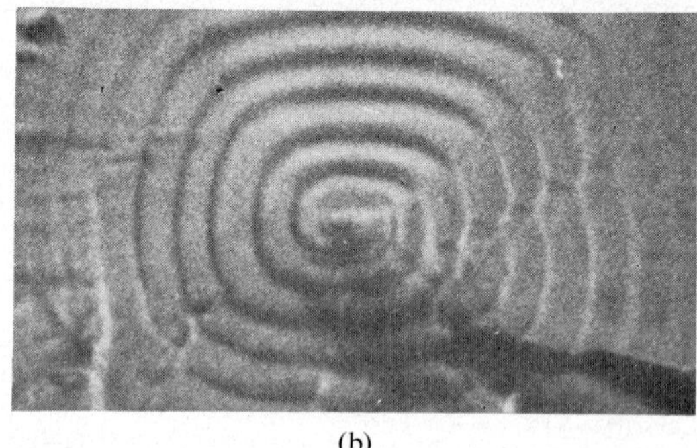

(b)

Fig. 2. (a) Spiral shock wave, (b) transforming into double-start spiral. ×165.

visibility and the spacings between successive loops are different. A system of class (b) consists of a series of closed loops which are equally spaced, while one of class (c) consists of a shock wave in the form of a spiral. In general, systems of class (a) were observed on hills with small slopes, while systems of the other two classes were seen on steep hills. The shock waves were difficult to resolve when they were close to the summits of the hills, becoming most clearly visible at a distance of 10–60 μm from it. In some instances the form of a shock wave system of type (b) or (c) remained unaltered for about an hour, but usually continual changes occurred. For example, sometimes each newly formed turn of a spiral was of a lower resolution then its predecessor leading to the apparent disappearance of the system; at other times a spiral shock wave split into two or more, usually fainter, shock waves at its centre, see figs 2a and 2b, with the subsequent development of a multi-start spiral.

Changes also occurred which led to the appearance of fresh shock wave systems. However, in general, fewer and fainter systems of shock waves were seen as growth proceeded on the cleavage face and after about 1 mm of bulk growth there were very few clearly visible systems.

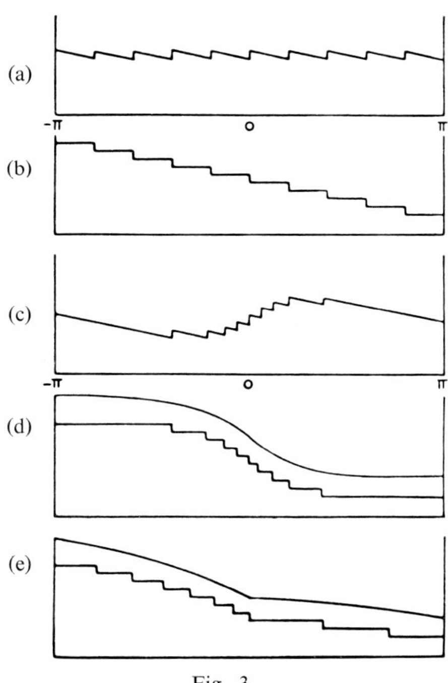

Fig. 3.

4. Discussion

The spiral forms of the shock waves of type (c) suggest that the growth sources were groups of screw dislocations emerging in the surfaces, and the behaviour described above suggest the following model.

Consider the growth hill originated by a group of n elementary dislocations of like sign grouped sufficiently closely together for all of them to act in cooperation as a growth source. The step pattern originated by the source consists of n similar concentric spirals. These steps could be equally spaced in orientation about the summit; then the height of the surface of the hill at any fixed distance from the summit and relative to a plane perpendicular to the axis of the hill would vary with orientation in the way shown in fig. 3a (for $n = 10$). Similarly the height of the surface in any fixed orientation would vary with distance from the summit in the way shown in fig. 3b. Alternatively the spirals could be bunched about one orientation, figs. 3c and 3d.

Suppose that the steps advance from the summit of the hill in such a manner as to form a surface profile similar to that in fig. 3d; then the kinetics of step movement are such that a kinematic shock wave would be developed, fig. 3e, and this would have a spiral shape. If step movement close to the summit were described by the same kinetics as movement far from it

the spiral shock wave would extend all the way to the dislocation group at its centre; also, it would slowly disappear[6]). However, the shock waves were found to be difficult to resolve near the summit. This suggests that within a distance of about 10 μm from the summit a different relationship between step spacing and velocity existed and profiles similar to that shown in figs. 3c and 3d could be formed. The smoothly concave regions in these profiles would become unstable as their distance from the summit exceeded 10 μm and they would gradually transform into shock waves which would become fully developed at a distance of about 50 μm from the summit. As they continued to advance down the growth hill the shock waves would slowly fade away[6]).

As growth proceeded the dislocations would extend into the freshly deposited crystal, a process which would be accompanied by changes in the spatial arrangement of the dislocations within the group. This in turn would change the relative orientations of the spiral steps; consequently the form of the shock wave system could be changed while the activity of the dislocation group and hence the average slope of the hill remained unaltered.

More drastic rearrangement of the dislocations during growth could result in a group splitting into two or more parts which could become so distantly spaced as to operate as independent growth sources, generally with reduced activities. Similar processes may exist whereby dislocation groups unite producing a source of higher activity. Rearrangements of this type would account for the changes in the activities of growth sources which were most common in the early stages of growth.

References

1) R. F. Strickland-Constable, *Crystallization* (Academic Press, London, 1968) p. 267.
2) P. H. Egli and S. Zerfoss, Discussions Faraday Soc. **5** (1949) 61; P. H. Egli and L. R. Johnson, in: *The Art and Science of Growing Crystals*, Ed. J. J. Gilman (Wiley, London, 1963) p. 194.
3) N. Albon and W. J. Dunning, Acta. Cryst. **12** (1959) 219; A. E. Morgan and W. J. Dunning, J. Crystal Growth **7** (1970) 179.
4) A. C. Terrell, The Microscope and Crystal Front **14** (1964) 174.
5) W. K. Burton, N. Cabrera and F. C. Frank, Phil. Trans. Roy. Soc. London A **243** (1951) 299.
6) F. C. Frank, in: *Growth and Perfection of Crystals* Eds. R. H. Doremus, B. W. Roberts and D. Turnbull (Wiley, New York, 1958) p. 411.

   Journal of Crystal Growth **13/14** (1972) 458–461 © *North-Holland Publishing Co.*

CROISSANCE DE L'HYDROXYDE DE NICKEL Ni(OH)$_2$ À PARTIR D'UN HYDROXYDE DE NICKEL TURBOSTRATIQUE

S. LE BIHAN

Laboratoire de Chimie des Solides, Faculté des Sciences, 9, Quai Saint Bernard, 75–Paris 5e, France

et

M. FIGLARZ

Laboratoire de Chimie des Solides Pulvérulents, Faculté des Sciences, 33 Rue St. Leu, 80–Amiens, France

The authors have studied the crystal growth of the nickel hydroxide Ni(OH)$_2$ from a turbostratic nickel hydroxide. The turbostratic hydroxide is obtained by the reaction of nickelous nitrate with ammonia followed by washing the precipitate and by repeated centrifugations. X-ray and infrared spectroscopy show that the structure of the turbostratic compound consists of a stacking of bidimensional parallel Ni(OH)$_2$ leaflets which are disoriented one from the other; the leaflets are separated by water molecules which are involved, in hydrogen bonding with the leaflet hydroxyl groups. When the turbostratic hydroxide is put in pure water this compound evolves slowly toward crystallization (oriented leaflets, absence of intercalary water); X-ray, infrared, electron microscopy and selected area diffraction studies, allow us to conclude that the mechanism of crystallization is a diphasic mechanism.

Par action de l'ammoniaque sur une solution de nitrate de nickel on obtient un précipité qui, suivant le mode de lavage employé, conduit soit à un composé amorphe (fig. 1, spectre 0) soit à un hydroxyde de nickel à structure turbostratique (figs. 1 et 2, spectre 1). Le composé amorphe est obtenu lorsque le lavage est effectué par une série de simples décantations. Lorsqu'on opère par une série de lavages et centrifugations on aboutit au composé turbostratique[1]. Ce composé présente un diagramme de diffraction X sur lequel on observe (spectre 1 des figures 1 et 2):

(a) deux raies larges dont les maximums correspondent à $d = 8.55$ Å et $d = 4.25$ Å;

(b) des bandes, parmi lesquelles seules les deux premières ont une intensité suffisante pour être examinées, leur maximum correspondant à $d = 2.65$ Å et $d = 1.55$ Å.

Ce type de spectre s'interprète en considérant que la structure du composé diffractant est formée d'un empilement de feuillets d'hydroxyde bidimensionnel parallèles et désorientés les uns par rapport aux autres. Une telle structure, qualifiée de turbostratique, donne un spectre formé de bandes hk et de raies $00l$.

En ce qui concerne ce composé, si nous admettons que l'arrangement des atomes dans les feuillets est le même que dans l'hydroxyde de nickel tridimensionnel, nous pouvons indexer les bandes hk comme des bandes

10 et 11. En revanche la première raie ne peut pas être indexée en considérant un simple empilement de feuillets désorientés d'hydroxyde: elle correspond en effet à des plans réticulaires distants d'environ 8.5 Å au lieu de 4.6 Å pour les plans (001) de Ni(OH)$_2$ cristallisé. La façon la plus simple de rendre compte de cet écart est de considérer qu'entre les feuillets d'hydroxyde s'intercalent des molécules; l'étude infrarouge nous montrera que ces molécules sont des molécules d'eau. Les premières raies du diagramme de diffraction correspondent alors aux raies 001 et 002 propres au composé turbostratique.

Nous avons étudié l'évolution de cet hydroxyde de nickel turbostratique, mis en suspension dans l'eau pure, vers la forme cristallisée Ni(OH)$_2$. On constate alors une évolution progressive des diagrammes de diffraction X, avec d'une part affaiblissement des bandes et des raies du composé turbostratique, et, d'autre part apparition des raies de l'hydroxyde cristallisé Ni(OH)$_2$ (spectres 2, 3 et 4 des figures 1 et 2). Il faut souligner que l'on n'observe aucun déplacement de la raie 001 du composé turbostratique, mais son affaiblissement au profit de 001 de l'hydroxyde tridimensionnel. Ces observations nous permettent de conclure que le passage au composé cristallisé (feuillets orientés, absence de molécules intercalaires) s'effectue par une réaction biphasique.

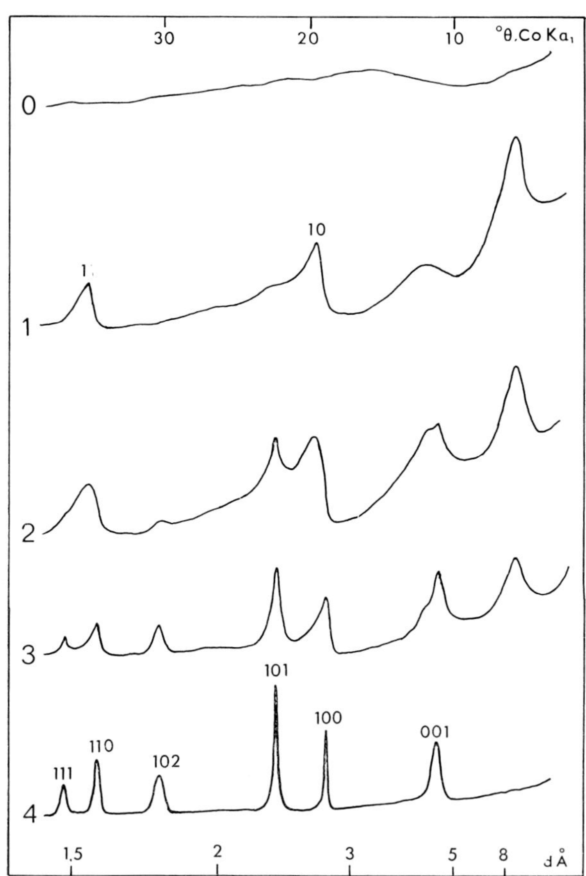

Fig. 1. Evolution des spectres de diffraction X au cours de la cristallogénèse de Ni(OH)$_2$ (diffractomètre à compteur, radiation CoKα$_1$).

placée par une bande large centrée vers 3500 cm^{-1}, et les bandes δ(OH), γ(OH) et ν(NiO) sont déplacées vers des plus courtes longueurs d'onde. Pour l'hydroxyde turbostratique les bandes situées à 3500 et 1600 cm^{-1} (figs. 3 et 4) indiquent la présence d'eau en assez grande quantité, dont une partie se trouve entre les feuillets. Le déplacement des bandes δ(OH), γ(OH) et ν(NiO) et la disparition de la bande étroite à 3650 cm^{-1} montrent qu'il s'établit les liaisons hydrogène, relativement fortes, entre les molécules d'eau interfeuillets et les hydroxyles des feuillets. La cohésion des empilements de feuillets désorientés dans l'hydroxyde turbostratique est par conséquent assurée par les liaisons hydrogène entre les molécules d'eau intercalaires et les hydroxyles des feuillets.

De l'examen des spectres d'absorption relatifs aux produits intermédiaires de la cristallisation (spectres 2, 3 et 4 de la figure 4), nous concluons, comme nous l'avons fait à la suite de l'étude radiocristallographique, que le passage du composé turbostratique au composé cristallisé s'effectue par une réaction biphasique. En effet on observe une diminution d'intensité des bandes du composé turbostratique et une augmentation d'intensité de celles de l'hydroxyde cristallisé.

Il faut enfin noter la présence et l'évolution des bandes dans la région du spectre comprise entre 700 et 1600 cm^{-1} (fig. 4). Elles s'interprètent [2]) comme dues à la présence de deux types d'ions nitrate adsorbés lors

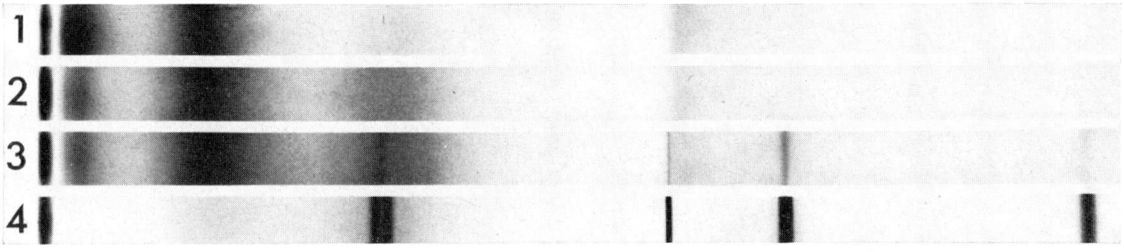

Fig. 2. Spectres de rayons X (Chambre Guinier de Wolff, radiation Cr Kα) de l'hydroxyde turbostratique, d'échantillons en cours de cristallisation et de Ni(OH)$_2$ cristallisé.

Nous avons complété cette étude radiocristallographique par une étude en spectroscopie infrarouge portant sur l'hydroxyde turbostratique et sur son évolution vers Ni(OH)$_2$ cristallisé[2]). Les spectres d'absorption de l'hydroxyde turbostratique et de l'hydroxyde cristallisé diffèrent notablement (fig. 3): la bande étroite ν(OH) à 3650 cm^{-1} caractéristique de la vibration de valence des OH libres de Ni(OH)$_2$ cristallisé est rem-

de la précipitation à partir de nitrate de nickel: des ions NO$_3^-$ faiblement liés (bande ν$_3$ (NO$_3^-$) à 1360 cm^{-1}) et des ions ONO$_2^-$ engagés dans des liaisons covalentes (bandes ν$_1$(ONO$_2^-$) à 1300 et ν$_4$(ONO$_2^-$) à 1500 cm^{-1}).

Le comportement thermique de l'hydroxyde turbostratique a également été étudié par microthermogravimétrie et microanalyse thermique différentielle. Cette étude montre que le composé turbostratique retient par

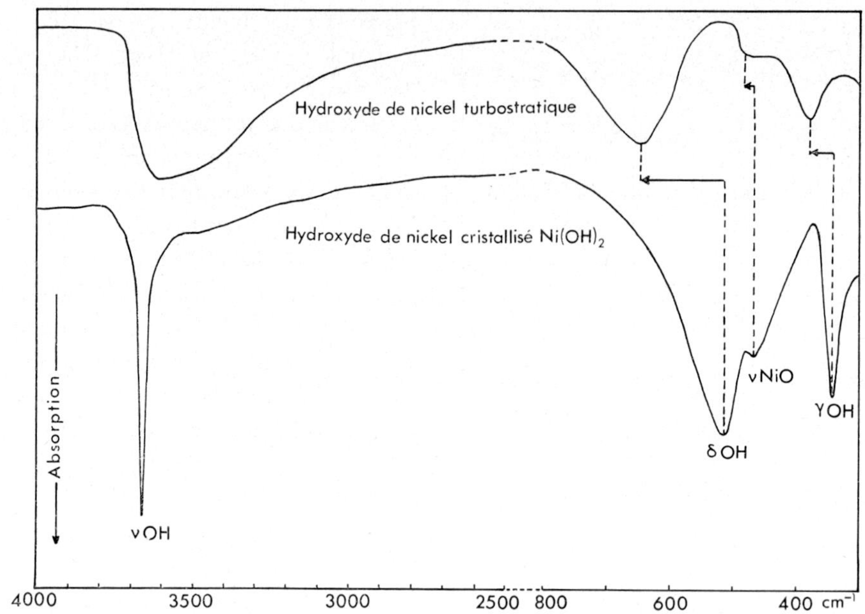

Fig. 3.   Spectres infrarouge de l'hydroxyde turbostratique et de l'hydroxyde cristallisé (Ni(OH)$_2$).

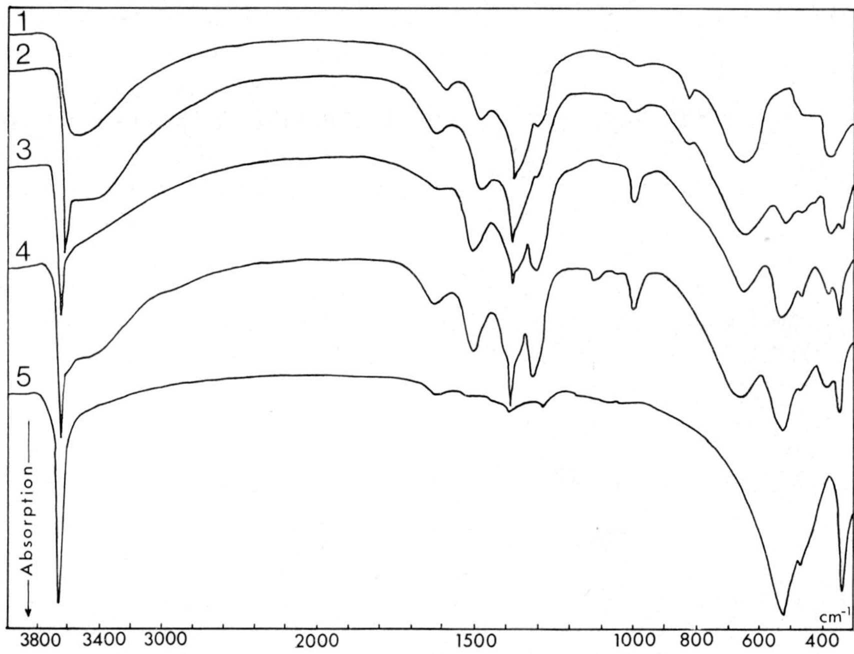

Fig. 4.   Evolution des spectres infrarouge au cours de la cristallisation de Ni(OH)$_2$ à partir de l'hydroxyde turbostratique.

adsorption des quantités d'eau importantes (20 % en poids environ) qui sont éliminées par chauffage à 150 °C sans modification de la structure turbostratique comme le prouvent des mesures radiocristallographiques et infrarouge. Au delà de 150 °C le produit se décompose en une seule étape pour donner l'oxyde NiO. On constate que les températures de décomposition des composés turbostratique, cristallisé et des mélanges intermédiaires sont très voisines et se situent vers 250 °C; toutefois les cinétiques de décomposition sont différentes. Cela confirme que les liaisons hydrogène entre l'eau intercalaire et les OH des feuillets d'hydroxyde

Fig. 5.   Hydroxyde de nickel turbostratique.

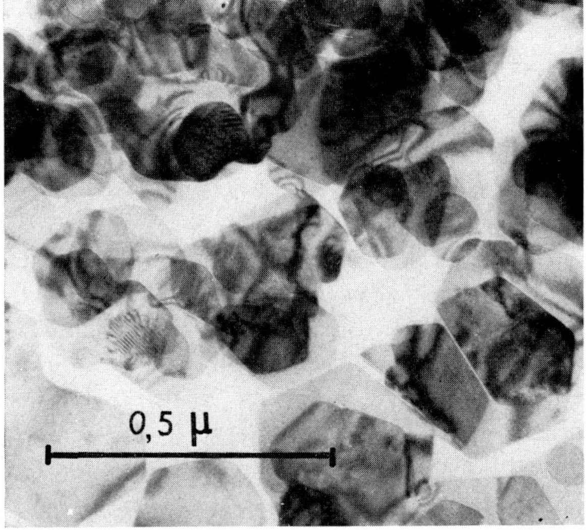

Fig. 6.   Hydroxyde de nickel cristallisé Ni(OH)$_2$.

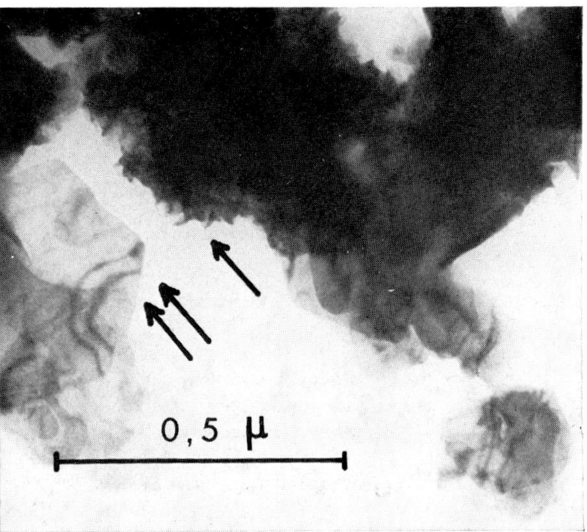

Fig. 7.   Echantillon en cours de cristallisation: → hydroxyde turbostratique, ⇉ hydroxyde cristallisé.

sont fortes. Les résultats de l'analyse thermogravimétrique permettent en plus de conclure qu'entre les feuillets de l'hydroxyde turbostratique on a une seule couche d'eau, la formule du composé pouvant s'écrire Ni(OH)$_2$, H$_2$O compte non tenu de l'eau d'adsorption.

Les différents échantillons ont été examinés en microscopie et microdiffraction électroniques. La microscopie montre que l'hydroxyde turbostratique se présente sous la forme de films très minces et très enchevêtrés qu'il est difficile de disperser (fig. 5). L'hydroxyde cristallisé apparaît comme formé de plaquettes hexagonales minces, monocristallines mosaïques (fig. 6). Quant aux échantillons prélevés au cours de la cristallisation, ils contiennent deux types distincts de particules caractéristiques de l'hydroxyde turbostratique et de Ni(OH)$_2$ cristallisé (fig. 7) ce qui confirme bien le caractère biphasique de la réaction de cristallisation.

Indiquons pour terminer les premiers résultats d'une étude magnétique réalisée sur le composé turbostratique[3]. L'étude magnétique comparée des composés turbostratique et cristallisé indique que la nature et la grandeur des interactions entre atomes de nickel situés dans les feuillets bidimensionnels ne sont sensiblement pas modifiées, ce qui confirme l'hypothèse avancée précédemment sur l'arrangement des atomes dans les feuillets. Par ailleurs le composé turbostratique présente des propriétés magnétiques originales justifiables selon les zones de température et de champ magnétique étudiées soit de la théorie des grains fins antiferromagnétiques, soit de la théorie des grains fins ferromagnétiques de Néel.

Bibliographie

1) S. Le Bihan, J. Guenot et M. Figlarz, Compt. Rend. (Paris) C **270** (1970) 2131.
2) M. Figlarz et S. Le Bihan, Compt. Rend. (Paris) C **272** (1971) 580.
3) J. Deportes, P. Mollard, J. Penelon, S. Le Bihan et M. Figlarz, Compt. Rend. (Paris) B **272** (1971) 449.

MÉTHODE DE CROISSANCE EN SOLUTION DES MONOCRISTAUX PAR L'ADDITION GRADUELLE D'UN RÉACTIF

I. F. NICOLAU*, M. ITTU et R. DABU

Institut de Physique Atomique, B.P. 35, Bucharest, Roumanie

The method generally applies to the growth of single crystals of some chemical compounds from a suitable solution. The supersaturation needed for growth is achieved by the chemical reaction of two reactants soluble in the choosen solvent one of which is added at the beginning to the solution bath, the second being gradually added during growth.

The supersaturation of the solution yielded by addition one reactant is calculated for the general case of an inorganic salt obtained as a result of the neutralization reaction between the acid and the salt base. The advantages presented by this method over methods that yield the supersaturation by evaporation or by gradual cooling are discussed.

A laboratory installation is described and the results obtained are presented for growth of KDP single crystals from aqueous solutions acidified with H_3PO_4 by gradually addition of KOH solution.

KDP single crystals useful for electrooptic devices have been grown by means of this method at an absolute supersaturation of 10^{-4} g/ml with growth rates along the $\langle 001 \rangle$ axis of 2.3 mm/day. The supersaturation achieved by one addition of KOH solution is equivalent to the supersaturation obtained by cooling the solution by 0.014 °C.

1. Introduction

Nous proposons une nouvelle méthode de croissance de monocristaux de KDP en solution aqueuse, considérant les applications importantes liées à l'utilisation de ces monocristaux en optique. La necessité de produire de tels cristaux, à grande échelle, aussi parfaits que possible a imposé l'amélioration des méthodes de croissance en solution aqueuse déja existantes. La méthode que nous présentons est susceptible d'être utilisable pour la préparation d'autres cristaux.

La sursaturation nécessaire à la croissance des monocristaux en solution est réalisée d'habitude en utilisant soit la méthode de l'évaporation graduelle du solvant[1-3], soit la méthode de la décroissance graduelle de la température de la solution[4-7].

La première méthode présente l'avantage d'effectuer la croissance des cristaux à une température constante, ce qui permet la croissance des monocristaux de n'importe quelle substance soluble, sans tenir compte de la forme de la courbe de solubilité en fonction de la température et permet aussi l'obtention de cristaux exempts des tensions provoquées par des variations de température. Cépendant la méthode présente le désavantage que la vitesse de cristallisation ne peut pas être con-

trolée et maintenue fixe avec une précision suffisante. Cela est dû à l'évaporation même, puisque d'abord la vitesse d'évaporation et, par suite, la sursaturation de la solution en surface depend à la fois de la convection du gaz au dessus la solution et de la pression partielle du solvant dans le gaz, ces deux paramètres étant difficiles à controler. Par ailleurs, la sursaturation qui est nécessaire à la croissance des cristaux est conditionnée par la diffusion et la convection du soluté de la surface de la solution vers l'intérieur de celle-ci, là encore ce phenomène est difficile à controler. De plus la sursaturation de la couche de solution près de la surface, même à des sursaturations faibles de la solution globale peut dépasser parfois la sursaturation limite conduisant à la formation des germes cristallins qui se répandent dans la solution.

La deuxième méthode est applicable seulement aux substances qui présentent une variation suffisante de la solubilité en fonction de la température. La vitesse de cristallisation peut être controlée et maintenue constante avec précision si on impose un certain programme de décroissance de la température avec le temps en fonction de la courbe de solubilité (supposée connue). Cependant les cristaux peuvent présenter des tensions dues aux variations de température. Cette méthode a trouvé beaucoup d'applications, mais elle nécessite la construction d'un cristalliseur dans lequel la

* Présente adresse: Physikalisches Institut der Universität Stuttgart, Wiederholdstr. 13, 7 Stuttgart 1, BRD.

IX – 6

température pourrait être controlée avec une précision de $\pm 0.002\ °C$ et qui permetrait une décroissance de température de $0.15\ °C$ à $0.015\ °C$ par jour, l'ensemble devant fonctionner sans interuption et sans dérive pendant plusieurs mois. La méthode exige parallèment un dispositif électronique complexe qui permet la réalisation d'un tel programme possédant la fiabilité voulue.

La méthode de croissance qui permet la sursaturation par l'addition graduelle d'un réactif a l'avantage de travailler à une température constante et de permettre le contrôle et le maintien d'une vitesse de croissance presque constante. Ainsi la sursaturation se trouve elle-même controlée à un taux aussi faible que nous le désirons, au moins équivalent (ou aux besoin plus petit) à la sursaturation correspondant à une décroissance de $0.015\ °C$ utilisée dans la méthode de décroissance graduelle de la température. La technique expérimentale se trouve nettement simplifiée par rapport à la méthode de décroissance graduelle de la température, le contrôle de la décroissance de $0.015\ °C$ étant remplacé par le contrôle de l'addition d'un volume de solution de l'ordre de 0.1 ml.

2. Description de la méthode

Considerons un cristalliseur dans lequel se trouve à une température constante, une solution saturée d'un sel $A_p B_q$ dans l'eau acidifiée par l'acide correspondant au sel, où nous désignons par A le radical acide, par B le métal de la base et par p et q la valence respectivement du métal de la base et du radical acide. Etant donnée W_0 la quantité initiale d'eau en g et connaissant C_0^S la concentration initiale de saturation la solution en sel, C^S la concentration de saturation en sel en absence de l'acide et la concentration initiale de l'acide C_0^A, tout les trois en $g/ml\ H_2O$ et supposant connue la courbe de solubilité limite de la solution en sel en fonction de la concentration de l'acide pour la température t, courbe approximée avec une droite conformément à la fig. 1, nous cherchons à déterminer la variation de la sursaturation absolue α_n de la solution en sel en $g/ml\ H_2O$ après l'addition de x ml de solution aqueuse basique dans laquelle la concentration de la base est y g/ml sol. La sursaturation de la solution est obtenue par addition de la solution basique, la base additionnée réagissant avec l'acide en excès conformément à la reaction de neutralisation

$$p\ AH_q + q\ B(OH)_p \rightarrow A_p B_q + pq\ H_2O\,. \tag{1}$$

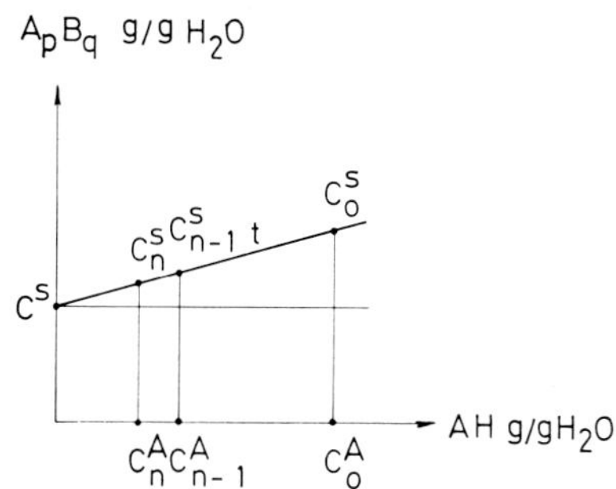

Fig. 1. Variation de la solubilité d'un sel en fonction de la concentration de l'acide.

A la n-ième addition on ajoute à la solution la quantité de base xy, ce qui engendre par réaction une quantité de sel xyM_S/qM_B et une quantité d'eau de neutralisation $xypM_W/M_B$ et on consomme une quantité d'acide $xypM_A/qM_B$. La quantité d'acide restant en solution est alors $C_0^A W_0 - npxyM_A/qM_B$. La quantité d'eau ajoutée avec la solution basique est $x(d-y)$, où M_A, M_B, M_S et M_W représentent respectivement les poids moléculaires de l'acide, de la base, du sel et de l'eau en g/mol, d étant la densité de la solution basique en g/ml. La quantité totale d'eau dans le cristalliseur après n additions sera:

$$\begin{aligned}
W_n &= W_{n-1} + x(d-y) + \frac{pxyM_W}{M_B} \\
&= W_0 + nx(d-y) + \frac{npxyM_W}{M_B},
\end{aligned} \tag{2}$$

et la concentration de l'acide en solution après la n-ième addition:

$$C_n^A = \frac{C_0^A W_0}{W_n} - \frac{npxyM_A}{W_n M_B}\,. \tag{3}$$

La sursaturation absolue du sel en solution α_n étant

$$\alpha_n = \frac{1}{W_n}\left(\frac{xyM_S}{qM_B} - C_n^S W_n + C_{n-1}^S W_{n-1}\right), \tag{4}$$

où C_n^S représente la concentration du sel en solution après la n-ième addition. Remplaçant dans la relation (4) C_n^S et C_{n-1}^S par leurs expressions en fonction de C_n^A et respectivement C_{n-1}^A en tenir compte de la relation linéaire ci-dessous:

$$C_n^S = C^S + \frac{C_0^S - C^S}{C_0^A}C_n^A, \tag{5}$$

remplaçant maintenant W_n et W_{n-1} par leurs expressions (2), C_n^A et C_{n-1}^A par leurs expressions (3) on obtient pour α_n l'expression suivante:

$$\alpha_n = \frac{1}{W_n}\left(\frac{xyM_S}{qM_B} + \frac{C_0^S - C^S}{C_0^A}\frac{xypM_A}{qM_B} - C^S x(d-y)\right.$$

$$\left. - C^S\frac{xypM_W}{M_B}\right) = \frac{1}{W_n}\left[\frac{1}{qM_B}\left(M_S + \frac{C_0^S - C^S}{C_0^A}pM_A\right.\right. \quad (6)$$

$$\left.\left. + C^S qM_B - C^S qM_B - C^S pqM_W\right)y - dC^S\right]x.$$

Ainsi on constate, dans le cas d'une courbe de solubilité linéaire, que la quantité de sel qui doit cristalliser, $\alpha_n W_n$, est une constante ne dépendant ni du rang de l'addition ni de la quantité initiale de solvant. Mais la sursaturation absolue α_n dépend tant du rang de l'addition que de la quantité initiale de solvant, à cause de l'augmentation graduelle de la quantité de solvant. En pratique nous choisissons des valeurs pour W_0 et y aussi grandes que possible afin de pouvoir considérer la quantité d'eau dans le cristalliseur pratiquement constante $W_n \cong W_0$. Dans l'expression de la sursaturation (6) apparaissent quatre termes: le premier, positif dû à la réaction chimique, le deuxième, positif ou négatif suivant que la solubilité du sel dans le milieu acide est supérieure ou inférieure à celle dans le milieu neutre, le troisième, négatif, provient de l'eau de la solution basique et le quatrième, négatif, est dû à l'eau de réaction.

La sursaturation α_n dépend des variables x et y. La sursaturation peut être faite aussi petite que nous désirons en diminuant x. La sursaturation peut être diminuée en diminuant y, si on ne peut pas diminuer x, mais pour qui l'en résulte une sursaturation positive, la valeur de y doit être comprise dans l'intervalle

$$y_{sat} \geq y > \frac{dqC^S M_B}{M_S + \dfrac{C_0^S - C^S}{C_0^A}pM_A + qC^S M_B - pqC^S M_W}.$$

Cette méthode de realisation de la sursaturation a été experimentée pour la croissance des monocristaux de KDP.

3. Description de l'installation

Une installation de cristallisation a été construite comme dans la fig. 2. La solution saturée de KDP

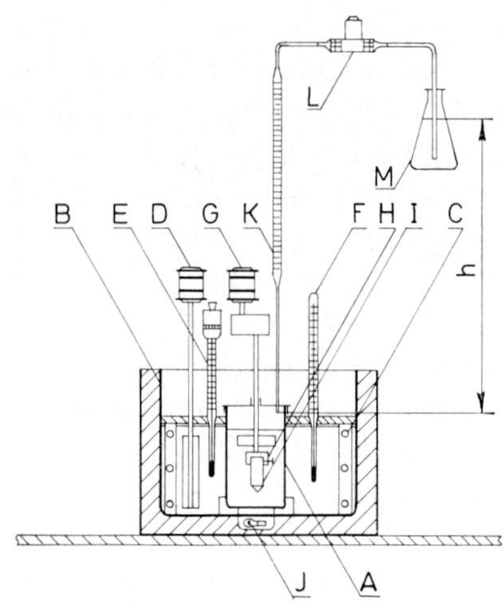

Fig. 2. Appareillage de cristallisation. (A) Becher en silice; (B) bac de verre; (C) résistance de CrNi; (D) agitateur du bain; (E) thermomètre de contact; (F) thermomètre de contrôle; (G) agiteur de la solution; (H) mandrin de teflon; (I) cristal; (J) ampoule; (K) burette en silice; (L) électrovanne; (M) ballon.

acidifiée avec H_3PO_4 est contenue dans un becher en verre de silice d'un litre (A). Le récipient repose sur le fond d'un bac de verre (B) en son milieu. Le bac contient une bain d'huile de 8 litres dans lequel sont plongés la résistance spirale (C) de 80 W, l'agitateur (D), le thermomètre de contact (E), et le thermomètre de contrôle (F), cet ensemble constituant un thermostat classique. Dans l'intérieur du bac se trouve plongé l'agitateur axial (G) en plexiglas, qui se termine à sa partie inférieure par un mandrin (H) de teflon qui tient le cristal (I). Les pertes de chaleur du récipient par le fond sont controlées par la mise en place d'une ampoule (J) de 3 W qui éclaire et chauffe continuellement. A l'intérieur de la burette en silice de 100 ml (K), se trouve la solution de KOH. La burette presente a son extremité inférieure un tube capilaire de silice qui passe à travers le couvercle du verre pénétrant dans le récipient juste au-dessus de la solution. A l'autre extrémité la burette est connectée par un tube de plastique à une électrovanne (L). L'autre extrémité de l'électrovanne est liée par un autre tube de plastique à un ballon (M) rempli d'huile. La partie supérieure de la burette, les tubes de plastique et l'électrovanne contiennent aussi de l'huile exempte de bulles d'air.

La thermostatage maintient avec une précision de $\pm 0.05\,°C$ (dans la bain d'huile) grace à l'utilisation d'un système de commutation avec transistors. L'addition de la solution de KOH se fait par l'ouverture de l'électrovanne. La quantité de solution ajoutée dépend du temps d'ouverture de l'électrovanne et de l'intervalle de temps entre deux ouvertures. Un montage électro-mecanique composé d'un moteur électrique maintenu en fonctionnement permanent couplé avec un système de démultiplication mécanique qui agit sur un interrupteur nous permet de choisir et de maintenir l'intervalle de temps entre deux ouvertures de l'électrovanne. Par la fermeture de l'interrupteur l'ouverture de l'électrovanne est commandée à l'aide d'un générateur et amplificateur d'impulsions transistorisé, la durée d'impulsions pouvant être en outre réglée. Le plus court temps d'ouverture de l'électrovenne, limité par son inertie mécanique, nous a permis l'addition d'une seule goutte, l'impulsion électrique d'ouverture correspondante étant de 0.25 sec. La vitesse de rotation de l'agitateur de la solution a été maintenue à 50 rot/min. L'alimentation électrique est branchée au secteur par l'intermédiaire de transformateurs et de redreseurs convenables. En cas de panne toutes les alimentations sont commutées sur une batterie d'accumulateurs. En utilisant dans les montages électroniques des pièces surdimensionées et en éliminant les relais, la fiabilité de toute l'installation a permis le fonctionnement continu pendant plusieurs mois, sans défections.

4. Résultats expérimentaux

Nous nous sommes attachés au cours de l'expérimentation de la méthode à réaliser; par addition d'un réactif, une sursaturation dans la solution équivalente à celle produite par la décroissance de température de 0.015 °C rapportée par Torgesen et al.[6]. Cette condition assure l'obtention de monocristaux sans inclusions. Nous avons travaille à la température de 40 ± 0.03 °C. Pour le KDP, en utilisant les données de solubilité de Stephen[8], on trouve une variation de la solubilité moyenne de 0.0075 g/g H_2O °C dans l'intervalle de température de 30 à 60 °C. La solution initiale a été faite par la dissolution jusqu'à la saturation de KDP p.a. trois fois recristallisé dans 800 ml d'eau contenant 40 g H_3PO_4. Une solution de KOH contenant $y = 0.4$ g KOH/ml sol. a été utilisée comme réactif. Un intervalle de temps entre deux ouvertures de l'électrovanne de 50 min a été établi qui correspond approximativement à 29 ouvertures par 24 heures. Nous avons utilisé l'impulsion électrique de 0.25 sec qui ajoute une seule goutte pour un contact. Nous avons ainsi ajouté

en 24 heures 1.9 ml solution de KOH. Une goutte correspond donc à $x = 0.0655$ ml sol. de KOH. Dans ces conditions en prenant pour le deuxième terme de la relation (6): $C_0^s = 0.292$ g/g H_2O[9] et $C^S = 0.246$ g/g H_2O[8] (toutes les deux pour 25 °C, la valeur pour C_0^s à 40 °C étant inconnue) et pour les autre termes $C^S = 0.335$ g/g H_2O à 40 °C[8]) et $d = 1.30$, nous obtenons pour la sursaturation absolue la valeur approximative $\alpha \cong 1.04 \times 10^{-4}$ g/ml et pour la quantité de KDP qui dépasse la saturation et doit cristalliser $\alpha W_0 = 0.0835$ g. Cette sursaturation correspond à une abaissement de température de 0.014 °C. Dans ces conditions de croissance, après calcul, on trouve une vitesse journalière de croissance des cristaux selon la direction $\langle 001 \rangle$ de 2.6 mm/jour. Plusieurs des cristaux ont été crûs pendant 3–4 semaines; des croissances le long de l'axe $\langle 001 \rangle$ de 5–7 cm ont été obtenues en utilisant comme germes des cristaux de section 20×20 mm^2; ceci nous a permis de déterminer une vitesse moyenne de croissance de 2.3 mm/jour. La différence est explicable puisque pendant la croissance, le cristal gagne aussi en section; à une croissance de 5 cm selon la direction $\langle 001 \rangle$ correspond une croissance de 1 mm selon la direction $\langle 100 \rangle$. Au cours des premières expériences nous avons travaillé avec un nombre d'additions inférieur à 29 par jour dans le but d'obtenir des faibles vitesses de croissance de 0.7–1.7 mm/jour[6]). En augmentant progressivement le nombre d'additions par jour jusqu'à 29, nous avons constaté que la perfection du cristal n'a pas souffert. Nous n'avons jamais constaté des inclusions de solution dans la masse cristalline sauf dans la région de la jonction du nouveau cristal avec le germe où des inclusions apparaissent systématiquement le long des arêtes de la pyramide du germe, en accord avec les observations de Brooks et al.[10]). Nous avons constaté un retrécissement de la section des cristaux semblable à celui de Byteva[11,12]) pour les cristaux d'ADP, seulement quand nous avons employé une eau désionisée avec une résistivité inférieure à 20 Mohm cm. Nous avons essayé d'obtenir des cristaux en utilisant une solution contenant un excès de KOH, mais les cristaux ainsi obtenus ont présenté un grand nombre de fissures et sont opalescents. La qualité des cristaux crûs dans milieu acide a été appreciée comme excellente par examination optique entre polariseurs croisés et par observation microscopique d'éventuelles discontinuités ou inclusions.

5. Considérations générales

La méthode peut être appliquée en général pour la croissance des monocristaux de tout sel si l'acide et la base sont solubles et stables en solution aqueuse. C'est le cas de la majorité des sels minéraux ou des métaux alcalins, des sels d'ammonium, et des sels des amines solubles dans l'eau. En particulier, la méthode permet la croissance des monocristaux de NaCl dans une solution aqueuse.

La méthode permet la croissance des monocristaux des halogénures alcalins uniformément dopés avec une certaine impureté, puisque le rapport entre la quantité d'impuretés et la quantité de substance reste pratiquement constant en solution pendant la cristallisation, ce qui n'arrive ni dans le cas de la méthode de cristallisation par la décroissance graduelle de la température, ni dans le cas de la méthode de l'évaporation graduelle du solvant, ni dans le cas de croissance en bain fondu.

En général la méthode peut être appliquée dans tous les cas où on peut trouver deux composés qui par réaction chimique forment le composé qui croît, et lorsque les deux composés et le produit résultant de leur réaction sont solubles dans le solvant choisi.

Bibliographie

1) H. E. Buckley, *Crystal Growth* (Wiley, New York, 1951) p. 43.
2) A. Smakula, *Einkristalle* (Springer, Berlin, 1962). p. 125.
3) P. H. Egli and L. R. Johnson, Ionic Salts in: *The Art and Science of Growing Crystals*, Ed. J. J. Gilman (Wiley, New York, 1963) p. 194.
4) A. N. Holden, Discussions Faraday Soc. **5** (1949) 312.
5) V. Sip and Vanicek, in: *Growth of Crystals*, Vol. 3 (Consultants Bureau, New York, 1962) p. 191.
6) J. L. Torgesen, A. T. Horton and C. P. Saylor, J. Res. NBS **670** (1963) 25.
7) J. Mastner and J. Janta, Czech. J. Phys. B **20** (1970) 230.
8) H. Stephen and T. Stephen, *Solubilities of Inorganic and Organic Compounds*, Vol. 1 (Pergamon, Oxford, 1963).
9) A. I. Krasilsicov, Izv. Fiz. Him. Anal. (1933) 159.
10) R. Brooks, A. T. Horton and J. L. Torgesen, J. Crystal Growth **2** (1968) 279.
11) I. M. Byteva, in: *Growth of Crystals*, Vol. 4 (Consultants Bureau, New York, 1966) p. 16.
12) I. M. Byteva, in: *Growth of Crystals*, Vol. 5B (Consultants Bureau, New York, 1968) p. 26.

Journal of Crystal Growth **13/14** (1972) 467–470 © *North-Holland Publishing Co.*

ZÜCHTUNG UND PHYSIKALISCHE EIGENSCHAFTEN VON MONOKLINEM ZINNDIFLUORID

E. ACKER

Mineralogisches Institut der Universität, Poppelsdorfer Schloß, 5300 Bonn 1, Germany

S. HAUSSÜHL

Institut für Kristallographie der Universität, Zülpicher Str. 49, 5000 Köln 41, Germany

and

K. RECKER

Mineralogisches Institut der Universität, Poppelsdorfer Schloß, 5300 Bonn 1, Germany

Large monocrystals of monoclinic SnF_2 with dimensions up to 4 cm were grown from water solutions for the examination of their physical properties. Two different techniques were used: growth by lowering temperature (50 to 25 °C) and growth by isothermal evaporation of the solvent (50 °C).
The crystals belong to space group C2/c. They are fully transparent from 0.240 to 11 µm. Their density is 4.87 g/cm³. The following further properties were measured: refractive indices for several wave lengths, indentation hardness, elastic constants, thermal expansion, and thermoelastic constants.

1. Einführung

Zinndifluorid (Raumgruppe C_{2h}^6–C2/c)[1]) besitzt eine Reihe sehr interessanter kristallographischer und kristall-physikalischer Eigenschaften, die aus den ungewöhnlichen Bindungsverhältnissen und der Isotopenreinheit hervorgehen. Die Kristalle sind insbesondere für Mößbauer- und Neutronenbeugungs-Experimente geeignet. Vorversuche zur Kristallisation zeigten, daß eine Züchtung aus wäßriger Lösung möglich ist und zu Kristallen extrem hoher Reinheit führen würde.

2. Züchtung

Zur Vervollständigung und Überprüfung der Literaturwerte[2,3]) der Löslichkeit in Wasser wurde diese zunächst neu bestimmt (Abb. 1). Dabei stellte sich heraus, daß geeignete Bedingungen für die Züchtung aus wäßriger Lösung im Temperaturbereich von ca. 60 °C bis 0 °C bestehen.

Die Züchtung wurde nach zwei Methoden durchgeführt: (1) nach dem Temperatursenkverfahren, (2) durch kontinuierlichen Entzug von Lösungsmittel bei konstanter Temperatur. Es wurde nur mit reinen Lösungen ohne Zusätze gearbeitet. Die benutzten Kristallisatoren sind in Abb. 2 dargestellt.

Die Wachstumsgefäße bestehen aus Plexiglas. Die Keimkristalle, gewonnen durch spontane Keimbildung,

wurden in einem Silikonschlauch am unteren Ende der Rührerachse befestigt. Es wurde gleichsinnig mit 40 U/min gerührt. Die erzielte Temperaturkonstanz betrug 0.004 °C.

Bei der Züchtung durch Temperaturänderung wurde die Temperatur pro Tag kontinuierlich um etwa 0.35 °C gesenkt. Größere oder kleinere Temperatursenkungen erwiesen sich als ungünstig. Bei dieser Übersättigungserzeugung kamen in Gefäßen von ca. 1 l Inhalt lineare Wachstumsgeschwindigkeiten von ca. 0.4 mm pro Tag zustande.

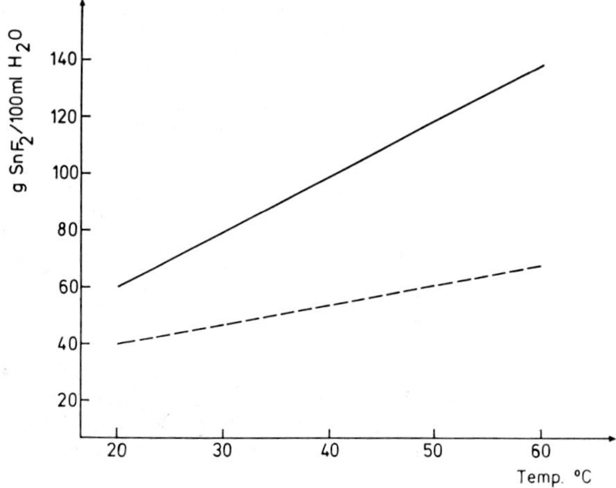

Abb. 1. Löslichkeit von SnF_2 in Wasser. (——) eigene Werte; (– – –) Werte nach Beal[2]).

Bei der Züchtung durch Lösungsmittelentzug wurde das Kondensat in einer Rinne des wassergekühlten Deckels gesammelt und mit einer Injektionsspritze kontinuierlich abgesaugt. Die tägliche Lösungsmittelentnahme betrug etwa 12 ml.

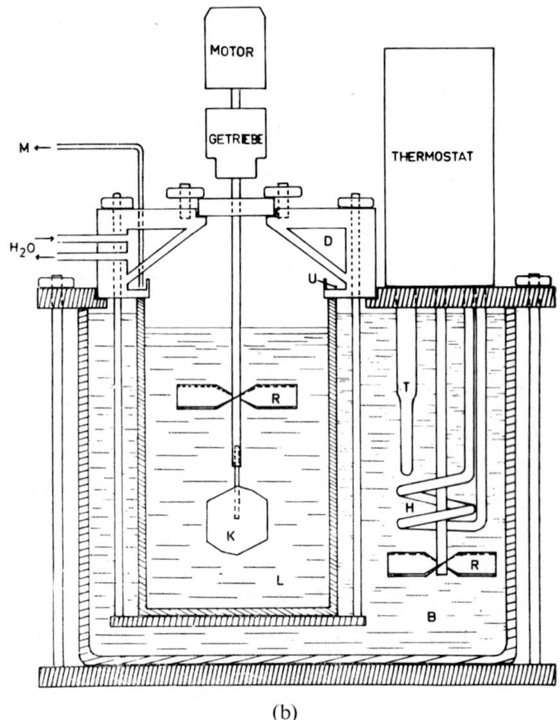

(a)   (b)

Abb. 2. Kristallisatoren zur SnF_2-Züchtung. (a) Temperatursenkverfahren, (b) Lösungsmittelentzug bei konstanter Temperatur (L = Lösung, B = Thermostatenbad, K = Kristall, R = Rührer, U = Überlaufrinne, M = Mikropumpe, D = Deckelkühlung, T = Kontaktthermometer, H = Heizung).

Abb. 3. SnF_2-Einkristalle, gezüchtet aus wäßriger Lösung.

IX – 7

TABELLE 1

Kristallographische und kristallphysikalische Eigenschaften von SnF_2

Eigenschaft	Dimension	Meßwerte	Verfahren	Maximaler relativer Fehler
Dichte	$g\ cm^{-3}$	4.8749	Auftriebsverfahren an großen Kristallen	0.2%
Metrik $a_1:a_2:a_3$ α_2		1.4086:1:2.7291 $109°\ 24'$	Morphologisch bestimmt mit Reflexionsgoniometer	
Gitterkonstanten [nach Bergerhoff[1])] a_i, α_2	Å	$a_1 = 13.46$ $a_2 = 4.92$ $a_3 = 13.86$ $\alpha_2 = 109.5°$		
Hauptbrechwerte $n_{i;\lambda_0}$ (λ_0 Vakuumwellenlänge in mm) $v_{\lambda_0} = \sphericalangle\,(\mathfrak{n}'_1, \mathfrak{a}_1)$ (zwischen $\mathfrak{a}_1$ und $-\mathfrak{a}_3$)		$n'_{1;625} = 1.8251$ $n'_{2;625} = 1.7945$ $n'_{3;625} = 1.8710$ $n'_{1;578} = 1.8335$ $n'_{2;578} = 1.8025$ $n'_{3;578} = 1.8816$ $n'_{1;546} = 1.8430$ $n'_{2;546} = 1.8098$ $n'_{3;546} = 1.8926$ $n'_{1;436} = 1.8890$ $n'_{2;436} = 1.8506$ $n'_{3;436} = 1.9506$ $v_{625} = \ 9°\ 55'$ $v_{578} = 10°\ 12'$ $v_{546} = 10°\ 33'$ $v_{436} = 11°\ 27'$	Prismenmethode	
Elastische Konstanten c_{ij}	$10^{11}\ dyn\ cm^{-2}$	$c_{11} = \ \ 4.789$ $c_{12} = \ \ 0.932$ $c_{33} = \ \ 3.355$ $c_{22} = \ \ 2.065$ $c_{13} = \ \ 0.528$ $c_{23} = \ \ 1.484$ $c_{44} = \ \ 1.735$ $c_{55} = \ \ 1.293$ $c_{66} = \ \ 1.441$ $c_{15} = -0.510$ $c_{35} = \ \ 0.655$ $c_{46} = \ \ 0.346$ $c_{25} = \ \ 0.314$	Beugung von monochromatischem Licht an stehenden Ultraschall-wellen in dicken Platten; Frequenz ca. 15 MHz	$c_{11}, c_{22}, c_{33}: 2°/_{00}$ $c_{12}, c_{13}, c_{23}: 2\%$ $c_{44}, c_{55}, c_{66}: 1\%$ die übrigen: 5%
Transparenzgrenzen (s. Abb. 5)	μm	UV-Kante: 0.240 IR-Kante: 11	Spektralphotometer	
Vickers-Härte	$kp\ mm^{-2}$	maximal auf (001): 75 minimal auf (010): 55	Leitz-Durimet	10%

Anmerkungen zur Tabelle:
Alle Indices (mit Ausnahme der Brechungsindizes) beziehen sich auf kartesische Achsen $\mathfrak{n}_i$, die wie folgt an die kristallographischen Achsen $\mathfrak{a}_i$ angeschlossen sind: $\mathfrak{n}_1 \parallel \mathfrak{a}_1, \mathfrak{n}_2 \parallel \mathfrak{a}_2, \mathfrak{n}_3 = \mathfrak{n}_1 \times \mathfrak{n}_2$.
Sämtliche Meßwerte gelten für 20 °C. Die Brechungsindizes beziehen sich auf die Schwingungsrichtungen $\mathfrak{n}'_i$, deren Lage zu den kristallographischen Achsen durch den Auslöschungswinkel v gekennzeichnet ist ($\mathfrak{n}'_2 = \mathfrak{n}_2$).

Beide Verfahren lieferten im Laufe von drei Monaten große, wasserklare Einkristalle bester optischer Qualität mit Abmessungen bis zu 4 cm (Abb. 3).

Die Morphologie (Abb. 4) der Kristalle ist durch folgende Rangfolge der Formen bestimmt: $\{100\}$, $\{001\}$, $\{111\}$, $\{011\}$, $\{\bar{1}22\}$, $\{\bar{1}11\}$, $\{\bar{3}22\}$, $\{\bar{1}0\bar{1}\}$, $\{\bar{2}0\bar{1}\}$.

IX – 7

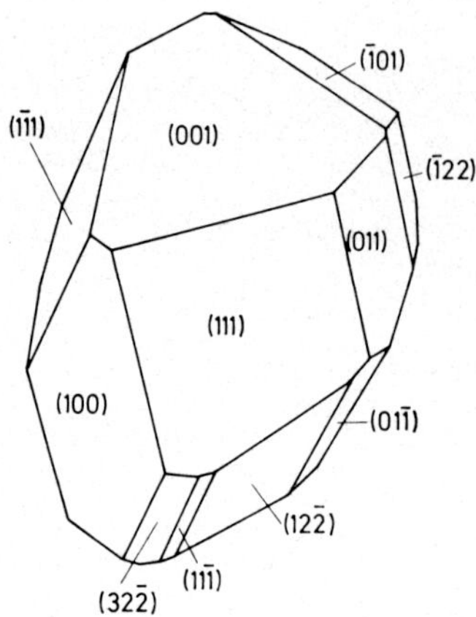

Abb. 4. Morphologie von SnF$_2$-Einkristallen.

3. Eigenschaften

Die Kristalle sind bei einer relativen Luftfeuchtigkeit unterhalb 60% nicht hygroskopisch. Sie bleiben bis zu einer Temperatur von ca. 150 °C (auch in längeren Zeiten) unzersetzt.

In Tab. 1 sind die von uns gemessenen weiteren Eigenschaften einschließlich der benutzten Methoden und Schranken für die relativen Fehler zusammengestellt.

Neben den in der Tabelle aufgeführten Eigenschaften wurden auch die thermische Ausdehnung und das thermo-elastische Verhalten gemessen. Ein Vergleich der Eigenschaften des SnF$_2$ mit denen anderer Fluoride, wie z. B. CaF$_2$, führt zu dem überraschenden Ergebnis, daß SnF$_2$ wesentlich schwächere Anteile an ionogener Bindung und stärkere Van-der-Waalssche Bindungsanteile aufweist als andere Fluoride. SnF$_2$ muß daher eher als Molekülkristall, weniger als Ionenkristall angesprochen werden. Dies kommt besonders deutlich zum Ausdruck in den viel kleineren elastischen Konstanten des SnF$_2$, gemessen am reziproken Molvolumen$^{4/3}$ [s. Haussühl[4])], sowie den viel größeren Koef-

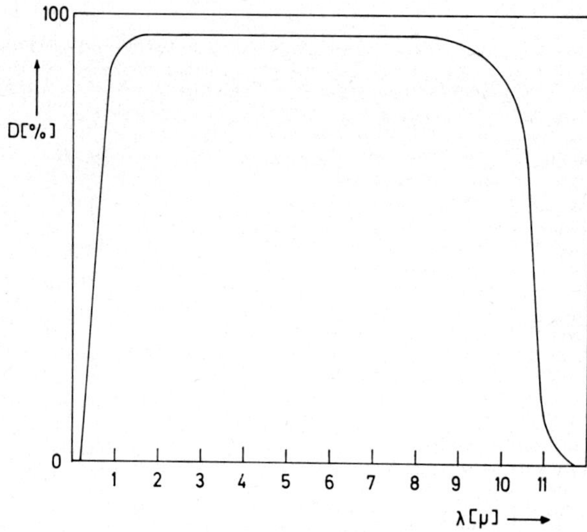

Abb. 5. Absorptionsspektrum von SnF$_2$ (D = Durchlässigkeit).

fizienten der thermischen Ausdehnung und der thermo-elastischen Konstanten, verglichen mit CaF$_2$.

Die Abweichungen von den Cauchy-Relationen – hier überwiegen alle elastischen Scherwiderstände über die zugehörigen Querkontraktionskoeffizienten – geben einen Hinweis auf starke intramolekulare kovalente Bindungsanteile.

Unerwartet groß sind auch die Anisotropie-Effekte, wie sie z.B. im elastischen und thermischen Verhalten sichtbar werden (z.B. $c_{11} \approx 2.4\, c_{22}$). Eine Interpretation dieser Anisotropie-Effekte aus der Struktur ist mangels einer vollständigen Strukturbestimmung noch nicht möglich.

Literatur

1) G. Bergerhoff, Acta Cryst. **15** (1962) 509.
2) J. B. Beal, Jr., Ph.D. Thesis, A. & M. College of Texas, 1962.
3) J. J. Hefferren, J. Pharm. Sci. **52** (1963) 1090.
4) S. Haussühl, Phys. Status Solidi **3** (1963) 1072.

Journal of Crystal Growth 13/14 (1972) 471–475 © *North-Holland Publishing Co.*

GROWTH KINETICS OF LEAD DENDRITES IN GELS

H. M. LIAW and J. W. FAUST, Jr.

University of South Ccarolina, College of Engineering, Columbia, S.C. 29208, U.S.A.

Kinetic studies were made on lead dendrites grown in gels under a variety of parameters. The parameters fell into two classes: (a) those associated with the gel, and (b) those associated with the reducing metal. The effect of these parameters on the growth rate is given. In all cases a parabolic growth law was observed.

1. Introduction

Crystal growth in gels has recently taken its place as a technique for preparation of special materials[1]. Besides the growth of crystals of slightly water soluble ionic compounds such as tartrates, sulfates, iodides, and calcite, metal dendrites have also been grown from gels as reported by Taft and Stareck[2]), Henisch, Vand and Hanoka[3]), and more recently by Liaw and Faust[4]). The principle of growing metal crystals in gels lies in the reduction of metal ions to the elemental state by a more active metal.

So far essentially nothing has been reported on the growth kinetics of metal dendrites in gels. Taft and Stareck[2]) have made several suggestions on the factors affecting the growth rate; however, quantitative information was not given. In preliminary studies on the growth of lead dendrites in gels, Liaw and Faust[4]) showed that the growth rates are affected both by factors associated with the reducing metals and with the gel. Examples of the latter are age of gels, concentration of lead ions, and pH of the gels. The purpose of the present paper is to give quantitative data on these parameters.

2. Experimental procedures

The silica gel (acidified sodium meta silicate) was used as the growth media throughout this study. Conventionally[5]), the gel solution was prepared by titrating 0.85 M sodium metasilicate into an equal volume of 2.0 M acetic acid. To this gel solution, a desired volume of 1.05 M lead acetate was added. The gel solution was then poured into growth cells and allowed to set. After the gel was set, the growth of dendrites was started by inserting a piece of the reducing metal into the surface of the gel. For the study of the effect of gel parameters and the nature of reducing metals, the following variables were adjusted.

2.1. The variation of gel parameters (under this study zinc was always used as the reducing metal)

(1) The variation of the age of gels. The gel was made by adding 1 volume of 0.85 M Na_2SiO_3 to 1 volume of 2.0 M HAc. (This will be referred to as the gel forming solution.) To 10 cm^3 of this gel forming solution, 6.0 cm^3 of 1.05 M $PbAc_2$ was added. The gel forming solution was then poured into cells and allowed to set in open air at room temperature for different times. (For other experiments, all of the gel forming solutions were allowed to set under saturated water vapor at room temperature.) After the gels had set, the zinc was inserted and the growth rates measured.

(2) The variation of the concentration of $PbAc_2$ in gels. Gels of various concentration of $PbAc_2$ were obtained by adding 7.5 cm^3 of various concentrations of $PbAc_2$ to each 10 cm^3 of the gel forming solution.

(3) Variation of the pH of gels. The pH values of gels were also varied by adding HCl. The 2.0 M HAc solutions for the preparation of the gels were first mixed with different amount of 10% HCl. To each of the mixed acid solutions an equal volume of 0.85 M Na_2SiO_3 was added. To 10 cm^3 of this gel forming

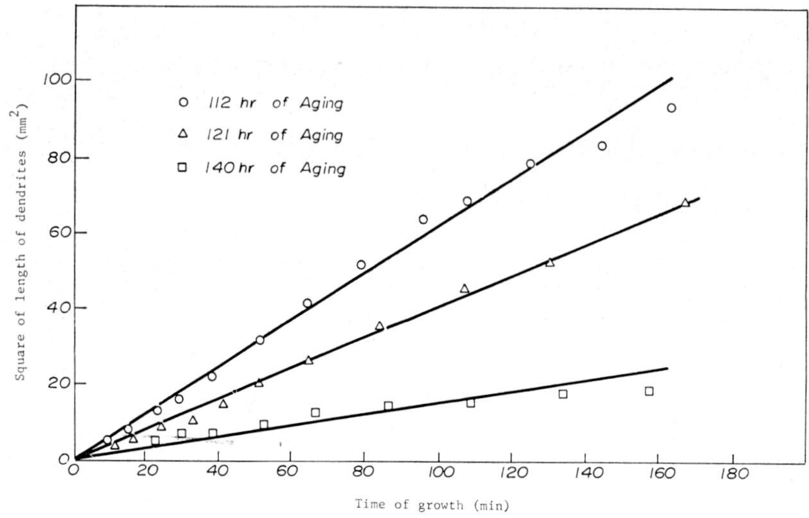

Fig. 1. Plot of the square of the length of dendrites versus time of growth for dendrites grown in gels of different ages.

solution, 2.0 cm³ of 1.05 M PbAc₂ was then added.

(4) Variation of Na_2SiO_3 concentration in gels. Gels for this study were made by adding different concentrations of Na_2SiO_3 into equal volumes of 2.0 M HAc. The pH of the gel forming solutions was adjusted to the same value by either adding a few drops of NaOH or HAc. To 10 cm³ of these gel forming solutions, 2.0 cm³ of 1.05 M lead acetate was added.

2.2. THE VARIATION OF REDUCING METALS

Several metals (Al, Cr, Co, Ni, Fe, Cd, In, Mg, Mn, Sn, Tl, Zn) of higher emf have been tried as a reducing agent for the growth of lead dendrites. In each case the same type of gel forming solution was used. The growth rates were measured whenever any reaction leading to dendritic formation was found.

The effect of the size of the reducing metal on the growth rate was studied by using zinc to react with the same type of gels. The growth rate of dendrites was also studied to see if the nature of the surface of the reducing metal affected it. For this purpose, two different types of surface treatment of zinc sheet were used.

TABLE 1

Growth rate constants and ages of the gels

Age of gel (hr)	Growth rate constant (mm²/min)
112	0.618
121	0.408
140	0.155

One was the original surface and the other was the sheet exposed to HAc vapors.

3. Experimental results and discussion

3.1. GROWTH KINETICS; AN EVALUATION OF GROWTH RATE CONSTANT

In order to be able to describe the growth kinetics of dendrites, a specific mathematical function should be chosen to express the relation between the length of dendrites and time of growth. In the plots of length of dendrites versus time of growth for dendrites reduced by any of reducing metals in any of gel medium, a parabolic relationship was always found. When the square of the length of dendrites was plotted against the time of growth for any of the parabolic curves, a good straight line was obtained. Their relationship can be expressed as $y^2 = Kt$, where y is the length of dendrites, t is time of growth, and K is a constant. In this study K is defined as the growth rate constant in units of mm²/min.

It has been shown that parabolic kinetics is a characteristic of one-dimensional diffusion controlled processes. This result agrees with the kinetic study of ionic crystals grown in gels by Henisch et al.[6]).

3.2. EFFECT OF GEL PARAMETERS ON GROWTH RATE OF DENDRITES

(1) The age of gels. For gels kept in air, it was found that the older the gel the slower the growth of dendrites.

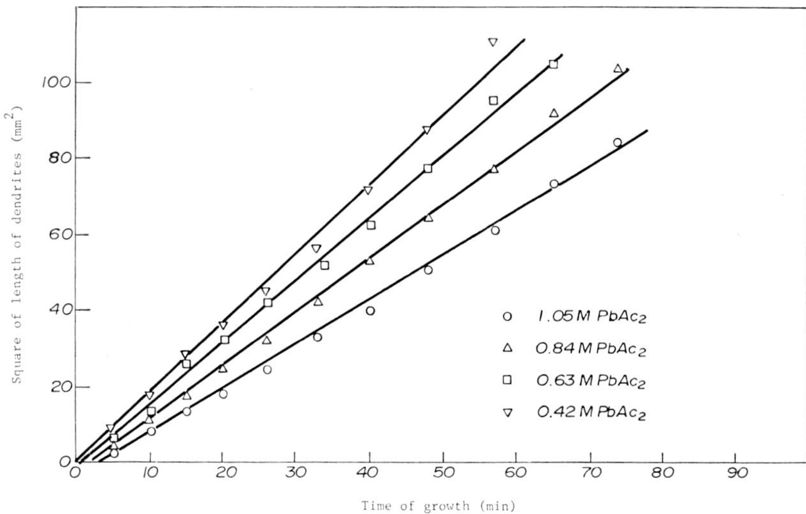

Fig. 2. Plot of the square of the length of dendrites versus time of growth for dendrites grown in gels of different PbAc$_2$ concentrations.

TABLE 2

Variation of PbAc$_2$ concentrations in gels, growth rate constants, pH, specific gravities of gel solution, density of gels, and electrode potentials

Concentration of PbAc$_2$ added (M)	Growth rate constant (mm^2/min)	pH	Specific gravity of gel solution	Density of gel	Cell potential referring to Pt (V)
1.05	1.07	5.55	1.0427	1.1607	0.56
0.84	1.35	5.60	1.0363	1.1174	0.89–0.62
0.64	1.61	5.55	1.0298	1.1354	0.72–0.65
0.42	1.82	5.55	1.0232	1.0941	0.78–0.76

Fig. 1 is the plot of the square of the length of the dendrites versus the time of growth in gels of various ages. The growth rate constants, determined from these plots, are tabulated in table 1.

It is evident that the longer a gel sets in room air the greater amount of water will evaporate out of the gel. The effect of water evaporation should be considered in two parts, namely before and after the formation of gel framework. Before the gel is set, evaporation of water will cause an increase in gel density, which in turn will decrease the diffusivity of ions in the gel and therefore decrease the growth rate. After the gel is set, evaporation of water causes not only the lack of ionic carriers in the channel of the gel framework, but also discontinuities in the channels due to the shrinkage of the gel. Both of these effects would adversely affect the diffusion of ions.

(2) Effect of lead ion concentration in gels. The effect of lead ion concentration in gels on growth rate was studied by adding the same volume of various PbAc$_2$ concentrations into a given volume of gel solution. It was found that as the PbAc$_2$ concentration in the gels was increased, the growth rate of the dendrites decreased. Fig. 2 shows the plots of the square of the length of dendrites versus time of growth in gels of various PbAc$_2$ concentrations. The growth rate constants taken from the slopes of these straight lines are listed in table 2.

pH, specific gravity of gel solutions and density of these gels were measured and also listed in table 2. It can be seen that the change of PbAc$_2$ concentrations in gels does not cause an appreciable change of pH. However, the specific gravity of gel solutions and density of gels varied regularly with the change of PbAc$_2$ concentration. It is, therefore, likely that the increase of growth rate by the decrease of PbAc$_2$ concentration is partly due to the decrease of gel density.

It has also been shown that the growth of lead dendrites in gels can be regarded as a type of voltaic cell reaction. The growth rate of dendrites will then be

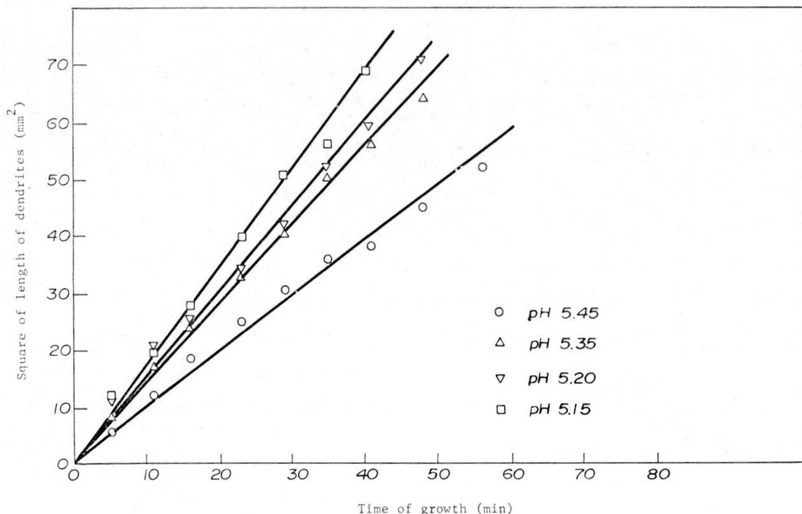

Fig. 3. Plot of the square of the length of dendrites versus time of growth for dendrites grown in gels of various pH values adjusted by HCl.

expected to be a function of the cell potential. It is also well known that the potential of the voltaic cell varies with the concentration of the electrolyte. The "cell potential" with various $PbAc_2$ concentrations was therefore increased. The method consisted of inserting a piece of Zn and Pt into the gels. The potential difference of two electrodes is the "cell potential" referred to a Pt potential in the gel. It was found that the cell potentials decay with time after growth of the lead on zinc. Table 2 lists the cell potentials of the gels read at the first and last stages. From these values it can be seen that the increase of the growth rates by the decrease of $PbAc_2$ concentrations is partly due to the increase of the cell potential, i.e., the driving force for crystallization.

(3) The effect of pH on growth rate of dendrites. The effect of the pH on the growth rate was studied by the change of pH without a change of gel composition and $PbAc_2$ concentration. It was achieved by replacing part of the HAc by HCl in the preparation of the gels. Table 3 lists the ratio of these mixed acids for gel preparation and pH of the gels. Fig. 3 is the plot of the square of the length of dendrites versus time of growth for dendrites grown in these gel mediums. The growth rate constants obtained from the plots are given in table 3. It shows that a decrease of pH increased the growth rate of dendrites.

This result is what was expected. Since the growth of dendrites is a reduction reaction occurring in an acid medium, an increase in the acidity of the medium might be expected to increase the ionization of the $PbAc_2$ and also the solubility of the reducing metals in gels. The growth rate would then also increase.

(4) The effect of Na_2SiO_3 concentration in gels. The usual way to make a gel was to mix 0.78 M Na_2SiO_3 with equal volume of 2.0 M HAc. In this study, however, several gels were made by mixing 2.0 M HAc with equal volume of various concentrations of Na_2SiO_3 (0.778 M, 0.637 M, 0.495 M, and 0.354 M). Within experimental errors, the growth rates of the dendrites

TABLE 3

List the variation of acid mixture for preparation of gels, pH of gel solution and growth rate constants

Mixture of acid		pH of Gel solution	Growth rate constant (mm²/min)
Vol. of 2 N HAc	Vol. of 10% HCl		
12.0cc	0.0	5.45	1.00
11.0	1.0	5.35	1.39
10.5	1.5	5.20	1.49
10.0	2.0	5.15	1.72

TABLE 4

Growth rate constants and type of reducing metals

Type of reducing metal	Growth rate constant (mm²/min)
Zn	1.210
Cd	0.392
Fe	0.170

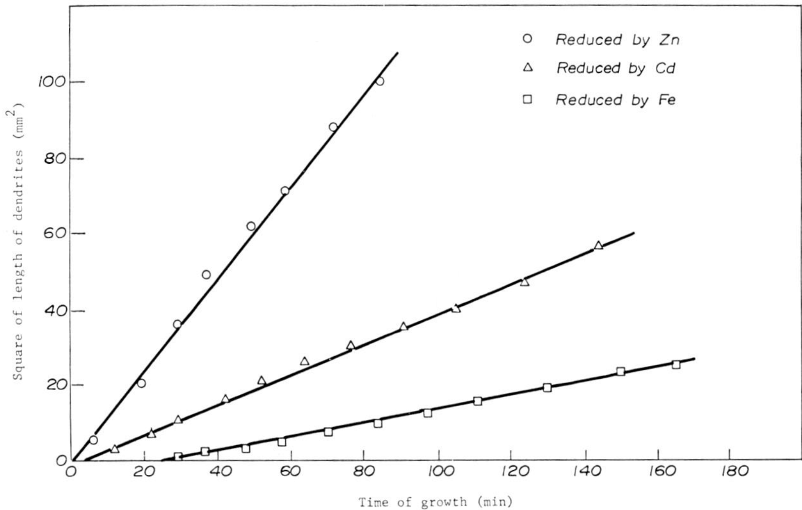

Fig. 4. Plot of the square of the length of dendrites versus time of growth for dendrites reduced by various metals in the gels.

were not found to vary with changes in the concentration of Na_2SiO_3 in the gels. This suggests that as far as the framework of the gel is concerned, the rate of diffusion of ions is independent of the composition of the Na_2SiO_3.

3.3. EFFECT OF THE CHOICE OF REDUCING METALS

It was found that not all metals whose emf is higher than lead are able to displace lead ions to give dendrites. Only Cd, Mn, Tl, Zn, or Fe of those investigated gave the desired effect.

At a low concentration of $PbAc_2$ in gels (i.e., 2 cm³ of 1.05 M $PbAc_2$ per 10 cm³ gel solution) the rate of growth of dendrites are in the same sequence as the emf of the reducing metals. Fig. 4 contains plots of the square of the length of dendrites versus time of growth for dendrites reduced by Zn, Cd, and Fe. The growth rate constants of these plots are listed in table 4. At a medium concentration of $PbAc_2$ in the gel (i.e., 4 cm³ to 6 cm³ of 1.05 M $PbAc_2$ per 10 cm³ gel solution), the same relation holds for metals of Mn, Zn, Fe and Cd. An exception was found for dendrites grown by Tl; these always grew faster. At a high concentration of $PbAc_2$ in gels (8 cm³ of 1.05 of 1.05 M $PbAc_2$ per 10 cm³ gel solution) no such relationship between the growth rate and emf of the reducing metals was found.

The size of a reducing metal (e.g. Zn) was not found to have any effect on the growth rate. The growth rate,

however, was affected by the surface preparation of the reducing metal. The growth rate of dendrites reduced by a zinc sheet exposed in HAc vapor was slower than that reduced by a zinc sheet without the acid vapor exposure.

4. Conclusion

The growth kinetics of lead dendrites in gels follow a parabolic law which is a characteristic of a one-dimensional diffusion controlled process. The factors which affect the growth rate of the dendrites are the driving potential of crystallization and diffusivity of the gel. The driving potential varies not only with emf of the reducing metals but also with the concentration of $PbAc_2$ in the gels. The emf of the reducing metals can be related to the driving force only when $PbAc_2$ concentration is not too high. The diffusivity of the gels is mainly governed by the age, density and pH of the gels.

References

1) H. K. Henisch, *Crystal Growth in Gels* (Pennsylvania State Univ. Press, University Park, 1970).
2) R. Taft and J. Stareck, J. Chem. Educ. 7 (1930) 1520.
3) H. K. Henisch, V. Vand and J. I. Hanoka, private communication.
4) H. M. Liaw and J. W. Faust, Jr., ACCG, Conf. on Crystal Growth, Gaithersburg, Maryland, 1969.
5) H. M. Liaw and J. W. Faust, Jr., J. Crystal Growth 8 (1971) 8.
6) H. K. Henisch, J. I. Hanoka and J. Denis, J. Electrochem. Soc. **112** (1965) 627.

     Journal of Crystal Growth **13/14** (1972) 476–480 © *North-Holland Publishing Co.*

BRUSHITE CRYSTALS GROWN BY DIFFUSION IN SILICA GEL AND IN SOLUTION

RACQUEL ZAPANTA LeGEROS

New York University, 339 E. 25th Street, New York, N.Y. 10010, U.S.A.

and

JOHN P. LeGEROS

Cooper Union College, Cooper Square, New York, N.Y., U.S.A.

Brushite, $CaHPO_4 \cdot 2H_2O$, occurs in phosphatic deposits and in biological systems as one of the components of human urinary and dental calculi. Because of its occurrence in biological systems; it is important that the conditions for the crystal growth of brushite and its relationship with the other biologically important calcium phosphates (e.g., octacalcium phosphate, $Ca_8H_2(PO_4)_6 \cdot 5H_2O$; whitlockite; β-$(Ca, Mg)_3(PO_4)_2$; and apatite $Ca_{10}(PO_4)_6(OH)_2$; are understood. Two crystals growth methods which utilize the slow diffusion of ions, and therefore best simulate the conditions under which biological crystals form, are reported here: (i) growth by diffusion in silica gel and (ii) growth by diffusion in dilute acid solution. A variation of the latter was made by growing brushite crystals on the surface of teeth suspended in phosphate solutions. Differences in growth morphology are observed as influenced by temperature, and presence of other ions (such as F^-, CO_3^{2-}, $P_2O_7^{4-}$, Sr^{2+}, Mg^{2+}). It is demonstrated that the gel system offers a convenient tool for observing effects of various ions and conditions (pH and temperature) on the growth morphology of brushite crystals and on its co-existence and relationship with octacalcium phosphate, whitlockite and apatite.

1. Introduction

Brushite, $CaHPO_4 \cdot 2H_2O$, occurs as crystals of varying sizes in natural phosphorite cavities; guano deposits and as incrustations on buried human and animal bones[1,2]). It is also found in biological systems as one of the components in human dental and urinary calculi[3,4]).

Because of its occurrence in biological systems, it is important that the conditions be identified for the growth of brushite and its relationship with the other biologically important calcium phosphates: octacalcium phosphate, $Ca_8H_2(PO_4)_6 \cdot 5H_2O$; whitlockite, β-$(Ca, Mg)_3(PO_4)_2$; and apatite, $Ca_{10}(PO_4)_6(OH)_2$.

To this end, an attempt has been made to simulate the biological conditions in which brushite growth occurs by employing systems which utilize the slow diffusion of ions: Two such systems are reported here: (i) growth by diffusion in silica gel[5,6]) and (ii) growth by diffusion in dilute acid solution. A variation of the latter was made by growing brushite on the surface of the teeth suspended in phosphate solutions.

2. Experimental

2.1. GROWTH OF BRUSHITE CRYSTALS BY SLOW DIFFUSION IN SILICA GEL

Silica gels were prepared using 10% solution of sodium metasilicate ($Na_2SiO_3 \cdot 9H_2O$, Mallinkrodt reagent) and acidified to the desired pH (4 to 8) by the addition of 3 M HAc. The PO_4^{3-} ions were incorporated in the gel (using either 0.1 M H_3PO_4 or 0.1 M NaH_2PO_4) and the Ca^{2+} ions [using 0.5 M $Ca(Ac)_2$] in the top solution. (The Ca/P molar ratio is about 1.) The gel tubes were kept in a water bath maintained at 37 to 60 °C, or at room temperature (25 °C), or refrigerated at 2 °C. The effects of impurities were studied by the incorporation of the anions (e.g., F^-, CO_3^{2-}, or $P_2O_7^{4-}$) in the gel with the PO^{3-} and of the cations (e.g., Mg^{2+} or Sr^{2+}) in the top solution replacing some of the Ca^{2+} ions. The crystals were harvested after 4 weeks. This gel method is a modification of that used by Blank[7a,b]).

2.2. GROWTH OF BRUSHITE CRYSTALS BY SLOW DIFFUSION IN ACID SOLUTION

Two general variations of this systems were employed. In one case, 900 ml of $Ca(NO_3)_2$ (0.61 M, pH adjusted to 4.0 with 0.01 M HNO_3) was contained in the inner jar and 640 ml of $NH_4H_2PO_4$ (0.85 M, pH adjusted to 4) was contained in the outer jar: the liquid levels of both inner and outer jars were below the lip of the inner jar and made equal. A layer of 0.1 mM

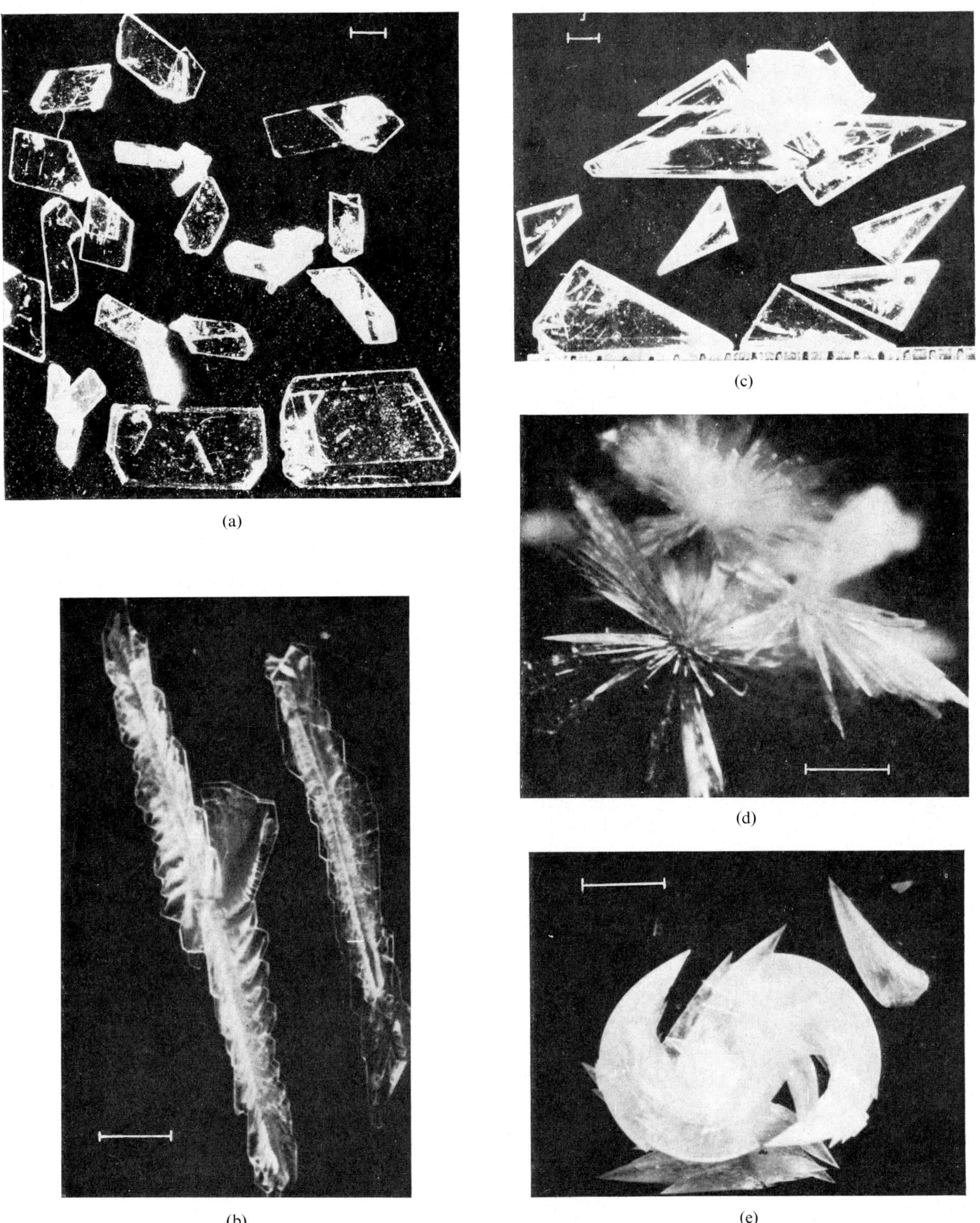

Fig. 1. Brushite crystals grown by slow diffusion in: (a) dilute acid, pH 4, 25° C, (b) silica gel, pH 6, 2 °C, (c) silica gel, pH 6, 37 °C, (d) and (e) have similar conditions as (c) except for the presence of pyrophosphate and strontium, respectively. The bar indicates 1 mm.

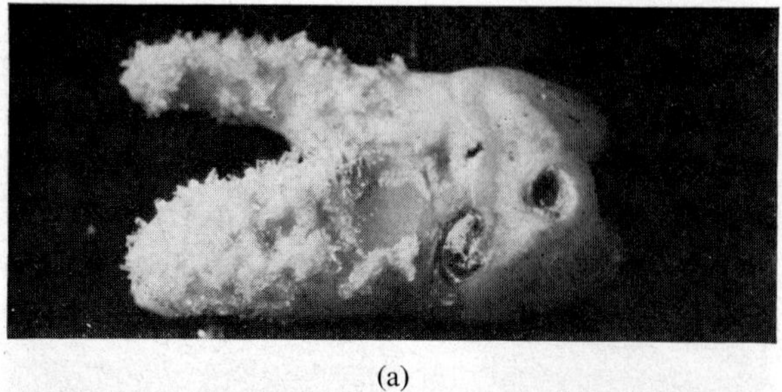

(a)

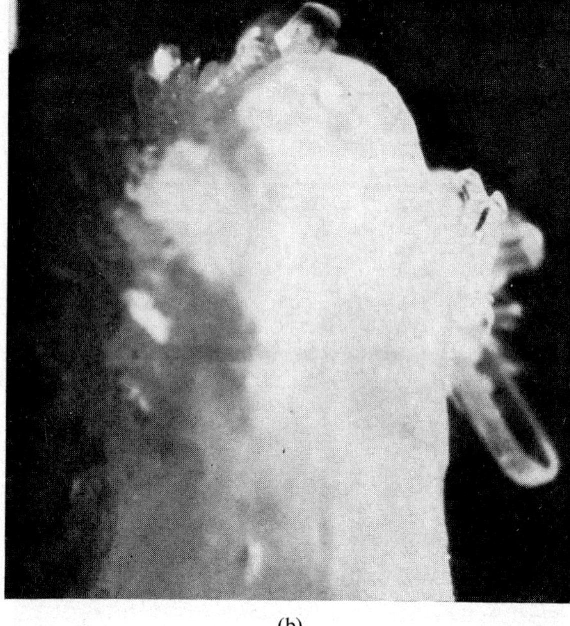

(b)

Fig. 2. Brushite crystals growing on tooth immersed in NaH₂-PO₄ solution, pH 4, 25 °C. (b) is a further magnification (8×) of a portion of the root (magnification of tooth 4×).

HNO₃ was carefully stratified above both solutions; thus serving as the diffusion medium for the calcium and phosphate ions from their respective reservoirs: Experiments of this type were conducted for periods of 1 to 4 months at 25 °C.

A variation of the system described above was made with cleaned extracted teeth which were suspended for 1 month at 25 °C in 50 ml NaH₂PO₄ (0.5 M) at pH levels 3 to 6.

2.3. ANALYSIS OF THE CRYSTALS

The crystals were usually platelets with surface parallel to (010). Angles between faces were determined with a Stoe optical goniometer and polarizing microscope equipped with rotating stage. The latter technique proved useful since the (h0l) faces are not developed on

Fig. 3. Gel tube pH 6, 37 °C, showing the co-existence of octacalcium phosphate spherules OCP and the brushite platelets B in the same medium.

these crystals. For indexing purposes; the unit-cell dimensions reported by Beevers[8]) ($a = 5.182$ Å; $b = 15.180$ Å; $c = 6.239$ Å; $\beta = 116°\ 25'$) were used: some of the crystals of modified habit were powdered and positive identification made using X-ray diffraction and infrared absorption spectroscopy.

3. Results

Brushite crystals grown by the slow diffusion in acid solution were first observed after 3 days. Most of these crystals were deposited on the rough edge of the inner jar. The crystals collected after 4 months ranged in sizes from 2 to 42 mm in length, 1 to 14 mm in width, and 0.2 to 6 mm in thickness (fig. 1a).

The brushite crystals which grew on the dentine surface of a tooth (fig. 2) were first observed after about 2 weeks. These crystals were smaller and the morphology similar to those grown by diffusion in acid solution.

Brushite crystals grown by the slow diffusion in silica gel were observed after 3 days in tubes kept at 25 to 37 °C; after 7 days in tubes kept at 2 °C; and grew in tubes kept at 60 °C for 2 days after it has cooled at room temperature. In the series of experiments at 37 °C, pH 5 to 8, octacalcium phosphate, OCP (fig. 3) was found to co-exist with brushiteplatelets.

The crystals grown by these methods were usually tabular on the (010) face. A perfect large crystal grown

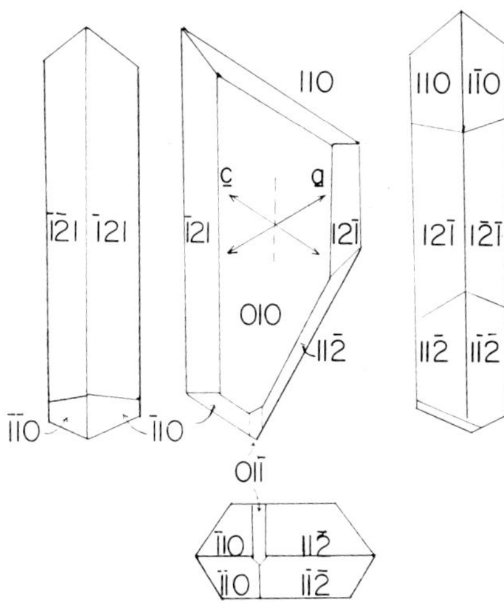

Fig. 4. Representation of single crystal of brushite obtained by diffusion in acid; with the respective interfacial angles (see tables 1 and 2).

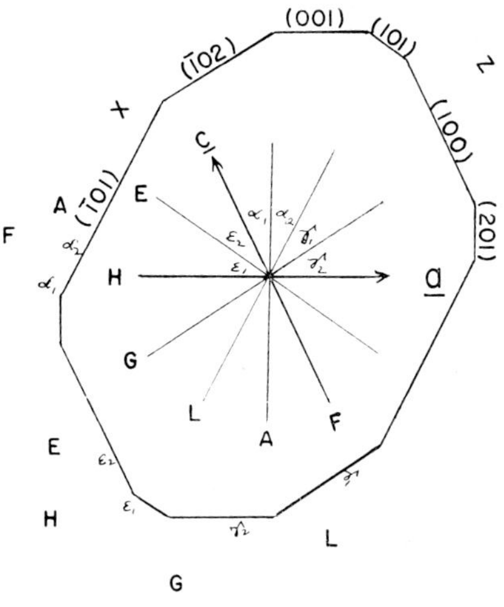

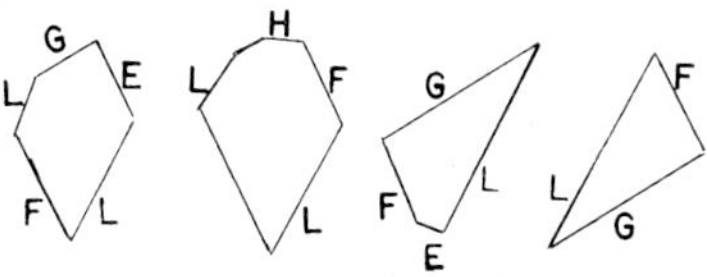

Fig. 5. Unit-cell net of brushite with platelet parallel to (010) face superimposed. Below at left; 2 platelets grown in dilute acid; at right, 2 platelets grown in silica gel. Approximate extinction directions: F to X–170°; X to E–13.5°; G to Z–10.5°; and Z to L–18.7°.

TABLE 1

Face angles in the (010) zone

$$
\begin{aligned}
(001) & \\
(101) & \;\; \varepsilon_1 = 33°05' \\
(100) & \;\; \varepsilon_2 = 30°30' \\
(20\bar{1}) & \;\; \alpha_1 = 27°45' \\
(10\bar{1}) & \;\; \alpha_2 = 27°11' \\
(10\bar{2}) & \;\; \gamma_1 = 29°12' \\
(00\bar{1}) & \;\; \gamma_2 = 32°16'
\end{aligned}
\Bigg\} \; \beta = 116°25'
$$

TABLE 2

Face angles between (010) and its bevels

Edge face	Angle
(011)	69°47′
(111)	78°12′
(110)	71°04′
(211)	79°15′
(121)	56°04′
(112)	78°28′

by diffusion in acid solution is shown in fig. 4. Since the monoclinic angle β is very near 120°, it is difficult to identify the crystal faces by visual inspection without exact measurement of the face angles in the [010] zone.

The crystal habit of brushite is affected by: (1) method of growth – whether in solution or in silica gel; (ii) temperature; and (iii) presence of impurities. The last two effects were studied only in the gel systems.

Growth in silica gel exhibited an inhibition of the (101) and (001) faces (fig. 5). In comparison with the clear crystals obtained in dilute acid solution, the crystals grown in silica gel at 2 °C (fig. 1b) and at 25 °C[76]. Crystal growth at 2 °C is considerably slowed down and showed much twinning. The presence of F^-, CO_7^{2-}, or Mg^{2+} was found to interfere with the growth of brushite: Only small amounts and very tiny brushite crystals were found in gels containing either F^- or Mg^{2+} and no growth was observed in gels containing CO_3^{2-}. The presence of $P_2O_3^{4-}$, reduced the amount and size of the brushite crystals and caused significant modification in the crystal habit. In the presence of $P_2O_7^{4-}$, aggregates of small needle-shaped crystals growing from a common center (fig. 1d) were found. [Brushite crystals incrusting bones dug from a 900-year old grave[2]) showed similar aggregates.] The presence of strontium caused an even more dramatic effect on the crystal habit. The triangular brushite crystals grew end to end forming a spiral aggregate (fig. 1e).

These preliminary studies demonstrate that the gel systems offer a convenient tool for growing large crystals and observing effects of various ions and conditions (pH and temperature) on the resulting morphology. It may also be concluded from this work that different calcium phosphates, such as brushite and octacalcium phosphate, can co-exist in the same medium. This is especially significant since different types of calcium phosphates: brushite, octacalcium phosphate, whitlockite, and apatite, are found to co-exist in human dental calculi and in phosphatic urinary calculi (kidney stones).

Acknowledgement

We gratefully acknowledge the helpful discussions with Drs. O. R. Trautz, Z. Blank and E. Klein; the technical cooperation of Dr. L. Reich; and the valuable photographic assistance of G. M. Domingo. The support of USPHS Research Grant DE-00159 is also acknowledged.

References

1) C. Palache, H. Berman and C. Frondel, *Dana's System of Mineralogy*, Vol. 2 (Wiley, New York, 1951) pp. 704–705.
2) H.-D. Mirau, D. R. Trautz and J. Vahl, Deut. Zahnaerzl. Z. **26** (1971) 37.
3) H. E. Schroeder, *Formation and Inhibition of Dental Calculus* (Hans Huber Publishers, Berne, 1969).
4) E. L. Prien and C. Frondel, J. Urol. **57** (1947) 949.
5) R. Z. LeGeros, L. Reich, O. R. Trautz and J. P. LeGeros, in: *Proc. 48th Gen. Meeting Intern. Assoc. Dent. Res.*, March 1970, p. 58.
6) A. J. Saffir and B. Rubin, in: *Proc. 6rd Ann. Scanning Electron Microscopy Symp.*, April 1970.
7) (a) Z. Blank, D. M. Speyer, W. Brenner and Y. Okamoto Nature **216** (1967) 1103;
   (b) Z. Blank, private communication.
8) C. A. Beevers, Acta Cryst. **11** (1958) 273.

Industrial bulk crystallization

N. J. J. HUIGE
H. A. C. THIJSSEN
A. W. NIENOW
P. D. B. BUJAC
J. W. MULLIN
E. G. DENK, Jr.
G. D. BOTSARIS

E. P. K. OTTENS
A. H. JANSE
E. J. DE JONG
H. GARABEDIAN
R. F. STRICKLAND-CONSTABLE
J. GARSIDE
C. GASKA

Journal of Crystal Growth **13/14** (1972) 483–487 © *North-Holland Publishing Co.*

PRODUCTION OF LARGE CRYSTALS BY CONTINUOUS RIPENING IN A STIRRER TANK

N. J. J. HUIGE and H. A. C. THIJSSEN

Laboratory for Physical Technology, Eindhoven University of Technology, Eindhoven, The Netherlands

A new bulk crystallisation process is described in which the supersaturation in the crystalliser is maintained by feeding very small subcritical crystals. Supercritical crystals that are present in the crystalliser grow at the expense of the dissolving subcritical crystals.

The process is described by means of a mathematical model. With this model the influences of crystal concentration, production rate, and size of the feed crystals upon the average product crystal size are calculated. The model predicts an increase of the mean size of the product crystals with a decrease of feed crystal size for small feed crystals. The size of the product crystals also increases with an increase in crystal concentration in the crystalliser.

The process has been studied experimentally for the growth of ice crystals from aqueous sucrose solutions. The theoretically predicted effects of crystal concentration and feed crystal size on product crystal size are confirmed by the experiments.

1. Introduction

The mean size of product crystals from a continuously operated crystallizer increases by decreasing the net nucleation rate, J_{net}. The net nucleation rate is defined as the number of product crystals obtained per unit time. A low net nucleation rate is commonly obtained by fines destruction or by suppressing the gross nucleation rate J_{gross}, which is the rate at which new crystals are formed. J_{gross} can be kept down by preventing high local supersaturations and by minimizing mechanical abrasion of the crystals.

In this paper a new process will be described by which, in contradiction to common practice, a low net nucleation rate is obtained as a result of a very high gross nucleation rate. At a high gross nucleation rate very small crystals are produced. When these small crystals are fed to a well mixed recrystallizer containing a suspension of large crystals, these large crystals will grow at the expense of the small feed crystals which dissolve. This so-called ripening effect is based upon the larger solubility of smaller crystals. The growth rate of the large crystals and the dissolution rate of the small crystals both increase with an increase in size difference between the large and the small crystals.

The new process will be described for the crystallization of ice from sucrose solutions. Since only few experiments at unsteady state conditions are available as

yet, a mathematical model will be used to predict the effects of crystal concentration and size of the feed crystals upon the product size at steady state conditions.

2. Experimental

The apparatus is schematically indicated in fig. 1. An aqueous solution of 35 wt% sucrose is fed continuously to a scraped surface heat exchanger used as a crystallizer. From this crystallizer small dendritic ice crystals with an effective diameter of 10–20 μm are produced. These crystals are fed to a 90 liter adiabatic recrystallizer containing a well-mixed suspension of larger crystals. The recrystallizer is provided with a filter for the recirculation of mother liquor to the entrance of the crystallizer. The mean residence time of the suspension

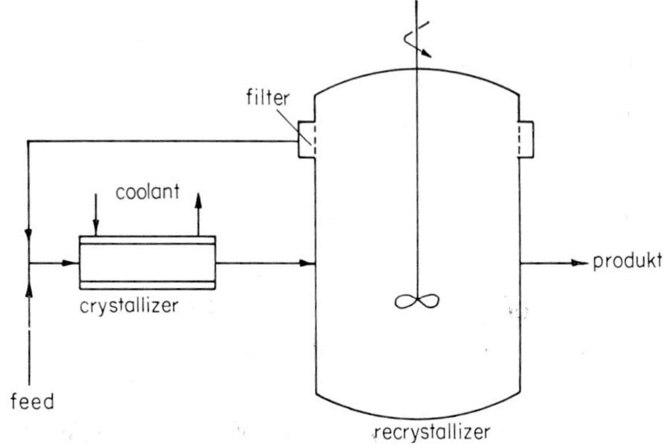

Fig. 1. Experimental set-up.

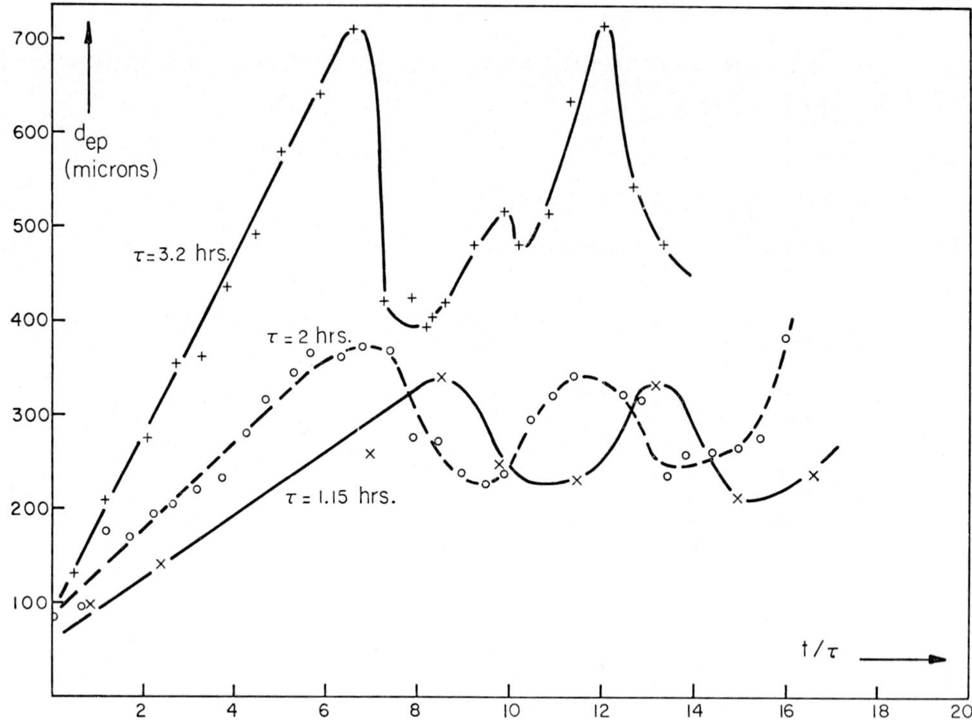

Fig. 2.   Effect of t/τ on effective diameter of product crystals.

in the crystallizer is 5 sec. This short residence time prevents the small nuclei that are formed from growing too large. In all runs the production rate was kept constant at a value of 6 kg ice/hr. The mean residence time τ of product crystals in the recrystallizer was varied at values of 1.15, 2.0, and 3.2 hr. The corresponding weight fractions X of ice in the recrystallizer are 0.069, 0.12, and 0.192. For every run X is constant with time.

The mean size of the product crystals can be characterized by means of an effective diameter d_{ep}. The effective diameter is calculated with the Ergun–Kozeny equation from the permeability and porosity of a bed of the crystals. Crystal size distributions are determined from photographs of the product crystals taken at one hour intervals. The crystals appeared to be nearly spherical.

The effect on d_{ep} of τ and of the time t elapsed after reaching a constant ice concentration is shown in fig. 2. For values of $t/\tau < 6$, fig. 2 shows that d_{ep} increases with τ and with t/τ. A constant value of d_{ep} was not obtained even at values of t/τ as large as 15.

The area averaged crystal size determined from the photographs appeared to be approximately equal to d_{ep}. It was remarkable to observe that before reaching the first maximum in d_{ep} the crystal size distributions were extremely narrow. When d_{ep} starts to decrease a large number of small crystals appear in the product. These small crystals increase in size and new crystals do not appear in the product until the next maximum in d_{ep}.

At a constant value of X the total crystal surface is inversely proportional to d_{ep}. Since d_{ep} increases with t/τ, the total crystal surface A decrases with time. With decreasing A the bulk supercooling ΔT increases until the point is reached where ΔT becomes so large that the corresponding critical diameter is smaller than that of the largest feed crystals. As a result these feed crystals will survive.

It can further be expected that when smaller crystals are fed to the crystallizer the critical supercooling will be reached later. The maximum value of d_{ep} will therefore increase with a decrease in feed crystal diameter. At steady state conditions this might result in a larger diameter of the product crystals.

The quantitative effect of process parameters on the steady state value of d_{ep} will be calculated from a mathematical model. For this model information on nucleation, ageing and growth rate is necessary.

3. Nucleation

Small crystals are obtained from the scraped surface heat exchanger at a high rate of secondary nucleation and a short residence time of the nuclei in the heat exchanger. The high nucleation rate is obtained by applying a high flux of heat withdrawal, and a high rate of mechanical agitation.

The net nucleation rate in the recrystallizer determines the effective product diameter. J_{net} equals the sum of the number of feed crystals per unit time and the gross nucleation rate in the recrystallizer minus the number of crystals that disappear per unit time through ageing.

Since the supercooling in the recrystallizer is very low only nucleation by crystal fragmentation can be expected. From the narrow product size distribution it can be concluded that at the applied stirrer speed of 200 rpm nucleation by this mechanism is negligible.

4. Ageing

A suspension of crystals of various sizes is not stable from a thermodynamical point of view. The Gibbs free energy of the system decreases by a decrease of the total crystal surface. This change toward a thermodynamic equilibrium is called ageing and can occur in several ways[1]:

(1) growth of the larger crystals at the expense of smaller crystals, the so-called Oswald ripening;
(2) agglomeration of small crystals followed by recrystallization;
(3) ageing of single crystals by a change in shape.

The difference in freezing point ΔT_{fd} between a large crystal and a small crystal with a diameter d is given by the Gibbs–Thomson equation:

$$\Delta T_{fd} = \frac{4\sigma T_i^*}{\Delta H \rho_s d}, \tag{1}$$

where σ is the surface free energy, T_i^* is the freezing point of the solution at the interface (°K), ΔH is the heat of fusion, and ρ_s is the specific weight of ice. From eq. (1) it follows that small crystals have a lower equilibrium freezing temperature than large crystals.

In an Oswald ripening process the bulk temperature T_b will be between the equilibrium temperature of the largest and that of the smallest crystals present. The size of a crystal that is in equilibrium at T_b is called the critical size d_{cr} belonging to that tempeature. Crystals with a size $d < d_{cr}$ are subcritical and will melt while crystals with $d > d_{cr}$ will grow.

Photographs of the product crystals show an almost uniform crystal size. From this it may be concluded that all the small feed crystals disappear as a consequence of Oswald ripening in the recrystallizer.

5. Growth rate

In the recrystallizer the solute molecules and the heat of fusion have to be transported away from the interface of the growing crystals. The heat and mass transfer coefficients can be calculated from the power input ε of the impeller per unit mass of suspension, the crystal diameter d, and the appropriate physical constants. From the experimental data reported by Brian and Sherwood[2], the following correlations have been obtained:

$$\mathrm{Nu} = 2 + 1.3 \left(\frac{\varepsilon d^4}{v^3}\right)^{0.17} \mathrm{Pr}^{0.25} \quad \text{for} \quad \left(\frac{\varepsilon d^4}{v^3}\right) < 10^6, \tag{2a}$$

and

$$\mathrm{Nu} = 2 + 0.4 \left(\frac{\varepsilon d^4}{v^3}\right)^{0.243} \mathrm{Pr}^{0.25} \quad \text{for} \quad \left(\frac{\varepsilon d^4}{v^3}\right) > 10^6, \tag{2b}$$

where Nu is the Nusselt number for heat or mass transfer, Pr is the Prandtl number for heat transfer or the Schmidt number for mass transfer and v is the kinematic viscosity. A heat balance yields:

$$v = h(T_i - T_b)/\Delta H \rho_s, \tag{3}$$

in which T_i is the interface temperature, v is the linear growth rate (cm/sec), and h is the heat transfer coefficient. The inbuilding kinetics can be described by the correlation[3]:

$$v = 0.27(T_{ed} - T_i)^{1.55}. \tag{4}$$

T_{ed} is here the equilibrium temperature of a crystal with size d. Since no solute molecules are built into the crystals a solute mass balance yields:

$$v C_i \rho_s / \rho = k(C_i - C_b), \tag{5}$$

where C_i is the solute concentration at the interface, C_b is the solute concentration in the bulk, ρ is the specific weight of the solution and k is the mass transfer coefficient. For known values d, ε, the equilibrium data, and the physical constants v can be calculated iteratively from eqs. (1) through (5).

6. Mathematical model

For steady state conditions a population balance over the recrystallizer yields:

$$\frac{\partial(nv)}{\partial d} = -\frac{n}{\tau} + \frac{n_f}{\tau},\tag{6}$$

where d is the crystal dimension and n is the population density in the recrystallizer defined as

$$n = \lim_{\Delta d \to 0} \frac{\Delta N}{\Delta d},$$

in which ΔN is the number of crystals of sizes between d and $d+\Delta d$. In eq. (6), n_f is the population density of the feed crystals.

For a given size distribution of feed crystals $n_f = n_f(d)$ and a known growth function $v = v(d, \Delta T(d))$ the size distribution of the product crystals $n = n(d)$ can be calculated numerically from eq. (6) and the boundary condition: $n_f(d_{cr}) = n(d_{cr})$. In this equation $n_f(d_{cr})$ is the density of feed crystals which at the bulk temperature of the recrystallizer have the critical size d_{cr}. Since both v and $n_f(d_{cr})$ are dependent on the supercooling an iterative procedure is necessary. In this procedure a value of the bulk temperature, which lies between the equilibrium temperatures of the largest and the smallest feed crystals, is estimated. At this value of the bulk temperature v can be calculated as a function of d. From the given feed size distribution and Tb, $n_f(d_{cr})$ can be determined. The iteration criterion follows from a mass balance over the recrystallizer.

For the size distribution of the feed crystals a Poisson distribution corresponding to 4 ideal mixers in series is chosen. Calculations were made for production rates of 6 and 12 kg ice/h and for average residence times of 1, 2, and 4 h. The influence of the effective diameter of the feed crystals d_{ef} upon the effective product diameter d_{ep} is shown in fig. 3 with τ as parameter. For effective feed crystal diameters <50 microns and a mean residence time of the crystals in the recrystallizer >1 hour d_{ep} increases with decreasing d_{ef}. This favorable trend has to be attributed to more effective ripening, which causes the absolute number of feed crystals that survive to decrease with decreasing d_{ef}. For values

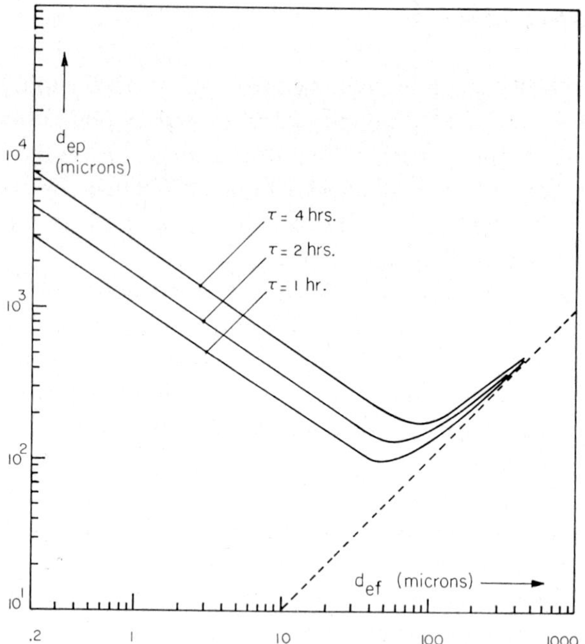

Fig. 3. Calculated effect of the diameter of feed crystals upon the diameter of product crystals.

of $d_{ef} > 100\ \mu m$ the driving force for the ripening process becomes very small and d_{ep} will therefore approach d_{ef}.

From microscopic observations the dendritic ice crystals in the feed to the recrystallizer were estimated to be approximately $40\ \mu m$ in length and $8\ \mu m$ in diameter. The smallest dimension is probably the most important one in determining the rate of dissolution.

TABLE 1

τ	d_{ep} (μm)	
	$d_{ef} = 10\ \mu m$	$d_{ef} = 20\ \mu m$
1.15	260	180
2	370	240
3.2	520	315

To get a general impression, however, values of d_{ef} of 10 and 20 μm will be taken for comparing the theoretical and experimental results. The theoretically predicted values of d_{ep} are presented in table 1. These theoretical values of d_{ep} are in reasonable agreement with the experimental data from fig. 2.

The increase of d_{ep} with τ can be explained in terms of a linear increase of X with τ at constant production rate and constant volume of the recrystallizer: with in-

X – 1

creasing X the bulk temperature increases, so that a larger fraction of feed crystals becomes subcritical, resulting in larger product crystals. From fig. 2 it can be seen that even though stationary conditions were not attained during the experiments, d_{ep} also increases with increasing τ.

Increasing the production rate at constant τ did not have any effect on the calculated value of d_{ep} since the ratio of X and the number of feed crystals remains constant.

References

1) A. E. Nielsen, *Kinetics of Precipitation* (Pergamon, Oxford, 1964).
2) P. L. T. Brian, H. B. Hales and T. K. Sherwood, A. I. Ch. E. J. **15** (1969) 727.
3) N. J. J. Huige and H. A. C. Thijssen, in: *Proc. Symp. Industrial Crystallization*, London, April 1969, p. 69.

SLIP VELOCITIES IN AGITATED VESSEL CRYSTALLISERS

A. W. NIENOW, P. D. B. BUJAC* and J. W. MULLIN

Department of Chemical Engineering, University College London, Torrington Place, London W.C. 1, England

Growth rates of ammonium alum crystals have been measured at different fluid velocities in a single crystal growth cell and a fluidised bed, and at different stirrer speeds in a turbine agitated vessel. To compare the latter with the former, it is necessary to know the particle–fluid slip velocity as a function of the impeller speed, and the method proposed by Hughmark, using Kolmogoroff's theory of isotopic turbulence to make this prediction, is found to be unsatisfactory in the present case.

An alternative method is now proposed in which agitated-vessel crystal dissolution data are used to predict the slip velocity by considering it to be related to the particle terminal velocity, but enhanced by the agitation. By this means, all the crystal growth data are correlated successfully with velocity, although growth rates in the agitated vessel are always somewhat higher. It is postulated that this is due to the effect of turbulence on the adsorption layer.

1. Introduction

Mullin and Garside[1,2] showed how the face growth rates of single crystals of potash alum could be used to predict overall crystal growth rates in fluidised bed crystallisers. Briefly, a single crystal was fixed on a wire in an upward flowing stream of supersaturated solution (called a growth cell) and growth rates were measured for a range of fluid velocities[1]. They found that under the same degree of supersaturation, crystals in a fluidised bed[2] would grow at the rate predicted from the face growth rates if the crystal–liquid relative velocity was the same as the absolute velocity in the single crystal growth cell. Methods have recently been put forward for predicting this relative velocity for a wide range of crystal sizes, densities and bed voidages[3]. A similar study has been made with ammonium alum[4] and the work was extended to include growth rate measurements in an agitated vessel.

2. Apparatus and technique

The work on fluidised beds and single crystal growth measurements was carried out in equipment, and using a technique, basically identical to that of Mullin and Garside[1,2].

The agitated vessel geometry used was suggested by the work of Nienow[5,6] on dissolution in agitated vessels. In particular, a large disc turbine impeller with a small clearance above the base was used (detailed di-

mensions are shown in fig. 1) at a rotational speed some 10% above that predicted[6] as being necessary to suspend the crystals. By this means, a slow rotational speed could be used which is desirable in agitated vessel crystallisers if secondary nucleation is to be prevented, and indeed it did not occur. A second speed, fixed as the highest obtainable before air became entrained in the system and approximately equal to twice the first, was also used. Again no nucleation occurred.

A known weight of seed crystals was added to the agitated vessel and growth rates were obtained by rapidly removing the crystals after a given time by means of a snugly fitting stainless-steel mesh liner of the type previously used in dissolution studies[5]. Further growth of the crystals was thus prevented and they could be dried and weighed. Runs were carried out at

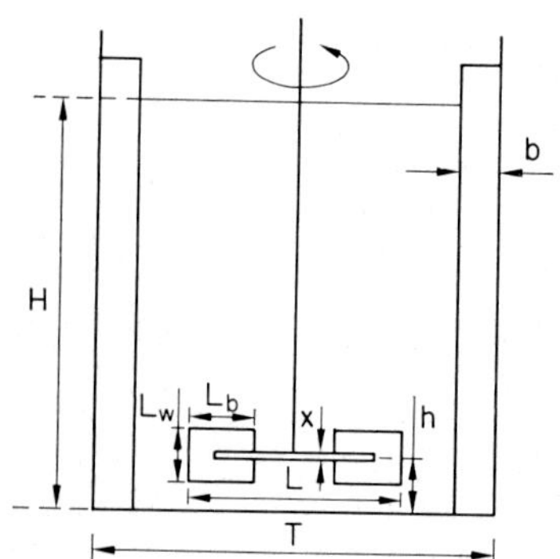

Vessel: H=T=0·14m h/H=1/7; L/T=1/2
4 baffles; b=0·1 T

Impeller: 6 blades; L_w/L = 1/5; L_b/L = 1/4
x/L_w = 1/3

Fig. 1. Agitated vessel geometry.

---

* Present address: I.C.I. Mond Division, Runcorn, Cheshire, England.

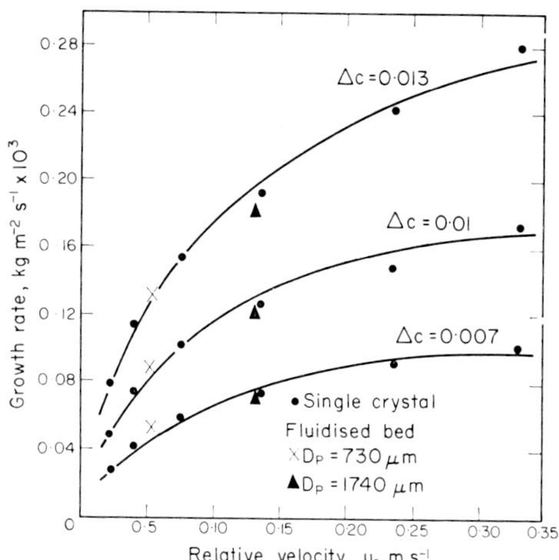

Fig. 2. Correlation of single crystal growth cell and fluidised bed growth rates for ammonium alum at 32 °C.

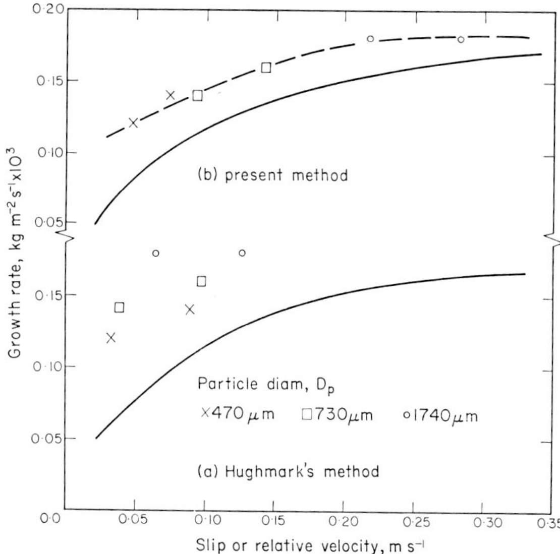

Fig. 3. Comparison of single crystal growth cell, fluidised bed and agitated vessel growth rates for ammonium alum at 32 °C. Lower: using slip velocities calculated by the Hughmark–Kolmogoroff method. Upper: using slip velocities calculated by the present work. (The solid line is the best line from the growth cell and fluidised bed data at $\Delta C = 0.01$ kg of hydrate per kg of water.)

32 °C for different degrees of supersaturation with 420 µm, 650 µm, and 1540 µm mean size seeds. Full details of the equipment, experimental technique and calculation of growth rates are given elsewhere[4]).

3. Results and discussion

Fig. 2 is a plot of growth rate against relative velocity for ammonium alum crystals grown in a single crystal growth cell and in a fluidised bed crystalliser. The results confirm the work of Mullin and Garside on potash alum in that the growth rates are similar for the two different systems at the same relative velocities. In both cases, of course, the relative velocity between the growing crystals and the supersaturated liquid can be easily and fairly accurately measured. However, in the complex flow system occurring in an agitated vessel, no single relative velocity exists, only some statistical mean; and even this cannot be measured. Therefore, unless some way of estimating this velocity can be found, agitated vessel growth rates cannot be estimated from single crystal growth cell data.

Hughmark[7]) suggested that this relative or (as it is generally called in agitated vessel work) slip velocity could be determined as follows. Firstly he assumed that the mass transfer equations of the Froessling type for fixed particles in a flow field would be applicable and then, by substituting measured agitated vessel mass transfer coefficients[8,9]) into these equations, the slip velocity could be calculated. Secondly he derived equations for the slip velocities using the concepts of Kol-

mogoroff's theory of isotropic turbulence and a comparison of the values predicted by the two methods showed good agreement. Hughmark's equation applicable to the present work was*

$$u_s = 0.055 \left(\frac{D_p}{v}\right)^{\frac{1}{3}} [L(P/G)]^{\frac{4}{9}} \left(\frac{v(H_2O, 25\ ^\circ C)}{v}\right)^{\frac{4}{9}} \left(\frac{\Delta\rho}{\rho}\right)^{\frac{2}{3}},$$

$$Re_p > 2. \tag{1}$$

The physical and geometrical data relevant to the present agitated vessel crystallisation study are shown at the top of table 1 together with the crystal size and the stirrer speed in the first and second columns respectively. Substitution of these data into eq. (1) enables slip velocities to be determined and these are shown in table 1 in the fourth column. The third column gives the experimental growth rates for a driving force of 0.01 kg of hydrate/kg of H_2O.

Fig. 3a is a plot of growth rate against slip velocity as predicted by Hughmark's equation; also shown is the equivalent line from fig. 2 for the fluidised bed and growth cell work (growth rates were also obtained at the other driving forces shown in fig. 2, but are omitted from fig. 3 for the sake of clarity). It is clear, therefore,

* See Nomenclature, at the end of this paper.

TABLE 1

Experimental growth rates and predicted slip velocities for ammonium alum crystals at 32 °C

Data: $L = 0.07$ m; $v = 1.1 \times 10^{-6}$ m^2 s^{-1}; $\rho = 1.06 \times 10^3$ kg m^{-3}; $\Delta\rho = 0.58 \times 10^3$ kg m^{-3}; $g = 9.81$ m s^{-2}; $N_p = 5$[10,11]); $G = 2.27$ kg; $v(\mathrm{H_2O}, 25\,°\mathrm{C}) = 0.895 \times 10^{-6}$ m^2 s^{-1}

Experimental data			Hughmark-Kolmogoroff theory	Present theory		
Mean particle size D_p (μm)	Agitator speed N (rev. s^{-1})	Crystal growth rate* (kg m^{-2} s^{-1})	Slip velocity (ref. 7) u_s (m s$^{-1} \times 10^2$)	Agitator suspension speed (ref. 6) N_{js} (rev s^{-1})	Enhancement factor (ref. 5) E	Slip velocity u_s (m s$^{-1} \times 10^2$)
470	3.17	0.12	3.05	2.83	1.2	4.70
	6.67	0.14	8.25			7.33
730	3.33	0.14	3.82	3.0	1.24	9.40
	6.67	0.16	9.56			14.2
1740	4.00	0.18	6.5	3.67	1.3	21.8
	6.67	0.18	12.7			28.2

* Growth rate for a driving force of $\Delta C = 0.01$ kg of hydrate per kg of water.

that agitated vessel growth rate data cannot be correlated with fluidised bed and growth cell data using the Hughmark–Kolmogoroff method.

A slightly different approach can be developed from the work of Harriott[9]) and Nienow[5]) on agitated vessel dissolution. Here it was suggested that when dissolving crystals were just suspended in an agitated vessel[6]), the mass transfer coefficient, k_{js}, should not be less than that, k_t, which would result from crystals falling at their terminal velocity under the influence of gravity. Furthermore, k_t could be determined by substituting the terminal velocity of the particles into a modified Froessling equation of the form

$$\mathrm{Sh_p} = 2 + \Phi \, (\mathrm{Re_p})^{\frac{1}{2}} (\mathrm{Sc})^{\frac{1}{3}} . \tag{2}$$

Experiments showed this hypothesis to be valid and it was found that

$$k_{js} = E k_t , \tag{3}$$

where E was a function of particle size and was considered[5]) to be a measure of the enhancement of the mass transfer coefficient due to the turbulence in the system.

Of course, the increase in k_{js} above k_t could also be considered to be due to the relative magnitudes of the slip and terminal velocities; in which case, E would be a measure of the actual slip velocity u_{js} as compared to the terminal velocity, u_t. From the Froessling equation, since the constant 2 has a small effect, for one solid–liquid pair,

$$k_t \propto u_t^{\frac{1}{2}} , \tag{4}$$

and

$$k_{js} \propto u_{js}^{\frac{1}{2}} , \tag{5}$$

where the constants of proportionality in relationships (4) and (5) are identical. Therefore, substituting into eq. (3)

$$u_{js} = E^2 u_t . \tag{6}$$

Developing this idea further, any increase in mass transfer coefficient above k_{js} due to increases in impeller speed could also be considered to be due to further increases in the slip velocity, i.e.

$$k_s \propto (u_s)^{0.5} , \tag{7}$$

where k_s is the mass transfer coefficient associated with any arbitrary slip velocity u_s provided the particle is suspended.

For the system geometry and solid–liquid pair used in the present work, Nienow found[5]) that

$$k_s \propto N^{0.4} , \tag{8}$$

where N was the stirrer speed causing particle suspension, and combining relationships (7) and (8) gives

$$k_s \propto (u_s)^{0.5} \propto (N)^{0.4} . \tag{9}$$

One limit of this relationship is when the particles are just suspended, i.e.

$$k_{js} \propto (u_{js})^{0.5} \propto (N_{js})^{0.4} . \tag{10}$$

Therefore from relationships (9) and (10):

$$u_s/u_{js} = (N/N_{js})^{0.8}. \tag{11}$$

In order to test the applicability of the above hypothesis to agitated vessel crystallisers, terminal velocities are determined from the equations[5])

$$u_t = \frac{0.153 \, g^{0.71} D_p^{1.14} \Delta\rho^{0.71}}{\rho^{0.29} \mu^{0.43}} \quad \text{for} \quad D_p < 500 \, \mu m, \tag{12}$$

and

$$u_t = \left(\frac{4 D_p g \, \Delta\rho}{3\rho}\right)^{\frac{1}{2}} \quad \text{for} \quad D_p > 1500 \, \mu m. \tag{13}$$

For crystals of intermediate size, an interpolated mean value is used. Relationships (6) and (11) can be then used to determine slip velocities at the impeller speeds employed.

Table 1 shows the calculated impeller speed[6]) which just suspends the particles in the fifth column and the experimental enhancement factor[5]) for the relevant particle size in the sixth column. Finally the seventh column gives the slip velocity calculated as outlined above; these values are used in fig. 3b which in all other respects is identical to fig. 3a.

It can be seen from fig. 3b that there is now a distinctly similar trend in the growth rate–slip velocity plot for the agitated vessel and the growth rate–relative velocity plot for the single crystal growth cell and fluidised bed. However, at all velocities, growth in the agitated vessel is faster than in the other systems. Though perhaps surprising in itself, the enhanced growth rate in the agitated vessel is consistent with the experimental results of this study as explained below.

It is considered that in the growth cell and fluidised bed, the growth of potash alum[1]) and ammonium alum[4]) crystals from supersaturated solutions is at least partially bulk-diffusion controlled at low relative velocities, becoming almost entirely surface integration controlled at high relative velocities. Hence the shape of the growth rate–relative velocity plot of fig. 2. When surface integration controls, the growth should be independent of velocity and the maximum obtainable for any one driving force.

The plot of growth rate against slip velocity (fig. 3b) for agitated vessel growth suggests the same two mechanisms are present, and for the largest particles where the slip velocity is highest, the growth rate again approaches a maximum, independent of velocity. However, the agitated vessel growth rates are always some 10 to 15% greater than those obtained in the other two systems. This fact suggests that the actual surface integration mechanism is enhanced in agitated vessels as compared to the other system. One possible explanation may lie in the effect of turbulence on the adsorption layer and the arrangement of growth units on the crystal face.

4. Conclusions

When crystal–fluid slip velocities are calculated by the method proposed by Hughmark, the growth rates of ammonium alum crystals in agitated vessels cannot be correlated with those in fluidised beds and single crystal growth cells. On the other hand, reasonably good agreement is obtained using the method outlined in this paper, developed from Nienow's work on dissolution. However, in general the growth rates obtained in the agitated vessel exceed those in the growth cell and fluidised bed, and it is considered that this might be due to the system turbulence increasing the rate of the surface integration step.

Acknowledgements

One of the authors (P. D. B. Bujac) gratefully acknowledges the award of an SRC postgraduate studentship.

Nomenclature

ΔC concentration difference between supersaturated and saturated solution (kg hydrate/kg water)

D_p particle size (m)

E enhancement factor (dimensionless)

G mass of fluid in vessel (kg)

g gravitational acceleration (m s^{-2})

k_s mass transfer coefficient when a particle is freely suspended at stirrer speed $N > N_{js}$ (arbitrary dimensions)

k_{js} mass transfer coefficient when a particle is just suspended† (arbitrary dimensions)

k_t mass transfer coefficient when a particle is at its terminal velocity (arbitrary dimensions)

L impeller size (m)

N impeller speed (rev s^{-1})

N_p power number defined by $N_p = P/\rho N^3 L^5$ (dimensionless)

N_{js} impeller speed to just suspend particle[†] (rev s^{-1})

P power imput (W)

Re_p particle Reynolds number (dimensionless)

Sc Schmidt number (dimensionless)

Sh_p particle Sherwood number (dimensionless)

u_r relative velocity between crystal and surrounding fluid (m s^{-1})

u_s slip velocity of crystal at an impeller speed N $(N > N_{js})$ (m s^{-1})

u_{js} slip velocity of crystal when just suspended (m s^{-1})

u_t terminal velocity of crystal (m s^{-1})

μ viscosity (kg m^{-1} s^{-1})

v kinematic viscosity (m^2 s^{-1})

$v(H_2O, 25\,°C)$ kinematic viscosity of water at 25 °C (m^2 s^{-1})

ρ density of liquid (kg m^{-3})

$\Delta\rho$ density difference between solid and liquid (kg m^{-3})

Φ constant in the Froessling equation (dimensionless)

L_b, L_w, h, T, H, b, x system dimensions, see fig. 1 (m)

[†] For a comlete description of the condition "just suspended". see ref. 6.

Reference

1) J. W. Mullin and J. Garside, Trans. Inst. Chem. Eng. **45** (1967) T285.

2) J. W. Mullin and J. Garside, Trans. Inst. Chem. Eng. **45** (1967) T291.

3) J. W. Mullin and J. Garside, Brit. Chem. Eng. **15** (1970) 773.

4) P. D. B. Bujac, Ph. D. Thesis, University of London, 1969.

5) A. W. Nienow, Can. J. Chem. Eng. **47** (1969) 248.

6) A. W. Nienow, Chem. Eng. Sci. **23** (1968) 1453.

7) G. A. Hughmark, Chem. Eng. Sci. **24** (1969) 291.

8) J. J. Barker and R. E. Treybal, Am. Inst. Chem. Eng. J. **6** (1960) 28.

9) P. Harriot, Am. Inst. Chem. Eng. J. **8** (1962) 93.

10) R. L. Bates, P. L. Fondy and R. R. Corpstein, Ind. Eng. Chem. (Proc. Des. and Dev.) **2** (1963) 318.

11) A. W. Nienow and D. Miles, Ind. Eng. Chem. (Proc. Des. and Dev.) **10** (1971) 41.

Journal of Crystal Growth **13/14** (1972) 493–499 © *North-Holland Publishing Co.*

FUNDAMENTAL STUDIES IN SECONDARY NUCLEATION FROM SOLUTION

EDWARD G. DENK, Jr.* and GREGORY D. BOTSARIS

Department of Chemical Engineering, Tufts University, Medford, Massachusetts 02155, U.S.A.

The nucleation behavior of seeded sodium chlorate solutions was studied experimentally. Because sodium chlorate crystallizes in two easily distinguished enantiomorphic forms, the origin of the secondary nuclei (whether the seed crystal or the solution) could be easily identified.

This work shows that secondary nucleation does not involve a singular mechanism but that different kinds of secondary nucleation can take place under different conditions of supersaturation, liquid velocity, and impurity concentration.

It was found that under certain conditions the resulting nuclei were all of the same enantiomorphic form as the seed. It was then concluded that these nuclei resulted from the growth and detachment of surface irregularities such as dendrites from the seed crystal.

In two other ranges of conditions, however, secondary nucleation produced both left and right handed crystals. In one case it was shown that the nucleation could have taken place in an impurity concentration gradient resulting from the rapid incorporation of dissolved impurities in the growing seed crystal. To explain the second case, a different mechanism (largely speculative) was proposed. This mechanism hypothesized that secondary nucleation could occur because of changes in the liquid layer surrounding the crystal. These changes in the liquid layer could be the result of an ordering of the solvent water molecules at the crystal–solution interface.

1. Introduction

Recently there has been considerable interest in the phenomenon of secondary nucleation – that is, nucleation which occurs in the presence of seed crystals of the material being crystallized. Strictly speaking the secondary nuclei do not have to come directly from the seed crystals, but the presence of the seeds is essential to the nucleation process because when no crystals are present, no nucleation occurs. Most of this recent interest has come about because it now appears that the nucleations which occur in the suspensions of crystals encountered in industrial crystallizers involve secondary nucleation mechanisms.

2. Theories of secondary nucleation

Strickland-Constable[13]) and Botsaris, Denk, Ersan et al.[2]) have outlined some of the ways in which secondary nucleation could occur.

In a recent paper Botsaris, Denk and Chua[1]) showed that secondary nucleation could occur in the boundary layer surrounding a seed crystal as the result of an impurity concentration gradient resulting from the incorporation of impurity into the growing crystal. Their

mechanism was based on the observation that in many crystallizing systems trace amounts of certain dissolved impurities can drastically suppress the rate of nucleation of crystals or even prevent their formation. Often the same impurities that were preventing nucleation were readily incorporated into the growing crystals[3,4]). In such a system a situation like that shown in fig. 1 could take place.

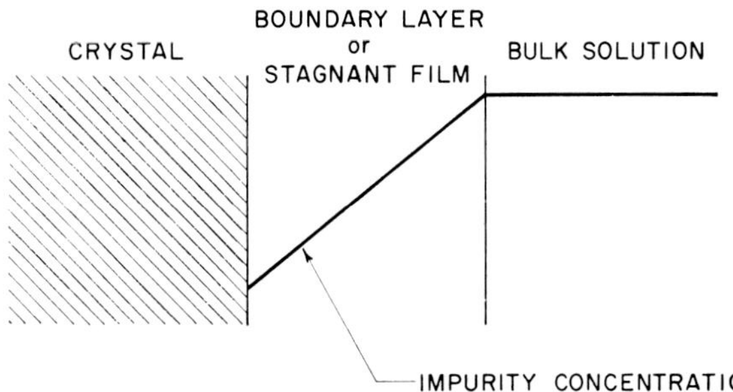

Fig. 1. Concentration gradient of impurity in a boundary layer.

As shown in this figure, the uptake of impurity by the growing seed crystal may lead to an impurity concentration gradient in the boundary layer surrounding the seed crystal. Because of this gradient, the impurity concentration near the crystal surface may be considerably lower than the bulk concentration. The Impurity Concentration Gradient (ICG) Nucleation Model says

*Present address: Polaroid Corporation, Emulsion Development Laboratory, Waltham, Massachusetts 02154, U.S.A.

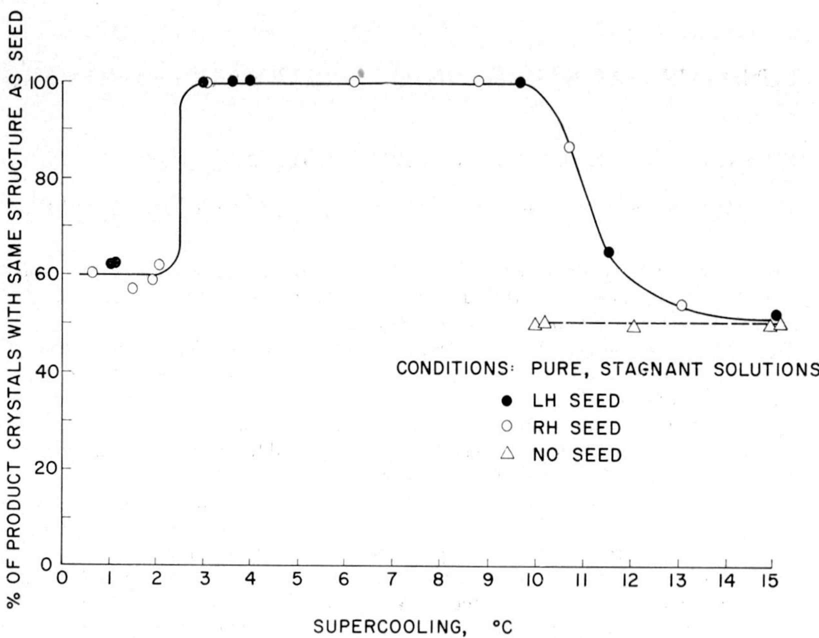

Fig. 2. Type of product crystals obtained in pure, stagnant solutions.

that if the conditions were such that the only reason that spontaneous nucleation did not occur in the bulk solution was because of the high impurity concentration that existed there, and if the impurity concentration near the crystal surface was reduced to a low enough level by the mechanism outlined in fig. 1, nucleation could occur in the boundary layer near the crystal surface.

3. Experimental approach

Much could be learned about the process of secondary nucleation if the origin of the secondary nuclei could be determined. If the source of the nuclei was known it would be possible to say, for example, whether these nuclei were created by breaking dendrites from a seed crystal or by nucleation that took place in the boundary layer because of an impurity concentration gradient.

With this in mind, experiments were conducted in which pretreated sodium chlorate seed crystals were added to aqueous $NaClO_3$ solutions under different conditions of supersaturation, liquid velocity, and impurity concentration. Sodium chlorate was chosen because it crystallizes in two enantiomorphic forms which can be easily distinguished by the direction in which they rotate polarized light[5]). With such a system the origin of the secondary nuclei (whether the seed crystal or the solution) could be identified. If, for example, a

left handed crystal was introduced into a supersaturated solution and both left and right handed crystals were later observed among the product crystals, it could be concluded that the secondary nuclei did not all come from the seed crystal.

The secondary nucleation experiments were not all conducted in the same apparatus, but instead the apparatus was changed as different types of secondary nucleation were studied. These experiments and the apparatus in which they were conducted are described below. (All the experiments were conducted at 30 °C.)

4. Results

4.1. PURE, STAGNANT SOLUTIONS

The experiments with pure, stagnant solutions were carried out in sealed, 2 liter Erlenmeyer flasks. To conduct an experiment, a mounted seed crystal was washed in double distilled water and then immediately inserted into the solution. While it was being added to the system, the seed crystal was rotated slightly to dissipate whatever water film might be adhering to its surfaces. The seeds were kept suspended in the solutions for a period of one week. At the end of this time, the seed crystal was removed from the solution, and the product crystals (if there were any) were then separated from the mother liquor by filtration, collected, and analyzed.

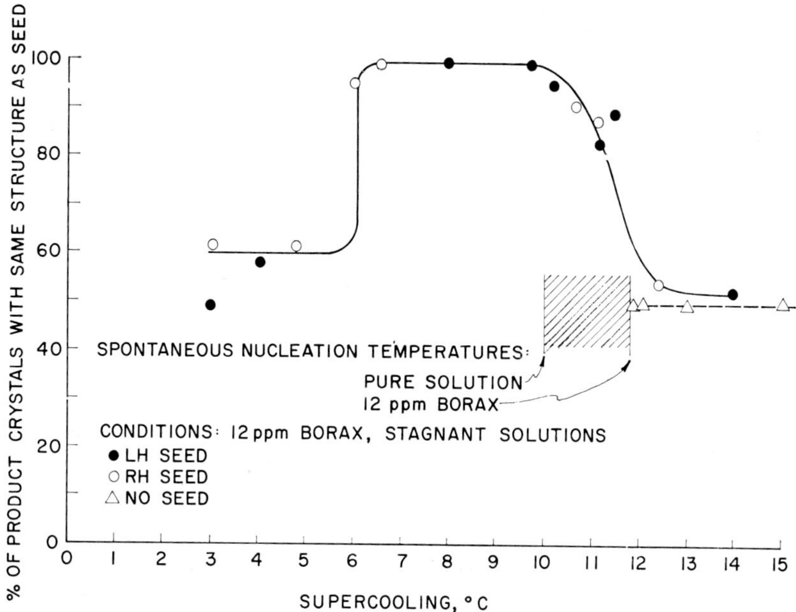

Fig. 3. Type of product crystals obtained in impure, stagnant solutions.

The results of these experiments are presented in fig. 2. This figure shows the type of product crystals that were obtained at different supercoolings. Note that there appear to be regions of supercooling in which different kinds of secondary nucleation took place. As shown in fig. 2, secondary nucleation did not take place until a certain minimum supersaturation was attained. Beyond this point secondary nucleation commenced and the first crop of product crystals obtained was such that 60% of the crystals had the same structure as the parent seed. At higher supersaturations, the mechanism of secondary nucleation (and growth) appeared to change and only crystals with the same structure as the original seed were ever observed in the product. This kind of nucleation could be associated with the appearance of dendrites and other visible growth irregularities on the surface of the seed crystal. At still higher supersaturations, spontaneous nucleation began to take place and the solutions nucleated whether or not a seed crystal was present. When spontaneous nucleation did occur (i.e. when the solutions were so highly supersaturated that they nucleated in the absence of a seed crystal) equal numbers of left and right handed crystals were always obtained.

The number of product crystals obtained in these experiments exhibited a relatively low order dependence (approximately first order) on the supersaturation[7]).

4.2. IMPURE, STAGNANT SOLUTIONS

A number of experiments were also carried out in solutions to which borax impurity had been added. The concentration of the dissolved impurity was equivalent to 12 moles of borax per million moles of sodium chlorate. The experiments were of interest because they offered a test of the impurity concentration gradient nucleation model discussed earlier. Except for the fact that impurity was present, the experiments used and procedure followed were exactly the same as those employed in the experiments with pure solutions.

With one major exception, the kind of secondary nucleation that took place in the impure solutions was qualitatively the same as that observed in the pure solutions. At very low supersaturations, no nucleation occurred, at low supersaturations the nature of the secondary nucleation process was such that 60% of the product crystals had the same structure as the seed, at moderate supersaturations only product crystals with the same structure as the seed were ever obtained, and at very high supersaturations a spontaneous nucleation yielding equal numbers of left and right handed crystals took place (fig. 3).

The number of product crystals obtained in these experiments exhibited the same kind of low order dependence on the supersaturation that was observed in

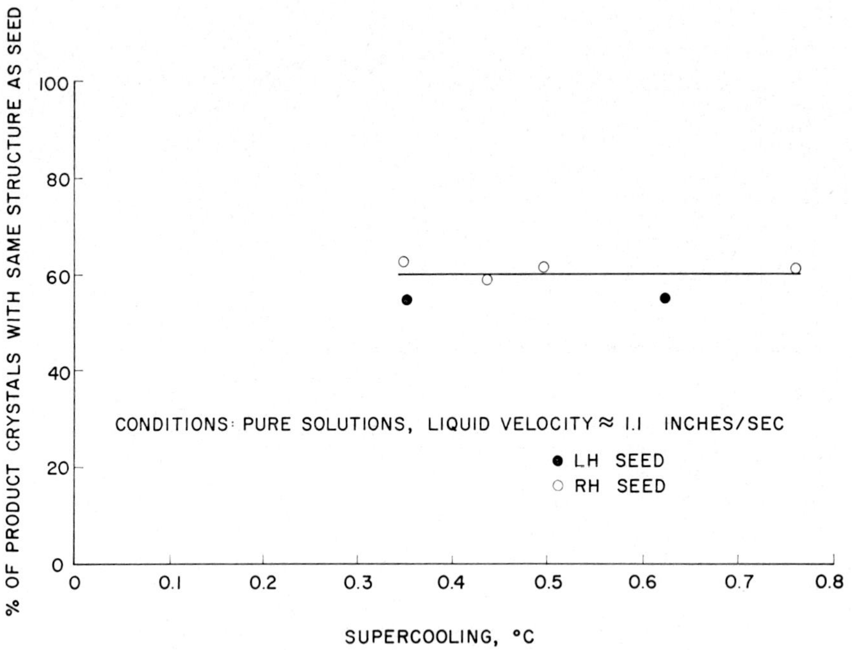

Fig. 4. Type of product crystals obtained in the flow system.

the pure solutions. However, far fewer crystals were obtained with the borax impurity present[7]).

The most important point about these experiments was that the borax impurity increased the spontaneous nucleation temperature to 11.8 °C supercooling. This meant that in the region from 10.0 °C (the spontaneous nucleation temperature for a pure solution) to 11.8 °C supercooling the only thing preventing spontaneous nucleation was the presence of the impurity. This, of course, had already been established as one of the requirements for impurity concentration gradient nucleation[1]). Therefore, if ICG nucleation was going to occur at all, it would have to take place in this region. As shown in fig. 3 a new mechanism of secondary nucleation, one yielding both left and right handed crystals did appear in this region.

4.3. FLOW SYSTEMS

A few experiments were conducted in which a stream of supersaturated sodium chlorate solution was slowly pumped past a stationary seed crystal. The advantage of a flow system such as this was that the supersaturation seen by the seed crystal remained constant throughout the course of a run. Even when secondary nucleation occurred and the supersaturation decreased, this decrease occurred downstream from the seed crystal.

The procedure followed in the flow experiments was simply to pump the solution (under nitrogen pressure) past the seed crystal, seal up the receiving flasks, and age the contents of the flasks until any product crystals that may have formed were large enough to analyze.

The results of these flow experiments are given in fig. 4. Note that these experiments were all conducted at low supersaturations. Note also that there was a certain minimum supersaturation below which nucleation did not take place and that once this supersaturation was reached and nucleation commenced, 60% of the product crystals that were obtained had the same structure as the seed crystal. The number of product crystals obtained in these experiments again exhibited a low order dependence on the supersaturation[7]).

Since the pumping apparatus did not employ any recycle of the solution, the flow experiments could not be conducted at high liquid velocities because the volume of solution available was quickly exhausted under these conditions.

4.4. STIRRED SOLUTIONS

To learn more about the kind of secondary nucleation that might take place at high liquid velocities, experiments were conducted in stirred solutions. The technique employed in these experiments was to in-

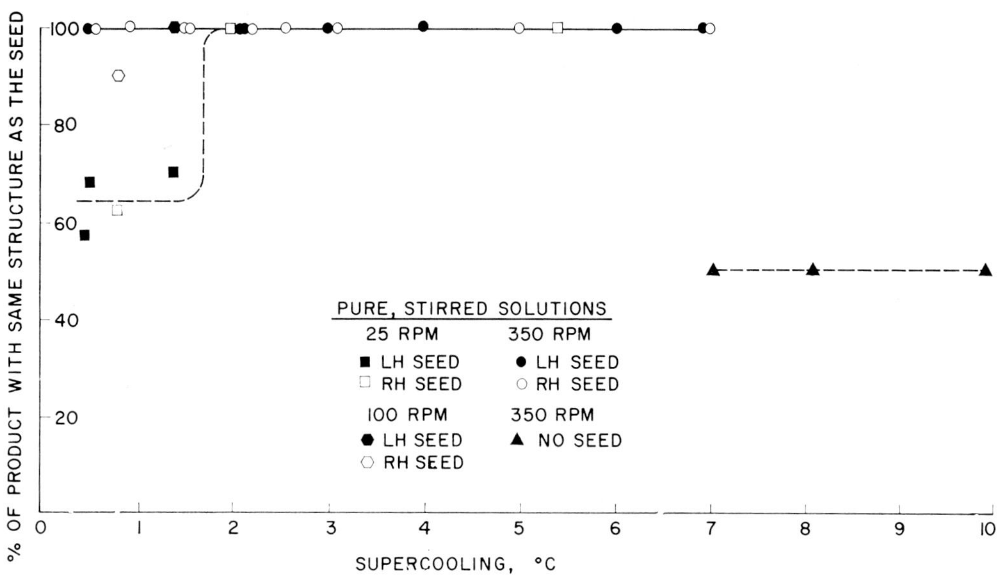

Fig. 5. Type of product crystals obtained in stirred solutions.

troduce a mounted and washed seed crystal into the solution, wait until product crystals were observed, and then shut off the stirrer and remove the seed. The product crystals were then aged until they became large enough to analyze.

At low stirring rates, the results were similar to those obtained in stagnant solutions (fig. 5). At very low supersaturations, no nucleation took place until a certain minimum supersaturation was attained. The first product crystals obtained (at low supersaturations) were such that 60% of the crystals had the same structure as the seed crystal. At higher supersaturations, the mechanism of secondary nucleation changed and only product crystals with the same structure as the parent seed were ever observed.

At high stirring rates there was also a region at very low supersaturations where secondary nucleation did not take place. However, when secondary nucleation did occur, only product crystals having the same structure as the seed were ever obtained regardless of the supersaturation (fig. 5). It should be noted that in all of the stirred solution experiments, no visible surface irregularities ever developed on the growing seed crystals.

4.5. BLANK SOLUTIONS

It was mentioned earlier that a number of experiments were conducted with blank solutions. Some of these solutions were true blanks (i.e. they contained no

seed crystal), while others contained a glass seed crystal (a piece of glass cut to the same size and shape as a normal sodium chloride seed crystal). These blank experiments were important because they showed that whatever nucleation took place in the seeded solutions was true secondary nucleation and not some kind of random spontaneous nucleation. These experiments showed that the solutions could be aged at supercoolings just 0.2 °C less than the spontaneous nucleation temperature for very long periods of time (weeks or more) without undergoing spontaneous nucleation. Because of this, it was concluded that the nuclcations observed in the seeded solutions were the result of secondary nucleation and not some kind of random, long term spontaneous nucleation.

5. Discussion of results

Cumulative these experiments with optically active seed crystals showed that secondary nucleation does not involve a singular mechanism but that different kinds of secondary nucleation can occur under different conditions of supersaturation, liquid velocity, and impurity concentration. The various secondary nucleation mechanisms observed in this study include:

5.1. IMPURITY CONCENTRATION GRADIENT (ICG) NUCLEATION

It was concluded that ICG nucleation took place in

the experiments in which the conditions were such that the only thing preventing spontaneous nucleation was the presence of the dissolved borax impurity (fig. 3). In these experiments, the addition of a seed crystal to the solution readily induced nucleation. Because both left and right handed crystals were obtained in the product, and because the experiments with the blank solutions had shown that spontaneous nucleation did not occur under these conditions, it was concluded that ICG nucleation was taking place here.

5.2. NUCLEATION CAUSED BY THE GROWTH AND DETACHMENT OF IRERGULARITIES ON THE SURFACE OF THE SEED CRYSTALS

Under certain conditions, the process of secondary nucleation could be associated with the appearance of visible irregularities (primarily dendrites) on the surface of the growing sodium chlorate seed crystals. Usually the product crystals obtained under these conditions all had the same structure as the parent seed (e.g. see figs. 2 and 3). It was postulated that these irregularities could have become detached from the main crystal, passed into the bulk solution, and served as secondary nuclei. Because all crystals contain surface irregularities of varying degrees (such as bunched steps and microscopic dendrites) it was further postulated that this mechanism could operate even when the surface irregularities were only microscopic in nature.

The dendrites could have become detached from the main crystal because of the liquid shear forces resulting from the flow of solution past the seed crystal (for the stirred solutions), because of inherent vibrational forces in the system, and may also have involved a phenomenon known as dendrite coarsening. The latter mechanism may have been the predominant mode of detachment in the *stagnant* solutions. Kliya[10]) and Saratovkin[12]) have taken photomicrographs showing the steps that took place during the coarsening of ammonium chloride dendrites. What they observed was that dendrites of relatively uniform cross sections appeared first. Soon the tip of each dendrite grew faster than the base and the cross section of each dendrite did not remain uniform. Later the dendrites began to dissolve away at the base while at the same time the tip and bulk crystal were still growing. Finally, some of the dendrites completely dissolved away at the base and passed into the bulk solution.

It is suggested here that the same kind of coarsening process could have taken place with the sodium chlorate seed crystals used in this study. If coarsening did occur and dendrites did dissolve away at the base and passed into the bulk solution, they could easily have served as secondary nuclei.

In a stagnant solution, the local supersaturation between coarse dendrites is likely to be lower near the base of the dendrites than near the tip. This is probably why the region near the tip of each dendrite grows faster than the base as dendrite coarsening takes place.

Once a growing dendrite assumes a non-uniform cross section, the base of each dendrite would have a higher solubility than the tip [because of the relationship between solubility and particle size[11])]. If these solubility effects became significant enough, it is possible to envision conditions under which the base of a dendrite might be dissolving away at the same time the tip and the bulk crystal are still growing. This could explain the dissolving away process that takes place in dendrite coarsening.

5.3. NUCLEATION CAUSED BY ORDERING OF THE WATER MOLECULES AT THE CRYSTAL SOLUTION INTERFACE

A secondary nucleation mechanism, operating at low supersaturation, low liquid velocities, and yielding both left and right handed crystals was also observed (figs. 2–5). Because this kind of secondary nucleation consistently yielded a product in which 60% of the crystals had the same structure as the seed, it was felt that two mechanisms were operating here – one yielding only crystals with the same structure as the seed, the other yielding equal numbers of left and right handed crystals. If 80% of the product crystals formed by the latter mechanism, and 20% by the former, a product in which 60% of the crystals would have the same structure as the parent seed would be obtained.

The former mechanism probably involved the kind of growth and detachment of surface irregularities described in section 5.2.

The mechanism yielding equal numbers of left and right handed crystals was not as obvious however. After a number of possible mechanisms were tested and rejected, it was hypothesized that this kind of secondary nucleation might be the result of an ordering of the solvent water molecules near the crystal solution interface. Drost-Hansen[8]) has presented a considerable amount of evidence which shows that the properties of water near interfaces are usually markedly different from the bulk properties. He reports that an ordering of the water molecules near the solid/liquid interface could account for these differences.

If the kind of ordering proposed by Drost-Hansen did occur the solubility of the crystallizing species in the ordered water structure could be less than the solubility in the bulk solution, and the local supersaturation in the ordered water region could be much higher than the bulk supersaturation. Should the local supersaturation near the crystal become high enough because of ordering of the water molecules, spontaneous nucleation might then occur.

6. Conclusions

The main conclusion that can be drawn from this study is that secondary nucleation does not involve a singular mechanism but that different kinds of secondary nucleation can take place under different conditions of supersaturation, liquid velocity, and impurity concentration.

Secondary nuclei can come from the parent seed crystal by (i) the growth and detachment of irregularities such as dendrites on the surface of the seed crystal.

Secondary nuclei can also come from the solution (ii) because of an impurity concentration gradient in the boundary layer resulting from the uptake of impurities by the growing seed crystal.

There is also a possibility (which is largely speculative) that secondary nuclei can come from the solution (iii) because the presence of the seed crystal leads to an ordering of the water molecules near the surface of the crystal and this in turn leads to a high local supersaturation.

Acknowledgement

This study was supported by the National Science Foundation through Grant GK-3715. The financial assistance is gratefully acknowledged.

References

1) G. D. Botsaris, E. G. Denk and J. O. Chua, paper presented at the Third Joint Meeting of the American Institute of Chemical Engineers and the Instituto Mexicano de Ingenieros Quimicos, Denver, Colorado, September 1970. [Published in Chem. Eng. Progr. Symp. Ser. No. 118 (1972)].
2) G. D. Botsaris, E. G. Denk, G. S. Ersan and others, Ind. Eng. Chem. **61** (10) (1969) 86.
3) G. D. Botsaris, E. A. Mason and R. C. Reid, J. Chem. Phys. **45** (1966) 1893.
4) G. D. Botsaris, E. A. Mason and R. C. Reid, A.I.Ch.E. J. **13** (1967) 764.
5) C. W. Bunn, *Chemical Crystallography* (Oxford Publ. Co., Oxford, 1961) p. 38.
6) N. Cabrera and D. A. Vermilyea, in: *Growth and Perfection of Crystals*, Eds. R. H. Doremus, B. W. Roberts and D. Turnbull (Wiley, New York, 1958) p. 393.
7) E. G. Denk, Jr., Fundamental Studies in Secondary Nucleation, Ph. D. Thesis, Dept of Chem. Engineering, Tufts Univ., 1971.
8) W. Drost-Hansen, Ind. Eng. Chem. **61** (11) (1969) 10.
9) F. C. Frank, in: *Growth and Perfection of Crystals*, Eds. R. H. Doremus, B. W. Roberts and D. Turnbull (Wiley, New York, 1958) p. 411.
10) M. O. Kliya, Soviet Phys.-Cryst. **1** (1956) 456.
11) W. Ostwald, Z. Physik. Chem. **34** (1900) 493.
12) D. D. Saratovkin, *Dendritic Crystallization* (Consultants Bureau, New York, 1959) p. 64.
13) R. F. Strickland-Constable, *Kinetics and Mechanism of Crystallization* (Academic Press, New York, 1968) p. 112.

500   *Journal of Crystal Growth* **13/14** (1972) 500–505 © *North-Holland Publishing Co.*

SECONDARY NUCLEATION IN A STIRRED VESSEL COOLING CRYSTALLIZER

E. P. K. OTTENS, A. H. JANSE and E. J. DE JONG

Technische Hogeschool Delft, Laboratorium Chemische Werktuigen, Mekelweg 2, Delft, The Netherlands

The effect of the crystal concentration, stirrer type and -speed, and the supersaturation on the total net nucleation rate were examined in a continuous stirred vessel cooling crystallizer. A model for nucleation in suspensions is proposed, which is based on collision breeding. This type of secondary nucleation is treated in the literature, but until now its mechanism has not been established. To evaluate the results of the experiments, all possible collisions of each crystal during its movement through the crystallizer are considered. The model shows a linear relationship between the net nucleation rate and the energy dissipated by the stirrer, which is verified by the experiments.

The experiments also show that the net nucleation rate is proportional to the crystal concentration.

From the model, it can be concluded that the nucleation rate is determined by the nucleation due to crystal-stirrer collisions.

Another result is an apparent low order dependence of supersaturation on the net nucleation rate.

1. Introduction

For continuous, bulk crystallization working with low relative supersaturation up to a few percent, collision breeding is mainly the source of secondary nuclei. This type of nucleation takes readily place, if a crystal is free to collide with moving surfaces, for instance a stirrer or a pump; stationary surfaces, for instance the vessel wall or with another crystal. Until now the mechanism of collision breeding is not clearly established.

Clontz and McCabe[1]) investigated collision breeding, which they call contact nucleation, with a single $MgSO_4 \cdot 7H_2O$ crystal, which was allowed to collide, in a well defined way, with a metal rod or another single crystal. Their data can be represented with eqs. (1) and (2):

Crystal-rod collisions:

$$\frac{N}{A} = k_1 \left(\frac{c_b - c_s}{c_s}\right) E_{imp} A^{-\frac{1}{2}}.$$ (1)

Crystal-crystal collisions:

$$\frac{N}{A} = k_2 \left(\frac{c_b - c_s}{c_s}\right) E_{imp}.$$ (2)

Identification of symbols is given at the end of this paper.

In both cases the total number of generated nuclei N is proportional to the impact energy at a collision E_{imp}.

Their experiments show, that the relative velocity of the solution past the crystal during a collision, has no

effect on the number of generated nuclei. However, the type of collision and the direction of approach just before a collision affects the nucleation considerably. In order to obtain an approximative expression for the dependence of the nucleation rate in a crystallizer on the process variables, such as supersaturation ΔC, crystal concentration M and the hydrodynamical conditions in the crystallizer, we will now introduce the following simplified statistical derivation.

For $n_0 d(d_p)$ crystals with size between d_p and $d_p + d(d_p)$ which collide with a frequency ω_{d_p}, where an impact energy E_{imp} is involved, the contribution to the total nucleation rate can be written, in analogy with the equations of Clontz and McCabe, as:

$$dJ_{c-i} = k_3 (\Delta C)^p \omega_{d_p} E_{imp} n_0 \, d(d_p).$$ (3)

The absolute value of the impact energy E_{imp} is assumed to be proportional to the kinetic energy of the colliding crystal relative to the other colliding surface. A part of this energy is available for creating nuclei. This part is included in constant k_3. It should be noted that no effect of the crystal surface area is encountered in eq. (3). Impacts on crystal surfaces can be thought to happen mostly on edges and corners. Such an impact area has no relation with the total surface area of a crystal face. Also the direction of approach of the two colliding surfaces is not encountered in eq. (3). The variations are infinite, but for a large number of crystals of a particular size, this effect is some statistical average, which is included in constant k_3.

X – 4

To find the total nucleation rate over the whole size interval eq. (3) must be integrated:

$$J_{c-i} = \int_{d_x}^{d_{max}} dJ_{c-i}$$

$$= \int_{d_x}^{d_{max}} k_3 (\Delta C)^p \, \omega_{d_p} E_{imp} n_0 \, d(d_p). \tag{4}$$

The crystal size d_x is the minimum size, which contributes to collision breeding. This size d_x is mainly determined by the density difference between the crystal and the solution and the flow conditions of the solution in the near vicinity of the crystal. We will now consider three possibilities for collisions: crystal–stirrer (c–ms); crystal–stationary surface (c–ss) and crystal–crystal (c–c) collisions.

2. Nucleation due to crystal-stirrer collision

The derivation is based on macro flow of the crystals through the crystallizer. The stirrer causes circulation of the suspension in the vessel. For a turbine- or propeller stirrer the pumping capacity may be written as:

$$\phi_{vp} = k_4 n_r D_r^{\,3} . \tag{5}$$

The time t_c, necessary for each volume element to pass through the stirrer area is:

$$t_c = V_c / \phi_{vp} . \tag{6}$$

If the suspension is ideally mixed, the individual collision frequency ω_{d_p} becomes:

$$\omega_{d_p} = k_5 / t_c . \tag{7}$$

For all crystals the impact energy is considered to be proportional to:

$$\bar{E}_{imp} \sim m_{d_p} v_t^2 . \tag{8}$$

The expression for the nucleation rate J_{c-ms} becomes:

$$J_{c-ms} = k_3 (\Delta C)^p \int_{d_x}^{d_{max}} \omega_{d_p} \bar{E}_{imp} \, n_0 \, d(d_p)$$

$$= k_3 (\Delta C)^p \int_{d_x}^{d_{max}} \frac{k_6 n_r D_r^3}{V_c}$$

$$\times k_7 k_v \rho_s d_p^3 (n_r \pi D_r)^2 n_0 \, d(d_p) \tag{9}$$

$$= k_8 (\Delta C)^p \frac{n_r^3 D_r^5}{V_c} \int_{d_x}^{d_{max}} k_v \rho_s d_p^3 n_0 \, d(d_p)$$

$$= k_8 (\Delta C)^p \frac{n_r^3 D_r^5}{V_c} M_x .$$

The energy dissipated by the stirrer can be written as:

$$P = P_0 \rho_{sl} n_r^3 D_r^5 , \tag{10}$$

$$P_0 = \text{constant for Re} > 10^4,$$

$$P_0 = 0.35 \text{ for the propeller stirrer,}$$

$$P_0 = 5.0 \text{ for the turbine stirrer.}$$

These power numbers are taken from Uhl and Gray[2]). The dissipated energy per unit mass of slurry ε is:

$$\varepsilon = \frac{P}{\rho_{sl} V_c} = P_0 \frac{n_r^3 D_r^5}{V_c} . \tag{11}$$

Eq. (9) can now be written as:

$$J_{c-ms} = k_9 (\Delta C)^p \varepsilon M_x . \tag{12}$$

Eq. (12) suggests that for all types of stirrer the nucleation rate is the same if the crystallizer operates with the same ΔC, ε, M_x. This is not true. In the derivation of the equations, we did not consider the absolute value of the constants k_4 and k_6. These constants depend on the stirrer type.

3. Nucleation due to crystal-stationary surface collision

The flow of the suspension past the vessel wall or cooling coils is maintained by the circulation capacity of the stirrer. Because the circulation capacity is dependent on the pumping capacity of the stirrer, the expressions for ω_{d_p} and E_{imp} can be assumed to be equivalent to eqs. (7) and (8). Consequently, the expression for the nucleation rate J_{c-ss} will be equivalent to eq. (12).

We are presently carrying out experiments to investigate the influence of stationary surfaces on the total nucleation rate.

4. Nucleation due to crystal-crystal collision

This derivation is based upon the relative motion of particles in suspension.

The impact energy will be much less in this case, but the total collision frequency may be so high, that the net effect on nucleation cannot be neglected. To determine the collision frequency, it is assumed that all crystals are spherical with an average size $\bar{d}_p$ according to:

$$N_{tx} \cdot \tfrac{1}{6} \pi \bar{d}_p^3 \rho_s = M_x . \tag{13}$$

The crystals with size $\bar{d}_p$ move under the influence of turbulent forces through the crystallizer. The theory of

TABLE 1

Equipment and experimental conditions

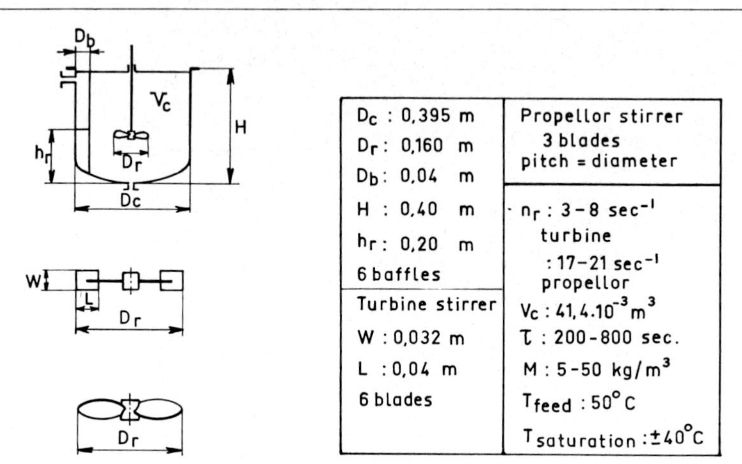

D_c : 0,395 m	Propellor stirrer 3 blades pitch = diameter
D_r : 0,160 m	
D_b : 0,04 m	
H : 0,40 m	n_r : 3–8 sec^{-1} turbine
h_r : 0,20 m	: 17–21 sec^{-1} propellor
6 baffles	V_c : 41,4.10^{-3} m^3
Turbine stirrer	τ : 200–800 sec.
W : 0,032 m	M : 5–50 kg/m^3
L : 0,04 m	T_{feed} : 50° C
6 blades	$T_{saturation}$: ±40° C

local, isotropic turbulence may be used to give information on the turbulent intensity in the small volume around the particle. Levich[3]) calculates the motion of a particle in such a turbulent field by means of a force balance. His derivation gives:

$$v_m \sim \left(\frac{\rho_s - \rho_l}{\rho_s}\right)^{\frac{1}{2}} \left(\frac{\rho_s}{\rho_l}\right)^{\frac{1}{3}} \left(\frac{1}{C_d}\right)^{\frac{1}{3}} \varepsilon^{\frac{1}{3}} d_p^{\frac{1}{3}}. \tag{14}$$

Calculation of the mean collision frequency $\bar{\omega}$ is as follows: The trajectory of a crystal relative to the solution per second is equal to v_m, which corresponds with a flow of solution of volume V_s:

$$V_s = \tfrac{1}{4}\pi \bar{d}_p^2 v_m. \tag{15}$$

In this volume $N_{tx} \cdot \tfrac{1}{4}\pi \bar{d}_p^2 v_m$ crystals are available for collisions, so:

$$\bar{\omega} = N_{tx}\tfrac{1}{4}\pi \bar{d}_p^2 v_m. \tag{16}$$

The impact energy $\bar{E}_{imp}$ is assumed to be proportional to the kinetic energy of the moving crystal:

$$\bar{E}_{imp} \sim E_{kin} = \tfrac{1}{2} \cdot \tfrac{1}{6}\pi \bar{d}_p^3 \rho_s v_m^2. \tag{17}$$

With eqs. (13), (14), (16) and (17) total nucleation rate becomes:

$$J_{c-c} = k_{10} (\Delta C)^p \varepsilon M_x^2. \tag{18}$$

In section 6 it will be shown that upon using the above eqs. (12) and (18), we are able to correlate in a satisfactory way the quantities J, ΔC, ε and M.

In fact M_x differs from the real crystal concentration M, which is caused by the fact that M_x represents the total mass of crystals above size d_x. It is assumed that d_x is relatively small. Consequently the total mass of crystals below size d_x can be regarded as relatively small compared to the total mass of crystals M.

5. Experimental apparatus and procedure

The experiments were carried out with potassium-alum in a simple, stirred vessel cooling crystallizer. Apparatus and process specifications are listed in table 1. The stirrer speed was always higher than that defined by Zwietering[4]). To check whether homogenity of the suspension was complete, two samples were taken in the top of the crystallizer and near the stirrer. These samples were exactly identical in crystal size distribution.

All samples were taken in the discharge of the crystallizer. A duplicate sample was taken one hour later. The crystals were filtered, dried and sieved. The crystal size distribution which results from a sieve analysis is given in a population density plot. Such a plot gives information about net nucleation rate and growth rate of the individual crystals[5]). To obtain accurate results, we used a least-square computer program.

Checks were made, to be sure that the crystallization process was at steady state.

6. Experimental data

Fig. 1 shows a characteristic population density plot for the crystallization of potassium alum in the stirred

X – 4

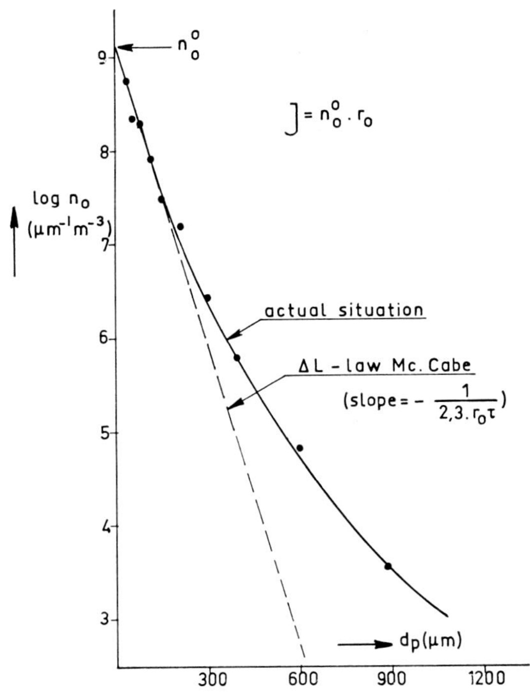

Fig. 1. A particle population density versus its size.

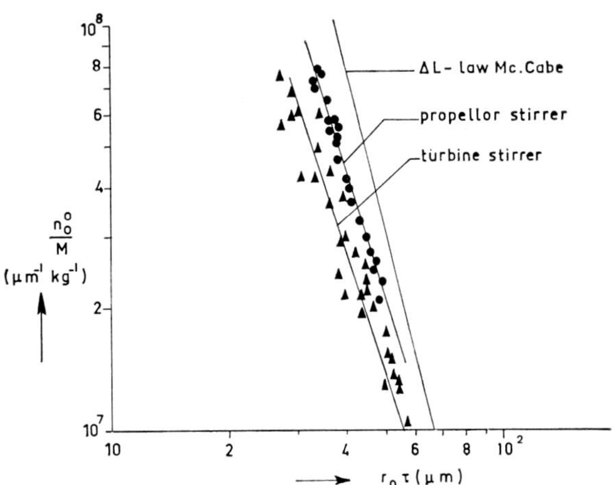

Fig. 2. Modified nuclei population density versus mean particle size.

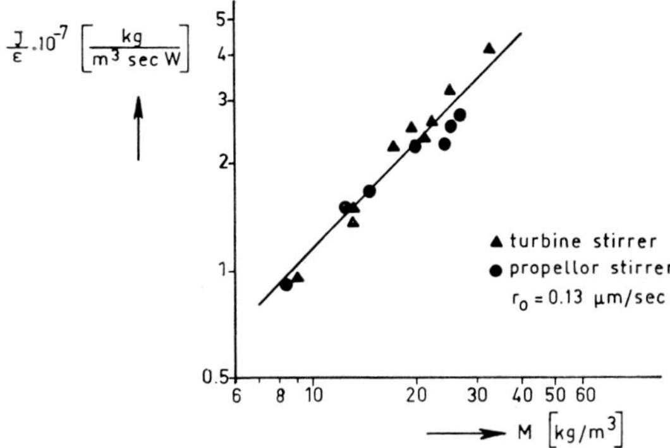

Fig. 3. A modified nucleation rate versus crystal concentration.

vessel cooling crystallizer. The dotted line in the curve indicates the case, if all crystals should grow with the same rate (ΔL-law of McCabe). The value $r_0\tau$ indicates the mean particle diameter on a length basis if the ΔL-law of McCabe is valid. In the case of potassium-alum where the ΔL-law is not valid, r_0 means the growth rate of the smallest crystals. The curvature in the plot can be explained by assuming non-ideal mixing or increasing growth rate with increasing size.

Non-ideal mixing was not measured or observed, and therefore can be ruled out. The experiments show that the curvature is mainly dependent on $r_0\tau$ and decreases when $r_0\tau$ increases. From this a transition to the ΔL-law of McCabe can be expected at high $r_0\tau$. For the interpretation of the nucleation rate data only those data were compared which had the same curvature in the population density plot.

From the above it can be concluded that the growth rate of the crystals increases with increasing size.

Identical observations were made on $CuSO_4 \cdot 5H_2O$ by McCabe and Stevens[6]) in a stirred vessel. If the mass transfer by convection from the bulk of the solution to the crystal interface would be rate determining during crystal growth, the growth rate should slightly decrease with increasing crystal size. Probably crystal collisions are more responsible for an increase in crystal growth.

Fig. 2 shows n_0^0/M as a function of $r_0\tau$, for our experiments and for the case that the ΔL-law of McCabe is valid. The data can be best represented by:

$$M = \alpha k_v \rho_s n_0^0 (r_0\tau)^{2.6}, \tag{19}$$

with:

$\alpha = 3 \times 10^{-15}$ for the turbine stirrer,

and

$\alpha = 2 \times 10^{-15}$ for the propeller stirrer.

The differences in the absolute value of constant α are caused by the difference in curvature in the population density plots. For the ΔL-law of McCabe we obtain:

$$M = 6 \times 10^{-18} k_v \rho_s n_0^0 (r_0\tau)^4. \tag{20}$$

In fig. 3, J/ε is plotted against M for the same growth rate r_0 for the turbine- and propeller stirrer. The slope

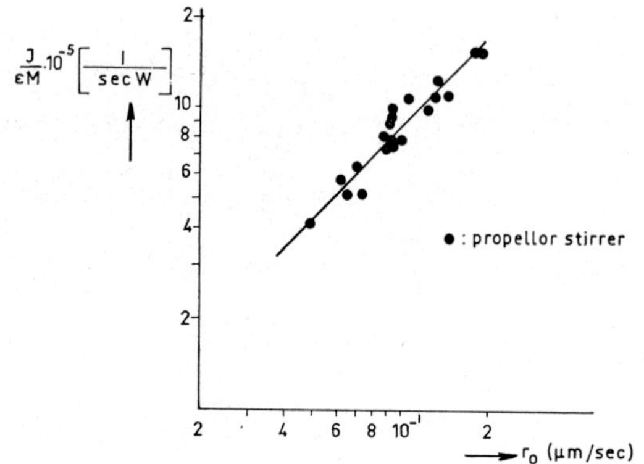

Fig. 4. A modified nucleation rate versus growth rate of the smallest crystals.

is unity. This is in good agreement with eq. (12). The variations in dissipated energy per unit mass of slurry were:

for the turbine stirrer: $0.4 \leq \varepsilon \leq 6.5$ W/kg slurry,

for the propeller stirrer: $4.3 \leq \varepsilon \leq 8.3$ W/kg slurry.

The growth rate r_0 of the smallest crystals is taken as a measure for the supersaturation ΔC. This is done because growth rate and supersaturation are always coupled together, for instance:

$$r_0 = k_g(\Delta C)^i.$$

In general i lies between one or two. From the data given by Mullin[7]), it may be concluded that $i = 1.25$ for small crystals. Fig. 4 shows a plot of $J/(\varepsilon M)$ against r_0 for the propeller stirrer. As it is seen $J/(\varepsilon M)$ is proportional to the growth rate r_0.

Upon comparing eqs. (12) and (18) we are able to distinguish crystal–crystal and crystal–stirrer collision breeding by checking the power of the crystal concentration M in the empirical correlation for the nucleation rate. From the experiments carried out with low crystal concentration (≤ 60 kg/m³), it can be concluded that the nucleation rate is determined by the nucleation due to crystal–stirrer collisions.

List of symbols

A	contact area at a collision	m²
c_b	bulk concentration of solute	kg/m³
c_s	saturation concentration	kg/m³
C_d	drag coefficient	
d_x	minimum crystal size which contributes to contact nucleation	m
d_{max}	maximum crystal size present in the crystallizer	m
d_p	particle size	m
$\bar{d}_p$	mean particle size according to eq. (13)	m
D_r	stirrer diameter	m
E_{imp}	impact energy	J
$\bar{E}_{imp}$	mean impact energy	J
E_{kin}	kinetic energy	J
J_{c-i}	contact nucleation rate due to collision of a crystal with an object ($i = ms$, moving surface; $i = ss$, stationary surface; $i = c$, crystal)	m⁻³ sec⁻¹
k_g	kinetic constant for growth of crystals	
k_v	volume factor	
$k_{1,2,\ldots10}$	(empirical) dimensional constant	
M	crystal concentration $\int_0^{d_{max}} n_0 k_v \rho_s d_p^3 \, \mathrm{d}(d_p)$	kg/m³
M_x	total crystal mass above size d_x: $\int_{d_x}^{d_{max}} n_0 k_v \rho_s d_p^3 \, \mathrm{d}(d_p)$	kg/m³
m_{d_p}	mass of crystal of size d_p	kg
n_0^0	population density of crystals with size d_p	m⁻³ μm⁻¹
n_0	nuclei population density	m⁻³ μm⁻¹
n_r	stirrer speed	sec⁻¹
N	number of nuclei generated by one collision	—
N_t	total number of crystals in crystallizer: $\int_0^{d_{max}} n_0 \, \mathrm{d}(d_p)$	m⁻³
N_{tx}	total number of crystals per unit volume in crystallizer above size d_x: $\int_{d_x}^{d_{max}} n_0 \, \mathrm{d}(d_p)$	m⁻³
P	dissipated energy by the stirrer	W
P_0	power number	—
r_0	growth rate of smallest crystals	μm/sec
Re	Reynolds number of stirrer	—
t_c	circulation time	sec

v_t	tip velocity of stirrer	m/sec
v_m	maximum slip velocity of particle to surrounding solution	m/sec
V_s	volume displaced by the particle per second	m^3/sec
V_c	volume of crystallizer	m^3
α	dimensional constant [eq. (16)]	
ε	dissipated energy per unit mass of slurry by the stirrer	W/kg
ρ_l	density of mother liquor	kg/m^3
ρ_s	density of crystal	kg/m^3
ρ_{sl}	density of slurry	kg/m^3
τ	mean residence time of solution in the crystallizer	sec
ϕ_{vp}	pumping capacity of stirrer	m^3/sec
ω_{d_p}	collision frequency of crystal with size d_p	sec^{-1}
$\bar{\omega}$	mean collision frequency of a crystal with size $\bar{d}_p$	sec^{-1}

References

1) N. A. Clontz and W. L. McCabe, Contact nucleation of magnesium sulphate heptahydrate, paper presented on the 62nd Annual Meeting of A.I.Ch.E., November 1969.
2) V. W. Uhl and J. B. Gray, *Mixing* (Academic Press, New York, 1966).
3) V. G. Levich, *Physicochemical Hydrodynamics* (Prentice Hall, Englewood Cliffs, N.J., 1962).
4) Th. N. Zwietering, Chem. Eng. Sci. **8** (1958) 244.
5) M. A. Larson and A. D. Randolph, Chem. Eng. Progr. Symp. Ser. **65** (1969) 95.
6) W. L. McCabe and R. P. Stevens, Chem. Eng. Progr. **47** (4) (1951) 168.
7) J. W. Mullin and J. Garside, Trans. Inst. Chem. Engrs. **45** (1967) T 285.

                    Journal of Crystal Growth **13/14** (1972) 506–509 © *North-Holland Publishing Co.*

COLLISION BREEDING OF CRYSTAL NUCLEI: SODIUM CHLORATE. I

H. GARABEDIAN and R. F. STRICKLAND-CONSTABLE

Imperial College, London, S.W.7, England

When crystals of $NaClO_3$ slide along a glass surface in a supersaturated solution large numbers of new crystals appear along the sliding path after the elapse of a certain time. The numbers increase rapidly with supersaturation. The results resemble closely earlier work with $Mg\,SO_4 \cdot 7H_2O$.

In explanation, it is assumed that contact of the crystal with the glass produces a large number of very small particles by mechanicl attrition: then the Gibbs-Thomson equation determines the proportion of these particles which survive – the higher the supersaturation, ΔC, the smaller the critical radius and hence the higher the proportion which survives, and vice versa. This interpretation is strongly supported by experiments in which the sliding process is carried out at a given ΔC_1, and the solution is immediately afterwards diluted to a ΔC_2; the number of crystals surviving is found to depend on ΔC_2 rather than on ΔC_1.

1. Introduction

A study of collision breeding has been published recently by Clontz and McCabe[1]). Inter alia the results obtained confirmed in a very satisfactory manner earlier work described by Strickland-Constable[2]), and carried out by Lal, Mason and Strickland-Constable[3]). Both these studies were with single crystals; but both used the same material, $MgSO_4 \cdot 7H_2O$: one of the objects of the present work was to extend the study to a second substance, for which $NaClO_3$ was chosen.

In the work of Lal, Mason and Strickland-Constable[3]) it was found that the number of nuclei produced by a given collision process was an increasing function of the supersaturation. This was explained in the following way: a given collision process produces the same number of nuclei, irrespective of the supersaturation*: but the particles are so small that many of them re-dissolve, and only those survive whose linear size is greater than r_0, where r_0 is given by the Gibbs–Thomson equation as**

$$\frac{C_0 - C_s}{C_s} \sim \frac{2\sigma V}{RT} \frac{1}{r_0}. \tag{1}$$

The present research provides substantial evidence in support of the theory, which will be referred to as the "survival" theory.

---

* If the particles are produced primarily by mechanical impact it is very unlikely that the number and size distribution will be substantially affected by small differences in supersaturation.
** See list of symbols at the end of this paper.

2. General experimental method

Sodium chlorate was the solute and distilled, de-ionised water the solvent. The experimental cell is shown in fig. 1. The main tube was 21 cm long, and $3\frac{1}{2}$ cm in diameter. It was held, initially, in a horizontal position, in a thermostat bath. A run was carried out as follows: $100\ cm^3$ of an aqueous solution of sodium chlorate of accurately known composition (corresponding to saturation temperatures above 35 °C) were introduced into the cell: the liquid level is shown by the hatching in fig. 1. To avoid initial breeding the cell was then heated to a temperature at least 3° above the saturation temperature, and a crystal of sodium chlorate (about $4 \times 2 \times 1$ mm in size) was introduced via neck A: the solution was then cooled to 35 °C, at which temperature all the sliding experiments took place. The treatment just described was shown to be adequate to avoid initial breeding.

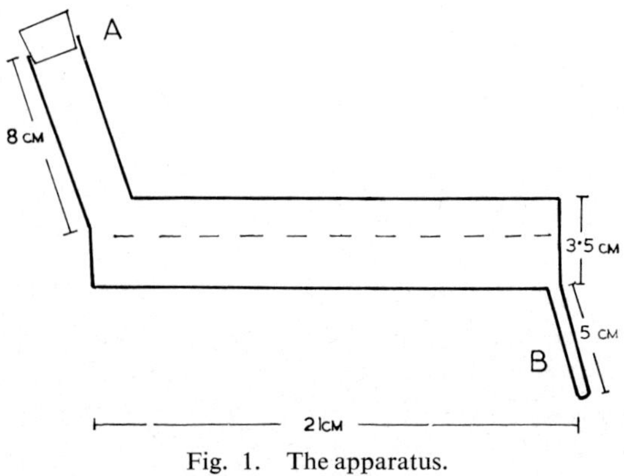

Fig. 1.   The apparatus.

The cell was next tilted through 65°, so that the crystal slid along the bottom of the cell and came to rest at the bottom of tube B: the cell was then again brought to the horizontal position.

The sliding of the crystal produced a number of nuclei along its path: these were at first too small to be visible. The cell was brightly illuminated and after a period of time, depending on the supersaturation, the nuclei grew and became visible as tiny points of light; the number increased at first, but finally became constant: the final number n was recorded.

Concentrations are all expressed as g-moles of sodium chlorate per cm^3 of solution. According to Seidl[4]) the solubility is given as 770×10^{-5} g-moles/cm^3 (total solution) at 40 °C and 723×10^{-5} g-moles/cm^3 (total solution) at 30 °C.
Density of solid: 2.49 g/cm^3. Molecular weight: 106.45.

3. Results

Series I: Several runs were carried out at each of a series of values of ΔC: the temperature was in all cases 35 °C, and this temperature was maintained during the whole run during which the nuclei grew to visible size and were counted. The results are shown in table 1.

TABLE 1

$10^5 \Delta C$	ΔT	n for individual runs	average n
33	7.0	580, 580, 660, 540	590
24	5.0	500, 500	500
18	4.0	280, 270	275
16	3.2	120, 100, 282 ,120	155
13	2.8	100, 110	105
9	1.7	56, 46	51
0	0	0	0

The values from table 1 are plotted in fig. 2, as n against ΔC.

The results call for the following comments:
(a) The number of crystals produced is very high considering the sliding time is only about 12 sec.
(b) The number increases rapidly with supersaturation.
(c) The curve is S-shaped, and in this way resembles the curves of Lal, Mason and Strickland-Constable[3]) with $MgSO_4 \cdot 7H_2O$.

Series II: This series differs from all the others in that the cell was tilted only just sufficiently to induce the crystal to slide; the total sliding time was approximately $1\frac{1}{2}$ min. At least 2 runs were carried out at

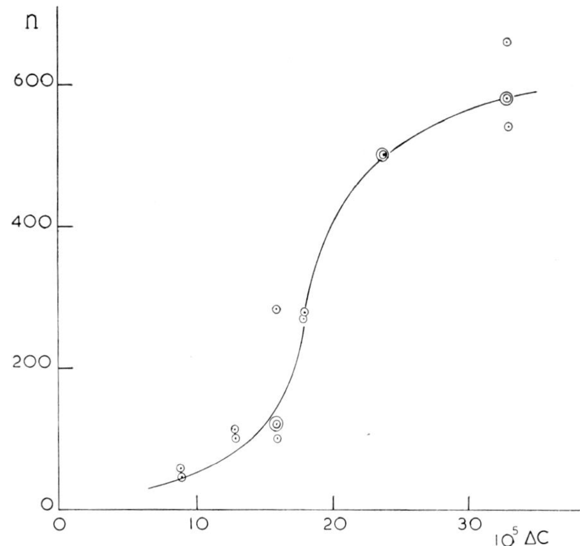

Fig. 2.   Series I. Number of crystals, n, against ΔC (moles/cm^3 solution).

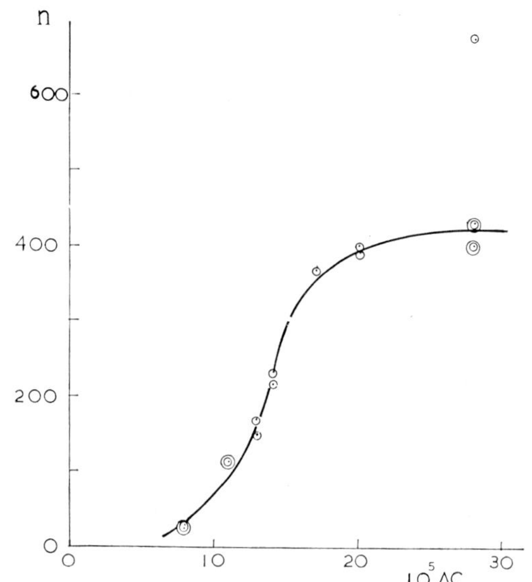

Fig. 3.   Series II. Number of crystals, n, against ΔC (moles/cm^3 solution).

each ΔC. The temperature was 35 °C. All other details as for Series I.

The results are plotted in fig. 3: they appear rather more self-consistent than Series I, which suggests that this method might be the more suitable for future work. The curves are of very similar shape, and the rates of nucleation do not differ greatly. The series confirms the conclusions of Series I.

Series III: It was at one time suggested that the small number of crystals produced at the lower values of ΔC

X – 5

might be due solely to the fact that insufficient supersaturation remained in the solution to develop to visible size the nuclei formed during the collision process. Experiments were designed to test this point. Sliding was carried out at 35 °C.

With a ΔC of 9×10^{-5} ($\Delta T = 1.7$): the crystals were allowed to develop as usual at the same ΔC until no more appeared (count no. 1). The supersaturation was then increased to a ΔC of 22.6×10^{-5} ($\Delta T = 4.7$) by dropping the temperature, and the number of crystals was counted periodically until no more appeared (count no. 2). Two experiments were carried out as shown in tables 2 and 3.

TABLE 2

1st count at $\Delta C = 9 \times 10^{-5}$, $\Delta T = 1.7$				
After (h):	$\frac{3}{4}$	$1\frac{3}{4}$	2	$2\frac{3}{4}$
Number of crystals:	45	47	47	47

2nd count at $\Delta C = 22.6 \times 10^{-5}$, $\Delta T = 4.7$					
After (min):	10	20	30	35	40
Number of crystals:	47	52	52	53	53

TABLE 3

1st count at $\Delta C = 9 \times 10^{-5}$, $\Delta T = 1.7$		
After (h):	$1\frac{1}{2}$	$2\frac{1}{2}$
Number of crystals:	58	62

2nd count at $\Delta C = 22.6 \times 10^{-5}$, $\Delta T = 4.7$				
After (min):	10	20	30	60
Number of crystals:	61	64	62	63

As will be seen, the differences between the 1st and 2nd counts are relatively small. And they can in fact be regarded as negligible when compared with the 480 crystals produced in Series I at the ΔC of 22.6×10^{-5} ($\Delta T = 4.7$). In other words, the procedure of Series I and II is capable of developing to visible size substantially all the nuclei produced, even at the lowest values of ΔC. A number of further runs gave similar results, and confirmed this general conclusion.

Series IV: This series was designed to check the validity of the survival theory, discussed above. The temperature was 35 °C throughout each run; the results are given in table 4. In each run the sliding was carried out at a ΔC_1 of 24×10^{-5} ($\Delta T = 5.0$), and in the first run the same supersaturation was maintained during the development period. In the second run the solution

was diluted as quickly as possible after the sliding operation to give a ΔC_2 of 16×10^{-5} ($\Delta T = 3.2$): the dilution process took only a few seconds; and this ΔC_2 was maintained throughout the development period. In the 3rd run the solution was diluted after sliding to a ΔC of 9×10^{-5} ($\Delta T = 1.7$). The number of crystals counted is given in the 3rd column; and the number of crystals obtained in Series I at the value of ΔC equal to the value of ΔC_2 in the present series is given in the 4th column.

TABLE 4

		Number of crystals counted	
$10^5 \Delta C_1$	$10^5 \Delta C_2$	Present Series	Series I
24	24	500	500
24	16	105	155
24	9	65	51

The results show that the number of crystals produced depends on the value of ΔC_2 and not on ΔC_1; that is to say, it depends on the conditions prevailing during the development period, and not during the sliding process.

4. Discussion

The results of Series IV are clearly what may be expected from the survival theory. The size distribution of nuclei produced by the collision process is assumed the same in all cases: the number of crystals finally counted will be only those crystals whose size exceeds the critical value given by eq. (1) at the dilution ΔC_2: at low supersaturations r_0 is large and nearly all crystals redissolve; and vice versa.

It is perhaps unexpected that the dilution can be changed rapidly enough to enable this effect to be observed. Further experiments concerning this point have been carried out, which it is hoped to report shortly.

A further conclusion from this work is that the primary particles produced by the collision process must all be extremely small (in accordance with eq. (1)), and it remains to suggest a mechanism which can give rise to such a result.

Acknowledgements

The authors wish to express their gratitude to the Atomic Energy Research Establishment, Harwell, for an apparatus grant covering this research; as also for a maintainence grant for H. Garabedian.

Symbols

C_0	concentration	moles/(cm^3 of total solution)
C_s	saturation concentration	moles/(cm^3 of total solution)
n	number of crystals in 1 run	
$\bar{n}$	average number of crystals in a series of runs at a given ΔC	
r	a linear dimension of a crystal	cm
r_0	critical value of r corresponding to concentration C_0	cm
R	gas constant	erg mole^{-1} K^{-1}
t	time	s
T	temperature	K
V	molar volume of solid	cm^3/mole
$\Delta C = C_0 - C_s$		moles/(cm^3 of total solution)
ΔT = Saturation temperature – actual temperature of solution		
σ	surface free energy	erg/cm^2

References

1) N. A. Clonz and W. L. McCabe, 62nd A.I.Ch.E. Annual Meeting, Washington, D.C. 1969.
2) R. F. Stickland-Constable, *Kinetics and Mechanism of Crystallization* (Academic Press, London, 1968).
3) D. P. Lal, R. E. A. Mason and R. F. Strickland-Constable, J. Crystal Growth **5** (1969) 1.
4) A. Seidl, *Solubilities of Inorganic and Metal Organic Compounds*, Vol. 1, 3rd ed. (1940) p. 1250.

Journal of Crystal Growth **13/14** (1972) 510–516 © *North-Holland Publishing Co.*

CRYSTAL GROWTH RATE STUDIES WITH POTASSIUM SULPHATE IN A FLUIDIZED BED CRYSTALLIZER

J. GARSIDE, C. GASKA and J. W. MULLIN

Department of Chemical Engineering, University College London, Torrington Place, London, W.C.1, England

Experiments on the nucleation and growth of potassium sulphate in a laboratory scale fluidized bed crystallizer are reported. The latent period of potassium sulphate solutions before desupersaturation is a function of the solution supersaturation and velocity, and the seed area available for growth. The amount of nucleation also depends on these variables and, in particular, is very sensitive to the total crystal seed area present in the crystallizer, a large area resulting in low nucleation.

Growth rate measurements in fluidized beds of low voidage (80–85 per cent) show good agreement with data previously obtained for high voidage (> 98 per cent) beds. During cooling crystallization processes, high yields are obtained when a large surface area of seed crystals is present. This ensures that sufficient surface on which growth can occur is available and so prevents excessive nucleation.

1. Introduction

The nucleation, crystal growth and dissolution characteristics of potassium sulphate were reported previously[1]. In the present paper, further data on nucleation and growth are presented. Desupersaturation rates of seeded and unseeded solutions have been measured, the growth rate of crystals in dense beds has been investigated and results of batch cooling crystallizations are recorded. All the runs reported here were carried out in a fluidized bed crystallizer. A description of this crystallizer and full details of the experimental procedure adopted have been given elsewhere[1-3].

2. Desupersaturation of potassium sulphate solutions

In Oslo-type crystallizers the solution is supersaturated into the metastable condition in one part of the equipment and then conveyed to another region where the supersaturation is partially released onto a mass of crystals fluidized by the upward-flowing solution. These crystallizers can be operated as either "circulating liquor" types, in which the fluidized bed of crystals is retained in the growth zone of the apparatus, or as "circulating magma" types[4]. In the latter, the solution circulation rate is increased to allow the crystals to be circulated with the liquor through the whole crystallizer. One of the factors to be taken into account in the choice between a circulating liquor or circulating magma crystallizer is the length of time a supersaturated solution can remain in the metastable condition with-

out nucleating. If the liquor transfer time between the "supersaturation" and "growth" zones is too long for the solution to remain metastable without nucleating, a circulating magma crystallizer is recommended.

In the following work desupersaturation rates were measured at a constant temperature of 30 °C (± 0.03 °C) under different conditions of supersaturation, circulation velocity and seed size. The change in solution concentration depends on the number of nuclei formed and their subsequent growth. Since it was impossible to make a continuous measure of the number and size of the nuclei produced in the crystallizer, the term "desupersaturation rate" is used here in preference to the term "nucleation rate". The desupersaturation rate is defined as the change in concentration per unit time per unit volume of solution.

2.1. EXPERIMENTAL PROCEDURE

A solution of potassium sulphate of known concentration was filtered through a sintered glass funnel (porosity 3, effective pore size 20–30 µm) before being charged into the crystallizer at 50 °C. The circulation rate through the crystallization section was adjusted to the desired value for a particular run, and the solution was cooled at a constant rate of 1 °C/min. Desupersaturation rate measurements were commenced when the solution temperature reached 30 °C, and the solution concentration was measured at suitable intervals of time by evaporating samples to dryness. The duration of all runs was fixed at 200 min.

In all the runs carried out there was a time lag or "induction period" after the solution reached 30 °C before any nuclei could be observed in the circulating solution. Nuclei were first noticed as scintillating particles when a beam of light was passed through the solution. In general induction periods could not be measured very accurately. Negligible change in solution concentration occurred until some time after the onset of nucleation; the number of nuclei then rapidly

2.2. EFFECT OF SUPERSATURATION

The effect of supersaturation on the desupersaturation rate was investigated for three solutions of initial concentration 15.1, 14.7 and 14.4 g K_2SO_4/100 g H_2O (saturation temperatures of 41.2, 38.6 and 37.3 °C respectively). At a working temperature of 30 °C these correspond to supersaturation ratios, S, of 1.15, 1.13 and 1.10 respectively. Both unseeded and seeded solu-

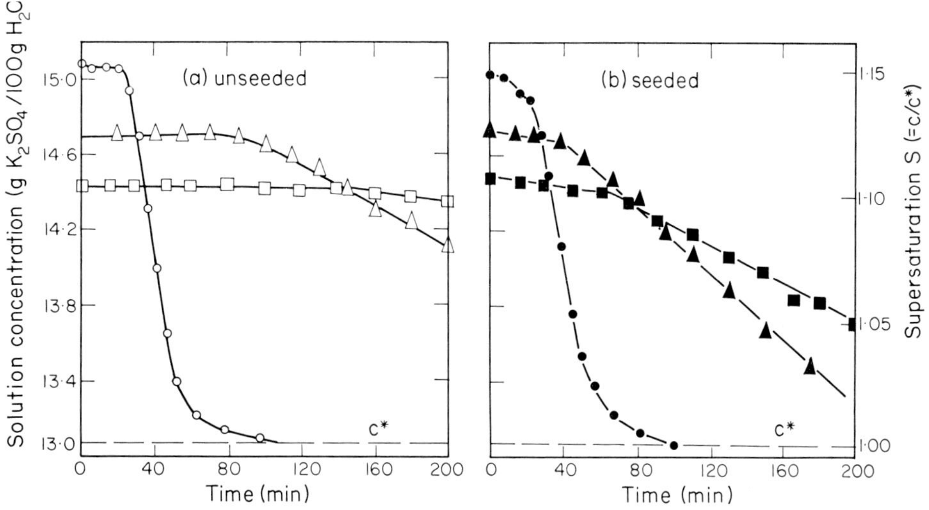

Fig. 1. Effect of supersaturation on the desupersaturation rate of potassium sulphate solutions. (Temperature = 30.0 °C; solution velocity = 10.7 cm/s; solutions seeded (b) with 0.875 g of 460 μm seeds.)

increased and the solution concentration rapidly decreased. The total time span up to this point will be referred to as the "latent period". Because this latent period was well defined and easily measured from the desupersaturation curve (see fig. 1), it was used to characterize the nucleation behaviour of the system.

The subsequent behaviour of the system depended mainly on the solution concentration. At low concentration the nuclei grew and coalesced into particles resembling snowflakes. At high concentration the solution rapidly turned "milky" and the coalescence of nuclei continued at a more rapid rate due to the greater number of nuclei formed.

An attempt was made to measure the maximum size of the visible "nuclei" by observing samples of solution rapidly removed from the crystallizer under a low-powered microscope fitted with an eyepiece graticule. No particles larger than about 50 μm were detected and most were much smaller than this.

tion were used. In the latter case seeds of mean size 460 μm (500–421 μm, between 30 and 36 mesh B.S. sieves) were used in all the runs and each seed charge had a total surface area of about 50 cm². The seeds were added as soon as the solution reached 30 °C and the solution velocity in the crystallization section was kept constant at 10.7 cm/s for all runs.

The effect of initial solution concentration on the latent period and desupersaturation rate is shown in figs. 1a and 1b for unseeded and seeded solutions respectively. The latent period increases as the solution concentration decreases. With unseeded solutions there is only a negligible change in solution concentration until a large number of nuclei are formed, and then the desupersaturation rate increases to a reasonably constant value which is maintained for a long period. At high supersaturation, rapid desupersaturation occurs as a result of the large number of nuclei formed and the high driving force causing rapid growth of these nuclei.

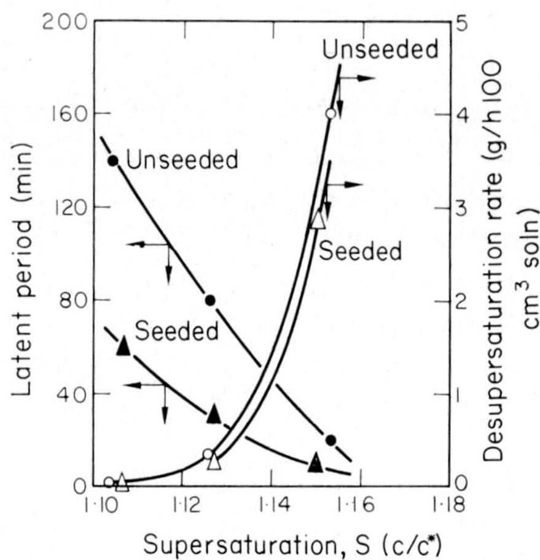

Fig. 2. Effect of supersaturation on the desupersaturation rate and latent period of potassium sulphate solutions. (Temperature = 30.0 °C; solution velocity = 10.7 cm/s; solutions seeded with 0.875 g of 460 μm seeds).

For the seeded solution (fig. 1b) the desupersaturation curves are of similar form to the unseeded ones but there are several differences. The two cases are summarized in fig. 2. The latent periods for the seeded solutions are shorter than for the unseeded case. At a given supersaturation the ratio of latent periods for unseeded and seeded solutions is constant at about 2.5 for the range of supersaturations investigated. This more rapid onset of nucleation with seeded solutions is probably due to the production of secondary nuclei.

The desupersaturation rate is also higher for the seeded solutions. This would be expected because solute is constantly being deposited on the added seeds as well as being precipitated as new nuclei. In order to calculate the desupersaturation rate due to nucleation alone, the depletion of solution concentration due to growth on the seeds was calculated using growth rate data that had previously been determined[1]). A stepwise calculation was used such that the fall in solution concentration for any one step did not exceed $0.2 \text{ g } K_2SO_4/100 \text{ g } H_2O$.

The resulting desupersaturation rate due to nucleation is plotted in fig. 2. There is little significant difference between the desupersaturation rates for the two cases.

2.3. EFFECT OF SOLUTION VELOCITY

It is well known that nucleation is enhanced by mechanical influences such as agitation and solution tur-

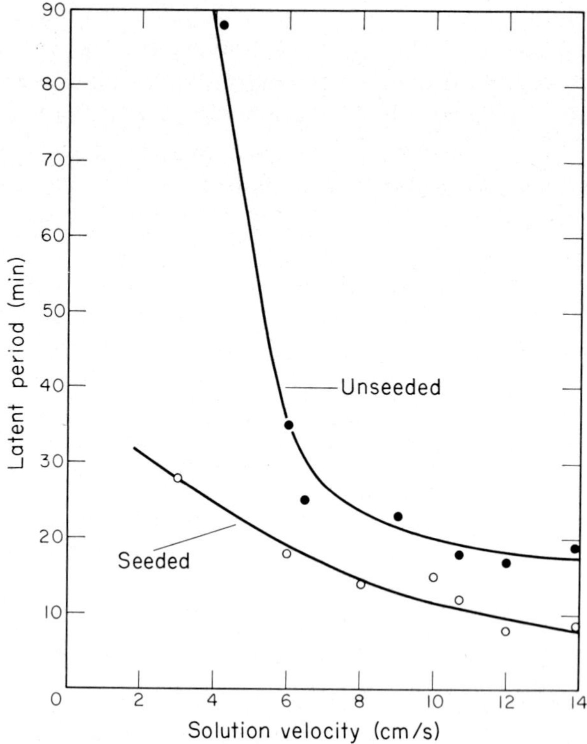

Fig. 3. Effect of solution velocity on the latent period of potassium sulphate solutions (temperature = 30.0 °C; solution concentration = 15.1 g K_2SO_4/100 g H_2O; seed area = 50 cm²).

bulence. In the crystallizer, the solution turbulence varies as it passes round the apparatus and is probably greatest in the pump. For convenience, however, the solution movement through the apparatus is recorded here in terms of the mean velocity in the main 7.6 cm diameter crystallization zone. This velocity was varied between 3 and 14 cm/s.

The variation of latent period with velocity is shown in fig. 3. For unseeded solutions the latent period decreases with increase in solution velocity and tends to a minimum value of about 17 min for a velocity of 14 cm/s. At 3 cm/s the induction period is in excess of 100 min. With seeded solutions the latent periods are much shorter; nuclei appear fairly quickly and very rapidly increase in number.

2.4. EFFECT OF SEED AREA

As the seed area increases so does the desupersaturation rate of the solution. The overall desupersaturation rate recorded represents the change in solution concentration due to both growth and nucleation.

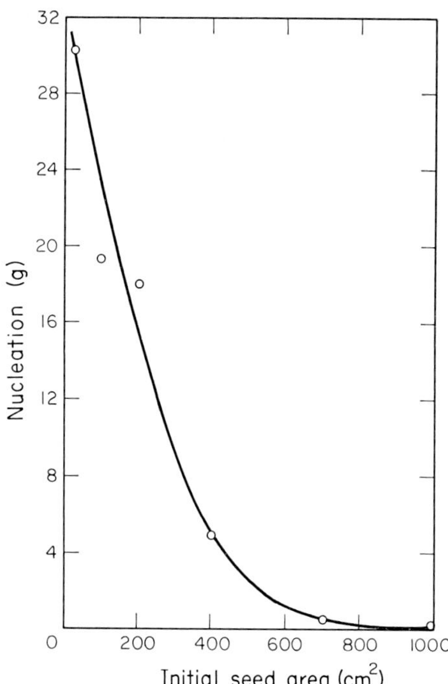

Fig. 4. Effect of seed area on the nucleation of potassium sulphate solutions (temperature = 30.0 °C; solution velocity = 10.7 cm/s; initial solution concentration = 14.4 ± 0.1 g K_2SO_4/ 100 g H_2O).

The effect of seed area on the nucleation is shown in fig. 4. The nucleation is defined as the mass of fines (crystals smaller than the initial added seeds) deposited during a run. This quantity was calculated from the desupersaturation curve by taking the difference between the calculated theoretical solute deposition for the given supersaturated solution, and the mass of grown crystals larger than the initial seed size removed from the crystallizer.

These data indicate that supersaturation was a much more important factor than attrition in producing nuclei. It is apparent that attrition between crystals is not here responsible for the nucleation because the amount of nucleation *decreases* as the seed area is increased. The effect of increasing seed area is to increase the rate of desupersaturation, thus keeping the level of supersaturation low enough to prevent excessive nucleation. A minimum seed surface area of about 1000 cm^2 (17.5 g/460 μm) is required to desupersaturate the solution completely in about 100 minutes from an initial concentration of 14.4 g K_2SO_4/100 g H_2O, with practically no nucleation occurring. An increase in the seed area also has a large effect on the habit and quality of

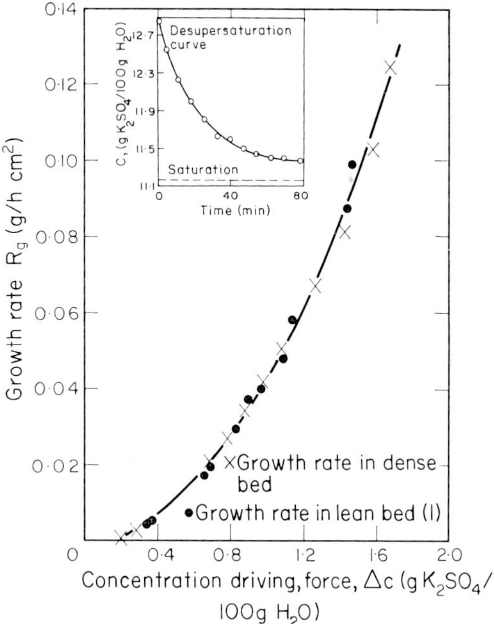

Fig. 5. Comparison of growth rates at 20 °C for dense and lean beds.

the product crystals, due to the level of supersaturation at which the crystals grow.

3. Growth rate measurements in dense beds

The growth rate measurements reported in the previous paper[1]) were carried out with relatively few crystals (5 g) in the fluidized bed (76 cm high, 7.6 cm diameter) giving a bed voidage, ε, greater than 0.98. Industrial crystallizers, however, are operated with much denser beds ($\varepsilon \sim 0.8$–0.85). It was clearly desirable to check if the growth rates measured in lean beds are applicable to dense beds. The experimental technique has already been described[1–3]). The supersaturated solution, kept at a constant temperature, was seeded with a given size and weight of crystals and the solution concentration was measured at selected intervals of time by evaporating samples to dryness. During the run the solution circulation velocity was increased to keep the growing crystals in the fluidized bed.

Figs. 5 and 6 show desupersaturation curves and crystal growth rates for solutions seeded with 200 g of 925 μm mean size crystals at 20 and 30 °C respectively. Crystal growth rates were calculated from these desupersaturation curves using the method outlined in the Appendix. The growth rates obtained at 20 °C with a 5 g/925 μm seed charge[1]) are also shown in fig. 5.

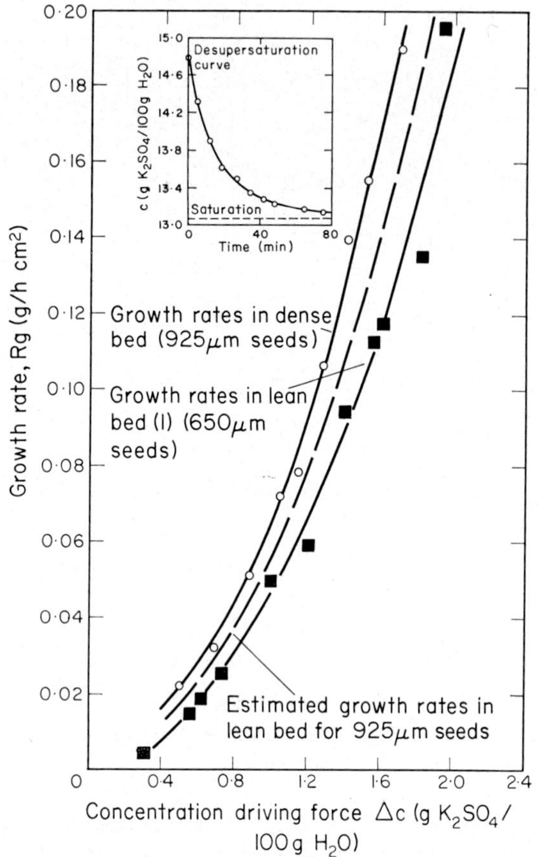

Fig. 6. Comparison of growth rates at 30 °C for dense and lean beds.

For the desupersaturation run the final weight of the fluidized crystal bed was over 390 g, corresponding to a bed voidage of about 0.85. The corresponding values for the 5 g bed charge were 15 g and 0.98. In both cases the mean crystal product size was about 1100 µm. The growth rates measured by the two methods show very good agreement and the data may be correlated by the equation*

$$R_g = 1.15 \times 10^{-5} (c - c^*)^2, \qquad (1)$$

where the units of R_g are g/cm² s (see ref. 1).

Calculated growth rates for a desupersaturation run at 30 °C are shown in fig. 6. No measurements were made for 925 µm seeds in lean beds at 30 °C, but lean-bed growth rates for smaller seeds (625 µm) at 30 °C[1] included in fig. 6 are about 50 % lower than those from the dense bed. Earlier work[1] indicated that, when grown in a fluidized bed, the growth rate of potassium

* See Nomenclature at the end of this paper.

sulphate crystals was proportional to $d^{0.6}$, where d is the crystal sieve aperture size. Applying this correction to the data for 650 µm seeds in fig. 6 gives the broken line. The effect of temperature on growth rate has also been determined[1] and using the appropriate temperature dependence, growth data for 925 µm seeds at 20 °C were extrapolated to 30 °C. This again gives results that lie on the broken line in fig. 6.

In contrast to the data for 20 °C, therefore, the results at 30 °C indicate that growth rates in the dense bed (minimum voidage $\varepsilon \sim 0.85$) are 15–20 % higher than in the lean bed (minimum voidage $\varepsilon \sim 0.98$). As with the data at 20 °C, however, both lean and dense bed results indicate that the growth rate is proportional to the square of the concentration driving force.

The speed with which data can be obtained from measurements in dense beds is a great advantage over the technique used with lean beds[1]. Only one desupersaturation run is required to calculate the growth rate for the whole range of solution concentration. It is essential, however, in using the desupersaturation method to ensure that sufficient seed crystals are employed so that all the desupersaturation is due to mass deposition on the seed crystals and no nucleation occurs.

4. Cooling crystallization

Many large-scale crystallization operations are carried out by batchwise cooling. A supersaturated solution is seeded with suitable crystals and the desired supersaturation level is maintained by a controlled rate of cooling. To obtain a uniformly sized crystal product it is essential that all the deposition of crystalline matter takes place on the seed crystals and that no spurious nucleation occurs. The temperature is therefore controlled so that the system is kept in the metastable state throughout the operation, and the rate of growth of the seeds is governed solely by the rate of cooling.

Two series of cooling crystallization runs were carried out to investigate the effect of (a) seed quantity and (b) seed size on the product yield. The effect of seed quantity was investigated using initial seed charges ranging from 25 to 200 g, for two seed sizes of 460 µm (500–421 µm or 30–36 mesh B.S. sieves) and 1100 µm (1200–1000 µm or 14–16 mesh B.S. sieves). The effect of seed size was investigated using 25 g seed charges which ranged in size from 325 to 1300 µm.

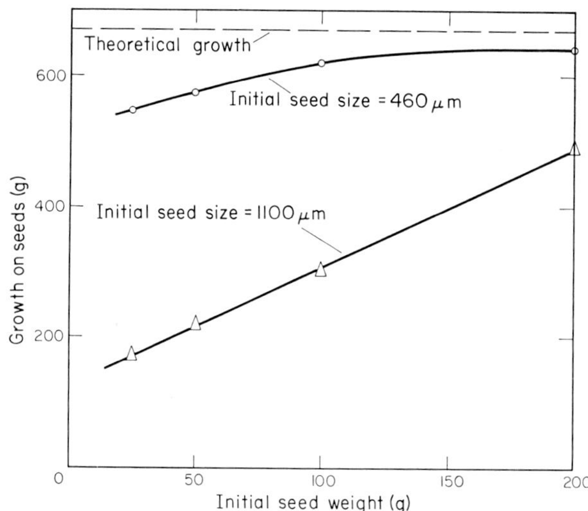

Fig. 7. Cooling crystallization – effect of seed size and quantity.

In all the cooling crystallization runs a constant volume of 13 litres of solution saturated at 50 °C was used. The solution was cooled to 45 °C before adding the seeds and then cooled from 45 °C to 20 °C, at a constant rate of 0.1 °C/min, by controlling the flow of water through the cooler. The solution velocity in the crystallization section was kept constant at 13.4 cm/s. At the end of the run the crystals were removed from the crystallizer, filtered, washed in methanol and dried.

Fig. 7 shows the growth on seed crystals as a function of initial seed weight for two seed sizes of 460 and 1100 μm. Because of the surface area effect, the seed size has a greater influence on the product yield than has the mass of seeds added. For example, using 25 g of seeds of mean size 460 and 1100 μm the product obtained was 570 and 195 g respectively. These yields are 86.5 and 26% respectively of the theoretical yield. The difference between the theoretical and actual yields may be taken as a measure of the nucleation that had occurred during the growth period.

It can be seen from fig. 7 that in order to obtain high product yield, the seed crystals should be small (e.g. 460 μm size) so as to provide sufficient area on which growth can occur. It can also be seen that a minimum quantity of about 125 g of 460 μm size seed crystals is required to obtain the maximum yield without any nucleation. This corresponds to 95% of the theoretical, the difference between the theoretical and actual yield being mainly due to the residual supersaturation of the solution at the end of the run.

TABLE 1

Mean product size and coefficient of variation for runs with 460 μm seeds

Seeds (g)	Product (g)	MS/CV (μm/percent)
25	571	1580/13.1
50	624	1230/10.2
100	740	1040/8.6
200	843	825/11.5

The effect of seed quantity on the product size distribution for the 460 μm seed runs is shown in table 1. The size distribution is characterised in terms of the median size, MS, and the coefficient of variation, CV, defined as[4])

$$CV = 100(d_{84} - d_{16})/2\,d_{50},\qquad(2)$$

where d is the equivalent sieve aperture size of the crystals.

5. Conclusions

(1) A decrease in the latent period of the system occurs as a result of an increase in supersaturation, the introduction of seeds or an increase in solution velocity.
(2) The nucleation decreases as the area of seed crystals is increased.
(3) Crystal growth rates in dense beds ($\varepsilon \sim 0.8$–0.85) are within about $+15\%$ of those obtained in lean beds ($\varepsilon > 0.98$).
(4) In order to obtain a high yield in cooling crystallizers, small seed crystals are required to provide a large surface area on which growth can occur. For example, the yield decreases from 95% of theoretical with 25 g of 328 μm seeds to less than 20% with the same mass of 1300 μm seeds.

Nomenclature

a	surface area of a single crystal
A	surface area of crystals
c	solution concentration
c^*	solution concentration at saturation
CV	coefficient of variation [eq. (2)]
d	equivalent sieve aperture size of crystal
$d_{16,50,84}$	equivalent sieve aperture sizes corresponding to 16, 50 and 84 cummulative per cent
f_s	surface shape factor ($= a/d^2$)
f_v	volume shape factor ($= v/d^3$)
F	overall shape factor ($= f_s/f_v$)

M	mass of crystals
MS	median size of crystals [eq. (2)]
n	number of crystals
R_g	crystal growth rate
S	supersaturation ratio ($= c/c^*$)
t	time
v	volume of a single crystal
W	mass of water in solution
ρ_c	crystal density

Subscripts

i	initial value
1	value after first time interval

Appendix

Calculation of crystal growth rates from solution desupersaturation curves

At time $t = 0$,

    initial seed size $= d_i$,

    initial mass of crystals $= M_i$,

    initial solution concentration $= c_i$.

Assume

(1) no nucleation occurs, hence the number of crystals in the crystallizer remains constant;

(2) volume and surface shape factors, f_v ($= v/d^3$) and f_s ($= a/d^2$), remain constant.

The product $nf_v\rho_c$ remains constant and is given by $nf_v\rho_c = M_i/d_i^3$.

At time $t = t_1$,

    solution concentration $= c_1$, and

    mass of solid deposited on crystals, $\Delta M_1 = W (c_i - c_1)$,

    mass of crystals in crystallizer, $M_1 = M_i + \Delta M_1$,

    mean size of crystals, $d_1 = M_1/(nf_v \rho_c)^{\frac{1}{3}}$,

    area of crystals, $A_1 = M_1 F/\rho_c d_1$.

Mean growth rate during period 0 to t_1,

$R_{g\,1} = \Delta M_1/A_1 t_1$.

Mean driving force at which this growth occurs $= \frac{1}{2}(c_i + c_1) - c^*$.

The above calculation is repeated for successive time intervals.

The change in solution concentration for any one interval is not greater than 0.1 g K_2SO_4/100 g H_2O.

References

1) J. W. Mullin, and C. Gaska, Can. J. Chem. Engng. **47** (1969) 483.
2) J. W. Mullin, J. Garside and C. Gaska, Chem. Ind. (1966) 1704.
3) C. Gaska, Ph.D. Thesis, University of London, 1966.
4) J. W. Mullin, *Crystallization* (Butterworths, London, 2nd. edition, 1972).

Section XI

Hydrothermal growth

Y. TOUDIC
A. REGRENY
M. PASSARET
R. AUMONT
J. F. BAYON
M. L. BARSUKOVA
V. A. KUZNETSOV
A. N. LOBACHEV
YU. V. SHALDIN

R. DHARMARAJAN
R. F. BELT
R. C. PUTTBACH
L. N. DEMIANETS
I. P. KUZMINA
E. N. EMELIANOVA
NGUYEN DUC CHINH
S. MROCZKOWSKI
J. ECKERT

Journal of Crystal Growth **13/14** (1972) 519–523 © *North-Holland Publishing Co.*

NOUVEAU TYPE D'APPAREILLAGE APPLIQUÉ À LA CROISSANCE HYDROTHERMALE DE MONOCRISTAUX DE CINABRE DE GRANDES DIMENSIONS

Y. TOUDIC, A. REGRENY, M. PASSARET, R. AUMONT et J. F. BAYON

Laboratoire CNET–CNRS, Centre National d'Etudes des Télécommunications,
Route de Trégastel, 22 – Lannion, France

A new apparatus for hydrothermal growth of cinnabar (α-HgS) single crystals has been designed and tested. Since corrosive Na_2S solutions were used as solvents at 260–275 °C, the first runs were performed in conventional vessels lined with PTFE and closed hermetically. The degree of filling needed in the experiments to have one liquid phase in the autoclave, under growing conditions, may induce important pressure variations even for small temperature variations which would modify equilibria in the system; moreover, conventional apparatus does not permit pressure measurements. To overcome these drawbacks, an apparatus has been developed for working in one liquid phase from the beginning of the experiment, while controlling the pressure. Single crystals of cinnabar up to 20 g were grown in this apparatus.

1. Introduction

Le cinabre (α-HgS) cristallise dans le système trigonal et possède des propriétés très intéressantes: grand pouvoir rotatoire, forte biréfringence, indices de réfraction élevés. La difficulté de se procurer des cristaux de cinabre naturel a amené de nombreux chercheurs[1-12] à mettre au point différentes méthodes d'obtention de ce matériau, les plus utilisées étant la méthode de transport en phase vapeur[1,2,4] et la méthode hydrothermale[5,6,8,10,12]. Pour notre part, l'intérêt du cinabre en optique non linéaire mis en évidence ces dernières années au CNET[13-15] et aux Etats-Unis[16,17] sur des échantillons de cinabre naturel, nous a conduit à rechercher une méthode permettant d'obtenir des cristaux synthétiques de bonne qualité et de taille suffisante utilisables à cet effet. Dans des publications précédentes[6,12] nous avons montré la possibilité d'obtenir des monocristaux de cinabre par voie hydrothermale en solution de sulfure de sodium (Na_2S) en utilisant la réversibilité de la réaction $HgS + S^{--} \rightleftarrows HgS_6^{--}$ (réfs. 18, 19). Cet article présente les résultats obtenus avec un nouveau type d'appareillage.

2. Etudes de solubilité

Des études de solubilité avaient été effectuées par Dickson[18,19] dans des solutions de Na_2S entre 50 °C et 250 °C, 1 et 1800 bar. Cependant, pour obtenir une solubilité importante[20] tout en opérant à des tempé-

ratures relativement basses (200 °C à 275 °C), nous avons utilisé des solutions de Na_2S à 0.5 M/l alors que les courbes publiées par Dickson étaient relatives à des domaines de concentrations nettement moins élevés. Les solubilités sont déterminées par la méthode de perte de poids dans des autoclaves de 25 cm^3 chemisés en PTFE et fermés hermétiquement par des obturateurs de même nature, la solution utilisée étant très corrosive. Cette méthode présente l'inconvénient de ne pas permettre la mesure de la pression interne de l'autoclave. Les résultats sont reportés sur les courbes des figures 1 et 2. Ces courbes montrent que, dans les conditions expérimentales, la solubilité augmente avec la température et diminue avec la pression.

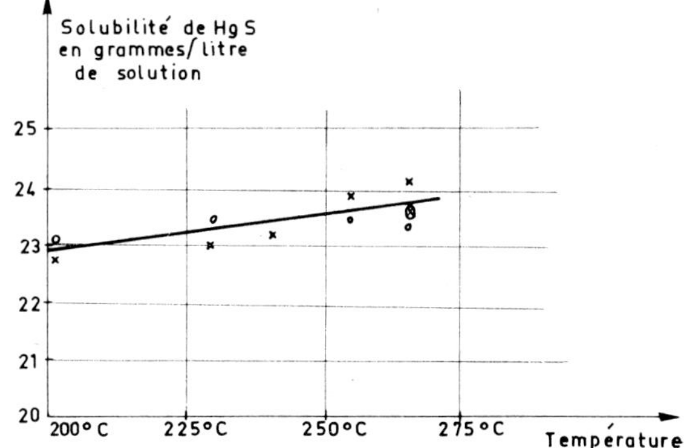

Fig. 1. Solubilité du cinabre en fonction de la température dans une solution de Na_2S à 0.5 Mole/litre, avec un remplissage initial de 87%. ($\times$) mesures faites en solution en dosant Hg par absorption atomique; ($\circ$) mesures donnant les variations de poids du cristal de départ.

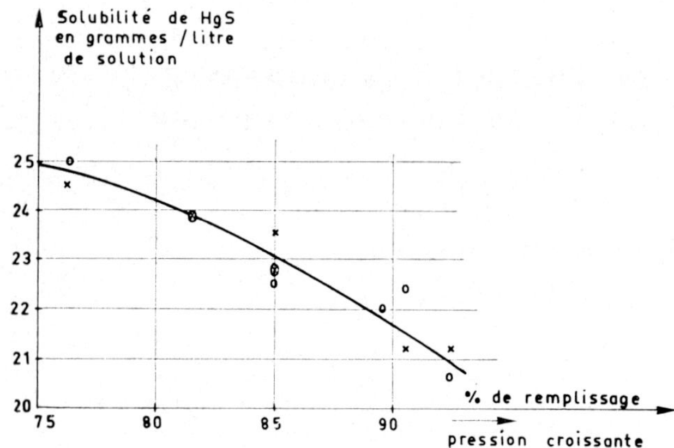

Fig. 2.   Solubilité du cinabre en fonction du remplissage dans une solution de Na_2S à 0.5 Mole/litre, à température constante: 265 °C. ($\times$) mesures faites en solution en dosant Hg par absorption atomique; ($\bigcirc$) mesures donnant les variations de poids du cristal de depart.

3. Croissance hydrothermale

Les premiers essais de croissance ont été effectués dans des autoclaves de 25 cm^3 et 250 cm^3 chemisés en PTFE et fermés hermétiquement afin de travailler dans des conditions très propres (type de fermeture: cône sur cône), mais ne permettant pas la mesure de la pression.

Dans ce système, classique en hydrothermal, le corps-mère utilisé est constitué de α-HgS obtenu par transport en phase vapeur à partir de poudre de vermillon, méthode donnant des agglomérats de petits cristaux plus purs que la poudre et d'une utilisation plus commode.

Les germes sont des cristaux de cinabre provenant d'essais antérieurs. Les expériences ont été réalisées en solution de Na_2S à 0.5 M/l à des températures variant de 260 °C à 275 °C avec un remplissage initial de l'ordre de 90%, le coefficient de passage de l'écran étant le plus souvent de 5%. Ces conditions ont permis l'obtention de cristaux de 900 mg dans les autoclaves de 25 cm^3 et de 8 g dans les autoclaves de 250 cm^3. Pour un gradient de température d'environ 5 °C, les vitesses de croissance sont alors de 0.1 à 0.2 mm/jour pour les faces {0001} et de 25 à 50 µm/jour pour les faces {10$\bar{1}$0}.

L'obligation d'avoir un remplissage important au départ afin d'obtenir une phase unique à l'intérieur de l'autoclave dans les conditions opératoires, peut entraîner des variations de pressions importantes si la température varie même faiblement et par conséquent modifier les équilibres à l'intérieur du système. D'autre part, ce type d'appareillage ne permet pas la mesure de la pression. Pour remédier à ces inconvénients, nous avons imaginé un dispositif permettant, d'une part de travailler en une seule phase liquide dès le début de l'essai et, d'autre part, de contrôler la pression.

3.1. Appareillage

Le dispositif utilisé est représenté par le schéma de la figure 3. L'ensemble comprend quatre parties principales:
– l'autoclave A,
– l'enceinte en PTFE: B,
– le réservoir de pression: C,
– la bouteille d'argon: D.

(1) L'autoclave proprement dit A, de volume intérieur 8 litres, possède une fermeture de type sphère sur cône et peut travailler à 300 bar jusqu'à la température de 400 °C. Le chauffage de l'autoclave est assuré par des colliers chauffants serrés sur une virole qui sert en outre à maintenir l'autoclave en place. La partie correspondant à la zone de dissolution est calorifugée par de la laine de roche, la partie correspondant à la zone de recristallisation étant laissée sans calorifuge afin d'en favoriser le refroidissement par l'air.

(2) L'enceinte B réalisée en PTFE, est constituée de 3 parties:
– un soufflet en PTFE susceptible de s'allonger d'environ la moitié de sa longueur, et qui forme la partie "dissolution";
– un tube en PTFE possédant une collerette à chaque extrémité et qui forme la partie "recristallisation"; le tube est séparé du soufflet par un écran convenablement percé;
– une fermeture en PTFE à la partie supérieure du tube.

Ces trois pièces sont reliées entre elles par des brides en acier inoxydable afin d'assurer une très bonne étanchéité en cours de fonctionnement. Le volume utile de l'enceinte ainsi définie est actuellement d'environ un litre. Une enceinte d'un volume utile de 2.5 litre est en cours de montage.

L'ensemble est porté par la tête d'autoclave à l'aide de tiges support. Le soufflet en PTFE est placé à l'intérieur d'un tube en acier inoxydable, ce qui permet d'obtenir des gradients de température importants (jusqu'à 20 °C) à l'intérieur de l'enceinte B alors que ce gradient serait beaucoup plus faible si l'on n'utilisait pas cet artifice.

Ce système permet de séparer l'enceinte B en deux

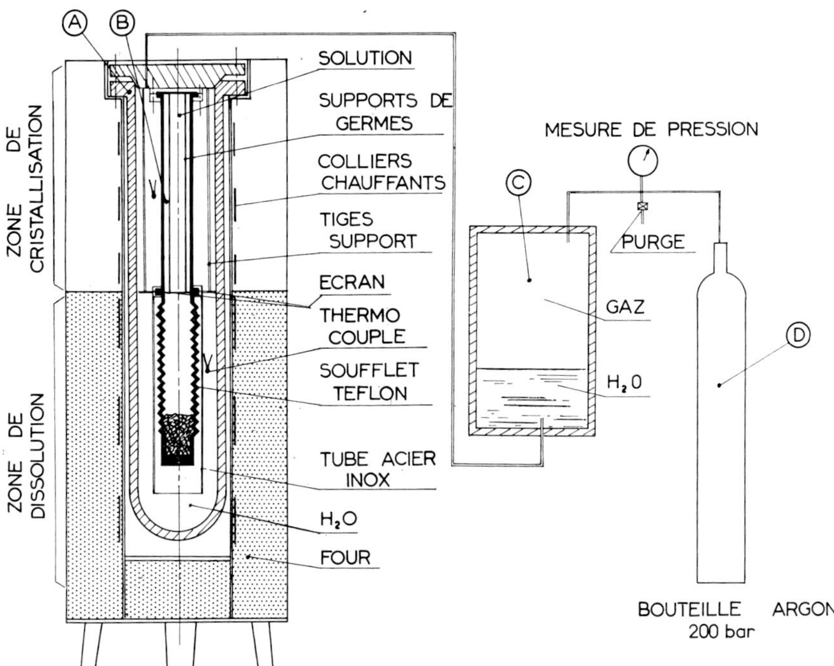

Fig. 3. Appareillage.

zones pratiquement isothermes séparées par un gradient de température abrupt au niveau de l'écran. Ce profil de température correspond à des conditions très favorables pour la recristalisation hydrothermale. Des thermocouples disposés le long de l'enceinte B dans les zones de solubilisation et de croissance permettent la mesure des températures à l'intérieur de l'autoclave. Un étalonnage préalable permet en outre de connaître les températures correspondantes à l'intérieur de l'enceinte B.

(3) La réserve C a un volume de 4 litres.

(4) La bouteille d'argon sert de générateur de pression. La mesure de pression se fait à l'aide d'un manomètre à tube de Bourdon.

3.2. Mode opératoire

Le corps-mère constitué par des agrégats de cristaux de sulfure mercurique (400 à 500 g) obtenus par transport en phase vapeur, est introduit dans l'enceinte B. Les germes sont accrochés sur leur support en PTFE, à l'aide de fils de même nature. Le soufflet est alors comprimé et l'enceinte B totalement remplie d'une solution de Na_2S à 0.5 M/l fraîchement préparée à partir de cristaux de Na_2S, $9H_2O$ "Merck" et d'eau distillée, est fermée hermétiquement.

L'enceinte B est introduite dans l'autoclave A. Celui-ci est rempli d'eau et mis en communication avec le réservoir C contenant environ un litre d'eau, surmonté par une pression d'argon fournie par la bouteille D, la pression étant appliquée à froid dès le début de la manipulation. Lors de la montée en température, la solution à l'intérieur de B se dilate et le soufflet se déforme évitant ainsi la rupture de l'enceinte. Le trop-plein de liquide dû à la déformation de B et à la dilatation de l'eau dans l'enceinte A passe dans le réservoir C. Cependant, le volume de la bouteille D assurant la mise en pression est suffisamment important pour que la variation de pression résultante reste faible au cours de la montée en température et pour qu'il n'y ait plus de variations une fois les conditions de manipulation atteintes.

3.3. Résultats

Les essais de recristalisation ont été effectués avec des températures vraies de l'ordre de 225 °C, les gradients variant de 10 °C à 20 °C. Le tableau 1 donne les résultats obtenus au cours de deux expériences.

Dans les autoclaves de type classique, les expériences ont été réalisées à des températures de recristalisation nettement supérieures (au-dessus de 250 °C) et des gra-

TABLEAU 1

Manipula-tion	Durée (jours)	P (bar)	Températures de recristallisation (°C)	Températures de dissolution (°C)	Gradient (°C)	Poids des germes (g)	Poids final (g)	Orientation des germes	Vitesse de croissance (µm/jour)
5 H	56	110	225	237.5	12.5	3.612	13.383		
						2.591	10.744		
						5.341	21.820		
6 H	40	185	225	237.5	12.5	435	3.880	‖ à {10$\bar{1}$0}	50
						727	6.830	‖ à {10$\bar{1}$0}	56
						1037	9.200	‖ à {10$\bar{1}$0}	62.5

Fig. 4.   Monocristaux de cinabre obtenus dans la manipulation 6H.   ×2.

dients de 5 °C à 7 °C. Les vitesses de croissance obtenues sont sensiblement identiques dans les deux cas, l'abaissement de la température de recristallisation dans le cas présent, étant compensée par une augmentation du gradient.

La figure 4 représente les cristaux obtenus dans la manipulation 6H.

Les études de caractérisation effectuées sur les échantillons de cinabre ont essentiellement porté sur:

(1) Des études optiques sur des coupes de cristaux qui ont permis de montrer la présence de nombreux défauts localisés à l'interface germe-cristal. Pour une même qualité de germes, ces défauts sont plus importants dans les cristaux obtenus dans les autoclaves de type classique (25 cm³ et 250 cm³) que dans les cristaux obtenus dans l'autoclave précédemment décrit. On peut donc penser que l'existence d'une phase unique pendant toute la durée de la manipulation permet d'éviter l'influence néfaste qu'auraient sur la qualité des cristaux, les fluctuations de pression dans le système.

(2) Des études cristallographiques de perfection cristalline par la méthode de Barraud–Lambot[21,22]) qui ont permis de montrer que les désorientations maximales obtenues sur les meilleurs échantillons sont de l'ordre de quelques secondes.

(3) Une détermination aussi complète que possible des impuretés présentes dans les échantillons par spectrophotométrie d'absorption atomique et par spectrométrie de masse, les impuretés principales étant des impuretés métalliques (Ca, Fe, Zn, Cu, Ag). La somme de ces impuretés est de l'ordre de 50 µg/g.

(4) Une étude de la transmission optique dans les spectres visible et infra-rouge.

4. Conclusion

Les conditions de croissance dans un appareillage de ce type semblent favorables à une bonne recristallisation de cristaux de cinabre de grandes dimensions et de bonne qualité. Il est possible d'envisager son utilisation pour la croissance d'autres cristaux nécessitant de faibles températures de recristallisation.

Cette étude a été rendue possible grâce à l'aide de la D.G.R.S.T. (Contrat n° 69.01.78).

Bibliographie

1) D. R. Hamilton, Brit. J. Appl. Phys. **9** (1958) 103.
2) O. L. Curtis, J. Appl. Phys. **33** (1962) 2461.
3) P. M. Koval's'kii, V. M. Melnik et A. D. Shneider, Ukr. Fiz. Zh. **11** (1966) 921.
4) E. H. Carlson, J. Crystal Growth **1** (1967) 271.

5) H. Rau et A. Rabenau, Solid State Commun. **5** (1967) 331.

6) Y. Toudic et R. Aumont, Compt. Rend. (Paris) **269** (1969) 74.

7) E. Cruceanu et N. Nistor, J. Crystal Growth **5** (1969) 206.

8) S. D. Scott et H. L. Barnes, Mater. Res. Bull. **4** (1969) 897.

9) A. F. Armington et J. J. O'Connor, J. Crystal Growth **6** (1970) 278.

10) A. Pajaczkowska, J. Crystal Growth **7** (1970) 93.

11) R. W. Garner et W. B. White, J. Crystal Growth **7** (1970) 343.

12) Y. Toudic et R. Aumont, J. Crystal Growth, à paraître.

13) J. Jerphagnon, E. Batifol, G. Tsoucaris et M. Sourbe, Compt. Rend. (Paris) B **265** (1967) 495.

14) J. P. Noblanc, J. Loudette et G. Duraffourg, Solid State Commun. **5** (1967) 803.

15) J. Jerphagnon, Etude CNET/PCM n° 851 (1967).

16) W. L. Bond, G. D. Boyd et H. L. Carter, J. Appl. Phys. **38** (1967) 4090.

17) G. D. Boyd, Th. J. Bridges et F. G. Burkhardt, J. Quantum Electron. *Q.E.* **4** (1968) 515.

18) F. W. Dickson, Econ. Geol. **59** (1964) 625.

19) F. W. Dickson, Bull. Volcanologique **29** (1966) 605.

20) R. A. Laudise et E. D. Kolb, Endeavour **28** (105) (1969) 114.

21) J. Barraud, Bull. Soc. Franc. Mineral. (1951) 223.

22) H. Lambot, L. Vassamillet et J. Dejace, Acta Met. **1** (1953) 711

Journal of Crystal Growth **13/14** (1972) 524–529 © *North-Holland Publishing Co.*

RECRISTALLISATION DE TiO_2, GeO_2, SiO_2, $Si_{1-x}Ge_xO_2$ EN SOLUTIONS HYDROTHERMALES FLUORÉES

M. PASSARET, Y. TOUDIC, A. REGRENY, R. AUMONT et J. F. BAYON

Laboratoire CNET–CNRS, Centre National d'Etudes des Télécommunications,
Route de Trégastel, 22 – Lannion, France

The crystallisation of titanium, germanium, and silicon oxides in KF, CsF, and RbF solutions was studied at 300–600 °C and 500–1500 atm. Crystallisation of mixed $Si_{1-x}Ge_xO_2$ crystals, with Ge atomic concentration up to 5%, is described. The spontaneous crystallisation of rutile, quartz and mixed $Si_{1-x}Ge_xO_2$ crystals and their growth on seed crystals was studied. The optimum concentration of solvent for growth of a particular oxide in the range 0.5–5 M alkali fluoride differs with the alkali. When the temperature is higher, germanates or fluorides are obtained.

1. Introduction

Depuis longtemps le quartz synthétique est fabriqué par voie hydrothermale dans des solutions de soude, cependant, nous savons qu'un tel solvant est impropre à la recristallisation de TiO_2 et GeO_2. Aussi, nous avons été amenés à utiliser des solutions de fluorures alcalins, qui peuvent se prêter particulièrement bien à la solubilisation et à la recristallisation de ces oxydes à haute température et sous pression. Nous avons pu, par cette méthode, obtenir des cristaux de rutile et de quartz, ainsi que des cristaux de quartz renfermant en substitution plusieurs atomes pour cent de germanium.

2. Etude de solubilité

Pour savoir s'il était envisageable de faire la synthèse hydrothermale de ces oxydes, on a mesuré la solubilité de TiO_2 et SiO_2 dans des solutions aqueuses de KF, CsF et éventuellement RbF.

Les mesures ont été effectuées par la méthode de perte de poids: des cristaux de quartz ou de rutile (Verneuil) sont introduits dans des capsules de métaux nobles en présence du solvant. La capsule, une fois scellée, est placée dans un autoclave de $10\ cm^3$ (Tem-Pres) avec une contre pression d'argon, puis portée à la température de travail et à la pression prédéterminée. L'expérience dure quatre jours, temps au bout duquel on ne décèle plus de variation de poids de l'échantillon de départ. L'autoclave est alors refroidi rapidement pour éviter une précipitation importante sur les cristaux. Dans toutes ces expériences, il se forme en effet au cours de la descente en température de fines aiguilles blanches d'anatase, en accord avec le diagramme TiO_2–H_2O publié par Osborn[1]).

Les courbes des figures 1 et 2, permettent de suivre les variations de poids des cristaux de rutile, exprimées en g/l de solution, en fonction de la température, dans des solutions de KF 2 N et 3 N à 1000 bar et CsF, RbF, 2 N et 5 N à 1000 et 1500 bar.

Nous constatons qu'à partir d'une certaine tempé-

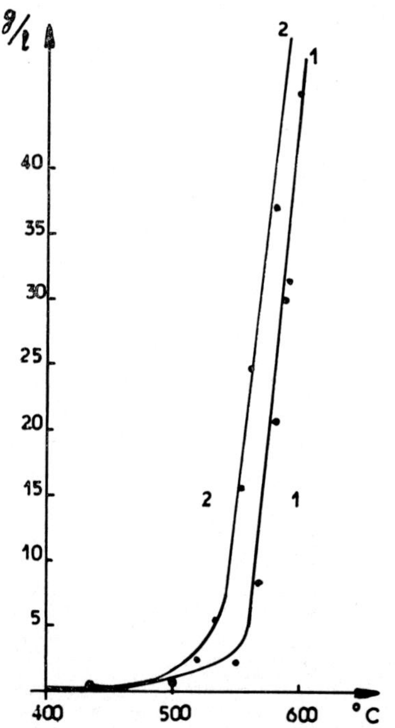

Fig. 1. Solubilité du rutile. (1) dans KF 2N, $P = 1000$ bar; (2) dans KF 3N, $P = 1000$ bar.

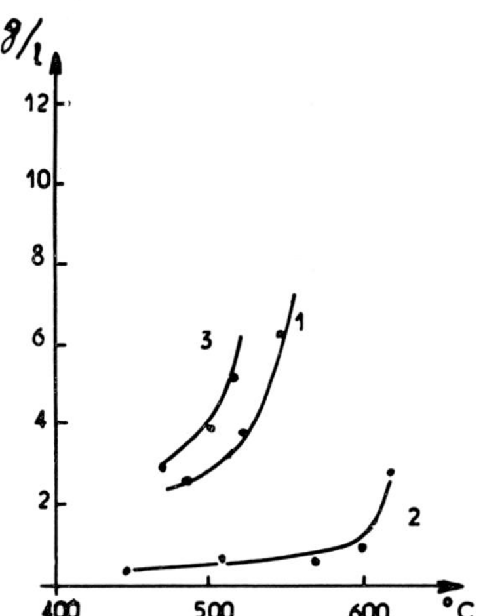

Fig. 2. Solubilité du rutile. (1) dans RbF 5N, $P = 1000$ bar;
(2) dans CsF 2N, $P = 1500$ bar; (3) dans CsF 5N, $P = 1000$ bar.

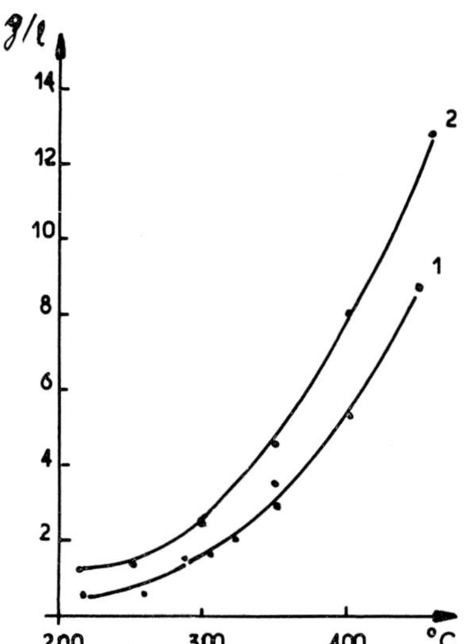

Fig. 3. Solubilité du quartz. (1) dans KF 2N, $P = 500$ bar;
(2) dans KF 3N, $P = 500$ bar.

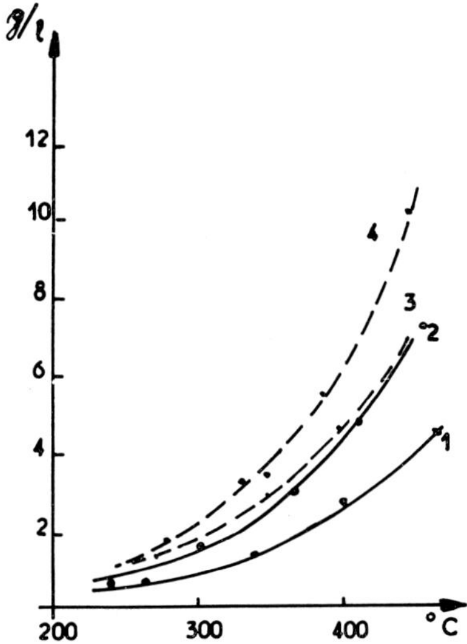

Fig. 4. Solubilité du quartz. (1) dans CsF 1N, $P = 500$ bar;
(2) dans CsF 2N, $P = 500$ bar; (3) dans CsF 1N, $P = 1000$ bar;
(4) dans CsF 2N, $P = 1000$ bar.

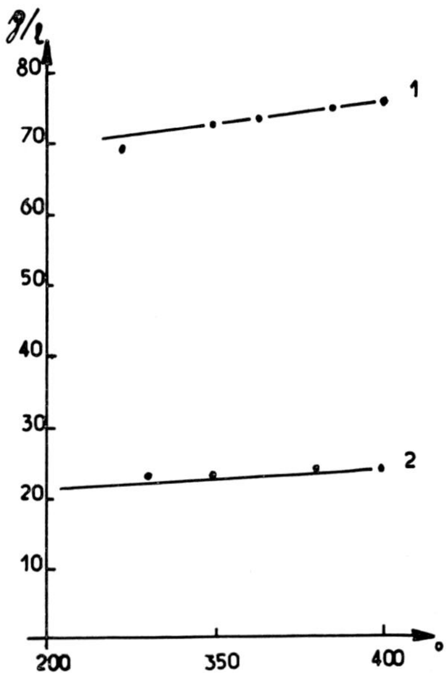

Fig. 5. Solubilité du quartz dans la soude. (1) dans NaOH 1N,
$R = 80\%$; (2) dans NaOH 0.5N, $R = 80\%$.

rature, d'autant plus basse que la concentration du solvant est élevée, la solubilité croît très rapidement. On note dans ces conditions l'apparition à température ambiante et pression atmosphérique de complexes du type TiF_6K_2, TiF_6Cs_2, etc., dont la nature est confirmée par diffraction de rayons X et analyses chimiques.

Si la pression est inférieure à 1000 bar, pour une même concentration du solvant, ce phénomène apparaît à des températures nettement plus basses où la solubilité est faible.

Les courbes des figures 3 et 4 relatives aux variations de poids du quartz dans des solutions de KF et CsF permettent de faire les mêmes remarques que précé-

demment, mais il sera possible dans ce cas de travailler à des pressions plus faibles (500 à 1000 bar) et des températures moins élevées (450 °C), domaine où les solubilités sont plus importantes que pour le rutile.

Par contre, les courbes de la figure 5, relatives à la solubilité du quartz dans la soude, montrent que le phénomène est alors linéaire, dans les domaines de températures et de pressions choisis.

Pour étudier la solubilité de GeO_2 par cette méthode,

TABLEAU 1

La solubilité et Ge/Si pour $GeO_2 + SiO_2$

Produits de départ (g)	Solvant	t (°C)	P (bar)	Ge en solution (atome g/ml)	Si en solution (atome g/ml)	Ge/Si rapport atomique
GeO_2 0.2655 SiO_2 0.232	CsF 1N 0.9 cm³	450	1000	58×10^{-6}	14×10^{-6}	4.1
GeO_2 1.4 SiO_2 1.6	CsF 0.5N 4 cm³	450	1000	34.2×10^{-6}	9.9×10^{-6}	3.4
GeO_2 1.25 SiO_2 1.1	CsF 3N 3.8 cm³	300	1000	21.3×10^{-6}	4.8×10^{-6}	4.5
GeO_2 1 SiO_2 1	KF 2N 3 cm³	450	1100	23×10^{-6}	6.8×10^{-6}	3.4
GeO_2 0.880 SiO_2 1	KF 3N 3 cm³	450	1100	37.5×10^{-6}	9.3×10^{-6}	4

TABLEAU 2

Recristallisation de TiO_2–SiO_2–GeO_2 dans des solution de fluorures alcalins

Produits de départ	Solvant	t (°C)	Pression (bar)	Poids du germe (mg)	Poids du cristal obtenu (mg)	Observations
TiO_2 poudre	KF 2N	570	1500			Transport de cristaux noirs, $l = 1$ mm
Rutile Verneuil	KF 2N	560	1500	122	246	Durée 6 jours
Rutile en flux	RbF 5N	520	1100			Cristaux noirs isométriques de rutile
Rutile Verneuil	CsF 5N	485	1750	85	343	Durée 4 j, $V[110] = 0.375$ mm/j
Rutile Verneuil	CsF 5N	470	500	85	225	Durée 5 j, $V[110] = 0.3$ mm/j
Quartz	KF 2N	450	1100	85	186	Durée 3 j, $V[110] = 0.45$ mm/j
Quartz	KF 2N	280	500	85	92	Durée 8 j
Quartz	KF 2N+HF, pH 3	280	500	82	31	Dissolution du germe
Quartz	CsF 3N	450	1200			Transport de petits cristaux de quartz
Quartz	CsF 1N+HF, pH 1	280	500	94	82	Dissolution du germe
GeO_2 hex	KF 3N	500	1200			Transport de GeO_2 quadratique
GeO_2 hex	KF 3N	500	800			Transport de GeO_2 quadratique $+ K_2Ge_4O_9$
GeO_2 hex	RbF 2N	350	870			Corps-mère transformé GeO_2 quadratique
GeO_2 quad	CsF 0.2N	400	1100			Transport de l'ensemble GeO_2 quadratique
GeO_2 hex	CsF 2N	400	1100			Transport de monocristaux de GeO_2 quadratique
GeO_2 hex	CsF 2N	400	1100			Transport de GeO_2 quadratique

il faudrait posséder des cristaux suffisamment importants. Désirant fabriquer des cristaux de quartz dopés au germanium, nous nous sommes bornés en partant d'un mélange GeO_2 hexagonal–SiO_2 quartz, à mesurer les quantités de Germanium et de Silicium dans des solutions de fluorures alcalins, ramenés à pression et température ambiante. La solubilité de GeO_2 dépend de sa forme polymorphique; nous vérifions par diffraction de rayons X, la structure hexagonale du produit de départ. Notons que dans les conditions opératoires utilisées (300 à 450 °C, 1000 à 1100 bar), même dans l'eau, le GeO_2 se transforme entièrement sous sa forme quadratique.

Le tableau 1 met en évidence la plus grande solubilité de GeO_2 dans les conditions données. Si la concentration du solvant devient élevée (KF, 5 N par exemple) un fluorure double de silicium et d'alcalin précipite ainsi qu'un germanate ($K_4Ge_9O_{20}$).

Cette étude permet donc d'envisager la recristallisation hydrothermale du quartz et de cristaux mixtes SiO_2–GeO_2, dans des solutions de CsF et KF à des températures de l'ordre de 450 °C et des pressions comprises entre 500 et 1000 bar.

Pour le rutile, il faudra généralement travailler à des températures et des pressions plus élevées (470–600 °C; 1000–1500 bar) afin de rester dans un domaine où les réactions sont réversibles. Cependant, en prenant des solutions concentrées de RbF et CsF (5 N) il est possi-

XI – 2

TABLEAU 3

Recristallisation de cristaux mixtes du type $Si_{1-x}Ge_xO_2$

Corps-mère (g)	Solvant	t (°C)	Pression (bar)	Résultats	Rapport atomique Ge/Si dans cristaux
Quartz 0.820 GeO$_2$ hex 0.800	KF 2N	450	1000	Epitaxie sur un germe de quartz, $\Delta p = 329.3 - 71 = 258.3$ mg	1/28 (croissance)
Quartz 1.113 GeO$_2$ hex 0.886	KF 3N	450	1100	Epitaxie sur un germe de quartz, $\Delta P = 177.3 - 85.7 = 91.6$ mg, nombreuses nucléations	1/18 (nucléations) 1/18 (croissance)
Quartz 1.132 GeO$_2$ hex 1.000	KF 3N	450	1000	Croissance sur germe de quartz Perturbation par les parois de la capsule, $V_z = 0.23$ mm/j, $\Delta p = 324 - 67 = 257$ mg, 8 j	
Quartz 1.100 GeO$_2$ hex 1.250	CsF 3N	450		Pas de germes au départ, nombreuses nucléations	1/18 (nucléations)
Quartz 0.232 GeO$_2$ hex 0.2655	CsF 1N	450	1000	Nombreuses nucléations, cristaux isométriques de 3 mm de long	1/20 (nucléations)
Quartz 0.400 GeO$_2$ hex 0.400	CsF 0.2N	450	1200	Nombreuses nucléations	1/17 (nucléations)
Quartz 0.830 GeO$_2$ hex 0.920	CsF 0.2N	450	1200	Croissance d'un germe, $\Delta p = 225.6$ mg	1/28 (croissance)
Quartz 1 GeO$_2$ hex 1	CsF 2N+ HF pH 5.5	280	500	Très légère croissance d'un germe, $\Delta p = 1.5$ mg	
Quartz 1 GeO$_2$ hex 1	CsF 2N+ HF pH 5.5	450	1100	Croissance d'un germe+nucléations, $\Delta p = 196.5 - 69.7 = 126.8$ g, nucléations 70 mg	1/20 (nucléations) 1/18 (cristal)
Quartz 1 GeO$_2$ hex 0.200	KF 2N	450	1000	Nucléations 42 mg	1/175 (cristal)

ble de recristalliser du rutile à des pressions beaucoup plus faibles (500 bar) (voir tableau 2).

3. Recristallisation de TiO_2, SiO_2, GeO_2

Les études de recristallisation ont été généralement faites dans les mêmes autoclaves que ceux utilisés pour les mesures de solubilité, en utilisant des capsules de métaux nobles plus longues, afin d'obtenir un gradient de température suffisant.

3.1. RUTILE

Les différentes manipulations faites dans des solutions de KF, CsF ou RbF dans les domaines de températures et de pressions décrits précédemment ont permis de recristalliser du rutile et de faire croître des germes obtenus en flux ou en fusion de Verneuil (voir tableau 2). L'axe c est dans l'axe du germe. Les faces de départ sont les faces {100} ou {110}.

On peut remarquer que les solutions de RbF et CsF nécessitent des températures et des pressions moins élevées que les solutions de KF; ainsi en milieu CsF, le rutile cristallise à partir de 470 °C, 500 bar alors qu'en milieu KF, il est nécessaire de travailler à 550 °C, 1000 bar.

En accord avec Kuznetsov et Panteleev[2,3]) nous pou-

vons conclure que le rutile recristallise correctement dans des solutions de fluorures alcalins à des vitesses de l'ordre de 0.2 à 0.3 mm/jour suivant [110]. Cependant, les cristaux obtenus sont toujours noirs et possèdent de nombreuses lacunes en oxygène que l'on a pu mettre en évidence en observant les variations de certaines propriétés physiques avant et après réoxygénation (constante diélectrique, tangente de l'angle de perte)[4]. Ces expériences ont permis de montrer la possibilité de recristalliser le rutile par voie hydrothermale malgré la faible solubilité de ce matériau dans les solvants utilisés (fig. 6).

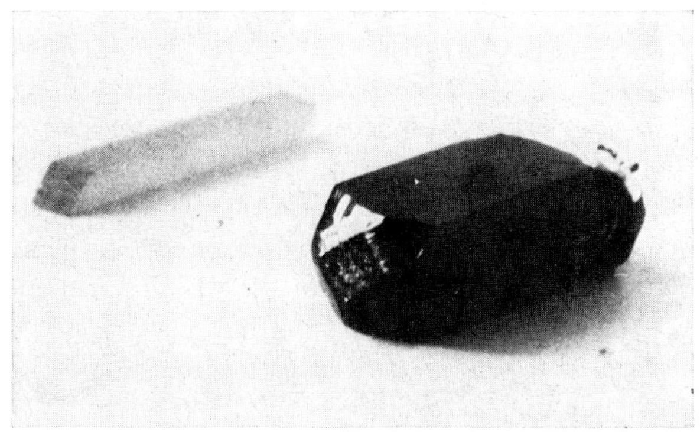

Fig. 6. Monocristal de rutile obtenu par voie hydrothermale. Grossissement 4.5×.

XI – 2

Fig. 7. Cristaux mixtes $Si_{1-x}Ge_xO_2$ obtenus dans un autoclave de 250 cm³, dans une solution de KF 2N à 400 °C, 500 bar. Gradient extérieur de température: 30 °C.

3.2. Quartz

Il est possible de recristalliser du quartz aussi bien dans des solutions de KF que de CsF à condition d'opérer à des températures suffisantes, dès 280 °C et des pressions supérieures à 500 bar. Les vitesses de croissance sont comparables à celles obtenues dans des solutions de soude[6]). La concentration du solvant doit, cependant, être inférieure à une certaine limite au delà de laquelle l'hydrolyse des complexes en solution est loin d'être complète.

Le tableau 2 nous présente des expériences de transport ou de croissance. On peut remarquer que les synthèses conduites en milieu acide (en ajoutant HF au solvant) dans des capsules en PTFE, ont toujours conduit au fluorure double, alors que dans des conditions peu différentes, certains auteurs[6]) ont obtenu du quartz.

3.3. GeO₂

Toutes les études faites dans des solutions de KF et CsF, dans des conditions identiques aux précédentes, en partant soit de GeO_2 quadratique soit de GeO_2 hexagonal, ont toujours conduit à un transport de GeO_2 quadratique sous forme de monocristaux de petites dimensions. Si la concentration en KF est supérieure à 3 M/l (à 450 °C, 1000 bar), on transporte un peu de germanate $K_2Ge_4O_9$ dans la partie froide de l'autoclave; notons qu'en solution de soude 0.5 N, on obtient uniquement des germanates dans les conditions de température et de pression citées précédemment.

4. Recristallisation de cristaux mixtes $Si_{1-x}Ge_xO_2$ et épitaxie sur du quartz

Quoique GeO_2 et SiO_2 ne recristallisent pas dans le même système cristallin, dans les conditions opératoires décrites plus haut, nous avons essayé, du fait des rayons ioniques voisins de Ge^{4+} (0.53 Å) et de Si^{4+} (0.42 Å), de substituer du silicium par du germanium dans le quartz.

Les expériences réalisées dans des capsules d'or, placées dans des autoclaves de 25 cm³ avec un rapport initial pondéral GeO_2/SiO_2 voisin de 1, ont permis d'obtenir aussi bien dans des solutions de KF que dans des solutions de CsF des germinations de "quartz" renfermant de 5 à 6 % de germanium en substitution. Soit $0.05 < x < 0.06$ pour des cristaux de formule $Si_{1-x}Ge_xO_2$.

Nous avons par ailleurs fait croître de tels cristaux sur des germes de quartz préalablement orientés. Les croissances ayant eu lieu dans des capsules d'or introduites dans des autoclaves de petites dimensions, les gradients de températures sont mal connus; les vitesses de croissances obtenues sont de l'ordre de 0.1 à 0.3 mm/jour suivant l'axe optique du quartz (fig. 7).

La méthode de Barraud–Lambot[7]) permet de mettre en évidence sur une face polie d'un cristal (plan YZ) une variation de paramètre très nette entre la partie germe et la partie croissance du cristal; la variation de $d_{11\overline{2}0}$ est de l'ordre de 3×10^{-3} Å pour une teneur en germanium de 5 % en atome.

Les figs. 8 et 9 permettent de repérer le germe et la

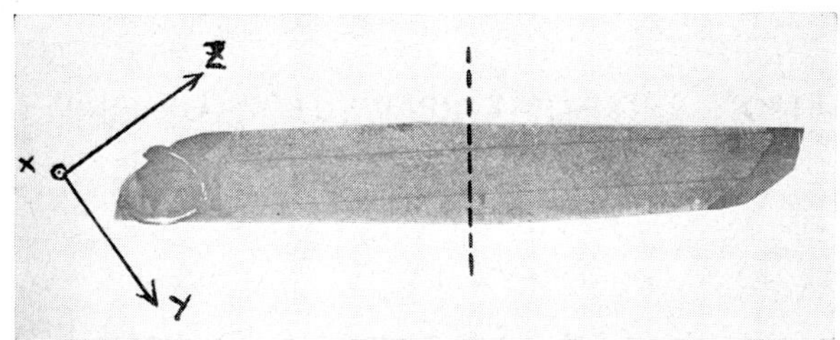

Fig. 8. Cristal de $Si_{1-x}Ge_xO_2$. Position du faisceau de rayons X sur l'échantillon au cours de la manipulation de Barraud–Lambot. Grossissement 4×.

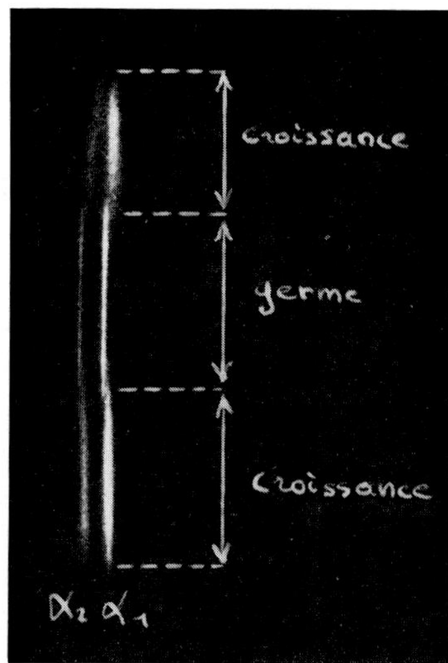

Fig. 9. Diagramme de Barraud–Lambot réalisé sur le cristal de la figure 8. $Cu\lambda K_{\alpha1} = 1.54051$ Å, $Cu\lambda K_{\alpha2} = 1.54433$ Å. Réflexion $(11\bar{2}0)$, $2\theta_{11\bar{2}0}$ / 36.58°.

partie croissance, on peut apprécier les variations de paramètres enregistrées.

Notons que dans l'ensemble de ces manipulations nous avons toujours utilisés des solutions sursaturées en SiO_2 et GeO_2 dans la partie la plus chaude de l'autoclave.

5. Conclusion

Si les essais de dopage du rutile par le germanium nous ont toujours conduit à la recristallisation de rutile et de GeO_2 quadratique séparément, il est démontré la possibilité de fabriquer du quartz et des cristaux mixtes $Si_{1-x}Ge_xO_2$ dans des solutions de fluorures alcalins par voie hydrothermale. Les cristaux mixtes obtenus ont une structure identique à celle du quartz et des paramètres très voisins.

Le silicium et le germanium sont dosés par spectrophotométrie d'absorption atomique et la substitution du Si par Ge a été confirmée par un diagramme effectué grâce à l'E.S.C.A. (Electron Spectroscopy for Chemical Analysis).

Remerciements

Les auteurs remercient tous ceux qui ont contribué aux différentes analyses nécessitées par cette étude et notamment Monsieur Auvray en ce qui concerne la partie cristallographie.

Bibliographie

1) E. F. Osborn, J. Am. Ceram. Soc. 36 (1953) 149.
2) V. A. Kuznetsov, J. Crystal Growth 3, 4 (1968) 405.
3) V. A. Kuznetsov et V. Panteleev, Soviet Phys.-Cryst. 10 (1965) 369.
4) Y. Toudic, Document C.N.E.T.-L.C.C./5 (1969).
5) R. A. Laudise et J. W. Nielsen, Solid State Phys. 12 (1971) 187.
6) V. S. Balitski et L. Tsinober, Dokl. Akad. Nauk SSSR 187 (1969) 1142.
7) H. Lambot, L. Vassamillet et J. Dejace, Acta Met. 1 (1953) 711.

     Journal of Crystal Growth **13/14** (1972) 530–534 © *North-Holland Publishing Co.*

HYDROTHERMAL CRYSTALLISATION AND SOME PROPERTIES OF BISMUTH TITANATES

M. L. BARSUKOVA, V. A. KUZNETSOV, A. N. LOBACHEV and Yu. V. SHALDIN

Institute of Crystallography, Academy of Sciences of the U.S.S.R., Lenin Pr., V-333, Moscow, U.S.S.R.

This article gives the results of investigations into hydrothermal crystallisation in the Bi_2O_3–TiO_2 system and a study of the non-linear optical properties of the synthesized crystals.

1. Introduction

The Bi_2O_3–TiO_2 system has been the subject of investigation more than once because of the unique characteristics of the bismuth-titanate series. Nevertheless until recently there have been differences of opinion on the number and composition of individual compounds in this system. According to ref. 1, there exist nine titanates of bismuth with high dielectric permittivities. Subsequent work[2,3] does not support this conclusion. The existence of only three titanates of bismuth is firmly established: $Bi_4Ti_3O_{12}$, $Bi_2Ti_4O_{11}$ and Ti–sillenite. More than one opinion exists about the composition of the latter: $Bi_{24}TiO_{38}$[2]), $Bi_{12}TiO_{20}$[4]) Bi_8TiO_{14}[3]), and in ref. 5 sillenite is considered as a solid solution with the ratio Bi_2O_3/TiO_2 (mol.) from 25:1 to 8:1. There are also indications of the existence of a phase intermediate in composition between $Bi_4Ti_3O_{12}$ and $Bi_2Ti_4O_{11}$, to which both the formulae $Bi_2Ti_2O_7$[6]) and $Bi_2Ti_3O_9$[2]) have been ascribed.

One of the reasons for the disagreements indicated above relating to phase-formation in the Bi_2O_3–TiO_2 system is, apparently, that previous investigations have been conducted by differential thermal and X-ray structual analysis without isolation of the individual bismuth titanates in the form of single-crystals (except $Bi_4Ti_3O_{12}$). One of the problems before us in the study of crystallisation in the Bi_2O_3–TiO_2 system was the necessity of isolating the bismuth titanates in the form of single-crystals, suitable for individual investigation by chemical and X-ray methods.

The second problem was to obtain sufficiently large and optically homogeneous crystals of bismuth titanates suitable for studying their non-linear properties. As is known, sillenite crystals (Si–sillenite, Ge–sillenite), which are characterised by high values of ionic ($\varepsilon - \eta_\infty^2 > 30$) and electronic ($\eta_\infty^2 > 4$) polarisability[4,7]) occupy a special place amongst non-linear cubic crystals. By analogy, high values of polarisability may be expected for Ti–sillenite. In the present work special attention was paid to the separate study of crystals of Ti–sillenite and $Bi_4Ti_3O_{12}$.

2. Experimental

The investigations were conducted in steel autoclaves of $150\ cm^3$ capacity. The starting charge was a carefully blended mix of the chemical reagents Bi_2O_3 and TiO_2. The ratio Bi_2O_3/TiO_2 (mol.) was varied from 20:1 to 1:6; with large TiO_2 content in the charge, the titanium oxide separated out as an independent phase. The temperature of the experiments was 400–600 °C, but below 450 °C the synthesis of bismuth titanate reaction proceeded very slowly, and, as a rule, only very small crystals formed which were unsuitable for study of their physical characteristics. In a similar way, the filling factor of the autoclaves was not less than 0.7–0.8.

The choice of solvents which will ensure recrystallisation and production of sufficiently large crystals of bismuth titanates presents considerable complications. On the basis of previous investigations into crystallisation of TiO_2 and titanates of lead[7]), we used KF solutions as solvents.

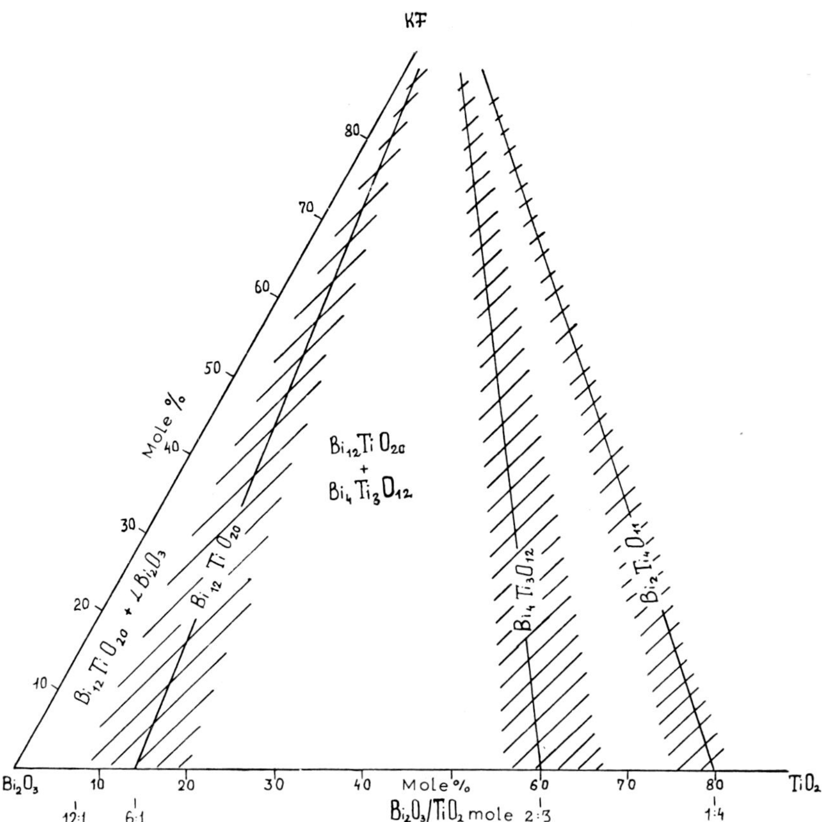

Fig. 1.   Phase relation in the Bi_2O_3–TiO_2–KF–H_2O system. Mono-mineral crystallisation is shown by the hatching.

3. Results

Only three individual compounds, $Bi_4Ti_3O_{12}$, $Bi_2Ti_4O_{11}$ and Ti–sillenite, were formed in the Bi_2O_3–TiO_2 system under the conditions employed. The identity of the synthesised compounds was confirmed by chemical, X-ray and optical methods. The region of crystallisation of the bismuth titanates is shown in fig. 1.

Ti–sillenite is formed in the part of the system rich in bismuth, being practically a single phase when the TiO_2 content in the starting charge is from 7.5 to 20 mole %. When the TiO_2 content is less pale-yellow elongated crystals of oxide bismuth α-Bi_2O_3, are formed together with Ti–sillenite.

The composition of the Ti–sillenite crystals corresponds to $Bi_{12}TiO_{20}$ (Bi_2O_3, 96.44 %; TiO_2, 2.16 %) and does not depend on the composition of the starting charge. These results do not confirm the formation of solid solutions on the basis of Ti–sillenite under the conditions of our experiments. The Ti–sillenite was in the form of isometric crystals of yellow, sometimes reddish colour. When charges enriched with Bi_2O_3 (TiO_2 content is 7.5–14.5 mole %) were used, the facets of tetrahedron (111), trigontritetrahedron (112) and rhombododecahedron (110) predominated on the crystals and the facets of cubes (100) were somewhat less strongly developed (fig. 2). With TiO_2 content 14.5–20 mole %, the facets of tetrahedron and trigontritetrahedron were most developed, and the facets of cube and rhombododecahedron practically disappeared.

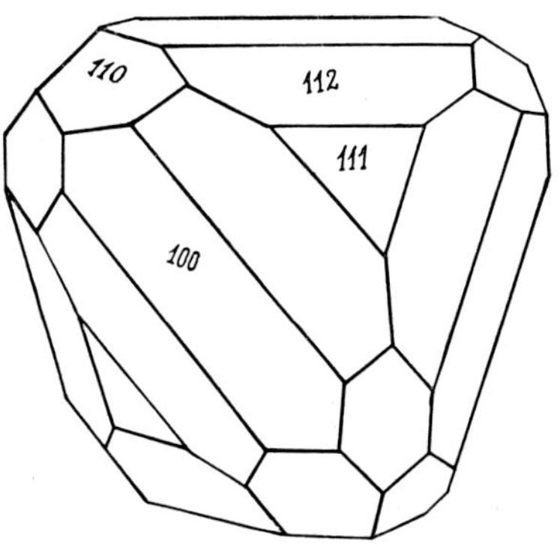

Fig. 2.   Crystal of Ti–sillenite.

The parameters of the elementary cell of the synthesised $Bi_{12}TiO_{20}$ were $a = b = c = 10.15$ Å. The crystals appear practically isotropic in polarised light. The refractive index is 2.65. The pycnometric density of the Ti–sillenite crystals is 9.1 g/cm³, the microhardness, determined on "PMT-3" equipment on artificial polished sections, is 380 kg/mm². The melting temperature of synthesised Ti–sillenite is 880 °C.

The relative dielectric permitivity of Ti–sillenite, measured on a "VM-271" instrument of the firm "Tesla" at a frequency 0.2 MHz, is of the order 50. The edge of the electron adsorption band, determined by the disappearance of double refraction induced by the electric field, is 400 nm and is displaced towards the longer wavelength part of the spectrum compared to the Ge– and Si–sillenites[9]).

$Bi_4Ti_3O_{12}$ (Bi_2O_3, 77.00 %; TiO_2, 20.80 %) begins to crystallise, together with Ti–sillenite, in the form of transparent thin, up to 0.1 mm, tetragonal platelets of light yellow colour when the TiO_2 content in the charge is greater than 20 mole %. As the TiO_2 in the charge is increased, the quantity of $Bi_4Ti_3O_{12}$ increases and it becomes practically single-phase in the 55–67 mole % TiO_2 range. Simultaneously, the crystals become thicker, up to 1 mm. Along with the pinacoid (001) a development of facets of a lateral band (101) is obtained on the crystals (fig. 3).

The parameters of the elementary cell of $Bi_4Ti_3O_{12}$ are $a = 5.41$ Å, $b = 5.45$ Å, $c = 32.83$ Å. The refractive index is 2.59; the pycnometric density is 7.8 g/cm. the mean microhardness is 313 kg/mm². A strongly expressed anisotropy of the microhardness from 272 to 364 kg/mm in two mutually perpendicular directions in the plane of the pinacoid is characteristic for $Bi_4Ti_3O_{12}$. The temperature of the phase transition of $Bi_4Ti_3O_{12}$ is 670 °C.

The edge of the electron absorption band lies in the region of 400 nm and in value is near to the absorption band in Ti–sillenite. The electrical resistivity is approximately 10^{12} ohm cm, and the value of the relative dielectric permitivity $\simeq 150$.

$Bi_2Ti_4O_{11}$ (Bi_2O_3, 58.30 %; TiO_2, 40.80 %) is the predominant phase in the range of values 75–82 mole % TiO_2 and is found as lime-yellow, often greenish transparent crystals of two morphological types. When there is an exess of Bi_2O_3 is the charge over the stoichiometric composition plate-like, strongly striated crystals

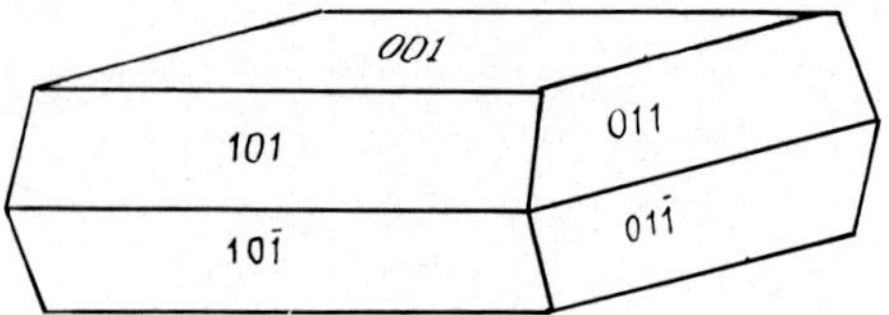

Fig. 3.   Crystal of $Bi_4Ti_3O_{12}$.

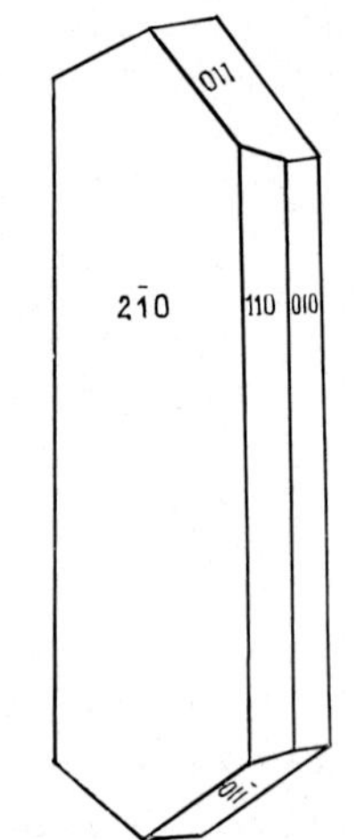

Fig. 4.   Platelike crystal of $Bi_2Ti_4O_{11}$.

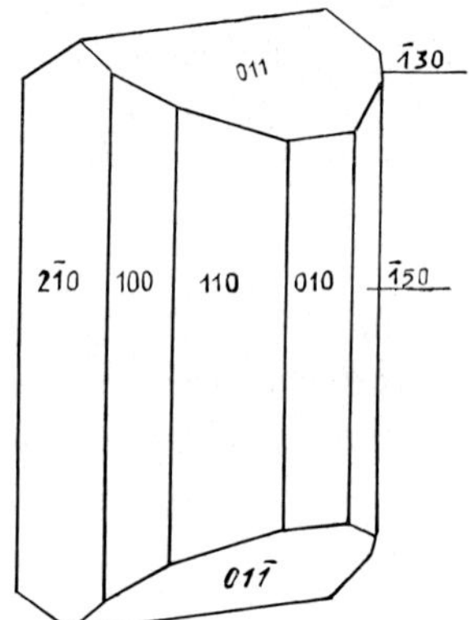

Fig. 5.   Prismatic crystal of $Bi_2Ti_4O_{11}$.

are formed (fig. 4). With $Bi_2O_3 : TiO_2 = 1:4$ in the charge, or with excess of TiO_2 the crystals are short-prismatic with well developed facets (fig. 5). The pycnometric density of the crystals if 6.1 g/cm³, the microhardness 758 kg/mm².

All the synthesised compounds below 500 °C are in the form of small crystals, less than 1 mm. Increase in the temperature promotes increase in their dimensions and in the 550–600 °C range the crystals attain a size of several mm. In the presence of a temperature gradient along the body of the autoclave in the 550–600 °C region simultaneously with synthesis recrystallisation of the material from the lower to the upper zone of the autoclave occurs; then crystals having maximum size are formed: Ti–sillenite, up to 8 to 10 mm; $Bi_4Ti_3O_{12}$, up to 6 mm; $Bi_2Ti_4O_{11}$, up to 3 mm. Reduction in the crystal size (simultaneously with a lowering of the amount of transport of the material) together with an increase of the TiO_2 content in the crystal provides evidence of a lower mobility of titanium oxide compared with bismuth oxide.

Non-linear properties of Ti–sillenite and $Bi_4Ti_3O_{12}$

To measure the non-linear change of the refractive index of Ti-sillenite which is produced by an electric field, an orientated specimen in the form of a rectangular platelet of dimensions $3.6 \times 4 \times 0.4$ mm was prepared from the single crystals. The electric voltage applied to previously deposited electrodes of finely dispersed silver, created a field along the $\langle 110 \rangle$, and the light beam was normal to the (110) surface. Measurement of the dispersion of "controlling" (half wave retardation) voltage $u_{\frac{1}{2}\lambda}$ and $\Delta\eta$ were made in circularly and linearly polarised light beams. Fig. 6 shows how the effective signal, the magnitude of which determines the values of $u_{\frac{1}{2}\lambda}$ and $\Delta\eta$ changes in both cases. The data from calculations based on the results of measurement in circularly and linearly polarised light are in good agreement with the value indicated above.

The dispersion of values of $u_{\frac{1}{2}\lambda}$ and $\Delta\eta$ in Ti–sillenite crystals was investigated in the region from 0.7 to 0.46 µm. A "SPM-1" monochrometer with a filament lamp "Cu8-200" was used as the source of monochromatic radiation. The dependencies $u_{\frac{1}{2}\lambda}(\lambda)$ and $\Delta\eta(\lambda)$ were constructed (fig. 7) from the experimental data with account of the relationship for $u_{\frac{1}{2}\lambda}{}^{10}$) (the latter is single-valued and determined by the value of the voltages applied to the crystal and of the alternating and constant voltages at the output of the photodetector). The curve of the dispersion (dotted) of $u_{\frac{1}{2}\lambda}$ for KH_2PO_4 crystals is also given for comparison. Noticeable deviation of the dependence $u_{\frac{1}{2}\lambda}$ from a linear law is observed as the absorption band is approached. Since photoconductivity occurs in melt grown crystals of Ge– and Si–sillenites[11], attempts were made to investigate the dependence of $u_{\frac{1}{2}\lambda}$ on the intensity of light and on time at the wavelength 500 nm, but no effects associated with photoconductivity were observed.

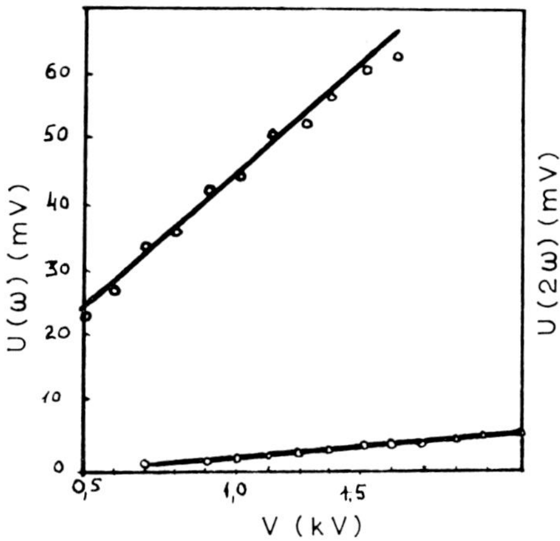

Fig. 6. The value of the effective signal at the output of the "FEU" (photomultiplier) as a function of the voltage frequency, ω, on a specimen of titanosillenite. Top line: Circularly polarised; and bottom line: linearly polarised light beam. Abscissa: voltage in kilovolts. Ordinate: the alternating component of the signal $u(\omega)$ and $u(2\omega)$ in millivolts.

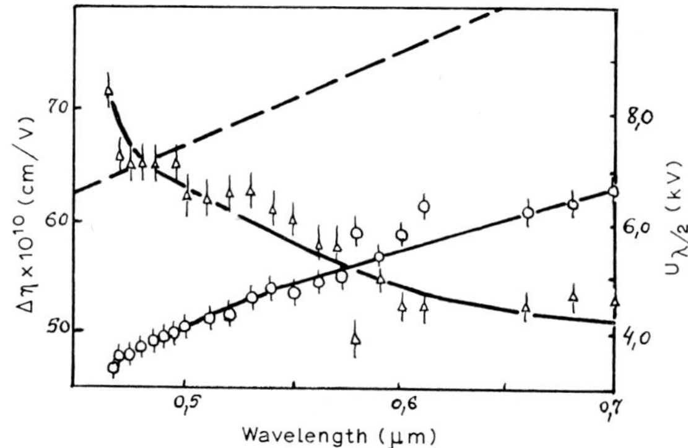

Fig. 7. Dispersion of the values of the controlling voltage $u_{\frac{1}{2}\lambda}(0)$, and of the non-linear refractive index $\Delta\eta$ ($\triangle$) in Ti–sillenite crystals. The dotted curve is the change of $u_{\frac{1}{2}\lambda}$ in KH_2PO_4 crystals. Abscissa: wavelength in microns. Ordinate: value of the controlling voltage $u_{\frac{1}{2}\lambda}$ in kilovolts and of the non-linear refractive index $\Delta\eta$ in units of 10^{10} cm/V.

The dispersion of $u_{\frac{1}{2}\lambda}$ and $\Delta\eta$ in dependence on frequency ω_2 was investigated on these same crystals.

The results of the experiment are presented in fig. 8. Weak increase of $\Delta\eta$ which exceeds the experimental error is observed as the frequency increases.

$Bi_4Ti_3O_{12}$. The non-linear refractive index $\Delta\eta$ was measured on specimens of $2.5 \times 2.5 \times 0.3$ mm dimensions. Electrodes to which the voltage was applied via needle contacts were deposited on one of the narrow faces. For circularly polarised light beam, passing along (010), $\Delta\eta = \eta_1^3 r_{11}^T - \eta_3^3 r_{31}^T$, where η_i is the refractive index along the i direction, r is the value of the coeffi-

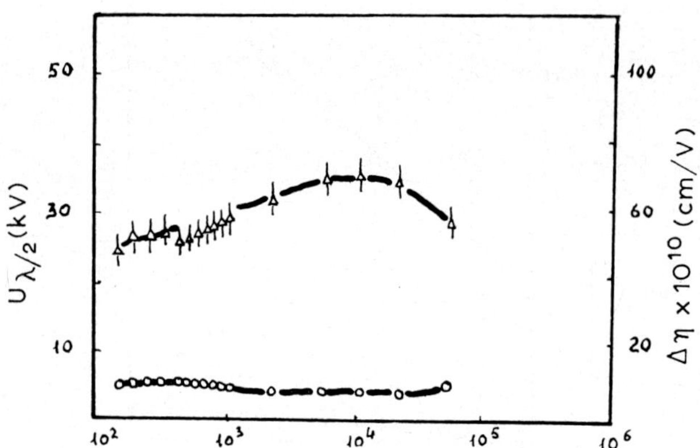

Fig. 8. The controlling voltage $u_{\frac{1}{2}\lambda}$ ($\bigcirc$) and the non-linear refractive index $\Delta\eta$ ($\triangle$) as a function of the applied field. Ordinate: the value of the controlling voltages $u_{\frac{1}{2}\lambda}$ in kilovolts and of the non-linear refractive index $\Delta\eta$ in 10^{10} cm/V. Abscissa: frequency of the electric field in Hz.

cients of the electro-optical constants. The measurements were obtained in circularly polarised light at 535 nm wavelength. The change of the effective signal as a function of the field in this case is linear and has a nature similar to that shown in fig. 6, but the tangent of the angle of slope is somewhat different and is 0.13×10^{-5}. The value obtained was corrected for the initial double refraction ($\gamma_0 \sim 10°$) of the crystal specimen and finally the value $\mathrm{tg}\,\alpha = 1.8 \times 10^{-5}$ was obtained. The value obtained for the controlling voltage $u_{\frac{1}{2}\lambda} = 5.0$ kV (correspondingly $\Delta\eta = 54 \times 10^{-10}$ cm/V), which in value is near to $u_{\frac{1}{2}\lambda}$ for titanosillenite.

4. Conclusions

(1) The hydrothermal method confirms that only three compounds: Ti–sillenite, $Bi_4Ti_3O_{12}$, $Bi_2Ti_4O_{11}$, are formed in the Bi_2O_3–TiO_2 system.

(2) A number of the physical properties were determined on Ti–sillenite, $Bi_4Ti_3O_{12}$ and $Bi_2Ti_4O_{11}$ crystals grown to sizes from 3 to 10 mm.

(3) The non-linear properties of Ti–sillenite and $Bi_4Ti_3O_{12}$ were investigated.

Acknowledgements

The authors express their thanks to O. G. Unanova for the chemical analysis of the titanites of bismuth and to T. N. Ivanova for determination of the physical constants.

References

1) G. I. Skanavi and A. I. Demeshina, Zh. Eksperim. i Teor. Fiz. **31** (1956) 565.
2) I. N. Belyaev, N. P. Smolyanov and N. R. Kal'nitskii, Zh. Neorgan Khim. **2** (1963) 348.
3) I. E. Speranskaya, I. S. Rez, L. V. Koslova and V. M. Skorikov, Izv. Akad. Nauk SSSR, Ser. Neorgan. Mater. **1** (1965) 232.
4) S. C. Abrahams, P. B. Jamieson and J. L. Bernstein, J. Chem. Phys. **47** (1967) 4034.
5) I. E. Speranskaya and V. M. Skorikov, Izv. Akad. Nauk USSR, Ser. Neorgan. Mater. **2** (1967) 345.
6) E. Aleshin and R. Roy, J. Am. Ceram Soc. **45** (1962) 18.
7) B. N. Litvin, Yu. Shaldin and I. E. Pitavranova, Kristallografiya **13** (1968) 1106.
8) V. A. Kuznetsov, J. Crystal Growth **3, 4** (1968) 405.
9) V. N. Batog, V. I. Burkov, V. A. Kizen' and V. M. Skorikov, Kristallografiya **13** (1968) 928.
10) Yu. V. Shaldin, Yu. V. Piscarevskii and Yu. S. Men'shikov, Zh. Prikl. Spektroscopii **3** (1965) 463.
11) P. V. Lenzo, E. F. Spenser and A. A. Ballman, Phys. Rev. Letters **19** (1967) 641.

Journal of Crystal Growth **13/14** (1972) 535–539 © *North-Holland Publishing Co.*

HYDROTHERMAL AND FLUX GROWTH OF ZIRCON CRYSTALS

R. DHARMARAJAN, R. F. BELT and R. C. PUTTBACH

Airtron Division of Litton Systems, Inc., 200 E. Hanover Avenue, Morris Plains, New Jersey 07950, U.S.A.

Attempts to grow single crystals of zircon ($ZrSiO_4$) by flux and hydrothermal techniques are described. Crystals grown from a lithium tungstate melt were colored yellow brown in corners indicating absorption of a growth impurity along preferred crystallographic planes. The impurity is suspected to be iron reported in the lithium tungstate flux. Hydrothermal zircon growth was accomplished using a 3 molar aqueous KF mineralizer at about 30 000 psi and 500 °C. The nutrient used was ZrO_2 and SiO_2 powders pressed at 30 000 psi and sintered at 1520 °C. Due to the relatively low solubility of zircon in various molten salts, the hydrothermal method appears to be more suitable than the flux method for growing large single crystals of zircon.

1. Introduction

Since published phase diagrams[1] indicate that zircon ($ZrSiO_4$) is incongruently melting, attempts to grow single crystals of zircon have been confined to hydrothermal and flux techniques. Available literature on flux growth of zircon include papers by Ballman and Laudise[2] and Chase and Osmer[3]. These authors obtained well-formed, macroscopic crystals using alkali molybdate or vanadate fluxes. Frondel and Collette[4] have reported zircon formation from mixed ZrO_2 gels at 150–700 °C and 71 to 14 700 psi. The authors indicate that the hydrothermal reaction kinetics are accelerated by trace additions of ZrF_4. Hydrothermal synthesis of zircon from $ZrO(OH)_2$ and silicic acid has been accomplished by Maurice[5].

Based on this past work, a program was initiated to produce zircon crystals up to 2 in. in diameter. Since the solubility of zircon in most known fluxes is under 5%, our approach is to obtain flux grown crystals of sufficient size to be used as seeds in a hydrothermal system. This paper describes work completed during the initial phase of the program. It includes results of flux and hydrothermal growth experiments, zircon solubility, and stability in a variety of molten salt solvents.

2. Experimental

Flux growth experiments were carried out using sealed 265 cm^3 platinum crucibles. The furnace used possessed a 3 in. diameter alumina core and was heated by six bar-type SiC heating elements. The furnace is capable of generating specimen temperatures up to 1400 °C. Associated control equipment maintained specimen temperatures to within ± 2 °C and a programmer allowing cooling rates up to 6 °C/hr.

Solubility of natural zircon in a variety of molten salt solvents was determined by the weight loss method. In these experiments natural zircon crystals were equilibrated with fluxes at temperature for 24 hr in sealed 0.25 in. diameter × 0.020 in. wall × 6 in. long platinum capsules. The stability of zircon in several molten salt systems was studied by reacting high purity ZrO_2 and SiO_2 with the flux in similar sealed capsules and cooling at 3 to 5 °C/hr.

Hydrothermal growth feasibility tests were designed to evaluate mineralizers at various pressure–temperature conditions for zircon growth. Studies were carried out in a four station Tem-Pres Research Hydrothermal Unit. Operating temperatures ranged up to 600 °C and pressures up to 30 000 psi. Reaction capsules were fabricated from 0.25 in. diameter × 0.010 in. wall platinum tubing. In the hydrothermal growth tests reported, a 0.80 in. diameter × 0.010 in. wall × 12 in. long reaction capsule was used along with a 1 in. diameter autoclave designed at Airtron. The growth furnace is a Nichrome resistance wire wound alumina tube. The associated equipment controls temperatures of the solution and growth zones and also the temperature differential between the zones. Temperature measurements are accomplished by chromel–alumel thermocouples attached to the outer surface of the autoclave.

Pressure is measured by a gauge connected directly to the internal cavity of the autoclave.

Sintering of ZrO_2 znd SiO_2 powders (~ 200 mesh) to produce an acceptable hydrothermal nutrient was performed in induction heated iridium crucibles. An equimolar mixture of the ZrO_2 and SiO_2 powders, was pressed at 30000 psi and sintered at 1520 to 1650 °C for 16 hr.

3. Discussion

3.1. FLUX GROWTH

The solubility of natural zircon in several molten salt fluxes is shown in table 1. The solubilities of zircon in the alkali molybdate fluxes shown in table 1 are in essential agreement with published values[2]). The solubility in $0.5 \, Li_2WO_4$–$0.5 \, WO_3$ flux is more than twice the solubility in $0.5 \, Li_2MoO_4$–$0.5 \, MoO_3$ at 1200 °C. As shown in table 1, PbO–PbF_2 based fluxes showed the

highest solubility for natural zircon in the 1100–1200 °C temperature range.

Results of zircon stability experiments using high purity ZrO_2 and SiO_2 powders are shown in table 2. With PbO or PbF_2 based fluxes, Pb-rich phases were observed. An additional difficulty was that wetting of

TABLE 1

Solubility of natural zircon in several molten salt fluxes

Flux no.	Flux composition	Natural zircon solubility as wt% of flux
1	$0.5 \, Li_2MoO_4$–$0.5 \, MoO_3$	1.55 at 1200 °C
2	$0.33 \, Na_2MoO_4$–$0.67 \, MoO_3$	1.71 at 1200 °C
3	$0.5 \, Li_2WO_4$–$0.5 \, WO_3$	3.71 at 1200 °C
4	$0.52 \, KHCO_3$–$0.48 \, WO_3$	0.92 at 1100 °C
5	$0.5 \, PbO$–$0.5 \, PbF_2$	18.72 at 774 °C
6	$0.5 \, PbO$–$0.5 \, PbF_2$	36.97 at 1100 °C
7	PbF_2	18.72 at 1108 °C
8	$0.4 \, PbO$–$0.4 \, PbF_2$–$0.2 \, B_2O_3$	24.40 at 1100 °C
9	$0.85 \, PbO$–$0.15 \, WO_3$	33.93 at 1100 °C
10	$0.66 \, PbO$–$0.34 \, B_2O_3$	26.98 at 1100 °C

TABLE 2

Zircon formation from ZrO_2 and SiO_2 powders in several molten salt systems

Test No.	Flux composition	Temp. range (°C)	Results
1	$0.5 \, PbO$–$0.5 \, PbF_2$	900–1295	Colorless glassy, Pb-rich phase
2	$0.4 \, PbO$–$0.4 \, PbF_2$–$0.2 \, B_2O_3$	900–1295	Small platelets
3	$0.85 \, PbO$–$0.15 \, WO_3$	900–1295	Glassy phase
4	$0.66 \, PbO$–$0.34 \, B_2O_3$	900–1295	Glassy phase
5	$0.5 \, Li_2WO_4$–$0.5 \, MoO_3$	878–1300	Colorless microcrystallites showing first order dipyramids and prisms characteristic of zircon
6	$0.5 \, V_2O_5$–$0.5 \, MoO_3$	878–1300	Blue zircon microcrystallites, up to 0.2 mm in size
7	$0.5 \, V_2O_5$–$0.5 \, Li_2WO_4$	878–1300	Blue zircon microcrystallites, up to 1 mm in size
8	$0.4 \, Li_2WO_4$–$0.3 \, MoO_3$–$0.3 \, V_2O_5$	878–1300	Colorless zircon microcrystallites, up to 0.2 mm in size
9	$0.5 \, Li_2WO$–$0.5 \, WO_3$	878–1300	Colorless zircon microcrystallites, up to 0.2 mm in size

TABLE 3

Results of zircon growth tests in molten salts

Flux composition	Solute (wt%)	Solution		Crystallization		Results
		Temp. (°C)	Time (hr)	Range (°C)	Rate (°C/hr)	
$0.25 \, Li_2O$–$0.75 \, MoO_3$	2.5	1204	24	792–1204	2.2	Cristobalite and tridymite silica with very few zircon crystallites in acicular morphology
$0.4 \, Li_2O$–$0.6 \, V_2O_5$	6.98	1366	67	907–1366	1.28	Zircon cristals up to 4 mm showing characteristic first order prism and dipyramid morphology; about 60% yield; all of the crystals colored dark blue
$0.5 \, Li_2WO_4$–$0.5 \, WO_3$	4.0	1374	89	969–1374	2.34	Colorless to yellow brown zircon crystals showing characteristic morphology; about 35% yield

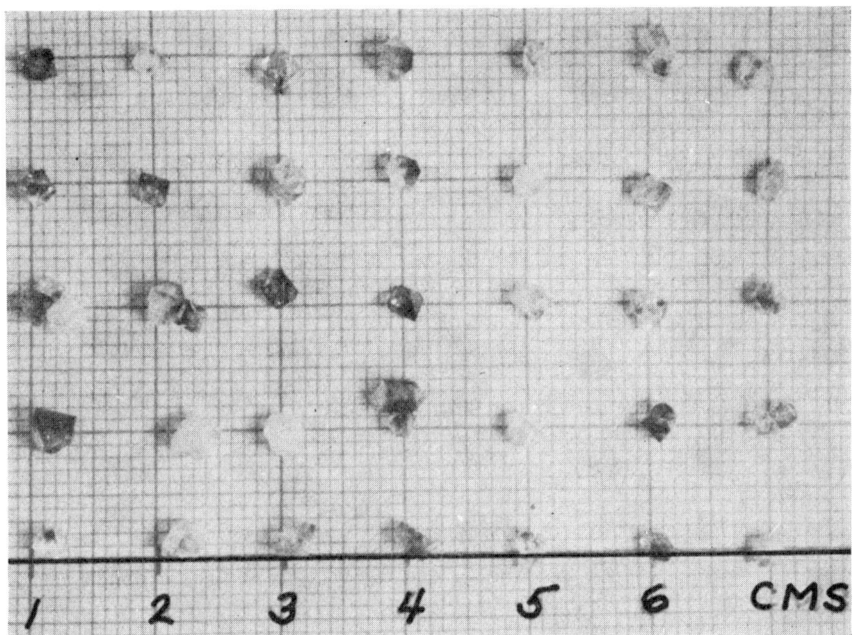

Fig. 1. Zircon single crystals grown from a lithium tungstate flux. $\times 1.7$.

the solute powders by the molten fluxes was difficult. However, zircon formation was confirmed in the alkali molybdate, vanadate and tungstage fluxes by optical microscopy and X-ray diffraction analyses.

Based on these data and published solubility values[2]) growth runs were conducted in 265 cm³ crucibles. The results are summarized in table 3. Results obtained using the Li₂O–3 MoO₃ flux are at variance with those reported by Ballman and Laudise[2]), who succeeded in growing zircon under conditions where we observed

mostly other phases. Apparently at temperatures below 900 °C, zircon decomposes and excess silica precipitation occurs in this system.

Dark blue crystals were obtained with the 0.4 Li₂O– 0.6 V₂O₅ flux and this result is consistent with published data[3]). With a Li₂WO₄–WO₃ flux, colorless to yellow brown crystals were obtained as shown in fig. 1. These crystals showed the first order prism and dipyramid morphology characteristic of zircon. The coloring was primarily at crystal corners indicating absorption

TABLE 4

Results of a one-inch hydrothermal zircon growth run

Seed No.	Seed weight (g)		Seed weight gain (%)	Seed width (cm)		Gain in width (%)	Seed thickness (cm)		Gain in thickness (%)
	Before	After		Before	After		Before	After	
1A	2.704	3.731	38.0	1.48	1.56	5.4	0.227	0.300	32.2
1B	0.495	0.776	56.8	0.61	0.91	49.1	0.393	0.416	5.8
2	3.915	5.235	33.4	1.75	1.80	2.8	0.230	0.282	22.6
3	3.035	4.007	34.3	1.71	1.75	2.3	0.224	0.282	25.8
4A	1.958	2.772	41.9	1.25	1.31	4.8	0.245	0.292	19.2
4B	0.308	0.443	43.8	0.56	0.70	25.0	0.277	0.310	11.9

Test conditions
Duration of test: 386 hr.
Mean temperature of top zone: 504 °C.
Mean temperature of nutrient zone: 530 °C.
Mean temperature of growth zone: 502 °C.

Mean ΔT: 28 °C.
Mean pressure: 29 000 psi.
Internal fill: 78%; external fill: 70%.
Capsule volume: before test: 90 cm³; after test: 87 cm³.
Weight of nutrient charged: 30 g.

XI – 4

of a growth impurity along preferred crystallographic planes. Tentatively, the growth impurity is identified to be the iron reported in the Li_2WO_4 used.

3.2. HYDROTHERMAL GROWTH

Since literature data[4]) indicated that fluoride ions accelerated hydrothermal zircon formation, several fluoride mineralizers were evaluated. These feasibility experiments showed that a 3 molar aqueous solution of KF is superior to HF, KHF_2 and ZrF_4 solutions. Optimum hydrothermal conditions were found to be 500–600 °C and 30000 psi.

Using sintered ZrO_2 and SiO_2 as nutrient, a growth run was performed in a 0.8 in. diameter × 12 in. long platinum reaction capsule. The duration of the test was 386 hr. Mean pressure inside the autoclave was 29000 psi and the mean temperature of the dissolution (nutrient) zone was 530 °C. A mean temperature differential of 28 °C was maintained between the dissolution and growth zones.

Results of this growth run are shown in table 4. All of the seeds used gained weight and the weight gain ranged from 33.4 to 56.8 %. Seeds 1B and 4B (table 4) were pebble-like colorless Australian zircon with essentially rounded off surfaces. These seeds showed the

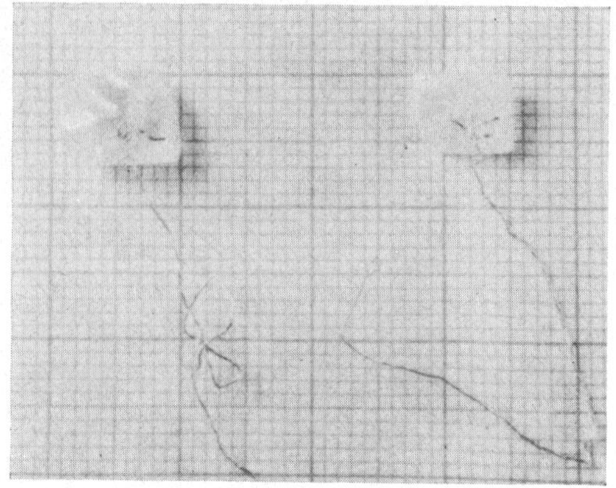

Fig. 2.   Hydrothermal growth on pebble-like Australian zircon seeds.

higher weight gains and the growth on these seeds was colorless, transparent and anisotropic (fig. 2). Seeds 1A, 2, 3 and 4A were cut from a crystal of dark brown Uganda zircon. The flat faces in these seeds (fig. 3) are parallel to the zircon (110) prismatic planes. The growth on these seeds was of high quality and colorless.

4. Conclusions

It has been demonstrated that $PbO–PbF_2$ based fluxes have high solubilities for natural zircon, but crystallization of other compounds occur upon cooling

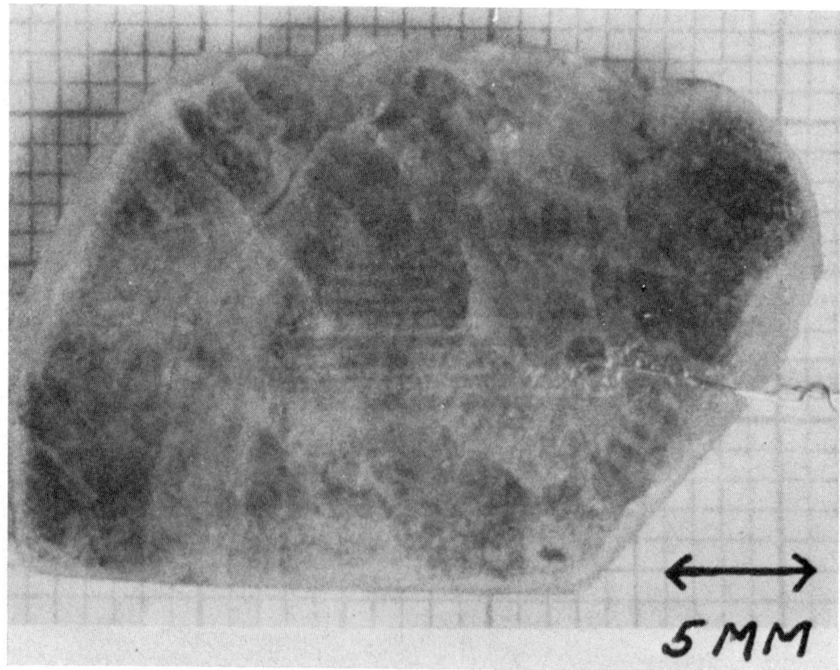

Fig. 3.   Hydrothermal growth on Uganda zircon seeds.

a mixture of these fluxes with ZrO_2 and SiO_2 powders. The solubility of natural zircon in 0.5 Li_2WO_4–0.5 WO_3 flux is twice that in 0.5 Li_2MoO_4–0.5 MoO_3 flux at 1200 °C. When high purity components are used, colorless crystals can be expected from the 0.5 Li_2WO_4–0.5 WO_3 flux system. High quality hydrothermal zircon growth has been accomplished using a 3 molar KF aqueous mineralizer at 530 °C and 30000 psi. An acceptable hydrothermal nutrient can be prepared by isostatically pressing an equimolar mixture of ZrO_2 and SiO_2 at 30000 psi and sintering the compact at about 1600 °C for 16 hr. Due to relatively low solubility of zircon in the fluxes investigated, the hydrothermal method appears to be more suitable than the flux method for the growth of large zircon single crystals.

Acknowledgement

This work was sponsored by the Office of Naval Research, Washington D.C. under contract No. N00014-70-C-0379.

References

1) E. M. Levin, C. R. Robbins and H. F. McMurdie, *Phase Diagrams for Ceramists* (American Ceramic Society, 1964 and 1969 Supplement, Columbus, Ohio).
2) A. A. Ballman and R. A. Laudise, J. Am. Ceram Soc. **48** (1965) 130.
3) A. B. Chase and J. A. Osmer, J. Electrochem. Soc. **113** (1966) 198.
4) C. Frondel and R. L. Collette, Am. Mineralogist **42** (1957) 759.
5) O. D. Maurice, Econ. Geol. **44** (1949) 721.

INVESTIGATION OF THE SOLUBILITY AND CRYSTALLIZATION KINETICS OF SODIUM-ZINC GERMANATE (Na_2ZnGeO_4) UNDER HYDROTHERMAL CONDITIONS

A. N. LOBACHEV, L. N. DEMIANETS, I. P. KUZMINA and E. N. EMELIANOVA

Institute of Crystallography, Academy of Sciences of the U.S.S.R., Lenin Pr., V-333, Moscow, U.S.S.R.

A study of the solubility of Na_2ZnGeO_4 under hydrothermal conditions was made. A change in the temperature coefficient of solubility was found at 18–22 wt% NaOH. The Van 't Hoff equation was obeyed and ΔH ranged from 1.1 to 3.0 kcal/mol. The growth rate on (011), (001) and (00$\bar{1}$) was studied and an activation energy of 6.6 kcal/mol was found. The solubility data were used to choose growth conditions.

1. Introduction

The investigation of the crystallization of divalent metal germanates under hydrothermal conditions resulted in some previously unknown complex germanates. Among these compounds the most interesting one is Na_2ZnGeO_4.

The X-ray data for Na_2ZnGeO_4 are: monoclinic cell, $a = 8.91$ Å, $b = 5.60$ Å, $c = 5.33$ Å, $\beta = 127°$, $Z = 2$, space group Pa. The main crystallographic planes of these crystals are (011), (01$\bar{1}$), (110), ($\bar{1}$10), (001), (00$\bar{1}$), (100), ($\bar{1}$00). Na_2ZnGeO_4 is luminescent under conditions of photo, cathode, electron and X-ray excitation when doped with Mn^{+2}. This compound is also characterized by triboluminescence and piezo-effect[1,2]. Sodium–zinc germanate is one of the main crystalline phases in the system Na_2O–ZnO–GeO_2–H_2O. Na_2ZnGeO_4 is crystallized from aqueous solutions in the temperature range of 200–500 °C, the NaOH concentration has to be in excess of 12 wt%[3].

2. Experimental

In order to determine the optimum conditions of Na_2ZnGeO_4 growth, the solubility was measured in an aqueous solution of NaOH (15–35 wt%) at temperatures from 200 to 350 °C and at % fill of 50–90%.

The solubility was measured by the weight loss method; the chemical composition of the solutions was determined for some of the experiments. Autoclaves with silver cans were used, the volume was 40–50 cm³. Na_2ZnGeO_4 monocrystals obtained previously by a hydrothermal method were used as seeds.

3. Results and discussion

The influence of temperature on solubility at constant NaOH concentration was investigated as well as the influence of concentration at constant T and that of pressure (or fill degree) at constant temperature and concentration.

3.1. EXPERIMENTAL DATA ON THE SOLUBILITY OF Na_2ZnGeO_4

(a) $S = f(T)_{C = \text{const.}}$. The solubility of Na_2ZnGeO_4 was determined in 15, 20, 25, 30 and 35 wt% NaOH solutions at temperatures of 200, 250, 300 and 350 °C. As shown in fig. 1, the character of the solubility changes with varying NaOH concentration. The transition from a negative temperature solubility coefficient to a positive one was observed in the investigated concentration interval. The solubility of Na_2ZnGeO_4 in 15% NaOH decreases with increasing T from 4.3 wt% (200 °C) to 3.3 wt% at 350 °C.

In contrast, the solubility in NaOH with a concentration above 25% increased with increasing T. At a NaOH concentration of 20%, the solubility of Na_2ZnGeO_4 is practically temperature-independent.

(b) $S = f(C)_{T = \text{const.}}$. The results of the determination of Na_2ZnGeO_4 solubility versus NaOH concentration are given in fig. 2. All isotherms intersect in the NaOH concentration range of 19–22 wt%. The 200 °C isotherm has a maximum at a concentration of nearly 25 wt%. The dependence of Na_2ZnGeO_4 solubility on T in the temperature interval of 250 to 350 °C is nearly linear. This dependence can be used for determining

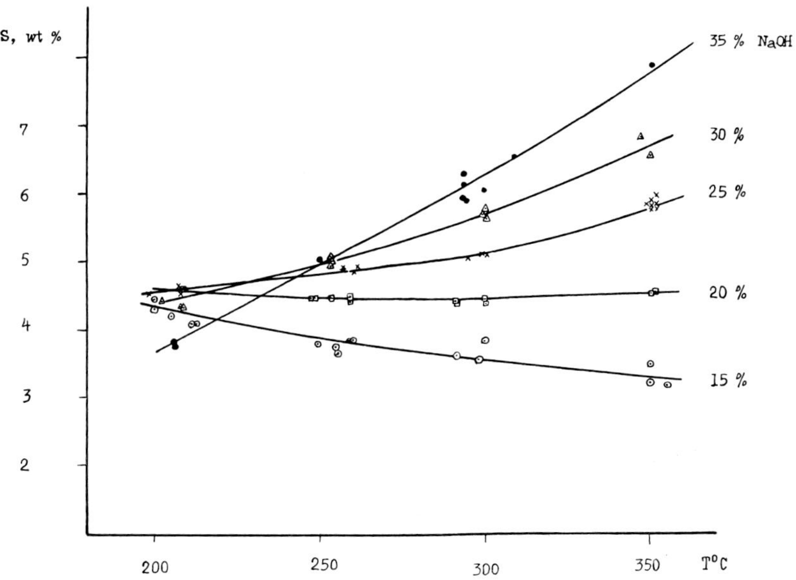

Fig. 1. Solubility of Na_2ZnGeO_4 as a function of temperature at constant concentration of NaOH (independent of % fill between 50–80%).

the supersaturation by the temperature difference method.

(c) $S = f(P)_{T,C\ =\ \mathrm{const.}}$ In the investigated range of fill degree (0.5–0.9), the Na_2ZnGeO_4 solubility is practically independent of pressure at temperatures between 200–300 °C within the error limits of the experiments.

4. Experimental data on the crystallization kinetics

The kinetics of hydrothermal crystallization was investigated in a pressure–temperature–concentration range where Na_2ZnGeO_4 was a stable solid phase and where the other solid crystal phases were absent. Considering the solubility data, these conditions are in conformity with a concentration of NaOH above 25 wt%.

The variable parameters were the following: temperature, temperature difference (supersaturation in the growth zone), concentration of solvent. Most of the experiments were performed in 30 wt% NaOH solutions. The growth rates of the best developed faces of the crystal (011), (001), (00$\bar{1}$) were determined.

(a) $V = f(T)_{C,P\ =\ \mathrm{const.}}$ The growth rates of the different faces as a function of temperature was determined experimentally as described below. The relation of

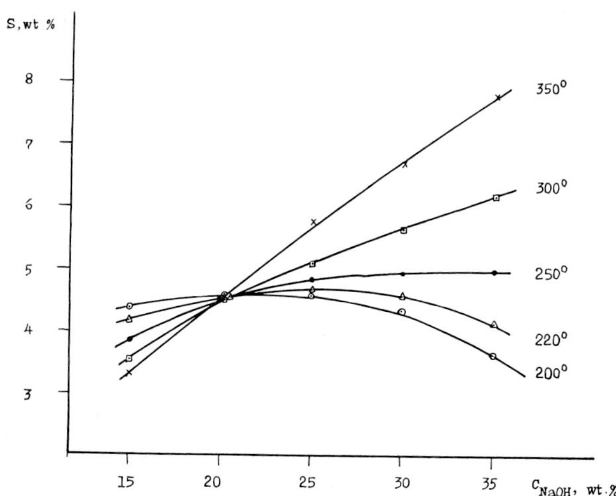

Fig. 2. The isotherms of Na_2ZnGeO_4 solubility in aqueous solutions of NaOH.

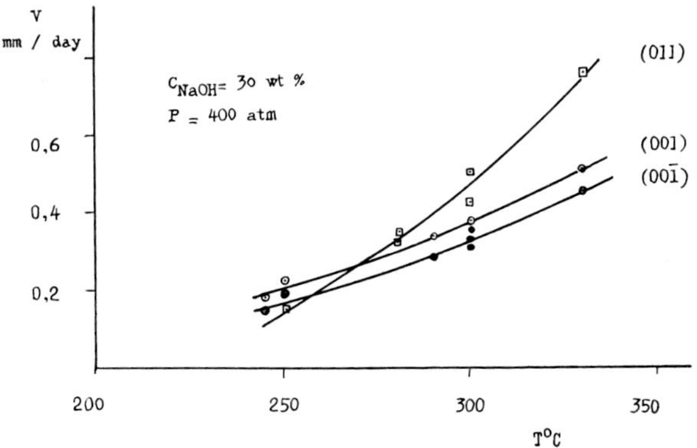

Fig. 3. Growth rates versus temperature in 30% NaOH at 400 atm.

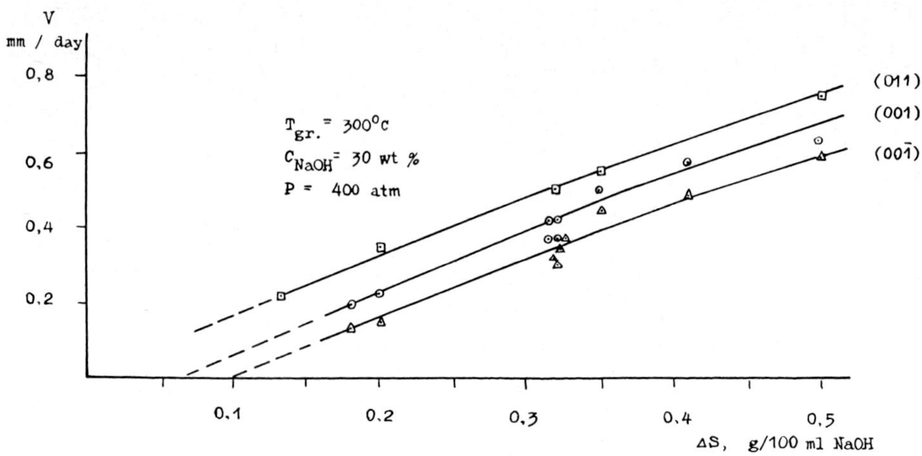

Fig. 4.   Growth rates versus supersaturation ΔS in 30% NaOH at a temperature of 300 °C and a pressure of 400 atm.

two simple planes (001)/(011) was chosen as a characteristics for the habit of the crystals.

$T < 260°$ C:

$$V_{(001)} > V_{(00\bar{1})} \gtrsim V_{(011)} > V_{(100)} > V_{(110)},$$

$$V_{(001)}/V_{(011)} > 1;$$

$260 < T < 300$ °C:

$$V_{(011)} > V_{(001)} > V_{(00\bar{1})} > V_{(100)} > V_{(110)},$$

$$V_{(001)}/V_{(011)} \gtrless 1;$$

$T > 300$ °C:

$$V_{(011)} > V_{(001)} > V_{(00\bar{1})} > V_{(100)} > V_{(110)},$$

$$V_{(001)}/V_{(011)} < 1.$$

Fig. 3 shows the dependence of the growth rate on temperature in a 30% NaOH solution. Rates increase by a factor of four to five as the temperature rises from 250 to 340 °C. The greatest increase was observed for the (011) faces.

(b) $V = f(\Delta S)_{T,P,C = \text{const.}}$. The dependence of the growth rates of (011), (001), (00$\bar{1}$) on the absolute supersaturation at a $T_{gr} = 300°$, $P = 400$ atm and $C_{NaOH} = 30$ wt% is shown in fig. 4.

The growth rates of the principal faces increase nearly linearly as the supersaturation increases. The extrapolation of these data determe the critical supersaturation, which amounts to 0.06–0.09 g per 100 ml solvent for (001), (00$\bar{1}$) faces, and less than 0.02 g per 100 ml solvent for (011) face.

(c) $V = f(C)_{T,P = \text{const.}}$. Increasing the concentration

of NaOH from 30 wt% to 45 wt% results in increases of growth rates of individual faces by 20–30%. A further increase of the concentration leads to crystals having gas–liquid inclusions and other crystalline imperfections.

Decreasing the concentration of the solvent to 20–22 wt% leads to a sharp decrease of the growth rates. It can thus be inferred that solubility in NaOH solutions of the specified concentration (according to the experimental data) becomes independent of temperature and the supersaturation in the growth zone practically equals zero. The crystals immersed in these solutions are subjected to a dynamic equilibrium between solution and growth.

5. Discussion

On the base of the Na_2ZnGeO_4 solubility data it can be concluded that NaOH solutions of a concentration of less than ~25 wt% are unsuitable for growing Na_2ZnGeO_4 by the temperature difference method. The properties of the system Na_2ZnGeO_4–NaOH–H_2O are most distinctly reflected in the dependence $S = f(C)_{T=\text{const.}}$. All the isotherms intersect in one region of the NaOH concentration (19–22 wt%). On the left of this region the system is characterized by a negative temperature coefficient of solubility (retrograde solubility), and on the right, by a positive one.

Apparently, such a solubility change is connected with a qualitative change of the solution structure and changes of relationship and composition of the main hydrate complexes in the solution.

The possibility of varying the complex composition

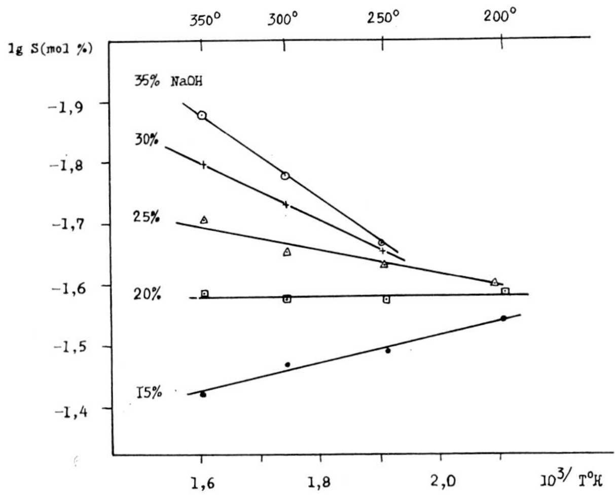

Fig. 5. Log S versus reciprocal absolute temperature.

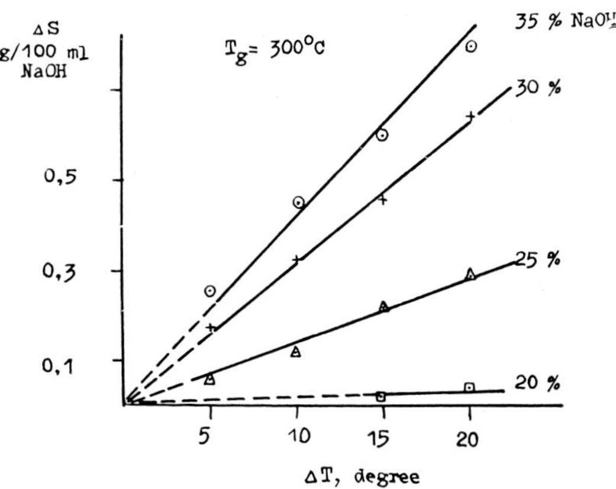

Fig. 6. Supersaturation versus temperature difference at $T_{gr} = 300\,°C$.

in alkaline solutions is linked to the individual properties of Ge and Na ions. Ge in solids, and probably also in liquids, can be characterized by the coordination number toward oxygen of 4 or 6[4]). The transition $Ge^{IV} \rightarrow Ge^{VI}$ in oxides under hydrothermal conditions[5]) is observed at a temperature of $\sim 185\,°C$ ($GeO_{2\,hex} \rightarrow GeO_{2\,tetr}$). Therefore, at higher T a coordination number of 6 can exist. In some germanates of alkali metals we have both Ge^{IV} and Ge^{VI}. In sodium-zinc-germanate we have only Ge^{IV} in tetrahedra[6]). Besides, Na in solution can have different coordination numbers varying from 6 in diluted solution through 5 (two sub-spheres) to 4. Such an alternation of Na-coordination is reflected in the structure of solid Na_2O with a strict coordination number 4 (anti-fluorite type).

From these facts it follows that in solutions there is dynamic equilibrium between the two main aqueous species:

$$Ge(OH)_6 \rightleftarrows Ge(OH)_4.$$

On the base of the structure of solid Na_2ZnGeO_4, it may be concluded that the region of the sharp change in the solubility character is connected with the transition to the dominant aqueous species: $Ge(OH)_4$ and $Na(OH)_4$. It is likely that in $\sim 20\%$ NaOH there is a stable equilibrium:

$$Na_xGe(OH)_6(OH)_y(H_2O)_z$$

$$\rightleftarrows Na_mGe(OH)_4(OH)_n(H_2O)_p.$$

In the system Na_2ZnGeO_4–NaOH–H_2O the temper-

ature dependence of solubility obeys Van 't Hoff's law (fig. 5) within the investigated temperature intervals:

$$d \ln S/dT = -\Delta H_s/RT^2,$$

where S is the equilibrium solubility, ΔH_s is the heat of solution.

The estimated heats of solution of Na_2ZnGeO_4 in 15, 25, 30, and 35 wt% NaOH are equal to -1.2, 0.9, 2.9, and 2.9 ± 0.3 kcal/mole, respectively.

The supersaturation in the growth zone must be about 0.1–0.2 wt% to allow the growth of a crystal at a rate 0.2–0.3 mm/day. This corresponds to a temperature difference of 3–8 °C at a temperature of 300 °C, depending on the solvent concentration (fig. 6).

The rate of crystal growth is nearly linear and increases with the supersaturation (ΔS); it can be connected with the dependence $V = \beta \Delta S$, where β is the velocity constant, and V is the growth rate of the face.

The temperature dependence is defined by the expression

$$d \ln \beta/dT = \Delta E_a/RT^2,$$

where E_a is the energy of activation under the constant ΔS, $\beta = V'$, where $V' = V/\Delta S$, and therefore ΔE_a can be estimated from the slope of this curve. Calculated in this manner ΔE_a was found to be 6.6 ± 0.5 kcal/mole for the (001) plane.

Our investigation of the solubility of Na_2ZnGeO_4 in aqueous solutions of NaOH at high temperatures and of the kinetics leads to the following conclusions:

(1) The system Na_2ZnGeO_4–NaOH–H_2O is characterized by a change of the temperature coefficient in the NaOH concentration range of 19–22 wt%.

(2) The heats of solution of Na_2ZnGeO_4 in aqueous solutions of NaOH with concentrations of 15, 25, 30, and 35 wt% were found to be $-$ 1.2, 0.9, 2.1, and 2.9 ± 0.3 kcal/mole.

(3) The growth rates of the principal crystallographic planes of the Na_2ZnGeO_4 crystals were determined to be from 0.2 mm/day to 0.8 mm/day depending on supersaturation, temperature and solvent concentration.

(4) From the temperature dependence of the growth rate of the (001) faces the activation energy was found to be 6.6 ± 0.5 kcal/mole.

(5) On the base of the obtained data the following conditions of Na_2ZnGeO_4 growth were found:

$$T_{gr} = 250\text{--}300\,°C, \Delta T = 3\text{--}10\,°C, C_{NaOH} = 30 \pm 5\,\text{wt}\%.$$

References

1) M. I. Tombak, I. P. Kuzmina and A. N. Lobachev, Zh. Prikl. Spektroskopii **13** (1970) 921.
2) I. P. Kuzmina, O. K. Melnikov and B. N. Litvin, in: *Hydrothermal Synthesis of Crystals*, Ed. A. N. Lobachev (Izd. Nauka, Moscow, 1968) (in Russian).
3) I. P. Kuzmina, B. N. Litvin and H. S. Kurajkovskaia, in: *Research in Crystallization Processes under Hydrothermal Conditions*, Ed. A. N. Lobachev (Iz. Nauka, Moscow, 1970) (in Russian).
4) I. V. Tananaiev and I. Ya. Shpirt, *Chemistry of Germanium* (Izd. Khimiya, 1967) (in Russian).
5) R. Schwarz and C. Huf, Z. Anorg. Chem. **203** (1932) 188.
6) A. N. Lobachev, I. P. Kuzmina, L. N. Demianets, V. V. Iljuchin and N. V. Belov, Kristall und Technik **5** (1970) 425.

Journal of Crystal Growth **13/14** (1972) 545–548 © *North-Holland Publishing Co.*

MÉTHODE DE DOPAGE AU CHROME DANS LA PRÉPARATION DU RUBIS HYDROTHERMAL

NGUYEN DUC CHINH

Centre National d'Etudes des Télécommunications, Groupement Physique Électronique et Composants, 196, rue de Paris, 92 – Bagneux, France

Hydrothermal growth of ruby single crystals, on $\{\bar{1}22\}$ oriented seeds, has been carried out, using silver cans in a 150 cm³ autoclave. Chromium doping using potassium dichromate in an aqueous solution of potassium carbonate with sapphire as source material, caused inhomogeneous chromium distribution with unequally colored bands in the crystal. The use of ruby both as nutrient and doping reagent yielded a more regular chromium distribution. Both the two preceding methods generally produced heavy chromium accumulation at the seed–crystal interfaces and chromium rich green deposits on seeds have often been observed; in the latter case no further growth occurred. In order to greatly reduce the initial chromium concentration in solution during the warm-up period, gibbsite was added to the ruby nutrient; in the same way, an experiment performed with sapphire instead of gibbsite was not satisfactory. First, the starting crystallisation temperature of ruby was determined, then the gibbsite solubility was determined, so that the useful amount of gibbsite could be correctly estimated. With the use of an appropriate gibbsite–ruby nutrient mixture, optically clear and uniformally doped ruby single crystals have been prepared.

1. Introduction

La croissance cristalline du corindon par voie hydrothermale a fait l'objet de nombreux articles[1–3]) et les conditions d'obtention des cristaux de rubis dont la désorientation maximum des plans réticulaires est inférieure à 8 minutes et la densité des dislocations suivant $\{111\}$ de l'ordre de $10^5/\text{cm}^2$, sont maintenant bien connues[4]). En revanche, l'obtention de cristaux de rubis uniformément dopés en chrome constitue un problème non encore résolu d'une façon satisfaisante, et auquel nous nous proposons d'apporter une solution.

Nous avons réalisé les essais de dopage dans un autoclave possédant les caractéristiques suivantes: capacité = 150 cm³, pression de service = 4000 bar, température de service = 500 °C. Les croissances ont été effectuées dans des capsules d'argent en employant comme germes des lames de rubis dopé à 500 ppm et orientées $\{\bar{1}22\}$.

2. Méthodes de dopage

2.1. PAR LE BICHROMATE DE POTASSIUM

Nous avons utilisé du corindon comme corps-mère, et ajouté du bichromate de potassium à la solution aqueuse de carbonate de potassium. Avec des solutions contenant 4, 1, 0.5, 0.25 g/l de chrome, nous avons

constamment obtenu des cristaux irrégulièrement colorés. Par exemple, en opérant avec une solution contenant 0.25 g/l de chrome, nous avons remarqué: une couche profonde rouge foncé riche en chrome, une zone périphérique rose contenant 3500 ppm de chrome et une région intermédiaire jaune pâle ayant 1500 ppm de chrome (fig. 1).

La répartition non homogène du dopant peut s'expliquer par l'abaissement de sa concentration au cours de la croissance cristalline, car la totalité du chrome se trouve en solution dès la mise en température de l'enceinte. Cette hypothèse de travail nous a conduit à

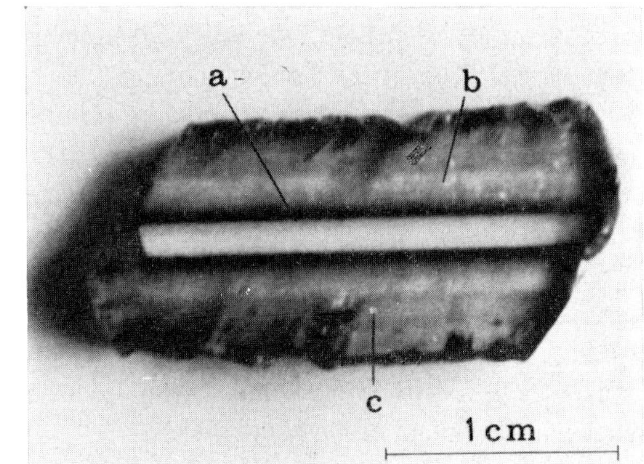

Fig. 1. Répartition non homogène de chrome: couche profonde rouge foncé (a); région intermédiaire jaune pâle (b); zone périphérique rose (c).

rechercher une source de chrome libérant progressivement cet élément pendant la formation du cristal.

2.2. Utilisation de rubis comme corps-mère

Nous avons employé du rubis d'origine Verneuil, concassé en grains de 5 mm et contenant 270, 500, 800, 1000, 3770, 12750 ppm de chrome. Dans les cristaux obtenus, la répartition du chrome n'est toutefois pas entièrement uniforme.

Dans tous les cristaux, nous avons constaté la présence à l'interface germe–cristal d'une bande très colorée, riche en chrome, le reste du cristal étant régulièrement coloré. En effet, au début de la croissance cristalline, la quantité de chrome provenant de la mise en solution saturée du corps-mère s'ajoute à celle libérée par la dissolution progressive de ce dernier, d'où l'accumulation de chrome à l'interface germe–cristal (fig. 2).

Certains cristaux présentent une zone d'accumulation de chrome parallèle aux faces du cristal: ceci semble provenir des diverses modifications du corps-mère durant l'essai: diminution de poids et volume, diminution du diamètre des grains, déplacement des grains, cristallisation parasite, etc. Ces modifications se répercutent sur le transport de matière dissoute et produisent finalement des fluctuations dans la concentration du chrome (fig. 3).

Parmi les essais préparés soit avec du bichromate en solution, soit avec du rubis comme corps-mère, nous avons parfois noté l'absence de croissance accompagnée de formation d'un dépôt vert sur les germes et sur le corps-mère. Les germes étaient faiblement dissous dans la plupart de ces cas; la concentration en chrome de ces dépôts analysés à la microsonde électronique variait entre 3.5 % et 6.8 % en poids. Le même phénomène a été signalé par Kuznetsov[5]) et Monchamp[6]). Il peut être expliqué par la formation de phases nouvelles, d'après le diagramme de miscibilité du système Al_2O_3–Cr_2O_3 étudié par Neuhaus[7]); la forte adhérence des dépôts verts est due probablement à la solubilité de Cr_2O_3 dans Al_2O_3 aux faibles concentrations de chrome[7]); l'absence de cristallisation ultérieure semble être liée à une différence notable des paramètres de réseau entre les deux composants, selon les valeurs établies par Newnham et De Haan[8]).

Afin d'éviter la formation du film vert à forte concentration de chrome, nous avons cherché à réduire la teneur initiale en chrome des solutions, en remplaçant partiellement le rubis par du corindon dont la vitesse de dissolution est supérieure à celle du rubis, d'après les données de Kuznetsov[5]).

2.3. Utilisation du mélange corindon–rubis comme corps-mère

Nous avons employé, pour ce faire, environ en poids $\frac{2}{3}$ de rubis et $\frac{1}{3}$ de corindon, ce dernier était disposé au fond de l'autoclave, c'est-à-dire dans la partie la plus chaude. Nous n'avons pas obtenu de croissance; l'expérience avec du corindon a été abandonnée. Nous avons opéré ensuite avec de la gibbsite $Al(OH)_3$ qui se dissolvait très rapidement.

2.4. Utilisation du mélange gibbsite–rubis

Dans ces essais, nous avons d'abord utilisé 18 et 25 g de gibbsite pour environ 110 cm³ de solution. Nous

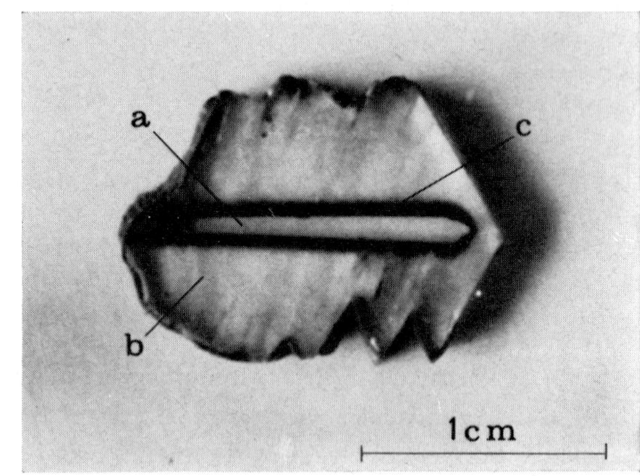

Fig. 2.   Germe (a); cristal (b); accumulation de chrome (c).

Fig. 3.   Accumulation de chrome à l'interface germe–cristal (a) et dans la masse du cristal (b).

TABLEAU 1

Croissance des germes rubis; résultats obtenus avec des mélanges gibbsite–rubis

Essais	Gibbsite (g)	Rubis à 500 ppm Cr (g)	$^*T_d$ (°C)	$^*T_c$ (°C)	Durée en température (heure)	Cristaux formés Variation de poids (g)	Variation d'épaisseur (mm)
1	25	28	500	475	140	+3.6	+5.3
2	18	30	500	470	168	+5.4	+6.3

* Les températures de dissolution T_d et de cristallisation T_c ont été mesurées dans la paroi de l'autoclave, respectivement aux niveaux du corps mère et des germes.

avons pu obtenir de cette manière le développement des cristaux sans la formation d'oxyde de chrome vert.

Les résultats du tableau 1, ainsi que beaucoup d'autres obtenus par la suite dans des essais semblables, nous ont permis de confirmer ce mode opératoire. De plus, dans les cristaux ainsi préparés, l'accumulation de chrome à l'interface germe–cristal a été fortement réduite, sinon éliminée (fig. 4).

L'emploi du mélange gibbsite–rubis dans les proportions indiquées au tableau 1 fournit cependant des cristaux peu limpides et à faible teneur en chrome. On peut alors penser que l'addition d'une trop grande quantité de gibbsite provoque une forte sursaturation dans la zone de croissance, empêche la dissolution superficielle du germe, augmente la vitesse de cristallisation, ce qui engendre des défauts (voiles, inclusions) et enfin perturbe la concentration en chrome des cristaux. Par conséquent, il est nécessaire de déterminer la température T_i de cristallisation commençante et la masse exacte de gibbsite servant à saturer la solution K_2CO_3.

(a) Pour mesurer la température T_i nous avons effectué des essais à différentes températures en observant les conditions générales suivantes:

– germes d'orientation $\{\bar{1}22\}$;

– corps-mère, environ 100 g $Al(OH)_3$ par litre de solution;

– solution aqueuse K_2CO_3, 4 M;

– coefficient de remplissage de l'autoclave $\rho = 85\%$ volume;

– gradient de température dans la paroi $\Delta T = 20$ °C.

Après chaque essai, nous avons examiné au microscope le facies cristallin des germes, dont les variations de poids et d'épaisseur étaient également contrôlées. Dans ces conditions la cristallisation commençante a été repérée autour de 370 °C (mesurée dans la paroi de l'autoclave).

(b) Pour évaluer la masse utile de gibbsite, nous

avons ensuite effectué une dissolution de α-Al_2O_3 avec les conditions indiquées (a) et en l'absence de germes et de gibbsite. Pour les températures du haut T_c et du bas T_d de l'autoclave respectivement égales à 370 °C et 390 °C, la solubilité était de 68.5 g Al_2O_3 par litre de solution, soit 104.6 g $Al(OH)_3$ par litre de solution.

Au cours de nos essais de cristallisation, effectués dans les conditions optimales, une fraction du corps-

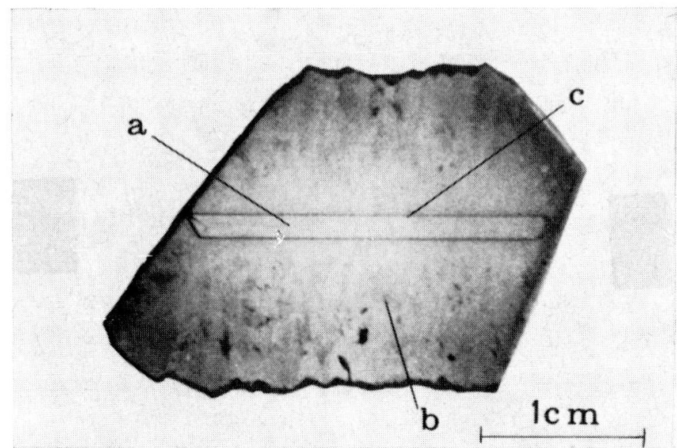

Fig. 4.   Germe (a); cristal (b) uniformément coloré; interface germe–cristal (c).

Fig. 5.   Echantillon de rubis hydrothermal: longueur 30 mm, épaisseur 4 mm.

TABLEAU 2

Concentrations respectives de chrome du corps-mère et du cristal

(Cr) du corps mère (ppm)	(Cr) du cristal obtenu (ppm)
270	95
500	370
800	430

mère rubis servait à saturer finalement la solution K_2CO_3 entre 370 °C et 480 °C. Afin de diminuer cet apport de rubis, dans la pratique, nous avons augmenté la masse de gibbsite jusqu'à la valeur utile de 107 g $Al(OH)_3$ par litre de solution, soit 13.7 g pour un autoclave de 150 cm^3 rempli à 85%.

A partir de cette valeur utile, nous avons fait croître des germes d'orientation $\{\bar{1}22\}$ découpés dans des barreaux de rubis fabriqués par le procédé de tirage Czochralski, en respectant les conditions optimales de cristallisation suivantes:

– corps-mère: rubis dopé à 500 ppm de chrome + gibbsite, 107 g/l;

– solution: K_2CO_3, 4 M; remplissage $\rho = 85\%$ volume;

– température de dissolution $T_d = 500$ °C et température de cristallisation $T_c = 480$ °C, mesurées dans la paroi de l'autoclave.

Des monocristaux de rubis d'une très grande limpidité ont été obtenus (fig. 5); avec un germe de 4.4 g et 1.5 mm d'épaisseur nous avons obtenu au bout de 640 heures un cristal de 31.7 g et 13 mm d'épaisseur.

Des essais de dopage à l'aide de rubis à différentes concentrations de chrome ont été faits. Les résultats sont indiqués dans le tableau 2.

3. Conclusion

Pour résumer les étapes successives de nos investigations, nous pouvons considérer que l'emploi du rubis est préférable aux solutions de bichromate pour réaliser une meilleure distribution des ions chrome dans les cristaux. L'utilisation du mélange gibbsite–rubis, en proportion convenable, constitue une méthode efficace pour éviter l'accumulation de chrome à l'interface germe–cristal et pour obtenir un dopage uniforme de ce dernier.

Remerciements

Je tiens à exprimer toute ma reconnaissance à Monsieur M. Marais, Ingénieur en Chef, pour ses précieux conseils et ses encouragements. Qu'il me soit également permis de remercier les Laboratoires d'analyse et de contrôle des matériaux du Groupement PEC pour leur fructueuse assistance.

Bibliographie

1) R. A. Laudise et A. A. Ballman, J. Am. Chem. Soc. **80** (1958) 2655.
2) J. Butcher et E. A. D. White, Mineral. Mag. **33** (1964) 974.
3) V. A. Kuznetsov, Kristallografiya **9** (1964) 123.
4) M. Marais, dans: *Croissance de Composés Minéraux Monocristallins*, Séminaires Chim. Etat Solide, No. 2 (Masson, Paris, 1967–68) p. 33.
5) V. A. Kuznetsov et A. A. Shternberg, Soviet Phys.-Cryst. **12** (1967) 280.
6) R. R. Monchamp, J. W. Nielsen and R. Puttbach, Project No. 8–132, contract AF33(657–1058) AIRTRON, AD-441224, Interim Report (1963).
7) A. Neuhaus, *Physics and Chemistry of High Pressures* (Soc. Chem. Industry, London, 1963) p. 237.
8) R. E. Newnham et Y. M. de Haan, Z. Krist. **117** (1962) 235.

Journal of Crystal Growth **13/14** (1972) 549–551 © *North-Holland Publishing Co.*

HYDROTHERMAL GROWTH OF SINGLE CRYSTALS OF HEAVY RARE EARTH HYDROXIDES*

S. MROCZKOWSKI and J. ECKERT**

Becton Center, Yale University, New Haven, Connecticut 06520, U.S.A.

The growth techniques for the lighter rare earth hydroxides proved unsuccessful for the heavier elements, Er–Lu. Solubility studies on $Er(OH)_3$ yielded data which could be applied to the modification of the earlier methods. The use of a large, specially cooled autoclave and the consequent large thermal gradients over the length of the growth region proved essential in the growth of macroscopic single crystals. Experimental details of the procedure are described.

1. Introduction

The hydrothermal growth of single crystals of the lighter rare earth hydroxides [$RE(OH)_3$] from lanthanum through holmium has been described earlier[1]. The optimal growth conditions were shown to include the use of 55 wt% NaOH as a mineralizer and temperatures as high as possible. This latter condition was determined by the point at which the $RE(OH)_3$ phase underwent a transition to the REOOH phase.

Attempts to grow the heavier rare earth hydroxides by a similar method were unsuccessful, resulting in the formation of microcrystalline specimens only. This is but one example of a number of striking differences in the characteristics of the two groups of hydroxides, including pronounced differences in the solubilities. In particular, Ivanov-Emin et al.[2] have shown that the solubility of the heavier hydroxides exhibit a maximum as a function of NaOH concentration under ambient conditions, whereas the lighter hydroxides show no such peak. As a first step it was therefore decided to extend our solubility measurements under hydrothermal conditions to the heavier hydroxides.

2. Solubility

A weight loss technique was used and the results were analyzed in terms of the adjustable parameters

for each run. [Most of the work was done on $Er(OH)_3$.] Full details will be published[3]), but for the present it is sufficient to quote those details which are relevant to the present work.

Fig. 1 shows the phase boundary for $Er(OH)_3$ and ErOOH. The two species can be distinguished not only by X-ray diffraction but also by their color. $Er(OH)_3$ is pale pink while ErOOH is significantly darker.

The data are in agreement with that of Klevtsov and Sheina[4]), although their investigation was limited to solutions with a 10% NaOH concentration and a pressure of 1000 atm.

For the range of temperatures covered in our study (up to 250 °C), no solubility maximum was found under hydrothermal conditions, despite the presence of such a maximum at ambient conditions. Rather, the solubility of $Er(OH)_3$ was found to increase approxi-

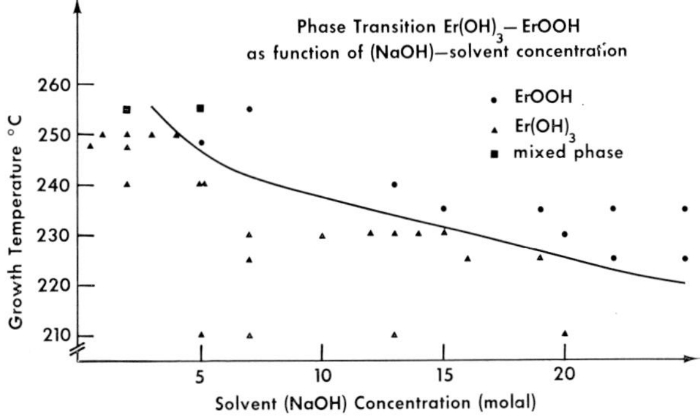

Fig. 1. Phase boundary for $Er(OH)_3$ and ErOOH as a function of solvent concentration. ErOOH is stable above the line while $Er(OH)_3$ is formed below the line.

* Work supported by the U.S. Atomic Energy Commission and the National Science Foundation.
** Present address: Princeton University, Engineering Quadrangle, Solid State Materials lab., Princeton, New Jersey 08540, U.S.A.

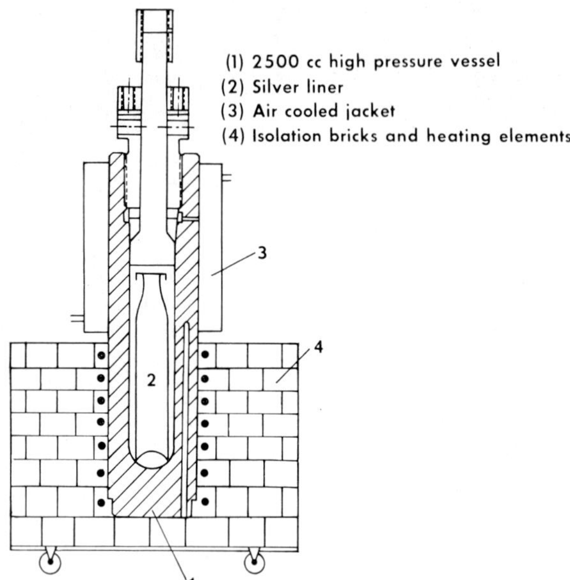

Fig. 2. Schematic diagram of the autoclave showing the air-cooled jacket used to produce the thermal gradient.

mately linearly with increasing concentrations of NaOH at fixed temperature, while for a fixed concentration it increased linearly with temperature. Moreover, no dependence on pressure was observed over the range studied (5000–12000 psi). This behavior is similar to that encountered for the lighter $RE(OH)_3$.

3. Experimental procedure and results

A large autoclave, 76 cm long, with a volume of 2500 ml was used. A silver tube containing the actual nutrient and solvent was placed inside the autoclave. The tube was 58 cm long and 5.7 cm in diameter. The bottom had a welded seal while the top was closed with a larger version of the mechanical seal described by Mroczkowski et al.[1]).

The relatively large size of these vessels was necessitated by the desirability of maintaining a thermal gradient of 40–60 °C (and sometimes as much as 90 °C) along the length of the growth region. This gradient proved critical for the success of the various runs.

For the growth of $Er(OH)_3$ the bottom of the autoclave was heated to a temperature of 245 °C. Although this is above the transition temperature to the monohydroxide phase, the large thermal gradient ensured that only $Er(OH)_3$ would be formed at the top of the silver tube. In addition, the large temperature at the bottom of the vessel makes it possible to take full

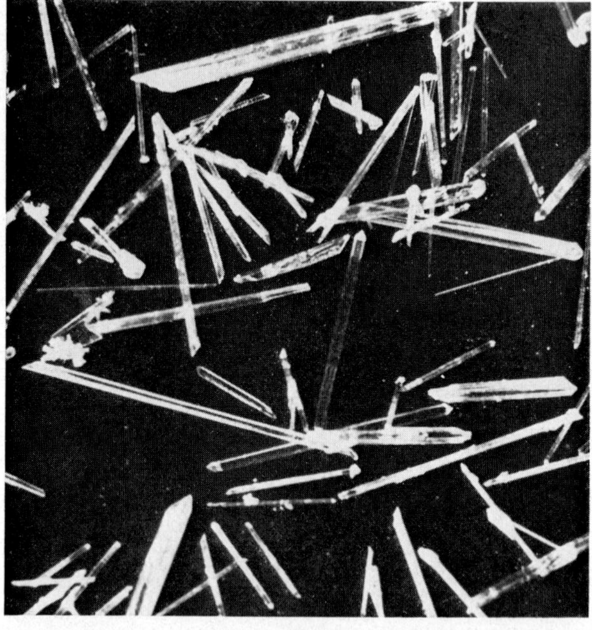

(a)

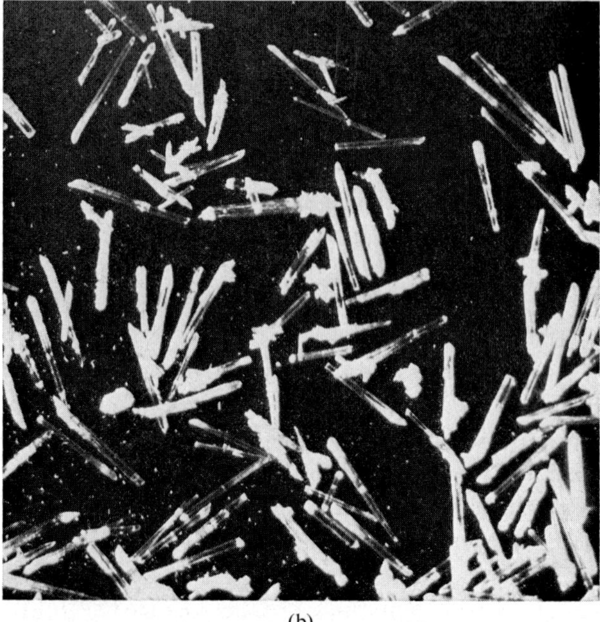

(b)

Fig. 3. Examples of crystals grown by the method described. (a) $Er(OH)_3$, and (b) $Yb(OH)_3$. Magnification 7.5×.

advantage of the increase in solubility with temperature.

Even when only half of the autoclave was heated, the desired temperature gradient could not be maintained. The autoclave was therefore placed in the configuration shown in fig. 2. The top half was encased in a tightly fitting aluminum jacket which was cooled by

TABLE 1

Lattice parameters and density of $Er(OH)_3$ and $Yb(OH)_3$ (error limits: lattice parameters ± 0.03 Å; pycnometric density ± 0.05 g/cm^3)

$RE((OH)_3$	Parameters (Å)		c/a	Density (g/cm³)	
	a	c		X-ray	pycnometer
$Er(OH)_3$	6.23	3.50	0.54	6.15	6.09
$Yb(OH)_3$	6.20	3.47	0.58	6.46	6.33

circulating air, the proper rate of flow having been determined from experience.

The nutrient consisted of 150 g of $Er(OH)_3$ in a precipitated powder form which had been prepared in the same manner as the lighter $RE(OH)_3$[1]. 1200 ml of 55 wt% NaOH was used as a mineralizer. The ingredients were placed in the large silver tube so that the fill factor was approximately 90%. A balanced pressure technique was employed, wherein the autoclave itself is filled with sufficient water to keep the inner tube from collapsing or bursting.

The results are shown in fig. 3. For $Er(OH)_3$, crystals 3–5 mm long and 0.3 mm thick were obtained. A similar procedure was used for $Yb(OH)_3$, except that the bottom of the autoclave was heated only to 230 °C. For this material, specimens 3 mm and 0.2 mm thick were produced.

4. Conclusions

The use of a substantially larger temperature gradient than that used for the growth of the lighter $Re(OH)_3$ proved to be essential for the successful growth of $Er(OH)_3$ and $Yb(OH)_3$. In principle there should be no problem obtaining equally good results for $Tb(OH)_3$, but the starting material is rather expensive. $Lu(OH)_3$ must be excluded from this approach since the transition to $LuOOH$ occurs around 60 °C. It is interesting to speculate, however, whether extremely large vessels and correspondingly large thermal gradients might not result in some success.

The crystals which were grown have been checked with X-ray diffraction and the gravimetric method reported earlier[1]. Details are given in table 1.

Acknowledgements

The authors would like to thank J. Sneider for his expert assistance in all of the mechanical phases of this work. They would also like to thank Prof. W. P. Wolf and Dr. H. Meissner for helpful discussion, R. L. Cone for assistance in the development of the aluminum jacket, and J. C. Doran for critical reading of the manuscript.

References

1) S. Mroczkowski, J. Eckert, H. Meissner and J. C. Doran, J. Crystal Growth **7** (1970) 333–342.
2) B. N. Ivanov-Emin, E. N. Siforowa, V. Mel'yado Kampos and E. Balestav Lafert, Russ. J. Inorg. Chem. **2** (1966) 1054.
3) S. Mroczkowski and J. Eckert, 9th Rare Earth Conference, October 1971.
4) P. V. Klevtsov and L. P. Sheina, Neorg. Mater. **1** (1965) 912.

Flux growth

D. ELWELL

R. D. DAWSON

H. J. SCHEEL

W. TOLKSDORF

F. WELZ

A. BROCHIER

P. COEURE

B. FERRAND

J. C. GAY

J. C. JOUBERT

J. MARESCHAL

J. C. VIGUIE

J. C. MARTIN-BINACHON

J. SPITZ

J. M. ROBERTSON

B. W. NEATE

M. SCHIEBER

A. GRILL

Y. AVIGAL

R. KRISHNAN

H. MAKRAM

L. TOURON

J. LORIERS

G. GARTON

S. H. SMITH

B. M. WANKLYN

M. VICHR

C. BÉLIN

J. J. BRISSOT

R. E. JESSE

L. VU TIEN

A. GRANDIN DE L'EPREVIER

V. GABIS

A. M. ANTHONY

J. W. GOODRUM

Journal of Crystal Growth **13/14** (1972) 555–559 © *North-Holland Publishing Co.*

555

THE VARIATION OF CRYSTAL GROWTH RATE WITH SUPERSATURATION IN FLUXED MELTS

D. ELWELL and R. D. DAWSON

Physics Department, Portsmouth Polytechnic, Park Road, Portsmouth, England

The growth rate of barium strontium niobate and nickel ferrite crystals from borate fluxes was determined by a thermogravimetric method. In the case of the $Ba_{0.5}Sr_{0.5}Nb_2O_6$ crystals the growth rate varies as the square of the supersaturation, as predicted by the BCF screw dislocation theory. The data for nickel ferrite shows a linear variation which is believed to be controlled by boundary layer diffusion.

1. Introduction

Flux growth is the least well understood of all the widely used crystal growth techniques. It is at present very difficult to give a quantitative discussion of such factors as nucleation, growth rate and the origin of imperfections. The main handicap to the formulation of such basic considerations is the lack of numerical data on fluxed melts.

A major omission is the absence of any measurement of the growth rate as a function of the supersaturation (or supercooling). Such data have been obtained in a large number of cases of growth from aqueous solutions and so a comparison may readily be made.

The aim of this investigation is to determine the variation of the growth rate of some materials of practical importance from fluxed melts in order to determine the rate controlling mechanism.

2. Theory of crystal growth from solution

In the generally accepted model of crystal growth, the growth units first diffuse through a boundary layer to a site on the surface of the crystal. The units then diffuse across the surface until they reach a step at the edge of an advancing layer. Integration into the crystal occurs at vacant sites or "kinks" in a step. In order to arrive at an expression for the rate of growth of a given material, it is frequently possible to neglect one or more of these stages as being extremely fast or extremely slow in comparison with the others.

In solution growth, Nernst[1]) assumed that volume diffusion of solute particles through the boundary layer is the rate controlling process. The growth rate will, by Fick's law, vary as the concentration gradient normal to the interface and the simplest solution to the diffusion equation gives a constant solute gradient across the boundary layer. A plane interface normal to the x axis at $x = 0$ will then have a linear growth rate v given by

$$v = \frac{D}{\rho}\left(\frac{\delta n}{\delta x}\right)_{x=0} = \frac{n_e D\sigma}{\rho\delta},\tag{1}$$

where n is the solute concentration, D its effective diffusion coefficient, n_e the equilibrium concentration, σ the relative supersaturation, ρ the density of the crystal and δ the thickness of the boundary layer.

Burton, Cabrera and Frank[2]) (BCF) included the effect of the interface kinetics but assumed that surface diffusion is negligible in solution growth. Their treatment leads to a relation between the growth rate and the supersaturation of the form

$$v = A\sigma^2\tag{2a}$$

for supersaturation values below some critical value, and to a linear relation

$$v = B\sigma\tag{2b}$$

for higher values of σ.

A similar treatment was used by Chernov[3]), who considered a layer rather than a spiral growth mechanism. His expression for the growth rate may be approximated over a wide range by a law of the form

XII – 1

$$v \propto \sigma^{1.65}. \qquad (3)$$

Bennema[4]) made careful measurements on the rate of growth of crystals in aqueous solutions and concluded that surface diffusion should not be neglected. His results for potassium aluminium alum and sodium chlorate are best described by the BCF surface diffusion equation

$$v = \frac{C\sigma^2}{\sigma_1} \tanh \frac{\sigma_1}{\sigma}, \qquad (4)$$

which was derived for growth from the vapour phase. Here, σ_1 is a critical value of the relative supersaturation. Eq. (4) may be approximated at low values of σ by a square law

$$v = \frac{C}{\sigma_1} \sigma^2 \qquad \text{(for } \sigma < \sigma_1\text{),} \qquad (5a)$$

and at higher supersaturations by a linear law

$$v = C\sigma \qquad \text{(for } \sigma \gg \sigma_1\text{).} \qquad (5b)$$

Recently Gilmer, Chez and Cabrera[5]) have given a formalism which includes simultaneous volume and surface diffusion. They arrived at a fairly simple expression for the growth rate which reduces to the forms given above when the appropriate approximations are made.

In flux growth, it is also possible in principle for the growth rate to be determined by the rate at which the heat of crystallization is removed from the interface. If heat is transported by conduction through a medium of thermal conductivity K, the growth rate will by given by

$$v = \frac{K}{\rho\phi} \frac{\Delta T}{\Delta x}, \qquad (6)$$

where ϕ is the heat of solution and $\Delta T/\Delta x$ the temperature gradient in the conducting medium. In an ideal solution where $n \propto \exp(-\phi/RT)$, $\sigma \; (= \Delta n/n) = \phi\Delta T/RT^2$ and so a linear relation between v and σ would be observed provided that the supercooling ΔT is small compared with T.

It is clear from the above discussion that a determination of the relationship between v and σ will not give an unambiguous indication of the rate controlling mechanism, but it is the most valuable single experiment and is particularly useful when combined with a meas-

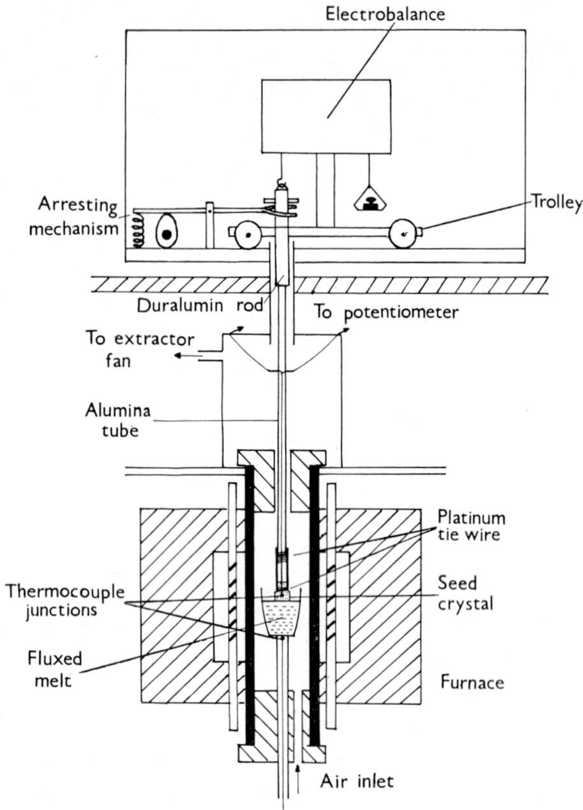

Fig. 1. Thermogravimetric apparatus (diagrammatic).

urement of the dependence of the growth rate on the degree of stirring in the solution.

3. Growth rate determination

Since continuous observation of the growing crystal in a fluxed melt is not possible because of the opacity of the melt, the growth rate was measured by a thermogravimetric method. The apparatus is based on that of Smith and Elwell[6]) and is shown diagrammatically in fig. 1.

The fluxed melt was contained in a 25 ml crucible inside a vertical tube furnace, in a temperature gradient with the base of the crucible hotter than the melt surface. A seed crystal was suspended with one habit face just below the surface of the melt from one arm of a C.I. Electronics Ltd. electrobalance, in an attempt to confine growth to this face.

The balance was supported on a trolley mounted on rails so that it could be displaced in order to facilitate adjustments to the suspension. Facilities were provided for accurate location of the balance in the weighing

XII – 1

position and for locking the wheels of the trolley. The platform was completely covered by a perspex box to protect the balance from dust and was well insulated to minimise heat transfer from the furnace.

The suspension consisted of a twin bore alumina tube, 2 mm in outside diameter and 40 cm long. The top end of this tube was firmly located in a duralumin rod, which was fitted with a hook for connection to the electrobalance, and with short horizontal arms by which the suspension could be arrested by means of a lever mechanism attached to the platform.

The seed crystal was attached to the lower end of the alumina tube by two platinum wires of 0.75 mm diameter which were bound tightly to the rod by fine platinum wire. The seed crystal was grooved parallel to its axis to incorporate the supporting wires and fine platinum wire was tightly bound round the grooved circumference of the seed. A hole was bored into the crystal along the axis using a jet of alumina powder so that a fine thermocouple could be inserted as closely as possible to the growing interface. The thermocouple leads passed through the alumina tube and emerged through holes bored in this tube above the level of the furnace. Extremely fine (0.05 mm) thermocouple wires were connected to these protruding leads in order to avoid interference with the balance mechanism.

The supercooling was measured as the difference between the temperature registered by the thermocouple in the seed crystal and that indicated by a second thermocouple pressed against the base of the crucible. The two thermocouples were calibrated against each other and the error in the measurement of the higher temperature was estimated from a determination of the temperature difference between thermocouples pressed against opposite sides of the crucible base.

With excess nutrient always present at the base of the crucible, a Rikadenki chart recorder was used to monitor the change in weight of the crystal as a function of time for a measured supercooling. The weight increase was normally found to be linear. This determination was repeated for various supercoolings with the seed temperature held constant.

Additional experiments were performed to determine the density of the fluxed melt and also the solubility curve of the crystalline phase so that the experimental data could be expressed as a plot of the linear growth rate versus the supersaturation.

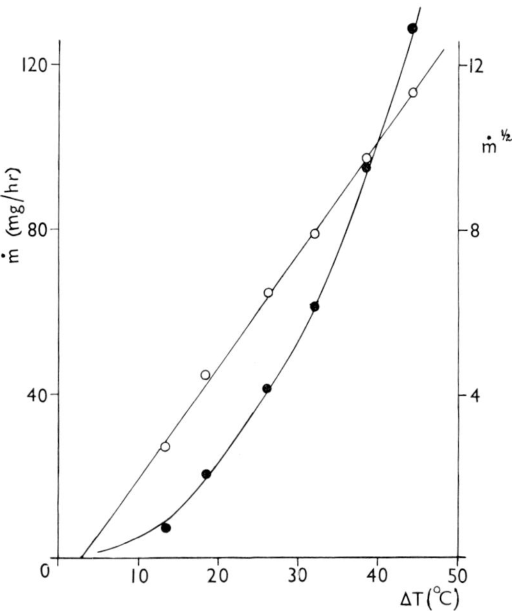

Fig. 2.   Rate of increase in weight $\dot{m}$ (●) and $\dot{m}^{\frac{1}{2}}$ (○) versus supercooling of barium strontium niobate at 1126 °C.

4. Results

The crystals studied were $Ba_{0.5}Sr_{0.5}Nb_2O_6$, a dielectric material with the tetragonal tungsten bronze structure, grown from a $Ba_{0.5}Sr_{0.5}B_4O_7$ flux, and nickel ferrite, $NiFe_2O_4$, grown from a flux of composition $BaO \cdot 0.62\, B_2O_3$. The borate fluxes were chosen mainly on account of their low volatility. The temperature of growth was over 300 °C below the melting point in the case of the niobate and over 500 °C below the melting point in the case of the ferrite.

Fig. 2 shows the variation of the rate of increase of crystal weight, $\dot{m}$, with the supercooling ΔT corrected for the temperature drop across the crucible base. The experimental data clearly do not follow a linear growth law but a plot of $\dot{m}^{\frac{1}{2}}$, also shown, is linear within experimental error. The extrapolation to $\dot{m}^{\frac{1}{2}} = 0$ does not pass through the origin but the intercept of about 3° on the ΔT axis may be a systematic error as it is extremely difficult to obtain absolute values of ΔT to a high accuracy because of the temperature drop across the crucible base. The relative values of ΔT are, however, reliable and rather large supercoolings have been used to minimise any errors due to uncertainty in the absolute values.

The values of $\dot{m}$ have been converted to give the linear growth rate (parallel to the c-axis) of the crystal

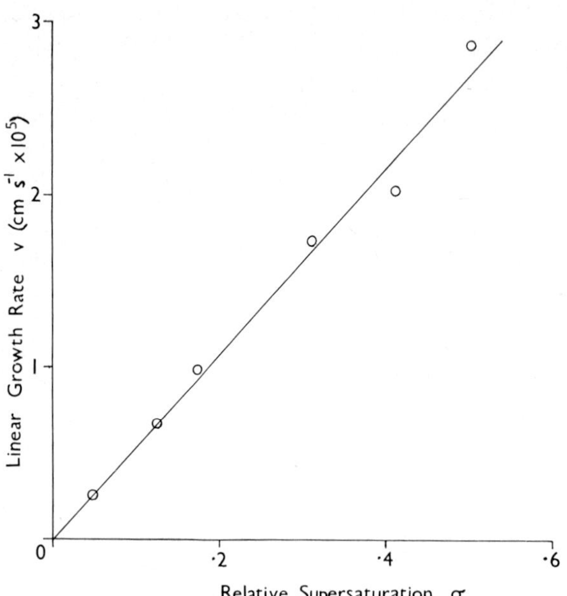

Fig. 3. Linear growth rate v of (111) face of nickel ferrite versus supersaturation at 1206 °C.

from the measured values of the solution and crystal densities and of the area of the growing interface. Due to surface tension, some growth inevitably occurs on the vertical sides of the crystal and, although a correction was made for the resulting increase in surface area, the absolute growth rates should be regarded as uncertain by about $\pm 25\%$. Since the increase in effective area due to surface tension did not change appreciably with temperature, this uncertainty does not affect the variation of apparent growth rate with supercooling.

The supercooling values were used to give the relative supersaturation σ using solubility data obtained using the same apparatus. The solubility versus temperature curve does not differ appreciably from linearity over the small temperature interval considered. The data of fig. 2 indicate that the growth rate is proportional to σ^2 in the range from $\sigma = 0.024$ to $\sigma = 0.081$. The corresponding values of the linear growth rate v are from 4 to 100×10^{-6} cm s^{-1}.

In contrast to the niobate system, the data for nickel ferrite shown in fig. 3 indicate a linear variation of growth rate with supersaturation. The graph in this case passes close to the origin. Again rather high values of supercooling (up to 53.7 °C) were used and the rate of growth of the (111) habit face of the crystal is similar to that of the niobate.

5. Discussion

The square growth law found for the barium strontium niobate may be explained either by eq. (2a) or by eq. (5a). Unfortunately, it has not yet been found possible to establish the dependence of the growth rate on the degree of agitation in the system and so distinguish between the BCF volume and surface diffusion theories.

A remarkable feature of the results is the persistence of the square law at supersaturations up to 0.081, corresponding to a 44.2 °C supercooling at 1126 °C. The critical supersaturation σ_1 for a transition to a linear law must have a value of at least 0.10 which is two orders of magnitude greater than the values reported by Bennema[4]) for crystallization from aqueous solutions. According to Bennema's analysis, a high value of σ_1 indicates a high activation energy for the entry of a growth unit into a kink. This suggestion is consistent with our observation that crystallization of this material by slow cooling normally yields tiny crystals, which indicates that nucleation occurs readily in comparison with growth on established crystallites. A high value of the activation energy for entry into a kink would account for this observation since growth units at the interface would have a high probability of resolution rather than growth.

If the growth process is in fact described by eq. (5a), the constant C/σ_1 has a value of 2×10^{-2}, so that C is also two orders of magnitude greater than Bennema's values. A quantitative interpretation is, however, not possible at present since many of the parameters which determine the value of C are unknown. The supersaturation values are, in fact, rather high for application of the BCF theory, but no other theory predicts a square growth law.

In the case of nickel ferrite, the linear relation between v and σ is believed to be due to boundary layer diffusion control. Although quantitative measurements were not made, it was found in crystal growth experiments[7]) that the rate of growth on rotating seeds was appreciably greater than that in static solutions. If it is assumed that eq. (1) is applicable to this system, the value of D/δ from the experimental data is 5.7×10^{-4} cm s^{-1}. The diffusion coefficient D for aqueous solutions and for ionic salts is normally of the order of 10^{-5} cm^2 s^{-1} and δ for aqueous solutions has been found to be approximately 10^{-2} cm. These values give $D/\delta \simeq 10^{-3}$ cm s^{-1} which is in reasonable agreement with our experimental value.

We therefore conclude that the rate of growth of barium strontium niobate crystals is determined by the BCF screw dislocation mechanism, although it is not possible to distinguish between the volume and surface diffusion theories. The growth rate of nickel ferrite is controlled by the volume diffusion stage.

Acknowledgements

We are grateful to Dr J. C. Brice for the barium strontium niobate seed crystals and to A. W. Morris for the construction of the thermobalance support. One of us (R.D.D.) wishes to thank the Science Research Council for the award of a CAPS research studentship.

References

1) W. Nernst, Z. Physik. Chem. **47** (1904) 52.
2) W. K. Burton, N. Cabrera and F. C. Frank, Phil. Trans. Roy. Soc. London A **243** (1951) 299.
3) A. A. Chernov, Soviet Phys.-Usp. **4** (1961) 129.
4) P. Bennema, J. Crystal Growth **5** (1969) 29.
5) G. H. Gilmer, R. Ghez and N. Cabrera, J. Crystal Growth **8** (1971) 79.
6) S. H. Smith and D. Elwell, J. Mater. Sci. **2** (1967) 297.
7) S. H. Smith and D. Elwell, J. Crystal Growth **3, 4** (1968) 471.

ACCELERATED CRUCIBLE ROTATION: A NOVEL STIRRING TECHNIQUE IN HIGH-TEMPERATURE SOLUTION GROWTH

H. J. SCHEEL

IBM Zurich Research Laboratory, 8803 Rüschlikon, Switzerland

The role of diffusion and convection in conventional flux growth is compared with the strong stirring effect of the accelerated crucible rotation technique (ACRT). ACRT allows fast solution flow rates at the growing crystal faces. By this and by homogenization of the solution many problems in flux growth are solved. Control of nucleation in a closed crucible is achieved by combination of ACRT with localized cooling. Large inclusion-free crystals with applications in solid state physics and technology have been grown.

1. Introduction

This paper describes a new technique of stirring and its application to crystal growth from high-temperature solutions[1].

In crystal growth from melts and from solutions, stirring and, to a lesser extent, convection, reduce the thickness of the boundary layer in front of the growing crystal[2–7]. In a well-stirred system, surface diffusion and particle integration (including the desolvation processes) dominate as rate-limiting growth processes over volume diffusion through the unstirred layer if supersaturation is high enough. Also fast motion of melt or solution relative to the growing crystal surface prevents the adverse effect of constitutional supercooling on the interface stability[8–10].

By stirring, the solution is homogenized with respect to temperature and concentration. Therefore further nucleation after the first nucleation or after insertion of a seed crystal can be minimized or prevented altogether.

For these reasons stirring is employed in crystal growth wherever possible. But in high-temperature solution growth stirring by seed crystals is extremely difficult because generally volatile solvents like PbO, PbF_2, Bi_2O_3, or mixtures of these with B_2O_3 are used. Rotating seed crystals were applied only in a few exceptional cases which are listed, with the obtained maximum stable growth rates in table 1.

In the following it will be shown that an adequate stirring effect and growth rates as listed in table 1 can be achieved in a closed crucible by accelerated crucible rotation.

Compared to diffusion and convection which are

TABLE 1

Maximum stable growth rates in high-temperature solution growth (from experiments)

Crystal	Solvent	Linear growth rate (Å/sec)	Remarks	Reference
$GdAlO_3$	$PbO–PbF_2–B_2O_3$	~ 200	ACRT	This work
$NiFe_2O_4$	$Na_2Fe_2O_4$	200	Pulling from solution	11
$NiFe_2O_4$	$BaO–B_2O_3$	~ 500	Pulling from solution	12
$NiFe_2O_4$	$PbO–PbF_2$	~ 260	Seeded growth from solution	13
YIG	$BaO–B_2O_3$	120	Seed crystal on stirrer	14
YIG	$BaO–B_2O_3$	150	Pulling from solution	15
YIG	$BaO–B_2O_3$	260	Garnet film growth in a temperature gradient	16
YIG	$BaO–B_2O_3$	~ 150	Pulling from solution	17
$Ba_2Zn_2Fe_{12}O_{22}$	$BaO–B_2O_3$	200	Pulling from solution	18

XII – 2

slow material transport mechanisms in conventional flux-growth experiments, the Accelerated Crucible Rotation Technique (ACRT) reduces the thickness of the stagnant boundary layer and therefore allows faster stable growth rates. In addition, homogenization of the solution by ACRT combined with localized cooling permits control of nucleation. These statements are supported by estimates of the boundary-layer thickness and the maximum stable growth rate, by simulation experiments, and by successful crystal-growth experiments.

2. The role of convection in high-temperature solution growth

In the conventional flux-growth practice by the slow-cooling technique one relies exclusively on natural convection as the mechanism for material transport and homogenization. By comparison, heat conductivity and diffusion are significant processes only if there are appreciable temperature or concentration gradients. Thus they contribute to heat and material transport at boundary layers, but only very little within the bulk of the solution where, by the crystallization process, by inhomogeneous heating and by flux evaporation, relatively small concentration and temperature gradients are constantly generated. Therefore convection is generally present in crystal-growth experiments*.

The flow velocity of thermal convection is low and amounts to the order of 1 to 5×10^{-2} cm sec^{-1} for typical flux-growth conditions[19]. Even for a low viscosity liquid of 1 centipoise and the high thermal expansion coefficient of 10^{-3} °C^{-1}, the flow velocity will reach only 10^{-1} cm sec^{-1} for a temperature difference $\Delta T = 10$ °C. The width of the metastable Ostwald–Miers region imposes a limit on the permitted temperature difference of roughly this order of magnitude. For these flow velocities the stable crystal growth rate is limited by volume diffusion through the relatively thick boundary layer of approximately 5×10^{-2} cm and lies in the order of 50 Å sec^{-1}. Because of constitutional supercooling only small inclusion-free crystals have been obtained so far.

By the presence of convection cells, pockets with high supersaturations are probably formed locally and cause nucleation of many crystals instead of one or two which then may grow to large sizes. Localized cooling and convection generated by asymmetric temperature distribution increased only slightly the size and quality of the flux-grown crystals. In their crystal-growth experiments with a 2 gallon crucible, Van Uitert et al.[20] combined crucible rotation with localized cooling at the bottom. Despite the expected improvement in convection the authors reported multinucleation. The resulting YIG and YAG crystals were very large when compared to crystals obtained from smaller crucibles, while the ratio of crystal-to-crucible size seems to have been comparable for large and small crucibles.

3. The accelerated crucible rotation technique (ACRT) in high-temperature-solution growth

Convection alone or in combination with uniform rotation is far from optimum as a material transport mechanism. In high-temperature solution growth a mechanism is required which agitates the solution in much the same way as conventional mechanical stirring in crystal growth from aqueous solutions. Accelerated and decelerated crucible rotation is proposed to achieve such a stirring effect in flux growth. Consider a crucible which is initially at rest. Then constantly accelerate the crucible to rotation about its vertical axis. The outer region of the solution follows the change in crucible motion with little delay. The center of the melt, however, continues to remain at rest due to inertia. Thus a shearing of solution rings is produced. Slowly the center of the solution begins to move, too, until, after a considerable time, it rotates at the same rate as the crucible.

By the laminar slip of the solution around the rotation axis a spiral is formed which, for typical flux-growth conditions and acceleration to 60 rpm, has the order of a thousand arms. Therefore any local variation in solute concentration or in temperature is distributed along this spiral. Since the spiral arms have a radial distance of less than a tenth of a millimeter, the concentration and temperature differences disappear by diffusion and heat conduction in a time of less than a second. The hydrodynamics of convection and of ACRT will be presented in the paper by Schulz-DuBois[21].

Alternating accelerated and decelerated crucible ro-

---

* Convection flow rates due to thermal and concentration differences are believed to be of the same order of magnitude generally, depending on the material parameters and the growth conditions.

TABLE 2

Comparison of the influence of diffusion, convection, and ACRT on thickness of boundary layer and on maximum linear growth rate for volume-diffusion-limited growth

	Mean velocity of solution (or particles) at the crystal surface (cm sec^{-1})	Thickness of boundary layer[a] (cm)	Maximum linear growth rate (Å sec^{-1})
Diffusion	(5×10^{-5})	(1.7)	(0.5–3)
Convection[b]	4×10^{-2} [c]	6×10^{-2}	20–80
ACRT	1	1.2×10^{-2}	100–500
ACRT	10	4×10^{-3}	$(300–1500)$[d]
ACRT	100	1.2×10^{-3}	$(1000–5000)$[d]

[a] Calculated from formulas of Carlson[4]) and of Bennema[6]).
[b] For $T = 1400\ °K$, $\Delta T = 10\ °C$, $l = 5$ cm, $\eta = 2$ cp, $\rho = 7.5$ g cm^{-3}, $c_p = 0.15$ cal g^{-1} deg^{-1}, $\beta = 8 \times 10^{-4}$ deg^{-1}, $K = 0.02$ cal cm^{-1} sec^{-1} deg^{-1}.
[c] After Cobb and Wallis[19]), we get the same value for our conditions.
[d] Surface reactions are rate-limiting processes.

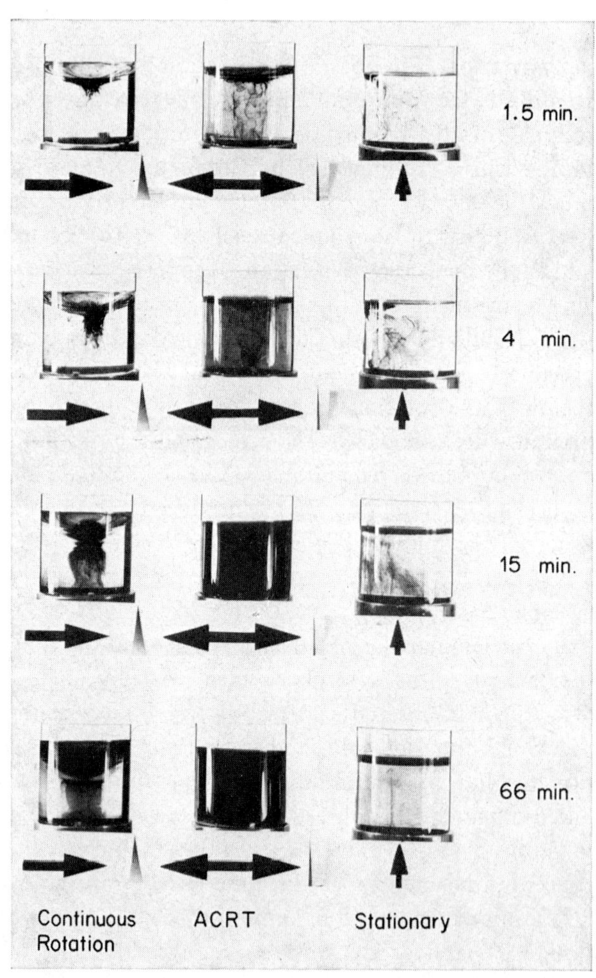

Fig. 1. Simulation experiment demonstrating the strong mixing effect of ACRT.

tation acts much the same way as a stirrer acting in the laminar regime. In addition to this rotational stirring effect, mixing in vertical and radial directions is provided by the Ekman flow and by the deformation of the liquid due to centrifugal forces, which is demonstrated by the parabolic surface at high rotation rates and the nearly flat surface of the liquid at rest. And, of course, there will be the relatively small mixing effect of the superimposed natural convection.

If the ACRT stirring technique is combined with localized cooling, control of nucleation is possible and only one or two nuclei are formed. Similarly, if a seed crystal is arranged in the crucible and all conditions are properly controlled, there is no additional spontaneous nucleation.

A forced flow of solution is produced along the crystal surface. The flow rate is determined by the properties of the solute-solvent system, by the crucible (and crystal) geometry and dimensions, and by acceleration and rotation rates. Flow velocities of 1 or even 100 cm sec^{-1} relative to the crystal are not difficult to

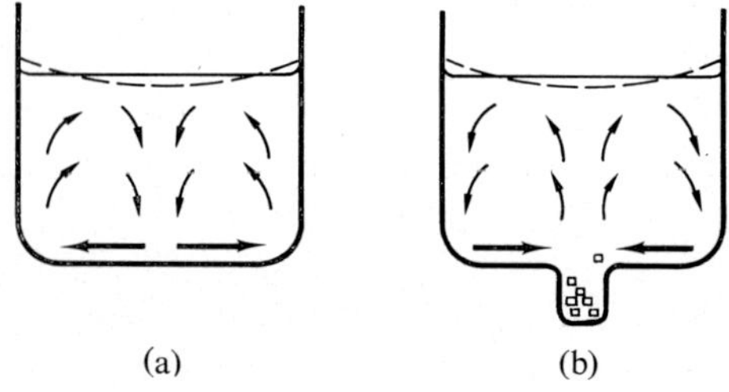

Fig. 2. (a) Direction of radial and vertical flow during acceleration due to centrifugal forces (Ekman flow). (b) Trap of crystallites during deceleration.

XII – 2

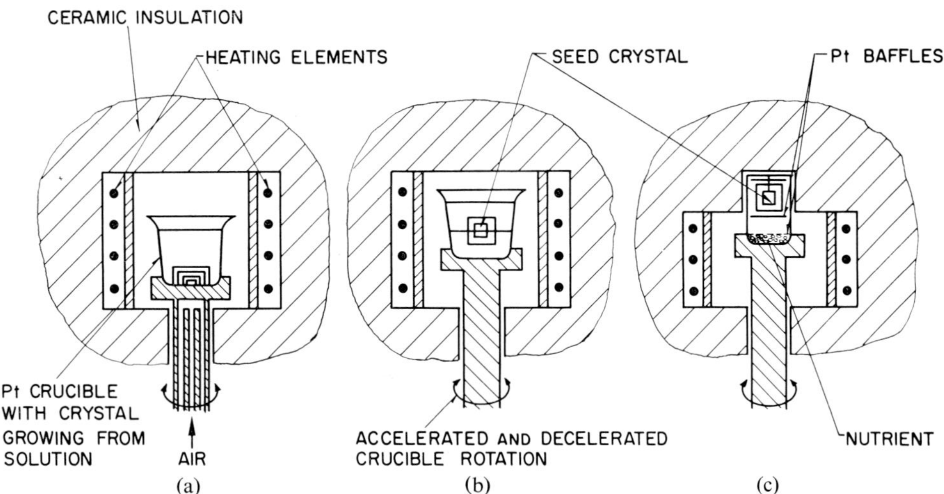

Fig. 3. (a) ACRT with localized cooling for nucleation control (slow cooling technique). (b) ACRT with a seed crystal under iothermal conditions (slow cooling technique). (c) ARCT with as seed crystal in constant temperature gradient.

realize. These reduce the thickness of the stationary boundary layer[4,6]) surrounding the crystal to the order of 10^{-2} to 10^{-3} cm, which, in turn, permits greater stable growth rates. In table 2 a comparison of the boundary layer thicknesses and the corresponding stable growth rates calculated for volume-diffusion-controlled growth are shown for the flow rates of convection and of ACRT, respectively. The data for pure diffusion are also shown and are somewhat hypothetical since one would expect convection to be present generally in flux-growth experiments. Crystal growth by convection requires an extremely low cooling rate in order to maintain a stable crystal growth rate of approximately 50 Å sec^{-1} only. In ACRT a flow rate of 1 cm sec^{-1} is adequate for a stable growth rate of 100 to 500 Å sec^{-1}. From table 1 it appears that no growth rates higher than this order of magnitude have ever been observed experimentally. This evidence suggests that by application of ACRT the diffusion through the boundary layer is no longer the main growth-rate limiting process.

4. Simulation experiments and crystal-growth results

The ACRT stirring effect can be easily demonstrated by a simulation experiment. Three glass beakers containing a liquid of a viscosity of 2 centipoise (H$_2$O + H$_2$SO$_4$) are heated from below in order to establish convection. Equal amounts of KMnO$_4$ as a coloring agent are placed at the center of the liquid surface where they float due to surface tension. The left beaker is continuously rotated, the right one remains at rest, while the center one is subjected to accelerated and decelerated rotation by a motor which is switched on and off for about a minute at a time. Fig. 1 shows the situation at times of 1.5, 4, 15, and 66 min, respectively, after the beginning of the experiment. In the center beaker the KMnO$_4$, as far as it is dissolved, is distributed homogeneously over the entire beaker volume in a time of one or two cycles of acceleration. After 15 min all the KMnO$_4$ is completely dissolved and a homogeneous solution is obtained.

In the uniformly rotated beaker there are stable unmixed regions with no coloration at all. This agrees with the observations by Carruthers and Nassau[22]). Due to Taylor–Proudman cells in the uniformly rotated beaker there seems to be even less mixing than in the stationary one.

In fig. 2 the directions of Ekman flow and the deformation of the liquid body due to centrifugal force are shown. The most simple experiment for demonstration of the Ekman flow is a cup of tea containing tea leaves. These tea leaves float to the circumference, when the tea is stirred, and back to the center when the stirring action is stopped.

Figs. 3a–3c show examples of experimental arrangements for flux growth by ACRT. To date, the greater part of our experience is with the method of fig. 3a. The 500 cm^3 platinum crucible with its special mechan-

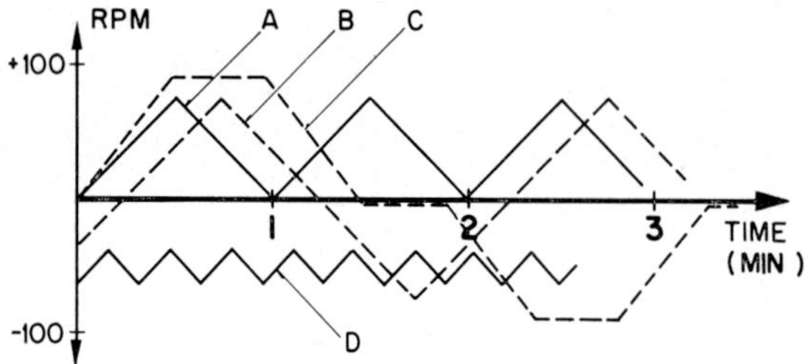

Fig. 4. Examples of cycles of accelerated crucible rotation.

ical stable shape is sealed by argon-arc welding, and is supported by a suitably shaped ceramic disk which in turn is placed on a ceramic tube. The latter is rotated about its vertical axis with periodically alternating acceleration and deceleration. Several useful cycles are illustrated in fig. 4. For flux growth, cycles A, B, C are appropriate. Cycle D may be useful in the Czochralski technique, when corresponding accelerated counter rotation of the crystal is used additionally.

Initial experience with the ACRT was gained by growing gadolinium aluminate, $GdAlO_3$, an orthorhombic perovskite-type compound, from $PbO-PbF_2-B_2O_3$ solution. Many experiments without ACRT in crucibles of 30 to 1000 cm^3 yielded crystals of $GdAlO_3$ up to 20 g weight with flux inclusions and inclusion-free crystals up to 8 mm in size (approx. 4 g). For ACRT experiments the mixture of the starting materials of highest purity was premelted in the 500 cm^3 platinum crucible, which was sealed and heated to 1300 °C for 15 hr. Then it was cooled to 900 °C at a rate which increased from 0.3 to 0.5 °C hr^{-1}. Faster cooling rates than these caused interface instability (flux inclusions) by stronger constitutional supercooling (large faces, higher viscosities, decreasing diffusion coefficients). At the end of the run the crucible with the one or two crystals formed was allowed to cool slowly to room temperature.

The first two runs yielded just one crystal of $GdAlO_3$ each, of 210 g and 104 g weight, respectively. Another run in a small 140 cm^3 crucible yielded just two crystals with a total weight of 44 g. Fig. 5 shows the above-mentioned crystal of 210 g weight in the interior of which the light from behind is reflected. The overall dimensions of the crystal are $3.5 \times 3 \times 2.5$ cm^3. Like other spontaneously nucleated crystals, this crystal shows a central region of dendritic growth. However the outer portions of these crystals of about 1 cm

Fig. 5. Gadolinium aluminate crystal of 210 g weight showing internal light reflection through clear portion.

thickness are optically clear and free of inclusions. In crystal growth from aqueous solutions dendritic growth can be prevented by use of seed crystals. First experiments with seed crystals and ACRT in crystal growth from high-temperature solutions are promising. Other work is directed to magnetic materials, and large inclusion-free garnets have been grown.

Subsequent to our preliminary description of ACRT at national crystal growth conferences[23]), other laboratories had similar success with crystal growth of magnetic garnets and of manganese telluride by ACRT[24]).

5. Conclusions

The accelerated crucible rotation technique (ACRT) has been demonstrated to have a strong stirring effect in crystal growth from high-temperature solutions by successful crystal-growth runs and by simulation experiments. Specifically, ACRT accomplishes control of nucleation, it suppresses the adverse effects of constitutional supercooling and convection, and it permits the realization of stable crystal-growth velocities in the range of several hundred Å sec^{-1}. The method is believed to be especially useful in the growth of large crystals for the magnetic bubble domain technology, for laser rods, and for electrooptic or nonlinear optical materials. Simulation and preliminary crystal-growth experiments indicate that ACRT is a valuable aid also for crystal growth from melts, from the vapor phase, and from hydrothermal solutions.

Acknowledgement

The author thanks Dr. D. Elwell, Dr. E. O. Schulz-DuBois, Dr. W. V. Smith, Prof. W. A. Tiller and Dr. E. A. D. White for valuable discussions.

References

1) H. J. Scheel and E. O. Schulz–DuBois, J. Crystal Growth **8** (1971) 304.
2) W. K. Burton, N. Cabrera and F. C. Frank, Phil. Trans. Roy. Soc. London A **243** (1951) 299.
3) J. A. Burton, R. C. Prim and W. P. Slichter, J. Chem. Phys. **21** (1953) 1987.
4) A. Carlson, in: *Growth and Perfection of Crystals*, Eds. R. H. Doremus, B. W. Roberts and D. Turnbull (Wiley, New York, 1958) p. 421.
5) J. C. Brice, J. Crystal Growth **1** (1967) 161.
6) P. Bennema, J. Crystal Growth **5** (1969) 29.
7) W. A. Tiller, J. Crystal Growth **2** (1968) 69.
8) W. A. Tiller, K. A. Jackson, J. W. Rutter and B. Chalmers, Acta Met. **1** (1953) 428.
9) H. J. Scheel and D. Elwell, J. Crystal Growth **12** (1972) 153.
10) S. O'Hara, L. A. Tarshis, W. A. Tiller and J. P. Hunt, J. Crystal Growth **3, 4** (1968) 555.
11) W. Kunnmann, A. Ferretti and A. Wold, J. Appl. Phys. **34** (1963) 1264.
12) S. H. Smith and D. Elwell, J. Crystal Growth **3, 4** (1968) 471.
13) J. Kvapil, V. Jon and M. Vichr, in: *Growth of Crystals*, Vol. 7, Ed. N. N. Sheftal, (Consultants Bureau, New York, 1969) p. 233.
14) R. A. Laudise, R. C. Linares and E. F. Dearborn, J. Appl. Phys. **33** (1962) 1362.
15) R. C. Linares, J. Appl. Phys. **35** (1964) 433.
16) R. C. Linares, R. B. McGraw and J. B. Schroeder, J. Appl. Phys. **36** (1965) 2884.
17) M. Kestigian, J. Am. Ceram. Soc. **50** (1967) 165.
18) T. R. AuCoin, R. O. Savage and A. Tauber, J. Appl. Phys. **37** (1966) 2908.
19) C. M. Cobb and E. B. Wallis, Report AD 655388 (1967).
20) W. H. Grodkiewicz, E. F. Dearborn and L. G. Van Uitert, in: *Crystal Growth*, Ed. H. S. Peiser (Pergamon, Oxford, 1967) p. 441.
21) E. O. Schulz–DuBois, J. Crystal Growth **12** (1972) 81.
22) J. R. Carruthers and K. Nassau, J. Appl. Phys. **39** (1968) 5205.
23) Talks at National Conferences on Crystal Growth at Bristol, U.K. (Sept. 11, 1970), Zurich (Sept. 22, 1970) and Munich (Oct. 16, 1970).
24) Dr. W. Tolksdorf (Philips Zentrallaboratorium, Hamburg), private communication; see also the papers of W. Tolksdorf and D. Mateika at this Conference.

THE EFFECT OF LOCAL COOLING AND ACCELERATED CRUCIBLE ROTATION ON THE QUALITY OF GARNET CRYSTALS

W. TOLKSDORF and F. WELZ

Philips Forschungslaboratorium Hamburg GmbH, 2 Hamburg 54, Germany

To characterize growth and properties of yttrium iron garnet single crystals a detailed description of growth conditions is given including temperature distribution in the fluxed melt in terms of isotherms. The influence of local cooling and the accelerated crucible-rotation technique, recently published by Scheel and Schulz-DuBois, are described. Growth features on crystal faces are discussed, and analytical data about impurities are summarized.

1. Introduction

In a paper recently published by Van Uitert et al.[1] the following statement can be read: "The dynamics of growth of rare earth iron garnets are somewhat unusual. It is quite possible to obtain orthoferrite where garnet is expected and vice versa. Nucleation conditions and growth rates are of major importance... It is of interest that there is only a 50–50 chance that pure yttrium iron garnet (YIG) will form in the course of usual flux run."

We must underline this statement. Sometimes we find single crystals of magnetoplumbite ($PbFe_{12}O_{19}$) or yttrium oxyfluoride ($Y_{20}O_{16}F_{28}$) besides or instead of YIG ($Y_3Fe_5O_{12}$). Because of this uncertainty it is the aim of this contribution to carefully describe growth conditions and properties of YIG single crystals.

2. Experimental

Our standard composition of the melt similar to the composition used by Grodkiewicz et al.[2] is: PbO 36.3; PbF_2 27.0; B_2O_3 5.4; CaO 0.1; Fe_2O_3 3.4 and Fe_2O_3 17.38 and Y_2O_3 10.42 (in mole%). The procedure we use has been described already by Tolksdorf[3].

A crucible in a top heated chamber furnace can be turned around its horizontal axis to separate the bottom grown crystals from the flux. Temperature stability is of the order of ± 0.01 °C at the bottom of the platinum crucible which is thermally isolated by an alumina-silicate ceramic fibre and an alumina ceramic support crucible. To localize nucleation an alumina ceramic pipe cooled by a pressure stabilized air stream is pressed on to the center of the bottom of the platinum crucible. To apply the accelerated crucible-rotation technique (ACRT) described by Scheel et al.[4], the support crucible is mounted on an alumina shaft. As long as the crucible is rotated around this vertical axis, the horizontal axis has to be withdrawn.

The temperature distribution was measured inside the melt in steps of 5 mm with three thermocouples; in the centre, in 16 mm and 32 mm distance from the centre, respectively. During the measurements the crucible could not be rotated. The isotherms obtained are given in fig. 1 assuming a symmetrical and steady distribution. The shape of the isotherm is very sensitive to the form of the crucible and its surroundings. There is no evidence of convection within the flux, because a filling with alumina granulate shows a similar temperature distribution. No temperature oscillations larger than the temperature stability of the order of ± 0.01 °C could be detected. The effect of the cooling finger at the bottom of the platinum crucible is limited to an area with a radius of about 15 mm around the centre. Outside of this zone there is still competitive nucleation. Fig. 2 shows a typical result of a 1100 g run. Note that nucleation is essentially of one crystal on the cooling finger. Nucleation occurs at 1100 ± 10 °C, as has been measured by many runs around this temperature; the undercooling is less than the given temperature range of ± 10 °C. The cooling rate was 1 °C/h, the cooling air stream was 160 l/h. At 1010 ± 10 °C the crystal was taken out of the flux by turning the

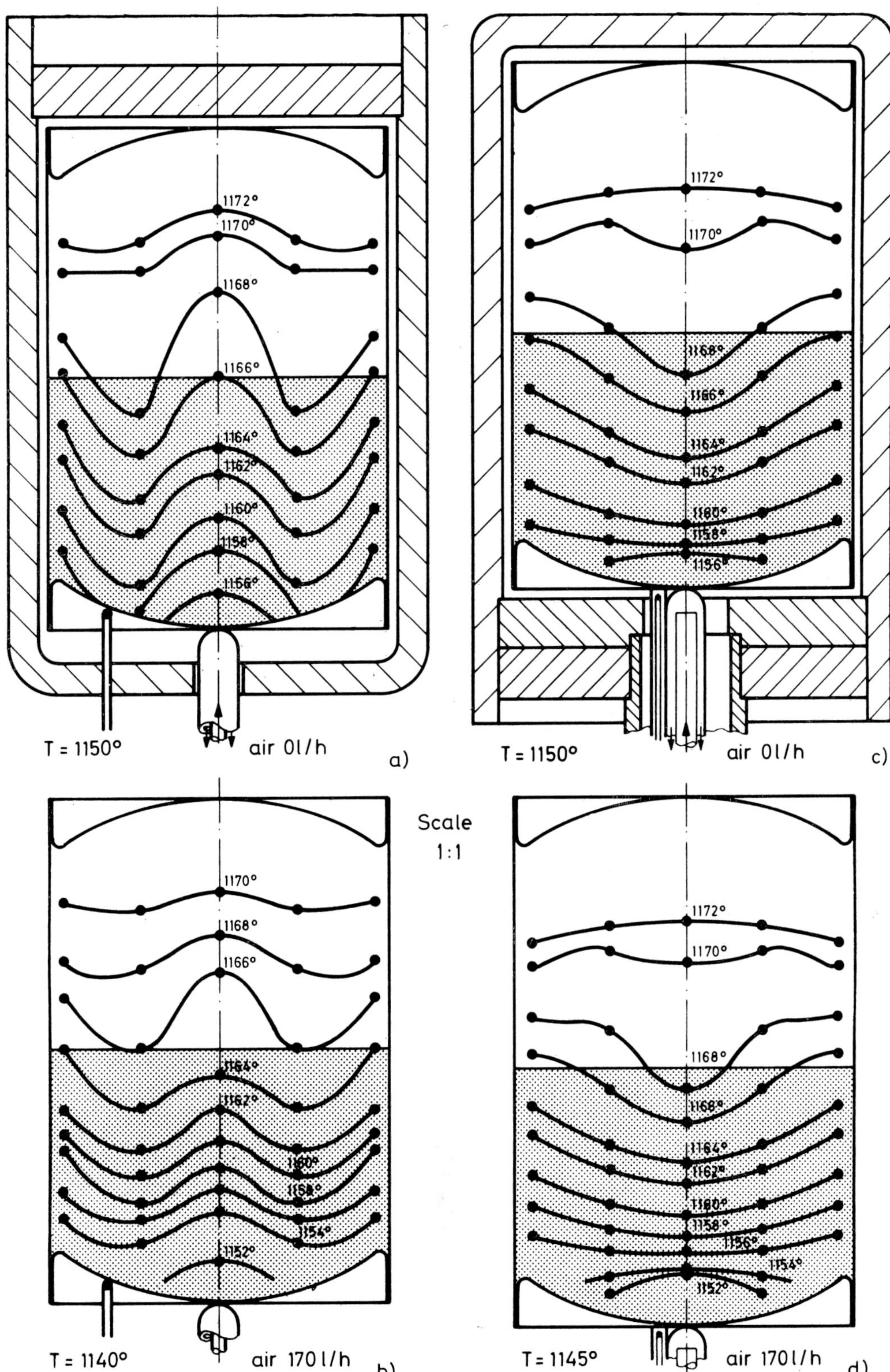

Fig. 1.   Temperature (°C) distribution in the melt. (a, b) standard mounting, (c, d) mounting for ACRT[4]).

Fig. 2. Central YIG crystal (scale in mm).

Fig. 3. Yield of YIG crystals grown by ACRT[4]) (scale in mm).

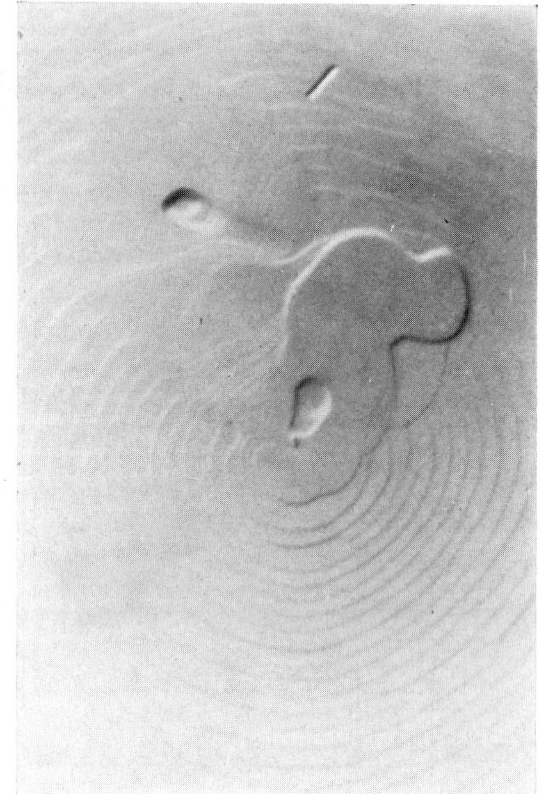

Fig. 4. Growth centre on a (110) face of YIG. Magnification ×1000.

crucible around its horizontal axis. In many runs with the given composition YIG was the only phase to crystallize in the temperature range mentioned. In one case, one magnetoplumbite single crystal of substantial size grew at the center of the bottom, while the other crystals were YIG. Sometimes surface crystals were present. This indicates that there is not always equilibrium in the melt as a whole without stirring. After cooling with a speed of 50 °C/h (this is normal speed after finishing crystal growth) an analysis of the solidified melt had shown different zones of composition.

To apply a stirring effect by ACRT[4]) the crucible was alternately rotated (50 rpm) and stopped around the vertical axis. In this case there is a strong tendency for nucleation at the edge of the bottom of the crucible, as shown in fig. 3, when no cooling air stream is applied. The crystals had fewer inclusions, the yield was larger (up to 45 wt% of the input) by this method, and no surface crystals or crystals of other phases were present.

3. Characterization

The growth features on the crystal faces are the same as in the unstirred case. As reported by Lefever and Chase[5]), Komatsu and Sunagawa[6]) and others, growth spirals can be seen on (110) and (211) faces of YIG crystals. Under our standard conditions often very few spirals, sometimes only one, cover (110) faces larger than 1 cm^2. The centre of such a growth figure is given in fig. 4 on a photomicrograph with differential interference contrast like that used by Nomarski[7]). The height of the steps varied between 50 and 150 Å. More regular were the growth features found on (110) faces of rare earth orthoferrite crystals (R.E.FeO$_3$) which

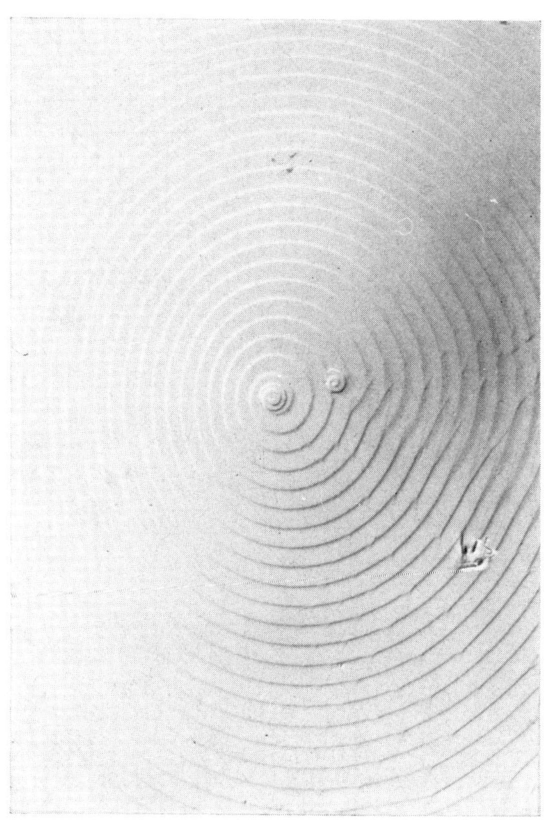

Fig. 5. Growth centre on a (110) face of $(Sm_{0.55}Tb_{0.45}FeO_3)$. Magnification $\times 180$.

were grown by the same technique (fig. 5) where the height of the steps varied from 250 Å near the centre to 150 Å near the edge of the face. The measurement was carried out by multibeam interferometry after Tolanski[8]). Very often the center of the growth spiral is near the edge of the face. This may be due to the fact that the slip direction of the dislocation differs mostly within about $5°$ from the normal of the face as we found by a photoelastic method in agreement with the results of Prescott and Basterfield[9]) on $Gd_3Ga_5O_{12}$ and $Sm_3Ga_5O_{12}$ garnets which were grown by the same technique as YIG. By etching it could be proved that there is always a dislocation – or a cluster of dislocations – in the center of the growth feature. Dislocation density was often found to be less than $10 \, cm^{-2}$.

Characterization of YIG-crystals often is only given by the low linewidth value of the magnetic resonance. In our crystals which have a very low linewidth (ΔH less than 0.5 Oe, ΔH_K less than 0.16 Oe measured at

9 GHz) we normally find the following impurities given in formula units

$$(Y^{III}_{3-a-b}Pb^{II}_a Ca^{II}_b Fe^{III}_{5-c-d} Fe^{II}_c Si^{IV}_d O_{12-e-x} F_e)$$

Pb: $a = 0.012$ to 0.019;
Ca: $b = 0.0008$ and less;
Fe^{2+}: $c = 0.005$ to 0.01;
Si: $d = 0.001$ to 0.003;
F: $e = 0.010$ to 0.016.

For analysis only optically clear specimens were used. The molar concentrations of lead and fluorine were in the same range. Fe^{2+}-values were measured by cerimetric analysis. Less than 0.00001 platinum per formula unit was found in the crystals. Values of Pb and Fe^{2+} decrease slightly after high temperature annealing (13 hr, 1300 °C in O_2) of small pieces (diameter ≈ 1 mm) of YIG.

The lattice constant a_0 for YIG was determined by a method described by Winkler et al.[10]). with an accuracy of $\pm 3 \times 10^{-4}$ Å at 22 °C. The values varied from run to run from 12.3745 Å to 12.3776 Å. After high temperature annealing of powdered crystals we found for all samples $a_0 = 12.3763 \pm 0.0005$ Å and the Fe^{2+}-content dropped under the detection limit of 0.002 per formula unit.

4. Conclusion

With ACRT equilibrium within the whole melt can be achieved, and no competitive phase crystallizes. Crystallization can be localized within a certain area by applying an air cooler. The influence of the cooler is diminished by the size of the growing crystal. Crystals grown by ACRT have very few flux inclusions and cracks. Cubes with (111), (110) and (211) faces as large as $12 \times 12 \times 12$ mm free of cracks and inclusions could be produced. There is no significant difference of growth features and impurities between crystals grown with and without ACRT. The low numbers of growth centers and the regular form of growth features on crystal faces indicate constant growth conditions.

References

1) L. G. Van Uitert, W. A. Bonner, W. H. Grodkiewicz, L. Pietroski and G. J. Zydzik, Mater. Res. Bull. **5** (1970) 825.
2) W. H. Grodkiewicz, E. F. Dearborn and L. G. Van Uitert, in: *Crystal Growth*, Ed. H. S. Peiser (Pergamon, Oxford, 1967) p. 441.

3) W. Tolksdorf, J. Crystal Growth **3, 4** (1968) 463.

4) H. J. Scheel and E. O. Schulz-DuBois, J. Crystal Growth **8** (1971) 304.

5) R. A. Lefever and A. B. Chase, J. Am. Ceram. Soc. **45** (1962) 32.

6) H. Komatsu and J. Sunagawa, Miner. J. (Tokyo) **4** (1964) 203.

7) W. Lang, Zeissinformationen **70** (1968) 114.

8) W. Müller, Leitz, Mitteilungen aus dem Labor, Anwendung Mikro, Nr. 27, 1968.

9) M. J. Prescott and J. Basterfield, J. Mater. Sci. **2** (1967) 583.

10) G. Winkler, P. Hansen and P. Holst, to be published in Philips Res. Rept. (1972).

Journal of Crystal Growth **13/14** (1972) 571–575 © *North-Holland Publishing Co.*

HÉTÉROÉPITAXIE DE COUCHES MINCES DE GRENAT DE FER–YTTRIUM PAR MÉTHODE DE FLUX ET SYNTHÈSE HYDROTHERMALE

A. BROCHIER, P. COEURE, B. FERRAND, J. C. GAY, J. C. JOUBERT, J. MARESCHAL, J. C. VIGUIE, J. C. MARTIN-BINACHON et J. SPITZ

L.E.T.I.–C.E.N.G., Centre d'Etudes Nucléaires de Grenoble, Cedex 85, 38–Grenoble–Gare, France

Epitaxial growth of thin monocristalized YIG films has been performed by flux and hydrothermal methods using flux grown GdGaG single crystals as seed plates. BaO–$0.6B_2O_3$ solvent and 1050–1110 °C deposition temperatures were used in flux method. 2M $NaOH$ solvant solutions and different nutrient materials were tested for hydrothermal synthesis; the best results were obtained by reacting $FeNaO_2$ and $Y(OH)_3$ at 500 °C and 2 kbar with a 30 °C temperature gradient and a 5 to 20% baffle opening. The deposited films are of quite good quality and uniformity, especially when seeds were cut parallel to (211) and (110) orientations; with a thickness of 2 to 3 μm, the films were transparent, with a yellow-green color. Observed under polarized light, some films show very regular band shaped domains. ΔH was also tested and was shown to be as low as 1 Oe in some cases.

1. Introduction

L'effort important entrepris depuis quelques années pour fabriquer des films minces monocristallins de ferrite à structure grenat trouve sa justification dans les applications envisageables en hyperfréquence – isolateurs à résonance[1]), transducteurs[2]), amplificateurs par ondes de surface[3]) – et dans des dispositifs de mémorisation magnéto-optique[4–6]) ou à déplacement de domaines[7–8]).

Quatre méthodes de préparation ont permis l'épitaxie de ferrite grenat:
- recristallisation de dépôt en phase solide[1–6,9,10]);
- recristallisation de projection d'échantillons polycristallins[5,6,9]);
- réaction chimique en phase vapeur[11]);
- croissance à partir d'une solution sursaturée[12]).

Les inhomogénéités résultant de la formation de phases étrangères dans les méthodes de recristallisation, l'apparition de tensions engendrant des modes parasites ou influençant la mobilité des domaines, les risques de modification des propriétés intrinsèques de la couche par diffusion du support dans les techniques hautes températures de flux et de phase vapeur, nous ont conduits à mettre au point une technique originale d'épitaxie: la synthèse hydrothermale. Parallèlement afin de tester les résultats obtenus, nous avons entrepris l'élaboration de couches minces par la méthode de flux. Ces deux techniques viennent de faire l'objet de deux communications[13,14]).

2. Le support

Quelle que soit la technique de croissance utilisée, le choix du support est essentiel et doit satisfaire aux impératifs habituels – inertie chimique vis à vis du milieu environnant dans les conditions de la réaction, arrangement structural et coefficients d'expansion thermique voisins de ceux du matériau déposé, bonne qualité cristalline – auxquels s'ajoute ici celui d'être dia- ou paramagnétique dans le domaine de température envisagé pour les applications.

Le grenat de gallium–gadolinium (ou GdGaG) a été reconnu comme le support le plus approprié au dépôt de films minces de grenat de fer–yttrium (ou YIG)[11,15]). Nous l'avons élaboré par une méthode de solvant dans un mélange de fluorure et oxyde de plomb et d'anhydride borique. L'ajustement des paramètres de croissance et en particulier du rapport molaire PbO/PbF_2 (0.8 à 0.9) nous a permis d'obtenir, dans des creusets de 200 cm³, des monocristaux exempts d'inclusions dans des volumes de plus d'un cm³.

Avant leur utilisation comme support, ces monocristaux ont été découpés en lamelles d'une épaisseur de 250 à 500 μm orientées parallèlement aux plans cristallographiques (100), (110), (111) ou (211) avec une précision de 2 à 5 minutes d'arc; les faces destinées à recevoir le dépôt de YIG ont été polies mécaniquement avec des poudres abrasives diamantées puis soumises à des recuits ou à des traitements chimiques (courant gazeux d'HCl à 1000 °C ou acide orthophosphorique

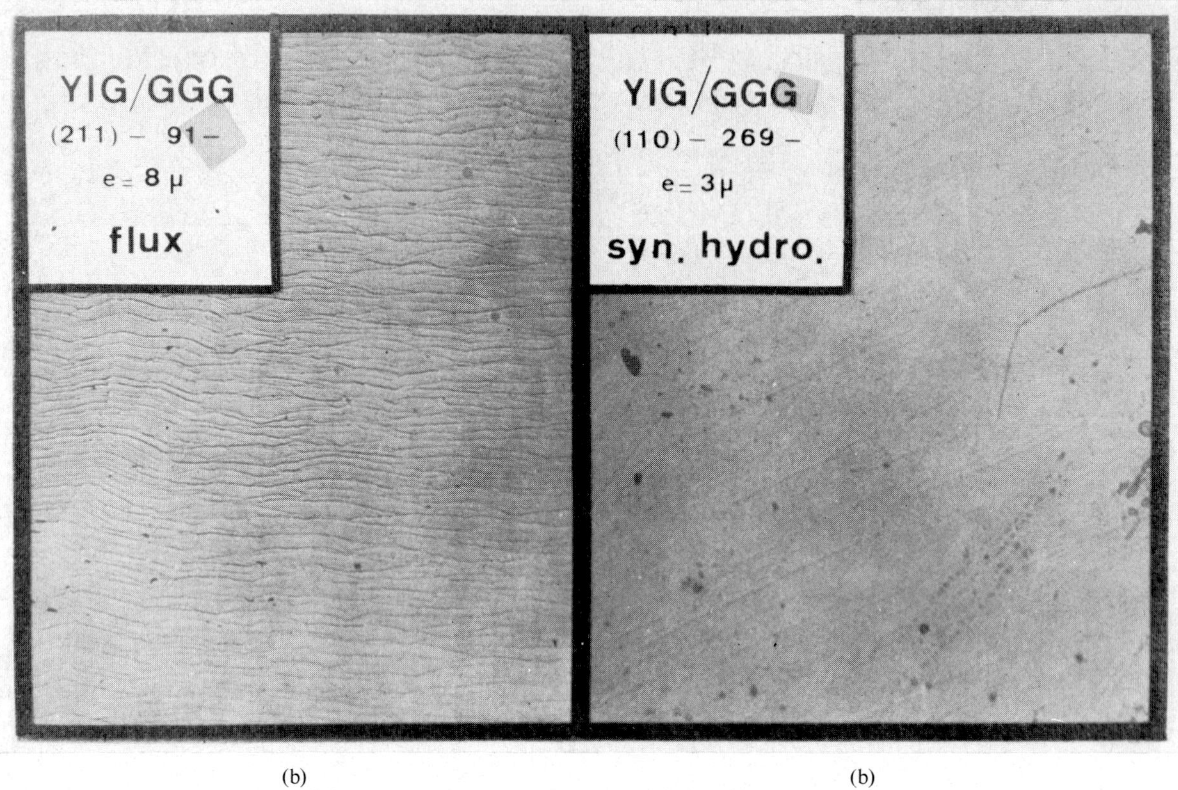

Fig. 1. Couches minces monocristallines de YIG/GdGaG obtenues par méthode de flux et synthèse hydrothermale; aspect de la surface ($\times$126).

à 270–280 °C) pour éliminer les tensions et défauts de polissage.

3. Epitaxie par méthode de flux

La méthode consiste à déposer une couche monocristalline de YIG sur le support choisi en mettant celui-ci en contact avec une solution sursaturée des oxydes constituants. Elle a été utilisée pour la première fois par Linares[12]) pour déposer du YIG sur du YAG, le contact puis la séparation bain-support, tous deux contenus dans un creuset, étant obtenus en basculant celui-ci d'un côté puis de l'autre.

Nous avons adopté un procédé sensiblement différent, dérivé de la méthode de tirage dans un flux introduite par Kestigian[16]) et développée dans nos laboratoires[17]): un creuset de platine de 60 mm de diamètre et de 120 mm de hauteur, contenant une solution de YIG dans un mélange B_2O_3–BaO, est placé dans l'enceinte d'un four à résistance qui maintient un gradient de température inverse, contrôlé et vertical de plusieurs dizaines de degrés sur la hauteur du creuset; la tempé-

rature du bain près de la surface est ajustée pour que la solution y soit juste saturée. Le support, rendu solidaire d'une broche refroidie par eau pouvant se déplacer verticalement, est amené au contact avec la solution pendant le temps nécessaire à l'épitaxie, puis retiré lentement du four. La broche est animée d'un mouvement alternatif de rotation à courte période destiné à détruire la sursaturation occasionnelle à l'interface de croissance. Le contact support-solution est repéré par la résistance ohmique de l'ensemble creuset–broche.

Les meilleures conditions d'élaboration de couches monocristallines sont les suivantes:
– composition de la solution: identique à celle indiquée par Linares[12]);
– température de dépôt: comprise entre 1050 et 1110 °C;
– durée de contact: variable entre 15 min et 15 h;
– mise en température du support: 2 à 6 h;
– refroidissement des films: 3 à 15 h.

Après refroidissement, le support et le film sont séparés du porte-échantillon par traitement dans l'acide nitrique dilué chaud.

4. Epitaxie par synthèse hydrothermale

Cette technique de croissance a été utilisée par plusieurs auteurs[18,19]) pour fabriquer du YIG massif. Nous l'avons adaptée à l'hétéroépitaxie de films minces.

Dans les conditions décrites par Laudise[16]) pour le YIG massif, aucun résultat n'a pu être obtenu, le solvant utilisé (NaOH ou KOH 20 M), trop corrosif, attaquant rapidement le support avant qu'aucun dépôt de YIG ne puisse s'effectuer. Le même effet corrosif a été observé en utilisant une solution de NH_4Cl à 7%[17]): un dépôt monocristallin de qualité moyenne a néanmoins été obtenu à partir d'une telle solution.

Pour remédier à ces inconvénients, nous avons été amenés à effectuer dans l'autoclave, en présence du support, une réaction de double décomposition du type:

$$5\ NaFeO_2 + 3\ Y(OH)_3 \rightarrow Fe_5Y_3O_{12}$$
$$+ 5\ NaOH + 2\ H_2O.$$

La charge de départ est un mélange de $NaFeO_2$ et de $Y(OH)_3$ en poudre dont la proportion Fe/Y varie de 5/3 à 9/3. Le solvant est de la soude. La réaction d'échange utilisée entrainant la formation de soude, la concentration initiale du solvant est calculée pour obtenir en fin de réaction une concentration voisine de 2 M. Les dépôts monocristallins s'effectuent à des températures voisines de 500 °C en utilisant des taux de remplissage de l'autoclave de 40 à 60% et un gradient de température de 40 à 60 °C.

Ainsi dans les conditions suivantes:
– température de dissolution: 500 °C;
– température de cristallisation: 450 °C;
– taux de remplissage: 60%;
– solvant: NaOH (2 M);
nous avons obtenu des films minces monocristallins de 2 à 5 µm d'épaisseur et de très bonne qualité (cf. fig. 1).

5. Résultats

Nous avons effectué une qualification systématique des films minces obtenus par l'une ou l'autre des techniques employées:
– mesure de l'épaisseur de la couche déposée par observation directe sous un microscope à fort grossissement;
– nature de la transmission optique dans le visible à l'aide d'un spectromètre à réseau; cette mesure permet aussi d'atteindre l'épaisseur du film avec une bonne

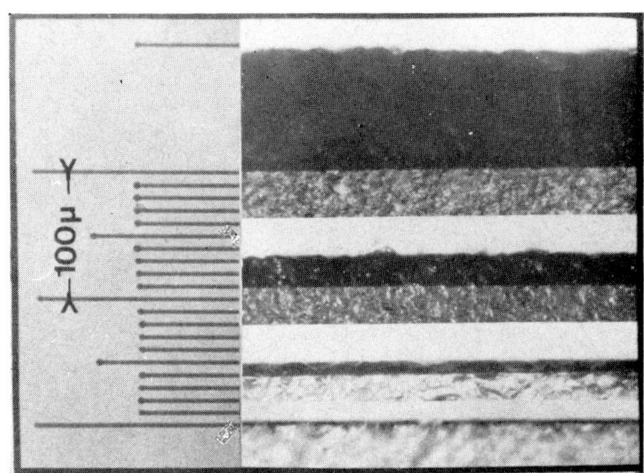

Fig. 2. Coupes de couches minces de YIG/GdGaG: détermination des épaisseurs.

précision lorsque ce dernier est de bonne qualité.
– mesure de la largeur de la raie de résonance ferrimagnétique ΔH en résonance perpendiculaire
– observation par effet Faraday des domaines magnétiques.

Les dimensions des films monocristallins obtenus par l'une ou l'autre des techniques de croissance ne sont limitées que par celles du matériau support utilisé; 7×7 mm^2 sont des dimensions courantes. Leur coloration évolue du jaune-vert au noir avec l'épaisseur de YIG déposée.

Des couches monocristallines non craquelées ont été déposées de façon reproductible, sur des épaisseurs allant de 2 µm à 100 µm par la méthode de flux et de 1 µm à 20 µm par la méthode hydrothermale (cf. fig. 2). Les variations d'épaisseur, vérifiées par étude topographique de la surface déposée à l'aide d'un "talysurf", sont dans tous les cas inférieures à 4% et le plus souvent comprises entre 1 et 2%.

La vitesse de croissance est un des facteurs déterminants de la qualité et de l'uniformité des films minces obtenus. Elle est influencée par les paramètres habituels qui régissent chaque technique: orientation cristallographique du support, température du dépôt, taux de remplissage, gradient de température; dans la méthode de flux, ce dernier paramètre agit de façon prépondérante sur l'allure de la loi de croissance en fonction du temps. Pour des conditions de dépôt identiques, les vitesses de croissance varient avec l'orientation cristallographique du support de la façon suivante:

$$V\,(110) \simeq V\,(211) < V\,(100) < V\,(111);$$

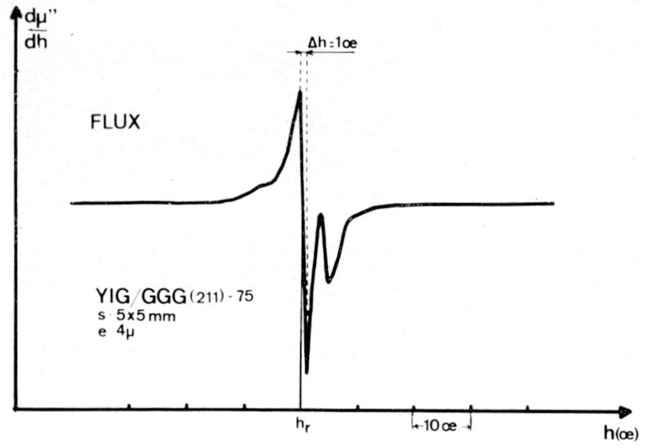

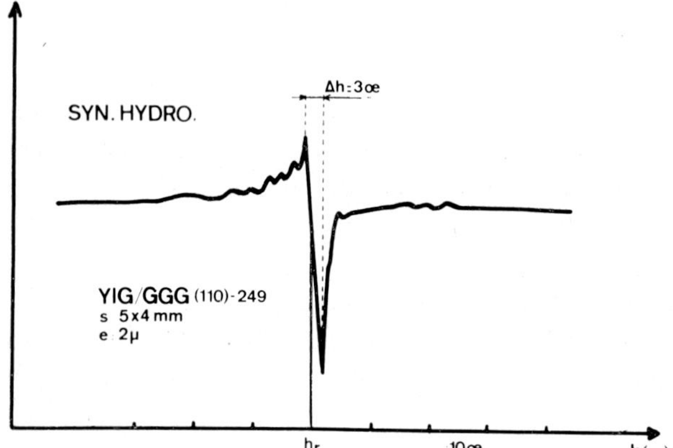

Fig. 3. Comparaison des largeurs de ligne ΔH des couches minces YIG/GdGaG obtenues par flux et synthèse hydrothermale

ces résultats sont en accord avec le faciès naturel observé dans les cristaux massifs. Pour certaines orientations, les vitesses de croissance des films sont abaissées par une faible désorientation du support par rapport aux directions cristallographiques (1 à 4°). Dans la gamme de températures utilisables, les vitesses de dépôt sont d'autant plus élevées que la température est plus basse.

Des vitesses de croissance de 1 à 3 μm/h fournissent des films parfaitement transparents et sans fissures déposés sur des supports de l'une des quatre orientations ci-dessus.

Des largeurs de raie de résonance de 1 Oe ont été mesurées sur les meilleurs échantillons obtenus par épitaxie dans un flux; cette valeur est beaucoup plus faible que celles signalées dans réf. 12. Celles des échantillons élaborés par synthèse hydrothermale sont de 3 Oe (cf. fig. 3); nous pensons que cette valeur élevée de ΔH est attribuable à la substitution d'ions O^{2-} par des ions

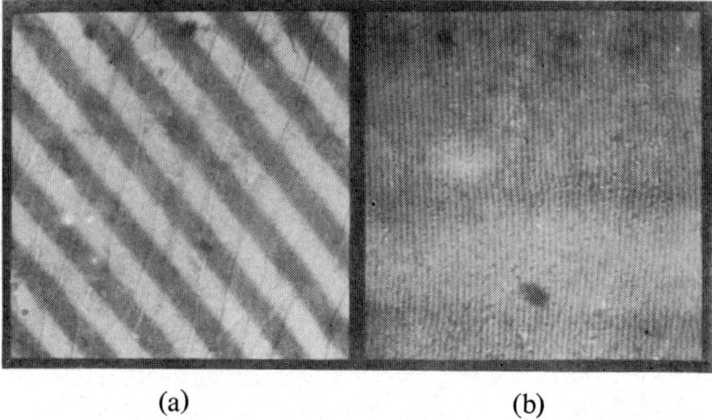

Fig. 4. Domaines magnétiques en bandes parallèles vus par effet Faraday (×270) sur des couches YIG/GdGaG brut de fabrication. (a) YIG/GdGaG (211)–91; $e = 13\ \mu m$, $s = 3 \times 4$ mm, flux. (b) YIG/GdGaG (110)–249; $e = 2\ \mu m$, $s = 4 \times 5$ mm, syn. hydro.

OH^- entrainant, par un mécanisme de compensation de charges, la formation d'ions Fe^{2+} qui élargissent la raie de résonance. Des essais sont en cours pour travailler en milieu plus oxydant.

Des domaines magnétiques en bandes ont été observés sur tous les échantillons élaborés par l'une ou l'autre des techniques; les dispositions régulières (domaines en bandes parallèles) existant sur certains échantillons bruts de croissance montrent que les contraintes magnétostrictives induites lors de l'élaboration restent faibles (cf. fig. 4).

6. Conclusion

Deux techniques ont été mises en œuvre pour réaliser l'hétéroépitaxie du YIG sur GdGaG: méthode de flux et synthèse hydrothermale; cette dernière est originale.

La qualité des couches obtenues est comparable au stade actuel de l'étude. La méthode de synthèse hydrothermale semble cependant la plus prometteuse, tant au point de vue qualitatif qu'en fonction des critères économiques de production.

Ce travail est financé en partie par la Direction des Recherches et Moyens d'Essais.

Bibliographie

1) W. L. Wade, T. Collins et W. J. Skudera, IEEE Trans. Magnetics, Mater. Pack. **MMP1** (1965) 3, 26.
2) J. E. Mee, F. A. Pizzarello, G. R. Pulliam et T. N. Hamilton, Appl. Phys. Letters **13** (1962) 3, 93.
3) Microwave IC's-Electronics **66** (11 Nov. 1968).
4) R. E. MacDonald, O. Voegeli et C. D. Mee, J. Appl. Phys. **38** (1967) 4101.

5) E. Sawatzky et E. Kay, IEEE Trans. Magnetics **3** (1967) 530.

6) P. Coeure, F. Forrat et J. C. Gay, Intermag. Conf., Amsterdam, 1969.

7) A. H. Bobeck, Bell Syst. Tech. J. **46** (1967) 1901.

8) P. C. Rossol, J. Appl. Phys. **40** (1969) 1082.

9) W. D. Westwood, H. K. Eastwood, R. G. Poulsen et A. G. Sadler, J. Am. Ceram. Sco. **50** (1967) 119.

10) M. Oron, I. Barlow et W. F. Traber, J. Mater. Sci. **4** (1969) 271.

11) J. E. Mee, G. R. Pulliam, J. L. Archer et P. J. Besser, IEEE Trans. Magnetics **MAG-5** (1969) 717.

12) R. C. Linares, J. Crystal Growth **3, 4** (1968) 443.

13) E. D. Kolb et R. A. Laudise, J. Appl. Phys. (1971).

14) L. K. Shick, J. W. Nielsen, A. H. Bobeck, A. J. Kurtzig, P. C. Michaelis et J. P. Reekstin, J. Appl Phys. (1971).

15) S. Geller, G. P. Espinosa et P. B. Crandall, J. Appl. Cryst. **2** (1969) 86.

16) M. Kestigian, J. Am. Ceram. Soc. **50** (1967) 765.

17) J. Mareschal, A. Brochier et J. C. Grenier, Marché DRME No. 68 34 744 00 480 75 01, Nov. 1970.

18) R. A. Laudise et E. D. Kolb, J. Am. Ceram. Soc. **45** (1962) 51.

19) B. V. Mill' et I. I. Naumova, Kristallografiya **6** (1961) 800 [English Transl. Soviet Phys.-Cryst. **6** (1961) 644].

SOME OBSERVATIONS ON THE GROWTH OF YIG UNDER OXYGEN PRESSURE BY THE "FLUXED MELT" TECHNIQUE

J. M. ROBERTSON* and B. W. NEATE

Physics Department, Portsmouth Polytechnic, Park Road, Portsmouth, England

The effects of an oxygen pressure on YIG single crystals grown from PbF_2 rich fluxes are: (a) larger crystals, (b) glossy appearance, (c) raised leading edges, and (d) more flux inclusions up to 2.2 kg cm^{-2} pressure after which they decrease in percentage to the normal values of air grown garnets.

1. Introduction

Following the work of Robertson et al.[1]) on the growth of $NiFe_2O_4$ under oxygen pressure, a study has been made of the mode of crystal growth of YIG from a $PbO–PbF_2–B_2O_3$ flux under oxygen pressure. Makram[2]) reported that the quality of YIG single crystals from lead solvents was improved by the application of a high oxygen gas pressure above the melt during growth. In the present work oxygen pressures up to 6 atm absolute have been employed and the quality of the crystals assessed in respect of the inclusions and stoichiometry.

2. Experimental

The furnace used was a modification of the type reported by Robertson et al.[1]) in which a mullite tube is supported on a brass water cooled base plate which locates into a flange which is glued to the tube with araldite. In fig. 1 (bottom) is shown the bottom end plate with the oxygen inlet and the pedestal for the crucible support. Fig. 1 (top) shows the upper end of the furnace tube, which has a pyrophyllite plug to prevent heat reaching the upper brassplate by radiation or convection. This upper end is supported by a spring loaded mechanism which prevents all the weight from resting on the tube. This design is used in preference to the earlier furnace design since the mullite tubes were subject to faults during manufacture making them

* Now at Philips Research Laboratories, N.V. Philips' Gloeilampenfabrieken, Eindhoven, The Netherlands.

unreliable for high pressure usage. Fig. 2 shows the complete furnace.

A non-destructive method by which the solvent inclusion weight percentages were assessed was based on a technique used to eliminate flux inclusions from solvent grown crystals using a travelling solvent [Elwell and Neate[3]]. The trapped solvent diffuses through the crystal when it is maintained in a temperature gradient until it resides on an outside surface where it may be removed by nitric acid. The crystal weight difference

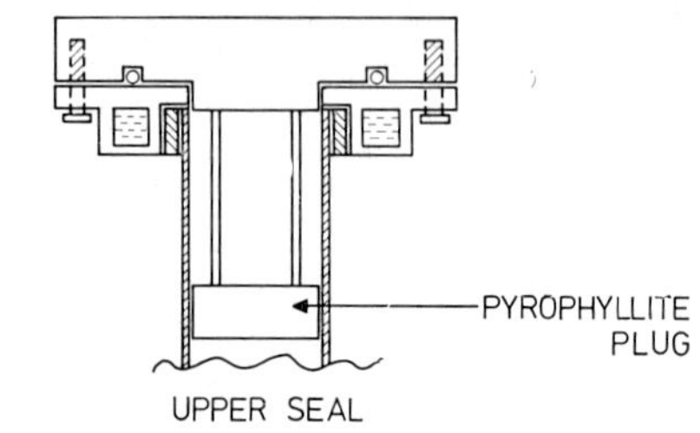

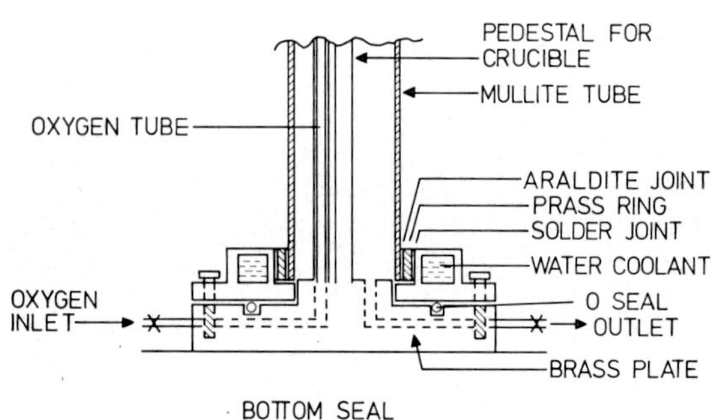

Fig. 1. Furnace. Top seal and bottom seal.

Fig. 2.   General view of furnace.

Fig. 3.   YIG crystal grown under 2 kg cm^{-2} oxygen pressure.

both before and after this treatment gives the amount of solvent present. Experiments have shown that greater than 90 % of the trapped solvent is eliminated from the crystal by this procedure.

3. Crystal growth

A typical crystal growth procedure involved a mix of 44.14 g stoichiometric YIG; 51.6 g PbF$_2$; 39.0 g PbO and 2.56 g B$_2$O$_3$ placed in a 50 cm^3 platinum crucible. This composition has been used since it has shown to give consistently good yields of YIG crystals grown in air. The air was flushed out of the system and the required pressure of oxygen applied. The furnace temperature was slowly raised to 1330 °C, the oxygen pressure being adjusted continuously to the required value. The furnace was held at this temperature for several hours and then slow cooled to 1060 °C at 0.75 °C hr^{-1}, after which point the furnace was allowed to cool naturally. During the cooling process no adjustment to

the oxygen pressure was made. Finally the crystals were leached from the flux with hot nitric acid.

4. Results

In general, the size of crystals obtained are twice the size of those produced at air pressure in the same furnace, the largest being 12 mm produced at 2.2 kg cm^{-2} oxygen pressure. This may be due to the fact lead fluoride has a high vapour pressure at the temperatures used, Robertson and Taylor[4]), and at atmospheric pressure the crystals normally grow by a combination of flux evaporation and temperature reduction. The former effect is now reduced by the application of the high gas pressure above the melt allowing the crystals to grow by temperature reduction only. Experiments are being set up to distinguish between the effects of oxygen and pressure on the evaporation of PbF$_2$ during growth. With this in mind there are several secondary effects associated with oxygen pressure which might influence the crystal size. Firstly, depression of the evaporation of solvent also leads to a smaller convection current effect in the liquid. Secondly, the initial nucleation critical nucleus size may be modified and a smaller nucleus in greater numbers may cause a larger temperature growth range on the solubility diagram [as in fact has been noted by Elwell et al.[5])], leading to larger crystals. Or perhaps the suppression of the Fe^{2+} ion content has an effect on the nucleus size since

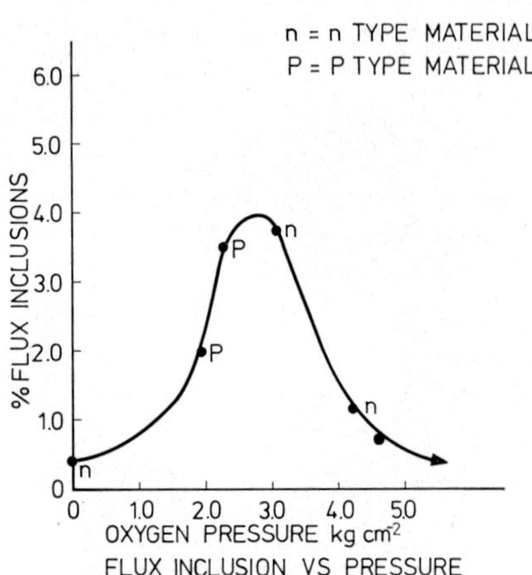

Fig. 4. Flux inclusions as a function of pressure.

the addition of small quantities of divalent calcium enhance the growth of garnet [van Uitert et al.[6])]. Thirdly, solvent impurities may be substituted onto lattice sites modifying the growth habit [Lefever et al.[7])].

YIG crystals grown under oxygen pressure, fig. 3, have a surface which appears glazed and the (211) faces have leading edges similar to the (111) faces of NiO reported by Hill and Wanklyn[8]). The crystals also have (211) faces modified by (110) faces similar to that reported by Chase and Osmer[9]) for YGG grown from PbF_2 rich melts whose composition was similar to those we have used. Titova[10]) has shown also that of the two dominant faces of YIG the (211) faces vanish in favour of the (110) under certain conditions when grown from lead fluxes. These effects have been summarized by Quon and Sadler[11]). It is possible that the leading edge may be formed by an edge becoming established so that the liquid system at the centre of the face then becomes depleted of solute in a manner similar to cellular impurity growth from the melt [Rutter and Chalmers[12])].

It is found that the quality of the garnets initially decreases up to 2.2 kg cm^{-2} oxygen pressure, the aver-

age percentage of inclusions reaching about $4\frac{1}{2}\%$ as shown in fig. 4. Crystals grown at below 2.2 kg cm^{-1} were preliminarily observed to have p-type conductivity, above this pressure n-type conductivity was observed. Further conductivity studies are being made. At higher oxygen pressures the inclusion percentage began to approach the value of air grown crystals and it may be assumed to reach the values of good quality YIG reported by Makram[2]) produced under an oxygen gas pressure of 15 kg cm^{-2}. The effect of the oxygen pressure on the flux inclusion percentages is not yet understood and at the present time work is being done to assess the effect of other gases on the flux inclusions to establish if it is a genuine oxygen effect or a pressure effect.

Throughout the whole pressure range the garnets exhibited low ferrimagnetic resonance linewidths $\ll 2$ oersteads. The p-type conductivity may be due to the presence of Fe^{4+} ions caused by the substitution of lead into the garnet structure. The presence of Fe^{4+} ions will be established by anisotropy measurements being carried out at Portsmouth at the present time.

References

1) J. M. Robertson, D. Elwell and S. H. Smith, J. Crystal Growth 5 (1969) 189.
2) H. Makram and K. Krishnan, J. Phys. Chem. Solids, Suppl. No. 1 (1964) 464.
3) D. Elwell and B. W. Neate, to be published.
4) J. M. Robertson and R. G. F. Taylor, J. Crystal Growth 2 (1967) 171.
5) D. Elwell, B. W. Neate and S. H. Smith, in: *Proc. Intern. Conf. Ferrites*, Japan, 1970 (to be published).
6) L. G. Van Uitert, W. A. Bonnier, W. H. Grodkiewicz, L. Pictroski and G. J. Zydzik, Mater. Res. Bull. 5 (1970) 825.
7) R. A. Lefever, A. B. Chase and J. W. Torpy, J. Am. Ceram. Soc. 44 (1961) 141.
8) G. J. Hill and B. M. Wanklyn, J. Crystal Growth 3, 4 (1968) 475.
9) A. B. Chase and J. A. Osmer, J. Crystal Growth 5 (1969) 239.
10) A. G. Titova, Soviet Phys.-Cryst. 8 (1964) 1714.
11) H. H. Quon and A. G. Sadler, J. Am. Ceram. Soc. 36 (1967) 33.
12) J. W. Rutter and B. Chalmers, Can. J. Phys. 31 (1953) 15.

Journal of Crystal Growth **13/14** (1972) 579–581 © *North-Holland Publishing Co.*

FLUX GROWTH AND MAGNETIC CHARACTERIZATION OF CALCIUM-SILICON SUBSTITUTED YTTERBIUM AND YTTRIUM IRON GARNETS*

M. SCHIEBER, A. GRILL and Y. AVIGAL

Department of Materials Science, School of Applied Science and Technology, Hebrew University of Jerusalem, Israel

Single crystals of $R_{2.9-x}Bi_{0.1}Ca_xFe_{5-x}Si_xO_{12}$ where $R = Y$ or Yb and $0.6 < x < 1.5$ have been grown from a $PbO–B_2O_3$ flux. The crystals have been characterized by measurements of the unit cell dimension, Curie temperature and magnetization measurements between liquid helium and room temperature.

1. Introduction

Mixed cation iron garnet crystals have been grown for a program of calibration materials for magnetic moments where ferrimagnetic compounds with weak magnetic moments are needed. Such compounds are also of interest in studying the strength of the magnetic interaction in diamagnetic substituted ferrimagnetic garnets[1]. Growth of multi-cation iron garnets is of great interest at present, especially for bubble domain devices. The crystallization of mixed rare earth garnets with application in magnetic domain bubbles[2] and the growth of $Y_{2.9-x}Bi_{0.1}Ca_xFe_{5-x}Si_xO_{12}$ up to $x = 1.1$ have been reported[3]. The present investigation extends the range of the Ca–Si–Bi substituted YIG (B–Y)† to $x = 1.5$ and reports at the same time Ca–Si substituted YbIG (B–Yb)†.

2. Growth procedure

Crystals of B–Y and B–Yb with x up to 1.1 have been grown from a flux, similar to that published elsewhere[3] for B–Y, i.e. $Y_{2.9-x}Bi_{0.1}Ca_xFe_{5-x}Si_xO_{12}$. The composition of the flux is: Fe_2O_3: 25, MoO_3: 5, PbO: 30, B_2O_3: 3, $(R_2O_3+Bi_2O_3)$: 12, $(CaO+SiO_2)$: 25 mole %, where R is Y or Yb. In order to vary the amount x in the substituted garnets, the composition of SiO_2 has been varied between 3 and 7 mole %, and the composi-

tion of R_2O_3 has been varied between 4 and 9 mole %. All the conditions of growth[3] have been similar for both Y_2O_3 and Yb_2O_3. The crystals thus obtained never exceeded the value of $x = 1.1$ due to the fact that high concentrations of CaO tended to form $CaFe_2O_4$ rather than increase the content of Ca in the garnet. Although it is supposed that the increase of the content of SiO_2 in the melt should increase the content of Si in the crystal this does not happen. The increased content of SiO_2 in the melt increases the viscosity of the melt and there is a tendency to form glasses. One way to lower the viscosity of the melt and still have a higher content of SiO_2 is to lower the content of Fe_2O_3. This was done by using the flux composed of Y_2O_3: 4, Fe_2O_3: 21, SiO_2: 11, $CaCO_3$: 22; Bi_2O_3: 1, MoO_3: 5, B_2O_3: 3, PbO: 30. From this flux B–Y garnets were obtained with $x = 1.5$. In this case no $CaFe_2O_4$ was formed but instead ternary iron calcium silicates insoluble in HNO_3 were crystalised together with the garnets. A similar modified flux for B–Yb yielded measurable crystals with $x = 1.25$.

Crystals with $x = 0.6$ have been grown from fluxes having a content of Yb_2O_3 of 7–9 mole %. Thus a slight increase or decrease of Yb_2O_3 in the $Yb_2O_3 + Bi_2O_3$ constant composition of 12 mole % does not change the overall composition of the garnet crystal.

3. Characterization of the crystals

The unit cell dimensions (a), Curie temperatures (T_c) and spontaneous magnetization M from liquid helium to room temperatures of the grown crystals were meas-

---

* Supported in part by the U.S. Government through a grant of the National Bureau of Standards, Washington D.C.
† Ca–Si substituted garnets have been termed B in our previous work[3].

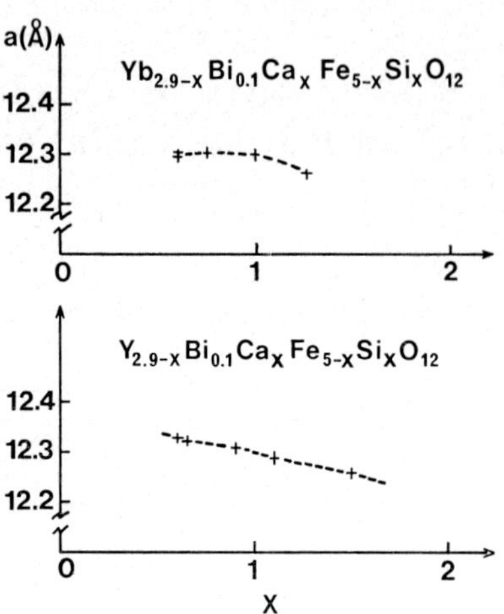

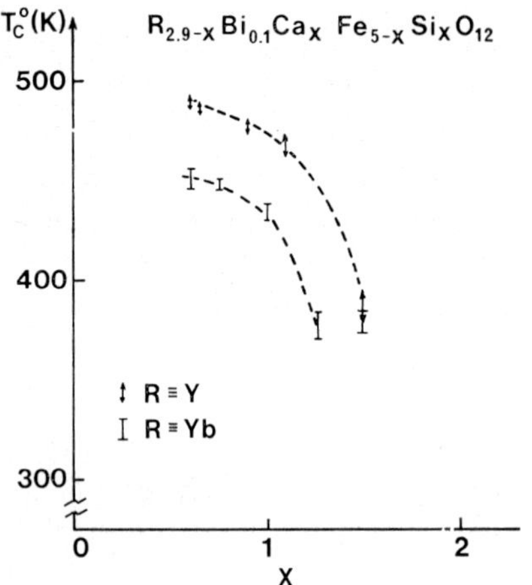

Fig. 1. Unit cell versus composition of $Y_{2.9-x}Bi_{0.1}Ca_xFe_{5-x}Si_xO_{12}$ and $Yb_{2.9-x}Bi_{0.1}Ca_xFe_{5-x}Si_xO_{12}$.

Fig. 2. Curie temperature versus composition of $Y_{2.9-x}Bi_{0.1}Ca_xFe_{5-x}Si_xO_{12}$ and $Yb_{2.9-x}Bi_{0.1}Ca_xFe_{5-x}Si_xO_{12}$.

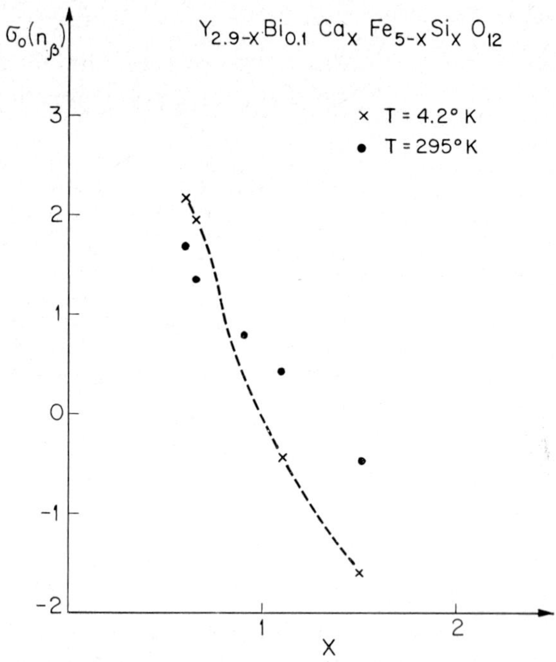

Fig. 3. Spontaneous magnetization versus composition at 4.2 °K and 295 °K of $Y_{2.9-x}Bi_{0.1}Ca_xFe_{5-x}Si_xO_{12}$.

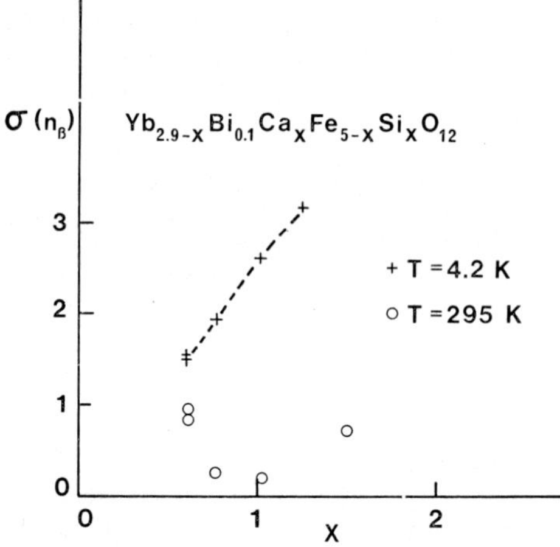

Fig. 4. Spontaneous magnetization versus composition at 4.2 °K and 295 °K of $Yb_{2.9-x}Bi_{0.1}Ca_xFe_{5-x}Si_xO_{12}$.

ured. The results are summarized in figs. 1–4. The value of x for B–Y and B–Yb is deduced from the value of M at 4.2 °K assuming Nowik's statistical theory[4]) which takes into account any canting of the substituted iron sublattices. A full discussion of the magnetic phenomena will be published elsewhere. Figs. 1 and 2 show that the values of a cannot be used for the characterization of x. However, the values of T_c are consistent with the value, dictated by the measurements of M. Only two compounds of B–Yb with $x = 0.60$ and 0.75 have zero magnetization at compensation temperatures of 33 and 55 °K respectively. Figs. 3 and 4 show the values of M versus x calculated from Nowik's

model and measured at 4.2 °K. Figs. 3 and 4 also show the values of M at 295 °K.

It should be pointed out that chemical analysis by X-ray fluorescence and atomic adsorption analysis for Y, Yb and Fe were much less reliable than the analysis by magnetic characterization as shown by a comparison of the scatter in results obtained by the various analytical methods. Production of polycrystalline samples with known values of x of both B–Y and B–Yb and the measurements of a and T_c was used to confirm the value of x assumed for the single crystals.

Acknowledgement

The technical assistance of Mrs. L. Goldenberg is appreciated.

References

1) A. Grill and M. Schieber, to be published in J. Phys. (Paris) February 1971.
2) L. G. Van Uitert, W. A. Bonner, W. H. Gradkievicz, L. Pictroski and G. J. Zydzik, Mater. Res. Bull. **5** (1970) 825.
3) M. Schieber, A. Grill and I. Shidlovsky, J. Crystal Growth **3, 4** (1968) 467.
4) I. Nowik, J. Appl. Phys. **40** (1969) 5184.

Journal of Crystal Growth **13/14** (1972) 582–584 © *North-Holland Publishing Co.*

SOME FEATURES OF THE GROWTH OF SINGLE CRYSTALS OF THE $Y_3Fe_{5-x}Cr_xO_{12}$ SYSTEM

R. KRISHNAN

Laboratoire de Magnétisme, C.N.R.S., 92–Bellevue, France

Single crystals of $Y_3Fe_{5-x}Cr_xO_{12}$ with $0 < x < 0.43$ have been grown by the fluxed melt technique using $PbO–PbF_2–B_2O_3$ as the flux. The molar ratio of PbO/PbF_2 was varied between 1.0 and 0.60 and it was found that at a given temperature the solubility of Cr_2O_3 increased with PbF_2 content. The flux with $PbO/PbF_2 = 0.70$ gave the best results. The effect of crucible size on garnet crystallisation was shown to be related to evaporation rate. When the Cr_2O_3 content in the melt exceeded 1 mole% no garnet phase could be crystallized for small crucibles (50 cm^3), whereas garnets crystallized when bigger crucibles (100 cm^3) were employed. Electron probe analysis of a section of a crystal did not reveal any gradient of Cr_2O_3 concentration between the center of the crystal and its surface.

1. Introduction

In order to understand the contribution of certain paramagnetic ions to the magnetic anisotropy and similar properties it is necessary to substitute these ions in suitable ferrimagnetic host lattices and then study their properties. In such a project we were led to investigate the preparation of yttrium iron garnet (YIG) crystals containing Cr_2O_3 and having the formula $Y_3Fe_{5-x}Cr_xO_{12}$. Previous work on polycrystalline samples has revealed that x can have a maximum value of about 0.5[1]). The only work relevant to ours appears to be that of Petrov et al.[2]) who have studied the preparation of $Y_3Fe_{5-x}Cr_xO_{12}$ crystals by the flux method but furnish barely any detail. Hence we were led to start a detailed investigation of this preparation which is described in this paper.

The fluxed melt technique was used and we chose $PbO–PbF_2$ as the flux due to its excellent characteristics for the growth of garnet crystals[3]). However, introducing Cr_2O_3 into the crystal presents several problems mainly to do with its poor solubility relative to the other oxides components of the crystal[4]). As seen from the work of Timofeeva[4]) who investigated the flux system $PbO–PbF_2–B_2O_3$, the solubility of Cr_2O_3 is very small (of the order of a few percent even at 1400 °C) and it becomes practically negligible at 1200 °C. This second aspect particularly gives rise to inhomogeneous distribution of Cr_2O_3 in the crystals. The reason is that at a higher temperature when the crystal begins to nucleate more Cr_2O_3 is in solution and thus it enters

the crystal at a certain concentration but as the crystallisation proceeds with a decrease in temperature there is less and less Cr_2O_3 in solution due to its decrease in solubility and consequently a lower concentration enters the crystal. Chemical analyses have shown that the Cr_2O_3 content is higher at the centre of the crystal than at the surface. As it is known that the solubility of an oxide and its temperature dependence are strong functions of flux composition, we thought that by using proper ratios of PbO/PbF_2 one might improve chromium homogeneity. Thus a search had to be made for a flux in which Cr_2O_3 solubility would be as high as possible.

2. Experimental details

The furnace used in this work was heated by SiC rods controlled by a "Eurotherm" regulator and programmer. The performance of the furnace can be characterised as follows: (1) Temperature stability better than ± 0.5 °C, (2) cooling rates continuously variable between 0.5 to 15 °C/h, and (3) a positive temperature gradient near the crucible of about 1 °C/cm. Platinum crucibles of two different sizes 50 and 100 cm^3 were used. The crucibles were fitted with tightly crimpled lids to minimise the loss by evaporation. The crystal growth experiment consisted of maintaining the crucible and its contents at a temperature in the range 1240 to 1330 °C for 4 to 12 h then cooling to about 1000 °C at 0.5 to 1 °C/h. The crystals were separated in the usual manner. The Cr_2O_3, Fe_2O_3 and Y_2O_3 contents of the crystals were determined by spectrophotometric

TABLE 1

Summary of results of various crystal growth runs

Expt.	PbO/PbF_2	PbO	PbF_2	B_2O_3	Fe_2O_3	Cr_2O_3	Y_2O_3	X in crystal	Remarks
					All expressed in moles				
				Crucible size 50 cm³					
YIGCr[III]	1.0	30	30	5	26	1	8	0.05	Garnets
YIGCr[I]	1.0	30	30	5	22	5	8		Orthoferrites
YIGCr[V]	0.70	34	48	4	8	2	4		Only orthoferrites $Y_3Fe_{0.8}Cr_{0.2}O_3$
YIGCr[IV]	0.70	28	40	2	18	4	8		Orthoferrites
YIGCr[VIB]	0.70	28	40	2	19	3	8		Orthoferrites
				Crucible size 100 cm³					
YIGCr[VIII]	0.70	28	40	2	21	1	8	0.14	Garnets
YIGCr[II]	0.70	28	40	2	20	2	8	0.27	Garnets
YIGCr[VI]	0.70	28	40	2	19	3	8	0.43	Garnets very few magnetoplombite
YIGCr[VII]	0.78	30	38	5	17	3	7	0.28	Garnets

methods. In certain cases, electron probe analysis was also carried out to determine the homogeneity of the composition in a crystal.

3. Results

The starting materials were: PbO, PbF_2, Fe_2O_3 and Y_2O_3 specpure (Johnson–Mathey); B_2O_3 and Cr_2O_3 of purity better than 99.9 % (E. Merck).

Four flux compositions with PbO/PbF_2 = 1.0, 0.78, 0.70 and 0.60 were tried. In some cases 2 to 5 mole % of B_2O_3 was added which apparently had only the effect of diminishing the loss by evaporation. In all twelve experiments were carried out. To start with 50 cm³ crucibles (containing about 100 g of mixture) were employed. It was found that when the Cr_2O_3 content in the melt exceeded 1 mole % no garnet phase could be crystallised over the range of flux compositions and experimental conditions investigated. The only phase obtained was large blocks of plate like crystals which were weakly magnetic. Chemical and X-ray analysis revealed them to be orthoferrite crystals corresponding to the formula $YFe_{1-y}Cr_yO_3$. However, when we used a larger crucible (100 cm³) containing about 150 g of mixture, surprisingly garnet crystals could be obtained for all the flux compositions that had failed earlier with the smaller crucible. It is well known that larger crucibles often improve the size of flux grown crystals but this, in our knowledge, is the first time an effet of crucible size on the stability of a

particular phase has been observed. The garnet crystals obtained had typical {211} and {110} faces with average overall length of 6 mm. Some of them attaining even 18 mm. The large crystals had inclusions and fine cracks, the contents of which upon analysis were found to be PbO and Cr_2O_3. A similar observation has been reported by Petrov and Tigova[2]). The undissolved Cr_2O_3 is apparently trapped during crystal growth. Table 1 shows a summary of the results obtained.

4. Discussion

The changes in phase stability with crucible size are probably caused by changes in flux composition. In the smaller crucibles where the ratio (surface/volume) is larger, evaporation of volatile components will be greater. Besides the effect of crucible size on the stability of the garnet phase, the following results should be noted. To get more Cr_2O_3 into the crystal it is reasonable to increase its content in the starting mixture and then to try to increase the amount dissolved. Though the latter can be achieved by increasing the temperature of the solution, it is not always possible to get a higher Cr_2O_3 content by this procedure in the crystal because crystal nucleation occurs at more or less the same temperature (around 1200 °C) and hence the increased quantity of Cr_2O_3 dissolved at a higher temperature precipitates out before 1200 °C. Consequently it is nor available in solution during crystallisation. On the other hand by suitably adjusting the flux composition it was

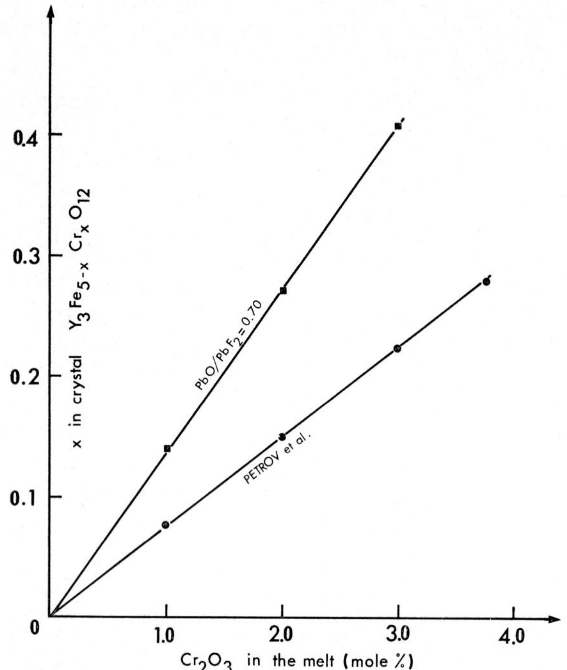

Fig. 1.   The relation between Cr_2O_3 content in the crystal and in the melt.

possible to increase the Cr_2O_3 solubility at a given temperature. This was achieved by making the flux richer in PbF_2 as illustrated by experiments YIGCr[VI] and YIGC[VII]. For the same Cr_2O_3 content in the melt this yielded crystals with higher Cr_2O_3. This point is also clear from fig. 1 which shows the dependence of Cr_2O_3 content (X) in the crystal on the amount (in mole %) of Cr_2O_3 in the starting mixture. The advantage of the flux with $PbO/PbF_2 = 0.70$ is evident. It is also seen that for this flux system there is a linear dependence between X of the crystal and the Cr_2O_3 content of the starting mixture. Also fig. 1 shows the results of Petrov et al.[2]). A comparison with our data reveals the suitability of our flux system which enabled

a higher value of X (= 0.43, practically the upper limit) to be obtained. However, decreasing the ratio PbO/PbF_2 to 0.6 is not found to be helpful because of a considerable increase in the volatility of the flux which causes a continuous change in the flux composition during crystal growth.

In order to check the homogeneity of Cr_2O_3 substitution in the crystals a rectangular plate $6 \times 4 \times 1$ mm was cut from the centre portion of a crystal with $X = 0.27$ and subjected to electron probe analysis. The electron beam was swept along two mutually perpendicular axes at a speed of about 100 µm/min. The total difference in Cr_2O_3 content across the length of the plate was found to be within $10 \pm 5\%$ (in terms of X the variation is 0.27 ± 0.03). It was not possible to infer conclusively that the concentration was higher at the centre. This small variation of Cr_2O_3 concentration across a length of 6 mm shows that in a spherical specimen of about 2 mm diameter (used for magnetic measurements) it is practically constant within the limits of experimental error thus enabling us to study and interpret the magnetic properties. Indeed, several samples from a single batch have identical magnetic properties[5]).

In conclusion, we have developed a flux composition well suited for preparing garnet crystals doped with Cr_2O_3 and shown the high degree of homogeneity of doping achieved by electron probe analysis. The effet of crucible size on the crystallisation of garnet phase has been observed for the first time.

References

1) G. Villers and J. Loriers, Compt. Rend. (Paris) **245** (1957) 2033.
2) R. A. Petrov and A. G. Tigova, Inorgan. Mater. **5** (1969) 120.
3) J. W. Nielsen and E. F. Dearborn, J. Phys. Chem. Solids **5** (1958) 202.
4) V. A. Timofeeva, J. Crystal Growth **3** (1968) 496.
5) R. Krishnan, to be published elsewhere.

Journal of Crystal Growth **13/14** (1972) 585–587 © *North-Holland Publishing Co.*

PHASE RELATIONS IN THE SYSTEM Fe_2O_3-B_2O_3 AND ITS APPLICATION IN SINGLE CRYSTAL GROWTH OF $FeBO_3$

HELMY MAKRAM, LOUIS TOURON and JEAN LORIERS

Laboratoire de Recherches sur les Terres Rares, C.N.R.S., 92–Bellevue, France

A study of the phase relations in the system Fe_2O_3–B_2O_3 to aid in finding the optimum conditions to grow single crystals of iron borate was made. The compositions of the system Fe_2O_3–B_2O_3 studied were with 20, 30, 40, 50, 60, 70 and 80 mole% of Fe_2O_3, at temperatures of 670, 800, 860 and 900 °C. At all compositions and at low temperatures (670–800 °C), $FeBO_3$ is the prevalent stable phase. $FeBO_3$ is stable up to 800 °C for $0.25 \leq Fe_2O_3/B_2O_3 \leq 2.6$. At 750 °C and for $1 \leq Fe_2O_3/B_2O_3 \leq 2.6$, two phases are present, $FeBO_3$ and Fe_3BO_6. When $Fe_2O_3/B_2O_3 > 2.6$, $FeBO_3$ disappears and Fe_3BO_6 and α-Fe_2O_3 are found. Thus, $FeBO_3$ is obtained at low temperatures and with an excess of B_2O_3. Single crystals of borate $FeBO_3$ with maximum dimensions of up to 8 mm are obtained from the flux with a mixture containing 1 Fe_2O_3–5 B_2O_3– 1 Bi_2O_3. To avoid the formation of α-Fe_2O_3 and Fe_3BO_6, the mixture was cooled rapidly between 1135 and 860 °C.

1. Introduction

Iron borate has been the object of considerable interest in recent research from both the practical and the theoretical viewpoint. This interest arises from the unusual magneto-optical properties. Thus, phase equilibria data necessary for the growth of good large crystals under controlled and reproducible conditions are needed, together with reliable growth methods. A study of the chemical changes which take place during crystallization is therefore necessary in understanding crystal growth of these materials. The first step in this investigation has been to carry out a study of the phase equilibrium relationships in the system Fe_2O_3–B_2O_3.

The partial information presently available on the system Fe_2O_3–B_2O_3 is given by Joubert et al.[1] for two compositions: 1 $H_3BO_3 + Fe_2O_3$ and $H_3BO_3 + \frac{3}{2}Fe_2O_3$ sintered respectively at 900 and 670 °C. However, the compounds of the general formula $Fe_{1-x}M_xBO_3$, where M stands for Ga^{3+}, Cr^{3+} or Ti^{3+}, were first prepared by Bernal et al.[2] in polycrystalline form and monocrystalline films, but they were unable to prepare either of the compounds in bulk, nor pure phase where $x = 1$.

The aim of the present investigation is to study the phase relations in the system Fe_2O_3–B_2O_3 and to use the results as a guide to determine the optimum conditions to grow single crystals of iron borate.

2. Experimental procedure

The starting materials used in the determination of the phase relations or for growing single crystals were of reagent grade. The samples were prepared by standard ceramic techniques and fired in air at the desired temperatures for the same interval of time (100 hr). The samples used for the study of reaction kinetics were prepared by careful mixing of the desired proportions of the oxides in an agate mortar, pressing the powders into pills placed in a small platinum microdish and firing at a given temperature. The accuracy of the temperature measurement in these runs is probably ± 2 °C. The samples were then quenched by dropping the product directly into water. The phases present at the temperature of equilibrium were then identified with a metallographic microscope and by X-ray diffraction techniques using FeK_α radiation.

The compositions of the system Fe_2O_3–B_2O_3 studied in the present investigation, are with 20, 30, 40, 50, 60, 70 and 80 mole% of Fe_2O_3. Each composition was reacted for 100 hr to achieve equilibrium at temperatures of 670, 800, 860 and 900 °C.

3. Results and discussion

The samples corresponding to each composition have been separated into four lots to show the effect of temperature. The results are listed in table 1.

3.1. TREATMENT AT 670 °C

For all compositions studied we have observed the existence of $FeBO_3$. From these results one may assume that $FeBO_3$ is a stable phase at low temperatures. How-

TABLE 1

Composition mole - ratio		Tempe-rature (100 hrs)	Phases Observed After Quenching.
Fe_2O_3	B_2O_3		
20	80	670 °C	$FeBO_3$
		800 °C	$FeBO_3 + Fe_2O_3 + Fe_3BO_6$
		860 °C	$Fe_3BO_6 + Fe_2O_3$
		900 °C	Fe_2O_3
30	70	670 °C	$FeBO_3$
		800 °C	$FeBO_3 + Fe_3BO_6 + Fe_2O_3$
		860 °C	$Fe_3BO_6 + Fe_2O_3$
		900 °C	Fe_2O_3
40	60	670 °C	$FeBO_3$
		800 °C	$FeBO_3 + Fe_3BO_6 + Fe_2O_3$
		860 °C	$Fe_3BO_6 + Fe_2O_3$
		900 °C	$Fe_2O_3 + Fe_3BO_6$
50	50	670 °C	$FeBO_3$
		800 °C	$FeBO_3 + Fe_3BO_6 + Fe_2O_3$
		860 °C	$Fe_3BO_6 + Fe_2O_3$
		900 °C	$Fe_3BO_6 + Fe_2O_3$
60	40	670 °C	$FeBO_3 + Fe_2O_3$
		800 °C	$Fe_3BO_6 + Fe_2O_3 + FeBO_3$
		860 °C	$Fe_3BO_6 + Fe_2O_3$
		900 °C	$Fe_2O_3 + Fe_3BO_6$
70	30	670 °C	$FeBO_3 + Fe_2O_3$
		800 °C	$Fe_3BO_6 + Fe_2O_3$
		860 °C	$Fe_2O_3 + Fe_3BO_6$
		900 °C	Fe_2O_3
80	20	670 °C	$FeBO_3 + Fe_2O_3$
		800 °C	$Fe_2O_3 + Fe_3BO_6$
		860 °C	$Fe_2O_3 + Fe_3BO_6$
		900 °C	Fe_2O_3

* The size of the letters in the last column indicates the relative proportion of the phases.

ever, for compositions with Fe_2O_3 equal or more than 60 mole%, we note on the X-ray pattern the presence of increasing intensity characteristic lines of Fe_2O_3.

3.2. TREATMENT AT 800 °C

In the range of 20–50 mole% Fe_2O_3, we observed the presence of $FeBO_3$ while with mixtures containing

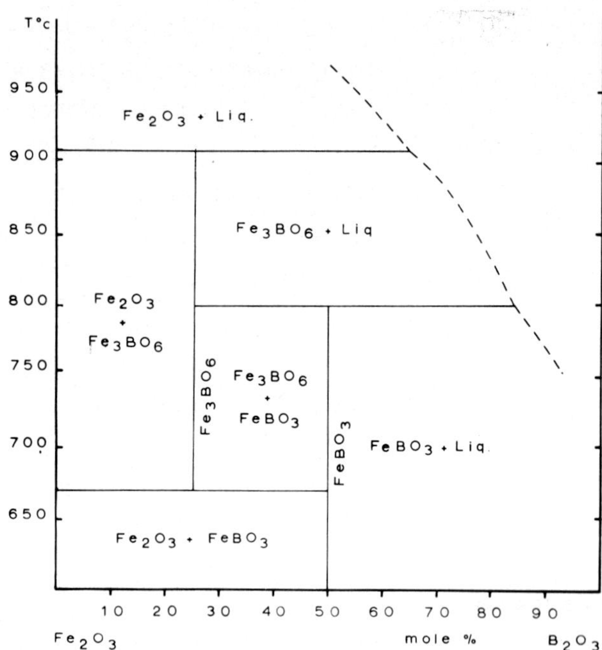

Fig. 1.   Phase diagram for the system Fe_2O_3–B_2O_3.

more than 50 mole% Fe_2O_3, both Fe_3BO_6 and Fe_2O_3 are present.

3.3. TREATMENT AT 860 °C

In another series of samples treated at 860 °C, we observed for all the compositions, the presence of iron orthoborate Fe_3BO_6 as a prevalent phase and Fe_2O_3. In compositions with 60 mole% or more Fe_2O_3, the predominant phase is α-Fe_2O_3.

3.4. TREATMENT AT 900 °C

When samples were heated at 900 °C, the X-ray pattern showed Fe_2O_3 as prevalent phase + Fe_3BO_6, for compositions where Fe_2O_3 exceeded 40 mole% in the starting mixture. Samples of composition equal to 40, 50 or 60 mole% Fe_2O_3, sintered at 900 °C, yielded a mixture of two phases Fe_2O_3 and Fe_3BO_6. Above 60 mole%, only Fe_2O_3 was present. It is evident, in this case, that Fe_3BO_6 is an unstable phase.

Fig. 1 shows the phase diagram deduced from the above experiments, which is in good agreement with the one predicted by Joubert et al.[2]).

The preceding information enabled us to determine the approximate conditions (temperature and composition) of $FeBO_3$ formation and to predict the crystallization path of this phase.

4. Synthesis of iron borate single crystals

All runs were made in $100\ cm^3$ platinum crucibles heated in an electric furnace at atmospheric pressure. Different compositions and different heat treatments have been tried. $FeBO_3$ cannot be grown from excess B_2O_3 without the addition of an additional component (flux) to decrease viscosity. Bi_2O_3 was used for this purpose. It appears that the most favourable conditions are to heat mixture using the mole ratio: 1 Fe_2O_3, 5 B_2O_3, 1 Bi_2O_3 at 1135 °C for a period of 6–8 hr to ensure complete dissolution of compounds followed by a rapid cooling to 850 °C to avoid the crystallization of Fe_2O_3 and Fe_3BO_6. The melt was then cooled slowly at a rate of 0.5 °C/hr to 600 °C. The $FeBO_3$ monocrystals were present as a single phase and the crystals were of good quality with maximum dimensions of up to 8 mm.

5. Conclusion

It is concluded that at 670 °C the approximate compositional limits, where the iron borate $FeBO_3$ exists as a single phase, are 20–50 mole% Fe_2O_3. The metaborate and the ferric oxide phases are present when the temperature and Fe_2O_3 were increased. Such information greatly expedites the growth of iron borate single crystals from the flux.

References

1) J. C. Joubert, T. Shirk, W. T. White and R. Roy, Mater. Res. Bull. **3** (1968) 671.
2) I. Bernal, C. W. Struck and J. G. White, Acta Cryst. **16** (1963) 849.

 Journal of Crystal Growth **13/14** (1972) 588–592 © *North-Holland Publishing Co.*

CRYSTAL GROWTH FROM THE FLUX SYSTEMS PbO-V$_2$O$_5$ AND Bi$_2$O$_3$-V$_2$O$_5$

G. GARTON, S. H. SMITH and BARBARA M. WANKLYN

The Clarendon Laboratory, Oxford, England

Fluxes of composition 2PbO · V$_2$O$_5$ and 4Bi$_2$O$_3$ · V$_2$O$_5$ have been found to be particularly suitable for crystal growth. Starting compositions and conditions for growth are given for Al$_2$O$_3$, Cr$_2$O$_3$, α-Fe$_2$O$_3$, β-Ga$_2$O$_3$, ThO$_2$, TiO$_2$, GaFeO$_3$, NiFe$_2$O$_4$, NiTiO$_3$, Fe$_2$TiO$_5$, TbNbO$_4$, ThPbV$_2$O$_8$ and RVO$_4$ (R = rare earth ion).

1. Introduction

The paper describes the application of lead and bismuth vanadate fluxes to the growth of a number of simple and complex oxides. The advantages of these fluxes have been described previously[1]); in particular the addition of V$_2$O$_5$ to the powerful solvents PbO and Bi$_2$O$_3$ greatly reduces the degree to which platinum is attacked. When these solvents are used alone, crucibles are prone to attack because the oxides are easily reduced and both lead and bismuth form low melting point alloys with platinum. On the addition of V$_2$O$_5$, however, any traces of free metal in the solvents are oxidized and the V$_2$O$_5$ is reduced only to a lower oxide. As a result prolonged experiments may be conducted, even at high temperatures (> 1350 °C). The vanadate fluxes are also nonvolatile, simple to prepare, readily soluble in dilute acids and permit the growth of a small number of large crystals by spontaneous nucleation.

2. Experimental

The crystals were grown in a muffle furnace which was heated by two vertical arrays of horizontal Morgan Crusilite elements, one on each side of the hearth. This arrangement produced a favourable temperature distribution for crystal growth and a small vertical temperature gradient which was sufficient to ensure that nucleation occurred at or near the bases of the crucibles.

The crucibles were provided with tightly fitting lids to minimize vaporization and the volumes of the melts were chosen so that each crucible was not more than half full. The latter measure enabled a hot-pouring technique to be used to facilitate the recovery of crystals.

In a typical experiment, several crucibles were embedded in a block of Morgan M.I.28 alumina brick and any spaces were packed with powdered alumina. The block was then heated in the furnace to such a temperature that complete dissolution of the solutes was assured and was maintained at this temperature for a predetermined period. At the end of the period, the furnace was cooled at a specified rate by means of a Eurotherm electronic programmer.

When the cooling programme had been completed, the block was withdrawn from the furnace, inverted and replaced. This sequence of operations was performed rapidly with the aid of a mechanical device so that the flux flowed into the lids of the crucibles before it could solidify. The furnace was then allowed to cool and the crucibles were removed from the block.

After a period of immersion in hot, dilute nitric acid, the lids were removed from the crucibles and each in turn was placed on a cone of perforated platinum foil which was mounted above a second, empty crucible. The assembly was then heated in the furnace to 1000 °C so that the flux melted and drained through the perforated cone into the lower crucible. By this means the flux was separated from the crystals, which remained either on the cone or on the inner surface of the upper crucible. The method had the advantage that crystals were recovered without recourse to tedious leaching procedures and, in the case of the rare earth vanadates,

TABLE 1

Starting compositions and experimental conditions for crystal growth from $2\,PbO \cdot V_2O_5$

Desired compound	Starting composition	Crucible volume (ml)	Initial temperature (°C)	Holding time (hr)	Final temperature (°C)	Cooling rate (°C hr^{-1})	Final product
(1) Fe_2TiO_5	16 g Fe_2O_3, 8 g TiO_2, 205 g PbO, 89 g V_2O_5	100	1330	1	950	1.7	Black plates up to 2.5 cm × 8 mm and rods
(2) $GaFeO_3$	2.5 g Ga_2O_3, 2.1 g Fe_2O_3 52 g PbO, 21 g V_2O_5	50	1300	14	900	5	Rods up to 1 cm × 0.3 mm × 0.3 mm and plates of β-Ga_2O_3 up to 4 mm in width
(3) $NiTiO_3$	5.6 g NiO, 6 g TiO_2 103 g PbO, 45 g V_2O_5	100	1320	14	950	1	Black rhombohedra up to 3 mm on edge and plates up to 1 cm in width
(4) α-Fe_2O_3	13 g Fe_2O_3, 202 g PbO, 94.5 g V_2O_5	100	1120	15	810	2.5	Thin basal plates up to 3 cm in width
α-Fe_2O_3	25 g Fe_2O_3, 202 g PbO, 94.5 g V_2O_5	100	1345	24	750	1.3	Thicker plates and rhombohedra
(5) TiO_2	4.2 g TiO_2, 51.4 g PbO, 24 g V_2O_5	50	1330	14	1000	1	Rods up to 2 cm × 1.5 mm × 1.5 mm
(6) ThO_2	14 g ThO_2, 95 g V_2O_5, 202 g PbO	100	1330	14	1000	1	Clear octahedra up to 2.5 mm on edge
(7) $ThPbV_2O_8$	14 g ThO_2, 112 g V_2O_5, 238 g PbO	100	1320	14	750	1.3	Thin, pale yellow plates up to 6 mm in width
(8) $DyVO_4$	9 g Dy_2O_3, 91 g V_2O_5, 214 g PbO	100	1330	14	950	1	Clear rods up to 2 cm × 1.5 mm × 1.5 mm

enabled the crystals to be recovered intact since these were prone to fracture if allowed to remain in contact with the flux until it solidified.

3. Results

3.1. CRYSTAL GROWTH FROM LEAD VANADATE

The composition $2\,PbO \cdot V_2O_5$ was found to be the most suitable flux for the growth of crystals and the following materials have been prepared from it in accordance with the experimental procedures which are summarized in table 1.

(1) Fe_2TiO_5: This material occurs in nature as pseudobrookite. Large crystals were obtained, as shown in fig. 1, and E.P.M.A. has shown that these contained less than 0.3% V.

(2) $GaFeO_3$ and β-Ga_2O_3: Rods of $GaFeO_3$ up to 1 cm in length were obtained from a 78 g melt. A number of semi-transparent, brown, pseudo-hexagonal crystals were also formed and these were identified as β-Ga_2O_3.

(3) $NiTiO_3$: $NiTiO_3$ was found to grow in the form of rhombohedral crystals at high temperatures and

basal plates at lower temperatures. Similar results were obtained for Cr_2O_3 and α-Fe_2O_3, which are structurally similar, and such behaviour has also been reported in the growth of the latter materials from PbO–$K_2B_4O_7$[2]).

(4) α-Fe_2O_3: When dilute solutions of Fe_2O_3 in $2\,PbO \cdot V_2O_5$ were cooled from below 1000 °C, large basal plates were formed. At higher temperatures the plates were thicker and rhombohedra were also produced. E.P.M.A. showed that the crystals contained less than 0.05% Pb and 1–2% V.

(5) TiO_2: Rods 2–3 cm in length were obtained of

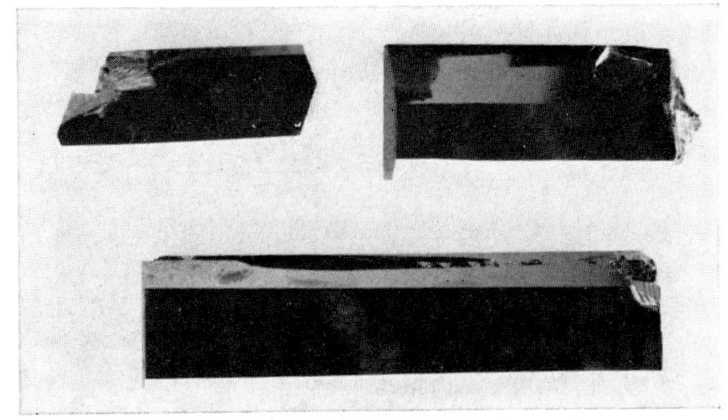

Fig. 1. Crystals of Fe_2TiO_5 (magnification 3×).

XII – 9

TABLE 2

X-ray powder pattern data for ThPb V_2O_8

hkl	I	ThPb V_2O_8	
		$\sin^2\theta$ (obs)	$\sin^2\theta$ (calc)
101	w	0.0199	0.0201
011	m	0.0253	0.0253
11$\bar{1}$	m	0.0314	0.0315
101	mw	0.0343	0.0341
111	m	0.0456	0.0455
200	s	0.0525	0.0526
120	vs	0.0582	0.0584
210	w	0.0634	0.0639
$\bar{1}$12	w	0.0660	0.0664
012	s	0.0674	0.0673
121	mw	0.0797	0.0794
211, $\bar{2}$12	mw	0.0907	0.0919
112	mw	0.0942	0.0944
220	vw	0.0969	0.0978
$\bar{1}$22	vw	0.0998	0.1004
130	mw	0.1151	0.1154
031	mw	0.1173	0.1158
31$\bar{1}$	mw	0.1213	0.1227
221, 22$\bar{2}$	mw	0.1249	0.1258
$\bar{1}$13	vw	0.1294	0.1298
131	vw	0.1348	0.1363
212	m	0.1471	0.1479
301	w	0.1531	0.1533

Unit cell dimensions for ThPb V_2O_8: $a = 6.96$ Å, $b = 7.25$ Å, $c = 6.75$ Å, $\beta = 105°$.

which the largest tended to contain flux inclusions. The crystals varied in colour from straw to deep red and some exhibited termination facets.

(6) ThO_2: Clear, colourless octahedra were obtained of which the larger possessed an internal dendritic structure and the (111) facets consisted of thin plates. When a melt was cooled below 1000 °C, resolution occurred and a previously unreported phase, $ThPbV_2O_8$ was formed. That platinum dissolved in the melt to an appreciable extent was evident from the fact that small platelets of the metal were subsequently found at the base of the crucible.

(7) $ThPbV_2O_8$: This previously unreported material crystallized in the form of transparent, yellow plates which exhibited extinction under the polarizing microscope. The X-ray powder pattern of the material was found to be similar to that of monazite, $LaVO_4$, and has been similarly indexed. The X-ray diffraction data and the lattice parameters are included in table 2. No superlattice was observed, so it would appear that Th^{4+} and Pb^{2+} are randomly disposed in the lattice and thus the formula may be written (Th, Pb)VO_4. The crystals were analyzed and were found to contain 32.8% Pb and 14.2% V, a result which is in good agreement with the theoretical values of 31% and 15% respectively.

Similar but colourless crystals were produced on cooling a solution of ThO_2 in $Pb_2P_2O_7$, and the X-ray powder pattern of the material was consistent with a monazite-type structure of formula (Th, Pb)PO_4. Both materials rapidly decomposed on heating and attempts

TABLE 3

Starting compositions and experimental conditions for crystal growth from $4Bi_2O_3 \cdot V_2O_5$

Desired compound	Starting composition	Crucible volume (ml)	Initial temperature (°C)	Holding time (hr)	Final temperature (°C)	Cooling rate (°C hr^{-1})	Final product
(1) Cr_2O_3	20.4 g Cr_2O_3, 240 g Bi_2O_3, 24 g V_2O_5	100	1345	24	750	1.3	Rods and plates up to 5 mm in width
(2) Al_2O_3	7 g Al_2O_3, 84 g Bi_2O_3, 8.5 g V_2O_5	50	1300	14	900	5	Rhombohedral platelets up to 5 mm on edge
(3) α-Fe_2O_3	12 g Fe_2O_3, 120 g Bi_2O_3, 12 g V_2O_5	50	1300	4	900	5	Rhombohedra up to 7 mm on edge
(4) β-Ga_2O_3	7 g Ga_2O_3, 84 g Bi_2O_3, 8.5 g V_2O_5	50	1300	14	900	5	Platelets up to 6 mm × 2 mm × 1 mm
(5) $NiFe_2O_4$	3.7 g NiO, 8 g Fe_2O_3, 40 g PbO, 4 g V_2O_5	50	1350	24	950	1	Octahedra up to 3 mm on edge
(6) $TbNbO_4$	20 g Tb_4O_7, 14 g Nb_2O_5, 200 g Bi_2O_3, 20 g V_2O_5	100	1330	14	900 500 then to	1 3	Bipyramids up to 5 mm in width
(7) $GdVO_4$	15 g Gd_2O_3, 180 g Bi_2O_3, 24 g V_2O_5	100	1320	14	950	1	Clear, yellow rods 7.5 mm × 2 mm × 2mm

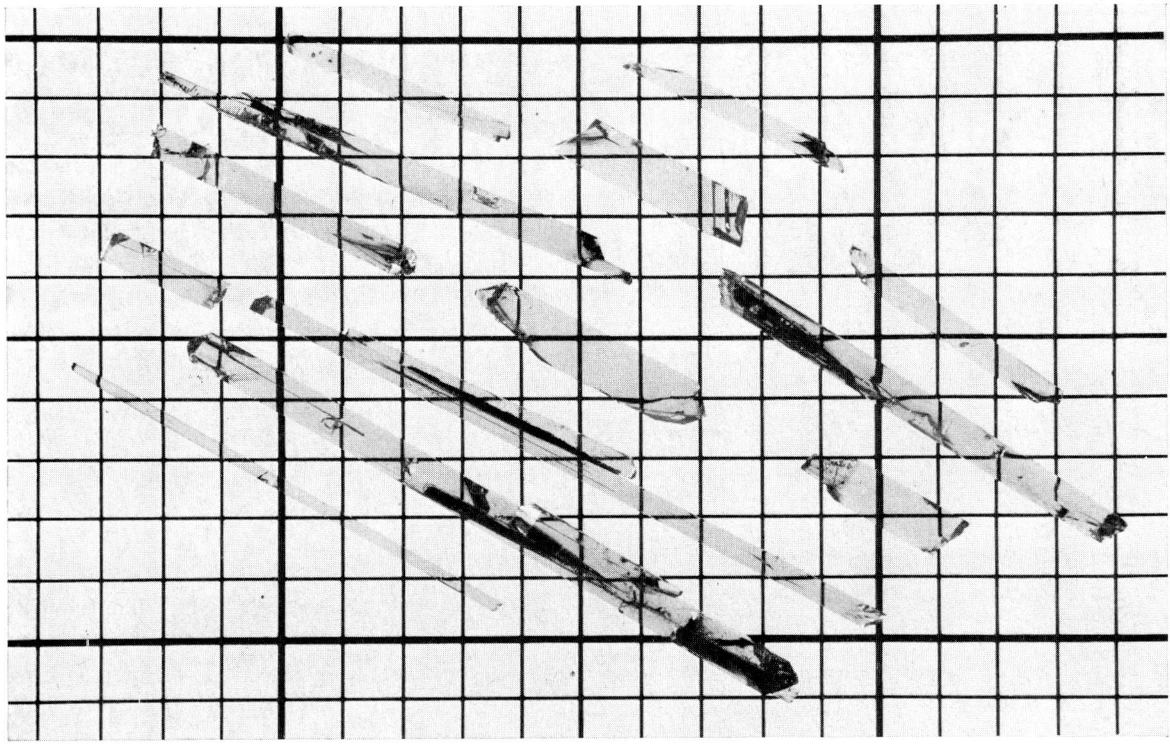

Fig. 2.   Crystals of $DyVO_4$ (2.5 mm grid).

to synthesize them from the constituent oxides were unsuccessful.

(8) RVO_4 (R = rare earth ion and yttrium): All of the RVO_4 crystals grew in the form of long, clear rods from $2PbO \cdot V_2O_5$ alone as shown in fig. 2. However, as was reported for $GdVO_4$[3]), the growth habit was dependent on the composition of the flux and the presence of additives. For example, the addition of $NaVO_3$ and $Na_2B_4O_7$ to the flux caused the crystals to grow in the form of plates or thick rods, while the addition of V_2O_5 or B_2O_3 caused the rods to be elongated. The $GdVO_4$ crystals were analysed and found to contain 0.03% Pb.

3.2. Crystal growth from bismuth vanadate

The composition $4Bi_2O_3 \cdot V_2O_5$ was found to be the most suitable flux and the following materials have been grown from it in accordance with the experimental procedures which are summarized in table 3. The preparation of the flux from an intimate mixture of its constituent oxides was shown to minimize damage to the crucibles.

(1) Cr_2O_3: As a result of the great difference in density between the flux and the solute, dissolution of the latter was usually incomplete and a layer of Cr_2O_3 remained at the surface of the melt. Despite the fact that this layer provided nucleation sites, large crystals grew at the base of the crucible, generally in the form of pseudo-hexagonal plates but also as rhombohedra. A hexagonal crystal in which grain boundaries are clearly defined is shown in fig. 3.

(2) Al_2O_3: As in the case of Cr_2O_3, nucleation occurred on undissolved material at the surface of the melt and also at the base of the crucible. The crystals

Fig. 3.   Crystal of Cr_2O_3 showing grain boundaries.

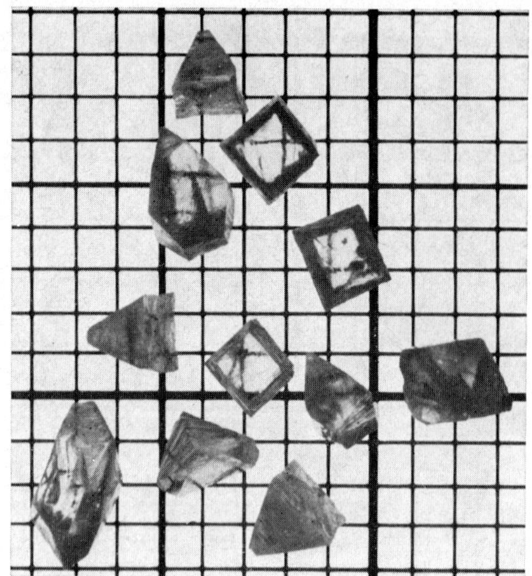

Fig. 4.   Crystals of TbNbO$_4$ (mm grid).

grew in the form of rhombohedral platelets and some attained 5 mm on edge.

(3) α-Fe$_2$O$_3$: Only rhombohedral crystals were obtained and it was concluded that the presence of Bi$_2$O$_3$ inhibited the formation of basal plates. Under conditions which promoted rapid growth dendrites were formed. The crystals contained less than 0.1 % Bi and less than 1 % V.

(4) β-Ga$_2$O$_3$: Transparent, colourless plates up to 6 mm edge were obtained from a 100 g melt.

(5) NiFe$_2$O$_4$: Octahedra up to 3 mm on edge were grown from a 56 g melt.

(6) TbNbO$_4$: Bismuth fluxes are normally to be avoided for rare earth compounds because Bi^{3+} is similar in both size and charge to R^{3+}, as a result of which partial substitution can occur. However, it was shown that rare earth iron and aluminium garnets could be produced, but the extent of the substitution may be appreciated from the fact that, in an attempt to grow crystals of Dy$_3$Fe$_5$O$_{12}$, crystals of formula Dy$_{2.46}$-Bi$_{0.52}$Fe$_{5.01}$O$_{12+\delta}$were obtained. In the case of RNbO$_4$

however, the only known alternative solvent to bismuth vanadate is bismuth borate[4]), and in experiments with RNbO$_4$–Bi$_2$O$_3$–B$_2$O$_3$ mixtures it was found that the crucibles were attacked so severely that none remained intact. When bismuth vanadate was used, the crucibles were not damaged and bipyramidal crystals of TbNbO$_4$ were obtained, as shown in fig. 4. The crystals contained about 4 % Bi and 2 % V.

(7) RVO$_4$: The rare earth vanadate rods which were grown from bismuth vanadate were considerably larger in cross-section than those which were grown from lead vanadate. The degree to which the crystals were contaminated by the flux was, however, considerably greater. Analysis of GdVO$_4$ crystals showed that these contained about 7 % Bi.

4. Conclusions

Fluxes of composition 2 PbO · V$_2$O$_5$ and 4 Bi$_2$O$_3$ · V$_2$O$_5$ have been found to be suitable for the growth of a variety of oxide crystals. The properties of these fluxes, and in particular their low volatility at high temperatures, favour their use in epitaxial deposition and crystal-pulling experiments. Future work will be directed towards these ends.

Acknowledgements

The authors are grateful to Dr. E. A. D. White for valuable discussions, to Mrs. J. Berrington for the X-ray study, to Philips Research Laboratory, Hamburg for the chemical analysis of the crystal dysprosium iron garnet, and to Dr. D. M. Poole of AERE, Harwell, for EPMA of α-Fe$_2$O$_3$ and TbNbO$_4$ crystals.

This work was supported in part by the Science Research Council.

References

1) B. M. Wanklyn, J. Crystal Growth **7** (1970) 368.
2) R. E. Banks and R. Roy, in: *Crystal Growth*, Ed. H. S. Peiser (Pergamon, Oxford, 1967) p. 497.
3) G. Garton and B. M. Wanklyn, in: *Proc. Conf. Les Eléments des Terres Rares*, Paris–Grenoble, 1969, pp. 343–348.
4) G. Garton and B. M. Wanklyn, J. Mater. Sci. **3** (1968) 395.

Journal of Crystal Growth **13/14** (1972) 593–596 © *North-Holland Publishing Co.*

593

SYNTHÈSE DE MONOCRISTAUX DE VANADATE D'YTTRIUM PAR LA MÉTHODE DE TIRAGE EN FLUX

JEAN LORIERS et MIROSLAV VICHR

Laboratoire de Recherches sur les Terres Rares, C.N.R.S., 92–Bellevue, France

Yttrium vanadate has been crystallised on a seed crystal from molten vanadium pentoxide by a modified pulling method. Transport in a temperature gradient was used: an excess of YVO_4 was placed in the lower, hotter, part of the crucible, and the seed was suspended in the upper, cooler, region on a rapidly rotating support. Crystalline YVO_4 nutrient and seed crystals were prepared by slow cooling of YVO_4–V_2O_5 melts. Different temperature gradients (30 °C/cm, 10 °C/cm, 1 °C/cm), pulling rates (5 mm/day, 20 μm/hr) and dissolving temperatures (900–1250 °C) were examined. Large yttrium vanadate single crystals were obtained with the following optimal conditions: dissolving temperature 1050 °C, temperature gradient 4 °C/cm, pulling rate in [001] direction 50 μm/hr. Fast pulling rates (5 mm/day), high temperature gradients (30 °C/cm) and low crystallization temperatures (900 °C) resulted in dendritic parallel growth.

1. Introduction

On sait que le vanadate d'yttrium est un matériau très intéressant pour la préparation de composés luminescents activés aux terres rares; dopé à l'europium, il a reçu des applications importantes dans l'industrie de la télévision en couleur ("phosphor" rouge des écrans) et de l'éclairage (lampes fluorescentes). YVO_4 est également une excellente matrice pour la fabrication de cristaux "laser" dopés au néodyme: cristallisant, comme tous les autres vanadates de terres rares à l'exception du lanthane, dans la structure quadratique du zircon (groupe d'espace D4h) il constitue un "cristal-hôte" favorable pour les ions trivalents des lanthanides, qui peuvent se substituer en toutes proportions à l'yttrium, et qui occupent un site cristallin de basse symétrie renforçant la possibilité des transitions radiatives. C'est ce qui explique que YVO_4: Nd soit un cristal laser[1,2]) d'une efficacité comparable à celle du grenat d'yttrium–aluminium.

L'obtention de monocristaux de vanadate d'yttrium de grandes dimensions est malheureusement peu aisée: la méthode de cristallisation en flux, la plus facile à mettre en œuvre, ne permet pas d'obtenir des cristaux ayant plus de quelques mm de coté[3-5]). C'est par croissance directe à partir du sel fondu que de grands cristaux (20 mm de long) ont pu être obtenus, soit par tirage suivant la méthode Czochralski[6,7]), soit par une technique dérivée de la zone flottante[8]); mais ces méthodes impliquent des températures élevées – de l'ordre de 1800–2000 °C – pour la fusion de YVO_4, et il est difficile dans ces conditions d'éviter les pertes d'oxygène par réduction de V^{5+}; les cristaux obtenus sont généralement très colorés ou même noirs, donc inutilisables comme laser, et cet inconvénient ne peut être évité qu'un opérant en atmosphère contrôlée (mélange d'argon et d'oxygène).

2. Choix de la méthode et mode opératoire

Les principales difficultés rencontrées jusqu'à présent étant dues à l'instabilité du vanadate fondu à haute température, nous avons pensé qu'il serait préférable de revenir au principe de la cristallisation en flux, qui peut s'effectuer à température beaucoup plus basse, et de combiner cette méthode avec la technique de tirage à partir d'un germe, qui devrait permettre l'obtention de cristaux de bonnes dimensions.

2.1. Choix du flux

Parmi les différents flux utilisables ($NaVO_3$, V_2O_5, PbO) le pentoxyde de vanadium V_2O_5 nous a paru le meilleur après quelques essais préliminaires: c'est celui qui donne les cristallites les plus développés, et il a l'avantage de ne pas introduire de cations étrangers dans le bain de tirage; en outre, le système YVO_4–V_2O_5 est un système eutectique simple[9]) (la température de cristallisation de YVO_4 passant de 1810 °C pour le sel pur à 672 °C pour l'eutectique à 96 % de V_2O_5) et la solubilité de YVO_4 dans V_2O_5 est importante, avec un coefficient de température fortement positif: 20 mole%

XII – 10

à 950 °C et 30 mole% à 1150 °C. Ces caractéristiques permettent d'obtenir la cristallisation unique de YVO_4 entre de larges limites de températures, et sont favorables à la technique de tirage isotherme avec gradient de température dans le creuset, qui est plus facile à réaliser qu'un tirage en refroidissement lent. Pour assurer une composition constante du bain fondu (déterminée par la température choisie) une réserve de cristaux de YVO_4 sera placée dans le creuset; les cristaux resteront au fond, car la densité de YVO_4 (4.31 g/cm^3) est supérieure à celle de V_2O_5 (3.357 g/cm^3). Ces conditions sont particulièrement favorables à la préparation de cristaux dopés homogènes.

Dans notre méthode, le cristal se développera donc par transport à travers le flux entre la réserve de YVO_4 non dissous et le germe monocristallin, grâce au gradient de température.

2.2. APPAREILLAGE

L'appareillage utilisé est d'une réalisation très simple. Il comprend seulement un four tubulaire vertical à chauffage électrique par résistance de Kanthal (température maximale 1250 °C), une régulation de température (précision ± 0.1 °C), et un dispositif mécanique de tirage avec rotation (déplacement vertical 0.5 à 10 mm/jour; rotation 6 à 60 tours/min). Le germe est fixé à l'extrêmité d'une tige d'alumine au moyen d'un petit tube de platine; la tige passe à travers un écran en céramique réfractaire percé d'un trou, qui recouvre le creuset; celui-ci est un récipient cylindrique en platine d'une contenance de 100 ml. Le gradient dans le bain peut être ajusté entre 1 °C/cm et 50 °C/cm en fixant la base du creuset (point le plus chaud) au niveau voulu dans le four.

2.3. MODE OPÉRATOIRE

Le creuset est chargé en plaçant dans le fond le vanadate d'yttrium (sous forme de poudre obtenue par frittage de $Y_2O_3 + V_2O_5$, ou de préférence sous forme de cristaux obtenus en flux) et par dessus le fondant V_2O_5. Les proportions sont choisies de telle façon qu'il subsiste un excès de YVO_4 pour alimenter le bain pendant toute la croissance du monocristal. Le creuset placé dans le four de tirage est porté à la température choisie pour l'opération et maintenu à cette température pendant 24 h pour atteindre l'équilibre de dissolution de YVO_4. Le germe est alors amené au contact du

Fig. 1.   Croissance dentritique de YVO_4. $\times 20$.

bain, et sa croissance commence dès que le système de rotation est mis en marche.

3. Résultats expérimentaux

Parmi les facteurs influençant la croissance du monocristal, la température et son gradient dans le creuset sont les plus importants, car c'est eux qui fixent la concentration de YVO_4 dans le flux. Nous avons donc effectué une série d'expériences d'exploration en fixant ces paramètres: gradients de 30 °C/cm, 10 °C/cm et 1 °C/cm, températures comprises entre 1250 °C et 900 °C; et en faisant varier les autres facteurs: vitesse de tirage et de rotation, géométrie des écrans du creuset, direction cristallographique de tirage.

Aux températures élevées (supérieures à 1200 °C) la nucléation spontanée sur les parois du creuset et à la surface du bain perturbe la croissance du germe; avec des températures plus basses et des gradients élevés (30 °C/cm), on observe des croissances dendritiques d'agglomérats de vanadate; les figures 1 et 2 montrent les cristaux obtenus respectivement à 1050 °C et 900 °C avec une vitesse de tirage de 5 mm/jour. De meilleurs résultats sont obtenus en diminuant le gradient de température: ainsi le premier monocristal de bonne taille (6 à 8 mm de diamètre, 15 mm de long) a été tiré à

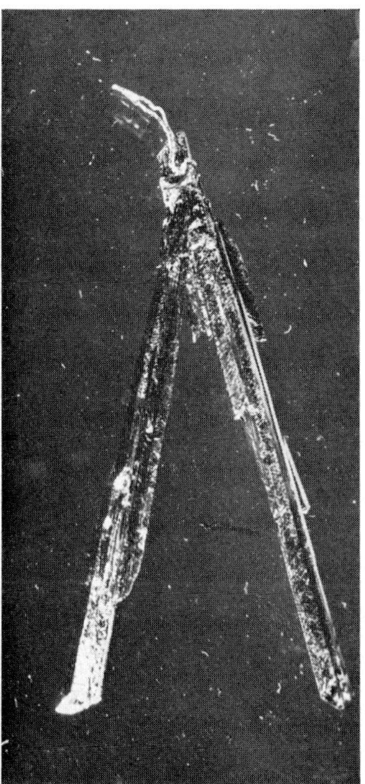

Fig. 2.   Cristaux aciculaires de YVO$_4$. $\times 1.3$.

Fig. 4.   Cristal de YVO$_4$ obtenu par tirage en flux suivant l'axe c. $\times 8$.

994 °C avec un gradient de 4 °C/cm et une vitesse de tirage de 50 µm/h (environ 1 mm/jour).

De faibles vitesses de tirage sont favorables à la formation de cristaux bien développés, mais la vitesse de croissance dépend beaucoup de la direction de tirage. Les figures 3 et 4 montrent des monocristaux obtenus à 1065 °C avec un gradient de 1 °C/mm, le premier ayant été tiré dans une direction perpendiculaire à l'axe c à une vitesse de 28 µm/h, le second parallèlement à cet axe c'est à dire dans la direction [001] à une vitesse de 98 µm/h. Il faut signaler que le cristal représenté figure 3 s'est développé dans une direction perpendiculaire à celle du tirage, pour suivre l'axe c; cet axe est donc un axe de croissance facile pour YVO$_4$.

On observe de façon générale une très nette tendance pour les cristaux à se développer en déviant de la direction de croissance choisie; cet inconvénient est probablement lié au gradient latéral de température qui existe dans notre creuset. Il faut noter d'autre part que nous avons amélioré sensiblement la qualité des cristaux en sélectionnant des germes taillés dans les meilleures parties des cristaux déjà obtenus.

Fig. 3.   Cristal de YVO$_4$ formé par croissance latéral. $\times 8$.

4. Conclusions

A ce stade de nos recherches, on peut dire que les conditions optimales de croissance sont réalisées à des températures moyennes, des gradients de température et des vitesses de tirage faibles. Du point de vue pratique, ces conditions ne sont pas avantageuses, car la durée de fabrication des cristaux est longue. Nous espérons les améliorer par l'emploi de plus hautes températures, mais elles ne seront applicables que dans un appareillage plus perfectionné (en particulier en ce qui concerne la géométrie du creuset et des écrans thermiques).

Notre étude, réalisée avec un appareillage peu onéreux, présente en tout cas l'intérêt de proposer une nouvelle méthode permettant la fabrication de monocristaux de composés à points de fusion élevés; elle pourrait être appliquée facilement au tirage de cristaux à fusion incongruente.

Ce travail a été effectué avec l'aide financière de la Direction des Recherches et Moyens d'Essais.

Bibliographie

1) J. R. O'Connor, Appl. Phys. Letters **9** (1966) 407.
2) K. S. Bagdasarov, A. A. Kaminskii, V. S. Krylov et V. I. Popov, Phys. Status Solidi **27** (1968) K 1.
3) L. G. Van Uitert, R. C. Linares, R. R. Soden et A. A. Ballman, J. Chem. Phys. **36** (1962) 702.
4) L. H. Brixner et E. Abramson, J. Electrochem. Soc. **112** (1965) 70.
5) R. Heindl et J. Loriers, Brevet France No. 1 528 089 du 28.4. 1967.
6) J. J. Rubin et L. G. Van Uitert, J. Appl. Phys. **37** (1966) 9920.
7) H. M. Dess et S. R. Bolin, Trans. Met. Soc. AIME **239** (1967) 359.
8) K. Muto et K. Awazu, Japan. J. Appl. Phys. **8** (1969) 1361.
9) E. M. Levin, J. Am. Ceram. Soc. **50** (1967) 38.

Journal of Crystal Growth **13/14** (1972) 597–600 © *North-Holland Publishing Co.*

THE GROWTH OF CALCITE SINGLE CRYSTALS BY TRAVELLING SOLVENT ZONE MELTING

C. BELIN, J. J. BRISSOT and R. E. JESSE*

Laboratoires d'Electronique et de Physique Appliquée, 94 – Limeil-Brévannes, France

Travelling Solvent Zone Melting has been used to grow calcite ($CaCO_3$) single crystals. In order to obtain an interface as flat as possible a heating strip is embedded in the molten solvent zone. Some small holes in the strip ensure the liquid contact on both sides. The solvent is driven along the solid by the displacement of the solid phases with respect to the strip. Some characteristics of the transport process in this zone are considered in more detail. Experiments show that the formation of single crystals is determined by the transport process more than by kinetics.

Some physical properties of the synthetically grown crystals are compared with those of the natural crystals.

1. Introduction

The birefringent properties of calcite ($CaCO_3$) with rhombohedral structure make this material interesting for the purpose of obtaining polarised light in a wide spectra range (0.2–3 μm). The growth of calcite from its pure melt is difficult because $CaCO_3$ at its melting point has a high equilibrium pressure (≈ 100 atm) of CO_2. Therefore several other methods have been tried to prepare calcite single crystals at lower temperatures and pressures.

One of these methods is the growth from a solvent melt (flux growth). Li_2CO_3 is a suitable solvent. Nester and Schroeder[1] have obtained calcite single crystals by slowly cooling down a solution of $CaCO_3$ in Li_2CO_3 rich in $CaCO_3$ with respect to the eutectic composition. However it appeared difficult to separate the single crystals obtained from the flux because the chemical properties of the flux and calcite are very similar.

This inconvenience can be overcome by application of the principle of Travelling Solvent Zone Melting (T.S.Z.M.) proposed initially by Pfann[2]. In this technique a thin layer of solvent is sandwiched between a polycrystalline feed and a seed and displaced along the feed.

In one realization of the T.S.Z.M. technique an external heater which may consist of a furnace keeps the flux in a molten state. However, for the growth of large diameter crystals a flat solid–liquid interface is

very useful. Therefore following Gasson[3] the external heater is replaced by a Pt strip which is immersed in the solvent itself. This Pt strip is heated by an alternating current. Some small holes drilled in its center allow for contact of the liquid at both sides and the transport of the solute from feed to seed.

With this set up, described previously[4], regular growth rates of 6×10^{-6} cm/sec (≈ 5 mm/day) could be obtained.

2. Some considerations on the transport process

A schematic representation of the phase diagram of the components A and B and of the experimental set-up is given in figs. 1 and 2. In the present paper A and B represent $CaCO_3$ and Li_2CO_3 respectively. The slopes of the liquidus lines, e.g.

$$m_\alpha = \left(\frac{\delta T}{\delta C}\right)_{C < C_E},$$

and of the solidus lines are taken to be linear. C represents the concentration of B and C_E that of B in the eutectic mixture which melts at T_E.

At the beginning of the experiment a thin solid layer of a eutectic composition of A and B is deposited on the Pt ribbon. The minimum temperature $T(0)_{min}$ of the heating strip must be such that liquid of eutectic composition with thickness l is in contact with α-phase only, i.e. the temperature at the solid–liquid interfaces is $T_E [T(0)_{min} > T_E]$.

$T(0)_{min}$ is determined by the original amount of eutectic flux, the thermal conductivity of both α-phases (K_α),

* On a year's leave from Philips Research Laboratories, Eindhoven, The Netherlands.

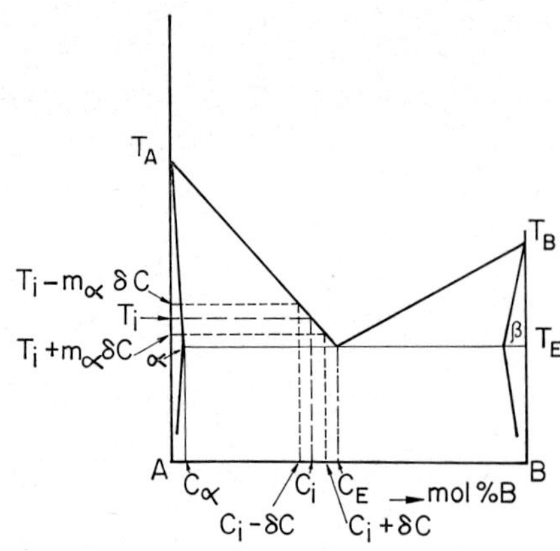

Fig. 1.   Phase diagram of components A and B.

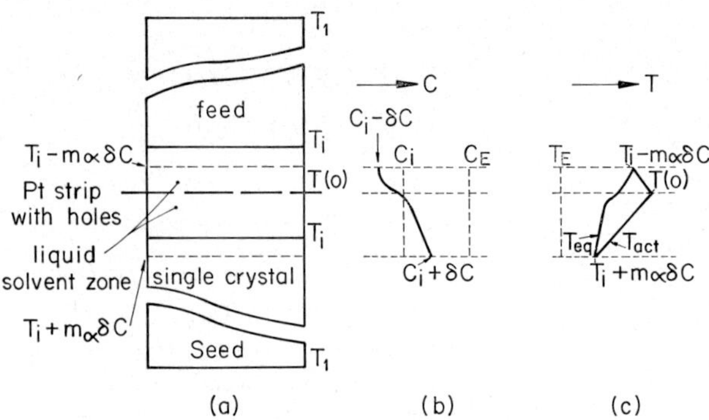

Fig. 2.   (a) Schematic representation of experimental set-up. (b) Concentration distribution in liquid zone in case of growth. (c) Equilibrium temperature distribution T_{eq} and the actual temperature distribution T_{act}.

and of the liquid phase (K_{liq}) and the temperature gradient in the solid phases.

If it is assumed that the thermal gradients on both sides of the heating strip are linear, that they are symmetrical, that the flow of heat is bidirectional from the liquid to both feed and seed and that it occurs by conduction only it follows that $T(0)_{min}$ is given by:

$$T(0)_{min} = T_E + \frac{K_\alpha}{K_{liq}} \frac{l}{L-l}(T_E - T_1),$$

where T_1 is the temperature of the α-phases at a distance $\frac{1}{2}L$ from the heating strip ($T_1 < T_E$).

If the temperature $T(0)$ of the strip is below $T(0)_{min}$ the liquid will be in contact with solid eutectic. In this case a solid of eutectic composition will be deposited onto the seed at the beginning of the growth process. If after a while the amount of B is so exhausted that only A will be deposited, a polycrystalline α-phase will grow. For the case of no growth the concentration in the liquid zone will be uniform. The concentration C_i will be inferior to C_E if $T(0) > T(0)_{min}$, a condition necessary for the growth of single crystals. For a concentration C_i there corresponds a length l_i ($l_i > 1$) and a temperature at the solid–liquid interfaces $T_i = T_E - m_\alpha (C_E - C_i)$.

In the case of growth the stationary state will be modified by two effects:

1) due to the growth process a concentration gradient has been built up in the solvent zone. At the growing interface the concentration of B has increased from C_i to a value nearer to C_E. At the melting interface the concentration of B has decreased with respect to C_i. As a consequence the equilibrium temperature of the grow-

ing interface will decrease, i.e. the interface will move away from the ribbon. Similarly, the melting interface will approach the ribbon. The shift, δz, of both interfaces is essentially determined by both the temperature gradient G near the ribbon and the concentration shift, $\pm \delta C$, at the interfaces with respect to C_i.

2) During growth, heat of fusion will be absorbed at the melting interface and liberated at the growing interface. This heat effect can be neglected since the involved heat flux is small as compared with the heat flux which already exist in the case of no growth.

The value of the concentration gradient in the liquid is determined by the velocity V with which the solvent moves and by the transport of material from the dissolving interface to the growth interface.

The material transport in this zone occurs generally by both diffusion and convection while it is hindered by the presence of the heating strip. Experimentally the thickness l_i of the molten zone with concentration C_i is about 6×10^{-2} cm. This value can be compared with that of the diffusion layer in the theory of Burton, Prim and Slichter[5]. The thickness of this diffusion layer is given by:

$$\delta \sim 1.6 \, D^{\frac{1}{3}} v^{\frac{1}{6}} \omega^{-\frac{1}{2}}.$$

D represents the diffusion coefficient, v the kinematic viscosity and ω the angular velocity of crystal rotation. Insertion of some realistic values, i.e., $D = 10^{-5}$ cm^2/sec, $\lambda = 1$ cm^2/sec and $\omega = 1/6$ sec^{-1} gives $\delta \sim 7 \times 10^{-2}$ cm.

Although the case described by Burton, Prim and Slichter is somewhat different from the T.S.Z.M. case described here, an analogy exists. It is considered therefore that, if a thin solvent zone is applied, material transport is governed essentially by diffusion.

Since the applied travelling rate is about 6×10^{-6} cm/sec, it follows that $(V/D)l_i \ll 1$. It can be easily verified from a simple diffusion consideration, that the concentration gradient in the liquid zone is practically linear. It must be realized that in fact, due to the presence of the heating strip, the concentration will change more rapidly in the neighbourhood of the strip than in the rest of the liquid zone. Apart from this local variation in linearity it can be seen that the equilibrium temperature gradient in the liquid zone will be linear. The actual temperature, however, increases linearly from the equilibrium value at the feed interface up to $T(0)$ and decreases from $T(0)$ down to the equilibrium value at the seed interface. It can be concluded therefore that in the present case constitutional supercooling will not exist (fig. 2).

If during the growth process the dissolving interface touches the heating strip, the actual temperature distribution and the equilibrium temperature distribution in the liquid are close to each other. In this case the process, which is then at the limit of constitutional supercooling, will stop because the solid feed cannot pass the heating strip.

An interesting conclusion is therefore that with this growth technique constitutional supercooling will not occur due to the fact that at the moment it starts the process will stop.

In order to estimate δz a purely diffusion model has been considered. If it is taken into account that $(V/D)l_i \ll 1$ and that the total amount of B in the liquid before growth started is practically equal to that during growth, it follows that the deviation δC at the growing interface and $-\delta C$ at the dissolving interface from C_i is given by:

$$\delta C \sim \tfrac{1}{2}(1 - k_\alpha) \frac{V}{D} l_i C_i,$$

where k_α is the distribution coefficient of B. The shift δz equals $m_\alpha \delta C/G$ where G is the temperature gradient in the liquid zone. It is possible therefore to estimate δz for the case of growth of calcite from a Li_2CO_3–$CaCO_3$ flux with near eutectic composition.

Taking $C_i = 60$ mol% Li_2CO_3, $l_i = 5 \times 10^{-2}$ cm and $k_\alpha = 0.01$ (in reality k_α is much smaller, since $C_\alpha \sim 3 \times 10^{-4}$) it follows that $\delta C \sim 7 \times 10^{-3}$. From the phase diagram[6]) it is seen that $m_\alpha = -11.3$ deg/mol% and from a rough measurement it was found that $G \sim 400$ deg/cm. Therefore $\delta z \sim 2 \times 10^{-2}$ cm.

3. Experimental results

The conclusions which follow from the foregoing are confirmed qualitatively by experiments. It has been found that $T(0)_{min}$ increases with the amount of eutectic flux. In order to get an idea of the change of $T(0)_{min}$ with the amount of flux two typical cases are considered. $T(0)_{min} = (690 \pm 15)$ deg if 110 mg flux of eutectic composition is placed between feed and seed. For 220 mg eutectic flux $T(0)_{min} = (750 \pm 15)$ deg C.

It has also been verified that the distance between the dissolving interface and the heating strip decreases on increasing V. A typical value in the case of no growth for both the distances between feed and strip and seed and strip is 3×10^{-2} cm. At $V = 2.8 \times 10^{-6}$ cm/sec, however, the former distance became about 2×10^{-2} cm while the latter distance became about 4×10^{-2} cm. At $V = 8.4 \times 10^{-6}$ cm/sec the process stopped due to the fact that the feed touched the strip.

The foregoing values were obtained for the case where the strip contained 3 holes with a diameter of 1 mm which were 8 mm apart and distributed on the tips of a regular triangle (crystal diameter 1 cm). If instead a strip is taken with 13 holes which are about 4 mm apart the value of V above which the process stopped was 1×10^{-5} cm/sec.

Single crystals have been grown with a diameter of 1 cm and a length of 2–3 cm (fig. 3). In case of growth along the [100] direction the crystal axis tends to deviate from the axis of the feed-seed system. Further the cross section of the synthetic crystal tends to become elliptic during growth. In the case of growth along the [111] direction the crystal axis stays coincident with the axis of the feed-seed system while the crystals are more cylindrical.

The synthetic crystals have been etched with a 30% formic acid in a water solution for 2 sec. It appears that the density of etch pits obtained is about 10^5 cm^{-2}. The natural seed crystals had a much lower density.

The refraction indices have been measured with the Na_D light ($\lambda = 5843$ Å). The dispersion has been measured for the wavelengths $\lambda = 4861$ Å (F) and $\lambda = 6563$ Å (C). The following indices were found with a precision of 0.0005:

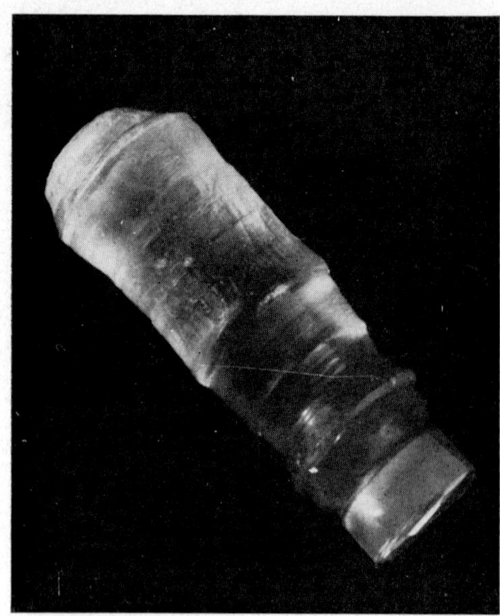

Fig. 3.   Synthetic single crystal of calcite. Magn. ×1.8.

$n_{0D} = 1.6581$ and $n_{0F} - n_{0C} = 0.014,$

$n_{eD} = 1.4864$ and $n_{e}F - n_{eC} = 0.006.$

These results are in good agreement with already published data. The images of birefringence of two types of crystal (natural and synthetic) are compared in fig. 4. Data on light absorption and chemical composition have already been given[4]).

4. Conclusion

The growth of single crystals of calcite with a diameter of 1 cm by the T.S.Z.M. technique is relatively simple. Melting of the solvent zone by a strip heater gives rise to flat solid–liquid interfaces. Such flat interfaces are convenient in order to grow crystals with a large diameter: a 15 mm diameter and 20 mm length single crystal has just been obtained.

The macroscopic aspects of the growth process by this method can be understood by a simplified consideration of the transport of heat and material in the liquid zone. It follows that constitutional supercooling

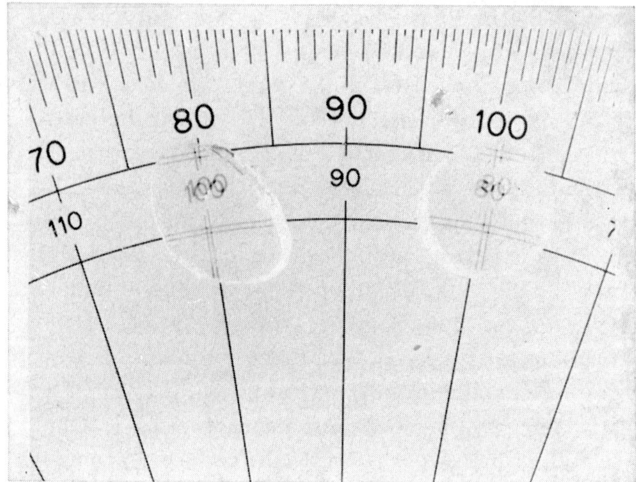

Fig. 4.   Birefringence images of natural and synthetic crystals.

will not occur if sufficiently thin zones are used. In principle the method also makes possible the growth of homogeneously doped crystals.

In this investigation the growth of calcite has been studied. However it is clear that the method can be applied to the growth of single crystals of a different species as well.

Acknowledgments

The authors are indebted to Mr. M. Pertus for his assistance during the construction of the crystal growth equipment and for his aid in carrying out the experiments. They also wish to express their gratitude to Mr. Vulmière of the Institut d'Optique Théorique et Appliquée for measurements of refraction indices.

References

1) J. F. Nester and J. B. Schroeder, Am. Mineralogist **52** (1967) 276.
2) W. G. Pfann, Trans. AIME **263** (1955) 961;
   W. G. Pfann, *Zone Melting*, 2nd ed. (Wiley, New York, 1966).
3) D. G. Gasson, J. Sci. Instr. **42** (1965) 114.
4) J. J. Brissot and C. Belin, J. Crystal Growth **8** (1971) 213.
5) J. A. Burton, R. C. Prim and W. P. Slichter, J. Chem. Phys. **21** (1953) 1987.
6) E. M. Levin, C. R. Robbins and H. F. McMurdie, *Phase Diagrams for Ceramists* (The Am. Ceramic Soc., 1964).

Journal of Crystal Growth **13/14** (1972) 601–603 © *North-Holland Publishing Co.*

AMÉLIORATION DE LA MÉTHODE DES FLUX, SYNTHÈSE ET PURIFICATION D'UN MINÉRAL NATUREL: LA FORSTÉRITE

L. VU TIEN, A. GRANDIN DE L'EPREVIER, V. GABIS et A. M. ANTHONY

Centre de Recherches sur la Physique des Hautes Températures (C.N.R.S.), Laboratoire de Géochimie et Minéralogie, Faculté des Sciences d'Orléans, 45–Orléans–02, France

It is well known that single crystals of iron-free olivine have numerous applications in geological studies as well as in ESR and infra-red spectroscopy. Previously, clear and stoichiometric single crystals of centimeter size have been grown only by Bridgman or flame-fusion technique with all the inconveniences of the methods.

We report growth using fluxes such as PbO–V_2O_5, MoO_3–Li_2O, WO_3–Li_2O. The solubility has been increased with fluxes containing MoO_3, Li_2O, V_2O_5 and larger crystals have grown. Centimeter-size single crystals of forsterite with uniform composition were obtained after slow-cooling. The morphology of these compounds was the same as native olivines.

Impurity concentration was measured by microprobe and optical emission spectrometry. Molybdenum was found only in inclusions. Some crystals doped with Gd^{3+} and Fe^{3+} were also studied by ESR techniques to locate the doping ions in the lattice.

Inclusions were completely eliminated by heating crystals in a hydrogen atmosphere.

1. Introduction

L'indigence des minéraux naturels ayant une composition homogène est notoire. En particulier, les études minéralogiques sont très mal pourvues en monocristaux de la famille des olivines $(Mg, Fe)_2SiO_4$. Les échantillons naturels contiennent une proportion toujours variable de Fe. Pour cette raison, nous avons entrepris la synthèse du pôle magnésien de la série, la forstérite Mg_2SiO_4.

La forstérite a déjà fait l'objet de synthèse par la méthode de Verneuil[1]). Les cristaux obtenus sont fissurés et non stoechiométriques. Par la méthode des flux, les essais ont conduit jusqu'alors à plusieurs phases – diopside, akermanite, wollastonite et forstérite – en petits cristaux[2]).

Bien que le diagramme de phase montre la possibilité de deux silicates de Mg, nous avons pu obtenir la seule phase forstérite en monocristaux utilisables en minéralogie, et en physique du solide (RPE, Raman, infrarouge).

2. Synthèse

Le solvant le plus adapté a été sélectionné en comparant la solubilité de la forstérite dans différents flux. Trois flux comparés dans les mêmes conditions ont donné les résultats indiqués dans le tableau 1.

Nous avons retenu MoO_3–Li_2O qui présente l'avantage d'être soluble dans l'eau. Ces 3 fondants cependant ont donné des cristaux de forstérite.

Le remplacement d'une partie de MoO_3 par V_2O_5 augmente légèrement la solubilité et réduit la nucléation, très intense avec MoO_3–Li_2O pur.

TABLEAU 1

Fondants	Solubilité de la forstérite	
	à 1200 °C	à 1400 °C
PbO–V_2O_5	1.8%	3.7%
MoO_3–Li_2O	5.5%	6.5%
WoO_3–Li_2O	3.3%	6.0%

Un excès de magnésie est nécessaire pour éviter la formation d'une phase vitreuse silicatée. Un rapport $Mg/Si = 2.25$ paraît parfaitement adapté.

Les poudres, fondues dans un creuset de platine de 40 cm^3, sont portées à 1500 °C pendant 20 heures, puis refroidies lentement à la vitesse de 5 °C par heure environ, jusqu'à 1100 °C.

Les cristaux sont toujours localisés au fond du creuset. Le rendement est compris entre 50 et 70% des produits de départ.

Les cristaux obtenus sont transparents. Certains possèdent des inclusions visibles à l'oeil, mais d'autres semblent être exempts d'inclusions.

Fig. 1.   Monocristal de forstérite. 3×.

On a pu ainsi synthétiser des cristaux dépassant 1 cm dans leur plus grande dimension (fig. 1).

Les expériences avec une fuite thermique contrôlée pour obtenir un seul germe de forstérite ont donné des résultats très encourageants. On espère ainsi améliorer la taille du monocristal obtenu.

3. Analyse

Les diagrammes Debye–Scherrer et de cristal tournant montrent un excellent accord entre les paramètres calculés pour nos cristaux et ceux de la fiche ASTM (No. 4-0770):

	ASTM	Nos cristaux
a (Å)	4.76	4.762
b (Å)	10.20	10.192
c (Å)	5.99	5. 99

Leur morphologie est la même que celle des cristaux automorphes d'olivine naturelle avec les faces (001), (010), (110), (101) et (021). Cependant, un excès de vanadium favorise la croissance des faces (110).

Les indices de réfraction ont été mesurés par la méthode du liseré de Becke. On a obtenu les valeurs suivantes:

ASTM	Nos cristaux
N_g 1.670	1.669 ± 0.001
N_m 1.651	1.651 ± 0.001
N_p 1.635	1.631 ± 0.001

L'analyse chimique effectuée à la microsonde de Castaing a montré une répartition homogène de Mg et Si. La figure 2 montre une cartographie du cristal sur une section de 200 μm de côté. Le molybdène et le vanadium ne sont décelés que dans les inclusions. La microsonde ne détecte pas d'autre impureté.

On a pu facilement doper les cristaux avec Gd, Fe,

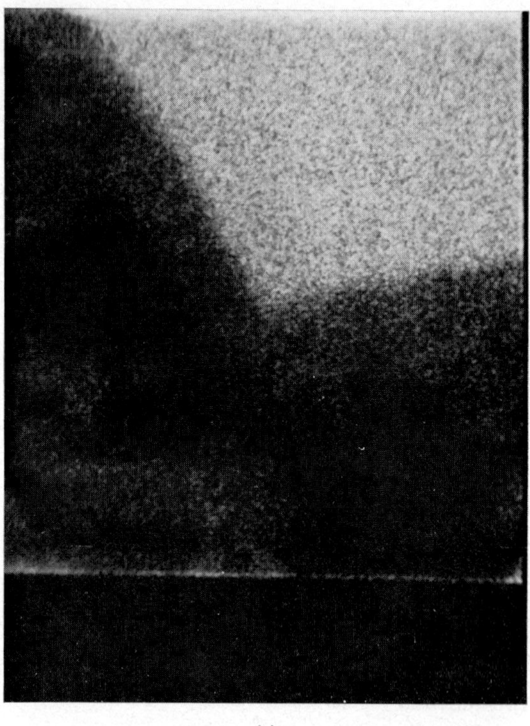

(a)

(b)

Fig. 2.   (a) Répartition homogène du Si à l'intérieur du cristal. (b) Présence de Mo sur une crête du cristal.

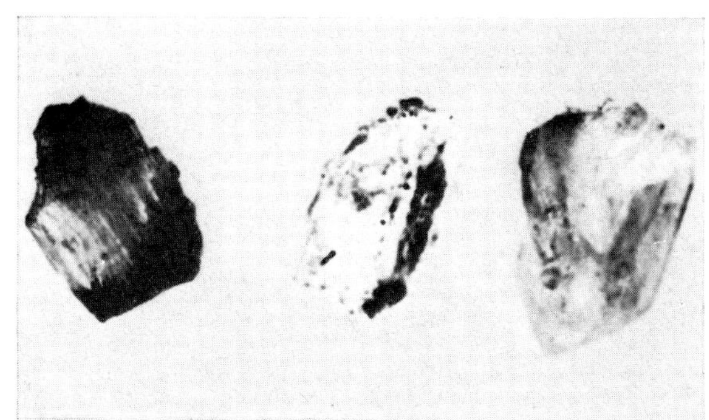

Fig. 3.   Purification: cristal initial (à gauche); cristal purifié et apparition de Mo et V à la surface; cristal purifié et lavé (à droite).

Ni. Ces ions étrangers ne gênent en rien la croissance et se localisent dans le réseau sur les sites Mg^{2+}.

4. Purification

Une étude thermodynamique nous a montré la possibilité de purifier les cristaux en traitant ceux-ci entre 700 et 1000 °C sous hydrogène. Comme le montrent les figures 3 et 4, on observe expérimentalement une migration des impuretés vers l'extérieur de l'échantillon.

Cette méthode est applicable aux impuretés de $MoO_3 \cdot V_2O_5$ dont les oxydes sont moins stables que les constituants du cristal.

Des dosages par spectrométrie optique d'émission ont donné des résultats suivants:
– cristal non purifié 3700 ppm de Mo,
– cristal purifié 840 ppm de Mo.
70 % du molybdène contenu dans le cristal a donc été éliminé.

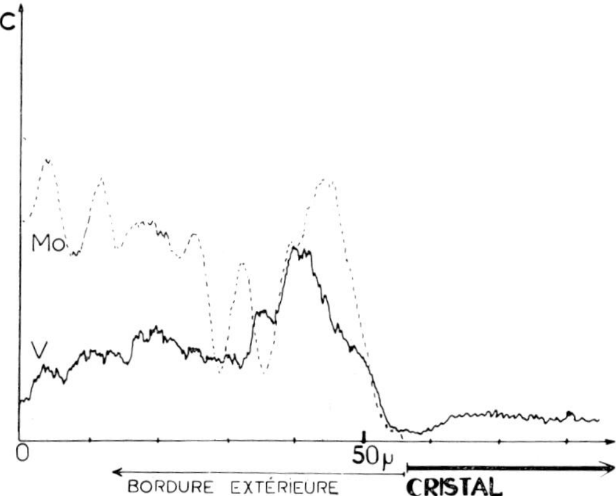

Fig. 4.   Analyse à la microsonde: présence de Mo et V à la surface du cristal.

5. Conclusion

Nous avons pu obtenir en flux des monocristaux de forstérite qui ont pu être analysés et utilisés dans trois laboratoires de recherche.

La méthode des sels de molybdate de lithium est bien adaptée pour préparer l'autre pôle de la série des olivines, la fayalite, ainsi que d'autres silicates recherchés par les Sciences de la Terre.

Cette synthèse contribue à enrichir et compléter la collection des cristaux naturels disponibles.

Bibliographie

1) Shankland et Hemmenway, Am. Mineralogist **48** (1963) 208.
2) W. H. Grodkiewicz et L. G. Van Uitert, J. Am. Ceram. Soc. **46** (1963) 356.

SOLUTION TOP-SEEDING: GROWTH OF GeO$_2$ POLYMORPHS

JOHN W. GOODRUM

Air Force Cambridge Research Laboratories (AFSC),
L. G. Hanscom Field, Bedford, Massachusetts 01730, U.S.A.

Single crystals of both tetragonal (rutile) GeO$_2$ and hexagonal (quartz) GeO$_2$ have been grown by top-seeding of appropriate molten salt solutions. Tetragonal GeO$_2$ is grown from a Na$_2$O solution (92.3 molar% GeO$_2$), and hexagonal crystals are prepared using a Li$_2$O · 2WO$_3$ solution (26.2 molar% GeO$_2$). The top-seeded method permits one to control solution growth so that only one crystal grows and the major crystal growth direction can be selected. Also, the growth process can be observed to some extent. Details of the apparatus and procedures for growing each of the GeO$_2$ structures are given. Variations of thermal geometry in the solution, stirring, and seed orientation are shown to markedly influence the top-seeded growth process. The crystals are characterized by X-ray, optical absorption, and physical property studies. Spectrographic analysis shows that impurity levels in the GeO$_2$ polymorphs are low.

1. Introduction

Historically, top-seeding may be considered a modification of Kyropoulos' method. In this method, slow cooling of the melt produces growth on a seed suspended on the top surface. In recent practice, solutions have been seeded in several ways. The seed may be immersed in the liquid so that growth can occur in any crystallographic direction. This method has been used to grow good quality YIG crystals, Laudise and Linares[1]), also, the seed may be suspended at the liquid surface by wires or rigidly attached to a seed rod and introduced to the liquid surface, Miller[2]). The latter method has been used in this work. "Top-Seeding" in this discussion refers to surface seeding of slow-cooled solutions, rather than pure melts. The top-seeded method permits one to control solution growth much more effectively than does spontaneous nucleation: growth can be limited to only one crystal and the major crystal growth direction can be selected. The growth process can be observed to some extent; thus, the immediate effect of different cooling rates, different thermal geometries in the solution, etc., can be seen.

The hexagonal, high temperature (quartz) phase of GeO$_2$ melts at 1115 °C, while the tetragonal (rutile), low temperature phase is thermodynamically stable below 1050 °C, as reported by Sarver and Hummel[3]). The phase study by Murthy et al.[4]) indicates that sodium oxide (Na$_2$O) can serve as solvent for GeO$_2$ and lower the liquidus temperature below the 1050 °C transition point. Viscosity measurements by Riebling[5]) also

indicate that the alkali oxides decrease the extremely high viscosity of a GeO$_2$ melt, an effect which would facilitate crystal growth. The initial mixture used to crystallize GeO$_2$ was chosen so that the mixture would have a composition of 0.3 wt% above the eutectic. This mixture was 92.30 molar% GeO$_2$ combined with Na$_2$O. The point is chosen to be near the eutectic in order to provide a low growth temperature. At this composition, a crystal of about 5 g could be grown from a 30 g mixture without crystallizing the eutectic composition. This solution also has the lowest viscosity since the viscosity decreases with increasing alkali oxide content, Riebling[5]). Hexagonal GeO$_2$ crystals were grown from the solvent, Li$_2$O · 2WO$_3$, described by Finch and Clark[6]). This solvent system yielded only hexagonal spontaneously nucleated crystals when slow-cooled over the temperature range 1050 °C to 950 °C in unseeded experiments by Finch and Clark[6]). For top-seeded growth, solutions containing 7 g GeO$_2$, 6 g Li$_2$O, and 88 g WO$_3$ were prepared. This composition corresponds to approximately 26.2 molar% GeO$_2$ in LiO$_2$ · 2WO$_3$.

Preliminary results of top-seeded growth of tetragonal GeO$_2$ have already been reported by Goodrum[7]). We will therefore emphasize the application of the top-seeding method to grow hexagonal GeO$_2$, and give here the details of the growth techniques.

2. Experimental

The materials used were electronic grade GeO$_2$ and reagent grades of alkali oxide and tungsten trioxide.

These are dry mixed for 15 hr, and melted at 1200 °C for 1 hr to promote additional mixing. Platinum crucibles are used for the growth experiments. RF heating within a glass-closed growth area is used in order to avoid irritating vapors which come from Na_2O–GeO_2 solutions. The equipment and configuration for top-seeded growth is shown in fig. 1. To eliminate the inhomogeneous heating, etc., caused by direct r.f. coupling to the platinum crucible containing the solution, an external susceptor is used in experiments. A thick-walled platinum cylinder (fig. 1) plus discs at the bottom and top form the "heat source". These elements were isolated from the solution crucible by alumina so that r.f.-induced thermal anomalies in the susceptor are not transmitted to the solution. In addition, this arrangement also reduces transient thermal variation in the solution caused by irregular r.f., generator output. If a small resistance (Kanthal wire wound) furnace is used for top-seeded growth, the various thermal gradients – surface radial and vertical solution gradients – are much less sharp. Thus to obtain good quality large crystals: (1) the exact temperature at which top-seeding is initiated is more critical with a resistance furnace and, (2) the rate of cooling should be slower than the rate used with r.f.-heated solutions. Gradients in a resistance-type furnace may be so small that top-seeding is not successful, and nucleation may occur at the surface–crucible interface, etc., as well as a simultaneous growth of 2 or 3 crystals from the seed surface. Therefore a r.f.-heated apparatus was generally used.

A similar procedure is followed for top-seeded growth of both the hexagonal and tetragonal forms of GeO_2. Initially, it is necessary to grow c-axis needles (tetragonal) and small hexagonal crystals for seeds. Seeds are grown by cooling a solution in the presence of a platinum wire which is attached to the seed rod. The seeds, of about 1.5 to 3 mm in diameter, are mounted in the end of the seed rod and dipped in the surface of the growth solutions. To determine the temperature where the solution is at equilibrium with the solid phase GeO_2 (i.e., the seed) proved to be a critical step in the top-seeded technique. A solution with slight excess of supersaturation, while not immediately precipitating out GeO_2 onto the seed, will nonetheless produce an excessive growth rate causing very poor quality crystals. The best technique is to start cooling from as high a temperature as the seed will withstand without dis-

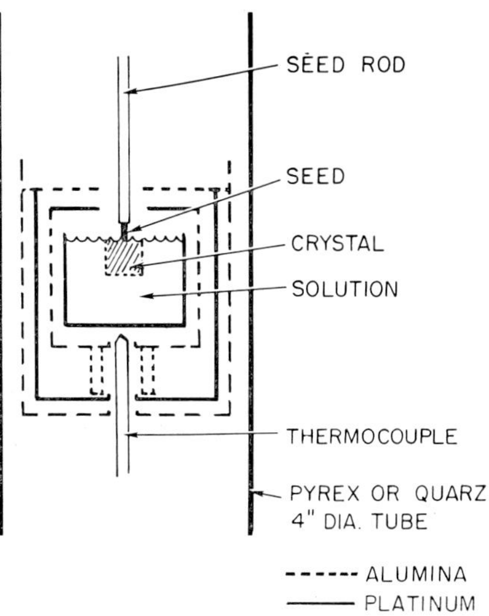

Fig. 1.   Furnace schematic: top-seed growth.

solving. This is empirically found by an incremental temperature increase process. The solution temperature is programmed down at a rate of 0.5–1 °C/hr, and the seed rod is rotated to improve mixing of the solution. All crystallization occurs on the seed as the temperature is lowered between approximately 1050 °C and 950 °C. As the solution is about to freeze at the eutectic, the tetragonal crystal is pulled out of the liquid. Hexagonal crystals are pulled clear of the solution at about 950 °C, mainly to separate the crystal from solvent residue. For both polymorphs, the crystals are cooled at about 200 °C/hr to room temperature.

Variations of thermal geometry in the solution, stirring, and seed orientation markedly influence the top-seeded growth of hexagonal and tetragonal GeO_2. It was found that stirring and the control of the thermal gradient were more important for the growth of tetragonal GeO_2 than for hexagonal GeO_2. Secondary nucleation during growth can be stopped by introducing radial thermal gradients which tend to localize only on the seed. Thermal gradients in the solution were optimized to encourage downward crystal growth, rather than the surface growth obtained in initial top-seeding experiments. Initial top-seeded crystals were grown using poorly insulated crucibles which were heated directly by r.f. where the solutions had sharp vertical and very flat radial thermal gradients. This configuration produced crystals which grew downward on the seed for a short period and then either by extension of the seed, or more often by random nucleation on the crystal surface, growth would tend to be only on the surface of the liquid. Crystals grown under these conditions were of very poor quality. To improve the thermal conditions, a disc of alumina ceramic with a center hole approximately 10 mm diameter was used to reduce thermal losses from the solution surfaces. Later a platinum disc was found more effective, since it coupled in the r.f. field and became a heat source, rather than a thermal insulator only.

As might be expected, if a crystal is to grow from the liquid surface into a solution, the liquid at the surface must be cooler than liquid at the bottom of the crucible. A vertical gradient of about 40 °C/cm was found appropriate; too small a gradient made extre-

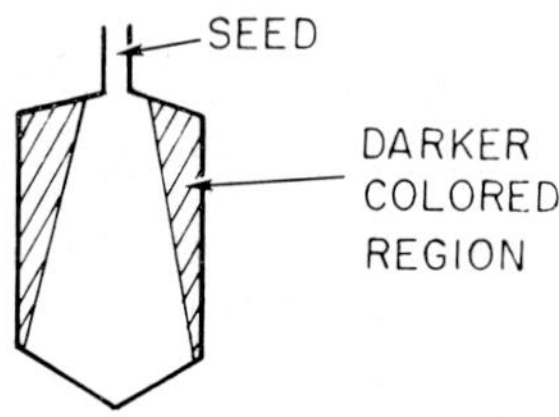

Fig. 2.   Cross-section of Na_2O-grown crystal (schematic). Crystal section through seed and parallel to primary growth direction.

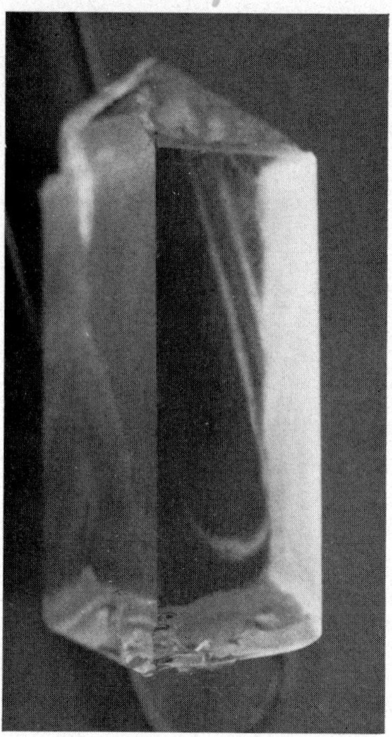

Fig. 3.   Sodium oxide grown tetragonal (GeO_2). Length ~ 1 cm.

mely slow cooling rates necessary while too large a gradient caused the formation of needles. Indeed the usual needle-like morphology of tetragonal GeO_2 could be altered by a suitable thermal gradient in the solution during growth. With an increase in the vertical temperature gradient, the length/diameter ratio of tetragonal polymorph crystals could be varied from 5:1 to 1:1. While with smaller vertical gradients, hexagonal and tetragonal crystals of GeO_2 were found to grow down into the solution more readily.

Various approaches were used to promote circulation, or stirring effects, in the solutions during top-seeded growth. Experience with solutions for growing GeO_2 polymorphs indicates that a combination of stirring by crystal rotation and by convection is most effective. Top-seeded growth without both seed rotation and convection induced by thermal gradients was found to yield only badly twinned crystals. As the rotating crystal grows down into the solution, it becomes a more effective stirrer. The greater bulk of the crystal also causes the diffusion path for the solute to be shortened.

Top-seeded growth of tetragonal GeO_2 crystals from

alkali oxide systems is more strongly influenced by the crystallographic orientation of the seed than is hexagonal GeO_2. It has been found best to orient the hexagonal and tetragonal seeds so that the preferred crystal growth direction is directed into the regions of greatest supersaturation, i.e., the fastest growth orientation should be pointed downward into the solution rather than in some horizontal direction.

Hexagonal GeO_2 crystals were grown using (001) seeds. Tetragonal GeO_2 grows best along the c-axis as needles. If the seed is oriented so that its c-axis is perpendicular to the solution surface, growth is primarily downward, with slower growth on the sides. If the thermal gradients in the solution have the geometry earlier discussed, and if the cooling rate is not excessive, untwinned unflawed growth is observed on the seed.

During growth of tetragonal GeO_2, there is a strong tendency toward secondary nucleation. If the seed is not oriented so that the c-axis is perpendicular to the surface of the solution, one obtains a clump of crystals, the largest of which usually follows seed orientation. As might be expected, good quality growth in c ends when a c-axis bar grows close to a crucible wall. It would appear that at that time, growth on the other

Fig. 4. Hexagonal GeO$_2$ crystal (mm scale).

crystal surfaces being relatively slow, large supersaturations develop in the solution, and extensive nucleation of c-axis needles occurs on the crystal surface.

3. Crystal properties

Crystals of tetragonal GeO$_2$ grown by top-seeding have a rectangular solid morphology with the cross-section perpendicular to the c-axis approximately square. The bottom of the crystal has a pyramidal morphology. (110) facets and (111) facets form the exterior of the crystals. The crystals appear to be slightly yellow. The yellow color of these crystals is predominantly outside a conical boundary whose apex is at the seed (fig. 2). Optical studies of crystals suggest the color is due to trapped electrons in lattice defects. The density of the crystals is 6.28 g/cm^3, refractive indices are 1.960 and 2.047, hardness is 1990 (Moh microhardness). Major impurities in the crystals are Al (80 ppm), Fe (40 ppm), Cr (30 ppm), Na (20 ppm).

Hexagonal GeO$_2$ crystals, colorless and of optical quality, up to 5 mm diameter have also been grown by top-seeding. The crystals were examined by the X-ray powder diffraction and Laue back-reflection techniques. 50 ppm (W) and 20 ppm Na are present as impurities in the crystals.

References

1) R. A. Laudise, R. C. Linares and E. F. Dearborn, J. Appl. Phys. **33** (1962) 1362.
2) C. E. Miller, J. Appl. Phys. **29** (1958) 233.
3) J. F. Sarver and F. A. Hummel, J. Am. Ceram. Soc. **43** (1960) 336.
4) M. K. Murthy and J. Aguayo, J. Am. Ceram. **47** (1964) 444.
5) J. F. Riebling, J. Chem. Phys. **39** (1963) 1889.
6) C. B. Finch and G. W. Clark, Am. Mineralogist **53** (1968) 1394.
7) J. W. Goodrum, J. Crystal Growth **7** (1970) 254.

Section XIII

Melt growth – basic studies and semiconductors

J. R. CARRUTHERS   B. W. STRAUGHAN
M. GRASSO   P. J. TUFTON
T. ARIZUMI   E. W. WILLIAMS
N. KOBAYASHI   A. R. VON NEIDA
K. J. GÄRTNER   L. J. OSTER
K. F. RITTINGHAUS   J. W. NIELSEN
A. SEEGER   A. R. PEAKER
W. UELHOFF   P. D. SUDLOW
D. F. O'KANE   A. MOTTRAM
T. W. KWAP   J. STEININGER
L. GULITZ   A. J. STRAUSS
A. L. BEDNOWITZ   E. ANDRÉ
J. B. MULLIN   J. M. LE DUC
W. R. MACEWAN   M. MAHIEU
C. H. HOLLIDAY   R. UEDA
A. E. V. WEBB   O. OHTSUKI
B. C. GRABMAIER   K. SHINOHARA
J. G. GRABMAIER   Y. UEDA
A. ROYLE

Journal of Crystal Growth **13/14** (1972) 611–614 © *North-Holland Publishing Co.*

THE STABILITIES OF FLOATING LIQUID ZONES IN SIMULATED ZERO GRAVITY

J. R. CARRUTHERS and M. GRASSO

Bell Telephone Laboratories, Incorporated, Murray Hill, New Jersey 07974, U.S.A.

The stabilities of floating liquid zones under conditions commonly used for crystal growth were studied in a simulated zero-gravity environment. Factors studied in detail were meniscus constraints at the solid–liquid interfaces, unequal end diameters and rotation. The behavior of nonrotated zones was shown to be predictable from Rayleigh's theory[1] of the stability of cylindrical liquid zones while the behavior of the rotated cylindrical zones agreed with the theory of Hocking[2] and Gillis[3]. Implications for the growth of real crystals in space are discussed.

1. Introduction

The availability of zero-gravity environment in the planned Manned Orbiting Laboratory of NASA has caused much speculation on the advantages of such an environment to crystal growth[4]. It appeared to us that the two principal advantages of zero gravity would be the absence of natural convection and the availability of much larger crucible-free systems than can be obtained by floating zone or levitation processes on earth. It should be remembered that it is possible to eliminate natural convection on earth by use of the Bridgman-Stockbarger geometry as well as magnetic fields in the case of conductive melts. Thus the chief advantage of zero gravity is that the gravitational constraint on the geometry of floating zones is removed. This constraint has been analyzed by Heywang[5] to be

$$l_{max} = 2.84\,(\sigma/\rho g)^{\frac{1}{2}},$$

when

$$R > 2.00\,(\sigma/\rho g)^{\frac{1}{2}},$$

where l_{max} is the maximum stable length of a cylindrical zone of radius R and σ, ρ are the liquid surface tension and density respectively. Consequently, we decided to study the stability of floating zones under conditions of interest in crystal growth and in a simulated zero-gravity environment. It became apparent to us that such zones were quite easy to simulate on the surface of the earth.

There are two simulation systems which one may use for such studies. Soap films, which are effectively weightless, have been used by Pfann and Hagelbarger[6] in a similar study. The suspension of one liquid inside another with which it is immiscible but precisely equal in density has been used originally in a remarkable series of experiments by Plateau[7] in 1859. The second method was chosen here because of its greater time of stability (soap films evaporate) as well as because we also wished to eventually study the forced convection flow patterns inside these zones.

Plateau determined that the maximum stable length of a uniform cylinder of liquid constrained by non-wetting solid surfaces was equal to its circumference under zero-gravity conditions ($l_{max} = 2\pi R$). This value was derived theoretically by Rayleigh[1] and more recently by Pfann and Hagelbarger[6]. Precise experimental verification has been provided recently by Mason[8]. None of these experiments considered the influence of wetting and contact angle at the end solid supports.

The stability of rotating liquid zones in the absence of gravity has not been studied experimentally. However, two separate theoretical analyses have been performed by Hocking[2] and Gillis[3] which show that rotation of a cylindrical zone reduces the maximum stable length according to

$$\frac{l_{max}}{R} = \frac{2\pi n}{[1+\rho R^3 \Omega^2/\sigma]^{\frac{1}{2}}}, \tag{1}$$

where ρ is the density, Ω the rotation rate and n is an integer defining the mode of instability; $n = l/\lambda$ where λ is the wavelength of the instability.

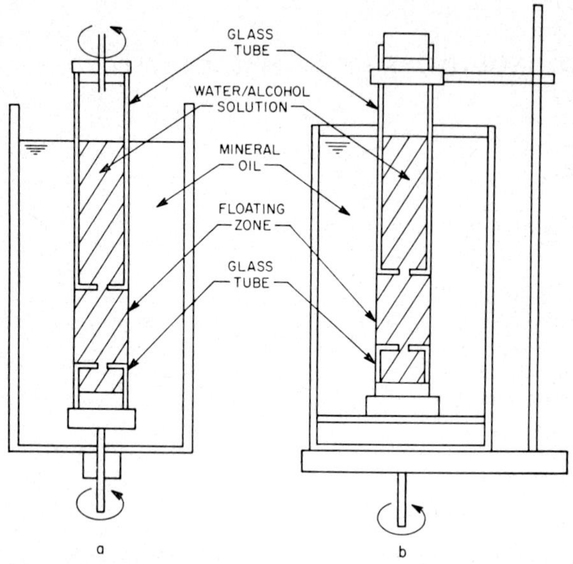

Fig. 1. Schematic drawings of experimental apparatus for simulating floating liquid zones in a zero gravity environment: (a) rotating glass supporting tubes, (b) rotation of whole assembly.

In this paper, we investigate the influences of the wetting conditions of the zone at the solid (end) supports, noncylindrical shapes and rotation on the stability of floating liquid zones in the absence of gravitational forces.

2. Experimental

The experimental apparatus is shown schematically in fig. 1. In fig. 1a only the glass supporting tubes were rotated but the outer nujol bath was stationary, whereas in fig. 1b the whole apparatus could be rotated. Both open-ended glass tubes as well as the closed-end tubes shown were used. The outer container was transparent to permit observation of the zone. We chose to use nujol, a high molecular weight mineral oil, and a water/ethanol solution as the two immiscible liquids because easy density balance could be achieved by adjusting the solution composition and because the materials were inexpensive.

The filling procedure is quite important. First, the outer container is filled with nujol. Then the lower pyrex tube is filled with alcohol/water solution of slightly greater density than the nujol (70 vol% ethanol) and carefully lowered into position. The upper tube is placed into the nujol and the tube position adjusted until the ends meet. Then the upper tube is carefully filled through the upper opening with an alcohol/water solution of slightly lower density than the nujol (75 vol% ethanol). The nujol inside the upper tube is displaced downward into the outer region through the opening where the two tubes meet. When the two alcohol/water regions meet, they are made continuous with a metal wire (to perforate the interface and stir the solution to make it uniform). Final density balance was achieved by adding appropriate amounts of water or ethanol and stirring. The density balance became more critical as the zone length increased. In all experiments described here, density balancing was performed at a zone length $l \approx 5R$ before other experiments were performed. An equal density of 0.8750 g/cm^3 was found at 72.0 vol% ethanol for the nujol used. It was found that the exact maximum zone length, $l_{max} = 2\pi R$ for cylindrical zones, could not be achieved because of nonuniformity of the alcohol/water solution, slight temperature gradients in the chamber and room temperature changes during the experiment. However, values of $l = 0.85\, l_{max}$ could be obtained easily at density differences of $\Delta\rho \approx 10^{-3}$ g/cm^3. The density balancing is discussed in more detail in work to be published elsewhere[9]). An appropriate dye was added to the zone to improve the visibility. The outer portions of the glass tubes are coated with stopcock grease which gives a wetted surface for the nujol and nonwetted surface for the alcohol/water. This was necessary to prevent capillary "run away" of the zone along the outside of the tubes.

We have also measured the liquid–liquid interfacial surface tension of the above combination by means of the Wilhelmy slide method[10]). In this method, it was necessary to use carefully cleaned glass slides and suspend them on a rigid steel wire from the balance arm. This method works only when the liquid–liquid interface is constrained to be nearly flat. Consequently, we measured the surface tensions of combinations on both sides of the equal density value and interpolated to find $\sigma = 8.0 \pm 0.5$ dyne/cm. The variation with composition was 0.156 dyne/cm vol% ethanol.

2.1. WETTING OF SOLID SUPPORTS BY ZONE LIQUID

The most important concept of this model is the constraint placed on the shape of the zone by the volume of liquid moved into the zone when the length is changed. Referring to fig. 1a, when the upper tube is moved up to increase the zone length, liquid is transferred from the inside of the tube to the zone. Since the zone diameter is equal to the tube outside diameter rather than the inside diameter, a slight volume decrease occurs which must be compensated by adding liquid to the zone. When the lower tube is moved down, the zone volume is decreased by the volume swept through by the tube since there is no free surface and excess liquid to compensate for the change. Hence all of this liquid must be externally added to the liquid to retain the same zone shape. When these volume constraints were taken into account, we observed no change in the shape of the zone as the zone length was changed when either open- or closed-end tubes were used. In addition, there was no detectable difference in the measured static maximum zone length, $l_{\max} = 2\pi R$ within the measurement errors.

The influence on zone stability of unequal diameters at each end of the zone was studied. The volume constraint on the shape of the zone by moving the upper tube is obvious for this case. If the upper tube is smaller in diameter, the zone volume decreases and conversely if the upper tube is larger in diameter. The particular case studied was the stability of a conical surface between the two tubes and liquid was added to or removed from the zone to achieve this condition as the zone length was increased. It was found empirically that the maximum zone length for a conical surface was obtained for the condition:

$$l_{\max} = 2\pi[R_1^2 - (R_1 - R_s)^2]^{\frac{1}{2}}, \qquad (2)$$

where R_1 and R_s are the larger and smaller tube radii, respectively. This result would probably be expected from an extension of the theoretical treatment for the equal diameter case [see Rayleigh[1]) or Pfann and Hagelbarger[2])].

2.2. ROTATION OF FLOATING ZONES

The influence of rotation on the stability of liquid zones was of particular interest since molten floating zones must be rotated to provide the radial temperature uniformity required for crystal growth. However, it is likely that forced convection will also be required in a zero-gravity environment to modify the steep thermal gradients arising from conductive heat transfer. Such steep gradients will be deleterious because of the associated adverse solid–liquid interface shapes.

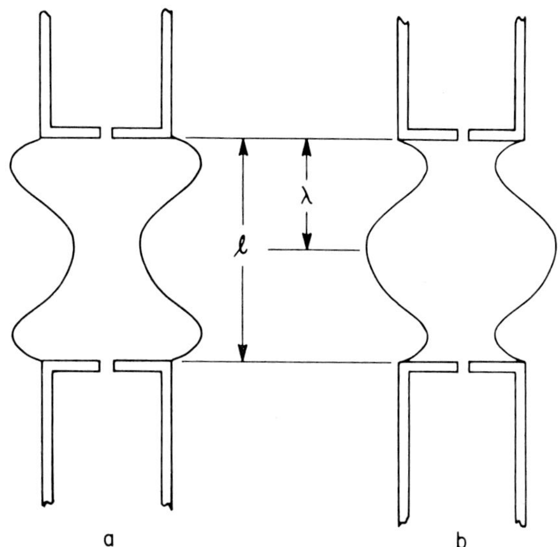

Fig. 2. Instability modes of initially cylindrical floating liquid zones for cases of differential rotation between the inner and outer liquids: (a) glass support tubes rotating, (b) outer nujol bath rotating.

It was found that the zone stability in our model depended quite sensitively on the rotating boundary conditions. When the outer nujol bath and glass tubes rotated at different velocities, the unstable shapes shown in fig. 2 developed from an initially cylindrical zone. Fig. 2a is the case for rotating tubes only and fig. 2b for rotating the outer nujol bath only. The unstable wavelength is basically $\lambda = \frac{1}{2}l$ for both cases. These two shapes are easily understood from centrifugal force considerations. When the tubes and container rotate simultaneously at the same velocity, the zone remains cylindrical and only becomes unstable at a value of l determined by eq. (1) with $n = 1$ (i.e., $\lambda = l$). It is quite important not to allow the upper surface to be free in these experiments or the radial pressure increase will not occur across the surface of the zone.

The rotational stability results for a typical zone radius $R = 1.43$ cm are shown in fig. 3 where the maximum (normalized) zone length is given for various rotation rates. It can be seen that there is a basic difference between open- and closed-end tubes and also between the rotating tubes and rotating nujol. The solid curves are given by eq. (1). The closed-end tubes show a consistent $n = 2$ behavior over the length and rotation ranges investigated while the open-end tubes show a transition between $n = 1$ and $n = 2$ behavior

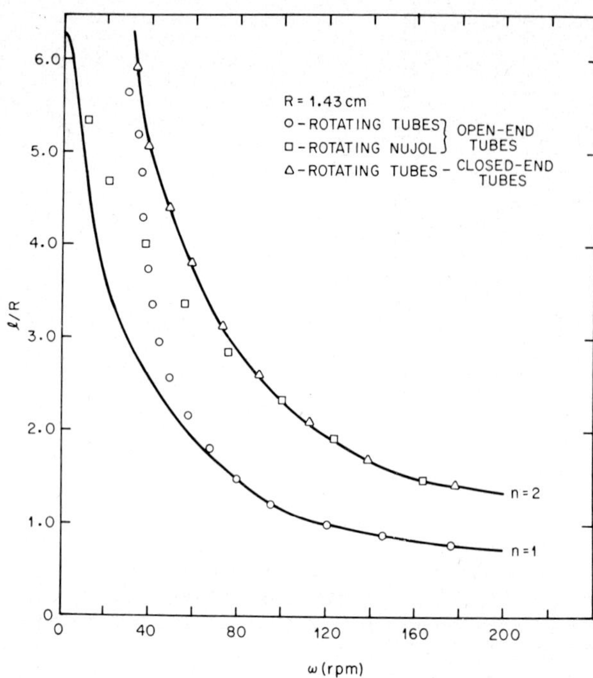

Fig. 3. Maximum stable zone length for a cylindrical liquid zone in the presence of rotation for the indicated boundary conditions. Solid curves are from the theoretical work of Hocking[2]) and Gillis[3]).

3. Discussion

We can gain some idea of the stability of real liquids used in crystal growth by substituting the following values into eq. (1): $R = 1.43$ cm, $\rho = 4.0$ g/cm^3, $\sigma = 500$ dyne/cm and $l/R = 3.3$ to find $\Omega = 9.5$ sec^{-1} or $\omega = 95$ rpm whereas, for our model, $\omega = 30$ rpm corresponded to the limit of stability for $n = 1$ at the same radius and zone length. It is not yet known to what extent forced convection will have to be used to provide heat transfer in the absence of normal thermal convection. Our results indicate that there will be a definite limit to the cylindrical zone length and radius because of the forced convection requirements. It should be possible to use conical zone shapes to relax the zone length requirements (assuming the growing crystal to be the smaller diameter). Forced convection and levitation from electromagnetic sources will overcome many of these difficulties for metallic melts.

Further studies to be reported elsewhere[9]) have been performed on the convective flow patterns in the floating zone caused by rotation of the end members. Other factors, which may be important in a zero-gravity environment, are the influences of surface tension variations and random accelerations on the stability of liquid floating zones.

which depends on the method of rotation. The $n = 1$ behavior occurs at large zone lengths for the outer nujol bath rotation because of the influence of the centrifugal radial pressure gradient in causing an overall diameter increase in the zone. The $n = 1$ behavior occurs at small zone lengths for the case of supporting tube rotation because the end effects of the tubes extend over the entire zone surface and again cause an axially uniform radial pressure gradient.

The rotational stability of the conical shapes resulting from different end tube diameters was studied in great detail and full results will be presented elsewhere[9]). It was found that different rotation velocities were required for each of the two end tubes to produce stability at any given zone length. In fact, the critical rotation rate for each tube corresponded to the value obtained for the same zone length in the equal diameter rotating cylindrical zone case.

Acknowledgment

We wish to thank R. A. Laudise for many stimulating discussions of this work.

References

1) Lord Rayleigh, *Scientific Papers*, Vol. 1 (Cambridge Univ. Press, London, 1899) pp. 377–401.
2) L. M. Hocking, Mathematika **7** (1960) 1.
3) J. Gillis, Proc. Cambridge Phil. Soc. **57** (1967) 152.
4) Space Processing and Manufacturing, National Aeronautics and Space Administration publication ME-69-1, October 12, 1969.
5) W. Heywang, Z. Naturforsch. **11a** (1956) 238.
6) W. G. Pfann and D. W. Hagelbarger, J. Appl. Phys. **27** (1956) 12.
7) J. A. F. Plateau, translated in Smithsonian Institution Annual Report of Board of Regents, pp. 250*ff* (1863).
8) G. Mason, J. Colloid Interface Sci. **32** (1970) 172.
9) J. R. Carruthers and M. Grasso, submitted to J. Appl. Phys.
10) W. D. Harkins and T. F. Anderson, J. Am. Chem. Soc. **59** (1937) 2189.

Journal of Crystal Growth **13/14** (1972) 615–618 © *North-Holland Publishing Co.*

THEORETICAL STUDIES OF THE TEMPERATURE DISTRIBUTION IN A CRYSTAL BEING GROWN BY THE CZOCHRALSKI METHOD

TETSUYA ARIZUMI

Faculty of Engineering, Nagoya University, Nagoya, Japan

and

NOBUYUKI KOBAYASHI

Faculty of Engineering, Toyama University, Takaoka, Toyama, Japan

The problem of heat transfer in a crystal being grown from a melt was solved computationally by considering the exchange of thermal radiation from the various surfaces at different temperatures and the gaseous convection. When a crystal is grown in a vacuum, the exchange of thermal radiation greatly influences the temperature distribution in the growing crystal; when it is grown in a gaseous atmosphere, the gaseous convection dominates.

1. Introduction

A knowledge of the heat transfer problem in a crystal being grown from a melt is important in order to predict the conditions for obtaining better quality crystals. Some attempts have been made to solve the heat transfer problem in the crystal growth process theoretically[1-8]. In the present paper, a previously postulated crystal growth model[6-8] is refined by considering additionally the gaseous convection and the interchanges of thermal radiation from the various surfaces at different temperatures. The resulting temperature distribution in the growing crystal is thus predicted theoretically.

2. Method of simulation

Previous work[6] has shown that the temperature distribution and hence the solid–liquid interface shape is considerably influenced by the heat radiated from the surface of the crystal and the melt. This means that the heat transfer processes should be treated more rigorously in order to simulate the practical conditions of crystal growth.

If a crystal growing system involves N surfaces with different emissivities at different temperatures, the mutual radiation interchange is adequately treated by introducing the concept of the geometrical view factor[9]. The net heat flux radiated from the first surface is expressed by

$$q_1^{net} = \varepsilon_1 \sigma \sum_{i=1}^{N} \varepsilon_i f_{1i}(T_1^4 - T_i^4),\tag{1}$$

when considering only the direct radiation interchange, where σ is the Stefan–Boltzmann constant, T_i and ε_i the temperature and the emissivity of the ith surface respectively and f_{1i} is the geometrical view factor viewed from a point on the first surface to the ith surface. If the temperature of the ith surface is not uniform, T_i in eq. (1) should be replaced by the equivalent average temperature $\overline{T}_i$ for $i \neq 1$.

The geometry of a crystal growth system is shown in fig. 1 and for simplicity the thickness of the crucible wall is assumed to be negligible. The net heat flux radiated from the side surface of the crystal is given by

$$q_s^{net} = \varepsilon_s \sigma [\varepsilon_l f_{sl}(T_s^4 - \overline{T}_l^4) + \varepsilon_c f_{sc}(T_s^4 - T_c^4) \\ + f_{sa}(T_s^4 - T_a^4)].\tag{2}$$

The first term of the right-hand side is the radiation interchange between the crystal surface and the melt free surface, the second term the interchange between the crystal and crucible inner surfaces and the last one is the heat radiation into the atmospheric environment.

The net heat flux radiated from the top surface of the crystal is given by

$$q_t^{net} = \varepsilon_s \sigma (T_s^4 - T_a^4).\tag{3}$$

Similarly the net heat flux radiated from the melt free surface is given by

$$q_l^{net} = \varepsilon_l\sigma[\varepsilon_s f_{ls}(T_l^4 - \overline{T}_s^4) + \varepsilon_c f_{lc}(T_l^4 - T_c^4)$$
$$+ f_{la}(T_l^4 - T_a^4)]. \tag{4}$$

The suffixes s, l, c and a in the above formulae stand for the crystal, the melt, the crucible and the atmosphere, respectively (cf. Appendix).

The heat flux transfered by the gaseous convection[10] is assumed to be expressed by

$$q_c = \alpha(T - T_a)^{\frac{5}{4}}, \tag{5}$$

where α is a parameter.

The heat transfer problem in the crystal growth is described as follows, using the notations and quantities tabulated in Appendix. In the crystal and the melt, the temperature is assumed to be governed by the Laplace equation,

$$\frac{\partial^2 T}{\partial r^2} + \frac{1}{r}\frac{\partial T}{\partial r} + \frac{\partial^2 T}{\partial z^2} = 0, \tag{6}$$

subject to the following boundary conditions:
On the crystal axis, by symmetry, we have

$$\frac{\partial T}{\partial r} = 0, \tag{7a}$$

on the crystal side surface,

$$-\lambda_s\left(\frac{\partial T}{\partial r}\right)_s = q_s^{net} + q_{con}, \tag{7b}$$

on the top surface of the crystal,

$$-\lambda_s\left(\frac{\partial T}{\partial z}\right)_s = q_t^{net} + q_{con}, \tag{7c}$$

on the melt free surface,

$$-\lambda_l\left(\frac{\partial T}{\partial z}\right)_l = q_l^{net} + q_{con}, \tag{7d}$$

on the crucible surface,

$$T = T_c, \tag{7e}$$

and on the solid–liquid interface we have

$$T = T_m \tag{7f}$$

and

$$-\lambda_s\left(\frac{\partial T}{\partial n}\right)_s = -\lambda_l\left(\frac{\partial T}{\partial n}\right)_l + LF_n. \tag{7g}$$

By solving the Laplace equation computationally,

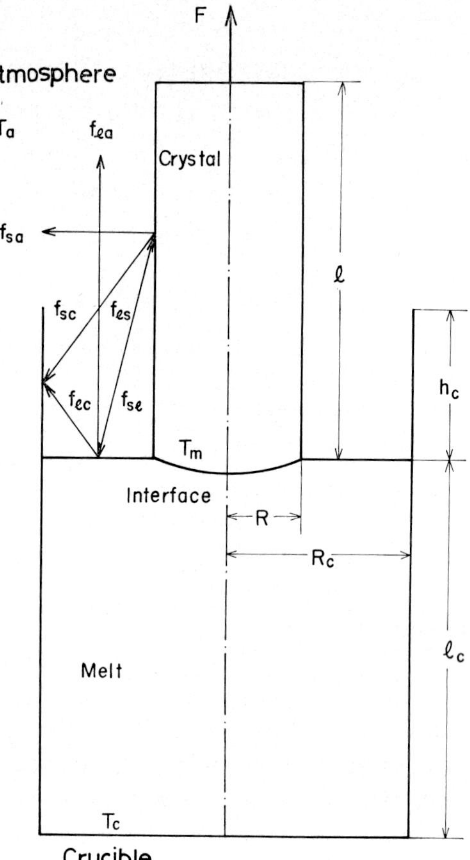

Fig. 1.　Crystal growth model.

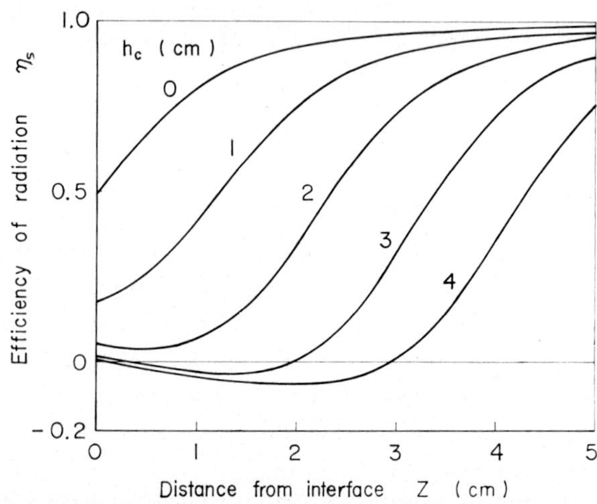

Fig. 2.　Efficiency of radiation on the crystal side surface.

the temperature distribution in the crystal and the solid–liquid interface shape is studied, concentrating mainly on the effects of the crucible inner wall height

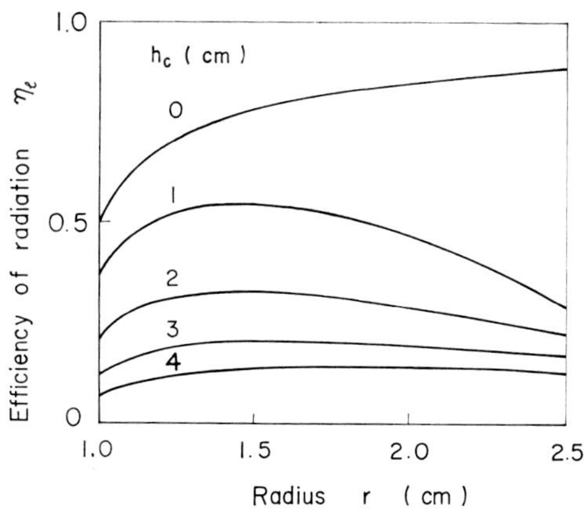

Fig. 3. Efficiency of radiation on the free melt surface.

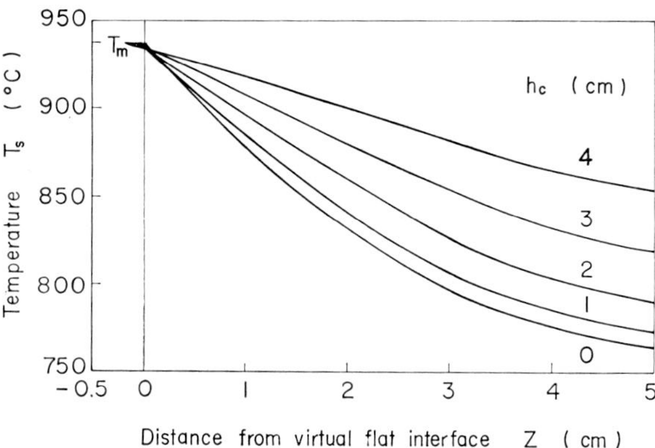

Fig. 4. Temperature distribution at the axis in a crystal.

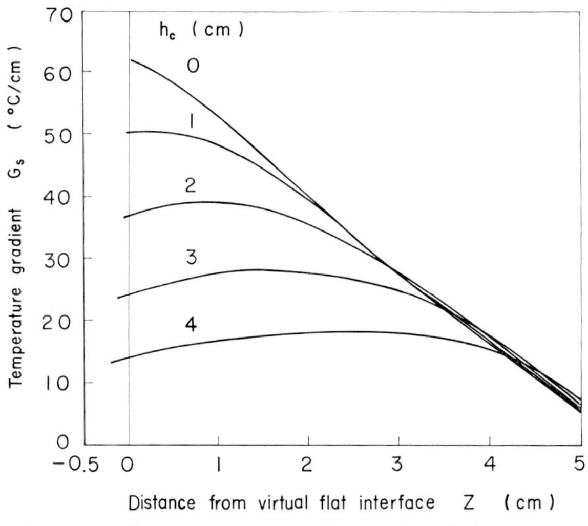

Fig. 5. Axial temperature gradient at the axis in a crystal.

above the melt free surface (h_c in fig. 1). The emissivity of the crucible is assumed unity and the atmospheric temperature is assumed to be 300 °K.

3. Results

3.1. IN A VACUUM

The heat radiated from the crystal, the melt and the crucible are strongly influenced by h_c. The radiation efficiency from a surface is generally defined by

$$\eta = q^{\text{net}}/\varepsilon\sigma T^4. \tag{8}$$

In fig. 2 is shown η_s, the radiation efficiency of the crystal side surface which strongly depends on the value of h_c; it becomes rapidly smaller as h_c increases. If h_c is larger than about 3 cm, η_s becomes negative at a distance of about 2 cm from the solid–liquid interface (curve 3). If the distance from the interface, Z in fig. 2 is smaller than 2 cm, the heat is no longer dissipated but absorbed from the crucible. On the other hand η_l decreases monotonically as h_c increases, as shown in fig. 3. The temperature distribution in a crystal is shown in fig. 4 with the parameter h_c. The axial temperature gradient clearly reflects the change of the thermal conditions in the crystal. In a lower part of the crystal, the axial temperature gradient decreases gradually to approach a value which is nearly constant with increasing h_c. The solid–liquid interface shape changes from concave to convex towards the melt as h_c increases.

3.2. IN A GASEOUS ATMOSPHERE

The heat dissipation by gaseous convection has just as the radiation a large influence on the temperature distribution in a growing crystal. The temperature distribution is solved by assuming $\alpha = 4 \times 10^{-4}$ W/cm^2 °K$^{5/4}$, a value in air for a crystal of a few centimeters in length. The temperature distribution and the axial temperature gradient of the growing crystal differs considerably from that predicted for the growth in vacuum; and under no conditions does the axial temperature gradient approach a nearly constant value, because the gaseous convection always compensates the heat gain by radiation. The solid–liquid interface shape is always concave towards the melt even though h_c increases (fig. 5).

4. Discussion

In a model previously mentioned, the axial temperature gradient at the interface is determined only by the length of the growing crystal; the solid–liquid interface shape is usually concave towards the melt. As a result of considering the radiation effects of the crucible and the melt surface, such a restriction can be eliminated. Some experimental results indicate that the temperature distribution near the interface is almost linear[11,12]), which is consistent with the present results. Brice[5]) predicted that the axial temperature gradient is maximum at a position of about 1 cm from the interface. Such a situation cannot be realized unless the heat is absorbed near the interface. His experimental measurements, however, satisfy the present model fortunately.

We used the simplified eq. (3) for the gaseous convection in atmosphere. More rigorously, however, the heat transfer problem involving the gaseous convection should be solved fluid-dynamically, as was done in the melt, but such a calculation is extremely complicated and the assumption (3) can approximate the practical case with enough accuracy.

Appendix*

Geometrical view factor	$f_{1i}, f_{sl}, f_{sc}, f_{sa}, f_{ls}, f_{lc}, f_{la}$
Normal component of growth rate	F_n
Crucible wall height	h_c
Crystal length	l (5 cm)
Crucible depth	l_c (5 cm)
Latent heat	L $(2.16 \times 10^3 \text{ J/cm}^3)$
Heat flux by convection	q_{con}
Heat flux by radiation	$q_1^{net}, q_s^{net}, q_l^{net}, q^{net}, q_t^{net}$
Radial coordinate	r
Crystal radius	R (1 cm)
Crucible radius	R_c (2.5 cm)
Temperature	T, T_s, T_l

* Suffix s, l, c, and α mean solid, liqued, crucible, and atmosphere, respectively.

Atmospheric temperature	T_a (300 °K)
Crucible temperature	T_c (943 °C)
Melting point	T_m (937 °C)
Temperature gradient at interface	$(\partial T/\partial n)_s, (\partial T/\partial n)_l$
Axial coordinate	z
Heat transfer coefficient	α
Emissivity	$\varepsilon, \varepsilon_1, \varepsilon_i, \varepsilon_s$ (0.2), ε_l (0.2), ε_c (1.0)
Efficiency of radiation	η, η_s, η_l
Thermal conductivity	λ_s (0.24 W/cm deg) λ_l (0.71 W/cm deg)
Stefan–Boltzmann constant	σ

Acknowledgement

The authors are grateful to Prof. Ichikawa for his helpful suggestions and to the staff of the Data Processing Centre, Kyoto University, for performing the numerical calculations.

References

1) E. Billig, Proc. Roy. Soc. (London) A **229** (1955) 346.
2) E. Billig, Proc. Roy. Soc. (London) A **235** (1955) 37.
3) W. R. Wilcox and F. L. Duty, J. Heat Transfer **88c** (1966) 45.
4) T. B. Reed, in: *Crystal Growth*, Ed. H. S. Peiser (Pergamon, Oxford, 1967) p. 39.
5) J. C. Brice, J. Crystal Growth **2** (1968) 395.
6) T. Arizumi and N. Kobayashi, Japan. J. Appl. Phys. **8** (1969) 1091.
7) N. Kobayashi and T. Arizumi, Japan. J. Appl. Phys. **9** (1970) 361.
8) N. Kobayashi and T. Arizumi, Japan. J. Appl. Phys. **9** (1970) 1255.
9) W. H. McAdams, *Heat Transmission* (McGraw-Hill, New York, 1954) p. 63.
10) W. H. McAdams, *Heat Transmission* (McGraw-Hill, New York, 1954) p. 172.
11) J. C. Brice and P. A. C. Whiffin, Solid-State Electron. **7** (1964) 183.
12) M. G. Mil'vidskii and V. V. Eremeev, Soviet Phys.-Solid State **6** (1965) 1549.

Journal of Crystal Growth **13/14** (1972) 619–623 © *North-Holland Publishing Co.*

AN ELECTRONIC DEVICE INCLUDING A TV-SYSTEM FOR CONTROLLING THE CRYSTAL DIAMETER DURING CZOCHRALSKI GROWTH

K. J. GÄRTNER, K. F. RITTINGHAUS and A. SEEGER

Zentrallabor für Elektronik der Kernforschungsanlage Jülich, 517 Jülich, Germany

and

W. UELHOFF

Institut für Festkörperforschung der Kernforschungsanlage Jülich, 517 Jülich, Germany

Almost in all cases of crystal growth by the Czochralski method it is very useful to control the crystal geometry, the crystal diameter being the reference input for the RF power control system. The crystal diameter is measured using a TV-camera, looking directly or under high vacuum conditions through a special mirror system, at the growing crystal. An electronic device gates the video signal of a preselectable line as close as possible to the phase boundary, and measures the crystal diameter at this position by digital methods. After digital-to-analog conversion the signal proportional to the diameter is fed into the reference input of the RF power controlled system. It is also possible to register the crystal profile automatically within a desired region without perturbing the geometry control. The resolution of this diameter measuring system is about 0.1 per cent.

1. Introduction

The necessity for geometrical control during crystal growth is shown from the following points:
(1) the avoidance of uncontrolled changes in diameter;
(2) the avoidance of uncontrolled thermal stresses;
(3) the avoidance of chemical inhomogeneities in doped melts arising from an uncontrollably changing distribution coefficient.

Control of the geometry through visual observation and manual adjustment of the heating energy is always incomplete. A better possibility for the avoidance of uncontrolled growth is offered by the control of the heating energy over empirically determined performance programmes[1]. One disadvantage in this procedure is uncontrolled changes in temperature, brought about by changing absorption coefficients of the container wall (e.g. by evaporation). This disadvantage can be avoided by means of control of the geometry of the growing crystal with the aid of a programme regulation of the melt temperature[2]. In this process the following disadvantages may occur:
(1) Impossibility of measuring the temperature because the melt temperature is too high, i.e. above approximately 1500 °C[3].
(2) Possibility of contaminating of the melt with the thermocouple material.
(3) Since the crystal diameter, as well as the temper-

ature, is still dependent upon the protective gas pressure and the pulling rate, it is necessary to keep the two previously mentioned parameters constant during the application of a temperature control, in order to obtain a constant diameter.

For these reasons it is useful in many cases to control the crystal geometry with the aid of an electronic diameter detector which is used to control the heating energy. Such a system is described in the following sections.

2. Construction of the system

2.1. Requirements and general construction

The TV-camera produces an intensity profile of the crystal during line scanning as shown in fig. 1. Horizontal synchronizing pulses H with a duration of 46 µs together with the video-signal of the crystal with its duration T can be seen. This duration T is proportional to the diameter and has to be measured with an optimum accuracy. The electronics now must fulfil the following duties:
(1) Selection of a scanning line of the television picture at each desired place, in particular evaluation of the picture content with regard to the crystal diameter at the interface with the highest possible degree of accuracy, for the purpose of regulation at a constant value.
(2) Diameter measurement in successive lines within a

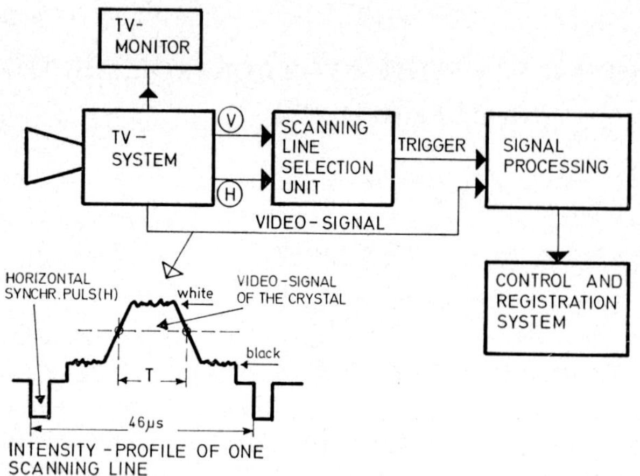

Fig. 1. Intensity profile of a line scan and a general block diagram. The vertical V and horizontal H synchronising pulses generate the trigger for signal-processing.

determined area which may be selected particularly for measuring the meniscus profile.

(3) Recording of all data occurring from the above measurements.

These requirements were realised in an electronic system according to fig. 1.

2.2. Television system

We use a black and white television system having 875 scanning lines and 50 frames per second (interlaced scanning). A highly resolving vidicon serves as camera-tube. The zoom objective is remote controlled with a focal distance of 50 mm to 250 mm, light intensity 1:5 and a horizontal aperture angle of 3° to 15°. According to the experimental conditions an ancillary lens and colour filter are connected in series. The camera looks either directly at the growing crystal or, in high vacuum operation, over a mirror system under an angle of 15° (see below). Visual control is provided by the monitor.

2.3. Line selection

In order to obtain the trigger impulse (fig. 2) for the signal processing, it is necessary to have:

(a) Selection of single lines. At the beginning of each odd-numbered frame of the television signal, there exists in the NAND-gate a coincidence whereby the preselected counter on a determined line is started for single line selection. It counts the line impulses which are at its disposal over the selector switch up to the preselected line and delivers impulses at 40 ms intervals to the NOR-gate.

(b) Profile measurement. When the "start" key is pressed the divider by 12 takes care that the profile counter, which is set up at a determined place in the picture, is only started every 480 ms. Simultaneously a switch-over command arrives at the selector switch which now passes along the line impulses for the dura-

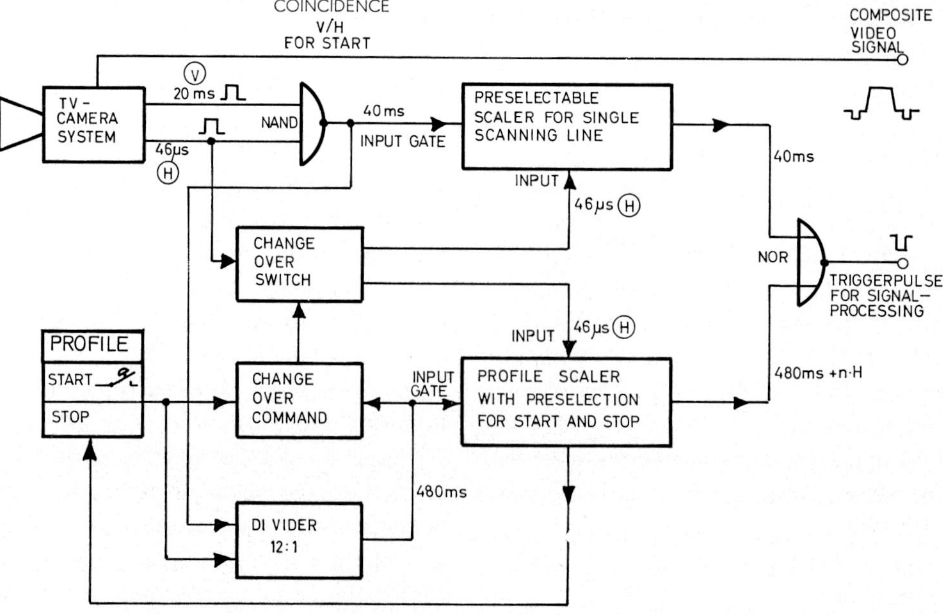

Fig. 2. Block-diagram of the trigger circuit.

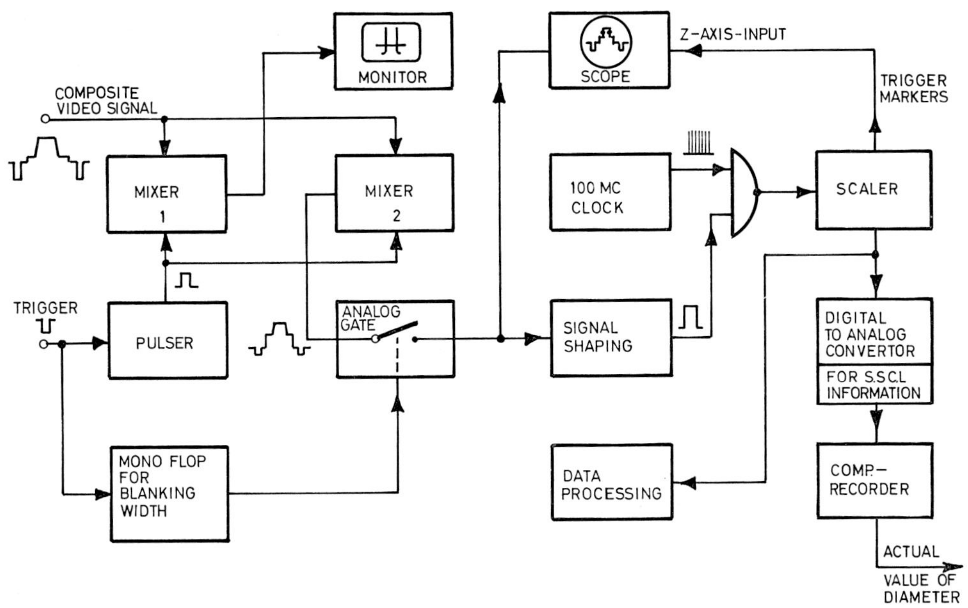

Fig. 3.   Block-diagram of the signal processing.

tion of a maximum 40 ms to the profile counter instead of to the counter for single line selection. The profile counter provides pulses at the NOR-gate at intervals of

$$480\,\text{ms} + n \times \text{H} = 480\,\text{ms} + n \times 46\,\mu\text{s}.$$

Here, n signifies the number of lines following on to each other in the preselected picture area. If the last line in this area is reached, the profile measurement automatically stops. Thus, only every 12th television frame is evaluated for profile measurement: thereby it is assured that the thickness regulation is not disturbed during the process of profile measurement.

2.4. SIGNAL PROCESSING

Fig. 3 shows the signal processing. The signal obtained in the way described triggers a pulser with adjustable delay τ_p, impulse width T_p and amplitude A_p. At first its output impulse is mixed in mixer 1 with the composite-video signal, and led to the TV-monitor. Then the adjusted line and also the picture area selected during profile measurement are distinguished by means of brightening on the screen. Then in mixer 2 the composite-video signal is again mixed with the pulser output and led to the analog-gate. A monoflop is triggered by the trigger impulse and this opens the analog-gate for the duration of the preselected line. At the output

of this gate then, only the videosignal of this one line is present, mixed with the pulser output.

The following signal shaper contains a threshold-amplifier with adjustable barrier and a comparator for the generation of a suitable gate pulse for the 100 MHz counter.

The width of this gate pulse is proportional to the diameter of the crystal at the selected place.

The adjustable delay τ_p, pulse duration T_p and pulse amplitude A_p of the pulser make it possible amongst other things to suppress potential light reflections in the field of view in order thereby to avoid erroneous measurements. This is made clear in fig. 4. Only the signal above the discriminator barrier is evaluated.

The selected video-signal is reproduced on the control scope. Trigger points at the time points t_1, t_2, are marked by unblanking.

The whole electronic data acquisition is connected to the counter and besides this a digital analog-convertor. At the output there is an analog-signal proportional to the diameter for use in analog constant diameter regulation and in recording.

3. Resolution of the system

In consideration of the accuracy of the system we have to discuss:

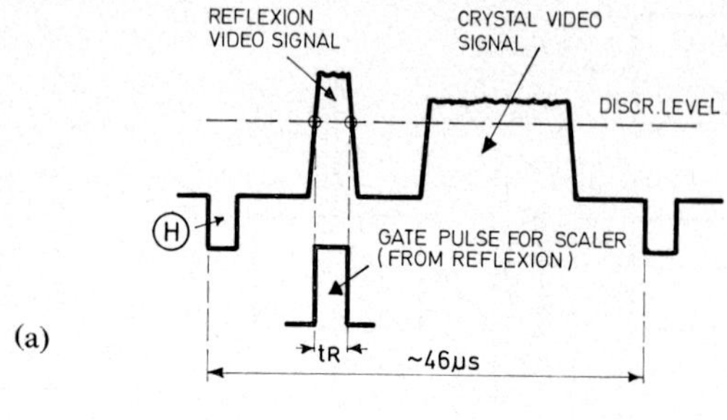

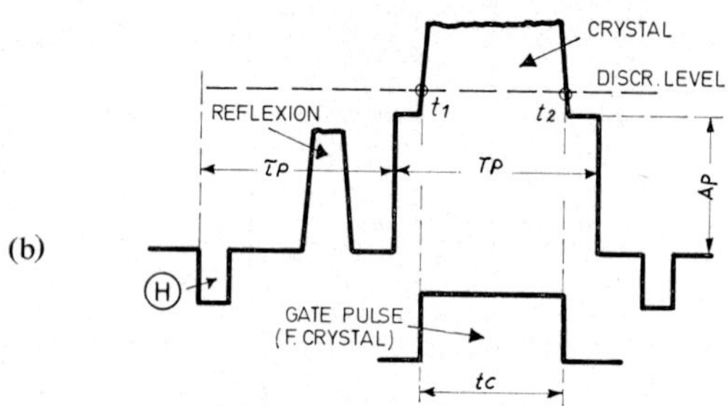

Fig. 4. False measurements by reflections (a) are avoided by a superposition of a suitable pulse (b).

(1) the measurement area at a given magnification
(2) the accuracy of the diameter measurement.

The measurement area is predetermined by the resolution of the television system, i.e. by the band width of vidicon and video amplifier. In our equipment it is 12 MHz. Thereby a black and white border develops with a rise time of $t_R = 250$ ns.

If one takes into consideration the fact that the diameter D of a crystal is scanned in the time t_D, whereby

$$t_D = D \times V/V_H,\qquad(1)$$

where V = linear optical magnification, V_H = horizontal deflection rate (0.273 µm/ns), there results then, on account of the finite rise time of the scanning, the fact that a crystal diameter of less than

$$D_{min} = \frac{V_H}{V} t_{D\,min},\qquad(2)$$

$$D_{min} = \frac{V_H}{V} 2t_R.\qquad(3)$$

The optical magnification is so selected that the maximum diameter D_{max} of a crystal which occurs during a process is produced approximately in a time of $t_{D\,max} = 34$ µs. This means that a diameter ratio of

$$\frac{D_{max}}{D_{min}} = \frac{t_{D\,max}}{t_{D\,min}} = \frac{34000\ \text{ns}}{500\ \text{ns}} = 68,\qquad(4)$$

is detectable without any change in the magnification. With $D_{max} = 10$ mm we obtain approximately 0.15 mm for D_{min}.

The accuracy of the diameter measurement is not dependent upon the band width but only upon the resolving power Δt_D in the time measurement, and this, with the counter used, is ± 10 ns. This means a degree of accuracy of

$$\frac{\pm \Delta t_D}{t_{D\,max}} = \frac{\pm 10\ \text{ns}}{34000\ \text{ns}} \times 100\% \approx \pm\, 0.3^0/_{00},\qquad(5)$$

for D_{max}, and

$$\frac{\pm \Delta t_D}{t_{D\,min}} = \frac{\pm 10\ \text{ns}}{500\ \text{ns}} \times 100\% = \pm 20^0/_{00},\qquad(6)$$

for D_{min}.

4. The contrast conditions in different methods of operation

Ideal operational conditions for the system are present when the crucible is completely filled with the melt, and the camera sight line is horizontal. The background (recipient) is darker than the crystal and this again is darker than the melt in the meniscus (for copper)[4].

In general, however, the crucible is not completely filled with the melt. In order to recognise the phase boundary, then, it is very advantageous to use a crucible with as large a diameter as possible. The crucibles we use in the growth of copper, silver and gold crystals have a diameter of 90 mm, so that with a crucible which is full in the beginning the melt sinks 1.25 cm when a crystal of 10 mm diameter and 100 mm length is pulled.

Nevertheless the measurement during the whole process is possible using an angle of inclination of the sight line of $B = 15$ degrees against the horizontal.

Since an even interface is reproduced on the vidicon screen as a half ellipse with an inclined sight line, the signal in the environment of the interface is more complex than with a horizontal insight. The adjusted picture section now shows the crystal and meniscus and as background the surface of the melt, that exhibits a higher luminous density than the meniscus because of its higher temperature. In this case also the system can be used in the way described by inversion (negative modulation) of the video signal.

In growth processes under a vacuum, a direct observation is only possible for a short time because of

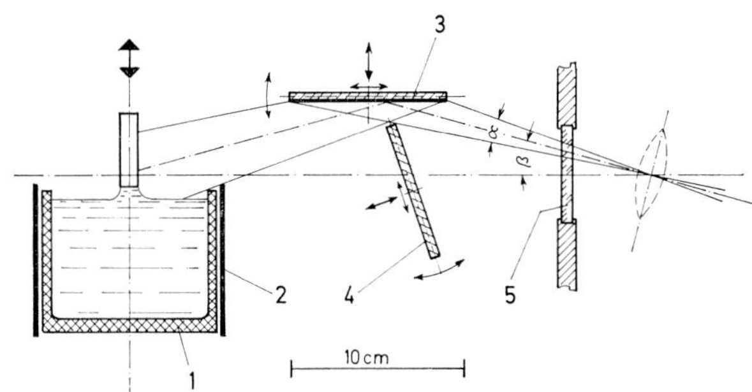

Fig. 5. Mirror system to avoid evaporation on the window in a high vacuum process (1) crucible; (2) induction coil; (3) mirror; (4) shield; (5) window.

evaporation on the window. This problem is avoided by means of a mirror system as in fig. 5, similar to the one used by King[5]). A quartz panel serves as a mirror here; it is already vapour deposited with the melt material before the crystal growth process.

5. Results

In the first experiments with the equipment, the following results were obtained during high-vacuum growth of Cu-crystals.

(1) The mirror system allows a process of diameter measurement lasting up to approximately 4 hr. During this time the video signal, with open objective stop (1:5) at the phase boundary at the beginning of the process has an amplitude of approximately $100 \, \mathrm{mV_{pp}}$ (= peak to peak voltage). After about 4 hr it becomes weakened to $20 \, \mathrm{mV_{pp}}$ through the fall off in reflecting power of the mirror.

(2) During crystal-growth under hydrogen the camera is looking directly through a window into the chamber onto the crystal. In this case we have with a light intensity of 1:8 a video-signal amplitude of $500 \, \mathrm{mV_{pp}}$, which means 50% of the maximum value.

(3) The RF-interference voltage in the video-signal, caused by the 12 kW RF-generator, could be reduced by excellent shielding and filtering to about $5 \, \mathrm{mV_{pp}}$.

(4) The noise of the TV-system is about $2.5 \, \mathrm{mV_{pp}}$. Together with the RF-interference therefore we have about $7.5 \, \mathrm{mV_{pp}}$, which means at the beginning of the process using the mirror system a signal-to-noise ratio of 13:1. After four hours we have about 2.7:1. Without the mirror system there is a signal-to-noise ratio of 67:1.

(5) The experimental accuracy of diameter measurement was approximately $\pm 0.3\%$, which signifies a deterioration from the theoretical value by a factor of 10, caused by the noise level, and a non-optimum optical magnification of only 60% of the maximum value.

(6) Tests made clear, that basically it is possible to regulate the diameter with the present system. It is important to avoid dead time in regulation by measuring the diameter exactly at the interface. In the current version the operator has to select the desired line at the interface by manual means.

Considerable improvement could be achieved by automatic line-scanning-selection at the interface. This should be possible especially for metals and semiconductors since a significant intensity difference occurs above and below the interface.

We thank Mr. P.-G. Manteuffel for discussions and Mr. D. Winkler for the co-operation in the construction of the apparatus.

References

1) K. J. Bachmann, H. J. Kirsch and K. J. Vetter, J. Crystal Growth **7** (1970) 290.
2) E. Kappler, W. Uelhoff, H. Fehmer and F. Abbink, *Herstellung von Kupfereinkristallen* (Westdeutscher Verlag, Opladen, 1971).
3) B. E. Walker, Rev. Sci. Instr. **36** (1965) 601.
4) E. Menzel and J. Stahlknecht, Z. Angew. Physik **26** (1969) 216.
5) G. D. King, J. Phys. E. **3** (1970) 730.

    Journal of Crystal Growth **13/14** (1972) 624–628 © *North-Holland Publishing Co.*

INFRARED TV SYSTEM OF COMPUTER CONTROLLED CZOCHRALSKI CRYSTAL GROWTH

D. F. O'KANE, T. W. KWAP, L. GULITZ and A. L. BEDNOWITZ

IBM Thomas J. Watson Research Center, Yorktown Heights, New York 10598, U.S.A.

An infrared television system of computer controlled Czochralski crystal growth is described. Crystal diameter measurements were made by computer analyses of the TV data, and the power to the crystal puller was controlled by the computer to regulate the diameter. Single crystals of $(KNbO_3)_{10}(LiNbO_3)_{35}(BaNb_2O_6)_{55}$ were grown near 1400 °C. Relative temperatures on the melt surface and growing crystal were obtained from the TV data.

1. Introduction

An infrared television system for computer controlled Czochralski crystal growth is described. The system is suitable for growth processes above 325 °C and was used to grow single crystals which melt near 1400 °C. Previous methods of automatic control of Czochralski crystal growth used a fixed sensor to monitor crystal diameter and computer regulation of the power and other variables[1], or programmed regulation of the power to the crystal puller to alter the crystal diameter[2]. The television system allows the crystal diameter to be measured continuously from the time of the seed dipping to the completion of crystal growth.

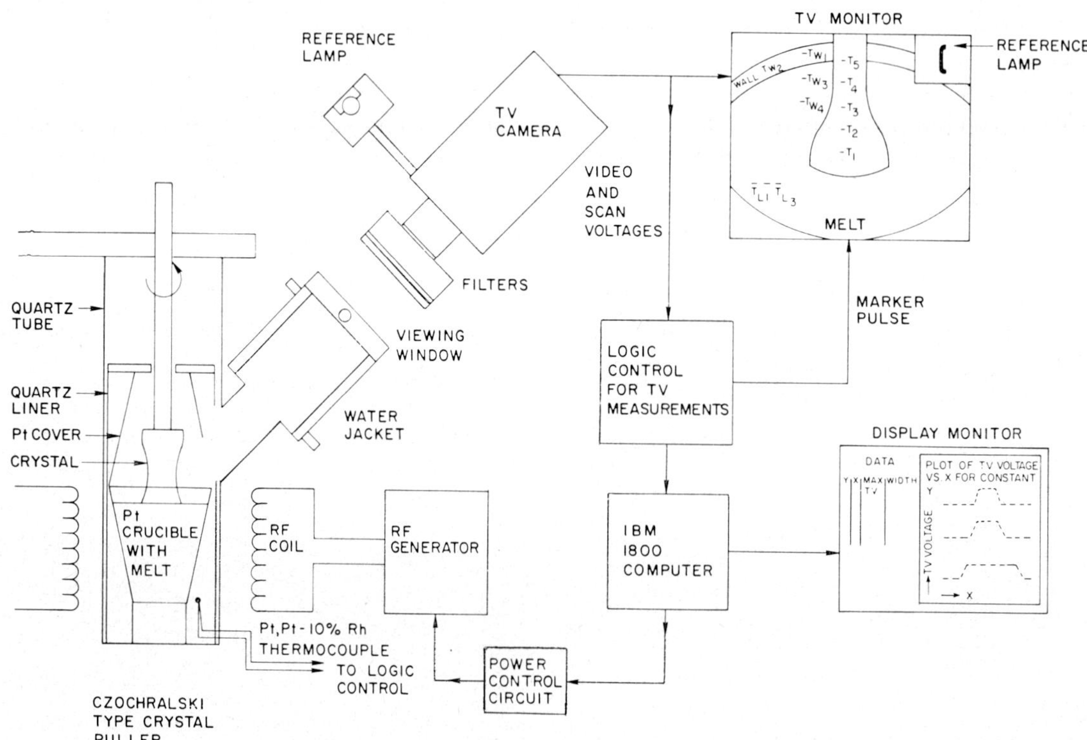

Fig. 1.   Infrared TV system of computer controlled Czochralski crystal growth.

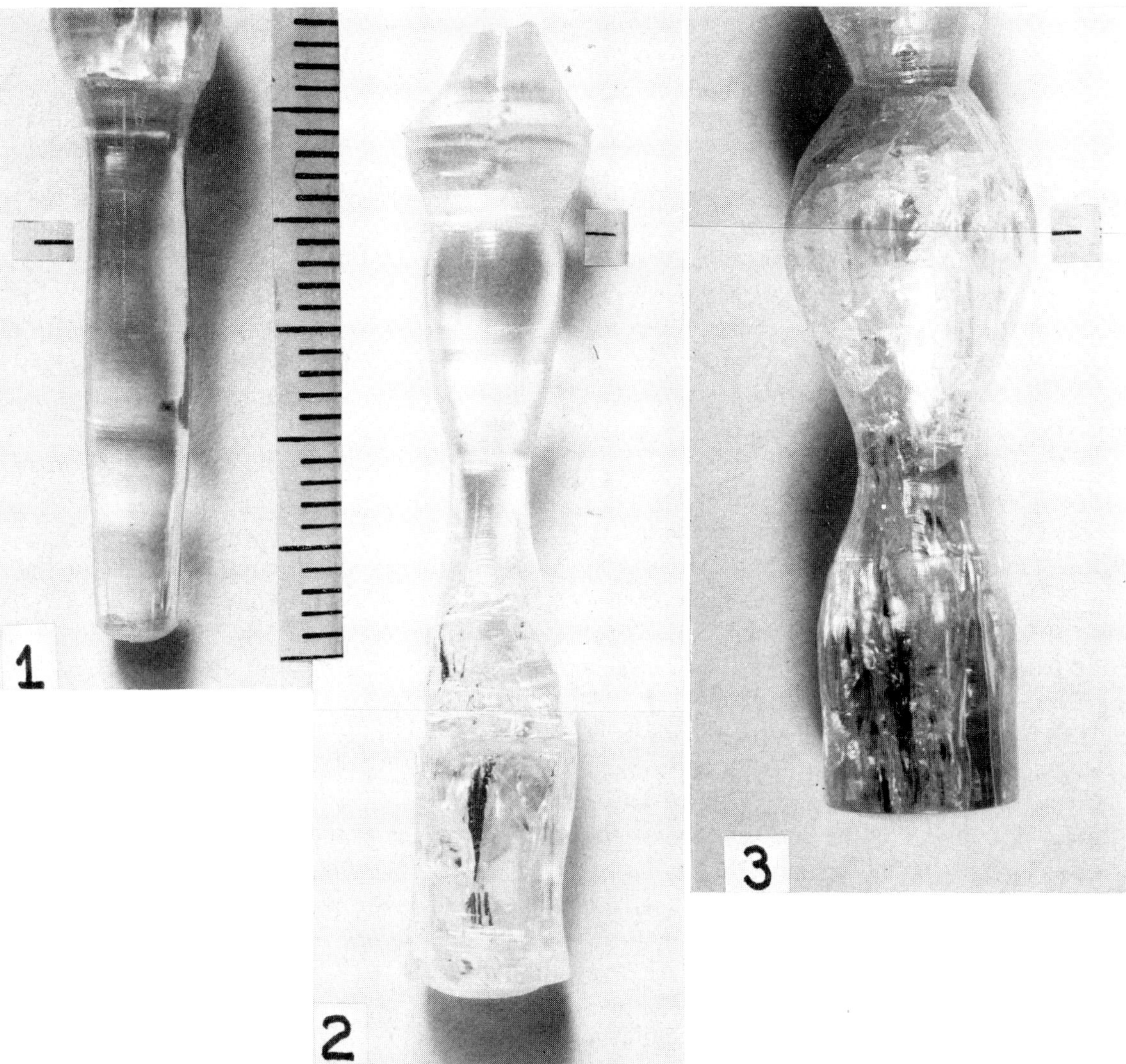

Fig. 2.   Single crystals of $(KNbO_3)_{10}(LiNbO_3)_{35}(BaNb_2O_6)_{55}$ grown under computer control. Black line near the top of each crystal indicates the start of computer control.

A computer analyzes the diameter data from the TV camera and makes power adjustments to regulate the crystal diameter to different sizes at various times in the growth process. Crystal diameter is reduced to remove dislocations[3,4]) and then increased to a larger size for the remainder of the growth. The TV system is capable of measuring relative temperature differences at points in the crystal puller which can be viewed by the TV camera. The TV camera has a temperature sensitivity near 4 mV/°C[5]). This is dependent on the camera lens filters and the emissivity of the surface being examined.

2. Experimental procedures

A crystal puller described previously[6]) for the growth of electrooptic crystals was used to grow single crystals of $(KNbO_3)_{10}(LiNbO_3)_{35}(BaNb_2O_6)_{55}$[7]) which melt near 1400 °C. The RF generator power supply to the crystal puller was modified to permit control with an IBM 1800 computer. An infrared TV camera with a Tivicon type silicon diode array tube[8]) was mounted at the viewing window of the puller. Infrared absorbing and neutral density filters were placed in front of the camera lens to reduce the light level and allow convection currents on the melt surface to be seen when the image was displayed on the TV monitor (fig. 1). The video and scan voltages from the TV camera were converted to digital form before being sent to the computer. This allowed TV voltage to be measured as a function of X–Y position. Data were collected at the rate of 30 points per second. The sensitivity of the TV voltage to temperature changes has been reported for the camera[5]). The radiant energy to the camera can be adjusted with filters to provide a temperature sensitivity of about 4 mV/°C over a 300 °C range, such as 1200 to 1500 °C. This permits relative temperature measurements to be made on the melt surface and the growing crystal.

A manually activated marker pulse on the TV monitor was used to instruct the computer on the points to be measured. In the automatic mode, the pulse showed the points being measured by the computer. A reference lamp was attached to the side of the TV camera and a magnified image of its hot filament was focused on a small portion of the TV camera. This computer controlled lamp provided a stable source of radiant energy which was used to check drifts in the TV system and as a reference source for TV voltage readings at other points in the field of view of the camera.

The crystal diameter was determined by analyzing the video voltage from a scan across the melt surface and growing crystal interface. An abrupt change in TV voltage indicated the position of the interface. The data were analyzed by the computer and scaled to give the actual crystal width. This width was compared with a stored value in the computer.

The computer, which was in a remote location, was programmed to analyze the TV data and provide information on crystal diameter and relative temperatures. These data were sent to a computer controlled storage display scope (Tetronix 611) near the crystal puller and shown in graphical and numerical form. Operator commands relating to the position of the points to be used for crystal diameter measurements, the control limits for crystal diameter, and the size of the increments of the power adjustments to be made by the computer during growth were sent by keyboard to the computer before the start of crystal growth. Multiplexed lines were provided to send information to the computer on thermocouple temperature near the crucible, power setting, rotation rate, angular position of the rotating

crystal, pull rate, TV voltage, and horizontal and vertical positions of the TV voltage reading.

Power to the crystal puller was obtained from a 20 kW output RF generator which was controlled by a magnetic amplifier. The output power of the generator was proportional to the supply voltage (0–20 V) to the magnetic amplifier. A digital-to-analog converter, which was controlled by the computer, supplied input power to the magnetic amplifier.

The crystal width readings were synchronized to the angular position of the rotating crystal by a reed contact which was mounted on the shaft holding the crystal. A shaft encoder will be added to provide a crystal width profile for cross sections other than circular.

3. Results

Three single crystals of $(KNbO_3)_{10}(LiNbO_3)_{35}$-$(BaNb_2O_6)_{55}$ which were grown with the TV system are shown in fig. 2. The black lines indicate the point at which computer control of the crystal growth was started. Crystal 1 was grown by computer regulation of the power to keep the diameter constant at 5 mm. Crystals 2 and 3 were grown by specifying a small diameter to produce the tapered diameter selection and then indicating the final diameter which was to be held constant. Cracking of crystals 2 and 3 occurred during cooling; however, the crystals had the desired taper followed by a constant diameter section. The crystals

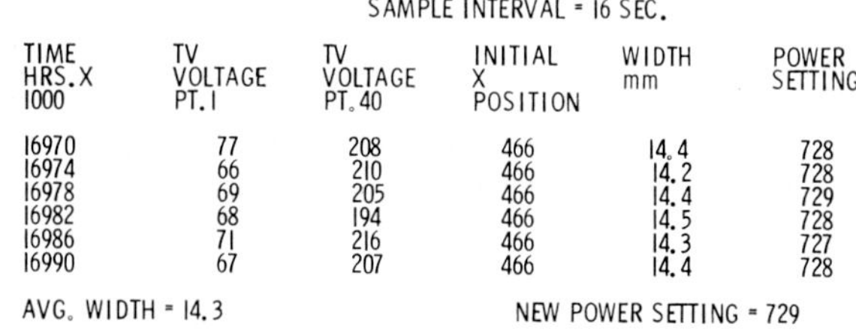

SAMPLE INTERVAL = 16 SEC.

TIME HRS.X 1000	TV VOLTAGE PT.1	TV VOLTAGE PT.40	INITIAL X POSITION	WIDTH mm	POWER SETTING
16970	77	208	466	14.4	728
16974	66	210	466	14.2	728
16978	69	205	466	14.4	729
16982	68	194	466	14.5	728
16986	71	216	466	14.3	727
16990	67	207	466	14.4	728

AVG. WIDTH = 14.3   NEW POWER SETTING = 729

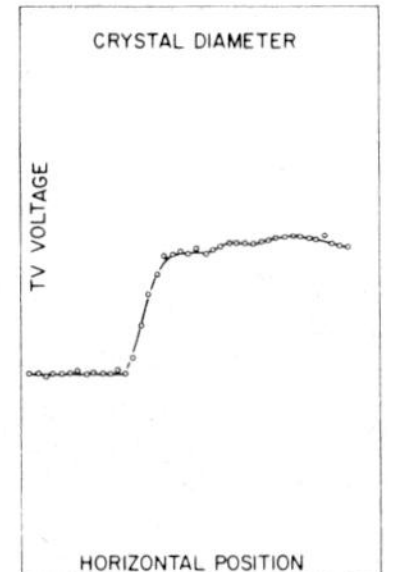

Fig. 3. Data displayed on the storage scope after each crystal diameter measurement.

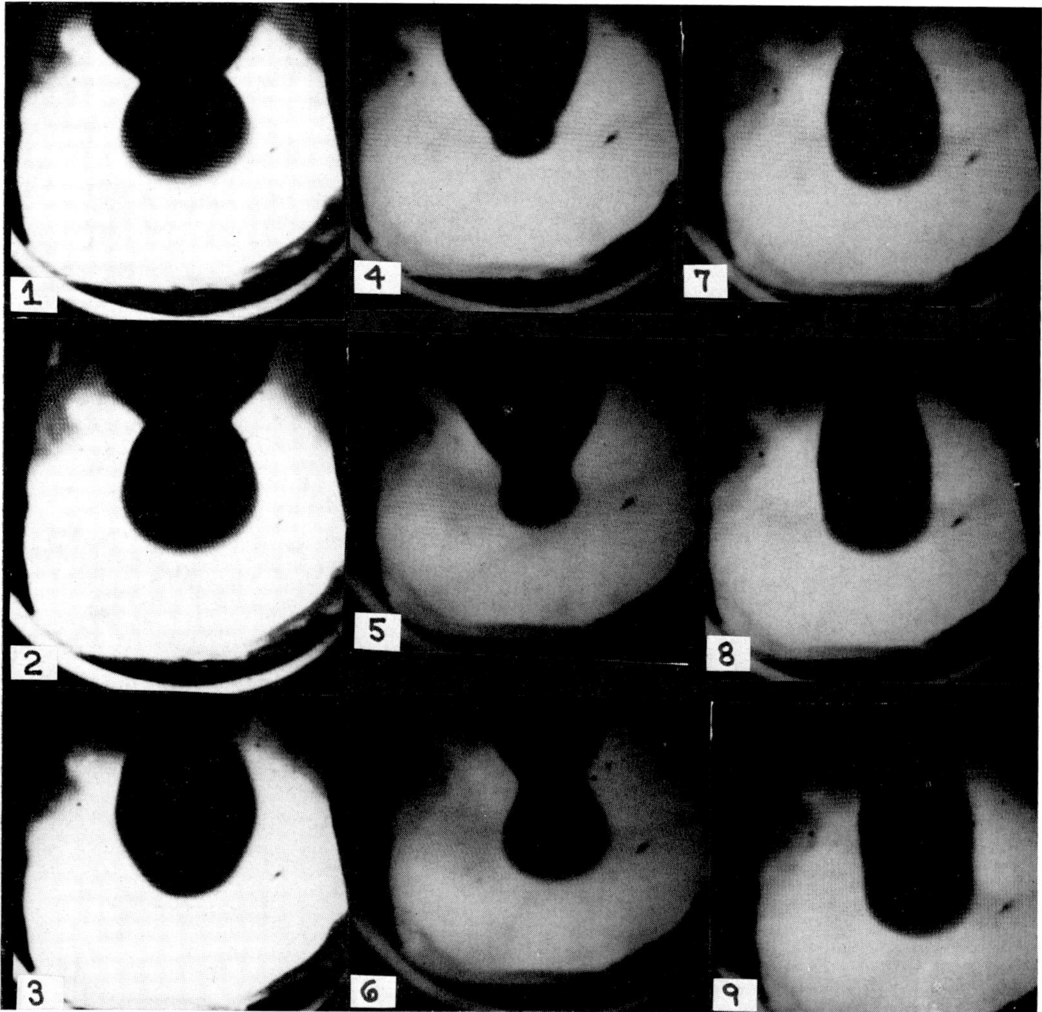

Fig. 4.   Crystal 3 viewed on the TV monitor at different time intervals from the start to the end of computer controlled growth.

were grown at 4 mm/hr and rotated at 80 rpm. The present system is dependent on power changes to control the crystal diameter. A time delay occurs between the change in power setting and the response of the crystal diameter to the power change. This response time will be reduced when the rotation and pull rate are controlled directly by the computer and can be used to help regulate the crystal diameter.

Fig. 3 shows the data displayed on the storage scope after a crystal diameter measurement. The TV voltage is plotted against horizontal position (X) for the scan across the melt surface and the growing crystal interface. The position of the left growth interface is determined from the sharp rise in the TV voltage. This position is compared with the initial crystal diameter to determine the new diameter. Crystal width can be read to 0.1 mm. Data on the time of the measurement, the TV voltage at the first and last point, the initial X position, the crystal width, and the power setting are tabulated on the storage display scope. Each line of data corresponds to one data plot. After 6 readings in 96 sec a statistical analysis is performed on the width data and the average value is compared with the required value. If the width is not within the control limits, the power is adjusted. In addition to the display scope near the crystal puller a permanent record of the process variables during crystal growth is typed at a terminal near the computer.

Another keyboard command can be used to provide a display scope plot of TV voltage versus horizontal position at a large number of vertical positions along the growing crystal. This provides information on the relative temperature gradient along the grown crystal.

Fig. 4 shows crystal 3 (fig. 2) at various times during the crystal growth. After the start of computer control in picture 1, the crystal diameter was reduced from 12 to 5 mm and then increased to 9 mm and held constant. The length of the crystal grown under computer control was 25 mm. Convection currents on the melt surface can be seen in pictures 5 and 6.

4. Summary

A new infrared TV system of computer controlled Czochralski crystal growth was used to grow single crystals at 1400 °C. A silicon diode array TV camera served as a non-contact sensor to measure crystal width and relative temperatures on the melt surface and the growing crystal when the temperature was above 325 °C. Data were taken from the camera at the rate of 30 points per second and analyzed by a computer. Crystal diameter was measured and controlled from the time the seed was dipped into the melt to the completion of crystal growth. Diameter was measured to 0.1 mm with the lens system used on the TV camera. Relative temperatures were measured over a 300 °C range before adjustment of lens filters was necessary.

Acknowledgements

The authors would like to thank R. A. Bochman, Dr. G. H. Schwuttke, and Dr. W. Westdorp of the IBM Components Division, East Fishkill, N.Y. for their financial support which made this research possible. Dr. L. M. Foster and M. A. Koblenz of the Research Division supported and encouraged the program.

R. V. Dobransky provided technical assistance on the interface to the computer and P. C. Yin assisted in the crystal growth.

References

1) E. J. Patzner, R. G. Dessauer and M. R. Poponiak, SCP and Solid State Technology, 25, October (1967).
2) K. J. Bachmann, H. J. Kirsch and K. J. Vetter, J. Crystal Growth, **7** (1970) 290.
3) W. C. Dash, J. Appl. Phys. **30** (1959) 459.
4) W. N. Borle, S. Tata and S. K. Varma, J. Crystal Growth **8** (1971) 223.
5) T. W. Kwap, D. F. O'Kane and L. Gulitz, 5th Symposium on Temperature, June (1971).
6) D. F. O'Kane, G. Burns, B. A. Scott and E. A. Giess, J. Electrochem. Soc. **115** (1968) 1081.
7) D. F. O'Kane, G. Burns, E. A. Giess and A. W. Smith, J. Electrochem. Soc. **117** (1970) 195C.
8) M. H. Crowell and E. F. Labuda, Bell System Tech. J. **48** (1969) 1481.

Journal of Crystal Growth **13/14** (1972) 629–634 © *North-Holland Publishing Co.*

PRESSURE BALANCING: A TECHNIQUE FOR SUPPRESSING DISSOCIATION DURING THE MELT-GROWTH OF COMPOUNDS

J. B. MULLIN, W. R. MACEWAN, C. H. HOLLIDAY and A. E. V. WEBB

Royal Radar Establishment, Malvern, Worcs, U.K.

The salient features of the techniques for the melt-growth of many of the compounds of current technological interest in the fields of semiconductor and electro-optic application such as the III–V's, II–VI's etc. are compared and some of the outstanding growth problems identified. Pressure balancing is suggested as a solution to some of these problems. Conceptually, pressure balancing involves dynamically-balancing the dissociation pressure with an equivalent pressure of inert gas. The essential features of a practical vertical pulling apparatus embodying this concept are analysed. They include a simple non-critical construction, a novel liquid seal which will allow free pull and rotation, a liquid manometer for sensing pressure differentials and a valve which will effect rapid pressure equilisation but prevent diffusive loss of volatile components. An apparatus which can be used both for the formation and growth of compounds is described together with experimental results on the preparation of III–V's.

1. Introduction

Many of the compounds of current technological interest in the fields of semiconductor and electro-optic application such as GaAs, InP, ZnTe, $ZnSiP_2$ etc, are difficult to prepare in bulk form from the melt.

The dominant difficulty is that of controlling the vapour pressure. If it is sufficiently high to cause evaporation then serious problems can occur during crystal growth. These problems range from loss of visibility to melt nonstoichiometry with concomitant effects due to constitutional supercooling. Crystal contamination by impurities from the crucible and growth environment is another intrinsic problem. Of course, these compounds can be prepared by methods other than by melt-growth, but these will not be considered here. The main purpose of the present paper is to analyse the principles of the current melt-growth techniques and show how a novel technique referred to as Pressure Balancing can be used to solve some of these outstanding problems.

2. Comparison of melt-growth techniques

It is clear that control of dissociation is the central problem encountered during the growth of the afore mentioned and like compounds. Without it single crystals can not be prepared from the melt. It is relevant to note that the use of a blanket gas at high pressures does not alleviate the previously-mentioned crystal growth problems. Certainly evaporative-loss can be re-

duced by pressures up to about 10 atm (but not beyond)[1]), but this reduction is not significant in the present context. The only known way of controlling dissociation is to allow the melt to equilibrate in a closed environment which is capable of withstanding the dissociation pressure. This principle is common to all the melt-growth techniques so far developed.

In table 1 a comparison is attempted of the salient features of the various techniques for the melt-growth of dissociable compounds. The comparisons are necessarily subjective; but they are an approximate reflection of the relative merits of the techniques.

The first technique noted in table 1 is exemplified by the Bridgman process. Here the compound is completely sealed generally in a silica tube. Dissociative loss is negligible. It is limited to that necessary to produce the equilibrium pressure in the free volume of the tube. The attractive features of the technique are its simplicity and the ease with which the crystal growth can be fully automated. Variants of this technique have been used for the preparation of gallium phosphide[2]).

In the second technique the melt and its crucible are enclosed in a sealed or semi-sealed chamber, the walls of which are sufficiently hot to prevent condensation of the volatile components. The technique has been developed in a variety of vertical pulling modifications[3,4]) the two most successful variants being the syringe puller (a) and the magnetic puller (b). Both techniques can in principle be used to synthesise compounds prior to crystal growth.

TABLE 1

Comparison of melt-grown techniques

Problem	1 Horizontal/ vertical Bridgman	2a Syringe puller	2b Magnetic puller	3 Liquid encapsulation	4 Pressure balancing
Dissociation pressure control	* * *	* *	* * *	* * *	* * *
Stoichiometry control	* *	* *	* *	†	* * *
Contamination	*	* *	*	*	* *
Physical perfection	*	* * *	* *	* * *	* * *
Synthesis	* * *	* * *	* *		* * *
Simplicity of apparatus	* * *	* *		* * *	* *

* The more asterisks the better.
† Stoichiometry can be maintained at a fixed level.

The Liquid Encapsulation technique[5,6]) is a considerable simplification over the other melt growth techniques. It effectively reduces the growth problems to those of conventional pulling in a pressure chamber. The main disadvantages of the technique are the need to use synthesised material and encapsulants which do not react with the melt and impair the encapsulants' transparency. Unfortunately the higher melting point II–VI compounds such as ZnSe and ZnTe do just this. They have not been grown successfully by liquid encapsulation.

A semi-sealed system based on the use of a liquid gallium seal was suggested by Richards[7]) in 1957 for the growth of gallium arsenide. This apparatus consisted essentially of a closed upper cylinder containing the seed and a lower cylinder which contained the melt. The upper cylinder could be rotated and lifted from inside an annular cup containing liquid gallium which surrounded the lower cylinder. The technique however proved of limited value because of its relative complexity and because the system was prone to seize-up due to the formation of crystallites of GaAs which form when arsenic saturates the gallium seal.

In the course of the present work an apparatus was developed for growing GaAs which used a simple silica bell operating in a seal of liquid B_2O_3.

Although it worked successfully its development was dropped in favour of the more convenient system described in section 3.

The fourth technique, Pressure Balancing, was developed in an attempt to overcome the limitations of the techniques discussed[1-3]) whilst retaining as far as possible all their advantageous features.

3. Principles of pressure balancing

In this technique the dissociation pressure of the compound is dynamically balanced by a pressure of inert gas. This balance can be achieved by using a liquid seal to sense and control means for eliminating the development of a pressure differential across the walls of the growth chamber. The essential features of pressure balancing technology comprise, an apparatus of relatively simple non-critical construction, a system which permits free pull and rotation in the case of a vertical pulling arrangement, a pressure sensor and a regulating valve which will effect rapid pressure equalisation but prevent diffusive loss of components. The role of these features can be explained with reference to the schematic diagram of a pulling system shown in fig. 1.

This shows a crystal 3 growing from a melt 4 in the inner growth chamber 2 which is supported within the main pressure chamber 1. The pull rod 5 passes through a normal seal in 1 and then through a *liquid seal* 7 contained in a well in the pull rod bearing seal block 6. The nub of the technique is the means for maintaining the liquid seal in position. It is prevented from running down the pull rod under the action of gravity by the screwing action of the pull rod against a female screw thread cut in the bearing seal block. Any liquid between the rod and the thread is screwed back into the well provided the rotation is in the "upward sense".

The liquid seal prevents diffusive loss of volatile species where the pull rod enters the growth chamber. Mechanical loss of volatile species by pressure differentials across the liquid seal is prevented by pressure

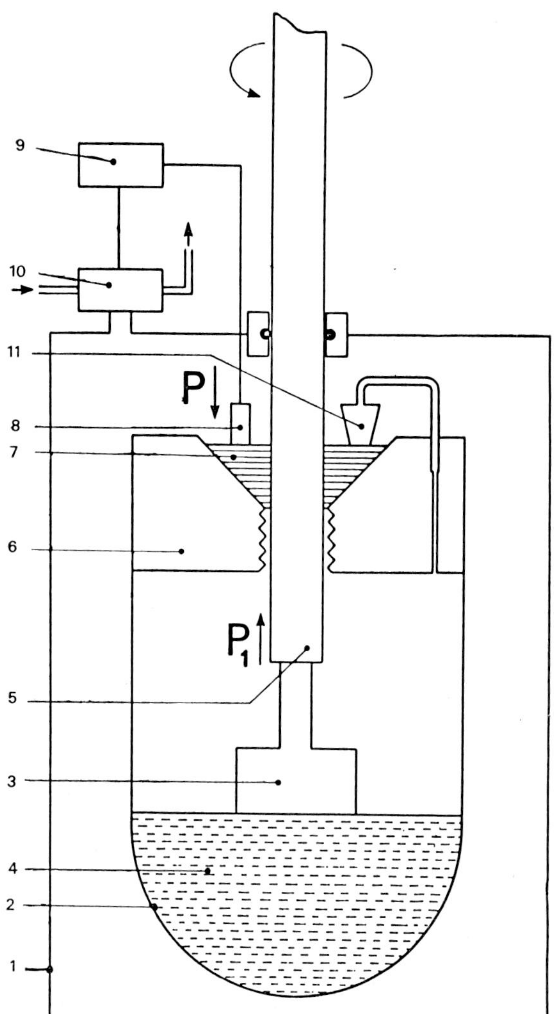

Fig. 1. Schematic diagram of a pulling apparatus using Pressure Balancing. (1) Main pressure chamber, (2) inner growth chamber, (3) (seed) crystal, (4) melt, (5) pull rod, (6) pull rod bearing seal block, (6a) peripheral heater, (6b) female thread, (7) liquid seal, (8) pressure balancing sensor, (9) out-of-balance relay, (10) pressure regulating valve, (11) pressure balancing bubbler, (12) auxiliary heater, (13) tie rods with nimonic springs, (14) growth chamber support bracket, (15) heater and thermocouple lead throughs, (16) susceptor, (17) RF coils, (18) inert gas supply. (Legend also applies to fig. 2.)

balancing. Pressure balancing can be effected in either of two ways, by inert gas pressure regulation or by partition balance.

In the first method involving pressure regulation the inert gas pressure P is controlled by the pressure regulating valve 10. This valve is actuated by the relay 9 via a signal from the pressure (liquid level) sensor 8. In practice using the pressure regulation mode both chambers are initially evacuated. Then, during melt-up as the vapour pressure of the compound increases inert gas is introduced into chamber 1 to maintain pressure balance. In principle pressure fluctuations can be controlled automatically without vapour loss by the controlled introduction and venting of inert gas.

In the second method of pressure balancing by partition balance, a two way pressure balancing bubbler 11 allows the excess pressure of vapour (or inert gas) to flow and partition itself between the two chambers 1 and 2. Clearly, the bubbler 11 can pass vapour from chamber 2 to chamber 1; but additionally it will work in the reverse direction if as a result of the higher pressure in 1 liquid flows into the suitably-shaped bell of 11 and lowers the level of liquid in the well sufficiently to uncover the end of the bell. Inert gas can then bubble through into 2. The bell must not overfill. One needs to maximise the liquid level height change relative to the volume of the bubbler (see experimental). In practice, prior to growth the pressure in both chambers 1 and 2 is raised to a pressure in excess of that which will be generated by dissociation of the melt. During melt-up as the vapour pressure of the compound increases a small amount of gas (inert gas plus the volatile component(s) of the compound) is forced through the pressure balancing bubbler until equilibrium is reached. The actual quantitative loss of the volatile component of the compound in this process is insignificant in terms of stoichiometry of the melt. Pressure differentials are generated generally only before and after the growth of the crystal as a result of heating or cooling. The crystal growth process itself generally takes place under steady-state conditions with no vapour loss. Whilst conceptually, pressure balancing represents a technique for suppressing dissociation of compounds or the loss of volatile additives during crystal growth, additionally one may use such an apparatus for the formation of compounds prior to growth.

4. Experimental

4.1. APPARATUS

A photograph of an experimental version of an apparatus for preparing crystals using pressure balancing is shown in fig. 3. Its construction can be understood with reference to figs. 1 and 2 and their detailed legends. One may consider the apparatus as a low pressure

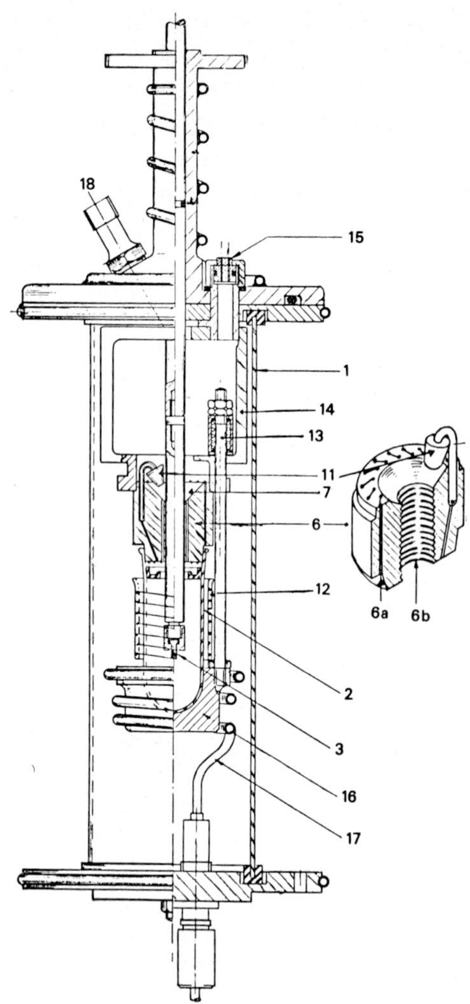

Fig. 2. Actual diagram of a pulling apparatus using Pressure Balancing. See fig. 1 for details.

system suitable for use under near atmospheric conditions in view of the use of a silica envelope as the main pressure chamber. The growth chamber 2 is a simple silica tube. It was secured against a machined taper on the bearing seal block by tie rods 13 fitted with nimonic springs to counteract expansion problems. The pull rod bearing seal block 6 was machined from two inch diameter boron nitride rod. It has a peripheral resistance heater 6a. This keeps volatile materials from condensing on the top of chamber 2 and it also keeps liquid seals such as boric oxide sufficiently fluid. The initial studies were carried out using a silica rod for the lower section of the pull rod; but a new silica rod was required with each run since freezing boric oxide cracks silica. Boron nitride rods were subsequently used since they

can be used repeatedly. The surface of boron nitride is remarkably resistant to damage by continual melting and freezing of boric oxide.

The liquid seal functioned exceptionally well and required only minor modifications during development. Neither the bore tolerance nor the thread pitch seem particularly critical. The shape but particularly the position of the pressure balancing bubbler are more critical. The design of bubbler system shown has worked but it is rather difficult to predict its height adjustment in relation to the liquid level. It was intended that the bubbler would work as a result of the excess pressure of inert gas forcing the liquid level down into the bore of the bearing seal block in order to expose the open end of the bubbler and re-establish equilibrium. In practice the liquid was forced more readily into the bubbler. This required a bubbler which could handle a none too predictable volume of liquid. This design

Fig. 3. Photograph of a low pressure version of a Pressure Balancing crystal puller used for growing InAs and GaAs.

problem can be avoided by using a bubbler working in a separate small area well at the top of the block. It proved to be more effective since it could be accurately preset and was particularly pressure sensitive.

It will be appreciated that the liquid seal itself can function as a one way pressure bubbler. Controlled experiments with InAs and GaAs showed that since the pressure rose during melting and then remained constant during growth the pressure balancing bubbler could be dispensed with on occasion. However it is anticipated that it will be needed as a safety measure and in high pressure applications.

4.2. CRYSTAL GROWTH

An attractive feature of the apparatus is the ease with which compounds can be prepared in situ prior to crystal growth. The procedure used with InAs or GaAs was to load both the group III element and the arsenic into the bottom of the growth chamber 2. The system was evacuated and the arsenic distilled up the silica tube by heating the susceptor. This treatment removed traces of As_2O_3. The B_2O_3 was then formed into the liquid seal by using the bearing seal block heater and rotating the pull rod. The auxiliary heater 12 was then employed to distill the arsenic into the group III metal. This procedure took about an hour. The growth of the crystal then followed the conventional pattern. It was necessary to avoid rapid movements of the pull rod (say > 10 cm/hr) which could extract B_2O_3 from the well. Nevertheless this was a useful expedient at the end of a run since rapid withdrawal freed the pull rod from the bearing seal block. The B_2O_3 could then easily be brushed off the rod when cold.

4.3. APPLICATIONS AND SCOPE OF PRESSURE BALANCING

Both InAs and GaAs have been prepared and grown as crystals (fig. 4) using the technique in order to develop the method and demonstrate its viability. In principle the technique can be used to grow highly dissociable compounds since the outer silica envelope can be replaced by a standard steel pressure vessel[5]). As with Liquid Encapsulation the hot inner growth chamber is not subject to pressure differentials at temperature which could distort it; only the cold outer vessel is subject to a mechanical force. It is interesting that it is possible to record and control pressures directly during crystal growth. This could eliminate the uncertainties in other techniques of interpreting a "cold point" in order to control a pressure. This may be useful in stoichiometry control during crystal growth and in measuring crystal/vapour equilibria. The relative simplicity and versatility of the technique could be

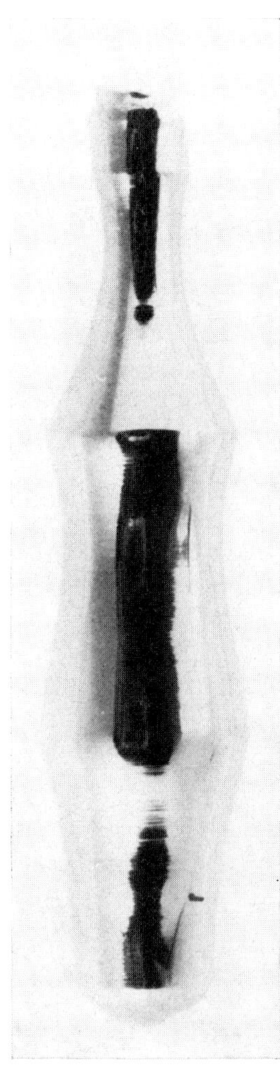

Fig. 4. As-grown crystal of InAs prepared and grown in the system shown in fig. 3.

a considerable asset in making and growing ternary compounds. The liquid seal need not be boric oxide; indeed it need not be transparent. It could be metallic; hence the technique offers scope in the growth of dissociable oxides and non-metallic compounds. Pressure Balancing could be also applied to the horizontal growth of compounds.

5. Summary

The advantages and limitations of established techniques including Liquid Encapsulation for growing dissociable compounds have been considered. It is proposed that a novel technique called Pressure Balancing offers all the significant advantages of all the other techniques. A low pressure pulling system has been demonstrated for the formation and growth of InAs and GaAs. The application of the technique to the preparation and crystal growth highly dissociable compounds is anticipated since the growth apparatus has been designed to fit a standard high pressure chamber[8].

Acknowledgements

The authors gratefully acknowledge the useful suggestions and advice of our colleagues particularly Drs. D. T. J. Hurle and P. Tufton. Contributed by permission of the Director of RRE. Copyright controller HMSO.

References

1) M. Chesswas, B. Cockayne, D. T. J. Hurle, E. Jakeman and J. B. Mullin, J. Crystal Growth **11** (1971) 225.
2) S. E. Blum and R. J. Chicotka, J. Electrochem. Soc. **115** (1968) 298.
3) R. Gremmelmaier, in: *Compound Semiconductors*, Vol. I, Preparation of III–V Compounds, Eds. R. K. Willardson and H. L. Goering (Reinhold, New York, 1962) p. 254.
4) A. G. Fischer, J. Electrochem. Soc. **117** (1970) 41C.
5) J. B. Mullin, R. J. Heritage, C. H. Holliday and B. W. Straughan, J. Crystal Growth **3, 4** (1968) 281.
6) S. J. Bass and P. E. Oliver, J. Crystal Growth **3, 4** (1968) 286.
7) J. L. Richards, J. Appl. Phys. **34** (1957) 289.
8) Metals Research, Cambridge, England.

Journal of Crystal Growth **13/14** (1972) 635–639 © *North-Holland Publishing Co.*

DISLOCATION-FREE GaAs BY THE LIQUID ENCAPSULATION TECHNIQUE

B. C. GRABMAIER and J. G. GRABMAIER

Research Laboratories of Siemens AG, Munich, Germany

Dislocation-free GaAs single crystals were grown by the liquid encapsulation technique by means of a necking-in procedure. It was found that the thinning of the crystal diameter is the main step for diminishing the dislocations. The stoichiometry of the melt composition is not very critical; it can be sufficiently maintained with regard to an adapted growing rate. A low temperature gradient near the solid–liquid interface is necessary and can be governed by the encapsulating B_2O_3 itself. No dislocations must be created by the B_2O_3 melt. Crystals were grown in [111]- as well as in [100]-orientation. The dislocations were examined by etching the (111)-Ga surface and (100) surface, respectively. It was found that all dislocations of the seed crystal have disappeared from the neck portion toward the crystal surface.

1. Introduction

Dislocation-free single crystals of GaAs are of considerable interest as a basic material for luminescence diodes, laser diodes[1]) and avalanche diodes, especially for photo-avalanche diodes[2]).

In general, for the production of high quality single crystal GaAs, the Gremmelmaier's method[3]) is used. In recent times, the liquid encapsulation technique[4]) has often been applied. This latter technique is much simpler in design and easier to control. Steinemann and Zimmerli[5]) first succeeded in growing dislocation-free single crystal GaAs using the Gremmelmaier's method by means of a necking-in procedure.

They found the following experimental conditions to be necessary:
(a) extremely severe control of the stoichiometric melt composition, i.e. constant partial pressure of As within the sealed system;
(b) a low temperature gradient in the solidified portion of the crystal near the solid–liquid interface;
(c) a horizontal interface, i.e. purely axial heat flow;
(d) a thin "neck" with a diameter of 1–2 mm over a length of 10–20 mm after seeding.

These requirements for dislocation-free crystal pulling however, seem to be in several respects unnecessary when using the liquid encapsulation technique[6]).

2. Experimental procedure

It was shown by Metz et al.[7]) that it is possible to suppress volatilization from a melt by covering it with a layer of an inert liquid. They also showed that it is possible to pull crystals through such a layer by Czochralski's method. More recently it was shown by Mullin et al.[4]) that this method of crystal growth could be applied to III–V compounds using molten B_2O_3 to prevent loss of the group V component. Bass and Oliver[8]) described in detail the usual apparatus for pulling GaAs by the liquid encapsulation technique.

The apparatus we used differed from the usual Czochralski-crystal puller: The motors for pulling and rotation of the crystal were situated not on the apparatus directly, but on a special support (fig. 1). The pulling and the rotation movements were transmitted by means of friction driven rubber connecting rings so no disturbing vibrations could influence the pulling system.

The quartz crucible, containing the molten GaAs which is covered with the molten B_2O_3, was held in an iridium susceptor inductively heated at 450 kc/s. The susceptor was surrounded by an alumina radiation shield with a hole in the bottom (fig. 2). The melt temperature was stabilized within ± 0.1 °C by a photo-element which was directed towards the bottom of the susceptor.

We used polycrystalline undoped GaAs fabricated by Mining and Chemical Products (carrier density: 10^{16} cm^{-3}; mobility: 4000 cm^2/V s) and boric oxide, "suprapur", fabricated by Merck AG. B_2O_3 must be dried thoroughly before using; a temperature of about 1000 °C, in a vacuum of about 10^{-5} torr, over several days is suitable. This B_2O_3 showed satisfactory results.

The B_2O_3 layer on the GaAs melt had a thickness of 8–10 mm. The Argon pressure was 1.5 atm.

The dislocations were made visible via the etch pit technique. Etch pits appear on the Ga (111)-plane by etching in a mixture of $HNO_3:H_2O = 2:3$ at a temperature of 60 °C[9]) and on the (100)-plane by etching in a KOH-melt at a temperature of 330 °C[10]).

3. Results

In the course of pulling several hundred GaAs single crystals we found the conditions for growing single crystals with a maximum etch pit density of about $1 \times 10^4 \, cm^{-2}$. These high quality crystals had a diameter of about 15 mm and a length of about 50 mm. In general the seeds were oriented in the [111]-direction, sometimes they were oriented in the [100]-direction. Contrary to other authors[5]) we never found a lower twinning probability as a result of using a seed with the (111) Ga-face towards the melt compared with a seed with the (111) As-face towards the melt.

These conditions are as follows:

(1) We have to use a high quality seed, because the faults of the seed always propagate into the growing crystal. The faults increase with the diameter and with the length of the growing crystal.

(2) We have to apply a suitable (low) pulling rate of the crystal. A high pulling rate always results in a reduction of the crystal quality because the pulling rate is connected with the heat flow through the growing

Fig. 1.   Crystal puller used for liquid encapsulated growth of high quality GaAs.

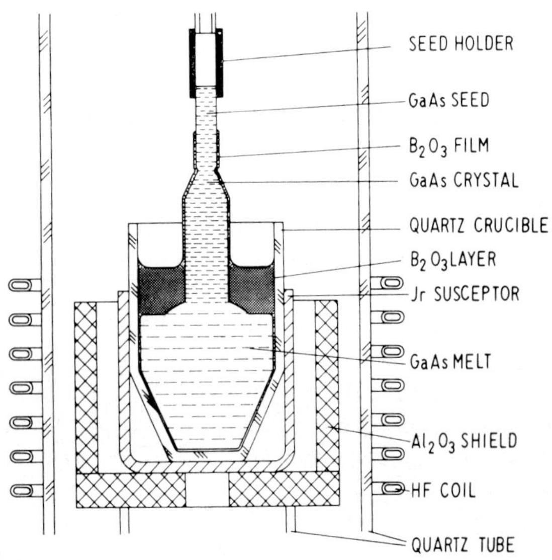

Fig. 2.   Furnace arrangement used for liquid encapsulated growth.

Fig. 3.   Dislocation-free [111]-crystal (left) and [100]-crystal (right), respectively, grown by the necking-in procedure.

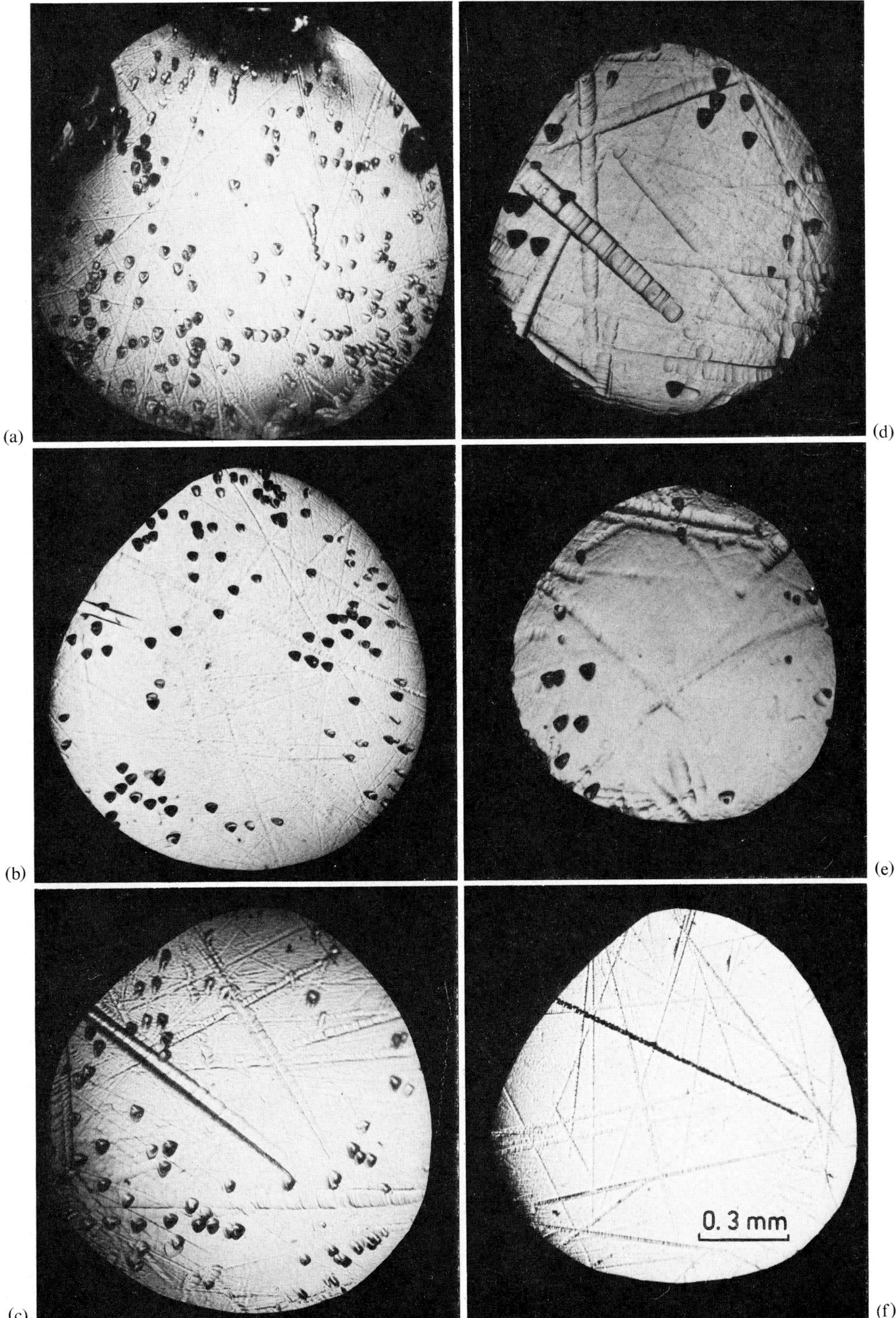

Fig. 4. Elimination of dislocations by necking in a [111]-crystal. Sequence of etched horizontal cross-sections Ga (111) of a neck.

TABLE 1

Diameter of the crystal (mm)	Pulling rate (mm/hr)
15	6
10	8
6	10

crystal. For growing high quality crystals we found the relationship as given in table 1 between the crystal diameter and the pulling rate to be a useful working rule.

(3) We have to reduce mechanical disturbances of the pulling system for example produced by the vibrations of the motors for pulling and rotation because they influence the perfection of the growing crystal.

(4) We have to avoid the sources causing a high twinning probability of GaAs. We found three sources for the twinning:

(a) areas of high etch pit density $\geq 10^6$ cm^{-2});

(b) solid particles at the solid/liquid interface;

(c) mechanical shocks.

By considering these conditions we are able to grow high quality crystals with a frequency of about 95%.

Being able to produce high quality crystals we then tried to grow dislocation-free crystals by the necking-in procedure. By paying attention to the conditions described above we found that three out of four necking-in experiments were successful. We grew [111] crystals as well as [100] crystals. In fig. 3 two crystals are shown which have been produced by the necking-in procedure. The left crystal is [111]-oriented, the right one is [100]-oriented. In general, crystals grown by the necking-in procedure exhibit marked facets. [111]-crystals have 3 facets and [100]-crystals four facets. Furthermore we observed the cross section of an (111)-crystal being more triangular, the cross-section of an (100)-crystal being more oval.

Examination of the dislocation density in the crystals was carried out by etching cross-sections of the crystals. Two successful experiments are illustrated in figs. 4 and 5. They show a sequence of etched (111) cross-sections and (100) cross-sections, respectively, of the neck. Across a length of 3–5 mm the etch pit density of these crystals is reduced from about 5×10^3 cm^{-2} to zero. The reduction of the density is not homogeneous across the cross-section.

By carefully increasing the diameter of the disloca-

tion-free crystal neck a dislocation-free crystal with a larger diameter can be grown. By the necking-in procedure we found that it is of advantage if the angle be-

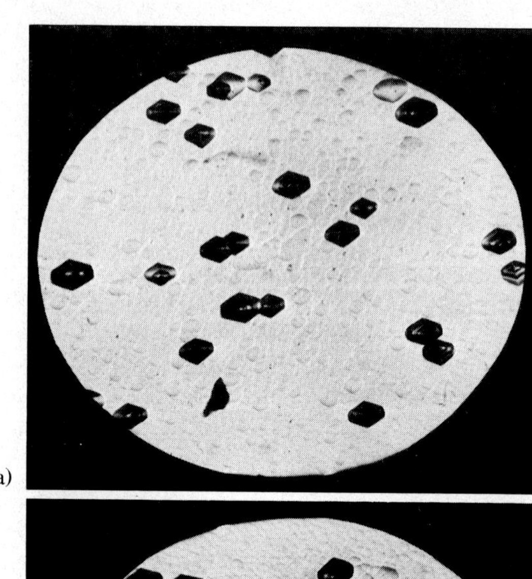

(a)

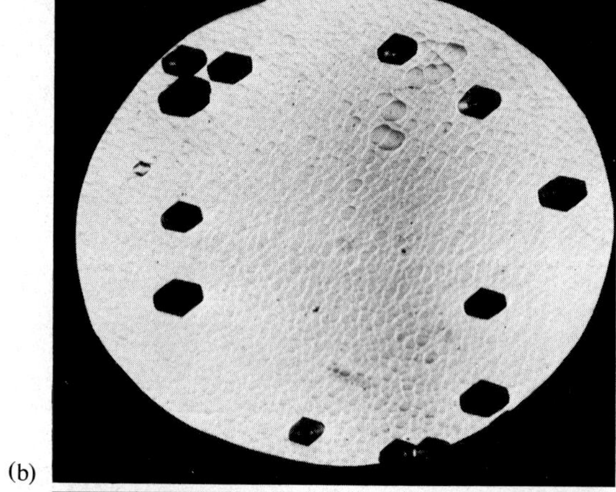

(b)

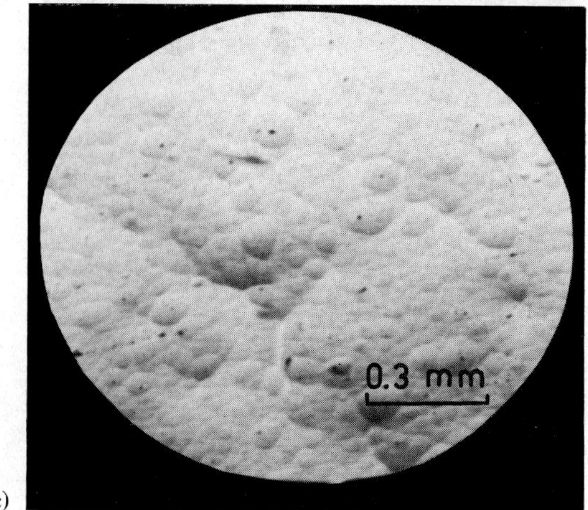

(c)

Fig. 5.   Elimination of dislocations by necking in a [100]-crystal. Etched horizontal cross-sections.

XIII – 6

tween the peripheral crystal surface and crystal axis is smaller than 25 degrees. New dislocations can be created on the surface of the crystal if the diameter after the necking-in procedure increases very rapidly.

Crystals successfully grown by the necking-in procedure are dislocation-free over their whole length. However, if in the last period of the pulling procedure after separation from the GaAs melt, the crystal is withdrawn with a rate which is higher than the normal growing rate, then on the bottom of the crystal over a length equal to the thickness of the B_2O_3 layer dislocations can be created by plastic deformations.

4. Discussion

Completely dislocation-free GaAs crystals can be grown from the melt by the liquid encapsulation technique. The conditions we found deviate in some points from the conditions Steinemann and Zimmerli[5] found for pulling dislocation-free crystals by the Gremmelmaier's method.

(1) The stoichiometric melt composition is not important if the pulling rate is low enough. From the binary system Ga-As the crystallization of GaAs materials with a stoichiometric composition and regular structure is possible. Ga inclusions occur if the pulling rate is too fast so that constitutional supercooling occurs.

(2) A low temperature gradient near the solid–liquid interface is important and is guaranteed by the molten encapsulant B_2O_3 itself. The molten B_2O_3 acts as an afterheater.

(3) The growing surface of the solid must be horizontal or convex. The shape of the growing interface can be influenced by the variation of the position of the quartz crucible-susceptor. Further precautions concerning the axial heat flow are not necessary.

(4) It was found that the thinning of the crystal diameter is the main step for diminishing the dislocations. The length of the subsequent neck is of less importance. Within the region of the crystal of decreasing diameter the dislocations grow in $\langle 111 \rangle$-type directions towards the crystal surface and disappear there.

References

1) K. Mettler, FSDERC, München, März 1971.
2) H. Kressel, RCA Rev. **28** (1967) 175.
3) R. Gremmelmaier, Z. Naturforsch. **11a** (1956) 511.
4) J. B. Mullin, B. W. Straughan and W. S. Brickel, J. Phys. Chem. Solids **26** (1965) 782.
5) A. Steinemann and U. Zimmerli, in: *Crystal Growth*, Ed. H. S. Peiser (Pergamon, Oxford, 1967).
6) A. Steinemann and U. Zimmerli, in: *Symp. on GaAs*, 1966 (The Institute of Physics and the Physical Society, London, 1966).
7) E. P. A. Metz, R. C. Miller and R. Mazelski, J. Appl. Phys. **33** (1962) 2016.
8) S. J. Bass and P. E. Oliver, in: *Symp. on GaAs*, 1966 (The Institute of Physics and the Physical Society, London, 1966).
9) H. A. Schell, Z. Metallk. **48** (1957) 158.
10) J. G. Grabmaier and B. C. Watson, Phys. Status Solidi **32** (1969) K 13.
11) W. A. Tiller, J. Appl. Phys. **29** (1958) 611.

CRYSTAL GROWTH AND PROPERTIES OF GROUP IV DOPED INDIUM PHOSPHIDE

J. B. MULLIN, A. ROYLE, B. W. STRAUGHAN, P. J. TUFTON and E. W. WILLIAMS

Royal Radar Establishment, Malvern, Worcs., U.K.

InP crystals have been grown from the melt by the high pressure liquid encapsulation technique. A study of the segregation of group IV elements based on measurements of free electron concentrations in doped crystals indicates that the mean effective distribution coefficients for Ge and Sn are 2.4×10^{-2} and 2.1×10^{-2} respectively. Whereas doping with Ge and Sn gave high quality and reproducible n-type crystals, controlled doping experiments using Si and Pb were not successful. The role of these latter elements together with that of C is discussed. The characteristic photoluminescence spectra of these doped crystals were measured at 77 °K, and in conjunction with the electrical and chemical evidence the peaks were assigned to specific impurity/vacancy interactions. The highest energy peak at about 1.41 eV is associated with donors and is sensitive to Ge and Sn doping, while the acceptor level appears to lie at 1.37 eV. The peaks at 1.32 eV and 1.35 eV are attributed to the effects of growth from B_2O_3 and copper doping respectively, and the broad peak between 1.12 and 1.20 eV may be associated with phosphorus vacancies.

1. Introduction

Although InP is a well-known semiconductor it has not received the systematic study accorded to other III–V's such as InSb and GaAs. For the most part this is due to the difficulty of growing single crystals from the melt of a material with a high vapour pressure, but the relative simplicity of the high pressure liquid encapsulation technique[1] which has recently been developed has now made the growth process reasonably straightforward. The present investigation concerns the problems involved in the preparation of single crystals of InP doped with group IV elements, together with an assessment of their photoluminescence properties in an attempt to understand the vacancy/impurity interactions.

Melt-growth of InP prior to 1962 has been reviewed by Richman[2]; most of the studies were concerned solely with the preparation and electrical properties of the material. However Folberth and Schillman[3] did look at group IV doping and reported the effect of Ge, Sn and Pb on the conductivity of various III–V compounds together with estimates of their distribution coefficients. Although they do not state specifically by which method they grew their crystals, it can be inferred that they used the zone melting technique previously described by Folberth and Weiss[4] which involves crystal growth under In-rich conditions[2]. These results and those of Rosztoczy et al.[5] who measured the distribution coefficients of Ge, Sn and Te during the liquid phase epitaxial growth of InP are discussed later.

2. Crystal growth

2.1. GENERAL PROCEDURES

The technique of growing crystals by high pressure liquid encapsulation has already been described when the characteristics of undoped and Cr-doped (semi-insulating) InP were reported[6]. The crystals were pulled from the melt through a liquid boric oxide seal backed by a gas pressure of twenty seven atmospheres of nitrogen. Thus dissociative loss of phosphorus was prevented and visual control maintained throughout.

When commercial[7] polycrystalline InP is pulled it yields a uniform stoichiometric material, and it was these prepulled crystals that were used as the starting charges for all the doping experiments to be described. Such material contains from 5–25 ppm of impurities as determined by a mass spectrometric analysis and the free carrier concentration ranges from about 10^{14} to 10^{17} carriers/cm^3. Direct interpretation of deliberate doping effects can therefore only be made on crystals having free carrier concentrations in excess of 10^{17} carriers/cm^3.

2.2. CARBON

The free carrier concentrations measured in undoped and Si-doped crystals were found to be significantly less than one would expect from the C contents of these crystals assuming that each atom produced one free electron. Hence no deliberate attempts were made to dope with C. It is significant that under similar conditions GaP crystals grown in silica crucibles[1] contained more C (130–230 ppma) than InP crystals grown in silica (5–13 ppma) and surprisingly even more C (25–40 ppma) than GaP grown in vitreous carbon. A probable explanation of these results is the reaction of the silica crucibles with the graphite susceptors used for coupling in power. The CO produced – the highest concentration occurring with the higher melting point compound in silica – would equilibrate with the melt through the B_2O_3.

2.3. SILICON

Controlled Si doping has been frustrated by the reaction of Si with B_2O_3. The reaction causes a scum to form on melting. This scum initiates twins and a polycrystal results. Although the scum disappears in the course of crystal growth it is probable that it merely becomes attached to the boule as it is withdrawn. Si-doped GaAs can be pulled from scum free melts provided that the amount of Si is limited to that yielding crystals with free carrier concentrations of about 10^{18} per cm^3, but the cause of scum formation is not understood and the surfaces of all the Si-doped InP melts were contaminated even when a reducing environment of forming gas replaced nitrogen in the pressure chamber. Furthermore the Si levels in the crystals were very low. Table 1 indicates that doping with a melt concentration of about 0.3 mg/g of charge, a concentration which produces in GaAs 10^{18} carriers per cm^3, gave a non-homogeneous but very low carrier concentration with a mass spectrographic analysis of about one ppm at. ($\equiv 4 \times 10^{16}$ atoms cm^{-3}). Thus, although the conventional effective distribution coefficient for Si in GaAs is $k = 0.1$, and that for the liquid encapsulation technique is $k = 0.02$, the value for Si in InP grown under liquid encapsulation seems to be more than an order for magnitude lower still (i.e. $\sim 8 \times 10^{-4}$).

2.4. GERMANIUM AND TIN

By contrast Ge and Sn showed well-behaved doping and segregation characteristics; reproducibly doped n-type crystals could be grown from clear scum-free melts. Conventional electrical measurements[2] showed

that doping in the range $\sim 10^{17}$ up to at least 10^{19} carriers per cubic cm was possible. As well as the free carrier (N_D–N_A) concentrations the individual ionised donor N_D and acceptor N_A concentrations were calculated from a knowledge of the mobility μ using a Brooks–Herring[2]) analysis. Whilst this analysis was not applicable above about 3×10^{18} carriers per cm^3 it was useful for giving at least a semi-quantitative picture of the acceptor behaviour in the material. The segregation behaviour of the parameters $N_D - N_A$, N_D and N_A were then analysed as a function of the mass fraction, g, of crystal pulled. Assuming an impurity of concentration C_0 in the melt is distributed as a concentration C_s in the crystal according to the normal freeze relationship

$$C_s/C_0 = k_{eff}(1-g)^{k_{eff}-1},$$

where k_{eff} is the effective distribution coefficient then a plot of log C_s/C_0 versus log $(1-g)$ should give a slope of $k_{eff}-1$ and an intercept of log k_{eff}. Equivalent plots were made for $N_D - N_A$, N_D and N_A for the majority of the crystals. The parameters $N_D - N_A$ and N_D showed approximate though not accurate normal freeze behaviour but it was possible to measure k_{eff} from the intercept. The acceptors however showed more capricious behaviour. This is considered along with photoluminescence evidence in section 3.

The results of the measurements of the effective distribution coefficients are listed together with relevant segregation data in the table. The distribtution coefficients have been estimated on two assumptions; firstly that no dope was lost in the encapsulant, and secondly that each doping atom gave rise to one conduction electron in the grown crystal.

2.5. LEAD

No evidence of Pb doping was obtained from the free carrier determinations made on sections taken from crystals pulled from even very heavily Pb-doped melts (5 and 10% Pb by weight). The 5 wt% crystal pulled at 0.5 cm/hr was initially single crystal.

3. Photoluminescence properties

3.1. AS-GROWN MATERIAL

Fig. 1 gives a comparison of the photoluminescence spectra of InP grown by four different methods but all

Table 1

Doping and segregation behaviour of group IV elements in InP

Crystal	Dopant	Distance from seed on mm	Carriers at 77 °K from half width of 1.41 eV line (cm^{-3})	(N_D-N_A) at 77 °K from electrical measurements (cm^{-3})	Concentration of dopant added to melt (cm^{-3})	(N_D-N_A) (297 °K) from electrical measurements (cm^{-3})	μ (297 °K) $(cm^2\ V^{-1}\ sec^{-1})$	N_D from calculations and B-H theory (cm^{-3})	$\bar{k}(N_D-N_A)$ and standard deviation	$\bar{k}(N_D)$ and standard deviation
									Number of	results
L101	Sn	46	7.0×10^{17}	$9.5\ \times10^{17}$	2.66×10^{19}	9.98×10^{17}	1860	1.24×10^{18}	$\bar{k}=2.1\times10^{-2}$	$\bar{k}=2.8\times10^{-2}$
		55	2.1×10^{18}	$1.4\ \times10^{18}$		1.49×10^{18}	1080	2.59×10^{18}		
L170	Sn	12	5.0×10^{17}	$6.0\ \times10^{17}$	2.77×10^{19}	6.01×10^{17}	2360	7.23×10^{17}	$S=6.6\times10^{-3}$	$S=9.4\times10^{-3}$
		62	1.5×10^{19}	$1.6\ \times10^{19}$		1.51×10^{18}	1000	—	from 10 crystals	
L190	Ge	28	2.5×10^{18}	$2.9\ \times10^{18}$	9.72×10^{19}	2.31×10^{18}	1020	3.85×10^{18}	$\bar{k}=2.4\times10^{-2}$	$\bar{k}=4.1\times10^{-2}$
		53	7.0×10^{18}	$7.5\ \times10^{18}$		7.91×10^{18}	820	—		
L199	Ge	23	2.5×10^{17}	$2.7\ \times10^{17}$	1.08×10^{19}	3.03×10^{17}	2320	4.94×10^{17}	$S=3.7\times10^{-3}$	$S=1.1\times10^{-2}$
		54	5.0×10^{18}	$2.5\ \times10^{18}$		2.40×10^{18}	1420	3.06×10^{18}	from 3 crystals	
L98	Si	6		1.67×10^{13}	3.25×10^{19}	7.58×10^{13}	1690	1.51×10^{17}	$\bar{k}=7.5\times10^{-4}$	$\bar{k}=6.3\times10^{-3}$
		28		5.39×10^{15}		1.03×10^{16}	2700	1.37×10^{17}		
L202	Si	25		2.13×10^{16}	1.64×10^{19}	3.03×10^{16}	3360	1.21×10^{17}	$S=8.5\times10^{-4}$	$S=2.3\times10^{-3}$
		56		1.81×10^{16}		2.65×10^{16}	3770	9.46×10^{16}	from 4 crystals	

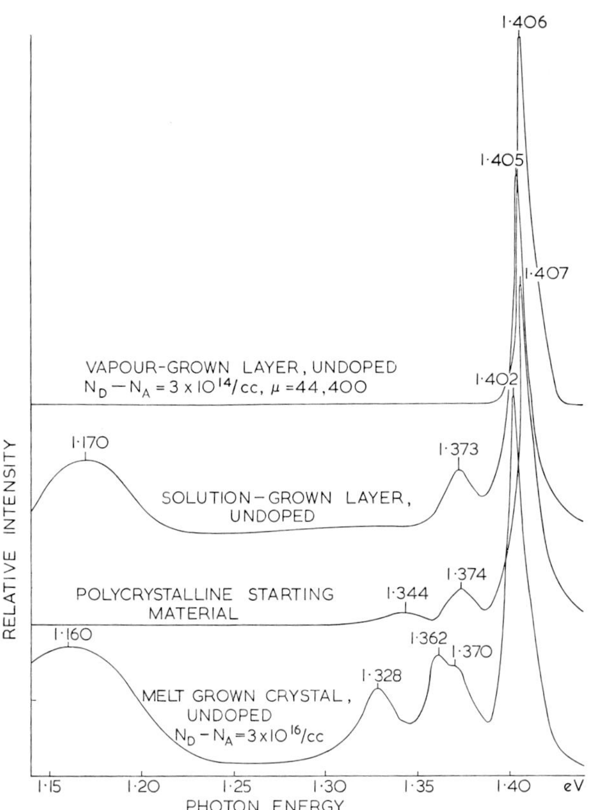

Fig. 1. Photoluminescence spectrum observed at 77 °K from a high purity vapour epitaxial layer of InP compared with similar spectra from a high purity solution grown layer, from a section of polycrystalline starting material and from a section of "undoped" melt-grown crystal.

without intentional doping. Here the peak energies are recorded to three places of decimals, but since these values shift slightly with background doping it is convenient to identify a characteristic energy level accurate only to two places of decimals. In this way we distinguish five major peaks in the spectra and can attempt correlations of the basic underlying energy transitions with impurity/vacancy phenomena.

The photoluminescence measurements were made at 77 °K, and at this temperature the energy gap of InP is 1.42 eV whereas the highest energy peak recorded in the n-type crystals was 1.41 eV. As this energy is ascribed to donor to valance band recombination the activation energy of the donor must be approximately 10 meV in good agreement with that determined recently by Eaves et al. from photoconductivity measurements[8]).

The highest purity InP yet produced, which is vapour epitaxial material grown from a PCl_3/In system, shows only the 1.41 eV peak. In the highest purity solution grown epitaxial InP on the other hand almost invariably a broad 1.17 eV peak is observed as well. The broadness and shape of this additional peak is similar to an equivalent one found in GaAs which is thought to be associated with vacancies[9,10]). Such vacancy associated luminescence is consistent with the strong evidence for high equilibrium vacancy concentrations in III–V crystals at their melting points; for example GaAs is estimated to have 10^{19} vacancies per cm^3 at 1237 °C[11]). As this low energy peak showed up in practically all of the three hundred odd solution grown layers investigated and was also observed in all the samples of InP taken from near the bottoms of crystals, even when no line appeared in samples taken from the tops of these same crystals, it can be inferred that the level is associated with a defect caused by growth from excess In. Indeed it is tempting to assume that the line is due solely to phosphorus vacancies. But there is contrary evidence. The level is not found in polycrystalline InP starting material which is In-rich, nor in a few of the solution grown layers. It seems more reasonable therefore to assume that the 1.17 eV line comes from impurity/P vacancy complexes, and this hypothesis is supported by the evidence from the bulk grown crystals as not only are the bottoms grown from slight In excess but the impurity concentration is higher. Nevertheless detailed quantitative studies are needed to establish unambiguously the cause of this peak.

It is relevant here to note that others have explained the 1.17 eV peak as due to Cu present in the lattice[12]). Not only have we absolutely no evidence that Cu is responsible but we have diffused Cu into a high purity vapour epitaxial layer and re-recorded the photoluminescence spectrum (fig. 2); no suggestion of a 1.17 eV peak is observed but a strong 1.35 eV level appears in addition to the 1.41 eV line.

The polycrystalline starting material is manufactured direct from high purity In and P. However it presumably picks up impurities during processing and the spectrum (fig. 1) showed a peak at 1.37 eV. This line was also observed in the melt and solution grown InP and has been attributed to acceptor impurities. Certainly Zn (fig. 3) and Cd[13]) which are acceptors in InP produce material which gives photoluminescence peaks at 1.37 eV, but the use of the spectra at low resolution due to a high background and a relatively high tem-

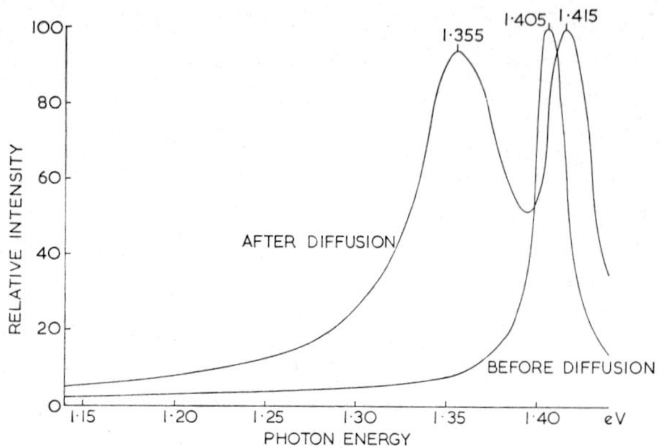

Fig. 2. Photoluminescence spectra observed at 77 °K from a vapour grown layer of InP into which Cu had been diffused.

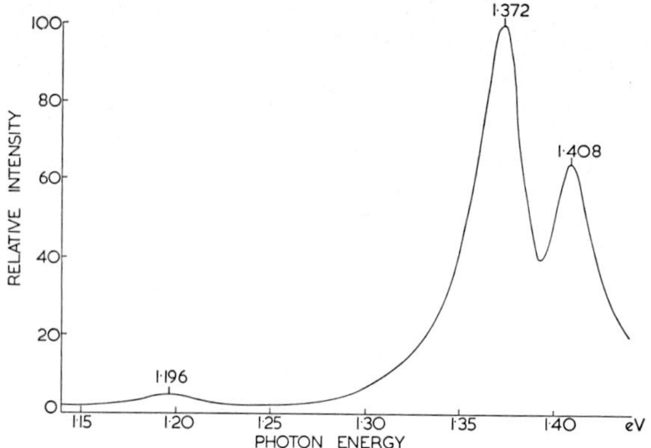

Fig. 3. Photoluminescence spectrum observed at 77 °K from a Zn-doped solution-grown layer of InP.

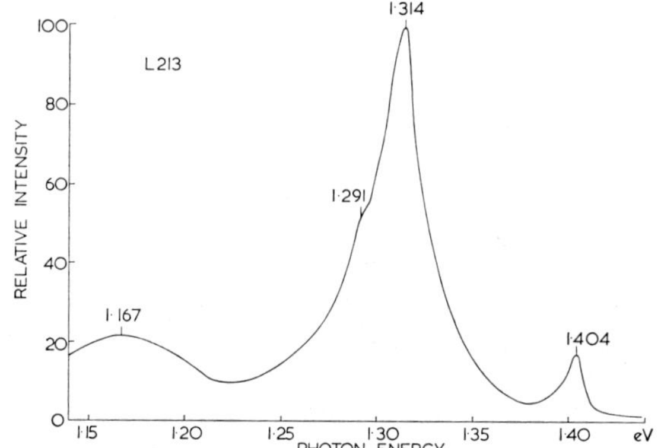

Fig. 4. Photoluminescence spectrum observed at 77 °K from a section of Al-doped melt-grown InP.

perature can rarely give unambiguous identification.

The pre-pulled undoped crystals used as starting charge for the doping experiments generally produced complex spectra with peaks at 1.41, 1.37, 1.35, 1.32 and 1.17 eV. The mass spectrographic analysis[14]) showed C (5–13 ppma), B (0.01–0.1 ppma), O_2 (1 ppma), S (0.15 ppma), Cu (0.02 ppma) and Zn (0.3–0.6 ppma) as the main impurities. The correlation of these peaks with specific impurities has not been attempted though, as has been stated above, we know that Zn enhances the 1.37 eV level and Cu the 1.35 eV level.

The peak at 1.32 eV is found in pulled crystals only and never in the polycrystalline starting material or in any of the many solution grown layers of InP. It is therefore associated with liquid encapsulated B_2O_3 growth, but whether the peak is directly due to elemental B is more problematical and the unequivocal verification of this hypothesis will require experimental work beyond the scope of this paper. There is nevertheless interesting indirect evidence. The intensity of the 1.32 eV peak in InP pulled from Al-doped melts (fig. 4) is considerably greater than that from an undoped melt. Indeed it is the dominant peak. Al would be expected to reduce boric oxide and release B. However the spectrum from a crystal pulled from a melt doped with 0.25 wt% Al is virtually identical to that from a melt containing 5 wt% Pb, and it is hardly conceivable that Pb could liberate sufficient amounts of B by direct reduction. Nevertheless the Pb disappeared into the boric oxide, next to none found its way into the crystal and the 1.32 eV peak was strong. The Si results were even more confusing. With Si having an intermediate reducing power one might expect equiv-

alent results. This, however, does not appear to be so. These Si-doped crystals gave no discernable 1.32 eV level (fig. 5a) but usually a 1.33 eV peak, and we believe these two to be distinguishable. As a scum invariably forms on a melt doped with Si, even under reducing conditions or when the Si has been previously dissolved in the starting charge, a different type of reaction may be involved. The B_2O_3 was quite clear with all the Pb and Al-doped crystal growths.

3.2. Si-DOPED CRYSTALS

Si-doping has produced InP from which the most complex spectra have been obtained (fig. 5a). Moreover the overall pattern of the peaks changed markedly even along the length of the crystals. Near the top of L98, the actual crystal from which fig. 5a was taken, where it was closely compensated but still contained 1.5 ppma of Si ($N_D - N_A \sim 10^{13}$ per cm^3) the 1.33 eV peak dominated the 1.41 eV peak. Further down the same crys-

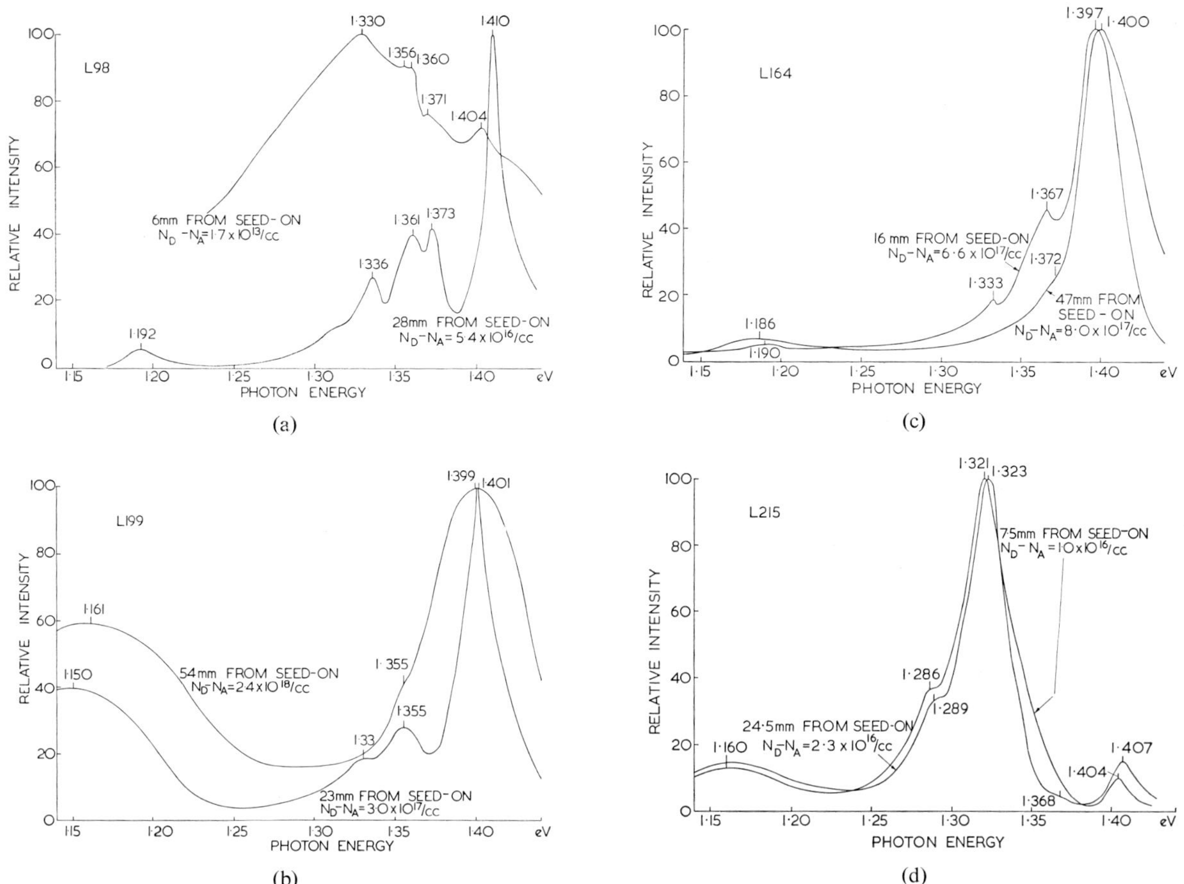

Fig. 5. Comparison of photoluminescence spectra of group IV-doped melt-grown InP observed at 77 °K: (a) Si-doped, (b) Ge-doped, (c) Sn-doped, (d) Pb-doped.

tal where there was less compensation ($N_D - N_A \sim 10^{16}$ per cm^3) the reverse situation held and the 1.41 eV peak was now the highest. In both spectra however an intermediate 1.37 eV line was observed together with other lines. Another Si-doped InP crystal, L142, which also was found to contain 0.9 ppma Si in the early stages of growth, a prominent 1.41 eV line was matched by a strong one at 1.16 eV, the 1.33 and the 1.37 eV peaks being rather weak. The capricious nature of these and other Si-doped crystals is a reflection of the reactivity of Si and the consequent unreliability of the doping technique.

3.3. Ge AND Sn-DOPED CRYSTALS

These elements were similar to each other in their results. Both enhanced the intensity and broadened the 1.41 eV peak. Further, as shown in the table, the carrier concentration as determined by Hall measurements correlated well with a carrier level inferred from half-width measurements made on the 1.41 eV peak[15]). A strong 1.37 eV peak was also observed in the Sn-doped specimens (fig. 5c). There would not appear to be sufficient Zn impurity present to account for the intensity of this acceptor peak, but clearly if Sn, in spite of its large size, was entering the lattice on the P site such good carrier concentration agreement would not be expected when only the linewidth of the donor peak was measured. The Ge spectra (for example fig. 5b) showed a small peak at 1.36 eV and a strong peak at 1.15/1.16 eV together, of course, with the enhanced 1.41 eV donor peak. It is considered significant that the lowest level peak, appearing from 1.14 eV to 1.20 eV, was particularly strong in InP from the ends of all the group IV (Si, Ge and Sn) doped crystals. Quite often this line completely dominated the spectra. One cannot but draw the conclusion that this peak is the result of a group IV/P vacancy complex.

3.4. Pb-DOPED CRYSTALS

The photoluminescence spectrum observed on Pb-doped crystals (fig. 5d) has already been discussed in relationship to the 1.32 eV peak. It is worth noting that the carrier concentration determined from the half width of the 1.41 eV line is in close agreement with the Hall measurements.

4. Conclusions

The effective distribution coefficients for Ge and Sn measured on crystals grown from stoichiometric melts are 2.4×10^{-2} and 2.1×10^{-2} respectively and are to be compared with the corresponding measurements of 1.1×10^{-2} and 2.2×10^{-3} obtained by Rosztoczy et al.[5] from liquid epitaxial growth experiments carried out in the 600–650 °C temperature range. The results of the Folberth and Schillman[3], 5.0×10^{-2} and 3.0×10^{-2}, although in general agreement are difficult to compare since they were obtained using melts of unknown stoichiometry. It is interesting that the effect of excess In in the growth of InP is to reduce k_{eff} for Sn by an order of magnitude while only halving the k_{eff} for Ge compared with the corresponding value of k_{eff} obtained by stoichiometric growth. The reason for this may be a consequence of the greater size of the Sn atom compared with the Ge atom. Sn might find it more difficult than Ge to accommodate in the lattice when the In vacancy concentration is being suppressed as a result of growth from excess In. The failure to incorporate Pb is almost certainly due to the large size of the Pb atom which is greater than that of the P or In atoms.

The electrical properties of the crystals have been used to interpret the pattern of donor and acceptor segregation in the course of crystal growth. This analysis is based, as mentioned earlier, on a Brooks–Herring treatment[6]; it gives a reasonable semi-quantitative picture of acceptor behaviour in Ge and Sn-doped crystals. It is clear from the plots of the log $(N_D - N_A)$ versus log $(1 - g)$ curves and the corresponding curves for N_D and N_A that the crystals are always closely compensated. The N_A plots for each crystal usually run approximately parallel to the corresponding plots for $N_D - N_A$ and N_D. Three Sn-doped crystals were however exceptional. In these the acceptor level increased much more rapidly towards the bottom of the crystal;

so much so that the interpreted k_{eff} from the N_A plot was negative. These results could be explained by a mechanism which generates acceptors, it does not seem possible from the segregation behaviour that the acceptors are due solely to impurities substituting for In. A more complex vacancy/impurity effect is indicated. It is probably relevant that in these crystals the 1.37 eV peak was prominent in the photoluminescence analysis. The results of k_{eff} for Ge and Sn which take account of the acceptor levels in the crystals are shown in the table. The role of the acceptors will need more detailed investigation before the full significance of these results can be interpreted.

Acknowledgements

The authors gratefully acknowledge the skilled experimental assistance of Mr. S. Benn and Mrs. M. Webb on the electrical and photoluminescent measurements respectively. Also, it is a pleasure to acknowledge the very thorough mass spectrographic studies of Mr. L. G. Harvey and Mr. G. W. Blackmore of the Admiralty Materials Laboratory. Contributed by permission of the Director, RRE. Copyright Controller HMSO.

References

1) J. B. Mullin, R. J. Heritage, C. H. Holliday and B. W. Straughan, J. Crystal Growth **3, 4** (1968) 281.
2) D. Richman, in: *Compound Semiconductors*, Vol. 1, Preparation of III–V Compounds, Eds. R. K. Willardson and H. L. Goering (Reinhold, New York, 1962) p. 214.
3) O. G. Folberth and E. Schillman, Z. Naturforsch. **12a** (1957) 943.
4) O. G. Folberth and H. Weiss, Z. Naturforsch. **10a** (1955) 615.
5) F. E. Rosztoczy, G. A. Antypas and C. J. Casau, in: *Intern. Symp. on Gallium Arsenide and Related Compounds*, Aachen, 1970 (Institute of Physics, London) p. 86.
6) J. B. Mullin, A. Royle and B. W. Straughan, in: *Intern. Symp. on Gallium Arsenide and Related Compounds*, Aachen, 1970 (Institute of Physics, London) p. 41.
7) Mining and Chemical Products Ltd., Wembley, Middlesex, U.K.
8) L. Eaves, R. A. Stradling, F. Askenazy, J. Leotin, J. C. Portal and J. P. Ulmet, J. Phys. C **3** (1970) L 42.
9) E. W. Williams, Phys. Rev. **168** (1968) 922.
10) C. J. Wang, Phys. Rev. **180** (1969) 827.
11) R. M. Logan and D. T. J. Hurle, J. Phys. Chem. Solids **32** (1971) 1739.
12) O. Röder, U. Heim and M. H. Pilkuhn, J. Phys. Chem. Solids **31** (1970) 2625.
13) E. W. Williams, unpublished results.
14) L. G. Harvey and G. W. Blackmore, Admiralty Materials Laboratory, Holton Heath, Poole, U.K., unpublished results.
15) B. D. Joyce and E. W. Williams, in: *Intern. Symp. on Gallium Arsenide and Related Compounds*, Aachen, 1970 (Institute of Physics, London) p. 57.

Reprinted from:
Journal of Crystal Growth **13/14** (1972) 647–650 © *North-Holland Publishing Co.*
Printed in the Netherlands

LIQUID ENCAPSULATED GROWTH OF GaP FROM NON-STOICHIOMETRIC MELTS

A. R. VON NEIDA, L. J. OSTER and J. W. NIELSEN

Bell Telephone Laboratories, Incorporated, Murray Hill, New Jersey 07974, U.S.A.

Crystals of GaP have been grown by the liquid encapsulation Czochralski (LEC) technique from non-stoichiometric melts at temperatures between 1430 °C and 1200 °C. Using melt compositions of 36 atom per cent phosphorus (36 at% P) and 30 at% P, satisfactory growth was obtained at 0.5–0.3 in./hr. These crystals which had a maximum diameter of 0.6 in. and weighed up to 25 g were free of second phase and contained lower dislocation densities (10^4 cm^{-2} and 10^2 cm^{-2} respectively) than GaP grown from stoichiometric melts. Crystals grown at rates between 0.3–0.1 in./hr from solutions containing 20 at% P and 10 at% P exhibited some cellular growth morphology and generally contained excess gallium. Correlated with decreasing growth temperature is a large increase in the red photoluminescence quantum efficiency reaching 0.8% for a crystal grown from a 10 at% P solution.

1. Introduction

Large area crystals of GaP have been prepared mainly by two methods, self nucleation from Ga rich solutions at around 1100 °C and the recently developed liquid encapsulated Czochralski (LEC) growth technique[1,2] from the melt at 1470 °C. Of principal interest is the use of this compound for light emitting diodes where large quantities of specifically oriented material are required. The LEC technique is superior for the production of large crystals, however, the efficiency of the light emitting process is much reduced in this material when compared with solution grown GaP[3]. To aid in understanding such differences a study was undertaken on the LEC growth of single crystals from Ga rich solutions employing growth temperatures between 1430 °C and 1200 °C. The technique for such growth as well as preliminary results of an examination of some physical and electrical properties of these crystals are described herein.

2. Experimental

The Czochralski puller used for the experiments was designed for growth to 100 atm and has been described previously[4]. All crystals, except where noted, were grown from 1.5 in. diameter silica crucibles using an initial charge of 12–16 cm^3. Crystals were seeded, with random face polarity, using $\langle 111 \rangle$ axis oriented crystals from standard LEC melts. The seeds rotating at 25 rpm were introduced into the solution at a temperature 6–10 °C higher than the dendrite formation point obtained on cooling. The encapsulant used to prevent phosphorus vapor loss from the GaP was B_2O_3. The properties of this material for encapsulation have been described[1,2]. Prebaking the B_2O_3 in a platinum crucible at 1500 °C for several hours provided excellent visibility at the B_2O_3–GaP interface both at the start of a run and as long as 12 hr later. The encapsulant thickness was 0.4 in. A pressure of 20 atm, well in excess of the dissociation pressure at the highest solution concentration employed, 36 atom per cent phosphorus (36 at% P), was used initially to insure no volatilization loss. Later, at this composition, it was found that growth could occur at reduced pressures in accordance with the vapor pressure measurements of phosphorus over GaP by Marina et al.[5]. At the melt composition 27 at% P about one atm N_2 pressure may be used over the encapsulated melt.

3. Crystals

Crystals were grown from four solution concentrations containing initially 36, 30, 20, and 10 at% P. These will hereafter be designated respectively as solutions A, B, C, and D. Some representative crystals are shown in fig. 1. Crystals from solution A could be conveniently pulled at 0.5 in./hr, although growth rates as high as 3 in./hr were used initially to aid in producing thin necks. With decreasing phosphorus solution concentration the growth velocity had to be reduced reaching about 0.1 in./hr for crystals from solution D (see

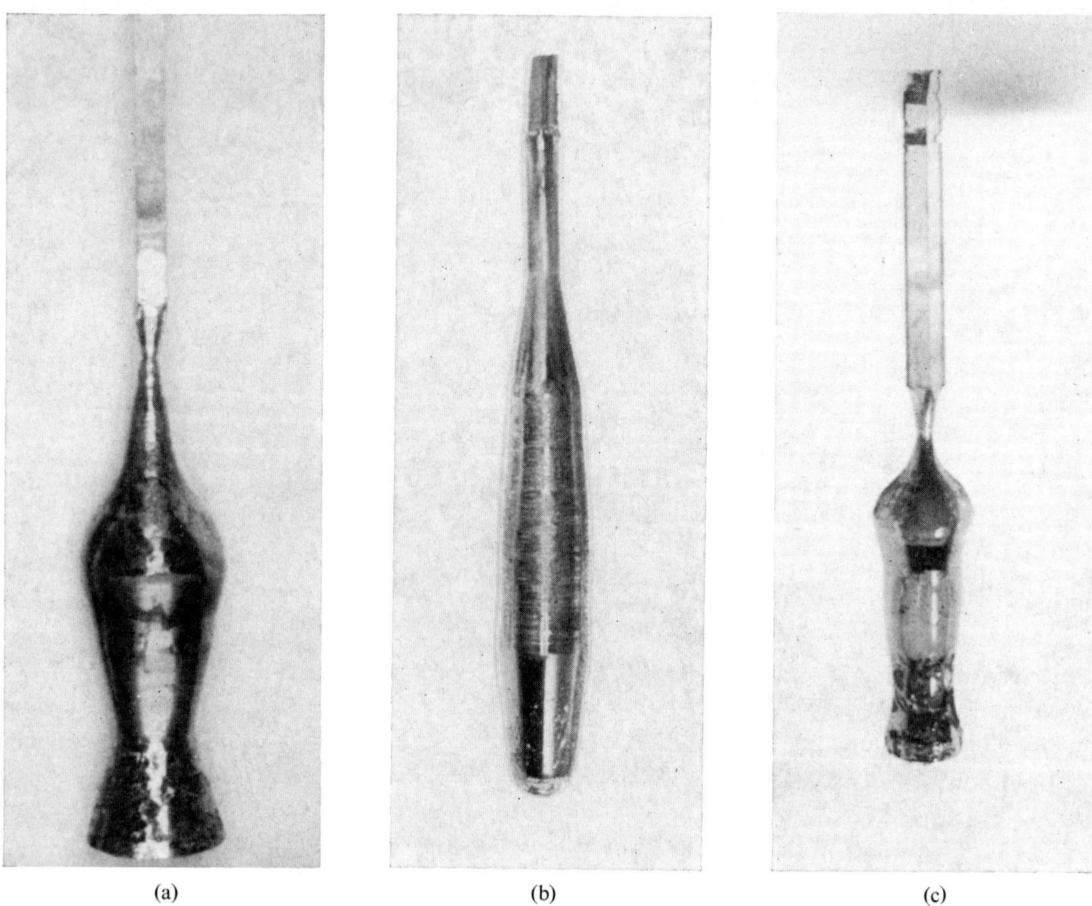

Fig. 1.   Crystals A, B, C, grown from 36, 30, and 20 at % P solution compositions, respectively. Surface cracking in lower sections of crystals A and C is due to B_2O_3 which adhered when crystals were suddenly raised from melt. ×1.2.

table 1). Crystals from the A and B solutions were usually single along the entire length. About one half of the eight crystals grown from the C and D solutions were entirely single, weighing from 2–8 g. The remainder either started poorly or became unstable at some point. Usually these crystals (C and D solutions) could not be grown much larger in diameter than about

0.35 in. One crystal from solution D, grown at 0.05 in./hr using an initial charge of 34 g GaP, 175 g Ga, in a 1.75 in. diameter crucible grew to a 0.5 in. diameter but the region in the center exhibited extensive cell formation and twinning. In order to achieve this diameter the temperature had to be lowered substantially probably resulting in considerable constitutional supercooling.

The macroscopic growth interface was always flat and most of the crystals (with the exception of those from solution D) showed extensive surface faceting with {211} plane faces. No effort was made to replenish the charge during growth so that each crystal was grown through a range of solution compositions. For example, crystals from the 36 at % P solution typically weighed about 25 g; with an initial charge of 45 g GaP and 24 g Ga the solution composition had shifted to near 29 at % P at the tail end of the crystal.

TABLE 1

Growth parameters and PL efficiency

Melt comp. (at% P)	Approx. growth (°C)	Nominal growth rate (in./hr)	PL efficiency (%)
50	1470	1	0.02
36	1430	0.5	0.04
30	1400	0.3	—
20	1320	0.3–0.1	0.1
10	1200	0.1–0.05	0.8

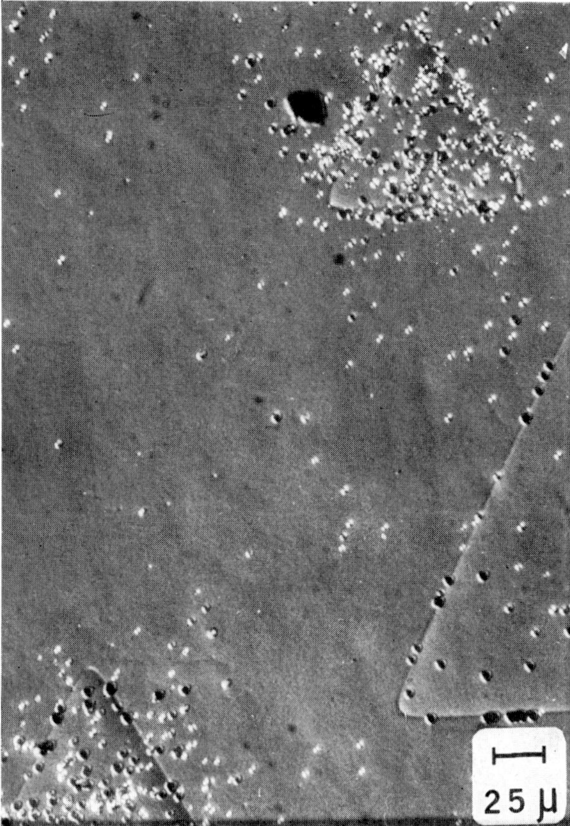

Fig. 2. Second phase inclusions present in a p-type crystal grown from a 10 at% P solution. Dislocation etch pits and impurity etch pits decorate the inclusions. Chemically polished in methanol-bromine and etched in AB solution. Orientation {111}. Contrast interference.

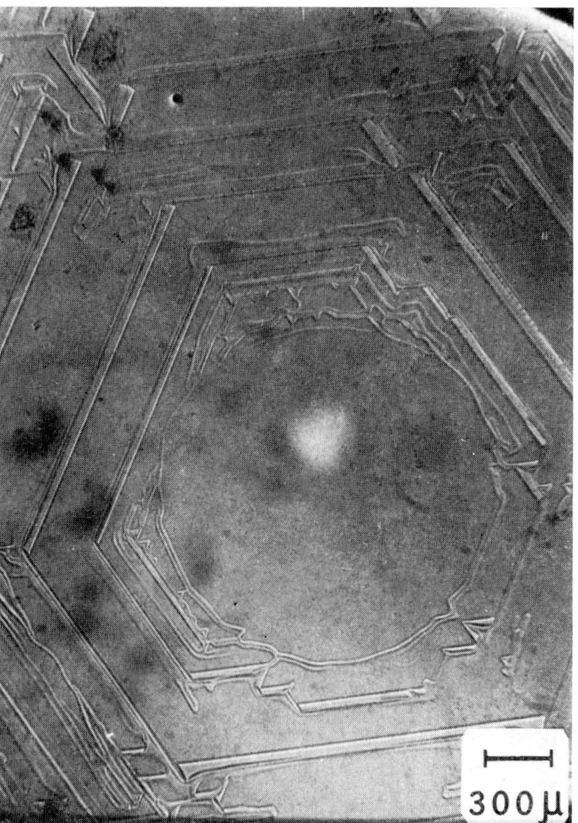

Fig. 3. Cellular morphology in crystal from D solution, mechanically polished and chemically etched in AB solution. Contrast interference.

4. Discussion

Although it was used mainly as a seal to prevent loss of the volatile specie, the B_2O_3 should also be considered for its role in promoting favorable crystal growth conditions. Use of the encapsulant provides greater thermal stability at the GaP surface reducing the chance of spurious nucleation. Chemical effects should also be noted since the melt surface will be cleaned by the B_2O_3 acting as a solvent for metal oxides. No attempt has been made to date to grow a crystal from an unencapsulated melt. However, Stambaugh et al.[6] have used the Czochralski technique (unencapsulated) for growth of GaP from solution at ≈ 1375 °C. Their ingots were largely polycrystalline.

Crystals from the A and B solutions showed little excess phase and no slip traces or twins. Occasionally some cellular structure was observed but the morphology was always fine and irregular. Crystals from the C and D solutions often showed narrow twin bands and had second phase inclusions (10–100 cm^{-2}) ranging in size from several μm to mm. Fig. 2 shows three triangularly shaped inclusions oriented along $\langle 211 \rangle$ in the {111} growth plane of the GaP. This type of inclusion results because Ga which is rejected by the freezing solid cannot be equilibrated (at the growth rates used) by diffusion with the main body of solution. Other results of constitutional supercooling, produced particularly in crystals from the D solutions, are shown in fig. 3. Here a coarse cellular morphology is developed in the {111} growth plane. Gallium or a Ga-rich phase once present in the cell boundaries has been removed by the chemical etch leaving the trace outlines.

It is expected with LEC solution growth that the concentration of structural defects will be reduced from that observed in crystals grown from the melt. Indeed, crystals from the 36 at% P solution had a dislocation density, $\rho = 10^4$–10^5 cm^{-2}, reduced by at least the factor 10 from material grown on stoichiometry. A 0.4 in. diameter section of one of these crystals showed an about equal dislocation distribution between core and edge, indicative of low radial strains associated with a small radial temperature gradient.

In crystals grown from initial solution concentrations of 30 at% P the defect densities decrease to much lower values. For some of these crystals dislocation etch pit densities <10 cm^{-2} have been observed[7]. It should also be noted that the impurity etch pit density as described by Iizuka[8] decreases in these crystals with a reduction pattern generally similiar to that discussed for dislocations. Further, it appeared that the dislocations were more mobile than those in stoichiometric melt grown material. In two crystals (Band C in fig. 1) a {211} X-ray diffraction topograph was used to study the structure in the neck region; it was observed that the dislocation network from the seed was immediately reduced in the initial growth region[9]. Such reduction may be related to a vacancy supersaturation causing dislocation climb to the interface. Following the initial decrease the dislocation network in the B crystal moved at a 30° angle to the growth direction, presumably along a {111} plane in a ⟨110⟩ direction disappearing at the surface. This crystal had a $\rho < 10^2$ cm^{-2}.

Crystals from the C and D solutions have a different defect configuration as shown in fig. 2. Both dislocations and impurity etch pits extensively decorate the second phase structure where the accommodation strains are high. In areas away from inclusions the total defect concentration is near zero.

It is evident that a decrease in the growth rate would be required to reduce second phase entrainment in crystals from the 20 at% P and 10 at% P solutions. Some improvement could be expected by steepening the thermal gradient at the GaP–B_2O_3 interface through changes in susceptor design and the use of thinner encapsulant layers. In fact, the lowest total defect densities (<10 cm^{-2}) have been obtained in crystals grown from the B solutions at higher pressures (60 atm); the effect of pressure is to increase the gradient at the GaP surface[4].

The red photoluminescence quantum (PL) efficiency of P type LEC stoichiometric melt grown crystals is typically of the order of 0.02% compared with a 4% efficiency for solution grown GaP[3]). Some of the present crystals were doped with zinc and oxygen in order to compare their PL efficiency with the above (see table 1). Low PL efficiency in GaP from stoichiometric melts is known to be associated with non-radiative centers which electrically shunt the luminescence path. Once present these complexes cannot be removed by any currently known technique. The origin of these centers is unknown at present but it is clear from the large increase in PL efficiency with decreasing growth temperature that the center concentration is principally determined by the highest thermal growth history of the material.

In summary, where higher photoluminescence efficiency and lower defect densities than those presently obtainable in LEC melt grown crystals are required, it is apparent that Czochralski growth from encapsulated solutions having compositions around 30 at% P can be used. Growth rates in this region allow reasonable size crystals to be produced in commensurate periods of time with much reduced over pressures. At lower phosphorous solution concentrations (20 at% P) and 10 at% P) the crystals exhibit cellular growth and entrained Ga.

Acknowledgments

The authors wish to thank the following people for assistance and helpful discussion: J. R. Carruthers, R. Caruso, M. DiDomenico, Jr., S. E. Haszko, T. Iizuka, G. Rozgonyi, and L. J. Varnerin.

References

1) J. B. Mullin, R. J. Heritage, C. H. Holliday and B. W. Straughan, J. Crystal Growth **3, 4** (1968) 281.
2) S. J. Bass and P. E. Oliver, J. Crystal Growth **3, 4** (1968) 286.
3) R. Caruso, M. DiDomenico, H. W. Verleur and A. R. Von Neida, J. Phys. Chem. Solids, to be published.
4) A. R. Von Neida and J. W. Nielsen, to be published.
5) L. I. Marina, A. Ya. Mashelskii, V. N. Vigdorovich and D. D. Bakonova, Russian J. Phys. Chem. **38** (1964) 296.
6) E. P. Stambaugh, J. F. Miller and R. C. Himes, *Metallurgy of Elemental and Compound Semiconductors* (Interscience, New York, 1960) p. 319.
7) G. A. Rozgonyi, A. R. Von Neida and O. C Miller, J. Elect. Mater., A. I. M. E., to be published.
8) T. Iizuka, J. Electrochem. Soc., to be published.
9) S. E. Haszko, private communication.

Reprinted from:
Journal of Crystal Growth **13/14** (1972) 651–656 © *North-Holland Publishing Co.*
Printed in the Netherlands

SOLUTION GROWTH OF GALLIUM PHOSPHIDE p-n JUNCTIONS BY LIQUID PHASE EPITAXY

A. R. PEAKER, P. D. SUDLOW and A. MOTTRAM

Ferranti Ltd., Chadderton, Lancashire, England

Layers of gallium phosphide have been grown by homo-epitaxy from gallium solution onto liquid-encapsulated Czochralski grown substrates. The impurities tellurium and zinc have been introduced to produce electroluminescent p-n junctions and electrical studies of these dopants in gallium phosphide have been used to investigate their incorporation into the crystal lattice near the p–n junction. During growth the distribution of dopants predicted from the values of effective and equilibrium distribution coefficients is disturbed by the diffusion of zinc from the doped gallium solution and from the p-type material into the n-type layer. Under normal growth conditions this latter effect has been found to be of greater significance than the redistribution of impurities associated with the back etching and regrowth of the n-type region.

1. Introduction

The liquid epitaxy process using gallium solution is almost universally employed for the growth of layers of gallium phosphide for light emitting diodes. Although the use of suitable impurities as dopants allows both red and green electroluminescent emission to be obtained, this paper is only concerned with the double liquid epitaxy structure in which tellurium is used in the n-type region and zinc and oxygen in the p-type layer. This combination is normally used to produce red emitting devices from gallium phosphide. The layers are grown separately in different sets of apparatus, the n-type layer being grown first so that the p–n junction is formed at the very beginning of the second growth cycle. This stage of the process is of paramount importance in studies of light emission by electroluminescence and although the added impurities have a dominant role in determining the light emission and electrical characteristics of the final device only the region within a few micrometres of the junction plays any part in the electroluminescent process. The necessity to establish accurately, concentrations of less than ten parts per million in a region only a few microns wide severely limits the number of available analytical techniques. Electrical methods have been developed to study free carrier concentrations in the regions of p–n or metal-semiconductor junctions[1–3] and this paper describes the use of these techniques and their subsequent interpretation in terms of impurity concentra-

tions to study the processes occurring during the early stages of the second growth cycle.

2. Growth method

The equipment used for the growth process is shown in fig. 1. The melt was held in a quartz crucible and the substrates on a flat vertical spade which could be withdrawn or inserted into the melt as desired. An atmosphere of palladium diffused hydrogen or hydrogen-nitrogen mixture was established in the furnace tube and the furnace arranged so that the temperature could be raised or lowered at a rate between 1 and

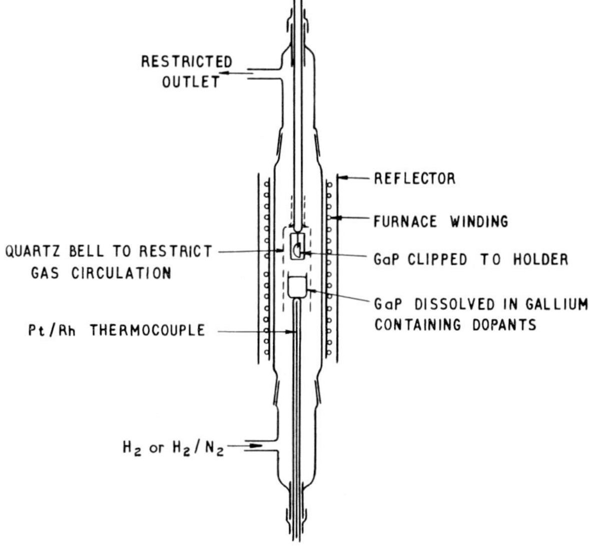

Fig. 1.   The liquid epitaxy apparatus.

20 °C min^{-1}. The epitaxial layers were grown on the (111)B faces of polished substrates cut from ingots grown by the liquid encapsulated Czochralski process. The tellurium doped layers were grown from a gallium melt consisting of approximately 10^{-2} at% of 99.999% pure tellurium and 8.4 mole% gallium phosphide, thus ensuring that excess gallium phosphide was present at all melt temperatures used. The melt was raised to 1060 °C and held there for several minutes until thermal equilibrium was reached. The spade was then inserted and the temperature raised slightly to back etch the substrate, this being necessary to ensure that the substrate surface was free from localised work damage and contamination. After a few seconds the controlled cooling cycle was initiatied and the temperature reduced at the rate of 10 °C min^{-1} until a temperature of 910 °C was reached, the layers were then withdrawn from the melt and cooled rapidly. The growth procedure for the zinc–oxygen doped layers was identical to that used for the tellurium doped material except that the substrates were withdrawn at a lower temperature, the doping being accomplished by adding approximately 0.09 mole% of ZnO and 0.04 mole% Ga_2O_3 to the melt. Previous work has shown that the oxygen is present in the grown layers only to the extent of 5×10^{16} atoms cm^{-3} due to its low solubility and low distribution coefficient from gallium solution[4,5]. Impurity profiles of zinc and zinc oxygen doped gallium phosphide with a net carrier concentration of 10^{18} cm^{-3} have shown that the effect of oxygen on the distribution of the major impurity is negligible and consequently in the subsequent discussion of impurity levels the presence of oxygen in the grown layer is neglected.

3. Measurement techniques

Two principal methods of measurement have been used, these are the determination of the capacitance–voltage relationship of the grown gallium phosphide diode which establishes the shape of the impurity profile in the immediate vicinity of the junction and a Schottky diode technique[3] which enables a more general picture of the impurity distribution throughout the layer to be obtained. Both these methods can only determine the free carrier concentration which is normally equal to the net value of electrically active impurity concentration, i.e. it represents the difference between ionised donor and acceptor concentrations,

the added tellurium acting as a donor and the zinc as an acceptor in the gallium phosphide. Although in the bulk of each individual layer the concentrations of donors and acceptors differ by two orders of magnitude, the levels near the p–n junction are much more closely matched resulting in a highly compensated region and consequently this plays a significant part in the interpretation of the results discussed in section 5. However the individual values of donor and acceptor concentrations can be related quite accurately to the total impurity concentration by radiotracer techniques[6] and over the range of values studied in this paper the two concentrations can be taken as being almost identical.

The Schottky diode technique consists of polishing a shallow bevel through the depth of the grown layer at an angle usually of somewhat less than 1° to the original surface. By evaporating a matrix of gold dots onto this face and measuring the change of capacitance with applied voltage for each diode in turn it is possible to determine the impurity profile at 5 μm intervals through the depth of the layer to an absolute accuracy of $\pm 10\%$ and a relative accuracy of $\pm 2\%$. Similarly the measurement of the change of capacitance with voltage at the p–n junction enables the shape of the doping profile to be estimated up to a distance of 0.3 μm from the junction. In cases where an abrupt transition in carrier concentration occurs at the junction the absolute value of $n_j p_j/(n_j + p_j)$ can be determined where p_j is the hole concentration in the p-type material and n_j the electron concentration in the n-type material at the junction.

The capacitance measurements were made using an autobalance RF bridge operating at 1 MHz with a signal amplitude of 5 mV. Electrical connections were made via a gold wire probe attached to a micro-manipulator head. The values of carrier concentration for individual dots in the matrix were determined from the gradients of the C–V curves at an appropriate value of capacitance.

4. Results

The doping gradients in the bulk of individual epitaxial layers have been analysed and discussed in detail elsewhere[3,7]. However, when a p–n junction is grown these profiles are significantly modified at considerable distances from the junction. Fig. 2 shows a carrier concontration profile on layers where the impurity level in

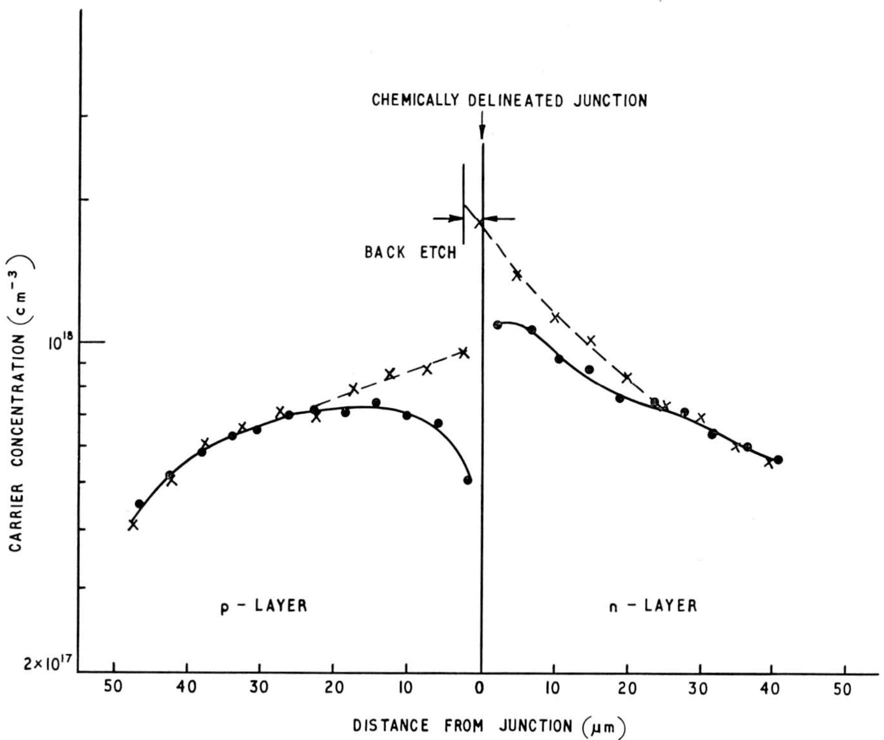

Fig. 2. Doping gradients for the case where $n > p$. The circles show the values at the p-n junction and the crosses the values obtained when the layers are grown individually.

the n-type region is somewhat higher than in the p-type material. The dots represent the values measured on a p–n junction slice using the Schottky technique and the crosses measurements on separate individual layers grown at the same time as the corresponding layer on the p–n junction slice but using a p-type substrate of comparable carrier concentration in the case of the p-type epitaxial layer and a n-type high resistivity substrate for the n-type layer. The values obtained enable the impurity levels throughout the p–n junction to be compared with the impurities present when the distribution is not disturbed by the proximity of a region of the opposite type. In the case shown in fig. 2 it can be seen that the carrier concentration of the p-type layer decreases rapidly towards the junction and that both profiles of the p–n junction material deviate significantly from the values obtained on individual layers even at distances up to 20 μm from the junction. The Schottky technique cannot be used to determine the carrier concentration at the actual junction but the value of $n_j p_j/(n_j + p_j)$ can be established in this region. In the case shown in fig. 2 the value was 3×10^{17} cm^{-3} indicating that the carrier concentration of one of the

layers continued to fall from the values that were measured some 3 μm from the junction. It seems most likely that the trends determined by the Schottky techniques will continue up to the junction so agreeing with the p–n junction value making the probable carrier concentrations of the p and n layers at the actual junction 4.3×10^{17} cm^{-3} and 1.0×10^{18} cm^{-3} respectively. The value ascribed to the thickness of the part of the layer which was back-etched, was obtained by measuring the difference in thickness of the n-liquid layer before and after growth of the p-region.

Fig. 3 illustrates a case where the doping level in the p-layer is higher than in the n-layer and compared to fig. 2 shows the larger deviation observed between n-type profiles before and after growth of the p-layer. Junction capacitance measurements on this layer have shown the value $n_j p_j/(n_j + p_j)$ to be 2×10^{17} cm^{-3}.

5. Discussion

The difference in the n-type carrier concentration profiles measured on the p–n junction layers and on the layers grown without a p-layer is very substantial. As the deviation extends 20 μm into the layer and com-

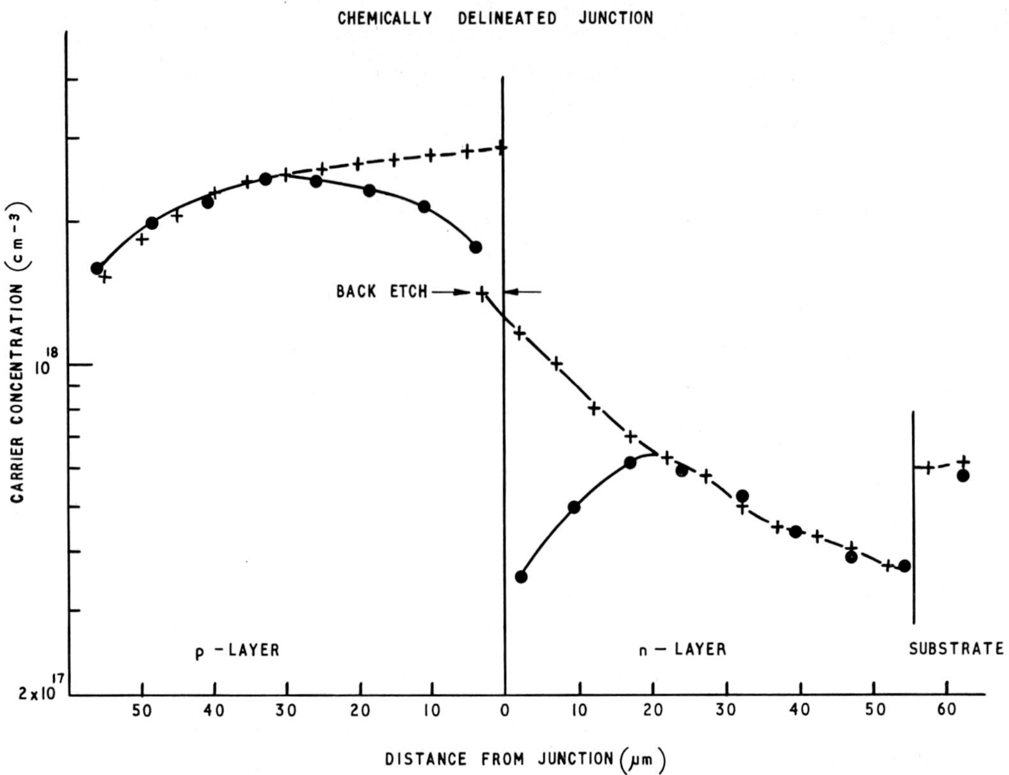

Fig. 3. Doping gradients for the case where $p > n$. The circles show the values at the p–n junction and the crosses the values obtained when the layers are grown individually.

pared to this the degree of back etching is very small, the variation can be accounted for by two possible mechanisms. The measurement is of free carrier concentration so the deviation can be due to the loss of the added donor atom or the addition of an acceptor atom. The rate of diffusion of tellurium in solid gallium phosphide is extremely low and so any change in distribution throughout the layer on the scale shown could only be attributed to an annealing effect. A study of the effects of annealing has shown that it does not influence either the overall carrier concentration level or the slope of the doping profile. Consequently it appears that the deviation between the two n-type profiles is due to the diffusion of zinc into the layer. The zinc could originate from either the p-type layer itself or from the gallium solution while the n-type layer was in contact with it, for example during the back etch process or more likely during the period between back etching and the start of growth. This period is normally kept as short as possible but inevitably there must be some time delay because of the thermal capacity of the system and because of the necessity to establish some

supersaturation. This period, referred to subsequently as the dwell time, is estimated to be approximately 2 min. The two diffusion processes are essentially different, the movement of zinc from the p-type grown layer approximates to a limited source diffusion whereas the gallium melt acts as a low concentration infinite source. The effective concentration of this source would be expected to be considerably less than the concentration in the melt because of the gallium–gallium phosphide interface acting as a barrier. In order to establish the zinc diffusion behaviour under these conditions several n-type Czochralski grown substrates with a carrier concentration of 5×10^{16} cm^{-3} were immersed in Ga/Zn solution at 1060 °C. After slight back etching the slices were left in contact with the gallium melt for varying periods, the temperature being maintained constant. On removing the slices it was found that a junction could only be detected on those which had been immersed longest (10 min) and this was only 2 μm from the surface. The carrier concentration at the surface was marginally less than that predicted from the equilibrium distribution coefficient and consequently it was

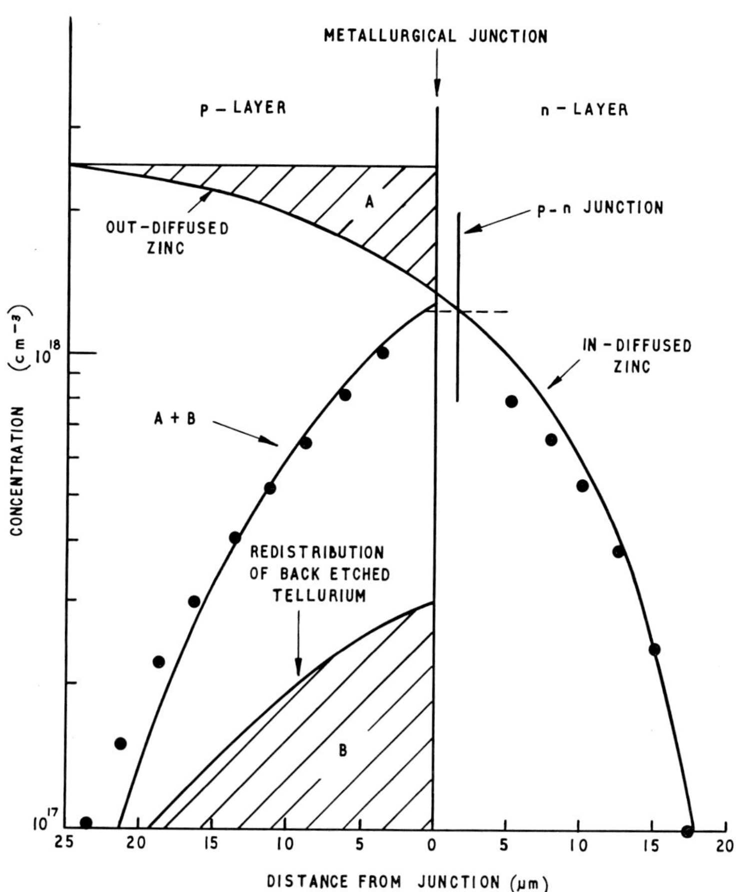

Fig. 4. Analysis of theoretical compensation levels for the condition where $p = 2.5 \times 10^{18}$ cm^{-3} and $n = 1.2 \times 10^{18}$ cm^{-3} using computed zinc diffusion data and taking k_0 for Te at 1040 °C (Ga$_L$ → GaP$_s$) to be 0.25. Theoretical values are shown as continuous lines and the measured values derived from the data represented in fig. 3 are shown as solid circles.

assumed that a thin layer had been grown isothermally and the contribution from the Ga → GaP diffusion of zinc was small enough to be neglected under these circumstances.

White[8]) has calculated the diffusion profiles of zinc in gallium phosphide under various conditions taking the value for the diffusion coefficient[9]), D, as:

$$D = 7.5 \times 10^{-8} C^{0.45} \exp(-2.5/kT) \text{ cm}^2 \text{ s}^{-1},$$

where C is the concentration of zinc in the solid in atoms cm^{-3}. For the conditions occurring during the growth of the gallium phosphide layers as shown in fig. 3, namely a temperature decrease at the rate of 10 °C/min and a limited source of 2.5×10^{18} cm^{-3} taken to be 25 μm thick, the profile extends into the n-type layer as shown in fig. 4.

The level of zinc diffusion into the layer for the case shown is higher than the level of tellurium in the layer, consequently the p–n junction moves during the diffusion process. It is, therefore, necessary to distinguish between the metallurgical junction and the p–n junction. The metallurgical junction is the point at which back etching stops and growth starts whereas the p–n junction is the electrical boundary. Where chemical etching is used to delineate the junction it is the p–n junction which will be most apparent and consequently the data presented in figs. 2 and 3 is given in relation to the p–n junction not the metallurgical junction. Although the p–n junction represents a change in material polarity a large change in carrier concentration should appear at the metallurgical junction, such a change can also be stained chemically in favourable circumstances and cases have been observed where a weaker line is stained close to the strong p–n junction demarcation.

This is true only where the overall carrier concentration of the n-layer is less than that of the p-layer and the separations observed have been between 1 and 5 μm. Calverley and Wight[10]) using the cathodoluminescent mode on a scanning electron microscope have observed "dead layers" between 0.1 and 10 μm wide at gallium phosphide junctions grown by techniques similar to those used in this investigation. Although the analysis of the cathodoluminescent behaviour is considerably more complex than the primary consideration of dominant impurity concentration in this work, it would be expected that a p-type region produced by zinc diffusion into tellurium doped gallium phosphide would not display red cathodoluminescent emission.

The calculated diffusion profile represents a somewhat idealised case, the p-type layer is assumed to grow faster than the diffusion front moves and although this is probably the case during the first 20 μm the growth is much slower in the later stages. This and the fact that the p–n junction presents a barrier to the zinc diffusion in a manner somewhat different to Chang's equation[9]) is though to account for the deviation between the calculated and measured profiles in the n-type region as shown in fig. 4.

In the p-type layer the carrier concentration profile is influenced by the loss or out diffusion of zinc and for the case in fig. 3 this is shown as area A in fig. 4. However, in addition, after back etching a local concentration of tellurium exists in the solution at 2×10^{18} cm^{-3}. As the diffusion of tellurium in gallium is slow this will be redistributed in the p-layer during the early stages of growth. Assuming that the distribution coefficient of tellurium ($k_0 = 0.25$ at 1040 °C) is unaffected by the presence of the zinc dopant the level of tellurium in the first grown region of the zinc doped layer will be of the order of 3×10^{17} cm^{-3}. An exact value for the diffusion coefficient of tellurium in gallium is not known but from previous work[7]) it is apparent that the rate of diffusion is comparable with the growth rate. The

profile of the redistributed tellurium has been calculated on this basis and the results shown in fig. 4 as area B.

When the loss of acceptor atoms is added to the gain of donor atoms the result should be equal to the numerical difference between the two p-type layer profiles as shown in fig. 3. As can be seen from fig. 4, this is in fact the case and agreement between the theoretical and measured values is apparent. A similar analysis of the compensation levels for the case where $n > p$ as shown in fig. 2 also gives good general agreement although not as precisely as the layers discussed above, probably due to the greater significance of the redistribution of the back etched tellurium, the profile of which is not accurately predictable due to uncertainties in the value of its diffusion coefficient in gallium.

Acknowledgements

We would like to thank Mr. M. Abbott for his assistance with the electrical measurements and Mr. A. Calverley and Dr. D. Wight of S.E.R.L. for helpful discussions. This work was done in part under a C.V.D. contract and is published by permission of the Ministry of Defence (Navy Department).

References

1) Y. F. Chang, Solid-State Electron. **10** (1967) 281.
2) W. Nuyts and R. J. Van Overstraeten, Electron. Letters **5** (1969) 54.
3) A. R. Peaker and B. L. Smith, Solid-State Electron. **13** (1970) 1407.
4) L. M. Foster and J. Scardefield, J. Electrochem. Soc. **116** (1969) 494.
5) A. R. Peaker, S. J. Fisk and A. Mottram, Electron. Letters **5** (1969) 186.
6) F. A. Trumbore, H. G. White, M. Kowalchik, R. A. Logan and C. L. Luke, J. Electrochem. Soc. **112** (1965) 782.
7) P. D. Sudlow, paper presented at B.A.C.G. Bristol 1970, to to published J. Mater. Sci.
8) T. White, private communication.
9) L. L. Chang and G. L. Pearson, J. Appl. Phys. **37** (1964) 2.
10) A. Calverley and D. Wight, Solid-State Electron. **13** (1970) 382.

Journal of Crystal Growth **13/14** (1972) 657–662 © *North-Holland Publishing Co.*

PHASE DIAGRAMS AND CRYSTAL GROWTH OF PSEUDOBINARY ALLOY SEMICONDUCTORS*

JACQUES STEININGER** and ALAN J. STRAUSS

Lincoln Laboratory, Massachusetts Institute of Technology, Lexington, Massachusetts 02173, U.S.A.

The ternary Zn–Cd–Te and pseudobinary CdTe–CdSe and PbTe–PbSe phase diagrams have been determined by thermal and X-ray analysis and analysed in terms of thermodynamic theories of liquid and solid solutions. The phase diagram data for two of the investigated systems has then been applied to the selection and application of two different techniques for growing pseudobinary alloy crystals: diffusional freezing and solution zoning.

1. Introduction

The development of techniques for growing alloy crystals from liquid solutions[1–4] has created a need for comprehensive and precise thermodynamic data, especially phase diagrams of liquid–solid equilibria in pseudobinary and ternary systems. The principal methods used for the determination of liquidus and solidus data (thermal analysis, metallographic or X-ray analysis of quenched samples, and determination of segregation coefficients) are non-equilibrium techniques and may be subject to significant experimental errors. These errors are usually difficult to detect from thermodynamic first principles because of our limited understanding of non-ideal solutions and because of the paucity of basic thermodynamic data for many of these systems.

This communication is concerned with the determination of the ternary Zn–Cd–Te and pseudobinary CdTe–CdSe and PbTe–PbSe composition–temperature phase diagrams by thermal and X-ray analysis. The resulting phase diagrams have been analyzed in terms of thermodynamic theories of liquid and solid solutions. This led in particular to the development of a new method for the calculation of the liquidus–solidus gap in homogeneous, monotonic alloy systems. The phase diagram data for two of the investigated systems, Zn–Cd–Te and CdTe–CdSe, have also been applied to

the selection and application of two different techniques for growing pseudobinary alloy crystals: diffusional freezing and solution zoning.

2. Phase diagrams

2.1. EXPERIMENTAL

The liquidus and solidus curves for the pseudobinary CdTe–ZnTe, CdTe–CdSe and PbTe–PbSe systems (fig. 1) and the liquidus surface for the ternary Zn–Cd–Te system (fig. 2) have been determined by thermal analysis of high purity liquid and solid alloy samples in a specially constructed D.T.A. apparatus[5–7]. Particular attention was paid to thorough homogenization of both the liquid and solid samples by prolonged annealing at high temperature for periods of up to two weeks. The liquidus or solidus temperatures for non-congruently melting compositions were determined exclusively from the initial thermal arrests either on cooling or heating curves at slow rates of 2 °C/min. The accuracy of the thermal arrests was estimated to be better than ±1 °C for the liquidus curves and about ±2 or 3 °C for the solidus curves.

The sub-solidus boundaries of the cubic and hexagonal phase fields in the CdTe–CdSe system have been investigated by X-ray analysis of annealed and quenched powder samples in the temperature range of 800 to 1050 °C[6].

2.2. RESULTS

The CdTe–ZnTe and PbTe–PbSe systems show typ-

* This work was sponsored by the Department of the Air Force.
** Present address: Arthur D. Little, Inc., Cambridge, Massachusetts 02140, U.S.A.

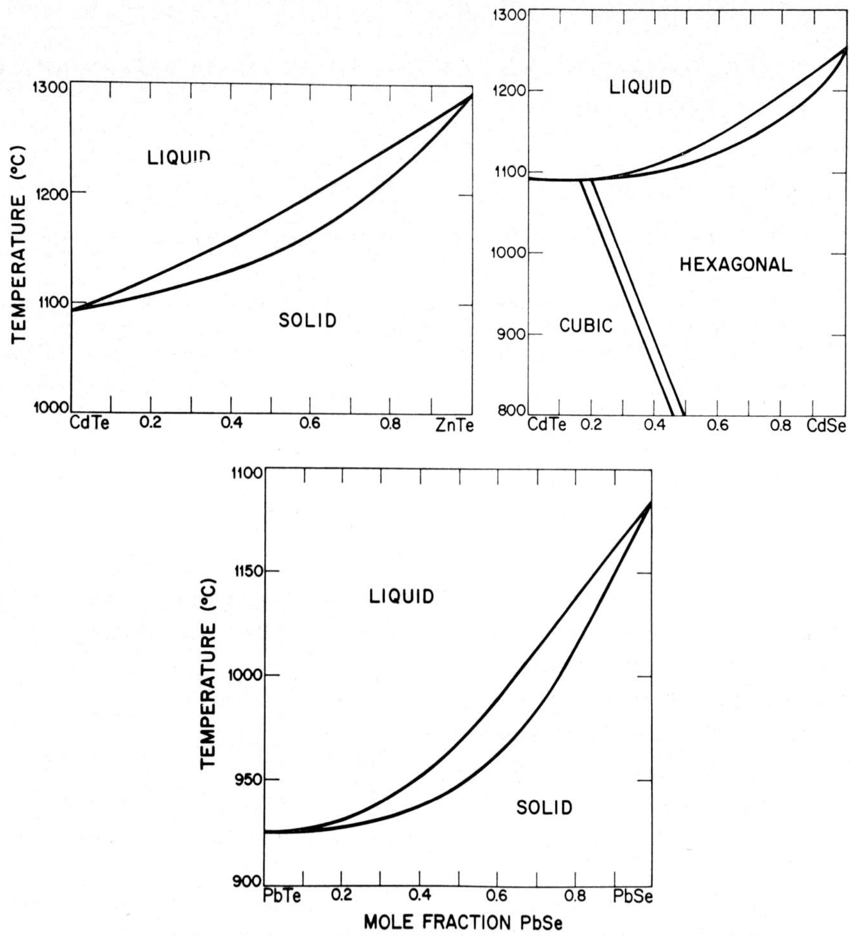

Fig. 1. Phase diagrams of the CdTe–ZnTe, CdTe–CdSe and PbTe–PbSe pseudobinary systems.

ical lens-shaped phase diagrams with sublinear variations in temperature with composition and relatively narrow liquidus–solidus gaps (fig. 1). The CdTe–CdSe system presents almost the same type of liquidus–soli-

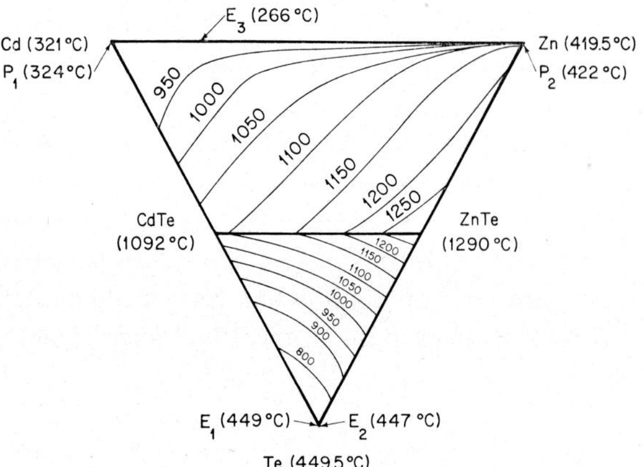

Fig. 2. Phase diagram of the Zn–Cd–Te ternary system.

dus phase diagram with a degenerate eutectic near 20 mole% CdSe. The narrow cubic–hexagonal two-phase region inside the solidus field shows increasing stability of the hexagonal phase with increasing temperature.

The ternary liquidus isotherms in the Zn–Cd–Te system (fig. 2) show a smooth variation in liquidus temperature between the binary Cd–Te and Zn–Te systems. The liquidus surface is strongly asymmetric with higher temperatures on the metal-rich side. The marked increase in temperature along the pseudobinary CdTe–ZnTe composition line is attributed to ordering in the liquid phase resulting from the strong interactions between chalcogen and metal atoms.

3. Thermodynamics

3.1. PSEUDOBINARY SYSTEMS

The experimental liquidus and solidus data for the

TABLE 1

Difference between partial excess free energies of mixing in solid and liquid phases for binary and pseudobinary alloy systems

System	Mean D (kcal/g-at)	Standard deviation (kcal/g-at)
CdTe–ZnTe	−0.040	0.252
CdTe–CdSe	+0.168	0.272
PbTe–PbSe	+0.099	0.486

three pseudobinary systems have been used to calculate the values of the difference D between the partial excess free energies of mixing in the two phases

$$D = \frac{\partial \Delta G_{e,m}^{s}}{\partial X_{A}^{s}} - \frac{\partial \Delta G_{e,m}^{l}}{\partial X_{A}^{l}},$$

from the general[8]) liquidus–solidus equation:

$$RT\left(\ln \frac{X_{A}^{s}}{X_{B}^{s}} - \ln \frac{X_{A}^{l}}{X_{B}^{l}}\right)$$
$$= \Delta H_{A}\left(1 - \frac{T}{T_{A}}\right) - \Delta H_{B}\left(1 - \frac{T}{T_{B}}\right) - D.$$

As shown in table 1, the values of D are remarkably small. This result has also been obtained for a large number of binary and pseudobinary systems which exhibit complete miscibility in the two phases and monotonic variations of the liquidus and solidus curves[8]). It is attributed to the relatively limited (but not negligible) deviations from ideality in these systems. For this type of system, the ideal liquidus–solidus equation where D is neglected can therefore give a good approximation of the relationship between equilibrium concentrations in the two phases. This ideal equation is of particular interest for the calculation of one of the boundaries of the two-phase region (solidus or liquidus) when the other one (liquidus or solidus) and the enthalpies of fusion of the terminal compounds are already known. Typical applications[8]) include the prediction of one of the two phase boundaries when experimental data are unavailable (such as for the ZnTe–ZnSe and HgTe–ZnTe systems) and the detection of thermodynamic inconsistencies in published phase diagrams (Cu–Ni, InAs–InP, HgTe–HgSe, PbTe–PbSe systems).

3.2. TERNARY SYSTEMS

The experimental determination of the ternary liquidus–solidus tie-lines, which are needed for the applica-

tion of crystal growth techniques from non-stoichiometric solutions, can be prohibitively time-consuming and prone to excessive experimental errors. Attempts have therefore been made to develop models of liquid solutions that can provide a better understanding of these systems and can be used to calculate the ternary phase diagram data from the more easily determined and more reliable thermodynamic data of binary and pseudobinary systems. Preliminary results[9]) however indicate that theoretical models based on physical or chemical concepts of the liquid phase (such as the regular, quasichemical, regular associated or non-random two-liquid solution models) fail to give a satisfactory representation of the strong physical interactions even in the binary systems. More consistent results however can be obtained with several empirical thermodynamic expressions[9]).

4. Crystal growth

The major difficulty in the growth of homogeneous alloy crystals results from the differences in the equilibrium compositions of the liquid and solid phases, which lead to segregation during solidification and compositional variations in the grown crystals.

4.1. DIRECTIONAL FREEZING

Directional freezing of a liquid alloy solution under equilibrium conditions with complete mixing in the liquid phase results in the formation of solid alloy solutions of continuously varying composition along the growth axis. The theoretical profile shown in fig. 3 demonstrates the distribution for a constant segregation coefficient of 0.7. In reality, mixing is never complete and solidification proceeds under near-equilibrium conditions with a tendency to form a solute-rich boundary layer near the solidifying interface. As a result, the initial concentration gradient in the solid is even higher than that shown by the theoretical curve. If mixing is completely suppressed, the regime then becomes purely diffusional and the composition profile[10]) takes the shape indicated by the curves 1, 2 or 3 corresponding to increasing values of the freezing rate V. As indicated in fig. 3, it should be noted that the curves 1–3 for diffusional directional freezing are exactly similar to the more familiar profiles obtained for zone melting with progressively narrower liquid zones. These profiles are characterized by an initial transient of steadily increasing solute concentration, a steady state region with a constant concentration equal to the initial melt composition and a final transient of rapidly increasing solute concentration.

The region of constant composition is of particular interest for the preparation of homogeneous alloy crystals. Its length can be increased at the expense of the two transient regions by increasing [10]) the rate of solidification V. In order to avoid constitutional supercooling however, it is necessary to maintain sufficiently high values of the temperature gradient G in the liquid ahead of the interface. The minimum values required for G in various alloy systems have been calculated from the phase diagram data using the familiar simplified relationship shown in table 2 for a convenient numerical value of $V/D = 1$ cm, where D is the diffusion constant in the liquid. For a typical value of $D = 5 \times 10^{-5}$ cm^2/sec, V is equal to about 4.3 cm/day which corresponds to commonly used rates of directional freezing. Table 2 shows that the minimum value of G is strongly dependent upon the values of the phase diagram parameters and can vary over more than three orders of magnitude for various systems.

In contrast with most other crystal growth techniques, which strive to achieve near-equilibrium conditions, the use of directional freezing to obtain the diffusional profiles shown in fig. 3 requires the establishment of non-equilibrium conditions which may be difficult to maintain. In the present study, homogeneous cm-size single crystals of CdTe–ZnTe alloys have been grown by directional freezing under conditions closely approx-

TABLE 2

Constitutional supercooling in alloy systems

$$\frac{G}{V} > \frac{mC_0}{D} \frac{1-k}{k}$$

Alloy system	C_s for $C_0 = 0.50$	k	m (°C/mole fraction)	G (°C/cm) for $V/D = 1$ cm
Ag–Au	0.49	0.98	90	0.9
SnTe–PbTe	0.40	0.80	110	13.7
CdTe–CdSe	0.40	0.80	186	23.2
CdTe–ZnTe	0.35	0.70	210	45.0
HgTe–CdTe	0.19	0.38	600	490.0
InSb–AlSb	0.05	0.10	270	1215.0

C_0 = initial melt composition,
C_s = solidus composition for $C_0 = 0.50$,
k = segregation coefficient,
m = liquidus slope,
D = diffusion constant in liquid,
V = solidification rate
G = temperature gradient in liquid ahead of interface.

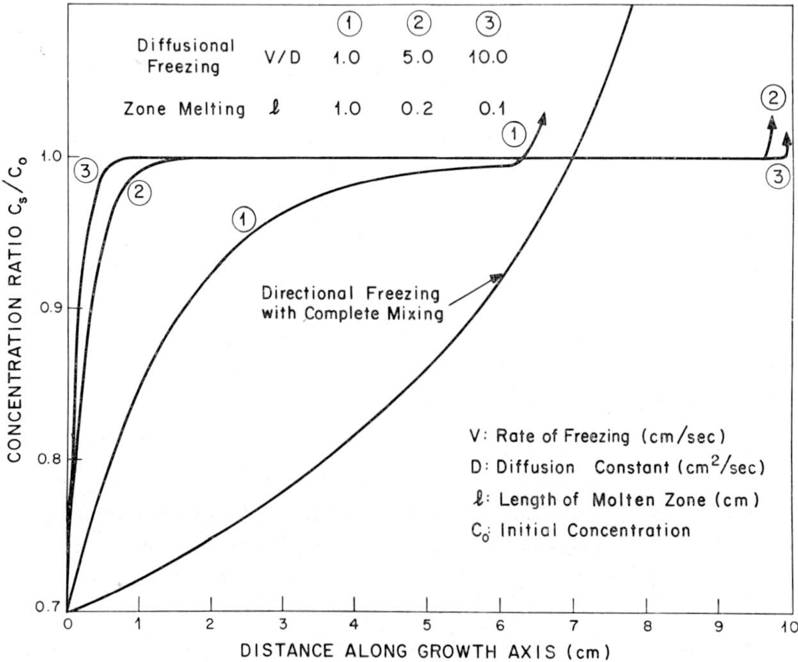

Fig. 3.   Theoretical composition profiles in directional freezing of alloys; $k = 0.7$.

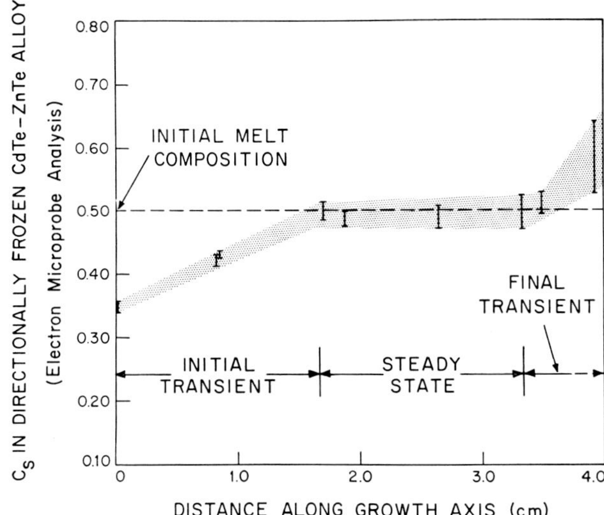

Fig. 4.   Composition profile of directionally frozen CdTe–ZnTe alloy.

imating a purely diffusional regime. To minimize mixing in the melt, we used resistance heating, temperature gradients limited to about 30 °C/cm and small vertical ampoules about 1 cm in diameter. Fig. 4 shows the result of an electron microprobe analysis of one such ingot, grown at a rate of 4 cm/day, which exhibits a typical diffusion-controlled profile, with a steady state region of constant concentration equal to the initial melt composition of 50 mole% ZnTe. The middle regions of the crystals were examined by electron microprobe and metallographic analysis and found to be homogeneous with no evidence of dendritic or cellular growth. The segregation coefficient for the first-to-freeze tip of this ingot (corresponding to the initial equilibrium freezing) had a value of 0.7, in excellent agreement with the phase diagram obtained by thermal analysis[6]). The amount of solute accumulated in the boundary layer was determined by integration along the initial transient and used to calculate the diffusion coefficient in the steady state region, giving a value of 5.6×10^{-5} cm²/sec. The minimum temperature gradient required to avoid constitutional supercooling was then calculated to be about 30 °C/cm in good agreement with the experimental evidence obtained by varying the freezing rate between 1 and 10 cm/day.

4.2. SOLUTION ZONING

The data of table 2 show that for certain alloy systems the thermal gradients required to avoid constitu-

tional supercooling at reasonable rates of growth can become quite significant. Relatively high and stable temperature gradients can be obtained by using various technical devices, such as the heat pipes[11]) which are to be discussed in another session of this Conference[12]). For certain systems however, it obviously becomes imperative to use a more suitable crystal growth technique. In the present study, the possibility of applying the temperature gradient solution zoning technique, previously used for the growth of II–VI compound crystals[4]), has been investigated for stoichiometric solutions in the CdTe–CdSe system and for non-stoichiometric solutions in the Zn–Cd–Te system. The experimental procedure used was the same as for compound crystals except that two separate charges of homogenized alloys were prepared: a zone charge about 1 cm long with a composition corresponding to the liquidus and a feed charge about 5 cm long with a composition corresponding to the solidus. A sealed quartz ampoule containing the two charges was placed in a vertical furnace with a small temperature gradient of about 10 °C/cm at the temperature corresponding to the selected equilibrium liquidus and solidus compositions.

In the case of the CdTe–CdSe alloys, the two charge compositions were selected to correspond to compositions on the pseudobinary phase diagram, which was accurately known. After a period of approximately 10 days, the zone charge was found to have migrated upward through the feed charge leaving behind crystals having the composition of the feed. This method is particularly attractive for pseudobinary systems with wide liquidus–solidus gaps. Its success depends on precise knowledge of the equilibrium liquidus and solidus compositions and on good control of the temperature and temperature gradient in the system.

The pseudobinary liquidus temperatures in the CdTe–ZnTe system are higher than in the CdTe–CdSe system, and attempts were made to grow pseudobinary crystals at lower temperatures by passage of a non-stoichiometric zone containing excess Te. The equilibrium liquidus temperatures were determined from the ternary liquidus surface. The ternary liquidus–solidus tie-lines are not known however. In order to determine the equilibrium composition for the feed charge it was rather arbitrarily assumed that there was no metal/metal segregation on solidification, i.e. that the tie-lines corresponded to lines of constant Cd/Zn ratio. This assumption is probably not warranted since the crystals grown by this technique were small and inhomogeneous, indicating the possibility of ternary constitutional supercooling. This result confirms the need for further studies of the phase diagram of this ternary system.

References

1) W. G. Pfann, *Zone Melting* (Wiley, New York, 1958).
2) W. M. Yim and J. P. Dismukes, in: *Crystal Growth*, Ed. H. S. Peiser (Pergamon, Oxford, 1967) p. 187.
3) G. A. Wolff, H. E. LaBelle, Jr. and B. N. Das, Trans. Met. Soc. AIME **242** (1968) 436.
4) J. Steininger and R. E. England, Trans. Met. Soc. AIME **242** (1968) 444.
5) J. Steininger, A. J. Strauss and R. F. Brebrick, J. Electrochem. Soc. **117** (1970) 1306.
6) A. J. Strauss and J. Steininger, J. Electrochem. Soc. **117** (1970) 1420.
7) J. Steininger, Met. Trans. **1** (1970) 2939.
8) J. Steininger, J. Appl. Phys. **41** (1970) 2713.
9) J. Steininger and R. F. Brebrick, Electrochem. Soc. Spring Conf. Washington, D. C. (1971).
10) V. G. Smith, W. A. Tiller and J. W. Rutter, Can. J. Phys. **33** (1955) 723.
11) G. Y. Eastman, Sci. Am. **218** (1968) 38.
12) J. Steininger and T. B. Reed, J. Crystal Growth **13/14** (1972) 106.

Journal of Crystal Growth **13/14** (1972) 663–667 © *North-Holland Publishing Co.*

LIQUID PHASE EPITAXY OF $Ga_{1-x}Al_xAs$

E. ANDRÉ, J. M. LeDUC and M. MAHIEU

*RTC, La Radiotechnique-Compelec, Centre Industriel de Caen,
Route de la Délivrande, B.P. 6025, 14–Caen, France*

In the liquid phase epitaxial growth of $Ga_{1-x}Al_xAs$, we have been faced with the problem of both strong absorption within the layers and inhomogeneities across their surfaces. A series of experiments have been carried out in the 700–970 °C temperature range with an aluminium concentration in the melt from 0.05% up to 0.8%. The layers we have obtained have been studied both by cathodoluminescence and electron microprobe analysis. From this study, we have been able to plot several curves which show how aluminium is incorporated into the ternary layer under the different conditions of growth.

1. Introduction

$Ga_{1-x}Al_xAs$ has been shown to be very convenient for electroluminescent devices[1–3]. Efficient red diodes ($\lambda \simeq 6700$ Å) have been made by zinc diffusion into $Ga_{1-x}Al_xAs$ layers grown by liquid phase epitaxy in our laboratory. Those diodes have an area of 3×10^{-3} cm^2 and are operated under 15 mA. They have been working under such conditions for a year and a half now without any noticeable decay in light power. This latter stands at an average value of 80 ft lambert A^{-1} cm^{-2}. However in an attempt to industrialise the process we had to deal with two main problems. Firstly, these diodes exhibited a strong surface absorption preventing light from coming out properly so that most of it was emitted at the edges of the crystal. Secondly, inhomogeneities across the surface were revealed both by cathodoluminescence and electroluminescence measurements. It was thought that these problems were linked with the aluminium concentration gradient through the layer, so a series of experiments controlling both the temperature of epitaxy and the concentration of aluminium in the melt as parameters has been made so that the part they play in the growth could be ascertained. Since the pn junction is formed near the surface of the layer most attention has been paid to the surface properties.

2. Experiments

The phase diagram of $Ga_{1-x}Al_xAs$ and its applications to liquid phase epitaxy have been studied by

several authors[4,5]), but most of their experiments have been carried out either at high temperatures (1000 °C) or at "high" aluminium concentrations in the melt ($C_l > 1\%$). The present work has been carried out in the 700–970 °C temperature range and with aluminium contents between 0.05 and 0.8 wt% in the gallium solution.

These experiments have been performed in a horizontal furnace by a technique similar to Nelson's[6]) but neither the furnace nor the crucible are tilted at any time during the process. A special crucible made of boron nitride has been designed with a sliding shutter (fig. 1)[7]. The substrate is set at the bottom of the crucible while the melt is kept apart from it by the shutter. The melt is composed of 25 g of pure gallium (AluSuisse 99.9999%) saturated with an excess of pure GaAs (grown in our laboratory by the Bridgman method) and the amount of aluminium needed to get the right bandgap. When the desired temperature is reached the shutter is pulled back from outside the furnace and the melt comes down onto the surface; then growth proceeds as the temperature of the furnace decreases at a rate of 1 °C/min. We have therefore two possibilities, either we let the growth proceed undisturbed to the end or we can push the shutter forward to the closed position before the growth has finished. In this latter case the results we can expect are strongly dependent on the clearance between the substrate and the shutter since this volume determines the amount of solution from which the growth proceeds; but this paper deals only with the results we have obtained

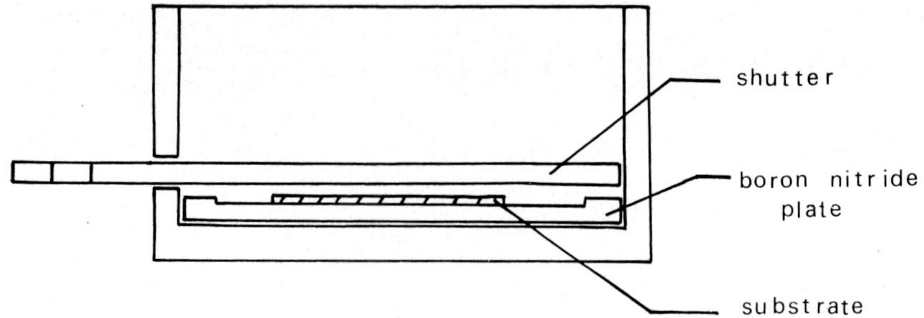

Fig. 1.   Crucible with shutter.

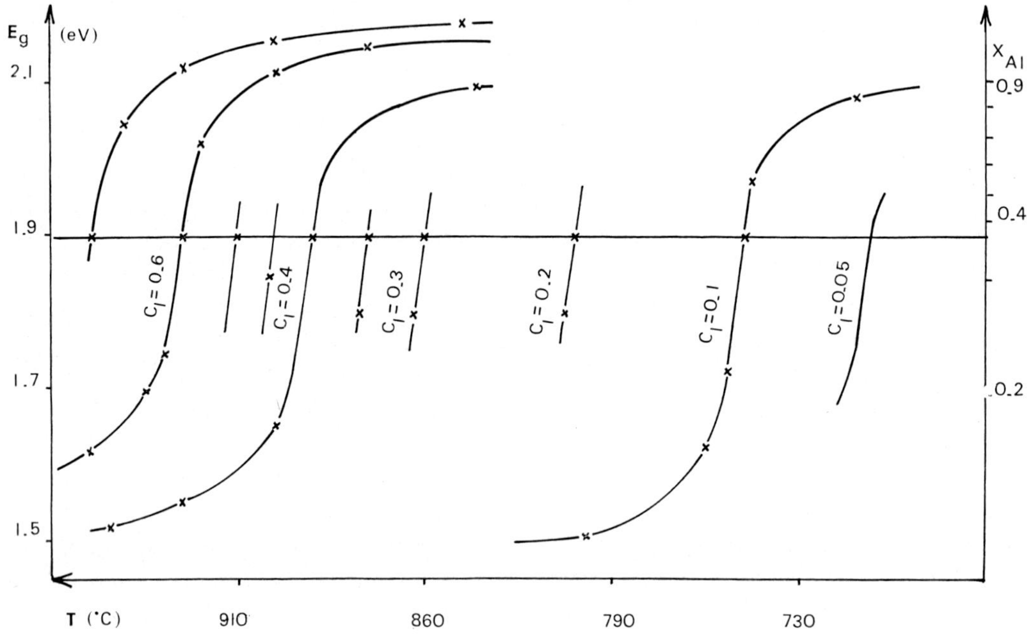

Fig. 2.   Band-gap at the surface of $Ga_{1-x}Al_xAs$ layers versus the starting temperature of growth at different C_i.

leaving the shutter open so that the aluminium concentration at the surface of the layers and its profile are determined by the cooling history of the material.

3. Results

Both cathodoluminescence and the electron microprobe are used to analyse the $Ga_{1-x}Al_xAs$ layers. Cathodoluminescence gives a good measure of the aluminium concentration at the surface of the layer and has proved to be a good way to sort out slices. Microprobe analysis is performed through the layer to determine the aluminium concentration profile. The electroluminescence of Zn-diffused diodes is in good agreement with these two measurements.

3.1. CATHODOLUMINESCENCE ASSESSMENTS – SURFACE STUDY

In fig. 2 we show the results of our studies on Aluminium concentration variations at the surface of layers versus the starting temperatures of epitaxy. For practical purposes we have plotted the band gap of the material linearly on the left hand side, the corresponding composition being shown on the right hand side.

As the material we are most interested in for devices has a visible light range which lies between 1.8 and 1.9 eV (cross-over point), we can only obtain it over a 3 °C temperature range for a given melt composition. The slope of C_s versus T changes abruptly around the

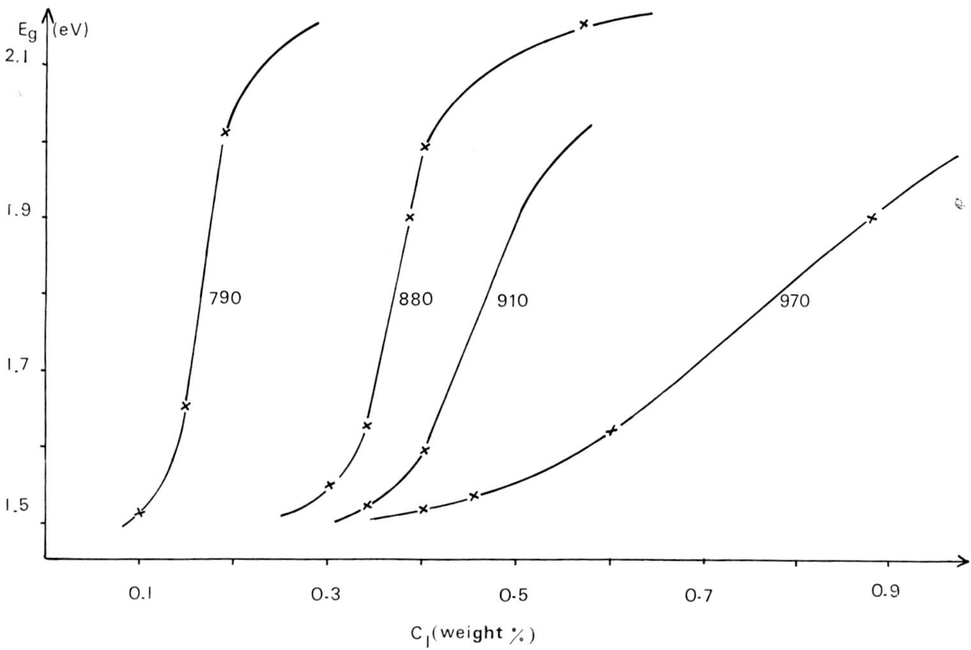

Fig. 3. Isothermal curves of the band-gap versus aluminium concentration in the melt.

temperature at which the material $Ga_{0.66}Al_{0.34}As$ is grown. Whatever the concentration of Al in the melt this change occurs at the cross-over point. Around this point each C_l has a temperature of its own from which either infra-red transition material or indirect material can be expected. Fig. 3 shows the variation of the band gap (E_y or $\times$) at different temperatures. It is clearly shown that the same 3 °C temperature range does not correspond to the same change in C_l. That set of curves points to the fact that until the content of aluminium in the melt reaches a certain level the aluminium content in the layer increases very slowly; then the AlAs concentration in the layer increases sharply over a very small change in aluminium content of the melt and finally saturates when the content of aluminium in the crystal reaches a very high level.

3.2. ELECTRON MICROPROBE ANALYSIS – ALUMINIUM CONCENTRATION PROFILE

Figs. 4 and 5 show the profile of the aluminium concentration through layers which have been grown from solutions with two different C_l and at different temperatures. The aluminium gradient increases with the temperature and the higher the temperature the lower the concentration both at the surface and the interface. From these curves it can be concluded that

the behaviour of the Al at the surface of the layers is the same at the interface but to a smaller extent. This is clearly shown on fig. 6 where the slope of $C_s = f(T)$ at the interface is 6 times less than the slope at the surface of the same layers grown from a solution with the same content of aluminium (0.1 %). Moreover the slope at the interface for a crystal grown from a solution with $C_l = 0.4\%$ is not the same as the slope at the interface for the layers grown from a melt with $C_l = 0.1\%$. This can be easily explained by the difference in temperature between the two sets of experiments, but it seems evident to us that whatever the concentration of aluminium in the melt and the temperature starting point of epitaxy the curves $C_s = f(T)$ at the surface of the layers always have the same slope. It could therefore account for the growth ending under the same conditions.

4. Comments

From our experimental study we have observed the following facts:

(1) the problem of chemical stability of mixed $Ga_{1-x}Al_xAs$ crystals has not been found as severe as can be expected from the literature. The surface of the layers were mirror-like and they withstood corrosion very well both in air and in aqueous solutions.

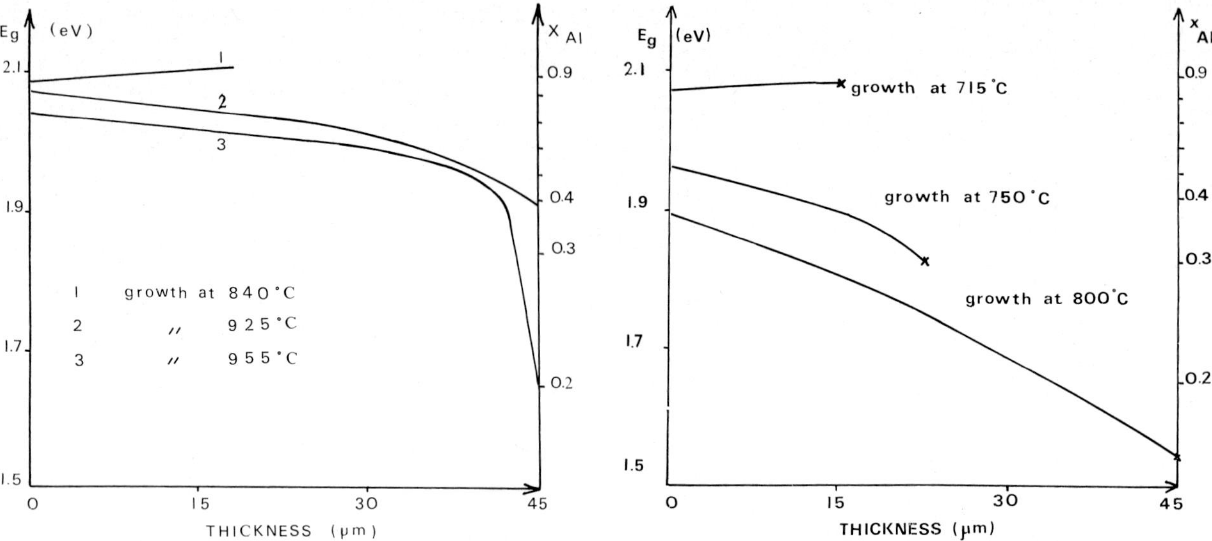

Fig. 4. Aluminium profiles measured by electron microprobe through three layers of $Ga_{1-x}Al_xAs$ grown at different temperatures from a solution with $C_l = 0.4\%$.

Fig. 5. Alumium profiles measured by electron microprobe through three layers of $Ga_{1-x}Al_xAs$ grown at different temperatures from a solution with $C_l = 0.1\%$.

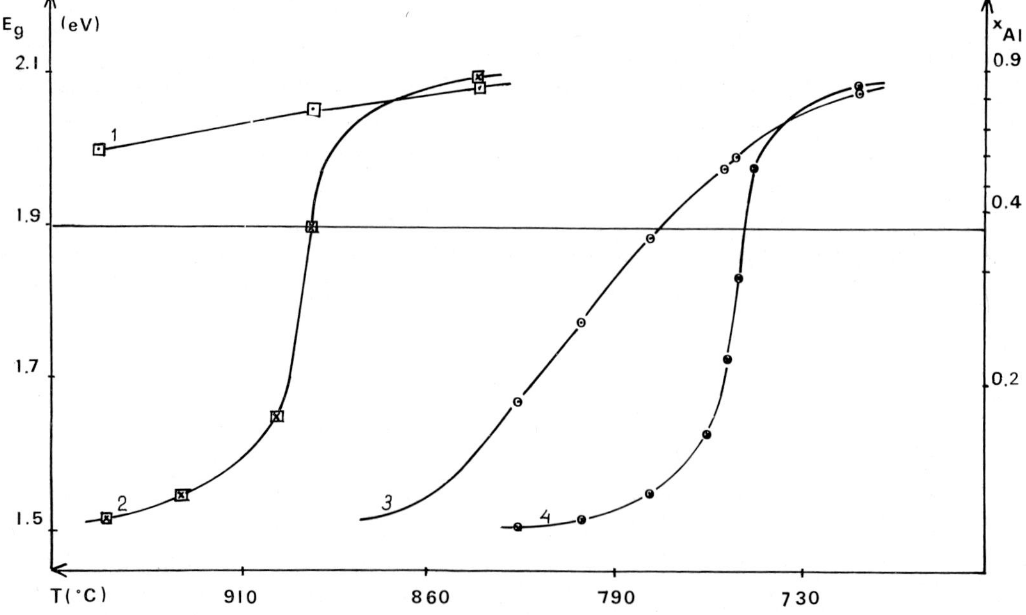

Fig. 6. Band-gap versus the starting temperature of growth: (1) at the interface, and (2) at the surface, of layers grown from a solution with $C_l = 0.4\%$; (3) at the interface, and (4) at the surface, of layers grown from a solution with $C_l = 0.1\%$.

(2) the distribution coefficient of aluminium is known to be very high and it varies with temperature. However the aluminium concentration profile of the layers is entirely controllable and reproducible, which is a very important point for device production.

(3) On the other hand, the inflection point of the curves C versus T occurs at the cross-over point which means that it is more difficult to obtain the correct composition.

(4) Liquidus and solidus curves in the $Ga_{1-x}Al_xAs$ system have been described theoretically by several authors[4,5] from a regular liquid solution and ideal solid

solution treatment. Their theory was found to give a good description of their experimental results. For instance, from Panish's work[4], it is concluded on the 1000 °C solid solubility isotherm that the composition of the ternary layer increases continuously with the melt composition. Our results at lower temperatures and lower aluminium contents of the melt could permit a more precise insight in this field. However, in the interpretation of our curves it must be borne in mind that for practical purposes we had to study only surface concentration of liquid epitaxial layers.

5. Conclusion

In this work we have looked more closely at the behaviour of aluminium during liquid epitaxy of $Ga_{1-x}Al_xAs$, at lower temperatures than those previously used. We have therefore been able to bear out that $Ga_{1-x}Al_xAs$ can be grown at the band-gap needed for the device purposes even at as low temperature as 700 °C at which it is easier to get layers of thickness around 10 μm. This could be achieved at high temperature but gaining mastery over the whole process is somewhat difficult, the more so since the surface of the substrate is damaged during the saturation process. At low temperatures and of course low aluminium concentrations, layers with a low C_s at the interface can be achieved more easily, so that the dislocation density is substantially reduced and the layers correspondingly improved. Since the thickness of the layer is thin enough the concentration gradient of aluminium is not as critical as it is when layers are thick. It has been observed that in any case the lower the temperature the better the surface of the layers. Moreover, at higher temperatures layers become more and more inhomogeneous, and cathodoluminescence makes it clear that aluminium is clustered either in patches or stripes which make device production very hazardous.

Acknowledgements

The authors wish to thank Mr. Bouley and Mr. Lefevre for the Cathodoluminescence measurements and Mr. Haroutounian for the electron microprobe analysis. Many thanks are due to Dr. Ilegems for his critical reading of the manuscript. This work has been made possible thanks to the Direction de la Recherche et des Moyens d'Essais which sponsors the study.

References

1) H. Rupprecht, J. M. Woodall and G. D. Pettit, Appl. Phys. Letters **11** (1967) 81.
2) K. J. Linden, J. Appl. Phys. **40** (1969) 2325.
3) J. M. Woodall, Electrochem. Soc. Meeting. Spring Meeting Los Angeles, May 10–15, 1970.
4) 4) M. B. Panish and S. Sumski, J. Phys. Chem. Solids **30** (1969) 129.
5) M. Ilegems, Thesis for the Ph.D. degree, Stanford University, October 1969.
6) H. Nelson, RCA Rev. **24** (1963) 603.
7) E. André, Patent No. H 0117/00//B 01 j–(31.12.1968).
8) G. A. Antypas and L. W. James, J. Appl. Phys. **41** (1970) 2165.

Journal of Crystal Growth **13/14** (1972) 668–671 © *North-Holland Publishing Co.*

CRYSTAL GROWTH OF $Hg_{1-x}Cd_xTe$ USING Te AS A SOLVENT

RYUITI UEDA, OSAMU OHTSUKI, KOUJI SHINOHARA and YOUITI UEDA

Fujitsu Laboratories, Ltd., 5 Wadayama-dori, 1,Chome, Hyoga-Ku, P.O. Box 1032, Kobe Central, Japan

A method of preparing homogeneous alloy crystals of $Hg_{1-x}Cd_xTe$ using a zone melting procedure which employs Te as a solvent is reported. The value of x in various ingots has been measured by atomic absorption analysis and electron probe microanalysis. The results show that the compositional uniformity of cross sections of ingots was found to significantly improve by reducing the volume of the liquid zone; this caused a change from a curved to a flat solid–liquid interface

1. Introduction

This paper describes a method for preparing homogeneous alloy crystals of $Hg_{1-x}Cd_xTe$ over a wide range of compositions.

Several problems arise in preparing these alloys by melt growth methods, and especially the most complicating factor is the effect of segregation of CdTe with respect to HgTe during crystal growth. The phase diagram (x, T) for these alloy shows a marked difference between the liquidus and the solidus curves[1–3]. Therefore, it is quite difficult to prepare homogeneous alloy crystals of $Hg_{1-x}Cd_xTe$ by conventional crystal growth methods. It is also quite difficult to maintain the melt stoichiometry because of the high vapour pressure of mercury over the melt. The segregation of excess Te can give rise to pronounced constitutional supercooling[4].

The preparation of homogeneous alloy crystals has been attempted by the rapid quenching method under appropriate conditions[5,6]. However, in this method, in order to prepare a perfect crystal, subsequent heat treatment is necessarily required for a long time at a high temperature with enough excess mercury. Moreover, it may be difficult to prepare a large size of single crystal. A different technique, the vertical zone-melting method[7] has been attempted but concentration fluctuations along the ingot due to the instability of hot zone are generally observed.

The growth of $Hg_{1-x}Cd_xTe$ epitaxial layers by various methods has been investigated (evaporation-diffusion mechanism[8,9]), interdiffusion between HgTe and $CdTe$[10,11]), however it is impossible to avoid the concentration gradient of composition along the crystal growth direction.

The deposition of $Hg_{1-x}Cd_xTe$ films on single crystal substrates by the cathodic sputtering has been investigated[12]), and also the formation of $Hg_{1-x}Cd_xTe$ by Hg-ion bombardment to CdTe has been carried out[13]), but an epitaxial growth layer has not been achieved.

In the present work we have investigated the possibility of the solution growth of $Hg_{1-x}Cd_xTe$ by means of the vertical zone-melting method using Te as a solvent. We at first prepared an alloy crystal of $Hg_{1-x}Cd_xTe$ in a 40 mm ID by 50 mm OD qualtz ampoule and investigated the relation between the shape of distribution of alloy compositions and the shape of solid–liquid interface during crystal growth. It is generally regarded that the shape of solid–liquid interface during crystal growth is nearly consistent with the shape which is plotted through the same x-value of the $Hg_{1-x}Cd_xTe$, and this has been confirmed by our investigation as shown in fig. 1. Therefore, in order to prepare homogeneous alloy crystals of $Hg_{1-x}Cd_xTe$ at least across a wafer cut from an ingot and especially cut perpendicular to the growth direction, we attempted to keep the shape of growth interface flat by decreasing the volume of liquid zone, because the less the volume of liquid zone, the flatter the shape of solid–liquid interface became. Our attempt to keep the shape of solid–liquid interface flat has been achieved by means

(a)

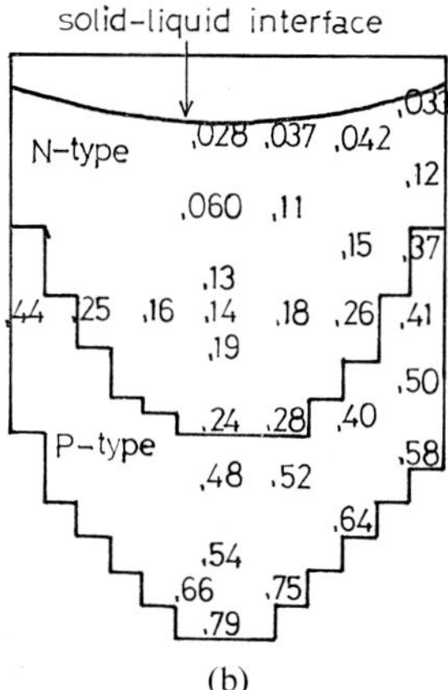

(b)

Fig. 1.   A comparison of the shape of the solid–liquid interface (a) with the distribution of x-values (b) in a $Hg_{1-x}Cd_xTe$ ingot.

of the vertical zone melting method using Te as a solvent[7,14]).

2. Experimental method

The method for preparing homogeneous alloy crystals of $Hg_{1-x}Cd_xTe$ presented in this paper is similar to the method by Dziuba[7]). The solid solutions of $Hg_{1-x}Cd_xTe$ were prepared from high purity HgTe and CdTe obtained previously. Appropriate quantities of HgTe and CdTe were placed in an 8–18 mm ID by 12–22 mm OD quatz ampoule coated with carbon, which was subsequently evacuated to a residual air pressure below 1×10^{-5} Torr and sealed off. The ampoule was put into a rocking furnace and the furnace temperature was slowly raised slightly above the melt-

ing temperature of $Hg_{1-x}Cd_xTe$. The HgTe/CdTe alloy when completely molten was mixed for about 24 hr in the rocking furnace; and then together with the rocking furnace, was turned to the vertical position, and air-quenched under appropriate conditions.

The ampoule which contained the quenched ingot was cut at one end, and a mixture of pure Te and $Hg_{1-x}Cd_xTe$ alloy was placed adjacent to it. This mixture forms a liquid zone during crystal growth. The ampoule was again evacuated below 1×10^{-5} Torr and sealed off with an empty space in one end of the ampoule. The apparatus which was used to prepare the alloy crystal of $Hg_{1-x}Cd_xTe$ is shown schematically in fig. 2a. The temperature profile of the furnace is shown in fig. 2b schematically. The ampoule was put into the furnace as shown in fig. 3a. After the mixture was completely molten, the ampoule was placed in the starting position as shown in fig. 3b and was kept as it was for a while till the thermal equilibrium between the quenched ingot and the molten materials was achieved, and then the ampoule was lowered through the hot zone at a rate of 0.2–0.3 mm/hour. The intermediate position in operation of the zone melting process is shown in fig. 3c. The end position of the zone melting process is shown in fig. 3d. After having finished this zone melting process, the ampoule was taken out from the furnace and inverted and then again put into the same furnace as shown in fig. 3e, and after a while lowered under the same condition mentioned above.

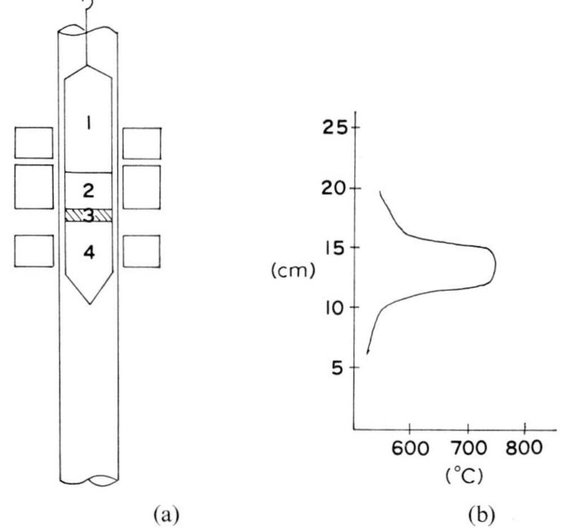

Fig. 2.   A schematic diagram showing the crystal growing furnace (a), together with its temperature profile (b). (1) quenched ingot, (2) vacant zone, (3) melt zone, (4) single crystal of HgCdTe.

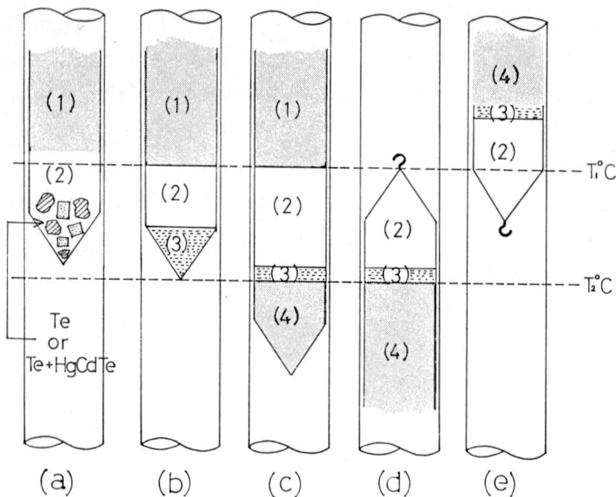

Fig. 3. A schematic diagram showing various stages in the zone melting process: (a) starting materials in sealed tube; (b) initial melt 3 in starting position; (c) growth position; (d) end of growth; (e) final inverted position for next zoning; (1) ingot, (2) vacant zone, (3) liquid zone.

Having finished these two processes, a homogeneous alloy crystal of $Hg_{1-x}Cd_xTe$ may be obtained.

3. Results and discussion

In attempting to grow an alloy crystal of $Hg_{1-x}Cd_xTe$ from Te-rich solution, we have investigated the phase diagram of ternary Hg–Cd–Te system by thermal analysis. Particular attention was paid to the Te-rich side of the ternary liquidus surface; we have especially investigated the liquidus at constant ratios of Hg/Cd on the Te-rich side, where in our case the Hg/Cd ratios were fixed at 0.95/0.05, 0.90/0.10, 0.85/0.15, 0.80/0.20. The results are presented in fig. 4.

In order to control the composition of the alloy crystal deposited from the solution, we have investigated the relationship between the x-values of $Hg_{1-x}Cd_xTe$ deposited and the compositions of the solutions. The x-values of $Hg_{1-x}Cd_xTe$ deposited from these solutions at the tip of crystal solidified by the normal Bridgman method have been determined by means of atomic absorption analysis of Cd. As a result, the x-value of $Hg_{1-x}Cd_xTe$ at the tip of the crystal deposited is about three times as large as that of the $Hg_{1-x}Cd_xTe$ which is contained in the solution; for example, if the x-value of $Hg_{1-x}Cd_xTe$ in the solution is 0.05, then the x-value of $Hg_{1-x}Cd_xTe$ deposited is about 0.15. Therefore, the temperature and the composition of the liquid zone is chosen appropriately from fig. 4.

The quenched ingot in the ampoule in this crystal growth method is transported through the following zone melting process. This starts at an initial condition as shown in fig. 3a. After being placed to the starting position, since there is a difference between the chemical potential of mercury at the surface region of the ingot and that of mercury in the liquid zone, mercury atoms at the surface of the ingot are vaporized into the empty zone and then absorbed into the liquid zone. At equilibrium, the vaporization of mercury at the surface of the ingot stops. But, whilst mercury at the surface of the ingot continues to vaporize, the surface composition of the ingot is abundant in Te; therefore, the melting temperature at the surface of the ingot is lowered causing a thin surface layer of the ingot to melt and form a small droplet which falls down into the liquid zone. When the ampoule is lowered as shown in fig. 3c, $Hg_{1-x}Cd_xTe$ alloy crystal begins to solidify in the liquid zone from the tip of the ampoule. Te is rejected into the melt in the solidification process, hence increasing its ability to absorb Hg vapour. A "steady state" process develops and the ingot is continuously transported.

The most important factor which contributes to the transportation of the material is the concentration of

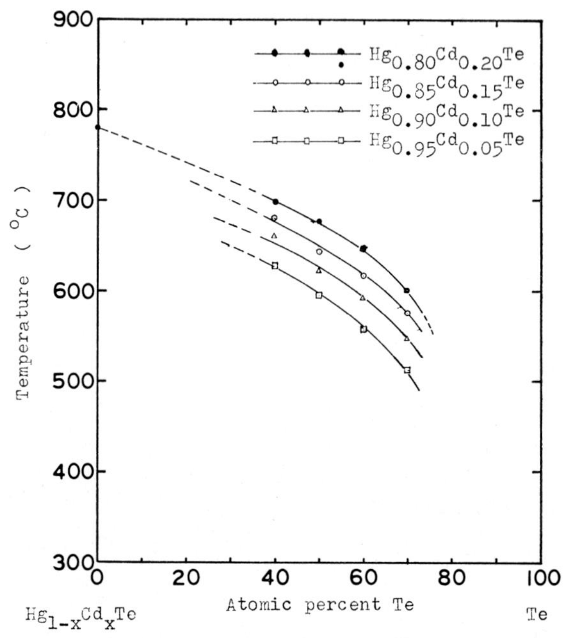

Fig. 4. Quasi-binary phase diagram at constant Hg/Cd ratios on the Te rich side of the diagram.

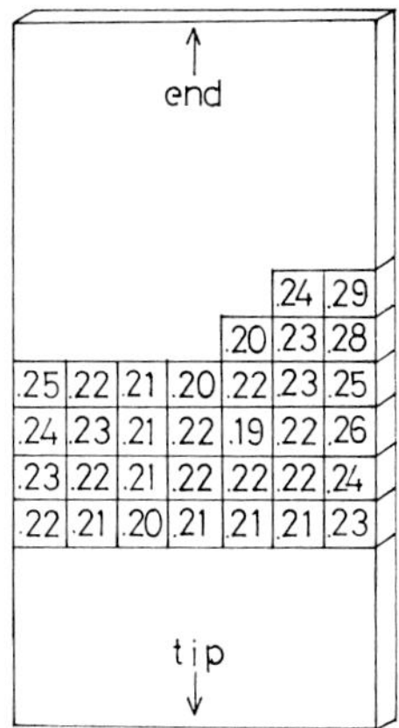

Fig. 5. Distribution of atomic concentration x (in the formula $Hg_{1-x}Cd_xTe$) as determined by atomic absorption analysis.

excess Te in the molten zone. We have shown in a separate experiment that the amount of Hg transported is a direct function of the Tellurium concentration in the liquid.

The uniformity of chemical composition in the grown ingots has been the subject of detailed investigation. For example, as shown in fig. 5, the distribution of the atomic concentration x in the formula $Hg_{1-x}Cd_xTe$ has been determined by atomic absorption analysis in a wafer cut parallel to the growth direction in the central part of an ingot. The result demonstrates the uniformity achievable by zone melting. The Bridgman technique shows marked non-uniformity in the growth direction.

The distribution of x-values in a wafer cut perpendicular to the growth direction in the central part of an ingot has been determined by electron probe microanalysis. In this case, the uniformity was also very good. A slight increase occurred at the edge, i.e. within about 10% of the width of the ingot. This increase was no more than 5–10% of the average value of x across the ingot.

The distribution of x-values in a wafer cut perpendicular to the growth direction which has been determined by atomic absorption analysis of Cd. In this case the depth of the liquid zone was kept up to 5 mm so that the shape of the solid–liquid interface became flatter than that of the two previous cases discussed. In this way, a homogeneous alloy crystal was obtained.

Approximately 40 determinations of the CdTe concentration across the slice of material showed a mean concentration of 0.075 ± 0.005 over 80% of the slice area.

4. Summary

The zone melting method described has the following:

(1) The amount of material to be transported from the ingot to the liquid zone is controlled by the amount of excess Te rejected in the liquid zone so that the volume of liquid zone is kept constant.

(2) The volume of the liquid zone can be kept so small that the depth of the liquid zone becomes thin and it is possible to form a flat liquid solid interface.

(3) The effect of purification by liquid Te (the change in segregation coefficient for most impurities) makes the crystal lower in carrier concentration in comparison to conventional zone melting methods[14].

(4) It is possible to prepare at will crystals with optional composition by casting an ingot which consists of the appropriate amount of HgTe and CdTe.

(5) Explosions caused by the vapor pressure of mercury do not occur at all during crystal growth.

References

1) B. Ray and P. M. Spencer, Phys. Status Solidi **22** (1967) 371.
2) T. C. Harman, in: *Physics and Chemistry of II–VI Compounds*, Eds. M. Aven and J. S. Prener (North-Holland Amsterdam, 1967) p. 784.
3) J. L. Schmit and C. J. Speerschneider, Infrared Phys. **8** (1968) 247.
4) B. E. Bartlett, J. Deans and P. C. Ellen, J. Mater. Sci. **4** (1969) 266.
5) L. N. Swink and M. J. Brau, Met. Trans. **1** (1970) 629.
6) E. L. Stelzer, J. L. Schmit and O. N. Tufte, IEEE Trans. Electron Devices **ED-16** (1969) 880.
7) E. Z. Dziuba, J. Electrochem. Soc. **116** (1969) 104.
8) O. N. Tufte and E. L. Stelzer, J. Appl. Phys. **40** (1969) 4559.
9) G. Cohen-Solal, Y. Marfaing, F. Bailly and M. Rodot, Compt. Rend. (Paris) **261** (1965) 31.
10) H. Rodto and J. Henoc, Compt. Rend. (Paris) **256** (1963) 1954.
11) F. Bailly, G. Cohen-Solal and Y. Marfaing, Compt. Rend. (Paris) **257** (1963) 103.
12) H. Kraus, S. G. Parker and J. P. Smith, J. Electrochem. Soc. **114** (1967) 616.
13) N. A. Foss, J. Appl. Phys. **39** (1968) 6029.
14) R. O. Bell, N. Hemmat and F. Wart, Phys. Status Solidi (a) **1** (1970) 375.

Section XIV

Melt growth – mainly oxides

J. AUBRÉE
M. TAKAHASHI
S. NANAMATSU
M. KIMURA
R. S. FEIGELSON
G. W. MARTIN
B. C. JOHNSON
J. BILLINGHAM
P. S. BELL
M. H. LEWIS
D. MATEIKA
J. W. NIELSEN
S. L. BLANK
J. DAVAL

D. CHALLETON
J. MARESCHAL
F. B. KHAMBATTA
P. J. GIELISSE
M. P. WILSON, Jr.
J. A. ADAMSKI
CH. SAHAGIAN
J. RICARD
A. CIOCCOLANI
R. FALCKENBERG
A. FREUND
P. GUINET
F. RUSTICHELLI
F. VANONI

Journal of Crystal Growth **13/14** (1972) 675–680 © *North-Holland Publishing Co.*

TIRAGE DU NIOBATE DE CALCIUM MONOCRISTALLIN DOPÉ AU NÉODYME

J. AUBRÉE

Département Cristallogenèse et Caractérisation des Matériaux du Centre National d'Études des Télécommunications, 92–Issy-les-Moulineaux, France

A method of preparing synthetic Nd-doped calcium niobate single crystals is described. The technique consists in pulling crystals from an RF-heated iridium crucible in an argon/oxygen atmosphere by the Czochralski method. The study deals with the preparation of the raw material, and the growth of doped variously-orientated single crystals. Rods machined from the single crystals exhibited CW laser operation at room temperature. The performance characteristics achieved were: emitted power better than 1 W for a pumping power of 1 kW, the threshold excitation pulsed or CW, being 1.09 J and 350 W, respectively.

1. Introduction

En vue d'applications éventuelles aux systèmes de transmission téléphonique à grande capacité, nous avons étudié la cristallogenèse de certains matériaux susceptibles de permettre la réalisation de laser à solide fluorescent émettant en régime continu à température ordinaire. Notre choix s'est limité à des cristaux dont les conditions de croissance paraissaient, à priori, plus favorables que celles du YAG et les performances comparables: faible seuil d'excitation et rendement énergétique élevé. C'est dans cette optique que nous avons entrepris l'étude de la préparation des monocristaux de niobate de calcium dopé au néodyme, après celle du tungstate de calcium et parallèlement à celle de la fluorapatite.

L'objectif de cette communication n'est donc pas de présenter un matériau nouveau ou de qualité exceptionnelle, mais simplement de faire connaître les solutions apportées aux problèmes techniques relatifs à la préparation des matières premières pures, la croissance des monocristaux de grandes dimensions, et surtout l'introduction d'une quantité suffisante de néodyme dans la matrice sans perturbation excessive de celle-ci.

Nous ne pouvons commencer cet exposé sans citer les quelques travaux déjà effectués dans ce domaine. Une importante étude sur la structure du matériau a été faite par Cummings et al.[1]) et par Whiston et al.[2]). De même Ibrahim et al.[3]) ont étudié le système $CaONb_2O_5$. Busch[4]) a déterminé par spectroscopie les niveaux d'énergie du néodyme dans le niobate de calcium. Rubin et al.[5]) ont élaboré des cristaux pour applications en optique non linéaire. Ballman et al.[6]) ont produit par la méthode Czochralski des cristaux dopés jusqu'à concurrence de 0.5% de néodyme avec charge de compensation. Cette compensation a été spécialement étudiée par Scott et al.[7]) et par Rooksby et al.[8]) qui ont examiné les niobates contenant des terres rares.

2. Préparation de la matière première

2.1. NIOBATE DE CALCIUM

Rappelons que le niobate de calcium $CaONb_2O_5$, à l'état naturel présente une structure comparable à celle de la fersmite. Il cristallise dans le système orthorhombique:

$$a = 5.73 \text{ Å}, \quad b = 14.94 \text{ Å}, \quad c = 5.22 \text{ Å},$$

et appartient à la classe de symétrie mmm. Son point de fusion est de 1560 °C. Il est préparé à partir des deux produits suivants:

oxyde de niobium Nb_2O_5 – pureté 99.95,
carbonate de calcium $CaCO_3$ – qualité RP.

Le carbonate de calcium est attaqué par l'acide nitrique de normalité 5 N environ. A la solution limpide de nitrate ainsi obtenue, on ajoute l'oxyde Nb_2O_5 en poudre. Le mélange est chauffé dans un creuset de platine, au bain de sable, puis calciné à 1000 °C pendant 4 heures. Les nitrates se décomposent et produisent un matériau qui est alors broyé et homogénéisé. Il serait possible de poursuivre la calcination à plus haute température pour achever la réaction, mais la fusion au moment du tirage aboutit au même résultat.

L'analyse aux rayons X montre que la formule du composé obtenu est mal définie: les diagrammes de poudres font apparaître plusieurs phases cristallines parasites qu'on retrouve dans les monocristaux sous forme de précipités d'aspect blanchâtre. Le produit contient 90 à 95% de $CaONb_2O_5$ de structure type fersmite, le reste étant constitué par:

(a) $2(CaO)Nb_2O_5$, monoclinique ($\beta = 98°25'$):

$a = 13.36$ Å, $b = 5.50$ Å, $c = 7.70$ Å;

(b) des oxydes de niobium dont les 2 principaux sont:
Nb_2O_5, monoclinique ($\beta = 119°48'$):

$a = 21.08$ Å, $b = 3.82$ Å, $c = 19.33$ Å;

Nb_2O_5, mononclinique ($\beta = 118°15'$):

$a = 22.10$ Å, $b = 7.63$ Å, $c = 19.52$ Å.

2.2. Dopant

Les différents dopants employés ont subi une préparation semblable à celle du niobate de calcium. Après cuisson à 1000 °C la poudre obtenue est traitée par de multiples broyages séparés par des calcinations à 1400 °C. Nous nous sommes inspirés des travaux de Rooksby et al.[8]) pour tenter une préparation sous forme de pastilles, mais les résultats atteints n'ont apporté aucune amélioration de l'homogénéité.

3. Tirage des monocristaux

3.1. Appareil de tirage

La figure 1 montre un schéma de l'enceinte et indique la nature des matériaux qui constituent l'installation. Le creuset (diamètre: 40 mm, hauteur: 40 mm, épaisseur: 2 mm) est posé sur un cylindre réfractaire fendu. Nous avons intercalé entre le tube interne froid et le cristal une double chemise en alumine pour diminuer le gradient thermique radial.

Le porte-germe est un demi-cylindre de platine fixé à un barreau d'alumine. L'ensemble, solidaire de la tige en acier inoxydable non refroidie, peut être animé de mouvements de rotation et de translation de vitesses variables.

Le chauffage est réalisé par induction directe dans le creuset. La température mesurée par le thermocouple est régulée à ± 0.5 °C. Le mélange argon–oxygène circule de haut en bas.

3.2. Tirage des cristaux non dopés

A partir du contenu d'un creuset on tire généralement trois cristaux: le premier est limpide, le second peut présenter quelques défauts, quant au troisième, il se termine par une partie opaque inutilisable. Les cristaux paraissant à l'oeil trop imparfaits sont purifiés une seconde fois. On constate par examen à la microsonde électronique une diminution progressive de la teneur en calcium (6%) entre le début du premier et la fin du troisième cristal.

La matière première non tirée est éliminée du creuset par le procédé suivant: on ajoute du tungstate de sodium – Na_2WO_4; l'ensemble est chauffé à 1200 °C, puis refroidi jusqu'à 200 °C. Une addition d'eau suffit alors de désagréger le matériau.

Les meilleurs barreaux sélectionnés sont employés à la fabrication de cristaux dopés. Le programme de refroidissement de 1560 °C à l'ambiante est exécuté en 3 heures.

Tous les tirages sont effectués sous balayage d'argon 92% et d'oxygène 8% avec un débit de 2 litre/min. La composition du gaz de balayage a été déterminée en vue d'éviter la formation de centres colorés bleu-noir résultant du manque d'oxygène. Cette coloration ne semble pas présenter d'inconvénient de fabrication, mais on peut en déduire qu'un recuit du barreau terminé, sous oxygène pur, ne pourra être que bénéfique. La condition idéale serait une cristallisation sous atmosphère d'oxygène pur, mais la détérioration du creuset provoquerait la contamination du bain. Pour chaque manipulation, la perte en poids du creuset est de 0.5 g environ. La contamination par l'iridium du niobate non dopé est évaluée par analyse radiochimique à 1.9×10^{-7} g/g.

3.3. Tirage des cristaux dopés

Cette étude a revêtu une importance particulière, la fragilité des cristaux posant alors un problème délicat. Les barreaux non dopés fabriqués pour la purification peuvent atteindre un diamètre de 10 mm, les fractures étant sans importance à ce stade. Par contre, pour les cristaux dopés, le maintien d'un petit diamètre et sa régularité sont essentiels.

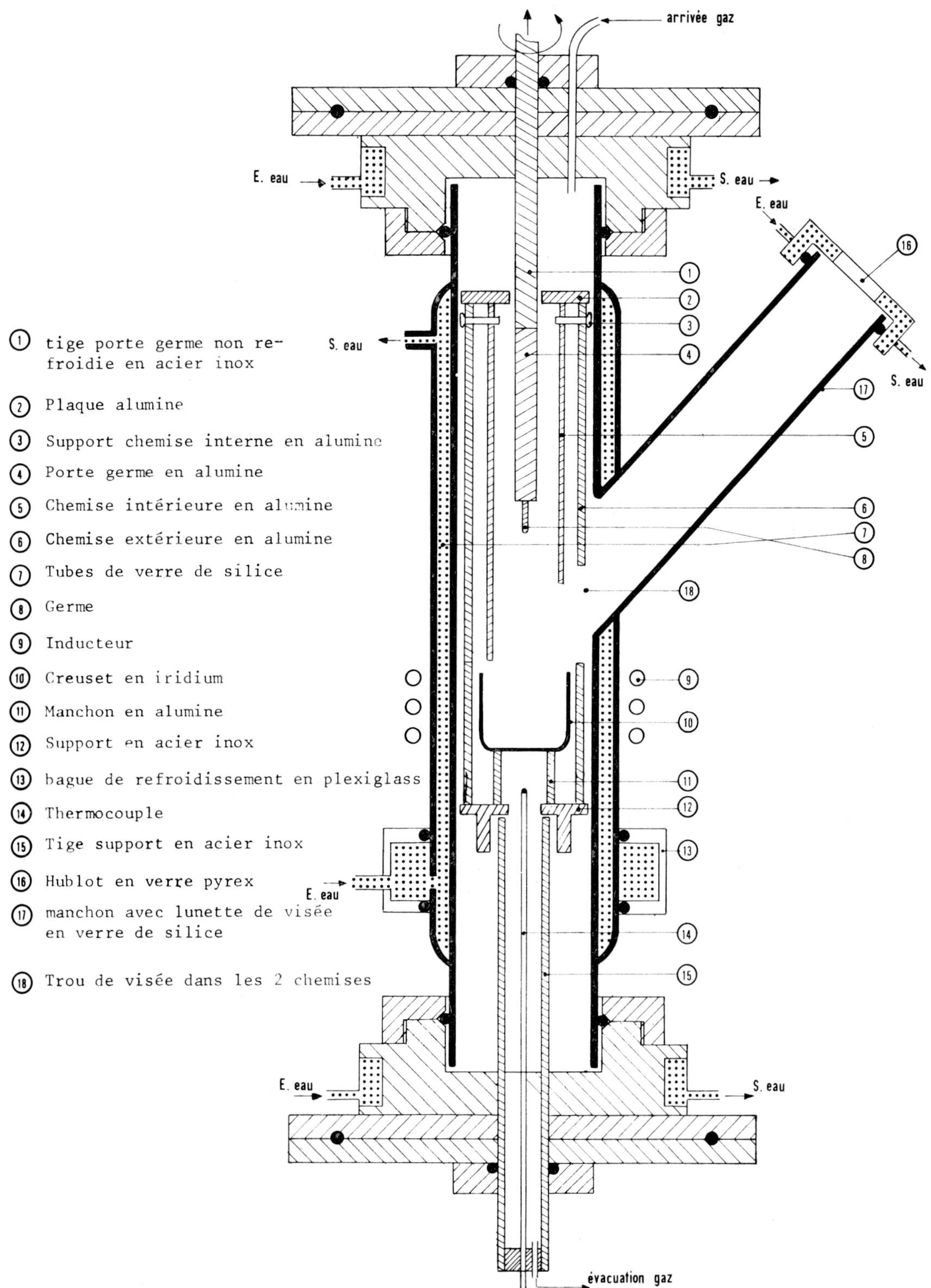

Fig. 1. Schéma de l'installation de tirage.

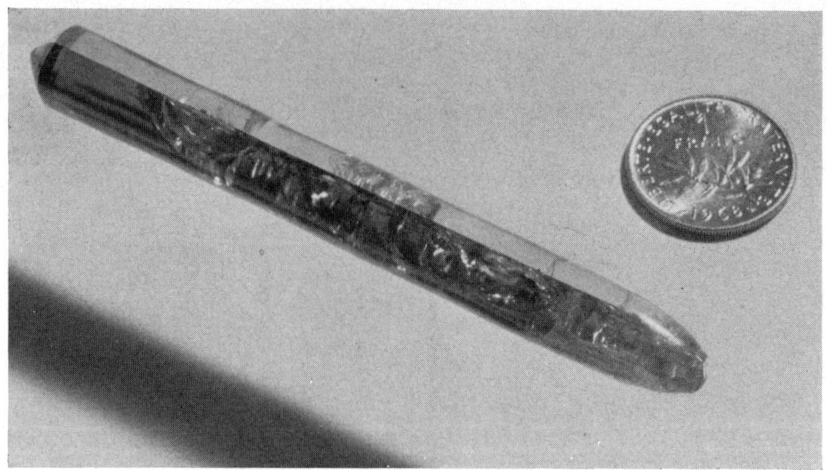

Fig. 2.   Cristal de niobate de calcium non dopé clivé suivant un plan médian.

3.3.1 *Influence de la direction de croissance*

La direction de tirage la plus courante est [001] ou axe c. Les plans de clivage (010) correspondent au plus grand paramètre cristallin.

Dans ce qui suit, la direction d'un cristal sera définie conventionnellement comme la normale au plan cristallographique indiqué.

Nous avons observé à plusieurs reprises qu'un cristal non dopé orienté suivant l'axe c clivait suivant un plan médian (fig. 2), l'une des parties respectant l'orientation du germe (coloration bleue) l'autre, incolore, orientée suivant (170). Nous avons tiré parti de cette orientation spontanée en exécutant une série de cristaux d'orientation (071), qui s'avéra moins fragile. L'orientation (071) a été choisie, d'une part par analogie à (170), les paramètres a et c étant très voisins; d'autre part, pour satisfaire aux conditions de fluorescence optimale. Ces dernières se trouvent être celles où le vecteur électrique est parallèle à l'axe a et contenu dans le plan de sortie du cristal, par conséquent, lorsque le plan perpendiculaire à l'axe de tirage est parallèle à l'axe a. C'est le cas de (071) dérivé de (170).

On a successivement utilisé différentes orientations à partir de (071), en vue de se rapprocher des axes optiques. Leurs orientations sont voisines de (032), (0$\bar{3}$2) et contenues dans le plan (yOz) à environ 35° de l'axe c. Le plan (071) ne diffractant pas aux rayons X nous avons abouti à (062), très proche de (071), et dont les plans réticulaires diffractent bien[9]). Des monocristaux ont été tirés suivant des orientations (071), (062), (131), (001). Des barreaux également cristallisés suivant (110)

et (131) ont été testés en fluorescence optique (fig. 3). On a remarqué que le pic du néodyme était sensiblement plus prononcé dans les orientations (071) et (062).

3.3.2. *Influence des dopants et du dopage*

En raison des travaux antérieurs, nous avons, tout d'abord, utilisé le composé $NaNd(NbO_3)_4$, concentration Nd/Ca: 1 % en atomes.

Nous rappelons que, théoriquement: l'addition d'un ion alcalin est nécessaire pour établir la compensation des charges, en admettant qu'à deux Ca^{++} peuvent se substituer un Na^+ et un Nd^{+++}. En effet dans ces conditions, il a été possible de tirer, un monocristal de longueur suffisante et non fêlé. Mais le deuxième barreau obtenu à partir du même bain ne peut s'élaborer: l'interface solide-liquide tend à devenir fortement concave et la croissance s'interrompt.

Des analyses radiochimiques montrent que si le néodyme, dans ce premier échantillon, est bien introduit dans la matrice avec un coefficient de ségrégation voisin de 1, le sodium, par contre ne s'y trouve qu'en quantité 10 fois moindre. Ce cristal a présenté l'effet laser et a fonctionné en régime continu à température ordinaire. On a obtenu des résultats identiques en utilisant un dopant où le sodium était remplacé par le potassium.

L'essai du dopant $K_2Nd\ (NbO_3)_5$ quadratique a été entrepris,

$$a = 12.49 \text{ Å}, \quad b = 12.49 \text{ Å}, \quad c = 3.92 \text{ Å}.$$

Les résultats ont été négatifs, par suite d'une fusion

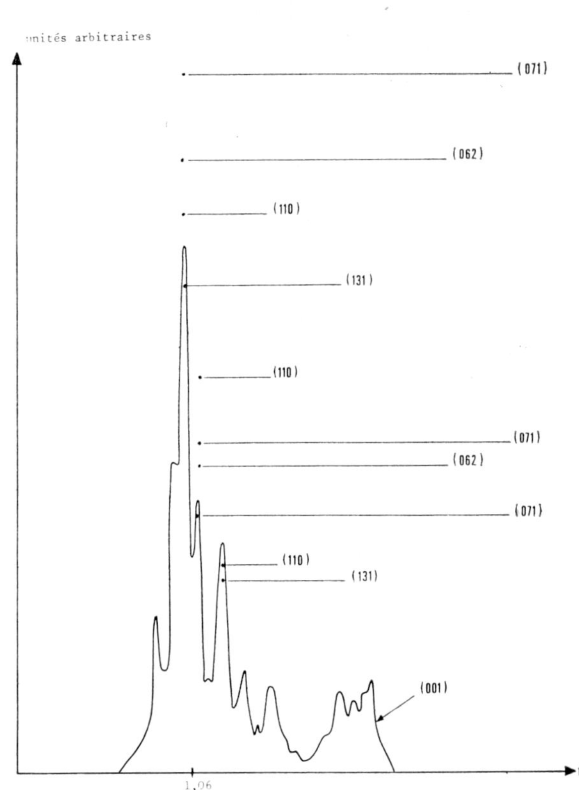

Fig. 3. Variation du spectre de fluorescence suivant les directions d'observation à 300 °K.

incongruente dûe à l'existence d'une phase encore solide à la température normale de cristallisation.

Pour tous les dopants utilisés, différentes conditions de tirage ont été examinées: variation des vitesses de tirage (2 à 5 mm h^{-1}) et de rotation du germe (5 à 180 tours min^{-1}). Après adaptation d'un four de recuit in situ, la fréquence de fracture des cristaux ne diminue pas.

La limpidité des cristaux est excellente, un barreau traversé par le faisceau d'un laser à hélium–néon ne laisse voir que de rares et faibles centres de diffusion optique.

Nous avons abandonné ce dernier dopant pour Nd (NbO$_3$)$_3$, quadratique: a et $b = 3.87$ Å, $c = 7.84$ Å, éliminant ainsi les éléments alcalins.

L'abaissement de la concentration du dopant à 0.3% diminue la fragilité. Quelques rares cristaux ont pu être élaborés et ont présenté l'effet laser. Le tirage en était facile, sans coupure spontanée entre le solide et le bain, mais la possibilité de réaliser des cristaux sans fracture et de longueur suffisante, restait problématique.

Nous avons alors recherché un dopant de paramètres cristallins se rapprochant le plus possible de ceux du niobate de calcium. C'est le cas de NdNbO$_4$, dont les paramètres dans le système monoclinique sont:

$$a = 5.45 \text{ Å}, \quad b = 11.24 \text{ Å}, \quad c = 5.13 \text{ Å}.$$

Au fur et à mesure de sa préparation (voir 2.2) la fraction de dopant monoclinique augmente progressivement jusqu'à 38%, la fraction restante: 62%, quadratique, est peu favorable.

On peut supposer que la proportion utile du dopant s'améliore au cours de la fusion. Ce dopant a finalement permis la fabrication d'excellents monocristaux, les fractures devenant extrêmement rares et souvent dues à de simples incidents de fabrication.

La concentration en dopant dans le bain a été portée progressivement de 0.3% à 1.7% sans provoquer de perturbation dans la réalisation des monocristaux.

Les derniers essais nous ont permis depréciser les conditions optimales de tirage pour la direction (001): vitesses de tirage de 9 mm h^{-1} et de rotation de 180 tours min^{-1}. L'opération s'effectue aisément, la seule précaution étant de respecter un diamètre faible et régulier.

La plupart de ces cristaux ont présenté l'effet laser et ont fonctionné en régime continu à la température ordinaire. Pour les quatre dopants essayés, les compensations de charge doivent se produire de la manière suivante:

Nd (NbO$_3$)$_3$ — compensation par vacance cationique,

Na Nd (NbO$_3$)$_4$ — compensation par un cation monovalent,

K$_2$ Nd (NbO$_3$)$_4$ — compensation par deux cations monovalents,

Nd NbO$_4$ — compensation par vacance ionique.

3.3.3. *Essais de recuit*

Le barreau fabriqué est usiné sous forme de cylindre de 50 mm de longueur et de 3 mm de diamètre, le poli final est donné par une meule diamantée de grain 800.

Les cristaux sont alors recuits sous oxygène pur à 1400 °C. La montée en température s'effectue en 7 heures. Après un palier de 15 heures, on rejoint la température ambiante à raison de 50 °C h^{-1}.

L'amélioration de la qualité cristallographique est

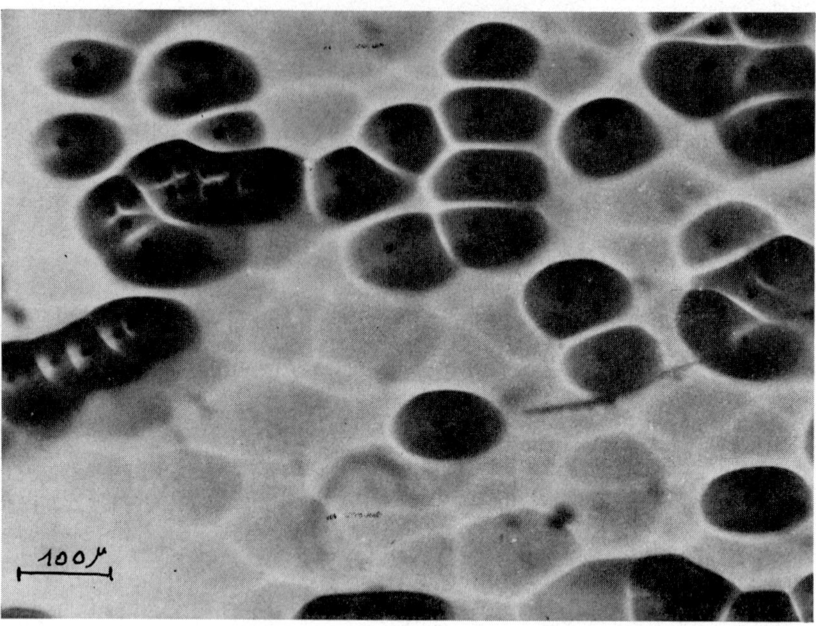

Fig. 4.   Dislocations avant recuit.

rendue évidente par l'examen des courbes de réflexion aux rayons X avant et après recuit. Dans certains cas l'effet laser n'apparaît qu'après ce traitement, l'amélioration des performances, pouvant se chiffrer à environ 30%.

3.3.4. *Dislocations*

La densité des dislocations révélées sur face (001) est de l'ordre de 5000 cm^{-2} au centre du barreau et 30000 cm^{-2} sur les bords (fig. 4).

4. Conclusion

Pour la clarté de l'exposé nous avons présenté séparément l'influence des différents paramètres sur la cristallogenèse du niobate de calcium. Dans la réalité, pour tout essai, nous avons recherché les circonstances favorables en variant la vitesse de tirage et de rotation, la teneur en dopant, le recuit in situ etc.

C'est à l'emploi d'un dopant à maille voisine de celle du niobate de calcium que nous devons d'avoir obtenu enfin des cristaux non fragiles. Il serait cependant logique, tout en conservant la formule du dernier dopant, d'ajouter à nouveau des alcalins, et de contrôler la réaction des cristaux.

La fabrication suivant (062) et (032) présenterait de l'intérêt et une amélioration sensible des performances est envisageable.

Les mesures laser effectuées sur nos cristaux dopés à 1% de Nd font apparaître une puissance émise supérieure à 1 W pour une puissance de pompage de 1 kW alternatif. Les seuils d'excitation comparables à ceux du YAG, en impulsions et en alternatif sont respectivement de 1.09 J et de 350 W.

Cette étude a été financée par la Direction des Recherches et Moyens d'Essais, à qui nous adressons nos remerciements. Nous remercions également les laboratoires de rayons X et de radiochimie, du département Cristallogenèse et Caractèrisation des Matériaux ainsi que ceux du département Physique Electronique et Solides pour les mesures laser.

Bibliographie

1) J. P. Cummings et S. H. Simonsen, Am. Mineralogist **55** (1970) 90.
2) C. D. Whiston et A. J. Smith, Acta Cryst. **23** (1967) 82.
3) M. Ibrahim, N. F. H. Bright et J. F. Rowland, J. Am. Ceram. Soc. **45** (1962) 329.
4) M. Busch, Document C.N.E.T., PEC **42** (1970).
5) J. J. Rubin, L. G. Van Uitert et H. J. Levinstein, J. Crystal Growth **1** (1967) 315.
6) A. A. Ballman, S. P. Porto et A. Yariv, J. Appl. Phys. **34** (1963) 3155.
7) B. A. Scott, E. A. Giess, G. Burns et D. F. O'Kane, Mater. Res. Bull. **3** (1968) 831.
8) H. P. Rooksby et E. A. D. White, Acta Cryst. **16** (1963) 888.
9) J. F. Rowland, N. F. H. Bright et A. Jongejan, in: *Advances in X-Ray Analysis*, Vol. 2 (Pitman, London, 1958) pp. 97–106.

Journal of Crystal Growth **13/14** (1972) 681–685 © *North-Holland Publishing Co.*

THE GROWTH OF FERROELECTRIC SINGLE CRYSTAL $Sr_2Nb_2O_7$ BY MEANS OF THE FLOATING ZONE TECHNIQUE

M. TAKAHASHI, S. NANAMATSU and M. KIMURA

Central Research Laboratories, Nippon Electric Co., Ltd., 1753 Shimonumabe, Kawasaki, Japan

A floating zone technique has been used for the synthesis of single crystals of new ferroelectric materials. The molten zone of the material was formed by radiant heating using a halogen lamp. The lamp and the ceramic rod of the material were arranged to be adjacent foci of an elliptical reflector. The temperature fluctuations in the molten zone are reduced by using a constant power supply to the lamp. Contamination by impurities is minimal. $Sr_2Nb_2O_7$ a newly-discovered ferroelectric, with m.p. 1700 °C, was synthesized and grown as pale yellow transparent single crystals. The colour of these crystals faded on annealing at 1000 °C for 2 hr. It is important that the starting material be stoichiometric to achieve good crystals. The composition of one of the crystals obtained was found to be $Sr_{2.00 \pm 0.03}Nb_{2.05 \pm 0.02}O_7$. Natural cleavage occurs parallel to the b-plane. The Laue patterns, conoscopic and orthoscopic figures, and the electromechanical coupling factor k_{33} before and after poling, were observed in order to determine the homogenity in the crystal. No internal stress or twins were found. The value of k_{33} before poling (0.26) was the same as that after poling. The single crystals appear to be homogeneous single domain material.

1. Introduction

In order to synthesize uniform single crystals of nickel ferrite and nickel–zinc ferrite materials, Akashi et al.[1] in our laboratories developed a new floating zone technique. The apparatus they used is described in their paper[1]. Its main advantages are the ability to supply a constant power into the molten zone, and the relative ease of minimizing impurity contamination. A halogen lamp is placed on one focal point, and a ceramic rod consisting of the material under investigation is placed on the other in an encapsulated elliptical reflector. The light power of the halogen lamp is controlled to keep the height and width of the molten zone constant. The zone can be observed by the enlarged image projected onto a screen by means of a lens. The molten zone is enclosed by a quartz tube. The reflector is cooled with water and the lamp with air. Nickel ferrite and nickel–zinc ferrite materials are black at room temperature and readily absorb radiant heat.

It was found possible, by suitably controlling the working conditions of the apparatus to synthesize single crystals of transparent materials having high melting points. In particular, we succeeded in preparing transparent single crystal of $Sr_2Nb_2O_7$ with m.p. 1700 °C. So far, the existence of the compound has been identi-

fied only by the synthesis of ceramics[2]. We found that the compound is ferroelectric, and has excellent piezoelectric and electro-optic properties similar to those of $LiTaO_3$ and $Ba_2NaNb_5O_{15}$[3].

We observed the orientation, conoscopic and orthoscopic figures, and determined the piezoelectric properties of the grown single crystal of $Sr_2Nb_2O_7$. The present paper deals with the results obtained on the melting point, conditions of preparation and the physical properties related to ferroelectric domain configuration of the crystal.

2. Experimental

2.1. SAMPLE PREPARATION

Powdered $SrCO_3$ (purity 99.8 %) and Nb_2O_5 (99.5 %) were used as raw materials. They were weighed and mixed by a ball mill for 24 hr. The mixed powder was pressed with a pressure 60 kg/cm² into a rod with diameter 8 mm and length 100 mm. The rod was sintered at 1300 °C for 3 hr. The resulting ceramic looked like a piece of chalk. Its density was 4.3 g/cm³ (X-ray density = 5.1 g/cm³).

So far, no report has been found on the melting points of compounds in the system $SrO–Nb_2O_5$. A melting point study using calcined powders having

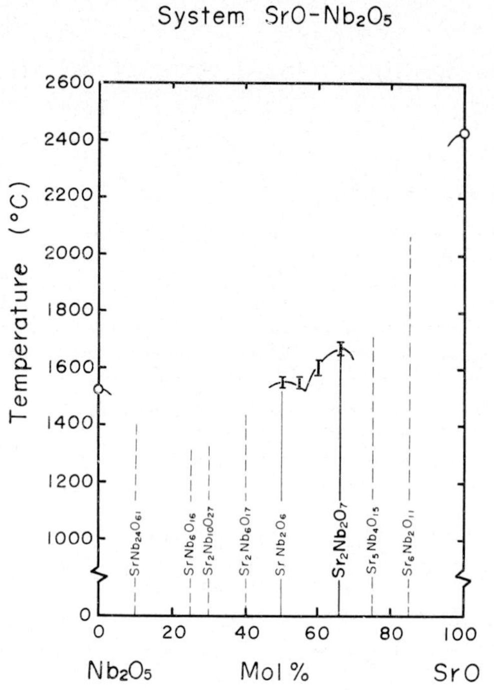

Fig. 1. Liquidus of the SrO–Nb₂O₅ system in the region of Sr₂Nb₂O₇.

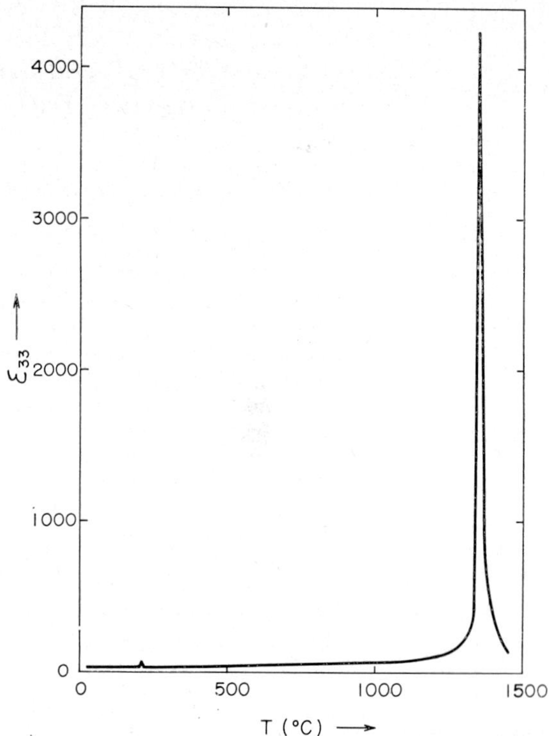

Fig. 3. Temperature variation of dielectric constant ε_{33}.

various compositions was therefore carried out. Small amounts of the powders were placed at various points in the furnace for which the temperature distribution had been determined using a thermocouple. Melting points were determined by observing the positions where powders began to melt. Stoichiometric $Sr_2Nb_2O_7$ was found to melt at approximately 1700 °C (fig. 1).

Recently Carruthers and Grasso[4]) reported a phase diagram on $SrO–Nb_2O_5$ which shows that $Sr_2Nb_2O_7$ single crystal can not be obtained from the material of stoichiometric composition. However, we found that single crystal material could be obtained from a congruently freezing melt. It should be noted that the values of the d-spacing given by Carruthers and Grasso were the same as those of calcined powder and the single crystal.

2.2. Crystal growth

A single crystal thickness 2 mm, width 5 mm and length 15 mm, was mounted as a seed. The ceramic rod and seed crystal were rotated at 30 rpm in opposite directions to each other, making their relative velocity 60 rpm. The rod was moved downward at the same time giving a zone speed of 10 mm/hr.

The single crystals obtained were pale yellow and transparent. The colour faded on annealing at 1000 °C for 2 hr (fig. 2). The average size of crystal was 6 mm in diameter and 40 mm in length. Crystals were grown

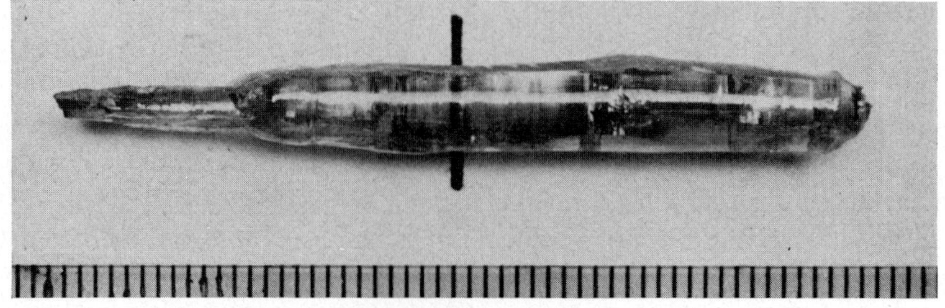

Fig. 2. A single crystal of $Sr_2Nb_2O_7$ grown by the floating zone technique. Scale in cm.

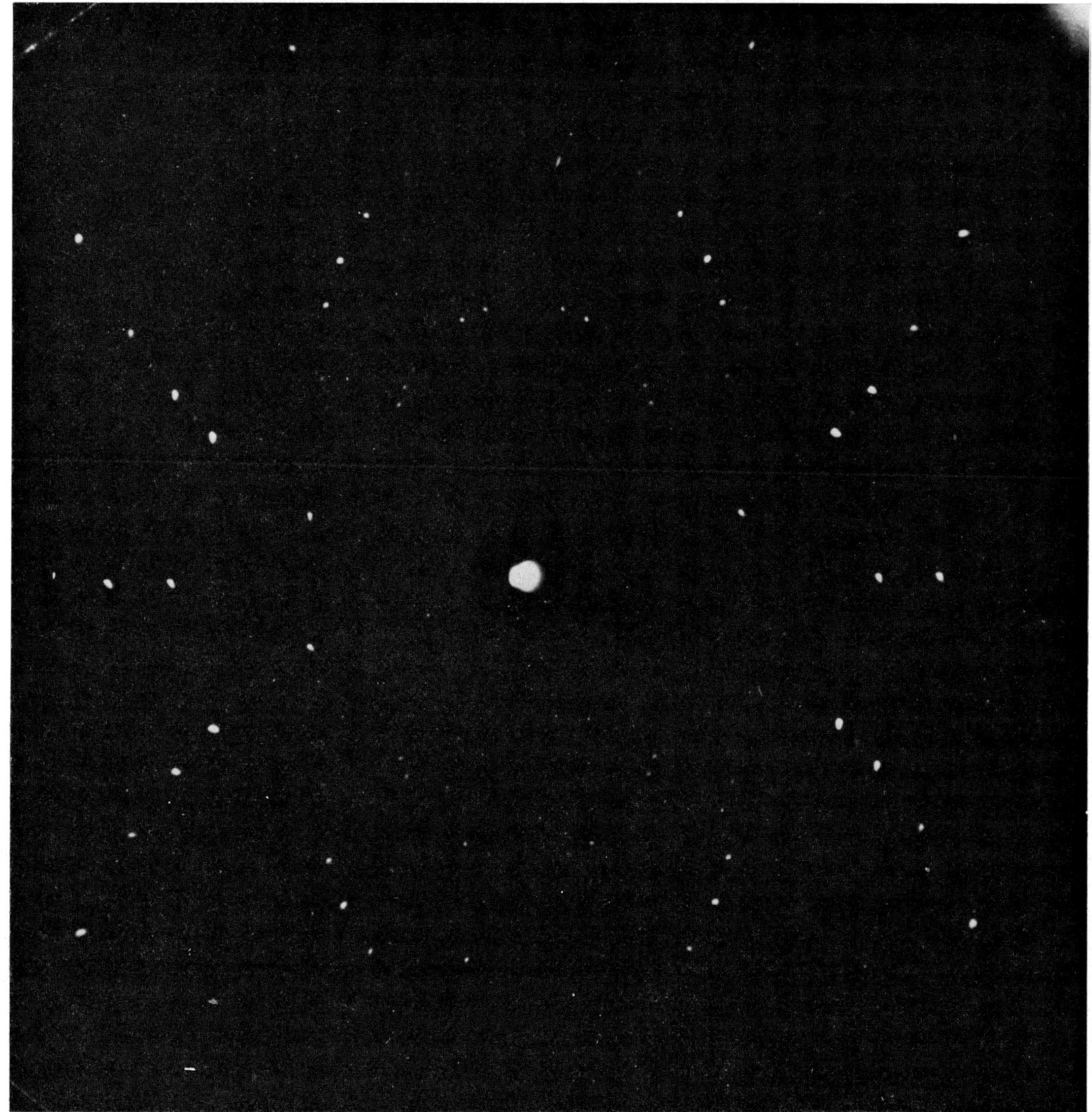

Fig. 4.   Laue patterns of $Sr_2Nb_2O_7$ single crystal. Incident beam is normal to the c-plane.

in a direction parallel to the b-plane, a plane in which natural cleavage occurs.

If the material contained 1 mol% Nb_2O_5 above stoichiometric composition, the molten zone became unstable. In this case a white "compound" segregated from the zone and made it more difficult to obtain homogeneous single crystal than in the case in which stoichiometric material was used.

If the material contained 1 mol% SrO above stoi-

chiometric composition, the viscosity of the molten zone increased and the synthesis of single crystal again became difficult. The crystal obtained was milk-white.

The composition of single crystals (obtained from the material corresponding to stoichiometric composition) was determined by means of a fluorescent X-ray method to be $Sr_{2.00\pm0.03}Nb_{2.05\pm0.02}O_7$. This is close to $Sr_2Nb_2O_7$. Single crystals were uniform in composition.

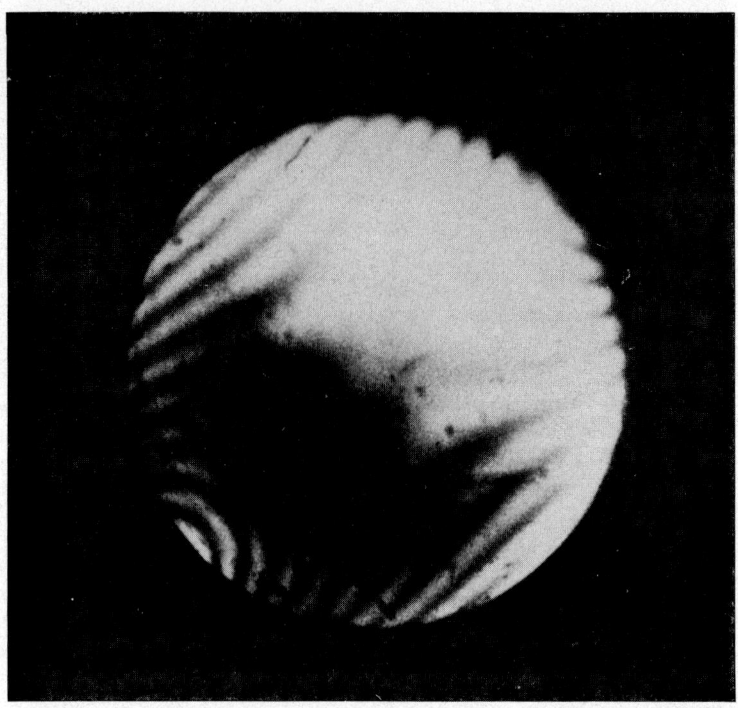

Fig. 5.   Conoscopic figure of $Sr_2Nb_2O_7$ single crystal using light of wavelength 5330 Å of mercury lamp.

2.3. PROPERTIES

The ferroelectric properties of $Sr_2Nb_2O_7$ single crystal were determined by measurements of (1) D–E hysteresis curve and (2) the Curie temperature. (1) was obtained using a Sawyer–Tower[5]) circuit in an ac field of 50 Hz and (2) by means of the temperature variation of dielectric constant at a frequency of 1 MHz.

The crystal orientation was determined by means of transmission Laue patterns. The spot size of the incident beam was chosen to be 1 mm in diameter. The homogenity was confirmed by the observation of conoscopic and orthoscopic figures.

The piezoelectric properties of the crystal were determined by means of electromechanical coupling factor in the longitudinal mode (k_{33}). The value of k_{33} was determined by resonance–antiresonance measurements in accordance with appropriate IRE Standards[6]).

3. Results and discussion

Hysteresis curve could be observed only along the c-axis. The spontaneous polarization at room temperature was 9 $\mu C/cm^2$ at a coercive field 6 kV/cm. The Curie temperature was determined to be 1350 °C (fig. 3).

Single crystals of ferroelectric materials similar to $Sr_2Nb_2O_7$, e.g. $LiNbO_3$ and $LiTaO_3$, are usually subjected to poling (application of a dc field) in order to produce single domain material. In the case of un-poled crystals multi-domain material is present in the crystal.

It should be noted that no evidence was found for the existence of domain walls even if the single crystal of $Sr_2Nb_2O_7$ was synthesized without poling. The single crystal seems to consist of single domain for the following reasons.

(1) Crystal structure. The most probable point group of the crystal was determined to be C_{2v} from the measurement of the Laue patterns. A typical result is given in fig. 4. The direction of the incident beam (fig. 4) was made to be normal to the c-plane (or parallel to the polarization axis) of the crystal. The pattern obtained suggests that the crystal is homogeneous irrespective of the twin or ferroelectric domain configuration produced by 90° domain walls.

(2) Conoscopic and orthoscopic figures. The single crystal was found to be homogeneous by the observation of conoscopic and orthoscopic figures. A typical conoscopic figure is given in fig. 5. The crystal seems to be free from internal stress and ferroelectric domain

configuration. This suggests that single crystal has no 90° domain wall.

(3) Piezoelectric properties. In order to examine the existence of 180° domain wall, piezoelectric properties of the single crystal were determined before and after poling. The value of k_{33} before poling (0.26) was the same as that after poling. This means that no 180° wall exists and the crystal consists of ferroelectric single domain.

Acknowledgements

The authors wish to thank Professor E. Tatsumoto and Dr. T. Tsuji for their encouragement. They are indebted to Dr. T. Kawamura, Mr. S. Matsushita and Mr. K. Doi for their help in the experiments.

References

1) T. Akashi, K. Matsumi, T. Okada and T. Mizutani, IEEE Trans. Magnetics **MAG-5** (1969) 285.
2) G. A. Smolenskii, V. A. Isupov and A. I. Agrranovskaya, Soviet Phys.-Dokl. **1** (1956) 300.
3) S. Nanamatsu, M. Kimura, K. Doi and M. Takahashi, J. Phys. Soc. Japan **30** (1971) 300.
4) J. R. Carruthers and M. Grasso, J. Electrochem. Soc. **117** (1970) 1426.
5) C. B. Sawyer and C. H. Tower, Phys. Rev. **35** (1930) 269.
6) Proc. IRE **49** (1961) 1161.

CRYSTAL GROWTH AND PROPERTIES OF SOME ALKALI METAL METAVANADATES

R. S. FEIGELSON, G. W. MARTIN and B. C. JOHNSON

Center for Materials Research, Stanford University, Stanford, California 94305, U.S.A.

Single crystals of $LiVO_3$, $NaVO_3$, and KVO_3 were grown from stoichiometric melts by the Czochralski technique utilizing slow growth rates 0.5–2 mm/hr. X-ray investigation showed $LiVO_3$ is most probably monoclinic with $a_0 = 9.524$, $b_0 = 8.406$, $c_0 = 5.874$, and $\beta = 93.33°$; $NaVO_3$ monoclinic, with $a_0 = 10.530$, $b_0 = 9.465$, $c_0 = 5.864$, and $\beta = 108.21°$; and KVO_3 orthorhombic, $a_0 = 5.70$, $b_0 = 10.82$, $c_0 = 5.22$. Crystals were optically transparent from 0.4 to 5 microns. Several strong, narrow linewidth Raman modes were detected in each of the crystals. Stimulated Raman scattering was successfully obtained from all three samples.

1. Introduction

Metavanadates of the alkali metals, XVO_3, exist over the entire range from lithium through cesium. Small crystals of all have been grown by slowly cooling stoichiometric melts. Large crystals, allowing extensive property evaluation were not produced, however, until recently, when Baughman and Farnum[1] produced sizeable crystals of KVO_3. The present investigation was aimed at the growth of large lithium, sodium, and potassium metavanadate single crystals of high optical quality for physical property evaluation.

The first study of solid–liquid equilibria in the system $Li_2O–V_2O_5$ was made by Canneri[2], who reported the formation of two intermediate compounds, one a 1:1 salt which melts congruently at about 618 °C. Reisman and Mineo[3] found other additional compounds and confirmed the existence of $LiVO_3$, as did also Kohlmuller and Martin[4]. Glazyrin[5] thought $LiVO_3$ to be orthorhombic based on optical measurements alone. Lithium metavanadate was reported to be very hydroscopic[6].

A number of compounds also exist in the $Na_2O–V_2O_5$ system including a composition at 50 mole% Na_2O. Sodium metavanadate, like $LiVO_3$, melts congruently[7], but at a slightly higher temperature (627–638 °C). Glazyrin and Fotiev[8] determined that $NaVO_3$ was monoclinic using optical techniques, which confirmed the earlier work of Sørum[9], who reported the space group to be $C2/c$ and the lattice constants $a = 10.14$ Å, $b = 9.45$ Å, $c = 5.86$ Å and $\beta = 69.58°$.

Lukacs and Strusievici[10] reported a polymorph of $NaVO_3$, prepared by dehydration of the crystal hydrate. This low temperature β form exhibits an irreversible transformation to the more common high temperature α form at 403–405 °C. Sodium vanadate was reported to be very soluble in water[11]. In 1950 Sawada and Nomura[12] reported that sodium metavanadate became a ferroelectric above room temperature and exhibited a domain structure and hysteresis in electric fields. Pulvari[13] and Miller et al.[14], however, conjectured that $NaVO_3$ may be antiferroelectric rather than ferroelectric above room temperature.

Potassium metavanadate has been more widely studied than any other metavanadate in this series, including highly refined crystal structure analysis[15–18] and optical studies[19]. Compound formation and phase diagrams for the $K_2O–V_2O_5$ system are also available[20–22]. From these studies it was evident that KVO_3 melts congruently at a temperature of 520 °C, and has an orthorhombic crystal structure (space group Pmab) with lattice constants $a_0 = 5.70$ Å, $b_0 = 10.82$ Å, $c_0 = 5.22$ Å.

The variation of refractive index and birefringence as a function of cation radius for the alkali metal metavanadate series was studied by Glazyrin[5]. A steep almost linear decline of birefringence was observed with increasing cation radius. The index of refraction also decreased with increasing cation radius, reaching a minimum, however, at Rb then increasing with Cs. Birefringence $n_e - n_0$ was 0.28 for $LiVO_2$, 0.225 for $NaVO_2$, and 0.142 for KVO_3.

2. Experimental procedure

2.1. MATERIAL SYNTHESIS

Metavanadates were prepared by reaction of alkali metal carbonates with vanadium pentoxide according to the following reaction*:

$$X_2CO_3 + V_2O_5 \rightarrow 2XVO_3 + CO_2\uparrow.$$

If caution was not exercised however, stoichiometric compositions yielded vanadium rich melts after synthesis, which had a reddish brown to black appearance. Two significant factors are the hydration or moisture adsorption of the carbonates, particularly sodium and potassium, or loss of the alkali metal oxide with CO_2 evolution during the decomposition reaction. To minimize the former problem all the carbonates were baked at a temperature of 200 °C for 2 hours in a dynamic vacuum of 10^{-5} torr prior to weighing. The dried powder was transferred in an evacuated tube to a controlled atmosphere chamber where it was then weighed and blended thoroughly with V_2O_5 under argon. Samples were slowly heated in platinum crucibles to a temperature just below the lowest eutectic temperature adjacent to the particular compound ($LiVO_3$: 545 °C, $NaVO_3$: 510 °C, KVO_3: 365 °C). These temperatures were determined from the respective phase diagrams. Samples were held at this temperature for 15 hr. The mixture was then slowly heated over a 7-hr period to just above the melting temperature and then rapidly heated to 1000 °C and held there for 2 hr. Finally, the melt was furnace cooled. Deviations from stoichiometry were determined by weight loss, color, chemical and DTA analysis.

2.2. GROWTH PROCESS

A simple Czochralski system utilizing a resistance furnace was used for the growth of all three vanadates. Since the melting temperatures of these materials is low, the crucible was placed high in the furnace and a funnel shaped quartz heat shield placed over the furnace opening to enhance both the axial and radial temperature gradients. The resulting axial gradient was 55 °C/cm directly above the melt surface. An ambient air atmosphere was maintained during growth. Temperature was controlled by a proportional two function controller in conjunction with a SCR power supply.

Seed crystals were selected initially from crystals embedded in the solidified melt surface and oriented so as to pull along their fastest growth direction. Most boules were grown along the [001]. The growth rate necessary for production of high quality crystals, even from nominally congruent melts was very low, less than 2 mm/hr, with the most successful rate about 0.5 mm/hr. Low diffusion rates of ionic or molecular species in the liquid phase at these low temperatures is probably a significant factor. Small deviations from an exact 1:1 composition could result in very rapid partitioning at the interface, which would lead to constitutional supercooling, interface instability, and then cellular type growth. This problem was most in evidence with KVO_3 melts.

Low rotation rates of 4–10 rpm were used.

2.3. X-RAY DIFFRACTION STUDY

A variety of X-ray diffraction methods were used to characterize the metavanadate crystals. Single crystals were examined with a Unicam rotation camera, a Weissenberg camera, and diffractometers. Powder diffractometer patterns were indexed with a computer program (see table 1). High temperature diffractometry data were taken with $NaVO_3$ and $LiVO_3$ from 23 °C to 600 °C.

3. Results and discussion

3.1. CRYSTAL PREPARATION

Of the three crystals, the easiest to grow was $NaVO_3$ where a 1.5 mm/hr growth rate yielded good quality, clear crystals. Potassium vanadate, because of its lower melting temperature and therefore poorer diffusion rates in the liquid phase, was difficult to grow without defects associated with constitutional supercooling. If rotation rates were increased from 10 to 25 rpm to enhance stirring near the boundary layer (pull rate 1.5 mm/hr) the band-like defect structure is converted into a narrow core defect. At higher rotation rates, however, massive breakdown occurs, giving a more extensive and uniform defect distribution. At 10 rpm, 0.5 mm/hr and along the [001], clear regions were obtained in straight sections of the boule. Whenever the boule diameter changed, however, the defects reappeared. This was also observed by Baughman and Far-

---

* Starting material purity was in the 99.99 per cent range with respect to cation impurities.

TABLE 1

Intensities I/I_1, interplanar d and hkl values for $LiVO_3$, α-$NaVO_3$, β-$NaVO_3$ and KVO_3

$LiVO_3$			α-$NaVO_3$			β-$NaVO_3$			KVO_3		
d(Å)	I/I_1	hkl*	d(Å)	I/I_1	hkl	d(Å)	I/I_1	hkl*	d(Å)	I/I_1	hkl
6.30	35	110	6.86	20	110	7.10	5	110	5.41	25	020
4.74	100	200	5.00	45	200	5.05	100	200	5.21	20	001
4.20	35	020	4.87	15	$\bar{1}11$	3.55	25	$\bar{2}11$	3.93	8	120
3.34	40	211	4.72	30	020	3.24	25	310	3.76	15	021
3.15	50	220	3.593	30	021	3.02	20	$\bar{2}21$	3.13	100	121
3.07	50	?	3.427	100	220	2.97	100	130	2.85	26	200
2.982	90	310	3.259	45	$\bar{2}21$	2.53	20	400	2.71	20	040
2.860	10	$\bar{3}01$	3.175	12	311	2.33	10	040	2.63	15	131
2.758	15	102	3.148	90	301	2.26	5	$\bar{3}02$	2.61	16	002
2.682	20	130	2.839	8	$\bar{2}02$	1.97	5		2.44	19	211
2.500	10	?	2.780	15	002	1.89	40		2.40	20	041
2.370	15	400	2.753	6	$\bar{1}31$	1.83	5		2.31	11	112
2.310	10	122	2.671	35	221	1.76	30		2.11	5	032
2.098	15	330	2.538	3	131	1.71	5		1.96	14	240
1.970	10		2.500	3	400	1.62	5		1.92	5	
1.920	5		2.445	10	311	1.59	10		1.89	3	
1.908	5		2.410	6	112	1.54	5		1.87	6	
1.854	15		2.362	6	040				1.80	7	
1.642	5		2.290	34	330				1.78	6	
			2.248	7	$\bar{4}02$				1.73	3	
			2.178	19	041				1.700	20	
			2.141	9	$\bar{1}32$				1.573	3	
			2.092	10	$\bar{2}41$				1.501	5	
			2.053	5	$\bar{5}11$				1.480	2	
			1.959	40	510				1.461	5	
			1.861	18	150				1.425	3	
			1.808	9	042				1.412	2	
			1.754	11	$\bar{4}41$				1.352	3	
			1.721	15	440				1.316	10	
			1.692	6	530						
			1.673	12	600						
			1.649	12	$\bar{1}33, \bar{5}33$						
			1.580	20	060						
			1.530	3	422						
			1.511	21	531						
			1.461	6	043						
			1.414	5	512, 710						
			1.375	20	550, $\bar{2}62$						
			1.360	10	$\bar{4}24$						
			1.333	9	$\bar{7}32$						
			1.253	3	641, 370						
			1.142	8	552, 750						

* Tentative fit with monoclinic cell.

num[1]). Defects were present even at the slowest pulling speeds when KVO_3 was grown along the [100].

When grown under conditions similar to those mentioned above, lithium vanadate presented some unique problems. During the initial part of the growth process the crystal was clear and single. As the crystal grew, however, the upper section became uniformly clouded. The defect structure moved uniformly downward as the crystal grew. The interface between clear and clouded regions was sharply defined. When the crystal was pulled from the melt surface, the interface moved through the remaining clear region leaving the entire crystal opaque. Visually the defect manifested itself as fractures along some of the prominent cleavage planes and in [001] growth was parallel to the growth axes. In [100] growth the fractures lay perpendicular to the growth direction. The effect was most pronounced in off-stoichiometric melts. DTA and chemical analysis experiments were

XIV – 3

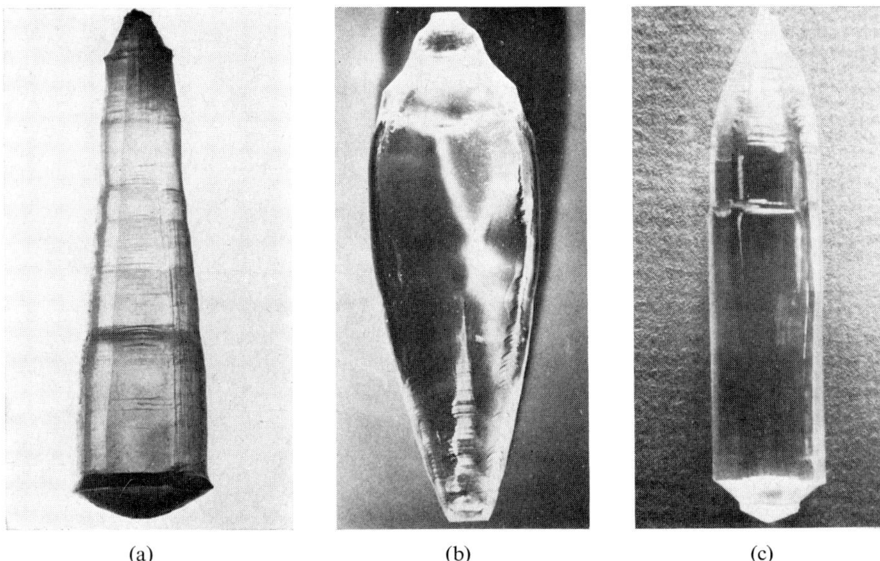

(a)   (b)   (c)

Fig. 1.   (a) $LiVO_3$, (b) KVO_3, (c) $NaVO_3$ ($\times 0.9$).

carried out to determine whether precipitation or a phase transition was involved. No evidence was found for either effect. It is possible that thermal stresses which develop as the crystal moves into colder regions of the furnace is responsible.

To minimize the problem, the melt was carefully prepared to insure stoichiometry. The melt was also lowered in the furnace by 2 cm, and the inner heat shield removed to give a lower overall axial gradient of 20 °C/cm. In addition, the crystal was slowly cooled to room temperature after growth, at a rate of about 20 °C/hr. Clear crystals were obtained by this technique at growth rates of 1.5 mm/hr and 5 rpm rotation rates. In uncracked transparent single crystals analyzed by X-ray techniques, microtwinning was observed.

Boules of 1.5 cm diameter and 6 cm in length were grown without difficulty. Three typical boules are shown in fig. 1.

Lithium vanadate in bulk single crystal form was not hydroscopic, as previously reported by Fotiev and co-workers[6]), unless the composition was off stoichiometry. Under these conditions surface deterioration was observed after several hours in an ambient air atmosphere. Sodium and potassium metavanadate crystals did not show this type of degradation at all. Lithium and potassium metavanadate were found to be roughly 10 times more soluble in H_2O than $NaVO_3$ at room temperature. Approximate solubility values at 28 °C for solution of anhydrous crystals crushed to fine particle size (~ 100 mesh) are $LiVO_3$ 13 g/100 cm^3; $NaVO_3$ 0.8 g/100 cm^3; KVO_3 9 g/100 cm^3. Solution of a KVO_3 single crystal occurs principally along the [001].

Sodium and potassium metavanadate crystals were pale yellow, potassium almost colorless; but lithium vanadate was a deep yellow brown or amber color, light yellow in thin sections. Thin crystals grown in a

TABLE 2

Chemical analysis*

Crystal	Alkali metal (%)		Vanadium (%)		Oxygen (%)		Permittivity ε'
	Calc.	Exp.	Calc.	Exp.	Calc.	Exp.**	
$LiVO_3$	6.554	6.57	48.113	48.75	45.334	44.68	(Z) 20 $\pm$ 10%
$NaVO_3$	18.855	18.67	41.779	41.50	39.366	39.83	(Z) 17 $\pm$ 10%
KVO_3	28.326	28.33	36.903	36.76	34.772	34.91	(Z) 9.5 $\pm$ 10%
$LiNbO_3$	—	—	—	—	—	—	(Y) 85

* $\pm 0.3\%$.

** Determined indirectly from mass balance.

XIV – 3

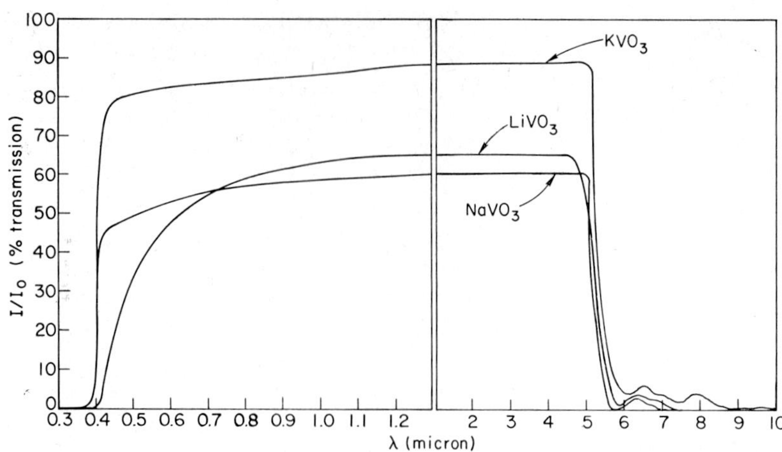

Fig. 2. Optical transmission for lithium, sodium and potassium meta vanadates. Measurement made parallel to the b-axis. Sample thickness: $LiVO_3 = 0.97$ mm, $NaVO_3 = 0.64$ mm, $KVO_3 = 0.66$ mm.

LiCl flux, however, were almost colorless. Emission spectrographic analysis showed similar impurities common to all three, including 100 ppm Fe, 50 ppm Ni, 15 ppm Al, 30 ppm Si and 2 ppm Mg (15 ppm Mg in KVO_3). The color of $LiVO_3$ may therefore be due to a slight deviation in stoichiometry. Results of wet chemical analysis are given in table 2, showing $LiVO_3$ to be slightly rich in V, and O_2 deficient. Growth of $LiVO_3$ in an O_2 atmosphere did not result in a color change nor did O_2 annealing at elevated temperatures. Absorption curves for these materials are given in fig. 2. These materials are transparent from about 0.4–5 μm.

Handling of all three crystals was difficult, due principally to the easy cleavage. Strained crystals fractured easily and it was generally troublesome to cut sections perpendicular to the [001].

3.2. CRYSTAL STRUCTURE

$LiVO_3$. High-temperature diffractometry indicated no change in $LiVO_3$ between 23 °C and 600 °C. Single crystal Weissenberg photographs suggested a monoclinic cell with $a_0 = 9.542$ Å, $b_0 = 8.406$ Å, $c_0 = 5.874$ Å, and $\beta = 93.33°$. However, since these values could not account for two of the lines in the powder pattern (see table 1) it is possible that the crystal could be triclinic. Crystal cleavage was observed on the {100} and {110} planes. Experimental density measurements based on Archimedes' Principle gave a density of 3.005 g/cm^3. The calculated X-ray density based on 8 molecules/unit cell is 3.00 g/cm^3.

α-$NaVO_3$. Detailed and precise measurements on α-$NaVO_3$ established that it was monoclinic with $a_0 = 10.530$ Å, $b_0 = 9.465$ Å, $c_0 = 5.864$ Å and $\beta = 108.21°$. The powder pattern was indexed, and showed excellent agreement with values obtained with single crystals. X-ray work reported by Sørum[9]) had indicated the space group to be C2/c (S.G. No. 15), but our Weissenberg photographs showed it could also be the non-centrosymmetric space group Cc (S.G. No. 9). Previous ferroelectric hysteresis measurements[12]) showed that the crystal was probably antiferroelectric, and therefore should belong to space group Cc. Our own preliminary hysteresis measurements also indicate antiferroelectric behavior. Crystals cleaved well on the {110} plane. The acute angle between {110} planes was 87°. The calculated X-ray density based on 8 molecules/unit cell was 2.91, which was in excellent agreement with the experimental value of 2.91.

β-$NaVO_3$. None of the low temperature β-$NaVO_3$ form was found to exist in the $NaVO_3$ boules grown at high temperatures. Some β-$NaVO_3$ was precipitated from solution at room temperature, and examined at elevated temperatures by diffractometry. Between 350 °C and 450 °C the beta form changed irreversibly to α-$NaVO_3$, as previously reported[10]). X-ray powder data for both forms are given in table 1.

KVO_3. Our powder diffraction pattern for KVO_3 was indexed using an orthorhombic unit cell of $a_0 = 5.70$ Å, $b_0 = 10.82$ Å and $c_0 = 5.22$ Å, in excellent agreement with previous research. A major crystal cleavage was the {010} plane. Assuming 4 molecules per unit cell, an X-ray density of 2.848 g/cm^3 was calculated compared to a value of 2.90 measured experimentally.

3.3. CRYSTAL PROPERTIES

The basic motivation for the development of large single crystal samples of the vanadate compounds is to attempt to gain some insight into the inherent relation between the elements which combine to form the crystals and the various macroscopic properties exhibited by them. Specific properties of interest in this case are: (1) nonlinear coefficients (e.g., Raman, parametric) which participate in frequency mixing, tunable optical and infrared generation, and (2) optical mode frequencies, linewidths, and scattering efficiencies.

The original consideration of the vanadate compounds as good candidates for nonlinear properties arose from studies conducted on two other well known materials, $LiNbO_3$ and $LiTaO_3$. Both of these crystals are in prominent use today for a variety of nonlinear applications (e.g., parametric oscillators, microwave acoustic devices, etc.). Both are ferroelectric at room temperature, and have exhibited stimulated polariton scattering which produces tunable optical and far infrared radiation[23]).

It was proposed that by substituting vanadium, the lightest element in Group V of the periodic table, for the Nb or Ta atom, together with Li (or other Group I cation) and three oxygens, that materials with interesting and useful bulk nonlinear properties might result. It is also well known that optical mode frequencies, v_1, follow a dependence $v_i \alpha_i (K_i/m_i)^{\frac{1}{2}}$, where K_i is the effective restoring force constant for the ith mode, and m_i the reduced mass of the nuclei participating in the vibration. Assuming that the K_i values do not vary significantly from compound to compound, introduction of the lighter vanadium atom for Nb or Ta should reduce m_i (at least for certain modes) and hence shift these optical modes to higher frequencies than corresponding modes in $LiNbO_3$ or $LiTaO_3$. When accompanied by high nonlinear coefficients, this result is of considerable importance for the generation of high power, coherent, tunable radiation in the far infrared[24]).

It is a general rule that materials with high refractive indices are likely to possess large nonlinear coefficients[25]. The fact that the vanadate compounds show index values comparable to those of $LiNbO_3$ and $LiTaO_3$ is a promising indication that nonlinear coeffi-

cients for the vanadates will prove to be large as well. Preliminary measurements have been undertaken to measure the coefficients.

Permittivity values for the vanadates as compared to $LiNbO_3$ are given in table 2. In order to determine the optical mode properties of the vanadates, Raman frequencies, linewidths, and scattering strengths of the three materials have been studied. An argon laser (5145 Å, 100 mW) was used in conjunction with a Jarrell–Ash double grating Raman spectrometer and a dry-ice cooled RCA–ERMA C31000F photomultiplier detector to observe 90° spontaneous Raman scattering from all three materials. Several strong, narrow linewidth Raman modes were detected in each of the crystals from the scattering configuration $X(ZZ)Y$, where X is the propagation direction in the crystal of the incident (laser) photons which are polarized along Z; scattered (Stokes) photons, shifted in frequency are also polarized along the crystal Z direction and propagate in the Y direction. Some results of the measurements are summarized in table 3, which presents mode frequencies and spontaneous linewidths for the three vanadate crystals.

TABLE 3

Raman frequencies and linewidths

Crystal	$LiVO_3$		$NaVO_3$		KVO_3	
	v_i	Γ_i	v_i	Γ_i	v_i	Γ_i
	527	12	178	20	213	9
	652	12	508	10	648	10
	903	6	919	4	938*	4
	955*	5	955*	4	955*	4

Mode frequencies, v_i, and spontaneous linewidths, Γ_i are specified in cm^{-1}.

* Indicates modes from which stimulated Raman scattering has been observed.

In a second series of experiments, the materials were pumped with a Q-switched ruby laser (1 MW, 20 nsec pulses), and stimulated Raman scattering was successfully obtained from all three samples. The high coherence of the Stokes radiation and substantial conversion efficiency from pump to shifted signal power indicated large values for the stimulated gain coefficients in the three crystals, thus reinforcing earlier predictions of large nonlinear coefficients in the vanadate compounds. It is of particular interest also to note that the mode frequencies observed in the stimulated effect are higher in frequency than *any* of the optical modes in either $LiNbO_3$ or $LiTaO_3$. This fact is important because it confirms the prediction that the introduction of the lighter vanadium atoms into the lattice would increase certain frequencies, based on a (K_i/m_i) dependence. In addition, since the shifts are identical irrespective of the cation (Li, Na, or K), it is reasonably certain that a vanadium–oxygen vibrational state is participating.

The high Raman scattering efficiency and narrow spontaneous linewidths of the modes indicate that the vanadium atom is favorably influencing the nonlinear properties of these materials.

Acknowledgments

We are indebted to Rosemarie Swanson, Ted Hopkins, and Marta de Rojas for assisting in X-ray analysis, to Robert Raymakers and Robert Griffin for Materials preparation, and to Wayne Kway for optical absorption measurements.

Work supported by the Advanced Research Projects Agency through the Center for Materials Research.

References

1) R. J. Baughman and E. H. Farnum, Mater. Res. Bull. **5** (1970) 993.
2) G. Canneri, Gazz. Chim. Ital. **58** (6) (1928).
3) A. Reisman and J. Mineo, J. Phys. Chem. **66** (1962) 1181.
4) R. Kohlmuller and J. Martin, Bull. Soc. Chim. Franc. (1961) 748.
5) M. P. Glazyrin, Kristallografiya **10** (1965) 761.
6) A. A. Fotiev, M. P. Glazyrin and N. V. Bausova, Russ. J. Inorgan. Chem. **13** (1968) 1007.
7) H. Flood and H. Sørum, Tidskr. Kjemi Bergvesen Met. **5** (1943) 57.
8) M. P. Glazyrin and A. A. Fotiev, Kristallografiya **92** (1964) 287.
9) H. Sørum, Kgl. Norske Videnskab. Selskabs, Forh. **16** (1943) 39.
10) I. Lukacs and C. Strusievici, Z. Anorg. Allgem. Chem. **315** (1962) 323.
11) D. J. McAdam, Jr. and C. A. Pierle, J. Am. Chem. Soc. **34** (1912) 604.
12) S. Sawada and S. Nomura, J. Phys. Soc. Japan **6** (1951) 192.
13) C. F. Pulvari, Phys. Rev. **120** (1960) 1670.
14) R. Miller, E. Wood, J. Remeika and A. Savage, J. Appl. Phys. **33** (1962) 1623.
15) H. Evans, Jr. and S. Block, Am. Mineralogist **39** (1954) 326.
16) Christ, J. Clark and H. Evans, Jr., Acta Cryst. **7** (1954) 407.
17) M. Petrasova, J. Madar and F. Hanic, Chem. Zvesti **12** (1958) 410.
18) H. T. Evans, Jr., Z. Krist. **114** (1960) 258.
19) M. P. Glazyrin and A. A. Fotiev, Kristallografiya **9** (1964) 506.
20) F. Holtzberg, A. Reisman, M. Berry and M. Berkenblit, J. Am. Chem. Soc. **78** (1956) 1536.
21) V. V. Illarinov, R. P. Ozerov and E. V. Kil'disheva, Zh. Neorgan. Khim. **1** (1956) 777.
22) M. P. Glazyrin and A. A. Fotiev, Russ. J. Phys. Chem. **42** (1968) 1288.
23) J. M. Yarborough, S. S. Sussman, H. E. Puthoff, R. H. Pantell and B. C. Johnson, Appl. Phys. Letters **15** (1969) 102.
24) C. H. Henry and C. G. B. Garrett, Phys. Rev. **171** (1968) 1058.
25) R. C. Miller, Appl. Phys. Letters **5** (1964) 17.

Journal of Crystal Growth **13/14** (1972) 693–697 © *North-Holland Publishing Co.*

THE CRYSTAL GROWTH OF TRANSITION METAL INTERSTITIAL COMPOUNDS BY A FLOATING ZONE TECHNIQUE

J. BILLINGHAM*, P. S. BELL and M. H. LEWIS

University of Warwick, Coventry, England

A floating zone technique for the preparation of large high purity crystals of vanadium carbide, oxide and nitride with controlled stoichiometry is described. An essential feature of the process is the provision of a facility which enables growth under positive ambient gas pressures (up to 20 atm) to be carried out. Inert gas atmospheres are used to reduce vanadium vaporisation losses in vanadium carbide and oxide and so enable closer compositional control to be achieved, while the nitrides are grown under similar pressures of nitrogen and helium to reduce nitrogen losses. Zone levelling has enabled large single crystals of vanadium monocarbide (VC) with differing and closely controllable compositions to be produced. Vanadium monoxide crystals were grown over the whole cubic stoichiometry range but in the vanadium nitrogen system single phase cubic vanadium mononitride (VN) could not be produced and the microstructure of the alloy of highest nitrogen content consisted of a eutectic of the cubic VN and the hexagonal divanadium nitride (V_2N) phases.
Finally, in addition to the usual chemical and optical microstructural examination, electron microscopy and electron diffraction have been used to characterise the type and extent of the defect structures which are present in all three alloy systems studied.

1. Introduction

The extremely high melting points and hardness of the transition metal–interstitial compounds make them candidates for future generations of high temperature structural materials. Recent high temperature deformation studies on the carbides[1-3] have indicated that promising mechanical properties can be obtained, but the associated low temperature brittleness probably means that in order to utilize the full potential of these materials careful design criteria or the possible use of composite structures will be required[1]. Non-stoichiometry is common in these materials and their phase diagrams are characterised by single phase fields with extremely wide homogeneity ranges, e.g. $VC_{0.72}$ to $VC_{0.90}$[4]) and $VO_{0.89}$ to $VO_{1.20}$[5]). Moreover recent electron and neutron diffraction studies[6,7]) have led to the discovery of ordered phases, which are difficult to detect by X-ray techniques, within these apparently single phase fields. The degree of non-stoichiometry and the presence of ordered phases have a profound influence on many physical properties of these materials and must be closely controlled and carefully measured in any such studies. In addition to high temperature properties considerable interest has been shown

in the electrical properties, especially the metal–semiconductor transitions, in vanadium oxygen alloys. (For a recent review in this field see ref. 8.)

The crystal growth programme was undertaken to prepare materials for an electron microscope study of the phases in the vanadium interstitial element systems, where ordered phases had previously been reported[5,6]. In studies of this type high purity material with known and variable degrees of non-stoichiometry are required. Preparation methods include growth from solution[9]), from the vapour[10]) and by reaction in the solid state[11]), but growth from the melt by the floating zone technique offers the best means for producing large crystals with high purity and controlled stoichiometry[12,13]). Although these materials are extremely stable at room temperature vaporisation of the constituents usually occurs at temperatures near the melting point. This vaporisation can be reduced by melting under a high pressure of inert gas. Thus under similar conditions Haggerty et al.[14]) have shown that the rate of boron evaporation from zirconium diboride is inversely proportional to $\sqrt{P}$ (where P is the gas pressure used). Precht and Hollox[12]) used this technique to prevent excessive vanadium loss in the growth of vanadium monocarbide single crystals.

No established equilibrium diagram exists for the vanadium nitrogen system although the existence of

* Present address, Fulmer Research Institute, Stoke Poges, Bucks., England.

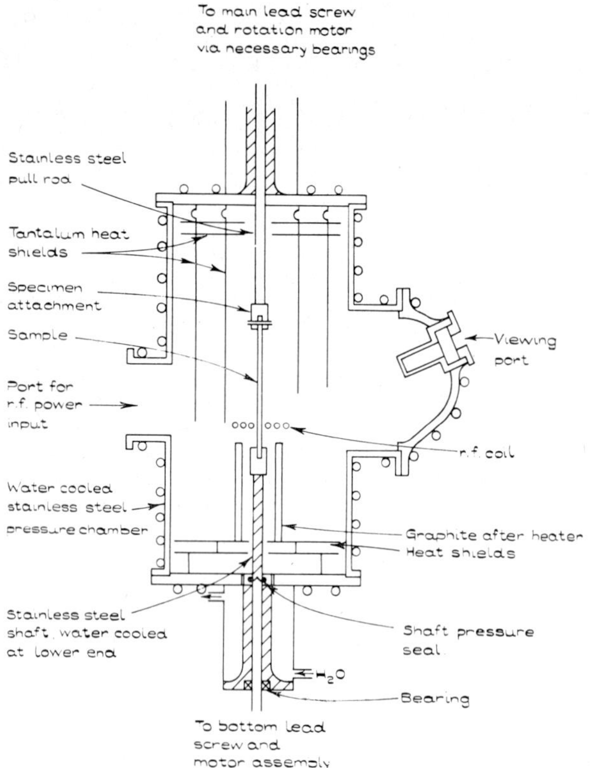

Fig. 1.   Schematic diagram of crystal growth apparatus.

the vanadium mononitride and divanadium nitride phases is well established[15]). These materials are difficult to prepare because of the ease of impurity contamination, especially by oxygen, and the instability of the phases at high temperatures. Because of these problems few attempts have been made to prepare these materials from the melt.

2. Experimental techniques

The starting materials were pure powders of vanadium and vanadium monocarbide, sesquioxide and mononitride of at least 99.5% purity. Powder mixtures of controlled stoichiometry were produced by blending the respective compounds with the required quantity of vanadium metal powder and these were isostatically pressed in rubber bags at room temperature at 80 kg/mm² to produce rods approximately 10 mm diameter and 150 mm long. Samples with high C/V and N/V atomic ratios required additions of 2% paraffin wax as a binder. By closely controlling the initial powder sizes and using wire mesh cages straight rods could be produced with a good surface finish thus ob-

viating the need for any additional difficult and expensive grinding operations. These rods were sintered to approximately 90% theoretical density (except for the oxides) by vacuum annealing at a temperature of approximately $0.6\,T_\mathrm{m}$ prior to zone melting.

The crystal growth furnace shown schematically in fig. 1 is designed to operate at ambient gas pressures up to 20 atm. In addition to the usual lead screw arrangement, an additional motor enables the bottom pull rod to be moved independently thus correcting for changes in length due to differences in density between the starting sintered material and the final fully dense zoned crystal. The chamber itself is made from stainless steel and is 200 mm diameter, 400 mm tall and 10 mm wall thickness. The zoning speed can be continuously varied over a wide range but the samples were usually zoned in an upwards direction at a speed of 5 to 10 mm/hr. These transition metal compounds are all electrical conductors so direct r.f. heating was used with a 30 kW induction generator operating at 450 kHz and controlled through a saturable reactor.

3. Results and discussion

3.1. Vanadium carbon system

Crystals were grown with compositions varying from $VC_{0.43}$ to $VC_{0.90}$. Single crystals were obtained at six compositions within the monocarbide phase field between $VC_{0.70}$ and $VC_{0.90}$. At compositions below $VC_{0.70}$ two phase microstructures containing both the cubic (VC) and hexagonal (V_2C) phases were produced. A typical VC crystal produced was 100 mm in length and 10 mm diameter. Single crystals were usually produced after approximately two zone lengths travel. Slow growth speeds were used to prevent cracking due to steep thermal gradients.

Chemical analysis, optical metallography, and Debye–Scherrer powder photography have been used to characterise the crystals produced. The results indicated that a condition of zone levelling is set up in these crystals whereby the final zone composition is similar to that predicted from the equilibrium diagram[4]) for the particular starting composition. By utilising the technique of zone levelling longer crystals of constant composition were produced. The alloys with high carbon content C/V > 0.84 lost vanadium on melting and increased their C/V ratios by approximately 0.02,

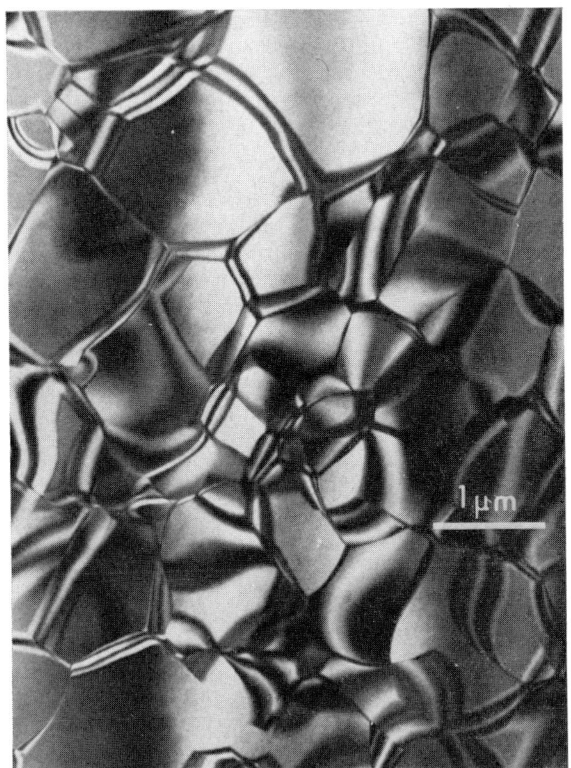

Fig. 2. Structure in $VC_{0.89}$ alloy. Diffraction data consistent with the cubic ordered compound V_8C_7 was obtained and a foam structure of anti-phase boundaries is shown.

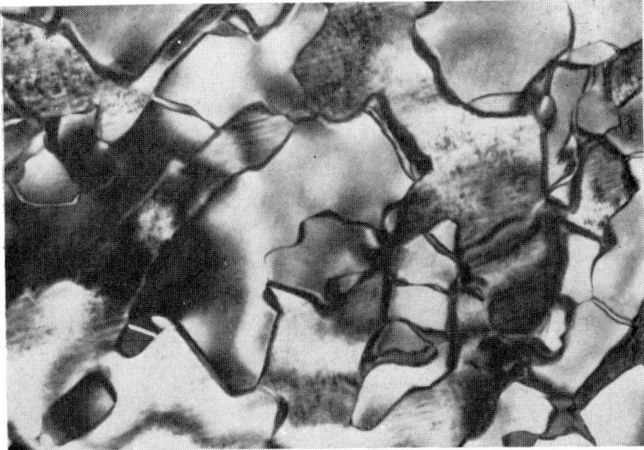

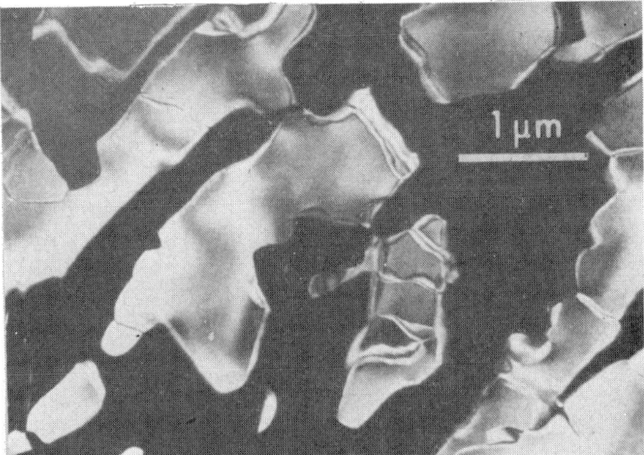

Fig. 3. Superlattice domains in $VC_{0.84}$ alloy arising from the presence of an ordered compound V_6C_5; (b) is a dark field micrograph illustrating how different domains are associated with different orientations of the superlattice unit cell with respect to the parent lattice.

whereas the lower carbon alloys e.g. $VC_{0.73}$ did not show any change in composition due to their much lower melting points.

Electron microscopy revealed some subgrain boundaries and dislocations in the monocarbide alloys but the microstructures were mainly characterised by the presence of substructure due to the formation of ordered compounds within this nominally cubic (rock-salt) phase field. Two ordered compounds V_8C_7 and V_6C_5 previously identified as $VC_{0.88}$[16]) and $VC_{0.84}$[6]) were found to exist over a range of composition: V_8C_7 from $VC_{0.87}$ to $VC_{0.90}$ (fig. 2) and V_6C_5 from $VC_{0.75}$ to $VC_{0.86}$[17]) (fig. 3). Between $VC_{0.70}$ and $VC_{0.75}$ diffuse banding in the electron diffraction patterns indicated a form of short range order in the alloys. Within this composition region and particularly below $VC_{0.73}$, large stacking faults are formed and these are nuclei for the transformation to the ζ-phase having a complex 12 metal atom layer stacking sequence which forms at lower carbon contents[18]).

3.2. VANADIUM OXYGEN SYSTEM

Crystals were grown over the composition range $VO_{0.80}$ to $VO_{1.35}$. Difficulty was experienced with this material in forming the first molten zone due to surface cracking and poor densification during sintering. This cracking prevented the outer skin of the rod from melting and increasing the power input to melt the outer skin caused overheating of the inner region resulting in loss of this material due to a sudden overrun. This could be avoided by heating very slowly near the melting point when the outer skin would melt by conduction from the inner region. Once a complete molten zone was formed the rod could be traversed satisfactorily.

No single crystals were obtained, but the grain size increased markedly with higher oxygen concentrations. Electron microscopy indicated the presence of the ordered tetragonal phase below 800 °C in the region

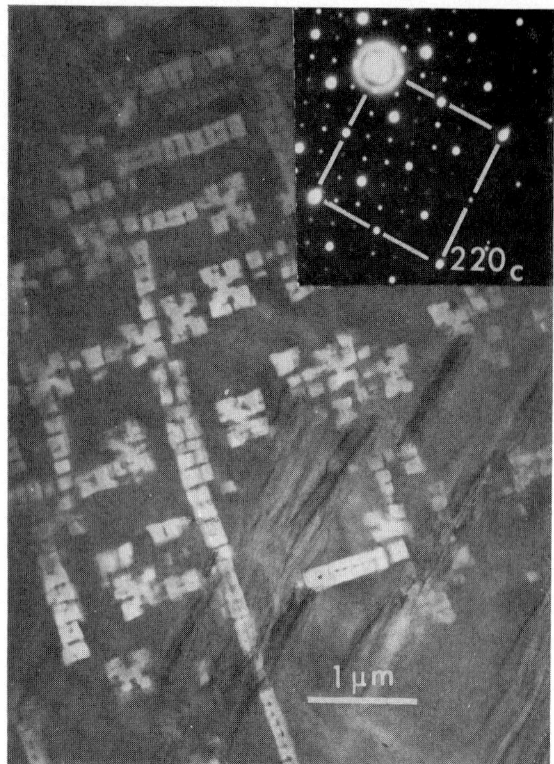

Fig. 4.   Electron micrograph from sample of $VO_{1.25}$ showing domains with tetragonal crystal symmetry formed by vanadium vacancy ordering. The beam direction is parallel to $\langle 100 \rangle$ cubic and to the tetragonal c-axis of the square shaped domains. Inset shows an electron diffraction pattern from a single domain.

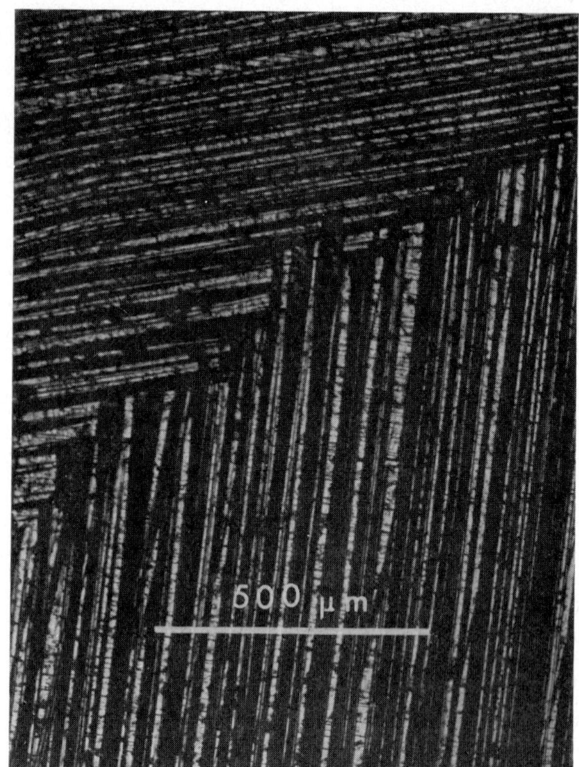

Fig. 5.   Optical micrograph from transverse section of $VN–V_2N$ eutectic.

$VO_{1.15}$ to $VO_{1.30}$. This phase has previously been reported with unit cells of composition $V_{204}O_{256}$[5]), and as $V_{52}O_{64}$[19]). However this present investigation has shown that the diffraction data is consistent with a larger unit cell structure of composition $V_{244}O_{320}$, as illustrated in fig. 4 (full details of structural analysis will be given in ref. 20). However, another previously unreported ordered phase at the composition $VO_{1.30}$ was formed below 600 °C after long annealing. Short range order, as evidenced by diffuse electron scattering, was present in samples annealed above the ordering temperature and was found to exist over the whole phase field.

3.3. VANADIUM NITROGEN SYSTEM

Crystals were grown under starting pressures of 0–10 atm of nitrogen, the total pressure being made up to 20 atm with helium. Contamination of the crystals by any residual oxygen in the gas mixture was mini-

mised by firstly melting an isolated zone at the top of the specimen well away from the area to be zoned.

Crystals whose initial compositions were within the mononitride phase field were grown under 0–2 atm nitrogen and consisted mainly of polycrystalline V_2N containing precipitates of bcc vanadium–nitrogen solid solution. Samples grown under 2–10 atm nitrogen all showed a large grained coarse eutectic microstructure (see fig. 5). Thus no single phase mononitride alloys could be produced. Higher nitrogen pressures could not be used without making modifications to the cooling of the apparatus. The constituents of the two phase alloys were found by electron microscopy to be hexagonal divanadium nitride V_2N and fcc vanadium mononitride VN, with overall chemical composition near to $VN_{0.65}$. The ordered ε-Fe_2N type structure was confirmed[21]) for the V_2N phase.

The formation of a eutectic between VN and V_2N is unexpected as in the Group VB transition metal carbon systems the M_2C phase melts peritectically. However the evidence gained from electron and optical micrographs shows that a eutectic is formed. The optical micrograph in fig. 5 shows a transverse section of $VN-V_2N$ eutectic containing a grain boundary. Within individual VN lamellae fine banding is visible. Electron micrographs confirmed that this banding was the same as the darker phase of fig. 5, i.e. V_2N, and was formed by solid state precipitation from the VN phase during cooling. The large interlamellar spacing is consistent with the similarity in chemical compositions of the two phases ($V_2N \sim VN_{0.52}$ and $VN \sim VN_{0.70}$) and low interfacial energy arising from the highly coherent interface relationship, i.e. $(111)_{VN} \parallel (0001)_{V_2N}$ and $\langle 110 \rangle_{VN} \parallel \langle 11\bar{2}0 \rangle_{V_2N}$ determined by electron diffraction.

References

1) G. E. Hollox, Mater Sci. Eng. **3** (1968) 121.
2) W. F. Brizes, University of Pittsburgh, Ph.D. Thesis, 1970.
3) D. W. Lee and J. S. Haggerty J. Am. Ceram. Soc. **52** (1969) 641.
4) L. M. Adelsberg and L. H. Cadoff, J. Am. Ceram. Soc. **51** (1968) 213.
5) S. Westman and C. Nordmark, Acta Chem. Scand. **14** (1960) 465.
6) J. D. Venables, D. Kahn and R. G. Lye, Phil. Mag. **18** (1968) 177.
7) S. Yamaguchi, J. Phys. Soc. Japan **24** (1968) 855.
8) C. N. R. Rao and G. V. Subba Rao, Phys. Status Solidi A **1** (1970) 597.
9) R. W. Bartlett and F. A. Halden, Stanford Res. Inst., NASA-49 (19), June 1967.
10) T. Takahaski, K. Sugiyama and H. Itoh, J. Electrochem. Soc. **117** (1970) 541.
11) L. R. Fleischer and J. M. Tobin, in: *Proc. Third Intern. Symp. on High Temperature Technology*, Asilomar, California 1967 (Butterworths, London, 1969).
12) W. Precht and G. E. Hollox, J. Crystal Growth **3, 4** (1968) 818.
13) J. F. Wenkus, J. S. Haggerty and D. W. Lee, Tech. Rep. AFML-TR-68-228, Sept. 1968.
14) J. S. Haggerty, J. L. O'Brien and J. F. Wenkus, J. Crystal Growth **3, 4** (1968) 291.
15) W. D. Schnell, University of Freiberg, Ph.D. Thesis 1960.
16) C. H. de Novion, R. Lorenzelli and P. Costa, Compt. Rend. (Paris) **263** (1966) 775.
17) J. Billingham and M. H. Lewis, in: *7th Congr. Intern. de Microscopie Electronique*, Grenoble, 1970, Vol. 2, p. 477.
18) J. Billingham and M. H. Lewis, Phil. Mag. **24** (1971) 231.
19) B. Andersson and J. Gjønnes, Acta Chem. Scand. **24** (1970) 2250.
20) P. S. Bell, University of Warwick Ph.D. Thesis, 1971.
21) H. Nowotny and F. Benesovsky, Planseeber Pulvermet. **16** (1968) 209.

GROWTH OF MnTe SINGLE CRYSTALS FROM NONSTOICHIOMETRIC MELTS BY LIQUID ENCAPSULATION

D. MATEIKA

Philips Forschungslaboratorium Hamburg GmbH, 2 Hamburg 54, Germany

MnTe single crystals have been grown from nonstoichiometric melts with tellurium excess in a graphite crucible. A tellurium excess of about 6 at% Te reduces the temperature of crystallization below the transformation point (1026 °C) of MnTe. A molten layer of boron oxide on the MnTe/Te melt prevents the evaporation of tellurium. The crucible is rotated during the growth period according to the accelerated crucible rotation technique. The temperature of the furnace is cooled down at a rate of 0.4 °C/h from 1018 °C till 920 °C. To prevent the peritectic reaction at 735 °C the MnTe single crystal is separated from the residual melt by turning the furnace.

MnTe single crystals with sizes up to 50 mm in length and 15 mm in diameter have been grown with this technique. The single crystalline character of the crystals was confirmed by Laue's back reflection method and by etching with dilute nitric acid.

1. Introduction

MnTe is an antiferromagnetic semiconductor. Crystals of MnTe have been grown from stoichiometric melts by the Stockbarger–Bridgman method[1-3]. The quality of this material was poor, due to the presence of grain boundaries, mosaic structure, cavities etc.

MnTe melts at about 1165 °C. It is polymorphic and crystallizes above 1026 °C in the cubic NaCl type structure, below this temperature in the hexagonal NiAs type structure[4,5]. This first order transformation at 1026 °C is probably the reason for the imperfection of the crystals finally obtained from stoichiometric melts.

In this paper a method is described for preparing MnTe single crystals from nonstoichiometric melts with tellurium excess. The tellurium excess reduces the temperature of crystallization of MnTe below the transformation point. A molten layer of boron oxide on the MnTe/Te melt prevents the evaporation of tellurium.

2. The MnTe–Te system

A knowledge of the exact liquidus curve and of the temperatures of polymorphic transformations in the MnTe–Te system is very important for single crystal growth of MnTe from nonstoichiometric melts.

In the MnTe–Te system there exist two compounds MnTe and $MnTe_2$. $MnTe_2$ forms by a peritectic reaction at 735 °C[5,6]. The hexagonal modification of MnTe transforms in the temperature range 990–1020 °C into the high temperature cubic form[7]. There is a slight miscibility between MnTe and Te at high temperatures which decreases at low temperatures[7-10].

In order to determine the exact liquidus curve in the region between the stoichiometric composition of MnTe and the peritectic of $MnTe_2$, thermal analysis was carried out in a commerical DTA apparatus (Netsch, Germany). For the preparation of the samples a homogeneous mixture of 1.5 g manganese (powdered, 99.995% pure) and tellurium (powdered, 99.999% pure) was placed into a graphitized quartz ampoule. In order to establish the same conditions as during crystal growth about 0.2 g boron oxide were placed on the mixture of MnTe/Te. The quartz ampoules were evacuated, sealed off and were heated and cooled several times in the DTA apparatus. The melting points were determined with an accuracy of ± 3 °C. The phase composition of the samples was analyzed by X-ray diffraction.

Fig. 1 shows the MnTe–Te phase diagram obtained from the results of the heating and cooling curves and from the X-ray analysis in the composition range 50–100 at% Te.

Manganese telluride melts at 1165 ± 3 °C. The polymorphic transformation occurs at 1026 ± 3 °C. Abrikosov et al.[7] and Johnston and Sestrich[5] observed two effects at 990 and 1020 °C on the cooling and heating curves. The nature of these effects was related in ref. 7 to the beginning and to the end of the polymorphic transformation. We also observed two effects at 995 and 1026 °C on the cooling and heating curves

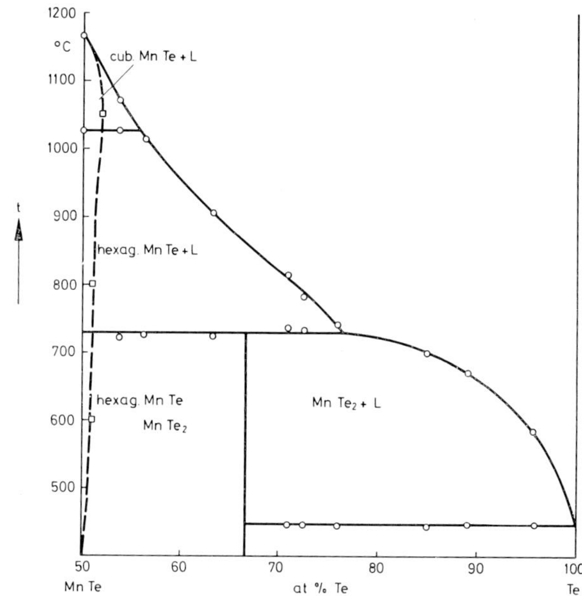

Fig. 1. MnTe–Te phase diagram; (○) measured values of the DTA analysis, (□) experimental values by Abrikosov et al.[7].

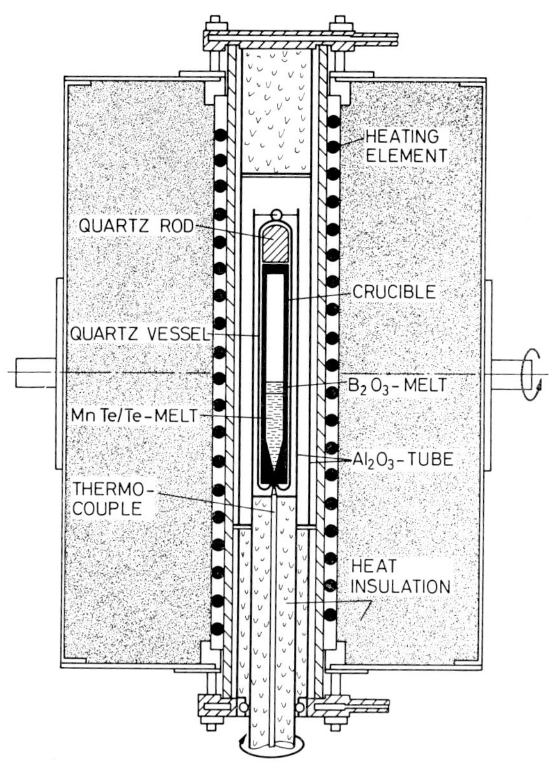

Fig. 2. Schematic drawing of the single crystal growth apparatus.

at a rate of 5 °C/min. But if we used a rate of 0.2 °C/min only one effect at 1026 °C was observed. Therefore we believe that only one defined transformation point exists at 1026 ± 3 °C.

From manganese telluride, the liquidus curve slopes steeply till the peritectic is reached at $730 + 10$ °C and about 77 at% Te. From there the slope of the curve is somewhat reduced up to a tellurium content of about 85 at%. The accuracy of the determination of the peritectic temperature is not too high, because the peritectic reaction occurs with retardation. Abrikosov et al.[7] noted that there is a degenerate eutectic on the Te-rich side with a melting point close to that of pure tellurium.

In order to prepare single crystals of manganese telluride which grow directly in the NiAs type structure, it is necessary to apply a tellurium excess of about

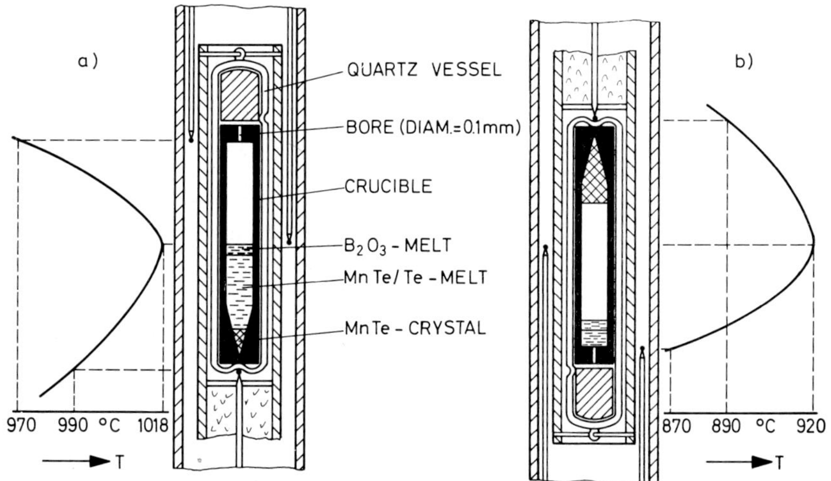

Fig. 3. Crystal growth phases of MnTe; (a) Beginning of crystal growth, (b) After turning of the furnace.

6 at %. Furthermore, the crystal growth must be stopped before the peritectic temperature is reached in order to prevent the formation of $MnTe_2$.

3. Crystal growth

A powdered mixture of manganese (43.6 at %) and tellurium (56.4 at %) is placed in a graphite crucible under a layer of boron oxide. To obtain a seed selection the crucible bottom tapers to a point with an angle of 25°. The top of the crucible is closed with a cover containing a hole of 0.1 mm bore. A quartz vessel containing the graphite crucible is evacuated for 20 h and sealed off under an argon atmosphere. The sealed quartz vessel is attached to a rotating Al_2O_3 tube inside the furnace (fig. 2). The tube of the furnace is closed with insulating material. The furnace can be turned around on its horizontal axis.

The temperature profile along the axis of the vertical furnace at the beginning of the crystal growth is shown in fig. 3a. In order to obtain a thoroughly mixed melt during crystal growth, the crucible is alternately rotated and stopped by means of a rotational mechanism at the bottom of the furnace [accelerated crucible rotation technique, ACRT[11])]. The furnace is cooled down at a rate of 0.4 °C/h until a temperature of 890 °C is reached at the crucible bottom. At this temperature the furnace is turned around on its horizontal axis by 180° (fig. 3b) and the residual melt is drained from the crystal. As the molten boron oxide layer is on the residual melt, tellurium cannot evaporate and react with the MnTe crystal to produce $MnTe_2$. This technique prevents the decomposition of the MnTe single crystal by the peritectic reaction at 730 °C (fig. 1).

The furnace is then cooled down to room temperature at a rate of 50 °C/h.

4. Results

Several MnTe single crystals have been grown with this technique. The size of the crystals is 15 mm in diameter and 50 mm in length. Fig. 4 shows such an MnTe crystal and some samples cut from a MnTe crystal.

The crystals according to X-ray analysis contained a trace of $MnTe_2$. In order to transform $MnTe_2$ into MnTe, the crystals were annealed several hours at 900 °C in vacuum. After annealing, X-ray analysis showed only the hexagonal NiAs type structure of

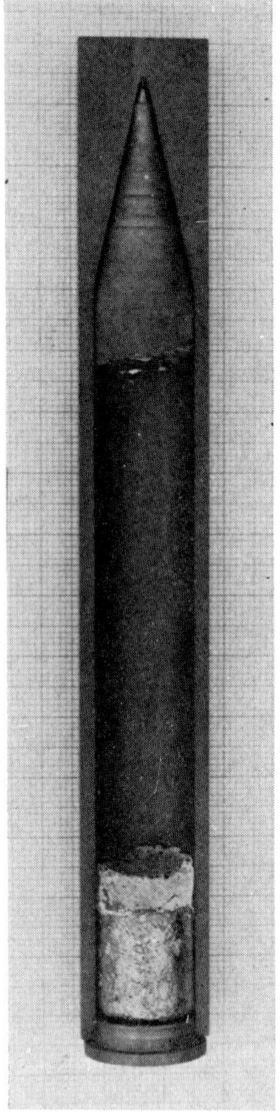

(a)

(b)

Fig. 4.   (a) Graphite crucible with MnTe single crystal and solid residual melt of MnTe/Te and B_2O_3; (b) MnTe single crystal and samples cut from such a crystal.

(a)

(b)

Fig. 5. Etch face of a MnTe single crystal sample; (a) Before heating, (b) after heating at 1100 °C. Size 1.5×2.2 mm.

MnTe. The probable reason for the formation of $MnTe_2$ is the reduced solubility of tellurium in solid manganese telluride with decreasing temperature.

The single-crystalline character of the MnTe crystals was confirmed by Laue's back reflection method and by etching with dilute nitric acid. Fig. 5a shows an etch face of a MnTe sample. The face is uniformly etched and does not show any grain boundaries.

An improvement in the quality of the MnTe single crystals was achieved by applying the accelerated crucible rotation technique. MnTe crystals grown without crucible rotation contained some cavities and cracks. The cavities and cracks in the MnTe crystals were reduced if the crucible was rotated during crystal growth. The density of the MnTe crystals increased from 5.96 ± 0.01 without rotation to 6.05 ± 0.01 g/cm^3 with rotation (theoretical density is 6.084 g/cm^3).

To study the behavior of a MnTe single crystal heated above the transformation temperature a sample of a MnTe single crystal was heated at 1100 °C and cooled to room temperature. Fig. 5 shows an etch face of this sample before and after heating. The sample was split up into several smaller crystals after heating.

The author expresses his thanks to Mr. H. Kohler and to Mr. G. Passig for their technical assistance.

References

1) T. Komatsubara, M. Murakami and E. Hirahara, J. Phys. Soc. Japan **18** (1963) 356.
2) G. Zanmarchi, J. Phys. Chem. Solids **28** (1967) 2123.
3) A. I. Zvyagin, L. V. Povstyanyi and R. M. Arefeva, Soviet Phys.-Solid State **11** (1970) 2759.
4) I. Oftedal, Z. Physik. Chem. **128** (1927) 135.
5) W. D. Johnston and D. E. Sestrich, J. Inorg. Nucl. Chem. **19** (1961) 229.
6) R. T. Delves and B. Lewis, J. Phys. Chem. Solids **24** (1963) 531.
7) N. Kh. Abrikosov, K. A. Dyul'dina and V. V. Zhdanova, Inorg. Mater. **4** (1968) 1638.
8) S. Furberg, Acta Chem. Scand. **7** (1953) 693.
9) M. A. Johansen, J. Inorg. Nucl. Chem. **6** (1958) 344.
10) J. van den Boomgaard, Philips Res. Rept. **24** (1969) 284.
11) H. J. Scheel and E. D. Schulz-Du Bois, J. Crystal Growth **8** (1971) 304.

CRYSTAL GROWTH AND PHASE EQUILIBRIUM STUDIES IN THE SYSTEM $(R.E.)_2O_3$–Fe_2O_3

J. W. NIELSEN and S. L. BLANK

Bell Telephone Laboratories, Incorporated, Murray Hill, New Jersey 07974, U.S.A.

The phase relationships near the composition 50 mole% Fe_2O_3 in the system $(R.E.)_2O_3$–Fe_2O_3 are very similar in the cases where the orthoferrites melt in the 1680–1730 °C range. On the other hand, the Y_2O_3–Fe_2O_3 system does not exhibit immiscible liquid formation in air up to 50% Y_2O_3. Further, $Y_3Fe_5O_{12}$ is stable in air when in equilibrium with liquids of 14–23 mole% Y_2O_3, but $Yb_3Fe_5O_{12}$ and $Dy_3Fe_5O_{12}$ exhibit no similar region of stability.

Rapidly quenched $(R.E.)_2O_3$–Fe_2O_3 melts prepared in air near the 50 mole% Fe_2O_3 composition contain a non-magnetic, previously unreported oxide of $(R.E.)^{3+}$, Fe^{3+} and Fe^{2+} whose general formula is $(R.E.)_{1+x}(Fe_{1-y}^{3+}Fe_y^{2+})_{1\pm x}O_{3-y(1\mp x)/2}$. The compound plays a major role in defect formation in ortho-ferrites.

High quality crystals of $YFeO_3$ can be grown by the Bridgman technique in one atmosphere of oxygen, and other orthoferrites could be grown in the same way provided that their melting points are low enough that platinum crucibles can be used and an oxygen overpressure near the equilibrium pressure can be applied.

1. Introduction

Recent requirements for crystals of rare earth ortho-ferrites, $(R.E.)FeO_3$, as media for bubble domain devices[1] have stimulated interest in crystal growth in the pseudo-binary systems $(R.E.)_2O_3$–Fe_2O_3. Since stoichiometric melt growth is generally faster and often purer than solution growth, the conditions of stability of the orthoferrites and their phase relationships in $(R.E.)_2O_3$–Fe_2O_3 melts were sought.

Three crystal growth techniques were examined: simple slow cooling, and the Czochralski and Bridgman methods. Czochralski-grown crystals a few millimeters in maximum extent were readily obtained but were defective. Larger and slightly less defective crystals were obtained by slow cooling. This result, followed by the work of Okada et al.[2] led us to the Bridgman technique which has yielded crystals suitable for bubble domain devices.

2. Experimental technique

Liquidus–solidus equilibrium conditions were determined using the same apparatus employed for Czochralski crystal growth. Similar growth apparatus is described in the literature[3,4]. The crucibles were iridium, 37 mm internal diameter and 37 mm high with 1.5 mm walls, fitted with lids containing an 18 mm diameter circular opening. Melts of about 110 g were prepared from 99.99% pure rare earth and ferric oxides pre-dried at 1000 °C. Runs were prepared in one of two ways. In some cases where the melting point of the orthoferrite was very high, such as $LaFeO_3$, the oxides were pre-sintered in stoichiometric proportions to form the compound; then, the powder was added slowly to the crucible at a temperature slightly above the liquidus. When the orthoferrite melting point was nearer 1700 °C it was found more convenient to add the mixed oxides carefully to the crucible held at a temperature just above the liquidus.

Temperatures were measured with a calibrated optical pyrometer which was further checked against a new Pt-versus-Pt–10% Rh thermocouple measuring the temperature of a $YbFeO_3$ melt held in a resistance furnace. Agreement was within 5 °C. Four measurements were made at each composition; usually two individuals made the measurements independently. Precision was about ± 8 °C. Agreement with the literature values for the liquidus temperatures of $LaFeO_3$[5] and Fe_2O_3[6] gave further assurance of the accuracy of the temperature measurements.

The Bridgman furnace and the specially designed crucibles have been previously described[7,8].

3. Phase studies

Preliminary experiments confirmed that all orthoferrites lose oxygen in air at their melting points. In order

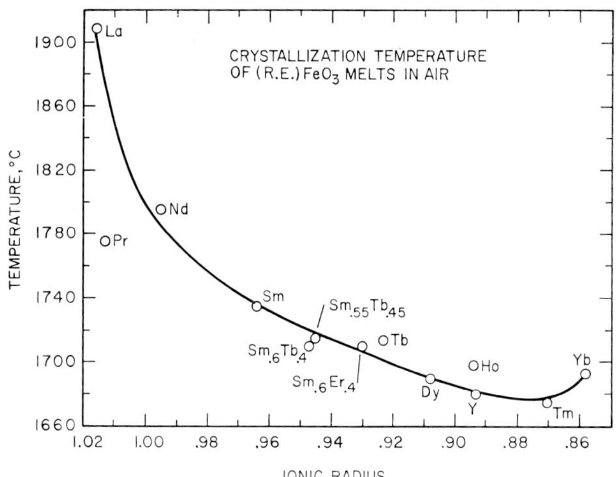

Fig. 1.   Crystallization temperatures of orthoferrite in air.

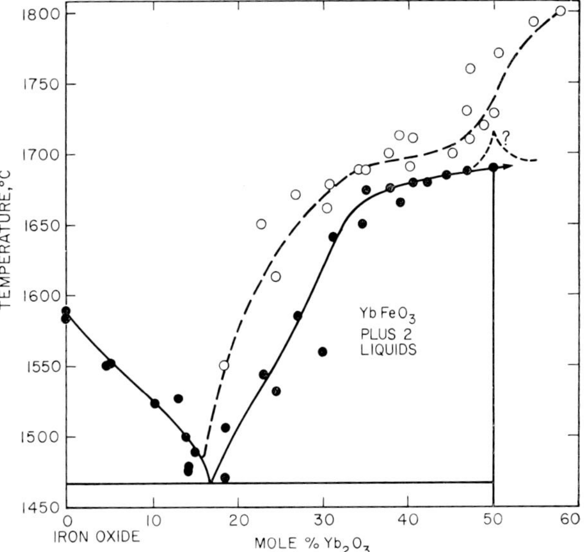

Fig. 2.   Phase relationships in the system Yb$_2$O$_3$–Fe$_2$O$_3$ in air.

to choose an orthoferrite melt which decomposed only a small amount, hence limiting the incorporation of Fe^{2+} in the solid, the crystallization temperatures of a large number of orthoferrites in air were measured. The results are shown in fig. 1 where it is seen that the crystallization temperatures decrease with rare earth ionic radius until a minimum is reached for TmFeO$_3$. (The points measured for the SmFeO$_3$ solid solutions are plotted using a weighted average for the R.E. ionic radius). Because of their relatively low crystallization points, and the view (held when this work began) that their magnetic properties were suitable for device applications, work was concentrated on YbFeO$_3$ and YFeO$_3$.

Van Hook[9]) has determined the phase relationships in the Y$_2$O$_3$–Fe$_2$O$_3$ system in air, O$_2$ and CO$_2$ in the stability region of the garnet phase Y$_3$Fe$_5$O$_{12}$. Our work in the system supported his findings. The Yb$_2$O$_3$–Fe$_2$O$_3$ system behaves differently, however, and a few of our measurements are shown in fig. 2. We stress that the figure is neither a binary phase diagram nor a section of the ternary diagram Fe$_2$O$_3$–FeO–Yb$_2$O$_3$. It is a flat projection of the isobaric compositions in equilibrium with 0.2 atm of O$_2$. We also emphasize that the two-liquid boundary is less accurate than the liquidus–solidus curves because the denser liquid was much more deficient in oxygen than the one from which orthoferrite crystallized, and since the latter covered the surface of the melt, equilibrium with air was difficult to achieve and maintain. Indeed, in the case of Fe$_2$O$_3$-rich melts it was possible to grow small crystals when

the system was supercooled with respect to separation of the second liquid.

Second liquid formation increased dramatically as the Yb$_2$O$_3$ concentration increased and at 52 mole% Yb$_2$O$_3$ the low emissivity of the less dense liquid prevented accurate pyrometer readings. At a concentration of 55 mole%, Yb$_2$O$_3$ was the primary solid phase.

In order to determine the composition of the liquid phases, quenched samples of the less dense liquid were obtained from melts originally 47.4 and 52.6 mole% Yb$_2$O$_3$ and then analyzed. The compositions, expressed as Yb$_2$O$_3$, were 46.81 and 52.22 respectively. Thus the two liquids formed near 50% Yb$_2$O$_3$ differed principally in their oxygen contents. This was confirmed by analyses of rapidly solidified entire melts quenched from the point of second liquid separation.

Solids crystallizing from the denser liquid were of interest principally because of the formation in quenched samples of large amounts of a previously unreported oxide of Yb^{2+}, Fe^{3+} and Fe^{2+}. X-ray powder patterns could not be identified with a known structure. The phase was non-magnetic down to 177 °K, and exhibited a solid solution range which included the compositions Yb$_{1.05}$Fe$^{3+}_{0.67}$Fe$^{2+}_{0.28}$O$_{2.86}$, Yb$_6$Fe$_6$O$_{17}$ or YbFe$^{3+}_{0.72}$-Fe$^{2+}_{0.28}$O$_{2.86}$, and Yb$_{0.667}$Fe^{3+}Fe$^{2+}_{0.33}$O$_{2.83}$. The compositions were found by analyses of single phase solid formed in quenched melts originally 52.6, 50 and 45 mole% Yb$_2$O$_3$ respectively. The unknown phase

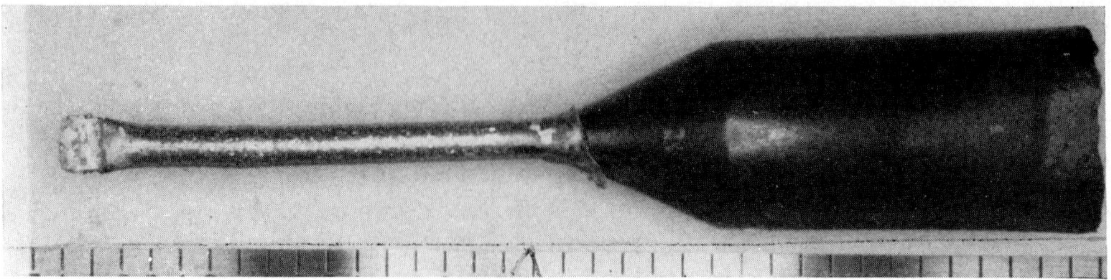

Fig. 3.   Single crystal of $YFeO_3$ grown by the seeded Bridgman method (magn. $1.5\times$).

appeared to be unstable at lower temperatures since none was observed in melts cooled slowly. Slow cooled melts contained only $YbFeO_3$ plus Yb_2O_3 or Fe_3O_4 depending on initial composition.

The dotted line in fig. 2 outlines a cusp in the liquidus near 50 mole % Yb_2O_3 which we believe exists in the true binary system. We noted that among the 65 melts which were prepared near 50 mole % Yb_2O_3, the *initial* crystallization points, which are too numerous to show with clarity, could be included within the dotted line. However, as equilibration with air proceeded and oxygen was lost, the temperature would descend to the heavy line. For stoichiometric melts second liquid formation was not observed until equilibrium with air had been reached.

In contrast to the Y_2O_3–Fe_2O_3 system, the garnet phase was not observed in any run made in air in the Yb_2O_3–Fe_2O_3 system (or the Dy_2O_3–Fe_2O_3 system). The garnet field does not extend far enough into the Fe_2O_3–FeO–Yb_2O_3 phase diagram to intersect the 0.2 atm O_2 isobar. Although the eutectic temperature and composition are nearly identical to those in the Fe_2O_3–$YFeO_3$, system no immiscible liquids were observed in the latter.

4. Crystal growth

The Czochralski method, in many attempts, yielded only small defective crystals of the orthoferrites of Dy, Sm, Tb, Tm, Y, and Yb. In the cases of Dy and Yb, excessive Fe^{2+} content and second liquid formation revealed in the phase studies were major problems. Their poor thermal conductivity and their low transmission made crystal pulling of all orthoferrites difficult. The phase studies revealed that slow cooling might improve crystal growth and some melts were simply cooled at 3 deg/hr until solidification was complete. Crystals of Dy, Yb, Tm, and $Sm_{0.55}Er_{0.45}$ orthoferrite about 0.5 cm^3 in volume were grown in this way,

but inclusions high in Fe^{2+} were still present. Thus the Bridgman method in an O_2 atmosphere was selected since it yielded better results. The successful use of this technique to grow 30 g crystals of $YFeO_3$ suitable for bubble domain device use has been reported[7]. A boule of $YFeO_3$ is shown in fig. 3.

5. Defects in orthoferrites

The defects most often observed in orthoferrite crystals grown in air were small inclusions of an opaque phase which were 2 to 3 microns in diameter and were frequently arranged in rows, presumably decorating dislocations. Upon annealing thin sections of crystals in 1 atm of O_2 at 1500 °C most, but not all of the precipitates disappeared[10]).

The formation and annealing behavior of the inclusions found in orthoferrite crystals can be explained if we assume the vertical section through the compositions (R.E.)/Fe = 1, in the ternary system Fe_2O_3–$(R.E.)_2O_3$–FeO, near the orthoferrite pole, is approximately as shown in fig. 4. (We ignore the two liquid region and exaggerate the solution ranges in the solids for clarity.) Crystals grown from liquid B, which we

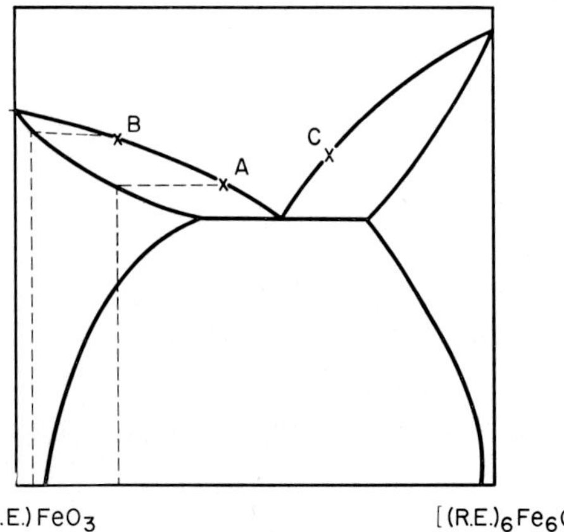

Fig. 4.   Schematic of part of the section R.E./Fe = 1, in the systems $(R.E.)_2O_3$–Fe_2O_3-FeO.

assume is the composition in 1 atm of O_2, will grow at equilibrium with a small Fe^{2+} content but no precipitates will form since the composition is stable at lower temperatures. Crystals grown from liquid A, in air, will, upon careful reduction of temperature, precipitate the phase $(R.E.)_6Fe_6O_{17}$. By annealing in 1 atm of O_2 the reduced phase can be oxidized to orthoferrite and apparently it will unite coherently with the lattice in most cases; hence, the precipitate disappears. However, the reduced phase is unstable and may decompose at equilibrium at low temperatures to $(R.E.)_2O_3$, $(R.E.)FeO_3$ and Fe_3O_4. When this occurs, as it does with a few precipitates, reoxidation is not sufficient to heal the crystal. Recombination occurs but does so imperfectly, leaving a trace of the original precipitates.

At equilibrium, A and B are invariant points until the liquid is solidified. Rapid growth, however, will cause the liquid composition to shift toward the liquidus of $(R.E.)_6Fe_6O_{17}$ to point C. If the interface rapidly advances in a Bridgman experiment, the build-up of Fe^{2+} ahead of it will eventually cause precipitation of $(R.E.)_6Fe_6O_{17}$. At that point the liquid locally is supersaturated with respect to O_2 and a bubble of O_2 is ejected. The bubble is included along with the precipitate by the rapid growth resulting in an inclusion pair.

Twinning can be avoided by preparing melts which are precisely 50 mole% Fe_2O_3. Since Yb_2O_3 was observed mixed with $YbFeO_3$ which had crystallized early during the solidification of a melt only 2.6 mole% rich in Yb_2O_3, the eutectic line between $YbFeO_3$ and Yb_2O_3 must closely approach the $Yb/Fe = 1$ vertical section in the ternary system. This may result in the formation of particles of Yb_2O_3 during growth as crystallization proceeds, especially if growth is rapid. Twinning can result as particles of Yb_2O_3 make contact with the growing interface.

Striations caused by temperature fluctuations were observed in all crystals grown in air, or in O_2 at rapid rates. The striations appeared as bands of higher optical absorption which was thought to arise from an increased Fe^{2+}/Fe^{3+} ratio in the solid. However, striations were only partially removed by annealing in O_2, so a complete explanation of their make-up will require further work.

Acknowledgments

We wish to thank P. V. Vittorio and L. K. Shick for technical assistance and Mrs. M. H. Read for X-ray work. We also thank L. J. Varnerin for discussions during the work and reviewing the manuscript.

References

1) A. H. Bobeck, R. F. Fischer, A. J. Perneski, J. P. Remeika and L. G. VanUitert, IEEE Trans. Magnetics **MAG-5** (1969) 544.
2) T. Okada, K. Matsumi and H. Makimo, Presented at the Intern. Ferrite Conf. Kyoto, Japan, 1970.
3) B. Cockayne, M. Chesswass and D. B. Gasson, J. Mater. Sci. **2** (1967) 7.
4) M. Kestigian and W. W. Holloway, Jr., in: *Crystal Growth*, Ed. H. S. Peiser (Pergamon, Oxford, 1967) p. 451.
5) V. L. Moruzzi and M. W. Shafer, J. Am. Ceram. Soc. **43** (1960) 369.
6) E. M. Levin, C. L. Robbins and H. F. McMurdie, Eds., *Phase Diagrams for Ceramists* (The American Ceramic Society, Columbus, Ohio, 1964) p. 39, No. 10.
7) S. Blank, L. K. Shick and J. W. Nielsen, J. Appl. Phys. **42** (1971) 1556.
8) J. W. Nielsen, Met. Trans. **2** (1971) 625.
9) H. J. Van Hook, J. Am. Ceram. Soc. **44** (1961) 213.
10) E. Heinlein, Bell Telephone Laboratories, unpublished work.

Journal of Crystal Growth **13/14** (1972) 706–709 © *North-Holland Publishing Co.*

CROISSANCE DE MONOCRISTAUX D'ORTHOFERRITE D'YTTRIUM PAR TIRAGE CZOCHRALSKI

J. DAVAL, D. CHALLETON et J. MARESCHAL

Cristallogénese, LETI–CEN–Grenoble, Cédex 85, 38–Grenoble-Gare, France

Single crystals of yttrium orthoferrites have been grown for the first time using the Czochralski technique. Starting from high purity oxides Y_2O_3 and Fe_2O_3 and using the congruent melting of $YFeO_3$ near 1700 °C in $Ar-O_2$ atmosphere, it is possible, from particular and well-oriented seeds, to obtain perfectly oriented [001] crystals of $YFeO_3$, 15 to 30 mm in diameter and several mm in length. The magnetostriction and magnetodynamic properties of such crystals, as well as the optical absorption spectra and the results of chemical analysis are presented and compared with those of flux grown crystals.

1. Introduction

Le développement des dispositifs de logique et mémorisation dits à "bulles magnétiques" qui utilisent les propriétés magnétiques des orthoferrites de terres rares ou d'yttrium[1]) est en particulier lié à la possibilité d'obtenir des monocristaux de ces composés qui soient à la fois suffisamment grands ($S > 1$ cm^2) et parfaits. Ceux qui ont été utilisés jusqu'ici sont obtenus par croissance naturelle à partir d'une solution sursaturée dans des solvants à base de sels de plomb[2–4]). Parmi les nombreux inconvénients inhérents à cette technique de croissance en solution nous ne citerons que ceux qui vont à l'encontre des applications envisagées:

– substitution des ions terres rares par des ions Pb^{2+} (ref. 4);
– inclusions macroscopiques de solvant qui gênent la propagation des domaines magnétiques;
– coût élevé de la fabrication provenant à la fois des faibles vitesses de croissance (0.5 à 1 mm/jour) et de la nécessité pour obtenir des cristaux de taille suffisante, d'utiliser des creusets en platine de grand volume.

Le diagramme de phase $Fe_2O_3-Y_2O_3$ a été étudié dans l'air[5]) et en fonction de la pression d'oxygène[6]). Ces deux études indiquent une fusion congruente de $YFeO_3$ à une temperature voisine de 1700 °C. Il nous a donc paru indiqué de faire croître les monocristaux d'orthoferrite d'yttrium par la méthode Czochralski de tirage vertical à partir du bain fondu, qui, tout en supprimant la plupart des inconvénients mentionnés plus haut, a l'avantage d'être mieux adaptée en vue d'un développement industriel.

Pendant le déroulement de notre étude, des monocristaux d'orthoferrites de Terres Rares et d'yttrium ont également été obtenus par solidification du bain fondu; ce sont:

– $ErFeO_3$ et $YbFeO_3$ par méthode Verneuil, par Nakazumi, Daido, Tsuboya et Saito[7]);
– $YFeO_3$ fusion de zone par Okada, Matsumi et Makino[8]);
– $YFeO_3$ par technique Bridgman par Blank, Shick et Nielsen[9]).

2. Appareillage

Nous avons utilisé une machine de tirage modulaire conçue dans notre laboratoire et adaptée au tirage Czochralski; elle autorise les mouvements habituels de rotation et de déplacement vertical du germe et du bain dans les gammes 0–60 tpm et 0–100 mm/h avec une reproductibilité de 2%. L'enceinte de travail, de 10 cm de diamètre et de 60 cm de hauteur, est en légère surpression d'un mélange gazeux argon–oxygène à 10–15% d'oxygène purifié par passage sur tamis moléculaires à basse température. La puissance de chauffage est fournie par un générateur H.F. de 35 kW, sous basse impédance, stabilisé en tension. Le chauffage se fait par couplage direct sur un creuset en iridium (dont les dimensions sont les suivantes: diamètre 45 mm, hauteur 60 mm, épaisseur 1 mm). Le porte-germe est une tige cylindrique d'iridium ou d'alumine. Une série d'écrans en zircone frittée limite le rayonnement vers l'extérieur.

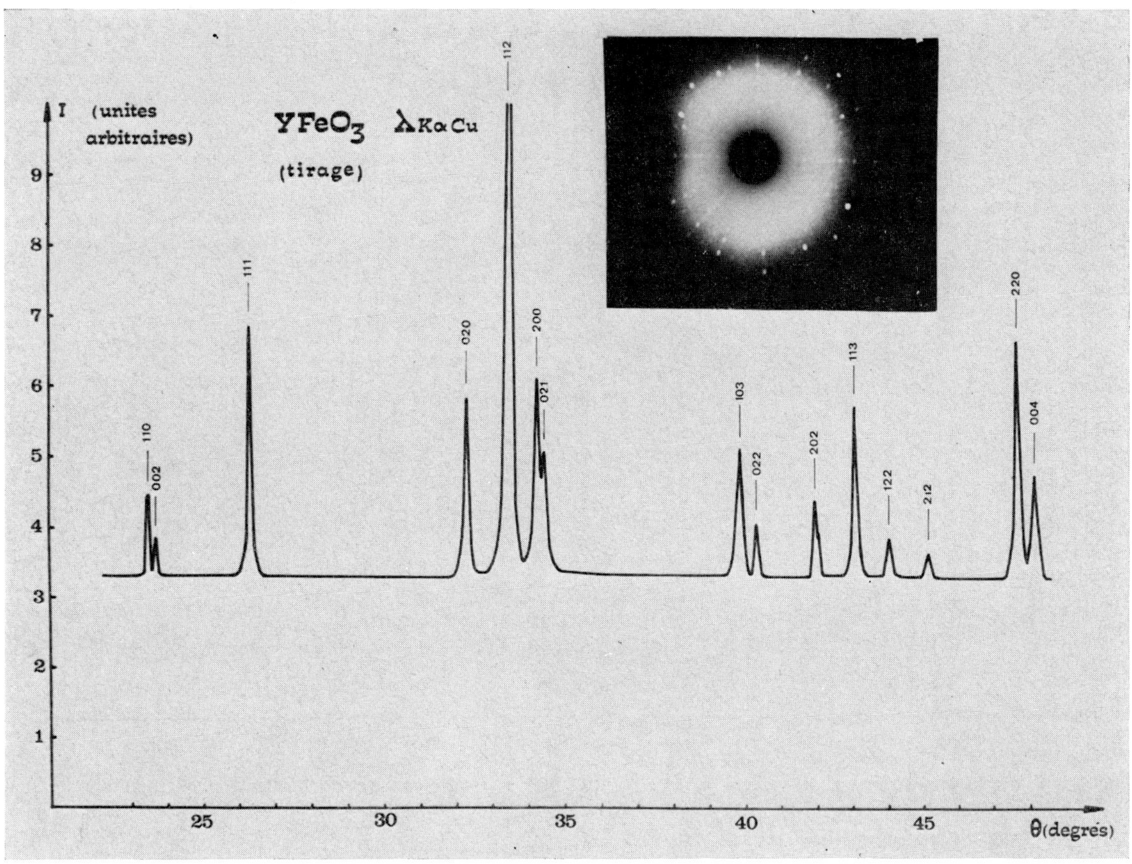

Fig. 1. Cliché de Laue et diagramme de poudre à partir d'un monocristal de YFeO₃ obtenu par tirage.

3. Conditions expérimentales de tirage et résultats

Nous avons utilisé dans nos essais des oxydes Y_2O_3 et Fe_2O_3 de haute pureté. Un traitement thermique à 1250 °C transforme un mélange stoechiométrique de ces oxydes en $YFeO_3$, préalablement à son introduction dans la machine de tirage. Le remplissage des creusets se fait en plusieurs fusions successives au-dessus de 1700 °C.

Les germes utilisés ont été obtenus par croissance naturelle dans un solvant PbO–PbF_2. Dans un tel système, pour certaines valeurs des rapports molaires solvant/soluté et PbO/PbF_2, les cristaux se développent sous la forme de parallélépipèdes de quelques mm^2 de section et de 10 à 20 mm de longueur dans la direction de l'axe c orthorhombique qui est aussi, dans la plupart des composés, la direction de faible ferromagnétisme. A partir des germes présentant cette orientation (la plus favorable pour l'utilisation envisagée) des monocristaux d'orthoferrite d'yttrium ont été élaborés sous forme de

"boules" de 15 à 30 mm de diamètre et plusieurs millimètres d'épaisseur.

Les vitesses de tirage utilisées varient entre 0.5 et 5 mm/h et les rotations du bain et du germe sont maintenues voisines de 10 tpm et 40 tpm respectivement.

Les meilleurs résultats ont été obtenus en limitant par un écran réflecteur en iridium de forme appropriée, placé au-dessus du bain, les courants de convection allant des parois vers le centre du creuset; une limitation de ces mêmes courants peut être réalisée en positionnant correctement le creuset par rapport à la spire de chauffage. Cet écran réflecteur permet d'effectuer le tirage en géométrie fermée et de maintenir ainsi au-dessus du bain liquide une congruence forcée.

4. Propriétés caractéristiques des monocristaux obtenus

4.1. Propriétés cristallographiques

Le caractère monocristallin des cristaux obtenus par tirage apparait bien sur le cliché de Laue (fig. 1)

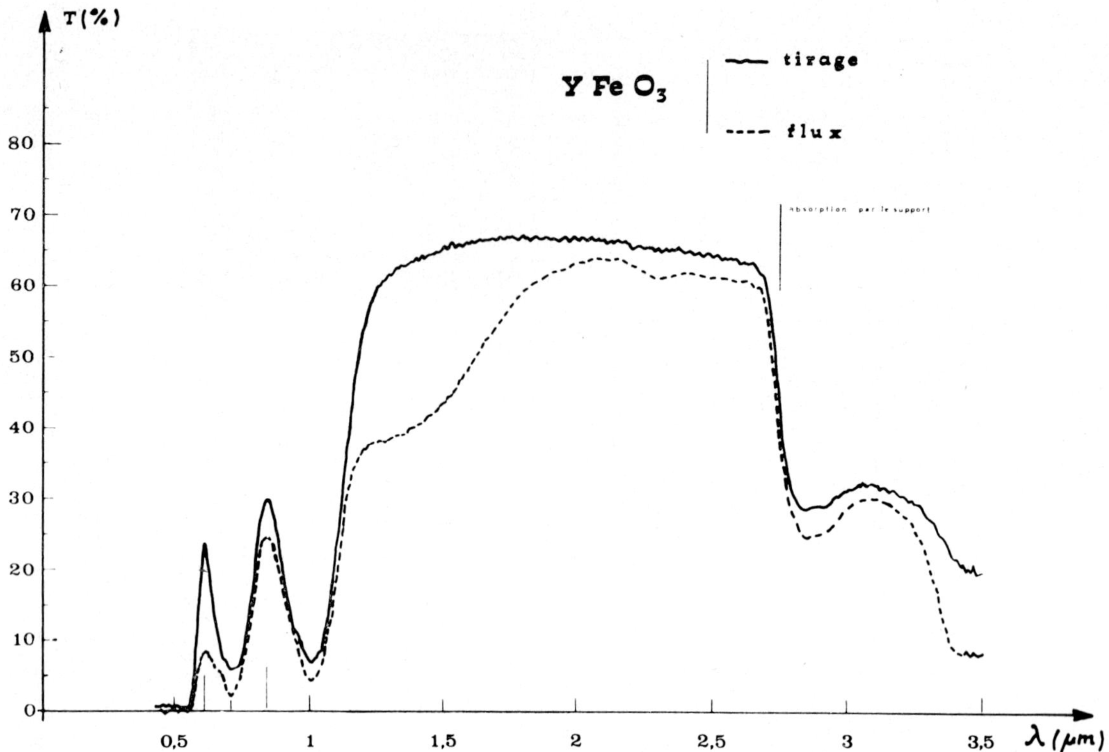

Fig. 2.   Transmission infra-rouge de monocristaux d'orthoferrites d'yttrium obtenus par tirage et par la méthode de flux.

pris suivant l'axe de tirage [001], axe de symétrie 2.

La composition chimique des cristaux peut être vérifiée sur un diagramme Debye–Scherrer (fig. 1) réalisée à partir d'un cristal pulvérisé sur lequel nous n'avons pu déceler la présence d'aucun autre composé ou constituent en excès.

4.2. Propriétés magnétiques

Les propriétés magnétiques statiques[10]) et dynamiques de ces monocristaux ont été observées conformes à la fois aux propriétés théoriques et à celles des échantillons de même composition obtenus par les méthodes habituelles de cristallisation en solution (tableau 1).

Les dernières valeurs de H_{rot} sont significatives d'une moins bonne homogénéité du matériau provenant du flux due à la présence de précipités non magnétiques qui sont autant de points d'ancrage pour les domaines magnétiques. Les observations microscopiques de surfaces et de récentes topographies par rayons X confirment ce résultat.

4.3. Propriétés optiques

La figure 2 présente, superposées, les courbes de

Tableau 1

Tirage	Flux
(I) *Mesure de* $4\pi M_s$	
(Echantillons massifs)	
$4\pi M_s = 103$ gauss	$4\pi M_s = 105$ gauss
(II) *Domaines magnétiques*	
Epaisseur des lames 47 µm (recuit)	
(a) Domaines en *bandes*	
largeur moyenne	
$d = 155$ µm	$d = 140$ µm
$\sigma_w = 1.68$ erg/cm²	$\sigma_w = 1.60$ erg/cm²
(b) Domaines *cylindriques* (bulles)	
diammètre moyen	
$2r = 78$ µm	$2r = 76$ µm
champ magnétique critique	
$H_a^c = 26$ Oe	$H_a^c = 25$ Oe
paramètre caractéristique	
$l_d = 19.4$ µm	$l_d = 19.1$ µm
soit $\sigma_w = 1.70$ erg/cm²	soit $\sigma_w = 1.67$ erg/cm²
(III) *Propagation des domaines*	

Motifs: disques de Permalloy, champ tournant: 25 kHz
Les domaines tournent régulièrement et à même vitesse autour du disque

$H_{rot} = 13$ Oe	$13 < H_{rot} < 16$ Oe

transmission du rayonnement visible et proche infrarouge de deux lames minces monocristallines de même orientation, taillées respectivement dans un cristal obtenu par tirage et dans un cristal obtenu par la méthode du flux.

Ces courbes d'allure identique présentent deux pics d'absorption à 7050 Å et 1 μm dûs aux transitions électroniques dans les ions Fe^{3+} en coordinance octaédrique[11,12]). Mais, toutes conditions égales par ailleurs (épaisseurs des lames, colle, support, largeur du faisceau) la transparence du cristal tiré est supérieure à celle du cristal provenant du flux, ceci sur toute l'étendue du spectre mais particulièrement manifeste à partir et en dessous de 2 μm. Nous attribuons cette différence essentielle à l'influence du plomb contenu dans les cristaux obtenus par la méthode du flux, alors que les cristaux tirés en sont exempts. La présence de plomb dans la matrice $TFeO_3$, à la place de la terre rare est en effet inhérente à la méthode de flux; elle provient des solvants PbO et PbF_2, et ce, en quantité non négligeable [Remeika et Kometani[3]) chiffrent cette quantité en moyenne à 0.29 % pour $YFeO_3$]. Une analyse par spectrométrie de masse a revelé la présence de 0.15 ppm de plomb dans les cristaux obtenus par tirage.

Il faut encore noter le gain de transparence à 6200 Å en faveur du cristal obtenu par tirage. Cette "fenêtre optique" est particulièrement appréciable lorsque l'expérience nécessite l'emploi de la longueur d'onde 6328 Å d'un faisceau laser.

Enfin la méthode de tirage ayant nécessité l'emploi d'un creuset en iridium, un dosage de cet élément dans les cristaux tirés fut effectué par spectrométrie de masse qui révéla la présence de 4 ppm atomiques d'iridium, quantité tout à fait négligeable.

Bibliographie

1) A. H. Bobeck, Bell Syst. Tech. J. **46** (1967) 1901.
2) J. P. Remeika, J. Am. Ceram. Soc. **78** (1956) 4259.
3) J. P. Remeika et T. Y. Kometani, Mater. Res. Bull. **3** (1968) 895.
4) B. M. Wanklyn, J. Crystal Growth **5** (1969) 323.
5) J. N. Nielsen et E. F. Dearborn, J. Phys. Chem. Solids **5** (1958) 202.
6) H. J. Van Hook, J. Am. Ceram. Soc. **44** (1961) 208.
7) Y. Nakazumi, K. Daido, I. Tsuboya et M. Saito, Intern. Ferrite Conf., Kyoto, Japan, 1970.
8) T. Okada, K. Matsumi et H. Makino, Intern. Ferrite Conf., Kyoto, Japan, 1970.
9) S. L. Blank, L. K. Shick et J. W. Nielsen, Intermag. 1971, Denver, Colorado.
10) E. F. Bertaut, J. Mareschal, P. Coeure et D. Challeton, NT LETI/DN/EPA No. 623, CEA-CEN-Grenoble, 1970.
11) A. V. Antonov, A. M. Balbaskov et A. Ya. Chervonenkis, Soviet Phys.-Solid State **12** (1970) 1363.
12) D. L. Wood, J. P. Remeika et E. D. Kolb, J. Appl. Phys. **41** (1970) 5315.

    Journal of Crystal Growth **13/14** (1972) 710–717 © *North-Holland Publishing Co.*

INITIAL THERMAL MODEL OF THE FLAME FUSION CRYSTAL GROWTH PROCESS

F. B. KHAMBATTA, P. J. GIELISSE and M. P. WILSON, Jr.

Department of Chemical Engineering, University of Rhode Island, Kingston, Rhode Island 02881, U.S.A.
and

J. A. ADAMSKI and CH. SAHAGIAN

Air Force Cambridge Research Lab., L. G. Hanscom Field, Bedford Massachusetts 01730, U.S.A.

A mathematical model of the flame fusion crystal growth process has been developed. Optimum conditions and growth rates have been calculated using the model with available data. The analysis indicates the importance of the muffle wall temperature distribution and the gas temperature in the annulus surrounding the boule. Outside the melt zone, radiative heat transfer is shown to be about an order of magnitude greater than the convective heat transfer rate. Experimental data for the average temperature of the crystal during the growth process will be presented along with the temperature distribution along the muffle wall and in the flame with the boule removed.

1. General

The flame fusion crystal growth process, though simple in principle, operation and mechanical design, is in reality complex as a result of many interacting process parameters. The thermodynamic aspects of the process are critical in formulating a model from which the effect of the variation of key parameters on the ultimate product may be determined. As a first step to a fuller understanding of the exact relationship between process parameters and crystal characteristics we have developed a generally applicable thermal model. The model allows calculation of optimum growth rates and growth conditions limited only by the availability of accurate input data. This type of data has been generated as part of this study. The present restrictions and limitations are discussed.

2. Thermodynamic model

The most important rate limiting step is the requirement that the latent heat of fusion be removed from the crystal–melt system. The intermediate crystallization zone, the dashed line in fig. 1, acts as a control volume, the thickness of which is reported to be of the order of approximately 20 μm[1,2]) and the radius of which is nearly equal to that of the crystal. Since it is in this volume that actual crystallization occurs, a thermal balance may be set up for the 20 μm layer selecting two points very close to it on either side, one in the melt and the other in the solid crystal, as shown in fig. 1. Latent heat + Heat transferred to the crystal from the melt = Heat conducted away from the interface, or

$$L\frac{\mathrm{d}m}{\mathrm{d}t}+k_1\left[\frac{\mathrm{d}T}{\mathrm{d}x}\right]_1 A_1 = k_2\left[\frac{\mathrm{d}T}{\mathrm{d}x}\right]_2 A_2, \tag{1}$$

where

L = latent heat of fusion (cal/g),
$\mathrm{d}m/\mathrm{d}t$ = amount of material crystallizing per unit of time (g/sec),
k_1 = thermal conductivity of the melt just above the crystallization layer (point 1) (cal/sec cm deg),
k_2 = thermal conductivity of the crystal just below the crystallization layer (point 2) (cal/sec cm deg),
$[\mathrm{d}T/\mathrm{d}x]_1$ = temperature gradient at point 1 (deg/cm),
$[\mathrm{d}T/\mathrm{d}x]_2$ = temperature gradient at point 2 (deg/cm),
A_1 = area of the isotherm through point 1 (cm^2),
A_2 = area of the isotherm through point 2 (cm^2).

The amount of material crystallizing per unit time may further be expressed as

$$\mathrm{d}m/\mathrm{d}t = V_\mathrm{g} A_\mathrm{c} \delta,$$

where

V_g = velocity of growth (cm/sec),
A_c = area of cross section of the crystal (cm^2),
δ = density of the solid phase near the melt–crystal
interface (g/cm^3).

Replacing subscripts (1) and (2) by L and S to denote liquid and solid phases respectively, eq. (1) can be rewritten as

$$LV_\mathrm{g}A_\mathrm{c}\delta + k_\mathrm{L}\left[\frac{\mathrm{d}T}{\mathrm{d}x}\right]_\mathrm{L} A_\mathrm{L} = k_\mathrm{S}\left[\frac{\mathrm{d}T}{\mathrm{d}x}\right]_\mathrm{S} A_\mathrm{S}. \tag{2}$$

If $A_\mathrm{L} = A_\mathrm{S} = A_\mathrm{c} = \pi r^2$, where r is the radius of the cylindrical portion of the crystal,

$$LV_\mathrm{g}A_\mathrm{c}\delta + k_\mathrm{L}\left[\frac{\mathrm{d}T}{\mathrm{d}x}\right]_\mathrm{L} A_\mathrm{c} = k_\mathrm{S}\left[\frac{\mathrm{d}T}{\mathrm{d}x}\right]_\mathrm{S} A_\mathrm{c}. \tag{3}$$

The right hand side of eq. (3) represents the amount of heat transferred *into* the crystal and must be equal to the heat lost *from* the crystal, assuming steady state conditions during cylindrical growth. The growing crystal suffers three types of heat losses, viz. (a) radiation loss (q_rad) considering the high temperatures prevalent in the oxy-hydrogen flame, (b) convection (conduction) loss due to the surrounding hot gases (q_gas) and (c) conduction loss (q_cond) due to the seed rod contact with the retraction system. Thus,

$$LV_\mathrm{g}A_\mathrm{c}\delta + k_\mathrm{L}\left[\frac{\mathrm{d}T}{\mathrm{d}x}\right]_\mathrm{L} A_\mathrm{c} = q_\mathrm{rad} + q_\mathrm{gas} + q_\mathrm{cond}. \tag{4}$$

If it is assumed that the crystal is essentially cylindrical and the radial temperature gradient within the crystal regarded negligible, i.e. $\mathrm{d}T/\mathrm{d}r = 0$ or $T = T(x)$ only, the crystal may be treated as a rod shaped extended surface or fin of uniform radius, r, protruding into a surrounding atmosphere of hot gases. See fig. 1. The conical lower portion is replaced by an equal volume cylindrical portion.

The heat-balance on a small element of the crystal of length $\mathrm{d}x$ (fig. 1) is, assuming steady state,
Rate of heat flow by conduction into the element at x = Rate of heat flow by conduction out of the element at $(x+\mathrm{d}x)$ + Rate of heat flow by convection from the surface between x and $(x+\mathrm{d}x)$,
or

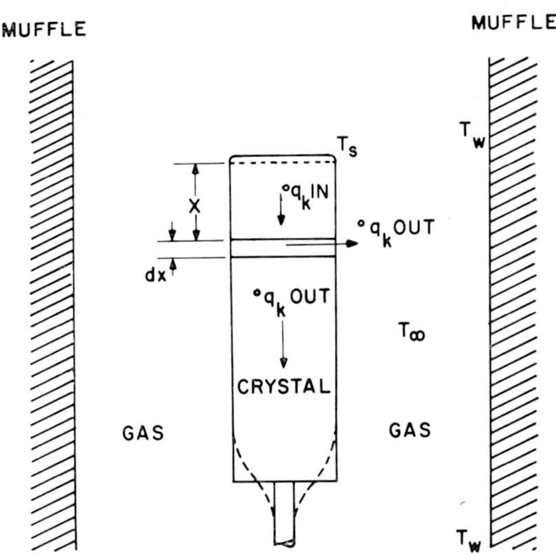

Fig. 1.   Assumed boule configuration as used in thermodynamic model (crystal as an extended cylindrical surface).

$$q_\mathrm{k}\,(\mathrm{in}) = q_\mathrm{k}\,(\mathrm{out}) + q_\mathrm{c}$$

$$q_\mathrm{k} = \left[q_\mathrm{k} + \frac{\partial q_\mathrm{k}}{\partial x}\,\mathrm{d}x\right] + q_\mathrm{c}$$

or

$$-kA\frac{\mathrm{d}T}{\mathrm{d}x} = -kA\frac{\mathrm{d}T}{\mathrm{d}x} + \frac{\mathrm{d}}{\mathrm{d}x}\left(-kA\frac{\mathrm{d}T}{\mathrm{d}x}\right)\mathrm{d}x$$
$$+ P\bar{h}(T - T_\infty)\,\mathrm{d}x,$$

from which

$$\frac{\mathrm{d}^2 T}{\mathrm{d}x^2} = \frac{P\bar{h}}{kA}(T - T_\infty) = m^2(T - T_\infty), \tag{5}$$

where

$\mathrm{d}T/\mathrm{d}x$ = longitudinal temperature gradient within the crystal (deg/cm),
k = thermal conductivity (average) of the crystal (cal/sec cm deg),
P = crystal perimeter = $2\pi r$ (cm),
$\bar{h}$ = heat transfer coefficient between the crystal surface and its surroundings (cal/sec cm^2 deg),
A = area of cross section of the crystal (cm^2),
T = crystal temperature at distance x from the source (deg),
T_∞ = temperature of surrounding fluid.

The general solution for $\mathrm{d}^2 T/\mathrm{d}x^2 = m^2(T - T_\infty)$ is,

$$T - T_\infty = C_1\,e^{mx} + C_2\,e^{-mx}, \tag{6}$$

where C_1 and C_2 are constants of integration whose values depend on the boundary conditions.

(a) First boundary condition

$$T = T_S \text{ at } x = 0,$$

i.e., at the crystallization layer where T_S is the temperature of the source or parent body.

$$\therefore T_S - T_\infty = C_1 e^{m(0)} + C_2 e^{-m(0)} = C_1 + C_2 .$$

(b) Second boundary condition

$$T = T_x \text{ at } x = X \text{ the total length of the crystal.}$$

In this treatment the heat loss by seed rod conduction is assumed to be negligible. The crystal looses the bulk of its heat by radiation and a lesser amount by gas convection both from its surface. Thus the bottom of the crystal would receive very little heat from the crystallizing zone as most of it is lost through the surface. The thermal conductivity of the crystal material is also very low; particularly at high temperatures[3]. Furthermore, the cross sectional area of the seed rod (0.125 inch diam.) is negligible by comparison to the surface area of the crystal. Hence, $[dT/dx] = O$ at $x = X$ where X is the total length of the crystal. Differentiating (6) and substituting, we have $C_2 = C_1 e^{2mX}$ from consideration of second boundary conditions. As a result, the complete solution of (5) becomes

$$T - T_\infty = (T_S - T_\infty) \left[\frac{e^{mx}}{1 + e^{2mX}} + \frac{e^{-mx}}{1 + e^{-2mX}} \right],$$

or

$$\frac{T - T_\infty}{T_S - T_\infty} = \frac{\cosh m(X - x)}{\cosh (mX)} . \tag{7}$$

Differentiation of (7) gives the temperature gradient within the crystal body at any distance x below the crystallization layer as follows:

$$\frac{dT}{dx} = (T_S - T_\infty) \left[\frac{m\, e^{mx}}{1 + e^{2mX}} - \frac{m\, e^{-mx}}{1 + e^{-2mX}} \right] . \tag{8}$$

It is obvious from eq. (8) that the temperature gradient within the crystal must be non-linear. The temperature gradient at the crystallizing layer, i.e., at $x = 0$ is,

$$\left[\frac{dT}{dx} \right]_{x=0} = (T_S - T_\infty)m \left[\frac{1}{1 + e^{2mx}} - \frac{1}{1 + e^{-2mX}} \right] . \tag{9}$$

The heat flow rate from the crystal is given by

$$q_{\text{crystal}} = -kA \left[\frac{dT}{dx} \right]_{x=0}$$

$$= (\bar{h}PkA)^{\frac{1}{2}} (T_S - T_\infty) \tanh (mX) . \tag{10}$$

Returning to the original eq. (3), we have

$$LV_g A_c \delta + k_L \left[\frac{dT}{dx} \right]_L A_c = q_{\text{crystal}} ,$$

which on substituting (10) becomes

$$LV_g A_c \delta + k_L \left[\frac{dT}{dx} \right]_L A_c$$

$$= (\bar{h}PkA)^{\frac{1}{2}} (T_S - T_\infty) \tanh (mX) .$$

As a first approximation we may assume the temperature gradient in the liquid to be $[dT/dx] = 0$ and the final form of the equation reduces to:

$$LV_g A_c \delta = (\bar{h}PkA)^{\frac{1}{2}} (T_S - T_\infty) \tanh (mX) , \tag{11}$$

where the terms have their previously defined meaning. Simplification of (11) by substituting $A_c = A = \pi r^2$ and $P = 2\pi r$, we have

$$LV_g r \delta = (2r\bar{h}k)^{\frac{1}{2}} (T_S - T_\infty) \tanh (mX) \tag{12}$$

$$V_g = \left(\frac{2\bar{h}k}{r} \right)^{\frac{1}{2}} \frac{(T_S - T_\infty)}{L\delta} \tanh (mX) \tag{13}$$

Rearranging and substituting ΔT for $(T_S - T_\infty)$ we have

$$V_g = \frac{\Delta T}{(r/2\bar{h}k)^{\frac{1}{2}}} \frac{\tanh (mX)}{L\delta} , \tag{14}$$

$$\text{Rate} = \frac{\text{Driving force}}{\text{Resistance}} \times \text{Process factor} .$$

Eq. (14) shows that the growth velocity is directly proportional to the driving force term ΔT and inversely proportional to the under-root of $r/2\bar{h}k$ comprising the resistance term, multiplied by a constant term. The importance of T_∞ is seen in the driving force as well as in the resistance term. T_∞ affects $\bar{h}$.

3. Evaluation of the thermal model

The successful application of eq. (14) is predicated on the availability of accurate thermal profiles in the crystal, the gas and the restricting muffle wall. This is particularly true for the precise determination of the value of the heat transfer coefficient, $\bar{h}$, which involves

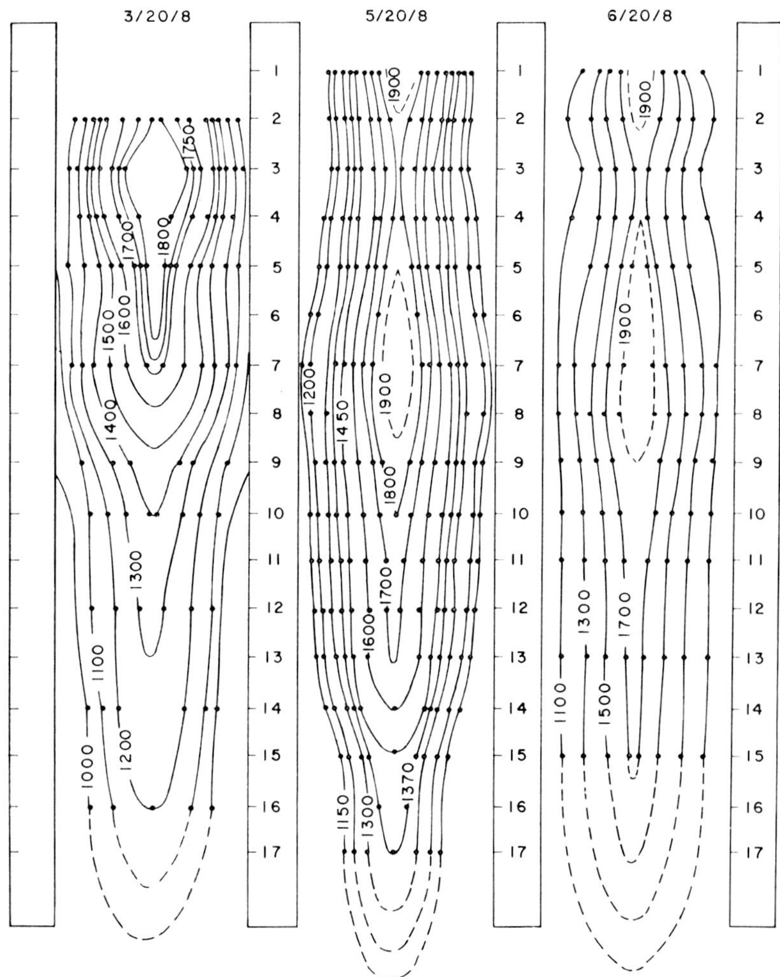

Fig. 2.   Flame thermal profiles at various flow rates (see text for explanation).

temperature dependent quantities such as the emissivity of the crystal and wall, the thermal conductivity and density of the crystal and the fluid properties of the gas. To this effect we have experimentally determined the flame thermal profiles for three different gas flow rates as shown in fig. 2, by direct thermocouple probing (Pt–Pt, 10% Rh). Probing during actual growth could not be done because of interference from the dropping powder with the thermocouple.

The thermal model clearly brings out the importance of T_∞. The gas envelope is, however, restricted by the muffle wall. Since the radiation component of the heat transfer coefficient is about an order of magnitude larger than the convection component (see below), T_∞ can be replaced by the average wall temperature T_w. Wall thermal profiles, determined with a standard optical

pyrometer are shown in fig. 3 for various flame conditions.

The value of certain crystal thermal properties requires a knowledge of the average crystal temperature. A thermocouple recorded the temperature at the base of the seed crystal. The arithmetic mean of the melt and base temperature was used as the crystal's average temperature assuming no thermal gradient within the seed rod. The results for various flame conditions are shown in fig. 4. Eq. (8) indicated a non-linear thermal gradient within the growing crystals. The solution requires knowledge of the exact wall temperature and assumes no heat loss by seed rod conduction. The nature of the crystal's thermal gradient may be approximated, however, by using the measured top and bottom temperatures and assuming linearity. The result is

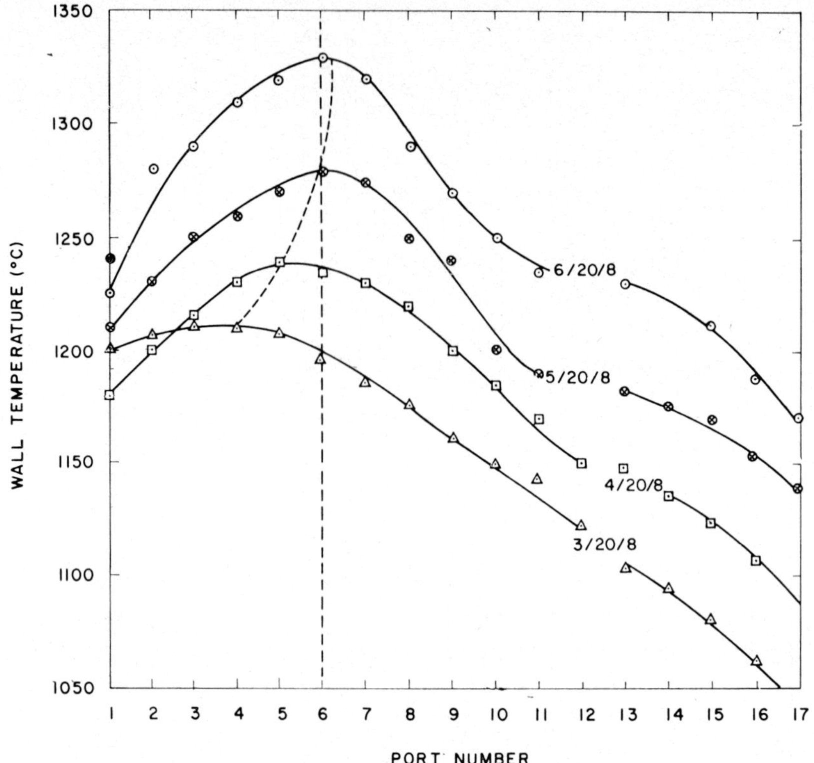

Fig. 3.   Muffle-bore wall thermal profiles.

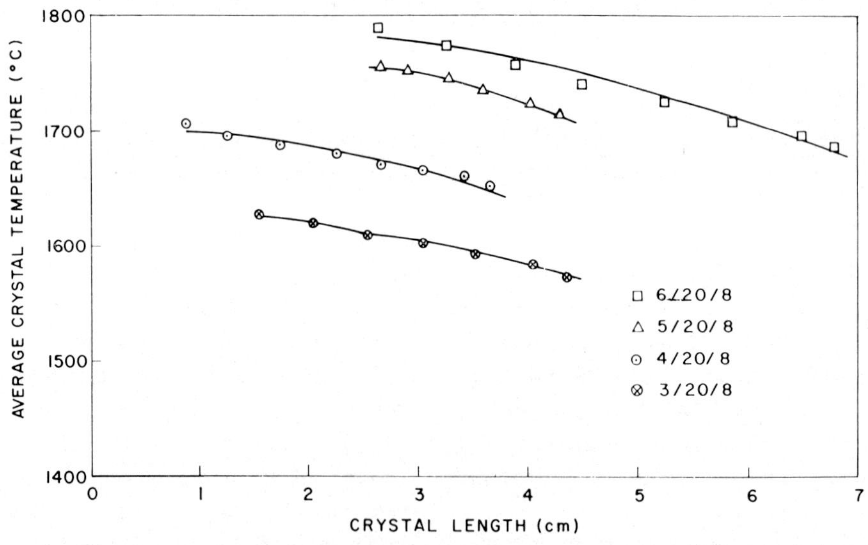

Fig. 4.   Average crystal temperatures as a function of crystal length for various gas-flow rates.

shown in fig. 5. Although the absolute values may be less significant, due to the assumptions, it may be noted that the curves indicate a rather shallow gradient after the first 3 to 4 inches of growth. This part of the boule is usually discarded and may be considered to represent the approach to relatively steady state conditions.

It is convenient to define a heat transfer coefficient in a thermal network involving convection and radiation processes as[4]),

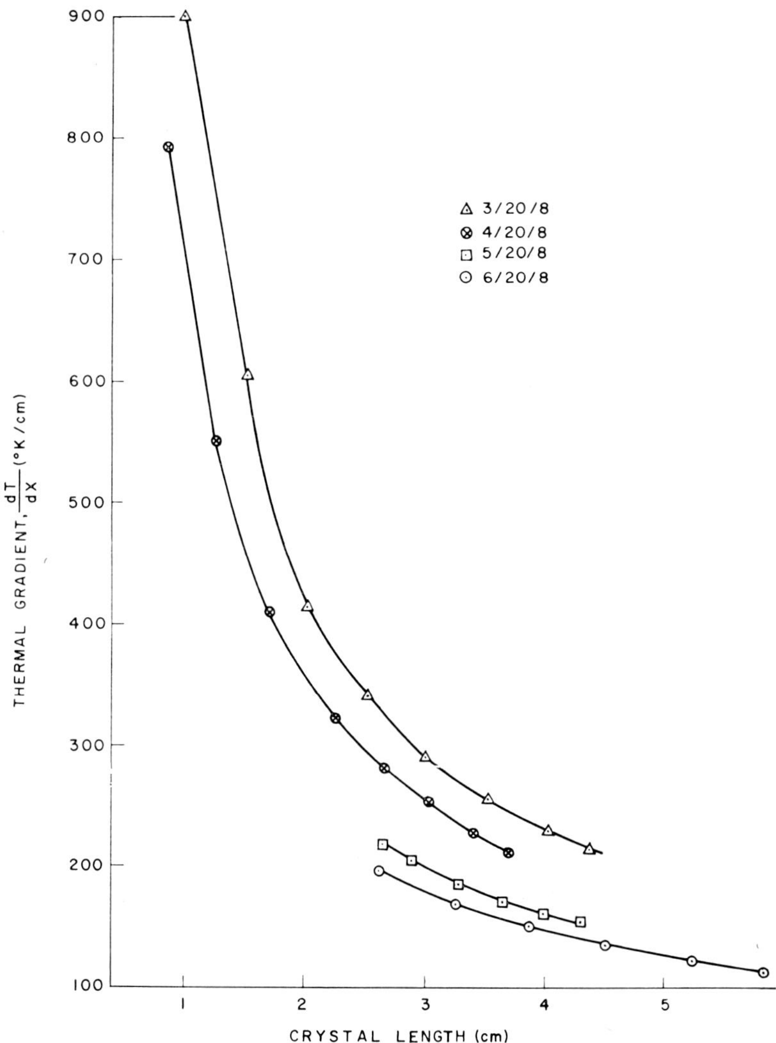

Fig. 5.   Crystal thermal gradient (longitudinal) for various gasflow rates.

$$\bar{h} = \bar{h}_c + \bar{h}_r.$$

The value of the convection coefficient may be determined from the expression for Nusselt's number, $N_u = \bar{h}_c D_H/k$, where D_H = hydraulic diameter of the annulus (diameter muffle-diameter boule) and k = thermal conductivity of the gas mixture. The value of the Nusselt's number $(R_e P_r D_H/L) \times 10^{-2}$ for any laminar flow may be found using Kay's plot[4]. This, in turn, requires a knowledge of the Reynold's number $R_e = D_H G/\mu$ and the Prandtl number $P_r = C_p \mu/k$, where

G = mass flow rate of the gases,

μ = fluid viscosity at the bulk temperature of the surrounding gases,

C_p = specific heat of the mixture,

L = crystal length.

Using the appropriate expressions[3] and interpolation[4] (H_2) to find component and mixture viscosities, the Reynold's number is found to have a value of 110. This low value indicates a definite laminar flow and absence of turbulent flow conditions. The convection heat transfer coefficient takes on a value of 1.13×10^{-3} cal/cm² sec deg.

The radiant heat transfer coefficient, $\bar{h}_r$, depends on the temperatures of the radiating surfaces, the respective emissivities, and a shape factor, and is given by

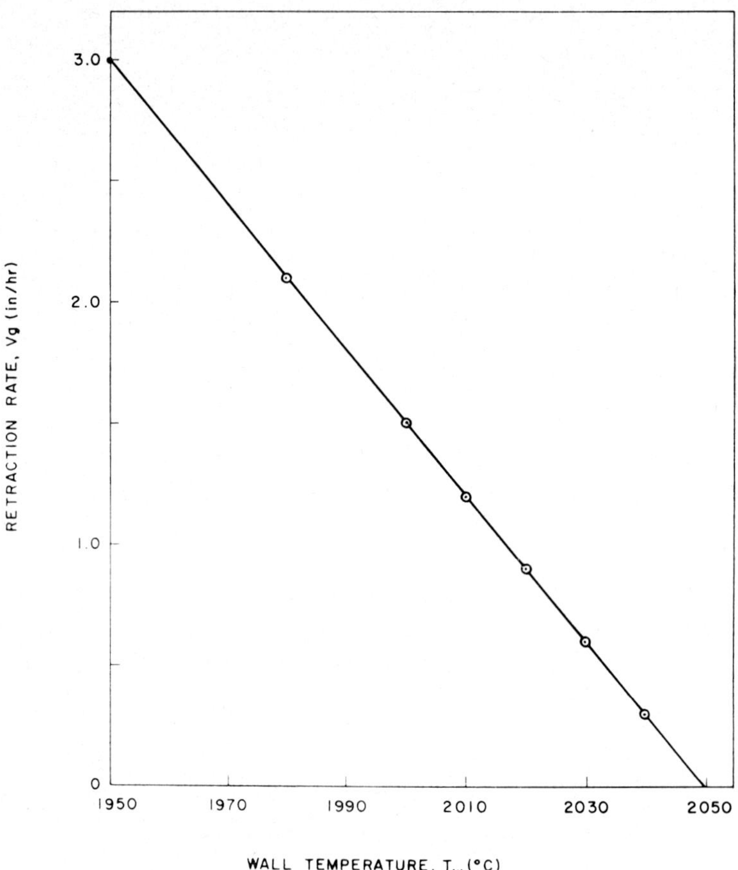

Fig. 6.   Retraction rate as a function of muffle wall temperature.

$$\bar{h}_r = F_{c-w}\left[\frac{\sigma(T_c^4 - T_w^4)}{T_c - T_w}\right],$$

where

F_{c-w} = grey body shape factor,

σ = Stefan-Boltzmann constant (cal/sec cm^2 °K^4),

T_c = average crystal temperature (°K),

T_w = average wall temperature (°K).

The grey body shape factor takes into account the respective emissivities and surface areas. Taking the value of the emissivity of the crystal[5]) as 0.4 and regarding the rough muffle wall as a black body, the radiation heat transfer coefficient $\bar{h}_r = 13.25 \times 10^{-3}$ cal/cm^2 sec °K, which is about an order of magnitude larger than the convection coefficient. The previously made assumption that radiation is the dominating phenomenon seems justified. The crystal gains heat from the gas and losses it to the wall. The overall heat trans-

fer coefficient will thus be $\bar{h} = \bar{h}_r - \bar{h}_c = 12.12 \times 10^{-3}$ cal/cm^2 sec °K.

Thermal conductivity of Al$_2$O$_3$ at the average crystal temperature (1767 °C) is not known. From the diffusivity[5]) the value of k has been determined as 0.011 cal/sec cm °K. Likewise, no information on the high temperature crystal density exists. From the relationship between thermal expansion and density and known values for a 60° inclined c-axis crystal[6]), the density was computed as $\delta = 3.8$ g/cm^3. The value of $\tanh(mx) = \tanh(7.43)$, varies little from unity for boule lengths between 1 and 3 inches. In this manner eq. (14) reduces, for the case of a $\frac{1}{2}$ inch diameter and $1\frac{1}{2}$ inch long boule in a 6/20/8 flame, to

$$V_g = 0.03(T_s - T_w) \text{ in./hr.}$$

The result is graphically shown in fig. 6. If perfection in crystal growth implies a growth rate controlled only by the rate at which the latent heat is removed from the crystallizing zone, than a $\Delta T = (T_s - T_w)$ of only $\sim 50\,^\circ\mathrm{C}$ would suffice for a growth rate of 1 in./hr as is common practice. Although present data do not allow the precise determination of ΔT, it is safe to assume that it is considerably larger than $50\,^\circ\mathrm{C}$ and much larger growth rates should be possible. An additional consideration seems more important, however. Eq. (14) was developed for the case where the temperature gradient in the liquid was considered negligible or no heat transfer between the crystal and the melt. If this term were not neglected, the growth rate from eq. (14) would be reduced by a term

$$\frac{k_1}{L\delta}\left(\frac{\mathrm{d}T}{\mathrm{d}x}\right)_1,$$

which, with the appropriate data, would have a value of $\sim 1.6 \times 10^{-2}\,(\mathrm{d}T/\mathrm{d}x)_1$ in./hr. A gradient of only 100 deg/cm would reduce the growth rate by 1.6 in./hr. It is likely that with the present method of intermittent non-regulated powder supply an appreciable gradient does exist across the liquid.

4. Summary

It appears possible, therefore, to control the growth rate in a flame fusion crystal growth process for any one flame condition, by monitoring the muffle thermal profiles and/or the liquid feed conditions. Appreciably higher rates of growth than presently used in practice should be possible. A continuous feed process, such as through fluidized bed type operation, could standardize the thermal conditions of the "liquid zone". Thermal conditions of the powder feed zone might have to be controlled beyond that which results from the combustion and gas-to-powder heat transfer process. Refinements of the process to the point that growth rates are continuously adjusted to account for time dependency of the heat transfer coefficient and physical characteristics of the growing crystal cannot be regarded fruitful until control of the primary parameters is fully realized. A computerized and adaptively controlled growth process appears necessary. Such a program is presently under development.

References

1) J. A. Adamski, R. C. Powell and J. L. Sampson, J. Crystal Growth **3, 4** (1968) 246.
2) J. G. Grabmaier, J. Crystal Growth **5** (1969) 105.
3) J. H. Perry, Ed., *Chemical Engineers Handbook* (McGraw-Hill, New York, 1963) 4th ed.
4) F. Kreith, *Principles of Heat Transfer* (Intern. Textbook Comp., Scranton, Pa., 1958).
5) Y. S. Touloukian, Ed., *Thermophysical Properties of High Temperature Solid Materials*, Vol. 4 (Thermal Properties Research Center, Purdue Univ. Press, 1966).
6) R. K. Kirby, National Bureau of Standards, Inorganic Materials Division, Washington, D. C., private communication.

CROISSANCE INDUSTRIELLE ET CARACTÉRISATION DE GRANDS MONOCRISTAUX DE SAPHIR PAR LA MÉTHODE DE VERNEUIL

J. RICARD

Ugine Kuhlmann, Centre de Recherches de Grenoble, 38–Jarrie, France

et

A. CIOCCOLANI

Le Rubis Synthétique des Alpes, 38–Jarrie, France

An industrial production of large white sapphire crystals reaching a few tons a year enables us to provide cylinders in the range 20 to 40 mm × 120 mm and now in the range 50 ± 5 mm × 80 mm. This relatively large scale production indicates that some characteristics of these sapphires are almost constant, e.g. the macromosaic structure, optical axis wandering, scattering of light and microbubbles and dislocations. It is to be noted that an increase in diameter from 30 to 55 mm results in some improvements in quality, because a new apparatus is used to obtain the larger diameters. This fact shows the perfectibility of the industrial Verneuil Method. It is difficult to compare our results with the available published data on Verneuil crystals, because of the large size of the sapphire crystals described here.

1. Fabrication industrielle du saphir blanc par la méthode Verneuil

Le corindon monocristallin non dopé, (saphir blanc) est fabriqué industriellement dans notre société par la méthode de Verneuil, avec des productions annuelles de plusieurs tonnes. Le contrôle de cette production importante nous permet de décrire les principales caractéristiques de ce matériau ainsi produit, après examen de milliers d'échantillons.

Notre appareillage Verneuil, construit dans nos ateliers, est semi-automatique. Il comporte un brûleur à deux gaz hydrogène-oxygène sans prémélange. En cristallisant à des vitesses linéaires de 6 à 12 mm/heure, nous travaillons habituellement sur des cycles de 24 à 48 heures.

Les monocristaux subissent, après la cristallisation, un traitement thermique pour réduire les tensions internes. Cette fabrication s'appuie sur un atelier de fabrication d'alumines[1,2]) dont la qualité constante est un facteur essentiel de succès.

Principales dimensions des saphirs blancs: Nos fabrications courantes comportent les catégories suivantes données dans le tableau 1 (fig. 1). Nous décrirons spécialement les monocristaux C et D.

2. Forme géométrique

Les anciennes "boules" sont maintenant des cylindres, avec des tolérances de ± 1.5 mm sur le diamètre moyen pour les saphirs de type C. Pour les saphirs D, nous observons parfois deux méplats opposés tout le long du cylindre (fig. 2). Il s'agit du plan (0001). Cette orientation 90° permet notamment de tailler des disques 0° avec un meilleur rendement.

TABLEAU 1

Type	Forme	Diametre (mm)	Longueur (mm)	Angle de l'axe optique par rapport à l'axe de croissance (deg)	Ordre de grandeur des poids (g)
A	baguette	2–4	300–400	0 à 90	
B	cylindre	16–24	100–150	60 à 90	80 à 300
C	cylindre	28–40	120–250	90	300 à 1000
D	cylindre	40–55	40–100	90	300 à 600

Fig. 1.

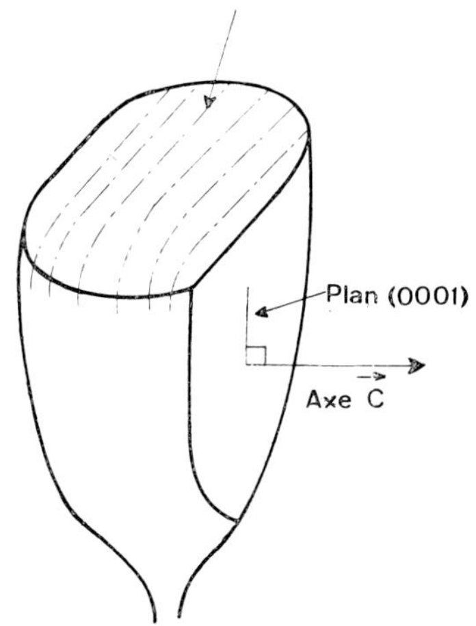

Fig. 2.   Saphir D à axe 90°.

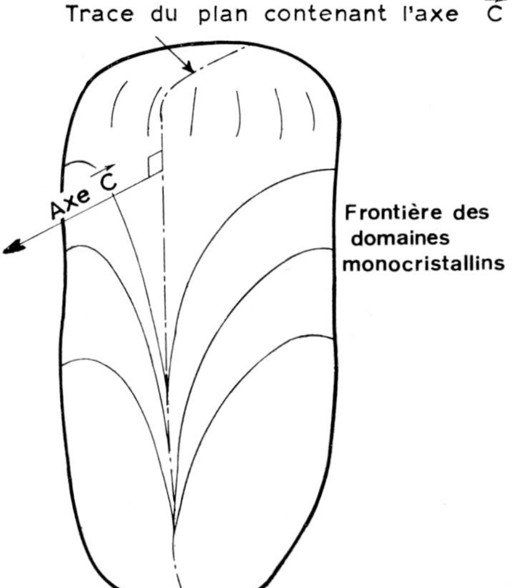

Fig. 3.   Structure mosaïque type des cylindres c à axe 90°.

3. Pureté chimique

Nous avons préféré le contrôle par spectrographie d'émission a celui par spectrographie de masse[3]), parce qu'il convient à la fois pour les poudres et les monocristaux. L'analyse type de nos saphirs est la suivante, en ppm par rapport à Al_2O_3 :

B	Ca	Co	Cr	Cu	Fe	Ga	Mg	Mn	Mo	Na
<5	≤3	<1	≤1	<0.3	≤3	<3	<1	<1	<3	<25

Ni	Pb	Si	Ti	V	Zn
<3	<1	2 à 5	10 ± 3	<3	<50

Le titane est un élément de dopage assez habituel. Les autres éléments ne sont pas détectés.

4. Caractéristiques cristallines

4.1. EXAMENS OPTIQUES

Les méthodes permettent l'examen de grands volumes.

Structure macromosaïque: Les agrégats de domaines monocristallins des saphirs Verneuil ont été souvent décrits[4,5]) mais, à l'exception des disques[6]), leur loi de répartition dans le volume des cristaux n'a pas été étudiée sur des populations importantes.

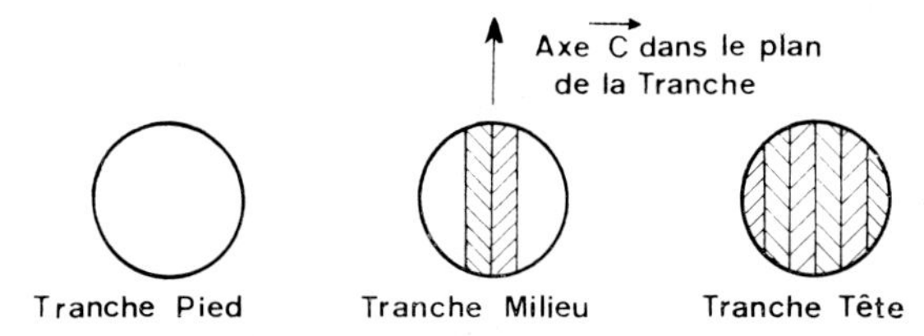

Fig. 4.   Structure macromosaïque des saphirs C en lumière polarisée.

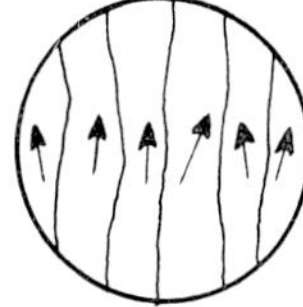

Fig. 5.   Variation relative des directions de l'axe optique en structure macromosaïque.

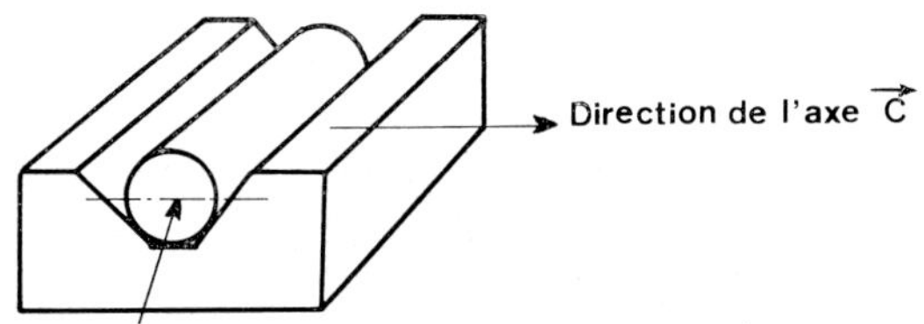

Fig. 6.   Berceau pour la mesure de la dérivé de l'axe optique.

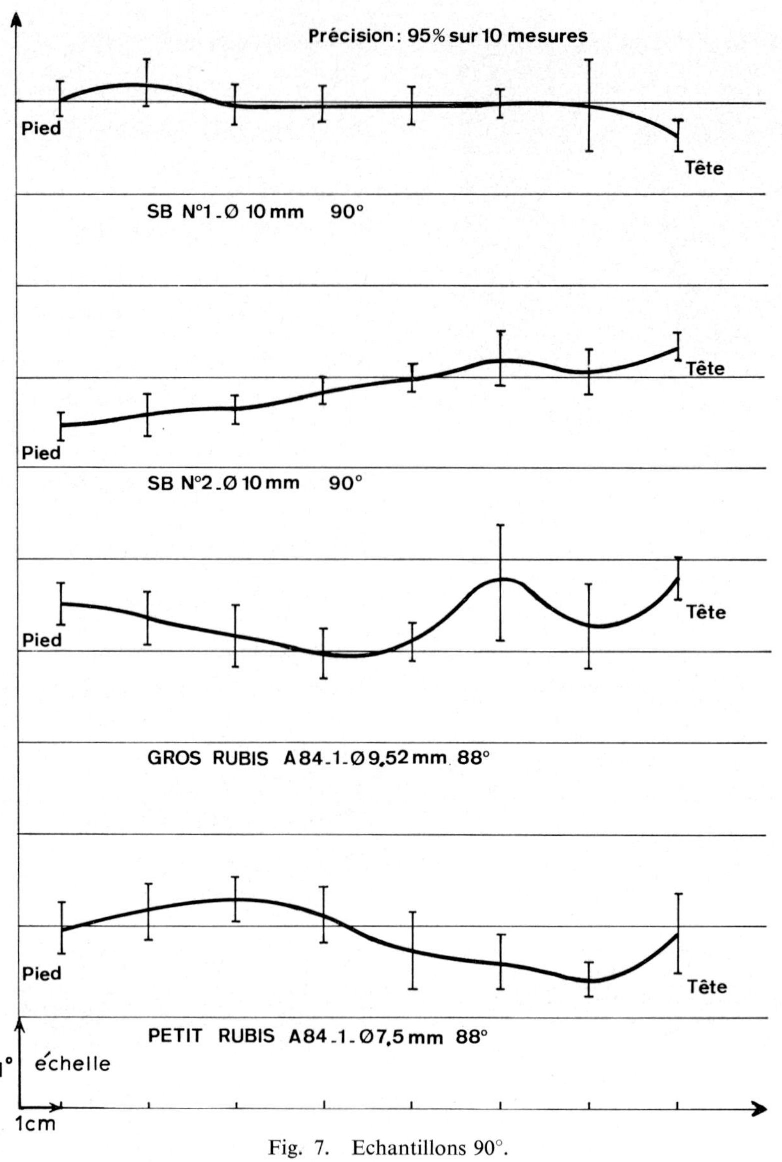

Fig. 7. Echantillons 90°.

Dans nos cylindres C (fig. 3), à environ 50 mm du pied, on voit apparaître des monocristaux accolés dont les nappes sont réparties de façon symétrique autour du plan contenant l'axe C[7]). En coupant des tranches perpendiculairement à l'axe de croissance, on obtient en lumière polarisée des domaines à frontières à peu près parallèles et légèrement désorientés entre eux (fig. 4). Ces frontières sont donc à peu près parallèles à l'axe C. Mesurée sur de nombreuses tranches, l'amplitude maximale de variation autour de la direction moyenne de l'axe C est de $\pm 2.5°$ (fig. 5).

Dans les saphirs D, le phénomène est plus symétrique, avec des frontières plus rectilignes. L'amplitude maximale de l'écart autour de l'axe C atteint seulement $\pm 1.5°$.

Mesure des dérivés de l'axe optique: Nous utilisons la méthode en lumière polarisée de Saucier[8]). En opérant avec un microscope et un berceau dans lequel le saphir est placé avec le plan contenant l'axe C horizontal, nous mesurons 10 extinctions par point le long de nos cylindres (fig. 6). Nous calculons alors l'écart type estimé pour chaque point avec 95 % d'intervalle de confiance. Nous arrivons à un intervalle moyen d'incertitude de $\pm 0.24°$ sur cylindre de 10 mm de diamètre.

Nous observons (fig. 7) que pour les cylindres C l'axe optique oscille autour de sa moyenne avec un écart

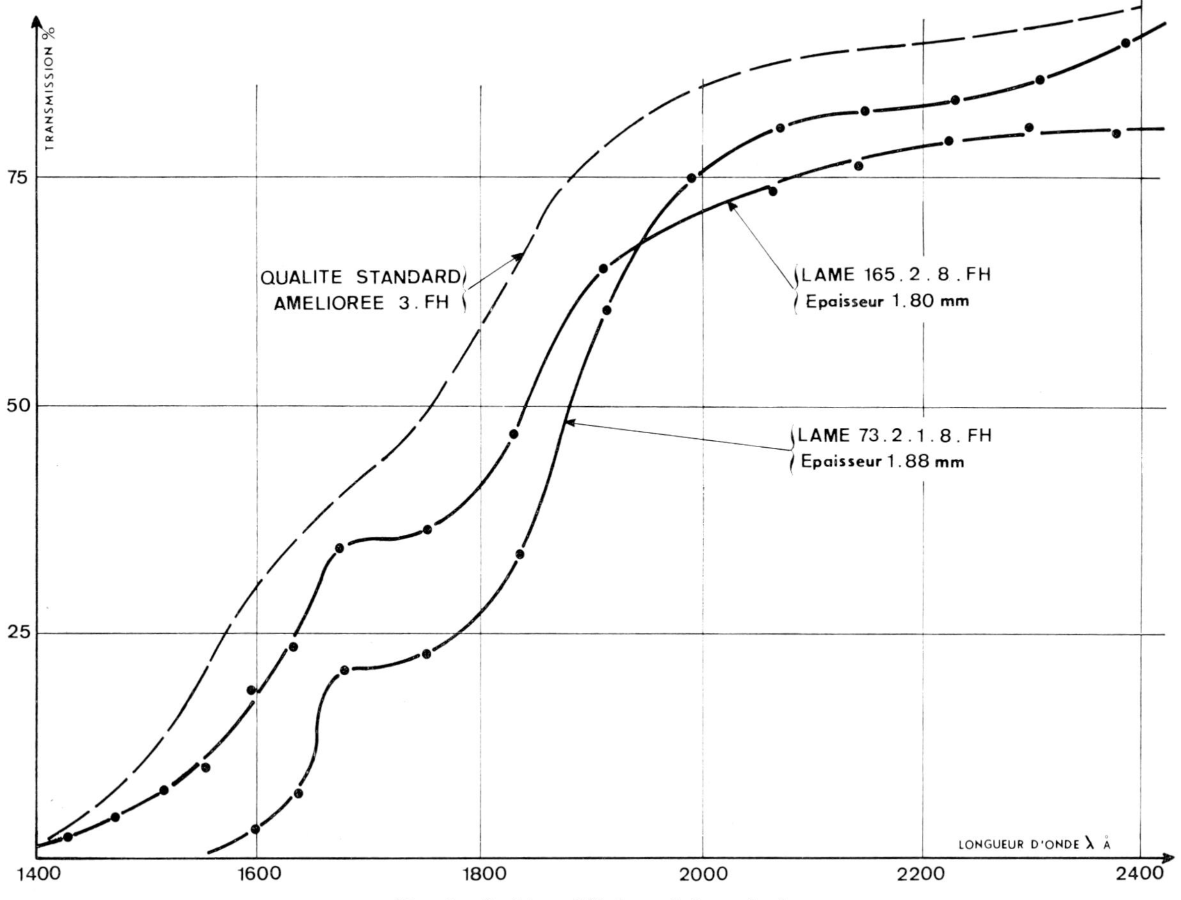

Fig. 8. Saphir stabilisé, qualité standard.

maximal de $\pm 2.5°$. Pour les cylindres et baguettes à axe $60°$, la dérive de l'axe C est constante, avec un gradient d'environ $0.2°/cm$. Ce phénomène a été signalé pour les baguettes[9]).

4.2. EXAMENS EN DIFFRACTION X

Compte tenu de la structure macromosaïque visible en lumière polarisée, nous avons peu employé la diffraction X. Pour une série de mesures, nous avons utilisé la méthode de Guignier-Tennevin avec le montage de Jan[11]) au laboratoire suisse de recherches horlogères. Nos saphirs C ont des désorientations pouvant atteindre $1.1°$, alors que nos saphirs D ne dépassent pas $0.2°$. Par comparaison, un échantillon de saphir Czochralski n'a présenté aucune désorientation mesurable.

4.3. DISLOCATIONS

Avec la méthode de révélation chimique de Scheuplein et Gibbs[12]) modifiée par Alford et Stephens[13]).

nous avons effectué de nombreuses mesures sur le plan (0001). Pour nos saphirs C, nous avons trouvé une moyenne de $5.0 \pm 1.10^4/cm^2$ de dislocations avec des écarts extrêmes 3.8×10^4 et 0.9×10^5. Pour nos saphirs D, la moyenne est de $1.4 \times 10^5/cm^2$ avec des écarts extrêmes $0.9 \times 10^4/cm^2$ à $5 \times 10^5/cm^2$.

Ces mesures nous ont montré que la densité de dislocation est répartie au hasard dans le volume cylindrique de nos saphirs, si l'on considère le plan (0001). A partir d'une cinquantaine de points de mesure, la répartition est en général de type gaussien.

5. Caractéristiques optiques

5.1. MICROBULLES ET CENTRES DIFFUSANTS LA LUMIÈRE

Ces microbulles mesurent de 1 µm à 1 mm dans les cas extrêmes. Elles se rassemblent parfois en filaments aux frontières des domaines monocristallins. Nous les comptons en effet Tyndall avec un laser He–Na. Pour les saphirs C, la densité oscille de 0.1 à 1 microbulle/cm^3. Elle est environ quatre fois plus faible pour les saphirs D.

5.2. SPECTRES D'ABSORPTION

En dessous de 2000 Å, nos spectres sont enregistrés par le laboratoire de technologie optique du C.N.R.S. à Marseille (fig. 8). Nous avons en préparation une qualité spéciale UV de transmission meilleure.

Dans le visible, ce saphir blanc n'a aucune bande d'absorption. Nous avons établi que la bande Ti^{+++} à 4900 Å[14]) n'est décelable qu'à partir de 25 ppm.

Dans l'infrarouge, nous obtenons 50 % de transmission à 6 µm sous 1 mm d'épaisseur. Sous 35 mm d'épaisseur nous avons mis en évidence une bande à 3.06 µm que nous attribuons à OH^- (réf. 15). Elle disparait au traitement thermique.

6. Caractéristiques mécaniques

6.1. APTITUDE AU SCIAGE

En opérant avec une scie diamantée, on travaille avec des vitesses linéaires de 30 m/sec sous refroidissement à l'eau. Un saphir de type C peut se tronçonner ainsi en 3 min. Les fractures au sciage sont très rares et sont souvent dues à des deviations de la scie.

6.2. APTITUDE AU POLISSAGE

Le plan le plus difficile à polir sans arrachement est le plan (0001). Nous opérons avec des métaux peu durs comme le cuivre et avec du diamant. Si l'on ne veut pas former de couches perturbées, il est nécessaire de terminer par un polissage à l'hydrogène[6]).

6.3. MESURES DE DURETÉ

Il est difficile d'éliminer la dispersion des mesures à cause des perturbations de surface créées par l'usinage.

Nous avons trouvé des duretés Knoop de 1600 à 2000 kg/mm^2, voisines de celles de Mangin[17]) et de Attinger[18]).

7. Conclusion

La fabrication industrielle du saphir permet de dégager des caractéristiques moyennes du matériau assez constantes. En augmentant les dimensions, et donc en modifiant la technique de cristallisation, on arrive à améliorer la qualité du saphir de manière significative et à étendre le champ des applications du saphir Verneuil.

Bibliographie

1) J. L. Henry et H. J. Kelly, J. Am. Ceram. Soc. **48** (1965) 217.
2) J. B. Schroeder, Investigation of the growth of optical crystals, Final Report 3, July 1962, AFCRL 62-593.
3) S. Stefani, A. Cornu, R. Bourguillot, A. Rubin et J. C. Brun, Chim. Analyt. **48** (1966) 252.
4) H. Conrad, Mechanical behavior of sapphire, 23 mars 1964, AD 435-976.
5) V. S. Doladugina et E. E. Berezina, Akad. Nauk SSSR Inst. Krist. **5** (1965) 401.
6) R. D. Olt et R. Rudness, Mater. Design Eng. (1963) 86.
7) L. M. Davies, Proc. Brit. Ceram. Soc. (1966) 1.
8) H. Saucier, Bull. Soc. Franç. Minéral. Crist. **76** (1953) 480.
9) Brevet français 938.492 du 9 décembre 1946.
10) cf. A. Guignier, *Theorie et Pratique de la Radiocristallographie* (Dunod, Paris, 1956) 2e éd., pp. 329–331.
11) J. P. Jan, Suisse d'Horlogerie No. 1–2 (1958) 1.
12) R. Scheuplein et P. Gibbs, J. Am. Ceram. Soc. **43** (1960) 458.
13) L. D. Stephens et W. J. Alford, J. Am. Ceram. Soc. **42** (1959) 81; **46** (1963) 193.
14) J. P. Jones, R. L. Coble et C. J. Mogab, J. Am. Ceram. soc. **52** (1969) 331.
16) J. L. Fraimbault, Thèse Grenoble 1967, Etude d'une structure épitaxiée silicium–corindon par hydrolyse du silane.
17) A. Mangin, Thèse Strasbourg 1960, Etude de certains systèmes binaires à base d'alumine: préparation, résistance mécanique dans l'état mono- et polycristallin.
18) C. Attinger, Bull. Soc. Suisse Chronométrie (1951).
19) D. A. Curtis et J. S. Thorp, Brit. J. Appl. Phys. **16** (1965) 734.

Journal of Crystal Growth **13/14** (1972) 723–725 © *North-Holland Publishing Co.*

GROWTH OF STOICHIOMETRIC Mg–Al SPINEL SINGLE CRYSTALS BY A MODIFIED VERNEUIL TECHNIQUE

R. FALCKENBERG

Forschungslaboratorien der Siemens AG, München 80, Germany

Stoichiometric Mg–Al spinel single crystals grown by the usual Verneuil technique crack during the cooling down period. The cause of the cracking is investigated. A modified technique is described by which mechanically stable crystals can be grown.

Mg–Al spinel single crystals are of increasing importance due to the advantages they offer as substrate for epitaxially grown silicon[1]). When used as a substrate they have to be of sufficient size (diameter ≥ 20 mm) and have to stay mechanically and thermally stable during all the processes which are necessary to fabricate integrated circuits. Such processes are sawing, grinding, polishing, epitaxial growth of silicon, diffusion and oxidation.

Because of the oxide nature of the spinel crystals and the high melting point of about 2000 °C the Verneuil technique[2]) is a suitable growth method. Fig. 1 shows the phase diagram[3]) $MgAl_2O_4$–Al_2O_3. The growth of crystals with a surplus of Al_2O_3[4]), for example of the composition $MgO : Al_2O_3 = 1 : 3.3$ is comparatively simple; these crystals are manufactured on a large scale to be used as gems and bearings. There is little application of Al_2O_3-rich crystals as substrate because of the precipitation[5]) of Al_2O_3 appearing when the crystals are thermally treated. Stoichiometric crystals of the composition $MgAl_2O_4$ are thermally stable; no precipitation of Al_2O_3 can be observed. Difficulties arise however in preparing crystals of adequate size and mechanical stability by the Verneuil process[3,6–11]).

To grow stoichiometric spinel crystals we used a Verneuil apparatus with a two-tube burner and without an after-heater. The first experiments in highly heat-insulating furnaces led to the result that the crystals fractured during the cooling down period. Frequently the plane of fracture went through the [100]-growth axis of the crystal. In order to study the cause of frac-

ture, spinel crystals with a surplus of Al_2O_3 have been grown. These crystals did not crack during cooling and could be used for photoelastic investigations without prior annealing. The crystals were cut perpendicular to the [100]-growth axis into pieces of 1–2 cm length. To suppress the isoclinics quarter-wave plates were put

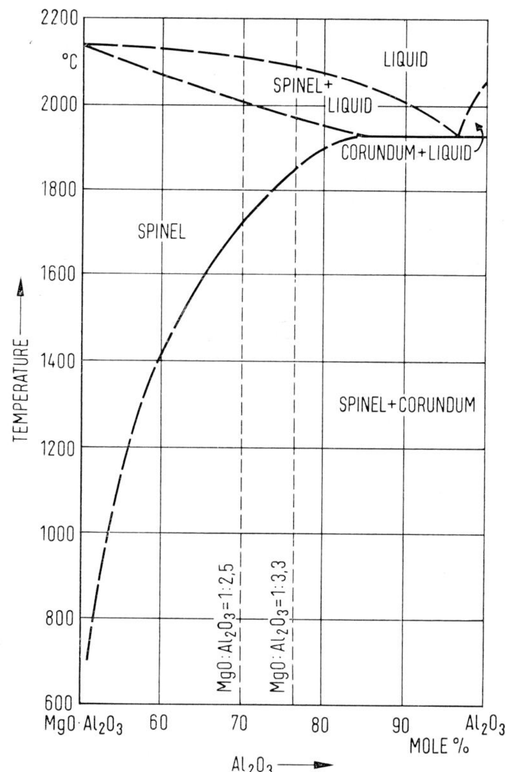

Fig. 1. Equilibrium diagram for the system $MgAl_2O_4$–Al_2O_3, after Arlett[3]).

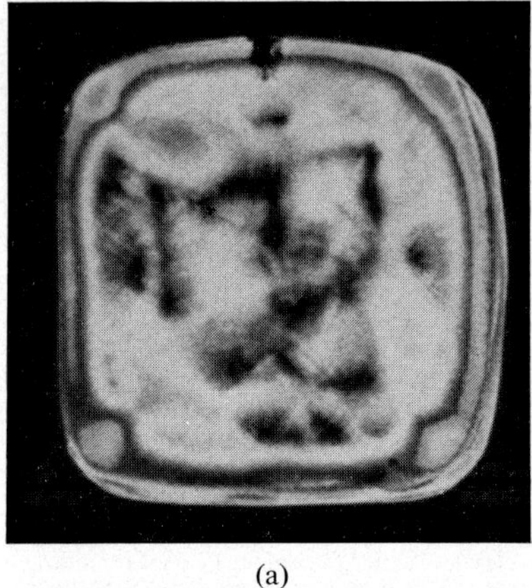

(a)

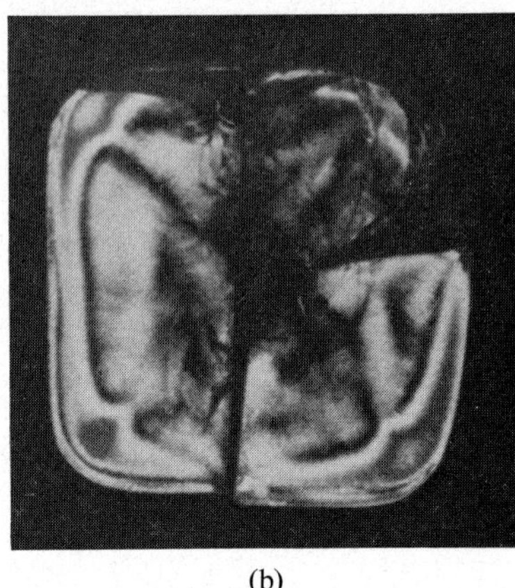

(b)

Fig. 2.   Isochromatic curves in an alumina rich Mg–Al spinel single crystal cut perpendicular to the growth axis ([100]); (a) before cleaving, (b) after cleaving. (See also footnote and explanation in the text.)

between the polarizers. Fig. 2a shows the pattern of isochromatic* curves typical for the crystals investigated. They run parallel to the side faces of the crystal in the marginal zone. As known along an isochromatic curve the state of stress is constant; therefore the stress gradient lies in a radial direction. Cleaving the crystal piece parallel to a plane through the growth axis changes the photoelastic picture (fig. 2b). The isochromatic curves are now confined to the corners of the crystal. This way of cleaving apparently leads to a drastic reduction of the stresses.

Crystal pieces cut parallel to the growth axis also have been investigated (fig. 3). No isochromatic curves

* Unfortunately it is not possible to reproduce coloured photographs. The black and white photographs shown do not allow to distinguish between different orders of isochromatic curves. The isochromatic curves referred to in the text show up as black lines.

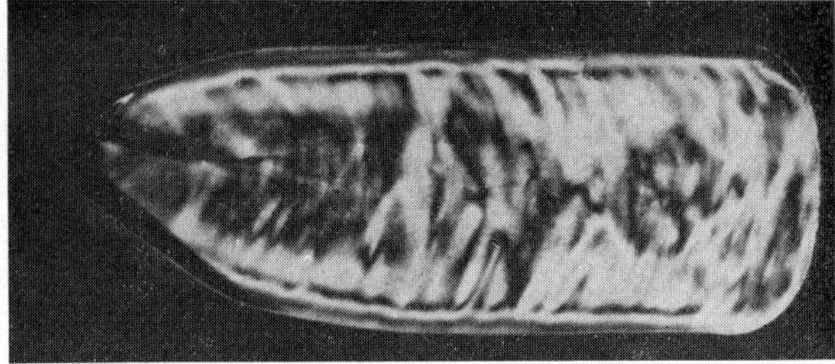

Fig. 3.   Isochromatic curves in an alumina rich Mg–Al spinel single crystal cut parallel to the growth axis.

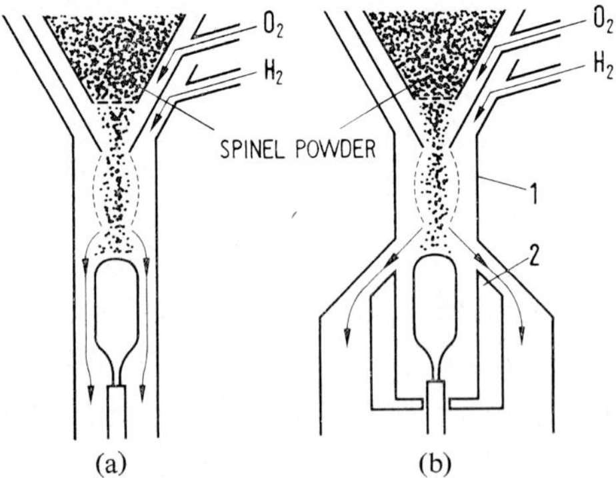

Fig. 4.   (a) Gas flow in a conventional Verneuil furnace. (b) Gas flow in the modified furnace.

could be found indicating differences of stress of comparable magnitude in the axial direction. These results lead to the following conception of the cooling down process: During the growth the hot flame gases flow closely around the crystal (fig. 4a). After the sudden extinction of the flame which has been found to be better than a slow cooling[12]) down, the temperature of the surface decreases faster than that of the core. A large radial temperature gradient develops. The differential contraction leads to the development of differences in the state of stress between the marginal zone and the core of the crystal. The differences can equalize by formation or rearrangement of dislocations as long as the core is plastically deformable. The limit of plastic deformation[13]) at the pressures present might be reached above 1900 °C. After the temperature of the core has decreased below this limit, the radial stresses then developing can not longer be reduced completely and may – as in the case of the stoichiometric spinel – lead to fracture. Fracture planes run through the growth axis.

Unlike these radial stresses, stress gradients arising as a consequence of an axial temperature gradient during the cooling down period mainly lead to elastic deformations, which decrease with decreasing temperature and do not contribute to the cracking of the crystal.

To avoid cracking one has therefore to take care that the radial stress gradients are reduced. Fig. 4b shows the scheme of a device which largely prevents the formation of differences in the state of stress in a radial direction during the cooling down period of the crystal. The flame gases are led as usual through a ceramic tube to the growth front and transfer the energy necessary to maintain the melt film on top of the crystal. Immediately underneath the growth front however the flame gases are deflected. This is done by a ceramic insert shaped like a crucible which encloses the growing crystal. The deflected flame gases heat the insert. Except for the growth front, the whole crystal is in thermal equilibrium with the bulk of the insert. Therefore the temperature shock is very small when the flame is quenched. The rate of cooling is largely determined by the thermal properties of the insert and can be adjusted in a way that fracture of the crystal is avoided.

In the axial direction the temperature gradient during the growth period is larger than with the usual Verneuil technique.

The bottom of the insert is closed by a plate (fig. 4b), which has just a small aperture for admitting the crystal holder. The plate effects the blocking of the gas stream through the insert; the diameter of the growth chamber can largely be adjusted to the special requirements. Generally the diameter was chosen 1 cm larger than the diameter of the crystals to be grown.

A crystal can be grown on a sintered cone or a seed of given orientation. A tube of sintered Al_2O_3 is used as a seed holder. The formation of the melt takes place between the insert and the lower end of the tube carrying the gas; it can be observed through lateral apertures. For hydrogen, the rate of flow may be 1500 l/h, for oxygen 400 l/h at the beginning of the growth process. The crystal can be broadened by increasing the flow rate of oxygen; with 700 l/h diameters above 20 mm can be reached. During the growth, the top of the crystal stays above the insert. The rate of growth is about 1 cm/h.

The further the crystal is drawn into the insert during the growth, the more heat it looses to the insert. If this loss is not compensated by increasing the flow rate of oxygen, the crystal tapers off. The growth is finished by interrupting the powder supply, by pulling the crystal down into the insert and by extinguishing the flame.

By this procedure we obtained large stoichiometric spinel single crystals which are mechanically and thermally stable and suitable as substrates. With an insert of 3 cm inner diameter crystals can be grown with a cross-section of about 20×20 mm^2 and a length of about 60 mm. In principle, the length of the crystals is not limited.

The spinel crystals grown had a growth axis parallel to the [100]. When the crystals are cut and polished as substrate wafers of a thickness of 0.3 mm, the yield is about 50%; by annealing the crystals at 1800 °C the yield reaches nearly 100%.

In a similar way the technique described can be used in cases where a conventional Verneuil furnace would fail because of the radial stresses or where a large axial temperature gradient is necessary to avoid for example precipitation during growth.

For helpful discussions I would like to thank H. Schlötterer and D. Schott; the latter also took the photoelastic pictures.

References

1) H. Schlötterer, Electronics **42** (1969) 113.
2) A. Verneuil, Ann. Chim. Phys. **3** (1904) 20.
3) R. H. Arlett, J. Am. Ceram. Soc. **45** (1962) 523.
4) W. F. Eppler, Z. Angew. Mineral. **4** (1943) 345.
5) H. Saalfeld and H. Jagodzinski, Z. Krist. **109** (1957) 2.
6) K. A. Wickersheim and R. A. Lefever, J. Opt. Soc. Am. **50** (1960) 831.
7) R. A. Lefever, Rev. Sci. Instr. **33** (1962) 1470.
8) R. S. Mitchell, Rev. Sci. Instr. **36** (1965) 1667.
9) R. H. Arlett and M. Robbins, J. Am. Ceram. Soc. **50** (1967) 273.
10) K. Nakano, H. Tabuta, H. Okuda and N. Kogyo, Gijutsu Shikensho Hokoku **17** (1968) 197.
11) C. C. Wang, J. Appl. Phys. **40** (1969) 3433.
12) W. Seifert, Dissertation, Universität München (1969).
13) J. Grabmaier and C. Watson, Phys. Status Solidi **25** (1968) K 7.

CRISTAUX À GRADIENT DE MAILLE[†]

A. FREUND[‡], P. GUINET[*], J. MARESCHAL[*], F. RUSTICHELLI[**] et F. VANONI[*]

Institut Max von Laue – Paul Langevin, Grenoble, France
Centre d'Etudes Nucléaires, Grenoble, France

This paper reports original results on a range of single crystals having a continuous lattice parameter variation. The main aim of these investigations is to replace conventional neutron monochromators by high perfection single crystals having a gradient in the lattice spacing. This is because it has been shown that such a type of crystal would considerably improve the neutron physics techniques. In order to obtain such crystals, the general idea followed in this work is to continuously change the relative concentration of a two-component single crystal solid solution by a suitable growing process.

The first experiments have been carried out on a CoO–NiO monocrystalline solid solution grown by the fame fusion process. Alkali halides have also been investigated using the KCl–KBr solid solution. Single crystals have been grown in a Bridgman crucible by the zone melting technique.

Finally using a special growth method, attention was focused on Cu–Ge solid solutions; these are of special interest in neutron diffraction. The Ge concentration was varied from $\simeq 2\%$ to $\simeq 10\%$, with a corresponding total relative variation of the lattice spacing of $\Delta d/d \simeq 1\%$. The mosaic spread of the crystal (dimensions, $h = 4$ cm, diam. $= 1$ cm) was found to lie between 35 and 120. The measurements were carried out by X-ray, γ-ray, neutron diffraction techniques and electron micro-analyser.

1. Introduction

On appelle "cristal à gradient de maille" (CGM) un monocristal dans lequel, par un artifice quelconque, le paramètre cristallin varie continuement. Le gradient est obtenu, dans cette étude, par une variation de concentration dans une solution solide.

Ces recherches sur les CGM ont débuté à Grenoble à partir d'une proposition de Maier-Leibnitz en vue de les utiliser comme monochromateurs de neutrons, en particulier pour le réacteur à haut flux en construction à Grenoble.

2. Intérêt des CGM en optique neutronique

Dans le cadre du développement des techniques modernes de physique neutronique, un effort particulier se porte sur l'amélioration des monochromateurs par diffraction, de manière à pouvoir mieux utiliser les sources de plus en plus puissantes de neutrons mises à la disposition des expérimentateurs[1]). On peut réaliser un nouveau type de monochromateur si on utilise des CGM ayant certaines caractéristiques qui seront définies plus loin.

Un faisceau monochromatique de neutrons, de largeur de bande donnée $\Delta\lambda$, peut être représenté sur un diagramme $(2d, \sin\theta)$ par la surface délimitée par les deux hyperboles représentant la loi de Bragg (fig. 1):

$$\lambda = 2d \sin\theta, \tag{1}$$

où d est la distance inter-réticulaire et θ l'angle de

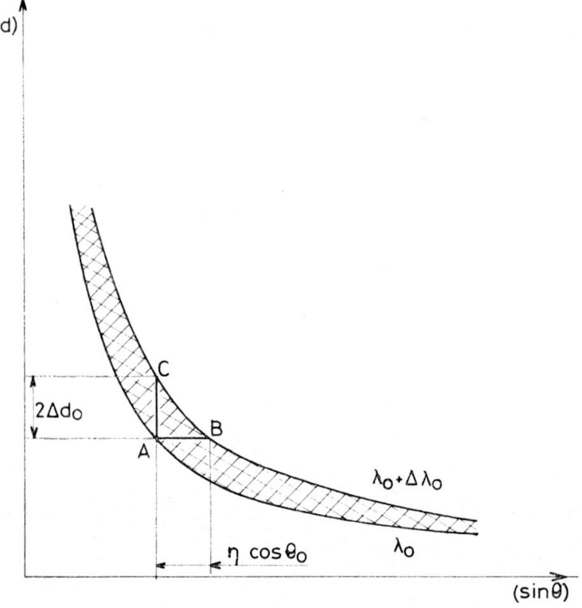

Fig. 1. Sélection d'une bande monochromatique de neutrons par un cristal mosaïque (segment AB) et par un cristal à gradient de maille (segment AC)[2].

[†] Dédié au Professeur H. Maier-Leibnitz, à l'occasion de son 60e anniversaire.
[‡] Institut Max von Laue – Paul Langevin.
 [*] Centre d'Etudes Nucléaires.
[**] Aussi C.C.R. Euratom, Ispra, Italie.

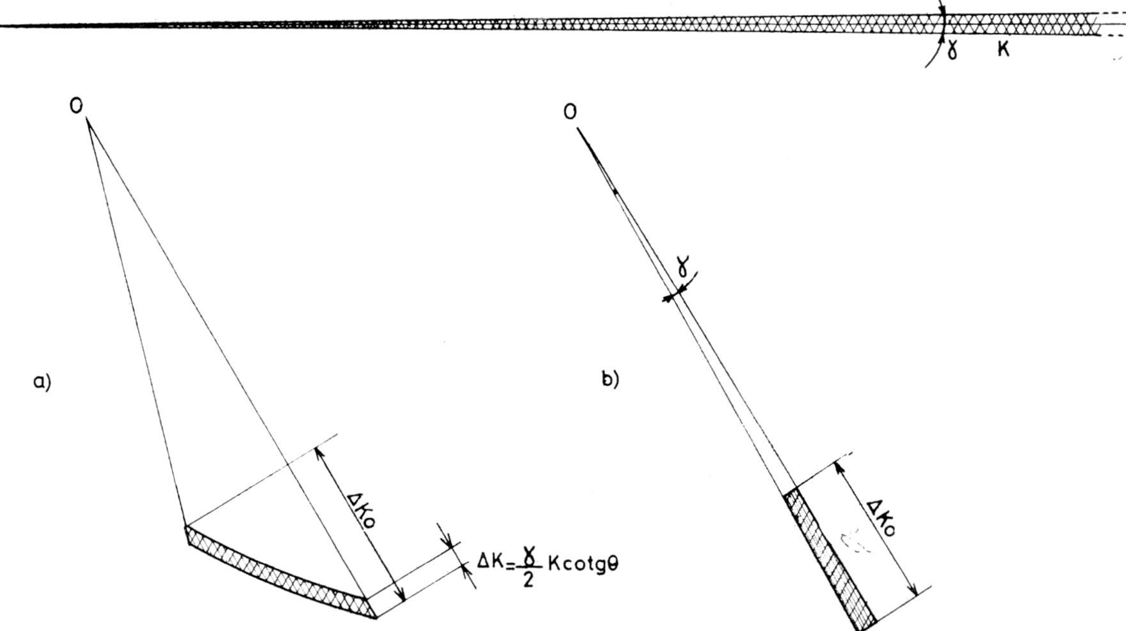

Fig. 2. Représentation dans l'espace des moments de la sélection d'une bande monochromatique de largeur ΔK_0: (a) par un cristal mosaïque, (b) par un cristal à gradient de maille[2].

Bragg. Actuellement la sélection de la bande de largeur $\Delta\lambda/\lambda$ est effectuée par des cristaux ayant une distribution mosaïque η et, dans la formule dérivée de l'équation (1),

$$\frac{\Delta\lambda}{\lambda} = \frac{\Delta d}{d} + \cot g\,\theta\,\Delta\theta, \qquad (2)$$

la valeur de $\Delta\lambda/\lambda$ est obtenue par la variation de θ due à la mosaïque avec $\Delta d/d = 0$. Une telle monochromation correspond au segment AB de la fig. 1.

Si par contre, on considère un cristal ayant une distribution mosaïque négligeable mais possédant un gradient de malle $\Delta d/\Delta x$ on obtient une sélection différente de $\Delta\lambda/\lambda$ qui est représentée sur la fig. 1 par le segment AC.

Les avantages originaux des CGM sont décrits en réfs. 2, 3. On rappelle le plus important, à savoir que le faisceau incident et le faisceau diffracté ont des divergences angulaires, identiques, même dans le cas de faisceaux parallèles, alors que les cristaux mosaïques donnent lieu à une dispersion angulaire (fig. 2).

3. Problèmes concernants les techniques de croissance

L'optimisation du gradient de distance inter-réticulaire dans le cas général et spécialement dans celui des systèmes monochromateurs composés a été traité dans réf. 2. Sur la base de cette étude, les problèmes à résoudre du point de vue croissance cristalline sont les suivants:

(a) établir un choix entre les différentes possibilités théoriques d'obtenir un gradient de maille, et en vérifier l'application pratique;

(b) obtenir une variation de maille de l'ordre de grandeur désiré $(\Delta d/d \sim 1\%)$[4];

(c) obtenir cette variation sur une longueur telle que le gradient de maille obtenu optimise les conditions d'extinction[2];

(d) pouvoir contrôler la valeur locale du gradient;

(e) obtenir les conditions précédentes avec des matériaux ayant des caractéristiques neutroniques convenables;

(f) obtenir des CGM ayant une dispersion mosaïque la plus réduite possible, inférieure à quelques minutes d'arc en général.

4. Premiers essais sur CoO–NiO

Après avoir écarté certaines techniques déjà réalisées dans d'autres domaines [diffusion d'un composant dans un cristal parfait, diffusion entre solides[5], implantation ionique], on s'est orienté sur l'utilisation de solu-

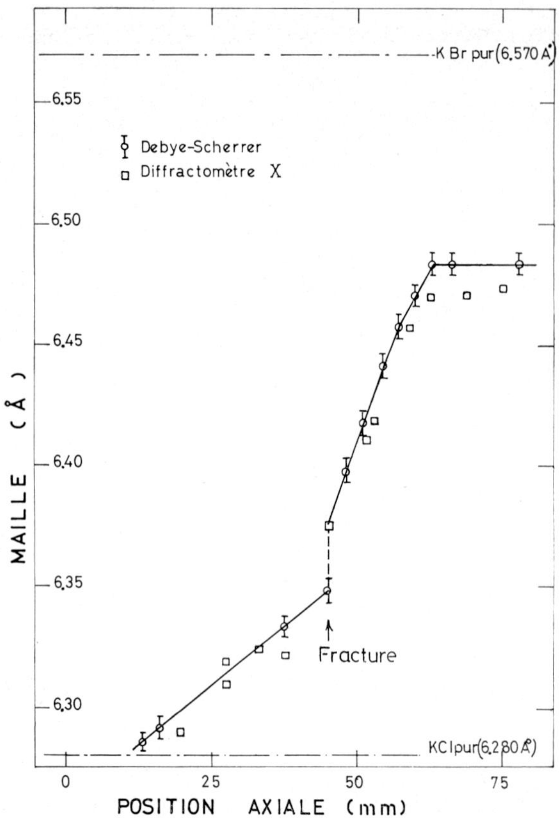

Fig. 3. Profil de variation de la maille dans le CGM KCl–KBr: les mesures Debye–Scherrer ont été effectuées sur des échantillons de cristal broyé, et les mesures en diffractométrie X sur des paillettes monocristallines obtenues par clivage.

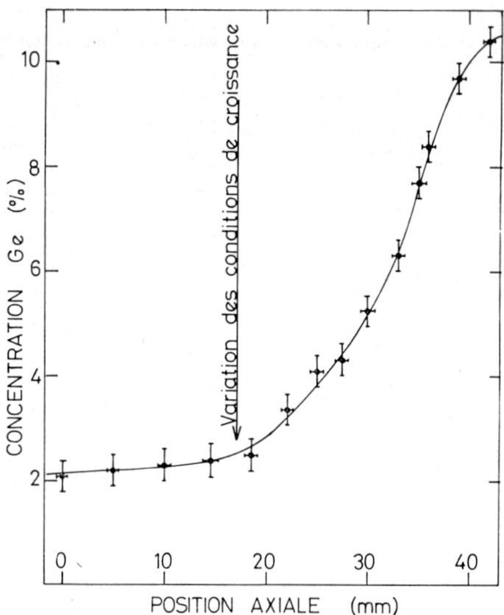

Fig. 4. Profil de concentration de Ge dans le CGM Cu–Ge le long de l'axe du cristal.

tions solides dans lesquelles le gradient de maille est obtenu par un gradient de concentration[6]).

Les premiers résultats positifs ont été obtenus avec les CGM CoO–NiO produits par la méthode Verneuil, avec une modification du système d'alimentation de la poudre: le distributeur est surmonté d'un cylindre de faible section dans lequel 10 zones de 5 g de poudre de concentration croissante en NiO ont été stratifiées, ce qui correspond à 10 zones de 3 mm pour le cristal formé. Le pied du cristal est formé de CoO pur et la tête de NiO pur. La température de flamme est continuement modifiée de 1935 à 1990 °C, de manière à correspondre à la température de fusion de la poudre à concentration variable, les débits de gaz étant de 800 l/h pour l'hydrogène et de 33 l/h pour l'oxygène. La vitesse de tirage est de 9 mm/h. En fin d'opération le monocristal a été recuit à 1900 °C, puis taillé sous forme de cylindre avec le gradient axial au cylindre, suivant la direction [100].

La cristal a été étudié en détail par diffraction neutronique et rayons X[7]). Une variation relative totale de maille de $\simeq 1.2\%$ correspondant à une variation de concentration en CoO de 20% à 90% a été mis en évidence. La distribution mosaïque était comprise entre $11'$ et $27'$.

5. Expériences sur les cristaux ioniques

Certains halogénures alcalins tels que des sels de rubidium ou les fluorures, présentent de bonnes propriétés neutroniques. Une première approche dans cette voie consiste à étudier la faisabilité d'un CGM de KCl–KBr pour mettre au point la cristallisation par fusion de zone en creuset Bridgman. De plus, le CGM KCl–KBr permet d'étudier d'autres propriétés de CGM en particulier les propriétés optiques (variations continues d'indices de réfraction).

Les résultats sont du même ordre de grandeur que ceux obtenus pour CoO–NiO, avec une variation de la maille de 3% sur 3 cm (fig. 3).

6. Expériences sur Cu-Ge

La solution solide Cu–Ge présente un grand intérêt comme monochromateur car Cu et Ge ont presque les mêmes caractéristiques neutroniques et peuvent être tirés sous forme de monocristaux parfaits.

Des solutions solides monocristallines presque parfaites[8]) ont d'abord été obtenues. Ensuite un gradient

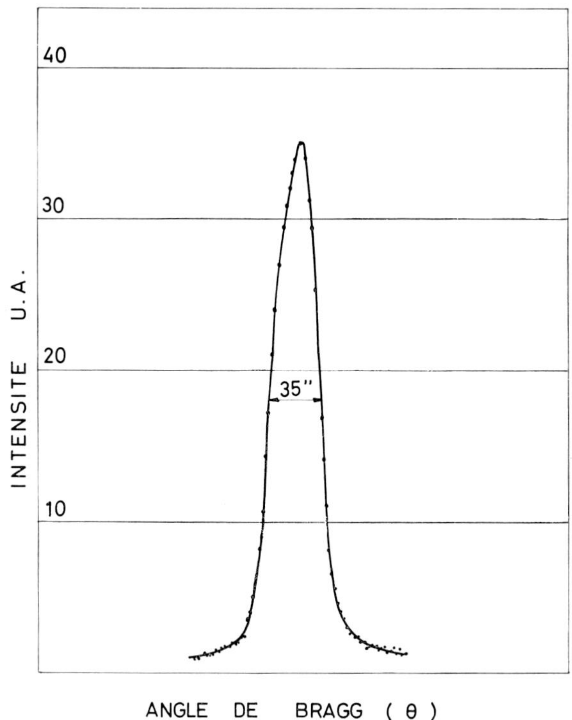

Fig. 5.   Profil de diffraction des rayons γ par le CGM Cu–Ge, obtenu par la radiation 412 keV d'une source d'or ($\lambda = 3.02 \times 10^{-2}$ Å, $\Delta\lambda/\lambda = 10^{-6}$) présentant une divergence angulaire de 10″.

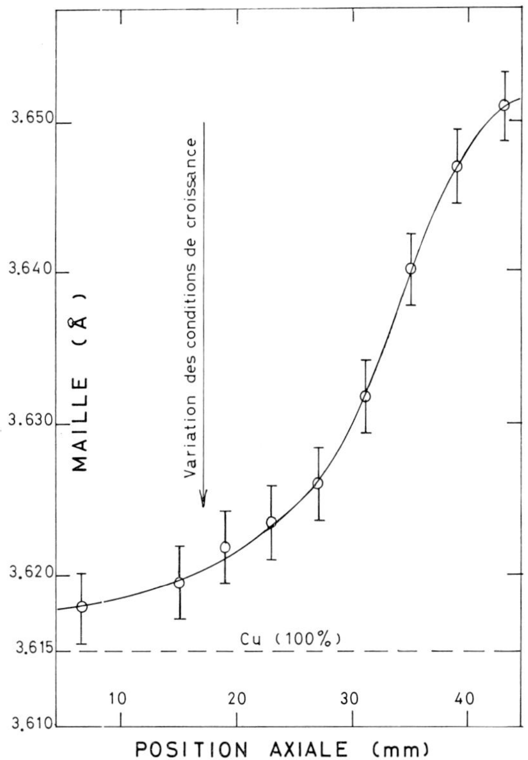

Fig. 6.   Profil de variation de maille du CGM Cu–Ge obtenu par clichés Debye–Scherrer.

de concentration a été introduit pendant la croissance en utilisant une nouvelle méthode de tirage qui fait l'objet d'une proposition de brevet.

Le cristal (diamètre $\simeq 1$ cm, $h \simeq 4$ cm) orienté sui-

vant [100] a été analysé à la microsonde électronique qui a mis en évidence (fig. 4) une variation de concentration de Ge de $\simeq 2\%$ à $\simeq 10\%$ (maximum de solubilité). La qualité cristalline a été étudiée par diffrac-

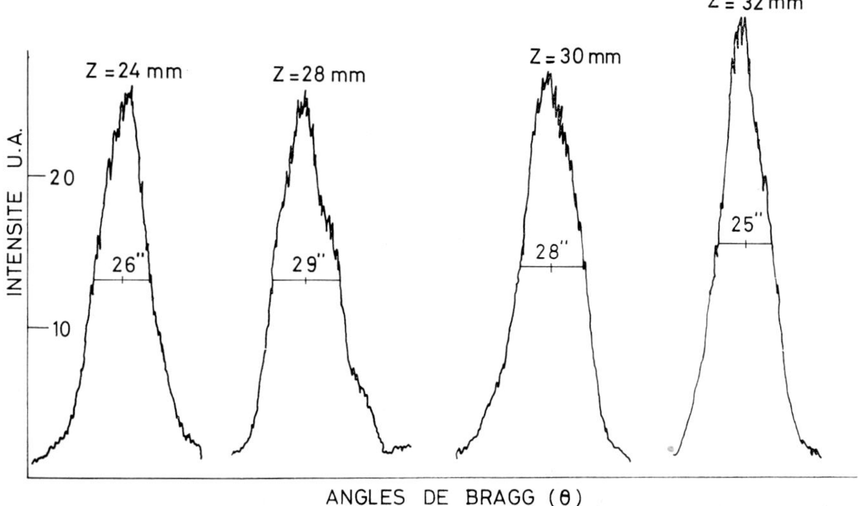

Fig. 7.   Profils de diffraction X obtenus en différentes positions axiales le long du CGM Cu–Ge en utilisant un spectromètre à double cristal: monochromateur Si (333) ($\theta_B = 19.9°$ pour la raie Mo Kα_1) et réflexion (400) sur Cu–Ge ($\theta_B = 23.1°$), surface illuminée 2×0.5 mm.

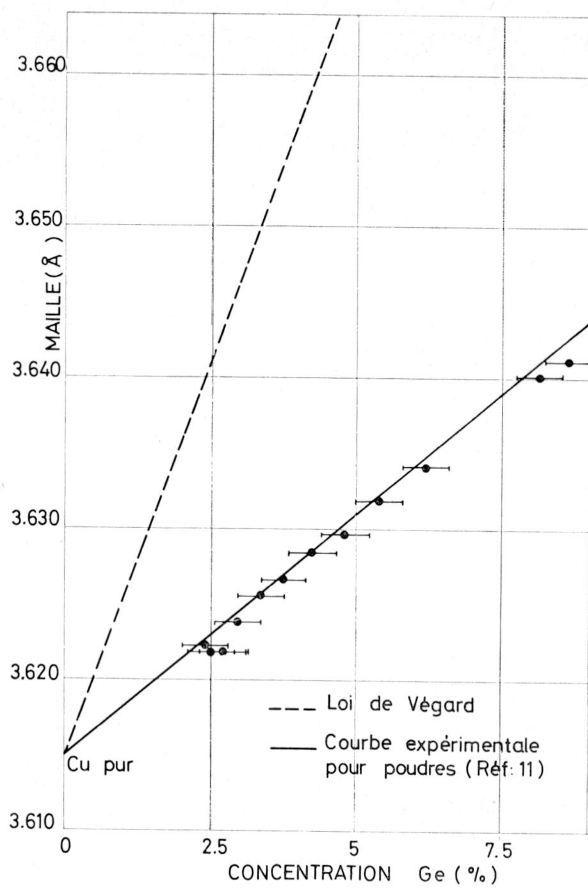

Fig. 8. Variations de paramètre cristallin en fonction des concentrations en Ge. (a) courbe obtenue dans des alliages polycristallins pulvérisés d'après Pearson[11]); (b) valeurs expérimentales obtenues sur le CGM monocristallin à partir de la méthode du spectromètre à double cristal; (c) droite correspondant à la loi de Vegard ($a_{Cu} = 3.61$ Å; $a_{Ge} = 5.65$ Å).

tion de rayons γ (faisceau de largeur 0.2 mm, normal à l'axe du cristal)[9]). La largeur des profils de diffraction varie entre 35″ (fig. 5) et 120″. La fig. 6 montre le profil de variation de maille obtenue par clichés Debye–Scherrer le long de l'axe: on remarque la correspondance avec le profil de concentration (fig. 4). La variation relative totale de maille est $\simeq 1\%$. Le spectromètre à double cristal a donné des profils de diffraction variant le long de l'axe du cristal depuis $\simeq 19″$ jusqu'à $\simeq 37″$ (fig. 7). Des valeurs précises de paramètre cristallin obtenues par la méthode de Bond[10]) ont été corrélées avec la concentration (fig. 4), reportées sur la fig. 8 et comparées à la courbe expérimentale obtenue sur les poudres[11]) et à la loi de Vegard.

7. Conclusion

Les résultats obtenus montrent qu'il est possible d'obtenir des monocristaux à gradient de maille possédant les caractéristiques neutroniques et cristallines nécessaires à leur utilisation en optique neutronique. L'étude sur Cu–Ge se poursuit en vue d'obtenir des cristaux de plus grandes dimensions.

D'autre part, l'intérêt des CGM ne se limite pas aux utilisations en optique neutronique. Toutes les branches de la physique des solides sont potentiellement intéressées depuis l'optique (variation d'indice) jusqu'a l'électronique (variation de largeur de bande interdite).

Enfin, en ce qui concerne les techniques d'élaboration, elles permettent une extension intéressante des techniques de croissance traditionnelles. En effet la technique utilisée pour Cu–Ge permet dans le tirage de monocristaux de solutions solides, de conserver, maigré l'influence de la ségrégation, les conditions d'un bain infini.

Remerciements

Nous remercions vivement Monsieur le Professeur H. Maier-Leibnitz qui a proposé ce sujet de recherche, MM. F. Forrat et B. Jacrot pour leurs intéressantes discussions, MM. A. Escoffier et R. Hustache pour leur efficace collaboration technique, MM. J. P. Morlevat et J. Roussignol pour les mesures à la microsonde électronique, et M. J. Schneider pour avoir mis à notre disposition son diffractomètre γ.

Bibliographie

1) H. Maier-Leibnitz, Ann. Acad. Acad. Sci. Fennicae, Phys. [VI] (1967) 267.
2) F. Rustichelli, Nucl. Instr. Methods **83** (1970) 124.
3) D. Marx, à paraître.
4) H. Maier-Leibnitz, communication privée.
5) Y. Marfaing, G. Cohen-Solal et T. Bailly, dans: *Crystal Growth*, Ed. H. S. Peiser (Pergamon, Oxford, 1967) 549.
6) F. Rustichelli, Report EUR 4414 i (1970).
7) J. Burgeat, J. Primot et F. Rustichelli, Z. Physik **238** (1970) 140.
8) A. Freund, F. Rustichelli et F. Vanoni, à paraître.
9) A. Freund et J. Schneider, J. Crystal Growth **13/14** (1972) 247.
10) W. L. Bond, Acta. Cryst. **13** (1960) 814.
11) W. B. Pearson, *A Handbook of Lattice Spacings and Structures of Metals and Alloys* (Pergamon, Oxford, 1958) pp. 578, 598.

Section XV

Eutectics, dendrites, inclusions

G. LESOULT
R. S. FIDLER
M. N. CROKER
R. W. SMITH
M. DURAND-CHARRE
F. DURAND
P. COMBRADE
M. TURPIN
J.-C. HUBERT
W. KURZ
B. LUX
A. T. CHAPMAN
R. J. GERDES

J. C. WILSON
G. W. CLARK
H. M. LIAW
J. W. FAUST, Jr.
J. CISSÉ
G. F. BOLLING
H. W. KERR
A. A. TZAVARAS
J. F. WALLACE
W. R. WILCOX
H. E. CLINE
T. R. ANTHONY

Journal of Crystal Growth **13/14** (1972) 733–738 © *North-Holland Publishing Co.*

733

THERMODYNAMIQUE DE LA SOLIDIFICATION EUTECTIQUE

G. LESOULT

Centre des Matériaux de l'Ecole des Mines de Paris, SNECMA, 91–Corbeil, France

The general evolution criterion derived by Glansdorff and Prigogine is applied to determine the steady state of controlled solidification.

Although difficulties with the expression of boundary conditions in the solid have not yet been solved rigorously, it is proposed to minimize $\frac{1}{2}\mathscr{P}+(w^0/T^0)\,G$ to find the stable steady state. $\mathscr{P}$ is the rate of entropy production, G the Gibbs free energy of the solid, w^0 the rate of solidification and T^0 the lowest temperature of the system. Only the rate w^0 as well as the upper and the lower temperatures of the system can be experimentally fixed.

This result expresses the competition which exists between the internal production of entropy and the storage of free energy in the solid as non-equilibrium structural components.

Coupled with approximate solutions of the chemical and thermal diffusion problems the new condition, when applied to the eutectic solidification problem, leads to relationships between λ and w^0 and between ΔT and w^0. Predicted values of $\mathscr{D}$ and $\gamma^{\alpha}/^{\beta}$ seem physically as meaningful as those obtained from other models. It is also shown that, for a given growth rate, the entropy production rate admits neither a minimum, nor a maximum in λ.

1. Introduction

Notre compréhension de la solidification diphasée a considérablement progressé depuis une dizaine d'années. Les approches théoriques actuelles ont le mérite de conduire à des lois en accord qualitatif avec l'expérience. Le seul point obscur de ces travaux est bien connu des métallurgistes: il existe mathématiquement une infinité de solutions stationnaires au problème de la croissance eutectique à vitesse imposée w et pourtant, expérimentalement, on n'observe le plus souvent qu'une valeur de la distance interlamellaire λ pour une valeur de w. La morphologie eutectique est donc contrôlée par des conditions plus restrictives que les conditions de stationnarité. L'objet principal de notre travail est de chercher si la thermodynamique peut nous guider dans le choix de ces conditions.

2. Les conditions d'optimisation

Nous pouvons regrouper les conditions utilisées dans le traitement des transformations eutectoïde ou eutectique en deux classes.

Depuis les travaux de Zener[1], on considère en général que l'énergie libre mobilisable pendant la réaction, ΔG_0, va se retrouver en partie sous forme d'énergie emmagasinée dans le produit de la réaction: énergie de volume quand la ségrégation est incomplète, ΔG_s, et énergie de surface des interfaces entre phases $(2\gamma/\lambda)V_{\mathrm{m}}$,

λ étant la distance caractéristique de la structure du solide supposée parfaitement périodique. Le complément ΔG_d aura été dissipé principalement dans des processus diffusionnels*:

$$\Delta G_0 = \Delta G_d + \Delta G_s + \frac{2\gamma}{\lambda}\,V_{\mathrm{m}}\,. \tag{1}$$

Suivant qu'on maintient constant l'écart à l'équilibre ΔG_0, comme lors de la transformation eutectoïde isotherme, ou la vitesse de transformation, comme lors de la solidification eutectique contrôlée, on choisit d'optimiser un des termes cités plutôt que l'autre. Pour la perlite, Zener[1] optimise la quantité $(\Delta G_d/\lambda)$ et Cahn[2] la quantité ΔG_d. Pour l'eutectique, la condition de Tiller[3] porte en fait sur ΔG_0.

Kirkaldy[4] puis Lesoult et Turpin[5] ont préféré s'inspirer des travaux de De Groot[6] et de Glansdorff et Prigogine[7,8] où il est établi que dans quelques cas précis un état stationnaire stable est caractérisé par un minimum de la vitesse de production d'entropie. En pratique, les conditions de validité des théorèmes invoqués ne sont pas remplies lors d'une solidification controlée.

Le présent travail est un nouvel essai dans le même sens: pour prévoir les caractéristiques du régime permanent le plus probable, nous avons cherché à exprimer que celui-ci était stable par rapport à des petites

* V. Liste des symboles.

XV – 1

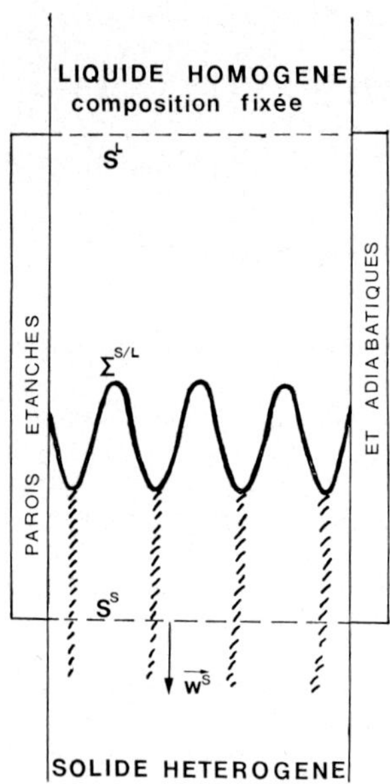

Fig. 1. Conditions aux limites: Sur la surface S^L: la composition du liquide et la température T^L. Sur la surface S^S: la vitesse w^S et la température T^S.

fluctuations de λ sans nous soucier du détail de la réalisation de ces fluctuations. Il est évident qu'une telle recherche est inutile si les mécanismes permettant de modifier λ en cours de croissance sont parfaitement bloqués: dans ce cas λ ne dépend uniquement que des conditions initiales de germination et non plus des caractéristiques du régime permanent établi.

3. Le critère d'évolution local

Nous présentons ici les étapes de notre raisonnement sur l'exemple simplifié de la solidification monophasée à plusieurs constituants. Il est raisonnable de penser que les frontières entre phases décrites parfois comme des surfaces mathématiques sont en réalité des couches de matière d'épaisseur finie bien que très petite au passage desquelles toutes les grandeurs physiques restent des fonctions continues et dérivables des coordonnées d'espace et de temps[9]). Nous pouvons donc admettre que les équations mécaniques et thermodynamiques locales connues en volume sont aussi valables dans ces zones de transition que dans les phases adjacentes et que le volume représenté sur la figure 1 enferme un milieu continu unique.

Dans le milieu continu unique ainsi défini, nous pouvons suivre la méthode proposée par Glansdorff et Prigogine[8]) pour exprimer la stabilité d'un système

ouvert. Le point de départ est la condition de stabilité thermodynamique d'un petit élément de volume déduite simplement du second principe[8]); quand les effets mécaniques peuvent être négligés, elle s'écrit:

$$\varphi = - \left\{ \sum_k \sum_i \frac{\partial(\mu_i^k/T)}{\partial t} \frac{\partial n_i^k}{\partial t} - \frac{\partial(1/T)}{\partial t} \frac{\partial(nu)}{\partial t} \right\} \leq 0. \quad (2)$$

En utilisant les équations de conservation des atomes et de l'énergie pour réécrire la forme φ, nous pouvons la présenter comme la somme d'un terme de flux et d'un terme de source lié à l'évolution interne du système.

4. Le critère d'évolution global

Composons les contributions des volumes élémentaires du milieu continu unique avec ses zones de transition:

$$\int_D \varphi \, d\omega = \Phi_{\text{flux}} + \Phi_{\text{source}} \leq 0, \quad (3)$$

$$\Phi_{\text{flux}} =$$

$$= - \int_S \left\{ \sum_k \sum_i \frac{\partial(\mu_i^k/T)}{\partial t} n_i^k w_i^k - \frac{\partial(1/T)}{\partial t} (\boldsymbol{\Psi} + nuw) \right\} \boldsymbol{n} \, dA,$$

$$\Phi_{\text{source}} = \int_D \left\{ - \sum_k \sum_i \text{grad} \left[\frac{\partial(\mu_i^k/T)}{\partial t} \right] \cdot \boldsymbol{J}_i^k + \right.$$

$$+ \text{grad} \left[\frac{\partial(1/T)}{\partial t} \right] \cdot \boldsymbol{\Psi} \right\} d\omega +$$

$$+ \int_{\Sigma^{L/s}} \sum_i \frac{\partial[(\mu_i^L - \mu_i^S)/T]}{\partial t} \chi_i^{L \to S} \, dA^{L/S}.$$

Nous reconnaissons dans le terme de source, Φ_{source}, la variation par unité de temps de la vitesse de production d'entropie due à la variation des forces thermodynamiques, notée $\partial_x \mathscr{P}/\partial t$ par Glansdorff et Prigogine[8]). Pour étudier les conditions d'intégrabilité de cette expression, nous devons y introduire les lois cinétiques usuelles. Quand le système est voisin d'un état stationnaire indicé par "0", il existe un potentiel local $\Phi(T^0, T, \mu_i^0, \mu_i)$, fonction du champ de température et des champs de potentiels chimiques dans l'état stationnaire et dans un état perturbé voisin. Si les conditions aux limites sur les surfaces extérieures étaient stationnaires, ce serait le minimum de ce potentiel qui caractériserait les régimes permanents avec les conditions subsidiaires:

$$T = T^0, \quad \mu_i = \mu_i^0. \tag{4}$$

Nous avons déjà indiqué que les potentiels chimiques ne pouvaient pas être fixés dans le solide lors d'un changement de phases dirigé. Le terme de flux va nous permettre de résoudre cette difficulté.

La ségrégation chimique lors d'une solidification est localisée près du front aussi admettrons nous que les flux de diffusion longitudinaux sont nuls sur les surfaces extérieures au domaine D. Par ailleurs, nous pouvons fixer la température dans les sections transversales d'entrée et de sortie, respectivement S^L et S^S, ainsi que la vitesse barycentrique molaire au niveau de S^S (fig. 1). Les frontières longitudinales du système étant pratiquement étanches et adiabatiques, il vient:

$$\frac{\partial(1/T)}{\partial t}(\boldsymbol{\Psi} + nu\boldsymbol{w})\boldsymbol{n} = 0 \quad \text{sur S}. \tag{5}$$

Si les volumes molaires partiels dans le solide sont égaux quels que soient les constituants, si la vitesse et la température sont constantes sur S^L d'une part et S^S d'autre part, le liquide initial étant homogène, le terme de flux se réduit à:

$$-\frac{\boldsymbol{n}^S \cdot \boldsymbol{w}^S}{T^S} \int_S \left. \sum_i \frac{\partial \mu_i^S}{\partial t} \right)_{p,T} n_i^S \, \mathrm{d}A^S. \tag{6}$$

En tentant d'exprimer ce terme de flux en fonction des variations de composition locale dans le voisinage d'un état stationnaire, nous nous heurtons provisoirement à une difficulté. Si dans l'expression (6) nous considérions les variations $\partial \mu_i^S/\partial t)_{p,T}$ au point $n = \{n_i\}$ plutôt qu'au point $n^0 = \{n_i^0\}$, nous la trouverions nulle d'après la relation de Gibbs–Duhem, le critère d'évolution se réduirait alors à:

$$\Phi_{\text{source}} = \frac{\partial \Phi}{\partial t} \leq 0. \tag{7}$$

Nous montrerons plus loin que l'application de ce résultat au cas de la solidification eutectique conduit à des conclusions en désaccord avec l'expérience. Cet échec peut provenir de l'inadéquation de notre modèle théorique à la description du phénomène physique. Par contre, si nous considérons les variations $\partial \mu_i^S/\partial t)_{p\,T}$ au point $n^0 = \{n_i^0\}$ pour calculer l'expression (6) nous trouvons:

$$\left. \sum_i n_i \frac{\partial \mu_i}{\partial t} \right)_{n=n} = \frac{1}{2} \sum_i \sum_j \frac{\partial^2 g}{\partial n_i \partial n_j} \frac{\partial}{\partial t}(\delta n_i \, \delta n_j). \tag{8}$$

Nous adopterons ici la dernière solution en indiquant que nous continuons à travailler ce point pour justifier notre choix. En supposant que la redistribution d'atomes se fait dans une couche parallèle à la surface de sortie et en intégrant sur cette surface, nous obtenons:

$$\int_S \left. \sum_i n_i^S \frac{\partial \mu_i^S}{\partial t} \right)_{n=n^0} \mathrm{d}A^S = \frac{\partial G^S}{\partial t}(n^0, n, T^S). \tag{9}$$

L'expression (9) représente la variation d'énergie libre par unité de volume de solide produit maintenu à la température de sortie quand le profil de concentration diffère légèrement du profil stationnaire, la composition moyenne étant maintenue constante.

Avec ces notations, le critère d'évolution global s'écrit dans le voisinage d'un état stationnaire:

$$\frac{\partial}{\partial t}\left\{ \Phi(T^0, T, \mu^0, \mu) + \frac{w^S}{T^S} G(n^0, n, T^S) \right\} \leq 0. \tag{10}$$

Les conséquences pratiques de ce critère nous encouragent à persévérer dans cette voie. Nous pensons que l'expression proposée ici dans un cas particulier est assez générale: pendant l'évolution d'un système macroscopique construit pour réaliser un changement de phases dirigé, la quantité $\mathrm{d}(\Phi + (w^S/T^S)G)$ est négative et s'annule quand le système atteint un régime permanent compatible avec les conditions aux limites.

5. La solidification eutectique

Nous avons insisté sur le fait que le résultat précédent n'était valable qu'au voisinage d'un état stationnaire supposé connu. Imaginons cependant qu'à propos d'un problème particulier nous sachions calculer les quantités Φ et G pour un ensemble d'états stationnaires fictifs dépendant d'un ou plusieurs paramètres continus. Les valeurs des paramètres correspondant à un minimum de la quantité $(\Phi + (w^S/T^S)G)$ caractérisent des régimes permanents stables par rapport aux fluctuations des paramètres en question. Par contre, nous ne savons pas si nous avons obtenu tous les régimes permanents stables par cette méthode.

L'étude de la solidification eutectique se prête bien à ce type d'approche. Le calcul du champ de concentration dans le liquide lors d'une solidification eutectique en régime stationnaire à vitesse fixée a été fait dans différentes approximations[10–13]. Nous avons déjà dit qu'il existe une infinité de solutions paramétrée par λ. Si expérimentalement on n'observe qu'une valeur de λ

pour une vitesse de croissance, c'est que seule une de ces solutions est stable par rapport à certaines fluctuations.

L'étude du potentiel Φ est très simplifiée si l'on admet d'une part que l'équilibre thermodynamique est établi le long du front de solidification, d'autre part que les transferts de masse et de chaleur sont découplés. Il vient:

$$\Phi = \tfrac{1}{2}(\mathscr{P}_{\mathrm{d}}+\mathscr{P}_{\mathrm{t}}). \tag{11}$$

Lesoult et Turpin[5]) ont trouvé pour première approximation de $\mathscr{P}_{\mathrm{d}}$, production d'entropie associée à la diffusion chimique dans le liquide:

$$\mathscr{P}_{\mathrm{d}} = \frac{R\xi}{(C_{\mathrm{E}}-1C_{\mathrm{E}})}\,\frac{(\Delta C_{\mathrm{E}})^2 P}{V_{\mathrm{m}}\mathscr{D}}\,\lambda w^2 A^{\mathrm{S}}. \tag{12}$$

Nous trouvons par ailleurs que la contribution à $\mathscr{P}_{\mathrm{t}}$ dépendant de la géométrie du solide, $\mathscr{P}_{\mathrm{t}}(\lambda)$, s'écrit:

$$\mathscr{P}_{\mathrm{t}}(\lambda) = w\left\{\frac{T_{\mathrm{E}}-T^{\mathrm{S}}}{T_{\mathrm{E}}T^{\mathrm{S}}}\,\overline{\Delta H} - \frac{\Delta S_{\mathrm{E}}}{V_{\mathrm{m}}}\frac{1}{T_{\mathrm{E}}}\,\overline{\Delta T}\right\} A^{\mathrm{S}}, \tag{13a}$$

où

$$\overline{\Delta H} = -\frac{2h^{\alpha/\beta}(T_{\mathrm{E}})}{\lambda} < 0, \tag{13b}$$

et

$$\overline{\Delta T} = (m^{\alpha}+m^{\beta})\frac{\Delta C_{\mathrm{E}}P}{\mathscr{D}}\,\lambda w + \frac{2}{\lambda}\,\Lambda, \tag{13c}$$

avec

$$\Lambda = -\frac{V^{\alpha}}{\Delta S^{\alpha}}\gamma^{L/\alpha}(T_{\mathrm{E}})\sin\theta^{L/\alpha}$$

$$+ \frac{V^{\beta}}{\Delta S^{\beta}}\gamma^{L/\beta}(T_{\mathrm{E}})\sin\theta^{L/\beta} > 0.$$

A ce point des calculs, il est très important de remarquer que le potentiel Φ est une fonction monotone croissante de λ (fig. 2). Il n'existe pas d'état tel que $\partial\Phi/\partial\lambda = 0$. Nous avons montré plus haut comment ce résultat troublant nous a incité à rechercher les états tels que:

$$\frac{\partial}{\partial\lambda}\left(\Phi + \frac{w^{\mathrm{S}}}{T^{\mathrm{S}}}G\right) = 0. \tag{14}$$

L'analogue de la fonction $G(n^0, n)$ pour un solide eutectique est, en première approximation, égal à:

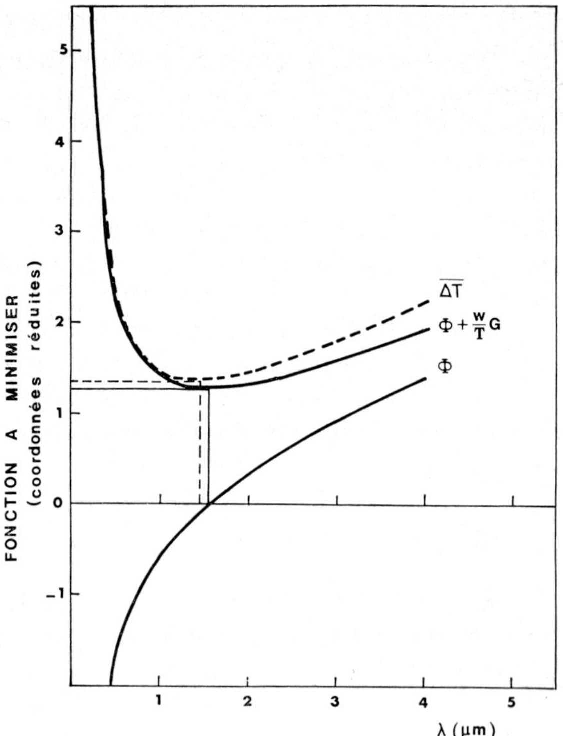

Fig. 2. Comparaison de trois critères d'optimisation sur l'exemple du système Pb–Sn. $\mathscr{D} = 1.8\times10^{-5}$ cm²/s, $\gamma = 47$ erg/cm², $w = 10$ µm/s.

$$G(\lambda) = \frac{2\gamma^{\alpha/\beta}(T_{\mathrm{E}})}{\lambda}\,A^{\mathrm{S}}.$$

Les calculs ayant été menés pour une diffusion chimique infiniment lente dans le solide, nous pouvons choisir la surface de sortie telle que $T_{\mathrm{E}}-T^{\mathrm{S}} \ll T_{\mathrm{E}}$; il vient alors:

$$\left(\Phi + \frac{w^{\mathrm{S}}}{T^{\mathrm{S}}}G(T^{\mathrm{S}})\right)_{\lambda} = \tfrac{1}{2}\left[\frac{R\xi\,\Delta C_{\mathrm{E}}}{C_{\mathrm{E}}(1-C_{\mathrm{E}})} - \frac{\Delta S_{\mathrm{E}}(m^{\alpha}+m^{\beta})}{T_{\mathrm{E}}}\right]\times$$

$$\times \frac{\Delta C_{\mathrm{E}}\,P}{V_{\mathrm{m}}}\,\lambda w^2 A^{\mathrm{S}} + \left[\frac{2\gamma^{\alpha/\beta}(T_{\mathrm{E}})}{T_{\mathrm{E}}} - \frac{\Delta S_{\mathrm{E}}}{V_{\mathrm{m}}}\frac{\Lambda}{T_{\mathrm{E}}}\right]\frac{w}{\lambda}\,A^{\mathrm{S}}. \tag{16}$$

Les deux quantités entre crochets étant toujours positives, l'expression (16) présente un minimum en λ. Par analogie avec le résultat avancé pour la solidification monophasée, nous prévoyons que l'état correspondant sera stationnaire et stable. La condition d'optimisation conduit à des relations $\lambda = f(w)$ et $\Delta T = g(w)$ analogues à celles obtenues par d'autres auteurs.

Nous avons rassemblé dans le tableau 1 les évalua-

TABLEAU 1

Tableau des évaluations de $\mathscr{D}$ et de $\gamma^{\alpha/\beta}$

Expériences		Modèles		
		Jackson et Hunt[11]	Tiller[3], modifié[5]	Présent travail
Hunt et Chilton[14]	$\mathscr{D}$ (cm²/s)	0.54×10^{-5}	0.88×10^{-5}	0.83×10^{-5}
	$\gamma^{\alpha/\beta}$ /erg/cm²)	90	157	150
Moore et Elliott[15]	$\mathscr{D}$ (cm²/s)	1.47×10^{-5}	2.4×10^{-5}	2.25×10^{-5}
	$\gamma^{\alpha/\beta}$ (erg/cm²)	24	41	39
Moore et Elliott[15],	$\mathscr{D}$ (cm²/s)	1.18×10^{-5}	1.9×10^{-5}	1.8×10^{-5}
Ha-Quac-Bao[16]	$\gamma^{\alpha/\beta}$ (erg/cm²)	29	51	49

tions de $\mathscr{D}$ et de $\gamma^{\alpha/\beta}$ (T_E) du système Pb–Sn faites suivant divers modèles. Les valeurs calculées par Jackson et Hunt[11] sont plus faibles que les nôtres. Nous avons aussi comparé sur la figure 2 les conditions d'optimisation sur $(\Phi + (w^S/T^S)G)$ et sur $\overline{\Delta T}$. Pour cette comparaison, nous avons utilisé la formule (13c) car elle seule est fondée thermodynamiquement[5] au contraire des formules proposées par Tiller[3] ou Jackson et Hunt[11]. L'accord entre les deux prédictions est remarquable étant donné la dispersion des valeurs expérimentales de λ et de $\overline{\Delta T}$ disponibles actuellement.

Des travaux sont en cours pour établir l'influence de la forme microscopique du front de solidification sur la forme des relations entre λ, w et $\overline{\Delta T}$.

6. Conclusions

(1) Sur l'exemple simple de la solidification monophasée, nous avons rappelé comment exprimer la stabilité locale d'un élément de volume fixe dans l'espace d'une part et le sens de l'évolution du système ouvert dans lequel est réalisée une transformation de phase dirigée d'autre part. La solution pratique que nous proposons ici reste à établir plus rigoureusement: lors de l'évolution d'un système macroscopique construit pour réaliser un changement de phases dirigé, la quantité $d(\Phi + (w^S/T^S)G)$ est négative et s'annule quand le système atteint un régime permanent compatible avec les conditions aux limites.

(2) La forme de la fonction que nous minimisons traduit la compétition qui s'instaure entre la production interne d'entropie et le stockage d'énergie dans les structures hors d'équilibre du solide fabriqué.

(3) Nous avons montré que la vitesse de production d'entropie n'admettait ni maximum ni minimum en λ dans le cas de la solidification eutectique à vitesse fixée. Le critère que nous proposons ici conduit à une loi analogue à celle trouvée par Tiller suivant une démarche plus intuitive.

(4) Ces résultats nous encouragent à appliquer notre méthode à l'étude de la taille des structures cellulaires et dendritiques en fonction des paramètres d'élaboration.

Liste des symboles

A^S Surface de la section transversale du solide

C_E Composition du liquide eutectique

ΔC_E Largeur du palier eutectique sur le diagramme d'équilibre

$\mathscr{D}$ Coefficient d'interdiffusion dans le liquide

$h^{\alpha/\beta}$ Enthalpie interfaciale par unité de surface

J_i^k Flux de diffusion du constituant (i, k)

m^α, m^β Pentes des liquidus comptées positives

n Nombre total d'atomes par unité de volume

n_i^k Nombre d'atomes du constituant i par unité de volume dans la phase k

$\boldsymbol{n}$ Normale intérieure à la surface frontière du domaine D

$$P = \sum_{n=1} \left(\frac{1}{n\pi}\right)^3 \sin^2\left(\frac{n\pi\lambda^\alpha}{\lambda^\alpha + \lambda^\beta}\right)$$

R Constante des gaz

ΔS_E Entropie de fusion de l'alliage eutectique

$\Delta S^\alpha, \Delta S^\beta$ Sont définis dans la réf. 5

$\partial/\partial t$ Dérivation locale par rapport au temps (repère fixe)

T Température absolue

T_E Température eutectique

u Energie interne par atome-gramme

V_m Volume molaire du solide

V^α, V^β Sont définis dans la réf. 5

w Vitesse de croissance imposée

$\boldsymbol{w}_i^k$ Vitesse du constituant (i, k)

XV – 1

w	Vitesse barycentrique molaire
γ	Energie libre de surface par unité de surface
$\gamma^{k/k'}$	Energie libre interfaciale de l'interface entre k et k'
$\theta^{L/\alpha}$, $\theta^{L/\beta}$	Sont définis dans la réf. 5
μ_i^k	Potentiel chimique du constituant i dans la phase k
ζ	Facteur thermodynamique dans le liquide
$\chi_i^{L \to S}$	Flux d'atomes du constituant i qui passent de l'état liquide à l'état solide par unité de temps et de surface
Ψ	Flux de chaleur

Bibliographie

1) C. Zener, Trans. AIME **167** (1946) 550.
2) J. W. Cahn, Acta Met. **7** (1959) 18.
3) W. A. Tiller, *Liquid Metals and Solidification* (Am. Soc. Metals, Cleveland, Ohio, 1958) p. 267.
4) J. S. Kirkaldy, Can. J. Phys. **37** (1959) 739.
5) G. Lesoult et M. Turpin, Mem. Sci. Rev. Met. **66** (1969) 619.
6) S. R. de Groot, *Thermodynamics of Irreversible Processes* (North-Holland, Amsterdam, 1952).
7) P. Glansdorff et I. Prigogine, Physica **20** (1954) 773.
8) P. Glansdorff et I. Prigogine, Physica **30** (1964) 351.
9) F. Fer, Thermodynamique des Systèmes Ouverts, ch. VI, dans: *Equations d'Interfaces* (Cours ENSMP, 1968).
10) W. H. Brandt, J. Appl. Phys. **16** (1945) 139.
11) K. A. Jackson et J. D. Hunt, Trans. AIME **236** (1966) 1129.
12) L. F. Donaghey et W. A. Tiller, Mater. Sci. Eng. **3** (1968) 231.
13) G. Lesoult, M. Colin et M. Turpin, Communication aux Journées d'Automne de la Société Française de Métallurgie (Paris, Sept. 1970).
14) J. D. Hunt et J. P. Chilton, J. Inst. Metals **92** (1963–64) 21.
15) A. Moore et R. Elliott, I.S.I. Publication **110** (Brighton Solidification Conf., 1968) p. 110.
16) Ha-Quac-Bao, Thèse d'Ingénieur Docteur, Grenoble (Oct. 1970).

Journal of Crystal Growth **13/14** (1972) 739–746 © *North-Holland Publishing Co.*

THE THERMODYNAMICS AND MORPHOLOGIES OF EUTECTICS CONTAINING COMPOUND PHASES

R. S. FIDLER, M. N. CROKER and R. W. SMITH

Department of Metallurgical Engineering, Queen's University, Kingston, Ontario, Canada

Following the successful correlation of eutectic morphologies with the entropies of solution of the constituent phases in simple binary eutectics, the correlation has been extended to include a further 14 systems. In each of these systems, one of the phases is a compound for which the calculation of the entropy of solution must be adapted from the treatment of simple eutectic phases.

Based upon the concepts of Hunt and Jackson's eutectic classification, the structures of all the eutectics studied have been shown to correlate well with the calculated entropies of solution. For the more complicated crystal structures of the compound phases, it is not possible to calculate roughness parameters which have the physical significance of those calculated for close packed structures. Thus, although the distinction between non-faceting and faceting behaviour cannot be defined exactly, the structures can be predictably classified by listing in the order of maximum entropies of solution per atom. From such a table the borderline between faceting and non-faceting behaviour is expected at $\Delta S_\alpha \simeq 5.5$ cal/g-atom °K.

1. Introduction

The Hunt and Jackson[1]) classification of eutectic structures has been shown[2]) to apply well to simple binary eutectics if the roughness parameters of the constituent eutectic phases are calculated at the eutectic temperature and composition. For a metal with a simple crystal structure, a roughness parameter, α, can be calculated according to Jackson's theory[3,4]) of rough and smooth interfaces. Following the suggestion by Kerr and Winegard[5]) the roughness parameters of eutectic phases formed under conditions corresponding to solution growth were shown[2,6]) to vary from the roughness parameters of the pure components at their melting points. Using α-values appropriate to the solidification conditions, the Hunt and Jackson classification[1]) predicts:

(1) for eutectics in which both phases have $\alpha \leq 2$, regular rod or lamellar structures; this is the non-faceted/non-faceted group;

(2) for eutectics in which one phase has $\alpha > 2$ and the other phase has $\alpha \leq 2$, the faceted/non-faceted group, the structure will be complex-regular, broken lamellar or irregular;

(3) for eutectics in which both phases have $\alpha > 2$, the faceted/faceted group, the structure will be irregular.

In this paper, the thermodynamic calculations are extended to treat eutectic systems in which one phase is a compound. Because the crystal structures of the compounds are complex, Jackson's simple models of solid-liquid interfaces are inadequate and roughness parameters therefore cease to provide satisfactory physical descriptions of interface structure. Consequently, it has been necessary to calculate the entropies of solution of the phases and then to correlate these with observed structures in an attempt to establish the range of values which characterize particular eutectic groups. As expected, on the basis of Jackson's simple theory[3,4]) and more recent general theories[7,8]) the entropy of solution per atom added has been found to govern the roughness of any solid–liquid interface during growth and thereby also to determine the structures of these eutectics.

2. Theory

Taylor, Fidler and Smith[2]) have previously described the calculation of the heat of solution of a solid binary phase in a liquid whose composition is given by the phase diagram as the tie-line liquidus composition. When a compound phase is considered, the calculation must be modified slightly to take into account (1) the heat of formation of the compound and (2) if the compound has a solid solubility range, the heat of formation of the solid solution, γ.

A solid solution, γ, of composition $x_B{}^*$, is conveniently thought of as a stoichiometric compound, A_pB_q,

---

* All symbols are defined in the Glossary at the end of the paper.

in which excess component A has mole fraction x'_A if the compound is considered as a single molecular species. x'_A is obtained from x_A. x_B by the equation:

$$x'_A = \frac{qx_A - px_B}{qx_A + (1-p)x_B}.$$

The heat of solution, L_γ, in liquid of composition x_l is the sum of the heat changes of the following reactions, described in the manner in which the calculation is carried out:

(1) The solid solution, γ, is divided into stoichiometric compound A_pB_q and excess of component A, with the absorption of the heat of solid solution. Assuming that within the solid solubility range of the compound, the partial heat of the stoichiometric compound is negligible, the heat of solid solution is $x'_A\overline{\Delta H_A}(\gamma)$. Writing $x = x_B$, the reaction can be written:

$$A_{(1-x)}B_x = x'_A\,A + (1-x'_A)\,A_pB_q,$$
$$\Delta H_1 = x'_A\,\overline{\Delta H_A}(\gamma).$$

$\overline{\Delta H_A}(\gamma)$ is calculated from the phase diagram solubility data as described previously[2]).

(2) The solid compound A_pB_q is divided into its pure solid components with the absorption of the heat formation:

$$A_pB_q = p\,A(s) + q\,B(s),$$
$$\Delta H_2 = (1-x'_A)\,\Delta H_f.$$

(3) The pure solid components are melted with the absorption of the heats of fusion:

$$\Delta H_3 = (1-x)L_A + xL_B.$$

(4) The pure liquid components are mixed into a large bath of liquid of the liquidus composition being

TABLE 1

Heats and entropies of solution calculated as functions of liquidus temperature and composition

System	Temp. (°C)	Liquidus composition (atomic fraction of component 2)		Solidus composition (atomic fraction of component 2		Heat of solution (kcal/mole)		Entropy of solution per atom (cal/g-atom °K)	
		Phase 1	Phase 2	Phase 1	Phase 2	Phase 1	Phase 2	Phase 1	Phase 2
Al–CuAl$_2$	548[3]	0.174	0.174	0.025	0.320	2.54	8.89	3.09	3.61
	562	0.160	0.200	0.022	0.322	2.54	9.09	3.05	3.63
	591[2]	0.101	0.328	0.015	0.328	2.56	9.53	2.96	3.68
	600	0.090		0.013		2.56		2.93	
	660[1]	0.000		0.000		2.57		2.75	
Bi–Au$_2$Bi	241[3]	0.189	0.189	0.000	0.333	2.61	3.85	5.08	2.50
	271[1]	0.000		0.000		2.60		4.82	
	300		0.264		0.333		3.90		1.30
Bi–Bi$_2$Mg$_3$	260[3]	0.043	0.043	0.000	0.600	2.59	37.11	4.86	13.93
	271[1]	0.000		0.000		2.60		4.78	
	300		0.060		0.600		37.15		12.97
	400		0.150		0.600		37.21		11.06
	500		0.300		0.600		37.25		9.64
	600		0.500		0.600		37.34		8.56
	700		0.560		0.600		37.48		7.70
	823[2]		0.600		0.600		37.46		6.84
Bi–Pb$_2$Bi	125[3]	0.437	0.437	0.005	0.580	2.32	2.95	5.82	2.47
	142	0.388	0.500	0.004	0.610	2.38	3.04	5.74	2.44
	184[2]	0.288	0.640	0.003	0.684	2.47	3.27	5.40	2.38
	250	0.100		0.001		2.58		4.91	
	271[1]	0.000		0.000		2.60		4.78	
Cd–CdSb	290[3]	0.070	0.070	0.000	0.500	1.53	6.61	2.72	5.87
	300	0.050	0.095	0.000	0.500	1.51	6.33	2.64	5.52
	321[1]	0.000		0.000		1.46		2.46	
	350		0.195		0.500		5.88		4.72
	400		0.306		0.500		5.89		4.38
	456[2]		0.500		0.500		6.12		4.20

XV – 2

TABLE 1 (continued)

System	Temp. (°C)	Liquidus composition (atomic fraction of component 2)		Solidus composition (atomic fraction of component 2)		Heat of solution (kcal/mole)		Entropy of solution per atom (cal/g-atom °K)	
		Phase 1	Phase 2	Phase 1	Phase 2	Phase 1	Phase 2	Phase 1	Phase 2
$Cu-Cu_2Mg$	722^3	0.219	0.219	0.070	0.317	2.01	6.52	2.02	2.18
	800	0.183	0.302	0.048	0.332	2.34	7.21	2.18	2.24
	819^2		0.333		0.333		7.27		2.22
	900	0.128		0.028		2.65		2.26	
	1000	0.065		0.013		2.88		2.26	
	1083^1	0.000		0.000		3.12		2.30	
$Mg-Mg_2Pb$	466^3	0.191	0.191	0.077	0.333	0.98	11.66	1.33	5.26
	500	0.170	0.220	0.063	0.333	1.18	11.88	1.53	5.12
	550^2	0.125	0.333	0.040	0.333	1.56	12.04	1.90	4.88
	600	0.070		0.020		1.90		2.18	
	650^1	0.000		0.000		2.14		2.32	
$Mg-Mg_2Sn$	561^3	0.105	0.105	0.033	0.333	1.63	15.70	1.95	6.27
	600	0.055	0.135	0.017	0.333	1.86	15.94	2.13	6.08
	650^1	0.000	0.170	0.000	0.333	2.14	15.71	2.32	5.67
	700		0.215		0.333		16.38		5.61
	778^2		0.333		0.333		16.58		5.26
$Pb-AuPb_2$	215^3	0.156	0.156	0.000	0.333	1.02	5.95	2.09	4.06
	233	0.131	0.200	0.000	0.333	1.05	5.99	2.07	3.95
	254^2	0.102	0.285	0.000	0.333	1.08	5.03	2.05	3.82
	300	0.038		0.000		1.12		1.96	
	327^1	0.000		0.000		1.14		1.90	
$Pb-Mg_2Pb$	253^3	0.157	0.157	0.059	0.667	1.06	11.06	2.02	7.01
	300		0.280		0.667		11.21		6.52
	327^1	0.000		0.000		1.14		1.90	
	400		0.430		0.667		11.57		5.73
	500		0.550		0.667		11.85		5.11
	550^2		0.667		0.667		12.04		4.88
$Sb-CdSb$	445^3	0.430	0.430	0.000	0.485	4.52	5.99	6.29	4.17
	456^2	0.408	0.500	0.000	0.500	4.53	6.12	6.21	4.20
	500	0.330		0.000		4.55		5.89	
	630.5^1	0.000		0.000		4.69		5.19	
$Sb-InSb$	500	0.317	0.317	0.000	0.500	4.47	8.66	5.79	5.60
	530	0.250	0.500	0.000	0.500	4.59	8.71	5.71	5.42
	556	0.200		0.000		4.63		5.58	
	597	0.100		0.000		4.67		5.37	
	630.5	0.000		0.000		4.69		5.19	
$Sb-ZnSb$	505^3	0.320	0.320	0.000	0.500	4.87	6.11	6.27	3.92
	525	0.269	0.398	0.000	0.500	4.83	6.07	6.05	3.80
	546^2	0.215	0.480	0.000	0.500	4.74	6.00	5.79	3.66
	575	0.140		0.000		4.70		5.55	
	600	0.080		0.000		4.71		5.40	
	630.5^1	0.000		0.000		4.69		5.19	
$Sn-Mg_2Sn$	203.5^3	0.096	0.096	0.000	0.667	1.67	14.69	3.50	10.28
	232^1	0.000	0.122	0.000	0.667	1.67	14.89	3.31	9.83
	318		0.200		0.667		14.94		8.42
	440		0.311		0.667		15.59		7.29
	666		0.517		0.667		16.31		5.79
	753		0.617		0.667		16.56		5.38
	770.5^2		0.667		0.667		16.58		5.26

Key to superscripts in Temperature column: 1 and 2: Melting points of components 1 and 2; 3: Eutectic temperature.

TABLE 2

Correlation of the entropies of solution of the constituent phases with the observed eutectic structures

System	C_E, atomic fraction of component 2	T_E (°C)	Roughness parameter α at C_E		Entropy of solution (cal/g-atom °K)		Observed eutectic morphology	Refs.
			Phase 1	Phase 2	Phase 1	Phase 2		
Cu–Cu$_2$Mg	0.219	722	0.78	–	2.02	2.18	rod	
Ag–Cu	0.399	779	0.71	0.90	2.80	3.56	lamellar	12
Al–CuAl$_2$	0.174	548	0.78	–	3.09	3.61	lamellar	13
Pb–AuPb$_2$	0.156	215	0.53	–	2.09	4.06	lamellar	
Cd–Zn	0.265	266	0.68	1.21	2.69	4.81	lamellar or rod	14
Bi–Au$_2$Bi	0.189	241	1.85	–	5.08	2.50	broken lamellar*	15
Mg–Mg$_2$Pb	0.191	466	.34	–	1.33	5.26	lamellar	
Cd–Pb	0.72	248	1.37	0.53	5.43	2.13	lamellar or rod	14, 16
Sb–InSb	0.317	500	1.95	–	5.79	5.60	broken lamellar	
Bi–Pb$_2$Bi	0.437	125	1.96	–	5.82	2.95	complex regular	17, 18
Cd–CdSb	0.07	290	.68	–	2.72	5.87	broken lamellar/irregular	
Bi–Sn	0.57	139	2.04	1.44	5.95	2.14	complex regular	17, 19
Bi–Cd	0.55	144	2.10	1.01	6.26	4.00	complex regular	19, 20
Mg–Mg$_2$Sn	0.105	561	0.49	–	1.95	6.27	complex regular	21
Sb–ZnSb	0.32	505	2.11	–	6.27	3.92	complex regular rod	
Sb–CdSb	0.189	241	2.12	–	6.29	4.21	broken lamellar	
Pb–Mg$_2$Pb	0.157	253	0.50	–	2.02	7.01	irregular	
Sn–Zn	0.15	198	1.23	1.91	3.67	7.60	broken lamellar	15, 17, 22
Pb–Sb	0.175	252	0.51	2.71	2.04	8.08	complex regular	12
Ag–Pb	0.953	304	2.23	0.49	8.87	1.96	broken lamellar	20
Ag–Bi	0.953	262	2.27	1.64	9.03	4.90	broken lamellar	6, 15
In–Zn	0.043	143.5	0.34	2.34	1.37	9.30	broken lamellar	23
Al–Ge	0.303	424	0.80	3.70	3.17	9.80	complex regular	24
Sn–Mg$_2$Sn	0.096	203.5	1.18	–	3.50	10.28	irregular	
Al–Sn	0.978	228.3	3.08	1.12	12.23	3.33	broken lamellar	19
Bi–Bi$_2$Mg$_3$	0.043	260	1.64	–	4.87	13.93	broken lamellar	

* Bi–Au$_2$Bi is strongly anomalous because here the faceting Au$_2$Bi phase[15] has $\Delta S_\gamma = 2.5$ cal/g-atom °K. In this case there is some doubt about the reliability of the thermodynamic data, which are scant.

considered, x_l. The heat change is given by:

$$\Delta H_4 = (1-x)\,\overline{\Delta H_A}(l) + x\,\overline{\Delta H_B}(l).$$

The heat of solution of the compound phase, L_γ is given by:

$$L_\gamma = \Delta H_1 + \Delta H_2 + \Delta H_3 + \Delta H_4.$$

The average entropy of solution per atom added to the compound phase is

$$\Delta S_\gamma = L_\gamma / T(p+q).$$

In all other respects, the calculation is the same as was described previously[2]. The required thermodynamic data were taken from Hansen et al.[9], Hultgren et al.[10], and Kubachewski et al.[11]) and, where possible the data used were corrected to the temperatures considered.

3. Results and discussion

Fourteen eutectic systems in which one phase is a compound have been examined. Al–CuAl$_2$, Bi–Au$_2$Bi, Bi–Bi$_2$Mg$_3$, Bi–Pb$_2$Bi, Cd–CdSb, Sb–CdSb, Cu$_2$Mg–Cu, Mg–Mg$_2$Pb, Pb–AuPb$_2$, Pb–Mg$_2$Pb, Mg–Mg$_2$Sn, Sn–Mg$_2$Sn, Sb–InSb and Sb–ZnSb. In table 1, the heats and entropies of solution of the two phases are tabulated as functions of liquidus temperature and composition. As described before[2]), we note that for a given phase considerable variation in the heats and entropies of solution occurs as the melt composition and temperature approach the eutectic point.

The roughness parameter, α, is the product of the entropy of solution term $(\Delta S_\alpha / R)$ and a crystallographic factor (n_1 / v). Thus, for eutectic systems in which each phase has a simple crystal structure, it is possible to calculate[2]) roughness parameters which give some physical significance to the predicted morphology of the solid–liquid interface. Unfortunately, however, the crystallographic factor cannot be defined easily for the crystal structure of a compound in which the two types

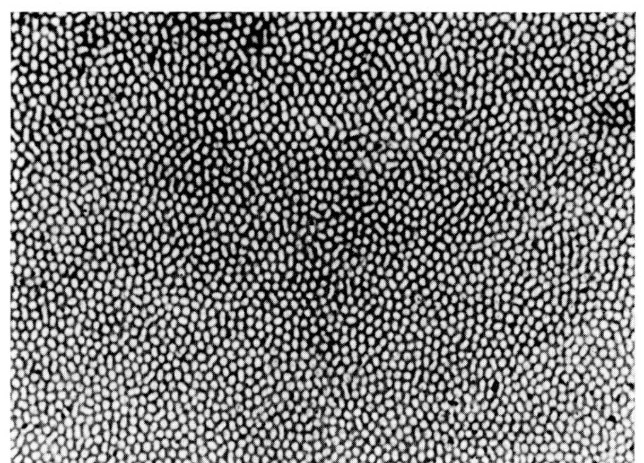

Fig. 1. Regular rod eutectic structure of Cu–Cu$_2$Mg. Transverse section. ×600.

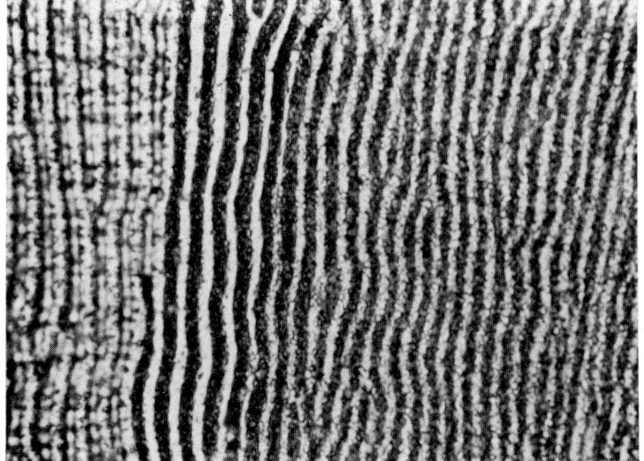

Fig. 2. Regular lamellar structure of unidirectionally solidified Mg–Mg$_2$Pb. Longitudinal section. ×520.

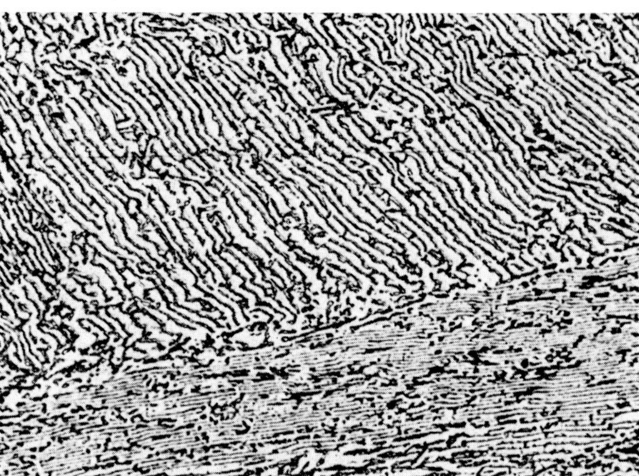

Fig. 3. Regular lamellar structure of unidirectionally solidified Pb–AuPb$_2$. Transverse section. ×400.

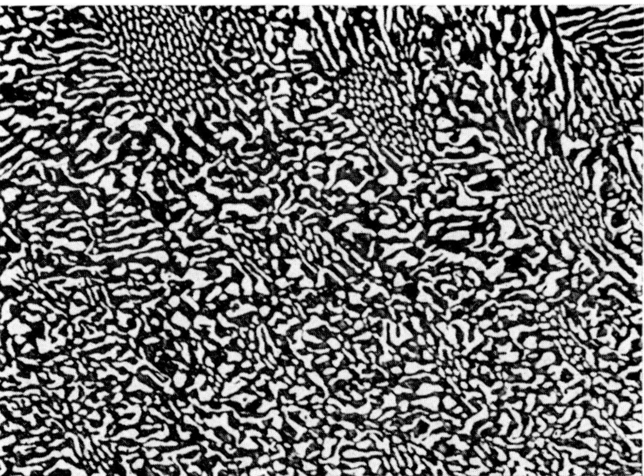

Fig. 4. Complex-regular array of rods in Sb–ZnSb. ×200.

of atoms have different coordination in the lattice. In the present results, therefore, roughness parameters do not appear for the compound phases.

Jackson's theories of solid–liquid interfaces[3,4,7,8] conclude that the entropy of fusion determines the growth forms of solid–liquid interfaces. Although it is unlikely that the transition from one type of growth to another is in reality as exactly defined as is implied by the change of α from ≤ 2 to > 2, it is desireable to estimate a value of the entropy of solution at which morphological changes are likely to become possible. Since eutectic structures are apparently sensitive to changes in the nature of the solid–liquid interface, the correlation of entropies of solution of the constituent phases with the eutectic structure provides a technique of estimating the transitional ranges.

In table 2, all the eutectics for which the entropies of solution have been calculated in this study and earlier[2]) at the eutectic temperature and melt composition, are tabulated in order of increasing entropy. The eutectic structures are listed for the purposes of correlation. Where the structures were not previously known, samples were prepared by weighing the pure components to the eutectic composition, melting and shaking under argon to facilitate mixing, and then slowly cooling to provide coarse grained boules which were sectioned and metallographically examined. In some cases, it was necessary to remelt the alloys and slowly solidify them unidirectionally to characterize the structure more accurately.

The structures which were not found in the literature are shown in figs. 1–10. Of these, the systems Cu–

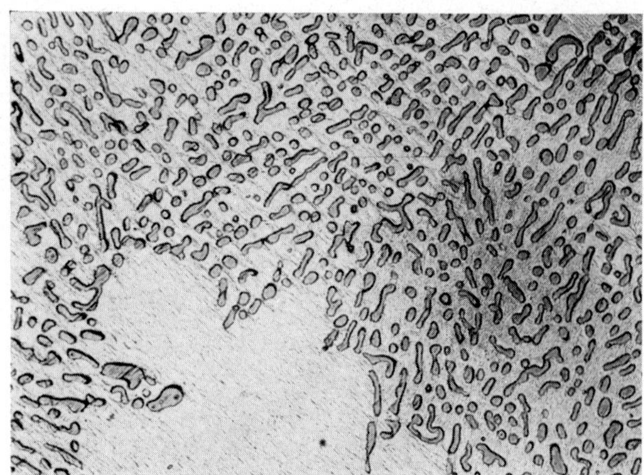

Fig. 5.   Broken lamellar structure of unidirectionally solidified Cd–CdSb. Transverse section. ×400.

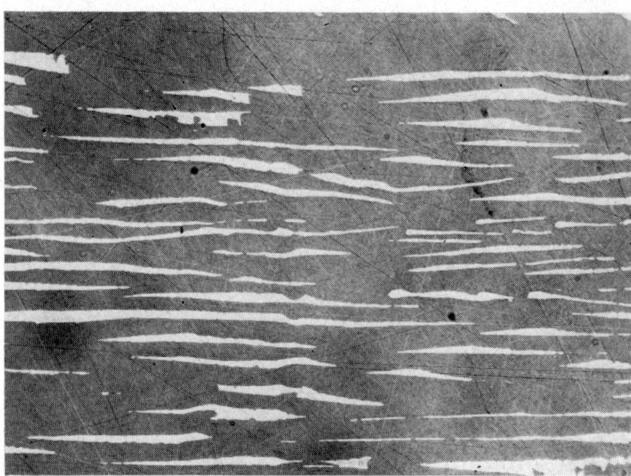

Fig. 6.   Broken lamellar structure of unidirectionally solidified Sb–CdSb. Longitudinal section. ×80.

Cu_2Mg, $Mg–Mg_2Pb$, $Pb–AuPb_2$ showed regular lamellar or rod structures typical of non-faceted/non-faceted eutectics. The Sb–ZnSb eutectic had a structure very similar to complex-regular Bi–Cd formed under the same growth conditions. $Sn–Mg_2Sn$ and $Pb–Mg_2Pb$ were acicular after the manner of Al–Si. These are observed forms of the faceted/non-faceted group of eutectics. Another morphology found in the faceted/non-

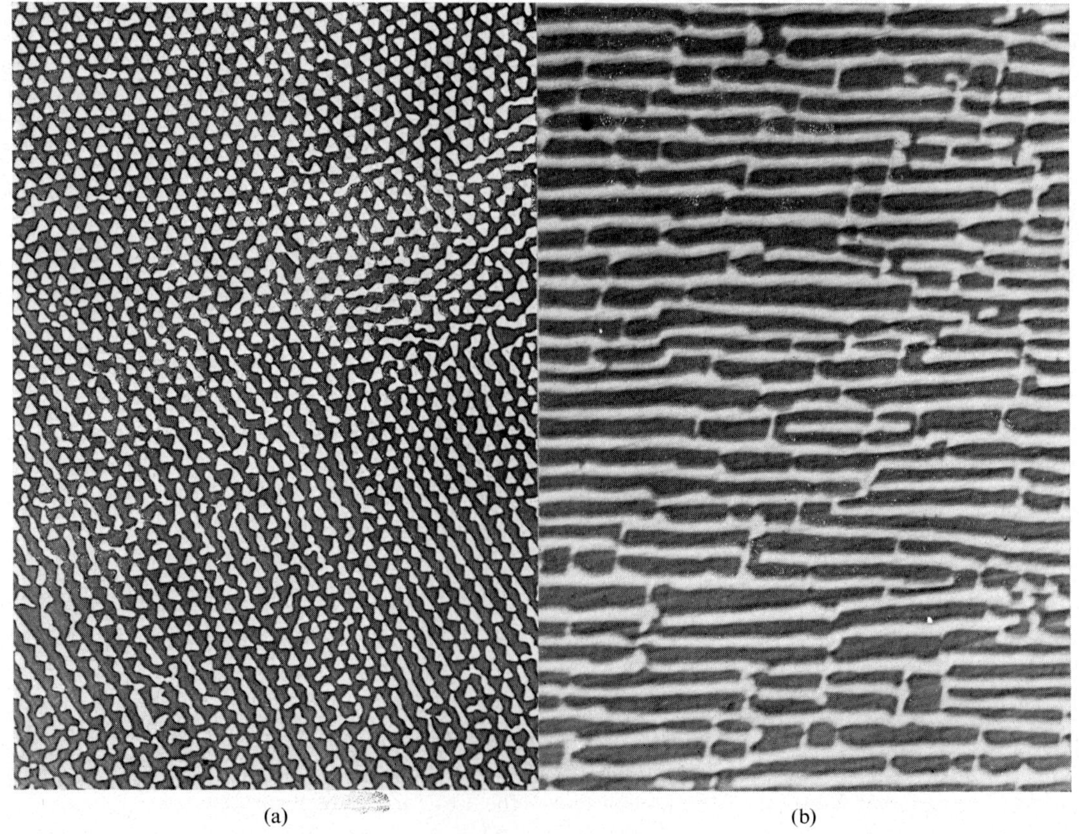

(a)   (b)

Fig. 7.   (a) Faceted rod structure and (b) broken lamellar structure of unidirectionally solidified Sb–InSb. Transverse section. (a) ×200, (b) ×310.

XV – 2

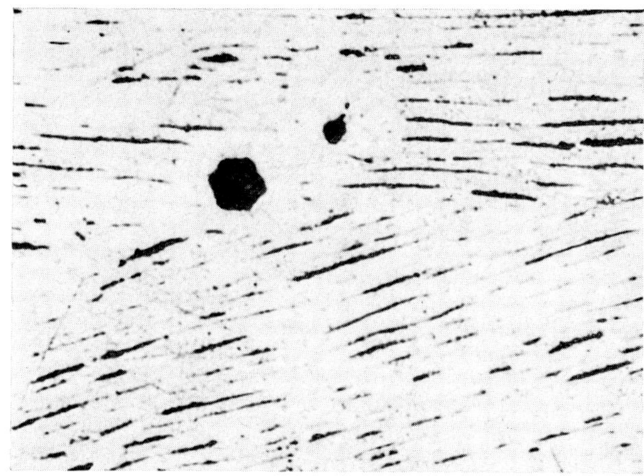

Fig. 8.   Broken lamellar structure of Bi–Bi$_2$Mg$_3$ eutectic with faceted pro-eutectic Bi$_2$Mg$_3$ particle. ×800.

Fig. 9.   Irregular structure of unidirectionally solidified Pb–Mg$_2$Pb. Longitudinal section. ×200.

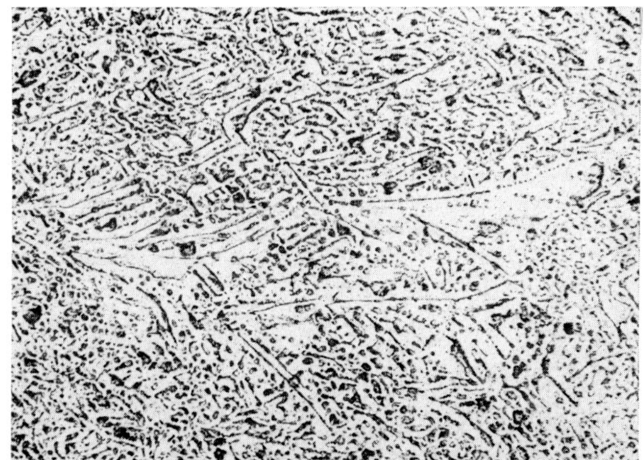

Fig. 10.   Irregular structure of unidirectionally solidified Sn–Mg$_2$Sn. Longitudinal section. ×140.

faceted group[2]), the broken lamellar structure, was observed in Cd–CdSb and Bi–Bi$_2$Mg$_3$. A variation of the broken lamellar structure is observed in Sb–CdSb and Sb–InSb in which the broken lamellar phase occupies a significant proportion of the volume. This structure is believed to be of the same type as that observed in Ni–22W by Cline et al.[25]). Indeed, the broken lamellar or plate-like structure in Sb–InSb can appear as a faceted rod form in the same way as Ni–22W[25]).

As has been observed previously[6,18]), where pro-eutectic phases appear in the microstructures of these eutectic systems, the shapes of the dendrites reflect the predicted faceted or non-faceted forms. Thus the appearance of angular pro-eutectic particles of Bi$_2$Mg$_3$ supports the relation of the broken lamellar structure with faceted/non-faceted behaviour[2]).

When table 2 is examined it is apparent that the demarcation of non-faceted/non-faceted behaviour from faceted/non-faceted growth occurs when the entropy of solution per atom of one of the phases exceeds a value of 5.5 cal/g-atom °K. Thus all eutectics, except Bi–Au$_2$Bi, for which the individual entropies of solution of the constituent phases are less then this value are regular, i.e., lamellar or rod-like whilst those eutectics in which at least one of the phases has an entropy of solution greater than this have a variety of forms such as complex-regular, broken lamellar or irregular.

The faceted/non-faceted eutectic group appears at first to possess a bewildering variety of growth forms. Many of these are growth rate dependent, as has been demonstrated recently by Cline et al.[25]). These workers report that the broken lamellar or plate-like structure may change to a faceted rod-like structure at lower growth rates. In faceted/non-faceted eutectics, factors other than the entropies of solution of the phases appear to have a determining influence on the eutectic structure. Of particular importance is the phasial volume ratio. This has been examined in detail by the present authors[26]) who have shown that other structures give way exclusively to the broken lamellar structure when the phasial volume ratio exceeds a value of 10:1. This appears to be independent of the entropy of solution providing the latter exceeds the value of 5.5 cal/g-atom °K, as previously pointed out.

4. Conclusions

While the solid–liquid interface structure of simple

close-packed solid phases may be described with reference to the roughness parameter, i.e., it will be atomically rough if $\alpha \leq 2$ and singular if $\alpha > 2$, it is seen that a similar demarcation for compound phases is given by an entropy of solution value of 5.5. cal/g-atom °K. In that this is a bulk thermodynamic property, its use is less subject to the criticisms that have been levelled at the Jackson interface model as representing a real interface configuration in solidification processes.

With the value of 5.5 cal/g-atom °K as the borderline between non-faceted and faceted growth, it is seen that the present results give further support to the Hunt and Jackson[1]) scheme for classification of eutectic structures.

While there is good correlation between the observed structures and those predicted by the entropies of solution per adatom of the phases, the simple mechanistic description of eutectic solidification is still not entirely satisfactory for structures other than the lamellar or rod structures. Thus, the way in which such a variety of structures can appear within the faceted/non-faceted class remains unclear.

5. Glossary of symbols

x_A and x_B atomic fractions of A and B respectively in solid solution α or γ.

$x \equiv x_B$

x'_A mole fraction of A in solid γ taking stoichiometric compound as a single species.

x_l liquidus composition in at. frac. B.

p and q number of atoms of A and B per compound molecule.

$\Delta H_A(\gamma)$ partial heat of component A in solid solution; γ.

ΔH_f molar heat of formation of compound.

$\overline{\Delta H_A}(l)$ and $\overline{\Delta H_B}(l)$ partial heats of mixing components A and B respectively in liquid, composition of x_l.

L_A and L_B heats of fusion of pure components A and B.

L_γ heat of solution of γ phase in liquid, composition x_l.

T liquidus temperature.

ΔS_γ average entropy of solution per atom for γ phase.

Acknowledgements

The work described here has been made possible via the financial support of the Ministry of Aviation Supply, U.K. Government and the National Research Council of Canada.

References

1) J. D. Hunt and K. A. Jackson, Trans. AIME **236** (1966) 843.
2) M. R. Taylor, R. S. Fidler and R. W. Smith, Met. Trans. **2** (1971) 1793.
3) K. A. Jackson, *Liquid Metals and Solidification* (Am. Soc. Metals, Cleveland, 1958).
4) K. A. Jackson, in: *Growth and Perfection of Crystals*, Eds. R. H. Doremus et al. (Wiley, New York, 1958).
5) H. W. Kerr and W. C. Winegard, in: *Crystal Growth*, Ed. H. A. Peiser (Pergamon, Oxford, 1967) p. 179.
6) M. R. Taylor, R. S. Fidler and R. W. Smith, J. Crystal Growth **3, 4** (1968) 666.
7) K. A. Jackson, J. Crystal Growth **3, 4** (1968) 507.
8) K. A. Jackson, J. Crystal Growth **5** (1969) 13.
9) M. Hansen and K. Anderko, *Constitution of Binary Alloys* (McGraw-Hill, Toronto, 1958).
10) R. Hultgren et al., *Selected Values of Thermodynamic Properties of Metals and Alloys* (Wiley, New York, 1963).
11) O. Kubachewski, E. Ll. Evans and C. B. Alcock, *Metallurgical Thermodynamics*, 4th Ed. (Pergamon, New York, 1967).
12) L. M. Hogan, J. Australian Inst. Metals **6** (1961) 279.
13) R. W. Kraft and D. L. Albright, Trans. AIME **221** (1961) 95.
14) J. D. Hunt and J. P. Chilton, J. Inst. Metals **91** (1962–3) 338.
15) R. S. Fidler, J. A. Spittle, M. R. Taylor and R. W. Smith, *The Solidification of Metals* (Iron and Steel Institute, London, 1968) p. 173.
16) G. A. Chadwick, J. Inst. Metals **92** (1963–4) 18.
17) H. W. Kerr and W. C. Winegard, Can. Met. Quart. **6** (1967) 67.
18) J. D. Hunt and D. T. J. Hurle, Trans. AIME **242** (1968) 1043.
19) E. P. Whelan and C. A. Haworth J. Australian Inst. Metals **12** (1967) 77.
20) W. A. Miller and G. A. Chadwick, *The Solidification of Metals* (Iron and Steel Institute, London, 1968) p. 49.
21) R. W. Kraft, Trans. AIME **227** (1963) 396.
22) P. J. Taylor, H. W. Kerr and W. C. Winegard, Can. Met. Quart. **3** (1964) 235.
23) G. Nishimura, R. S. Fidler, M. R. Taylor and R. W. Smith, Can. Met. Quart. **8** (1969) 319.
24) A. Hellawell, Trans. AIME **239** (1967) 1049.
25) H. E. Cline, J. L. Walter, E. Lifshin and R. R. Russell, Met. Trans. **2** (1971) 189.
26) M. N. Croker, R. S. Fidler and R. W. Smith (to be published).

Journal of Crystal Growth **13/14** (1972) 747–750 *North-Holland Publishing Co.*

EFFECTS OF GROWTH RATE ON THE MORPHOLOGY OF MONOVARIANT EUTECTICS: MnSb–(Sb,Bi) AND MnSb–(Sb,Sn)

M. DURAND-CHARRE and F. DURAND

Laboratoire de Thermodynamique et Physicochimie Métallurgiques, E.N.S.E.E.G. 18, rue Hoche, 38-Grenoble, France

MnSb–(Sb,Bi) and MnSb–(Sb,Sn) samples of monovariant eutectic composition were unidirectionnally solidified at growth rates increasing from 0.6 to 12.6 cm/hr in a temperature gradient of $G = 50\,°C/cm$. The microstructure has been found to be fibrous at growth rates (V) in the range $0.6 < V < 2$ cm/hr, changing to cellular then dendritic at growth rates greater than 2 cm/hr. The crystallographic orientation of the microstructure has been investigated. The longitudinal and transverse solute distribution inside the cells was studied using the electron probe microanalysis technique. It was found that the concentration at the tip of the cells decreases as an inverse function of the growth rate.

1. Introduction

Unidirectional solidification gives the Sb–MnSb eutectic a well-known morphology with MnSb fibers embedded in a Sb matrix[1]). Adding Bi (up to 6 at%) or Sn (up to 4 at%) does not disturb the MnSb fibers, but gives a solid solution with Sb: the eutectic equilibrium is then monovariant. The consequence of this is constitutional supercooling, which may change the morphology. This effect is examined on several unidirectionally solidified MnSb–(Sb,Bi) and MnSb–(Sb,Sn) samples. By increasing the rate of solidification a breakdown of the planar front to cellular morphology occurs, followed by a dendritic morphology. The crystallographic orientation of the growth axis is given. An investigation of longitudinal and transverse solute distribution inside the cells was made using an electron microprobe. The concentration at the tip of the cells decreases with growth rate, as predicted by Flemings.

2. Experimental procedure

MnSb–(Sb,Bi) and MnSb–(Sb,Sn) alloys of nearly monovariant eutectic composition were prepared in a vacuum sealed quartz tube of 10 mm inside diameter, 170 mm long. These ingots were unidirectionally solidified in a horizontal moving furnace, at solidification rates V ranging from 0.7 cm/hr to 12.6 cm/hr. Temperature gradients G in the melt were measured by a chromel–alumel thermocouple in direct contact with the melt. G could be varied between 35 °C/cm and

65 °C/cm, but was set at 50 °C/cm for all the experiments described.

After solidification, the microstructure appeared to be uniform in the central part of the ingot, for approximately 7 cm. Several compositions were processed. The following alloys were used:

MnSb–(Sb,Bi): 4.4 at% Bi, 14.8 at% Mn;
MnSb–(Sb,Sn): 3 at% Sn, 15 at% Mn.

3. Morphologies

For increasing growth rates the following morphologies were observed:

(a) At slow growth rates ($V = 0.7$ cm/hr) the microstructure was fibrous, with a plane interface (fig. 1).

(b) The transition from plane front to cells occurred for $V \cong 0.9$ cm/hr ($G/V \cong 2 \times 10^5\,°C\,s\,cm^{-2}$). Fig. 2 shows the early stage of the breakdown. On the cross-section, zones with a lower fiber density indicate the depressed regions of the growth front.

(c) Fig. 3 shows a longitudinal section of the quenched interface profile for $V = 2$ cm/hr. In the corresponding cross-section some cells are rectangular and others show trigonal symmetry.

(d) At higher growth rates, all the cells are of the latter shape (fig. 4).

(e) On alloys grown at rates exceeding about 4 cm/hr, periodic secondary arms may be seen in the cells (fig. 5). This microstructure looks like dendrites in solid solutions; by analogy, we shall name it "eutectic dendrites".

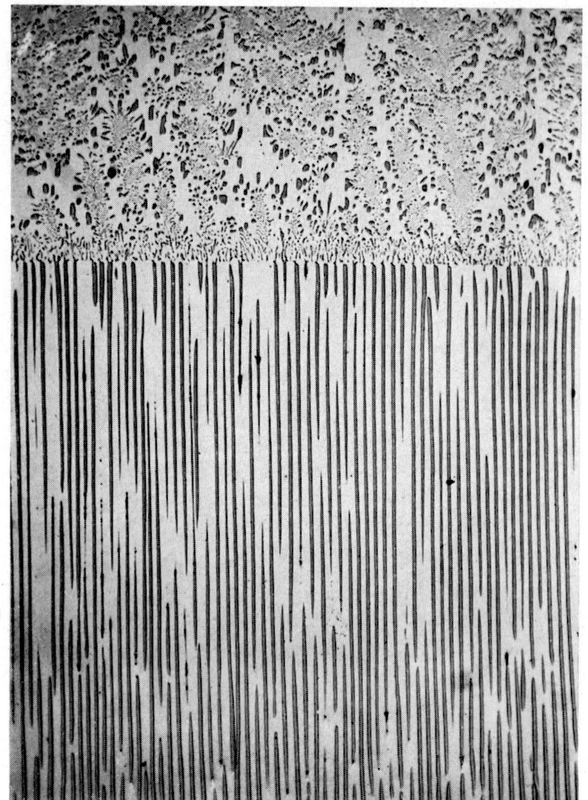

Fig. 1. MnSb–(Sb,Bi) eutectic quenched interface; $V = 0.7$ cm/hr; magnification 108 ×.

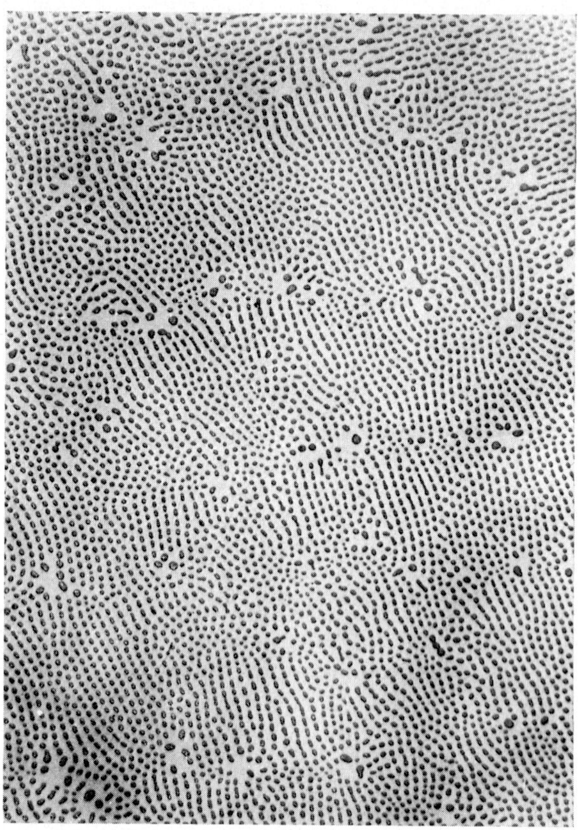

Fig. 2. MnSb–(Sb,Bi) transition to cells; cross-section, $V = 0.7$ cm/hr, $G \simeq 40$ °C/cm; magnification 108 ×.

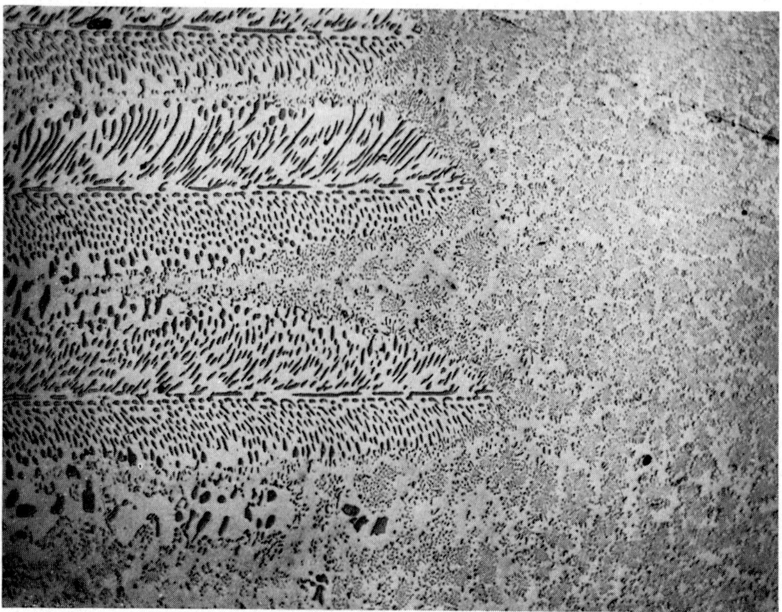

Fig. 3. Cellular MnSb–(Sb,Bi), quenched interface; longitudinal section; $V = 2$ cm/hr; magnification 108 ×.

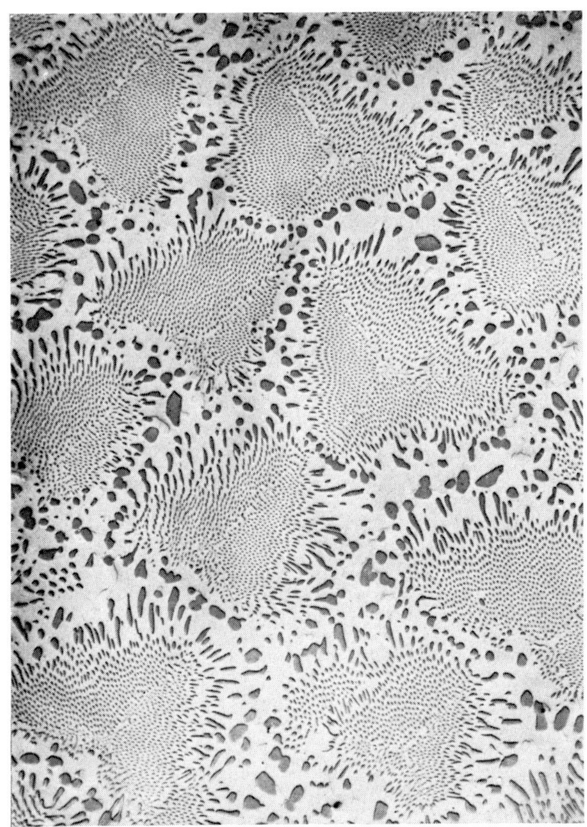

Fig. 4. Cellular MnSb–(Sb, Sn); cross-section showing the trigonal axis of the cells; $V = 6$ cm/hr; magnification $108\times$.

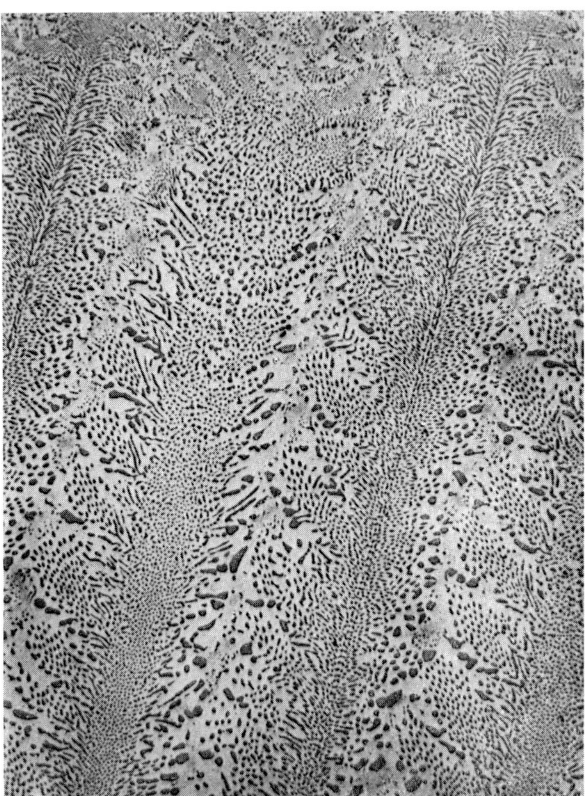

Fig. 5. Dendritic MnSb–(Sb,Bi); quenched interface; longitudinal section along the growth axis; $V = 6$ cm/hr; magnification $55\times$.

It should be noted that fast-grown alloys ($V > 6$ cm/hr) show two main orientations, one with the dendrite axis along the thermal gradient, and the other, in an adjacent grain, at about 20 degrees from the gradient axis.

When the solidification rate exceeds 12.6 cm/hr, the morphology resembles that of the quenched liquid in figs. 1 or 3.

4. Crystallographic orientation

The crystallographic orientation of the growth direction was found by X-ray determination (texture and Laue method): for low growth rates ($V \cong 0.7$ cm/hr), it is parallel to [0001] MnSb and parallel to [11$\bar{2}$0] (Sb,Si); for faster growth rates, the cell (or dendrite) axis is parallel to [0001] MnSb and to [0001] (Sb,Si). It is interesting to note that the present results agree with those found for (Sb,Bi) solid solutions[2,3].

5. Solute distribution in cellular and dendritic ingots

Solute distribution was measured in cellular and den-

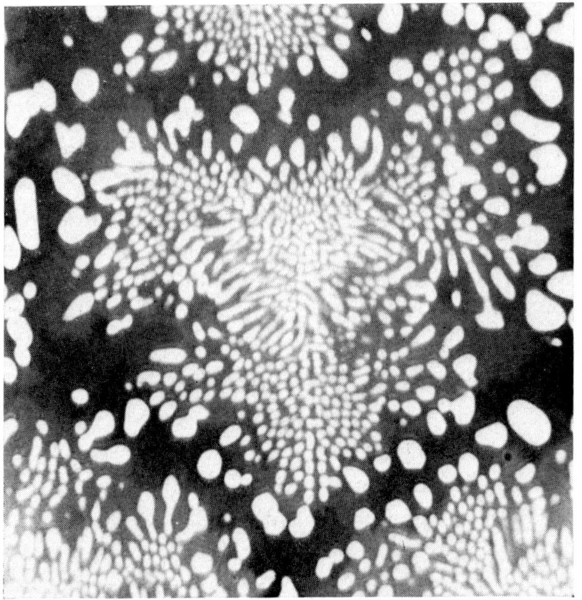

Fig. 6. Cell of MnSb–(Sb,Bi); cross-section; secondary electrons micrograph; $V = 7.8$ cm/hr; magnification $245\times$.

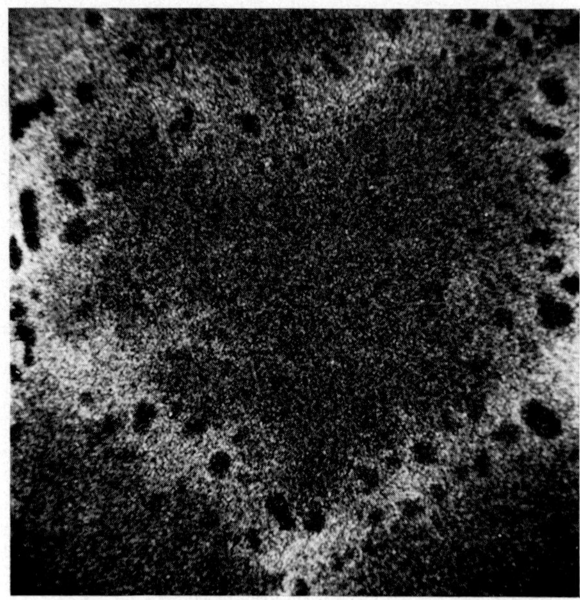

Fig. 7.   X-ray micrograph of Bi; cell as for fig. 6.

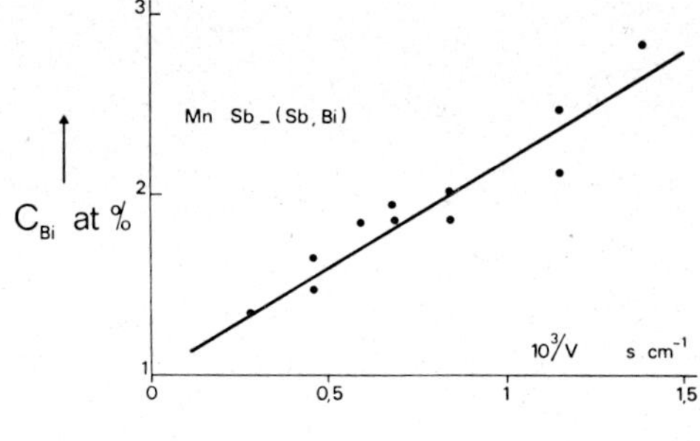

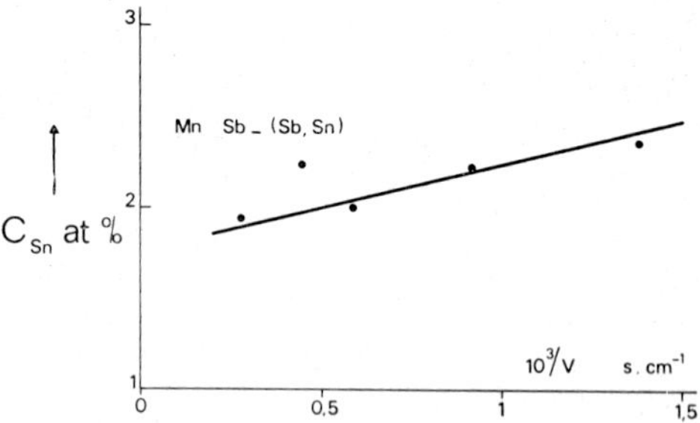

Fig. 8.   Solute concentration as an inverse function of growth rate V.

dritic ingots, using a Cameca MS 46 microprobe. Secondary electrons and X-ray micrographs (figs. 6 and 7) give a meaningful account of the solute distribution.

Measurements were made at different positions in the matrix, which is a (Sb,Bi) or (Sb,Sn) solid solution, without Mn. The spot was as thin as possible, approximately 1 μm wide. In order to avoid fibers, a spot position corresponding to a minimum manganese count was chosen.

Three main points of interest emerge from the above experiments:

(1) The cell cores are depleted and the intercellular regions enriched with solute (Bi or Sn). Some much richer grains (figs. 6 and 7) with up to three times the concentration in the middle of the cell can be seen in intercellular grooves. These discontinuities are similar to those observed by Brown and Heumann for (Sb,Bi) solid solutions[4]).

(2) The cell cores, i.e. the regions where the fibers seem identical and evenly spaced when observed in the cross-section, are nearly homogeneous.

(3) The composition C_t near the cell axis is constant along the ingot. It varies with V (fig. 8) according to Flemings[5]):

$$C_t - C^0 = - \frac{DG_L}{V} \frac{\partial C_{eq}}{\partial T}.$$

D is the diffusion coefficient of the solute, for which we assumed $D = 3 \times 10^{-5}$ cm^2/s, as Roberts and Hellawell[3]) did for (Sb,Bi) solutions. $\partial C_{eq}/\partial T$ is the variation of equilibrium concentration for one degree of temperature rise; from the slope in the diagrams, its value is estimated to be -0.8 at%/°C for MnSb–(Sb,Bi) and -0.24 at%/°C for MnSb–(Sb,Sn). The fact that the second value is lower seems to be related to the low solubility of Sn in Sb.

Acknowledgements

The authors are grateful to Dr. G. Massart for his microprobe analysis, to Dr. Caillet for his texture analysis and to Mr. Marmeggi for his Laue X-ray determinations.

References

1) M. R. Jackson, R. N. Tauber and R. W. Kraft, J. Appl. Phys. **39** (1968) 4452.
2) W. M. Yim and J. P. Dismukes, in: *Crystal Growth*, Ed. H. S. Peiser (Pergamon, Oxford, 1967) pp. 187–196.
3) P. A. Roberts and A. Hellawell, J. Crystal Growth **8** (1971) 104.
4) D. M. Brown and F. K. Heumann, J. Appl. Phys. **35** (1964) 1947.
5) T. F. Bower, H. D. Brody and M. C. Flemings, Trans. AIME **236** (1966) 624.

Journal of Crystal Growth 13/14 (1972) 751–756 © *North-Holland Publishing Co.*

SOLIDIFICATION MONOVARIANTE D'ALLIAGES TERNAIRES; APPLICATION À LA FUSION DE ZONE

P. COMBRADE et M. TURPIN

Centre des Matériaux de l'École Nationale Supérieure des Mines de Paris, 91 – Corbeil–Essonne, France

Many authors have studied the unidirectional solidification of many-component two-phase alloys. They encountered a serious problem of segregation which we have investigated in the simple case of monovariant eutectics. A general formulation of the problem of the macrosegregation allows complete study of regular two phase growth, qualitative conclusion on irregular growth and comparison between two techniques: solidification of a melt and zone melting. The most important result is that, in both cases, the compositions of the liquid bulk and of the growing solid do not follow the eutectic line. So we show, in good agreement with the experiments, that for the solidification of a melt the conditions required for regular growth are very difficult to obtain. However, zone melting which leads to a steady state whatever the diffusional flow, is particularly well suited for the growth of regular two phase composites. However, and in contradiction with a common point of view, the study of the evolution of the crystallization paths with the number of passes shows that this method is not suitable for determining monovariant eutectic lines in ternary alloys. These results can probably be extended to non-invariant eutectic reactions in *n*-component systems.

1. Introduction

Pour présenter un intérêt pratique, les alliages diphasés anisotropes doivent contenir plus de deux constituants[1,3]). Leur solidification pose alors des problèmes d'homogénéité et de régularité des structures[4,5]), qui ont pour origine le rejet de soluté à l'interface, caractéristique des réactions non-invariantes. Les solutions techniques de ces problèmes passent par la connaissance de la ségrégation majeure lors des solidifications dirigées. Nous examinons ici le cas simple des systèmes ternaires possédant une vallée eutectique et nous appliquons les raisonnements à la réaction eutectique monovariante.

L'étude de la ségrégation suppose l'écriture et la résolution de deux systèmes d'équations:
– le transfert de matière dans le liquide;
– le bilan massique global du système en évolution.

Pour ce qui nous concerne nous pouvons supposer que le transfert de matière devant l'interface se fait par diffusion chimique dans une couche relativement mince au-delà de laquelle l'homogénéité du liquide est assurée par la convection. Comme on le fait généralement[6]), nous supposons le régime de diffusion quasi-stationnaire et les coefficients constants.

La conservation de la masse dans la couche de diffusion s'écrit:

$$\operatorname{div}(\boldsymbol{J}_i + \boldsymbol{v}C_i) = 0, \quad i = \mathrm{B}, \mathrm{C}; \tag{1}$$

avec:

$$\boldsymbol{J}_\mathrm{B} = -D_\mathrm{BB}\,\operatorname{grad} C_\mathrm{B} - D_\mathrm{BC}\,\operatorname{grad} C_\mathrm{C}, \tag{2a}$$

$$\boldsymbol{J}_\mathrm{C} = -D_\mathrm{CB}\,\operatorname{grad} C_\mathrm{B} - D_\mathrm{CC}\,\operatorname{grad} C_\mathrm{C}. \tag{2b}$$

Prenons des axes liés à l'interface, z étant parallèle à la direction de la solidification et son vecteur unitaire $\boldsymbol{n}$ dirigé vers le liquide. Si l'on suppose égaux les volumes molaires des phases liquide et solides, l'intégration des équations (1) sur des sections S perpendiculaires à l'axe z conduit à:

$$\int_S \left(\frac{\partial(\boldsymbol{J}_i \cdot \boldsymbol{n})}{\partial z} - v\,\frac{\partial C_i}{\partial z} \right) \mathrm{d}x\,\mathrm{d}y = 0, \quad i = \mathrm{B}, \mathrm{C}. \tag{3}$$

Ce qui peut généralement se mettre sous la forme:

$$D_\mathrm{BB}\,\frac{\mathrm{d}^2 \bar{C}_\mathrm{B}}{\mathrm{d}z^2} + D_\mathrm{BC}\,\frac{\mathrm{d}^2 \bar{C}_\mathrm{C}}{\mathrm{d}z^2} + v\,\frac{\mathrm{d}\bar{C}_\mathrm{B}}{\mathrm{d}z} = 0, \tag{4a}$$

$$D_\mathrm{CB}\,\frac{\mathrm{d}^2 \bar{C}_\mathrm{B}}{\mathrm{d}z^2} + D_\mathrm{CC}\,\frac{\mathrm{d}^2 \bar{C}_\mathrm{C}}{\mathrm{d}z^2} + v\,\frac{\mathrm{d}\bar{C}_\mathrm{C}}{\mathrm{d}z} = 0, \tag{4b}$$

où $\boldsymbol{C}(z) = (\bar{C}_\mathrm{B}(z), \bar{C}_\mathrm{C}(z))$, est la concentration moyenne sur une section S:

$$\bar{C}_i(z) = \frac{1}{S} \int_S C_i(x, y, z)\,\mathrm{d}x\,\mathrm{d}y, \quad i = \mathrm{B}, \mathrm{C}. \tag{5}$$

Le système (4) a la même forme que le système obtenu pour une interface plane, mais ici la signification des variables $\bar{C}_B$ et $\bar{C}_C$ est tout à fait différente de celle des variables locales C_B et C_C utilisées dans le cas d'une interface monophasée plane.

Les conditions aux limites à l'interface traduisent la conservation de la matière à travers la zone interfaciale définie sur la figure 1: elles ne contiennent aucune hypothèse sur les rejets de soluté vers le liquide:

$$D_{BB} \frac{d\bar{C}_B}{dz}\bigg)_{z=0} + D_{BC} \frac{d\bar{C}_C}{dz}\bigg)_{z=0} + v(\bar{C}_B^I - \bar{C}_B^S) = 0, \quad (6a)$$

$$D_{CB} \frac{d\bar{C}_B}{dz}\bigg)_{z=0} + D_{CC} \frac{d\bar{C}_C}{dz}\bigg)_{z=0} + v(\bar{C}_C^I - \bar{C}_C^S) = 0, \quad (6b)$$

où

$$C^I = (\bar{C}_B^I, \bar{C}_C^I) \quad \text{et} \quad C^S = (\bar{C}_B^S, \bar{C}_C^S)$$

désignent les compositions moyennes de part et d'autre de la zone interfaciale.

Ecrivons également que la couche de diffusion est limitée:

$$C_i(\delta) = C_i^L. \tag{7}$$

La solution du système (3) assorti des conditions aux limites (6) et (7) est connue[7] et permet d'obtenir entre les compositions C^I et C^S et la composition du bain C^L les deux relations:

$$[(D_{BB}+D_{CB})\, r(+)+v]\, [(\bar{C}_B^L - \bar{C}_B^S)-(\bar{C}_B^I - \bar{C}_B^S)\, e^{r(-)\delta}]$$
$$+ [(D_{CC}+D_{BC})\, r(+)+v]\, [(\bar{C}_C^L - \bar{C}_C^S) \tag{8a}$$
$$- (\bar{C}_C^I - \bar{C}_C^S)\, e^{r(+)\delta}] = 0,$$

$$[(D_{BB}+D_{CB})\, r(-)+v]\, [(\bar{C}_B^L - \bar{C}_B^S)-(\bar{C}_B^I - \bar{C}_B^S)\, e^{r(+)\delta}]$$
$$+ [(D_{CC}+D_{BC})\, r(-)+v]\, [(\bar{C}_C^L - \bar{C}_C^S) \tag{8b}$$
$$- (\bar{C}_C^I - \bar{C}_C^S)\, e^{r(+)\delta}] = 0,$$

où

$$r(\pm) = -\frac{v}{D_{ij}}\{D_{BB}+D_{CC} \mp [(D_{BB}+D_{CC})^2 - 4D_{ij}]^{\frac{1}{2}}\}, \tag{9a}$$

$$D_{ij} = D_{BB}D_{CC} - D_{BC}D_{CB}. \tag{9b}$$

Si l'on ajoute aux relations (8) les équations de bilan global on obtient une formulation très générale pour la ségrégation majeure, uniquement paramétrée par le rejet moyen de soluté à l'interface (C^I–C^S). Ce dernier dépend naturellement du nombre et de la nature des phases déposées et de la morphologie du front de solidification.

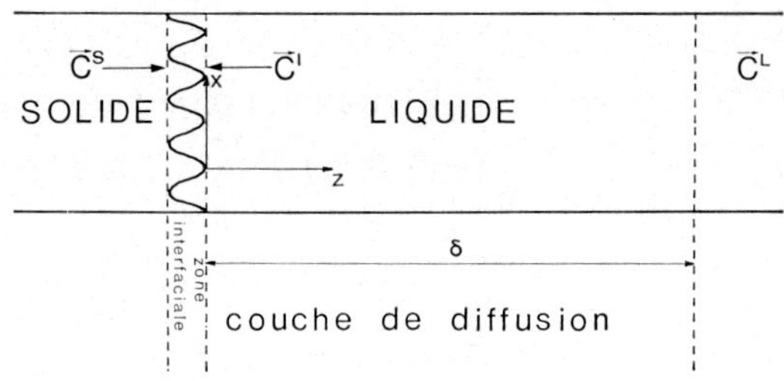

Fig. 1.   Schématisation de la zone de réaction et de transfert de matière dans le système en cours de solidification.

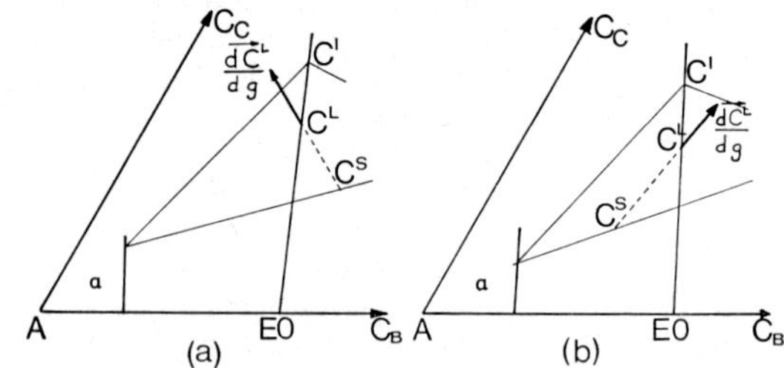

Fig. 2.   Solidification dirigée: composition d'un solide déposé par réaction eutectique monovariante: (a) $A < 0$; (b) $A > 0$.

2. Ségrégation majeure au cours de solidifications dirigées

Le bilan global s'écrit:

$$C^L - C^S = (1-g)\, dC^L/dg \tag{10}$$

(g = fraction solidifiée).

Sans perdre en généralité nous pouvons nous limiter au cas d'une vallée eutectique rectiligne, reliant un eutectique binaire E0 (C^{E0}) à une eutectique ternaire E3 (C^{E3}):

$$C_B = k_1 C_C + C_B^{E0}, \quad C_C \leq C_C^{E3}. \tag{11}$$

Les équations (8) et (10) donnent:

$$(1-g) \frac{d\Delta C^L}{dg} = (C_B^L - \bar{C}_B^S) - k_1(C_C^L - \bar{C}_C^S), \tag{12a}$$

$$(1-g) \frac{d\Delta C^L}{dg} = \frac{v(C_C^L - \bar{C}_C^I)}{D_{ij}[r(+) - r(-)]}\, AH \tag{12b}$$
$$+ K(\Delta \bar{C}^I - \Delta C^L),$$

avec:

$$\Delta \bar{C} = \bar{C}_B - k_1 \bar{C}_C - C_B^{E0}$$

$$= \text{écart de la composition } C \text{ à la vallée eutectique,}$$

$$K = \frac{e^{r(+)\delta}}{1 - e^{r(+)\delta}} + H\, \frac{r(-)\, [(D_{BB} - k_1 D_{CB})r(+) + v]}{v[r(+) - r(-)]},$$

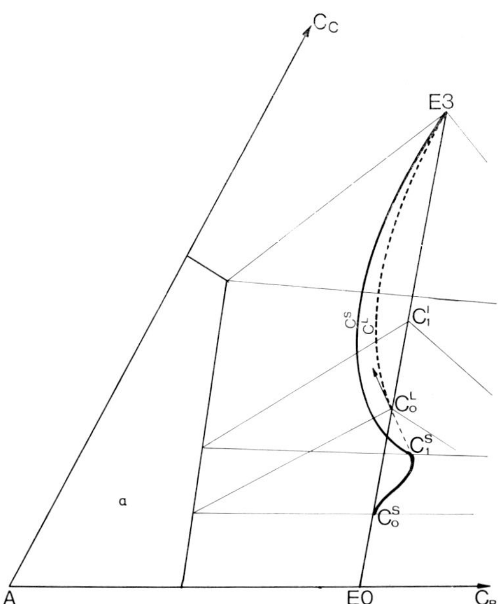

Fig. 3. Solidification dirigée: exemple de chemin de cristallisation d'un alliage dont la composition est initialement sur la vallée eutectique.

$$H = \frac{e^{r(+)\delta}}{1-e^{r(+)\delta}} - \frac{e^{r(-)\delta}}{1-e^{r(-)\delta}},$$

$$A = k_1^2 D_{CB} - k_1(D_{BB} - D_{CC}) - D_{BC}.$$

Ce dernier terme A ne dépend que des caractéristiques du système et ne s'annule, en pratique, que si les coefficients croisés sont nuls, puisque les coefficients directs deviennent alors égaux. Comme les termes H, K et $(\bar{C}_C^L - \bar{C}_C^I)$ ne sont généralement pas nuls, il devient évident d'après (12b) qu'un liquide initialement dans la vallée eutectique s'en écartera, quel que soit le mode de croissance.

Pour pousser l'analyse, considérons le cas de la croissance diphasée que nous qualifierons de régulière où l'interface est macroscopiquement plane et la structure du solide lamellaire ou fibreuse. Une solution de l'équation de la diffusion dans ce cas a une forme analogue à celle de la solution de Jackson et Hunt pour les eutectiques binaires et montre que l'on peut écrire des relations d'équilibre entre la composition moyenne du solide et celle du liquide à l'interface. D'après (10), (12b) et le diagramme d'état, C^I, C^L, C^S sont alors reliés par la construction de la figure 2. Si la croissance reste régulière nous pouvons tracer l'allure des chemins de cristallisation qui aboutissent toujours à l'eutectique ternaire. La figure 3 représente un cas possible où nous

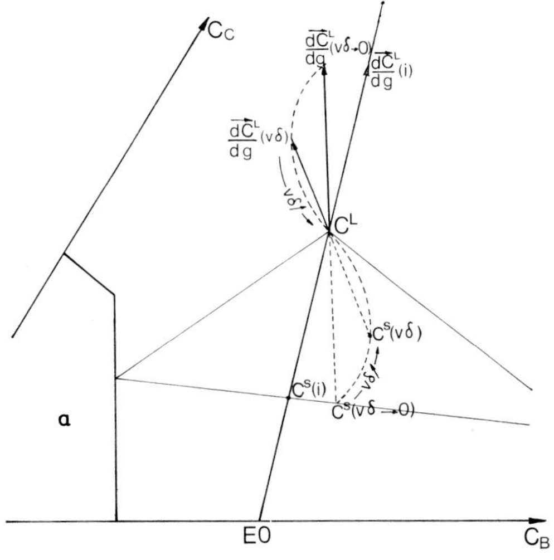

Fig. 4. Solidification dirigée: influence des conditions expérimentales sur la composition d'un solide déposé par réaction eutectique monovariante.

avons tenu compte du régime transitoire initial de diffusion. Un calcul d'ordre de grandeur pour le système Ni–Nb–Cr laisse prévoir des écarts à la vallée de plusieurs pour cent. Il est probable que la croissance cessera d'être régulière dans ces conditions. Un liquide dont la composition est initialement sur la vallée eutectique donnera généralement naissance à des dendrites. Malgré cette instabilité, l'allure générale des chemins de cristallisation n'est pas modifiée. Considérons en effet la direction d'évolution de la composition du liquide, caractérisée par:

$$\frac{d\Delta C^L/dg}{|dC^L/dg|} = \frac{(C_B^L - \bar{C}_B^S) - k_1(C_C^L - \bar{C}_C^S)}{[(C_B^L - \bar{C}_B^S)^2 + (C_C^L - \bar{C}_C^S)^2]^{\frac{1}{2}}}. \tag{13}$$

Lorsque le liquide est eutectique, cette grandeur ne dépend pas des relations liquide–solide de part et d'autre de la zone interfaciale puisque d'après (8), elle reste alors fonction du produit $v\delta$ et du diagramme des phases.

La présence de dendrites ne peut affecter que l'intensité de la ségrégation $(C^L - C^S)$. Les chemins de cristallisation sont alors plus près de la vallée eutectique mais ils restent du même côté.

3. Influence des conditions expérimentales

Elles agissent sur la ségrégation par le produit $v\delta$. La figure 4 montre d'abord que le modèle idéal où l'on suppose l'homogénéité parfaite du liquide ne constitue pas la limite des cas réels et ne peut donc être utilisé.

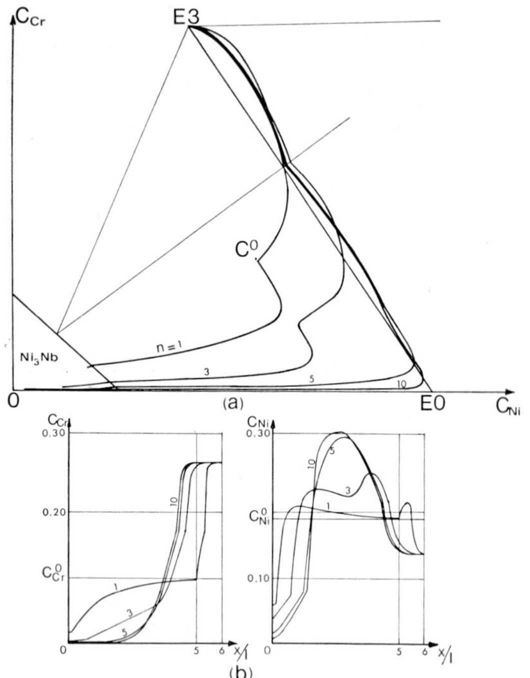

Fig. 5. Fusion de zone: Système Ni–Nb–Cr, (a) chemins de cristallisation et (b) profils de concentration.

La seule façon de rendre eutectique la composition du solide qui se dépose est donc de se rapprocher d'un régime stationnaire de diffusion en augmentant $v\delta$. Mais pour maintenir la stabilité de croissance il faut un rapport G/v important (G étant le gradient thermique à l'interface). Augmenter v impose d'augmenter G: or la réalisation de forts gradients est généralement liée à une forte convection. En pratique il est donc difficile d'augmenter v sans diminuer δ sauf peut être en bloquant les courants de convection par un champ magnétique.

En accord avec ces précisions nos expériences ont montré que, malgrè un régime très proche d'un état stationnaire, nous n'avons jamais obtenu, par cette méthode, plus de 5 cm de composite régulier (Ni–Ni$_3$Nb) + Cr. Nous sommes amenés à conclure que la solidification unidirectionnelle de barreaux fondus est une technique peu adaptée à l'obtention de grandes quantités de structures régulières par réaction eutectique monovariante.

4. Fusion de zone

L'apport de matière au bain liquide modifie les équations de bilan global en:

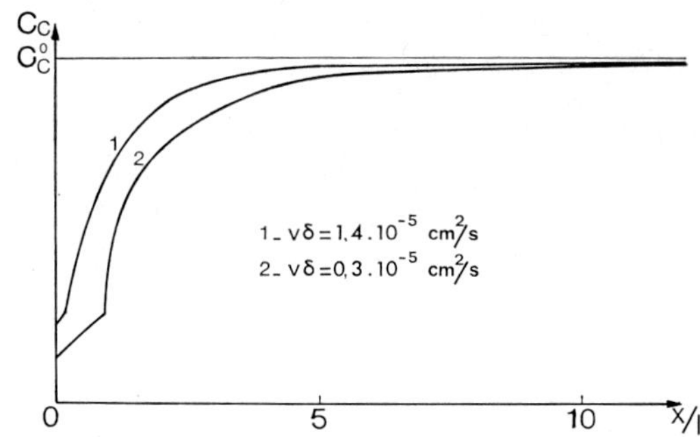

Fig. 6. Fusion de zone: système Ni–Nb–Cr: influence des conditions expérimentales sur l'établissement du régime stationnaire lors de la première passe.

$$0 \leq x \leq (Z-1)l,$$

$$C^n(x) - C^{n-1}(x+l) + l\,\frac{\mathrm{d}C^L}{\mathrm{d}x} = 0, \qquad (14a)$$

$$(Z-1)l \leq x \leq Zl,$$

$$C^n(x) - C^L(x) + (Zl-x)\,\frac{\mathrm{d}C^L}{\mathrm{d}x} = 0, \qquad (14b)$$

où $C^n(x)$ est la composition du solide déposé en x lors de la $n$ième passe; l est la longueur de la zone, Zl celle du barreau. La relation d'évolution du système, analogue à (12b) s'écrit alors:

$$l\,\frac{\mathrm{d}\Delta C^L}{\mathrm{d}x} = \frac{v(C_C^L - \bar{C}_C^I)AH}{D_{ij}[r(+)-r(-)]} - (K+1)\Delta\bar{C}^L + K\Delta C^I \\ + \Delta\bar{C}^{n-1}(x+l). \qquad (15)$$

L'analyse de cette relation d'évolution est la même que dans le cas de la solidification dirigée. Par contre l'équation (14) montre que dans ce cas, un régime stationnaire peut s'établir si le barreau est suffisant long; il est alors possible d'obtenir une structure régulière et homogène à partir d'un alliage de composition adéquate puisque le solide qui se dépose à la même composition que celui qui fond.

Puisque nous savons que la morphologie du front de solidification n'affecte pas la forme générale des chemins de cristallisation et en particulier ne change pas leur position par rapport à la vallée eutectique, nous avons cherché à préciser leur allure en retenant deux hypothèses supplémentaires:
– Manquant d'informations sur les valeurs des coefficients croisés de diffusion, nous supposons que, lorsque la composition du liquide est proche de la vallée eutectique, le constituant C ne distingue pas A de B c'est-à-dire:

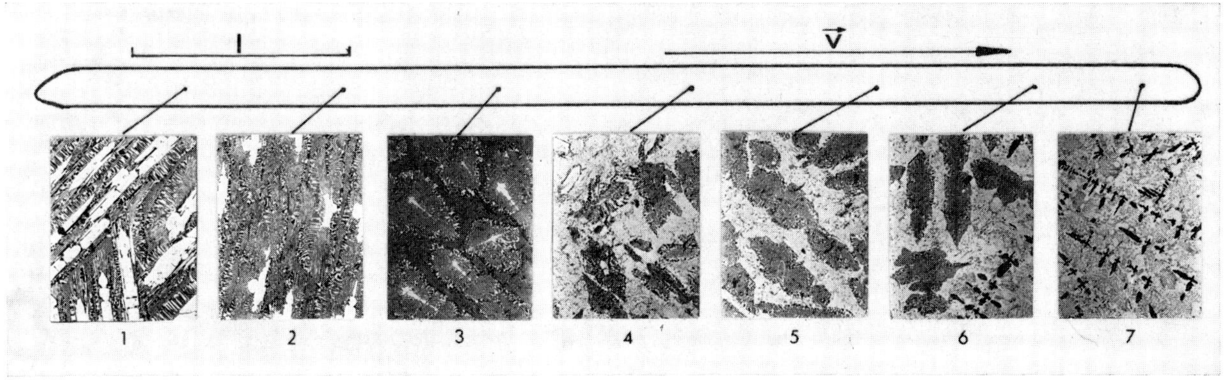

Fig. 7. Fusion de zone: système Ni–Nb–Cr: structure après 10 passes de fusion de zone d'un barreau initialement riche en niobium: (1) cristaux primaires Ni_3Nb+colonies diphasées+ eutectique ternaire; (2, 3, 4) cristaux primaires Ni_3Nb+dendrites diphasées+eutectique ternaire; (5) dendrites diphasées+eutectique ternaire; (6) cristaux primaires Ni(Cr)+dendrites diphasées +eutectique ternaire; (7) cristaux primaires Ni(Cr)+eutectique ternaire.

$$D_{CB} = 0, \quad J_B - J_A = -D_{BB} \, \text{grad} \, (C_B - C_A). \qquad (16)$$

– Nous écrivons en outre qu'il y a équilibre entre les compositions moyennes C^l et C^s.

Les chemins de cristallisation et les profils de concentration sont ainsi calculés en fonction du nombre de passes (fig. 5). On remarque surtout:

– que les chemins de cristallisation tendent non pas vers la vallée eutectique mais vers un chemin qui peut en être éloigné de plusieurs pour cent.

– qu'une forte agitation de bain et une faible vitesse les rapprochent de la vallée mais retardent l'établissement du régime stationnaire lors de la première passe (fig. 6).

La fusion de zone est donc une méthode adéquate pour la fabrication de composites par réaction eutectique monovariante: la longueur du barreau où le solide est irrégulier ou simplement inhomogène peut être réduite en diminuant la longueur de la zone et en augmentant $v\delta$ (voir à ce sujet les remarques faites précédemment). Par contre cette technique ne permet pas d'atteindre les vallées eutectiques.

Sur un barreau d'alliage Ni–Nb–Cr, riche en niobium et contenant 20% de chrome, nous avons fait 10 passes de fusion de zone, en nacelle horizontale. La vitesse de translation était de 6 mm/h et l'agitation électromagnétique du bain très importante. Pour éviter tout effet cumulatif de la gravité, le barreau était retourné sens dessus-dessous, après chaque passe. Ainsi après 10 passes, la structure observée sur des sections transversales est toujours homogène.

La figure 7 montre l'évolution de la structure le long du barreau. Bien qu'il soit plus riche en niobium que ceux pour lesquels nous avons fait le calcul, on peut constater un accord avec la théorie, il y a pratiquement partout des dendrites monophasées et la présence de cristaux primaires de nickel à la fin du barreau montre que la vallée eutectique a été franchie par les chemins de cristallisation: ceci confirme des observations de Yue et Clark[10]), sur le système Mg–Al–Zn.

Par ailleurs sur un barreau initialement riche en nickel, les chemins de cristallisation ne franchissent ni même n'atteignent jamais la vallée. De même, sur le système (Co–TiC)+Cr, nous n'avons jamais pu approcher la vallée par cette technique.

5. Conclusion générale

Nous avons montré, que, dans le cas général, les compositions des liquides et des solides ne suivent pas les vallées eutectiques dans les solidifications dirigées. Il est toutefois possible d'obtenir des structures régulières par réaction eutectique monovariante si l'on réalise un régime permanent de croissance: la technique la plus commode est la fusion de zone dont on améliore le rendement par une faible longueur de zone et un faible brassage de bain.

Enfin, contrairement à ce qui l'on pourrait penser à priori, la fusion de zone, méthode rapide et efficace pour la détermination des eutectiques invariants, ne peut être employée pour celle des lignes eutectiques monovariantes.

Tous ces résultats peuvent vraisemblablement être

généralisés aux réactions eutectiques non invariantes dans les systèmes *n*-aires.

Bibliographie

1) H. Bibring, M. Rabinovitch et G. Seibel, Compt. Rend. (Paris) C **268** (1969) 1666.
2) W. Kurz, J. C. Hubert et B. Lux, Journées d'Automne de la Métallurgie Paris 1970.
3) H. Cline et J. W. Walter, General Electric Report n° 70C–165.
4) H. E. Bates, F. Wald et M. Weinstein, J. Mater. Sci. **4** (1969) 25.
5) P. Combrade, Thèse, Nancy, à publier.
6) J. A. Burton, R. C. Prim et W. P. Schilchter, J. Chem. Phys. **21** (1953) 1987.
7) C. Potard et P. Desre, Mem. Sci. Rev. Met. **62** (1970) 229.
8) J. D. Hunt et K. A. Jackson, Trans. AIME **236** (1966) 1129.
9) Ha-Quac-Bao, Thèse, Grenoble, 1970.
10) H. S. Yue et J. B. Clark, Trans. AIME **221** (1961) p. 383.

Journal of Crystal Growth **13/14** (1972) 757–764 © *North-Holland Publishing Co.*

CROISSANCE EN SOLIDIFICATION DIRIGÉE DE L'EUTECTIQUE QUASI-BINAIRE Ni_3Al–Ni_3Ta

J.-C. HUBERT, W. KURZ et B. LUX

Institut Battelle, Centre de Recherche de Genève, 1227 Carouge/Genève, Suisse

The eutectic Ni_3Al–Ni_3Ta of the quasi-binary system Ni_3Al–Ni_3Ta, the phase diagram of which is not known, has been isolated as part of work directed towards the development of eutectic alloys for high temperature applications.

After a brief description of the experimental procedures, including the localization of the composition by the method of zone melting, the structures of this eutectic are described in terms of the solidification parameters.

The eutectic alloy, having a composition 64.35 wt% Ni, 30.2 wt% Ta, 5.45 wt% Al, has a fibrous structure of Ni_3Al in a matrix of Ni_3Ta, the fibres representing about 35% of the volume.

The two following morphological anomalies: – high interfibre spacing λ, and relatively high volume fraction of fibres – are explained by analogy with other eutectic systems showing a similar, but more marked anomaly, such as Co–CoAl and Ni–Cr, and also with the help of the growth law: $\lambda^2 R = $ constant.

This information, combined with electron probe microanalyses, gives an indication of the basic form of the phase diagram.

In conclusion, some tensile and creep properties at high temperature are presented.

1. Introduction

Une récente étude[1]) sur les alliages eutectiques montre que les deux tiers des microstructures des alliages eutectiques orientés connus sont soit fibreuses, soit lamellaires, soit les deux à la fois. Ce sont ces structures qui sont à l'origine de nombreux travaux sur les alliages eutectiques en vue d'applications diverses.

Dans le cadre de la recherche d'alliages à haute résistance à chaud, les alliages eutectiques à structure orientée Ni_3Al–Ni_3Nb[2]), Co–TaC[3,4]), Co–Cr–C[5]), etc. ont permis d'envisager des progrès substantiels. Dans cette classe d'alliages, l'eutectique Ni_3Al–Ni_3Ta a été isolé et étudié. Du point de vue des propriétés mécaniques, cet alliage est très voisin de l'eutectique Ni_3Al–Ni_3Nb, mais sa structure est fibreuse, ce qui est un avantage pour les applications dynamiques avec des contraintes triaxiales.

2. Préparation des alliages

Les alliages primaires sont fondus par chauffage à induction à haute fréquence dans une nacelle en argent refroidie à l'eau, sous pression statique d'argon purifié. Au préalable, un vide supérieur à 10^{-3} Torr est réalisé dans l'enceinte en quartz. Les traces d'oxygène et d'azote sont piégées par du titane chauffé à 1000 °C environ. Cette technique permet l'élaboration de lingots très propres.

Le tableau 1 indique la pureté des éléments utilisés pour l'élaboration des alliages.

La solidification orientée sous pression statique d'argon est réalisée dans un four vertical à résistance de molybdène à six zones de chauffage. Ce dispositif permet d'ajuster la valeur du gradient de température au front de solidification et de limiter la convection dans le liquide. Ce four fonctionne sans régulation de température. Une alimentation stabilisée à 0.1% permet d'avoir des variations de température inférieures à 0.1 °C à 1300 °C.

Les alliages sont fondus puis solidifiés dans des creusets d'alumine de 5 mm de diamètre intérieur, intro-

TABLEAU 1

Analyse des éléments de base (impurités en ppm; analyse spectroscopique donné par les fournisseurs)

Métal	Forme	Pureté	Ag	Al	C	Ca	Cu	Fe	H_2	Mg	Mo	Ni	O_2	Si	Ta	Ti
Al	barre	5N	—	reste	—	—	0.4	1	—	0.2	—	—	—	1	—	—
Ni	poudre	5N	<1	<1	—	<1	<1	5	—	2	—	reste	—	3	—	—
Ta	barre	3N5	—	—	30	—	—	50	7	—	30	—	50	—	reste	20

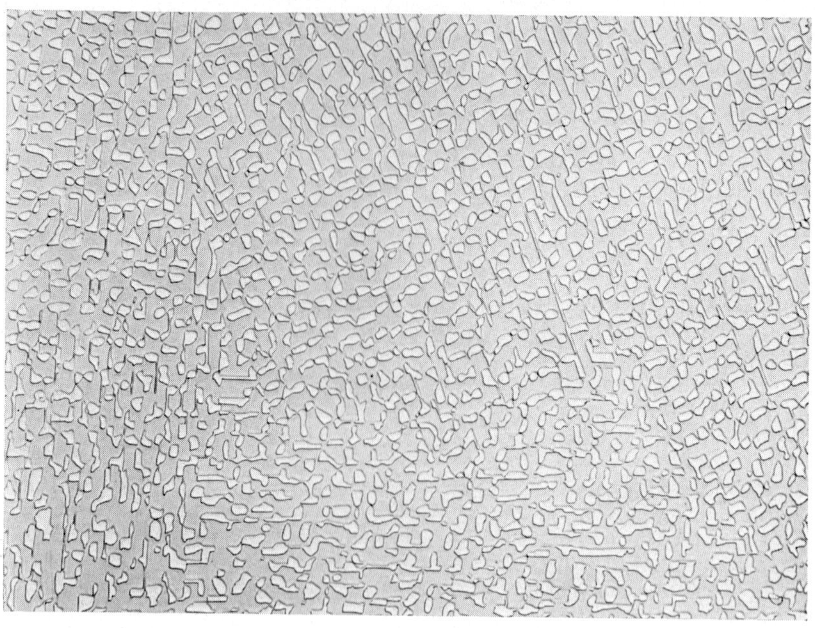

Fig. 1.   Eutectique orienté Ni$_3$Al–Ni$_3$Ta obtenu après un passage en zone flottante (coupe transversale: ×200).

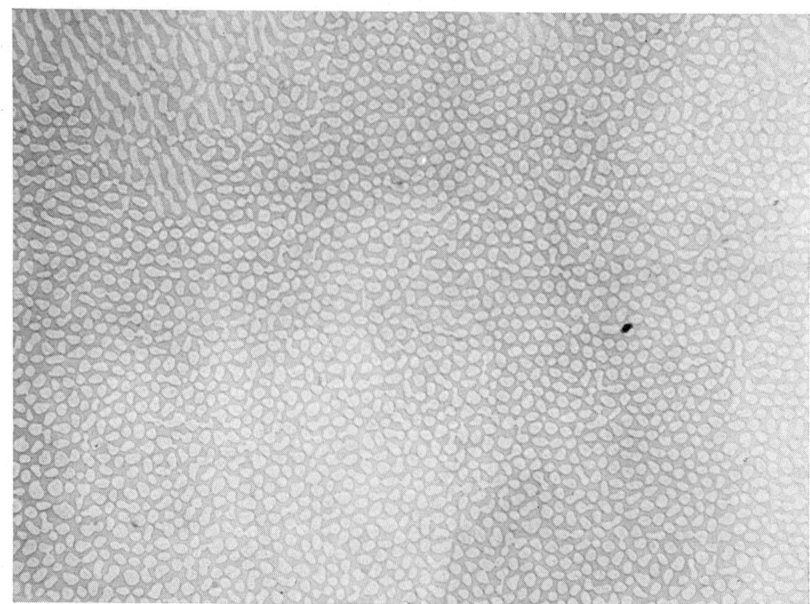

Fig. 2.   Eutectique orienté Ni$_3$Al–Ni$_3$Ta obtenu par solidification orientée ($R = 5.6 \times 10^{-4}$ cm/s, $G = 90$ °C/cm): fibres Ni$_3$Al – matrice Ni$_3$Ta (coupe transversale: ×100).

duits par la partie supérieure du four et déplacés à vitesse constante vers le bas. Chaque barreau ainsi solidifié a 5 mm de diamètre et 10 à 15 cm de long.

Les examens métallographiques ont été faits sur des coupes réalisées au milieu des barreaux, afin d'observer les structures correspondant à une vitesse de solidifica-tion pratiquement égale à la vitesse de déplacement du creuset.

3. Recherche de l'eutectique Ni$_3$Al–Ni$_3$Ta

Du fait que le système quasi-binaire Ni$_3$Al–Ni$_3$Nb[6]) possède un eutectique, l'analogie entre les composés

Ni$_3$Nb et Ni$_3$Ta permettait de supposer que le système Ni$_3$Al–Ni$_3$Ta pouvait aussi avoir un eutectique.

Dans ce cas, la nappe du liquidus du système Ni–Al–Ta était entièrement inconnue. L'eutectique du système quasi-binaire Ni$_3$Al–Ni$_3$Ta fut isolé par fusion en zone flottante, méthode décrite par Yue et Clark[7]).

Après une passe en zone flottante (vitesse de déplacement de la zone: 8.3×10^{-4} cm/s, longueur du barreau: 80 mm, longueur de la zone: 5.6 mm, diamètre du barreau: 5 mm) d'un alliage 64.0% Ni, 30.48% Ta, 5.52% Al en poids, une structure eutectique typique fut obtenue en queue du barreau (fig. 1) pour la composition suivante: 64.35% Ni, 30.2% Ta, 5.45% Al en poids, très proche de la composition eutectique estimée à 64.1% Ni, 31.0% Ta, 4.9% Al en poids.

4. Structure et morphologie

En général, nous avons trouvé une structure fibreuse très dense (fig. 2) avec des fibres de Ni$_3$Al dans une matrice de Ni$_3$Ta ($R = 5.6 \times 10^{-4}$ cm/s, $G = 90\,°$C/cm). La régularité de cette morphologie fibreuse est mise en évidence au microscope électronique à balayage (fig. 3) malgré la difficulté de réaliser une attaque différentielle entre les deux phases qui, comme nous le verrons, ont des compositions voisines (tableau 2). A la composition eutectique, le pourcentage en volume des fibres de Ni$_3$Al est de l'ordre de 35%, ce qui est élevé pour une structure fibreuse.

Dans certains cas, de légers écarts de composition ont donné, dans des conditions de croissance identiques, des dendrites de Ni$_3$Al et un eutectique interdendritique à structure fibreuse très régulière (fig. 4). Dans d'autres cas, une structure quasi-lamellaire composée de rubans ramifiés (fig. 5) a pu être observée pour une vitesse de croissance de 3.7×10^{-4} cm/s et un gradient de température au front de solidification de $90\,°$C/cm.

Enfin, pour des vitesses de solidification supérieures ou égales à 1.44×10^{-3} cm/s et des gradients de tempé-

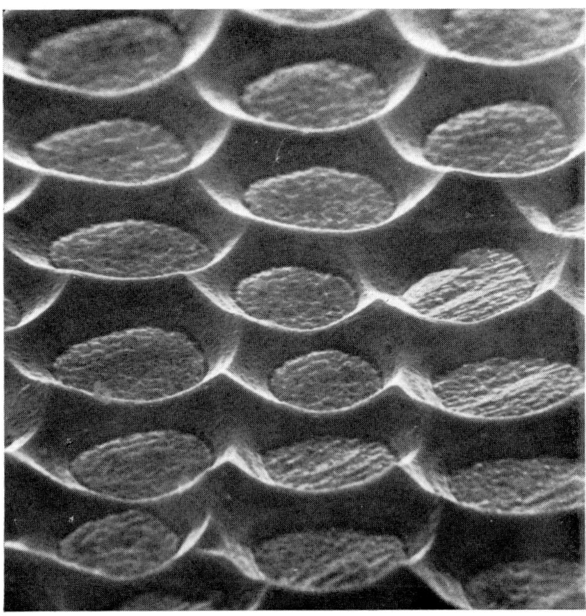

Fig. 3. Eutectique orienté Ni$_3$Al–Ni$_3$Ta: disposition régulière des fibres de Ni$_3$Al dans la matrice de Ni$_3$Ta après attaque profonde (microscope électronique à balayage: ×1920).

rature au front de solidification de l'ordre de $90\,°$C/cm, aucune structure régulière n'a été obtenue.

Comme l'ont décrit Mollard et Flemings[8]) pour le système eutectique Pb–Sn, nous avons constaté que pour des écarts de composition inférieurs à 2%at. de Ni$_3$Ta et pour des rapports G/R inférieurs à 0.34×10^6 °C s/cm^2, la structure composite est conservée avec apparition de lamelles.

5. Loi de croissance

Les mesures de la distance moyenne interfibre λ séparant les axes des fibres, montrent que celle-ci est fonction de la vitesse R. Ainsi, la courbe représentative de λ en fonction de R (fig. 6) dans un système logarithmique, peut être assimilée à une droite respectant la loi de croissance des eutectiques binaires proposée par

TABLEAU 2

Analyse à la microsonde électronique des phases de l'eutectique Ni$_3$Al–Ni$_3$Ta

Phase	Ni		Ta		Al		Formule stoéchiométrique approximative
	%pds	%at.	%pds	%at.	%pds	%at.	
Fibres	72.3	79.4	22.4	8.0	5.3	12.6	Ni$_3$(Al$_{0.6}$Ta$_{0.7}$)
Matrice	58.7	75.0	38.0	15.8	3.3	9.2	Ni$_3$(Ta$_{0.63}$Al$_{0.37}$)

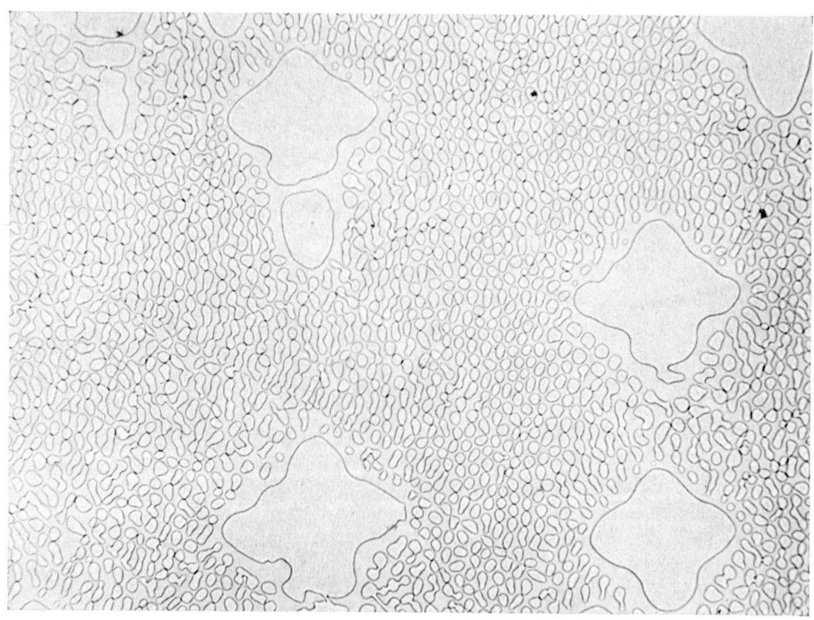

Fig. 4. Alliage hypoeutectique Ni$_3$Al–Ni$_3$Ta orienté: dendrites de Ni$_3$Al, eutectique interdendritique Ni$_3$Al–Ni$_3$Ta (coupe transversale: ×100).

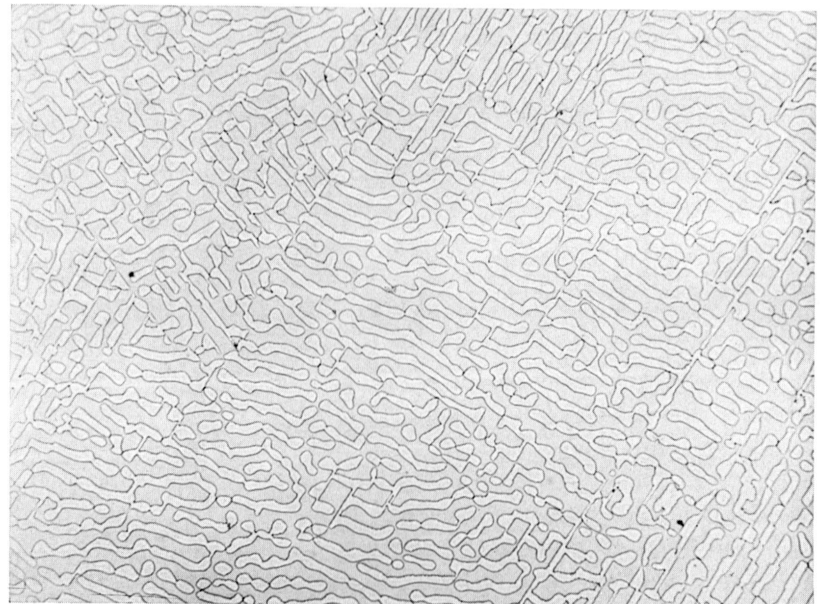

Fig. 5. Structure quasi-lamellaire de rubans ramifiés pour une composition proche de l'eutectique Ni$_3$Al–Ni$_3$Ta (coupe transversale: ×100).

Jackson et Hunt[9]:

$$\lambda^2 R = \text{constante}.$$

Toutefois, pour l'eutectique Ni$_3$Al–Ni$_3$Ta, on constate que ces distances λ sont sensiblement supérieures à celles mesurées dans la plupart des systèmes eutectiques connus[10]), lesquelles se situent dans la zone hachurée de ce graphique (fig. 6). Un certain nombre de systèmes semblent présenter ainsi des anomalies en comparaison des valeurs de λ habituellement mesurées, tel est le cas des eutectiques Ni–Cr[11]) et surtout Co–CoAl[12]).

Si l'on regarde de façon plus détaillée la loi de crois-

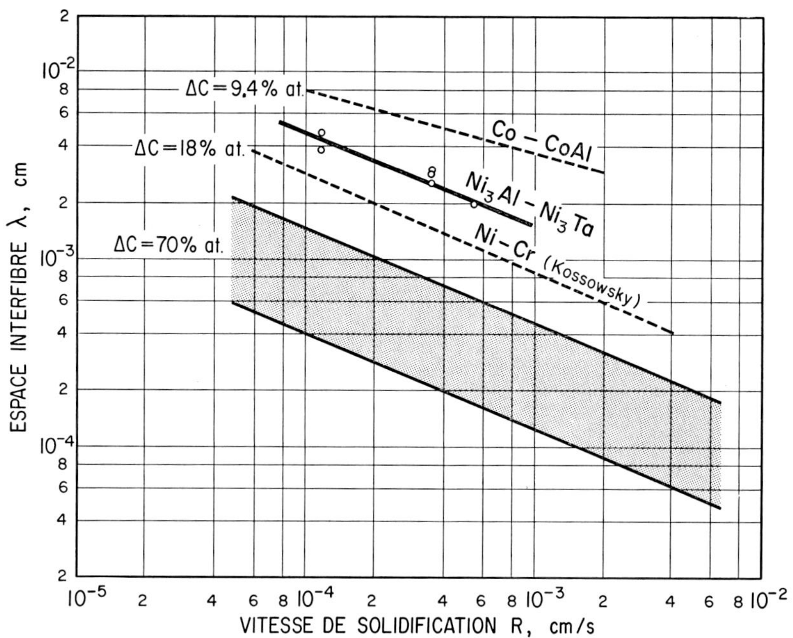

Fig. 6.   Influence de la différence de composition ΔC des deux phases d'un eutectique orienté sur l'espace interfibre λ en fonction de la vitesse de solidification R.

sance telle qu'elle est proposée par Jackson et Hunt[9]), nous avons:

$$\lambda^2 R = D \,\frac{1}{P}\,\frac{1}{\Delta C}\left(\frac{a_\alpha}{m_\alpha}\,v_\beta + \frac{a_\beta}{m_\beta}\,v_\alpha\right),$$

où

D = coefficient de diffusion dans le liquide,

P = une fonction du volume des deux phases de l'eutectique,

ΔC = $C_\alpha - C_\beta$; différence entre les compositions atomiques des phases α et β à la température eutectique*,

m_α, m_β = pentes des branches de liquidus à l'eutectique,

a_α, a_β = coefficients fonction de la géométrie de l'interface et des énergies interfaciales des phases α et β avec le liquide, respectivement,

v_α, v_β = fractions volumiques des phases α et β.

On remarque ainsi que

$$\lambda^2 R = f(1/\Delta C).$$

Pour tous les systèmes eutectiques dont la fonction $\lambda^2 R$ = constante se situe dans la zone hachurée, la dif-

* Cette différence doit être ramenée au seul système eutectique considéré qui peut être le tout ou la partie d'un diagramme de phases donné (par exemple, Co–CoAl).

férence ΔC est supérieure à 70 %, alors que pour les systèmes Ni–Cr et Co–CoAl, les différences ΔC valent respectivement 18 et 9.4 %at. D'autre part, l'étude des systèmes eutectiques à réaction eutectique binaire univariante, tels que (Ni, Co)–(Ni, Co)Al où ΔC varie régulièrement, a permis d'établir une corrélation entre ΔC et les valeurs de λ[12]).

Ces éléments nous indiquent que, pour l'eutectique Ni₃Al–Ni₃Ta, la longueur du palier eutectique, fixée par ΔC, est comprise entre 10 et 20 %at.

L'analyse des phases à la microsonde électronique et l'examen détaillé de la structure doivent nous permettre d'ébaucher le diagramme de phases Ni₃Al–Ni₃Ta encore inconnu.

6. Eléments du diagramme de phases Ni₃Al-Ni₃Ta

Le diagramme de phases Ni₃Al–Ni₃Ta correspond à une section quasi-binaire du diagramme ternaire Ni–Al–Ta comprise entre la phase γ' (Ni₃Al) du diagramme binaire Ni–Al[13]) et le composé intermétallique Ni₃Ta du diagramme binaire Ni–Ta[13]).

L'analyse à la microsonde électronique des deux phases (fibres et matrice) est donné au tableau 2 avec les équivalences en pourcentage atomique. Il est possible de situer ainsi les limites de solubilité des deux phases Ni₃Al et Ni₃Ta à l'état solide. On remarque

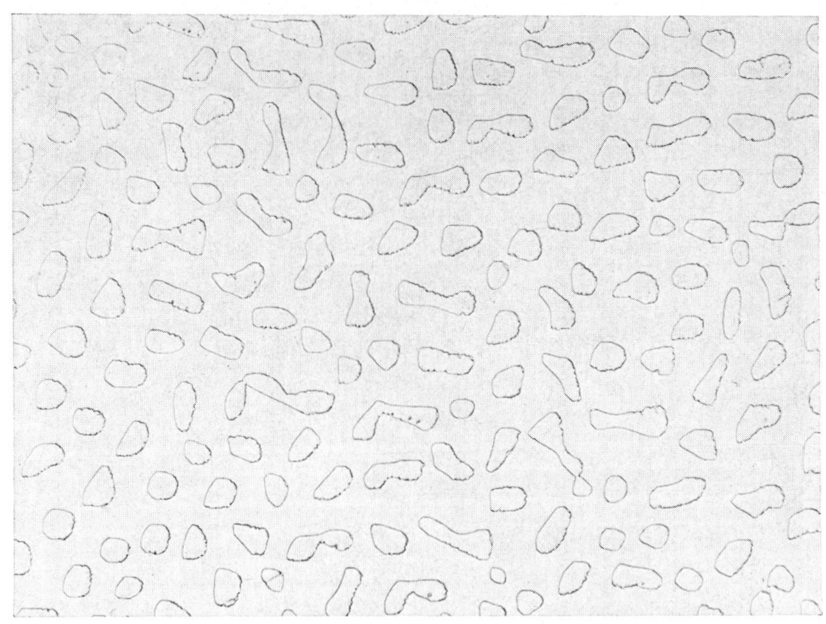

Fig. 7.   Coupe transversale de l'eutectique Ni$_3$Al–Ni$_3$Ta orienté, près du front de solidification: 26% en volume de fibres ($\times$200).

que la phase Ni$_3$Al (γ') dissout une quantité élevée de tantale, et la phase Ni$_3$Ta dissout de son côté une quantité appréciable d'aluminium.

L'examen des coupes métallographiques réalisées après trempe, près du front de solidification dans la partie orientée montre des fibres avec un bord dentelé caractéristique d'une diffusion à l'état solide (fig. 7). De plus, le pourcentage en volume des fibres de Ni$_3$Al n'est que de 26%. Cette valeur, étant inférieure à la fraction volumique critique établie par Jackson et Hunt[8]) (32%), permet d'expliquer que la structure déposée à la température eutectique ($\sim$1360 °C) soit fibreuse.

L'augmentation du pourcentage volumique par diffusion à l'état solide est d'autant plus importante que la courbure de l'un des solvus est forte. Ce phénomène a été particulièrement bien mis en évidence dans le système Co–CoAl[12,14]) où le pourcentage en volume des fibres de cobalt dans la matrice de CoAl est exceptionnellement élevé (70% en volume).

Ces quelques remarques sur la croissance de l'eutectique nous ont permis d'ébaucher le diagramme de phases quasi-binaire Ni$_3$Al–Ni$_3$Ta (fig. 8). En raison de l'analogie entre les phases Ni$_3$Ta et Ni$_3$Nb, nous avons superposé en trait fin le diagramme Ni$_3$Al–Ni$_3$Nb[6]).

6. Principales propriétés mécaniques

Si l'on compare l'eutectique Ni$_3$Al–Ni$_3$Ta aux su-

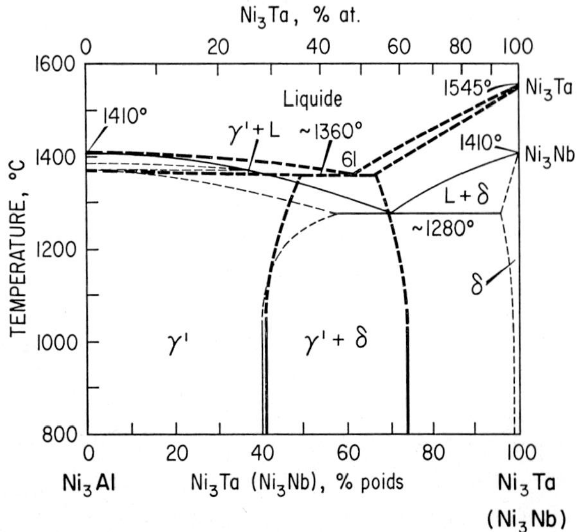

Fig. 8.   Diagramme de phases partiel du système Ni$_3$Al–Ni$_3$Ta superposé au diagramme de phases du système Ni$_3$Al–Ni$_3$Nb[7]).

TABLLAU 3

Comparaison du comportement au fluage de l'eutectique Ni$_3$Al–Ni$_3$Ta à celui du superalliage IN-100

Température (°C)	Contrainte (hbar)	Durée pour 1% d'allongement (h)	
		Ni$_3$Al–Ni$_3$Ta	IN-100
1038	10	1000	65
1000	14	640	58

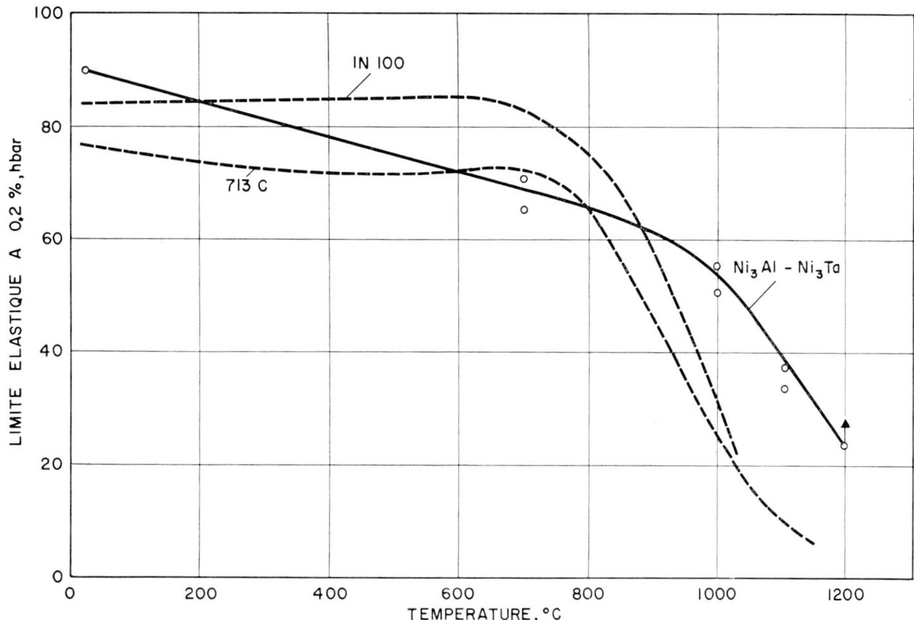

Fig. 9.   Limite élastique à 0.2% en fonction de la température de l'eutectique orienté Ni₃Al–Ni₃Ta comparée à celle des super-alliages IN 100 et 713 C.

peralliages actuellement utilisés (IN 100, 713 C), on constate un gain substantiel de la résistance à la rupture au-dessus de 1000 °C, et de la limite élastique à 0.2% (fig. 9)[15]) dès 900 °C. Pour un allongement de 1% en fluage, les temps de fluage à haute température comparés à ceux du superalliage IN 100 sont nettement améliorés (tableau 3): Ces quelques données sur les propriétés mécaniques mettent en évidence les possibilités des alliages eutectiques orientés.

7. Conclusions

Grâce à la technique de la fusion de zone et par analogie avec l'eutectique Ni₃Al–Ni₃Nb, l'eutectique a pu être isolé.

Cet eutectique présente une structure fibreuse. La distance interfibre est fonction de la vitesse de solidification et suit approximativement la loi de croissance:

$$\lambda^2 R = \text{constante.}$$

La structure présente deux anomalies morphologiques qui sont:

– une distance interfibre λ élevée en comparaison de celles mesurées pour de nombreux systèmes eutectiques binaires,

– un pourcentage en volume de fibres supérieur au pourcentage critique communément retenu.

L'analyse des paramètres de la loi de croissance et l'examen métallographique de l'évolution de la structure permettent d'expliquer ces anomalies.

Cette étude de la croissance de l'eutectique Ni₃Al–Ni₃Ta nous a permis d'ébaucher un diagramme de phases qui présente une analogie avec le diagramme du système Ni₃Al–Ni₃Nb.

Cet eutectique possède des caractéristiques mécaniques intéressantes pour son utilisation à très haute température.

Remerciements

Cette étude a été réalisée à l'Institut Batelle à Genève dans le cadre d'un contrat de recherche de la Direction des Recherches et Moyens d'Essais – Paris (contrat D.R.M.E. no. 69/537), à laquelle les auteurs expriment leurs vifs remerciements.

Bibliographie

1) L. M. Hogan, R. W. Kraft et F. D. Lemkey, Advan. Mater. Res., à paraître.
2) E. R. Thompson et F. D. Lemkey, Trans. ASM Quart. **62** (1969) 140.
3) H. Bibring et G. Seibel, Compt. Rend. (Paris) C **268** (1969) 144.
4) H. Bibring, M. Rabinovitch et G. Seibel, Compt. Rend. (Paris) C **268** (1969) 1666.
5) E. R. Thompson et F. D. Lemkey, Met. Trans. **1** (1970) 2799.

6) R. S. Mints, G. F. Byelyayeva et Y. S. Malkov, Zh. Neorg. Khim. **7** (1962) 2382.

7) A. S. Yue et J. B. Clark, Trans. Met. Soc. AIME **221** (1961) 383.

8) F. R. Mollard et M. C. Flemings, Trans. Met. Soc. AIME **239** (1967) 1534.

9) K. A. Jackson et J. D. Hunt, Trans. Met. Soc. AIME **236** (1966) 1129.

10) W. A. Tiller, dans: *Recent Research on Cast Iron*, Proc. Seminar, in Detroit, Michigan, 16–18 juin 1964, Ed. H. D. Merchant (1968), p. 129.

11) R. Kossowsky, W. C. Johnston et B. J. Shaw, Trans. Met. Soc. AIME **245** (1969) 1219.

12) J.-C. Hubert, W. Kurz et B. Lux, Morphologie de l'eutectique Co–CoAl et influence d'une addition de nickel, Communication aux Journées d'Automne, Paris 1970.

13) M. Hansen et K. Anderko, *Constitution of Binary Alloys*, 2e Ed. (McGraw-Hill, New York, 1958).

14) H. E. Cline, Trans. Met. Soc. AIME **239** (1967) 1906.

15) J.-C. Hubert, W. Kurz et B. Lux, Structure et propriétés d'un nouvel alliage eutectique à haute résistance à chaud, Communication aux Journées d'Automne, Paris 1970.

Journal of Crystal Growth **13/14** (1972) 765–771 © *North-Holland Publishing Co.*

UNIDIRECTIONAL SOLIDIFICATION BEHAVIOR IN REFRACTORY OXIDE-METAL SYSTEMS*†

A. T. CHAPMAN and R. J. GERDES

Georgia Institute of Technology, Atlanta, Georgia 30332, U.S.A.

and

J. C. WILSON and G. W. CLARK

Oak Ridge National Laboratory, Oak Ridge, Tennessee 37831, U.S.A.

Eutectic growth has been observed in a number of refractory oxide–metal systems. During controlled solidification of UO_2–W mixtures, structures consisting of tungsten fibers uniformly embedded in the oxide matrix are formed. Tungsten fiber diameters between 0.3 and 1.0 μm and densities between 4 and 25 million per cm^2 have been obtained. Scanning electron micrographs of the various UO_2–W composite structures, after selective chemical etching to expose the metallic fibers, are employed to analyze the structures. A preliminary interpretation of the solidification process leading to the different UO_2–W microstructures is presented using the starting oxide–metal ratio, a proposed phase diagram and growth parameters.

1. Introduction

Since the initial description of unidirectional solidification in the systems UO_2–W[1] and stabilized ZrO_2–W[2], there has been no published information on the parameters controlling the types of eutectic structures obtainable during coupled growth in refractory oxide–metal systems. Several distinctively different type UO_2–W composite structures have been observed and these are analyzed using a proposed phase diagram, oxide–metal ratios and the experimental growth conditions.

The technique used to grow this unique class of materials (cermets) will be briefly reviewed, prior to describing the solidification behavior and presenting scanning electron micrographs of the different types of structures obtained in the UO_2–W system. Recently eutectic growth was found in isolated areas of additional oxide–metal combinations[3].

2. Experimental techniques

2.1. GROWTH

A modified floating-zone technique, named the In-ternal Centrifugal Zone Growth (ICZG)[4], has been used to grow single crystals of refractory oxides and is employed to grow the oxide–metal composites. In this technique pressed rods of the oxide–metal mixture are sintered inside rf heated molybdenum tubes in an inert atmosphere to densify and preheat the material. After the initial heating, which also serves to increase the electrical conductivity of the oxides, the molybdenum tube heaters are separated to expose approximately 2 cm of the rod to an rf field of 3 to 30 MHz, depending on the material to be melted. The concurrent increase of temperature, electrical conductivity, and resistance heating continues until the interior of the rod melts at temperatures up to 3000 °C. The high radiant heat loss from the surface and the inherent low thermal conductivity of the oxides maintains the skin of the rod well below the eutectic temperature of the mixture. Composite growth is obtained by moving the molten zone up through the rod. In practice a cavity is generated in the molten zone because of the difference in density between the initial polycrystalline rod and solidified composite, and the oxide and metal melt from the roof of this cavity and solidify at the base.

During the growth of pure oxides it was found advantageous to rotate the rods at speeds above 300 rpm. During rotation the molten material is centrifugally cast against the sides of the cavity; thus, the liquid-solid interface is modified so that crystallization in the center of the rod leads solidification near the circum-

* This research was supported by the Advanced Research Projects Agency of the Department of Defense and was monitored by the U.S. Army Missile Command under Contract Number DAAHOI-70-C-1157.

† The views and conclusions contained in this document are those of the authors and should not be interpreted as necessarily representing the official policies, either expressed or implied, of the Advanced Research Projects Agency or the U.S. Government.

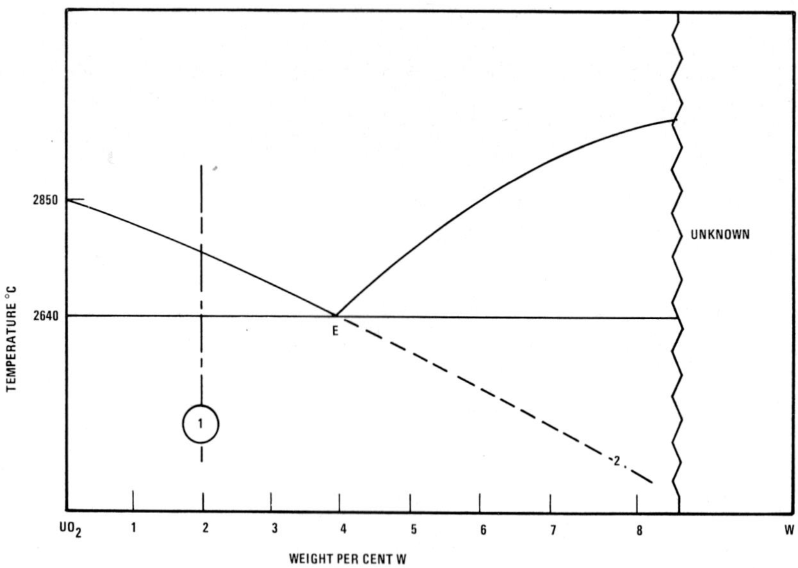

Fig. 1.   Proposed phase diagram for the system UO_2–W.

ference of the rod and promotes the uniform growth of one crystallographic orientation across the entire melt zone. The need for rotation during the growth of oxide–metal composites is uncertain at present because similar structures have been successfully obtained both with and without rotation. The occurrence of cell or grain boundaries across the composite structure, combined with the increased thermal conductivity due to the metallic fibers, may minimize the need for the liquid–solid interface geometry produced by rotation.

2.2. CHEMICAL ETCHING

After growth the UO_2–W composites were selectively chemically etched to remove a layer of oxide and expose the tungsten fibers for examination in the scanning electron microscope (SEM). Before etching the samples were mechanically polished to a bright finish using successively finer grinding paper and finishing with one micron diamond paste on a nylon cloth. Reflected light examination was used to check the surface prior to chemical etching. An etching solution[5]) was prepared consisting of:

(1) 20 ml saturated chromic acid,
(2) 10 ml glacial acetic acid,
(3) 7 ml concentrated nitric acid,
(4) 5 ml 48 % hydrofluoric acid.

Polished samples usually in the shape of wafers were clipped to a shaft rotating at between 20 to 30 rmp and immersed in the etching solution for a predetermined time, depending on the fiber length desired. After etching the samples were washed by rotation in a water bath prior to the SEM analysis.

2.3. SCANNING ELECTRON MICROSCOPE

The chemically etched samples were examined in a Stereoscan scanning electron microscope. Specimen preparation, such as plating with a conductive layer of gold palladium, was not necessary with the UO_2–W composites. The secondary emission mode of the SEM was normally used during these studies. The samples were viewed at tilt angles between 35° and 80° using incident beam voltages of 10 to 20 kV; thus, contrast due to differences in topography was recorded in the scanning electron micrographs.

3. Solidification behavior of the oxide–metal system UO_2–W

Since there is no existing phase diagram for the system UO_2–W, a tentative diagram is proposed (fig. 1) which is consistent with a number of experimental observations. This diagram, along with the scanning electron micrographs of various UO_2–W composite structures, is employed to interpret the controlled solidification leading to coupled growth in the refractory oxide–metal system UO_2–W.

The formation of eutectic structures from UO_2–W

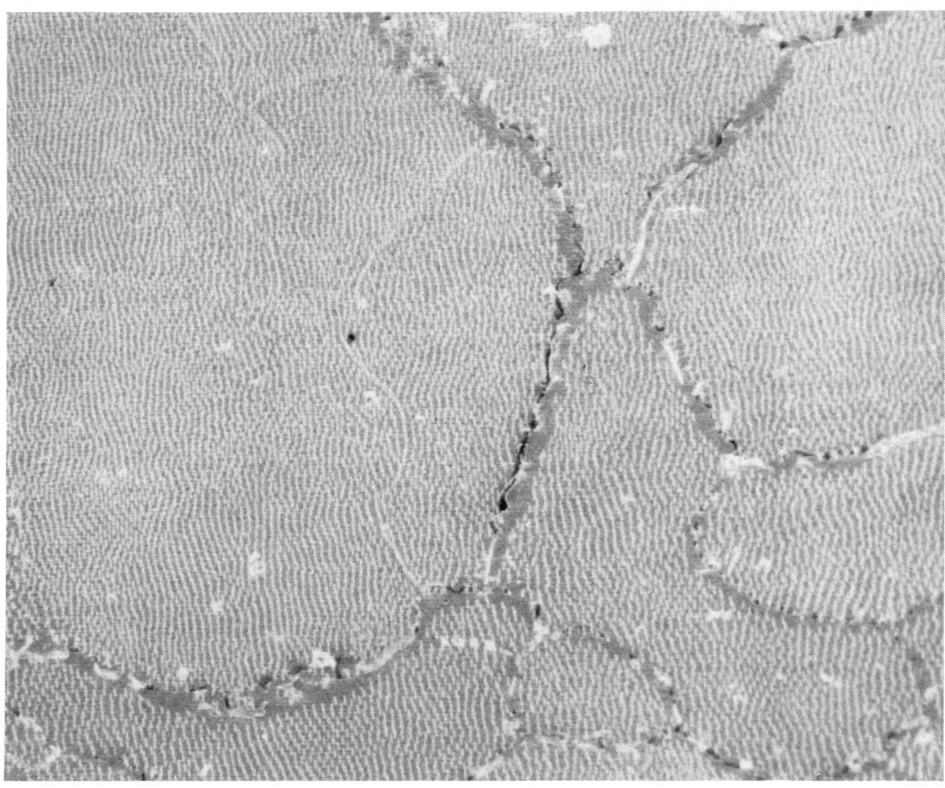

Fig. 2.   Scanning electron micrograph of UO_2–W composite displaying uniform W fiber distribution, emissive mode of SEM. ×220.

mixtures indicated these components form a simple eutectic system and, hence, justifies the general configuration of fig. 1. Very limited solid solubility is proposed because X-ray information showed lattice parameters very close to the standard values for pure UO_2 and W[6]). The eutectic temperature cannot be more than two or three hundred degrees below the melting point of UO_2, because melting pure UO_2 or UO_2–W mixtures yielded very similar sample surface temperatures. Several preliminary experiments[7]) have explored melting UO_2 crystals on a tungsten ribbon filament. The reaction (eutectic temperature) between the oxide and metal occurred at 2640 ± 30 °C. The eutectic composition is estimated, from analysis of SEM pictures, to occur between 3 to 5 wt% tungsten.

Initially it should be pointed out that simply using the starting oxide–metal ratio for the composition in the UO_2–W phase diagram is inaccurate, because transport processes are very rapid at the extremely high temperatures needed to melt these mixtures; and significant quantities of the metal are lost from the molten zone, probably by vaporization or a diffusion mechanism.

During the examination of the UO_2–W samples, two basically different types of oxide–metal structures have been observed. The first type is shown in fig. 2 and is characterized by the growth of W fibers essentially uniformly distributed across the melted zone. This growth pattern is interrupted by cell or colony boundaries where major changes in oxide and metal orientation occur. The second growth type is shown in fig. 3 and has circular areas of the oxide bounded by the composite (oxide–metal eutectic) structure. The larger grain or cell structure shown in the previous figure is also present in this type of sample.

In the case of the uniform fiber growth, fig. 2, no conclusion can be drawn at present as to the existence of a relationship between grain size, shape and orientation. The shapes of the grains are highly irregular, and the diameters vary from a few microns to a few millimeters. In the structures containing the circular oxide areas (fig. 3), the location of the cell boundaries also appears to be random, although they traverse the composite region more often than not.

Figs. 4a, b, c and d present higher magnification

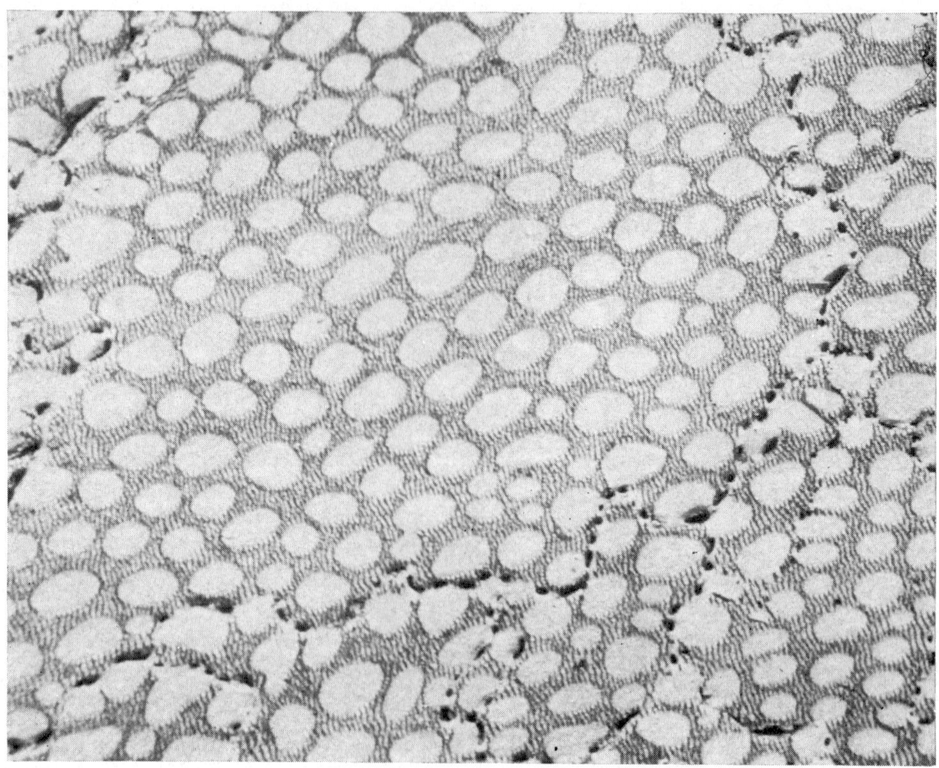

Fig. 3. Scanning electron micrograph of UO_2–W composite displaying circular primary phase areas of UO_2 plus eutectic growth, reflective mode of SEM. $\times 220$.

views of these types of structures, and the growth information for these samples is given in table 1. The W fiber diameter, density, and eutectic composition are average values obtained by examining scanning electron micrographs of many growth regions in these and other similar samples. The eutectic composition was determined by a simple area count of tungsten fibers and converting this information to weight percent metal using the fiber diameters and the theoretical density values for the UO_2 and tungsten.

TABLE 1

Growth and structural information for UO_2–W composites shown in fig. 4

	(a)	(b)	(c)	(d)
Starting composition (wt%)	12.5	10	10	5
Growth rate (mm/hr)	15	50	15	60
Fiber diameter (µm)	0.75–0.94	0.33	0.90	0.42
Fiber density (eutectic areas only) $\times 10^6$ per cm^2	4.5	25	6.5	56
Eutectic composition (wt% W)	3.6–5.3	3.8	7.1	12.8

Examination of many areas in samples displaying uniform growth (figs. 4a and b) indicated that the eutectic composition in the System UO_2–W occurred between 3.6 and 5.3 wt% W. (This analysis is the justification for the eutectic composition shown in the proposed phase diagram drawn in fig. 1.) Similar examinations of *only* the eutectic areas in the samples containing the circular primary areas of UO_2 (figs. 4c and d) indicated a eutectic composition between 7.1 to 12.8 wt% W. An explanation for this apparent discrepancy follows the analysis of Lundquist et al.[8]). They showed that solidifying near eutectic compositions in metal-metal systems produced continuous fine "eutectic" structures, whereas solidifying off-eutectic mixtures lead to primary phase regions distributed throughout the eutectic structure.

During cooling of oxide-rich UO_2–W mixtures, represented in general by composition 1 in the phase diagram, the primary phase nucleated at a temperature higher than the eutectic temperature; and, as cooling proceeded, the composition of the remaining liquid moved down the liquidus line toward the eutectic com-

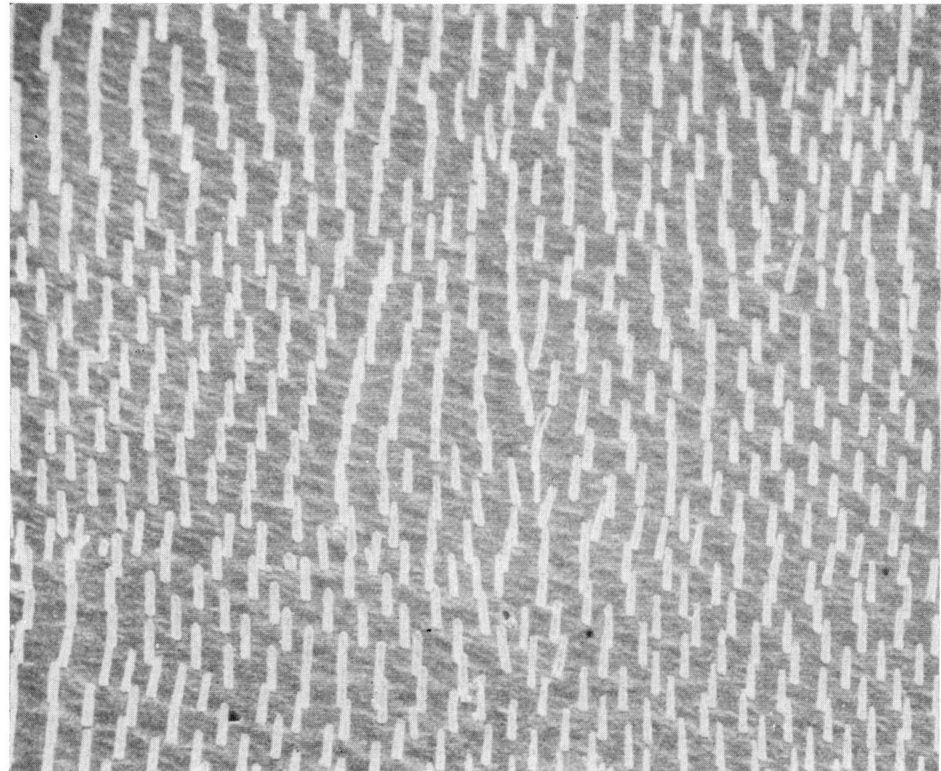

Fig. 4a

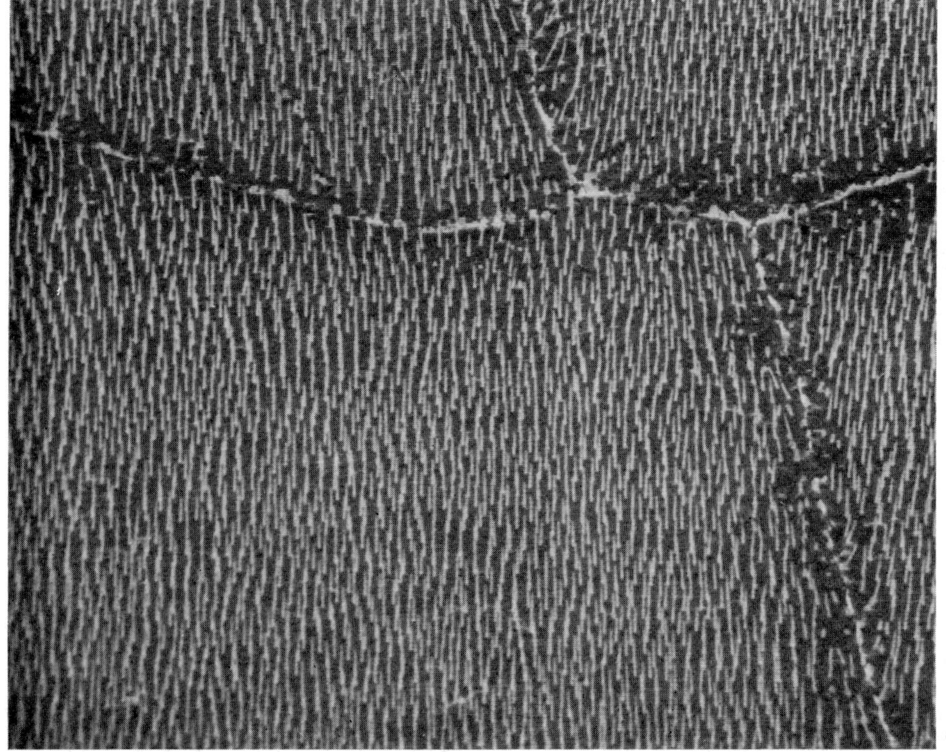

Fig. 4 b

XV – 6

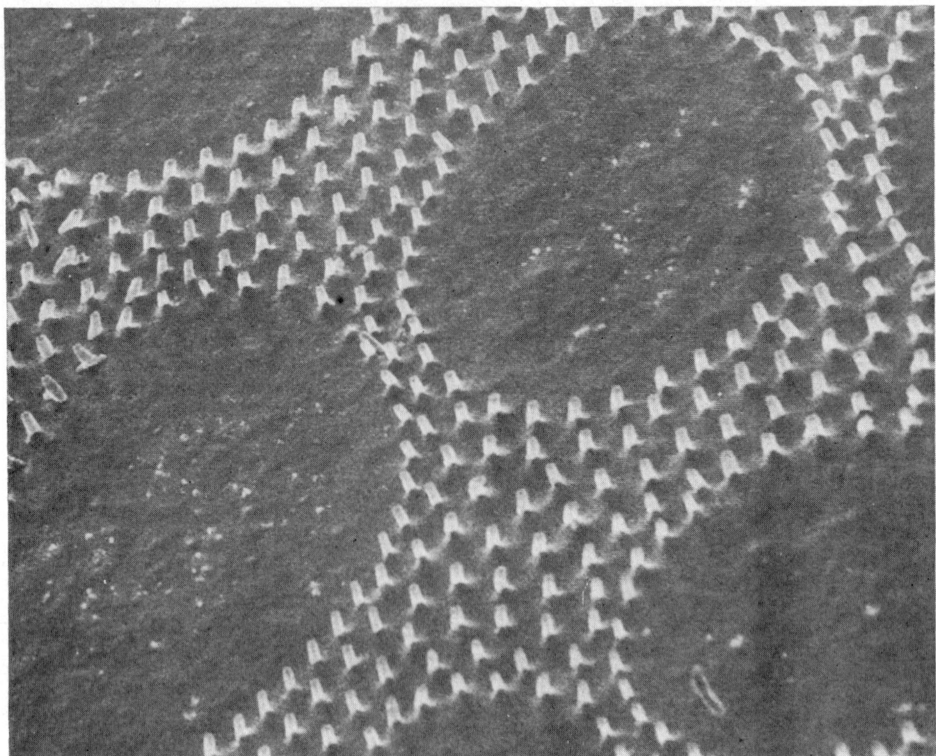

Fig. 4c

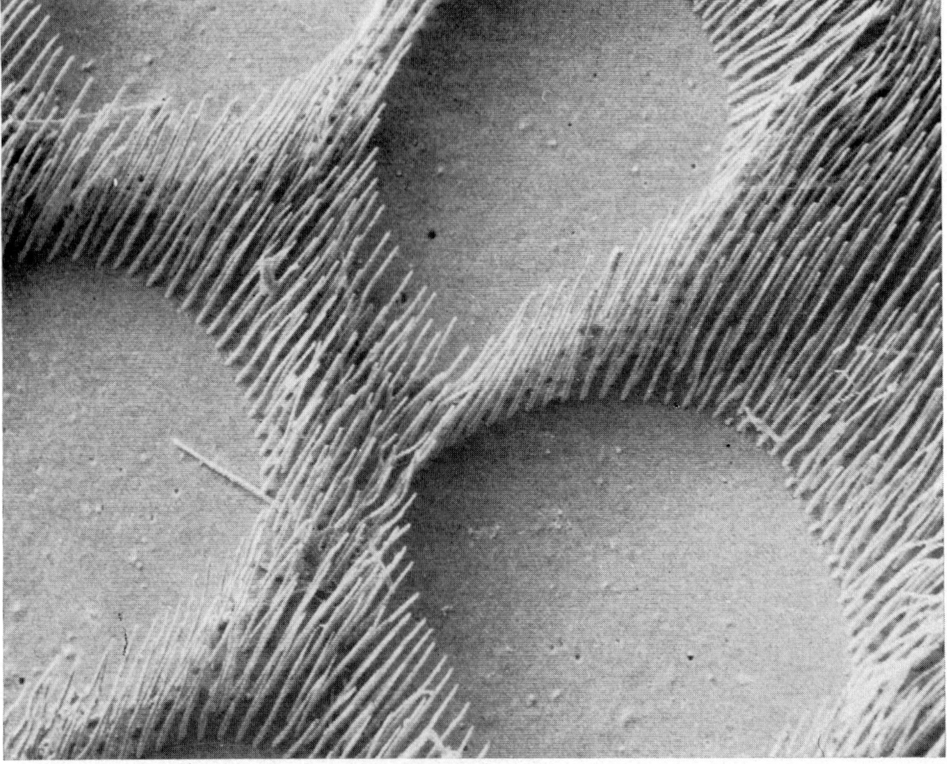

Fig. 4d

Fig. 4. Scanning electron micrographs of UO_2–W composites showing influence of oxide–metal ratio and growth rate on the composite structure and W fiber diameter and density. Samples chemically etched to expose the fibers, emissive mode of SEM. (a) Near eutectic composition, slow growth rate, $\times 1000$. (b) Near eutectic composition, fast growth rate, $\times 1100$. (c) Oxide-rich composition, slow growth rate, $\times 1600$. (d) Oxide-rich composition, fast growth rate, $\times 1500$.

position. Some undercooling is expected, and the liquid follows the metastable extension of the liquidus curve from E towards 2, and became richer in tungsten; and when sufficient undercooling was available to initiate nucleation, the tungsten content of the liquid approached 7 to 12 wt%. Thus, supercooling can account for the higher W concentration in the eutectic areas of samples (figs. 4c and d) containing the primary UO_2 regions. Also, it is noteworthy that increasing the growth rate should increase the undercooling and, hence, the amount of W in the eutectic areas of the faster grown samples. This behavior was observed for samples (c) and (d) (table 1). Since the melting point of UO_2 and the proposed eutectic temperature have a small spread, the liquidus line has a comparatively small slope; and the amount of undercooling required to change the metal–oxide ratios from 4 to 10% tungsten would be only 200 to 400 °C – not unreasonable values considering the extremely high eutectic temperature in this system.

The influence growth rate had on the fiber density and diameter can be readily seen by comparing the structures shown in fig. 4 and the data in table 1. The slower grown samples produced large, widely spaced fibers; whereas increasing the growth rate decreased fiber size and spacing. This behavior is typical of the eutectic solidification found in metal–metal systems, although the structures are usually lamellar rather than the fiber or rod type.

The analysis and controlled growth of a variety of UO_2–W mixtures is continuing. The initial attempts to melt tungsten rich mixtures using the induction heating technique have been unsuccessful, for reasons not clear at present.

4. Conclusions

The development of a technique suitable for the unidirectional solidification of refractory oxide–metal mixtures to form composites (cermets) provides a unique class of materials. The very stable nature of the oxides, the excellent bonding between metal and metal oxide, and the formation of structures containing many millions of less than 1 µm diameter metallic fibers per square centimeter, uniformly distributed in a refractory oxide matrix, yields a material with very anisotropic behavior. The growth and analysis of these composites has progressed to the point where the different types of microstructures can be related to the oxide–metal ratios, the phase diagram, and growth parameters. In the system UO_2–W the W fiber density can readily be varied between 3 to 5 and 25 to 30 million per cm^2 with corresponding changes in fiber diameter and spacing.

A number of potential uses for this class of composites are being considered. The ability to selectively etch the oxide and expose the ordered array of metallic fibers produces a structure of interest for electron emission. A material containing many small metallic fibers in a semi-conducting or insulating matrix may be of value in high resolution electro-optic applications. The electrical or thermal paths provided by the metallic fibers suggest oxide–metal composites may also be useful as substrates for integrated circuits or interconnecting media for sandwich type solid state structures.

References

1) A. T. Chapman et al., J. Am. Ceram. Soc. **53** (1970) 60.
2) M. D. Watson et al., J. Am. Ceram. Soc. **53** (1970) 112.
3) R. P. Nelson and J. J. Rasmussen, J. Am. Ceram. Soc. **53** (1970) 527.
4) A. T. Chapman et al., J. Am. Ceram. Soc. **53** (1970) 46.
5) A. J. Manly, J. Nucl. Mater. **15** (1965) 143.
6) R. J. Gerdes et al., Science **167** (1970) 979.
7) Personal communication, Benjamin Oliver, consultant to the Metals and Ceramics Division, Oak Ridge National Laboratory, Oak Ridge, Tennessee.
8) B. E. Lundquist et al., J. Metals **91** (1962–63) 204.

EFFECT OF MODES OF CRYSTALLIZATION ON HABIT AND MORPHOLOGY OF COPPER DENDRITES

H. M. LIAW and J. W. FAUST, Jr.

College of Engineering, University of South Carolina, Columbia, South Carolina 29208, U.S.A.

Copper dendrites were grown from molten thallium solutions by three modes of crystallization. The crystals were examined for facet planes and growth directions. Changes in habit and morphology are reported.

1. Introduction

The growth habit and morphology of some crystals can be modified by the presence of specific impurities. Much of the literature on this subject prior to 1956 has been taken up by Buckley[1]). Most of these studies were on ionic crystals grown from aqueous solutions. Studies on habit modification of other materials have been reported since then. For example, Faust and John[2]), and Faust, John and Pritchard[3]) reported on studies of semiconductors. The effect of poisons on the habit and morphology of copper dendrites was recently reported by Liaw and Faust[4]). However, relatively little work has been concerned with the study of the change of crystal habit or morphology by varying the growth parameters.

In their study of solution growth by the twin plane re-entrant edge (TPRE) mechanism, Faust and John[5]) reported that the cooling rate and the solute concentration are factors which affect growth habit and morphology. For fcc metals and semiconductors of the zinc blend structure slow cooling and dilute solution favor the growth of twin crystals which most often grow in $\langle 211 \rangle$ directions, while rapid cooling favors the growth of single crystals growing in $\langle 100 \rangle$ directions. In the growth of copper dendrites from thallium solutions Liaw and Faust[4]) also found that cooling rates affected crystal habit and morphology.

The purpose of this investigation is to report on studies undertaken to determine whether crystal habit and morphology can be changed by changes in crys-

tallization methods other than cooling rates. Particular attention was given to conditions under which the TPRE mechanism operates for the growth of dendrites.

2. Experimental procedures

The Cu–Tl system was chosen for this study because Cu could be crystallized from thallium solutions by three different methods: namely, (i) cooling of the solution, (ii) evaporation of the solvent, Tl, and (iii) seeding of the saturated solution.

The apparatus and the materials for the preparation of solutions have been described elsewhere[4]). Fig. 1 shows the phase diagram of Cu–Tl system with the three methods of crystallization sketched in.

2.1. Nucleation and growth by cooling of solution

The desired composition of Cu–Tl, AC of fig. 1, was sealed in a quartz ampoule under a vacuum. The ampoule was then placed in the hot zone of the furnace for complete dissolution. At point A it was then cooled either linearly or non-linearly along ABC to the solidification temperature C. The copper was removed from the thallium by phosphoric acid.

2.2. Nucleation by evaporation of the solvent

An alternative way to nucleate Cu from the solution at point A of fig. 1 is to evaporate Tl along path AD. The sample for this study was held in a quartz crucible which was sealed in a quartz ampoule whose length was 5–10 cm. The ampoule was placed in the process

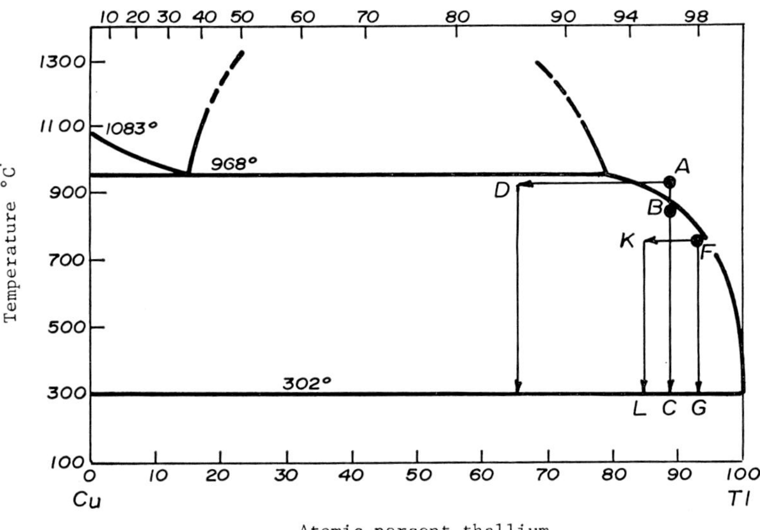

Fig. 1.   A sketch of phase diagram of Cu–Tl system.

tube of the preheated furnace in such a way that one end of the ampoule was hotter than the other end. The sample was kept at this temperature for a period of time to allow an appreciable amount of Tl to evaporate from the crucible and condense at the other end of the ampoule. After the evaporation path AD of fig. 1 had passed through the liquidus curve, the furnace was cooled down linearly to the solidification temperature of the melt. The amount of Tl evaporated, or the composition of solution D of fig. 1 was determined by either measuring the weight loss of the crucible or the of Tl condensed on the ampoule. The evaporation rate of Tl was determined by the amount of Tl evaporated and the time for the evaporation.

2.3. Growth of crystals by seeding

In this study seeding was not in the usual way of placing seeds into the solution but was a process of incomplete dissolution of the solute at a relatively low temperature. The desired composition of the sample (0.5 to 4.0 wt% Cu) was sealed in an ampoule and heated to a temperature at point F of fig. 1, which is a few degrees below its liquidus temperature. The nuclei at point F of fig. 1 were grown either by cooling along the path FG or by combination of evaporation and cooling along the path FKL. The procedures for the cooling and evaporation were the same as described in the previous paragraphs.

3. Experimental results

3.1. Growth habits obtained by cooling of solutions

Two types of cooling rates, linear and non-linear, have been used. The results agree with those published previously[4]). In short, the 3-dimensional [100] dendrites were obtained when the solutions were linearly cooled to their solidification temperatures. [110] acicular dendrites resulted if the solution was cooled be-

Fig. 2.   A stem of 3-dimensional primary ⟨110⟩ copper dendrites.

TABLE 1

Habit and morphology of crystals nucleated at various rates of evaporation; linear cooling rate was taken after the evaporation process

Initial composition (wt% Cu)	Final composition (wt% Cu)	Evaporation temperature (°C)	Evaporation rate (mg/hr)	Cooling rate (°C min)	Habit and morphology obtained
3.97	7.28	930	16.1	3.0	3D P ⟨110⟩ dendrites stalk and branches are formed by alignment of octahedra
3.95	100.0	930	13.0	3.0	Polycrystalline aggregates facets of crystallites were not well defined
3.95	5.75	940	17.0	3.0	3D P ⟨110⟩ dendrites tips of ⟨110⟩ dendrites split to ⟨100⟩ dendrites

low the liquidus temperature and then held at that temperature for a period of time before further cooling.

3.2. Habits of crystals nucleated by evaporation of the solvent

Crystal habits obtained by this mode of nucleation were essentially 3-dimensional primary ⟨110⟩ dendrites. The stems of the dendrites had the appearance of stacked octahedra as shown in fig. 2. Occasionally the tips of dendrites split into ⟨100⟩ dendrites. The facets of dendrites were bounded by {111} planes. When evaporation of Tl was completed, Cu crystals obtained were polycrystalline aggregates. The facets of aggregated crystals were not well defined. Table 1 lists the crystal habits and morphology obtained by several runs of this mode of nucleation.

3.3. Habits and morphologies of crystals nucleated by seeding and grown by linear cooling of the solution

Seeding was achieved by incomplete dissolution of the solute by heating up to a temperature just below its liquidus temperature. It was followed by linear cooling in the range of 0.75 °C to 10 °C/min. One run was performed by quenching. The habit of crystals obtained by most of the runs were either in the form of a rhom-

Fig. 3. A spear-like copper dendrites growth in a ⟨211⟩ direction.

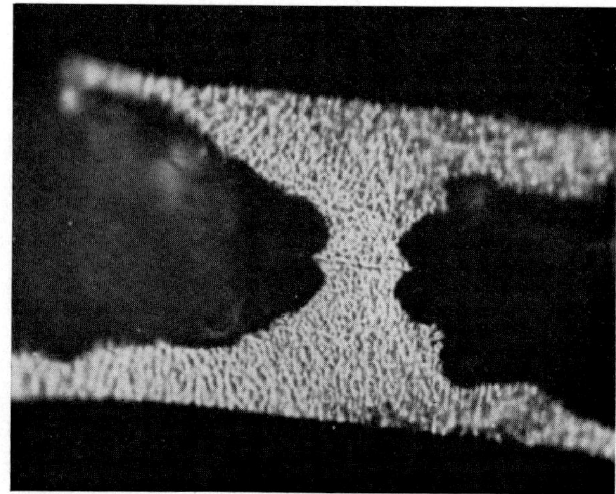

Fig. 4. An etched cross-section of a platelet dendrite showing a twin lamella in the core of H-arm like structure.

TABLE 2

Habits and morphologies of crystals nucleated by seeding and grown by linear cooling

Rate of linear cooling (°C min)	Wt% of Cu in solution	Max. temp. reached (°C)	Main type of habit and morphology obtained	Other type of habit and morphology
0.75	1.96	765	Only 2 massive 3D P $\langle 100 \rangle$ dendrites faceted by $\{111\}$	None
1.5	2.28	750	Rhombohedral platelets, $\{111\}$ main faces, growth in $\langle 211 \rangle$ direction	None
2.0	3.65	750	Either rhombohedral, butterfly or spear-like dendrites, $\{111\}$ facets growth in $\langle 211 \rangle$ direction	None
2.8	2.02	760	Either rhombohedral or butterfly-like platelets, $\{111\}$ facets growth in $\langle 211 \rangle$ direction	None
3.0	4.0	885	3D P $\langle 110 \rangle$ dendrites, tips of dendrites split in $\langle 100 \rangle$ direction	None
4.0	0.49	789	Only 3 hollow octahedra were obtained	None
10.0	0.47	550	Rhombohedral platelets, growth in $\langle 211 \rangle$ direction, $\{111\}$ facets	Octahedral crystallites and spongy
Quenched to room temp. by cold water	0.55	550	Tiny octahedral crystallites	Spongy copper

bus, butterfly, or spear-like platelets growing in the $\langle 211 \rangle$ directions as shown in fig. 3. Exceptions to this morphology were obtained when (i) The cooling rate was very slow (0.75 °C/min) in which case 3-dimensional primary $\langle 100 \rangle$ dendrites were formed. Stems of these dendrites were also formed by stacking of octahedra and faceted by $\{111\}$ planes. (ii) The composition of the solute was high (4.0 wt%) in which case 3-dimensional primary $\langle 110 \rangle$ dendrites were obtained. The tips of the $\langle 110 \rangle$ dendrites split into $\langle 100 \rangle$ dendrites. (iii) Very high cooling rate (>4.0 °C/min) or quenching resulted in tiny octahedra and spongy copper. Table 2 lists the habits and morphologies obtained by several different runs of this mode of nucleation.

3.4. Crystals nucleated by seeding and grown by the evaporation of the solvent

The path FKL of fig. 1 has been tried for the growth of copper dendrites. Due to the low vapor pressure of Tl at point F, which was taken as 760 °C in this study, no appreciable amount of Tl evaporated, thus the results are exactly the same as performed along the path FG in fig. 1.

3.5. Twinning in the solution grown copper dendrites

The different habits of the solution grown copper dendrites have been examined for twinning. The spear-like and butterfly-like platelet dendrites growing in the $\langle 211 \rangle$ directions were found to contain $\langle 111 \rangle$ 180° rotation twin planes. Fig. 4 shows an etched surface of a cross-sectioned platelet dendrite. A twin lamella is clearly seen across the core of H-arm like structure. Twin planes have not been observed in the arms of the copper dendrites as in the case of electrodeposited lead dendrites[6].

4. Discussions and conclusion

From the above results, it can be concluded that habit and morphology of copper dendrites can be modified by the choice of the crystallization methods.

The crystals nucleated by evaporation and grown by linear cooling give the morphology of $\langle 110 \rangle$ dendrites, which split to $\langle 100 \rangle$ dendrites at the later stage of growth. This result agrees with the previous study[4], which showed that crystals obtained by linear cooling were $\langle 100 \rangle$ dendrites.

The reasons that dendrites grow in the $\langle 110 \rangle$ directions instead of $\langle 100 \rangle$ directions have been discussed[4]. In short, this study confirms the previous study that the $\langle 100 \rangle$ directions of growth will be preferred if nucleation or growth has taken place with rapid heat dissipation, while the $\langle 110 \rangle$ direction is preferred for slow heat dissipation.

Twinned crystals obtained in this study by the incomplete dissolution of the solute requires further discussion. Since it has been shown[7] that most electrodeposited copper dendrites contain twin planes, it is believed that the twin platelets obtained in this study are due to the growth of twinned nuclei. (Since electro-deposited copper was used for the preparation of Cu–Tl solutions). However, some experiments have also been performed under the same mode of growth except with a higher concentration of copper in the solution. The results of these experiments showed that dendrites obtained are single crystals growing in the $\langle 100 \rangle$ direction. This gives further evidence that twin grooves are very effective sites for nucleation only if the supersaturation is not too high.

References

1) H. E. Buckley, *Crystal Growth* (Wiley, New York, 1956).
2) J. W. Faust, Jr. and H. F. John, J. Phys. Chem. Solids **25** (1964) 1407.
3) J. W. Faust, Jr., H. F. John and C. Pritchard, J. Crystal Growth **3, 4** (1968) 321.
4) H. M. Liaw and J. W. Faust, Jr., J. Crystal Growth **10** (1971) 302.
5) J. W. Faust, Jr. and H. F. John, Trans. AIME **233** (1965) 230.
6) H. M. Liaw and J. W. Faust, Jr., to be published.
7) J. W. Faust, Jr. and H. F. John, J. Electrochem. Soc. **110** (1963) 463.

Journal of Crystal Growth **13/14** (1972) 777–781 © *North-Holland Publishing Co.*

777

CRYSTALLOGRAPHIC ORIENTATIONS BETWEEN ALUMINUM GROWN FROM THE MELT AND VARIOUS TITANIUM COMPOUNDS*

J. CISSÉ and G. F. BOLLING

Scientific Research Staff, Ford Motor Company, P.O. Box 2053, Dearborn, Michigan 48121, U.S.A.

and

H. W. KERR

University of Waterloo, Waterloo, Ontario, Canada

The nucleation of aluminum by titanium has been attributed to many different reactions, but the only relationship thoroughly investigated seems to be that between Al and the peritectic compound Al_3Ti. We have now studied both the heterogeneous nucleation of Al by TiC and the solidification of Al in contact with massive pieces of TiC. The relationship, $(001)_{Al}//[001]_{TiC}$ and $[001]_{Al}//[001]_{TiC}$, seems to hold well even under conditions of varying carbide pre-oxidation. Such momplete epitaxy allows only one nucleus variant per compound particle in contrast to a greater number of variants in relationships (of partial epitaxy) accepted for Al_3Ti. With this result one can discuss the changing nucleation effectiveness of titanium at various concentrations and discover some important features influencing nucleation: In particular, we propose two apparently new and very simple rules that must apply every time that crystallographic relationships have to be considered in nucleation problems.

1. Introduction

The role of epitaxy in the phenomenon of heterogeneous nucleation has been discussed for years[1-3], but informative quantitative application either in practice or theory is quite limited in the study of growth from the melt. Heterogeneous nucleation of aluminum is, however, a practical reality. It is known, for example, that aluminum can be grain-refined by titanium and, perhaps more importantly, that for titanium addition above the peritectic composition the nucleant is (at least under definite conditions) the peritectic phase Al_3Ti[4,5]. Recent publications also show that there are definable crystallographic orientation relationships between this phase and aluminum[6,7] when Al_3Ti acts as an effective nucleant.

However, titanium compositions far below the peritectic have been experimentally shown to reduce liquidus undercooling and thus promote grain refinement[8,9]. This lower composition effect was attributed by Cibula[10] to the formation of titanium carbide nuclei. But to our knowledge, no intentional work on the possible influence of crystallographic orientation between TiC and Al has been done. The purpose of this work will first be to demonstrate that epitaxy occurs between TiC and Al and then to show how there must be some particular rules differentiating the effectiveness of nuclei. These rules can be established from the relationships now made available for TiC and Al_3Ti in practical refinement problems.

2. Experiments

Many procedures were used, covering the possible range of couples comprising a substrate of TiC upon which a charge consisting of a few grams of Al was melted. Melting was done in air and vacuum in a graphite crucible; the heating was either by electron beam or in a platinum furnace; the TiC was monocrystalline or polycrystalline, and was variously air cleaved, air cut and polished, air cut and polished close to a (111) surface, kept clean, or sometimes oxidized to form a significant oxide layer on the TiC (more than just the possible adsorbed surface that might be obtained in spite of due cleanliness precautions). In all cases a small temperature gradient was kept on the system during cooling, forcing solidification to start and continue preferentially from the substrate.

The solidified couples were polished in a plane perpendicular to the interface, and our study followed by micro-examinaton, etch pit orientation measure-

---

XV – 8

Fig. 1.   Cross section of a TiC/Al interface couple ($\times 20$). The TiC crystal at the bottom is covered by one single crystal of Al.

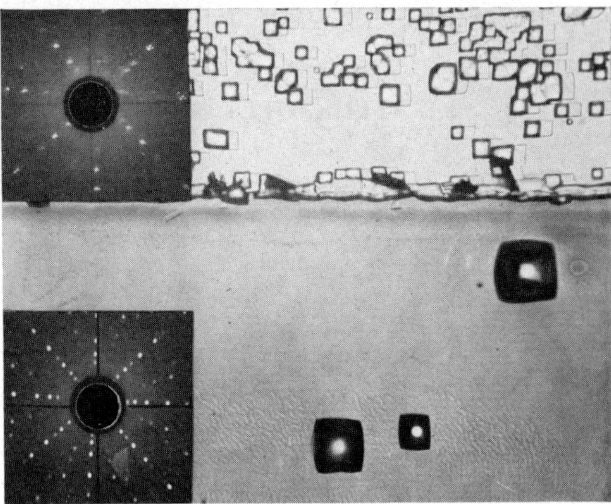

Fig. 2.   TiC/Al couple sectioned normal to a positioned {100} TiC cleavage-plane interface ($\times 350$). Laue diagrams and in-situ etch pits show the correspondence of {100} orientations in the Al (top) and the TiC (bottom).

ments, microprobe analysis, and X-ray measurements.

Fig. 1 shows a macroscopic section of a TiC/Al couple where only one aluminum crystal can be seen at the interface. Two representative X-ray images and a set of etch pit figures are shown in fig. 2. Studies on several surfaces revealed that the initial (and maintained) orientation of the Al on the TiC was completely epitaxial as, $(001)_{Al} \parallel (001)_{TiC}$; $[001]_{Al} \parallel [001]_{TiC}$. Preliminary studies were all performed by thorough X-ray analysis, but much of the subsequent investigation was pursued by using the relationship which could be established by the much simpler etch pit surface study.

The whole melting and freezing sequence was generally very short, and microprobe analyses through both the TiC and the Al revealed little or no interdiffusion of Ti and Al. We were, however, surprised by the presence of aluminum oxide needles in the Al and an aluminum carbide near the Al/TiC interface, providing reason for concern with surface reaction and oxides. Aluminum carbide is rarely mentioned[11]) and a complex ternary carbide of the type Ti_xAlC was more likely to be expected[12]). However, results from the microprobe (with correction to take into account background and absorption by Al) were within one percent of Al_4C_3 without the detection of any Ti. This particular result, due to the interaction between TiC and Al, seems to be at first view a contradiction to the hypothesis that TiC catalyzes (i.e., is an epitaxial nucleant for) the growth of Al. Fortunately, we know that the TiC is generally prone to carbon depletion[13]), and must,

therefore, be subject to dissolutions in which the carbon reacts with much faster kinetics than the titanium. The Al_4C_3 is thus a fine example of a compound that exists in the metastable situation where there is no initial equilibrium either of the Ti in the Al or vice versa. Since the presence of Al_4C_3 apparently had no effect on the nucleation of Al by TiC, the result will not be discussed further here.

As far as the oxide surface layer and the presence of oxide needles, we can say little except that they did not influence the observed orientation relationship. That is, either the oxide (on the TiC surface) is itself epitaxial and can also nucleate Al, or it is epitaxial and always nucleates the Al. The oxide needles really seemed to bear no relationship to the nucleation process at all, but may be related to the carbide formation[11]).

3. Discussion

A comparison between crystal structure and lattice parameter for TiC and Al shows the maximum misfit between the two species is 6.5% at room temperature[14]). The role of misfit has not yet proved to be absolutely clear[15,16]); but even so, this value is well below the 15% limit established by Royer[17]). Furthermore, the carbon depletion of TiC already mentioned directs the misfit factor to be even lower ($\cong 5.8\%$).

This favorable complete epitaxy indicates that any developed face of a TiC crystal will nucleate Al and that the derived Al crystal surrounding the parent TiC crystal will be identical to this crystal in all directions. This can only yield a maximum one-to-one ratio* between the number of TiC crystals and the number of Al grains nucleated if care is taken to avoid crystal multiplication by other processes[18]). In terms of a nucleation effectiveness, one can say that each TiC particle will be a potential nucleant for aluminum and that each single crystal particle will not nucleate more than one grain of aluminum.

It is now convenient for us to use the term "partial epitaxy" to describe the crystallographic relationship between Al_3Ti and Al as given by Davies et al.[7]); (but here other relationships could as well be used for illustration); namely, $(\bar{1}11)_{Al} \parallel (\bar{1}10)_{Al_3Ti}$; $[0\bar{1}1]_{Al} \parallel [001]_{Al_3Ti}$. In order to be effective, crystals of this Al_3Ti compound will have to exhibit a face having the $(\bar{1}10)$ orientation or one of the three other planes (110), $(1\bar{1}0)$, $(\bar{1}\bar{1}0)$ (but not the $\{011\}$ planes of this tetragonal compound). Let us assume now that this crystal has such faces. One can now position Al grains growing from these faces in such a way that from one to a maximum of four Al grains can be found for each crystal of Al_3Ti[19]), if one neglects the possibility of twin related Al nuclei. Crystallographic considerations show that all the surrounding aluminum grains must have between themselves definite orientation relationships. One can see that everything else being equal, Al_3Ti can nucleate more Al grains or thus will have a higher nucleation potential than TiC because of this partial epitaxy. One might expect, however, that the maximum should rarely be attained since this requires both that all $\{110\}$ planes develop on Al_3Ti, and that each such plane nucleates an Al crystal of each possible orientation (i.e., two for each $\{110\}$).

It should now be obvious that two very simple rules apply every time crystallographic relationships have to be considered in nucleation problems:
(i) The ratio (ideally) developed between the number of nucleated crystals and the number of nucleant crystals is either less than or equal to unity (ideally equal to unity) for complete epitaxy, and the ratio will generally be greater than unity for partial epitaxy.
(ii) In the case of partial epitaxy, there must be particular faces exposed or the nucleant will be ineffective;

* The rare instance of twin related Al nuclei has been seen on one massive piece of TiC.

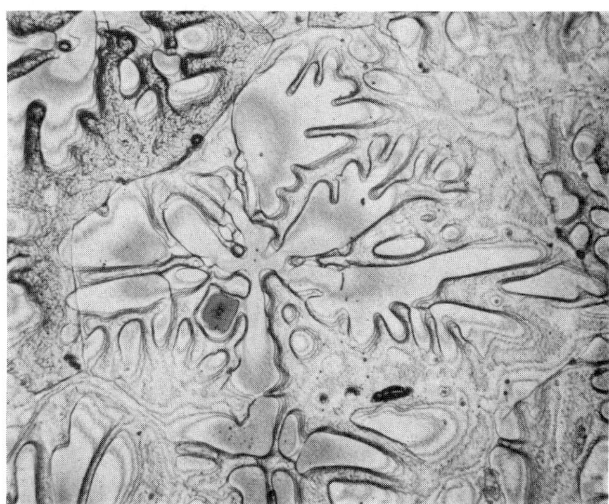

Fig. 3. Al – 0.5 wt% Ti cast under conditions inhibiting the nucleation efficiency of Al_3Ti[19]) ($\times$105, interferometric colors). A microscopic nucleant centers the dendrite giving one grain at a nucleation efficiency of 1/1, unaffected by the Al_3Ti particle trapped by a dendrite arm.

if these faces are exposed, the nucleation probability, being such a steep function of supercooling (supersaturation), ensures the separate operation of nearly all faces and high nucleation effectiveness.

These two rules, available from a very simple reasoning, seem to have been completely ignored until the present.

4. The comparative effectiveness of TiC and Al_3Ti

We have examined different titanium alloys in an investigative series that is still in progress. At this time we are *at the least* able to substantiate our reasoning as just presented with firm quantitative results for TiC and Al_3Ti. Most of this work will be presented elsewhere in due course, and so only a few examples are given here along with our positive statements.

In hypo-peritectic alloys when TiC is believed to be the active nucleant, we find TiC particles as the origin of dendrites. Special micrographic and microprobe analyses show us that nucleation (that is, we examine as do all other investigators, the earliest visible growth) starts in a central point located approximately at the center of each grain where the TiC particle is situated.

For hyper-peritectic alloys, under conditions for which Al_3Ti has been shown to be the nucleant, these particles can be found either in the center of grains, at grain boundaries, or at the intersection of more than

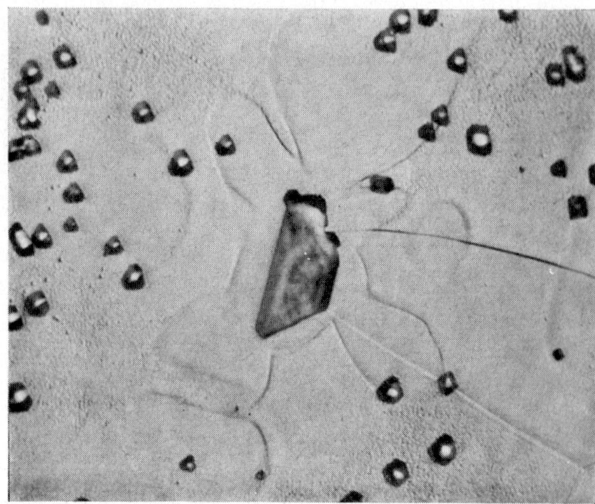

Fig. 4.   Al–0.5 wt% Ti normally cast with one Al_3Ti particle and at least four Al grains seen in this plane ($\times 700$). Etch pits show the particular Al orientations nucleated by the Al_3Ti in its defined shape, contrasting to the inactive, more irregular, trapped particle of fig. 3.

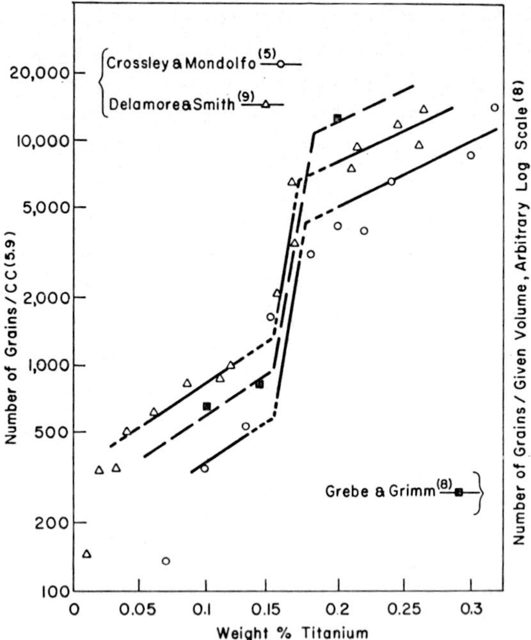

Fig. 5.   Grains per unit volume nucleated in Al by Ti as shown from Crossley and Mondolfo[5]) and Delamore and Smith[2]). Grebe and Grimm[2]) give arbitrary counts for the same Ti concentrations and their ordinate has been adjusted for best fit to Delamore and Smith[9]).

two grains. In the last two cases we find by different micrographic techniques that more than one dendrite originates at a single Al_3Ti particle. Orientation measurements of aluminum grains surrounding these particles show that our theoretical constructions are correct and that viable relationships do exist between different aluminum grains nucleated from one single particle. We were indeed able to find some casting conditions which inhibit Al_3Ti as a nucleant. In this case the microparticles are trapped by growing dendrites. Comparative examination of microscopic shapes of both active Al_3Ti nucleant and non-active nucleant particles show enough difference to enable us to emphasize shape and face orientation effects on nucleation potential.

Figs. 3 and 4 provide two examples of our statements. They are general observations.

We reproduce an important observation given by Grebe and Grimm[8]), Crossley and Mondolfo[5]), and Delamore and Smith[9]), the latter in a study where the concentration of Ti in Al melts was progressively increased and the resulting number of grains was counted (fig. 5). Since we find carbon sufficient to form TiC present in all our 99.99% purity aluminum, and since we find that Al can nucleate on TiC in agreement with the arguments by Cibula[10]), we presume that TiC acts as the lower concentration nucleant. Thus, for hypo-peritectics, there must be some kind of direct relation-

ship between number of grains and the Ti level since more Ti should mean more carbides. Above the peritectic reaction, since both TiC and Al_3Ti may nucleate the Al [regardless of whether or not we concern ourselves that TiC may nucleate Al_3Ti[9,19])] there should be a drastic nucleation frequency change. As shown, this is so; the volume effectiveness is 4 to 6 times higher when the peritectic is passed. The agreement with our hypothesis contrasting TiC and Al_3Ti is so good that we do not think it essential to mention the possibility of any other nucleation factors.

References

1) B. E. Sundquist and L. F. Mondolfo, Trans. Met. Soc. AIME **221** (1961) 607.
2) H. Sang and W. A. Miller, J. Crystal Growth **6** (1970) 303.
3) B. L. Bramfitt, Met. Trans. **1** (1970) 1987.
4) W. E. Sicha and R. C. Boehm, Trans. AFS **56** (1948) 398.
5) F. A. Crossley and L. F. Mondolfo, Trans. AIME **191** (1951) 1143.
6) J. A. Marcantonio and L. F. Mondolfo, J. Inst. Metals **98** (1970) 23.
7) I. G. Davies, J. M. Dennis and A. Hellawell, Met. Trans. **1** (1970) 275.

8) W. Grebe and G. P. Grimm, Aluminum **43** (1967) 673.
9) G. W. Delamore and R. W. Smith, 1970, private communication.
10) A. Cibula, J. Inst. Metals **76** (1949) 50, 321.
11) J. J. Trillat, L. Tertian and M. Bonnet-Gros, Mem. Sci. Rev. Met. **57** (1969) 845.
12) W. Jeitschko, H. Nowotny and F. Benesovsky, J. Less-Common Metals **7** (1964) 133.
13) I. Cadoff and J. P. Nielsen, Trans. AIME **197** (1953) 248.
14) W. B. Pearson, *Handbook of Lattice Spacings and Structures of Metals*, Vol. 2 (Pergamon, Oxford, 1967).
15) D. W. Pashley, Advan. Phys. **5** (1956) 173; **14** (1965) 327.
16) A. K. Green, J. Dancy and E. Bauer, J. Vacuum Sci. Technol. **7** (1970) 159.
17) L. Royer, Bull. Soc. Franc. Mineral Crist. **51** (1928) 7.
18) G. S. Cole and G. F. Bolling, Trans. Met. Soc. AIME **245** (1969) 725.
19) J. Cissé, to be published.

    Journal of Crystal Growth **13/14** (1972) 782–786 © *North-Holland Publishing Co.*

CAST STEEL STRUCTURES GROWN UNDER FLUID FLOW

A. A. TZAVARAS

Republic Steel Corporation, Research Center, 6801 Brecksville Road, Independence, Ohio 44121, U.S.A.
and

J. F. WALLACE

Department of Metallurgy, Case Institute, Case Western Reserve University, Cleveland, Ohio, U.S.A.

In a series of experiments the effect of induced fluid flow on the solidification structure of unidirectionally solidifying high strength, low alloy steel has been studied, It has been established that fluid flow influences the growth mode and the growth rate of the solidifying steel to the degree that morphologically new solidification structures can be produced by inducing fluid flow of a speed substantially above the speed of the thermally induced convection. The effect of fluid flow speed on the growth rate has been estimated and a relationship between fluid flow speed and solidification structures has been established.

1. Introduction

The purpose of this investigation was to establish the effect on the morphology of the steel growth structures, of induced fluid flow of a magnitude substantially above that encountered from thermally induced convection in normal castings.

In recent years the work of Chalmers[1]), Cole and Bolling[2]), and Tiller and Johnson[3]) has shed some light on the effect fluid flow has on solidification structures by identifying and describing the processes occurring when fluid flow interacts with solid growth. The systems studied have been almost exclusively non-metallic or non-ferrous and, although a great deal has been learned from the study of these systems, little was known[4]) on the effect of fluid flow on ferrous structures when this study was initiated. The advent of continuous casting in the steel industry made such a study more relevant. The AISI 4335 alloy was selected because of its general applicability and the existence of considerable comparative information.

2. Experimental technique

The apparatus used in the investigation consisted of a modified induction furnace containing an alumina crucible. The main characteristic of this furnace was its ability to rotate around a horizontal axis. A water-cooled copper chill machined to fit the mouth of the crucible could be fastened tightly with a simple fastening device when the furnace was in the upright position. The chill would remain fastened to the mouth of the

crucible when the crucible was turned upside down by rotating the furnace by 180°, thus becoming the chilled bottom part of a mold.

A charge consisting of an 8 kg ingot of 4335 steel was remelted in the crucible and brought to the desired casting temperature after which the chill was placed on the crucible. The furnace was rotated by 180° and unidirectional freezing of the melt occurred under the desired conditions. Thermocouples inserted into the crucible at various distances from the chill monitored the temperature.

During the unidirectional freezing, fluid flow was induced electromagnetically in the molten part of the ingot. The intensity of the fluid flow sweeping the solid liquid interface was set to the desired level for the desired length of time. This power-time parameter was selected on the basis of a pre-established relation between power input and fluid flow intensity. The structures produced under the pre-determined solidification and flow conditions were studied both with the conventional microscope and a scanning electron microscope. Details of the apparatus and the experimental procedures are given elsewhere[5]).

3. Results and discussion

This apparatus produced columnar structures extending at least up to 5 cm from the chill when no induced flow was applied. It is known from previous work[6]) that immobilizing the liquid in a solidifying melt enhances the columnar growth. However, it is evident that some flow prevailed in the melt under considera-

XV – 9

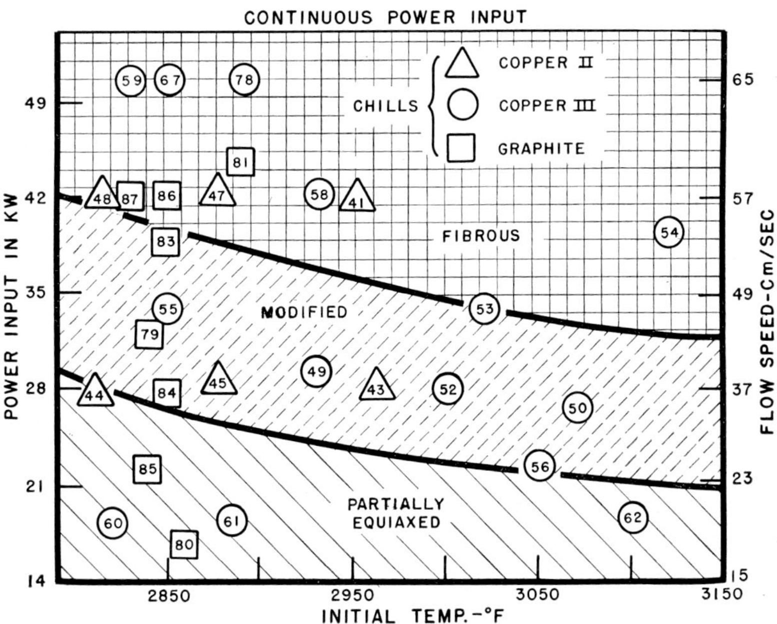

Fig. 1. Schematic description of the solidification conditions for various ingots depicting also the resulting microstructure.

tion partly because of the rotation of the apparatus and partly by thermal convection. This flow speed was estimated to be no higher than a few cm/sec[7]), whereas the induced flow was measured to be substantially higher.

The application of induced flow during freezing substantially reduced the length of the columnar zone, but the columnar structure was not eliminated until the applied fluid flow had reached a speed in excess of 23 cm/sec. For fluid flow speeds higher than 23 and up to 50 cm/sec, the ingots produced had a completely equiaxed structure regardless of the amount of superheat or the type of chill used, as indicated in fig. 1. At speeds exceeding 50 cm/sec, a columnar microstructure appeared anew but with substantially different morphological characteristics from that of the conventional unidirectional microstructure obtained when no fluid flow was induced in the molten part of the ingot during freezing.

Careful microscopic examination of the above structures revealed that fluid flow reduces the length of the columnar structure by fragmenting the columnar dendrites, i.e., by a mechanism suggested elsewhere[8]). The fragmented dendrites showed all the morphological characteristics of a dendritic structure in steel with a spine or primary arm from which secondary and, in some cases, tertiary arms were growing at square angles.

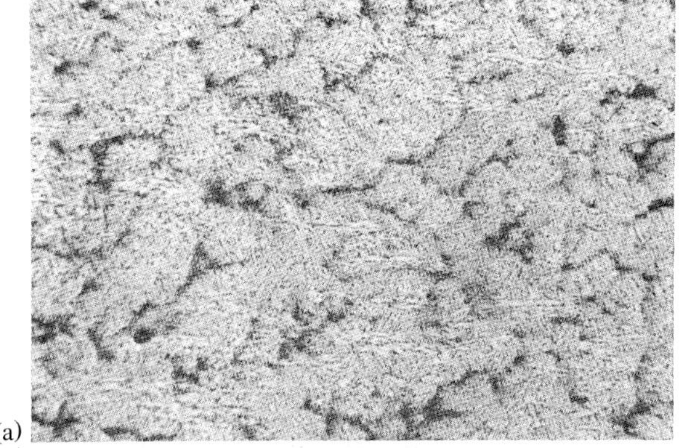

(a)

→ Growth direction

(b)

Fig. 2. (a) Flow modified microstructure of an ingot solidified unidirectionally. (b) Fibrous microstructure at 52 × magnification.

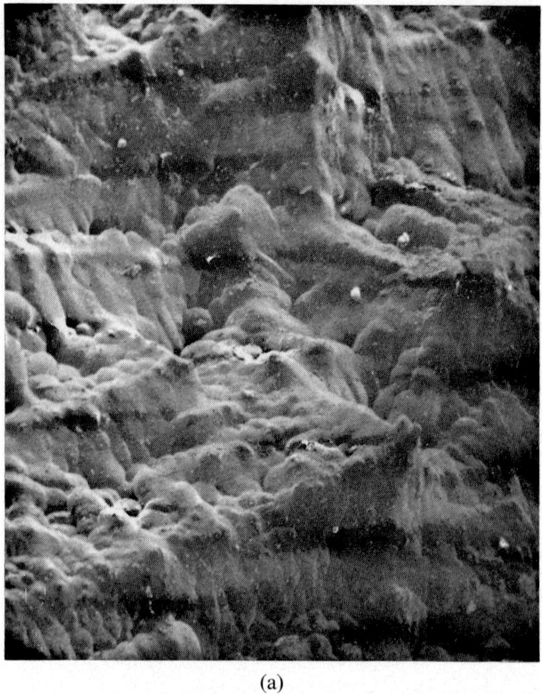

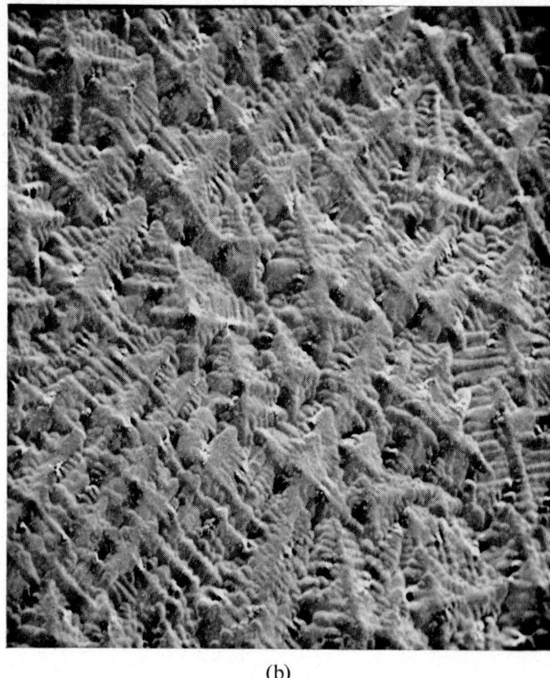

(a)   (b)

Fig. 3. Scanning electron microscope photographs of the solid–liquid interface at 90 × magnification of: (a) a columnar dendritic structure, (b) a structure rendered equiaxed by fluid flow.

The microstructure of ingots rendered completely equiaxed by fluid flow, however, had a substantially different morphology from that of the conventional dendritic equiaxed microstructure, as indicated in fig. 2a. This microstructure lacks any degree of geometrical symmetry. The primary or secondary dendritic arms are totally absent and replaced by irregular aggregates of round-shaped elements.

The columnar microstructure which appeared when the fluid flow speed exceeds 50 cm/sec is shown in fig. 2b. In this microstructure, primary arms are evident but no secondary arms occur. However, secondary arms did appear when the induced flow was stopped. In other experiments, mixed structures were produced by controlling the power input in the apparatus or the speed of the fluid flow during growth. In all these experiments no noticeable temperature rise was recorded in the melt.

Since it is obvious that fluid flow during solidification alters the growth mode by changing the growth conditions on the solid–liquid interface, a series of decanting experiments was made to produce interface specimens. These samples were studied by a scanning electron microscope; under no flow conditions, i.e.,

when columnar growth prevails, the interface is rough, as indicated in fig. 3a with the peaks or tips of the columnar dendrites extending deep into the liquid phase. Under flow of sufficient speed to break the columnar dendrites, the appearance of the interface changes substantially as indicated in fig. 3b. The orientation of the dendrites is completely random, indicating deposition rather than local growth of outward growing dendrites; the growth of the dendrites appears to take place in a direction primarily parallel to the fluid flow vector rather than perpendicularly to the solid liquid interface.

Under more intense flow (over 25 cm/sec) corresponding to that necessary for the formation of completely equiaxed ingots, the interface appears substantially different, as indicated in fig. 4a. It is evident in this figure that the dendrites have lost their symmetry and they are scarcely recognizable. This type of structure has been termed flow modified dendritic. This interface is substantially smoother with very few, if any, projections or holes as is the case with the columnar dendritic or even equiaxed dendritic solid–liquid interface. At even higher flow speeds (higher than 50 cm/sec), corresponding to the second columnar structure men-

(a)

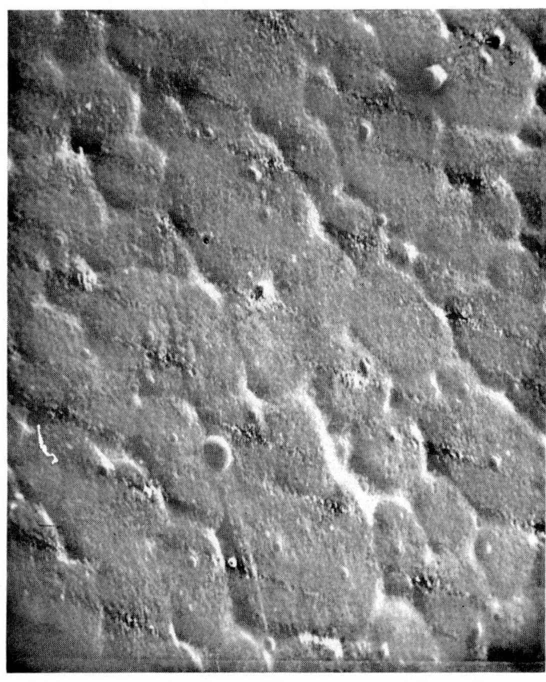

(b)

Fig. 4. Scanning electron microscope photographs of the solid–liquid interface at 180× magnification of: (a) a flow modified structure, (b) a fibrous structure.

tioned above, a very smooth interface forms as indicated in fig. 4b. This interface resembles that of a cellular structure; however, the cells are neither uniform or symmetrical. Etched sections cut parallel to the growth directions shown in fig. 2 indicate that the cells of this type structure continuously converge or diverge, appearing as knitted fibers. This fibrous structure is grown under thermodynamic conditions similar to these necessary for the formation of cellular structure; however, because of their differences in morphology, the two should be distinguished.

The result of this study supports the view that the reduction of the length of the columnar zone by flow is achieved by a fragmentation mechanism[8]) involving shearing off and transporting dendrites or fragments of dendrites. The modified structure, however, appears to form by a different mechanism involving growth of dendrites in situ. It is well known that dendrites "bend" or "turn" into the flow when subjected to the effect of laminar flow during growth. In the experimental system under consideration it has been observed that at about 20 to 25 cm/sec the flow becomes turbulent; such flow, considering the relative size of the dendrites and the eddies, would tend to "bend" the dendrite protrusions

at the interface in all directions at random, thus destroying the symmetry of the dendrites and producing the random round shapes shown in figs. 2a and 4a. At speeds higher than 50 cm/sec, the combined thermal and mechanical effect of flow on the interface increases the temperature gradient, G, in the liquid phase and decreases the growth rate, R, of the solid to the degree that the ratio G/R becomes as high or higher than that necessary for the production of cellular structures.

The morphological differences between the conventional cellular structure and the fibrous structure produced under high speed turbulent flow can be explained as follows. Hurle has shown[9]) that the cell size is proportional to the quantity $(GR)^{-1}$ under conditions of strong convective mixing. Under intense turbulent flow the quantity (GR) is expected to fluctuate locally because of the influence of flow on the momentum and therefore the concentration boundary layer[9])

$$\delta_m \sim (D/V)^{\frac{1}{3}} \delta_f .$$

It can be assumed then that the concentration boundary layer under conditions of high speed turbulent flow consists of traveling waves. These conditions would tend to change the cell size or cross section of a cell

XV – 9

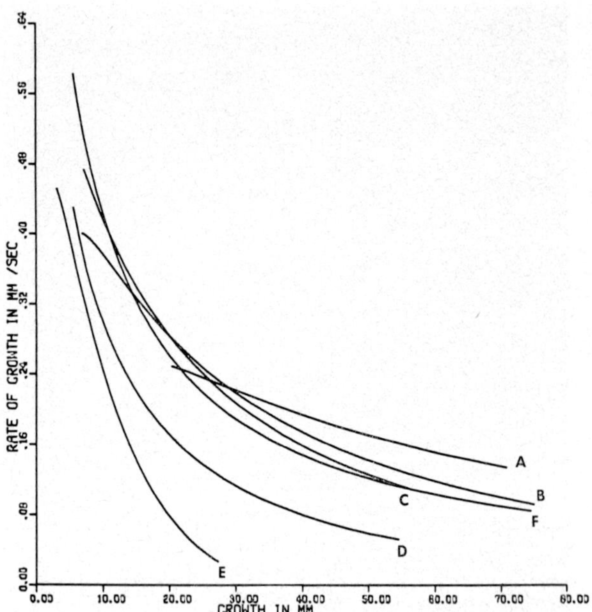

Fig. 5. Variation of growth rate as a function of growth for different flow conditions: (A) for induced flow speed $U = 20$ cm/sec; (B) $U = 36$ cm/sec; (C) $U = 45$ cm/sec; (D) $U = 52$ cm/sec; (E) $U = 65$ cm/sec; (F) curve produced for comparison on the basis of data in ref. 12 for static unidirectional growth.

continuously in a certain location during growth, thus producing the fibrous structure described previously.

To illustrate the effect of fluid flow on the critical quantity G/R, it was estimated that a 1.2 cm distance from the chill for 65 cm/sec flow (which normally would produce fibrous structure)

$$G/R = 2.38 \times 10^6 \; ^\circ\mathrm{C} \, \mathrm{cm}^{-2} \, \mathrm{sec},$$

or

$$G/R^{\frac{1}{2}} = 3.35 \times 10^5 \; ^\circ\mathrm{C} \, \mathrm{cm}^{-\frac{3}{2}} \, \mathrm{sec}^{\frac{1}{2}}.$$

This value is substantially higher than the

$$G/R = 10^4 \; ^\circ\mathrm{C} \, \mathrm{cm}^{-2} \, \mathrm{sec},$$

reported recently[10]) as critical value for the transition from cellular dendritic to cellular growth mode for an iron–8% nickel alloy.

The reduction of the growth rate by fluid flow, as estimated during this investigation for various growth conditions, is indicated in fig. 5. Reduction of growth rate by flow in unidirectional growth has also been reported recently by Russian investigators[11]). Their report indicates that the growth rate of steel solidifying unidirectionally under flow is 50% of the normal growth rate of steel in static ingots.

4. Conclusions

Induced fluid flow considerably higher than that of normal thermally induced convection influences both the growth and the growth mode of solidifying steel substantially. This is caused by the combined thermal and mechanical effect, the high speed, and particularly turbulent, fluid flow has on the solid–liquid interface. By controlling the speed of fluid flow it is possible to produce equiaxed, modified, equiaxed or fibrous structures in unidirectionally solidifying steel.

References

1) S. Wojciechowski and B. Chalmers, Trans. AIME **242** (1968) 690.
2) G. C. Cole and G. F. Bolling, Trans. AIME **236** (1966) 1366.
3) W. A. Johnston et al., Trans. AIME **233** (1965) 1865.
4) G. Pestel et al., U.S. Patent 2963758.
5) A. Tzavaras and J. F. Wallace, The Interaction of Intense Fluid Flow Inclusions and Growing Steel Structures, to be published.
6) D. Uhlmann et al., Trans. AIME **236** (1966) 527.
7) V. Khlynov and V. Gornovoi, Izv. Vysshikh Uchebn. Zavedenii Chernaya Met. **11** (1968) 40; H. Brutcher, translation 7688.
8) B. Chalmers, *Principles of Solidification* (Wiley, New York, 1964).
9) D. T. J. Hurle, J. Crystal Growth **5** (1969) 162.
10) G. R. Purdy, Progr. Rept. on Steel Solidification Res., to Am. Iron and Steel Inst., McMaster Univ., April 1970.
11) Gruzin et al., Izv. Vysshikh Uchebn. Zavedenii Chernaya Met. **13** (1970) 56.
12) M. C. Flemings et al., U.S. Army Materials Research Agency, Watertown, Mass., Tech. Rept. AMRA-CR-63-04/4.

Journal of Crystal Growth **13/14** (1972) 787–789 © *North-Holland Publishing Co.*

MOVEMENT OF LIQUID INCLUSIONS BY CENTRIFUGATION

WILLIAM R. WILCOX

Chemical Engineering and Materials Science Departments, University of Southern California, Los Angeles, California 90007, U.S.A.

A theory is developed for movement of solvent inclusions in crystals by a centrifugal field. It is necessary to account for the electric field generated by the differential migration rates of the different ions. The theory predicts a centripetal movement rate which increases as the solubility increases, which usually means as the operating temperature increases. Experimentally, the rate of movement of aqueous inclusions in potassium iodide decreases with increasing temperature and becomes centrifugal above about 30 °C. The origin of this discrepancy between experiment and theory is presently unknown.

When crystals grow from solution they almost invariably trap some of the solution as inclusions. One method to remove these inclusions is to place the crystal in a temperature gradient, which causes the inclusions to move[1]). Recent experiments have shown that a centrifugal field can similarly cause the inclusions to move through the crystal[2,3]). Comparison of the rates of movement with theoretical expressions yields information on the interfacial kinetics of growth and dissolution.

The theoretical expressions for inclusion movement in the ultracentrifuge are given in refs. 2 and 3 as, respectively

$$V = \frac{\omega^2 rD}{RT} \left\{ \frac{\dfrac{M - v_f \rho_f}{v} + \Delta_c^f v \rho_f}{\left(\dfrac{C_c}{C_f} - \dfrac{\rho_c}{\rho_{cf}}\right) + \dfrac{D}{LC_f}\left(\dfrac{1}{\mu_L} + \dfrac{1}{\mu_0}\right)} \right\}, \tag{1}$$

$$V = \frac{C_f D}{(C_c - C_f)(RT)} [(M - \rho_f v_c)g - K/L], \tag{2}$$

where ω is the angular rotation rate of the sample, r is the radial position of the inclusion, g is the acceleration field at the inclusion $(= \omega^2 r)$, D is the diffusion coefficient, R is the gas constant, T is the absolute temperature, C_f is the concentration of crystalline material in the inclusion (solubility), C_c is the solute concentration in the crystal, M is the molecular weight of the same species for which D and C apply, ρ_f and ρ_c are fluid and crystal densities, ρ_{cf} is the density the crystal would have if its components possessed their partial molal

volumes in the solution, $\Delta_c^f v$ is the change of partial molal volume upon solution, v_c is the molal volume of the solid $(= v_f - \Delta_c^f v)$, v_f is the partial molal volume in the fluid, v is the number of ions into which each molecule of solute dissociates in solution $(= 2$ for alkali halides in water), L is the length of the inclusion, and μ_L, μ_0 and K are all corrections for finite interface kinetics. It may not be immediately obvious, but these expressions are nearly identical for infinite interface attachment kinetics $(K = 0, \mu_L = \mu_0 = \infty)$. Nevertheless, both equations are incorrect. The correct derivation is given below. It is seen that the error in eq. (2) arose from neglecting dissociation into ions, while in eq. (1) dissociation was considered only in the sedimentation flux (first term in the numerator) and neglected in the diffusion flux (second term in the numerator).

When an ionized solute undergoes any form of mass transfer, the different ions tend to migrate at different rates. This generates an electric field which in turn influences the migration of the charged ions. Thus we must consider separately the migration of the different ions and connect the migration rates by the electric field and the net fluxes. The net flux of ion i into the solid is equal to the convective flux in the liquid plus the migration due to the gradient of electrochemical potential, η_{fi}, or

$$VC_{ci} = VC_{fi}\frac{\rho_c}{\rho_{cf}} + B_i C_{fi}\frac{\Delta\eta_{fi}}{L}, \tag{3}$$

where B_i is the mobility and ρ_c/ρ_{cf} represents the change in convective flow in the inclusion from that in the solid due to the partial molal volume differences[4]). The change in electrochemical potential in the liquid across the inclusion is equal to the change in the crystal minus the energy necessary to drive interfacial processes, $\Delta\eta_{Ki}$, or

$$\frac{\Delta\eta_{fi}}{L} = \frac{\Delta\eta_{ci} - \Delta\eta_k}{L} = (M_i - \rho_f v_{ci})g + \frac{z_i F \Delta\phi}{L} - \frac{\Delta\eta_{ki}}{L}, \quad (4)$$

where z_i is the charge on the ith ion and F is the Faraday constant. Strain effects in the solid are neglected for the simple reason that we are not certain how to take them into account.

For convenience we consider a binary solute $(S_1)_{v_1}(S_2)_{v_2}$ which dissociates completely in the solvent to $v_1 S^{z_1}$ and $v_2 S^{z_2}$. Since the solid must be stoichiometric, the liquid is likewise stoichiometric, so that

$$C = C_1/v_1 = C_2/v_2. \quad (5)$$

Substituting eqs. (4) and (5) into eq. (3) and writing the result for S^{z_1} and S^{z_2} we find that

$$\frac{\Delta\phi}{L} = \left\{ B_2 \left[(M_2 - \rho_f v_{c2})g - \frac{\Delta\eta_{k2}}{L} \right] \right.$$
$$\left. - B_1 \left[(M_1 - \rho_f v_{c1})g - \frac{\Delta\eta_{k1}}{L} \right] \right\} \quad (6)$$
$$\times [F(B_1 z_1 - B_2 z_2)]^{-1}.$$

and

$$V = \left\{ z_1 \left[(M_2 - \rho_f v_{c2})g - \frac{\Delta\eta_{k2}}{L} \right] \right.$$

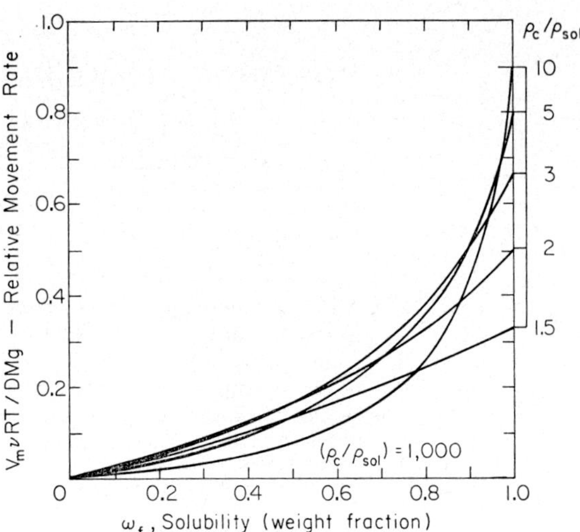

Fig. 1. Relative maximum inclusion movement rate as a function of solubility and relative densities of crystal and solvent, according to (approximate) eq. (20).

$$\left. - z_2 \left[(M_1 - \rho_f v_{c1})g - \frac{\Delta\eta_{k1}}{L} \right] \right\}$$
$$\times \left[\left(\frac{C_c}{C_f} - \frac{\rho_c}{\rho_{cf}} \right) \left(\frac{z_1}{B_2} - \frac{z_2}{B_1} \right) \right]^{-1}. \quad (7)$$

This may be made more useful by noting that

$$v = v_1 + v_2, \quad (8)$$
$$v_1 z_1 = -v_2 z_2, \quad (9)$$
$$M = v_1 M_1 + v_2 M_2, \quad (10)$$
$$v_c = v_1 v_{c1} + v_2 v_{c2}, \quad (11)$$

TABLE 1

Inclusion movement rates (centripetal)

Crystal	M	ρ_c (g/cm³)	T (°C)	w_f	v_f (cm³/m)	$D \times 10^5$ (cm²/sec)	$\dfrac{\partial \log \gamma}{\partial \log C}$	g	V (mm/hr) Calc.	Exp.	Exp. Ref.
AgNO₃	170	4.35	20	0.687	32.9	(1)	−0.503	210,000	0.14		
KCl	75	1.98	20	0.256	30.9	1.7	−0.164	53,600	0.0042	≤0.0018	3
KCl	75	1.98	40	0.287	(30.9)	(2.6)	(−0.164)	210,000	0.028	0.022	2
KBr	119	2.75	20	0.394	37.7	1.8	−0.099	210,000	0.043		
KI	166	3.13	5	0.567	(47.9)	(1.4)	(−0.065)	144,000	0.05	0.05	2
KI	166	3.13	24.3	0.597	47.9	2.5	−0.065	210,000	0.14	0.06	2
KI	166	3.13	40	0.615	(47.9)	3.5	(−0.065)	210,000	0.19	−0.06	2
NH₄Cl	54	1.53	20	0.272	39.2	2	(−0.08)	210,000	0.013		
NaCl	59	2.17	20	0.264	21.2	1.4	−0.055	210,000	0.0081		
PbI₂	461	6.16	20	0.001	(63.7)	(1)	(−0.01)	210,000	0.00006		
Pb(NO₃)₂	331	4.53	20	0.343	55.7	(1)	(−0.04)	210,000	0.028		
Sucrose	342	1.58	20	0.670	215	0.15	(0)	210,000	0.031		
Urea	60	1.32	20	0.517	44.2	1.1	(0)	210,000	0.020		

Data from refs. 8–10. Values in parentheses are author's estimates.

XV – 10

$$\Delta\eta_k \equiv K = v_1\Delta\eta_{k1} + v_2\Delta\eta_{k2}, \qquad (12)$$

and that the conventional tabulated diffusion coefficient may be written as[5,6])

$$D = \left[\frac{1}{z_1} - \frac{1}{z_2}\right]\left[1 + \left(\frac{\partial \ln \gamma}{\partial \ln C}\right)_{T,P}\right]RT$$

$$\times \left[\frac{1}{B_1 z_1} - \frac{1}{B_2 z_2}\right]^{-1} \qquad (13)$$

$$= z_1 v\left[1 + \left(\frac{\partial \ln \gamma}{\partial \ln C}\right)_{T,P}\right]RT$$

$$\times \left\{v_2\left[\frac{z_1}{B_2} - \frac{z_2}{B_1}\right]\right\}^{-1}.$$

Substituting eqs. (9)–(13), eq. (7) becomes

$$V = D[(M - \rho_f v_c)g - \Delta\eta_k/L]$$

$$\times \left\{vRT\left(\frac{C_c}{C_f} - \frac{\rho_c}{\rho_{cf}}\right)\left[1 + \left(\frac{\partial \ln \gamma}{\partial \ln C}\right)_{T,P}\right]\right\}^{-1}. \qquad (14)$$

When the crystal contains no solvent we find further that

$$\rho_c/\rho_{cf} = v_f/v_c, \qquad (15)$$

$$C_c = 1/v_c = \rho_c/M, \qquad (16)$$

and

$$C_f = F_f/v_f = w_f\rho_f/M, \qquad (17)$$

where F_f and w_f are the volume fraction and weight fraction of solute in the occluded solution. With these, eq. (14) becomes

$$V = \frac{D[M(1 - \rho_f/\rho_c) - \Delta\eta_k/L]}{vRT(\rho_c/w_f\rho_f)(1 - F_f)[1 + (\partial \ln \gamma/\partial \ln C)]}. \qquad (18)$$

In order to further elucidate the influences of solubility and crystal density we estimate the maximum movement rate V_m for an ideal solution with $v_c = v_f$ (no volume change on dissolution) and

$$\rho_f \approx \left(\frac{w_f}{\rho_c} + \frac{1 - w_f}{\rho_{sol}}\right)^{-1}, \qquad (19)$$

where ρ_{sol} is the density of the solvent. This yields the

equation

$$\frac{V_m vRT}{DMg} \approx \left(\frac{\rho_c}{\rho_{sol}} - 1\right)$$

$$\times \left[\frac{\rho_c}{\rho_{sol}}\left(1 + \frac{1 - w_f}{w_f}\frac{\rho_c}{\rho_{sol}}\right)\right]^{-1}. \qquad (20)$$

which is plotted in fig. 1. Note that the movement rate is influenced mostly by the solubility, increasing as the solubility increases. For $0 \le w_f \le 0.5$, the movement rate is maximum for $\rho_c \approx 2\rho_{sol}$. For $0.5 \lesssim w_f \lesssim 0.9$ the maximum movement rate occurs for $\rho_c \approx 3\rho_{sol}$.

Experimental results are also shown in table 1 for KI and KCl. The fact that the experimental results are lower than the theoretical results indicates that finite interface kinetics reduced the rate. However, the data of ref. 2 on KI indicate both a decrease of travel rate with increasing temperature and a reversal to centrifugal travel above about 30 °C. These observations are contrary to the predictions of the present theory, i.e., a centripetal movement which increases in velocity as the temperature increases. Possible explanations might involve strain in the solid, free convection, or sedimentation across a gas bubble (similar to the results of ref. 7).

Acknowledgement

This research was supported by the Advanced Research Projects Agency of the U.S. Department of Defense under Grant No. DAHC 15-70-G-14.

References

1) W. R. Wilcox, Ind. Eng. Chem. **60** (1968) 13.
2) W. R. Wilcox and P. J. Shlichta, J. Appl. Phys. **42** (1971) 1823.
3) T. R. Anthony and H. E. Cline, Phil. Mag., **22** (1970) 893.
4) W. R. Wilcox, J. Crystal Growth, in press.
5) Haskell, Phys. Rev. **21** (1908) 149.
6) J. R. Vinograd and J. W. McBain, J. Am. Ceram. Soc. **63** (1941) 2008.
7) W. R. Wilcox, Ind. Eng. Chem. **61** (1969) 76.
8) *International Critical Tables* (McGraw-Hill, New York, 1928)
9) Landolt-Börnstein, *Zahlenwerte und Funktionen*, 6th ed. Springer, Berlin, 1960).
10) *Handbook of Chemistry and Physics*, 51st ed. (Chemical Rubber Publ. Co., Cleveland, 1970–1971).

      Journal of Crystal Growth **13/14** (1972) 790–794 © *North-Holland Publishing Co.*

THE MIGRATION OF LIQUID DROPLETS IN SOLIDS

H. E. CLINE and T. R. ANTHONY

General Electric Corporate Research, Schenectady, New York, U.S.A.

The migration of liquid H_2O droplets in transparent KCl crystals was directly observed under various driving forces including thermal gradients, accelerational fields, and grain boundary and solid-liquid surface tensions. In all cases, it was found that the kinetics of attachment and detachment of atoms at the solid–liquid interface of the droplet is the factor which determines the migration rate and shape change of the liquid droplets. For example, with thermal gradients and accelerational fields, no droplet motion was observed below a critical droplet driving force because of these interface kinetics.

The influence of interface kinetics was also demonstrated by examining shape changes of liquid droplets in KCl in two independent experiments. In the first experiment, it was shown that interface kinetics produce a flattening of the droplet perpendicular to its direction of motion. In the second experiment, already extended droplets were allowed to relax to their equilibrium shapes. Interface kinetics were found to retard the relaxation rate and to prevent large droplets from ever attaining an equiaxed shape.

Finally, droplets were migrated into grain boundaries in KCl. From the critical thermal gradient or acceleration field required to free the liquid droplets from the grain boundary, the grain boundary energy of KCl was determined.

1. Introduction

The migration of a liquid droplet through a solid occurs by the dissolution and deposition of the solid at the forward and rear faces of the droplet, respectively[1]). For a given driving force[2-6]) on the droplet, the droplet migration rate is limited either by diffusive transport in the liquid or by the kinetics of attachment and detachment of atoms at the solid–liquid interface of the droplet. A variety of driving forces including thermal gradients[1,3,6-11]) accelerational fields[12-14]) solid–liquid surface tensions[15]), grain boundary tensions[16]) and electric fields[17]) have been used in studies of droplet migration through solids. This paper reports a series of experiments on the transparent system brine (H_2O saturated with KCl) KCl in which all of the above driving forces, with the exception of an electric field, have been used to study droplet migration in solids.

2. Droplet driving forces

The migration velocity V of a droplet through a solid is given by the product of the driving force on the droplet F and the droplet mobility M:

$$V = MF. \tag{1}$$

In the experiments discussed in this paper, the droplet mobility M is proportional to the volume diffusion rate in the liquid of the droplet as given in table 1. The total driving force F is the rate of change of the free energy[18])

of the system with respect to a virtual displacement of the droplet minus any frictional force produced by

TABLE 1

Driving forces and migration rates

Mobility	$M = \dfrac{C_1}{C_s}\dfrac{D}{(RT)}\dfrac{\bar{V}_s}{V_D}$ (volume diffusion)
Acceleration force	$F_g = V_D g(\rho_s - \rho_1)$
Interface kinetic friction	$F_K = \dfrac{K}{L}\dfrac{\bar{V}_s}{V_D}$
Thermal force	$F_T = \dfrac{V_D}{V_s} RT\left[\dfrac{1}{C_E}\left(\dfrac{\partial C_E}{\partial T}\right) + \sigma\right]G_1$
Grain boundary tension	$F_{GB} = 4X\,\gamma_{GE}$
Velocity of migration in an acceleration field	$V = M(F_g - F_K)$
Velocity in thermal gradient	$V = M(F_T - F_K)$

C_1 = concentration of salt in brine droplet
C_E = equilibrium concentration of salt in brine in contact with salt
C_s = concentration of salt in solid salt
D = diffusivity of salt in brine
R = gas constant
T = absolute temperature
K = kinetic potential
γ = grain boundary tension
L = dimension of droplet $\parallel$ to thermal gradient
X = dimension of droplet $\perp$ to thermal gradient
$\bar{V}_s$ = molar volume of solid salt
V_D = $X^2 L$ = volume of droplet
G_1 = temperature gradient in the brine droplet
σ = Soret coefficient of salt in brine
g = acceleration
ρ_s = density of solid
ρ_1 = density of liquid

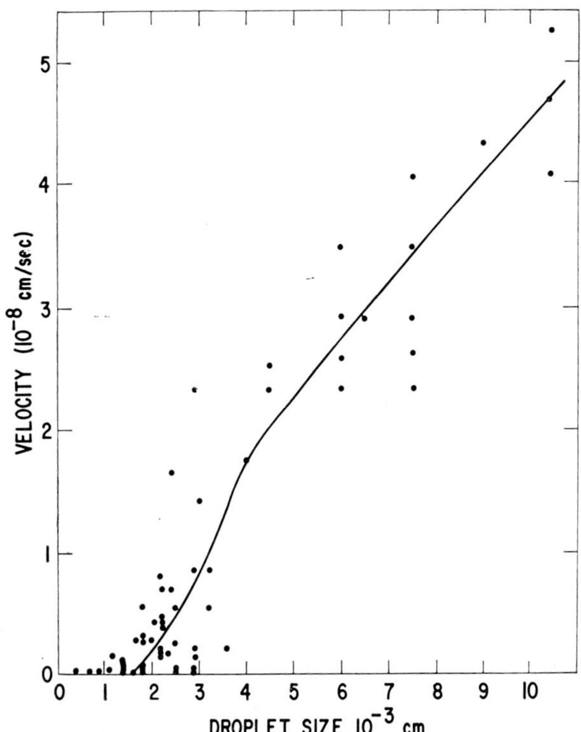

Fig. 1. The droplet velocity versus droplet dimension in an accelerational field of 53 600 times gravity.

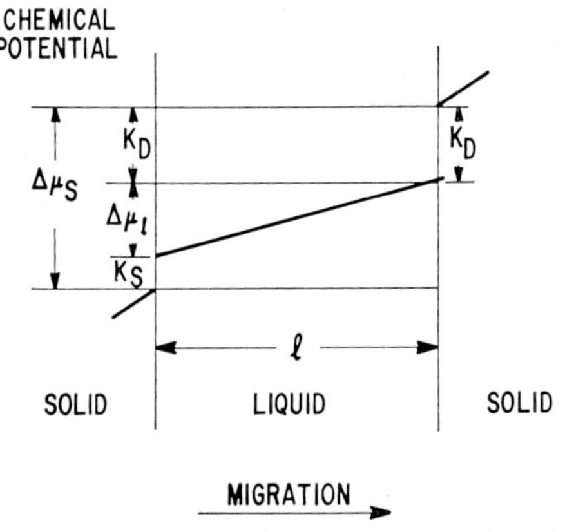

Fig. 2. The chemical potential across the liquid droplet, illustrating the kinetic potentials of dissolution K_d and solidification K_s at the solid–liquid interface that result in lowering the chemical potential gradient in the liquid. Also this figure illustrates the relation of the chemical potential gradient in the liquid to the droplet size.

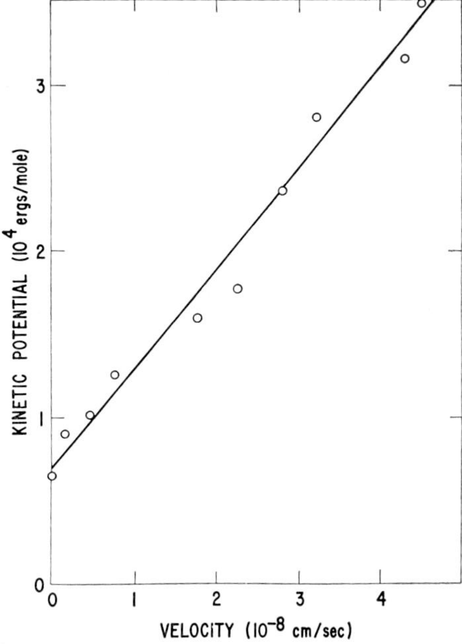

Fig. 3. The derived kinetic potential K versus the droplet velocity using average velocity data for different droplet sizes.

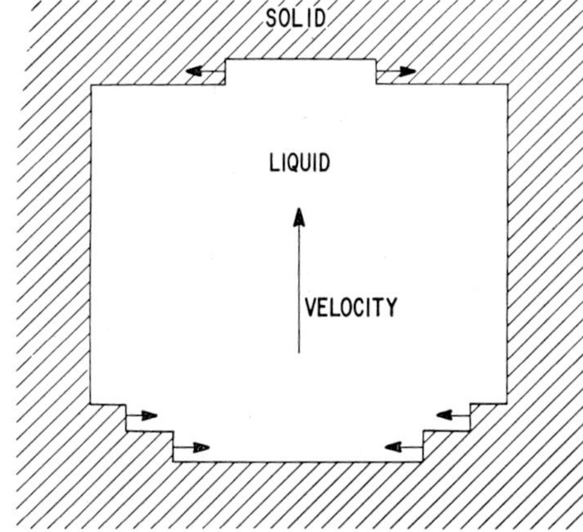

Fig. 4. Schematic diagram of the layer growth mechanism applied to droplet motion. The corners of the solidifying interface provide a continuous source of ledges for layer growth. In contrast, layers must be independently nucleated on the dissolving interface. Therefore, it is expected that interface kinetics will be largest on the dissolving interface.

irreversible processes associated with the transfer of atoms across the solid–liquid interface.

2.1. ACCELERATIONAL FIELD

In an accelerational field, a body force (see table 1) will be exerted on a liquid droplet in a solid if the density of the liquid and the solid are not equal. Because brine is less dense than H_2O, a buoyant force is produced on brine droplets in KCl in an accelerational field. Fig. 1 shows the migration velocities of brine droplets of various sizes in an accelerational field 53 600 times that of earth's gravity. Three important

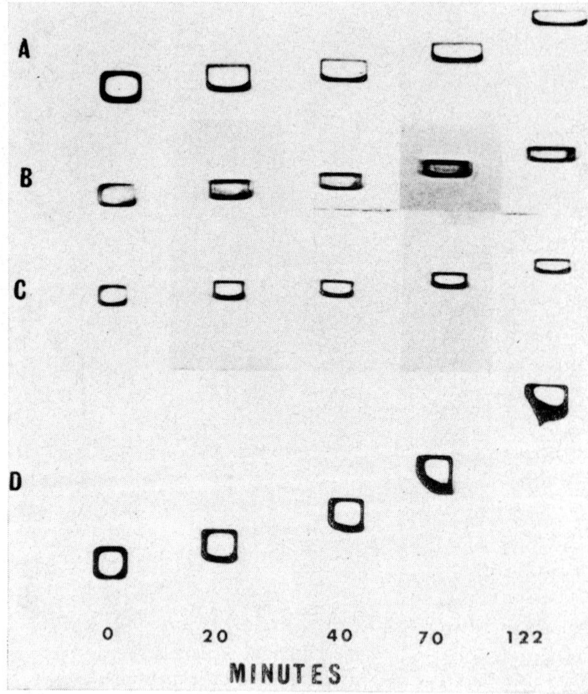

Fig. 5.   The shapes of brine droplets A, B, C and D after 0, 20, 40, 70 and 122 min of thermomigration, respectively. 200×.

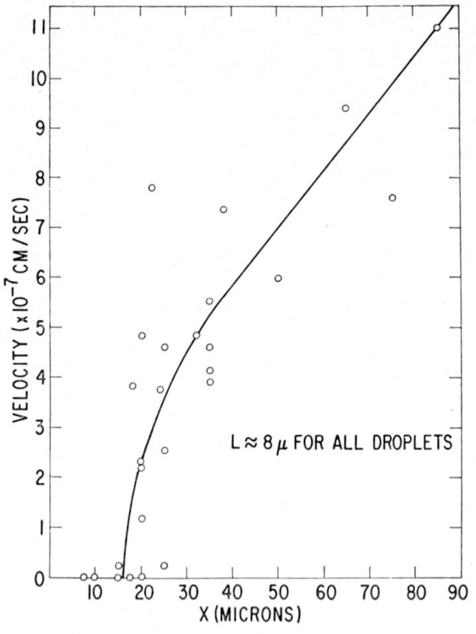

Fig. 6.   The steady-state velocities of brine droplets versus droplet size in a temperature gradient of 22 °C/cm.

features of fig. 1 should be noted. First, the droplet velocity decreases with decreasing droplet size. Secondly, below a critical size, there is no detectable droplet migration. Finally, the general magnitude of the observed droplet velocities is significantly smaller than that calculated by multiplying the buoyant forces on the droplets times their mobilities.

2.2. FRICTIONAL FORCE OF INTERFACE KINETICS

In order to understand the results of the accelerational field experiment shown in fig. 1 as well as the remaining experiments to be discussed in this paper, one must consider the effect of the irreversible transfer of atoms across the solid–liquid interface of the droplet on the velocity of the droplet[2,4,5]. As shown in fig. 2, this irreversible process results in a discontinuous change of the chemical potential of the salt on going across the solid–liquid interface at both the dissolving and solidifying interfaces of the droplet[5]. Thus, interface kinetics cause a decrease in the chemical potential gradient of salt in the liquid, which in turn reduces the droplet migration rate. As shown in table 1, the effect

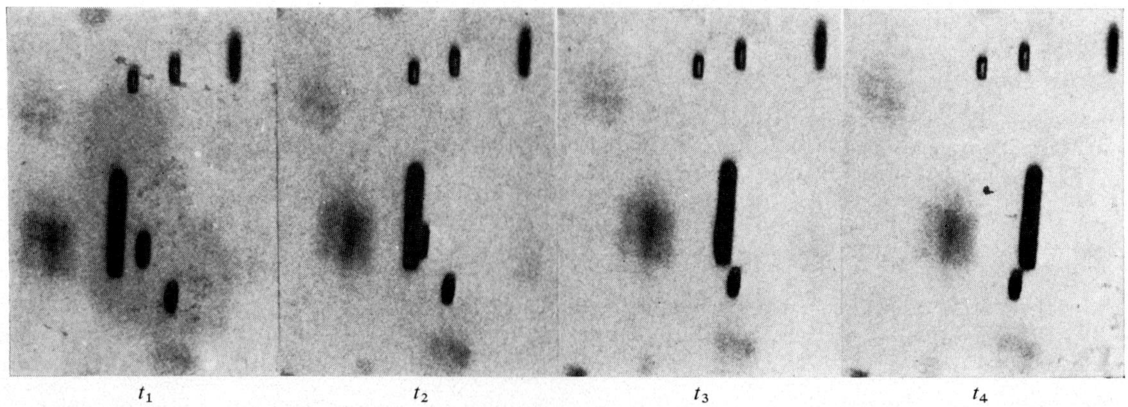

Fig. 7.   The collision and coalescence of two brine droplets migrating in a thermal gradient in KCl. The time interval between each sequence is 45 min. 100×.

of interface kinetics may be represented as a frictional force that depends on droplet size. By adding this frictional force to the buoyant force of the accelerational field experiments, the pertinent features of fig. 1 can be explained. Since the buoyant force decreases while the frictional force increases with decreasing droplet size (see table 1), it is apparent that the droplet velocity will decrease with decreasing droplet size and below a critical size, droplets will not move, as is observed (fig. 1). From the accelerational field experiments, the sum K of the chemical potential changes at the dissolving and solidifying interfaces and the variation of K with velocity was determined as shown in fig. 3.

A brine droplet in KCl migrating along a $\langle 100 \rangle$ direction assumes a rectangular faceted shape because crystal growth and deposition of KCl in an aqueous solution occur by the movement of ledges across (100) planes as illustrated in fig. 4[4,5]). Because of geometry, ledges are always available for deposition on the rear face of the droplet. In contrast, ledges must be either continually nucleated or be present as the result of dislocations for dissolution to occur on the front interface of the droplet. Consequently, the frictional force of interface kinetics is greater on the forward dissolving than on the rear depositing interface. One consequence of this inequality of frictional forces is that the more mobile rear interface moves faster than the front interface during the initial stages of droplet migration and causes a flattening of the droplet perpendicular to its direction of motion[11]).

2.3. THERMAL FORCES

A composition gradient in a liquid droplet in a solid in a temperature gradient is generated because of both the change of solubility of the solid in the liquid with temperature and the Soret effect[19,20]). This composition gradient produces a transport of the solid atoms through the droplet liquid from the front to the rear of the droplet and causes the droplet to migrate. Experiments in the brine–KCl system[11]) show that initially cubic droplets become flattened in a plane perpendicular to their direction of motion because of the greater frictional force of interface kinetics on the front as compared to the rear droplet face (fig. 5). After this initial shape change, the droplets migrate at a constant velocity which depends on the droplet size as shown in fig. 6. The larger droplets move more rapidly than the

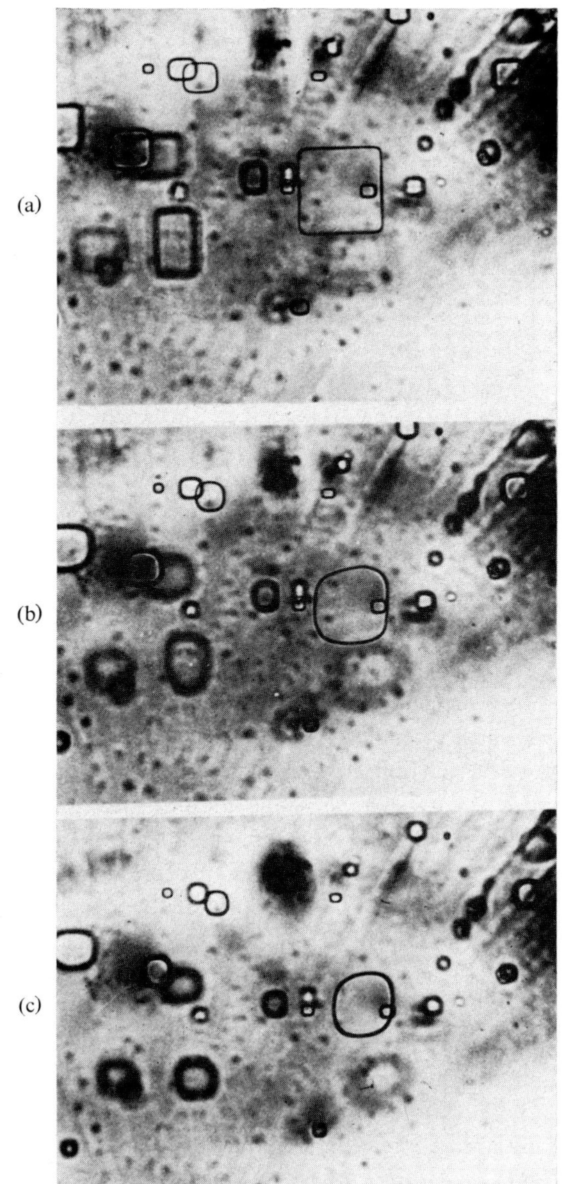

Fig. 8. Photomicrograph of a brine platelet perpendicular to the square face of the platelet at (a) zero, (b) 8 hr and (c) 48 hr after removal from a thermal gradient. 300×.

smaller droplets because the greater shape distortion of the larger droplets induces a larger thermal gradient in them[11]). This variation of droplet velocity with droplet size leads to an increased rate of collision and coalescence of droplets, an example of which is shown in fig. 7.

2.4. SURFACE TENSION FORCES

During thermomigration, brine droplets in KCl flatten into thin square platelets in a plane normal to the

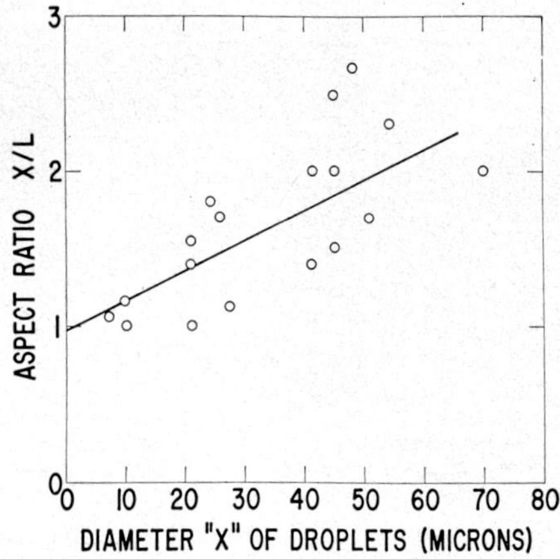

Fig. 9. Platelet aspect ratio X/L versus platelet diameter X after shape relaxation of the brine platelet has ceased. Without interface kinetics, all aspect ratios would be unity.

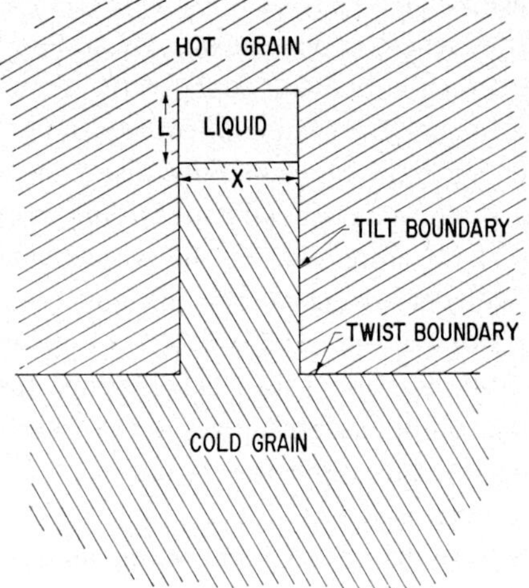

Fig. 10. A twist boundary of a KCl bicrystal after penetration by a migrating brine platelet. The grain boundary formed behind the migrating platelet is a square, columnar, pure-tilt boundary.

thermal gradient[11]) (fig. 5). After the thermal gradient is removed, the droplet relaxes under the driving force of surface tension towards a cubic shape of minimum surface energy[15]). During relaxation, the corners of the depositing faces become rounded (fig. 8) because of the easy nucleation of deposition ledges at these points while the dissolving faces remain flat and faceted because of the corresponding difficulty of nucleating dissolution ledges in a droplet (see fig. 4). Since the driving force of surface tension decreases with decreasing droplet distortion and increasing droplet size, a continuous series of distorted droplet sizes exist above which surface tension forces are unable to overcome the frictional force of interface kinetics[15]). Thus droplets above a certain size will never relax to an equiaxed shape of minimum surface energy as is shown in fig. 9 for a number of brine droplets in KCl relaxed for two months, a more than adequate time for diffusive transport to occur.

2.5. Grain boundary tension forces

After passing through a grain boundary, a liquid droplet is pulled back towards the grain boundary by the grain boundary tension as illustrated in fig. 10. From the critical thermal gradient force required to balance the grain boundary tension force (table 1) the grain boundary energy of a 15° tilt boundary in KCl was found to be 33 erg/cm^2 (ref. 16). A similar experiment was also carried out substituting an accelerational field for the thermal gradient to give another absolute determination of the grain boundary tension of KCl[14]).

References

1) W. G. Whitman, Am. J. Sci. [5] **11** (1926) 126.
2) P. G. Shewmon, Trans. AIME **230** (1964) 1134.
3) P. Hoekstra, T. E. Osterkamp and W. F. Weeks, J. Geophys. Res. **70** (1965) 5035.
4) W. A. Tiller, J. Appl. Phys. **34** (1963) 2757; J. Crystal Growth **6** (1969) 77.
5) W. Oldfield and A. J. Markworth, Mater. Sci. Eng. **4** (1969) 353.
6) J. H. Wernick, Trans. AIME **209** (1957) 1169; J. Chem. Phys. **25** (1956) 47.
7) W. D. Kingery and W. H. Goodnow, in: *Intern. Conf. on Glaciology, Ice and Snow, Process, Properties and Applications*, Ed. W. D. Kingery (MIT Press, Cambridge, Mass., 1963) p. 237.
8) J. D. Harrison, unpublished work.
9) R. E. Carter, J. S. Nadeau and J. H. Rosolowski, to be published.
10) W. D. Kingery and W. H. Goodnow, in: *Intern. Conf. on Glaciology, Ice and Snow, Properties, Processes and Applications*, Ed. W. D. Kingery (MIT Press, Cambridge, Mass., 1963) p. 237.
11) T. R. Anthony and H. E. Cline, J. Appl. Phys. **42** (1971) 3380.
12) T. R. Anthony and H. E. Cline, Phil. Mag. **22** (1970) 893.
13) W. R. Wilcox and P. J. Shlichta, J. Appl. Phys. **42** (1971) 1823.
14) T. R. Anthony and H. E. Cline, Phil. Mag. **24** (1971) 695.
15) H. E. Cline and T. R. Anthony, Acta Met. **19** (1971) 175.
16) H. E. Cline and T. R. Anthony, Acta Met. **19** (1971) 491.
17) Y. E. Geguzin, S. S. Simeonov and E. A. Eivazov, Soviet Phys.-Solid State **9** (1967) 1123.
18) J. W. Gibbs, *The Collected Works of J. Williard Gibbs*, Vol. 1 (Yale Univ. Press, New Haven, Conn., 1948).
19) S. R. De Groot, *Thermodynamics of Irreversible Processes* (North-Holland, Amsterdam 1951).
20) K. G. Denbigh, *Thermodynamics of the Steady State* (Wiley, New York, 1951).

Section XVI

Segregation and defects

A. S. YUE
J. T. YUE
C. POTARD
B. D. WARR
R. W. SMITH
J. DELOCHE

Y. MALMEJAC
B. SCHAUB
H. C. CASEY, Jr.
M. B. PANISH
P. BOTTELIER
F. BOUILLON

Journal of Crystal Growth **13/14** (1972) 797–803 © *North-Holland Publishing Co.*

THE EFFECT OF CONCENTRATION AND TEMPERATURE GRADIENTS ON SOLUTE REDISTRIBUTION*

A. S. YUE and J. T. YUE**

Materials Department, University of California, Los Angeles, California 90024, U.S.A.

A partial differential equation describing the process of solute redistribution in the melt of a dilute binary alloy is formulated by including contributions of concentration and temperature gradients in the liquid. The time-dependent solution of the partial differential equation is carried out for the case of a semi-infinite sample in which growing is initiated at one end and at time $t = 0$, and proceeds unidirectionally at a constant freezing rate R. By the use of a Laplace transform, the partial differential equation can be transformed into an ordinary differential equation which can be solved exactly by assuming a constant temperature gradient in the liquid ahead of the advancing solid–liquid interface. The exact solution of the Laplace equation is used as a zeroth order trial solution for solving the realistic partial differential equation containing a varying temperature gradient in the liquid by the iteration technique. The first order solution is found to converge rapidly with the temperature expansion parameter $(T_1 - T_0)D'/k$, where T_1 and T_0 are the temperatures of the heat source and the interface; k is the thermal diffusivity in the liquid; and D' is the thermal diffusion coefficient in the liquid.

1. Introduction

Solute redistribution in the melt has been discussed many times in recent years. Tiller et al.[1] assumed that solute diffusion in the liquid is incomplete. Wagner[2] included both convection and diffusion in the liquid. Pohl[3], Hulme[4], and Smith et al.[5] considered the time-dependent diffusion and various boundary conditions at the moving solid–liquid interface. They obtained slightly different solutions in which the solute redistribution in the initial transient solid is given as an infinite series. Although the above-mentioned authors have considered the various aspects of the solidification problem, they have not considered the significant contribution of the temperature gradient in the liquid ahead of the interface.

Recently, many investigators have succeeded experimentally in converting a cellular solid–liquid interface to a planar interface in dilute binary alloys[6,7], and in eutectic alloys[8,9] by increasing the temperature gradient in the liquid ahead of the interface. Thus, the inclusion of both concentration and temperature gradients in the theoretical formulation of solute redistribution in the melt is necessary. In this paper we shall solve a partial differential equation describing the so-lute redistribution in the melt of a dilute alloy in a coordinate system moving with the interface by including contributions of concentration and temperature gradients in the liquid by an iteration technique. The effect of convection current in the liquid is neglected in this treatment.

2. Theoretical analysis

According to Yue[7] a partial differential equation describing the process of solute redistribution in the melt of a dilute binary alloy by including contributions of concentration and temperature gradients in the liquid is formulated as

$$\frac{\partial C_L}{\partial t} = D \frac{\partial^2 C_L}{\partial x^2} + R \frac{\partial C_L}{\partial x} + D'\alpha \frac{\partial}{\partial x}\left[\exp\left(-\frac{Rx}{k}\right)C_L\right], \tag{1}$$

where D is the diffusion coefficient; R is the constant freezing rate; D' is the thermal diffusion coefficient; k is the thermal diffusivity; x is the distance into the melt measured from the moving interface;

$$\alpha = (R/k)(T_1 - T_0),$$

where T_1 and T_0 are the temperatures of the heat source and the interface; and $\partial C_L/\partial x$ is the concentration gradient of the liquid. The meaning of eq. (1) is illus-

* Research supported by the U.S. Army Research Office – Durham, under Grant No. DA-ARO-D-31-124-71-G61.
** Physics Department, Stanford University.

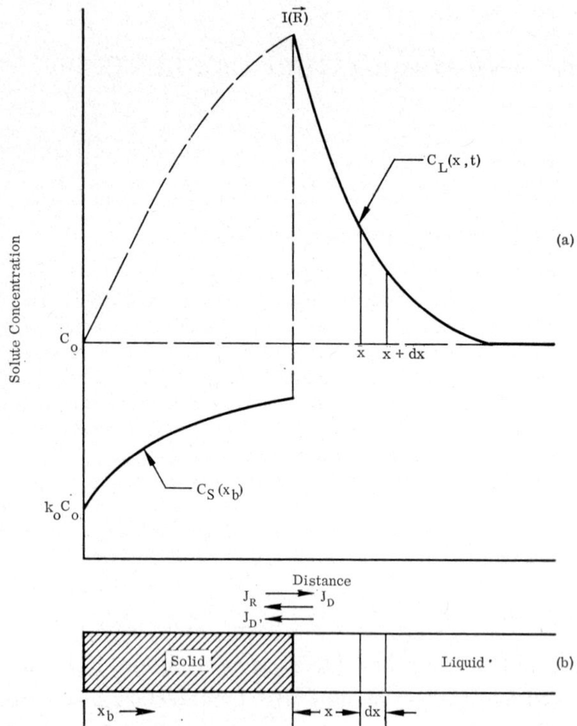

Fig. 1. (a) Distribution of solute in the liquid, $C_L(x, t)$, and in the solid, $C_S(x_b)$; (b) the frozen solid and the remaining liquid. x is the distance measured from the interface; x_b is the distance measured from the end of the boat.

trated in fig. 1, in which the three types of fluxes are indicated. J_D is the diffusive flux due to concentration gradient; J_R is the flux due to the movement of the interface; and J_D is the diffusive flux due to temperature gradient in the liquid.

The initial condition for equation (1) is

$$C_L(x, 0) = c_0 \quad \text{for} \quad x > 0, \qquad (2)$$

and the boundary conditions for eq. (1) are

$$C_L(\infty, t) = c_0 \quad \text{for} \quad t > 0, \qquad (3)$$

and

$$R(1 - k_0)C_L(0, t)$$
$$= -D \frac{\partial C_L(0, t)}{\partial x} - D'C_L \frac{\partial T(0, T)}{\partial x} \qquad (4)$$
$$\text{at } x = 0 \quad \text{for} \quad t > 0,$$

where c_0 is the initial solute concentration in the liquid and k_0 is the equilibrium distribution coefficient, and $\partial T(0, t)/\partial x$ is the temperature gradient at the solid–liquid interface and is equal to α.

The solution of eq. (1) will be carried out for the case of a semi-infinite sample in which growing is initiated at one end and at time $t = 0$, and proceeds at constant freezing rate R.

By the suse of a Laplace transform, eq. (1) can be transformed from a partial differential equation into an ordinary differential equation. This is obtained by replacing $C_L(x, t)$ by

$$L[C_L(x, t)] = \bar{C}_L(x, s) = \int_0^\infty e^{-st} C_L(x, t)\, \mathrm{d}t . \qquad (5)$$

Eq. (1) becomes

$$\frac{\mathrm{d}^2\bar{C}_L(x, s)}{\mathrm{d}x^2} + \frac{R}{D}\frac{\mathrm{d}\bar{C}_L(x, s)}{\mathrm{d}x} - \frac{s}{D}\bar{C}_L(x, s)$$
$$= -\frac{c_0}{D} - \frac{D'\alpha}{D}\frac{\mathrm{d}}{\mathrm{d}x}\left[\exp\left(-\frac{Rx}{k}\right)\bar{C}_L(x, s)\right]. \qquad (6)$$

Similarly, the transformed boundary conditions for eq. (6) are

$$\bar{C}_L(\infty, s) = c_0/s \quad \text{at} \quad x = \infty, \qquad (7)$$

and

$$\mathrm{d}C_L(0, s)/\mathrm{d}x = H\bar{C}_L(0, s) \quad \text{at} \quad x = 0, \qquad (8)$$

where

$$H = [R(k_0 - 1) - D'\alpha]/D .$$

3. Constant temperature gradient

Now the differential equation to be solved is the transformed eq. (6). To get an idea of what the temperature term

$$\frac{\alpha D'}{D}\frac{\mathrm{d}}{\mathrm{d}x}\left[\exp\left(-\frac{Rx}{k}\right)C_L(x, s)\right]$$

will do to the solution of eq. (6) and also to illustrate the technique we will use, let us start with a system with a temperature gradient described simply by $\mathrm{d}T(x, t)/\mathrm{d}x = \alpha$ instead of the more realistic gradient $\mathrm{d}T/\mathrm{d}x = \alpha \exp(-Rx/k)$. Eq. (6) then becomes

$$\frac{\mathrm{d}^2\bar{C}_L(x, s)}{\mathrm{d}x^2} + \frac{R}{D}\frac{\mathrm{d}\bar{C}_L(x, s)}{\mathrm{d}x} - \frac{s}{D}\bar{C}_L(x, s)$$
$$= -\frac{c_0}{D} - \frac{D'\alpha}{D}\frac{\mathrm{d}\bar{C}_L(x, s)}{\mathrm{d}x} . \qquad (9)$$

It is claimed here that the Laplace inverse of the

solution of eq. (9) would give the maximum effects due to a temperature gradient, imposed on the system, as the gradient $dT/dx = \alpha$ is always greater than $dT/dx = \alpha \exp(-Rx/k)$ for all x. Since the temperature effect on the solution of eq. (1) is a maximum for a system with a maximum temperature gradient throughout the liquid phase, the claim is justified.

Eq. (9) can be solved to give

$$\bar{C}_L(x, s) = \frac{c_0}{s} \left\{ 1 + \frac{q(R+\alpha D')}{\sqrt{D}} \exp\left(-\frac{(R+\alpha D')x}{2D}\right) \right.$$

$$\times \exp\left[-\frac{x}{\sqrt{D}}\left(s + \frac{(R+\alpha D')^2}{4D}\right)^{\frac{1}{2}}\right]$$

$$\left[(\tfrac{1}{2}-q)\left(\frac{R+\alpha D'}{\sqrt{D}}\right) + \left(s + \frac{(R+\alpha D')^2}{4D}\right)^{\frac{1}{2}}\right]^{-1} \right\}, \tag{10}$$

where $q = 1 - k_0$.

The Laplace inverse of eq. (10) can be performed exactly by using the tables of transform in Campbell and Forster[10]). We shall however, perform the inversion by a contour integration. This technique will be used in section 4 of the paper in which exact transforms are not available. The prescription for the Laplace inverse is

$$C_L(x, t) = \frac{1}{2\pi i} \int_{\gamma - i\infty}^{\gamma + i\infty} \bar{C}_L(x, s)\, e^{st}\, ds. \tag{11}$$

The analytic property of $\bar{C}_L(x, s)$ of eq. (10) is as follows: There is a simple pole at $s = 0$, a branch cut at $s = -(R+\alpha D')^2/4D$, and a singularity at $s = \bar{s} = -Rk_0[(1-k_0)R + \alpha D']/D$. For $0 < k_0 < 1$, the latter singularity is located between the branch point and the origin. We will assume for the rest of the paper that k_0 satisfies such a restriction. The nature of singularity at $\bar{s}$ can be shown to be of a simple pole. By Taylor expanding about $\bar{s}$, we obtain

$$\frac{1}{(\tfrac{1}{2}-q)\left(\dfrac{R+\alpha D'}{\sqrt{D}}\right) - \left(s + \dfrac{(R+\alpha D')^2}{4D}\right)^{\frac{1}{2}}} \tag{12}$$

$$\underset{s \to \bar{s}}{=} -\frac{1}{(s-\bar{s})}[(1-2k_0)R + \alpha D'].$$

The contour chosen is shown in fig. 2.

$C_L(x, t)$ can be evaluated by the Cauchy residue theorem,

$$C_L(x, t) = \sum (\text{residues at poles}). \tag{13}$$

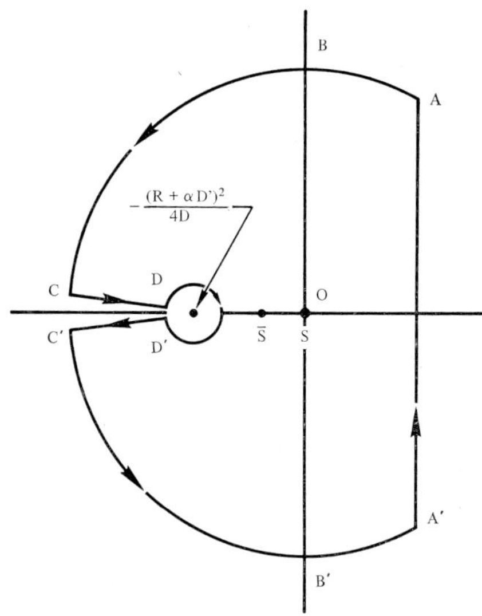

Fig. 2.   Contour used for evaluation of the inversion integral, with poles indicated for eq. (10).

In eq. (13), we have entirely neglected the contributions along the branch cut[3]). The validity of such an approximation can be seen at the end of the calculation by a comparison with the exact Laplace inversion. The contribution about the branch point is exactly zero.

Thus, we get

$$C_L(x, t) = c_0 \left\{ 1 + \left(\frac{1-k_0}{k_0}\right) \exp\left[-\frac{x}{D}(R+\alpha D')\right] \right.$$

$$- \left(\frac{1-2k_0}{k_0}\right) \tag{14}$$

$$\times \exp\left[-\frac{[(1-k_0)(R+\alpha D')][x + k_0(R+\alpha D')t]}{D}\right]\right\}.$$

How much confidence can we put into eq. (14)? Eq. (14) reduces to the solution[3,5]) when there is no temperature gradient, i.e.,

$$C_L(x, t) \underset{\substack{\alpha \to 0 \\ k \to \infty}}{\longrightarrow} c_0 \left\{ 1 + \left(\frac{1-k_0}{k_0}\right) \exp\left(-\frac{R}{D}x\right) \right.$$

$$- \frac{(1-2k_0)}{k_0} \exp\left[-\frac{R(1-k_0)(x + Rk_0 t)}{D}\right]\right\}. \tag{15}$$

Furthermore, eq. (14) satisfies the following conditions:

(i) $C_L(x, t = \infty)$

$$= c_0 \left\{ 1 + \left(\frac{1-k_0}{k_0}\right) \exp\left[-\frac{(R+\alpha D')x}{D}\right]\right\},$$

which is the steady state solution obtained by solving the steady state equation with a constant temperature gradient.

(ii) The exact Laplace inverse of eq. (10) is

$$
\begin{aligned}
C_L(x, t) = c_0 \Bigg\{ &1 + \frac{q}{2k_0} \exp\left[-\frac{x}{D}(R+\alpha D')\right] \\
&\times \operatorname{erfc}\left[\frac{1}{2\sqrt{(Dt)}}(x-(R+\alpha D')t)\right] \\
&- \tfrac{1}{2}\operatorname{erfc}\left[\frac{1}{2\sqrt{(Dt)}}(x+(R+\alpha D')t)\right] \\
&+ \frac{q}{2}\left(\frac{1}{q}-\frac{1}{k_0}\right) \\
&\times \exp\left[-q\frac{(R+\alpha D')}{D}(x+k_0(R+\alpha D')t)\right] \\
&\times \operatorname{erfc}\left[\frac{1}{2\sqrt{(Dt)}}(x+Rk_0-1)(R+\alpha D')t)\right]\Bigg\},
\end{aligned}
\tag{16}
$$

which reduces exactly to eq. (15) provided that $k_0 < \tfrac{1}{2}$, and

$$
\frac{x-(R+\alpha D')t}{2\sqrt{(Dt)}} \ll -1.
\tag{17}
$$

This can be achieved by looking at a region in the melt which is not too far away from the freezing interface. This criterion assures the negative sign. To satisfy the inequality, we choose $R \sim 10^{-3}$ cm/sec and $D \sim 10^{-5}$ cm^2/sec. These values are highly typical for a metallic system.

Thus, eq. (14) is an excellent solution for the constant temperature gradient problem. The neglect of the contributions along the branch cuts during the Laplace inversion process can be justified. We will use this approximation procedure in section 4.

The physics of the temperature gradient problem is already made clear by eq. (15). The solute distribution in the liquid ahead of the interface damps out faster by $\alpha D'/R$. The value of the solute concentration at the interface is enhanced over the zero temperature gradient values. Both of these effects will increase the concentration of the solute in the solid, in agreement with the more realistic calculations carried out in section 4.

4. Actual temperature gradient

We are now ready to solve the realistic problem described by eq. (6). Rewriting it, we obtain

$$
\begin{aligned}
\frac{d^2\bar{C}_L(x, s)}{dx^2} &+ \left(\frac{R}{D}+\frac{\alpha D'}{D}\right)\frac{d}{dx}\left[\bar{C}_L(x, s)\right] - \frac{s}{D}\bar{C}_L(x, s) \\
&= -\frac{c_0}{D} + \frac{\alpha D'R}{Dk}\exp\left(-\frac{Rx}{k}\right)\bar{C}_L(x, s) \\
&+ \frac{\alpha D'}{D}\left[1-\exp\left(-\frac{Rx}{k}\right)\right]\frac{d}{dx}\left[\bar{C}_L(x, s)\right].
\end{aligned}
\tag{18}
$$

The exact solution of eq. (18) is very complicated. It has been shown that it is a complicated function of a confluent hypergeometric function[11]) the Laplace inverse of which has been found impossible to perform. We will propose here an iterative procedure using eq. (10) as a zeroth order trial solution for the right-hand side (R.H.S.) of eq. (18).

If we neglect the term

$$
\frac{\alpha D'}{D}\left[1-\exp\left(-\frac{Rx}{k}\right)\right]\frac{d}{dx}\left[\bar{C}_L(x, s)\right],
$$

eq. (18) becomes

$$
\begin{aligned}
\frac{d\bar{C}^2_L(x, s)}{dx^2} &+ \left(\frac{R}{D}+\frac{\alpha D'}{D}\right)\frac{d}{dx}\bar{C}_L(x, s) \\
&- \frac{s}{D}\bar{C}_L(x, s) = -\frac{c_0}{D}+\frac{\alpha D'R}{Dk}\exp\left[-\frac{Rx}{k}\right]\bar{C}_L(x, s)
\end{aligned}
\tag{19}
$$

For small Rx/k, the term

$$
\frac{\alpha D'}{D}\left[1-\exp\left(-\frac{Rx}{k}\right)\right]\frac{d}{dx}\bar{C}_L(x, s)
$$

goes as

$$
\frac{\alpha D'}{D}\frac{Rx}{k}\frac{d}{dx}\bar{C}_L(x, s),
$$

which is negligible when compared to

$$
\frac{\alpha D'}{D}\frac{d}{dx}\bar{C}_L(x, s).
$$

Typically if $R \sim 10^{-3}$ cm/sec and $k \sim 10^{-1}$ cm^2/sec, then for $x \simeq 10$ cm, $Rx/k \sim 0.1$. So if we restrict our attention to a distance within 10 cm of the moving interface, the neglect of the above term is justified. Furthermore, this term can only alter the exponential damping rate of the solution, not the strength. In so far as Rx/k is small, the physics of the solution cannot change too much.

Our problem is now to solve eq. (19). Since $\alpha D'R/Dk \ll 1$, the iteration procedure is guaranteed to converge rapidly. Only a 1st order iteration will be necessary.

Inserting eq. (10) for $C_L(x, s)$ on the R.H.S. of eq. (19), we get the first iterated equation,

$$\frac{\mathrm{d}^2 \bar{C}_L(x, s)}{\mathrm{d}x^2} + \left(\frac{R+\alpha D'}{D}\right)\frac{\mathrm{d}\bar{C}_L(x, s)}{\mathrm{d}x} - \frac{s}{D}\bar{C}_L(x, s) \tag{20}$$

$$= -\frac{c_0}{D} + \frac{\alpha D'R}{Dk}g(x, s),$$

where

$$g(x, s) = c_0 \exp\left(-\frac{R}{k}x\right)\frac{1}{s}$$

$$\times \left\{1 + \frac{q(R+\alpha D')}{\sqrt{D}}\exp\left(\alpha^-(s)x\right)\right.$$

$$\left. \times \left[\left(\tfrac{1}{2}-q\right)\left(\frac{R+\alpha D'}{\sqrt{D}}\right) + \left(s+\frac{(R+\alpha D')^2}{4D}\right)^{\frac{1}{2}}\right]^{-1}\right\}.$$

We solve eq. (20) by the method of variation of parameters. The independent solutions of the homogeneous equation are $\exp\left(\alpha^+(s)\right)$ and $\exp\left(\alpha^-(s)\right)$ where

$$\alpha^{\pm}(s) = -\tfrac{1}{2}\left(\frac{R}{D}+\frac{\alpha D'}{D}\right) \pm \left[\tfrac{1}{4}\left(\frac{R}{D}+\frac{\alpha D'}{D}\right)^2 + \frac{s}{D}\right]^{\frac{1}{2}}. \tag{21}$$

A particular solution of eq. (20) is

$$Y^*(x, s) = V_1(x)\exp\left(\alpha^+(s)x\right) + V_2(x)\exp\left(\alpha^-(s)x\right)$$

where

$$V_1(x) = \frac{1}{\alpha^+(s)-\alpha^-(s)}\int^x \mathrm{d}x'\exp\left(-\alpha^+(s)x'\right)$$

$$\times \left[-\frac{c_0}{D}+\frac{\alpha D'R}{Dk}g(x', s)\right],$$

and

$$V_2(x) = \frac{1}{\alpha^-(s)-\alpha^+(s)}\int^x \mathrm{d}x'\exp\left(-\alpha^-(s)x'\right)$$

$$\times \left[-\frac{c_0}{D}+\frac{\alpha D'R}{Dk}g(x', s)\right],$$

After doing the necessary calculations, we get

$$Y^*(x, s) = \frac{c_0}{s}\left\{1-\frac{\alpha D'R}{Dk}\exp\left(-\frac{R}{k}x\right)\right.$$

$$\times \left[\frac{s}{D}+\frac{R}{k}\frac{(R+\alpha D')}{D}-\left(\frac{R}{k}\right)^2\right]^{-1}$$

$$+\frac{\alpha D'q(R+\alpha D')}{D\sqrt{D}}\exp\left[-\frac{R}{k}x+\alpha^-(s)x\right]$$

$$\times \left[\left(\tfrac{1}{2}-q\right)\left(\frac{R+\alpha D'}{D}\right)\left(s+\frac{(R+\alpha D')^2}{4D}\right)^{\frac{1}{2}}\right]^{-1} \tag{22}$$

$$\times \left[\frac{R}{k}+\frac{2}{\sqrt{D}}\left(s+\frac{(R+\alpha D')^2}{4D}\right)^{\frac{1}{2}}\right]^{-1}\right\}.$$

The general solution of eq. (20) in the Laplace transform space is

$$\bar{C}_L(x, s) = A_2 \exp\left(\alpha^-(s)x\right) + Y^*(x, s). \tag{23}$$

The arbitrary constant A_2 is determined by the boundary condition [eq. (8)]. We get

$$A_2 = \frac{1}{[\alpha^-(s)-H]}\left\{HY^*(0, s) - \frac{\mathrm{d}}{\mathrm{d}x}Y^*(0, s)\right\}.$$

Hence

$$\bar{C}_L(x, s) = \frac{\exp(\alpha^-(s)x)}{[\alpha^-(s)-H]}\left\{\frac{Hc_0}{s} + \frac{c_0}{s}\frac{\alpha D'R}{Dk}\right.$$

$$\times \left[\frac{\left(H+\dfrac{R}{k}\right)}{\left(-\dfrac{R}{D}+\dfrac{\alpha D'}{D}\right)\dfrac{R}{k}+\dfrac{R^2}{k^2}-\dfrac{s}{D}}\right.$$

$$\left.\left. +\frac{H\left(H+\dfrac{R}{k}-\alpha^-(s)\right)}{[\alpha^-(s)-H]\dfrac{R}{k}\left\{\dfrac{R}{k}+2\left[\tfrac{1}{4}\left(\dfrac{R}{D}+\dfrac{\alpha D'}{D}\right)^2+\dfrac{s}{D}\right]^{\frac{1}{2}}\right\}}\right]\right\}$$

$$+ Y^*(x, s). \tag{24}$$

This is the exact solution of eq. (20). The Laplace inverse of eq. (24) gives the first iterated solution to the problem as follows:

$$C_L(x, t) = C_0\left\{1+\left[\frac{(1-k_0)R+\alpha D'}{Rk_0}\right]\right.$$

$$\times \exp\left[-\left(\frac{R+\alpha D'}{D}\right)x\right] - \left[\frac{(1-k_0)R+\alpha D'}{Rk_0}\right]$$

$$\times \exp\left[-\frac{(1-k_0)R+\alpha D'}{D}(x+Rk_0 t)\right]$$

$$-\frac{\alpha D'}{D}\frac{(1-k_0)}{k_0}\exp\left[-\frac{(R+\alpha D')}{D}x\right]$$

$$\times \left[1 - \exp\left[\frac{-R(R+\alpha D')}{k} t \right] \right]$$

$$- \frac{\alpha D'}{R} \left(\frac{1-k_0}{k_0} \right) \left[\exp\left[-\frac{(R+\alpha D')}{D} x \right] \right.$$

$$- \exp\left[-(1-k_0)\frac{(R+\alpha D')}{D}(x + k_0(R+\alpha D')t) \right]$$

$$- \frac{\alpha D'}{R} \exp\left(-\frac{R}{k} x \right) \left[1 - \exp\left(-\frac{R(R+\alpha D')}{k} t \right) \right]$$

$$+ \frac{\alpha D'}{R} \frac{(1+k_0)}{k_0} \exp\left(-\frac{R}{k} x \right) \exp\left(-\frac{(R+\alpha D')}{D} x \right)$$

$$- \frac{\alpha D'}{R} \left(\frac{1-k_0}{k_0} \right) \exp\left(-\frac{R}{k} x \right)$$

$$\times \exp\left[-(1-k_0)\frac{(R+\alpha D')}{D} \right] \left[x + (R+\alpha D')k_0 t \right] \Big\} .$$

$$(25)$$

In eq. (25), we have assumed $1-k_0 \sim 1-2k_0$ and $1-k_0 \sim 1$, which is good for $k_0 \ll 1$. We have also neglected terms of order $D/k \ll 1$, $R/k \ll 1$, and have only kept terms with strength through $\alpha D'/R$.

The following checks have been performed on eq. (25) to assure that it is a valid first order iterated solution of the differential eq. (1).

(i) Eq. (25) satisfied eq. (1) through first order in $\alpha D'/k$, in so far as $D/k \ll 1$, $R/k \ll 1$ and the term

$$\frac{\alpha D'}{D} \left[1 - \exp\left(-\frac{R}{k} x \right) \right] \frac{\mathrm{d}}{\mathrm{d}x} C_L(x, t)$$

can be neglected.

(ii) At $t \leftarrow \infty$, the steady state solution obtained from eq. (25) is

$$C_L^{ST}(x) = c_0 \left\{ 1 + \left[\frac{(1-k_0)R + \alpha D'}{R k_0} \right] \right.$$

$$\times \exp\left[-\frac{(R+\alpha D')}{D} x \right]$$

$$- \frac{\alpha D'}{R} \frac{(1-k_0)}{k_0} \exp\left[-\frac{(R+\alpha D')}{D} x \right]$$

$$- \frac{\alpha D'}{R} \frac{(1-k_0)}{k_0} \exp\left[-\frac{(R+\alpha D')}{D} x \right]$$

$$- \frac{\alpha D'}{R} \exp\left(-\frac{R}{k} x \right)$$

$$+ \frac{\alpha D'}{R} \frac{(1-k_0)}{k_0} \exp\left(-\left[\frac{R}{k} + \frac{(R+\alpha D')}{D} \right] x \right) \right\} ,$$

$$(26)$$

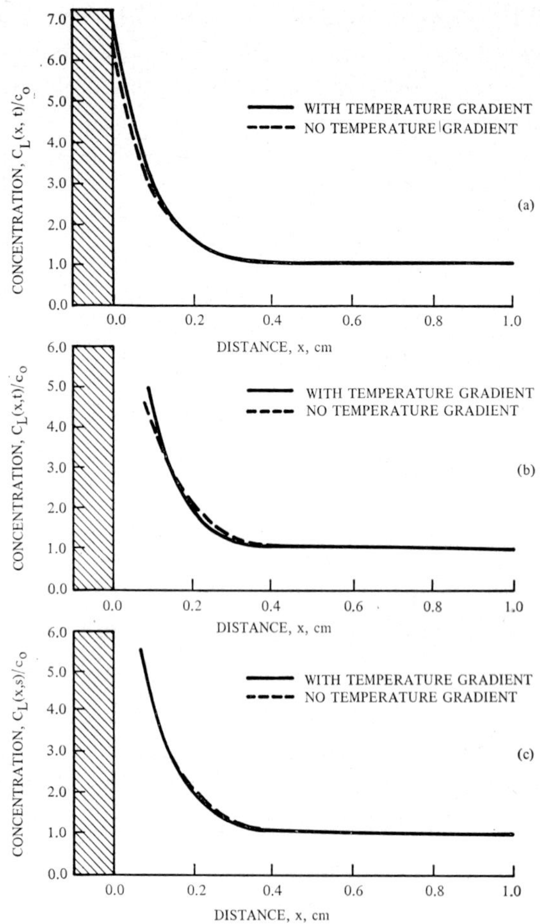

Fig. 3. Nonsteady state (transient) solute distribution in the liquid, $C_L(x, t)/c_0$, for $Rt = 1$ cm. (a) $k_0 = 0.1$; (b) $k_0 = 0.01$; and (c) $k_0 = 0.001$.

which agrees with the first order steady state solution in $\alpha D'/R$ (ref. 7).

The concentration of the solute at the interface is obtained by setting $x = 0$,

$$C_L(x = 0, t) \Big|_{\text{interface}} = c_0 \left\{ 1 + \left(\frac{1-k_0}{k_0} \right) \right.$$

$$\times \left[1 - \exp\left(-\left[\frac{(1-k_0)R + \alpha D'}{D} \right] k_0 Rt \right) \right]$$

$$- \frac{\alpha D'}{R} \left[\exp\left[-(1-k_0)\frac{(R+\alpha D')^2}{DR} k_0 Rt \right] \right.$$

$$\times \exp\left[-\left(\frac{R+\alpha D'}{k} \right) Rt \right] \Big] \Big\} .$$

$$(27)$$

Eq. (27) is in agreement with the conclusion reached at the end of section 3, that the value of the solute

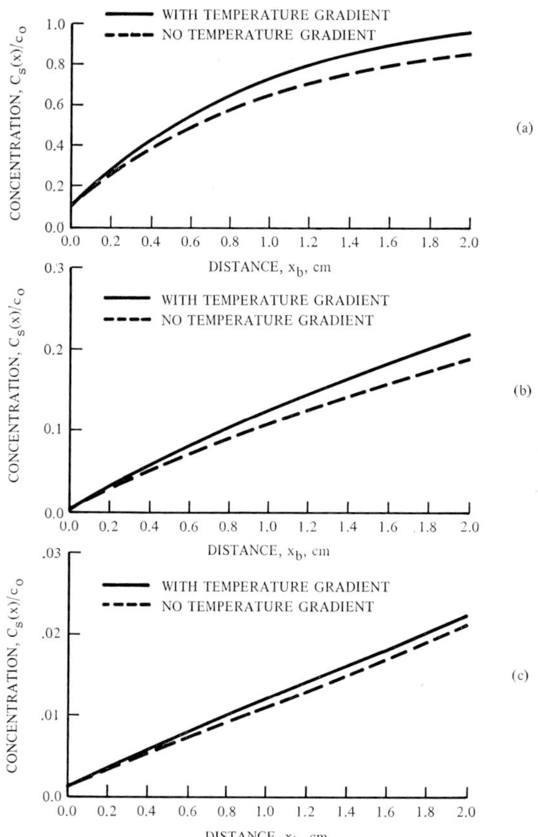

Fig. 4. Steady state solute distribution in the solid, $C_s(x)/c_0$, (a) $k_0 = 0.1$; (b) $k_0 = 0.01$; (c) $k_0 = 0.001$.

concentration at the interface is greater than the value for a zero temperature gradient situation.

The distribution of the solute in the solid is obtained by multiplying eq. (27) by k_0 and setting $Rt = x_b$, where x_b is the distance measured with respect to the boat

$$
\begin{aligned}
C_s(x) = k_0 c_0 \Bigg\{ &1 + \frac{(1-k_0)}{k_0} \\
&\times \left[1 - \exp\left[-\left(\frac{(1-k_0)R + \alpha D'}{D} \right) k_0 x_b \right] \right] \\
&- \frac{\alpha D'}{R} \left[\exp\left[-(1-k_0)\frac{(R+\alpha D')^2}{DR} k_0 x_b \right] \right. \\
&\left. - \exp\left[-\left(\frac{R+\alpha D'}{k} \right) x_b \right] \right] \Bigg\}.
\end{aligned}
\tag{28}
$$

In figs. 3 and 4, we plot the curves for $C_L(x, t)/c_0$ and $C_s(x)/c_0$ against distances along the lengths of the liquid and the solid. In all these plots, we assume $D = 10^{-4}\,\mathrm{cm^2/sec}$, $R = 10^{-3}\,\mathrm{cm/sec}$, and $D' = 3 \times 10^{-5}$

$\mathrm{cm^2/sec\,°C}$, $T' - T = 300\,°\mathrm{C}$, and $k = 0.1\,\mathrm{cm^2/sec}$. If $D' = 3 \times 10^{-7}\,\mathrm{cm^2/sec\,°C}$ is chosen and the other values remain the same, these plots would be different. However, if eq. (14) (constant temperature gradient) is used for these plots instead of eq. (25) (varying temperature gradient), the difference in solute distribution in the frozen solid would be 20% larger. The dashed curves represent the equivalent behavior when no temperature gradient is considered.

5. Conclusion

The major effect of a temperature gradient on the solute redistribution in the melt of a binary alloy results in an increase of the solute concentration in the solid (fig. 4) and a corresponding decrease in the melt for the steady state case. For the case when there is a temperature gradient in the liquid, the transient solute concentration near the interface is higher than the case when there is no temperature gradient in the melt. However, this transient solute concentration under-shoots the dashed curves (fig. 3) as x increases. No thermodynamic consideration of the delicate freezing process was considered. A fixed freezing rate R was assumed in addition to particular sets of constant parameter, k_0, D, D', $T_1 - T_0$, independent of temperature and concentration of the solute. We have arrived at the above conclusion purely from a kinetic point of view. The increased behavior in the solute concentration can be physically understood by referring to fig. 1. A temperature gradient present in the melt will create an additional solute flux toward the moving interface in the solute redistribution process so that as the freezing proceeds, more solute particles will be trapped in the solid than the situation when no such temperature gradient exists.

References

1) W. A. Tiller, K. A. Jackson, J. W. Rutter and B. Chalmers, Acta Met. **1** (1953) 428.
2) C. Wagner, Trans. AIME **200** (1954) 154.
3) R. G. Pohl, J. Appl. Phys. **25** (1954) 170.
4) K. F. Hulme, Proc. Phys. Soc. (London) B **68** (1955) 393.
5) V. G. Smith, W. A. Tiller and J. W. Rutter, Can. J. Phys. **33** (1955) 723.
6) W. A. Tiller and J. W. Rutter, Can. J. Phys. **34** (1956) 96.
7) A. S. Yue, J. Phys. Chem. Solids Suppl. B **28** (1967) 197.
8) H. W. Weart and D. J. Mack, Trans. AIME **212** (1958) 664.
9) A. S. Yue, Trans. AIME **227** (1963) 64.
10) G. A. Campbell and R. M. Foster, *Integrals to Practical Application* (Van Nostrand, 1948).
11) A. S. Yue, unpublished data.

CALCUL DES CONDITIONS DE CROISSANCE DE CRISTAUX EN SOLUTION

C. POTARD

Département de Métallurgie, Centre d'Etudes Nucléaires de Grenoble, C.E.A.–D.M.E.C.N., Cedex 85, 38 – Grenoble-Gare, France

The mass balance integral method is used to determine the transient growth rate and the critical thermal gradient for diffusion mass transport solution growth. Two general methods of growing crystals were studied. (1) Growth of thick crystals by controlled displacement of the presupposed crystal–solution equilibrium at the interface: The theoretical result was applied to cadmium-rich solution growth of Cd–Te crystals. The rate shows a maximum for relatively short time before a slowly decreasing variation take place. The critical thermal gradient shows a continuously increasing variation with time. (2) Liquid phase epitaxy under transient thermal conditions: This method was used by R. H. Deitch for the growth of Ga–As layer in Ga rich solution. The experimental results allow the numerical adjustment of two unknown parameters: the "thermal" parameters which depend on the experimental conditions and the arsenic liquid diffusion coefficient. The latter one has too low a value showing that interfacial phenomenon kinetics slow down the rate of growth and must be taken into account in theoretical treatment.

1. Introduction

La préparation de cristaux massifs ou en couche mince par cristallisation en solution, exige des conditions opératoires pour lesquelles l'interface plane en progression reste stable. Ceci impose la connaissance de la vitesse d'avancement de l'interface et du gradient thermique critique à l'interface qui en découle[1]).

La difficulté de ce problème réside dans la résolution rigoureuse, généralement impossible, d'une équation de diffusion dont les conditions aux limites sont variables avec le temps. L'utilisation d'une méthode intégrale due à Pohlhausen[2]), que nous développons ici dans le domaine de la cristallisation, permet souvent de trouver une solution satisfaisante à ce problème.

2. Méthode intégrale et hypothèses de base

La croissance unidirectionnelle d'un cristal en solution s'opère, en régime non stationnaire, à une vitesse $V(t)$ limitée par le transport purement diffusionnel des constituants.

Dans un alliage binaire, les grandeurs suivantes sont définies:

$S(t)$ = abscisse de l'interface mesurée par rapport à un système d'axes lié au laboratoire;

$\delta(t)$ = distance au-delà de laquelle le liquide n'est plus perturbé par la diffusion;

C_∞ = concentration en soluté dans le liquide à $x \geq \delta(t)$ (mole cm^{-3});

$C_e(t)$ = concentration en soluté dans le liquide en $S(t)$ (mole cm^{-3}). C'est une fonction du temps définie par les conditions expérimentales.

Dans le milieu liquide $S < x < \infty$, l'équation de diffusion classique:

$$D \frac{\partial^2 C}{\partial x^2} = \frac{\partial C}{\partial t} \tag{1}$$

(D = coefficient d'interdiffusion du soluté), s'applique car les axes de référence sont choisis fixes par rapport au liquide.

En outre, à l'interface, le bilan de matière permet d'écrire:

$$\frac{dS}{dt}\left(\frac{1}{\rho} - C_e(t)\right) = D\left(\frac{dC}{dx}\right)_s, \tag{2}$$

ρ = volume molaire du soluté.

L'équation intégrale à résoudre est obtenue par intégration de la relation (1) entre $S(t)$ et $\delta(t)$:

$$D \int_s^\delta \frac{\partial^2 C}{\partial x^2}\, dx = \int_s^\delta \frac{\partial C}{\partial t}\, dx. \tag{3}$$

L'équation de diffusion (1) ne sera ainsi satisfaite que dans sa valeur moyenne.

Puisqu'en δ le flux de diffusion est nul entraînant

$$(\partial C/\partial x)_\delta = 0, \tag{4}$$

le premier membre de l'équation intégrale est égal à:

$$- D(\partial C/\partial x)_s. \tag{5}$$

La règle de Leibnitz sur la dérivation des intégrales dépendant d'un paramètre, appliquée au second membre, permet d'écrire l'équation intégrale, en tenant compte des définitions citées plus haut sous la forme:

$$\frac{d}{dt} \int_S^\delta C \, dx - C_\infty \frac{d\delta}{dt} + C_e(t) \frac{dS}{dt} = -D \left(\frac{\partial C}{\partial x}\right)_S. \tag{6}$$

En choisissant une forme de solution de l'équation intégrale du type parabolique:

$$C(x, t) = a(t) + b(t)x + C(t)x^2, \tag{7}$$

Goodman[3]) a montré que la précision obtenue dans les problèmes de conduction de la chaleur était remarquable. Un exemple traité ci-après en donnera un aperçu.

Si les conditions aux limites du problème permettent d'expliciter (7) en fonction de S et de δ, son insertion dans (6) fournit alors une première équation différentielle dont les fonctions inconnues sont $S(t)$ et $\delta(t)$.

Une deuxième équation est nécessaire et sera naturellement obtenus en explicitant la relation (2) à l'aide de l'expression parabolique (7).

La méthode de calcul décrite a été appliquée à deux méthodes générales de cristallisation en solution.

3. Croissance par déplacement controlé de l'équilibre cristal–solution

La croissance de cristaux en couche mince[4,5]) ou sous forme massive, peut être réalisée en déplaçant lentement l'équilibre germe–solution par un refroidissement contrôlé du système.

Une analyse mathématique complète a été faite par Tiller et Kang[6]) dans le cas de la croissance en couche mince. Les conditions sont telles que l'épaisseur du cristal peut être négligée devant l'épaisseur de la couche liquide elle-même considérée comme un milieu fini. D'autre part, la vitesse de croissance est supposée très faible vis-à-vis de la cinétique de diffusion et ce terme n'apparaît pas dans l'expression du flux. Minden[7]) a traité le problème avec les mêmes hypothèses, mais en considérant également le cas d'une solution semi-infinie.

Dans le cas de la croissance de cristaux épais à partir de solutions concentrées, il est difficile de négliger l'épaisseur des dépôts et l'expression des flux de matière doit tenir compte de la vitesse d'avancement de l'interface. Le problème qui devient alors insoluble par les méthodes analytiques exactes, est bien adapté, à un traitement par la méthode intégrale.

3.1. CALCUL DE LA VITESSE D'AVANCEMENT DE L'INTERFACE

Le système germe-solution initialement en équilibre thermodynamique et dont l'interface est à la température T_0, est soumis à un refroidissement suivant une loi linéaire fixée par l'opérateur:

$$T_e = T_0 - \beta t, \tag{8}$$

T_e = température de l'interface, β = constante.

Au cours de ce refroidissement, en supposant la surfusion cinétique nulle, l'équilibre thermodynamique est conservé à l'interface.

En appliquant la loi de solubilité très générale: $\ln X = A - B/T_e$ la variation de la fraction atomique X en fonction du temps devient:

$$\ln X = A - B/(T_0 - \beta t). \tag{9}$$

Pendant une première phase de la croissance correspondant à la condition $\beta t \ll T_0$, un développement de (9) par rapport au temps limité au premier ordre conduit à une concentration à interface:

$$C_e(t) = C_\infty \, e^{-\alpha t}, \quad \text{avec} \quad \alpha = B\beta/T_0^2. \tag{10}$$

Les conditions aux limites peuvent alors s'écrire:

$t = 0:$

$$S(0) = 0, \quad C_e(0) = C_\infty,$$
$$\delta(0) = 0,$$
$$x > 0, \quad C(x, 0) = C_\infty; \tag{11a}$$

$t \neq 0:$

$$x = S(t), \quad C_e(t) = C_\infty \, e^{-\alpha t},$$
$$\frac{dS}{dt} \left(\frac{1}{\rho} - C_\infty \, e^{-\alpha t}\right) = D \left(\frac{\partial C}{\partial x}\right)_S,$$
$$x = \delta(t), \quad C(\delta, t) = C_\infty, \quad \left(\frac{\partial C}{\partial x}\right)_\delta = 0,$$

ρ = volume molaire du solute. $\tag{11b}$

En tenant compte de ces conditions, l'expression parabolique $C(x, t)$ devient:

$$C(x, t) - C_\infty = C_\infty \, (e^{-\alpha t} - 1) \left(\frac{\delta - x}{\delta - S}\right)^2. \tag{12}$$

La combinaison de cette expression et des équations (6)

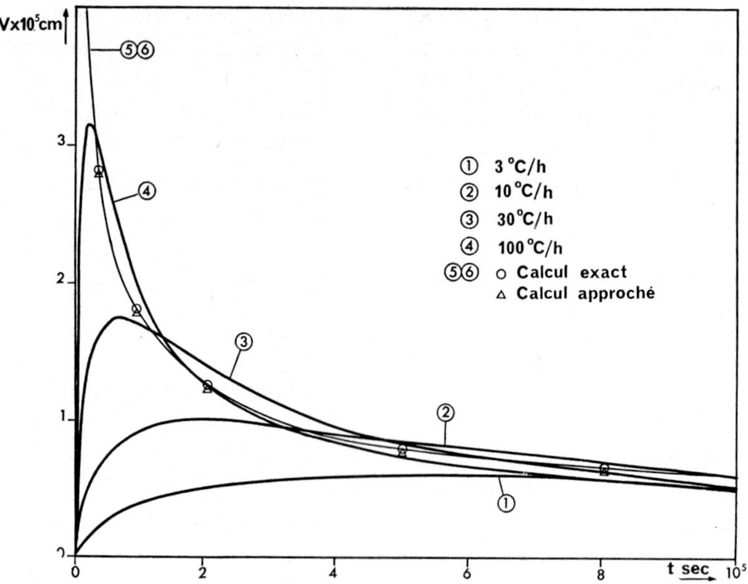

Fig. 1. Cristallisation du composé Cd–Te: vitesses de croissance théoriques pour différentes allures de refroidissement du four.

et (11) conduit, tous calculs effectués, à l'expression de la vitesse d'avancement de l'interface:

$$\frac{\mathrm{d}S}{\mathrm{d}t} = \left(\frac{D\alpha}{3}\right)^{\frac{1}{2}} (1-\mathrm{e}^{-\alpha t})^2 \, (b-\mathrm{e}^{-\alpha t})^{-1}$$

$$\times \left\{ \frac{b-1}{b}\left((1-b)^2 \ln \frac{b-\mathrm{e}^{-\alpha t}}{b-1} \right. \right. \tag{13}$$

$$\left. \left. -b(1-\mathrm{e}^{-\alpha t})+\alpha t\right)\right\}^{-\frac{1}{2}}$$

où $b = 1/\rho C_\infty$.

3.2. SURFUSION CONSTITUTIONNELLE

Le gradient critique G_C au-dessous duquel apparaît le phénomène de surfusion constitutionnelle se déduit de la connaissance du gradient de concentration instantané à l'interface et du liquidus dans la région du diagramme de phase concernée.

Le gradient des températures d'équilibre correspondant à la distribution spatiale des concentrations dans le liquide à l'interface peut s'écrire:

$$G_{\mathrm{Cr}}^{\mathrm{L}} = \left(\frac{\partial T_\mathrm{e}}{\partial x}\right)_S = \frac{\mathrm{d}T_\mathrm{e}}{\mathrm{d}C}\left(\frac{\partial C}{\partial x}\right)_S. \tag{14}$$

Dans cette expression le terme $\mathrm{d}T_\mathrm{e}/\mathrm{d}C$ s'exprime aisément en tenant compte de (8) et (10) par

$$\frac{\mathrm{d}T_\mathrm{e}}{\mathrm{d}C} = \frac{\beta}{\alpha C_\infty \, \mathrm{e}^{-\alpha t}}. \tag{15}$$

Le deuxième terme $(\partial C/\partial x)_S$ s'obtient à partir de l'équation (11) dans laquelle $\mathrm{d}S/\mathrm{d}t$ est remplacé par son expression (13). Le gradient thermique critique s'écrit alors:

$$G_{\mathrm{Cr}}^{\mathrm{L}} = \frac{\beta}{(3\alpha D)^{\frac{1}{2}}} (1-\mathrm{e}^{-\alpha t})^2$$

$$\times \left\{ \mathrm{e}^{-\alpha t}\left[\frac{b-1}{b}\left((1-b)^2 \ln \frac{b-\mathrm{e}^{-\alpha t}}{b-1}\right. \right. \right. \tag{16}$$

$$\left. \left. \left. -b(1-\mathrm{e}^{-\alpha t})-\mathrm{e}^{-\alpha t}\right)\right]^{\frac{1}{2}}\right\}^{-1}.$$

3.3. APPLICATION AU SYSTÈME Cd–Te

Ces résultats ont été appliqués à un exemple fictif de croissance du composé Cd–Te à partir d'une solution riche en cadmium. Pour une composition de 90%at. de cadmium, la température d'équilibre des phases est: $T_0 = 1168\,^\circ$K. Dans cette région du diagramme de phases, le liquidus suit la loi:

$$\ln X_{\mathrm{Te}} = 14.70 - 18.25 \times 10^3/T_\mathrm{e}.$$

En prenant: $D_{\mathrm{Te}} = 5 \times 10^{-5}\,\mathrm{cm}^2/\mathrm{sec}$, les vitesses de cristallisation théoriques correspondant à des allures de refroidissement du four de 100, 30, 10 et 3 °C/heure, ont été calculées (fig. 1).

Afin de s'assurer de la validité de la méthode utilisée, le calcul de la vitesse de cristallisation dans le cas simplifié d'une sursaturation constante de la solution a été effectué: le résultat obtenu est comparé sur la figure 1 à celui déduit d'une méthode de résolution exacte de ce problème particulier[5]). La superposition des courbes (5) et (6) atteste l'accord excellent entre les deux méthodes de calcul.

La figure 2 représente les variations du gradient thermique critique, G_{Cr}^{L} en fonction du temps, pour les précédentes allures de refroidissement du four. G_{Cr}^{L} est continuellement croissant et prend rapidement des valeurs élevées difficilement accessibles expérimentalement.

4. Croissance épitaxique sous conditions thermiques transitoires

Cette méthode, décrite par Deitch[8]) peut se schématiser comme suit:

Un germe de faible masse, à la température T_0 est plongé instantanément dans une solution saturée de grand volume, placée à une température T_∞ légèrement supérieure à celle du substrat.

Le four délivrant une puissance constante, la température de la solution n'est perturbée que dans le voisinage du substrat. En effet, la température de la première couche liquide en contact avec le germe froid suit tout d'abord une décroissance très rapide jusqu'à un minimum et dans un intervalle de temps très court. Succédant à cette phase presque instantanée du fait des conductibilités thermiques importantes du milieu métallique, une période plus longue de remontée de la température intervient jusqu'à atteinte de l'équilibre thermique.

Pendant ces périodes, les couches liquides refroidies sont sursaturées et un dépôt de soluté a lieu sur le germe. L'avantage de cette méthode est de conserver des gradients thermiques élevés à l'interface pendant la croissance, notamment pour les solutions métalliques.

L'épaisseur du dépôt est contrôlée par la durée d'immersion lorsque toutes les autres conditions sont fixées.

4.1. Vitesse d'avancement de l'interface

L'expérience a montré que la durée de la première phase d'évolution de la température peut être négligée

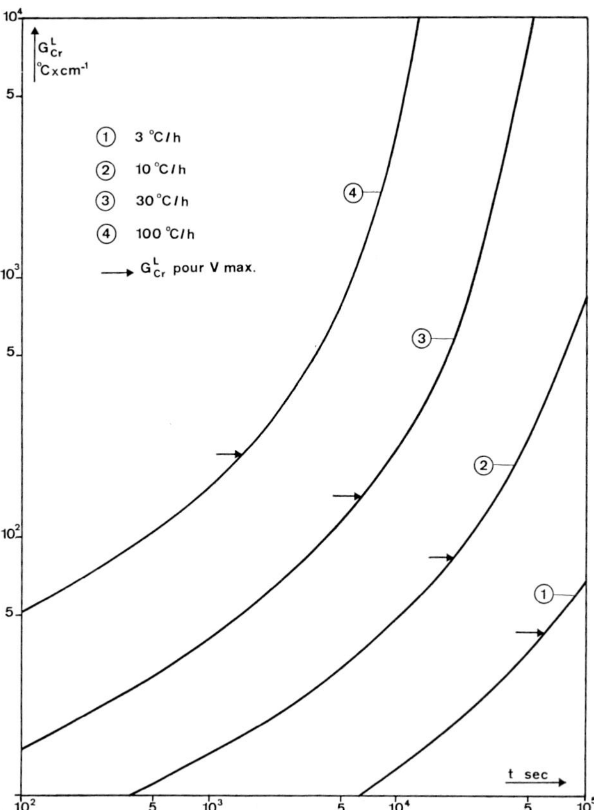

Fig. 2. Cristallisation du composé Cd–Te: gradients thermiques critiques théoriques.

devant la durée de la phase suivante. Ce point peut être mis à profit dans le calcul de la vitesse de croissance. En outre, il est permis de supposer que le minimum de température atteint est voisin de la température initiale du substrat. Dans ces conditions, la variation de la température d'interface suit une loi exponentielle[9]) du type suivant:

$$T_e = T_\infty - (T_\infty - T_0)\, e^{-\gamma t}, \tag{17}$$

γ = constante déterminée par les propriétés thermiques du système. Dans cette méthode, la différence de température $(T_\infty - T_0)$ n'excède pas 50 °C; il est alors raisonnable de considérer la partie du liquidus du diagramme de phases concerné comme un segment de droite.

La variation des concentrations d'interface en fonction du temps peut donc s'écrire:

$$C_e = C_\infty - (C_\infty - C_0)\, e^{-\gamma t}, \tag{18}$$

C = concentration en mole cm⁻³. Dans les couches

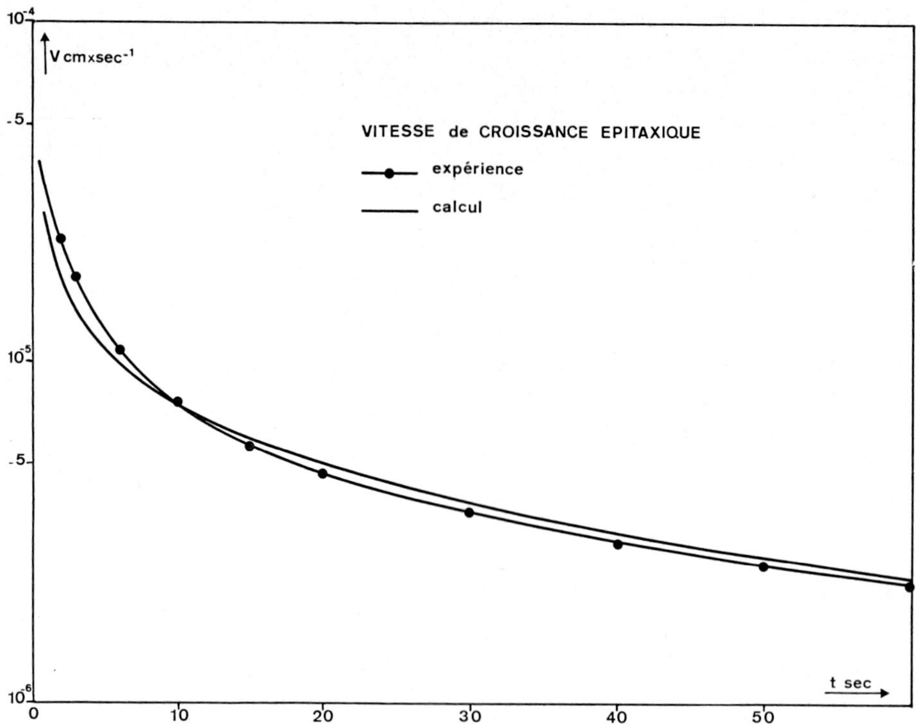

Fig. 3.   Cristallisation en film du composé Ga–As; comparaison entre les vitesses de croissance calculée et mesurée.

liquides sursaturées le mode de transport du soluté peut être considéré comme diffusionnel; les conditions aux limites du problème s'écrivent alors:

$t = 0$:

$\quad x = 0, \quad C(0, 0) = C_0,$

$\quad x \neq 0, \quad C(x, 0) = C_\infty;$ (19a)

$t \neq 0$:

$\quad x = S,$

$\quad C(S, t) = C_e(t) = C_\infty - (C_\infty - C_0)\, e^{-\gamma t},$

$$\left(\frac{\partial C}{\partial x}\right)_S = \frac{1}{D}\left[\frac{1}{\rho} - C_e(t)\right]\frac{\mathrm{d}S}{\mathrm{d}t},$$ (19b)

$\quad x = \delta, \quad C(\delta, t) = C_\infty,$

$$\left(\frac{\partial C}{\partial x}\right)_\delta = 0.$$

La distribution du soluté dans le liquide, comme dans le cas précédent, peut être représentée par une forme parabolique du type (7) qui devient, en appliquant les conditions aux limites ci-dessus:

$$C(t, x) = C_\infty - (C_\infty - C_0)\, e^{-\gamma t}\left(\frac{\delta - x}{\delta - S}\right)^2.$$ (20)

Une procédure de résolution identique à celle utilisée dans le cas précédent conduit à l'expression de la vitesse de croissance:

$$\frac{\mathrm{d}S}{\mathrm{d}t} = b(\tfrac{1}{3}\gamma D)^{\frac{1}{2}}\,(b + e^{-\gamma t})^{-1}$$ (21)

$$\times \left[b\frac{e^{-\gamma t} - 1}{e^{-\gamma t}} - \tfrac{1}{2}b^2\frac{e^{-2\gamma t} - 1}{e^{-2\gamma t}} - \ln\frac{(b+1)e^{-\gamma t}}{b + e^{-\gamma t}}\right]^{-\frac{1}{2}},$$

avec

$$b = \frac{(1/\rho) - C_\infty}{C_\infty - C_0}.$$

L'intégration numérique de cette fonction fournit l'épaisseur de cristal obtenu au bout d'un temps donné.

4.2. Surfusion constitutionnelle

L'expression de la surfusion constitutionnelle peut être obtenue à partir de l'équation (14) dans laquelle $\mathrm{d}T_e/\mathrm{d}C$ est une constante propre au diagramme de phase dans le domaine étroit des températures considérées.

$(\partial C/\partial x)_S$ est tiré de (19) en remplacant $\mathrm{d}S/\mathrm{d}t$ par son expression (21).

XVI – 2

4.3. APPLICATION AU SYSTÈME Ga–As

La croissance épitaxique de GaAs en solvant gallium, a été réalisée par Deitch qui a, en outre, mesuré l'épaisseur du depôt en fonction du temps pour les conditions expérimentales suivantes:

$$T_\infty = 800 \ ^\circ C, \quad X_{\infty As} = 6.30\% at,$$
$$T_0 = 750 \ ^\circ C, \quad X_{0 As} = 5.32\% at. \tag{22}$$

Dans cette expérience, la capacité calorifique du substrat n'étant pas négligeable devant celle de la solution, une décroissance sensible de la température du bain, T_∞ est enregistrée. Malgré cette différence entre l'expérience et notre traitement théorique, une comparaison a pu être tentée.

En effectuant l'interpolation numérique des valeurs expérimentales suivante:

$$S = 1.227 \times 10^{-4} \ t^{0.3059} \tag{23}$$

(facteur de correlation $= 0.994$), la dérivation par rapport au temps de cette expression conduit à la vitesse de croissance:

$$dS/dt = 3.750 \times 10^{-5} \ t^{-0.6941}. \tag{24}$$

Dans le demi-diagramme Ga–GaAs, les conditions (22) jointes à la valeur de $\rho_{Ga-As} = 3.7 \times 10^{-2}$ donnent $b = 23.1$.

Les paramètres inconnus γ et D_{As}^{L} sont alors ajustés sur les valeurs expérimentales; il vient:

$$\gamma = 10^{-2} \ sec^{-1}, \quad D_{As}^{L} = 10^{-6} \ cm^2 \ sec^{-1}:$$

Les courbes de vitesse expérimentale et théorique sont comparées sur la figure 3.

La valeur de la constante de temps γ ainsi obtenue est vraisemblable, car la différence $T_\infty - T_e$ qui en résulte et la vitesse de croissance expérimentale tendent simultanément vers zéro.

Par contre, la valeur de D_{As}^{L} obtenue est trop faible par rapport aux valeurs habituellement rencontrées dans les alliages liquides. Cette constatation permet de supposer que la vitesse de croissance n'est pas contrôlée par la diffusion seule, mais également par les cinétiques des phénomènes d'interface.

5. Conclusion

Le calcul des conditions de croissance de cristaux en solution se heurte aux difficultés de résolution rigoureuse de l'équation des flux de diffusion dans des conditions aux limites fonctions du temps.

L'utilisation d'une méthode mathématique intégrale nous a permis de calculer, sans hypothèses simplificatrices contraignantes, les vitesses de croissance en régime de transport de masse purement diffusionnel, ainsi que les gradients thermiques minimaux assurant la progression d'une interface plane.

Deux méthodes générales de croissance ont été étudiées:

(1) Croissance de cristaux épais par déplacement contrôlé de l'équilibre cristal-solution: le cas d'une vitesse de refroidissement de l'alliage constante a été examiné et le résultat appliqué au système Cd–Te. La vitesse de croissance présente un maximum tandis que le gradient thermique critique est continuellement croissant et atteint rapidement des valeurs élevées difficilement accessibles expérimentalement.

(2) Croissance épitaxique sous conditions thermiques transitoires: dans ce cas, la vitesse de cristallisation est rapidement décroissante. Nos résultats théoriques comparés à une expérience de cristallisation du composé GaAs en solvant gallium ont permis l'ajustement du paramètre "thermique" γ, contrôlant les flux de chaleur et du coefficient d'interdiffusion de l'arsenic en solution D_{As}^{L}. La valeur trop faible trouvée pour ce dernier, permet de supposer que les cinétiques des phénomènes d'interface ralentissent la vitesse de croissance et doivent être considérées dans un tel calcul.

Etude effectuée grace à l'appui de la D.G.R.S.T.

Bibliographie

1) W. A. Tiller, J. Crystal Growth **2** (1968) 69.
2) K. Pohlhausen, Z. Angew. Math. Mech. **1** (1921) 252.
3) T. R. Goodman, Trans. ASME **80** (1958) 335.
4) L. R. Dawson et J. M. Whelan, Bull. Am. Phys. Soc. **13** (1968) 375.
5) H. Nelson, RCA Rev. **24** (1963) 603.
6) W. A. Tiller et C. Kang, J. Crystal Growth **2** (1968) 345.
7) H. T. Minden, J. Crystal Growth **6** (1970) 228.
8) R. H. Deitch, J. Crystal Growth **7** (1970) 69.
9) H. S. Carslaw et J. C. Jaeger, *Conduction of Heat in Solids* (Oxford, 1959).

    Journal of Crystal Growth **13/14** (1972) 810–813 © *North-Holland Publishing Co.*

THE ORIENTATION-DEPENDENCE OF THE PARTITION COEFFICENT IN ZINC-BASE ALLOY SINGLE CRYSTALS

B. D. WARR and R. W. SMITH

Department of Metallurgical Engineering, Queen's University, Kingston, Ontario, Canada

The macrosegregation of cadmium occurring in dilute zinc–cadmium single crystals grown horizontally by the normal freeze technique from a zinc melt containing 100 ppm cadmium has been examined as a function of growth rate and orientation. Atomic absorption and autoradiographic analytical techniques were used. It was found that the partition coefficient is orientation dependent, being greatest when the basal plane of the crystal lies in the solid–liquid interface.

1. Introduction

When a dilute binary alloy liquid charge freezes, solute partitions in a variety of ways depending upon the manner in which freezing proceeds. The degree of segregation depends in the first instance on the ratio of the tie-line compositions at the equilibrium liquidus temperature of the freezing liquid. This ratio is defined as the equilibrium solute partition coefficient, k_0, and may have a value less or greater than unity. When a horizontal liquid charge is progressively frozen from one end, it is convenient to describe the solute partition taking place in terms of an effective partition coefficient, k_{eff}, i.e., by the ratio of the compositions of the solid near to, but not in, the solid–liquid interface and that of the bulk melt. In doing this, note is taken of the fact that a boundary layer may exist at the solid–liquid interface because solute mixing is incomplete in the bulk melt. Yet a further partition coefficient has been distinguished namely k^*, the dynamic partition coefficient. The value of this normally approaches that of k_0 but it takes note of the fact that freezing is a dynamic process and solute partitioning may well be dependent upon such variables as the rate of growth and the orientation of the crystal interface.

Whilst there is little reported evidence of anisotropic solute incorporation in metal systems, many cases of this have been reported in semiconductors[1–5]. The earliest of these was made by Hall[1]), who grew a bundle of three oriented seed crystals from a common antimony-doped germanium melt. He observed that above a certain growth rate the value of the partition coefficient was orientation dependent. Anomalous segregation, analogous with Hall's observation, has also been found in semiconductor crystals which exhibit a close-packet facet on an otherwise curved interface. Several investigations have shown, for alloys of germanium[4]), silicon[5]) and indium antimonide[3]), that the ratio of k_{eff} on a facet to that off the facet is almost without exception greater than unity.

For metals, Bocek et al.[6]) observed during the cellular growth of a dilute zinc alloy containing cadmium, that the cell size and transition from elongated cells to regular cells was dependent on crystal orientation. This has been attributed[7]) to a variation of solid–liquid interfacial energy with orientation. However, from a study of the factors influencing the stability of a planar solid–liquid interface in non-metals, Spittle et al.[8,9]) concluded that these anomalies[6]) might have been due to an orientation-dependence of solute partition. Accordingly, they grew dilute zinc–cadmium crystals by a normal-freeze technique using radio-tracer cadmium and applying the Pfann analysis[10]) to determine the k_{eff} for a particular crystal, and so examine the orientation dependence of the partition coefficient. The results obtained suggested that the partition coefficient was least when the basal plane was parallel to the solid–liquid interface. Since this result differed significantly from those reported earlier, where solute incorporation was always greater across the closest packed plane, it was decided to re-examine the segregation of cadmium in zinc using atomic absorption analysis and autoradio-

XVI – 3

graphy. This would allow considerably greater precision in determining composition than the techniques available to Spittle[9]).

2. Experimental procedure

The Pfann analysis used by Spittle has a number of limitations. In particular, it assumes that k_{eff} is constant during the freezing of the liquid charge and also that it is independent of melt composition. Whilst the latter assumption may well be valid for very dilute alloys, the effective stirring occurring in the horizontal liquid charge decreases as the liquid is consumed. Consequently, the interface boundary layer, and therefore k_{eff}, changes progressively as the interface moves along the charge. In view of this it was decided to adopt the following procedure to determine k_{eff}.

The apparatus consisted of a horizontal graphite boat containing three grooves approximately 5 mm wide, 1 cm deep and 25 cm long. One end of the graphite boat was water cooled to preserve the zinc alloy seed crystals which were placed at that end. Freezing was allowed to proceed along the entire liquid charge of zinc containing approximately 100 ppm cadmium under a reduced argon atmosphere. Following this, a section 5 mm long was removed from each crystal rod at a point 5 cm from the start end. The seed crystals were discarded whilst the remainder of each crystal rod was melted, stirred and chill cast. The 5 mm section and a small part of the chill cast material were then analysed for cadmium using an atomic absorption spectrophotometer. The ratio of these compositions was recorded as k_{eff}, for the particular growth conditions of that experiment. This was repeated for a range of growth rates. During any experiment, three crystal orientations were used. These were $\alpha = 90°$, $45°$ and $0°$, where α is the angle that the basal plane of the seed crystal made to the normal through the solid–liquid interface. In fact, it had been found from preliminary experiments that the plane of the interface lay approximately ten degrees from the vertical. The accuracy of any measurement of k_{eff} was assessed to be $< \pm 2\%$.

It was considered that the above method of determining k_{eff} would give a more reliable value than the application of the Pfann analysis. To check that the method used was not rendered unreliable because of the presence of microfluctuations in solute concentration, a number of crystals were grown containing the

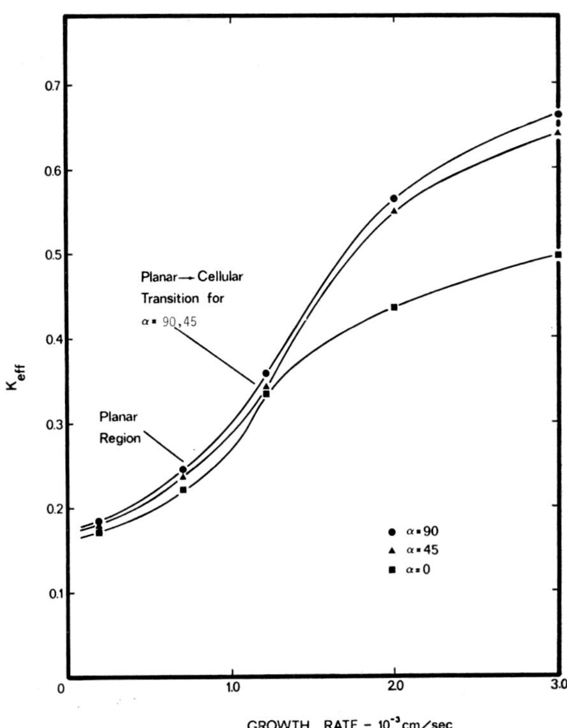

Fig. 1. Effective distribution coefficient as a function of growth rate and orientation for zinc containing 100 ppm cadmium. α is the angle between the basal plane and the normal to the plane of the interface.

radio tracer cadmium[115]. Subsequent autoradiographic exposure on fine grain nuclear emulsion did not reveal any observable microfluctuations in photographic density and therefore in cadmium content along the length of the crystals.

It was originally intended in these studies to examine the orientation dependence of the partition coefficient in bicrystals. Unfortunately, all attempts to grow bicrystals consisting of one crystal with its basal plane in the interface and another with the basal plane at 45° or perpendicular to the interface failed. However, it was found that when the seed crystals were moved from one groove to another and the molten charge solidified under the same general conditions the measured k_{eff} for any particular crystal orientation was unchanged.

3. Results

Fig. 1 summarises the change in k_{eff} as a function of growth rate and crystal orientation. It is seen that there is a small but consistent variation in k_{eff} between the various orientations up to a growth rate of approximately 1.3×10^{-3} cm/sec, above which the curve for $\alpha = 0$ parts company from the other two. This consistency was not present in the results of Spittle et al. who reported that the k_{eff} value for $\alpha = 90°$ was marginally less than those for other orientations.

4. Discussion

The major disadvantage of the experimental technique used in the present study when compared to that used by Spittle et al., is that higher cadmium concentrations are required in order to carry out accurate atomic absorption analysis. As a result of this, cellular growth was observed to occur in the crystals oriented with $\alpha = 45°$ and $90°$, when the growth rate exceeded approximately 1.3×10^{-3} cm/sec. Consequently, k_{eff} rises sharply with growth rate for these crystals at higher growth rates. However, at lower growth rates, planar growth ensued and permitted an accurate estimate of k_{eff} to be made.

The most significant feature of these results is that solute incorporation was greater when the basal plane lay in the interface. Since the basal plane is the close-packed plane of zinc, these results are not in conflict with those obtained with semiconductors, in that for these materials k_{eff} was always greatest across a close-packed solid–liquid interface.

The mechanisms suggested to account for anomalous solute partition in semiconductors invoke a change in the interface distribution coefficient k^* with orientation. This is necessary in order to explain for certain $k_0 < 1$ systems, a change in the value of k_{eff}, through unity to a value greater than unity. According to the theory of Burton et al.[11], if k^* were independent of orientation, the limiting value of k_{eff}, for the conditions of no stirring and complete stirring in the melt, would be unity and k^* respectively.

Hall[1] accounted for this observation in terms of the trapping of absorbed impurity atoms, the concentration of which, he assumed, varied with interface orientation. Alternative models for a possible exchange process have been examined by Thurmond[12] and Brice[13].

It now seems likely that the results obtained by Hall were caused by faceting associated with certain growth orientations. Any mechanism explaining the enhanced incorporation of solute on a facet could, therefore, also be used to interpret Hall's results. Since the presence

of a facet is synonymous with growth difficulty normal to the faceting plane, it follows that a certain supercooling is required to perpetuate faceted interface motion. Mullin[14] and Dikhoff[4] have both attempted to explain the change in k_{eff} value with orientation by postulating a dependence of solute incorporation on local kinetic supercooling. However, neither hypothesis explains all the experimental observations.

The most plausible mechanism for the reported results of anomalous solute incorporation in semiconductors appears to be in terms of the growth kinetics of a facet[3,15]. If growth normal to a facet occurs by the rapid lateral propagation of sheets of solid, this may cause enhanced incorporation of solute by the trapping of an adsorbed layer of impurity. Normal growth on non-faceted surfaces would probably be such as to allow solute to be incorporated with the equilibrium concentration.

The formation of a facet on a solid–liquid interface is usually indicative of the existence of a singular, i.e., atomically smooth, surface in contact with the melt. The conditions under which such a singular surface may exist has been examined by Jackson[16]. He showed that whereas the closest-packed planes of covalent materials in contact with their melts were likely to be faceted, all metal crystal interfaces were likely to be atomically rough and so be able to advance normal to themselves without requiring the nucleation of growth steps or the existence of a re-entrant corner site.

The experiments of Brice and Whiffin[17] support Jackson's contention in that the undercoolings at the interface of germanium crystals, grown in $\langle 111 \rangle$, $\langle 110 \rangle$ and $\langle 100 \rangle$ directions at 2.7×10^{-3} cm/sec were 1.8, 1.2 and 0.1°C, respectively. Since macrofacets have been observed on both (111) and (110) interfaces, it appears that these are atomically smooth surfaces while the (100) surfaces are rough.

Since Spittle et al.[9] were unable to detect supercooling for any of the zinc crystzls grown, it must be presumed that zinc always freezes with an atomically rough surface whatever the orientation of the plane of the interface, as Jackson's α factor would predict. In view of this, it would seem that whatever the cause of the variation in partition coefficient between the various crystal orientations used in the present experiments, an explanation in terms of faceted growth is not appropriate.

XVI – 3

It should be noted, however, that there is considerable unpublished evidence[18]) of an orientation dependence of solute incorporation in semiconductor crystals growing without a faceted solid–liquid interface.

In searching for an explanation of the observed variation in solute incorporation between crystal orientations, some significance is placed upon the fact that it proved impossible to grow bicrystals of zinc for which the basal plane of one of the crystals lay in the interface. This is interpreted as implying that the undercooling required to drive an interface containing a basal plane is larger than that required in the case of other orientations. On this assumption an argument of the type used by Dikhoff[4]) may be applicable.

Thus for a given growth rate, the interface temperature of a faceted surface would be lower than that of an atomically rough surface. Hence, in a system for which $k_0 < 1$, more solute will be incorporated in the faceted surface. In this case the difference in concentration between the crystal surface growing with a facet and without a facet will depend upon slope of the solidus curve and hence the equilibrium distribution coefficient, being larger when k_0 is larger. Since some undercooling is necessary for growth to occur at all, at a solid–liquid interface therefore, should this undercooling be greater for the advance of a zinc interface containing the basal plane than for an interface containing a less closely packed plane, then more solute would be incorporated in the basal plane and hence a small but finite orientation dependence of the partition coefficient should be observed.

It should be pointed out that whilst the observed variations in k_{eff} are small, they are appreciably greater than the experimental error which was estimated to be $< \pm 2\%$.

More recently it has been reported[19]) that when dilute zinc–cadmium alloys are frozen very rapidly, the compositions of the resulting solid are found to be larger than the stable or metastable solidus composition at any temperature and therefore a departure from local equilibrium at the liquid–solid interface occurs in the solid. This was interpreted as indicating that the cadmium solute atoms tended to concentrate in the solid–liquid interface in such a manner that, even though the tie-line composition might exist on either side of the interface at very low growth rates, as the growth rate increased so these localised atoms tended to be incorporated in the solid.

5. Conclusions

It has been observed that a small orientation dependence of the partition coefficient exists in zinc alloys containing small amounts of cadmium. This dependence varies with the rate of growth and is such that k_{eff} is greatest when the basal plane of the zinc crystal lies in the solid–liquid interface. It is suggested that this orientation dependence stems from the greater undercooling required to drive the solidification on a close-packed plane than on a less closely packed plane.

Acknowledgements

The above work was supported by the National Research Council of Canada.

References

1) R. N. Hall, Phys. Rev. **88** (1952) 139; J. Phys. Chem. **57** (1953) 836.
2) J. R. O'Connor, J. Electrochem. Soc. **110** (1963) 338.
3) J. B. Mullin and K. F. Hulme, J. Phys. Chem. Solids **17** (1960) 1.
4) J. A. M. Dikhoff, Solid-State Electron. **1** (1960) 202.
5) M. G. Mil'vidskii and A. V. Berkova, Soviet Phys.-Solid State **5** (1963) 517.
6) M. Bocek, P. Kratochvil and M. Valouch, Czech. J. Phys. **8** (1958) 557.
7) G. F. Bolling and W. A. Tiller, J. Appl. Phys. **31** (1960) 2040.
8) J. A. Spittle, Ph.D. Thesis, Birmingham University, 1966.
9) J. A. Spittle, M. D. Hunt and R. W. Smith, J. Crystal Growth **3, 4** (1968) 647.
10) W. G. Pfann, Acta Met. **1** (1953) 763.
11) J. A. Burton, R. C. Prim and W. P. Slichter, J. Chem. Phys. **21** (1953) 1987.
12) C. D. Thurmond, in: *Control of Composition in Semiconductors by Freezing Methods*, Ed. N. B. Hannay (Reinhold, New York, 1959).
13) J. C. Brice, *The Growth of Crystals from the Melt*, Selected Topics in Solid State Physics, Vol. 5, Ed. E. P. Wohlfarth (North-Holland, Amsterdam, 1965).
14) J. B. Mullin, in: *Compound Semiconductors*, Vol. 1, Eds. R. K. Williardson and H. L. Goering (Reinhold, New York, 1962).
15) A. Trainor and B. E. Bartlett, Solid-State Electron. **2** (1961) 106.
16) K. A. Jackson, in: *Liquid Metals and Solidification* (A. S. M., Cleveland, 1958).
17) J. C. Brice and P. A. C. Whiffin, Solid State Electron. **7** (1964) 183.
18) J. B. Mullin, RRE, Malvern, U.K., private communication.
19) J. C. Baker and J. W. Cahn, Acta Met. **17** (1969) 575.

  Journal of Crystal Growth **13/14** (1972) 814–817 © *North-Holland Publishing Co.*

ETUDE DES CONDITIONS DE CROISSANCE DE MONOCRISTAUX DE BERYLLIUM

J. DELOCHE, Y. MALMEJAC et B. SCHAUB

French Atomic Energy Commission, Metallurgy Department, Grenoble, France

The Czochralski technique has been used to prepare single crystals of beryllium. Various pulling parameters have been studied in terms of the perfection of the produced single crystals: those results are connected with the specific physical properties of the beryllium melt, chiefly its viscosity, its surface tension, its density, and its thermal diffusivity. The axial and radial distributions of various impurities (aluminium, copper, manganese) have been measured by a radioactivation analysis technique. They are related to the growth parameters. The crystalline quality of such crystals has been measured by neutron diffraction (Rocking Curves), and X-ray topography (Berg-Barrett and Lang methods). Disorientations of about 10 minutes were observed.

1. Introduction

Nous avons été amenés à étudier les conditions de croissance de monocristaux de beryllium en vue de leur utilisation en monochromateurs de neutrons: Celle-ci impose une bonne perfection cristalline et un contrôle des gradients de concentration en solutés métalliques.

2. Directions de croissance

De nombreuses expériences de cristallisation unidirectionnelle de beryllium (fusion de zone et croissance Czochralski), nous ont permis de constater la très forte anisotropie de croissance de ce métal.

Les cristaux se développent selon des directions préférentielles qui ne sont pas des directions cristallographiques simples; la direction de croissance, assimilée à l'axe moyen du barreau, se trouve soit à 8° environ de l'axe [10$\bar{1}$0], soit à 16° environ de l'axe [1$\bar{2}$10] de la structure hexagonale.

Le beryllium, à température ordinaire, est hexagonal compact; or on trouve dans la littérature[1,2] mention de l'existence, au voisinage du point de fusion, d'une phase cubique centrée; cette phase n'a jamais pu être isolée, mais ses paramètres cristallins ont été déterminés. On peut alors penser que les directions préférentielles de croissance observées sont dues à la transformation cubique centrée → hexagonale compacte lors du refroidissement.

On peut essayer, par une méthode graphique rappelant celle employée dans les cas d'épitaxie, de décrire la transformation du réseau cubique en réseau hexagonal.

Les cristaux cubiques ont deux directions de croissance privilégiées, [100] ou [111]; dans notre cas nous éliminons la possibilité de croissance suivant [100] car il n'apparaît pas de recouvrement possible avec les plans de la phase hexagonale.

Nous avons donc admis au départ que la croissance de la phase cubique se faisait dans la direction [111], et nous avons cherché comment pouvait se faire la transformation des plans (111) en plans (10$\bar{1}$0) ou (1$\bar{2}$10).

2.1. DONNÉES CRISTALLOGRAPHIQUES

Paramètres du réseau:

Be cubique: a_0 = 2.548 Å (entre 1283 et 1270 °C);

Be hexagonal: a = 2.3435 Å, c = 3.6585 Å (au voisinage du point de transformation) (tableau 1).

TABLEAU 1

Structure	Plan	Distances inter-atomiques
Cubique centrée	(111)	3.6034 suivant [$\bar{1}$10]
		3.1206 suivant [11$\bar{2}$]
Hexagonale compacte	(10$\bar{1}$0)	2.3435 suivant [1$\bar{2}$10]
		3.6585 suivant [0001]
Hexagonale compacte	(11$\bar{2}$0)	4.05906 suivant [10$\bar{1}$0]
		3.6585 suivant [0001]

2.2. TRANSFORMATION (111) → (10$\bar{1}$0)

Les répartitions atomiques dans les plans considérés, montrent que les réseaux coïncident le mieux lorsque la direction [$\bar{1}$10] du système cubique est confondue avec la direction [0001] du système hexagonal, et la direction [11$\bar{2}$] du système cubique confondue avec la direction [1$\bar{2}$10] du système hexagonal (fig. 1a).

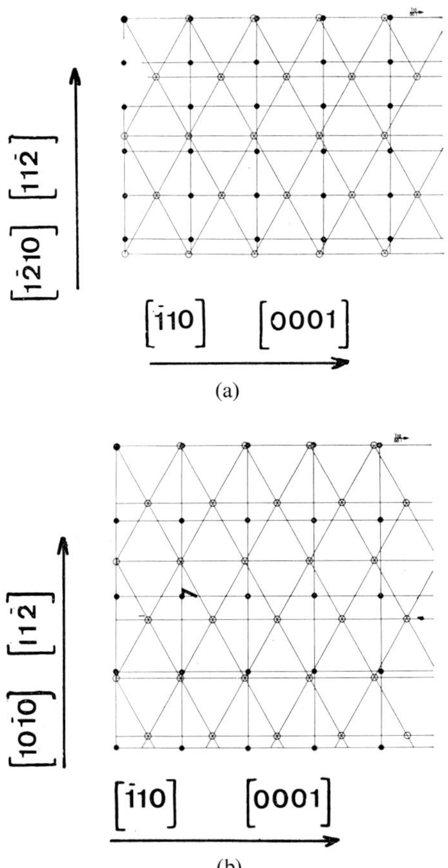

(a)

(b)

Fig. 1. (a) Transformation (111) → (10$\bar{1}$0). (b) Transformation (111) → (11$\bar{2}$0). (○) Atomes du système cubique; (●) atomes du système hexagonal.

On remarque que la coïncidence des réseaux s'améliore encore pour un angle de 9° 40′ entre les axes [$\bar{1}$10] et [0001] et un angle de 2° 50′ entre les axes [11$\bar{2}$] et [1$\bar{2}$10].

Cette double désorientation correspond à un angle entre les normales à chacun des 2 plans cubique centré et hexagonal compact égal à 9° 55′.

2.3. Transformation (111) → (11$\bar{2}$0)

Par un procédé analogue au précédent on trouve que la coïncidence optimale s'obtient pour un angle de 9° 40′ entre les axes [$\bar{1}$10] et [0001], et pour un angle de 11° 50′ entre les axes [11$\bar{2}$] et [10$\bar{1}$0] (fig. 1b).

Cette double désorientation correspond à un angle entre les normales à chacun des 2 plans, cubique centré et hexagonal compact, égal à 15° 20′.

La concordance des angles expérimentaux et théoriques paraît satisfaisante, d'autant plus que, pour la détermination expérimentale des angles, on se réfère à une direction géométrique mal définie qui est la "direction de croissance" assimilée à l'axe général du barreau. Il convient aussi de noter que les probabilités d'obtention de l'une ou l'autre des désorientations semblent équivalentes.

Aussi dans la suite de cette étude nous avons taillé nos germes soit à 8° de [10$\bar{1}$0] soit à 16° de l'axe [1$\bar{2}$10].

3. Technique de tirage

Nous utilisons un creuset en glucine pouvant contenir 700 g de béryllium. L'hélium purifié sur tamis moléculaires sert d'atmosphère protectrice. La vitesse de tirage est de 43 mm/h. Les vitesses de rotation du creuset et de la broche porte germe sont chacune égales à 10 t/min et de sens inverses.

Une programmation de température est nécessaire pour compenser les pertes thermiques et maintenir ainsi le diamètre du cristal constant. La surface libre du bain est maintenue à une cote constante par relevage progressif du creuset.

Il nous est ainsi possible de tirer des cristaux jusqu'a 55 mm de diamètre.

Nous avons analysé la qualité de ces monocristaux sous deux aspects:
– Répartition des impuretés principales;
– Perfection cristalline.

4. Distribution des solutés

Nous nous sommes particulièrement intéressés aux distributions axiales de l'aluminium, du cuivre et du manganèse, en fonction de la fraction solidifiée g. Le béryllium de départ (qualité CR Péchiney) contient 47 ppm d'aluminium, 20 ppm de manganèse, 7.5 ppm de cuivre.

Les résultats expérimentaux ont été lissés sur une fonction de la forme.

$$\log C_\mathrm{S} = b + m \log (1-g). \tag{1}$$

On trouve (fig. 2):
pour l'aluminium:
$\log C_\mathrm{S} \times 1.02696 - 0.61050 \log (1-g)$;
pour le manganèse:
$\log C_\mathrm{S} \times 0.94518 - 0.30510 \log (1-g)$;
pour le cuivre
$\log C_\mathrm{S} \times 0.84306 - 0.08328 \log (1-g)$;

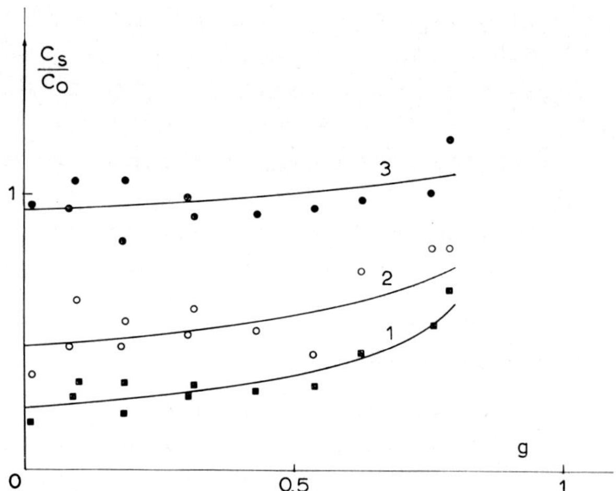

Fig. 2.   Repartitions longitudinales des impuretés Al, Mn, Cu.
(1) Al, ■, $C_0 = 47$ ppm; (2) Mn, ○, $C_0 = 19.6$ ppm; (3) Cu, ●,
$C_0 = 7.4$ ppm. g = fraction solidifiée.

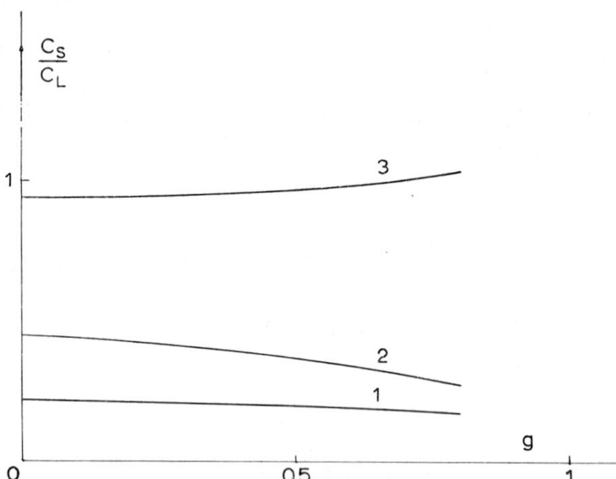

Fig. 3.   Coefficients de distribution effectifs des impuretés (1)
Al, (2) Mn, et (3) Cu. g = fraction solidifiée.

ce qui donne pour $g = 0$:

C_S/C_0 (aluminium) $= 0.226$,
C_S/C_0 (manganèse) $= 0.449$,
C_S/C_0 (cuivre) $= 0.941$.

A l'issue de chaque tirage on a effectué des analyses
dans le liquide résiduel et on a vérifié à mieux que 1%
près que:

$$\int_0^{g^*} C_S + C_L^*(1 - g^*) = C_0, \qquad (2)$$

g^* étant la valeur de la fraction solidifiée à l'issue du
tirage, C_L^* la concentration dans le liquide résiduel,

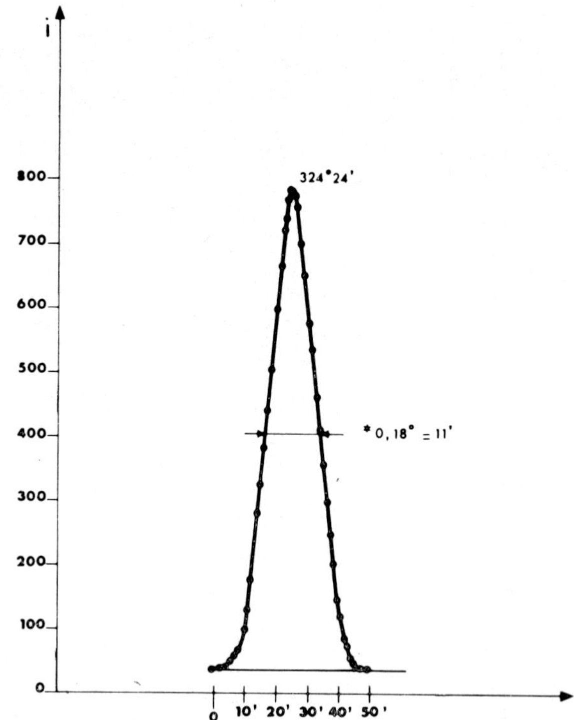

Fig. 4.   "Profil de diffraction" d'un monocristal de beryllium.

$\int_0^{g^*} C_S$ ayant été calculeé à partir de la relation (1).

On en déduit qu'aucune des trois impuretés ne
s'élimine par volatilisation sélective ce qui est conforme
aux données thermodynamiques relatives aux systèmes
Be–Al, Be–Mn, Be–Cu.

On peut donc calculer à partir de la relation (2) la
quantité de soluté dans le liquide à tout moment et en
déduire en fonction de g un coefficient de partage
effectif défini comme étant égal à C_S/C_L.

On trouve que les coefficients de partage effectifs
évoluent très faiblement, à partir des valeurs données
précédemment pour C_S/C_0, à $g = 0$ (fig. 3).

Il ressort de cette analyse que les impuretés principales cuivre, aluminium, manganèse n'introduisent
pas de gradient de concentration susceptible de nuire
à la qualité du monocristal tant que la fraction solidifiée,
g, ne dépasse pas 0.7. Les mêmes résultats ont été
récemment obtenus en ce qui concerne le fer et le
nickel.

5. Qualité des cristaux

Les "profils de diffraction" obtenus au double
spectromètre à neutrons montrent une largeur à

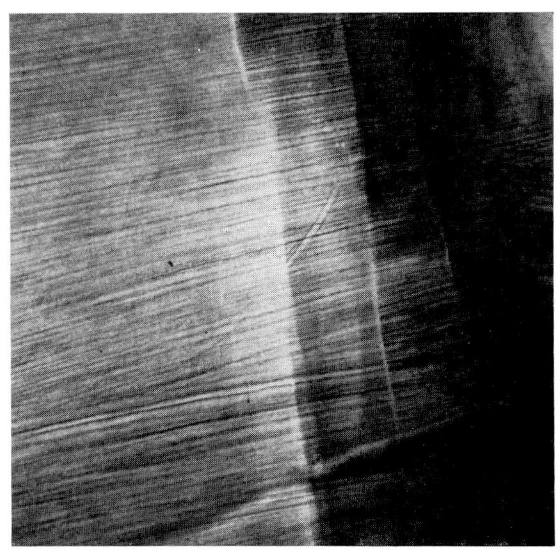

Fig. 5. Topographie par la méthode de Berg–Barrett d'un monocristal de beryllium.

mi-hauteur de 11′ et un pouvoir réflecteur élevé (fig. 4).

Des topographies par rayons X [Berg-Barrett: définition de 10′ sur le plan $(10\bar{1}3)$], confirment ce résultat (fig. 5).

On a par contre pu vérifier qu'en dehors des deux directions préférentielles dont nous avons fait état, il n'était pas possible d'obtenir de véritables monocristaux.

La direction de croissance la plus défavorable est celle de l'axe sénaire pour laquelle nous avons observé des polygonisations atteignant 2 degrés.

6. Conclusion

On a montré que les directions de croissance du béryllium étaient favorables au voisinage des axes binaires, et non pas comme on aurait pu le croire dans la direction de l'axe senaire.

On a montré que ceci était dû à l'existence d'une phase intermédiaire cubique centrée.

Par ailleurs le monocristal ne présentant pas de gradients de concentration importants en impuretés métalliques principales, sa perfection cristalline ne varie que peu au cours du tirage, dans la mesure où les conditions thermiques sont maintenues constantes.

Nous remercions M. Rustichelli et M. Tochetti de l'Institut Laue–Langevin à Grenoble pour l'aide qu'ils nous ont apportée dans l'examen neutronique des cristaux. La Section d'Application des Radioéléments du CEN-G a assuré l'analyse des solutés. M. Blet de Thomson CSF a réalisé la topographie présentée dans cet article.

Bibliographie

1) A. J. Martin et A. Moore, J. Less-Common Metals **1** (1959) 85.
2) C. Potard, C. Bienvenu et B. Schaub, Diagramme de phases du système Be–Ca, dans: *Thermodynamics of Nuclear Materials* (IAEA, Vienne, 1967).

THE INFLUENCE OF THE ELECTRICAL PROPERTIES OF THE SOLID PHASE ON IMPURITY INCORPORATION DURING CRYSTAL GROWTH

H. C. CASEY, Jr. and M. B. PANISH

Bell Telephone Laboratories, Incorporated, Murray Hill, New Jersey 07974, U.S.A.

Control of impurity incorporation during crystal growth requires consideration of the relationships between the concentration of the impurity in the liquid phase and its concentration in the growing solid phase. At the solid–liquid interface, the surface space-charge layer in the solid has been found to dominate the incorporation of slowly diffusing impurities. For extrinsic conditions, the amount of slowly diffusing impurity incorporated in the solid varies linearly with the amount in the liquid. Detailed analysis in terms of the surface-space charge layer is given for GaAs liquid-phase epitaxial layers doped with radioactive Te^{129m}. Not only is this linear dependence observed for all the group VI donors S, Se, and Te, but also for the slowly diffusing acceptor Ge. For Te in GaAs, the distribution coefficient goes from greater than unity at low temperatures to less than unity at high temperatures. Rapidly diffusing impurities are not influenced by the surface space-charge layer, and for singly ionized substitutional acceptors or donors the amount of impurity incorporated into the extrinsic solid is expected to vary as the square root of the amount in the liquid. This square-root behavior has been observed for the acceptor Zn which diffuses rapidly in GaAs.

1. Introduction

In the preparation and growth of materials for electronic components and devices, a primary consideration is the removal or the controlled introduction of dilute amounts of impurities into a host material. Control of impurity incorporation requires consideration of the relationships between the concentration of the impurity in the liquid phase and its concentration in the growing solid phase. This paper describes recent studies in the III–V compound semiconductor GaAs which permit the derivation of fundamental concepts for the manner in which the electrical properties of the solid phase influence impurity incorporation.

Treatment of impurity incorporation in semiconductors requires consideration of the properties of both the liquid and solid phase. It has been shown by Thurmond[1]) and Thurmond and Kowalchik[2]) that the binary liquid phase of an impurity element with Ge or Si may be described by regular solution activity coefficients. Jordan[3]) extended these regular solution calculations to the ternary liquidus isotherms for Ga–As–Zn and Ga–P–Zn. These thermodynamic treatments adequately describe the experimental liquidus data. The retrograde solidus for the binary Ge–impurity and Si–impurity systems was explained by Thurmond and Struthers[4]) in terms of regular solution theory. The concepts utilized to describe the liquidus and retrograde solidus are strictly thermodynamic and are not unique to semiconductors.

It was shown by Reiss and Fuller[5]) that the ionization of impurities in semiconductors led to departures from the solid solubility expected on the basis of the simple thermodynamic concepts described above. They demonstrated that the solubility of Li in Si was also dependent on the position of the Fermi level in the semiconductor bulk, where the Fermi level is the electron (or hole) chemical potential. Extension of the Reiss and Fuller concepts to the incorporation of singly ionized substitutional impurities in extrinsic GaAs leads to a prediction of a square-root dependence of the amount of impurity in the solid on the amount in the liquid, which agrees with the experimentally observed behavior of Zn in GaAs[6]) but disagrees with the observed linear dependence for Te in GaAs[7]).

Analysis by Zschauer and Vogel[8]) shows that the impurity diffusivity in the solid divided by the width of the surface space-charge region must exceed the growth rate in order for the liquid phase to be in equilibrium with the semiconductor bulk. This criterion is apparently met for the rapidly diffusing Zn in GaAs but not for the slowly diffusing Te in GaAs. In this latter case, it was shown by Casey et al.[7]) that the position of the Fermi level at the surface dominates the incorporation of Te in GaAs. The position of the Fermi level at the surface may be obtained by treating the solid–liquid interface as a metal–semiconductor Schottky barrier. The important result described by the above descussion is that the position of the Fermi level does influence the incorporation of impurities in semiconductors, as suggested by Reiss and Fuller[5] However, it is the impurity diffusivity in the solid that determines whether it is the Fermi level at the surface or in the semiconductor bulk that enters the solubility equilibrium relationships.

An additional consideration occurs at high impurity concentrations in the solid. The interactions of the free charge carriers at high concentrations affects the position of the Fermi level. For Zn in GaAs, the diffusivity in the solid is sufficiently rapid to ensure equilibrium between the liquid and the semiconductor bulk. However, the effects of the high impurity density influence the solubility equilibria through the decrease in the hole activity coefficient. These two examples, Te in GaAs and Zn in GaAs, demonstrate the manner in which the electrical properties of the solid may be included in the equilibrium relationships for the impurity incorporation during crystal growth.

2. The surface space-charge layer

Analysis by Zschauer and Vogel[8] has shown that $D\lambda$ (diffusivity D and reciprocal intrinsic Debye length λ) must exceed the growth rate v for the semiconductor bulk to be in equilibrium with the liquid phase. When $D\lambda < v$, the liquid is in equilibrium with the surface, while for growth rates between these limits, the impurity incorporation is growth rate dependent. The reciprocal intrinsic Debye length is given as

$$\lambda = (q^2 n_i / \varepsilon k T)^{\frac{1}{2}}. \tag{1}$$

With $n_i = 5 \times 10^{17}$ cm^{-3} (ref. 9), $\varepsilon = 1.1 \times 10^{-12}$ farad/cm, λ at 1000 °C is $\sim 10^6$ cm^{-1}. The actual width of the space-charge layer depends on both temperature and impurity density in the solid. Therefore, it is not entirely clear at the present time whether or not the intrinsic Debye length is the correct quantity to use for the space-charge layer width when the concentration in the solid is varied under extrinsic conditions.

The surface space-charge layer at the solid–liquid interface for a heavily doped n-type semiconductor is shown in fig. 1. This space-charge layer at the surface of the solid is due to surface states that result from disruption of the lattice periodicity by the surface and

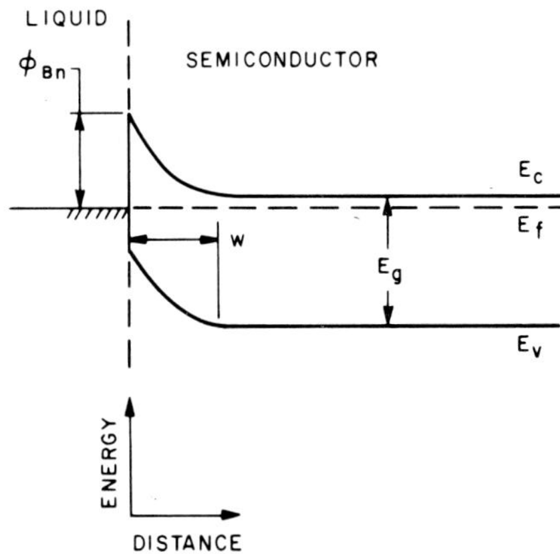

Fig. 1. Energy-band diagram for a liquid n-type semiconductor interface with a surface space-charge layer width w. The separation between the valance band E_v and the conduction band E_c is the energy gap E_g. The Fermi level is E_f and the barrier height ϕ_{Bn} is the position of E_f at the liquid–semiconductor interface.

the unsaturated or dangling bonds of the surface atoms[10]). The surface states result in levels within the energy gap which are characteristic of the surface, and they tend to compensate the donors (or acceptors) and control the position of the Fermi level at the surface. The significant quantities are the barrier height ϕ_{Bn}, which is the separation in energy of the Fermi-level and the conduction band edge at the surface, and the width w of the space-charge layer. Because of the high density of conduction electrons in the liquid, this interface is considered to behave as a metal–semiconductor Schottky barrier. It has been observed[11,12]) that the position of the Fermi level at the surface remains at a fixed energy above the valence band as the temperature varies:

$$E_g(T) - \phi_{Bn}(T) = \text{constant}, \tag{2}$$

where $E_g(T)$ is the temperature dependent energy gap.

3. Tellurium in GaAs

For the incorporation of Te in GaAs at 1000 °C by liquid-phase epitaxy, $D \approx 10^{-13}$ cm^2/sec (ref. 13), $\lambda \approx 10^6$/cm, and $v \approx 10^{-6}$ cm/sec. Therefore, $D\lambda/v$ is approximately 0.1, and the liquid should be in equilibrium with the surface rather than the semiconductor bulk. The reaction for Te incorporation during liquid-phase epitaxial growth may then be described by[7]):

$$\text{Te}(\ell) + \text{V}_{As} \rightleftarrows \text{Te}(s)^+ + e^-. \tag{3}$$

In this reaction, Te in the liquid Te (ℓ) is taken to react

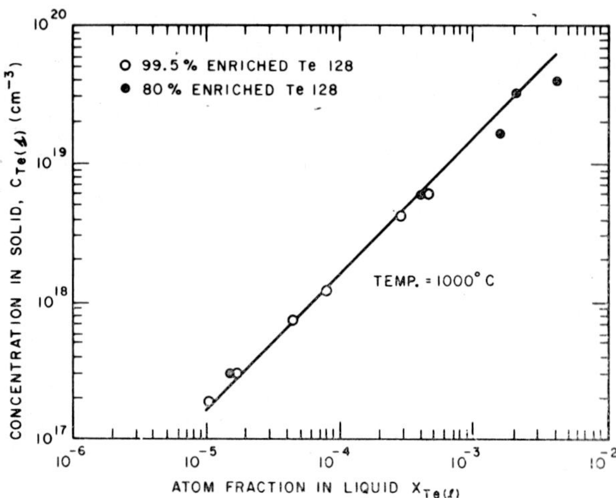

Fig. 2. A portion of the 1000 °C solid–solubility isotherm for Te in GaAs.

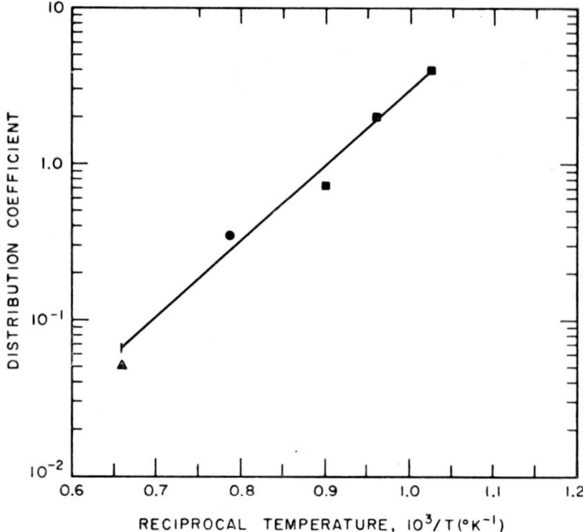

Fig. 3. The distribution coefficient of Te in GaAs as a function of temperature. (●) Ref. 7; (■) Ref. 16; (▲) Ref. 17.

with an As vacancy V_{As} to give an ionized substitutional donor on an As site $\mathrm{Te}(_{s})^{+}$ and a free carrier electron e^{-}. Since the As vacancy concentration is proportional to the activity of Ga in the liquid $\gamma_{Ga}X_{Ga}(\ell)$, the equilibrium relationship for eq. (3) may be written as

$$K_{1}(T) = \frac{\gamma_{\mathrm{Te}(s)}C_{\mathrm{Te}(s)}\gamma_{n}n}{\gamma_{\mathrm{Te}(\ell)}X_{\mathrm{Te}(\ell)}\gamma_{\mathrm{Ga}(\ell)}X_{\mathrm{Ga}(\ell)}}, \tag{4}$$

where γ_{n} is the activity coefficient of electrons in the solid, n is the electron concentration, and $\gamma_{\mathrm{Te}(s)}$ and $C_{\mathrm{Te}(s)}$ are the activity coefficient and concentration of $\mathrm{Te}(_{s})^{+}$ on As sites. The quantities $\gamma_{\mathrm{Te}(\ell)}$ and $\gamma_{\mathrm{Ga}(\ell)}$ are the activity coefficients for Te and Ga in the liquid. It is convenient to express $C_{\mathrm{Te}(s)}$ in units of atoms/cm^3 and the concentrations of the composition in the liquid, $X_{\mathrm{Te}(\ell)}$ and $X_{\mathrm{Ga}(\ell)}$, as atom fraction.

For discussion of the incorporation reaction, it is useful to simplify eq. (4). For the experimental data to be considered here, the solid solutions of Te in GaAs are dilute with maximum concentrations of about 5×10^{19} cm^{-3} (0.1 at %); therefore, Henry's law should be obeyed and $\gamma_{\mathrm{Te}(s)}$ is taken as constant. Also, the Te concentrations in the liquid are quite low, and $X_{\mathrm{Ga}(\ell)}$ is essentially constant at 0.88 atom fraction. Therefore, $\gamma_{\mathrm{Te}(\ell)}$ and $\gamma_{\mathrm{Ga}(\ell)}X_{\mathrm{Ga}(\ell)}$ may also be taken as constant. Treatment of the electron activity coefficient by Hwang and Brews[14]) shows that γ_{n} is relatively constant at a value of approximately 0.4 at 1000 °C in the concentration range considered here. Eq. (4) reduces to the simple expression

$$K_{2}(T) = C_{\mathrm{Te}(s)}n/X_{\mathrm{Te}(\ell)} \tag{5}$$

for no complex formation occurring at the growth temperature, and with the Te concentration in excess of the intrinsic carrier concentration n_{i}. For the liquid phase in equilibrium with the semiconductor surface, n is a function of temperature only and is given by[15])

$$n = N_{c} \exp\left(-\phi_{Bn}/kT\right), \tag{6}$$

where N_{c} is the effective density of states in the conduction band. Eq. (5) may now be written as

$$C_{\mathrm{Te}(s)} = K_{2}(T)\,X_{\mathrm{Te}(\ell)}/N_{c}\exp\left(-\phi_{Bn}/kT\right), \tag{7}$$

and hence $C_{\mathrm{Te}(s)}$ varies linearly with $X_{\mathrm{Te}(\ell)}$ at a given temperature. This predicted linear dependence is demonstrated by a portion of the 1000 °C solid solubility isotherm shown in fig. 2. The data shown in this figure were obtained from liquid-phase epitaxial layers grown in a closed system and were doped with radioactive Te^{129m}. The agreement between eq. (7) and the experimental data verify that when $D\lambda < v$, it is the position of the Fermi level at the surface rather than in the semiconductor bulk that dominates the impurity incorporation.

Because of the linear dependence of Te, a distribution coefficient may be assigned at each temperature. Kang and Greene[16]) determined the distribution coefficient in an open system between 700 and 850 °C by measuring the electron concentration and assuming that $C_{\mathrm{Te}(\mathit{s})} = n$. Milvidskii and Pelevin[17]) determined the distribution coefficient at the melting point by chemical analysis. Fig. 3 is a plot of data from refs. 7, 16, and 17 and illustrates the interesting result that the distribution coefficient increases at lower temperature. Note that the distribution coefficient goes from greater than unity at low temperature to less than unity at high temperature.

It should be noted that a linear dependence between the amount of impurity in the liquid and in the solid has also been observed for the group VI donors S (ref. 17) and Se (refs. 17 and 18). The linear dependence is not unique to donors and has been observed for the group IV acceptor Ge in GaAs[19]). The common property of these impurities in GaAs is that they all diffuse slowly.

4. Zinc in GaAs

Zinc in GaAs is an example of the incorporation of a rapidly diffusing impurity. At Zn concentrations in excess of 10^{18} cm^{-3} at 1000 °C, $D > 10^{-11}$ cm^2/sec (ref. 20). Therefore, with $\lambda \approx 10^6$ cm^{-1} and $v \approx 10^{-6}$ cm/sec, $D\lambda/v > 10$, and the liquid phase should be in equilibrium with the semiconductor bulk. The incorporation of Zn in the liquid Zn(ℓ) into a Ga vacancy V_{Ga} as a singly ionized substitutional acceptor $\mathrm{Zn}(\mathit{s})^{-1}$ on a Ga site plus a hole e^+ is represented by[6])

$$\mathrm{Zn}(\ell) + V_{\mathrm{Ga}} \rightleftarrows \mathrm{Zn}(\mathit{s})^- + \mathrm{e}^+ . \tag{8}$$

Since the Ga vacancy concentration is proportional to $\gamma_{\mathrm{As}(\ell)} X_{\mathrm{As}(\ell)}$, the equilibrium relation for eq. (8) may be written as

$$K_3(T) = \frac{\gamma_{\mathrm{Zn}(\mathit{s})} C_{\mathrm{Zn}(\mathit{s})} \gamma_{\mathrm{p}} p}{\gamma_{\mathrm{Zn}(\ell)} X_{\mathrm{Zn}(\ell)} \gamma_{\mathrm{As}(\ell)} X_{\mathrm{As}(\ell)}}, \tag{9}$$

where γ_{p} is the activity coefficient of holes in the solid, p is the hole concentration, and $\gamma_{\mathrm{Zn}(\mathit{s})}$ and $C_{\mathrm{Zn}(\mathit{s})}$ are the activity coefficient and concentration of $\mathrm{Zn}(\mathit{s})^{-1}$ on Ga sites. The quantity $\gamma_{\mathrm{Zn}(\ell)}$ and $\gamma_{\mathrm{As}(\ell)}$ are the activity coefficients for Zn and Ga in the liquid.

Eq. (9) may be simplified by assuming Henry's law applies for Zn in the solid so that $\gamma_{\mathrm{Zn}(\mathit{s})}$ may be taken as constant and included in $K_3(T)$ as $K_3'(T)$. For extrinsic conditions and fully ionized Zn in the solid, the condition of electrical neutrality may be expressed by $C_{\mathrm{Zn}(\mathit{s})} = p$. The solid solubility of Zn is then given by

$$C_{\mathrm{Zn}(\mathit{s})} = [K_3'(T) \gamma_{\mathrm{Zn}(\ell)} X_{\mathrm{Zn}(\ell)} \gamma_{\mathrm{As}(\ell)} X_{\mathrm{As}(\ell)} / \gamma_{\mathrm{p}}]^{\frac{1}{2}}. \tag{10}$$

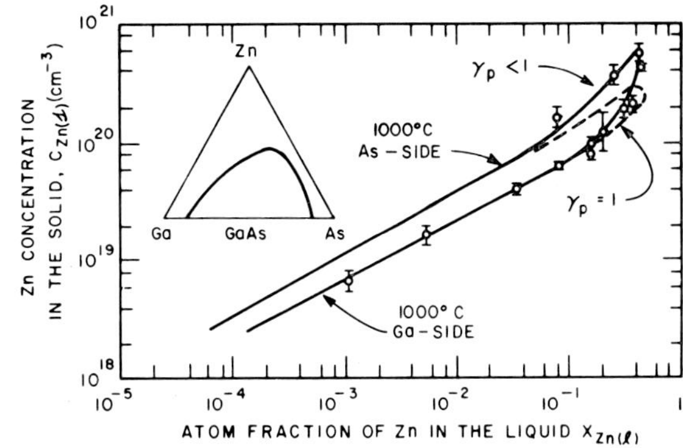

Fig. 4. The Zn concentration in the solid versus the atom fraction of Zn in the liquid along the Ga–As–Zn 1000 °C liquidus isotherm. The 1000 °C liquidus isotherm is shown by the inset.

At a given temperature, the variation of $X_{\mathrm{As}(\ell)}$ with $X_{\mathrm{Zn}(\ell)}$ may be obtained from the liquidus isotherm of the Ga–As–Zn ternary phase diagram[21]) which is shown for 1000 °C by the inset in fig. 4. Jordan's[3]) analysis of the Zn and arsenic pressure measurements by Shih et al.[22]) permits evaluation of $\gamma_{\mathrm{Zn}(\ell)}$ and $\gamma_{\mathrm{As}(\ell)}$. The complete solid solubility along the 1000 °C liquidus isotherm, as calculated by eq. (10) with $\gamma_{\mathrm{p}} = 1$, is given by the dashed line in fig. 4. The double valued solid solubility curve results from the fact that the liquidus isotherm is also double valued in Zn composition (see the insert of fig. 4). At high concentrations, the experimental solid solubility exceeds $C_{\mathrm{Zn}(\mathit{s})}$ obtained from eq. (10) with $\gamma_{\mathrm{p}} = 1$ and indicates that γ_{p} is less than unity in this region.

The hole activity coefficient is a convenient notation that includes all the effects of the charged carrier interactions which perturb the semiconductor band structure at high impurity concentrations. It is related to the Fermi level E_{f} by the usual thermodynamic relation for chemical potential

$$\mu(p) = -E_{\mathrm{f}} = \mu^0 + kT \ln (\gamma_{\mathrm{p}} p / N_{\mathrm{v}}), \tag{11}$$

where μ^0 is the reference potential and N_{v} is the effective density of states in the valence band. The reference potential may be shown to be the valence band edge E_{v} for dilute impurity concentrations. When the values of γ_{p} from ref. 20 are used in eq. (10), the solid line shown in fig. 4 is obtained and is in agreement with the experimental data. It should also be noted that

$$\gamma_{\mathrm{Zn}(\ell)} \gamma_{\mathrm{As}(\ell)} X_{\mathrm{As}(\ell)} / \gamma_{\mathrm{p}}$$

is constant for $X_{\mathrm{Zn}(\ell)} < 0.1$ atom fraction and the

the concentration dependence of the amount of Zn in the solid on the amount in the liquid is given by

$$C_{Zn(s)} = [K''_3(T)\, X_{Zn(\ell)}]^{\frac{1}{2}}, \tag{12}$$

which is a square-root dependence of the Zn concentration for equilibrium between the liquid and the semiconductor bulk.

5. Conclusions

Depending on the diffusivity of an impurity in the solid, the liquid phase is in equilibrium with either the semiconductor surface or bulk. Equilibrium between the liquid and semiconductor surface occurs for a slow impurity diffusivity in the solid and results in a linear dependence of the amount of singly ionized impurity in the solid on the amount in the liquid. This linear dependence was demonstrated by Te in GaAs. Equilibrium between the liquid and semiconductor bulk requires rapid impurity diffusivity in the solid and results in a square-root dependence of the amount of a singly ionized substitutional impurity in the solid on the amount in the liquid. In addition, high impurity concentrations result in interaction between the charge carriers and in enhanced solubility. These effects were demonstrated by Zn in GaAs. To treat impurity incorporation in semiconductors during crystal growth, it is necessary to not only consider the activity coefficients in the liquid phase, but to also include the electrical properties of the solid.

References

1) C. D. Thurmond, J. Phys. Chem. **57** (1953) 827.
2) C. D. Thurmond and M. Kowalchik, Bell Syst. Tech. J. **39** (1960) 169.
3) A. S. Jordan, Met. Trans. **2** (1971) 1965.
4) C. D. Thurmond and J. D. Struthers, J. Phys. Chem. **57** (1953) 831.
5) H. Reiss and C. S. Fuller, J. Metals **8** (1956) 276.
6) M. B. Panish and H. C. Casey, Jr., J. Phys. Chem. Solids **28** (1967) 1673.
7) H. C. Casey, Jr., M. B. Panish and K. B. Wolfstirn, J. Phys. Chem. Solids **32** (1971) 571.
8) K.-H. Zschauer and A. Vogel, in: *GaAs: 1970 Symp. Proc.* (Inst. of Phys., London, 1971) p. 100.
9) H. C. Casey, Jr., in: *Atomic Diffusion in Semiconductors*, Ed. D. Shaw (Plenum, London, in press).
10) A. Many, Y. Goldstein and N. B. Grover, *Semiconductor Surfaces* (North-Holland, Amsterdam, 1965) p. 165.
11) S. M. Sze, *Physics of Semiconductor Devices* (Wiley, New York, 1969) p. 363.
12) Y. Nannichi and G. L. Pearson, Solid-State Electron. **12** (1969) 341.
13) J. F. Osborne, K. G. Heinen and H. Riser, unpublished.
14) C. J. Hwang and J. R. Brews, J. Phys. Chem. Solids **32** (1971) 837.
15) Ref. 11, p. 429.
16) C. S. Kang and P. E. Greene, in: *GaAs: 1968 Symp. Proc.* (Inst. of Phys. and Phys. Soc., London, 1969) p. 18.
17) M. G. Milvidskii and O. V. Pelevin, Inorg. Mater. **3** (1967) 1024; translated from Izv. Akad. Nauk SSSR, Neorg. Mater. **3** (1967) 1159.
18) L. J. Vieland and I. Kudman, J. Phys. Chem. Solids **24** (1963) 437.
19) F. E. Rosztoczy and K. B. Wolfstirn, J. Appl. Phys. **42** (1971) 426.
20) H. C. Casey, Jr., M. B. Panish and L. L. Chang, Phys. Rev. **162** (1967) 660.
21) M. B. Panish, J. Electrochem. Soc. **113** (1966) 861.
22) K. K. Shih, J. W. Allen and G. L. Pearson, J. Phys. Chem. Solids **29** (1968) 367.

Journal of Crystal Growth **13/14** (1972) 823–826 © *North-Holland Publishing Co.*

STRUCTURE DE MONOCRISTAUX D'ALLIAGE CUIVRE–NICKEL FABRIQUÉS PAR LA TECHNIQUE DE BRIDGMAN

P. BOTTELIER et F. BOUILLON

Laboratoire de Chimie Minérale et Analytique, Université Libre de Bruxelles, Bruxelles, Belgium

Copper–nickel single crystals, in which the concentration C_0 of nickel ranges between 10^2 and 10^5 ppm, are grown from the melt by the Bridgman technique. Selective oxidation of the nickel in those crystals reveals a characteristic cellular structure, which can be related to constitutional supercooling. This is quite similar to many other alloys previously studied, in which the solute lowers the melting point of the solvent. According to Tiller's criteria, a plane interface should be stable for concentrations of nickel lower than 10^3 ppm. By means of selective oxidation in alloys containing 10^2 ppm of solute, no cellular segregation can be detected, though this technique is more sensitive than others used in this work, such as the autoradiography of ^{63}Ni, the electron microprobe and the electron scanning microscope.

1. Introduction

L'alliage cuivre–nickel constitue un matériau valable pour l'étude des réactions d'oxydation[1]). L'interprétation des mécanismes fondamentaux du processus n'est possible qu'à partir de résultats obtenus sur des échantillons monocristallins de structure bien connue. C'est pour cette raison que nous avons fabriqué quelques cristaux à faible teneur de nickel ($< 2\%$) marqué (^{63}Ni).

L'autoradiographie a permis de montrer que tous les échantillons dont la concentration en nickel est comprise entre 0.5 et 2% présentaient une structure cellulaire très stable (figs. 1–3)[2]), dont nous avons essayé d'interpréter l'origine dans le cadre des mécanismes modernes de la solidification unidirectionnelle.

2. Préparation des cristaux

Nous avons préparé deux séries de cristaux d'alliages à partir de cuivre ASARCO de pureté 99.999% et de nickel Koch-Light de pureté 99.995% fondus et solidifiés sous une pression de 10^{-5} torr dans des moules en graphite pur. Les premiers, fabriqués par technique Bridgman horizontale, avaient 150 mm de long, 20 mm de large et 2 mm d'épaisseur; les autres, cylindriques et réalisés par technique Bridgman verticale, avaient 10 mm de diamètre et 100 mm de long.

3. Stabilité de la structure

La figure 3 représente la structure régulière de l'alliage; elle est obtenue par autoradiographie de la face principale d'une lame plane monocristalline; elle est constituée d'alignements de cellules dont le coeur est suffisamment enrichi en nickel pour empêcher une homogénéisation ultérieure des alliages.

En effet, ni les recuits à temperature élevée, ni les traitements mécaniques et thermiques, ni la recristallisation sous contrainte ne changent la distribution générale du nickel. La structure observée est donc très stable.

4. Orientation des cristaux obtenus

Il n'apparaît aucune relation simple entre l'orientation des cristaux d'alliages et la teneur en soluté.

Il faut cependant noter que des conditions de croissance qui donnent de façon reproductible des lames d'orientation proche de (001) pour le cuivre pur, deviennent inopérantes après introduction de 1% de

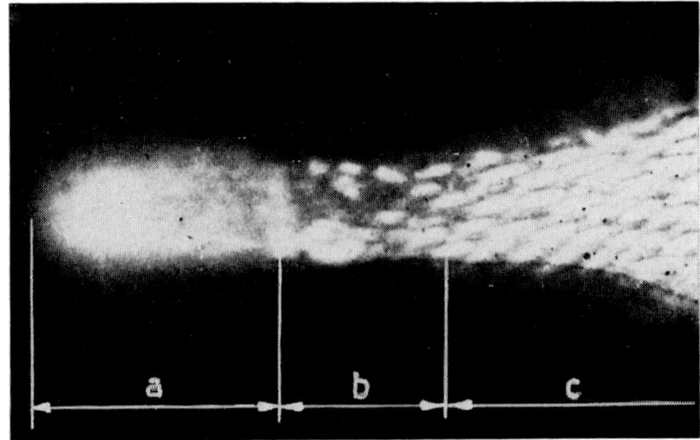

Fig. 1. Autoradiographie de la pointe d'une lame de 2 mm d'épaisseur; Cu–^{63}Ni 2.1%. Grossissement $\sim 5.5\times$.

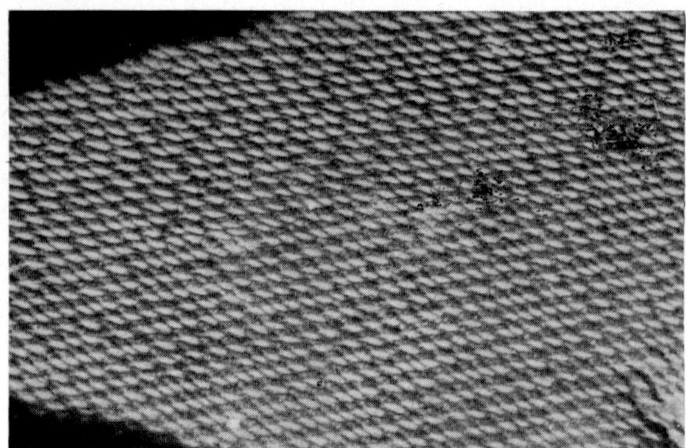

Fig. 2.　Autoradiographie de la région initiale de la même lame. Grossissement ~2.7×.

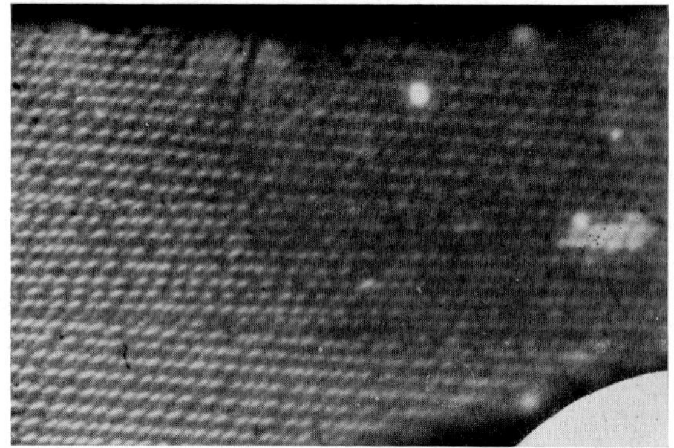

Fig. 3.　Autoradiographie de la région terminale: même lame. Grossissement ~2.7×.

Fig. 4.　Section suivant plan (001) dans un barreau cylindrique de 2 cm de diamètre, attaqué pendant 3 min dans $S_2O_8(NH_4)_2$ en solution aqueuse à 10% et additionnée de 25% de NH_4OH concentré; température ambiante; Cu–Ni 1.5%. Grossissement ~65×.

nickel puisque l'orientation bascule jusqu'à proximité du plan (332) à l'autre extrémité du triangle stéréographique. Rosenberg et Tiller[3]) ont observé des effets

similaires par introduction d'argent dans le plomb de zone fondue.

5. Mise en évidence et description de la structure cellulaire

5.1. Autoradiographie

La figure 1 est un cliché positif de la plaque autoradiographique d'un échantillon à 2.1% ^{63}Ni (vitesse de croissance $R = 2.5 \times 10^{-3}$ cm/sec). La pointe du monocristal (a) apparaît exempte de structure, les cellules ne s'édifient que progressivement dans la partie (b) et forment finalement un réseau régulier. L'extrémité (a) de la pointe se révèle, d'autre part, enrichie en nickel, ce qui est normal puisque le premier solide formé est enrichi en soluté (le coefficient de partage k_0 est supérieur à 1); immédiatement au delà, en (b) le cristal est appauvri en nickel. On atteint ensuite un état stationnaire et la lame a une concentration moyenne en nickel identique à la concentration de départ.

La figure 2 montre que les cellules riches en nickel s'ordonnent en alignements parallèles entre eux, selon une direction proche de l'axe du cristal. L'autoradiographie montre même deux régions légèrement désorientées l'une par rapport à l'autre (4° et 7° 30′ par rapport à l'axe de croissance). Dans cet échantillon, l'alignement des cellules se fait selon la direction $\langle 110 \rangle$, alors que la surface de la lame a une orientation située entre (332) et (221). Nous avons obtenu des résultats identiques pour d'autres lames cristallines avec d'autres vitesses de croissance, mais le ralentissement ($R = 0.4 \times 10^{-3}$ cm sec^{-1}) favorise l'alignement dans la direction de croissance.

Enfin, la réduction de la masse du liquide à l'extrémité du creuset rompt l'état stationnaire précédent et se manifeste par un appauvrissement en soluté (fig. 3). L'autoradiographie permet donc de visualiser la structure de l'alliage, mais l'activité réduite du ^{63}Ni ne permet pas de descendre en dessous de 0.5% de Ni.

5.2. Attaque chimique

Nous avons utilisé des solutions de persulfate d'ammonium-ammoniaque; elles permettent de révéler la structure interne des monocristaux telle qu'elle apparaît à la surface du cristal (pour les conditions d'attaque voir fig. 4). Comme les régions riches en nickel se dissolvent plus rapidement que les zones riches en cuivre, les cellules apparaissent comme des dépressions bordées d'un joint en relief. Grâce au tronçonnage chimique, il est possible d'examiner l'aspect de la structure cellulaire dans les différents plans cristallographiques.

La figure 4 est caractéristique du plan (100) d'un barreau à 1.5% de nickel obtenu par solidification verticale. L'ensemble des résultats montre que les cellules s'orientent dans une direction $\langle 100 \rangle$ plutôt que dans la direction $\langle 111 \rangle$ généralement admise pour une structure cubique à faces centrée.

5.3. Oxydation sélective

C'est certainement la méthode la plus originale que nous ayons utilisée pour étudier la structure cellulaire: il est en effet possible en se plaçant dans les conditions adéquates de n'oxyder que le nickel de l'alliage.

La présence, après oxydation, des cristaux de NiO (fig. 5) en surface de l'échantillon révèle la distribution du nickel dans l'alliage. La figure 6 montre les résultats d'une telle expérience; les cristaux d'oxyde de nickel y apparaissent clairs sur le fond noir correspondant au métal inaltéré; cette image présente une similitude totale avec les autoradiographies montrées antérieurement (figs. 1, 2, 3). La technique est cependant plus sensible que l'autoradiographie, puisqu'elle nous a permis de suivre la distribution du nickel jusqu'à la concentration de 0.1%. Les résultats obtenus sur les faces simples (100), (110) et (111) confirment l'existence de cellules quasi cylindriques enrichies en nickel (fig. 5).

5.4. Microscopie électronique a balayage*

La figure 7 représente une face (100) d'un alliage à 0.2% de nickel attaqué au persulfate d'ammonium; l'angle de vue est de 50° par rapport à la normale. L'image correspond au joint entre trois cellules contiguës; l'ensemble est constitué d'un faisceau de prismes alignés parmi lesquels ceux qui constituent le joint paraissent plus épais. Le sommet des prismes est constitué d'une pyramide à quatre faces.

5.5. Microsonde de castaing

La concentration locale en nickel se trouve à la limite de détection de l'appareil, mais un balayage de l'échan-

* Ces examens ont été effectués au Laboratoire de Géochimie grâce à l'amabilité de Monsieur Jedwab.

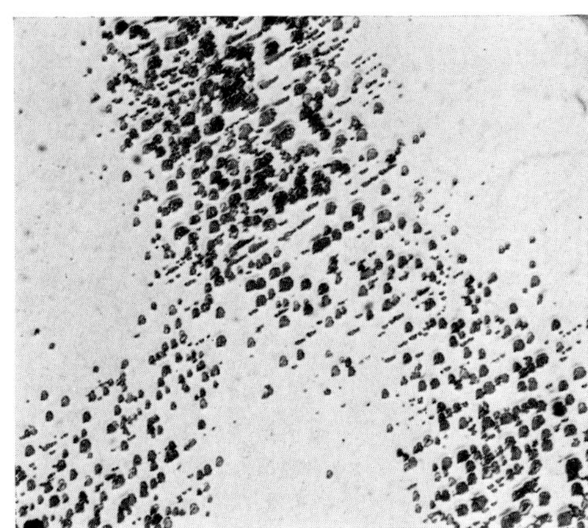

Fig. 5.   Alliage Cu–Ni 0.5%, oxydé sélectivement en NiO sous une pression de 2×10^{-4} Torr d'oxygène à 970 °C pendant 30 min. Grossissement $\sim 200 \times$.

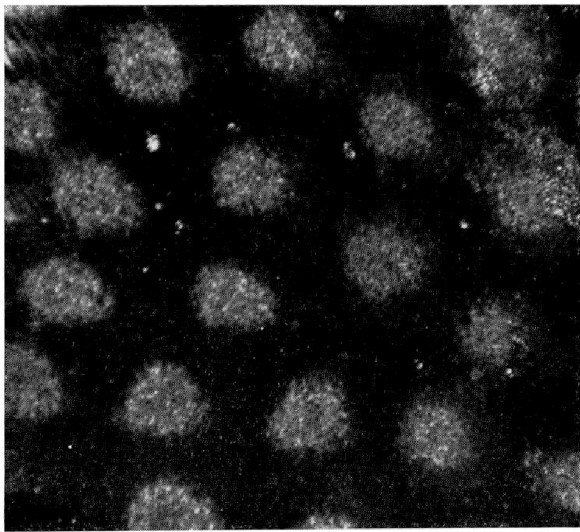

Fig. 6.   Alliage Cu–Ni 0.2%; orientation (001); oxydation sélective de Ni sous une pression de 3×10^{-4} Torr d'oxygène à 1.018 °C pendant 60 min. Grossissement $\sim 50 \times$.

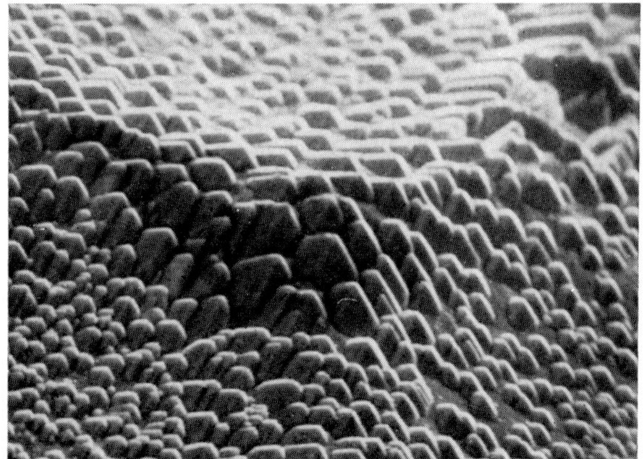

Fig. 7.   Aspect au Microscope Electronique à balayage d'une face (001) après légère attaque chimique, conditions d'attaque: voir fig. 4. Alliage à 0.2% Ni. Grossissement $\sim 450 \times$. Azimuth: 50°.

TABLEAU 1

C_0(Ni)	$D_{Cu}{}^{Ni}$ (réf. 8)	$\dfrac{1-k_0}{k_0}$	m	G/R $G = 25\,^\circ C\,\mathrm{cm}^{-1}$		$\dfrac{mC_0}{D}\left(\dfrac{1-k_0}{k_0}\right)$	Interface	Structure
1×10^{-1}	1.5×10^{-3}	0.56	1.052	5×10^3	$<$	39×10^3	Instable	cellules
2.5×10^{-2}	6×10^{-4}	0.73	1.103	5×10^3	$<$	33×10^3	Instable	Cellules
1×10^{-2}	3.5×10^{-4}	~ 1	1.109	5×10^3	$<$	37×10^3	Instable	Cellules
1×10^{-3}	1.1×10^{-4}	~ 1	~ 1.120	5×10^3	$<$	10×10^3	Instable	Cellules
1×10^{-4}	5×10^{-5}	~ 1	~ 1.120	5×10^3	$>$	2×10^3	Stable	Pas de cellules

tillon permet d'enregistrer en même temps la concentration en cuivre et en nickel. Le tracé du cuivre est très uniforme, ce qui paraît normal puisque l'échantillon en contient en moyenne 99.5 %. La concentration en nickel varie selon un tracé périodique présentant une alternance de pics et de dépressions, et la distance de centre à centre entre les cellules que l'on peut en déduire est d'environ 300 microns en moyenne.

L'ensemble des techniques confirme donc la structure de l'alliage monocristallin; l'oxydation sélective est la méthode la plus sensible et elle seule montre l'existence d'un gradient de concentration en nickel depuis le centre de la cellule jusqu'à la région intercellulaire.

6. Discussion

Il semble logique d'attribuer la structure observée au phénomène de "surfusion structurale" qui a fait l'objet de travaux importants[4]).

Cependant, fort peu d'alliages dont le soluté relève le point de fusion du solvant ont été étudiés jusqu'ici (voir par exemple réf. 5 pour Sn–Pb); mais tous les auteurs[4]) sont d'accord pour admettre que les conséquences de la "surfusion structurale" sont identiques à celles observées pour les systèmes où il abaisse le point de fusion du métal allié.

Le domaine d'existence de la structure cellulaire peut donc être déterminé en appliquant la relation de Tiller[6]) qui prévoit son existence lorsque

$$\frac{G}{R} < \frac{mC_0}{D}\left(\frac{1-k_0}{k_0}\right),$$

où G est le gradient de température, R est la vitesse de solidification, m est la pente du liquidus, k_0 est le coefficient de partage, C_0 est la concentration en soluté par gramme de solution, et D est le coefficient de diffusion dans le métal fondu.

En effet, l'interface solide–liquide n'est plane et stable que lorsque G/R est suffisamment grand par rapport à C_0. En tenant compte de nos conditions expérimentales, en calculant k_0 d'après les valeurs de Guertler et Tammann[7]) et m par la relation classique, nous obtenons finalement le tableau 1.

Nous devons donc nous attendre, dans les conditions expérimentales existantes, à avoir une structure cellulaire dans tous nos alliages, sauf dans ceux à 0.01 % de nickel, pour lesquels la stabilité de l'interface est atteinte.

Malheureusement, aucune des techniques expérimentales utilisées ne permet actuellement de le vérifier.

Nous remercions particulièrement Mademoiselle de Brouckère pour l'intérêt qu'elle n'a cessé de porter à nos travaux et Messieurs Binst et Delcambe pour leur collaboration de chaque instant.

Le Fonds National de la Recherche Scientifique a subventionné notre laboratoire, qui a aussi bénéficié de crédits Euratom et O.T.A.N.; que ces organismes soient assurés de notre reconnaissance.

Bibliographie

1) J. Bénard et al., *L'Oxidation des Métaux*, Vol. 1 (Gauthier-Villars, Paris, 1964).
2) P. Bottelier, F. Bouillon et M. Jardinier-Offergeld, Mém. Sci. Rev. Mét. **62** (1965) 86.
3) A. Rosenberg et W. A. Tiller, Acta Met. **5** (1957) 565.
4) W. C. Winegard, *The Solidification of Metals* (Inst. of Metals, Monogr. 29, 1964).
   B. Chalmers, *Principles of Solidification* (Wiley, New York, 1964).
5) T. Z. Plaskett et W. C. Winegard, Can. J. Phys. **37** (1959) 1555;
   H. Biloni, G. F. Bolling et G. S. Cole, Trans. Met. Soc. AIME **236** (1966) 930.
6) W. A. Tiller, K. A. Jackson, J. W. Rutter et B. Chalmers, Acta Met. **1** (1953) 428.
7) W. Guertler et G. Tammann, Z. Anorg. Chem. **52** (1907) 25.
8) S. Gerlach et B. Leidel, Z. Naturforsch. **22** (1967) 58.

Section XVII

Solid state transformations

Y. MONFORT

A. MAISSEU

G. ALLAIS

A. DESCHANVRES

F. BLANIÉ

J. LE HÉRICY

E. GROCHOWSKI

W. BRENNER

Journal of Crystal Growth **13/14** (1972) 829–833 © *North-Holland Publishing Co.*

CROISSANCE ORIENTÉE DES SOUS OXYDES DE NIOBIUM DANS LA MATRICE DE NIOBIUM

Y. MONFORT, A. MAISSEU, G. ALLAIS et A. DESCHANVRES

Groupe de Cristallographie et Chimie du Solide, Faculté des Sciences de Caen, Caen, France

Tetragonal niobium suboxides NbO_x (Nb_8O) and NbO_z (Nb_3O) have been grown by solid state oxidation of metallic niobium at 360 and 390 °C respectively. Electron microscopy and diffraction studies have been used to identify the lattice correspondence between the suboxides and the niobium solid solution matrix. The twinned structure of NbO_x has allowed to determine the interface energy of the twin planes which have been found to be in the range of 22–28 erg/cm².

1. Introduction

L'oxydation du niobium est caractérisée par la formation d'oxydes inférieurs avant que ne se forment les oxydes supérieurs NbO_2 et Nb_2O_5. L'existence de ces oxydes inférieurs a été mise en évidence par des études de cinétique d'oxydation. Les travaux de Hurlen[1]) résument ces résultats; il signale un oxyde NbO_x quadratique, deux NbO_y l'un quadratique, l'autre orthorombique, et NbO_z quadratique.

L'existence de sous oxydes est encore mise en évidence par Brauer et ses collaborateurs[2]); Norman[3]) observe la formation de NbO_z en métallographie, Terao[4]) en oxydant des films minces de niobium retrouve NbO_x et NbO_z.

Ces sous oxydes sont aussi responsables des contrastes observés en microscopie électronique par Van Landuyt[5]), Van Torne et Thomas[6]), Steeb et Renner[7]).

Le but du présent travail est d'analyser la formation des sous oxydes NbO_x et NbO_z et de rendre compte de leur morphologie. Nous n'avons jamais mis en évidence l'oxyde NbO_y.

2. Préparation des échantillons

L'étude a été essentiellement faite au microscope électronique. Les échantillons sont préparés à partir de feuilles minces de niobium très pur amincies électrolytiquement. Elles sont ensuite oxydées, mélangées à de la poudre de niobium pour que la réaction soit plus homogène. On peut ainsi contrôler l'apparition des sous-oxydes par diffraction de rayons X. L'oxydation a lieu dans un courant d'argon saturé de vapeur d'eau[8]). Entre 350 et 360 °C apparaît le sous oxyde NbO_x, entre 380 et 390 °C, apparaît d'abord la phase NbO_z, et une oxydation prolongée conduit à NbO_2. Les mesures pondérales indiquent une composition voisine de Nb_8O pour NbO_x et de Nb_3O pour NbO_z.

3. NbO_x

Nous avons identifié la phase NbO_x par les diagrammes de diffraction électronique: elle a une maille quadratique, M de paramètres $a = b = 3.38$ Å, $c = 3.276$ Å, $c/a = 0.968^{2-4}$). En fait la diffraction des électrons montre que le réseau de NbO_x est plus complexe: les réflexions correspondant au réseau réciproque construit à partir de la maille quadratique M sont les réflexions les plus intenses, et sont entourées de satellites plus faibles. Toutes les raies visibles de diffraction X peuvent être indexées dans le réseau construit sur la sous maille M, que nous conserverons par la suite.

Les paramètres de la sous maille utilisée sont très voisins de ceux de la maille de niobium (c.c., $a = 3.30$ Å). Nous avons pu vérifier que, dans les domaines qui n'ont pas été entièrement transformés en NbO_x, les directions [100], [010], [001] sont pratiquement les mêmes dans la matrice de niobium et dans le sous oxyde. Géométriquement l'oxydation de la matrice s'est traduite par une légère déformation.

En fait la phase NbO_x apparaît sous forme de domaines parallèles quasi-périodiques (figs. 2 et 4),

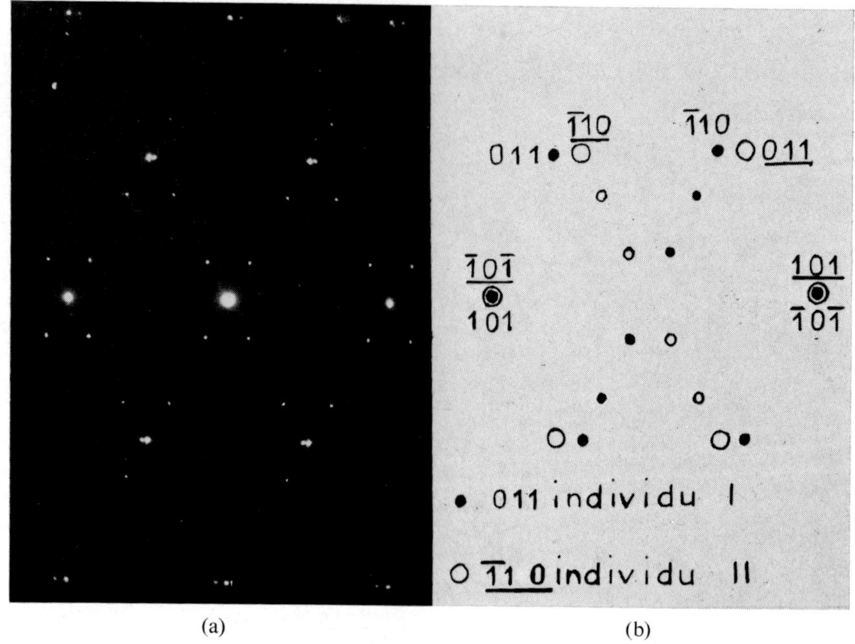

(a)   (b)

Fig. 1.   Diagrammes de diffraction de macle de NbO_x: plan commun $11\bar{1}$.

déjà observés par Van Landuyt[5]) avec des conditions de préparation différentes.

Les diagrammes de diffraction montrent que dans une zone contenant une seule famille de bandes, NbO_x se présente sous deux orientations distinctes I et II. La zone est formée d'un empilement de lamelles ayant alternativement ces deux orientations. La surface de séparation entre deux lamelles successives est un plan parallèle au plan réticulaire (101) dont la direction est la même dans les orientations I et II, qui se déduisent l'une de l'autre par une symétrie par rapport à ce plan. La correspondance entre les deux domaines monocristallins situés dans deux lamelles adjacentes est une macle par pseudo symétrie de plan (101). L'orientation relative des deux individus de la macle est visible sur les figures 1a et 1b qui montrent les plans réciproques $(11\bar{1})$, communs aux deux individus de la macle. (Les réflexions satellites y sont visibles de part et d'autre des réflexions intenses.)

Enfin nous avons pu observer que le plan de macle est parallèle à l'une ou l'autre des six directions | 101 | du niobium.

4. Interprétation de la structure maclée de NbO_x

NbO_x présente toujours des macles quasi-périodiques par rapport au plan (101). Nous allons montrer que cette structure peut s'expliquer en étudiant la croissance de NbO_x au sein de la matrice de niobium. Considérons un monocristal de niobium se présentant sous la forme d'une feuille mince parallèle au plan (001) et supposons qu'une zone carrée D de côté L s'est trans-

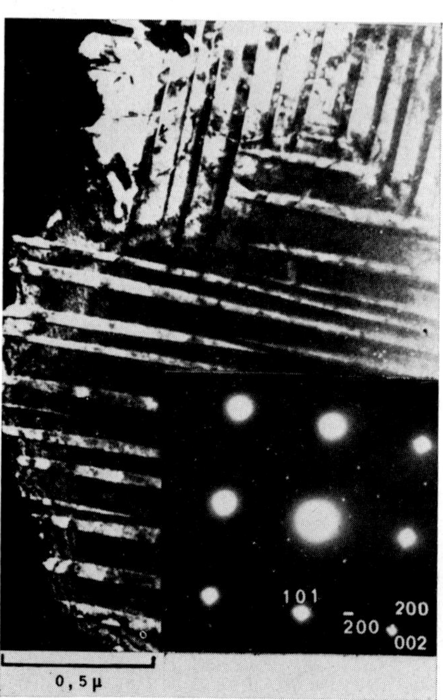

Fig. 2.   Orientation des domaines de NbO_x.

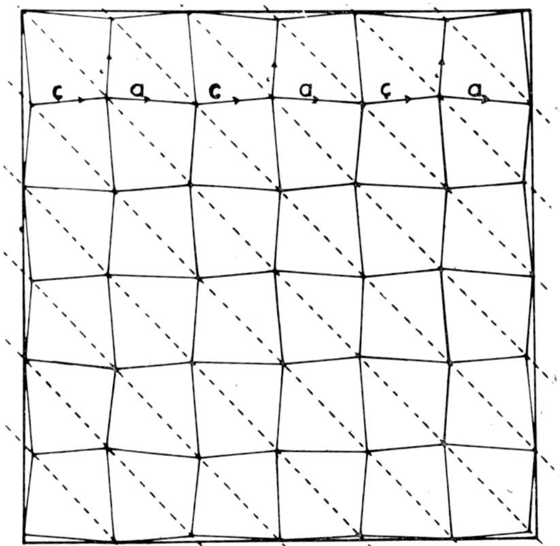

Fig. 3.   Représentation de macles périodiques de NbO$_x$; (– – –) paroi de macle.

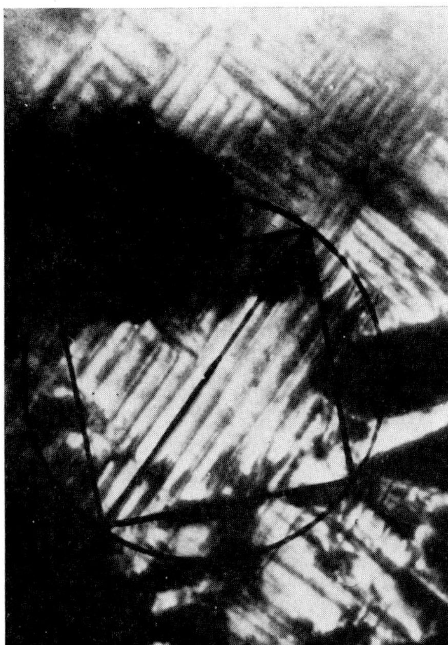

Fig. 4.   Domaines de NbO$_x$. Evaluation de l'énergie de paroi: $e^2/L = 26$ Å.

formée en NbO$_x$. Le réseau de l'oxyde est voisin, mais différent, de celui du métal pur : la formation de NbO$_x$ s'accompagne d'une déformation de la matrice de niobium et on peut calculer l'énergie de déformation de ce système en supposant que les constantes élastiques des phases Nb et NbO$_x$ sont les mêmes. Si le carré se transforme en un rectangle de côté $L(1+\alpha)$ et $L(1+\beta)$,

l'énergie élastique, si la déformation est petite, est proportionnelle à $(\alpha^2+\beta^2)L^2$.

Considérons trois possibilités:

(a) Le domaine D transformé en NbO$_x$ est monocristallin, et c est perpendiculaire à la lame: $\alpha^2+\beta^2 = 8.8 \times 10^{-4}$.

(b) Le domaine D est monocristallin; a et c sont dans le plan de la lame, et parallèles aux côtés du carré: $\alpha^2+\beta^2 = 5.4 \times 10^{-4}$.

(c) Le domaine D, ayant la direction b perpendiculaire au plan de la lame, est divisé en bandes parallèles; au passage d'une bande à l'autre, il y a échange entre a et c (figs. 3 et 4). Parallèlement à un côté L, il y a alternance de contractions et de dilatations qui se compensent partiellement. Globalement on a une dilatation de 0.48%, et $\alpha^2+\beta^2 = 0.46 \times 10^{-4}$.

La structure observée correspond donc à une énergie de déformation plus faible.

5. Evaluation de l'énergie de paroi

Dès que le nombre de bandes est un peu grand, la dilatation globale lors de la formation de NbO$_x$ ne dépend pas du nombre de macles. Nous nous proposons de montrer qu'une analyse plus fine des déformations à la limite entre Nb et NbO$_x$ permet d'évaluer à partir de la largeur des bandes l'énergie de paroi des macles de NbO$_x$.

En effet lorsque le nombre de bandes varie dans le domaine D, l'énergie du système comprend trois termes dont l'un, l'énergie W_1 dûe à la dilatation est constant. Les deux autres contributions à l'énergie sont:

(a) l'énergie W_2 de paroi, qui est proportionnelle à la surface des plans de contact, et qui augmente lorsque le nombre de bandes augmente,

(b) l'énergie W_3 de déformation du système au voisinage de la limite entre NbO$_x$ et la matrice de niobium.

Si le domaine D est un carré dont les côtés de longueur L sont parallèles à a et c, les parois de macles sont parallèles à une diagonale du carré (fig. 3). Outre la dilatation globale, il y a au voisinage des limites du domaine D, une déformation de cisaillements périodiques. Cette déformation est d'autant plus grande que la distance entre parois est grande. W_3 diminue lorsque le nombre de bandes augmente. Nous admettons que la largeur des bandes est telle que la somme W_2+W_3 soit minimum. [Des hypothèses similaires sont utilisées pour le calcul d'énergie de défauts d'empilement, à partir de l'énergie élastique dûe aux dislocations[9,10]).]

6. Calcul de W_2 et W_3

Nous supposons que l'épaisseur de la lame est égale à 1. La déformation au voisinage du joint Nb–NbO$_x$ est constituée par une suite périodique de cisaillements (fig. 3). On calcule l'énergie de déformation en déterminant l'énergie élastique associée à la même déformation, provoquée par une série de dislocations coins, perpendiculaires à la lame, situées à la limite Nb–NbO$_x$. On considère que le long d'un seul cisaillement de longueur $2l$ et d'angle $\phi = (1-c/a)$, se trouve $2(n-1)$ dislocations équidistantes de $a = l/n$; $(n-1)$ dislocations auront un vecteur de Burgers $\boldsymbol{b} = \phi\boldsymbol{a}$, et les $(n-1)$ suivantes, $-\phi\boldsymbol{a}$. L'énergie de déformation est l'énergie associée à la création de ces $(n-1)$ paires de dislocations. On calcule d'abord l'énergie de formation de la première paire, la distance entre les deux éléments étant $2(l-a)$, puis de la seconde paire, en tenant compte de l'interaction avec la première, la distance entre chaque élément étant de $2(l-2a)$, et ainsi de suite jusqu'à la $(n-1)$ème paire de dislocations. L'énergie obtenue s'exprime sous la forme:

$$W_{1c} = + \frac{\mu b^2}{2\pi(1-v)} \left(\sum_{i=1}^{n} \log \frac{2(n+1-i)a}{\alpha b} \right.$$
$$\left. + 2 \sum_{i=2}^{n} \sum_{j=0}^{i-2} \log \frac{2n+1-i-j}{i-j-1} \right),$$

où αb, représente la distance autour de chaque dislocation au-dessous de laquelle les expressions d'élasticité ne sont plus valables, μ est le module de cisaillement, et v est le coefficient de Poisson.

Le premier terme de la parenthèse représente l'énergie de formation des $(n-1)$ paires de dislocations indépendantes et le second, l'énergie d'interaction. Pour que le modèle soit plus proche de la réalité, on a fait tendre le nombre n vers l'infini, et l'énergie trouvée tend alors vers:

$$W_{1c} = \frac{\mu\phi^2 l^2}{\pi(1-v)} \log 2 .$$

On calcule ensuite l'énergie nécessaire à la création des cisaillements successifs le long d'un côté du domaine de NbO$_x$, en tenant compte des interactions avec les cisaillements déja existants. Ainsi l'énergie de création

de m cisaillements successifs sera évaluée à:

$$W_{mc} = \frac{m\mu\phi^2 l^2}{\pi(1-v)} \log 2 \left(1 - \frac{\log 2m}{8m \log 2} \right).$$

Le deuxième terme de la parenthèse devient rapidement négligeable lorsque m croît. On n'en tiendra pas compte.

Enfin on supposera que les côtés des domaines transformés en NbO$_x$ sont suffisamment grands pour que les cisaillements situés le long de deux côtés différents du domaine aient une énergie d'interaction négligeable. Dans ce cas, l'énergie W_3 s'exprime:

$$W_3 = \frac{4m\mu\phi^2 l^2}{\pi(1-v)} \log 2 ,$$

ou en fonction de L, longueur du côté, et de c/a:

$$W_3 = \frac{\mu(1-c/a)^2}{\pi(1-v)} \frac{L^2}{m} \log 2 .$$

L'énergie W_2 est proportionnelle à la surface des parois de contact entre lamelles maclées. Si γ est l'énergie de paroi par unité de surface:

$$W_2 = 2mL \sqrt{2}\,\gamma .$$

Calculant la valeur qui minimise $W_2 + W_3$, on aboutit à la relation

$$\gamma = A\,e^2/L ,$$

e étant l'épaisseur des bandes et:

$$A = \frac{\mu(1-a/c)^2\,2\sqrt{2} \log 2}{\pi(1-v)} .$$

A est voisin de 0.93×10^8 dyne/cm². Les valeurs obtenues pour γ d'après divers échantillons sont comprises entre 22 et 28 erg/cm².

7. Orientations de NbO$_x$

On a observé des domaines monocristallins de niobium partiellement transformés en NbO$_x$. Il a une maille quadratique, $a = b = 6.64$ Å, $c = 4.80$ Å.

L'oxyde prend l'une ou l'autre des deux orientations suivantes par rapport à la matrice.

Dans la première, déjà observée par diffraction de

rayons X^3) les directions [100], [010], [001] sont parallèles aux directions correspondantes de niobium, et on observe de bonnes concordances réticulaires dans le plan (001).

Ce type de croissance orientée est observé lorsque l'oxyde croît à la surface du niobium.

Lorsque l'oxyde NbO_z croît à l'intérieur d'un grain, il prend une orientation différente dans laquelle les rangées $\langle 110 \rangle$, $\langle 001 \rangle$, $\langle 1\bar{1}0 \rangle$ sont respectivement parallèles aux rangées $\langle 110 \rangle$, $\langle 1\bar{1}0 \rangle$ et $\langle 001 \rangle$ du niobium. On constate, parallèlement à ces trois rangées, de bonnes concordances réticulaires.

Bibliographie

1) T. Hurlen, J. Inst. Metals **89** (1960–1961) 273.
2) G. Brauer, H. Muller et G. Kuhner, J. Less Common Metals **4** (1962) 553.
3) N. Norman, P. Kofstad et D. J. Krudtaa, J. Less Common Metals **4** (1962) 52, 124.
4) N. Terao, Japan. J. Appl. Phys. **2** (1963) 156.
5) J. Van Landuyt, Phys. Status Solidi **6** (1964) 957.
6) L. I. Van Torne et G. Thomas, Acta Met. **19** (1964) 601.
7) S. Steeb et J. Renner, Z. Metallk. **56** (1965) 531.
8) A. Maisseu, Y. Monfort, G. Allais et A. Deschanvres, à paraitre, Rev. Chim. Minérale.
9) G. Saada, *Microscopie des Lames Minces* (Masson, Paris, 1966) p. 295.
10) B. J. Thomas, Metaux Corr. Ind. (nov. 1969) 363.

CROISSANCE, PAR RECRISTALLISATION SECONDAIRE, DE GROS CRISTAUX EXEMPTS DE MACLES DANS LE CUIVRE PURIFIÉ PAR FUSION DE ZONE

F. BLANIÉ et J. LE HÉRICY

Centre d'Etudes de Chimie Métallurgique (C.N.R.S.), 15, rue Georges Urbain, 94 – Vitry-sur-Seine, France

Only zone refined copper, rolled 91%, at −196 °C, and annealed at 1050 °C under high vacuum (10^{-7} Torr) yields large twin-free crystals by secondary recrystallization. The textures of the crystals are of the types: (110) [$1\bar{1}4$] and (751) [$\bar{1}\bar{2}\,7\,49$]. In the case of high purity copper (99.999%), small grains are obtained, while with OFHC copper (99.99%), the crystals show a lot of large twins. Each type of secondary grain is related to two principal components of the primary recrystallization texture by two rotations about two common $\langle 111 \rangle$ poles. The observed relationships are high angle boundaries with coincidence lattice sites (Kronberg-Wilson boundaries).

1. Introduction

La préparation de gros cristaux de cuivre par recristallisation secondaire nécessite de laminer très fortement les échantillons, à la température ambiante, de façon à produire par recristallisation primaire une structure à grains fins, de texture cubique (100) [$00\bar{1}$] très prononcée. Celle-ci s'oppose au grossissement normal du grain lors d'un recuit des éprouvettes près du point de fusion du cuivre (1000–1050 °C). Ce recuit provoque par recristallisation secondaire le développement de quelques gros cristaux, en nombre d'autant plus limité que la texture primaire est plus prononcée. Les cristaux secondaires présentent des relations d'orientations préférentielles avec la texture primaire[1,2]). Cette technique a été appliquée avec succès dans le cas du cuivre électrolytique de pureté courante ("tough pitch copper"), contenant de fortes teneurs en oxygène ($O_2 > 100 \times 10^{-6}$ poids), ou du cuivre OFHC ayant encore des concentrations importantes de métalloïdes $O_2 < 50 \times 10^{-6}$, $S < 50 \times 10^{-6}$)[2]). Cette méthode utilisée sur le cuivre de haute pureté 99.999% et le cuivre de zone fondue conduit à des cristaux de 5 mm environ présentant des macles[3]). La sélection de croissance est très forte dans le premier cas et plus faible dans le second. Cependant la texture primaire, moins prononcée sur ces deux types de cuivre[4]) que sur le cuivre ordinaire, pourrait être la cause de la taille relativement limitée des grains secondaires observés.

C'est pourquoi, il nous a semblé intéressant d'étudier la recristallisation secondaire du cuivre très pur sur des éprouvettes présentant des grains d'orientations différentes de la texture cubique. Ces grains peuvent en effet être obtenus par recuit des échantillons déformés à basse température (77 K) au lieu de la température ambiante. Dans ces conditions la texture d'écrouissage est très différente (type laiton)[5−7]), et il en est par suite de même de la recristallisation primaire[7,8]). Dans nos premières expériences[9]), du cuivre de zone fondue ainsi laminé et recuit ensuite à haute température avait conduit à de gros cristaux de 5 mm de diamètre environ, exempts de macles, et d'orientation (011) [$31\bar{1}$].

Le principal avantage de cette technique est donc l'obtention de cristaux sans macles sur le cuivre de pureté élevée. Comparée à la méthode classique de préparation de monocristaux par solidification directionnelle dans un moule de graphite, elle évite les pollutions souvent constatées avec ce matériau, qui n'est pas toujours aussi pur qu'on le désirerait. Par ailleurs, les monocristaux préparés par ce dernier procédé contiennent, à de rares exceptions[10]), une sous-structure, et ont généralement une moins bonne perfection cristalline que les cristaux obtenus dans l'état solide.

2. Pureté des différents lingots de cuivre utilisés et préparation des échantillons

Le tableau 1 indique les symboles que nous avons utilisés pour désigner les différentes nuances de cuivre étudiées.

Le tableau 2 en donne les analyses types. La pureté globale des échantillons peut être comparée suivant

TABLEAU 1

Liste des symboles utilisés

Notre symbole	Symbole commercial	Explication	Pureté conventionelle
O	OFHC	"Oxygen-free high conductivity"	99.98–99.99%
Z		Cuivre de zone fondue obtenu à partir du cuivre O[11])	Voir tableau 2 pour l'analyse des impuretés
S	Haute impureté	American smelting and Refining Company	99.999%
ZF		Cuivre de zone fondue flottante obtenu à partir du cuivre S[11])	Voir tableau 2 pour l'analyse des impurités

TABLEAU 2

Impuretés des divers échantillons (10^{-6} poids)

Cu	O (99.98–99.99%)	Z	S (99.999%)	ZF
As	1.1	0.04	0.006–0.008	<0.0001
Sb	7–8	0.0007	0.01–0.025	<0.0001
Ag	9–10	2	0.04–0.1	0.01–0.02
Au	0.03	0.003	0.0008	<0.00001
Hg	0.07	0.1	0.05	0.04
Na	<0.001	<0.001	0.02–0.04	<0.001
Zn	<0.05	<0.05	0.1–0.2	<0.05
Fe	2.7–2.9	2.9	0.4	0.5
Cr	0.04	<0.01	0.004	0.002
Mn	0.05–0.06	0.04	0.0001	<0.0001
Se	0.09	<0.001	<0.001	<0.001
S	5–8	0.4	0.17–0.2	<0.01
P	0.98	<0.01	0.07	0.01–0.03

Eléments $\ll 10^{-7}$: K, Sc, Co, Ga, Rb, Zr, Mo, In, Te, Cs, T.R., Hf, W. Th. U.

TABLEAU 3

Rapport de résistivité électrique $\rho_{He} = \rho_{4.2\,K}/\rho_{294\,K}$ des divers échantillons de cuivre utilisés (mesures effectuées sur des fils recristallisés, de 1 mm de diamètre)

Cu	O	Z	S	ZF
$\rho_{He} \times 10^4$	130–140	46	5	1.8

leur origine, d'après les valeurs du rapport des résistivités électriques à la température de l'hélium liquide (4.2 K) et à la température ambiante (294 K), que nous avons consignées dans le tableau 3.

Les échantillons sont mis en forme de parallélépipède d'épaisseur 6.5 mm, par un léger laminage à la température ambiante, puis recuits à des températures différentes suivant leur pureté pour obtenir, soit une mise en solution solide des impuretés (cuivre O, à 815 °C sous 10^{-7} Torr), soit des tailles de grains sensiblement comparables (cuivres S et Z à 1020–1050 °C, sous 10^{-7} Torr, et ZF à 600 °C sous H_2). Le diamètre des grains varie ainsi de 0.5 à 2 ou 3 mm, exceptionnellement 6 mm pour un cristal de l'échantillon Z.

Les échantillons subissent une réduction d'épaisseur de 91 % par passes unidirectionnelles dans un laminoir, dont les rouleaux sont immergés dans un bain d'azote liquide (77 K). L'épaisseur finale est de 55 à 60/100 mm, Les éprouvettes sont découpées aux ciseaux dans l'azote liquide, et gardées à cette température jusqu'à leur utilisation.

3. Recuits de recristallisation secondaire; résultats

Chaque échantillon d'une pureté donnée est recuit indépendamment des autres. Porté à la température ambiante, il est alors le siège d'une recristallisation primaire qui peut être complète en quelques jours (à l'exception du cuivre O, qui recristallise entre 80 et 140 °C). Après un décapage à l'acide nitrique, qui réduit son épaisseur à 0.5 mm environ, l'échantillon est placé sur un support de silice ultra pure, à l'intérieur d'un tube de silice transparente monté sur une installation de recuit sous ultravide à joints métalliques de cuivre. Quand le vide atteint 10^{-8} Torr après dégazage de l'enceinte, l'échantillon est porté lentement à 1050 °C dans un four à résistance, et subit un recuit de 15 h sous 10^{-7}–5×10^{-8} Torr, suivi d'un refroidissement lent jusqu'à la température ambiante. A la sortie du tube, une attaque macrographique à l'acide nitrique dilué révèle la taille des grains.

Nous avons déterminé l'orientation des grains secondaires par diffraction des rayons X (anticathode de cuivre) suivant la méthode de Laue en retour. Nous avons effectué des pointés individuels ou des pointés portant sur quelques cristaux (joint de grains, point triple…). Ces derniers nous ont permis d'obtenir les orientations de grains moins importants, ou de macles résiduelles.

3.1. CUIVRE DE ZONE FONDUE ZF

D'après la figure 1, montrant l'ensemble de quatre éprouvettes de différentes puretés, attaquées macro-

XVII–2

graphiquement, on peut voir que le cuivre de zone fondue ZF présente les plus gros cristaux de recristallisation secondaire. A une extrémité de cet échantillon, un grain s'étend sur toute la largeur, c'est-à-dire, sur 10 mm. Ces cristaux sont pour ainsi dire exemptes de macles. La figure 2 donne la figure de pôles $\langle 111 \rangle$ des orientations trouvées, indépendamment de l'importance des grains. Seules les orientations situées dans les cercles de dispersion de rayon 5° correspondent aux orientations idéales de grains importants. On peut remarquer que les cristaux secondaires présentent une texture dans laquelle on peut distinguer deux composantes:

(a) *Une composante majeure* (*A*) comprenant les orientations (110) [1$\bar{1}$3] et (110) [1$\bar{1}$4]. Ces grains sont particulièrement développés: le plus gros atteint 6 mm. La surface qu'ils occupent représente environ la moitié de l'échantillon. Les macles sont peu fréquentes et quand elles existent sont très peu développées; elles sont le plus souvent localisées au point triple de joints de grains. Les pôles de ces macles (111) sont situés sur le cercle stéréographique (110), l'un près de la direction de laminage (DL), l'autre de la direction transverse (DT). Le plan de macle est donc paralèle à une section droite de l'éprouvette.

(b) *Une composante mineure* (*B*) du type (751) [$\bar{1}$2 7 49] (Cette orientation idéale a été définie en tenant compte de tous les grains de ce type observés sur les quatre variétés de cuivre). Ces grains occupent environ 30 % de la surface de l'échantillon. Le plus gros s'étend sur toute la largeur de l'éprouvette, soit 10 mm, et représente à lui seul la moitié du pourcentage précédent. Nous avons pu déceler quelques macles résiduelles correspondant aux deux pôles $\langle 111 \rangle$ situés près de la direction transverse, ainsi qu'au pôle $\langle 111 \rangle$ le plus près de la normale à l'échantillon. Cette dernière macle, dont la trace est perpendiculaire à la direction de laminage est caractéristique de cette composante et permet, quand elle existe une identification rapide et simple d'un cristal de type B.

Les autres orientations sont intermédiaires entre celles de type A et de type B, quelques unes sont de type (110) [001]. Signalons deux cristaux de taille moyenne, d'orientation type (112) [$\bar{3}$11].

On remarquera que le cuivre ZF se caractérise, par rapport aux autres échantillons, par la présence de nombreux joints de grains rectilignes. La trace de ces

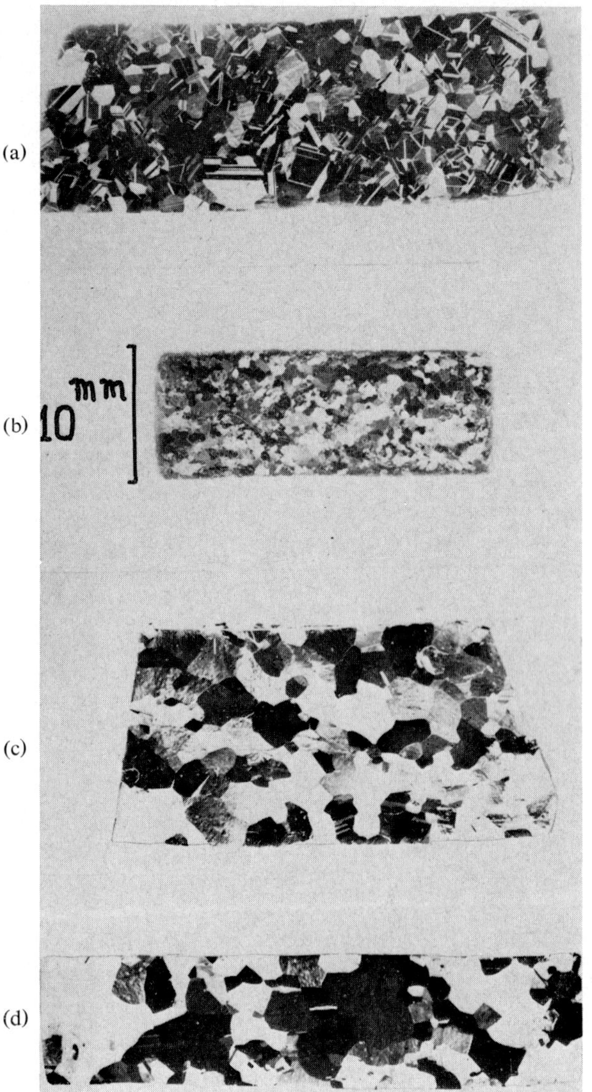

Fig. 1. Cristaux de recristallisation secondaires sur divers échantillons de différentes puretés, laminés de 91% à 77 K et recuits pendant 15 h à 1050 °C sous 10^{-7} Torr. Attaque macrographique à l'acide nitrique légèrement dilué. (a) Cu O, 99.99 %; (b) Cu S, 99.999 %; (c) Cu Z, fusion de zone à partir de Cu O; (d) Cu ZF, fusion de zone à partir de Cu S.

joints de grains à la surface de l'éprouvette correspond le plus souvent à la trace d'un plan de grande densité atomique de même type dans chaque grain [plans (111), (100), (110)], et plus rarement à des plans d'indices différents.

3.2. Cuivre de zone fondue Z

On peut voir sur la figure 1, que la taille moyenne du grain est quelque peu inférieure à celle du cuivre ZF

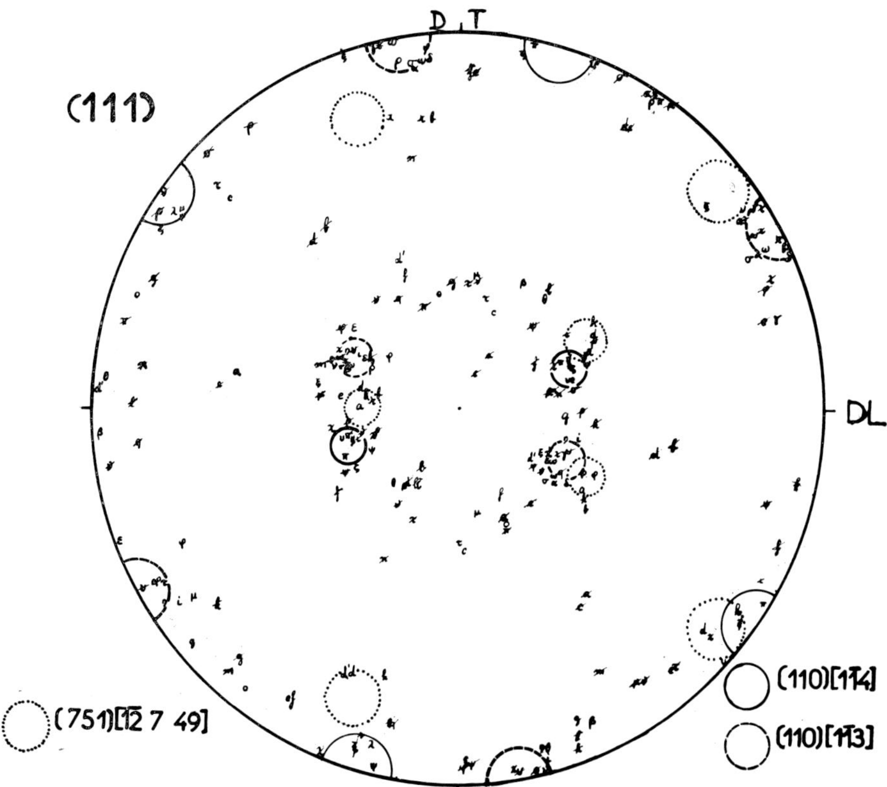

Fig. 2. Figures de pôles ⟨111⟩ des critaux secondaires du cuivre ZF. Les cercles correspondent à une dispersion de 5° en diamètre autour des orientations idéales indiquées. Cu ZF, pureté laminé de 91% à 77 °K, et recuits pendant 15 h à 1050 °C sous 10^{-7} torr.

et que les macles sont pratiquement inexistantes. Les plus gros cristaux pointés individuellement aux rayons X correspondent aux orientations du type A (110) [1̄14] et du type B.

3.3. CUIVRE DE HAUTE PURETÉ S (99.999 %)

Bien que les macles soient aussi rares sur ce cuivre, les grains sont cependant de taille réduite et les plus gros atteignent environ 1 mm (fig. 1). Ces derniers répondent aux orientations du type A (110) [1̄14] et du type B.

3.4. CUIVRE 0 (99.99 %)

La taille de grains est plus hétérogène sur cette nuance de cuivre et il est possible d'observer quelques gros cristaux (fig. 1). Néanmoins la caractéristique de ce cuivre est de présenter de nombreuses macles. Les plus gros cristaux étudiés aux rayons X ont des orientations du type A, s'étendant de (110) [1̄13] à (110) [1̄15], et du type B. Les orientations trouvées en dehors de ces 2 types correspondent aux macles de ces cristaux.

Il y a lieu de préciser que la croissance de gros cristaux débute à une température d'autant plus basse que le cuivre étudié est plus pur. Ainsi dès un recuit de $1\frac{1}{2}$ h à 250 °C, 30 a 50 % de la surface d'un échantillon de cuivre de zone fondue est envahie par les cristaux secondaires. La figure 3 montre la croissance de gros cristaux de même orientation dans une matrice à petits grains primaires. Les cristaux secondaires peuvent être amenés à se macler au cours de leur croissance, mais ces macles disparaissent ultérieurement en cours de recuit, ainsi que nous avons pu l'observer en continu dans le microscope à émission électronique (métioscope KE₃ de la Société Balzers* à Trübach, Suisse): dans la gamme de températures de 400 à 700 °C où nous avons travaillé, nous avons constaté la migration extrêmement aisée des joints incohérent des macles des cristaux secondaires dans le cas du cuivre 99.999 % et du cuivre de zone fondue. Cette migration se fait dans le sens

* Nous tenons à remercier ici MM. L. Wegmann, H. D. Dannöhl et M. Gribi de la Société Balzers, qui nous ont permis d'effectuer ces observations dans le métioscope KE₃.

Fig. 3. Cuivre ZF. Croissance de cristaux secondaires dans une matrice à petits grains par recuit de $1\frac{1}{2}$ h à 250 °C. Attaque dans une solution ammoniacale de H_2S: l'épaisseur du film épitaxique de sulfure de cuivre (et par suite sa teinte d'interférence) est fonction de l'orientation du cristal sous-jacent. Groississement $150\times$.

d'une diminution de la surface de la macle allant jusqu'à la disparition complète de celle-ci. Il semblerait par contre que les impuretés bloquent (ou ralentissent considérablement) les joints incohérents des macles du cuivre O, puisque celles-ci subsistent après des recuits à 1050 °C.

4. Discussion des résultats

4.1. Influence de la pureté du cuivre

D'après la figure 1, on voit que l'on peut définir deux seuils de pureté intervenant dans la croissance des gros cristaux sans macle:

(a) *entre 99.99% et 99.999%*: seuil au-dessous duquel les cristaux secondaires sont pratiquement exempts de macles;

(b) *entre 99.999% et la pureté "zone fondue"*: seuil au-dessous duquel les cristaux secondaires sans macles ont des vitesses de croissance notablement accrues.

La pureté globale du cuivre peut être un facteur déterminant, mais il semble probable que des impuretés spécifiques soient responsables de ces phénomènes de croissance. En effet, le cuivre Z conduit à des gros cristaux, alors que sa teneur en impuretés est très supérieure à celle du cuivre S (99.999%) ou du cuivre ZF

(voir tableau d'analyse 1). La taille de grain du cuivre O est, de même, supérieure à celle du cuivre S. On peut cependant affirmer que les traces de fer n'ont pas d'influence sur le phénomène. Il est fort possible, par contre, que des impuretés spécifiques soient responsables de la présence des macles, et que d'autres le soient du blocage des joints de grains secondaires, comme semblerait le prouver le comportement du cuivre O qui présente des cristaux maclés, dont la taille est supérieure à celle des cristaux du cuivre S (99.999%), ces derniers étant cependant exempts de macles.

Il y a lieu de rappeler d'autre part que, selon certains auteurs[6,13], les impuretés ont une influence sur la texture d'écrouissage à 77 K: la texture du cuivre 99.999% serait semblable à celle obtenue à la température ambiante (type cuivre) alors que celle du cuivre impur serait du type laiton. Selon Lücke[7,12], par contre, les cuivres 99.99% et 99.999% présentent une texture de transition. Le cuivre O déformé au-delà de 50% à 77 K montre de fines macles mécaniques et tend vers une texture laiton[14]).

Nous signalerons simplement, par ailleurs, que les impuretés du cuive O déplacent le domaine de température de recristallisation primaire vers 80–140 °C (avec des grains de 8 à 10 μm), en comparaison des autres

Fig. 4. Cuivre ZF. Cristal secondaire du type A envahi par un cristal d'orientation voisine à 250 °C. La migration du joint se fait en s'éloignant de son centre de courbure (flèches) indiquant que la force motrice correspond à une diminution d'énergie superficielle. Même attaque que sur la figure 3. Grossissement 200×.

nuances de cuivre étudiées ici, qui recristallisent entre 0 et 50 °C avec une taille de grain de 50 à 100 μm [8]).

4.2. FORCE MOTRICE DE LA RECRISTALLISATION SECONDAIRE

La force motrice de la croissance des cristaux secondaires dans la matrice primaire à petits grains est fournie par la diminution de l'énergie interfaciale des joints de grains. Cependant, nous avons constaté que dès 250 °C, les cristaux secondaires contigus et d'orientations voisines entraient en compétition. La migration du joint commun peut s'effectuer alors en s'éloignant de son centre de courbure, ce qui est caractéristique d'une migration induite par une diminution d'énergie superficielle[17]). La figure 4 illustre ce cas en montrant la disparition d'un cristal de type A (110) [$\bar{1}$14] au profit d'un autre cristal d'orientation voisine. Les flèches indiquent la progression du joint s'éloignant de son centre de courbure. On peut penser que par ce mécanisme seuls subsisteraient les cristaux secondaires du type (110) de plus faible énergie superficielle. Les joints cohérents des macles du cristal A, bien que de faible énergie interfaciale (3 à 5 % de l'énergie d'un joint quelconque), peuvent intervenir dans l'énergie motrice du joint du gros cristal secondaire, mais on remarquera que les macles m_1 et m_2 de premier ordre ont respectivement des orientations voisines de (110) [$\bar{1}$10] et (110) [$1\bar{1}1$]. Une diminution d'énergie superficielle entre ces macles et le cristal en expension pourrait expliquer la courbure et la migration du joint loin de son centre de courbure, suivant les flèches. On pourrait invoquer le même mécanisme pour expliquer l'élimination des macles des cristaux secondaires en cours de recuit, à condition que la pureté du métal (ou de l'atmosphère) soit suffisante (on a vu en effet que le cuivre O présentait toujours de nombreuses macles). Il est difficile par ailleurs de préciser dans quelle mesure l'énergie superficielle intervient, en même temps que l'énergie interfaciale, lors de la croissance des secondaires dans les primaires et quel rôle elle peut jouer dans la sélection des orientations secondaires qui vont se développer: en effet, la composante principale p_1–p_1' de la texture primaire (voir ci-dessous 4.3) et deux de ses orientations de macle ont une orientation (110) en surface, ainsi que les cristaux secondaires de type A.

4.3. ESSAI D'INTERPRÉTATION DE LA CROISSANCE DES GROS CRISTAUX

Avec les réserves que nous avons faites sur le rôle éventuel de l'énergie superficielle, il semble possible d'interpréter la croissance des cristaux secondaires suivant la théorie de la croissance sélective. On peut en effet, trouver des relations d'orientation bien déterminées entre cristaux primaires et secondaires. En appliquant au cas de la recristallisation secondaire les principes de Dillamore[18]), établis pour la recristallisation primaire des métaux c.f.c. laminés, on peut penser que les cristaux secondaires les plus développés seront ceux qui seront susceptibles de croître dans plusieurs composantes de la texture primaire et échanger avec chacune d'elle une relation d'orientation favorable pour la croissance orientée ; un centre de croissance devra donc avoir un pôle $\langle 111 \rangle$ commun à 10° près avec chaque composante primaire et leurs réseaux cristallins devront se correspondre par des rotations de 20 à 50° autour de ces axes.

Une étude micrographique des orientations des grains primaires à l'aide d'un réactif formant une pellicule épitaxique de sulfure de cuivre sur les cristaux (ref. 16), nous a montré que les composantes de la texture primaire sont par ordre d'importance :

$$\begin{cases} (110)\ [\overline{4}\,\overline{4}\overline{1}]\ p_1, \\ (110)\ [\overline{4}41]\ p_1', \end{cases}$$
$$(35\overline{1})\ [7\overline{3}6]\ p_2,$$
$$(140)\ [001]\ p_3,$$

et sont conformes à la figure de pôle $\langle 111 \rangle$ de la recristallisation primaire établie par Lücke[7,15]) sur du cuivre 99.99 % traité dans les mêmes conditions que les nôtres (avec cependant un laminage de 96 %).

Les orientations secondaires dans la figure de pôle $\langle 111 \rangle$ de Lücke se situent dans des zones de faible intensité de diffraction des rayons X. Les petits cristaux qui constituent des centres de croissance secondaire du type $(110)\ [\overline{1}14]$ et $(110)\ [1\overline{1}4]$ sont à environ 8° de deux orientations de macle de la composante primaire $(110)\ [\overline{4}\,\overline{4}\overline{1}]$.

(a) *Cristaux secondaires du type A* $(110)\ [\overline{1}14]$: en considérant l'orientation moyenne des deux orientations voisines p_1 en p_1' de la composante primaire du type $(110)\ [\overline{4}\,\overline{4}\overline{1}]$, et la composante primaire p_2 $(35\overline{1})$ $[7\overline{3}6]$, on trouve un couple de pôles $\langle 111 \rangle$ (un pour chaque composante) qui sont distants d'environ 70° (fig. 5). Ces deux pôles pourront donc appartenir à un cristal secondaire, qui aura ainsi un axe $\langle 111 \rangle$ commun avec chacune des deux composantes primaires. Le premier pôle sera placé en P_1, à 10° des deux pôles

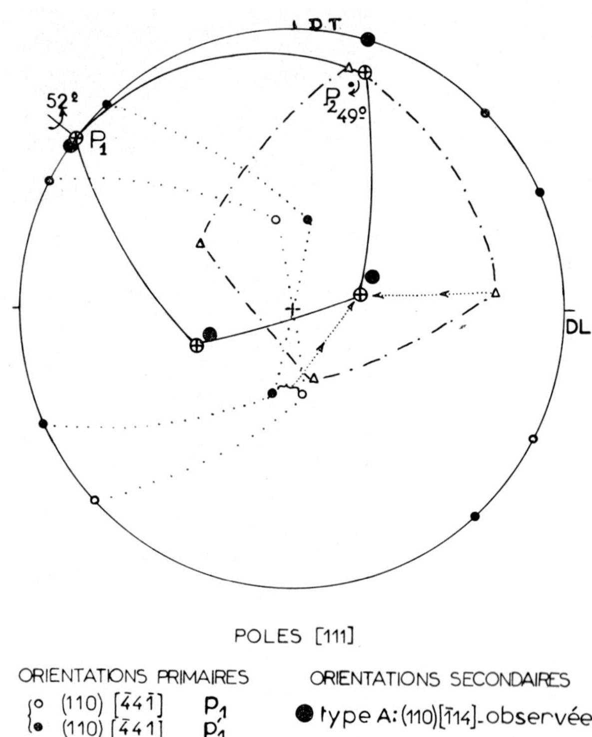

Fig. 5. Interprétation de la croissance des cristaux secondaires du type A : (110) [$\overline{1}$14].

$\langle 111 \rangle$ des deux orientations voisines p_1 et p_1' ; le deuxième sera placé à 70° du premier et près d'un pôle $\langle 111 \rangle$ de la composante p_2 (au voisinage du point P_2). Les deux autres pôles $\langle 111 \rangle$ du cristal secondaire se déterminent alors par construction géométrique. L'axe de rotation commun à l'orientation moyenne p_1–p_1' et au cristal secondaire est situé en P_1, l'angle de rotation étant de 52°. L'axe de rotation commun à la composante p_2 et au cristal secondaire a été déterminé par construction stéréographique et placé en P_2, l'angle de rotation étant de 49°. On voit que l'orientation secondaire déduite des deux composantes primaires est très proche de celle effectivement observée de type A (110) [$\overline{1}$14]. On peut remarquer que les deux relations d'orientations établies sont voisines des relations de Kronberg–Wilson : $\langle 111 \rangle$, 46°8, densité de sites en coïncidence : 1 sur 19.

(b) *Cristaux secondaires du type B* : (751) [$\overline{1}\overline{2}$ 7 49] : en procédant de la même façon avec les deux orientations de la composante p_1–p_1' et la composante p_3 (140) [001], on voit sur la figure 6, que l'on peut placer

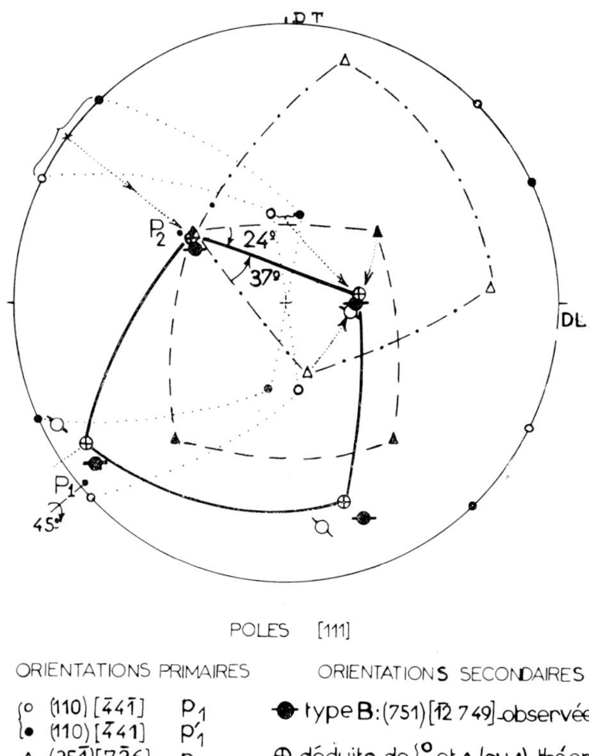

Fig. 6.   Interprétation de la croissance des cristaux secondaires du type B: (751) [1̄2 7 49].

un pôle $\langle 111 \rangle$ du cristal secondaire à 10° environ des pôles $\langle 111 \rangle$ de p_1 et p_1' (région de la figure marquée P_1), un autre pôle $\langle 111 \rangle$ étant situé près du pôle $\langle 111 \rangle$ de p_3 dans la région marquée P_2. L'orientation du cristal secondaire, construit sur ces deux pôles, se déduit de l'orientation moyenne de p_1–p_1' par une rotation d'axe P_1 proche de $\langle 111 \rangle$ et d'angle 45°. La relation observée est donc proche d'une relation de Kronberg–Wilson du type précédent. Par rapport à la composante p_3, la relation d'orientation est une rotation d'axe P_2 proche d'un axe $\langle 111 \rangle$ et d'angle 24°, qui correspond à une relation de Kronberg–Wilson du type $\langle 111 \rangle$, 22°, densité de coïncidence 1 sur 7. On remarquera que ce cristal secondaire théorique est aussi favorablement orienté pour croitre dans la composante primaire p_2, avec laquelle il échange un pôle $\langle 111 \rangle$, la rotation étant dans ce cas de 37°, de sens inverse à la précédente, et correspondant à une relation de Kronberg–Wilson $\langle 111 \rangle$ 38°, 1 sur 7. Les deux dernières rotations de 24° et 37° autour d'un même pôle peuvent s'expliquer par

le fait que les composantes p_2 et p_3 sont presque en orientation de macle. La figure 6 donne aussi l'orientation qu'aurait le cristal secondaire se développant uniquement dans les composantes p_2 et p_3 (pôles indiqués par ○). On voit donc que ces orientations théoriques sont très proches de l'orientation observée pour les cristaux du type B (dont les pôles $\langle 111 \rangle$ sont indiqués par ●).

5. Conclusion

Nous avons décrit une méthode de préparation d'échantillons de cuivre à gros grains, pour ainsi dire exempts de macles. Cette particularité est très intéressante pour de nombreuses expériences où la présence de macles n'est pas recherchée. Le mode opératoire en est relativement simple et consiste à laminer le cuivre à un taux de réduction de l'ordre de 90% à la température de l'azote liquide et à recuire l'échantillon pendant quelques heures à haute température. Toutefois, seul le cuivre de zone fondue permet d'obtenir de gros cristaux, les impuretés pouvant empêcher le grossissement du grain ou entraîner la persistance des macles. La force motrice de la recristallisation secondaire est la diminution d'énergie interfaciale de joints de grains, mais l'énergie superficielle des cristaux doit aussi intervenir dans la croissance des grains secondaires. Il est possible cependant d'interpréter les orientations de ces cristaux par la théorie de la croissance sélective, et l'application des principes de Dillamore sur la recristallisation des métaux c.f.c. permet d'expliquer les deux types d'orientations rencontrées (110) [1̄14] et (751) [1̄2 7 49], faisant intervenir des joints de grand angle du type Kronberg–Wilson.

Bibliographie

1) K. T. Aust, dans: *The Art and Science of Growing Crystals*, Ed. J. J. Gilman (Wiley, New York, 1965) p. 452.
2) M. L. Kronberg et F. H. Wilson, Metals Trans. **185** (1959) 501.
3) J. Le Héricy, Compt. Rend. (Paris) C **265** (1967) 620.
4) J. Le Héricy, Compt. Rend. (Paris) **253** (1961) 1687.
5) H. Muller, Oesterr. Akad. Wiss. Math. Naturw. Kl. Sitz. Ber. **7** (1958) 117.
6) Hsun Hu, R. S. Cline et S. R. Goodman, dans: *Recrystallization, Grain Growth and Textures* (Am. Soc. Met. Seminar, 1965) p. 286.
7) H. D. Mengelberg, M. Meixner et K. Lücke, Acta Met. **13** (1965) 835.
8) J. Le Héricy, Compt. Rend. (Paris) C **264** (1967) 1814.
9) J. Le Héricy, Compt. Rend. (Paris) C **268** (1969) 1932;

J. Le Héricy, communication présentée au "Fall Meeting of the Metallurgical Soc.", Cleveland, Ohio, 1967 [Abstract Bull. 2(2) 116].

10) F. W. Young, Jr. et J. R. Savage, J. Appl. Phys. **35** (1964) 1917.

11) J. Le Héricy, Ann. Chim. **1** (1966) 129.

12) H. Perlwitz, K. Lücke et W. Pitsch, Acta Met. **17** (1969) 1183.

13) R. E. Smallman et D. Green, Acta Met. **12** (1964) 145.

14) T. Leffers et A. Grum-Jensen, Trans. AIME **242** (1968) 314.

15) K. Lücke, dans: *Septiéme Colloque de Métallurgie: Ecrouissage, Restauration, Recristallisation*, C.E.N., Saclay, France, 1963, pp. 1–16.

16) P. A. Jacquet, Rev. Met. 42 (1944–45) 133.

17) C. G. Dunn et J. L. Walter, voir réf. 6, p. 461.

18) I. L. Dillamore, Acta Met. **12** (1964) 1005.

Journal of Crystal Growth **13/14** (1972) 843–847 © *North-Holland Publishing Co.*

THE EFFECTS OF SELENIDE MOLECULAR TRANSFORMATIONS ON ELECTRICAL RESISTIVITY

EDWARD GROCHOWSKI* and WALTER BRENNER

School of Engineering and Science, New York University, Bronx, New York 10453, U.S.A.

An experimental investigation of selenium and amorphous compositions of selenium containing either arsenic or thallium has been undertaken to ascertain the relationship between molecular transformations characteristic of these systems and electrical resistivity. Annealing amorphous selenium below 217 °C produces a complete transformation to the hexagonal polycrystalline phase with elimination of the Se_8 ring species. A sharp lowering in resistivity of five orders of magnitude accompanies such a reversible phase change. This purely thermally generated phenomenon is considered analogous to the memory effect observed in certain amorphous semiconductors. The addition of arsenic, a selenium molecular chain branching element, and thallium a chain cleaver was found to significantly modify these characteristics, including both amorphous and crystalline resistivities as well as crystallization rates. A figure of merit defined as the ratio of these resistivity values is proposed to evaluate amorphous selenides for device applications.

1. Introduction

The morphological transformations of selenium and various selenium containing compositions have been extensively investigated in the past with particular emphasis on crystallization behavior from melts[1,2]). Recent interest in the semiconducting properties of selenium and certain amorphous compositions in which selenium is a principal constituent, has prompted an experimental investigation on the relationship between certain morphological transformations and the electrical resistivity of such selenium containing systems in order to develop a better understanding of the role of molecular structure on semiconductive behavior.

It is well known that liquid selenium exists in two basic molecular configurations, namely eight-membered selenium rings (Se_8) and "polymeric" spiral types of chains with widely varying lengths of selenium atoms[3]). The presence and relative concentrations of these two species have been shown to affect crystallization characteristics both from the melt and from solid phase amorphous selenium and selenium containing compositions[2,5]). However, no relationships have as yet been developed on the effect of various morphological transformations on electrical resistivity although such information would be of considerable interest for obtaining much needed additional insight on the semi-

* Present address: IBM, Components Division, Hopewell Junction, New York 12533, U.S.A.

conductive properties of this element, especially for possible application in noncrystalline devices. The viscosity of selenium melts has been studied for this purpose. This physical parameter not only appears to be somewhat limited in response to certain molecular changes occurring in selenium melts but presents experimental difficulties for precise measurements at elevated temperatures. It was hoped that electrical resistivity would be a more sensitive monitoring parameter for the study of such morphological changes involving both selenium melts and the solid phase element per se, and in the presence of known amounts of impurity elements such as arsenic and thallium.

2. Discussion

2.1. PURE SELENIUM

A knowledge of the presence and relative concentrations of ring and chain configurations of selenium in the melt is essential for the investigation of its crystallization behavior and can be determined analytically employing the differences in solubility characteristics of the rings and chains of quickly quenched samples. Thus the eight membered selenium rings are soluble in solvents such as carbon disulfide at ambient temperatures. Since the chains are insoluble in the same solvent, their separation is possible and can be accurately accomplished.

This procedure was carried out with amorphous

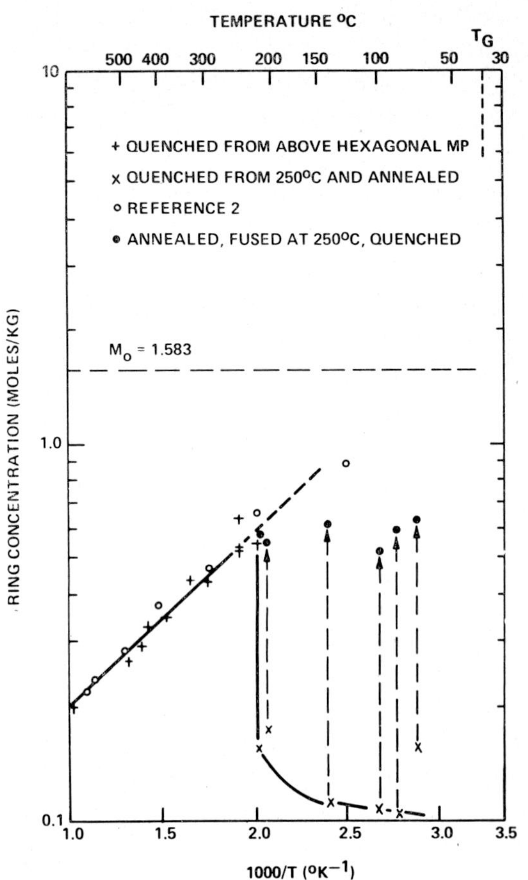

Fig. 1.   Variation of ring concentration of pure selenium with temperature.

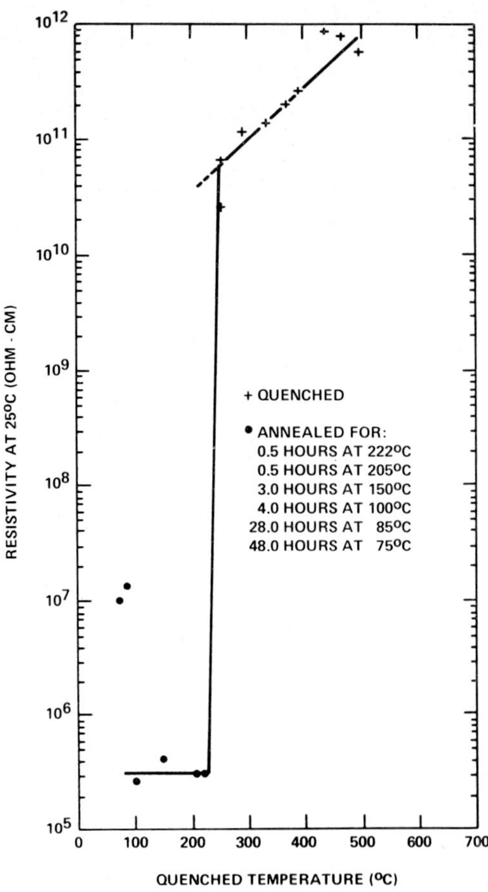

Fig. 2.   Restitivity variation with quenched /anneal temperature for pure selenium.

selenium samples heated to temperatures above 250 °C and quenched quickly to ambient temperatures. Fig. 1 shows the selenium ring concentration as a function of the temperature from which the samples were quenched. The plot of ring concentration versus reciprocal absolute temperature indicates a monotonic exponential dependency which has previously been employed for determining the equilibrium constant assuming a Van 't Hoff relationship[3]). It can be seen from this plot that amorphous selenium samples annealed below 218 °C for time periods long enough to attain thermodynamic equilibrium contain little if any Se_8 type ring molecules. Rearrangement of the rings to chain type structures must therefore occur below that temperature. This behavior can be related to the formation of hexagonal polycrystalline selenium upon cooling to below 217–218 °C, its melting temperature.

As indicated by the dotted lines in fig. 1 this crystalli-

zation process is completely reversible. Thus crystallized selenium samples which were subsequently fused at elevated temperatures and then quenched from 250 °C exhibit ring concentrations identical to those found for quenched selenium samples with no prior crystallization history. The termination of the monotonic dependency of the ring concentration-temperature relationship by the advent of crystallization can be viewed as a departure from polymerization behavior for selenium. If this process were to be continued, an increasing ring concentration would be observed at progressively lower temperatures until the total monomer concentration reached a maximum value (1.583 moles/kg) at the glass transition temperature.

The electrical resistivity of variously treated selenium samples whose ring/chain concentrations were known from the experiments which resulted in the development of the data shown in Fig. 1 was subsequently

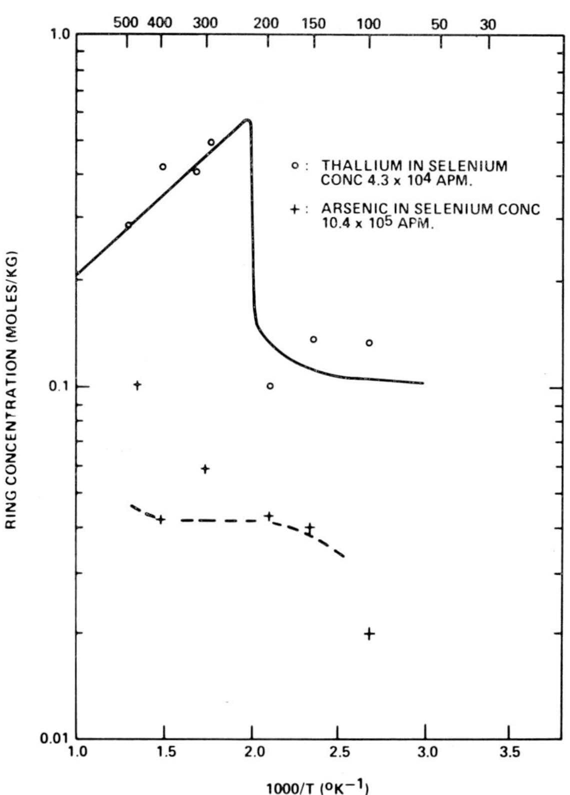

Fig. 3.   Restitivity variation with quenched/anneal temperature for selenium alloys.

measured employing the previously described experimental techniques[4]). Fig. 2 shows the ambient temperature resistivities as a function of quench temperature for these samples. Two regions indicative of selenium's behavior were observed. Above 250 °C the quenched liquid may be viewed as a true amorphous substance comprised of various equilibrium concentrations of rings and chains. Its resistivity corresponds to the expression

$$\rho = A \exp(BT), \quad 250 \le T \le 500 \; ^\circ C, \qquad (1)$$

where $A = 5.21 \times 10^9$ ohm-cm, $B = 1.03 \times 10^{-2} \; ^\circ C^{-1}$.

Liquid samples quenched from 250 °C and then annealed for varying predetermined times at temperatures below 218 °C were found to have quite substantially lower resistivities. As note previously this region corresponds to hexagonal phase type crystallinity and absence of essentially ring molecules. Completeness of crystallization was promoted by relatively short annealing periods at 200 °C, i.e. 30 min, or longer

annealing times at lower temperatures, i.e. 48 hr at 80 °C. The completeness of the growth of hexagonal crystallites resulted in a resistivity decrease of more than five orders of magnitude. The absence of ring molecules suggests that this growth occurs not only by enlarging the short range order of chain molecules as indicated by a process such as

$$X[\mathrm{Se_8}] \rightarrow [\mathrm{Se_8}]nX$$
(polymeric)   (crystalline)

but also by ring cleavage and subsequent nucleation of the resulting fragments.

2.2. SELENIDE ALLOYS

Given the development of an analytical technique of appropriate accuracy for the determination of the ring/chain concentrations of pure selenium, the possibilities of altering the ring/chain equilibrium by incorporation of selected impurity concentrations was experimentally explored. Specifically the effects of adding two elements, namely arsenic and thallium, on the relationship between certain morphological transformations and the resistivity of selenium was investigated.

It has been reported that additions of arsenic to selenium increases the glass transition temperature thus yielding a more temperature stable glass[5]). Branching of the selenium chains by the arsenic atoms has also been proposed. This has been shown to effect a moderate viscosity increase when compared with pure selenium[6]). Conversely, thallium has been reported to act as chain cleaver resulting in a significantly lower melt viscosity when added to selenium[1,6]).

In this investigation, selenium containing 10 wt% arsenic was found to be quite insoluble in carbon disulfide regardless of prior thermal history of the samples as shown in fig. 3. This phenomenon is indicative of the near complete absence of the soluble rings presumably in favor of branched chains. It is also evident from this figure that selenium containing 10 wt% thallium exhibits similar ring/chain concentration temperature relationships as pure selenium.

Fig. 4 shows the influence of both arsenic and thallium on selenium resistivity dependence with varying quench anneal temperatures. The resistivity of selenium alloys with 10 wt% arsenic was found to be greater than 10^{14} ohm-cm, and essentially independent of

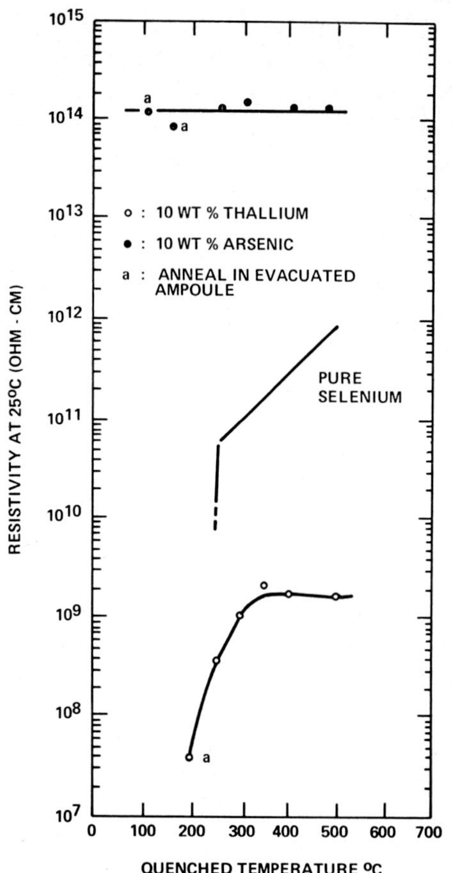

Fig. 4. Variation of ring concentration of 10 wt% selenium alloys with temperature.

quench or annealing temperatures in the range of 100–500 °C. However the resistivity of quenched selenium-thallium melts was typically 10^9 ohm-cm with near complete temperature independence.

Annealing of thallium–selenium alloys produced a resistivity decrease similar to pure selenium. With arsenic containing alloys this phenomenon was totally absent at this concentration.

X-ray diffraction patterns of selenium and selenide alloys were observed to characterize the structural nature of this material both after quenching and annealing. Copper Kα radiation was used filtered with aluminum foils. Quenched liquid selenium as well as arsenic or thallium alloys similarly prepared exhibited diffuse diffraction patterns with few discernible, regular lines. This behavior was attributed to the absence of crystallinity to any measurable degree at this stage of processing. Annealed selenium produced diffraction lines which are precisely characteristic of the hexagonal

crystalline phase. Annealed selenium with 10 wt% thallium also exhibited a crystalline diffraction pattern with lines attributed to both hexagonal selenium and tetragonal thallium selenide, TlSe. This behavior demonstrates that a diphasic crystallization of both hexagonal selenium and TlSe is prevalent from this amorphous alloy. No crystallization pattern was detected after the annealing of amorphous selenium containing 10 wt% arsenic.

The influence of both thallium and arsenic in various smaller concentrations down to 1 apm (atoms per million) was similarly studied to ascertain the concentration limits of these impurity elements in the resistivity-structure relationships. A quenched temperature of 350 °C was selected to prepare the amorphous alloys.

Arsenic was observed to reduce the ring concentration from that of pure selenium quenched from 350 °C starting at levels near 10^4 apm. The effect of thallium on ring concentration is minimal throughout the concentration range investigated, $1–10^5$ apm.

An increasing arsenic concentration was also observed to elevate the resistivity level above that of pure amorphous selenium to near 10^{14} ohm-cm. Beyond 10^4 apm. the resistivity is insensitive to the arsenic content. Since this concentration corresponds to the onset of Se_8 ring disappearance, ring dissociation to arsenic branched chains is an adequate explanation here.

Thallium also raises the resistivity of amorphous selenium to near $\sim 10^{14}$ ohm-cm at a concentration $\sim 10^2$ apm, the monotectic concentration. Following a concentration insensitive region, a rapid decrease in resistivity was observed with increasing thallium content due to a second phase formation.

3. Crystallization

Resistivity as a function of the degree of crystallization was determined by annealing at selected convenient temperatures pellets representing quenched, amorphous selenium and selenium alloys which were prepared as described earlier[4]). A temperature of 80 °C was found to be optimum to observe the crystallization phenomenon in terms of readily measurable time-resistivity profiles.

Typical time–resistivity relationships were observed for pure selenium as well as for the selenide alloys. These corresponded well with sigmoidal curves relating

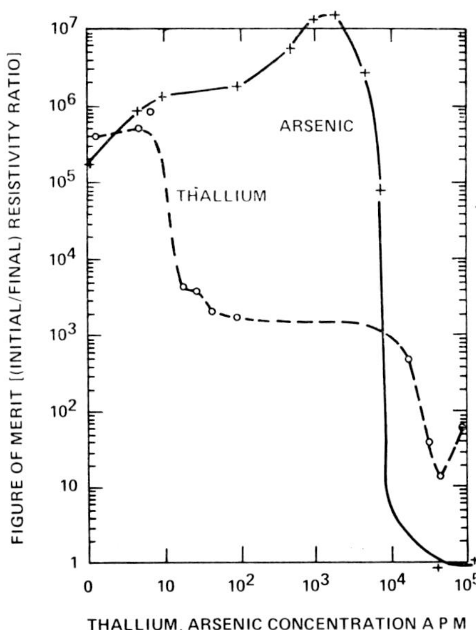

Fig. 5. Amorphous/crystalline resistivities (figure of merit) for selenium alloys.

the fraction of the crystallized phase to the crystallization time. The exception to this was the arsenic rich alloys with concentrations above 10^4 apm. These did not crystallize as previously discussed.

The parameter expressing the initial quenched resistivity of the amorphous phase to the final annealed resistivity of the crystalline phase was termed a "figure of merit" for a particular amorphous composition. Figures of merit are presented with varying impurity content in fig. 5. Thallium is observed to minimize this parameter with increasing content due to low amorphous resistivities. Arsenic yields a maximum value near the range 10^3–10^4 apm due to the magnitudes of both amorphous and crystalline phases.

This ratio would describe the electrical current levels in the "on" and "off" states in an amorphous semiconductive memory device as

$$\frac{\rho_A}{\rho_X} = \frac{R_A}{R_X} = \frac{I_X}{I_A} \tag{2}$$

at constant voltage, where $\rho_{A,X}$ are the resistivities of amorphous and crystalline phases, respectively; $R_{A,X}$ are the resistances; $I_{A,X}$ are the current levels conducted through the device. In addition

$$\frac{I_X}{I_A} = \frac{I_{on}}{I_{off}}. \tag{3}$$

4. Conclusion

Morphological transormations of selenium and selenium compositions with known impurity concentrations can be related to both molecular structure and electrical resistivity. The large and reversible changes in electrical resistivity can be associated with a crystallization transformation induced by specific thermal cycling.

These changes are of interest for the study and evaluation of amorphous semiconductor materials.

Crystallization and its resulting resistivity changes are influenced by addition of structure-modifying impurities. The presence of 10^4 apm of arsenic reduces the concentration of selenium ring molecules and obviates any crystallization phase changes. Conversely, thallium is considered to act as a chain cleaving agent; it increases the crystallization rate significantly.

The figure of merit values for selenium containing either of these two elements reflects this behavior. Large figures of merit, 10^5–10^7 are characterized by longer crystallization times. This behavior is prevalent with selenium containing traces of arsenic. Conversily, thallium yields figures of low merit at rapid crystallization times. Finally, this investigation has demonstrated that certain impurity elements can be added to amorphous selenium to obtain specific resistivity values as well as changes in resistivity values. These changes, promoted by a reversible crystallization phase change, are influenced by molecular structure in the amorphous material. Chain length is proposed as determining electrical resistivity by influencing carrier mobility in the amorphous and crystalline phases.

References

1) R. C. Keezer and M. W. Bailey, Mater, Res. Bull. **2** (1967) 185.
2) S. U. Dzhalilov, Russian J. Phys. Chem. **39** (1965) 10.
3) A. Eisenberg and A. V. Tobolsky, J. Polymer Sci. **46** (1960) 19.
4) E. Grochowski and W. Brenner, J. Non-Crystalline Solids **6** (1971) 83.
5) A. T. Ward and M. B. Myers, J. Phys. Chem. **73** (1969) 5.
6) W. C. Cooper, Ed., *The Physics of Selenium and Tellurium* (Proc. Intern. Symp., Montreal, 1967) (Pergamon, Oxford, 1969).

Journal of Crystal Growth **13/14** (1972) 849–850 © *North-Holland Publishing Co.*

REPORT ON THE ACTIVITIES OF THE ICCG-3 GENERAL ASSEMBLY (MARSEILLE) AND THE INTERNATIONAL ORGANIZATION OF CRYSTAL GROWTH (IOCG)

1. Constitution

In order to implement a decision made at ICCG-2 Birmingham in 1968, two draft constitutions – a long version of American origin and a short version of European origin – were submitted by postal ballot to participants of ICCG-1 Boston and ICCG-2 Birmingham. The vote was concluded in April 1971 by M. Schieber and A. D. McQuillan with the following results:

Long version – 146 votes,
Short version – 129 votes

In view of a decision reached by IOCG (formerly Comité International de Croissance Cristalline) at Zurich in September 1970, the long version was adopted until the close of ICCG-3 Marseille. In order to ameliorate difficulties presented by the strong divided opinions of the supporters of the two different versions of the Constitution, a compromise was reached at Marseille by which a revised version of the long Constitution was declared valid and in force until the closure of ICCG-4. Briefly, the revised version deletes all mention of dues or fees to be paid to the IOCG, but maintains most other concepts. Also, it was decided that a new Subcommittee for Constitution be appointed to search for further improvements (see paragraph 4 of this report).

2. Council of the International Organization of Crystal Growth (IOCG)

Based on this revised Constitution, a Council was elected having the following composition:

Officials (by Constitution)
M. Schieber, Past Chairman, ICCG-1 Boston
B. Chalmers, Past Chairman, ICCG-1 Boston
A. D. McQuillan, Past Chairman, ICCG-2 Birmingham
W. Bardsley, Past Chairman, ICCG-2 Birmingham
R. Kern, Co-chairman, ICCG-3 Marseille
W. Dekeyser, Co-chairman, ICCG-3 Marseille
R. R. Hasiguti, Chairman, ICCG-4 Tokyo
F. C. Frank, President, IOCG
C. S. Sahagian, Treasurer, IOCG

Delegates of National Associations

France–Belgium	A. Authier
	B. Mutaftschiev
	A.-M. Vergnoux
German Federal Republic	S. Haussühl
	R. Nitsche
Israel	L. Ben-Dor
Japan	N. Kato
	I. Sunagawa
Netherlands	P. Hartman
Spain	M. Font-Altaba
Switzerland	E. Kaldis
	(to be announced)
United Kingdom	C. H. L. Goodman
	E. A. D. White
U.S.A.	K. A. Jackson
	R. A. Laudise
	J. Wenckus

Delegate of IU Cryst

R. F. Strickland-Constable

Direct Representatives for Countries without National Associations

Bulgaria	E. Budevski
Czechoslovakia	B. Manec
German Democratic Republic	H. Bethge
India	P. Krishna
Poland	T. Niemyski
U.S.S.R.	K. S. Bagdasarov

The Council's tenure of office is until the close of ICCG-4 Tokyo, March 1974.

3. Executive Committee

The following officers of IOCG were re-elected to the Executive Committee:

F. C. Frank, FRS, President
M. Schieber, Secretary-General
C. S. Sahagian, Treasurer
W. Bardsley, Past President

No Vice-President and President-Elect were appointed in order not to bind the new Subcommittee for Constitution. The additional members of the Executive Committee are:

W. Dekeyser
S. Haussühl
R. Kern
R. A. Laudise
R. R. Hasiguti, Chairman ICCG-4, ex officio.

The tenure of office of the Executive Committee is until the close of ICCG-4, March 1974.

4. Subcommittee for Constitution

In order to search for further improvements and to minimize the differences of opinions, the Executive Committee appointed the following Subcommittee for Constitution:

S. Haussühl
K. A. Jackson
R. A. Laudise
A. D. McQuillan
B. Mutaftschiev

The agreed new version of the Constitution will be submitted for postal approval to all registrants of ICCG-1, 2 and 3. Changes to the Constitution would then be valid after ICCG-4.

5. Financial Report of IOCG

The books of IOCG were audited as of July 1, 1971 and the Treasurer's Report was presented at the General Assembly as follows:

Balance at Close of ICCG-2 . . . . . $ 1,550.49

Receipts:

3/70 Surplus – ICCG-2	$ 7,840.31	
11/70 Loan Repayment – Zurich Committee	250.00	
6/70 Interest – IOCG Bank Account	141.66	
	$ 8,231.97	

Disbursements:

9/68 Checking Accounts Costs	4.61	
2/70 Loan to Zurich Conference (repaid)	250.00	
2/70 Accountant Fees & Supplies	42.50	
4/70 Loan to ICCG-3 Committee	3,000.00	
9/70 Printing Costs	141.00	
9/70 Permanent Transfer to BACG	1,000.00	
1/71 Tax Liability to British Government	569.64	
2/71 Printing Costs	192,61	
	$ 5,200.36	

Balance . . . . $ 4,582.10

6. ICCG-4, March 1974, Tokyo, Japan

ICCG-4 will be held in March 1974 in Tokyo, Japan. The principal officers of ICCG-4 are as follows:

R. R. Hasiguti, Chairman, Organizing Committee
T. Arizumi, Secretary General
I. Sunagawa, Chairman, Program Committee
N. Kato, Chairman, Publication Committee
T. Sugano, Local Arrangements Committee

LIST OF AUTHORS QUOTED IN REFERENCES

Abbink, F. 257, 619
Abraham, F. F. 144, 148
Abrahams, M. S. 276
Abrahams, S. C. 27, 530
Abramowitz, M. 113
Abramson, E. 593
Abrikosov, N. Kh. 698
Achiwa, N. 282
Adamski, J. A. 710
Adelsberg, L. M. 693
Adlington, R. E. 39
Afromowitz, M. 365
Agranovskaya, A. I. 681
Aguayo, J. 604
Ahmed, A. M. 417
Aigrain, P. 365
Akasaki, I. 315
Akashi, T. 681
Albinet, G. 164
Albon, N. 454
Albright, D. L. 739
Alcock, C. B. 365, 739
Alefeld, B. 247
Aleshim, E. 530
Alexander, E. 285, 292
Alexander, J. A. 97
Alford, W. J. 718
Alix, J. E. 306
Allais, G. 829
Allen, F. G. 302
Allen, J. W. 818
Allpress, J. G. 48, 212
Amberger, E. 350
Amick, J. A. 306, 331, 342
Anderko, K. 739, 757
Anderson, J. S. 78
Anderson, R. L. 198, 315
Anderson, T. F. 611
Andersson, B. 693
Andersson, S. 78
André, E. 663
Andriech, A. 385
Andrade, E. N. da C. 57
Anthony, T. R. 787, 790
Antonov, A. V. 706
Antypas, G. A. 640, 663
Arakawa, K. 113

Aramaki, S. 78
Archer, J. L. 27, 571
Arefeva, R. M. 693
Arends, J. 276
Arizumi, T. 39, 315, 346, 615
Arlett, R. H. 365, 723
Armington, A. F. 519
Asbrink, S. 78
Ashby, M. 57
Askenazy, F. 640
Asti, G. 27
Attinger, C. 718
Aucoin, T. R. 560
Aumont, R. 519
Aust, K. T. 57, 834
Austin, J. B. 306
Authier, A. 34, 272
Aven, M. 375, 668
Avignon, M. 113
Avrami, M. 131
Awazu, K. 593

Bachmann, K. J. 619, 624
Badami, D. V. 97
Badiali, J.-P. 417
Badoz, J. 389
Bagdasarov, K. S. 593
Bagley, B. G. 48, 212
Bailey, J. E. 57
Bailey, M. W. 843
Bailey, S. M. 360
Bailly, F. 668
Bailly, T. 726
Baker, A. J. 97
Baker, J. C. 810
Bakonova, D. D. 647
Balbaskov, A. M. 706
Baldwin, T. O. 257, 427
Balitski, V. S. 524
Ball, D. J. 164, 167
Ballman, A. A. 530, 535, 545, 593, 675
Ban, V. S. 365
Banks, E. 433
Banks, R. E. 588
Baralis, G. 39
Baranov, B. V. 346
Bardeen, J. 252

Barker, J. J. 488
Barlow, I. 571
Barnes, H. L. 519
Barnes, R. S. 57
Barnes, S. C. 421
Barnet, F. R. 97
Barns, R. L. 27
Barraud, J. 519
Barsch, W. O. 106
Barthel, J. 39
Bartholomew, R. F. 78
Bartlett, B. E. 302, 668, 810
Bartlett, J. T. 241
Bartlett, R. W. 693
Basinski, Z. S. 380
Baskin, Y. 159
Bass, S. J. 629, 635, 647
Bassett, D. W. 167
Bassett, G. A. 19, 62
Basterfield, J. 268, 566
Bates, D. R. 174
Bates, H. E. 751
Bates, R. L. 488
Batifol, E. 519
Batog, V. N. 530
Batterman, B. W. 167
Baudelet, B. 34, 252
Bauer, E. 777
Baughman, R. J. 686
Baule, B. 236
Bausova, N. V. 686
Beal, Jr., J. B. 467
Beal, R. J. 365
Bean, K. E. 302
Beaumont, D. W. R. 97
Beaumont, P. 97
Beck, P. A. 57
Becker, R. 3, 44
Beevers, C. A. 476
Beingessner, C. J. 57
Belin, C. 597
Bell, P. S. 693
Bell, R. O. 668
Belov, N. V. 540
Belyaev, I. N. 530
Benacerraf, A. 389
Bénard, J. 823

851

SUBJECT INDEX

Each page number refers to the first page of a given reference

COMPOUND INDEX

Each page number refers to the first page of a given reference